Membrane
Fusion

5

CELL SURFACE REVIEWS
VOLUME 5

NORTH-HOLLAND PUBLISHING COMPANY
AMSTERDAM · NEW YORK · OXFORD

MEMBRANE FUSION

Edited by

GEORGE POSTE

Department of Experimental Pathology
Roswell Park Memorial Institute, Buffalo, New York

and

GARTH L. NICOLSON

Department of Developmental and Cell Biology
University of California, Irvine, California

1978

NORTH-HOLLAND PUBLISHING COMPANY
AMSTERDAM · NEW YORK · OXFORD

PUBLISHED BY:
Elsevier/North-Holland Biomedical Press
Jan Van Galenstraat 335, P.O. Box 211
Amsterdam, The Netherlands

SOLE DISTRIBUTORS FOR THE U.S.A. AND CANADA:
Elsevier North-Holland, Inc.
52 Vanderbilt Avenue
New York, New York 10017

Library of Congress Cataloguing in Publication Data

Main entry under title:

Membrane fusion.

  (Cell surface reviews; v. 5)
  Bibliography: p.
  Includes index.
  1. Membrane fusion. I. Poste, George.
II. Nicolson, Garth L. III. Series.
[DNLM: 1. Cell membrane. W1 GE1283 v. 5 / QH601
M5325]
QH601.M467   574.8'75   78-5321
ISBN 0-444-00262-6

MANUFACTURED IN THE UNITED STATES OF AMERICA

General preface

Research on membranes and cell surfaces today occupies center stage in many areas of biology and medicine. This dominant position reflects the growing awareness that many important biological processes in animal and plant cells and in microorganisms are mediated by these structures. The extraordinary and unprecedented expansion of knowledge in molecular biology, genetics, biochemistry, cell biology, microbiology and immunology over the last fifteen years has resulted in dramatic advances in our understanding of the properties of the cell surface and heightened our appreciation of the subtle, yet complex, nature of cell surface organization.

The rapid growth of interest in all facets of research on cell membranes and surfaces owes much to the convergence of ideas and results from seemingly disparate disciplines. This, together with the recognition of common patterns of biological organization in membranes from highly different forms of life, has led to a situation in which the sharp boundaries between the classical biological disciplines are rapidly disappearing. The investigator interested in cell surfaces must be at home in many fields, ranging from the detailed biochemical and biophysical properties of the molecules and macromolecules found in membranes to morphological and phenomenological descriptions of cellular structure and cell-to-cell interactions. Given the broad front on which research on cell surfaces in being pursued, it is not surprising that the relevant literature is scattered in a diverse range of journals and books, making it increasingly difficult for the active investigator to collate material from several areas of research. Thus, while scientists are becoming increasingly specialized in their techniques, and in the nature of the problems they study, they must interpret their results against an intellectual and conceptual background of rapidly expanding dimension. It is with these conflicting demands and needs in mind that this series, to be known under the collective title of CELL SURFACE REVIEWS, was conceived.

CELL SURFACE REVIEWS will present up-to-date surveys of recent advances in our understanding of membranes and cell surfaces. Each volume will contain authoritative and topical reviews by investigators who have contributed to progress in their respective research fields. While individual reviews will provide comprehensive coverage of specialized topics, all of the reviews published within each volume will be related to an overall common theme. This format represents a departure from that adopted by most of the existing series of "review" publications which usually provide heterogeneous collections of reviews on unrelated topics. While this latter format is considerably more convenient from an editorial standpoint, we feel that publication together of a number of related reviews will better serve the stated aims of this series—to bridge the information and specialization "gap" among investigators in related areas. Each volume will therefore present a fairly complete and critical survey of the more important and recent advances in well defined topics in biology and medicine. The level will be advanced, directed primarily to the needs of the research worker and graduate students.

Editorial policy will be to impose as few restrictions as possible on contributors. This is appropriate since the volumes published in this series will represent collections of review articles and will not be definitive monographs dealing with all aspects of the selected subject. Contributors will be encouraged, however, to provide comprehensive, critical reviews that attempt to integrate the available data into a broad conceptual framework. Emphasis will also be given to identification of major problems demanding further study and the possible avenues by which these might be investigated. Scope will also be offered for the presentation of new and challenging ideas and hypotheses for which complete evidence is still lacking.

The first four volumes of this series will be published within one year, after which volumes will appear at approximately one year intervals.

GEORGE POSTE
GARTH L. NICOLSON
Editors

Contents of previous volumes

Contents of forthcoming volumes

Preface

This volume, the fifth in the *Cell Surface Reviews* series, provides a comprehensive survey of our present understanding of events in membrane fusion and is the first publication to summarize the available information on this topic within a single book. Despite the large amount of published information dealing with various aspects of membrane fusion behavior, insight into the molecular mechanism(s) underlying this phenomenon is still limited. Most of our present knowledge has come from indirect observations of factors that control the diverse cellular and subcellular phenomena in which fusion occurs and fewer studies have been done with the specific aim of identifying the mechanism of fusion and the factors that regulate this process. This situation is changing, however, as more investigators from a broad spectrum of scientific disciplines become interested in defining the molecular events involved in fusion. The last few years have seen a dramatic increase in the number of publications concerned with membrane fusion and the growing interest in this phenomenon is also reflected in the increasingly diverse experimental strategies being used to study different examples of membrane fusion in both cellular and subcellular systems. We thus anticipate that knowledge of the molecular events in membrane fusion will expand substantially over the next few years. It is essential, however, that this growth should take place with a full appreciation of the complex nature of the various cellular phenomena in which fusion occurs. With this in mind, we have sought in this volume to strike a balance between chapters that describe the biological phenomena in which membrane fusion occurs and chapters that discuss experiments done with the specific goal of elucidating the mechanism(s) underlying fusion.

Membrane fusion cannot be considered in isolation from other aspects of membrane activity and analysis of the mechanism(s) of membrane fusion must take into account the basic properties of the membranes involved and identify the relationship between events in membrane fusion and other aspects of membrane function. In this way, the experimental study of membrane fusion can also provide useful information on basic membrane organization. At this level, the study of membrane fusion begins to overlap other major areas of contemporary membrane research concerned with the dynamic properties of membranes and their role in specialized cellular functions. This perspective is reflected in the fifteen chapters in this volume.

We thank the contributors to this volume and express our appreciation of their willingness to accept editorial suggestions. We again thank Molly Terhaar, Linda Bernstein and Adele Brodginski for their untiring help during the editing of the manuscripts and Margaret Quinlin and Susan Koscielniak of Elsevier North-Holland for their excellent help throughout the production stages.

August, 1977

GEORGE POSTE
Buffalo, New York

GARTH L. NICOLSON
Irvine, California

List of contributors

The numbers in parenthesis indicate the page on which the author's contribution beings.

Richard D. ALLEN (657), Pacific Biomedical Research Center and Department of Microbiology, University of Hawaii, Honolulu, Hawaii 96822, U.S.A.

J. Michael BEDFORD (65), Department of Obstetrics and Gynecology, Cornell University Medical College, New York, New York 10021, U.S.A.

Richard BISCHOFF (127), Department of Anatomy and Neurobiology, Washington University School of Medicine, St. Louis, Missouri 63110, U.S.A.

N. BORGESE (509), CNR Center of Cytopharmacology and Department of Pharmacology, University of Milan, Milan, Italy.

P. DE CAMILLI (509), CNR Center of Cytopharmacology and Department of Pharmacology, University of Milan, Milan, Italy.

Michael J. CARLILE (219), Department of Biochemistry, Imperial College of Science and Technology, London SW7 2A2, England.

B. CECCARELLI (509), CNR Center of Cytopharmacology and Department of Pharmacology, University of Milan, Milan, Italy.

E. C. COCKING (369), Department of Botany, University of Nottingham, Nottingham NG7 2RD, England.

Zanvil A. COHN (387), The Rockefeller University, New York, New York 10021, U.S.A.

George W. COOPER (65), Department of Anatomy, Cornell University Medical College, New York, New York 10021, U.S.A.

Paul J. EDELSON* (387), The Rockefeller University, New York, New York 10021, U.S.A.

David EPEL (1), Hopkins Marine Station of Stanford University, Pacific Grove, California 93750, U.S.A.

P. K. EVANS (369), Department of Biology, The University, Southampton SO9 5NH, England.

David GINGELL (791), Department of Biology as Applied to Medicine, The Middlesex Hospital Medical School, London WIP 6DB, England.

Lionel GINSBERG (791), Department of Biology as Applied to Medicine, The Middlesex Hospital Medical School, London W1P 6DB, England.

M. H. GINSBERG (407), Scripps Clinic and Research Foundation, San Diego, California 92037, U.S.A.

Graham W. GOODAY (219), Department of Microbiology, Marischal College, University of Aberdeen, Aberdeen AB9 1AS, Scotland.

Peter M. HENSON (407), Department of Pediatrics, National Jewish Hospital and Research Center, Denver, Colorado 80206, U.S.A.

Jack A. LUCY (267), Department of Biochemistry and Chemistry, Royal Free Hospital School of Medicine, University of London, London SCIN 1BP, England.

J. MELDOLESI (509), CNR Center of Cytopharmacology and Department of Pharmacology, University of Milan, Milan, Italy.

D. C. MORRISON (407), Scripps Clinic and Research Foundation, San Diego, California 92037, U.S.A.

Lelio ORCI (629), Institute of Histology and Embryology, University of Geneva Medical School, 1211 Geneva 4, Switzerland.

J. M. PAPADIMITRIOU (181), Department of Pathology, University of Western Australia, Perth Medical Centre, Western Australia.

Demetrious PAPAHADJOPOULOS (765), Department of Experimental Pathology, Roswell Park Memorial Institute, Buffalo, New York 14263, U.S.A.

Charles A. PASTERNAK (305), Department of Biochemistry, St. Georges Hospital Medical School, University of London, London SW17 0QT, England.

Alain PERRELET (629), Institute of Histology and Embryology, University of Geneva Medical School, 1211 Geneva 4, Switzerland.

George POSTE (305), Department of Experimental Pathology, Roswell Park Memorial Institute, Buffalo, New York 14263, U.S.A.

J. B. POWER (369), Department of Botany, University of Nottingham, Nottingham NG7 2RD, England.

Victor VACQUIER (1), Department of Zoology, University of California, Davis, California 95616, U.S.A.

*Dr. Edelson's present address is: Children's Hospital Medical Center, Boston, Massachusetts 02115, U.S.A.

Contents

1 Membrane fusion events during invertebrate fertilization, by D. Epel and V. D. Vacquier

2   *Membrane fusion events in the fertilization of vertebrate eggs,*
    by J. M. Bedford and G. W. Cooper

3   *Myoblast fusion, by R. Bischoff*

4   Macrophage fusion in vivo and in vitro: a review, by J. M. Papadimitriou

5   Cell fusion in myxomycetes and fungi, by M. J. Carlile and G. W. Gooday

6   *Mechanisms of chemically induced cell fusion, by J. A. Lucy*

7   *Virus-induced cell fusion, by G. Poste and C. A. Pasternak*

8   *Fusion of plant protoplasts, by J. B. Power, P. K. Evans and E. C. Cocking*

12   Ultrastructural aspects of exocytotic membrane fusion, by L. Orci and A. Perrelet

13   Membranes of ciliates: ultrastructure, biochemistry and fusion, by R. D. Allen

14   *Calcium-induced phase changes and fusion in natural
      and model membranes, by D. Papahadjopoulos*

15   *Problems in the physical interpretation of membrane interaction and fusion,
      by D. Gingell and L. Ginsberg*

xxii

MEMBRANE FUSION
5

Membrane fusion events during invertebrate fertilization

1

David EPEL and Victor D. VACQUIER

Contents

G. Poste & G. L. Nicolson (eds.) Membrane Fusion, pp. 1–63.
© *Elsevier/North-Holland Biomedical Press, 1978.*

2

1. Overview of fertilization

Fertilization can be viewed as a sequence of four coordinated membrane fusions which result in the triggering of embryonic development. The mechanism of these fusions, which is presumably similar in all eukaryote cells, can be easily studied in invertebrate gametes and has led to a recent burgeoning in our knowledge about fertilization. Much of the earlier work has been summarized in recent years (Allen, 1958; Runnström et al., 1959; Monroy, 1965; Tyler and Tyler, 1966; Colwin and Colwin, 1967; Dan, 1967; Metz, 1967; Giudice, 1973; Longo, 1973; Stearns, 1974; Austin, 1975; Epel, 1975). Our purpose here is to review the most recent work on the interaction and fusion of invertebrate gametes and to raise questions which may be amenable to experimental attack.

The first fusion associated with fertilization occurs in the sperm as it approaches the egg and is referred to as the acrosome reaction. The second, the fusion one primarily thinks of in relation to fertilization, is the actual fusion of the membranes of the sperm and egg. The third, which occurs in eggs of many, but not all, species is referred to as the cortical reaction and is a massive secretory response of the egg following sperm-egg fusion. These three fusions all occur at the plasma membrane of either the sperm or the egg. The fourth fusion is intracellular and occurs between sperm and egg pronuclei.

This chapter examines the nature of these fusions, especially the three plasma membrane events associated with fertilization. Two of the fusions, the acrosome and cortical reactions, are secretory or exocytotic events. The acrosome reaction also exposes the new membrane that will ultimately fuse with the egg surface and exposes the binding protein which acts as a ligand between sperm and egg. Cortical exocytosis involves thousands of 0.5 to 1 μm granules that are embedded in the egg cortex and fuse with the egg surface in response to fertilization. Fusion begins near the site of sperm-egg fusion and rapidly propagates around the egg. This exocytosis results in a dramatic alteration of the egg surface and in the elevation of the fertilization envelope. The latter event is related to preventing further sperm-egg fusions as part of the "block to polyspermy."

The fusion between sperm and egg at the plasma membrane is, of course, the most important kind of fusion in fertilization. Even though this fusion occurs over an extremely small portion of the egg surface, much is known about its regulation. The egg normally "admits" only one sperm and the nature of this block to polyspermy has thrown much light on fusion phenomena in general.

1.1. Experimental material

Eggs and sperm of marine invertebrates provide excellent material for fertilization studies. The best studied gametes are those of echinoids (e.g., sea urchins and sand dollars), starfish, molluscs, certain annelids, arthropods, and some urochordates. In optimal cases the organisms produce large amounts of gametes, and the breeding seasons are sufficiently long so that material is available for most of the year. In many organisms the gametes are readily extracted from the adult, are stable for at least several hours, and fertilization can be easily effected in vitro by simply mixing the sperm and eggs together. The best studied forms are sea urchin gametes and thus much of the emphasis in this chapter is on fertilization in these organisms.

1.2. General strategy of fertilization

The strategy of sexual reproduction in multicellular organisms utilizes gametes of dissimilar size (i.e., anisogamy) as opposed to unicellular organisms, where the gametes are similar in size (i.e., isogamy). The general strategy in the metazoans is to have a small motile male gamete, the sperm, and a large nonmotile female gamete, the egg. By complex behavioral mechanisms, the sperm is deposited in close proximity to the egg so that the two gametes can become juxtaposed and effect fusion of the two cells.

1.2.1. The egg
The external morphology of the egg differs with the species, the most obvious variables being size, pigment, extracellular coats, and whether the egg will exhibit a cortical reaction. The size of the egg is related to whether it will develop

4

directly into a miniature adult (i.e., direct development) or whether it will become an intermediate and feeding larval form (i.e., indirect development). Generally, the direct developing form has large amounts of yolk and tends to be opaque, whereas the indirect developing form tends to have less yolk and to be more transparent.

The major variant in egg morphology other than size relates to the extracellular coats or coverings of the egg. Moving in toward the egg surface, many eggs are surrounded by an extracellular polysaccharide referred to as the jelly layer. As is described further on, the components of this layer induce changes in the behavior of the sperm and also trigger the acrosome reaction.

The next extracellular layer might be a chorion or vitelline layer. In some forms, such as the keyhole limpet *Megathura,* the chorion is an extracellular envelope several hundred microns away from the egg surface and probably functions in a protective role. In other forms, such as the squid, the chorion is closer to the egg surface and possesses a specialized area for sperm entry, the micropyle. In sea urchins no chorion is apparent in the unfertilized egg, but a precursor "vitelline layer" is present on the egg surface. This layer, which also contains sperm receptors, is tightly apposed to the plasma membrane. After fertilization, the layer elevates away from the egg surface and becomes transformed into the activation calyx or fertilization envelope (section 6). In this position the fertilization envelope is more analogous to a chorion and probably serves a protective role and also functions in the block to polyspermy.

The plasma membrane of the egg is also specialized. In many, perhaps in all, cases eggs have prominent microvilli, and in the sea urchin these form a typical pattern on the egg surface that is constant for each species. These microvillar patterns are embossed in the overlying vitelline layer and the representative patterns of several species are shown in Fig. 1. The role of these microvilli may be as sites of sperm-egg fusion. The plasma membrane must also be specialized for sperm-egg reception and for triggering or activating development following successful interaction between sperm and egg.

1.2.2. The sperm

The sperm exhibits considerable morphological variation, and it is probably easier to distinguish species by their sperm than by their eggs (Afzelius, 1973; Bacchetti and Afzelius, 1976). The electron micrograph of an echinoid sperm (Fig. 2) illustrates the principal features of the sperm that are of interest in terms of sperm-egg fusion, the acrosome vesicle and the periacrosomal space (Dan, 1970; Tilney et al., 1973; Summers et al., 1975). The acrosomal vesicle contains lytic enzymes (sperm lysins) necessary for penetration of the sperm through the outer egg coats and binding proteins (bindins) for attaching the sperm to the

Fig. 1. The surface of the vitelline layers of eggs of five species of sea urchins. (a) *Strongylocentrotus franciscanus,* ×8,400; (b) *Echinometra mathei,* ×7,500; (c) *Eucidaris metularia,* ×8,100; (d) *Arbacia punctulata,* ×7,700; (e) *Allocentrotus fragilis,* ×8,100; (f) the fertilization envelope of *A. fragilis.* (Reproduced with permission of The Wistar Press from Tegner and Epel, 1977.)

5

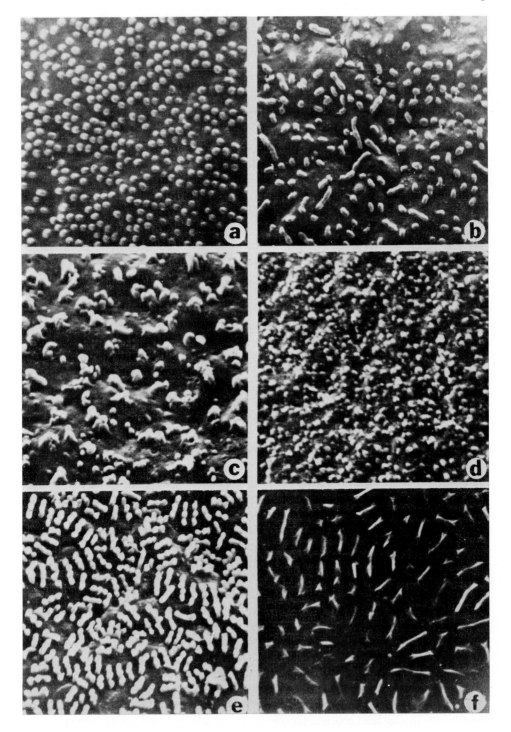

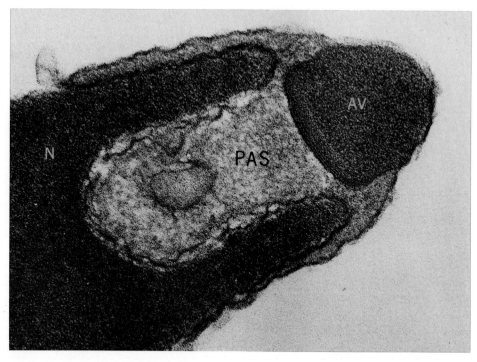

Fig. 2. Head of sperm of the sea urchin *Strongylocentrotus purpuratus* before the acrosome reaction. AV, acrosome vesicle; N, nucleus; PAS, periacrosomal space. ×111,850. (Reproduced with permission from Vacquier and Moy, 1977.)

egg. The periacrosomal region contains unpolymerized actin and probably plasma membrane precursors that are converted into a new plasma membrane during the acrosome reaction (Sardet and Tilney, 1977). As part of the acrosomal reaction, the actin polymerizes and pushes the precursor plasma membrane in front of it to form an extension of the sperm plasma membrane. This is referred to as the acrosomal process. The bindins become incorporated onto the surface of this new process, and the membrane at the tip of the process ultimately fuses with the egg plasma membrane (Fig. 3).

1.2.3. Sperm-egg interactions
The general strategy in fertilization is to retain the acrosome in an intact condition until the sperm approaches the egg. Components of the egg, the jelly layer or the vitelline layer, then act as inducers of the acrosome reaction with the resultant exposure of lysins and formation of the acrosomal process with associated bindins. Attachment to the egg surface then leads to sperm-egg fusion, incorporation of the sperm, and activation of development.

The responses of the egg to fertilization have been best investigated extensively in sea urchins, and these studies reveal that three major effects from sperm-egg fusion lead to incorporation of the sperm into the egg, the exclusion

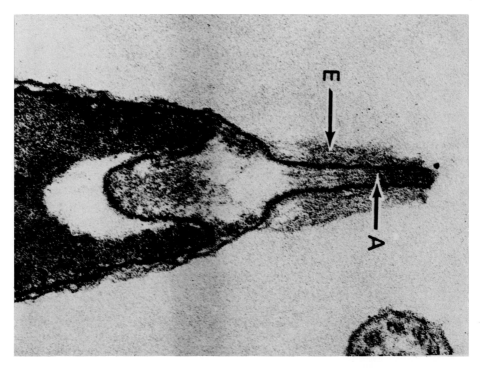

Fig. 3. Acrosome-reacted sperm of the sea urchin *Lytechinus pictus*. A, acrosomal process; E, extracellular material coating the acrosomal process composed of the protein bindin. ×102,600. (Reproduced with permission of Academic Press from Collins and Epel, 1977.)

of superfluous sperm, and the activation of development. The first of these changes is a rapid alteration in membrane potential. The resting potential of the unfertilized egg is between −60 and −10 mV, depending on species (Steinhardt et al., 1971; Ito and Yoshioka, 1972; Jaffe, 1976). Within three seconds of insemination the membrane depolarizes, and the potential rapidly increases to around +10 mV. It remains at this level for about 60 seconds, then drops back near 0 mV and, beginning at 5 minutes, gradually decreases to about −60 mV. As will be noted further on, this initial potential membrane change appears to be related to preventing further sperm from fusing with the egg as a part of the block to polyspermy. The second controlling change in fertilization is a release of intracellular calcium from cytoplasmic stores. This begins 20 to 25 seconds after fertilization, and measurements with aequorin indicate that calcium remains high for the next 2 to 3 minutes (Ridgeway et al., 1977; Steinhardt et al., 1977). This Ca^{2+} release appears to be related to the induction of cortical exocytosis (Vacquier, 1976; and section 6.2.1). The third change in fertilization, which probably is triggered by the calcium increase, is a dramatic change of membrane permeability with a resultant sodium-hydrogen exchange in which an influx of sodium is coupled to an efflux of hydrogen. The result is a large increase in in-

8

tracellular pH that seems to be necessary for metabolic activation (Johnson et al., 1976).

2. The acrosome reaction

As noted earlier, the acrosome reaction results from the exocytosis of the acrosome vesicle. Ultrastructural studies (Fig. 4a) indicate that exocytosis involves initial vesiculation of the plasma membrane overlying the vesicle (Summers et al., 1975). In the sperm of the sand dollar, vesiculation appears to be restricted to the rim of the vesicle, so that at the completion of vesiculation the upper part of this vesicle pops off almost like a cap (Fig. 4b). The result of vesiculation is the externalization of the vesicle contents and the exposure of a new membrane that previously lay beneath the vesicle. Exocytosis also exposes lytic enzymes involved in getting the sperm through the outer egg coats (Dan, 1967), exposes substances involved in initial stages of sperm-egg adhesion (see section 3.3), and exposes the plasma membrane which will ultimately fuse with the egg. Changes are also initiated that result in the polymerization of actin and the extrusion of the acrosomal process.

Considerable variation exists among species in the mode of the acrosome reaction and in the length of the acrosomal process. Some incredible engineering feats are seen in the reaction of starfish, sea cucumbers, and horseshoe crab sperm. Tilney and his colleagues, for example, have described the formation of a

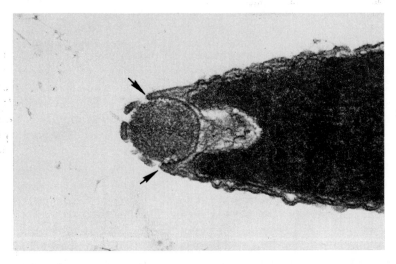

Fig. 4(a). Initiation of the acrosome reaction in the *Lytechinus variegatus* spermatozoon. The apical portion of the acrosomal vesicle membrane has fused with the overlying plasma membrane at multiple sites to form a series of membranous vesicles. The rim of dehiscence, which represents the site of continuity between acrosomal vesicle and plasma membranes, is indicated by arrows. ×67,000. (Micrograph courtesy of B. L. Hylander and R. G. Summers, unpublished.)

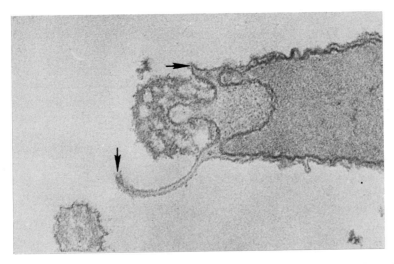

Fig. 4(b). An early stage of the acrosome reaction in the *Echinarachnius parma* spermatozoon in which the plasmalemma and acrosomal vesicle membranes have partically fused. The fused edges are indicated by arrows. The acrosomal tubule has begun to elongate by eversion of the remaining (basal) portion of the acrosomal vesicle membrane. Acrosomal vesicle contents have become altered in their morphology. ×61,500. (From R. Summers and B. Hylander, 1974.)

60 to 90 μm long process in the sperm of the sea cucumber and have elegantly documented the role of actin polymerization in the extension of the process (Tilney et al., 1973). Another remarkable example is the acrosome process in the horseshoe crab *Limulus*, where a major problem is the generation of new plasma membrane. In these forms it appears to involve packaging of preformed membrane and assembly of this material around the elongating acrosome process (Tilney, 1975; Sardet and Tilney, 1977).

2.1 Triggers of the acrosome reaction

2.1.1 Egg jelly
Most workers assume that some component of the egg jelly triggers the acrosome reaction since soluble egg jelly is a powerful inducer of this reaction (Dan, 1967; Metz, 1967). Also, transmission electron micrographs of sperm as they traverse the jelly layer show that many of the sperm embedded within this layer have undergone an acrosome reaction (Summers and Hylander, 1975).

The exact nature of the component in egg jelly that induces exocytosis is not yet known. Work of Hotta and co-workers (1973) and Lorenzi and Hedrick (1973) indicate that there are several distinct macromolecules in the jelly, but it is not known which one or combination of these is required to induce the reaction. Also, the egg jelly induces gross changes in motility, behavior, and respiration of the sperm and these individual responses may be related to the individual components of the jelly (Metz, 1967; Ohtake, 1976).

2.1.2. Other egg components

It is also possible that some other component of the egg surface acts as the inducer of the acrosome reaction. This is suggested partly by the recent finding of Decker and collaborators (1976) that *Arbacia* egg jelly is ineffective at physiological pH and also by an early observation that eggs will induce an acrosome reaction after the egg jelly has been removed. One possibility is that the vitelline layer itself may function as an inducer of the acrosome reaction. Another alternative is that the eggs are continually secreting an inducer that happens to be entrapped in the jelly. Both hypotheses are consistent with the recent observation of Aketa and Ohta (1977), who stripped intact jelly layer hulls from eggs and added sperm to these hulls; the sperm passed through the hulls without undergoing an acrosome reaction. Thus either the isolated jelly hulls lost the acrosome-inducing activity, or the inducing activity is on the egg surface.

It would be extremely interesting to purify the factors responsible for inducing the acrosome reaction, to study the nature of receptors on the sperm plasma membrane and how they effect the acrosomal exocytosis reaction.

2.2. Mechanism of induction of the acrosome reaction

2.2.1. Calcium

The acrosomal inducer probably acts by altering the permeability of the plasma membrane to calcium. This hypothesis was first suggested by Dan (1954), who observed that calcium was required in the extracellular medium for jelly to induce the acrosome reaction. The critical testing of this hypothesis was made possible with the advent of ionophores such as A23187 for divalent cations. Collins (1976) was the first to use ionophore A23187 for this purpose and found that in the absence of jelly the ionophore was a very potent inducer of the acrosome reaction of sea urchin sperm. The induction required calcium in the outside medium, supporting the idea that an influx of calcium triggers the reaction. This observation has since been confirmed and extended considerably by work from several other laboratories (Decker et al., 1976; Talbot et al., 1976; Collins and Epel, 1977). Another indication that calcium is an inducer of the acrosome reaction is the finding that simply increasing the concentration of calcium in the extracellular medium will also induce an acrosome reaction (Fig. 5). The induction of the reaction is not specific for calcium; barium and strontium, but not magnesium, are also effective.

Additional evidence shows that the locus of action is via affecting calcium metabolism because inhibitors of calcium transport will also prevent the acrosome reaction. Thus, incubation of sperm in lanthanum inhibits the induction of the acrosome reaction by egg jelly and the acrosome reaction can be prevented with local anesthetics such as procaine. The inhibition by procaine is antagonized by raising the concentration of calcium in the extracellular medium (Collins and Epel, 1977).

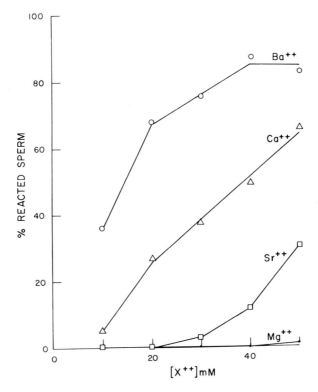

Fig. 5. Induction of the acrosome reaction by increased extracellular concentrations of Ca^{2+}, Ba^{2+}, Sr^{2+}: *L. pictus* sperm were suspended in Ca^{2+}-free sea water (SW) with 10 mM Tris buffer to which 0–50 mM $CaCl_2$, $BaCl_2$, $SrCl_2$, or $MgCl_2$ had been added (pH adjusted to 8.0). After 60 sec, sperm were fixed and assayed. Note that Ca^{2+}-free SW contains 55 mM Mg^{2+} before any additions. (Reproduced with permission of Academic Press from Collins and Epel, 1977.)

2.2.2. Intracellular pH

Elevating the pH can also induce the acrosome reaction, but only in the presence of extracellular calcium. Also, the pH dependence for the acrosome reaction is considerably lowered in the presence of egg jelly. These results support the concept that the binding of jelly alters the calcium channel, so that calcium now enters at physiological pH. As high pH per se can induce the acrosome reaction, increased alkalinity may act by altering calcium permeability channels directly.

An alternative possibility is that the activation of the sperm is similar to that of the egg, in which Ca^{2+} is a primary trigger but acts secondarily to induce a cytoplasmic pH change, which then directly causes a number of metabolic activations. This possibility is suggested by the finding that H^+ ion is released from sperm when they are activated with jelly (Collins and Epel, 1977), similar to the release of H^+ ion from eggs when activated by sperm (Johnson et al., 1976).

Tilney and his colleagues (1978) have recently studied the acid release from sperm and have presented compelling evidence that proton release is related to the initiation and regulation of actin polymerization in the acrosomal process. They reached this conclusion from their earlier work, which indicated that (1) actin was in a nonfilamentous state in the acrosomal region, (2) polymerization into filaments was prevented by an inhibitory protein, and (3) that this inhibitory protein was removed (and polymerization occurred) by increasing the pH from 6.4 to 8.0 (Tilney, 1976). Next, these investigators wondered whether polymerization in vivo was also induced by an increase in pH. They found that ionophores such as A23187 and X537A induced release of acid and concomitant actin polymerization. In the presence of Ca^{2+}, membrane fusion also occurred, whereas in the absence of Ca^{2+} only actin polymerization took place. Their results are consistent with the hypothesis that changes in Ca^{2+} initiate acrosomal exocytosis and that proton efflux initiates actin polymerization. The exact sequence of events (e.g., Ca^{2+} first, followed by a change in proton permeability) is not yet known.

Additional evidence shows that the calcium requirement is for membrane fusion because if egg jelly is added to sperm in the absence of calcium, no changes in the sperm membranes can be detected with the transmission electron microscope (Collins and Epel, 1977). This suggests that calcium is required for membrane fusion and that fusion might be a prerequisite for the subsequent steps of the acrosome reaction. The later reorganization of the apical region of the sperm might ensue entirely from the subsequent actin polymerization. This polymerization and the resulting exposure of membrane, lysins, and bindins might be programmed so that all components are in position for the next phase of fertilization, the attachment of sperm to egg.

2.2.3. Other changes induced by egg jelly

Although not related to fusion per se, the interaction between sperm and jelly also leads to profound alterations in behavior (Collins, 1976) and respiration (Ohtake, 1976) of the sperm. The behavioral alteration results in an apparent agglutination which is actually an aggregation of the sperm to some common locus (Collins, 1976). This behavioral change does not require exocytosis as it can occur in low concentrations of Ca^{2+}, which prevent the acrosome reaction but which appear to affect the behavioral change. Also, exocytosis can be induced as for example with A23187 without producing the behavioral change.

Incubation of sperm in egg jelly also results in a large increase in the synthesis of cyclic nucleotides (Garbers and Hardman, 1976). It is not known whether this is related to the exocytosis of the acrosome vesicle, the behavior alteration, or some other parameter.

3. Sperm-egg binding

Attachment or binding of the sperm to the egg surface or to some outer layer of the egg is a prerequisite for fertilization. The exact binding site will vary with the organism, and the nature of the site is also related to the mechanisms of incorporation of the sperm into the egg. One extreme example is seen in the sea cucumber, whose sperm attach to an outer layer 60 to 90 μm from the egg surface. The elongated acrosomal filament passes through this layer and ultimately makes contact with the egg; the contraction of this filament then draws the sperm toward the egg as a part of the fusion process (Colwin et al., 1973). A similar phenomenon, on a smaller scale, is seen in the eggs of *Urechis;* here sperm attach to the vitelline layer or chorion and send their acrosomal process through the 5 to 10 μm perivitelline space to the egg surface; incorporation involves an extension of egg cytoplasm (the fertilization cone) which envelopes the sperm akin to phagocytosis (Gould-Somero et al., 1977). The other extreme is seen in the sea urchin egg: the sperm in this form binds to the vitelline layer, which is tightly apposed to the egg's plasma membrane.

Irrespective of these variations, the first preliminary to fusion is the attachment or binding of the sperm to an outer egg coat. This binding exhibits a high degree of species specificity (Summers and Hylander, 1976). Also, this binding is very stable; sperm remain bound even though they continue to gyrate around their point of attachment. Thus the binding reaction has both specificity and stability, and studies on this binding might provide important insights into the general mechanisms of intercellular adhesion. The gamete adhesion system is also ideal since gram quantities of sperm and eggs can be obtained.

3.1. Characteristics of sperm-egg binding

3.1.1. Evidence for two types of binding

Electron micrograph studies on sperm binding generally reveal that sperm attached to eggs have undergone an acrosome reaction. In echinoid sperm, attachment appears to be via a protein that coats the acrosomal process and is referred to as bindin because of its probable role in sperm-egg binding (Glabe and Vacquier, 1977b; Vacquier and Moy, 1977). Occasionally however, one sees sea urchin sperm that have not undergone extrusion of the acrosome process, but have experienced the exocytotic release of the acrosome vesicle and are attached to eggs (Fig. 6). This apparent binding might relate to the aforementioned possibility that the induction of the acrosome reaction might also occur at the surface of the vitelline layer; in these cases the binding of the sperm would be a preliminary to extrusion of the acrosome process.

These observations suggest that sperm have two binding sites which can interact with components of the egg surface. One site is on the unreacted sperm surface, and its interaction with the vitelline layer would lead to the acrosome reaction. The nature and role of these receptors on sperm and egg are completely unknown but might account for the puzzling fact that eggs whose jelly has been removed can still be fertilized. The second binding site, between bindin and vitelline layer, is the one most often encountered in fertilization and the following sections are concerned with this type of binding.

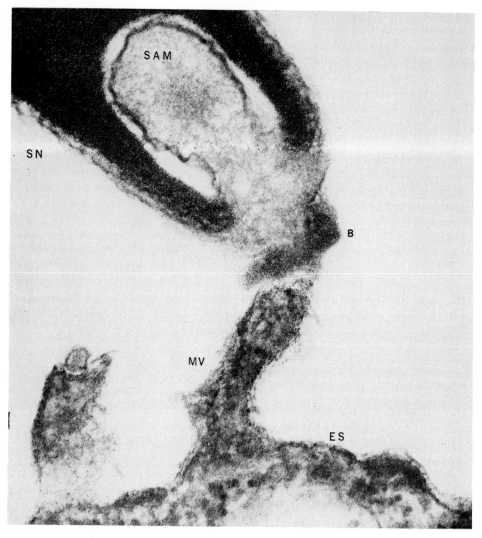

Fig. 6. Sperm of *S. purpuratus* possibly fused with the egg before extrusion of the acrosome process. Subsequent extrusion of the process would "inject" the egg with a bundle of actin filaments which may function in incorporation of the sperm by the egg. Fixation occurred at 7 sec after insemination. SAM, subacrosome material; SN, chromatin of sperm nucleus; B, bindin of obliquely sectioned mass of bindin from opened acrosome vesicle; MV, microvillus of egg, ES = egg surface. ×93,100. (Micrograph courtesy of Grudin and Vacquier, unpublished.)

3.1.2. The kinetics of sperm binding

Observations of fertilization in most species of invertebrates reveal that a great number of sperm attach to the outer envelope of the egg. The most dramatic example is that of the horseshoe crab *Limulus polyphemus,* where 10^6 sperm can attach to each egg (Brown, 1976).

At least several thousand sperm can attach to the sea urchin egg (Tegner and Epel, 1973; Vacquier and Payne, 1973) and to the egg of the Palolo worm *Tylorrhynchus heterochaetus* (Osanai, 1976). Thus sperm binding to an outer layer or vitelline layer is a general phenomenon of fertilization found in diverse animal groups.

A simple technique has been developed to quantitate the binding of sea urchin sperm to the vitelline layer as a function of time following the addition of sperm to eggs (Vacquier and Payne, 1973). The results of a sperm binding determination using this method are presented in Fig. 7. Sperm bind to eggs from 0 to 25 seconds after insemination. During that time (most probably by 3 sec after insemination; see Baker and Presley, 1969; Steinhardt et al., 1971; Jaffe, 1976) an interaction (fusion?) between the fertilizing sperm and egg results in a rapid membrane depolarization. The cortical granule reaction begins at 25 seconds and continues to 50 seconds, during which the cortical granules of the egg undergo exocytosis, releasing proteases that digest the bonds between the nonfertilizing sperm and the vitelline envelope (Vacquier et al., 1973; Carroll and Epel,

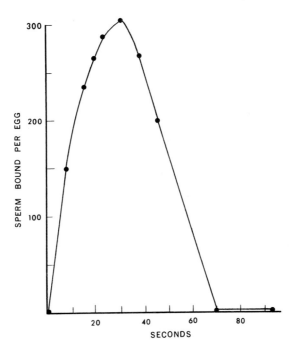

Fig. 7. Quantitation of sperm adhesion of sea urchin eggs. Sperm were mixed with eggs at time zero, samples were removed at various times, fixed in formaldehyde-sea water, and the number of sperm remaining bound to the eggs was determined. Sperm binding is fairly linear until cortical granule exocytosis begins at 30 sec. From 30 to 60 sec the protease released from the cortical granules digests the sperm receptor sites on the vitelline layers, resulting in the release of bound sperm (for method of quantitation, see Vacquier and Payne, 1973).

1975). Thus the kinetics reflect the binding phase (0–25 sec) and the detachment phase (25–50 sec).

The number of sperm bound per egg as a function of available sperm shows saturation kinetics, suggesting that a distinct number of sperm binding sites (sperm receptors) may be present on the vitelline layer. The number of sperm bound per *Strongylocentrotus purpuratus* egg varied from 1,048 to 3,160 (mean, 1744, 13 determinations). Values for *Lytechinus pictus* ranged from 1,018 to 1,745 (mean, 1,373, 11 determinations). For an *L. pictus* egg, this would mean that at saturation there is one sperm bound per 28 μm^2. Because the sperm heads are only 1.3 μm in diameter, if the entire surface of the vitelline layer were equally receptive to sperm binding, one would expect a much greater number of bound sperm than is consistently observed (Vacquier and Payne, 1973). The reason sperm binding exhibits saturation kinetics is unclear. One possibility is a limited number of receptors. Alternative explanations are the loss of the substance that induces the acrosome reaction (such as egg jelly), release of enzymes or other substances from sperm, depletion of oxygen, and steric hindrance of sperm attachment by the sperm already attached.

Summers and Hylander (1975, 1976) have quantitated the number of echinoid sperm bound per egg in 11 attempted inseminations between species where cross-fertilization does not occur. They found that although the acrosome reaction occurs in the presence of foreign egg jelly, the attachment of the acrosome process to a foreign vitelline layer does not occur. This failure in sperm attachment is presumably caused by the lack of recognition of cell surface macromolecules between heterologous sperm and egg.

Where cross-fertilization does occur, there is heterologous sperm binding to the foreign eggs. For example, sperm of *L. pictus* bind in substantial numbers to eggs of *S. purpuratus* and 20% fertilization can be obtained (Glabe, unpublished). This low percentage of fertilization in the face of high sperm binding suggests that gamete fusion requires more than mere attachment. There may be barriers to gamete fusion in the mechanism of penetration of the vitelline layer; since interphyletic hybrids are possible (e.g., mollusk and echinoid) (Afzelius, 1972), there is probably little variation in the mechanisms of the membrane fusion reaction.

3.1.3. Conditions for binding

Using the quantitative gamete binding procedure, Vacquier and Payne (1973) also investigated various parameters of sperm-egg binding. The optimum pH for sperm attachment was 8 and very little sperm binding occurred at pH 7 and 9. Failure of sperm binding at pH 7 may reflect the failure of sperm to undergo an acrosome reaction (Dan, 1967). Failure of binding at pH 9 does not result from effects on the acrosome reaction, since this reaction is actually promoted at this pH (Decker et al., 1976; Gregg and Metz, 1976; Collins and Epel, 1977). Rather, the effects on binding probably may reflect effects of high pH on the binding molecules themselves. Binding also requires Ca^{2+}, and between 1 to 10 mM Ca^{2+} the rate of binding and final number of sperm bound is directly proportional to the Ca^{2+} concentration. It is unclear, however, whether the Ca^{2+} requirement is one for binding, for the acrosome reaction or the binding of jelly to sperm receptors.

3.2. The vitelline layer

The early work on the vitelline layer (VL) of sea urchin eggs has been reviewed by Runnström (1966). The VL is a thin (100–300 Å) glycoprotein layer intimately bonded to the egg plasma membrane. Tegner and Epel (1973, 1976) have described the morphology of the VL as seen with the SEM on the intact egg of several species (see Fig. 1). In S. purpuratus the VL exhibits a dense array of projections that are casts of the egg microvilli which underlie the VL (Fig. 1).

3.2.1. Isolation and characterization of the vitelline layer
A procedure has recently been developed for the isolation of VLs from unfertilized eggs (Glabe and Vacquier, 1977a) in which the isolated VLs retain their native morphology and arrangement of casts of the cytoplasmic microvilli. When living sperm are mixed with such isolated VLs, the sperm attach by their acrosome process to the outer surface of the VL even though both inner and outer VL surfaces are accessible to sperm. This indicates that the outer surface may have sperm receptor sites not present on the inner surface (Glabe and Vacquier, 1977a). Sperm of the Palolo worm will also attach only to the outer surface of VLs isolated from Palolo eggs (Osanai, 1976).

Polyacrylamide-gel electrophoresis of isolated sea urchin VLs dissolved in sodium dodecyl sulfate resolves the VLs into numerous proteins ranging from 12,000 to 200,000 daltons. A similar complex pattern has been seen in VLs isolated from Xenopus eggs (Wolf et al., 1976). Only two of the sea urchin VL components stain for carbohydrate, and then only faintly.

Isolated VLs are 90 to 95% protein and 3.5% carbohydrate. There is nothing unusual about their amino acid composition, and carbohydrate analysis yields fucose, mannose, galactose, glucose, xylose, glucosamine, galactosamine, and sialic acid (Glabe and Vacquier, 1977a). Vitelline layers isolated from eggs of the limpet Megathura are composed of 37% protein and 63% carbohydrate (Heller and Raftery, 1976a). Sixty-one mol percent of the amino acids are threonine residues. The carbohydrates are galactosamine, galactose, and fucose, and almost every threonine residue is linked to a hexose moiety. The VL of the snail Tegula has also been isolated and shown to be a glycoprotein rich in threonine (Haino and Kigawa, 1966).

3.2.2. Evidence for sperm receptors on the vitelline layer
Evidence for the presence of sperm receptors on the vitelline layer comes from the following observations:

1. Trypsinization of eggs decreases their fertilizability (Aketa et al., 1972). This effect may result from removal of a sperm receptor glycoprotein (Schmell et al., 1977; Vacquier and Moy, 1977).

2. Unfertilized eggs treated with cortical granule protease do not bind sperm and are incapable of being fertilized until the protease-altered vitelline layer is removed (Vacquier et al., 1973; Carroll and Epel, 1975).

3. Sperm attach only to the outer surfaces of isolated vitelline layers (Glabe and Vacquier, 1977a).

4. Sperm binding is species-specific (in most cases; Afzelius, 1972; Summers and Hylander, 1975, 1976).

5. Sperm binding exhibits apparent saturation kinetics (Vacquier and Payne, 1973).

6. Membranous fractions from lysates of unfertilized eggs compete with eggs for sperm and thus inhibit egg fertilization. This inhibition is species-specific (Aketa, 1973, 1975; Schmell et al., 1977).

7. Fertilization can be prevented by low concentrations of concanavalin A (Schmell et al., 1977; Veron and Shapiro, 1977).

Observation 1 involves the digestion of the vitelline surface of unfertilized eggs with pancreatic trypsin. Aketa and co-workers (1972) noted that treated eggs required a much greater number of sperm to obtain the same percentage of fertilization observed in untreated eggs. This has recently been repeated by Schmell and associates (1977), who treated eggs with 0.1 mg/ml trypsin for 60 minutes.

Observation 2 arises from the discovery that a protease activity is extruded from fertilizing sea urchin eggs, which functions in establishing the late block against polyspermy (Vacquier et al., 1972a, 1972b, 1973; Schuel et al., 1973; Carroll and Epel, 1975; Fodor et al., 1975). When unfertilized eggs are treated with this cortical granule exudate, they rapidly become incapable of binding sperm and of being fertilized (Vacquier et al., 1973; Carroll and Epel, 1975). This protease activity appears to alter the VL surface so that it cannot bind sperm, but the VL is still capable of elevation when the eggs are activated with ionophore A23187 (Carroll and Epel, 1975). It is hypothesized that this altera-tion involves the specific removal of sperm receptors from the outer surface of the vitelline layer.

We have previously discussed observations 3, 4, and 5 and will not recount them here. Observation 6 provides excellent evidence for the existence of sperm receptors on the egg surface. Schmell and collaborators (1977) found that fertili-zation of *Arbacia punctulata* eggs is quantitatively inhibited by the addition of a membranous fraction prepared from *A. punctulata* eggs. The same is true for *Strongylocentrotus purpuratus* eggs mixed with membranes prepared from *S. pur-puratus* eggs. However, when *A. punctulata* eggs are fertilized in the presence of *S. purpuratus* membranes (or *S. purpuratus* eggs fertilized in the presence of *A. punctulata* membranes), no inhibition of fertilization is observed. The results of these competition experiments provide strong evidence for the existence of species-specific sperm receptors on the egg surface. These workers also found that receptor activity could be removed by trypsin treatment and that the recep-tor binds to Con A. These data are consistent with the hypothesis that the sperm receptor molecule may be a surface glycoprotein (Aketa et al., 1972; Aketa, 1973; Schmell et al., 1977; Tsuzuki et al., 1977).

Observation 7 is a quantitative study of the effects of Con A on the inhibition of fertilization (Veron and Shapiro, 1977). A Scatchard plot revealed two binding sites, a high affinity site ($K_a = 8 \times 10^{-7}$ M) on the vitelline layer and a low affinity site ($K_a = 4 \times 10^{-6}$ M) on the plasma membrane. The loss of fertilizability was associated with binding to the vitelline layer (high affinity site). These observations correlate well with the finding of Schmell and co-workers (1977) that the sperm receptor binds to Con A.

A problem with the experiments of both Schmell and colleagues (1977) and Veron and Shapiro (1977) is that it is not clear whether the effects observed are on sperm binding or the triggering of the acrosome reaction. As noted earlier, the vitelline layer may be the normal inducer of the acrosome reaction in eggs of some species. Also, Aketa (1975) treated sperm with soluble VL protein of its own species and found it lost fertilizing capacity without losing its motility and without undergoing an acrosome reaction. It would be interesting to induce the acrosome reaction to see whether fertilization was still inhibited by the membrane fraction or by Con A.

3.3 Egg-binding proteins on the sperm surface

In both invertebrate and vertebrate sperm the acrosome reaction probably occurs before the sperm bind to the egg surface (Dan, 1967; Hartman et al., 1972; Inoue and Wolfe, 1975; Summers et al., 1975). In marine invertebrates the reaction involves the rupture of the acrosomal vesicle and the extrusion of the acrosomal process. After the acrosome reaction, the contents of the acrosomal vesicle form a coating on the surface of the acrosome process. For several years it has been hypothesized that this acrosomal granule material may be the substance which bonds the acrosome process to the vitelline layer (Colwin and Colwin, 1967; Dan, 1967; Metz, 1967; Summers et al., 1975; Collins, 1976; Decker et al., 1976).

3.3.1. Isolation of egg-binding proteins (bindins) from sperm

To study the problem of gamete recognition in greater depth, a method was developed to isolate the acrosome vesicles of sea urchin sperm. The isolation method yields fairly pure preparations of acrosome vesicles devoid of their limiting vesicle membrane, as shown in Figs. 8 and 9 (Vacquier and Moy, 1977). This granular matrix material, which appears to be the major component of the acrosome vesicle, when dissolved in SDS and subjected to polyacrylamide gel electrophoresis (Weber et al., 1972) migrates as a single major component of 30,500 daltons (Fig. 10). This band stains strongly for protein, but is completely negative for carbohydrate when stained with PAS (Fairbanks et al., 1971). The isolated vesicles (Figs. 8, 9) also fail to show any measurable carbohydrate content as measured with the anthrone reaction (Spiro, 1966) and phenol-sulfuric methods for determination of sugars (Dubois et al., 1956).

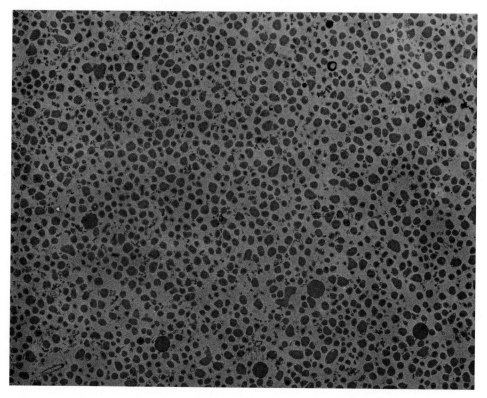

Fig. 8. Thin section of final pellet of isolated acrosome vesicles from *S. purpuratus* sperm. Larger circular objects are vesicles which have fused. ×6350. (From Vacquier and Moy, 1977.)

3.3.2. Proof of bindin involvement in sperm-egg adhesion

To determine whether this protein is involved in the sperm-egg bond, peroxidase-conjugated antibody to the protein was prepared and its location before and after the acrosome reaction was determined by the localization of diaminobenzidine precipitated on the cell surfaces (Vacquier and Moy, 1977; Moy et al., 1977). Before the acrosome reaction this protein appears to be localized in the acrosome vesicle (Fig. 11). Sperm were mixed with eggs, and after 15 seconds gamete interaction was stopped by formaldehyde fixation (Vacquier and Payne, 1973). The eggs, with sperm bound to their vitelline layers, were then reacted with the peroxidase-conjugated antibody. The diaminobenzidine precipitate appears on the vitelline surface of the egg, the acrosome process, and the anterior regions of the sperm plasma membrane (Fig. 12). Following the acrosome reaction and sperm-egg attachment, the acrosome vesicle protein not only coats the acrosome process but is also transferred to the area of the vitelline layer surrounding the point of contact of acrosome process and vitelline layer. This observation suggests this protein is involved in bonding the sperm to the vitelline surface and thus it has been proposed that the protein be named bindin (Vacquier and Moy, 1977).

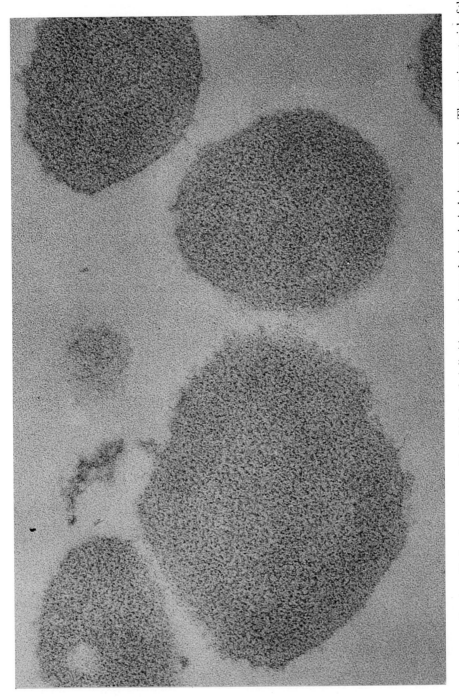

Fig. 9. High magnification of isolated acrosome vesicles which lose their limiting membrane during the isolation procedure. The matrix material of the vesicles is composed of a granular, insoluble material. ×192,000. (From Vacquier and Moy, 1977.)

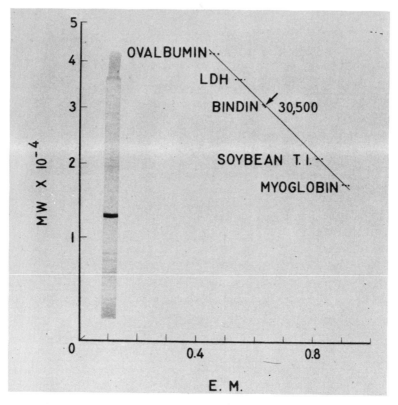

Fig. 10. Electrophoresis in 12.5% polyacrylamide gel containing 0.1% sodium dodecyl sulfate reveals the isolated acrosome vesicle matrix material to be a single component of apparently 30,500 daltons. On a dry weight basis this material is 100% protein. The name bindin has been tentatively proposed for this protein. Molecular weight standards are ovalbumin 43,000, lactic dehydrogenase 36,000, soybean trypsin inhibitor 23,000, and myoglobin 17,000. (Courtesy of V. D. Vacquier, unpublished.)

Is there other evidence that bindin mediates sperm attachment to sea urchin eggs? If bindin is the adhesive that binds sperm to the vitelline layer, one might expect bindin would agglutinate eggs. Furthermore, because sea urchin sperm binding is species-specific, one would predict that bindin would agglutinate only eggs of its own species. Bindin was isolated from *S. purpuratus* and *S. franciscanus* sperm. Eggs of these two species were then mixed with bindin and checked for agglutination after 5 minutes. The results of these experiments (Fig. 13) show that bindin agglutinates eggs and that the agglutination is species-specific (Glabe and Vacquier, 1977b). The proposal is made that bindin must recognize a species-specific receptor on the surface of the vitelline layer.

Other supporting evidence that bindin mediates sperm-egg adhesion comes from the following experiments. During cortical granule exocytosis, a protease (vitelline delaminase) cleaves the bonds between the vitelline layer and egg plasma membrane, allowing the vitelline layer to elevate and transform into the

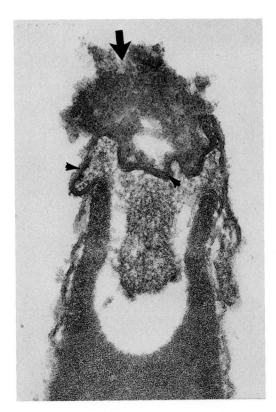

Fig. 11. Localization of peroxidase-conjugated rabbit antibindin in sea urchin sperm. Sperm treated with antibody after disruption of the plasma membrane and acrosome vesicle membrane by fixation in 3% formaldehyde in 50% glycerol. Bindin is exposed to peroxidase-conjugated antibody, and diaminobenzidine precipitate develops on the main mass of bindin (large arrow) and also on the surface of the inner acrosome membrane (small arrows). ×70,000. (Courtesy of Moy, Friend, and Vacquier, unpublished.)

fertilization envelope. A second protease (sperm receptor hydrolase) destroys the sperm receptor sites on the vitelline layer (Carroll and Epel, 1975). One would therefore expect that after the complete elevation of the vitelline envelope the bindin receptors would not be present, and that fertilized eggs would not agglutinate upon the addition of bindin. This is exactly the result that is obtained; when fertilized eggs are mixed with bindin, the eggs do not agglutinate. Also, bindin will agglutinate isolated vitelline layers. Trypsin-generated glycopeptides from egg surfaces destroy the ability of bindin to agglutinate eggs, and eggs that have been trypsinized are not agglutinated by bindin (Vacquier and Moy, 1977).

The finding that glycopeptides from egg surfaces inhibit egg-egg agglutination by bindin suggests the bindin receptor may be a carbohydrate moiety on the vitelline layer. Also consistent with this idea is that the ability of eggs to bind sperm is destroyed after periodate oxidation. This is best demonstrated on glutaraldehyde-fixed eggs, which retain their ability to bind sperm. When such eggs are treated with 0.1 M $NaIO_4$, pH 5.0, for 4 hours, they completely lose the ability to be agglutinated by bindin and also to bind living sperm (Vacquier, unpublished observations). The findings are also consistent with the work of Schmell and co-workers (1977) that the sperm receptor of the egg surface was trypsin-sensitive and species-specific, and that Con A bound to it.

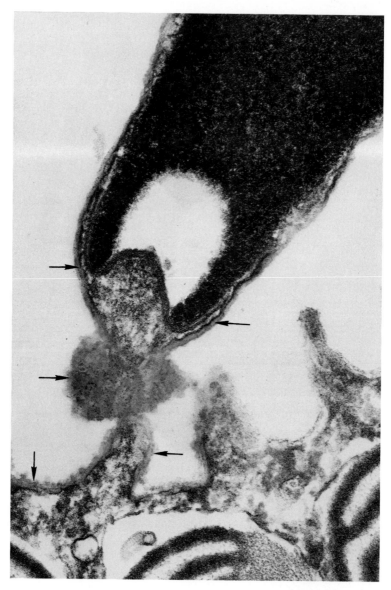

Fig. 12. Localization of bindin after formation of the sperm-egg bond. The peroxidase-conjugated antibody localizes bindin on the anterior membrane of the sperm and the egg microvillus to which the sperm is bound. Arrows point to diaminobenzidine precipitate on the cell surfaces. Note the microvillus on the right has no diaminobenzidine deposit. ×66,000. (Courtesy of Moy, Friend, and Vacquier, unpublished.)

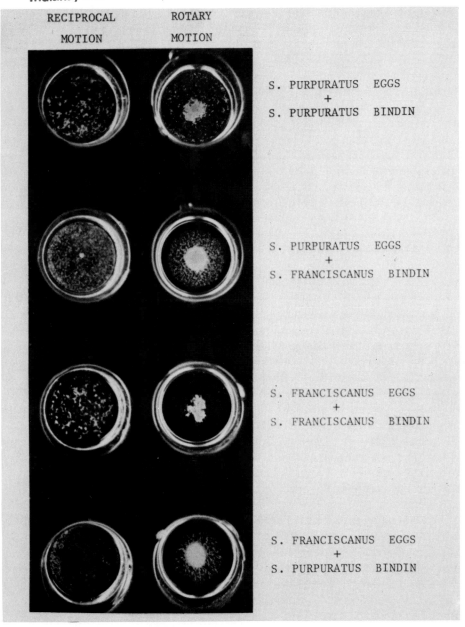

RECIPROCAL   ROTARY
MOTION   MOTION

S. PURPURATUS EGGS
+
S. PURPURATUS BINDIN

S. PURPURATUS EGGS
+
S. FRANCISCANUS BINDIN

S. FRANCISCANUS EGGS
+
S. FRANCISCANUS BINDIN

S. FRANCISCANUS EGGS
+
S. PURPURATUS BINDIN

Fig. 13. Species-specific agglutination of sea urchin eggs by bindin. Living, fresh, unfertilized eggs of *Strongylocentrotus purpuratus* and *S. franciscanus* were placed in wells of a microtiter plate containing a suspension of acrosome vesicles (bindin). Two different agitation motions were employed, rotary and reciprocal. After 3 to 5 min the plates were photographed using indirect illumination. Bindin agglutinated only its own species of eggs. This is evidence that species-specific receptors for bindin may exist on the vitelline layer of the egg (see Glabe and Vacquier, 1977b, for details; reproduced with permission of Macmillan Journals, Ltd).

Because sperm binding to the egg investment is a general phenomenon of fertilization it has been proposed that sperm bindins may be widely distributed throughout the animal kingdom (Vacquier and Moy, 1977). The immunological similarities of sea urchin bindin are now being studied by measuring the amount of cross-reaction between labeled anti-sea urchin bindin and sperm of various animal groups. To date it has been found that anti-sea urchin bindin will only cross-react with sperm of other echinoid species (Moy and Vacquier, work in progress); closely related echinoderms (e.g., starfish) do not cross-react.

3.3.3. Relevance to cell-cell adhesion studies

The evidence presented in the preceding section leads to the hypothesis that sea urchin sperm-egg interaction may be another example of cell-cell adhesion mediated by a protein-carbohydrate interaction. In this case the protein is bindin, which is released from the acrosome vesicle by the acrosome reaction and coats the acrosome process and the anterior plasma membrane of the sperm. The carbohydrate is probably the glycopeptide portion of a vitelline layer glycoprotein that appears to be present only on the outer vitelline surface (Glabe and Vacquier, 1977a; Tsuzuki et al., 1977).

Several other examples are known where intercellular recognition and adhesion are based on the interactions of cell surface proteins and carbohydrates. In yeast, a glycoprotein from one mating type agglutinates cells of the opposite mating type (Crandall and Brock, 1968; Yen and Ballou, 1974). In *Chlamydomonas,* interaction between + and − mating types depends on the presence of a glycoprotein and the interaction is blocked by proteases, Con A, or specific glycosidases (Wiese and Wiese, 1975). Recent work on plant fertilization shows that the reception of pollen by the surface of the stigma also depends on a protein-carbohydrate interaction (Knox et al., 1976).

In some protein-carbohydrate interactions involved in cellular adhesion, the protein has lectin-like activity and binds to specific terminal saccharides on cell surface glycoproteins. Examples include the aggregation factor from certain sponges that is specifically inhibited by uronic acid (Turner and Burger, 1973). In slime molds, proteins that bind specific carbohydrates appear to mediate the adhesion of aggregating amoebae (Rosen et al., 1974, 1975). Lectins are present on the plasma membranes of liver cells (Bowels and Kauss, 1976) and in the fusion-competent chick myoblasts in tissue culture (Nowak et al., 1976). Given these examples, sea urchin sperm-egg adhesion might also be a lectin-like interaction. To date, however, attempts to inhibit bindin-induced egg agglutination by the addition of common saccharides and saccharide derivatives have failed.

A very interesting case of gamete recognition, which must surely involve specific surface macromolecules, occurs in some hermaphroditic ascidians where the sperm from one individual do not fertilize the eggs of that same individual, that is, each individual is self-sterile (review, Metz, 1967). The block to gamete fusion is the chorion, since the egg can be fertilized by sperm from the same individual if the chorion is removed (Morgan, 1923). One possible explanation for this phenomenon is that each individual produces a component of the egg chorion which inhibits interaction with its own sperm. This component must vary enough among individuals so that it has no effect on sperm from a different individual. Given the microanalytical procedures now available, this interesting problem in gamete interaction may be amenable to experimental attack.

3.4. Passage of sperm through the vitelline layer

Once the sperm has bound to the vitelline layer, it must pass through it to become apposed to the plasma membrane as a preliminary to the fusion process. The nature of this "passage" is poorly understood. One possibility is that the sperm acrosomal process might literally bore through it if the momentum of the sperm were sufficiently great. Alternatively, sperm might attach to the VL before the acrosomal reaction; the force of the subsequent acrosomal process formation might then push it through the vitelline layer and perhaps against the plasma membrane as a part of the fusion process. Finally, the passage may be mediated by lysins. Levine and co-workers (1977) have recently reported a membrane-bound trypsin-like protease in sea urchin sperm. This enzyme is only seen after the acrosome reaction has occurred.

The presence of sperm lysins in molluscan sperm was demonstrated many years ago, and some of these enzymes have been purified (Hauschka, 1963; Heller and Raftery, 1973). The molecular action of one group of lysins from the keyhole limpet *Megathura* on isolated vitelline layers of that species has also been studied (Heller and Raftery, 1976b).

3.5. The role of sperm binding in fertilization

Although sperm-egg binding is a part of the fertilization process, it may not be a prerequisite for it. This is suggested by (1) studies on fertilization when the vitelline layer is removed or altered by proteolysis, and (2) studies on the role of the vitelline layer in preventing interspecies hybridization.

3.5.1. Effects of removing the vitelline layer
When eggs are treated with trypsin or other wide-spectrum proteolytic enzymes, the vitelline layer proteins are sufficiently digested that it does not elevate as the fertilization envelope following fertilization or activation (Berg, 1967; Epel, 1970). As noted, this treatment also removes the sperm receptors. Yet these treated eggs can be fertilized, suggesting that receptors are not required (Hultin, 1948; Bohus-Jensen, 1953).

Seemingly contradictory to the results above, eggs cannot be fertilized if the sperm receptors are specifically altered with the cortical granule proteases (Vacquier et al., 1973; Carroll and Epel, 1975). As noted earlier, there are trypsin-like enzymes contained within the cortical granules and released at fertilization. One (sperm receptor hydrolase) specifically alters the sperm receptors so that sperm cannot attach to the vitelline layer. The other (vitelline layer delaminase) detaches the vitelline layer from the plasma membrane as part of the process of vitelline layer-fertilization envelope formation (Carroll and Epel, 1975). The proteolysis carried out by these proteases is highly modulated, since the structural integrity of the vitelline layer is unimpaired as regards elevation and transformation into the fertilization envelope. Thus, if eggs treated with these proteases are activated with ionophore A23187, an apparently normal fertilization envelope elevates (Carroll and Epel, 1975). Consistent with these findings are results of studies with I^{125}-labeled egg surfaces. Treatment with trypsin or pronase removes 90% of these counts. Treatment with cortical granule proteases removes less than 5% of the surface counts (Carroll et al., 1977).

To recap these seemingly conflicting results, treatment with trypsin (1) removes sperm receptors and (2) removes the vitelline layer; such eggs *can* be fertilized. Treatment with cortical granule proteases (1) removes sperm receptors, but (2) does not remove the vitelline layer; such eggs *cannot* be fertilized. Fertilizability can be restored, however, if the overlying vitelline layer is removed with pronase or dithiothreitol.

One interpretation of these results is that the plasma membrane of the egg is a highly fusible membrane protected from promiscuous fusion by the VL. For fertilization to occur, therefore, sperm must first pass through the overlying vitelline layer and a prerequisite for this passage is binding or attachment. This binding allows proteases or lysins associated with the acrosomal process to digest through the layer and thus gain access to the plasma membrane. When eggs are treated with trypsin or dithiothreitol, the vitelline layer barrier is removed and fertilization can occur. When eggs are treated with cortical granule protease, the vitelline layer remains as a barrier; sperm cannot bind and therefore cannot get through. When the layer is removed, the plasma membrane surface is available and fertilization can now be effected.

3.5.2. Role of the vitelline layer in preventing hybridization

Studies on interspecies hybridization also support this concept and suggest that one role of the sperm receptor-VL barrier is as a bar to hybridization. In general, there appear to be formidable blocks to interspecies hybridization. If the species coexist in environments that overlap geographically, the barriers to hybridization are high. If the two species are from geographically different environments, there seem to be fewer barriers to hybridization. For example, the sand dollar *Dendraster excentricus* from San Diego, California hybridizes with the sand dollar *Encope grandis* from La Paz, Baja California, Mexico, and both possible gamete combinations result in advanced hybrid larvae (Vacquier and Epel, unpublished observations). In the former situation sperm do not attach to the vitelline layer, and it therefore appears that the lesion in fertilization is sperm-egg adhesion. In the latter situation, sperm can often attach to and fertilize the foreign egg.

This block to hybridization can be partially removed by treating the eggs with proteases (Hultin, 1948; Bohus-Jensen, 1953). This observation, which has been seen in a number of species, suggests that the vitelline layer is a filter at the level of sperm-egg binding and that species-specific fertilization is achieved through specificity of binding of sperm to the layer. If binding does not occur, fertilization will not ensue; if this barrier is removed or altered (as by proteases) fusion can take place.

In conclusion, binding to the egg vitelline layer is normally a prerequisite to fertilization, probably as part of the mechanism for the lysin-mediated passage through the vitelline layer. Thus the sperm receptors function as promoters of species-specific fertilization and barriers to interspecies hybridization.

3.6. Other roles of the vitelline layer

The vitelline layer may have other roles beside preventing interspecies hybridization. One such role may be ordering the patterns of microvilli on the egg surface. This is suggested by SEM observations of egg surfaces after removal of the VL; the microvilli of these eggs become disarrayed and the patterning is lost (Tegner, 1974; Eddy and Shapiro, 1976).

Eggs whose VLs have been removed also become flaccid, suggesting that another role of this layer is to maintain the structural rigidity of the egg. This effect is especially apparent in eggs of *Spisula*, which can actually become ameboid when the VL is removed (Rebhun, 1962; Ziomek and Epel, 1975).

3.7. Role of sperm motility in binding

Sperm motility appears to be required for binding to the egg surface. When motility is prevented, as by incubation in respiratory inhibitors or by removal of the sperm tails, binding is severely reduced. This reduction in binding is seen even when the acrosome reaction is induced by egg jelly prior to adding the metabolic inhibitors; thus the loss of binding does not result from effects on the acrosome reaction.

The reason for decreased binding is unclear. One possibility is that motility is required to overcome an "energy barrier" in the apposition of bindin to the egg surface. For example, one might imagine that in order to bind the bindin must hit the surface with a minimum impact. A related possibility is that the continued motion of the sperm pushes the bindin into the maximum available space so that the binding strength is maximized. Finally, motility may be required to insure that the acrosomal process is pointed in the proper direction relative to the microvilli of the egg surface.

4. Sperm-egg fusion

The incorporation of the sperm into the egg is obviously a multistep phenomenon. The preliminaries, discussed in the previous sections, include the induction of the acrosome reaction in the sperm and attachment to and passage through the VL to the egg surface. The fusion process itself is probably rapid and is then followed by the uptake of the sperm into the egg. The former might involve protein-lipid interactions in the microenvironment of the juxtaposed sperm and egg plasma membranes. The subsequent uptake probably involves a response of the egg to fusion and includes the mobilization of subcortical actin and surface microvilli to pull the sperm into the egg or to envelop and engulf the sperm.

The dissection of fusion is difficult to study in eggs of annelids and enteropneusts, since electron microscope studies indicate that fusion occurs within 4 seconds after sperm addition (Colwin and Colwin, 1967). The exact time of fusion in the sea urchin egg is not known, but electron microscope studies do not reveal any fusion until 10 to 20 seconds after insemination (Franklin, 1965). Such studies, however, are suspect since it is difficult to find the fertilizing sperm among the many sperm that attach to the egg, but behavioral studies also suggest that fusion is late relative to attachment. Cinemicrophotographic studies of Epel and co-workers (1977) show that the behavior of the fertilizing sperm cannot be initially differentiated from the nonsuccessful sperm until about 20 seconds after attachment. At this time the fertilizing sperm suddenly stops its movement and literally stands up perpendicular to the egg's surface, whereas the other unsuccessful suitors continue their original behavior. This suggests that fusion had not occurred until this time.

Experimental evidence for the multi-step nature of sperm incorporation in sea urchins has been provided by Baker and Presley (1969) and Presley and Baker (1970). They analyzed the effects of several fertilization inhibitors and found that these compounds arrested sperm incorporation at different times between the addition of sperm and the cortical reaction. For example, detergents stopped sperm-egg incorporation very early, whereas high potassium concentration stopped sperm-egg incorporation relatively late.

In the following sections, we first examine the fusibility of sperm and eggs, sites of fusion, and factors affecting sperm-egg fusion, including natural factors (the block to polyspermy, ions, and enzymes) and experimental factors (effects of inhibitors or promoters of fusion). We close with a consideration of the incorporation process.

4.1. Specific sites for fusion

If there were specific sites associated with fertilization, microscopic and chemical analyses of these specialized areas could throw light on the mechanism of fusion. The apical end of the acrosomal process appears to be specialized for fusion, and studies on this membrane could be interesting. Unfortunately, most eggs do not possess such specific areas. Although fusion occurs just under the micropyle in fish eggs (Yamamoto, 1961), it can occur anywhere on the surface of these eggs if the chorion is removed. In some eggs, such as those of tunicates, fusion tends to occur in the vegetal pole (Monroy et al., 1973). Lectin binding in these eggs is uniform, however, and so far no qualitative differences have been observed between vegetal and animal halves.

As noted earlier, eggs tend to have microvilli and fusion may be facilitated at the microvilli. This seems to be the case in the sea urchin (preliminary observations of F. Collins; unpublished work of David Chase), but this observation needs further examination. This would be consistent with the concept that fusion occurs in areas that minimize energy barriers (Poste and Allison, 1973).

4.2. Sperm and egg as fusible cells

We now consider whether the sperm and the egg are inherently fusible cells. Normally cells do not fuse with each other, and adjacent cells in tissues usually remain juxtaposed to each other; they might communicate ionically via intercellular connections but with no direct intermingling of their respective cytoplasms. Thus it appears quite difficult to fuse cells, and special techniques and treatments must be used to effect fusion.

The sperm and the egg are exceptions to the generalization above and suggest that these two cells might be specialized for rapid fusion. If this viewpoint is correct, then there must be controls on sperm-egg fusion that prevent fertilization of eggs by foreign sperm (i.e., a block to interspecies hybridization) and inhibit fusion of more than one sperm with the egg (i.e., a block to polyspermy). Both types of blocks exist, but their existence per se does not prove that gametes are fusible. However, several other lines of evidence suggest the validity of this viewpoint, as follows:

1. Can sperm fuse with each other?
    Sperm that have undergone an acrosome reaction will adhere to each other by their bindins and will occasionally fuse with each other. Transmission electron micrographs show that such sperm have actually fused between the acrosomal process and the midpiece (Fig. 14; Collins, 1976).
2. Can sperm fuse with eggs of their own species?
    This may seem like a foolish question, but the answer is interesting. Fusion is limited; generally only one sperm fuses with an egg. As is described in detail further on (section 4.3), there is a rapid alteration in sperm receptivity that prevents or blocks sperm entry as soon as a sperm has fused with an egg.
3. Can sperm fuse with eggs of other species?
    As noted earlier, there are formidable blocks to interspecific hybridization; the mechanism of this block appears to be at the level of the vitelline layer. Sperm cannot bind to the layer, and hence cannot pass through it to effect contact (fusion) with the plasma membrane. If the VL is removed, so that the egg plasma membrane is now exposed, fusion of foreign sperm can occur.

These studies suggest that the sperm and the egg, unlike other cells, are highly fusible. In light of this conclusion, the structures and behavior of the sperm and the egg can be seen to have evolved to prevent these fusible cells from fusing precociously. For example, the acrosomal process with its fusible membrane is not exposed until the sperm is quite near the egg. Similarly, the egg is covered

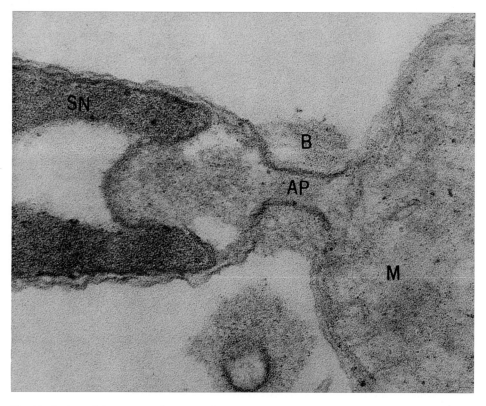

Fig. 14. Fusion of the acrosomal process of an *L. pictus* sperm with the midpiece mitochondrion of another sperm. AP, acrosomal process; SN, sperm nucleus; B, bindin; M, midpiece mitochondrion. ×168,000. (From Collins, 1976; reproduced with permission of Academic Press.)

with the VL to prevent random fusion with sperm or any other cells the egg might enounter.

4.3. The block to polyspermy—a natural regulator of fusion

One of the earliest responses of the egg to fertilization is a very rapid alteration in receptivity of the egg to the sperm, referred to as the block to polyspermy. This block is necessary to preserve the diploid genome of the organism, and therefore powerful mechanisms have evolved to ensure that only one sperm enters. It would be interesting to know how the egg prevents this additional fusion; such information could also provide insights into the normal fusion process.

4.3.1. Kinetics of the block to polyspermy
The standard means for determining the time of the block to polyspermy was developed by Rothschild and Swann (1952) and involves inseminating eggs with a light dose of sperm, followed by an extremely heavy dose of sperm at various

intervals thereafter (the heavy dosage being one which would induce polyspermy if added to eggs initially). One then determines exactly when the eggs no longer become polyspermic.

An example of such a study, from the work of Ziomek and Epel (1975) on eggs of the clam, *Spisula,* is shown in Fig. 15. This figure shows an extremely rapid loss of receptivity—50% reduced by 5 seconds and completely refractive to the addition of sperm by 15 seconds. Similar results have been obtained in *Urechis* eggs (Paul, 1975).

With sea urchin eggs, the original kinetic studies of Rothschild and Swann also indicated a rapid loss in receptivity, referred to as the "fast block to polyspermy." However, this block was incomplete and some sperm entered the eggs until the cortical reaction was complete. At this time sperm entry ceased completely and the cortical block, which is a mechanical barrier, is referred to as the "late block of polyspermy."

Byrd and Collins (1975) reexamined the kinetics of sperm-egg fusion using extremely high concentrations of sperm so that up to ten sperm could fuse with each egg before the cortical reaction. The rate of sperm entry was linear for both the fertilizing and the supernumerary bound sperm, suggesting that there was no

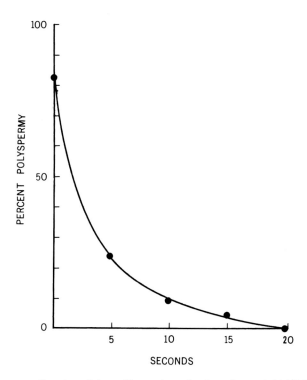

Fig. 15. Polyspermy as a function of time of heavy insemination after an initial light insemination at 0 in *Spisula solidissima* eggs. The eggs have a very rapid block to polyspermy. (Reproduced with permission of Science from Ziomek and Epel, 1975.)

fast block to polyspermy but rather that sperm-egg fusion was modulated and that polyspermy was not normally seen since very few sperm could fuse with the egg.

This particular viewpoint has since been reexamined and reinterpreted by Collins, who found that if sufficiently dilute sperm suspensions are added to eggs, sperm attach to and fertilize the eggs over a range of time so that the eggs undergo the cortical reaction asynchronously. As the sperm concentration is increased, more eggs exhibit a cortical reaction soon after sperm addition, until a sperm concentration is reached at which 100% of the fertilizable eggs begin the cortical reaction simultaneously and in the least possible time after sperm addition (about 20 sec, in *S. purpuratus*). At this sperm concentration the eggs may still be monospermic or slightly polyspermic. Therefore any increase in the sperm concentration should produce polyspermy, if there is no change in receptivity to sperm before the cortical reaction (i.e., no fast block). However, Collins (manuscript in preparation) found that the sperm concentration which just gave 100% cortical reaction in the minimum time possible could be increased by at least a factor of 20 without inducing additional polyspermy. These results suggest there may actually be a rapid alteration in sperm receptivity, but that this alteration or "early block to polyspermy" can be overwhelmed at high sperm-egg ratios.

4.3.2. Mechanism of the early block to polyspermy in sea urchins
The important work of Jaffe (1976) indicates that the rapid loss in sperm receptivity (i.e., the "fast block") results from depolarization of the plasma membrane. Using microelectrodes, she observed an extremely rapid depolarization (which began within 3 sec after adding sperm) during which the membrane potential went from -60 to between -10 and $+10$. The rate of depolarization varied among different eggs *(S. purpuratus)*. Eggs that depolarized rapidly were not polyspermic; eggs that depolarized more slowly became polyspermic. In these eggs the membrane potential might go to -10, remain there for several seconds, and then exhibit a small depolarization to about $+10$. This suggested to Jaffe that membrane depolarization was preventing sperm from entering the egg, but that it had to get to at least $+10$ to completely block out sperm. If it remained around -10, a second sperm could fuse with the egg, cause the egg to become polyspermic, and also account for the secondary rise in membrane potential.

Jaffe (1976) tested this hypothesis by applying a voltage clamp to the eggs. If the membrane potential was kept above zero, sperm would attach to the egg but would not fuse with it. If the potential was dropped below zero, the attached sperm rapidly fused with the egg and elicited a normal cortical reaction. These elegant experiments provide strong evidence that the change in membrane potential is responsible for the fast block to polyspermy. The rise in membrane potential in the eggs of *Lytechinus pictus* is not as rapid. It is interesting that these eggs are much more liable to polyspermy. The mechanism by which the membrane depolarization prevents additional sperm incorporation is unclear. Referring to the aforementioned idea that there are multiple steps to sperm entry, a possibility pointed out by Paul (1970) is that the fertilizing sperm must pass through a certain number of these steps before it will be incorporated by the egg.

The important question is the nature of this electrically sensitive step. One possibility is that the membrane depolarization triggers a change in subcortical microfilaments. The idea here might be that all attached sperm have fused with the egg but that further incorporation of the sperm is precluded, for example, by interfering with recruitment of microfilaments under the attached sperm. Another possibility is that the depolarization results in a conformational change in the receptors so that sperm can no longer bind to these receptors (Fig. 16). We might imagine that there are electrically coupled receptors, perhaps related to the receptors involved in sperm-egg adhesion, and that binding of a sperm to these receptors initiates cortical events leading to the incorporation of the sperm into the egg. The first sperm to bind is destined to be incorporated, but this binding also triggers membrane depolarization so that no additional sperm can attach to the receptor in a manner that will lead to sperm-egg incorporation (Fig. 16). A final mechanism might be that the membrane depolarization simply results in an

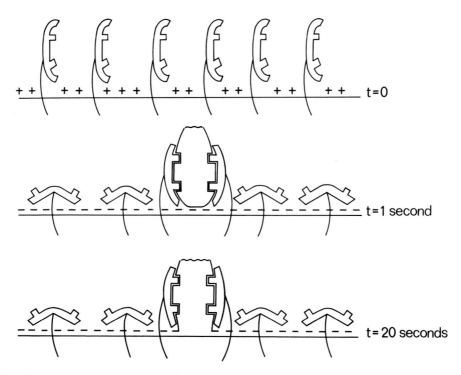

Fig. 16. A model for how altered surface charges might change sperm receptors to effect a block to polyspermy. At zero time (t = 0) all receptors are in a functional state capable of interacting with a sperm. After membrane depolarization (t = 1) those receptors that have not attached to a sperm are altered, because of the depolarization and charge difference, into a nonfunctional configuration. The sperm, which has previously interacted with the receptors, undergoes further changes and at some later time (t = 20 seconds) fuses with the egg (Courtesy of Epel, unpublished.)

altered surface charge which results in an electrostatic barrier to the sperm and egg membrane coming close enough together to effect fusion.

4.3.3. Mechanism of polyspermy prevention in Spisula

As noted earlier, there is a rapid block to polyspermy in eggs of *Spisula;* however, the mechanism of preventing polyspermy in these forms appears to be quite different from that in the sea urchin. In *Spisula*, the block to polyspermy is prevented by low doses (0.25 μg/ml) of cytochalasin B (Ziomek and Epel, 1975). This suggests that some membrane process sensitive to cytochalasin is involved in the fast block. An obvious possibility is that the passage of the first sperm into the egg triggers an alteration in subcortical microfilaments that precludes other sperm from entering. When this alteration is inhibited with cytochalasin B, sperm continue to enter the egg. The situation is therefore quite different from sea urchin and *Urechis* eggs, where sperm incorporation is actually prevented by cytochalasin (section 5).

This apparent variation in mechanism may at first seem disconcerting. Also, the egg of *Spisula* differs in requiring extracellular calcium to be activated by ionophore (Schuetz, 1975), as opposed to the eggs of other organisms where intracellular reserves are used (Steinhardt et al., 1974). One possibility is that the fertilizing sperm initiates a rapid change in calcium permeability, which then acts directly on microfilaments to effect a block to polyspermy. This would differ from eggs of other organisms, where the fertilizing sperm initiates a change in permeability to sodium to effect membrane depolarization; calcium changes occur later and result from release from intracellular stores rather than from increased uptake from the extracellular medium.

4.3.4. Mechanism of the late block to polyspermy

In the sea urchin the early block to polyspermy resulting from membrane depolarization does not completely stop polyspermy; kinetic studies show that the establishment of a complete block is coincidental with the cortical reaction (Rothschild and Swann, 1952). The mechanism of this late block to polyspermy is related to the elevation of the vitelline layer as it transforms into the fertilization envelope. This transformation is mediated through products released from the cortical granules, including two trypsin-like proteases (Carroll and Epel, 1975), an ovoperoxidase catalyzing protein cross-linking (Foerder et al., 1977) a β-1, 3-glucanase (Epel et al., 1969) and structural proteins (Bryan, 1970). The trypsin-like enzymes detach the vitelline layer from the egg surface and this action leads to the elevation of the VL; the ovoperoxidase catalyzes the formation of tyrosine dimers, which results in the hardening of the fertilization envelope. The function of the β-1, 3-glucanase is unknown at this time. A critical part of the polyspermy-preventing action is the destruction of sperm receptors of the vitelline layer by a specific sperm receptor hydrolase, which is one of the trypsin-like enzymes. The nature of these enzymes and reactions has been reviewed elsewhere (Epel, 1978) and will not be considered in further detail here.

4.4. Experimental means of promoting fusion

Another approach to understanding sperm-egg fusion is to study agents that promote fusion. As the sperm and the egg are probably highly fusible cells, the mechanism of these agents might be to inhibit membrane depolarization or to prevent the important consequences of depolarization that result in the block to polyspermy. Thus, these agents might not necessarily act to promote fusion but may act to keep the egg in its fusible state. We call these agents inducers of polyspermy, but their mode of action may be related to preventing the block to polyspermy.

The best agents for inducing polyspermy, such as nicotine, ammonia and the amine anesthetics, share the common property of having amine groups. Studies on the kinetics of sperm entry in nicotine and ammonia indicate that the rate of sperm-egg fusion is enhanced and that sperm continue to enter the egg at a linear rate until the time of the cortical reaction (Collins and Byrd, 1978). Nicotine might be expected to have effects on electrical properties of the cells, but the concentrations used are extremely high (mM) and hence their action may be nonspecific. An important clue as to how these agents might act has come from recent studies of Paul and collaborators (1976) and Johnson and co-workers (1976), who have found that the use of nicotine and ammonia results in a rapid and large increase in cytoplasmic pH (from 6.6 to > 7). This increased pH is a normal consequence of fertilization, but usually begins about 60 seconds after fertilization and takes about 4 minutes for the change to be complete (Johnson et al., 1976). Concentrations of ammonia and nicotine that induce polyspermy, however, result in an almost immediate pH change.

One hypothesis is that the pH change induced by these agents leads to polyspermy by inhibiting the operation of the normal polyspermy-preventing mechanism. For example, if the membrane depolarization alters the charge on sperm receptors, then the increased cytoplasmic pH might in effect neutralize this difference in charge. Alternatively, if the membrane depolarization triggers some secondary change (e.g., a change in microfilaments), this change might be precluded by the high pH. Nevertheless, it is important to note that normally the pH change does not begin until after the cortical reaction has taken place. Thus, if the normal cytoplasmic pH change throws the egg back into a sperm-receptive condition, it would make no difference, since the cortical reaction would have occurred and the elevation of the fertilization envelope around the egg would act as a barrier to any additional sperm.

Another obvious hypothesis is that these agents act to prevent membrane depolarization. This theory could be tested by examining the effects of the various agents on membrane depolarization.

The mechanism of action of the amine anesthetics, such as procaine, is more complex. These drugs induce the pH change, but because of their effects on Ca^{2+} metabolism they also prevent the cortical reaction. The result is that sperm enter the egg continuously, as there is no cortical reaction (Collins and Byrd,

1978). The rate of sperm entry, however, is not enhanced over normal eggs, even though the pH change has occurred in the procaine-treated eggs. This might indicate that a pH increase does not necessarily lead to increased fusion; however, procaine also affects the acrosome reaction so it is difficult to sort out these two effects from each other.

4.5. Fertilization inhibitors

Another approach to understanding fusion is to study the effects of agents that block fertilization (Metz, 1961). Agents reported to prevent fusion include detergents (Hagstrom and Hagstrom, 1954), high potassium concentrations (Baker and Presley, 1969), absence of magnesium (Sano and Mohri, 1976), uranyl ions (Presley and Baker, 1970) cytochalasin B (Gould-Somero et al., 1977; Longo et al., 1977), fluorescein dyes (Carroll and Levitan, 1977), lectins (Veron and Shapiro, 1977), and algal extracts and dermal secretions from sea urchins (Metz, 1967). The latter three agents probably affect sperm-egg binding and are of no interest in membrane fusion. Cytochalasin B affects incorporation and is discussed in detail in section 5. The locus of action of the ions is unknown; detergents probably affect unbound sperm; once sperm have attached or fused with the egg the detergents have no effect.

The inhibition of fertilization by fluorescein dye has just been reported by Carroll and Levitan (1977). This inhibitions is reversible, does not appear to affect the sperm, but causes a hyperpolarization of the egg plasma membrane. Although sperm binding is decreased, it is also probable that the dye is acting at some later stage of the sperm-egg fusion process. The exact mechanism of inhibition is still unclear, and further studies of this drug should be very interesting.

4.6. Ionic requirements for fertilization

Insights into the fusion process might be obtained from studies on the role of ions in fertilization. The major cations in sea water are sodium, potassium, calcium, and magnesium and the major anions are chloride, sulfate, and bicarbonate. Experiments on the role of these ions for fertilization can be assessed by inseminating in the absence of the ion, as long as sperm motility or the acrosome reaction is not interfered with. For example, it is difficult to effect fertilization in the absence of sodium because sperm motility is inhibited; therefore one cannot test the role of this ion in fusion.

Calcium is required for the acrosome reaction, but Takahashi and Sugiyama (1973) have circumvented this problem by initiating the acrosome reaction in normal sea water and then quickly adding the sperm to eggs in calcium-free sea water. Fertilization takes place, indicating that calcium is not required for sperm-egg fusion. This is an extremely interesting result because it is one of the few situations where extracellular calcium is not needed for cell-cell fusion. Intracellular stores of calcium are released within the egg when fusion is probably occurring, and perhaps these stores are adequate (Steinhardt and Epel, 1974; Steinhardt et al., 1977).

The role of magnesium has been investigated by Sano and Mohri (1976), who found that this ion was required at a yet undefined point between sperm-egg binding and the cortical reaction. They postulate an involvement in the fusion process. It will be interesting to determine the role of this ion in fertilization.

4.7. Enzymes involved in fusion

A final approach to the study of fusion would be to examine sperm and egg for enzymes possibly involved in promoting fusion. Conway and Metz (1976) observed a phospholipase in sperm, which was only apparent shortly after the acrosome reaction had been elicited. The activity was transient and unstable. A number of workers have proposed that phospholipases are involved in destabilizing the membrane as a part of the fusion process (Bach, 1974), and the discovery of a phospholipase in sperm is therefore of great interest.

Eggs also contain glycosyl transferases (Schneider and Lennarz, 1976). Whether there are similar transferases in sperm and whether they are involved in sperm-egg binding or sperm-egg fusion is still unknown.

5. Incorporation of the fused sperm into the egg

In this section we consider how the sperm is incorporated into the egg. The electron micrograph in Fig. 17a, which depicts a fusion at a microvillus, illustrates the problem of incorporation of this fused sperm (Fig. 17b) into the egg. As can be seen the acrosomal process is much smaller than the sperm diameter (Fig. 17a). How, then, does the sperm enter the egg? Is it pushed or pulled through this process, or is entry by phagocytosis, or combinations of these and other processes?

Cinemicrophotographic studies indicate that sperm motility per se is not involved in incorporation (Epel et al., 1977). These show that the fertilizing sperm ceases its flagellar motion just before it begins to enter the egg (Fig. 17b). The rate of movement into the egg is linear at 6 to 11 mm per minute. Since the flagellum is not involved in sperm entry, these observations suggest that either the cell cortex and/or acrosomal process provide the motive force for sperm entry.

5.1. Cell cortex and sperm entry

It is now generally accepted that the cell cortex contains actin-like microfilaments. This idea was presaged by the work of Mitchison and Swann (1954), who developed an elastimeter to measure the force required to deform the surface of the sea urchin egg. They calculated that the cell was not merely bounded by a plasma membrane but that this membrane must also be attached to an elastic wall

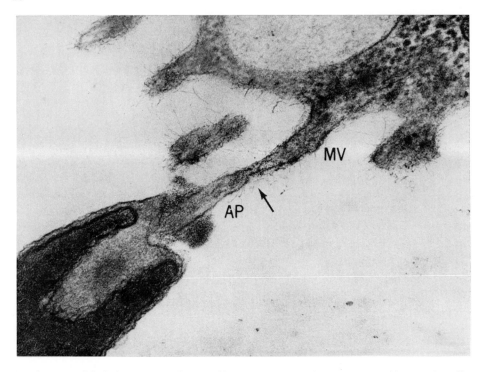

Fig. 17(a). Possible fusion (arrow) of sea urchin sperm acrosomal process (AP) with egg microvillus (MV). ×49,200. (Unpublished photograph courtesy of Dr. Frank Collins.)

1.5 μm in diameter, equivalent to the cell cortex. The rheological properties of the sea urchin egg have since been extensively studied by Hiramoto (1970), who found that the egg maintains an internal hydrostatic pressure slightly higher than that of the external medium and that the viscoelastic properties of the cortex change during fertilization and cytokinesis. The change at fertilization is an apparent stiffening of the cortex, as if it were undergoing a contraction (Hiramoto, 1974). The wave of increasing stiffness of the egg cortex begins at insemination, reaches a maximum at 1.5 minutes (just after completion of the cortical granule reaction), and then decreases to a minimum by 4 minutes after insemination. Another wave of increasing stiffness (contraction) reoccurs at cytokinesis (Hiramoto, 1974; Sawai and Yoneda, 1974).

The increasing stiffness of cell cortex during the period of sperm entry and the cortical granule reaction can be interpreted as a contraction of the cortex. Visual evidence for such a contraction has been described in several types of eggs by many workers (Kille, 1960; review, Runnström, 1966). For example, from 20 to 40 seconds after insemination in many species of sea urchin eggs the egg surface becomes concave or flat instead of remaining spherical, and in the center of this concavity the fertilizing sperm can be seen, over which the vitelline layer is elevating to form the fertilization envelope (Epel et al., 1977). The concave and

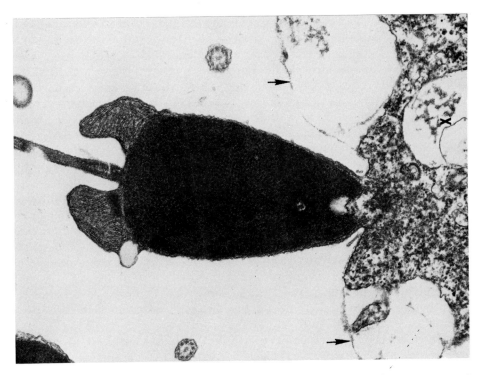

Fig. 17(b). Gamete membrane fusion in *Lytechinus variegatus*. Cytoplasmic confluence has been established by fusion of the acrosomal tubule membrane and the oolemma. The elevating fertilization envelope is indicated by arrows and a dehisced cortical granule is labeled X. Magnification ×29,300. (Micrograph courtesy of Drs. B. L. Hylander and R. G. Summers, unpublished).

flattened surface can extend to cover as much as half of the egg surface. After completion of the cortical reaction and the elevation of the vitelline layer, the egg returns (relaxes?) to a spherical shape. It is difficult to explain this consistently observed deformation of the cell surface without invoking a contractile mechanism (Runnström, 1966).

Sakai and co-workers (review, Mabuchi and Sakai, 1972) have isolated the egg cortex at various times throughout the cell cycle and have characterized several cortical proteins. The finding of mechanical changes in the cortex would encourage one to assume the presence of contractile proteins. Myosin (Mabuchi, 1974), actin (Miki-Noumura and Kondo, 1970), and an elastic protein similar to connectin (Maruyama et al., 1976) have all been purified from isolated egg cortices. An ATPase (Mabuchi, 1973) and a protein kinase (Murofushi, 1974) that show cyclic changes corresponding to phases of the cell cycle have also been found in the egg cortex. Transmission (Mabuchi and Sakai, 1972) and scanning (Vacquier, 1975; Schatten and Mazia, 1976, 1977) electron micrographs of isolated egg cortices show the presence of fibrous material that may be composed of bundles of these contractile proteins.

Given that there are contractile proteins and the above evidence for changes in cortex rigidity and contraction, are these proteins involved in incorporation of the sperm into the egg? One line of evidence suggesting this is that cytochalasin B (CB) prevents sperm incorporation in both sea urchins (Perry and Byrd, unpublished observations; Longo, 1977) and *Urechis* (Gould-Somero et al., 1977). When insemination is carried out in the presence of CB, egg activation occurs in both cases, as evidenced by a cortical reaction and respiratory burst (sea urchin) or vitelline layer elevation and germinal vesicle breakdown *(Urechis)*. In *Urechis,* the mechanism of inhibition is to prevent formation of the fertilization cone. This is a cytoplasmic upwelling that surrounds the sperm, and formation of this process appears sensitive to CB (Gould-Somero et al., 1977).

Another line of evidence, suggesting an involvement of contractile surface proteins, comes from two recent SEM (scanning electron micrograph) studies. Tegner and Epel (1976) and Schatten and Mazia (1977) have observed thin filaments emanating from the egg surface and surrounding the attached sperm. This association suggests some role in anchoring or incorporating the sperm. Even more interesting is the appearance of microvilli around the base of what is apparently the fertilizing sperm, and the subsequent complete covering of the sperm by the egg surface. It appears that the cell surface envelops the sperm (Schatten and Mazia, 1977).

An interesting hypothesis about sperm incorporation has been advanced by Schatten and Mazia (1976, 1977). They have observed sperm entry with the SEM into pieces of cortex, looking at the cortex from the inside. They suggest that the sperm is rapidly drawn in between the plasma membrane and a subcortical band of microfilaments, and that the sperm head then passes through this band into the egg cytoplasm. A problem with this idea is that there have been no TEM (transmission electron micrograph) observations of a sperm head lying laterally in the cortex, as is suggested by the authors' SEM observations. They therefore infer an extremely rapid entry process (< 1 second), but this idea seems contradictory to the aforementioned motion picture analysis, which shows that the sperm appears to enter linearly and over a period of 20 seconds. One obvious possibility is that the motion pictures were not of the sperm but of microvilli elongation at the site of sperm entry.

Regardless of the exact details, all these studies suggest that incorporation of the fused sperm involves an active process of the cortex in which the cell surface and associated structures are mobilized to surround and engulf the sperm. These observations also suggest that hypotheses regarding the fate of the sperm membrane at fusion should be reexamined. The Colwins (1967) originally postulated that the sperm plasma membrane remains as a patch on the egg surface, since their micrographs suggested that egg cytoplasm was forced up between the sperm plasma membrane and the sperm nucleus.

5.2. Acrosomal process and sperm entry

The final possibility is that the actin in the acrosomal process is involved in pulling the sperm into the egg. One mechanism might be that this actin acts as nucleation centers for the further polymerization and assembly of an actin (and myosin) complex, which would then contract to pull the sperm into the egg. Related to this hypothesis is the finding mentioned earlier that sea urchin eggs contain actin and actin binding proteins which appear to regulate the formation of bundles of actin filaments (Kane, 1975, 1976).

When intact sea urchin eggs are fixed by standard methods that preserve microfilaments in other cells, 50 to 70 Å cortical microfilaments are not observed. However, if the vitelline layer is removed by brief treatment in 1 M glycerol at pH 10, and the eggs are then washed into normal sea water, fertilized and fixed, a distinct band of cortical microfilaments is visible (Fig. 18). Such eggs become extremely polyspermic (Mann et al., 1976). Close examination shows that the cortical granules are no longer resting flush against the plasma membrane but are now located about 1 μm from the membrane. It is possible that in normal, untreated eggs the band of microfilaments is compacted and so closely associated with the plasma membrane-vitelline layer complex that it cannot be discerned in thin section (Schatten and Mazia, 1976). The harsh treatment of eggs with alkaline glycerol to remove the vitelline layer might also loosen the attachment of the microfilament band to the plasma membrane so that it can be seen readily. An alternative hypothesis is that the treatment to remove the vitelline layer, or the fact that the vitelline layer is no longer present, may somehow induce the formation of the microfilament band. Finally, the glycerol treatment might improve penetration of fixatives and so preserve the microfilaments.

Another indication of an interaction of the acrosomal process with the cortex seen in these glycerol-treated eggs is a clustering of tubular structures under the fused process (Fig. 19). These "microtubes" can be found with each fusing process within one minute of insemination (Mann et al., 1976), but it is not known whether they arise from the membrane of the sperm acrosome process or from the egg cortex. Does the egg sense the fusing sperm and instantly form microtubes to aid in sperm entry? To our knowledge the only other example of such structures associated with sperm fusion is seen in fertilization of a decapod crab, where they appear to arise from the membrane of the acrosome process (Brown, 1966). It is possible that these microtubes have not previously been observed in sea urchin eggs because removal of the vitelline layer greatly enhances the quality of fixation of these cells (D. Friend, unpublished observations) or, as noted above, that these processes are a manifestation of the harsh treatment.

6. Exocytosis of cortical granules

6.1. Introduction

The eggs of many diverse animal groups such as mammals, amphibians, fish, echinoderms, annelids, mollusks, and coelenterates contain a peripheral band of membrane-limited vesicles known as cortical granules (CGs). These vesicles are

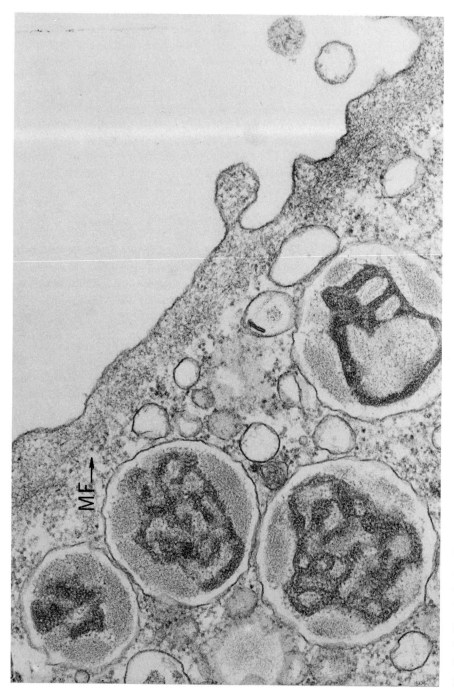

Fig. 18. Cortical band of microfilaments (MF) is preserved in *L. pictus* eggs after removal of the vitelline layer by treatment with glycerol at high pH. Note that the cortical granules have pulled away from the cell membrane. ×49,500. (Unpublished photograph courtesy of Dr. Daniel Friend.)

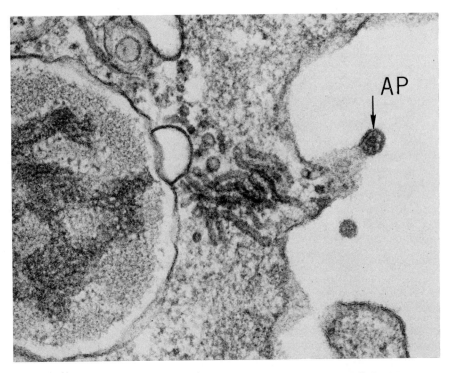

Fig. 19. Microtubes in cortex of alkaline glycerol-treated egg in response to fused sperm. A piece of the sperm acrosome process (AP) is the only part of the sperm seen in this section. Microtubes are thought to be composed of biomembranes. Are they formed by the egg in response to fusion, or are they remnants of the membrane of the acrosome process? ×69,000. (Unpublished photograph courtesy of Dr. Daniel Friend.)

produced by Golgi bodies of the developing oocyte and move to the periphery of the cell as oogenesis progresses (Anderson, 1968; Selman and Anderson, 1975). In sea urchin eggs CGs, about 1 μm in diameter, are attached to the inner surface of the plasma membrane (Fig. 20). Eggs of *S. purpuratus* contain approximately 15,000 CGs, which comprise 6.4% of the cell volume (Vacquier, 1975). As previously noted, one of the major consequences of fertilization is the fusion of the CG membrane with the egg cell membrane and the exocytosis of the CG's, which spreads as a wave across the cell surface. This secretion, termed the cortical granule reaction (CGR), begins 20 to 25 seconds after insemination and finishes by 40 to 50 seconds after insemination (Paul and Epel, 1971). As is discussed below, the CGR is one of the most massive and synchronous examples of natural membrane fusion known. Much work exists on CGs and the CGR that has been thoroughly reviewed by others (Allen, 1958; Runnström, 1966; Epel and Johnson, 1976). In this review we discuss the more recent studies of the CGR and the potential value of the CGR as a model system for studying the triggering mechanism of exocytosis and membrane fusion.

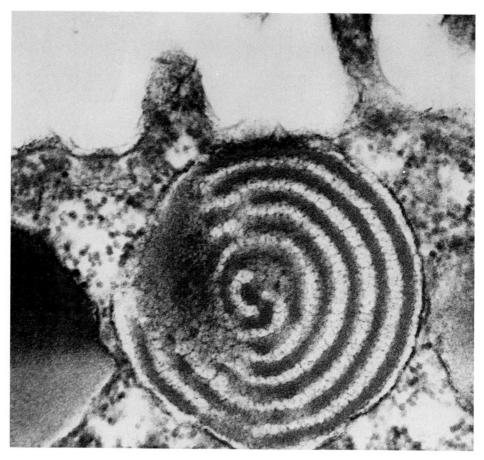

Fig. 20. Cortical granule of *S. purpuratus*. The membrane of the CG is in intimate contact with the plasma membrane of the cell. CGs have a species-specific ultrastructure. ×78,000. (Micrograph courtesy of V. D. Vacquier, unpublished.)

6.2. Initiation of the cortical granule reaction

Attachment and probably fusion of the acrosome process membrane with the egg plasma membrane initiates the CGR. This was beautifully demonstrated by Goodrich (1920), who succeeded in pulling the fertilizing sperm away from the surface of a *Nereis* egg. The CGR occurred normally but the egg did not divide. The CGR can also be initiated by artificial agents, such as pricking the egg with a needle, treatment with butyric acid, hypotonic shock, and detergents, numerous other agents (Loeb, 1913), and electric shock (Millonig, 1969).

The chain of reactions accompanying sperm-egg membrane fusion involves a change in electrical potential of the egg membrane, the elevation of the intracellular concentration of Ca^{2+}, and the possible contraction of the egg cortex underlying the cell membrane. As noted earlier, the resting potential of the sea urchin egg increases to about $+10$ mV at approximately 3 seconds after insemination (Steinhardt et al., 1971; Ito and Yoshioka, 1972; Jaffe, 1976), probably from an inrush of Na^+ from 3 to 25 seconds after insemination (Steinhardt et al., 1971). However, this potential change per se does not initiate the CGR since voltage-clamped eggs do not exhibit a CGR when their potential is raised to $+10$ mV (Jaffe, 1976).

The increase in Ca^{2+}, as measured by aequorin luminescence, does not begin until the beginning of the cortical exocytosis (Ridgeway et al., 1977; Steinhardt et al., 1977), and the Ca^{2+} probably comes from an intracellular store (Mazia, 1937; Nakamura and Yasumasu, 1974; Steinhardt and Epel, 1974; Ridgeway et al., 1977). The hypothesis we stress for the sea urchin egg is that approximately 25 seconds after insemination the intracellular Ca^{2+} increases above a certain threshold, which then triggers the exocytotic mechanism. Once initiated, the CGR becomes self-propagating and sweeps over the cell as a wave in every direction from the point of fusion (Ridgeway et al., 1977).

6.2.1. Evidence that Ca^{2+} triggers the cortical granule reaction

Evidence that Ca^{2+} initiates the CGR comes from four recent observations. First, divalent cation-transporting ionophores such as A23187 (Eli Lilly) and X537A (Hoffmann-LaRoche) invoke the CGR in the presence or absence of extracellular Ca^{2+} (Chambers et al., 1974; Steinhardt and Epel, 1974). Since extracellular Ca^{2+} is not required, the active Ca^{2+} presumably comes from an intracellular source. Second, when eggs of the Japanese Medaka fish are injected with the Ca^{2+}-indicating phosphorescent pigment, aequorin, and then fertilized, a dramatic increase in phosphorescence occurs which (upon quantitation) indicates that the free, ionic Ca^{2+} in the cell has increased at least 100-fold (Ridgeway et al., 1977). The timing of this increase in Ca^{2+} is concomitant with the CGR. Similar observations have been made about sea urchin eggs (Steinhardt et al., 1977). Third, direct injection of Ca^{2+} into amphibian eggs triggers the CGR and the metabolic activation of the egg (Hollinger and Schuetz, 1976). Finally, isolated CGs from sea urchin eggs fuse with each other and release their contents when exposed to Ca^{2+} greater than 0.2 mM (Vacquier, 1976).

6.2.2. Interaction of isolated cortical granules with calcium

The CGR of the sea urchin egg may be ideal for studying the biochemistry of exocytosis. A simple method is available for isolating "lawns" of intact sea urchin CGs that are bound to the plasma membrane-vitelline layer complex of the egg (Vacquier, 1975). The method is based on the attachment of eggs to a positively charged surface followed by lysing the cells, so that the plasma membrane and granules remain attached to the surface. The attachment of eggs is based on the fact that the egg surface has a net negative charge probably due to the presence of sialic acid (Glabe and Vacquier, 1977a). When flat surfaces such as plastic culture dishes are exposed to a solution of protamine sulfate, they become positively charged and the eggs electrostatically bond to such surfaces in an irreversible manner. If a thick suspension of eggs is poured into these protamine-coated dishes, the entire surface can be covered with a confluent layer of bound eggs. If a jet of isosmotic medium containing divalent cation chelators such as ethylenediaminetetraacetate (EDTA) or ethyleneglycoltetra-acetic acid (EGTA) is directed at the dish surface, the egg cytoplasm is sheared away and a dense population of cortical granules remains bound to the dish attached to the inner surface of the plasma membrane. Such isolated CG lawns (Figs. 21 and 22) appear free of cytoplasmic contamination. This is essentially a one-step isolation procedure that results in the exposure of intact CGs bound to a solid supporting surface such as a dish, coverslip, or microscope slide. The isolated CGs of *S. pupuratus* are 1.3 μm in diameter and appear to be tethered to the plasma membrane by 1000 to 2000 Å strands of unknown nature. The density of CGs can be up to 74 per 100 μm^2 of surface area.

When Ca^{2+} greater than 0.2 mM is added to these lawns of CGs, the granules undergo an instantaneous breakdown reaction. The CGs appear to swell, resulting in adjacent CG membranes making contact and fusing with each other (Fig. 23). The fused mass of membranes then contracts and, at the edges of the hull-like remains of the cell, pulls away from the plasma membrane-vitelline layer complex. As is true with other exocytotic vesicles, CG fusion can also be induced by Sr^{2+} and Ba^{2+} but not Mg^{2+}.

Because this membrane fusion induced by Ca^{2+} is primarily among CG membranes, and not CG and plasma membrane, caution must be used in terming this true exocytosis. However, it represents a fusion of CG membranes and release of CG contents. When the isolated CGs are viewed in cross section before (Fig. 24) and after (Fig. 25) the additions of Ca^{2+}, it appears that the contents of the granules are completely released and only fused membranes remain. A β-1, 3-glucanase and a characteristic Ca^{2+}-precipitable structural protein, which are markers for CG release, are both found in the supernatant after Ca^{2+}-induced CG fusion has occurred (Vacquier, 1975).

CGs can be removed from the plasma membrane by mild trypsinization. When Ca^{2+} is added to suspensions of these free CGs, the individual CGs explode and what remains appears to be their limiting membranes (Vacquier, 1975). The above analysis suggests that the direct trigger for the CGR is increased cytoplasmic Ca^{2+}. In the intact egg the CGR begins at the point of sperm fusion and radiates around the egg as a circular wave. Rothschild and Swann (1949) timed the rate of propagation of the CGR and concluded that the triggering stimulus diffuses through the cortex of the cell. A reanalysis of these data by Kacser (1955) rejected the idea of simple diffusion of the stimulus for the adoption of an autocatalytic propagation of the stimulus, which he supposed may be the release of Ca^{2+} from exocytosing CGs. Allen (1954) and Runnström and Kriszat (1953) also favored the hypothesis that the CGR was a self-propagating reaction.

Isolated CG lawns can be used to demonstrate that the Ca^{2+} actually arises from the cortical granules and that the self-propagating reaction depends on the release of Ca^{2+} from the fusing granules. The experiment is to remove the Ca^{2+} chelator from the solution over the lawns and replace it with Ca^{2+}-free isosmotic

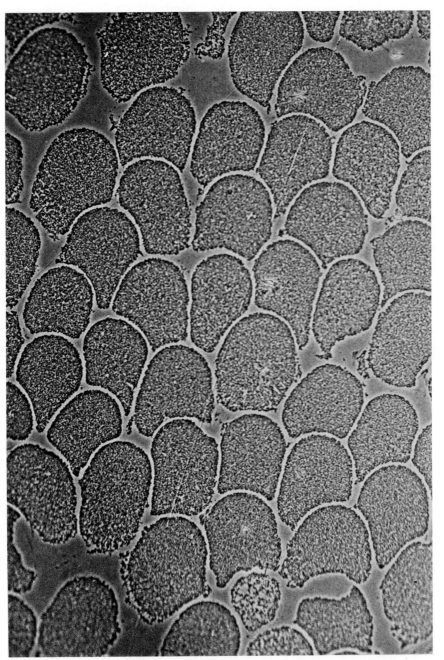

Fig. 21. Phase-contrast micrograph of cortical hulls of sea urchin eggs bound to a plastic culture dish. Eggs were electrostatically bound to a protamine-coated dish and the cytoplasm sheared away by a jet of isosmotic medium containing divalent cation chelator. Such hulls are composed almost entirely of intact cortical granules. ×241. (Reproduced with permission of Academic Press from Vacquier, 1975.)

50

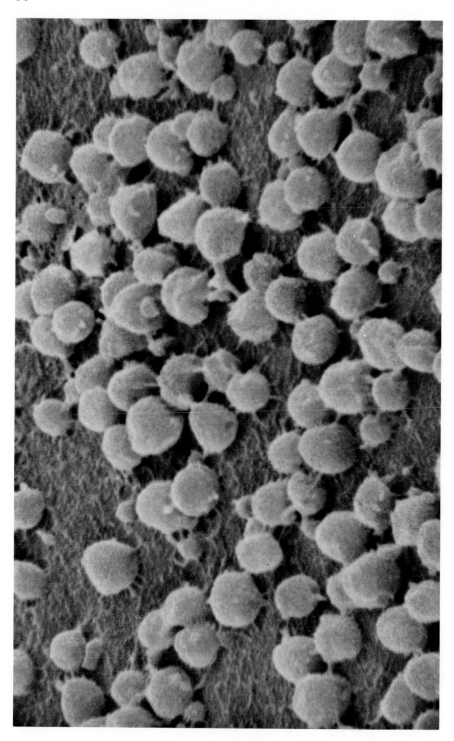

Fig. 22. Scanning electron micrograph (SEM) of the cortical hulls show the individual CGs (avg. diam. 1.3 μm) bound to the inner surface of the egg plasma membrane. In this preparation the jet of isolation medium removed many of the CGs, thus affording a direct view of the texture of the inner surface of the plasma membrane. The CGs appear to be tethered to the plasma membrane by their struts 100 nm diameter. ×11,500. (From Vacquier, 1975; reproduced with permission of Academic Press.)

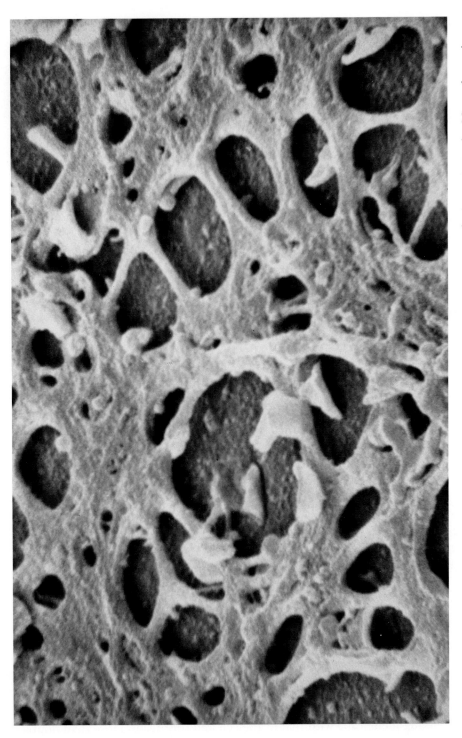

Fig. 23. SEM of fused mass of cortical granule membranes. When isolated CG lawns are exposed to Ca²⁺ above 0.2mM, the CG membranes instantaneously fuse with each other releasing the CG contents. ×14,600. (Reproduced with permission of Academic Press from Vacquier, 1975.)

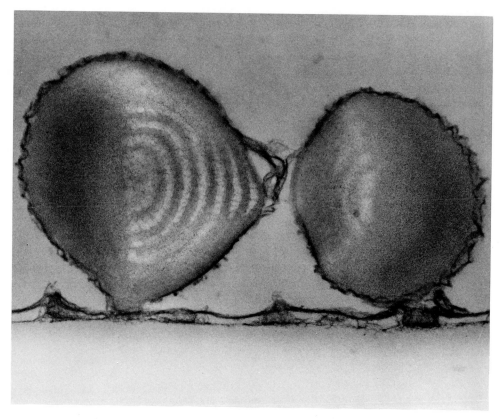

Fig. 24. Transmission electron micrograph (TEM) of isolated CGs bound to a plastic dish seen in cross-section. After embedding in epon, the dish was fragmented and sectioned using a diamond knife. The attachment of the CGs to the inner surface of the plasma membrane is clearly visible. ×61,500. (Reproduced with permission of Academic Press from Vacquier, 1975.)

media. When the edge of the lawn is scratched with a needle to disrupt some of the CGs, a wave of CG fusion observable with the naked eye propagates across the lawn in a few seconds.

Additional evidence for the localization of a Ca^{2+} triggering mechanism in the CG membrane is that the local anesthetic procaine, which is known to occupy Ca^{2+} binding sites on biomembranes (Blaustein and Goldman, 1966; Kuperman et al., 1968) will reversibly prevent the fusion reaction of isolated CGs by Ca^{2+}. Also, normal eggs fertilized in the presence of 5 to 10 mM procaine fail to undergo a CGR. Finally, La^{3+}, which has a high affinity for Ca^{2+} binding sites on membranes (Lehninger and Carafoli, 1971), inhibits Ca^{2+}-mediated CG fusion.

6.3. Membrane fusion and cell surface rearrangement—morphology

From 25 to 50 seconds after insemination of the sea urchin egg, the CGs fuse their limiting membrane with that of the cell and this fusion more than doubles

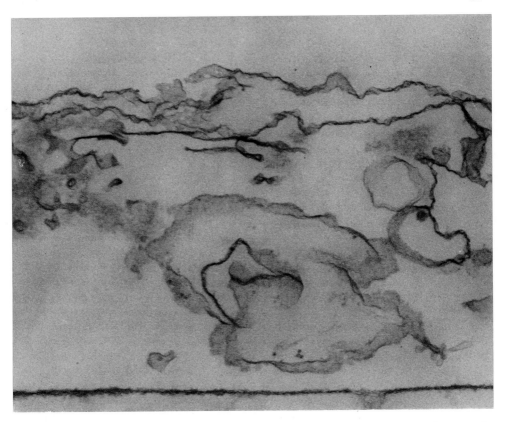

Fig. 25. TEM cross-section of CG membranes fused by the addition of Ca²⁺. It appears that most of the CG contents are released when the membrane fusion reaction occurs and only the limiting CG membranes remain. ×52,500. (Reproduced with permission of Alan R. Liss, Inc. from Vacquier, 1976.)

the area of the cell membrane (Eddy and Shapiro, 1976). However, the egg does not increase in diameter following the CGR, and the added surface membrane may be used in the lengthening of the cytoplasmic microvilli that extend from the cell surface. If the vitelline layer is removed by treatment with dithiothreitol before fertilization, the exocytosis of the CGs and the changing topography of the cell surface can be viewed directly by scanning electron microscopy (Eddy and Shapiro, 1976). Before fertilization, the naked egg surface displays a dense and evenly distributed array of microvilli (Eddy and Shapiro, 1976; Tegner and Epel, 1976). Observations of the cell surface in the area surrounding the fusing sperm show the opening of the fusion point between CG and the cell membrane, the presence of a pit after the expulsion of the contents of the CGs, and then the disappearance of the pits as the CG membrane comes up to the cell surface to form smooth areas devoid of microvilli. The cell membrane immediately after the CGR is a mosaic composed of patches of microvilli-containing membrane from the original plasma membrane and patches of smooth membrane derived from the cortical granules. The surface then rearranges by means of an increase in length of the microvilli. The final stage in surface rearrangement is seen as ridges of interconnecting microvilli that become regularly interspersed with smooth patches of membrane to form patterns resembling a honeycomb (Eddy and Shapiro, 1976).

Probably the most unusual and extensive examples of CG exocytosis and surface rearrangement are found in eggs of the cnidarian *Bunodosoma cavernata* (Dewel and Clark, 1974) and the prawn *Penaeus* (Clark et al., 1974). The surface of *B. cavernata* eggs possesses radial projections, each of which contains a dense core of 50 to 70 Å microfilaments extending down into the cell cortex in the form of a rootlet of microfilaments. These structures, termed cytospines (Dewel and Clark, 1974), closely resemble the microvilli of the intestinal brush border (Mooseker, 1976). The cortex of the egg contains a dense population of CGs that are not only flush against the plasma membrane but also exist as 4 to 6 rows of CGs extending down into the cortex. During the CGR, the uppermost row of CGs fuses with the plasma membrane while the lower rows all fuse with each other. A massive exocytotic release of CG contents ensues. Not only are the CG contents released, but some mitochondria and endoplasmic reticulum that get caught between rows of fusing CGs are also ejected from the cell. It is remarkable that the cell can survive such a violent perturbation and partial lysis of its surface. From the egg diameters before and after the CGR (Dewel and Clark, 1974), one can compute that approximately 10% of the cell volume is lost during the CGR.

Electron microscopic observations of the cortex of *Penaeus* eggs reveal the presence of rod-shaped bodies 6×9 μm in dimension, bounded by a limiting membrane. Shortly before ovulation these cortical rods move to the cell periphery where fusion of the membranes of the rods with the cell membrane occurs. At ovulation, when the egg contacts sea water, the rods are expelled into the cell surface. This rod-produced layer then disperses and a second CGR occurs that is involved in the formation of a "hatching membrane" which surrounds the cell. In this cell the CGR is such a massive exocytotic event that an approximate 30% reduction in cell diameter is seen after completion of the CGR (Clark et al., 1974; Lynn and Clark, 1975).

6.4. Membrane fusion and cell surface rearrangement—molecular changes

Both the lipid and protein composition of the cell surface changes after the CGR. Extracts of isolated membranes of sea urchin eggs before and after the CGR differ in amount of neutral and polar lipids (Barber and Mead, 1975). After the CGR the percentage of triglycerides and phosphatidylserine are increased while phosphatidylcholine decreases. Fatty acids of unfertilized membrane lipids are of longer chain and more unsaturated than those from membranes after fertilization. These data indicate that real biochemical differences may exist between the composition of the CG membranes and the membrane of the unfertilized egg.

The surface proteins of sea urchin eggs are also altered after fertilization (Shapiro, 1975). Surface proteins of unfertilized eggs were radioiodinated by the lactoperoxidase method. Polyacrylamide-gel electrophoresis in sodium dodecyl sulfate of labeled eggs showed 75% of the label to be greater than 130,000 daltons, with 5% so large that it was excluded from the gel. Fertilization of the labeled eggs caused a dramatic shift in the distribution of size classes of iodinated cell surface protein. After fertilization the amount of label excluded from the gel increased to 35 to 40%. A decrease in the ~ 130,000 dalton components occurred and new labeled components of from 20,000 to 30,000 daltons appeared. These changes do not occur if the eggs are fertilized in sea water containing soybean trypsin inhibitor, which indicates that the CG protease expelled from the cell during the CGR (Vacquier et al., 1972a, 1973) partially digests some cell surface protein.

Two additional points of interest come from this study. First, the proteolysis of membrane proteins is not needed for normal development, since monospermic eggs fertilized in soybean trypsin inhibitor will develop to the larval stage. Second, the labeled surface proteins appear to be conserved during embryogenesis, and after the limited proteolysis accompanying the CGR no further size changes or loss of label occur up through the larval stage. This indicates that those surface proteins of the egg that are labeled appear to be conserved during embryogenesis and are not turning over (Shapiro, 1975).

Cortical exocytosis is also required for turning on a Na^+-dependent amino acid transport system, the activity of which can increase several hundred fold after fertilization (Epel, 1972). If eggs are fertilized in the presence of inhibitors of the CGR very little transport develops (Epel and Johnson, 1976). The exocytosis requirement is partly for the proteolytic modification of the surface. Proteolysis alone is insufficient, however, suggesting that the new surface exposed by the exocytosis is also necessary. The insertion of amino acid carriers is sensitive to cytochalasin B, but transport per se is unaffected by this inhibitor. This suggests that insertion of carriers involves microfilaments, probably in the cell cortex. Interestingly, increased K^+ permeability does not require a cortical reaction and is insensitive to cytochalasin B. It appears that the membrane is a functional mosaic in regard to transport, with domains of K^+ permeability (not dependent on the CGR) and Na^+-dependent amino acid transport (dependent on the CGR; Epel and Johnson, 1976).

6.5. Isolated cortical granules as a model for studying exocytosis

Biochemical studies of the exocytosis mechanism are in their earliest stages. Most of the work to date has been done on the exocytotic release of histamine from secretory vesicles of mast cells (Goth and Johnson, 1975), mucocysts of protozoa (Satir et al., 1973), synaptosomes of neurons (Heuser and Reese, 1973), parotid gland vesicles (Castle et al., 1975), zymogen granules of pancreas (Rutten et al., 1975), and bovine chromaffin granules (Smith and Winkler, 1967). These systems may be less than advantageous for biochemical studies of the Ca^{2+} triggering and membrane fusion mechanisms because of inconvenience in obtaining large quantities of secretory vesicles and lack of synchrony of the exocytotic events within the population of secretory vesicles.

As an ideal model system for studying exocytosis a cell type should be used that permits the mass isolation of secretory vesicles free from cytoplasmic contamination. It would also be desirable if the secretory vesicles retained their physiological activity and underwent exocytotic release when given the proper stimulus. The cortical granules of invertebrate eggs may be ideal vesicles for both the biochemical study of the Ca^{2+} triggering mechanism and the membrane fusion reaction. This system of Ca^{2+}-mediated membrane fusion may have unique advantages that would permit a detailed biochemical analysis. Regarding the amount of secretory membrane, a circle of confluent cell hulls 2 cm in diameter would contain about 1.6×10^8 CGs with an average diameter of 1.3 μm. Each CG contains 5.3 μm^2 of membrane. Such an area would thus contain 8.3 cm² of secretory granule membrane (Vacquier, 1976).

The technique of sticking cells down and then washing away the cytoplasm to expose the inner surface of the plasma membrane has also been used to observe membrane-associated actin filaments of slime mold amoebae (Clarke et al., 1975) and polymorphonuclear leukocytes (Boyles and Bainton, 1970). It would be interesting to apply this technique to lymphocytes to determine the exact correspondence between patches and caps of con A receptors on the outside of the cell and the placement of cytoskeletal microfilaments and microtubules on the inner surface of the membrane (Albertini and Clark, 1975; Poste et al., 1975; Yahara and Edelman, 1975). For example, one could patch and cap lymphocytes with fluorescent con A, stick the cells to a locater grid (Clarke et al., 1975), shear away the cytoplasm, fix the cell remains and, by fluorescent microscopy, locate those cell hulls with distinct patches or caps. One could then examine the grid with the electron microscope to determine the distribution, orientation, and organization of membrane-associated cytoskeletal filaments.

Acknowledgments

During the writing of this review D.E. was a Fellow of the John S. Guggenheim Foundation and an Overseas Fellow of Churchill College, Cambridge University. Discussions with Drs. John Gurdon, Martin Johnson, and Michael Edidin are gratefully acknowledged. Drs. Frank Collins, Robert Summers, Bonnie Hylander, Daniel Friend, and Mia Tegner are thanked for providing photographs. We are also grateful to Dr. B. Brandriff for reading the manuscript.

Much of the work discussed in this review was supported by National Science Foundation grant BMS 73-07003-A2 (D.E.) and National Institutes of Health grant HD-08645 (V.D.V.).

References

Afzelius, B. A. (1972) Reactions of the sea urchin oocyte to foreign spermatozoa. Exp. Cell Res. 72, 25–33.

Afzelius, B. A. (1973) The Functional Anatomy of the Spermatozoon. Pergamon Press, London.

Aketa, K. (1973) Physiological studies on the sperm surface component responsible for sperm-egg bonding in sea urchin fertilization. I. Effect of sperm-binding protein on the fertilizing capacity of sperm. Exp. Cell Res. 80, 439–441.

Aketa, K. (1975) Physiological studies on the sperm surface components responsible for sperm-egg bonding in sea urchin fertilization. II. Effect of concanavalin A on the fertilizing capacity of sperm. Exp. Cell Res. 90, 56–62.

Aketa, K. and Ohta, (1977) Dev. Biol. In press.

Aketa, K., Onitake, K. and Tsuzuki, H. (1972) Tryptic disruption of sperm binding site of sea urchin surface. Exp. Cell Res. 71, 27–32.

Albertini, D. F. and Clark, J. I. (1975) Membrane-microtubule interactions: concanavalin A capping induced redistribution of cytoplasmic microtubules and colchicine binding proteins. Proc. Nat. Acad. Sci. U.S.A. 72, 4976–4980.

Allen, R. D. (1954) Fertilization and activation of sea urchin eggs in glass capillaries. I. Membrane elevation and nuclear movements in totally and partially fertilized eggs. Exp. Cell Res. 6, 403–424.

Allen, R. D. (1958) The initiation of development. In: The Chemical Basis of Development (McElroy, W. D. and Glass, B., eds.) pp. 17–67, Johns Hopkins Press, Baltimore.

Anderson, E. (1968) Oocyte differentiation in the sea urchin Arbacia punctulata, with particular reference to the origin of cortical granules and their participation in the cortical reaction. J. Cell Biol. 37, 514–539.

Austin, C. R. (1975) Membrane fusion events in fertilization. J. Reprod. Fert. 44, 155–166.

Bacchetti, B. and Afzelius, B. A. (1976) The biology of the sperm cell. In: Monographs in Developmental Biology, Vol. 10. S. Karger and Co.

Bach, M. K. (1974) A molecular theory to explain the mechanism of allergic histamine release. J. Theor. Biol. 45, 131–151.

Baker, P. F. and Presley, R. (1969) Kinetic evidence for an intermediate stage in the fertilization of the sea urchin egg. Nature 221, 488–490.

Barber, M. S. and Mead, J. F. (1975) Comparison of lipids of sea urchin egg ghosts prepared before and after fertilization. Wilhelm Roux' Arch. 177, 19–27.

Berg, W. E. (1967) Some experimental techniques for eggs and embryos of marine invertebrates. In: Methods in Developmental Biology (Wilt, F. H. and Wessells, N. K., eds.) pp. 767–776, Crowell, New York.

Blaustein, M. P. and Goldman, D. E. (1966) Competitive action of calcium and procaine on lobster axon. J. Gen. Physiol. 49, 1043–1063.

Bohus-Jensen, A. (1953) The effect of trypsin on the cross fertilizability of sea urchin eggs. Exp. Cell Res. 5, 325–336.

Bowles, D. J. and Kauss, H. (1976) Isolation of a lectin from liver plasma membrane and its binding to cellular membrane receptors in vitro. FEBS Lett. 66, 16–19.

Boyles, J. and Bainton, D. F. (1970) SEM of filament bundles on the cytoplasmic surface to PMN plasma membrane during phagocytosis. J. Cell Biol. 67, 40a.

Brown, G. G. (1966) Ultrastructural studies of sperm morphology and sperm-egg interaction in the decapod crab Callinectes sapidus. J. Ultrastruct. Res. 14, 425–440.

Brown, G. G. (1976) SEM and other observations of sperm fertilization reactions in Limulus polyphemus. J. Cell Sci. 22, 547–562.

Bryan, J. (1970) The isolation of a major structural element of the sea urchin fertilization membrane. J. Cell Biol. 44, 635–644.

Byrd, E. W., Jr. and Collins, F. D. (1975) Absence of fast block to polyspermy in eggs of sea urchin Strongylocentrotus purpuratus. Nature 257, 675–677.

58

Carroll, E. J., and Epel, D. (1975) Isolation and biological activity of the proteases released by sea urchin eggs following fertilization. Dev. Biol. 44, 22–32.

Carroll, E. J., Jr. and Levitan, H. (1977) J. Cell Biol. In press.

Carroll, E. J., Jr., Byrd, E. W., Jr. and Epel, D. (1977) A novel procedure for obtaining denuded sea urchin eggs and observations on the role of the vitelline layer in sperm reception and egg activation. Exp. Cell Res. 108, 365–374.

Castle, J. D., Jamieson, J. D. and Palade, G. E. (1975) Secretion of the rabbit parotid gland: isolation, subfractionation and characterization of the membrane and content subfractions. J. Cell Biol. 64, 182–210.

Chambers, E. L., Pressman, B. C. and Rose, B. (1974) The activation of sea urchin eggs by the divalent ionophores A23187 and X537A. Biochem. Biophys. Res. Commun. 60, 126–132.

Clark, W. H., Persyn, H. and Yudin, A. I. (1974) A microscopic analysis of the cortical specializations and their activation in the egg of the prawn, Penaeus sp. Am. Zool. 14, 1251.

Clarke, M., Schatten, G., Mazia, D. and Spudich, J. A. (1975) Visualization of actin fibers associated with the cell membrane in amoebae of Dictyostelium discoideum. Proc. Nat. Acad. Sci. U.S.A. 72, 1758–1762.

Collins, F. (1976) A reevaluation of the fertilizin hypothesis of sperm agglutination and the description of a novel form of sperm adhesion. Dev. Biol. 49, 381–394.

Collins, F. and Byrd, E. W., Jr. (1978) In preparation.

Collins, F. and Epel, D. (1977) The role of calcium ions in the acrosome reaction of sea urchin sperm: regulation of exocytosis. Exp. Cell Res. 106, 211–222.

Colwin, A. L., Colwin, L. H. and Summers, R. G. (1973) The acrosomal region and the beginning of fertilization in the holothurian Thyone briareus. In: The Functional Anatomy of the Spermatozoan (Afzelius, B. A., ed.) pp. 27–38, Proc. Second Werner-Gren Symp., Stockholm.

Colwin, L. H. and Colwin, A. L. (1967) Membrane fusion in relation to sperm-egg association. In: Fertilization (Metz, C. B. and Monroy, A., eds.) pp. 295–367, Academic Press, New York.

Conway, A. F. and Metz, C. B. (1976) Phospholipase activity of sea urchin sperm: its possible involvement in membrane fusion. J. Exp. Zool. 198, 39–48.

Crandall, M. A. and Brock, T. D. (1968) Molecular basis of mating in the Yeast Hansenula wingei. Bac. Rev. 32, 139–163.

Dan, J. C. (1954) Studies on the acrosome. III. Effect of calcium deficiency. Biol. Bull. 107, 335–349.

Dan, J. C. (1967) Acrosome reaction and lysins. In: Fertilization (Metz, C. B. and Monroy, A., eds.) pp. 237–293. Academic Press, New York.

Dan, J. C. (1970) Morphogenetic aspects of the acrosome formation and reaction. Adv. Morphogen. 8, 1–39.

Dan, J. C. and Hagiwara, Y. (1967) Studies on the acrosome reaction. IX. Course of acrosome reaction in the starfish. J. Ultrastruct. Res. 18, 562–579.

Decker, G. L., Joseph, D. B. and Lennarz, W. J. (1976) A study of factors involved in induction of the acrosomal reaction in sperm of the sea urchin Arbacia punctulata. Dev. Biol. 53, 115–125.

Dewel, W. C. and Clark, W. H. (1974) A fine structural investigation of surface specializations and the cortical reaction in eggs of the cnidarian Bunodosoma cavernata. J. Cell Biol. 60, 78–91.

Dubois, M., Gilles, K. A., Hamilton, J. K., Rebers, P. A. and Smith, F. (1956) Colorimetric method for determination of sugars and related substances. Anal. Chem. 28, 350–356.

Eddy, E. M. and Shapiro, B. M. (1976) Changes in the topography of the sea urchin after fertilization. J. Cell Biol. 71, 35–48.

Epel, D. (1970) Methods for removal of the vitelline membrane of sea urchin eggs. II. Controlled exposure to trypsin to eliminate postfertilization clumping of embryos. Exp. Cell Res. 61, 69–70.

Epel, D. (1972) Activation of an Na^+ dependent amino acid transport system upon fertilization of sea urchin eggs. Exp. Cell Res. 72, 74–89.

Epel, D. (1975) The program of and mechanisms of fertilization in the echinoderm egg. Am. Zool. 15, 507–522.

Epel, D. (1978) The mechanism of activation of sperm and egg during fertilization of sea urchin gametes. Curr. Top. Dev. Biol. In press.

Epel, D. and Johnson, J. D. (1976) Reorganization of the sea urchin egg surface at fertilization and its relation to the activation of development. In: Biogenesis and Turnover of Membrane Macromolecules (Cook, J. S., ed.) pp. 105–120, Raven Press, New York.

Epel, D., Cross, N. L. and Epel, N. (1977) Flagellar motility is not involved in the incorporation of the sperm into the egg at fertilization. Develop. Growth Differ. 19, 15–21.

Epel, D., Weaver, A. M., Muchmore, A. V. and Schimke, R. T. (1969) Beta-glucanase of sea urchin eggs: release from particles at fertilization. Science 163, 294–296.

Fairbanks, G., Steck, T. L. and Wallach, D. F. H. (1971) Electrophoretic analysis of the major polypeptides of the human erythrocyte membrane. Biochemistry 10, 2606–2617.

Fodor, E. J., Ako, H. and Walsh, K. A. (1975) Isolation of a protease from sea urchin eggs before and after fertilization. Biochemistry 14, 4923–4927.

Foerder, C., Eddy, E. M. and Shapiro, B. M. (1977) The fertilization membrane of the sea urchin is hardened by ovoperoxidase catalyzed formation of tyrosine crosslinks. Fed. Proc. 36, 926.

Franklin, L. E. (1965) Morphology of gamete fusion and of sperm entry into oocytes of the sea urchin. J. Cell Biol. 25, 81–100.

Garbers, D. L. and Hardman, J. G. (1976) Effects of egg factors on cyclic nucleotide metabolism in sea urchin sperm. J. Cyclic Nucl. Res. 2, 59–70.

Giudice, G. (1973) Developmental Biology of the Sea Urchin Embryo. Academic Press, New York.

Glabe, C. G. and Vacquier, V. D. (1977a) The isolation and characterization of the vitelline layer of sea urchin eggs. J. Cell Biol. 75, 410–421.

Glabe, C. G. and Vacquier, V. D. (1977b) Species-specific agglutination of eggs by bindin isolated from sea urchin sperm. Nature 267, 836–838.

Goodrich, H. B. (1920) Rapidity of activation in the fertilization of Nereis. Biol. Bull. 38, 196–201.

Goth, A. and Johnson, A. R. (1975) Current concepts on the secretory function of mast cells. Life Sci. 16, 1201–1214.

Gould-Somero, M., Holland, L. and Paul, M. (1977) Cytochalasin B inhibits sperm penetration into eggs of Urechis campo (Echiura). Dev. Biol. 58, 11–23.

Gregg, K. W. and Metz, C. B. (1976) Physiological parameters of the sea urchin acrosome reaction. Biol. Reprod. 14, 405–411.

Hagstrom, B. and Hagstrom, B. (1954) A method of determining the fertilization rate in sea urchins. Exp. Cell Res. 6, 479–484.

Haino, K. and Kigawa, M. (1966) Studies on the egg membrane lysin of Tegula pfeifferi: Isolation and chemical analysis of the egg membrane. Exp. Cell Res. 42, 625–633.

Hartman, J. F., Gwatkin, R. B. L. and Hutchison, C. F. (1972) Early contact interactions between mammalian gametes in vitro: evidence that the vitellus influences adherence between sperm and zona pellucida. Proc. Nat. Acad. Sci. U.S.A. 69, 2767–2770.

Hauschka, S. D. (1963) Purification and characterization of Mytilus egg membrane lysin. Biol. Bull. 125, 363–370.

Heller, E. and Raftery, M. A. (1973) Isolation and purification of three egg membrane lysins from sperm of the marine invertebrate Megathura crenulata. Biochemistry 12, 4106–4113.

Heller, E. and Raftery, M. A. (1976a) The vitelline envelope of eggs from the giant keyhole limpet Megathura crenulata. I. Chemical composition and structural studies. Biochemistry 15, 1149–1198.

Heller, E. and Raftery, M. A. (1976b) The vitelline envelope of eggs from the giant keyhole limpet Megathura crenulata II: products formed by lysis with sperm enzymes and dithiothreitol. Biochemistry 15, 1199–1203.

Heuser, J. E. and Reese, T. S. (1973) Evidence for recycling of synaptic vesicle membrane during transmitter release at the frog neuromuscular junction. J. Cell Biol. 57, 315–344.

Hiramoto, Y. (1970) Rheological properties of sea urchin eggs. Biorheology 6, 201–234.

Hiramoto, Y. (1974) Mechanical properties of the surface of the sea urchin egg at fertilization and during cleavage. Exp. Cell Res. 89, 320–326.

Hollinger, T. G. and Schultz, A. W. (1976) Cleavage and cortical granule breakdown in Rana pipiens oocytes induced by direct microinjection of calcium. J. Cell Biol. 71, 395–401.

Hotta, K., Kurokawa, M. and Isaka, S. (1973) A novel sialic acid and fucose-containing disaccharide isolated from the jelly coat of sea urchin eggs. J. Biol. Chem. 248, 629–631.

Hultin, T. (1948) Species specificity in fertilization reaction. II. Influence of certain factors on the cross fertilization capacity of Arbacia lixula. Ark. Zool. 40 1–12.

Inoue, M. and Wolf, D. P. (1975) Sperm binding characteristics of the murine zona pellucida. Biol. Reprod. 13, 340–346.

Ito, S. and Yoshioka, K. (1972) Real activation potential observed in sea urchin egg during fertilization. Exp. Cell Res. 72, 547–551.

Jaffe, L. A. (1976) Fast block to polyspermy in sea urchin eggs is electrically mediated. Nature 261, 68–71.

Johnson, J. D., Epel, D. and Paul, M. (1976) Intracellular pH and activation of sea urchin eggs after fertilization. Nature 262, 661–664.

Kacser, H. (1955) The cortical changes on fertilization in the sea urchin egg. J. Exp. Biol. 32, 451–467.

Kane, R. E. (1975) Preparation and purification of polymerized actin from sea urchin egg extracts. J. Cell Biol. 66, 305–315.

Kane, R. E. (1976) Actin polymerization and interaction with other proteins in temperature-induced gelation of sea urchin egg extracts. J. Cell Biol. 71, 704–714.

Kille, R. A. (1960) Fertilization of the lamprey egg. Exp. Cell Res. 20, 12–27.

Knox, R. B., Clarke, A., Harrison, S., Smith, P. and Marchalonis, J. J. (1976) Cell recognition in plants: determinants of the stigma surface and their pollen interactions. Proc. Nat. Acad. Sci. U.S.A. 73, 2788–2792.

Kuperman, A. S., Altura, B. T. and Chezar, J. A. (1968) Action of procaine on calcium efflux from frog nerve and muscle. Nature 217, 673–675.

Lehninger, A. L. and Carafoli, E. (1971) The interaction of La^{3+} with mitochondria in relation to respiration-coupled Ca^{2+} transport. Arch. Biochem. Biophys. 143, 506–515.

Levine, A. E., Fodor, E. J. B. and Walsh, K. A. (1977) An acrosin-like enzyme in sea urchin sperm. Fed. Proc. 36, 811.

Loeb, J. (1913) Artificial Parthenogenesis and Fertilization. Univ. of Chicago Press, Illinois.

Longo, F. J. (1973) Fertilization: a comparative ultrastructural review. Biol. Reprod. 9, 149–215.

Longo, F. J. (1977) J. Cell Biol. 75, 44a.

Lorenzi, M. and Hedrick, J. L. (1973) On the macromolecular composition of the jelly coat from *Strongylocentrotus purpuratus*. Exp. Cell Res. 79, 417–422.

Lynn, J. W. and Clark. W. H. (1975) A Mg^{2+} dependent cortical reaction in the egg of Penaeid shrimp. J. Cell Biol. 62, 251a.

Mabuchi, I. (1973) ATPase in the cortical layer of sea urchin egg, its properties and interaction with cortex protein. Biochim. Biophys. Acta 297, 317–332.

Mabuchi, I. (1974) A myosin-like protein in the cortical layer of cleaving starfish eggs. J. Biochem. 76, 47–55.

Mabuchi, I. and Sakai, J. (1972) Cortex protein of sea urchin eggs. I. Its purification and other protein components of the cortex. Dev. Growth Differ. 14, 247–261.

Mann, S., Schatten, G., Steinhardt, R. and Friend, D. S. (1976) Sea urchin sperm-oocyte interaction. J. Cell Biol. 70, 110a.

Maruyama, K., Mabuchi, I., Matsubara, S. and Ohashi, K. (1976) An elastic protein from the cortical layer of the sea urchin egg. Biochim. Biophys. Acta 446, 321–324.

Mazia, D. (1937) The release of calcium in *Arbacia* eggs on fertilization. J. Cell Comp. Physiol. 10, 291–308.

Metz, C. B. (1961) Use of inhibiting agents in studies on fertilization mechanisms. Int. Rev. Cytol. 11, 219–253.

Metz, C. B. (1967) Gamete surface components and their role in fertilization. In: Fertilization (Metz, C. B. and Monroy, A., eds.) pp. 163–237, Academic Press, New York.

Miki-Noumura, T. and Kondo, H. (1970) Polymerization of actin from sea urchin eggs. Exp. Cell Res. 61, 31–41.

Millonig, G. (1969) Fine structure analysis of the cortical reaction in the sea urchin egg: after normal fertilization and after electric induction. J. Submicr. Cytol. 1, 69–84.

Mitchison, J. M. and Swann, M. M. (1954) The mechanical properties of the cell surface. II. The unfertilized sea urchin egg. J. Exp. Biol. 31, 461–472.

Monroy, A. (1965) The Chemistry and Physiology of Fertilization. Holt, Rinehart & Winston, New York.

Monroy, A., Ortolani, G., O'Dell, D. and Millonig, G. (1973) Binding of concanavalin A to the surface of unfertilized and fertilized ascidian eggs. Nature 242, 409–410.

Mooseker, M. S. (1976) Brush border motility. Microvillar contraction in triton-treated brush borders isolated from intestinal epithelium. J. Cell Biol. 71, 417–432.

Morgan, T. H. (1923) Removal of the block to self-fertilization in the ascidian Ciona. Proc. Nat. Acad. Sci. U.S.A. 9, 170–174.

Moy, G. W., Friend, D. S. and Vacquier, V. D. (1977) Localization of bindin before and after the acrosome reaction of sea urchin sperm. Submitted for publication.

Murofushi, H. (1974) Protein kinase in the sea urchin egg cortices. Biochim. Biophys. Acta 364, 260–271.

Nakamura, M. and Yasumasu, I. (1974) Mechanism for increase in intracellular concentration of free calcium in fertilized sea urchin eggs. J. Gen. Physiol. 63, 374–388.

Nowak, T. P., Haywood, P. L. and Barondes, S. H. (1976) Developmentally regulated lectin in embryonic chick muscle and a myogenic cell line. Biochem. Biophys. Res. Commun. 68, 650–657.

Ohtake, H. (1976) Respiratory behavior of sea urchin spermatozoa. II. Sperm activating substance obtained from jelly coat of sea urchin eggs. J. Exp. Zool. 198, 313–322.

Osanai, K. (1976) Egg membrane-sperm binding in the Japanese Palolo eggs. Bull. Mar. Biol. Sta. Asamushi 15, 147–155.

Paul, M. (1970) Fertilization-associated changes in the eggs of Strongylocentrotus purpuratus and Urechis caupo. Ph.D. Thesis, Stanford University.

Paul. M. (1975) The polyspermy block in eggs of Urechis caupo: evidence for a fast block. Exp. Cell Res. 90, 137–142.

Paul, M. and Epel, D. (1971) Fertilization-associated light-scattering changes in eggs of the sea urchin Strongylocentrotus purpuratus. Exp. Cell Res. 65, 281–288.

Paul, M. and Gould-Somero, M. (1975) Evidence for a polyspermy block at the level of sperm-egg plasma membrane fusion in Urechis caupo. J. Exp. Zool. 196, 105–112.

Paul, M., Johnson, J. D., and Epel, D. (1976) Fertilization acid release of sea urchin eggs is not a consequence of cortical granule exocytosis. J. Exp. Zool. 197, 127–133.

Poste, G. and Allison, A. C. (1973) Membrane fusion. Biochim. Biophys. Acta 300, 421–465.

Poste, G., Papahadjopoulos, D. and Nicolson, G. L. (1975) Local anesthetics affect transmembrane cytoskeletal control of mobility and distribution of cell surface receptors. Proc. Nat. Acad. Sci. U.S.A. 72, 4430–4434.

Presley, R. and Baker, P. F. (1970) Kinetics of fertilization in the sea urchin: a comparison of methods. J. Exp. Biol. 52, 455–468.

Rebhun, L. I. (1962) Electron microscopic studies on the vitelline membrane of the surf clam Spisula solidissima. J. Ultrastruct. Res. 6, 107–123.

Ridgeway, E. B., Gilkey, J. C. and Jaffe, L. F. (1977) Free calcium increases explosively in activating medaka eggs. Proc. Nat. Acad. Sci. U.S.A. 74, 623–627.

Rosen, S. D., Simpson, D. L., Rose, J. E. and Barondes, S. H. (1974) Carbohydrate-binding protein from Polysphondylium pallidum implicated in intercellular adhesion. Nature 252, 128–150.

Rosen, S. D., Reitherman, R. W. and Barondes, S. H. (1975) Distinct lectin activities from six species of cellular slime molds. Exp. Cell Res. 95, 159–166.

Rothschild, L. and Swann, M. M. (1949) The fertilization reaction in the sea urchin egg. A note on diffusion considerations. J. Exp. Biol. 26, 177–181.

Rothschild, L. and Swann, M. M. (1952) The fertilization reaction in the sea urchin. The block to polyspermy. J. Exp. Biol. 29, 469–483.

Runnström, J. (1966) The vitelline membrane and cortical particles in sea urchin eggs and their function in maturation and fertilization. Adv. Morphogen. 5, 221–325.

Runnström, J. and Kriszat, G. (1953) The cortical propagation of the activation impulse in the sea urchin egg. Exp. Cell Res. 3, 419–426.

Runnström, J., Hagström, B. E. and Perlmann, P. (1959) Fertilization. In: The Cell (Brachet, J. and

Mirsky, A., eds.) vol. 1, pp. 327–398. Academic Press, New York.

Rutten, W. J., De Pont, J. J. H. H. M., Bonting, S. L. and Daemon, F. J. M. (1975) Lysophospholipids in pig pancreatic zymogen granules in relation to exocytosis. Eur. J. Biochem. 54, 259–265.

Sano, K. and Mohri, H. (1976) Fertilization of sea urchin needs Mg^{2+} in seawater. Science 192, 1339–1340.

Sardet, C. and Tilney, L. G. (1977) Origin of the membrane from the acrosomal process: is actin complex with membrane precursors? Cell Biol. Int. Repts 1, 193–200.

Satir, B., Schooley, C. and Satir, P. (1973) Membrane fusion in a model system. Mucocyst secretion in *Tetrahymena.* J. Cell Biol. 56, 153–177.

Sawai, T. and Yoneda, M. (1974) Wave of stiffness propagating along the surface of the newt egg during cleavage. J. Cell Biol. 60, 1–7.

Schatten, G. and Mazia, D. (1976) The penetration of the spermatozoon through the sea urchin egg surface at fertilization. Exp. Cell Res. 98, 325–337.

Schatten, G. and Mazia, D. (1977) The surface events at fertilization: The movements of the spermatozoon through the sea urchin egg surface the roles of the surface layers. J. Supramolec. Struct. 5, 343–369.

Schmell, E., Earles, B. J., Breaux, C. and Lennarz, W. J. (1977) Identification of a sperm receptor on the surface of the eggs of the sea urchin *Arbacia punctulata.* J. Cell Biol. 72, 35–46.

Schneider, G. E. and Lennarz, W. J. (1976) Glycosyl transferases of eggs and embryos of *Arbacia punctulata.* Dev. Biol. 53, 10–20.

Schuel, H., Wilson, W. L., Chen, K. and Lorand, L. (1973) A trypsin-like proteinase localized in cortical granules isolated from unfertilized sea urchin eggs by zonal centrifugation. Role of the enzyme in fertilization. Dev. Biol. 34, 175–186.

Schuetz, A. (1975) Induction of nuclear breakdown and meiosis in *Spisula solidissima* oocytes by calcium ionophores. J. Exp. Zool. 191, 443–446.

Selman, K. and Anderson, E. (1975) The formation and cytochemical characterization of cortical granules in ovarian oocytes of the golden hamster. J. Morphol. 147, 251–274.

Shapiro, B. M. (1975) Limited proteolysis of some egg surface components is an early event following fertilization of the sea urchin, *Strongylocentrotus purpuratus.* Dev. Biol. 46, 88–102.

Smith, A. D. and Winkler, H. (1967) A simple method for the isolation of adrenal chromaffin granules on a large scale. Biochem. J. 103, 480–492.

Spiro, R. G. (1966) Analysis of sugar found in glycoproteins. Methods Enzymol. 8, 3–26.

Stearns, L. W. (1974) Sea Urchin Development, Cellular and Molecular Aspects. Dowden, Hutchinson and Ross, Stroudsburg, Pennsylvania.

Steinhardt, R. A. and Epel, D. (1974) Activation of sea urchin eggs by a Ca^{2+} ionophore. Proc. Nat. Acad. Sci. U.S.A. 71, 1915–1919.

Steinhardt, R. A., Epel, D., Carroll, E. J. and Yanagimachi, R. (1974) Is Ca^{2+} ionophore a universal activator of unfertilized eggs? Nature 252, 41–43.

Steinhardt, R. A., Ludin, L. and Mazia, D. (1971) Bioelectric responses of the echinoderm egg to fertilization. Proc. Nat. Acad. Sci. U.S.A. 68, 2426–2430.

Steinhardt, R. A., Zucker, R. and Schatten, G. (1977) Intracellular calcium release at fertilization in the sea urchin egg. Dev. Biol. 58, 185–196.

Summers, R. G. and Hylander, B. L. (1974) An ultrastructural analysis of fertilization in the sand dollar *Echinarachinus parma.* Cell Tissue Res. 150, 343–368.

Summers, R. G. and Hylander, B. L. (1975) Species specificity of acrosome reaction and primary gamete binding in echinoids. Exp. Cell Res. 96, 63–68.

Summers, R. G. and Hylander, B. L. (1976) Primary gamete binding, quantitative determination of its specificity in echinoid fertilization. Exp. Cell Res. 100, 190–194.

Summers, R. G., Hylander, B. L., Colwin, L. H. and Colwin, A. L. (1975) The functional anatomy of the echinoderm spermatozoan and its interaction with the egg at fertilization. Am. Zool. 15, 523–551.

Takahashi, Y. M. and Sugiyama, M. (1973) Relation between the acrosome reaction and fertiliztion in the sea urchin I. Fertilization in Ca-free sea water with egg-water-treated spermatozoa. Dev. Growth Differ. 15, 261–267.

Talbot, P., Summers, R. G., Hylander, B. L., Keough, E. M. and Franklin, L. E. (1976) The role of calcium in the acrosome reaction: an analysis using ionophore A23187. J. Exp. Zool. 198, 383–392.

Tegner, M. J. (1974) Sea urchin sperm-egg interactions and the block against polyspermy: a scanning electron microscope and experimental study. Ph.D. Thesis, University of California, San Diego.

Tegner, M. J. and Epel, D. (1973) Sea urchin sperm-egg interactions studied with the scanning electron microscope. Science 179, 685–688.

Tegner, M. J. and Epel, D. (1976) Scanning electron microscope studies of sea urchin fertilization. I. Eggs with vitelline layers. J. Exp. Zool. 197, 31–58.

Tilney, L. G. (1975) Actin filaments in the acrosomal reaction of *Limulus* sperm: Motion generated by alterations in the packing of the filaments. J. Cell Biol. 64, 289–310.

Tilney, L. G. (1976) The polymerization of actin III. Aggregates of nonfilamentous actin and its associated proteins: a storage form of actin. J. Cell Biol. 69, 73–89.

Tilney, L. G., Hatano, S., Ishikawa, H. and Mooseker, M. S. (1973) The polymerization of actin: its role in the generation of the acrosomal process of certain echinoderm sperm. J. Cell Biol. 59, 109–126.

Tilney, L. G., Kiehart, D. B., Sardet, C. and Tilney, M. (1978) The polymerization of actin IV. The role of Ca^{2+} and H^+ in the assembly of actin and in membrane fusion in the acrosomal reaction in echinoderm sperm. J. Cell Biol. In press.

Tsuzuki, H., Yoshida, M., Onitake, K. and Aketa, K. (1977) Purification of the sperm-binding factor from the egg of the sea urchin *Hemicentrotus pulcherrimus*. Biochem. Biophys. Res. Comun. 76, 502–511.

Turner, R. S. and Burger, M. M. (1973) Involvement of a carbohydrate group in the active site for surface guided reassociation of animal cells. Nature 244, 509–510.

Tyler, A. and Tyler, B. S. (1966) Physiology of fertilization and early development. In: Physiology of Echinodermata (Boolootian, R. A., ed.) pp. 683–741. John Wiley, New York.

Vacquier, V. D. (1975) The isolation of intact cortical granules from sea urchin eggs: calcium ions trigger granule discharge. Dev. Biol. 43, 62–74.

Vacquier, V. D. (1976) Isolated cortical granules: a model system for studying membrane fusion and calcium-mediated exocytosis. J. Supramolec. Struct. 5, 27–35.

Vacquier, V. D. and Moy, G. W. (1977) Isolation of bindin: the protein responsible for adhesion of sperm to sea urchin eggs. Proc. Nat. Acad. Sci. U.S.A. 74, 2456–2460.

Vacquier, V. D. and Payne, J. E. (1973) Methods for quantitating sea urchin sperm-egg binding. Exp. Cell Res. 82, 227–235.

Vacquier, V. D., Epel, D. and Douglas, L. A. (1972a) Sea urchin eggs release protease activity at fertilization. Nature 237, 34–36.

Vacquier, V. D., Tegner, M. J. and Epel, D. (1972b) Protease activity establishes the block against polyspermy in sea urchin eggs. Nature 240, 352–353.

Vacquier, V. D., Tegner, M. J. and Epel, D. (1973) Protease released from sea urchin eggs at fertilization alters the vitelline layer and aids in preventing polyspermy. Exp. Cell Res. 80, 111–119.

Veron, M. and Shapiro, B. M. (1977) Binding of concanavalin A to the surface of sea urchin eggs and its alteration upon fertilization. J. Biol. Chem. 252, 1286–1292.

Weber, K., Pringle, J. R. and Osborn, M. (1972) Measurement of molecular weights by electrophoresis on SDS-acrylamide gel. Methods Enzymol. 26, 3–27.

Wiese, L. and Wiese, W. (1975) On sexual agglutination and mating type substance in isogamous dioecious chlamydomonads. Dev. Biol. 43, 264–276.

Wolf, D. P., Nishihara, T., West, D. M., Wyrick, R. E. and Hedrick, J. L. (1976) Isolation, physiocochemical properties, and the macromolecular composition of the vitelline and fertilization envelopes from *Xenopus laevis* eggs. Biochemistry 15, 3671–3678.

Yahara, I. and Edelman, G. M. (1975) Electron microscopic analysis of the modulation of lymphocyte receptor mobility. Exp. Cell Res. 91, 125–142.

Yamamoto, R. (1961) Physiology of fertilization in fish eggs. Int. Rev. Cytol. 12, 361–405.

Yen, P. H. and Ballou, C. E. (1974) Partial characterization of the sexual agglutination factor from *Hansenula wingei* V–2340 Type 5 Cells. Biochemistry 13, 2428–2437.

Ziomek, C. A. and Epel, D. (1975) Polyspermy block of *Spisula* eggs is prevented by cytochalasin B. Science 189, 139–141.

Membrane fusion events in the fertilization of vertebrate eggs

2

J. Michael BEDFORD and George W. COOPER

Contents

G. Poste & G. L. Nicolson (eds.) Membrane Fusion, pp. 65–125.

1. Introduction

This volume attests to the common role of membrane fusion as a fundamental device in many life processes. Nowhere, however, is this more important than at its inception. For it is apparent now that membrane fusion, both within the spermatozoon and between sperm and egg, is a critical feature of successive steps of the fertilization process culminating in syngamy. While this fact has become increasingly clear over the last twenty years, the place of membrane fusion during gamete interaction was something of an open question as recently as the late 1950s. Not only were the mechanisms involved in the acrosome reaction and cortical granule dehiscence scarcely studied, but the central event, incorporation of the sperm by the egg, was still seen by some solely as a phagocytic phenomenon (Tyler, 1959). Nothing was known then of the details of these events in any vertebrate class. Painstaking search with the electron microscope has revealed much about the behavior of various gamete organelles during fertilization in a variety of invertebrates (see chapter by Epel and Vacquier in this volume, p. 1) and in a few vertebrates where, with occasional exceptions, such efforts have been confined to a small number of mammals. In nonmammalian vertebrates there appears to be no relevant evidence about the nature of the sperm surface and little about that of the egg, the limited information that exists on the characteristics of vertebrate gamete surfaces again being restricted almost entirely to mammals. Thus, rather surprisingly, we are forced to limit the present discussion largely to consideration of this single vertebrate class.

It is common even now in general texts (e.g., Quinn, 1976) to find the incorrect assumption that the mammalian pattern of gamete fusion resembles that of invertebrates. Actually there are distinct differences, though membrane fusion appears to be an essential event in both groups at several of the sequential steps in gamete union. These steps include the sperm acrosome reaction, fusion of the spermatozoon with the vitelline surface, incorporation of the spermatozoon into the egg, cortical granule exocytosis triggered by sperm/egg fusion, and possible fusion of smooth endoplasmic reticulum to form the envelopes of the pronuclei.

The first visible indication of the acrosome reaction, involving point fusions between the sperm plasma membrane and the outer membrane of the acrosome, releases much of the enzyme content of the acrosome and is an essential prerequisite for sperm penetration through the zona pellucida. Association of the mammalian spermatozoon with the egg surface, a central event, involves initial fusion between the midregion of the sperm head and the oolemma. Unlike the situation described in several invertebrate groups, fusion of the sperm with the mammalian egg does not involve the persisting inner acrosomal membrane lying over the anterior part of the sperm head. For, as a final stage in the incorporation of the sperm head into the egg, egg membrane fuses with egg membrane during closure of a phagocytic vesicle which envelops the acrosomal region of the sperm head in the ingestion phase. Subsequently, as the organelles of the sperm tail are drawn in after the head, the tail plasmalemma is lost through a process of fusion with the oolemma. Interaction of the fertilizing spermatozoon with the egg surface also serves to stimulate the onset of a fusion reaction within the egg itself—that between the envelope of the cortical granules and the oolemma. The eruption of these granules at the vitelline surface is believed to bring about a block to polyspermy, probably by the action of their contents mediated either at the vitelline surface and/or at the zona pellucida, according to species. The envelope of the fertilizing sperm nucleus is lost during entry and subsequent reorganization of the nucleus within the mammalian oocyte. At this time, the now diffuse chromatin becomes limited by a pronuclear membrane which is constituted de novo during the first hours after sperm entry. A few hours after migration of both pronuclei to the center of the egg, disappearance of locally apposed areas of the pronuclear membranes allows the mutual approach and eventual association of the homologous chromatids of the male and female genome. This constitutes syngamy, the final event in the fertilization process.

The present discussion considers in some detail the fusion-associated events of gamete union outlined above. Notwithstanding recent studies of the interaction of mammalian gametes with somatic cells, it is apparent that normal fusion of gametes requires specialized characteristics of their respective surfaces. Moreover, it is evident that fusion can occur only when the membranes in question have been brought to an appropriate state of maturation—a process occurring in the spermatozoon as it passes through the epididymis and female tract. Therefore we will discuss not only the events of membrane fusion and the factors influencing them but also the changing character of the gamete membranes themselves. The relevance of current concepts about the organization of biological membranes to those of the spermatozoon in particular has been outlined recently by Johnson (1975). The present discussion focuses specifically on the known features of and current speculation about gamete membranes, little consideration being given to experimental or theoretical aspects of other membrane systems since these are covered widely elsewhere in this and previous volumes of the series.

2. The plasmalemma of the mammalian spermatozoon

2.1. General considerations

The character of the limiting membrane of a fertile spermatozoon is a product of the moieties determined by its genome and also of modifications, additive or otherwise, ordained by the milieus of the male and female reproductive tracts.

From the viewpoint of the student of somatic cell membranes, the mammalian sperm surface displays unusual properties in at least two respects, which we will discuss in more detail further on. In most mammals studied, change(s) in the chemical character of the spermatozoon surface have been detected during its ontogeny in the testis and also in its passage through the epididymis after release from the grasp of the Sertoli cell. Surface change is probably also an important feature of the subsequent phase of maturation in the female tract as a facet of *capacitation*, a process which finally enables the ejaculated spermatozoon to penetrate the ovum. A second unusual attribute of the mammalian spermatozoon, which is enveloped by a continuous plasma membrane, is an often sharply defined regional differentiation of its surface character. The existence of restricted domains is expressed clearly by its surface, and at least four distinct regions are recognizable on the basis of their different surface interactions and/or their affinities for surface ligands.

The relative stability of the plasmalemmal organization also merits brief consideration in an introduction of the idiosyncrasies of the spermatozoon surface. Observation of a turnover of surface components of somatic cells, the extent of which varies according to circumstance, fits well the current concept of the plasmalemma as being a fluid mosaic membrane. However, the genetically based transformation of cell surfaces by viruses and other agents, the replacement of some cell surface components after "capping," and to a lesser degree the normal repair and maintenance of the surface in various cell types, appear to involve a replacement of surface membrane components that probably is not possible in spermatozoa. For the spermatozoon is a highly differentiated genetically-repressed cell, irrevocably committed as such, in which the cytoplasmic machinery for protein synthesis, assembly, and insertion of new protein/lipid complexes into the plasma membrane appears to be minimal and perhaps absent. Thus, membrane modifications occurring in differentiated spermatozoa probably must be ordained mainly by external factors in the environment of the male and, subsequently, of the female tract.

Finally, it should be remembered that despite the probable absence of repair mechanisms, the mature mammalian spermatozoon and its plasmalemma can remain functional in the cauda epididymidis for several weeks, and that the spermatozoa of some bats may remain so in the female tract for several months. In a cell that probably does not possess the ability to repair its plasmalemma, in which the precisely defined domains of its plasmalemma imply the existence of a less than fluid arrangement, and which displays the ability to undergo sudden regulated breakdown or fusion with other membranes, it is tempting to attribute unusual properties not seen in other cell types to the limiting membrane of the spermatozoon.

2.2. Biochemical character

The chemical characteristics of the plasmalemma of mature mammalian spermatozoa are poorly defined. There is no mention of this aspect in the classical treatise of Mann (1964), and little information has been obtained since then. The isolation of a pure preparation of sperm plasma membrane presents singular problems which have not been entirely overcome. Even the most careful procedures to date result in a contamination of plasma membrane preparations with acrosomal membrane to a level of 20% or more (Esbenshade and Clegg, 1976). Thus, in the latter study, the possibility can not be ruled out that some of the 11 to 13 entities resolved as plasma membrane proteins were in fact contaminating components. The lipid composition of whole spermatozoa has been recorded for some mammals (Johnson et al., 1972; Poulos et al., 1973; Snider and Clegg, 1975; Darin-Bennett et al., 1976), but there appear to have been no lipid analyses of sperm plasma membrane preparations. This area of investigation is complicated further by the discouraging problem of the regional specialization of sperm topography, since this may make it difficult to interpret the functional significance of any one component isolated from sperm plasma membranes unless it can be assigned to a particular surface domain (Fig. 1).

Lectin-binding to the surface of spermatozoa implies the existence of carbohydrate moieties as surface components linked to intrinsic membrane proteins or perhaps, in some cases, glycolipids. Species variations in the pattern of lectin-binding, or those obtained with certain other surfaces markers, make it unlikely that there is a common molecular configuration which characterizes the surface of spermatozoa in different mammals. The claims that neuraminidase treatment alters the polarographic qualities (Rosado et al., 1973) and electrophoretic mobility of spermatozoa (Bey, 1965) suggest that sialic acid is a component of the sperm surface in several species at least. This is also implied by reports that myxoviruses have an affinity for a variety of vertebrate spermatozoa (Peleg and Ianconescu, 1966) and that Sendai viruses agglutinate rabbit spermatozoa head-to-head (Ericsson et al., 1971), these viruses being bound selectively to the periacrosomal surface of the sperm head (Buthala et al., 1971).

Sialic acid is reported to be present in the epididymal milieu (Bose et al., 1966; Fournier, 1968; Laporte, 1970) and Prasad and co-workers have concluded that the levels of sialic acid-associated components on spermatozoa vary according to their maturity (Arora et al., 1975; Bose et al., 1975; Rajalakshmi et al., 1976); the amounts bound to spermatozoa from the caput epididymidis being significantly higher than those bound to cauda spermatozoa. This is puzzling because, at low pH where sialic acid-contributed negative charges would still be expressed, the binding capacity of monkey spermatozoa and those of other mammals for cationized ferric colloids *increases* very significantly with epididymal passage (Cooper and Bedford, 1971a; Bedford et al., 1972, 1973; Yanagimachi et al., 1972; Bedford, 1974a), as does the expression of a sialic acid containing glycoprotein on rat spermatozoa (Olson and Hamilton, 1978). The latter observations are also consistent with the fact that Sendai viruses, which bind to cells via receptors terminated by sialic acid residues, do not agglutinate rabbit spermatozoa isolated from the caput epididymidis (Buthala et al., 1971), though their lack of motility might account for this. Although neuraminidase removes surface anionic charges detectable on erythrocytes at pH 1.8 (Chien et al., 1974), none of the negative charge expressed at the surface of spermatozoa at pH 1.8 is removed by neuraminidase (Yanagimachi et al., 1972; Koehler and Perkins, 1974;

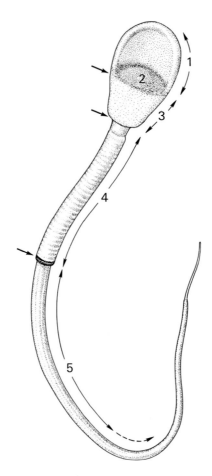

Fig. 1. Generalized diagram of a mammalian spermatozoon. Regions 1, 2, 3, 4, and 5 delineate the acrosomal, equatorial, postacrosomal, tail midpiece, and tail principal piece surface domains, respectively. Arrows pinpoint the major zones where the changing relationship of the plasmalemma to underlying structures may act to restrict lateral movement of surface components, thereby delineating the different surface domains. Although distinguished as a distinct zone by virtue of the underlying equatorial segment, to date no conclusive evidence exists that the properties of the surface overlying the equatorial region (2) differ from that over the remainder of the acrosome in the unreacted spermatozoon.

Fléchon, 1975; Fléchon and Morstin, 1975), even after mild hydrolysis of glutaraldehyde-stabilized spermatozoa with KOH (Cooper and Bedford, unpublished observations; cf. Cooper and Bedford, 1971b). The affinity of myxoviruses for the surface of the head rather than the tail of spermatozoa may be an indication that the negative charges expressed consistently by the tail at low pH are not contributed primarily by sialic acid, but by other groups which remain partially ionized at this low pH (e.g. sulfate). This seems questionable, however, as ferric colloid binding at pH 1.8 to human and rabbit spermatozoa can be

completely abolished by prior acid hydrolysis of glutaraldehyde-stabilized spermatozoa while retaining the ultrastructural integrity of the plasma membrane
(Cooper, unpublished observations; see Chien et al., 1974 for references to selective hydrolysis of sialic acid from cells).

There is a suggestion that the sperm plasmalemma has an unusually high
complement of -SH groups, varying somewhat between species (Mercado et al.,
1976). As judged by the ready solubility of the plasmalemma in a variety of detergents, however, the concentration of -SH groups, and thus the potential for
stabilization by their covalent crosslinking, must be small compared with that
seen in the outer mitochondrial membranes of mammalian spermatozoa (Wooding, 1973; Bedford and Calvin, 1974; Temple-Smith and Bedford, 1976). The
sperm surface has a glycocalyx (Fléchon, 1975), demonstrable also in guinea pig
spermatozoa by tannic acid fixation (D. Friend, unpublished observation) and
containing a variety of ATPase activities (see section 5.2.2.). These observations
constitute the extent of our knowledge of integral sperm plasmalemmal enzymes
and glycoproteins.

2.3. Regional differentiation

The character of the sperm surface reflected in its physical, immunologic, and
biochemical properties differs markedly over its various anatomic regions (Figs.
1–4). It appears, moreover, that the surface overlying the acrosomal region ex-
presses further localized differences not delineated by the boundaries of underlying organelles (Figs. 5, 12).

The difference between the head and tail surface of spermatozoa could be
foreseen in the observations of head-to-head and tail-to-tail agglutination dependent solely on adhesive properties of the surface (Yamane, 1921; Smith,
1949; Bedford, 1965a), on the orientation of spermatozoa in an electric field
(Bangham, 1961; Nevo et al., 1961), and on head-head and tail-tail agglutination
produced by antisperm antibody (Henle et al., 1938). The observations by Henle
and co-workers implying regional expression of certain antigens on the sperm
surface have since been confirmed and refined using visual markers for more
specific antigenic determinants. However, some of the recent reports which
make claims to this effect are difficult to evaluate. The visual evidence they provide is often not critical enough, perhaps because there may be relatively few determinants per unit area and also because of a heterogenity of the sperm population in this respect. A membrane glycoprotein isolated by O'Rand and Metz
(1976) is apparently present over all regions of the sperm surface, but other antigens are confined to distinct regions of the head. Among the latter that appear
most clearcut is periacrosomal localization of the T autoantigen in guinea pig
spermatozoa (Toullet et al., 1973), and of a similarly restricted location of an autoantigen in mice (Fellous et al., 1974). This distribution may agree with the
localized disruption of the periacrosomal plasmalemma of rabbit spermatozoa by
the complement-dependent autoantibody present in serum (Bedford, 1969b).
There is a clear local expression of the H-Y antigen over the acrosome of the

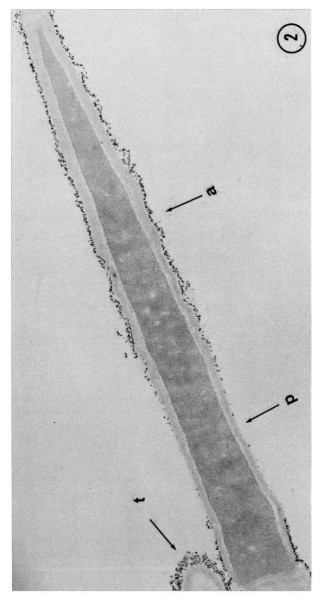

Fig. 2. Electron micrograph of the unstained head of a mature spermatozoon from the cauda epididymidis of the armadillo, *Dasypus novemcinctus*, exposed at pH 2.8 to cationic ferric oxide hydrosols according to a modification of the method of Gasic et al. (1968). The consistent absence of regular binding of colloid to the postacrosomal region (p) (region 3 in Fig. 1) clearly illustrates its different surface character compared with that over the acrosome (a) and tail (t). ×30,000.

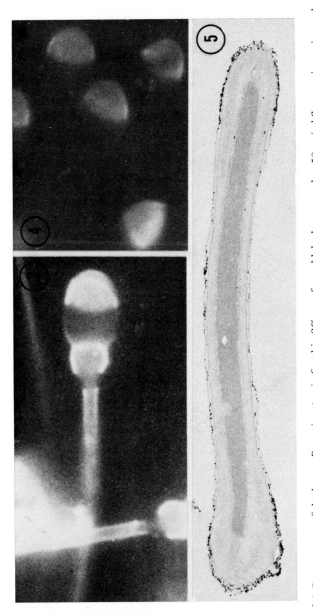

Figure 3 and 4. Spermatozoa of the hyrax, *Procavia capensis*, fixed in 2% paraformaldehyde, exposed to 50 μg/ml fluorescein-conjugated wheat germ agglutinin (WGA; Miles-Yeda) and photographed under oil with a Leitz incident-light fluorescent microscope. The immature spermatozoon from the head of the epididymis (Fig. 3) possesses receptors for WGA over the acrosome, cytoplasmic droplet, tail midpiece, and principal piece. After passage through the epididymis, the distribution of WGA receptors on spermatozoa from the vas deferens (Fig. 4) has become restricted to the periacrosomal surface.

Fig. 5. Electron micrograph of a transverse section through the more rostral region of the acrosome of a sperm from the cauda epididymidis of the rabbit. This unstained preparation was exposed to cationic ferric colloid at pH 2.6 after glutaraldehyde fixation and before postosmication. The conditions of the reaction were chosen to demonstrate the greater density of negative charge over the periacrosomal plasmalemma at the periphery of the acrosome, reflected here in the binding of cationic ferric colloid particles. ×35,000.

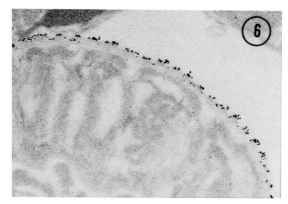

Fig. 6. Unstained cross-section of the tail midpiece of a mature spermatozoon of the opossum, *Didelphis virginianus*. This, exposed to cationic ferric colloid at pH 1.8 before postosmication, displays a regular interrupted pattern of colloid binding at the surface coincident with regular, spaced aggregations of material lying beneath the plasmalemma (see also Olson et al., 1977). ×58,000.

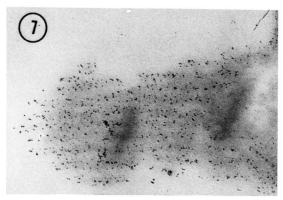

Fig. 7. A glancing tangential section of the surface of an opossum sperm tail midpiece prepared as in Fig. 6. This illustrates, in planar view, the spaced linearity of binding of colloid particles at the surface of the midpiece in this species. Unstained. ×54,000.

mouse (Koo et al., 1973) and of a primitive teratocarcinoma antigen on the post-acrosomal region of mouse and human spermatozoa (Fellous et al., 1974). Strikingly, however, the latter two studies and sex-ratio manipulation experiments with anti- H-Y antisera (Bennett and Boyse, 1973; Mathieson, 1976) suggest that less than 20% of the sperm population express the H-Y antigen at a detectable level. A final caveat in this type of question must be the possibility of adsorption (cf. Erickson et al., 1975), or 'leakage' to the surface of subsurface antigen (e.g., Hjört and Hansen, 1971; Husted, 1975; also see Johnson and Edidin, 1972). Indeed, the H-2 antigens on mouse spermatozoa appear to be located at the surface of the outer acrosomal membrane rather than within the plasmalemma (Vojtíšková et al., 1974).

The regionally differentiated state of the sperm plasmalemma has been revealed precisely and perhaps with greater regularity through a variety of other methods which, reflecting surface charge densities (Gasic et al., 1968), the presence of glycoproteins (Fléchon, 1975), terminal oligosaccharides (Edelman and Millette, 1971; Nicolson and Yanagimachi, 1974), and membrane architecture (Plattner, 1971; Koehler, 1972; Friend and Fawcell, 1974; Fawcett, 1975; Olson et al., 1977), allow distinctions to be made between one or more of the regions overlying the acrosomal and postacrosomal domains of the head, and the midpiece and principal piece of the tail. Regional distribution of some membrane enzymes also occurs, the ATPase expressed on rabbit spermatozoa at pH 7.0 being evident only within the periacrosomal region of the plasmalemma and not on the postacrosomal or flagellar surface (Gordon and Dandekar, 1977).

We have already alluded to the finding that the surface can express some heterogeneity within one anatomically defined domain. For instance, only the periphery of the rabbit sperm acrosome binds colloidal iron particles to any degree when exposed to ferric colloids at pH 2.6, the more posterior region of the periacrosomal surface remaining essentially free of bound colloidal particles at this pH (Bedford, 1974b; see Fig. 5). It is in this rostral region that the plasmalemma often appears to have a relatively closer relationship with the underlying acrosome, and in which lesions induced by antibody and complement first appear (Russo and Metz, 1974). There is also a suggestion that incubation of spermatozoa in the female tract results in a loss of Concanavalin A (Con A) binding ability over this restricted region of the acrosome (Gordon et al., 1974). A rabbit antiserum prepared against relatively small numbers of mouse spermatozoa appears to bind only to the swollen spine or periphery of the acrosome, the remainder of the acrosomal surface being devoid of fluorescence (P. Higgins, unpublished observations; and Fig. 12). In a similar vein, the periphery of the guinea pig acrosome is distinguished by its indifference to an antibody which does bind to the remainder of the periacrosomal surface (Koehler and Perkins, 1974). A different and unusual pattern of surface regional specialization is seen on the sperm head of *Suncus murinus,* a musk shrew. In addition to the usual regional differentiation over the long axis, an intriguing dorso-ventral asymmetry of surface character is expressed on all spermatozoa (Cooper and Bedford, 1976). The significance of this asymmetry for sperm/egg interaction in *Suncus* is unclear and is currently under investigation.

Not surprisingly, different techniques for examination of surface character do not always distinguish the same regional patterns on the sperm surface. For instance, regionality expressed in negative-charge density, as judged by ferric colloid binding, is not necessarily paralleled by the patterns obtained with lectin markers. Moreover, under given conditions, the ferric colloid binding patterns and those obtained with lectins over a particular region (see below) tend to vary widely among species, pointing to an individual molecular nature of the exposed surface components in different mammals. The exception in this respect has been the surface of the mainpiece of the tail whose character, as reflected in negative-charge density under given conditions, seems remarkably constant among many species (Cooper and Bedford, unpublished observations).

Among the least easily distinguished is the surface regionality of the human spermatozoon. Unlike those of most eutheria, human spermatozoa do not undergo head-to-head autoagglutination in serum-containing media, an indifference shared also by the spermatozoa of anthropoid monkeys (Bedford, unpublished observations). Indeed, comparative studies of sperm/egg interaction suggest that human spermatozoa have become unusually restricted in their sur-

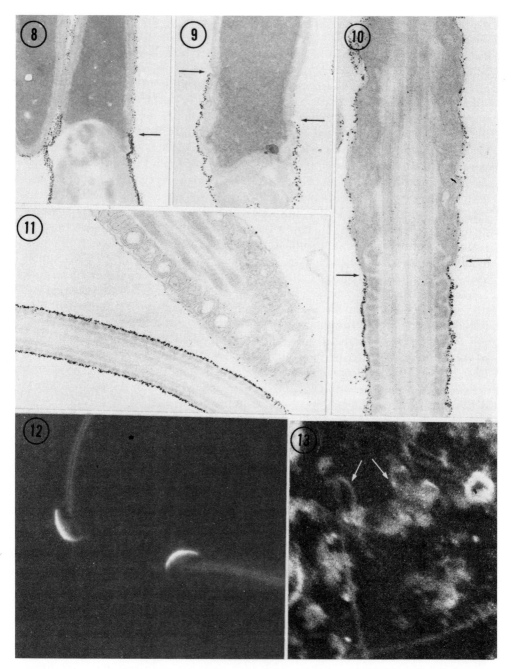

Figs. 8 and 9. Transmission electron micrographs of the head-neck region of the spermatozoon of the tree shrew, *Tupaia glis,* exposed to cationic ferric colloid at pH 3.0 (Fig. 8) and of the mouse at pH 1.8 (Fig. 9). In both cases a distinct change in the pattern of colloid binding (arrows) coincides with the location of the posterior ring, as it does in the rabbit (Fig. 10). Whether this relationship of the surface properties to the posterior ring is ubiquitous among mammals is not yet clear (e.g., Fig. 2). Some binding is seen on the postacrosomal region of the *Tupaia* sperm at pH 3.0, but at pH 1.8 there is none on this region in the mouse. Fig. 8 ×27,900. Fig. 9:×33,300.

face affinities (Bedford, 1977). Distinct binding patterns that distinguish the major part of the acrosome, its equatorial segment, the postacrosomal region, and the tail surface have been demonstrated on human spermatozoa with labelled antibodies from patients screened in a fertility clinic (Hjört and Hansen, 1971; Husted, 1975). The fact that the method of preparation of these spermatozoa involved air-drying makes it uncertain whether all the fluorescence seen really reflects surface antigens only, or the exposure also of subsurface antigens from within the cell. In our hands, regional patterns were not revealed on ejaculated human spermatozoa exposed to fluorescein-labeled wheat germ agglutinin or Con A. Unlike rabbit, bull, and ram spermatozoa, human spermatozoa do not orient in an electrophoretic field (Bedford, unpublished observations), nor do they display any regionality of binding when exposed to cationic ferric colloids between pH 1.8 and 3.0 (Bedford et al., 1973; Bedford, 1974a), suggesting that regional surface negative-charge differences on human spermatozoa are minimal, if they exist at all. Unfortunately, the picture of their surface character gained from polarographic measurement by Velázquez and Rosado (1972) does not resolve this point since this technique does not reveal the regionality of the sperm surface. Furthermore, their claim of a direct relationship of surface negative-charge density with the ionic strength of the medium, rather than the expected inverse one recorded in other studies of spermatozoa and somatic cells (Heard and Seaman, 1960; Bangham, 1961), as well as the very large calcium-binding values (cf. Blank et al., 1974), suggests that such polarographic data be viewed with some caution until confirmed.

The question of the relative fluidity of the membrane and the mobility of membrane components in the sperm plasmalemma has not been studied sufficiently, and the available evidence seems to be conflicting. Visible clumping of fluorescein-labeled antibody over the acrosome in ejaculated rabbit spermatozoa has been interpreted as an indication that the surface antigen—a single

Figs. 10–13. Photographs on opposite page.

Fig. 10. Section of the tail of an ejaculated rabbit spermatozoon, reacted after glutaraldehyde fixation with cationic ferric colloid at pH 3.5. At this pH the surface of the principal piece binds colloid avidly, whereas that of the midpiece shows only light sparse binding. The point of change from one pattern to the other coincides precisely with the end of the annulus lying beneath the plasmalemma (arrows). ×36,900.

Fig. 11. Sections of midpiece and mainpiece of the tails of ejaculated rabbit spermatozoa prepared without staining after reaction with cationic ferric colloid at pH 1.8. At this lower pH the different quality of the surface of the two adjacent regions of the tail is even more striking than that shown in Fig. 10. ×31,500.

Figs. 12 and 13. Fig. 12 shows mouse vas deferens spermatozoa exposed to rabbit antibody prepared against mouse vas deferens spermatozoa, and after washing to fluorescein-conjugated goat anti-rabbit IgG. The vas deferens spermatozoa stain specifically in a narrow band which delineates the crest of the acrosome only. By contrast, in Fig. 13, mouse testicular spermatozoa (arrows) treated similarly show no specific binding of anti-mouse sperm antibody. This type of experiment suggests that new antigen concentrated over a limited region of the periacrosomal plasmalemma, is acquired or is exposed on this surface as mouse spermatozoa pass from the testis through the epididymis to the vas deferens. (Photographs courtesy of Dr. Paul Higgins.)

membrane glycoprotein—is able to move laterally within the plane of the plasmalemma in this region (O'Rand, 1977a). On the other hand, Nicolson and Yanagimachi (1974) were unable to detect any redistribution of components in the periacrosomal membrane having an affinity for ferritin-labeled *Ricinus communis* agglutinin. They concluded that there is minimal or no lateral mobility of sperm membrane components in the acrosomal (and tail) region, but their observation of an apparent redistribution of labeled membrane components over the postacrosomal surface provoked the suggestion that this domain has a lability related to its possible role as the site of initial fusion with the egg. Nicolson and Yanagimachi (1974) failed, surprisingly, to draw attention to the fact that this appears true only for spermatozoa released from the caput epididymidis and not for those from the cauda epididymidis. Caput spermatozoa undergo marked changes in their surface properties, as well as lipid composition, during epididymal passage, and are not yet functionally competent in several other respects. One should be wary also of making functional interpretations of observations on epididymal or ejaculated spermatozoa in this respect, since important changes may occur in their membranes during passage through the female tract (see section 4.2).

The molecular features that ordain the abrupt regionality of the sperm plasmalemma have yet to be identified. Thin sections generally give no suggestion of an internal skeleton to the plasmalemma that might act to restrain transmembrane proteins over the acrosomal region or over the tail. The plasmalemma is closely related to underlying structures at several points along its axis, however, and these points often coincide with a discontinuity of surface properties between adjacent regions of the plasmalemma (Figs. 1–10). Thus, the plasmalemma over the head forms a sudden, more intimate, association with the underlying perinuclear material of the poastacrosomal region at the posterior border of the acrosome (Fawcett and Ito, 1965), and a "weld" between the plasmalemma and the nuclear membranes is seen at the posterior nuclear ring immediately rostral to the neck (Friend and Fawcett, 1974; Fawcett, 1975; see Figs. 8, 9). In a number of spermatozoa both of these sites delineate the limits of sperm surface domains. The more subtle regionality of the plasmalemma overlying the periphery of the acrosome (Figs. 4, 12) cannot easily be explained in this way, though it is true that the cell membrane is unusually closely applied to the underlying acrosome at this point. A similar discontinuity in surface character is sometimes evident over the tail at the annulus (Fig. 10), the terminus of the mitochondrial midpiece, over which the plasmalemma is closely applied (Fawcett, 1970; also see Fig. 11).

A rare example of an obvious relationship between surface characteristics and the distribution of subplasmalemmal components is seen in the sperm tail of the opossum, *Didelphis virginianus* (Olson et al., 1977; Cooper and Bedford, unpublished observation). As can be seen in Figs. 6 and 7, there is a linear arrangement of the negative charges expressed at pH 1.8 on the glutaraldehyde-stabilized midpiece of the tail and this is coincident with the regular spaced arrangement of electron-dense material which lies immediately subjacent to the plasmalemma.

3. The membrane of the acrosome

Little is known to date about the specific character of the acrosome membrane. It displays a structural asymmetry in which the outer lamina appears thicker than the inner (Bedford and Nicander, 1971; Roomans, 1975), and has a regionality reflected in a differential stability of its outer, equatorial, and inner portions respectively. Most of the outer membrane of the acrosome is easily disrupted by physical (Zahler and Doak, 1975) or chemical treatment (Hartree and Srivastava, 1965; Multamäki, 1973), by the action of complement-dependent antibody (Bedford, 1969a; Russo and Metz, 1974) and selectively during the acrosome reaction itself. The equatorial segment seems very stable by comparison (e.g., Srivastava et al., 1974), and the inner membrane of the acrosome even more so. The stability of the equatorial segment, which often has a septate appearance in the egg (Barros and Franklin, 1968), may be related to an unusual pentalaminar appearance of its membrane (Roomans, 1975) and to a particular character evident also in surface replicas (Phillips, 1977). The inner membrane of the acrosome, which is resistant to Hyamine (Hartree and Srivastava, 1965; Wooding, 1975) and persists for a period even after decondensation of the sperm nucleus within the egg, does not display special structural characteristics, at least in electronmicrographs. Consideration should be given to the possibility that its robust quality (and that of the postacrosomal plasmalemma) may be determined by the perinuclear material beneath it, which, in mature spermatozoa, is stabilized by -S-S- crosslinks (Calvin and Bedford, 1971).

The isolation of pure preparations of acrosomal membrane appears difficult, though the approach of Gall and Ohsumi (1976) may prove useful in this respect for certain species. Zahler and Doak (1975) have obtained outer acrosomal membrane preparations following physical disruption, but although the membrane pellet had only low acid phosphatase activity, no detailed consideration was given to possible contamination by head plasma membrane in that study, and the same comment may be made about the results of Multamäki (1973).

4. Maturation of sperm membranes

4.1. Changes in the male tract

Emphasis generally has been on membrane changes detected in the post-testicular phase of sperm maturation, but the first significant modification of the antigenic character of the surface of male germ cells begins in meiosis. Using complement-dependent lysis as an end point, O'Rand and Romrell (1977) have shown that new surface antigen(s) appear on spermatogenic cells as they reach the pachytene stage of meiosis. Other studies employing separated cell populations with immunofluorescent microscopy, complement-mediated cytotoxicity, and quantitation of surface immunoglobulin receptors confirm this finding and show further that the antigenic components that first appear on pachytene primary spermatocytes remain on round spermatids, residual bodies, and on mature spermatozoa, in which they tend to be confined to the acrosomal region and midpiece (Millette and Bellvé, 1977). Experiments employing ferritin-labeled antibody against a rabbit sperm membrane glycoprotein indicate that there is lateral mobility of this antigen in the plasmalemma of early round spermatids. However, restraints appear to be imposed at later stages of differentiation, as indicated by the absence of aggregation and patching of labeled antibody on both the sperm leaving the testis and the residual bodies which remain (Romrell and O'Rand, unpublished observations).

Of particular interest as a surface phenomenon is the phase of spermiation in which there is a programmed dissociation between the closely apposed plasma membranes of the Sertoli cell and the presumptive spermatozoon. In some species this is heralded by a disorganization of the periacrosomal cisternal complex within the Sertoli cell (Cooper and Bedford, 1976) and may be a response to specific change in the surface of the Sertoli cell membrane and/or that of the late spermatid (Bedford and Nicander, 1971), which results in an alteration of junctional specializations between the two (Ross, 1976). Recent observations by Russell and Clermont (1976) suggest that spermiation in the rat, monkey and perhaps other mammals involves rather more than a separation of apposing surfaces, however, and may require scission of fine spermatid surface projections held within deep recesses of the Sertoli cell.

Modification in the sperm surface character during epididymal transit was first demonstrated directly in rabbits, where a dissimilarity of the surface in caput and cauda spermatozoa was revealed by their electrophoretic behaviour. A majority of cooled, living rabbit spermatozoa from the caput epididymidis adopt a head-anode orientation (50V, 1.0 amp in 0.45M NaCl), whereas under the same conditions essentially all released from the cauda display a tail-anode orientation; this and the overall electrophoretic mobility being more rapid than that of spermatozoa from the caput (Bedford, 1963). The simple observation that significant numbers of moving rabbit spermatozoa from the caput epididymidis undergo tail-to-tail autoagglutination in a medium containing normal serum, whereas mature spermatozoa released from the cauda region agglutinate head-to-head (Bedford, 1965a), also suggested some regional modification of the properties of the surface during epididymal transit. Progressively developing associations between groups of spermatozoa within the epididymis of animals such as the guinea pig (Fawcett and Hollenberg, 1963; Fawcett, 1975), the flying squirrel (Martan and Hruban, 1970; Martan et al., 1971), and New World marsupials (Biggers, 1966) perhaps also reflect a concomitant alteration of such characteristics there.

At the molecular level, change in the sperm surface is reflected in modifications of the activity of a specific membrane-associated enzyme, and in the chemistry of the plasma membrane. Chulavatnatol and Yindepit (1976) have reported changes in the activity of a rat sperm-surface enzyme which has the character of a Na^+/K^+-dependent ATPase. Its specific activity decreases significantly during passage from caput to cauda epididymidis, being reduced then from 6.1 to 2.2

nM/min/10^6 spermatozoa. Whether this reduction is due to true disappearance of the enzyme or to masking of its activity by epididymal secretory products is unclear. Although this oubain-sensitive enzyme is possibly homologous with that detected cytochemically in mature epididymal rabbit spermatozoa (Gordon, 1973), the activity of the latter does not appear to change with epididymal maturation (Gordon and Dandekar, 1977). Species differences may be involved here, but the variability of findings in these and other studies (see section 5.2.3) makes it impossible at present to derive any firm conclusions about the role of ATPases in sperm membrane function.

Detection of alterations in sperm surface chemistry has been refined by the use of techniques that employ visual surface markers and in one recent study by isolation of a membrane glycoprotein labeled by a galactose oxidase-sodium metaperiodate probe (Olson and Hamilton, 1978). The binding patterns of cationized ferric oxide hydrosols on glutaraldehyde-stabilized spermatozoa at different pHs reveal changes in specific regions of the sperm surface during epididymal transit in a variety of mammals (Cooper and Bedford, 1971a; Bedford et al., 1972, 1973; Yanagimachi et al., 1972; Bedford, 1974a; Fléchon, 1975; Temple-Smith and Bedford, 1976) with one exception, the musk shrew *Suncus murinus* (Cooper and Bedford, 1976). The validity of this method of Gasic and co-workers (1968) has been questioned by Fawcett (1975) because of the low pHs at which this colloid binding must be carried out. This technique cannot be used to quantitate surface charge, does not measure net charge, and is not a substitute for cell electrophoresis. However, because of prior glutaraldehyde stabilization, this approach does not produce artefactual results for the questions asked (i.e., are there regional differences over the sperm surface, where are their limits, and how do these surface domains change with maturation in the male excurrent duct and during capacitation?). The ferric colloid-binding patterns on rabbit spermatozoa from the caput epididymidis and the changes in these with epididymal passage (Cooper and Bedford, 1971a; Yanagimachi et al., 1972) are consistent with their behavior in the cell electrophoresis system (Bedford, 1963). Moreover, in cells where negative surface charge is attributable to sialic acid (i.e. is neuraminidase sensitive), cell electrophoretic mobility and the amount of colloid bound at low pH are proportional to the surface complement of N-acetylneuraminic acid measured chemically (Chien et al., 1974). Furthermore, as the pH of the reaction is raised, and thus as the pK of different surface components is approached or exceeded, the binding of ferric colloid becomes denser.

Fléchon (1975) has reported a change in the glycoprotein at the surface of rabbit spermatozoa that results in a reduced staining of the surface by the phosphotungstic acid-hydrochloric acid technique (Pease, 1970). Although not yet localized to any specific region of the spermatozoon, Olson and Hamilton (1978) show in an elegant study the appearance in rat cauda epididymal spermatozoa of a membrane glycoprotein of 37,000 daltons not detectable on caput spermatozoa. Similarly revealing has been the visualization of changing lectin patterns by use of fluorescein isothiocyanate or ferritin-conjugated preparations

(Edelman and Millette, 1971; Nicolson, 1974; and Figs. 3, 4). The use of fluoresceinated lectins has revealed changes in the distribution of their receptors at the surface of spermatozoa passing through the epididymis of the rabbit (Nicolson, 1974), the Australian possum (Temple-Smith and Bedford, 1976), and the testicondid hyrax and armadillo (Bedford and Millar, 1978). A precise delineation of the regionality of terminal oligosaccharides associated with surface glycoproteins, or possibly glycolipids (Boldt et al., 1977), can also be obtained with ferritin-labeled lectins in the electron microscope (Nicolson, 1974; Nicolson and Yanagimachi, 1974).

In wild-type or T/t locus mutant mice also there is a change in the specific lectin (Con A) binding pattern on spermatozoa as they traverse the epididymis. While only 9% of caput spermatozoa show intense acrosomal and postacrosomal binding, 74% exhibit this in populations recovered from the vas deferens. However, spermatozoa recovered from the vas deferens of heterozygotes for three different alleles (t^{w2}, t^{w5}, and t^{w18}) show a significantly lower proportion with the "mature" pattern of binding (Dooher and Bennett, unpublished observations). This suggests that the process of maturation may often be retarded in such heterozygotes and indicates the possible influence of the sperm genome on the accomplishment of surface maturation changes.

Immunologic markers demonstrate the same principle in mice, since change can be revealed in the character of the antigens expressed over the periacrosomal region of the plasmalemma of mouse spermatozoa as they tranverse the epididymis. Spermatozoa from the testis, caput, and cauda of various strains of mice were exposed to an antiserum prepared in rabbits against whole mouse vas deferens spermatozoa, followed by fluorescein-conjugated anti-rabbit IgG. No fluorescence appeared on spermatozoa from testis and caput epididymidis, but a distinct specific fluorescent crest persisted over the greater curvature of the acrosome of washed spermatozoa from the cauda and vas deferens (P. Higgins, unpublished observations; and Figs. 12, 13).

Although all of these results suggest regional changes in the molecular species of the glycoproteins and, possibly, glycolipids represented at the outer face of the sperm plasmalemma, we are unable to be more specific. The developing presence of ionized groups at low pH with maturation suggests the acquisition of sialic acid-associated moieties which seem to be present in the epididymal environment (Fournier, 1968; Laporte, 1970; Rajalakshmi et al., 1976). Their presence at the sperm surface also becomes credible in light of the claim of Bey (1965) that neuraminidase treatment will modify the electrophoretic mobility of bull spermatozoa and of the observation that the glycoprotein acquired by rat cauda spermatozoa probably possesses terminal sialic residues on its oligosaccharide chains (Olson and Hamilton, 1978). Surprisingly, however, it has not been possible to modify by neuraminidase treatment, with or without prior KOH hydrolysis, the charge distribution reflected at the sperm surface with the cationic ferric colloid technique (see section 2.2). Changes occur in the total phospholipid composition of spermatozoa with epididymal maturation (Poulos et al., 1973), but how these alterations relate to the sperm membranes is unknown. Since the concept of fusion is linked to the notion of increased fluidity in inherently stable bilayers, it is perhaps relevant that the periacrosomal plasmalemma in mature epididymal spermatozoa often seems more easily distorted by preparation for electron microscopy than that in immature spermatozoa from the caput epididymidis (Bedford, 1975). Whether this signifies some necessary reduction of its stability or rigidity with epididymal maturation remains to be established. Apart from this increasing looseness or flexibility of the periacrosomal plasmalemma, we are unaware of additional examples of modifications in it that are expressed visibly. However, the surface changes occurring at this time in the spermatozoa of the marsupial possum *Trichosurus vulpecula,* are visually more dramatic. With maturation-associated reduction in the surface area of its acrosome, vesicles of the overlying plasmalemma are budded off to the exterior, presumably as a means of disposing of excess membrane. In the tail also, obvious invaginations appear in the plasmalemma overlying the midpiece of the flagellum as the spermatozoa reach the lower corpus epididymidis (Harding et al., 1975, 1976; Temple-Smith and Bedford, 1976). These have been likened to endocytotic vesicles by Harding and co-workers, but we believe that they remain as invaginations only, maintaining continuity with the surface, possibly to facilitate exchange between the mitochondria and the surface, across which there is an unusually wide separation in this species (Temple-Smith and Bedford, 1976).

In conclusion, we do not know whether the changes occurring in sperm surface character during the epididymal passage result from the acquisition of epididymal secretory components. At present it is also difficult to interpret the biological significance and role of these changes for the future economy of the spermatozoon, particularly in the fusion-associated events that will follow at fertilization. Preliminary observation of various other classes suggests that such changes are not a common characteristic of vertebrates other than the metatherian and eutherian mammals (Bedford, unpublished observation). Though Campanella (personal communication) suggests that the amphibian *Discoglossus pictus* has "seminal vesicles", during storage within which its spermatozoa acquire the ability to fertilize. Thus, together with the need for capacitation, these changes may represent a response to evolutionary change in the physiology of the female tract and/or the ovum of these mammals.

4.2. Changes in the female tract

It might appear that spermatozoa in the vas deferens have completed maturation of their membranes because they now display the potential, as free cells, to fertilize an egg. There is, however, a mass of evidence (mostly indirect) to indicate the need for further integral modification of their membranes before the acrosome reaction in order for the spermatozoon to be able to penetrate the egg. Clearly, the passage of spermatozoa through the female tract is accompanied over a period of hours by a functional change (capacitation) preceding fertilization (Austin, 1951, 1952; Chang, 1951). The ultrastructural findings in experimental manipulation of eggs and spermatozoa suggest that one facet of capacitation prepares the spermatozoon to be able to undergo the membrane fusion of the acrosome reaction (Bedford, 1969b). This has encouraged the idea that capacitation probably involves some change in the potential for fusion of the sperm plasmalemma and/or in the nature of the components at its surface. It is also possible that an observable change in the character of spermatozoa motility with capacitation (Yanagimachi, 1970) depends on a changed permeability of the tail plasmalemma, though there is no evidence to support this hypothesis.

It is disappointing that the various studies of this question do not yet allow even a general definition of the membrane modifications that may represent a key feature of capacitation. In part this is a consequence of incomplete experimental design that has made functional interpretation of many of the reported findings difficult. In contrast to many other systems used currently in cell biology, the sperm cells being studied appear as a heterogeneous population; some are dead, while the living spermatozoa are possibly undergoing capacitation at different rates. Especially limiting is the fact that capacitation is still an operational term, and is defined with certainty only as the functional change which allows an ejaculated/epididymal spermatozoon to penetrate an intact ovum. Within the framework of this definition, capacitation of the spermatozoa of a variety of mammals has now been accomplished in vitro, occasionally in defined media (Brackett and Oliphant, 1975 and Gwatkin, 1976 for literature). However, the conditions required for capacitation in vivo seem somewhat more demanding; often there is no cross-species compatibility; there is a differing potential of various compartments in the female tract, and their activity is modifiable by endocrine change. Moroever, rabbit spermatozoa subjected to the in vitro "capacitation" regime of Brackett and Oliphant (1975) will not fertilize in vivo unless also capacitated in vivo (Viriypanich and Bedford—unpublished observations). Thus, whether the steps or mechanisms by which "capacitation" is achieved in vitro really mimic the process as it occurs in vivo is questionable.

Indirect evidence for the idea of sperm surface changes as a feature of capacitation is seen in the interaction of leukocytes and rabbit spermatozoa. Leukocytes obtained from the uterus or other parts of the body quickly ingest damaged but not intact spermatozoa in vitro, and failure to ingest intact sperm within a period of several hours was also shown in the pseudopregnant uterus and pleural cavity, neither of which support capacitation. Only when the spermatozoon has resided in the (capacitating) environment of the estrous uterus for several hours does its surface become sufficiently attractive to leukocytes to stimulate ingestion (Bedford, 1965b). An "unmasking" change in the sperm surface as a concomitant of capacitation is also suggested indirectly by the observation that an unusual and potent sperm head agglutinating component in seminal plasma from a fertile rabbit immediately immobilized its own or other vigorously swimming spermatozoa after they had been in the uterus for 12 hours, but did not affect the motility of epididymal or ejaculated spermatozoa (Bedford, 1970a). In addition, the obvious sensitivity of rabbit spermatozoa to guinea-pig globulins only after uterine incubation (O'Rand and Metz, 1976) and the susceptibility to disruption by follicular fluid shown by the head membranes of uterine but not epididymal or seminal-plasma treated spermatozoa (Bedford, 1969b; Oliphant, 1976) could be interpreted as the result of unmasking components at the sperm surface as a consequence of uterine incubation. Functional interpretation of these membrane-dependent phenomena or other relevant studies (e.g., Gordon et al., 1974, 1975a; O'Rand, 1977), and of the "assay" for capacitation based on removal of seminal plasma sperm-coating antigen (Oliphant and Brackett, 1973a) would be strengthened, however, if consideration were given to controls incubated in a progesterone-dominated uterus or at other sites where capacitation fails to occur. The suggestion (Vaidya et al., 1971) that capacitation involves a reduction in electrophoretic mobility and thus reduction in net surface negative charge cannot be regarded seriously on present evidence because of the absence of such controls, and particularly because the measurements at 24 °C in such experiments would be confounded by the intrinsic motility of living spermatozoa; immobile spermatozoa so measured at this temperature are dead.

Much attention has been paid to the notion that the capacitated state must involve prior loss of material from the sperm surface. This seems a reasonable assumption in light of the reports suggesting a concomitant modification of the sperm surface (Johnson and Hunter, 1972; Oliphant and Brackett, 1973a, b; Koehler, 1976), though appropriate controls that allow these phenomona to be interpreted specifically as capacitation changes have not been employed (Vaidya et al., 1969). Undoubtedly there is a seminal plasma component, probably a glycoprotein (Davis, 1971), which inhibits the fertilizing ability of capacitated spermatozoa (Chang, 1957; Bedford and Chang, 1962a; Dukelow et al., 1967; Aonuma et al., 1973) and which probably is removed from spermatozoa in the uterus (Johnson and Hunter, 1972; Brackett and Oliphant, 1975). Other coating antigens such as blood group glycoproteins and lactoferrins (Boettcher, 1965; Weil, 1967; Hekman and Rümke, 1969; Roberts and Boettcher, 1969; Isojima et al., 1974) may be involved. Notably, however, epididymal spermatozoa devoid of such antigens require the same time period and conditions for capacitation and are equally as functional as ejaculated spermatozoa. Thus, although ejaculated spermatozoa are inevitably exposed to seminal plasma, removal of seminal components from their surface may be a trivial feature of the capacitation process involving no fundamental modification of the integral character of the plasma membrane. For this reason we suggest that in future biochemical or structural analyses of the molecular basis for capacitation consideration be given to the use of epididymal rather than ejaculated spermatozoa. Much of the future emphasis, as in the past, will probably be focused on the plasmalemma of the spermatozoon since this is the element exposed to the capacitating environment. One wonders whether the properties of the outer membrane of the acrosome, itself an integral partner that must suddenly acquiesce to the fusion events of the acrosome reaction, are not also modified by capacitation. It has been claimed that the integral proteins of the human acrosomal membrane change from a β to an α or random coil configuration in a capacitating situation (Delgado et al., 1976), but there is no indication that these authors' methods allowed them to obtain even reasonably pure acrosomal membrane preparations for their measurements.

5. Sperm membrane behavior during penetration

5.1. Interaction with egg vestments

Although the recent past has produced several electron microscopic and physiological studies of the initial interaction of spermatozoa with eggs, this phase of the fertilization process remains vague. Excepting human spermatozoa, which are unusually specific in this respect (Bedford, 1977), those of most experimental mammals readily adhere to the zona pellucida of homologous or heterologous oocytes in vitro though actual penetration occurs only between gametes of the same or closely related species. The reports of Hartmann and colleagues (1971, 1974), have drawn attention to the fact that capacitated hamster spermatozoa adhering to the zona pellucida in vitro appear to develop a binding relationship over a period of about 30 minutes with "receptors" at the zona surface, which themselves may permeate the zona (Gwatkin and Williams, 1977). Whatever the nature of the receptors of the zona pellucida, it is not clear which specific components of the spermatozoon surface must bind to them. Electron microscopic pictures of rabbit spermatozoa close to the zona pellucida in the interstices of the granulosa cell mass show them in different stages of reaction: the heads of some are intact, others have undergone the acrosome reaction (see below) and still others, whose heads have lost the acrosomal remnants, are bounded by the inner acrosomal membrane (Bedford, 1972). Unfortunately, the static nature of electron microscopy prohibits our knowing which of these categories of spermatozoa "caught" close to the egg surface can go on to penetrate the zona pellucida. Probably the determinants on the inner acrosomal membrane revealed after loss of the acrosome are different from those at the plasmalemmal surface. Moreover, one must wonder whether the membrane fusion of the acrosome reaction also modifies the components expressed at the surface of the plasmalemma which, now intermittent, forms the outer layer of the vesicles enveloping the sperm. It is unclear whether the plasma membrane of a fertilizing spermatozoon must remain intact until the zona surface is reached, or whether spermatozoa can first undergo the acrosome reaction and perhaps even shed the vesiculated membranes and acrosome content without prejudicing their ability to bind to and penetrate the zona pellucida. The reacted vesicles are left adhering to the zona by the penetrating spermatozoon, and it is possible that they must form a strong attachment to the zona surface after which the spermatozoon is able to intrude itself into the zona substance. If true, this would preclude any spermatozoon that has lost the acrosome before zona contact (e.g., Fig. 3 in Bedford, 1968) from penetrating the zona pellucida. The mechanisms involved in penetration of the zona substance are also unclear, as is the exact basis of the species-specificity of this stage of sperm entry, but it is inappropriate to consider these further here.

5.2. The acrosome reaction

5.2.1. Morphology

Speculation about the nature of the mammalian acrosome reaction (Colwin and Colwin, 1967) was resolved by ultrastructural studies of rabbit and hamster spermatozoa during fertilization, which showed that this is heralded or initiated by a series of point fusions between the plasma membrane and the outer membrane of the acrosome underlying it (Barros et al., 1967). Limited fusion between these adjacent membranes (Figs. 14, 15) results in the appearance of numerous gaps through which, presumably, the enzymic contect of the acrosome can escape. As can be seen in Fig. 14, the fusion reaction is limited strictly to that surface which overlies the more rostral region of the acrosome. Neither the narrow stable "equatorial segment" of the acrosome nor the postacrosomal surface of the head play any part in membrane fusion events at this stage.

As noted earlier, membranes of the equatorial segment have a special character (Roomans, 1975; Phillips, 1977) and we suggest that the failure of the equatorial segment to take part in the acrosome reaction may be a reflection of this, and perhaps its content, rather than of any particular attribute of the plasmalemma overlying it. Whether that portion of the plasma membrane overlying the equatorial segment has surface properties that distinguish it from the adjacent rostral plasmalemma is not known. A suggestion to this effect (in man, Hjört and Hansen, 1971), does not rule out the possibility that antigens expressed locally in the equatorial region emanate from subsurface structures.

It has not yet been possible to follow the acrosome fusion sequence closely from its inception. According to Talbot and Franklin (1976), fusion may be preceded by swelling of the apex of the acrosome and by an increase in membrane permeability. The plasmalemma often appears to have a more intimate association with the apex of the acrosome, and it has been suggested that this may be the initial site of fusion from which the acrosome reaction is propagated (Bedford, 1974b; also see Dan et al., 1974). As noted elsewhere (section 5.2.2), this is also the point at which antibody-induced rupture of the sperm head first occurs (Russo and Metz, 1974).

The fusion described above seems to be a consistent event in the pig (Szöllösi and Hunter, 1973), man (Soupart and Strong, 1975), rodent, and rabbit, and it is reasonable to anticipate that this may be common to most mammals. There are reports of vesiculation and separate loss of plasma and acrosomal membranes in human (Roomans and Afzelius, 1975) and pig spermatozoa (Jones, 1973), but unless this phenomenon is demonstrated with any consistency in fertilization situations it seems wise to consider with caution the suggestion of this mode as a physiological event. There is no doubt, however, that head membranes are often lost as separate entities as a result of damage, and these considerations raise a question about the essential role of the point fusion reaction for sperm function. Is this precise mechanism for loss of the rostral membranes still important for penetration where the egg is devoid of cumulus and corona cells, or is it needed

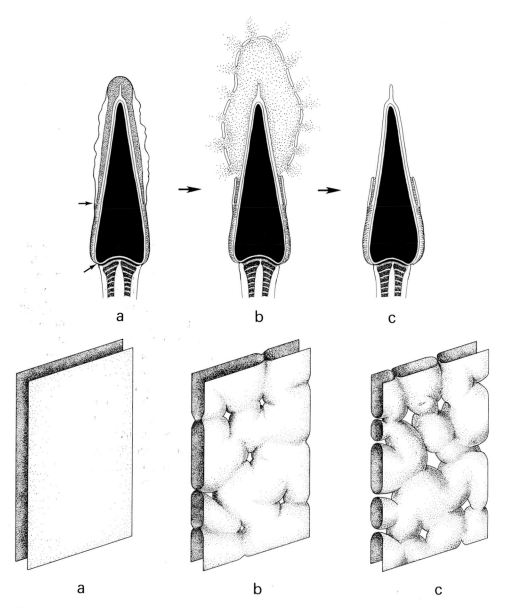

Figs. 14 and 15. These diagrams illustrate the fusion events in the spermatozoon during the acrosome reaction. The reaction involves a series of point fusions between the contiguous plasma and outer acrosomal membranes (Fig. 15), giving the impression of a string of associated vesicles arrayed around the rostral portion of the head in thin section (Fig. 14b). The resulting "ports" (Figs. 15b, c) are believed to allow the content of the acrosome to escape without loss of continuity of this surface of the spermatozoon. Although the vesicles remain around spermatozoa attached to the zona surface, and may be important for such attachment, they are lost before penetration of the zona, the form of the penetrating sperm being that shown in Fig. 14c. Swelling of the peri-acrosomal plasma membrane in 14a is a commonly seen artifact of fixation. (Adapted from Barros et al., 1967.)

only where controlled loss of acrosomal content is required for passage through the granulosa cell mass? As noted elsewhere (section 5.1), it is unclear whether a fertilizing spermatozoon can bind to the zona surface only by the agency of ligands integral to the plasmalemmal surface or whether the spermatozoon devoid of the acrosome before contact can go on to penetrate the zona. If the former is true, the importance of the point fusion reaction may lie in the way that such vesiculation allows release of the acrosomal content while maintaining intact the apparatus required to bind the sperm to the zona. Whether the important feature for the continued integrity of the tail compartment in the reacted spermatozoon is the "spot weld" of the posterior ring on the sperm head (Figs. 8, 9) or continuity provided by fusion of the plasmalemma with the rostral border of the equatorial segment (Fig. 14) is not clear. The latter becomes a possibility for consideration, however, in light of the consistent observation that tail movement ceases as the fertilizing spermatozoon begins to fuse with the oolemma.

There is a suggestion that the equatorial segment is involved in a subsequent fusion reaction as the spermatozoon penetrates the zona pellucida and that this is a necessary adjunct of this stage of fertilization (Yanagimachi and Noda, 1970b). This does not occur in the rabbit (Bedford, 1972), but observation of fixed hamster spermatozoa within the zona that have a vesiculating equatorial segment has provoked the suggestion that a delayed vesiculation in the equatorial region of the sperm penetrating the zona may release some essential zona lysin (Yanagimachi and Noda, 1970b; Yanagimachi, 1973). None of the hamster spermatozoa referred to in this context, however, had in fact passed through the zona. There is also an apparent correlation between equatorial segment disintegration in culture and loss of penetrating ability of hamster spermatozoa in vitro (Barros et al., 1973) and Barros and Herrera (1977) claim that the equatorial segment is lost before guinea pig spermatozoa penetrate zona-free hamster egg in vitro. We have attempted to resolve this question by examination of spermatozoa in the perivitelline space which have therefore penetrated the zona pellucida (Moore and Bedford, 1978). This has been achieved by placing immature follicular oocytes, which display no block to polyspermy, in the Fallopian tube of mated hamsters. In the oocytes recovered 6 to 8 hours later some of the spermatozoa present in the perivitelline space had an intact equatorial segment (Fig. 16a, b), and thus it is clear that vesiculation of this region is not a necessary adjunct of zona penetration. The receptivity of the vitelline surface during tubal incubation of the follicular oocytes also allowed penetration of some spermatozoa into the egg itself (Fig. 19). In these the persistence of the characteristic pentalaminar structure of both membranes of the equatorial segment (Figs. 20, 21, 22, 23) indicates that it remains intact even after penetration of the vitellus (Moore and Bedford, 1978; see also section 7.1).

It is quite possible, though not yet directly proven, that the acrosome reaction also propagates a change in the intact membrane overlying the equatorial segment or beyond and that this change is essential for sperm fusion with the egg.

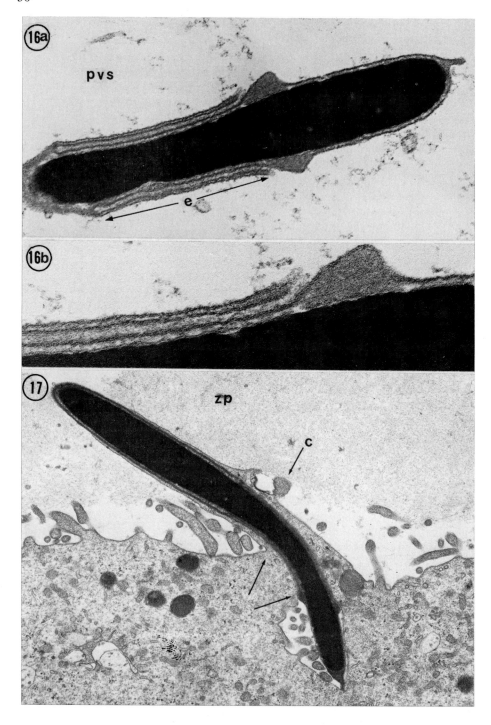

Initial observations by Yanagimachi and Noda (1970c) that hamster spermatozoa would not attach to the vitellus of zona-free eggs unless first capacitated, gave rise to the idea that capacitation itself may encompass some change in the postacrosomal surface. However, notwithstanding the contrary claim of Soupart and Strong (1975), the total impression from subsequent observation now suggests that even in zona-free eggs capacitated spermatozoa fuse with the vitelline membrane only after the acrosome reaction has occurred (Noda and Yanagimachi, 1976).

5.2.2. Induction

The point fusions between the outer acrosomal and plasma membranes that constitute the initial event of the acrosome reaction do not occur in normal physiological circumstances unless the spermatozoa have undergone capacitation (Bedford, 1970b). As noted earlier, this implied some capacitation-associated changes in the sperm head membranes which, under appropriate conditions, predispose to the fusion reaction. A primary question yet to be answered in mammals, however, concerns the nature of the stimulus or factor that initiates the reaction in vivo. Two alternatives, at first glance, are either a specific factor analogous in its function to the "fertilizin" of the echinoderm system, or the spontaneous response of sperm membranes to general (ionic) factors in the environment as they reach a final point in the capacitation process. The suspicion that different cells may become functionally capacitated at different rates, and that once this is achieved fertilizing ability may be lost relatively quickly, means that only a proportion of spermatozoa would be expected to react at any one time in either case.

Within the framework of present concepts, it seems illogical that the bulk of the acrosome content is lost long before the egg is reached; thus the reaction should occur close to the vestments of the egg. In fact, the few observations of the fine morphology of spermatozoa during mammalian fertilization in vivo

---

Figs. 16 and 17. Photographs on opposite page.
Fig. 16. Electron micrographs of a hamster spermatozoon head present in the perivitelline space (pvs) after it has penetrated (in vivo) the zona pellucida of a vesicular oocyte recovered from a secondary follicle. Clearly, the equatorial segment (e in 16a) is intact—a demonstration that vesiculation of this region is not required for zona penetration. Fig. 16b shows, at higher magnification, the arrangement of the membranes of the equatorial segment after completion of the acrosome reaction (see also face a, Fig. 23). In this, the content of the equatorial segment does not yet display the septate character commonly seen (Figs. 20, 21) after the spermatozoon has entered the vitellus. Fig. 16a: ×46,800. Fig. 16b: ×108,000.
Fig. 17. Hamster spermatozoon in an in vitro fertilization system at an early stage of its incorporation into the vitellus. This micrograph is included to illustrate the restricted region of initial association over the equatorial segment (arrows) and the fact that the ooplasm then appears to flow from there into the postacrosomal compartment. Note the cortical granule (c) in the process of eruption. zp, zona pellucida. ×18,900. (Figure courtesy of Dr. L. E. Franklin.)

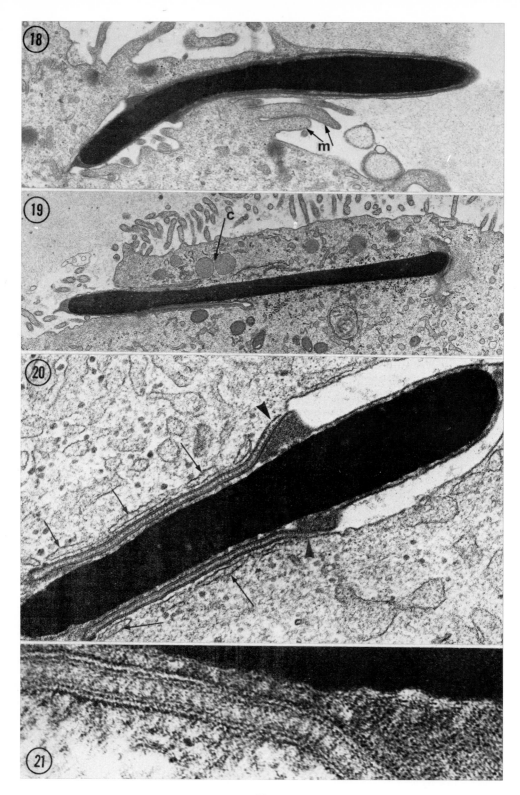

suggest that the acrosome tends to react in the close vicinity of the egg, but motile rabbit spermatozoa can undergo the acrosome reaction before ovulation and before they reach the tubal ampulla (Overstreet and Cooper, 1978). The long-standing precedent in invertebrates, in many of which the egg releases an inducing agent, "fertilizin" (Dan, 1967), and that in the anurans where the jelly coat is important in activating spermatozoa (Katagiri, 1974), raise the possibility that a comparable factor emanating from the egg or its vestments may operate in the mammalian situation. However, fertilization of washed ova will take place readily in vivo in the absence of follicular cells or other products of ovulation in the rabbit (Harper, 1970; Overstreet and Bedford, 1974) and hamster (Moore and Bedford, 1977) and in vitro in rabbit, hamster, and mouse (Bedford and Chang, 1962b; Frazer et al., 1971; Iwamatsu and Chang, 1972; Miyamoto and Chang, 1972; Pavlok and McLaren, 1972). Thus it becomes very apparent that neither granulosa cells nor follicular fluid are essential for induction of the acrosome reaction, even in vivo. The reaction occurs in fertile spermatozoa incubated with eggs denuded also of the zona pellucida (Yanagimachi and Noda, 1970c; Wolf et al., 1976; Yanagimachi et al., 1976). It can therefore be argued that if such a factor is involved, it must also be associated with the egg itself. On the basis of the idea that the fertilizin may be either a sialo-protein or residual progesterone, rabbit eggs were pretreated with neuraminidase, trypsin, chymotrypsin or anti-progesterone antibody, but this failed to modify the rate of sperm penetration through the zona pellucida, and so presumably did not affect the ability of capacitated spermatozoa in the vicinity of the ova to undergo the acrosome reaction (Overstreet and Bedford, 1975). It does not appear necessary that the oocytes should be alive for an acrosome reaction to occur, since the zonas of

---

Figs. 18–21. Photographs on opposite page.

Fig. 18. Another section of the hamster spermatoon shown in Fig. 17. This particular section illustrates well the impression that the microvilli (m) are relatively nonlabile structures, and that these are "squeezed" to one side rather than being the site of initial fusion contact with the equatorial segment of the spermatozoon. ×18,900. (Figure courtesy of Dr. L. E. Franklin.)

Fig. 19. Hamster spermatozoon head that has penetrated the zona pellucida and fused with the oolemma of a vesicular oocyte, in vivo. This was recovered 16 hours after instillation into the fallopian tube of a mated hamster. Although this fusion must have occurred some time earlier, there has been no exocytosis of cortical granules (c), nor has the nucleus undergone any decondensation. Note particularly the array of microvilli that adorn the ooplasmic fold enveloping the posterior part of the nucleus. ×12,600.

Fig. 20. Hamster sperm head in a similar situation to that in Fig. 19, which has, however, been incorporated completely by the vitellus. This illustrates the flattened vesicles (arrows) lying coaxially with the equatorial segment (just discernible in Fig. 19) and the impression of a septate organization of the equatorial segment content. The vesicles may be products of fusion of the equatorial sperm plasmalemma and the oolemma (see legend, Fig. 23). Arrow heads indicate the rostral border of the equatorial segment. ×50,400.

Fig. 21. High magnification of the equatorial segment of another hamster spermatozoon within the ovum in a similar situation to that in Fig. 20. This illustrates the similar pentalaminar appearance of both inner and outer membranes of the equatorial segment in spermatozoa that have fused with and entered the vitellus. This pentalaminar design typifies the membranes of the equatorial segment in the hamster (Moore and Bedford, 1978) and in man (Roomans, 1975). ×306,000.

94

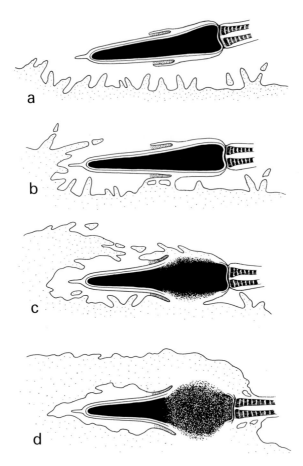

Fig. 22. These diagrams show the sequence of events through which the fertilizing spermatozoon fuses with and is believed to be incorporated by the vitellus. (a) the form of the spermatozoon with intact equatorial segment, immediately prior to contact with the oolemma. (b) the initial phase of fusion shows that the spermatozoon is first associated with the oolemma only by the defined region of the equatorial segment. This initial stage is then followed (c) by a peeling back of the membrane of the postacrosomal surface, giving the ooplasm access to the sperm nucleus, and by the beginnings of envelopment of the rostral inner acrosomal membrane-limited region of the sperm head by the egg surface. When this last phase is completed and the flagellar surface has been partially removed by fusion with oolemma (d), the sperm within the ovum is seen typically to be associated with a vesicle over its anterior surface; this is comprised by inner acrosome membrane and by endocytosed oolemma. By contrast, the posterior region of the nucleus is devoid of membrane and is the first to undergo the decondensation which presages the formation of the male pronucleus.

human cadaver oocytes taken from ovarian follicles 36 hours after death were penetrated in vitro by spermatozoa which had undergone an acrosome reaction (Overstreet and Hembree, 1976). Thus, evidence does not exist at present for the participation of a specific fertilizin in induction of the mammalian acrosome reaction.

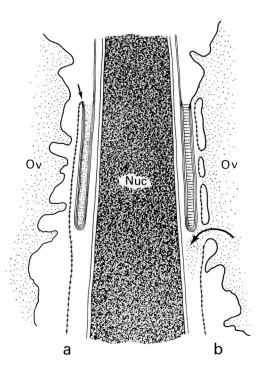

Fig. 23. Diagram using two faces of the spermatozoon to depict in greater detail the mode of what is believed to be the initial phase of fusion with the oolemma. Before fusion (face a), the plasmalemma of the spermatozoon (×××××) extends rostrally to the anterior border of the equatorial segment (arrow); this point forms the anterior boundary of the anticipated fusion with the oolemma. After this fusion has occurred (face b), vesicles are often seen initially along the line of the equatorial segment (Fig. 20). These vesicles may be a product of the fusion of the sperm plasmalemma over the equatorial region and the oolemma facing it and are depicted as such in the diagram. Elucidation of this point demands the use of surface markers to establish the surface origin of the vesicle components. Posteriorly, the fused plasmalemma seems to peel back to allow ooplasm to flow into the post-acrosomal region (curved arrow). The equatorial segment does not usually exhibit a septate organization of its content very clearly in intact spermatozoa before fusion, but the septate appearance depicted on face b is commonly seen in well-preserved sperm heads soon after fusion has occurred. Nuc, sperm nucleus; Ov, oocyte cytoplasm.

In the absence of ova, spermatozoa can be stimulated to undergo an acrosome reaction in vitro resembling that occurring in vivo. It may be dangerous to assume that the same end result, fusion between plasma and acrosomal membranes now appearing as a string of vesicles in thin section, necessarily implies the operation of common mechanisms in vivo and in vitro. There is no doubt that the need for capacitation can be overcome in vitro, and that the acrosome reaction itself can be induced by in vitro conditions which differ from those existing in vivo. The accomplishment of capacitation in vitro by incubation of spermatozoa in higher than physiological ionic strength media (mouse and rabbit—Oliphant and Brackett, 1973b; Brackett and Oliphant, 1975), with adrenal gland extracts (Bavister et al., 1976) or in a defined medium (guinea pig,

Yanagimachi, 1972), lends the impression that the environmental requirements for capacitation and the acrosome reaction are simple and non-specific. In vitro culture has considerable value in constructing experimental situations that may help in elucidating the physiology of prefertilization events, but the factors operating in vivo that bring about capacitation and, subsequently, the acrosome reaction may well prove to have an essential complexity and specificity far outweighing that of the defined elements used for this in vitro. For instance, rabbit spermatozoa treated according to the method of Brackett and Oliphant (1975) for in vitro capacitation cannot fertilize in vivo unless first capacitated completely in vivo, and hold no advantage in this respect over Tyrode's treated spermatozoa (Viriypanich and Bedford, unpublished observations). It is perhaps germane also that complete capacitation is not achieved readily in a foreign tract where the evolutionary distance between two species ensures that cross-fertilization cannot occur (Bedford, 1967b); neither is it supported to any extensive degree in several extra-uterine sites in vivo (isolated vagina, progestational uterus, anterior chamber of the eye, joint capsules—Bedford, 1967b; Hamner and Sojka, 1967) nor in the rodent uterus alone (Hunter, 1969), although all these milieus support good motility. Despite the wealth of experimental reports in which spermatozoa of various species have been 'capacitated' in vitro, the in vivo findings quoted above and the suggestion of a utero/tubal synergism (Bedford, 1969a; Hunter and Hall, 1974) indicate that functional capacitation in vivo may nonetheless require specific attributes of the environment found in the fallopian tube and/or uterus according to the species.

5.2.3. Mechanisms

Present attempts to explain the mechanism of fusion of the acrosome reaction in any detail must be seen as premature. Even if the factors triggering it were clearly defined, insufficient information exists about the molecular basis of fusion reactions in general and that of sperm membranes in particular.

In concentrations as low as 0.01%, Hyamine—a reagent that probably tends to disrupt alkali-labile and ionic bonds—induces an eruption in a significant proportion of bovine spermatozoa which ultrastructurally resembles the acrosome reaction (Wooding, 1975), and similar treatment of guinea pig spermatozoa with 0.003% Hyamine seems to bypass their requirement for capacitation (Yanagimachi, 1975). These observations and the nature of the other reagents producing such effects in the latter study suggest the need for perturbation of the molecular associations within the sperm head membranes for the fusion reaction to be able to occur, and it is possible that this may happen during capacitation in vivo. To date, however, substantial evidence is lacking for this last suggestion.

Nicolson and Yanagimachi (1974) have claimed that the integral proteins of the periacrosomal plasmalemma lack lateral freedom of movement, but this report means little in functional terms. As indicated previously (section 2.3), these observations were made on caput spermatozoa. O'Rand (1977a), on the basis of observation of fluorescent labels, has claimed that certain antigenic components of the periacrosomal plasmalemma possess the potential for lateral mobility in the plane of the membrane. This mobility is reported to be lost, however, after incubation of spermatozoa in the uterus, probably as a concomitant of capacitation. It is possible that capacitation involves a reordering of intramembrane molecular relationships whereby areas of now-rigid membrane are interspersed by more fluid regions, which would allow the point fusions of the acrosome reaction (see section 5.2.1). Yet, until such a mosaic has been demonstrated, the interpretation by O'Rand that the integral proteins of the periacrosomal plasmalemma display a reduced lateral fluidity after capacitation seems to be in conflict with the contemporary notion that the intrinsic proteins of a fusing membrane should be free to move laterally in a fluid bilayer, and perhaps, according to Ahkong and colleagues (1975), to migrate away from the immediate locus of lipid-lipid interaction.

In current postulates (Poste and Allison, 1973), considerable attention is paid to the possibility of a stabilizing role in membranes for calcium ions and, conversely, an increased freedom of lipid molecules within the bilayer following Ca^{2+} withdrawal. Allied to this is the modulating influence of the surface zeta potential apparently governing the closeness of approach to a distance ($-10Å$) that allows apposing membranes to be influenced by short-range attractive forces and to establish stable linkages between macromolecules of apposed membranes. Calcium ions bind to the sperm head membranes. Fixed under physiological conditions, the acrosomal membranes of rabbit and monkey epididymal spermatozoa display an asymmetry created by the greater thickness of the outer leaflet of the bilayer (Bedford and Nicander, 1971), an effect which, in ejaculated human spermatozoa, is reported to depend on the presence of calcium possibly interacting with integral proteins in this outer lamina (Roomans, 1975). Evidence now indicates that Ca^{2+} is an essential requisite for initiation of the mammalian acrosome reaction (Barros, 1974; Yanagimachi and Usui, 1974). Moreover, similar time-related effects of a calcium ionophore and the bearing of the calcium concentration (Summers et al., 1976; Talbot et al., 1976) lend some credence to the idea that capacitation finally involves a change in the Ca^{2+} permeability of the sperm head membranes. Thus the influx of calcium across the sperm surface, controlled perhaps by Mg^{2+} (Talbot, 1975; Rogers and Yanagimachi, 1976) may be an important element in initiating the acrosome reaction. But it is not known whether Ca^{2+} acts in this respect as an intermembrane ligand and/or as an activator for membrane ATPase (Poste and Allison, 1973), or induces lateral phase separations in membrane lipids which render the membranes unstable and susceptible to fusion (see, for example, the chapter by Papahadjopoulos in this volume, p. 765). The sperm plasmalemma contains a Na^+/K^+-dependent enzyme (Gordon, 1973; Chulavatnatol and Yindepit, 1976), while the Ca^{2+}-dependent ATPase of the acrosomal membrane is not inhibited by Mg^{2+} (Gordon, 1973), and the location of the Ca^{2+}/Mg^{2+}-dependent ATPase detected by biochemical methods on bull spermatozoa (Lindahl, 1973) is uncertain.

In considering the mode of action of calcium in this system, it may be relevant to refer to recent observations on the fusion of erythrocyte ghosts (Zakai et al.,

1976, 1977). For manipulation of their fusion by modification of the environmental conditions employed shows that agglutination or the operation of a ligand function is not sufficient for fusion. With an optimal pH of about 7.5, this requires a nascent calcium phosphate complex rather than the availability of free Ca^{2+} ions. Whether such a phosphate complex plays a key role in the induction of the acrosome reaction and in the subsequent fusion between the fertilizing spermatozoon and the oolemma should not be difficult to test in an in vitro situation.

"Reacted" spermatozoa are commonly seen within the granulosa cell matrix around the rabbit egg in the ampulla of the oviduct, but whether ampullary Ca^{2+} or Mg^{2+} levels differ significantly from those in the uterus is unclear (Hamner, 1971, 1973). Although ultrastructurally "reacted" spermatozoa are seen only rarely in the uterus, however, rabbit eggs are readily fertilized there (Chang, 1955; Bedford, 1969a). Thus, unless the egg is able to create a local ionic gradient around itself, it is difficult to believe that Ca^{2+}, whether complexed with phosphate or another counter ion, or even a particular Ca^{2+}/Mg^{2+} ratio, can be the only factor which ordains the onset of the acrosome reaction in fully capacitated spermatozoa. As discussed earlier, it seems desirable for the fertilizing spermatozoon to undergo the acrosome reaction in the vicinity of or at the surface of the zona pellucida.

6. The oolemma

6.1 General characteristics

6.1.1. Submammalian vertebrates
The topography of the vertebrate egg surface displays considerable species variation and also, in some submammalian species, a regional differentiation (Campanella, 1975; Monroy and Baccetti, 1975). Mature *Xenopus* eggs apparently have many fewer microvilli per unit surface area over the animal than the vegetal hemisphere (Monroy and Baccetti, 1975), but unfortunately the surface studies extant (Brummett and Dumont, 1976; Stenuit et al., 1977) add little further about the surface character of either pole. In several anurans, sperm/egg fusion occurs preferentially over the animal hemisphere (Elinson, 1975), but in certain species (e.g., *Discoglossus*) sperm access is confined to a restricted regional dimple (Campanella, 1975).

Surprisingly, the ultrastructure of the acrosome reaction of the anuran spermatozoon and that of its fusion with the egg has not been elucidated. We have no clue, moreover, about the role that regional differences in egg surface topography and chemistry may play for fertilization in this group. *Discoglossus* seems to be a useful system in which to study these early events in anurans and to correlate them with egg surface characteristics. We draw attention to the fact, however, that the spermatozoa of this group show marked species differences in the morphology of the sperm head and the structure of the acrosome. The head and acrosome configuration of the *Discoglossus* spermatozoon (Sandoz, 1975) is similar in morphology to that of the river lamprey (Nicander and Sjödén, 1971), and may therefore fuse with eggs via an acrosomal tubule, as do lamprey spermatozoa. The form of the acrosome of some anurans, on the other hand, is similar to that in mammalian spermatozoa (Burgos and Fawcett, 1956; Nicander, 1970; Poirier and Spink, 1971; Reed and Stanley, 1972). Since nothing is known of the ultrastructural response of anuran sperm membranes at fertilization or of the surface chemistry of anuran sperm and eggs, it is not possible to dwell further on the mechanism of fertilization in this group.

6.1.2. Eutherian mammals

By contrast with the situation in anurans, spermatozoa may fuse with any region of the mammalian egg surface other than that devoid of microvilli which overlies the meiotic spindle. How the surface of this "bald" region differs from the remainder is not known. An impression gained from light microscopy that it has a lower affinity for Con A (Johnson et al., 1975) has been shown to depend on the absence of microvilli and thus a reduced surface area there (Eager et al., 1976). This serves to illustrate the import of an awareness of topography when the intensity of fluorescence is used as an indicator of differences in surface chemistry. No systematic study of egg surface topography comparable to that in sea urchins (Eddy and Shapiro, 1976) has yet been performed on mammalian eggs. The few incidental observations on the latter suggest marked species variation, since that of the unfertilized rat egg appears relatively smooth (Szöllösi, 1967), while hamster and mouse eggs bristle with microvilli (Yanagimachi and Noda, 1972; Eager et al., 1976; Wassarman et al., 1977).

The species specificity of the oolemma for spermatozoa is less marked than that of the zona pellucida, but this does not mean that liberation of the egg from the zona pellucida necessarily removes the barrier to interaction between foreign gametes. The vitelline surface of the hamster ovum is highly receptive, and the spermatozoa of all of seven species tested to date have been shown to penetrate it. On the other hand, that of the rabbit seems compatible with mouse and rat but not with hamster and guinea pig spermatozoa, and the oolemma of the mouse is penetrated only infrequently by foreign spermatozoa (see Hanada and Chang, 1977 for literature). A minor note of caution should perhaps be sounded in interpretation of the experiments reported so far. In most the zona pellucida was removed by enzymes, carrying some risk of modifying the natural state of the vitelline surface. Actually, removal of the zona pellucida of the mouse egg by brief digestion in proteolytic enzyme solutions markedly reduces the rate of sperm penetration into the vitellus as well as the percentage of eggs showing polyspermy (Wolf et al., 1976). Hamster and rabbit eggs, however, do not appear to be sensitive to such enzyme treatment. Yanagimachi and Nicolson (1976) were unable to detect any change in lectin-binding properties of the hamster egg surface after prolonged treatment with 0.1% pronase, 0.1% trypsin, or 0.05 M 2-mercaptoethanol—reagents used routinely to lyze the zona pellucida of mammalian eggs—nor was the penetrability of rabbit eggs modified by nontoxic levels of neuraminidase, trypsin, or chymotrypsin (Overstreet and Bedford, 1975).

The significance for fertilization of egg surface glycoproteins, whose existence is implied by lectin affinities, is not understood. So far, appreciation of the nature of the surface components important for sperm/egg fusion has not been advanced by lectin-binding studies, except perhaps in a negative sense. Two lines of evidence indicate that receptor sites of vitelline surface components specific for a variety of lectins are not involved directly in the sperm recognition necessary for fertilization in vivo and in vitro. Firstly, the receptivity of the hamster vitellus cannot be related to demonstrated change in lectin-binding properties during its preovulatory maturation. Although *Ricinus communis* agglutinin I (RCA$_I$) and wheat germ agglutinin (WGA) bind more avidly to the vitelline surface of the ovulated ovum than to that of the primary oocyte (Yanagimachi and Nicolson, 1976), intact primary oocytes of the hamster are readily penetrated in vivo (see section 6.2; Moore and Bedford, 1978) as are zona-free oocytes in vitro (Usui and Yanagimachi, 1976). Thus the receptor sites on the oolemma to which these lectins bind seem unlikely to be involved as critical elements in vitelline penetrability. The second line of negative evidence comes from unpublished observations (cited in Nicolson et al., 1975), in which exposure of zona-free oocytes of the hamster to Con A, *Ricinus communis* agglutinin I (RCA$_I$) and *Dolichos biflorus* lectins at concentrations of up to 250 μg/ml failed to prevent their penetration by spermatozoa, and from preliminary results in the rabbit where exposure of zona-free eggs to as much as 100 μg/ml of Con A or (WGA) did not seem to prevent subsequent penetration in vitro (Gordon and Dandekar, 1976). Although lectins induce cortical exocytosis in the hamster (Gwatkin et al., 1976), the persistent penetrability of its lectin-treated zona-free egg is not unexpected in this species where the receptivity of the vitellus seems largely unchanged during luteotropic stimulation of egg maturation and after fertilization. We know nothing of the molecular characteristics of the oolemma that make it a suitable complementary surface for fusion with the spermatozoon. Apparently, however, the terminal sugars of oolemmal glycoproteins bound by the several lectins used are not important as recognition sites; and since such lectins induce clustering of egg surface glycoproteins (Nicolson et al., 1975), the spatial configuration of integral membrane proteins may also be insignificant for initiation of fusion in this system. In concluding, it should be recognized that these questions arise largely from observations in the hamster, the oolemma of which has unusually few restrictions to penetration in a variety of circumstances (Hanada and Chang, 1977). Other species must be studied before general conclusions can be drawn.

6.2 Acquisition of penetrability

In most vertebrates, viable oocytes in secondary follicles exist in a resting (often dictyate or diplotene) state (Baker, 1972). When destined to ovulate, however, they resume meiosis to reach metaphase II, where they remain until fertilized. Probably, the vestments of dictyate oocytes are penetrable in most mammals, but it has appeared in light microscopic studies that the vitelline surface of the primary oocyte is not receptive to motile spermatozoa which have penetrated the zona pellucida (rabbit: Overstreet and Bedford, 1974; rat and mouse: Iwamatus and Chang, 1972, Niwa and Chang, 1975; hamster: Barros and Muñoz, 1973, 1974). However, spermatozoa sometimes enter the vesicular oocyte of the pig (Polge and Dziuk, 1965), and not only will capacitated spermatozoa fuse with the oolemma of zona-free hamster oocytes (Usui and Yanagimachi, 1976), but they have been seen ultrastructurally to penetrate readily into the vitellus of the intact diplotene oocyte in vivo in both hamster (Moore and Bedford, 1978) and rabbit (Berrios and Bedford, unpublished observations). It seems particularly significant in the latter studies in the hamster and rabbit that such entry did not stimulate the cortical granule response which is not yet "coupled" functionally to the oolemma in immature oocytes (see section 6.3.2). In the immature rabbit oocyte, moreover, the oolemma seems unable in most instances to envelop the acrosomal region within a phagocytic vesicle; the sperm merely sinks into the egg leaving the acrosome membrane at the surface, which itself is unusually devoid of microvilli compared with the ovulated oocyte (Berrios and Bedford, unpublished). In a sense, therefore, the immature oolemma of the rabbit is less responsive to the stimulus of the fertilizing spermatozoon than is the mature egg surface. A slightly different situation is seen in the dog oocyte, which is ovulated and is fertilized while the germinal vesicle is intact (Van der Stricht, 1923), the first meiotic division being delayed for several days after ovulation (Phemister et al., 1973). Mahi and Yanagimachi (1976) also found that capacited spermatozoa can penetrate preovulatory dog oocytes but that, unlike any mammalian eggs studied earlier, the dog oocyte can induce decondensation of the fertilizing sperm nucleus before breakdown of the germinal vesicle.

There is, therefore, some variation among mammals in the state of the vitelline surface in the immature oocyte. We and Dr. Nicholas Cross, using cationic ferric colloids, fluoresceinated Con A or WGA, and cationized ferritin, have been unable to demonstrate any significant difference in the oolemmal surface between immature and ovulated oocytes in the rabbit which might correlate with its unresponsiveness alluded to above. Unfortunately, where a surface development has been demonstrated, this has not correlated with functional change in it. For although the concentration of receptors for WGA on the hamster oolemma increases as meiosis resumes (Yanagimachi and Nicolson, 1976), this surface responds normally to spermatozoa before meiosis is resumed (Usui and Yanagimachi, 1976; Moore and Bedford, 1978; Figs. 19, 20). The changes occuring in the vitelline surface as a result of fertilization are hardly clearer. Although there is no change in the Con A binding characteristics of the mouse oolemma as a result of fertilization (Pienkowski, 1974; Konwinski et al., 1977) a block to polyspermy does appear to occur there (as well as in the zona) after entry of the fertilizing spermatozoon (Braden et al., 1954). In the rabbit, the increase after fertilization in the concentration of oolemmal Con A receptors (Gordon et al., 1975b), and cationic binding sites attributable to n-acetyl-o-diacetyl neuraminic acid and perhaps other similar residues (Cooper and Bedford, 1971b), appears to correlate with the well-known vitelline block to polyspermy that follows fertilization in this animal. But whether these visible manifestations directly reflect the mechanism of the block remains to be proven. There is a pressing need for more detailed investigation of the nature of the egg surface before and after fertilization, especially since so few species have been examined. It may be useful in this respect to reexamine with modern techniques the types of inhibitors used by Parkes and associates (1954), which prevented interaction of spermatozoa with the vitelline surface in the rabbit.

Little more can be said about the egg surface in other vertebrates. Immature anuran coelomic oocytes are penetrable by spermatozoa, although their further development is not activated by sperm/egg fusion until a later stage of the first meiotic division is reached (Elinson, 1973, 1977; Katagiri, 1974; Katagiri and Moriya, 1976). Unfortunately the surface properties of such eggs also are virtually uncharted (Brummett and Dumont, 1976). It becomes apparent that we know nothing of the specific surface properties of the vertebrate eggs which allow sperm/egg fusion nor how its surface differs from that of somatic cells, which generally will not fuse with the sperm plasmalemma in the absence of fusogens (see section 9).

6.3. Exocytosis of cortical granules

The eggs of many vertebrates display an array of membrane-bound granules subjacent to the oolemma (Kemp and Istock, 1967; Szöllösi, 1967), and with certain exceptions that lack these, such as urodeles (Austin, 1968), fusion of the spermatozoon with this surface stimulates their rapid discharge, if the egg is in a coupled state (see section 6.3.2). Such exocytosis is thought to encompass a propagated series of events with sequential discharge of cortical granules occurring radially over the egg surface from the stimulus point (Grey et al., 1974). When induced chemically in anurans, this occurs simultaneously over the whole egg surface (Wolf, 1974), but after the point stimulus of pricking, which perhaps mimics the stimulus of the spermatozoon, completion of circumferential propagation of exocytosis around the large anuran egg requires about 90 to 180 seconds (Kemp and Istock, 1967; Wolf, 1974).

The pattern of cortical granule release in mammalian eggs and the time required for its completion are not known in detail. Granule discharge appears slower in mammals (Barros and Yanagimachi, 1971; Bedford, 1972) than in other vertebrates or most invertebrates (cf. Gould-Somero and Holland, 1975)—kinetics that can perhaps be correlated with the fact that mammalian ova are normally exposed to relatively few spermatozoa and do not need a rapid block to polyspermy (see Hunter and Léglise, 1971a, b, for disucssion of increase in polyspermy after tubal resection).

Ovulated mammalian eggs can have at least two populations of cortical granules (Gulyas, 1976), one lying immediately subjacent to the oolemma, the other tending to be scattered below the egg cortex. The former are probably the first to be discharged but we know nothing of the kinetics of exocytosis of granules located several microns from the egg surface or whether they are all discharged. Efforts to quantitate this in anuran ova suggest that about 1 to 2% are retained normally (Kotani et al., 1973; Campanella and Andreuccetti, 1977). This percentage is considereably greater in artificially activated eggs (Steinhardt et al., 1974; Wolf, 1974; Gulyas, 1976).

6.3.1. Mechanics of exocytosis

No detailed information about the mechanics of fusion of the cortical granule membranes has yet appeared from studies of the ultrastructure of exocytosis in nonmammalian eggs or from studies of fertilization or artificial activation of mammalian eggs. In the anuran egg the chance of catching the leading edge of the wave of exocytosis is small, and when seen by Grey et al. (1974), the authors were not interested specifically in the mechanics of membrane fusion, consequently their micrographs are not of sufficient resolution to reveal intermediate states of membrane apposition and fusion. In the smaller mammalian egg timing of sperm-stimulated activation is difficult, and following electric shock (Gulyas, 1976) exocytosis was not completed in 30 minutes. Other agents of egg activation that induce cortical granule release may result in different kinetics for exocytosis (Steinhardt et al., 1974), and this should be kept in mind when investigating their effects at the ultrastructural level.

Although there have been no detailed observations so far, the general notion we have of the mechanics of mammalian cortical granule exocytosis—right or wrong—is that of apposition and fusion of the limiting membrane of the granules with the oolemma at a single locus (Fig. 24). Possibly, lateral fusion between adjacent cortical granules may sometimes occur to produce cavernae which contain the contents of many granules prior to release to the perivitelline space (Szöllözi, 1967). Exocytosis in the sea urchin, by contrast, when induced electrically, apparently involves multiple sites of apposition and fusion between the oolemma and the limiting membrane of each granule (Millonig, 1969). Preceding this fusion, the minimal gap of cytoplasm separating the cortical granules from the oolemma must be reduced to less than 0.5 nm—an event which first occurs over the egg surface oriented toward the anode. How this is achieved is unknown. Reduction of the gap could involve the following: (1) deformation of the shape of the granules, as suggested for the secretion of parotid gland granules (Schramm et al., 1973); (2) formation of microblebs from the granule-limiting membrane, as seen during the fusion of lysosomes in vitro (Raz and Goldman, 1974); (3) possible involvement of the endoplasmic reticulum as an intermediate between the granules and the oolemma (Campanella and Andreuccetti, 1977).

Current notions about the structural mechanics of membrane fusion (Palade and Bruns, 1968; Satir, 1974; Ahkong et al., 1975; Pinto da Silva and Nogueira, 1977) suggest that a pentalaminar membrane complex occurs initially at points of apposition between exocytosing granules and the plasmalemma. Where such complexes have been looked for in mast cells (Lagunoff, 1973), they were not seen at the cell surface but were observed between the lateral cytoplasmic faces of mast cell granules. As a second step, a transitory trilaminar complex or bilayer

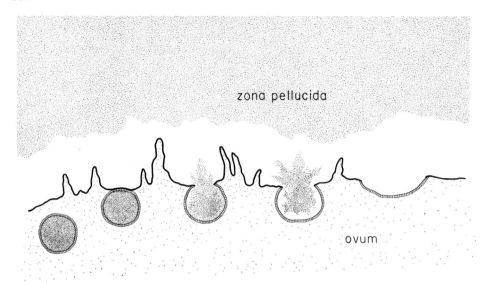

Fig. 24. Diagram to illustrate the general mode of cortical granule exocytosis in mammalian ova. After a mature ovum is penetrated by a spermatozoon, the cortical granules approach the oolemma and fusion occurs between the latter and the limiting membrane of each granule. This results (1) in release of the content of the granule, and (2) in insertion of its limiting membrane into the oolemma. To produce the block to polyspermy, the content of the granule may act only on the zona pellucida (e.g., in the hamster), or it may also contribute to the change induced by the fertilizing spermatozoon at the surface of the oolemma (this seems likely in the rabbit).

diaphragm can be seen where fusion has occurred between the inner layer of the plasmalemma and the outer layer of the exocytosing body's limiting membrane (Palade and Bruns, 1968; Pinto da Silva and Nogueira, 1977). Completion of this fusion, namely, establishment of membrane continuity, has not been visualized in eggs or in other cells, but it appears to be a cataclysmic event involving poorly understood rearrangement of membrane components. The concept of layered fusion during the release of secretory granules has been verified in several cells types (Palade and Bruns, 1968; Satir et al., 1973; Tandler and Poulsen, 1976; Lawson et al., 1977; Pinto da Silva and Nogueira, 1977), some of these having been studied by combined electron microscopy of thin sections and of freeze-fracture/freeze-etch replicas. However, there has yet to be a comparable study of cortical granule rupture in vertebrate eggs.

6.3.2. Coupling of the exocytosis response
In the mature egg, where the oolemma is fully penetrable, its stimulation by a fertilizing sperm or other active agent induces an accompanying exocytosis of cortical granules, whose contents effect the block to polyspermy. There are several circumstances, however, where these events are uncoupled or can be separated experimentally.

6.3.2.1. Anuran eggs. Although the vitellus of the immature anuran egg is penetrable (Elinson, 1973, 1975; Katagiri, 1974), the ability to release cortical granules in response to egg penetration does not appear until after initiation of the first meiotic division. Prior to this, the anuran egg is not able to achieve exocytosis even in response to mechanical stimulation (Skoblina, 1969; Smith and Ecker, 1969; Belanger and Schuetz, 1975; Katagiri and Moriya, 1976). The presence of a nucleus is not essential for development of the coupled state. For experiments with primary oocytes of *Bufo* (Katagiri and Moriya, 1976) and *Rana* (Smith and Ecker, 1969) show that coupling of the potential for exocytosis with penetration can occur when the germinal vesicle is removed and the enucleated eggs are cultured in progesterone-supplemented media. Exocytosis can also be induced before such coupling by microinjection of Ca^{2+} into *Xenopus* eggs as soon as cortical granules appear (Hollinger et al., 1976). But physiological coupling appears to require a change in egg surface properties, since it is inhibited by theophylline yet is independent of intraovular levels of cAMP (O'Conner and Smith, 1976). Once attained, the coupled state can later be uncoupled; as, for example, when mature *Xenopus* oocytes are incubated in vitro (de Roeper and Barry, 1976).

A variety of stimuli induce cortical granule exocytosis (Grey et al., 1974; Steinhardt et al., 1974; Wolf, 1974), yet each anuran species may have different requirements for this induction. In contrast to *Xenopus,* a presumed increase in intraovular calcium did not induce exocytosis in *Rana* primary oocytes until 2 to 3 hours after breakdown of the germinal vesicle. This interval coincides with the retraction of the egg surface from the vitelline envelope, but precedes by 12 hours the moment at which exocytosis can first be induced by pricking (Belanger and Schuetz, 1975).

6.3.2.2. Mammalian eggs. Mammalian oocytes generally have a full complement of cortical granules prior to the first meiotic division. An exception is the mouse, in which a full complement is elaborated during the 2 to 3 hours between ovulation and the onset of sperm transport to the ampulla (Stefanini et al., 1969; Zamboni et al., 1976). Coupling of the potential for cortical granule exocytosis with sperm/egg fusion has not been studied carefully in mamals. Hamster or rabbit primary oocytes would provide ideal material for such investigations since they are fully penetrable in vivo, with a high frequency of polyspermy and retention of cortical granules (Moore and Bedford 1978; Berrios and Bedford, unpublished observations; and see Fig. 19).

A variety of chemical, enzymic, and physical treatments will induce exocytosis in the eggs of mammals (Gwatkin and Williams, 1974; Steinhardt et al., 1974; Gulyas, 1976; Gwatkin et al., 1976; Zamboni et al., 1976; Uehara and Yanagimachi, 1977). For instance, as in mast cells where the events of membrane apposition and granule fusion have been studied by freeze-fracture techniques (Lawson et al., 1977), Con A induces exocytosis in hamster eggs (Gwatkin et al., 1976). These various agents offer means of studying the ultrastructural basis for coupling of the potential of the exocytosis response to the stimulus of sperm/egg fusion.

Mammalian eggs can be used in other ways to study this coupling. For example, development of rabbit eggs can be initiated without release of cortical granules by incubation at low temperatures (Gulyas, 1974; Longo, 1975), during which such eggs retain their penetrability (Fléchon et al., 1975); they may also become activated in response to sperm products (enzymes?) without releasing their cortical granules (Dandekar and Gordon, 1975). Similarly, mouse eggs activated by hyaluronidase (Graham, 1974), followed by osmotic shock (Graham and Deussen, 1974), show cortical granule retention (Solter et al., 1974). Such treatments probably activate eggs by changing the permeability of the oolemma, but why they fail to induce exocytosis is unknown. Possibly they act to uncouple the potential for exocytosis, as appears to be the case when hamster eggs heated to 40°C retained their capacity to be penetrated but failed to release their cortical granules, so becoming polyspermic (Gwatkin and Williams, 1974).

7. Sperm-egg fusion in mammals

7.1 General features

The studies of the Colwins (1967) in several invertebrate species first demonstrated that in these lower forms the remaining inner membrane of the reacted acrosome fuses with the oolemma. With the exception of one study in the river lamprey, where fusion with the egg also is effected via an acrosomal filament in a way resembling that in invertebrates (Nicander and Sjödén, 1971), gamete interaction has not been analyzed with profit in any vertebrate group except the eutherian mammals. By contrast with the lamprey and the invertebrates studied, initial fusion with the mammalian egg does not involve the inner acrosomal membrane. The moment of the first contact has not yet been visualized with sufficient resolution, but we believe that this must entrain the sperm plasma membrane overlying the mid or equatorial region of the spermatozoon. This region persists in rabbit spermatozoa in the perivitelline space, and the evidence of an intact equatorial segment in hamster spermatozoa in the perivitelline space and in those which have penetrated the vitellus in vivo (Figs. 16, 20, 21) provides strong reason to think that the plasmalemma over the equatorial segment fuses first with the oolemma. We thus believe that the zone of the spermatozoon involved in the initial events of fusion is limited anteriorly by the rostral border of the equatorial segment, and that it extends caudally as far as the postacrosomal region (Figs. 17, 22b, 23). As the postacrosomal plasmalemma, now continuous with the oolemma, is reflected back, the ooplasm gains access first to the perinuclear material and then to the dense chromatin of the nucleus itself. Subsequently, as surface movements of the egg cause the spermatozoon to sink into the ooplasm, the flagellum is also incorporated by a process which seems to involve a reaction of the oolemma to the tail surface and then fusion of the tail plasalemma with the oolemma (Bedford, 1972; Noda and Yanagimachi, 1976).

As shown diagrammatically in Fig. 23, the membranes of the acrosome that form the equatorial segment do not participate in the fusion reaction and, as has been clear in all the studies performed so far, neither does the persisting inner membrane of the acrosome take part in a fusion reaction during mammalian fertilization. The front or rostral part of the spermatozoon is brought into the egg as a result of envelopment of the acrosomal surface of the spermatozoon by pseudopodial folds thrown up around it at the egg surface (Fig. 22c). Its incorporation is complete when the borders of these folds fuse with each other, thus forming around the rostral part of the sperm a vesicle comprised by acrosome membrane and by endocytosed oolemma (Figs. 18, 20, 22d). In contrast to the mature ovum, the surface of the immature oocyte is unable to form such a vesicle (Berrios and Bedford, unpublished observation).

While a portion of the oolemma clearly is taken into the egg by the spermatozoon, the possibility has been recognized for some years that during fertilization some part of the plasma membrane of the spermatozoon is inserted into and constitutes, at least temporarily, a component of the egg surface. A suggestion to this effect is seen in the postfusion change in the colloid-binding character of what appears to be the postacrosomal surface (Yanagimachi et al., 1973), but the nonspecific nature of that marker does not rule out the possibility that this change is inherent in the sperm membrane and is not dependent on admixing with oolemmal components. Firmer evidence for the idea that sperm membrane components are incorporated into the egg surface comes from the observation that fertilized rabbit eggs were lysed when exposed to an isoantiserum prepared against whole sperm, whereas unfertilized eggs were unaffected by this. This complement-dependent ovo-toxic reaction was shown to be specific for testis associated antigens since prior absorption by testis, but not ovarian or other tissue homogenate, annulled the lytic effect of the antiserum (O'Rand, 1977b).

Although the broad morphological aspects of the union of sperm and egg have been defined and the place of different sperm organelles in this is implied, major conceptual questions nonetheless remain. It is not know precisely why the equatorial segment is excluded from the fusion events of the acrosome reaction. Indeed, it is not known why such a posterior "collar" has evolved in the acrosome of eutherian mammals since it is not seen in marsupials, and ultrastructural observation of the spermatozoa of a monotreme (echidna) and of several birds and reptiles gives no indication of its existence in lower species (Bedford, unpublished observation). The content of the equatorial segment seems different from the remainder of the acrosome because it may often display a ladderlike or septate organization within the egg (Barros and Franklin, 1968) or when exposed to Triton-X (Wooding, 1973). It may be that the organization of this content and the special (pentalaminar) structure of the membrane are the features which preclude its fusion with the overlying plasmalemma, and thus allow the preservation of this segment. As indicated earlier (section 5.2.1), the surface of this localized region does not seem competent to take part in fusion until the acrosome reaction propagates a change that renders it so. We now suggest, however, that the equatorial segment constitutes a device to ensure the availability of an intact region of head surface membrane, unencumbered by associated-S-S-stabilized material lying beneath, through which fusion with the vitellus can take place (Bedford and Moore, in preparation).

7.2. Role of microvilli

Cell-to-cell contact and adhesion are often mediated first via microspikes or other specialized surface projections. Vertebrate eggs bristle with microvilli in varying degrees depending on the species, and Austin (1968) has proposed that the microvilli of the mammalian egg function in sperm/egg fusion in a manner analogous to the acrosomal tubule of invertebrate spermatozoa. This view of the mechanics of sperm/egg fusion, one reiterated most recently by Gwatkin (1976), is based on a limited number of ultrastructural observations of mammalian eggs in fertilization situations and on theoretical proposals that cell-to-cell contacts are mediated via surface projections having a radius of curvature of 0.1 μm or less (Pethica, 1961; Bangham, 1964). In our opinion critical evidence for the notion that sperm/egg fusion first occurs over the egg microvilli has yet to be established, and the suggestion that the tips of the microvilli are the sites of fusion should be entertained with caution. Early electron micrographs depicting the ultrastructure of fertilization in the rat (Pikó and Tyler, 1964; also see Pikó, 1969) were interpreted as showing multiple sites of fusion between the egg surface and the postacrosomal region of the sperm plasmalemma. More recent studies on hamster sperm/egg interaction in vitro show that during the first step in sperm contact the microvilli of the egg appear to be in close apposition with the anterior region of the reacted sperm head, the inner acrosomal membrane (Yanagimachi and Noda, 1970a). The type of view of sperm/egg fusion in Fig. 17 taken by Dr. L. E. Franklin (unpublished observation), indicates that initially the region of fusion is confined to that overlying the equatorial segment, and other electron micrographs published to date probably represent later stages. Nevertheless, Fig. 17 and published evidence for sperm/egg fusion do not reveal whether the microvilli are the sites where fusion occurs. In this context we draw attention to the arrangement of the microvilli in Fig. 18, where the disappearance of the intervillous regions implies that the microvilli are rigid structures. As will be discussed, we believe that the intervillous regions of the oolemma may possess the ability to fuse with spermatozoa and that fusion is determined by the properties of these portions of the oolemma. Support for this viewpoint is provided also by our finding (Berrios and Bedford, unpublished observations) that spermatozoa fuse readily with the surface of the immature rabbit oocyte which displays only a very few widely dispersed microvilli.

Early scanning electron microscopy (SEM) studies of the intial sequence of contact in vitro between hamster eggs and spermatozoa (Yanagimachi and Noda, 1972) appeared to demonstrate a subsequent step in fusion, namely, the embrace of microvilli around spermatozoa during egg penetration. This suggests that the microvilli of the egg fold around and fuse with the remainder of the postacrosomal region of the sperm head after initial fusion. Since the material on which this report is based was air-dried, further SEM studies using critical point drying might profitably be undertaken to determine the early steps of interaction during fertilization. Transmission electron micrographs of hamster sperm/egg fusion often show the sperm head surrounded by what appears to be the egg surface, with rows of closely spaced microvilli arched over the sperm head. This and the microvillous spacing in Fig. 18 again suggest that the intervillous regions have the capacity to flow and fuse with the sperm plasmalemma. We believe that once fusion occurs the oolemma displaces the sperm plasmalemma from around the head, peeling back the sperm membrane and thereby bringing the oolemma, bearing microvilli, over the sperm (Fig. 19).

The viewpoint that mammalian spermatozoa probably fuse first with the egg surface derived from the intervillous regions is also supported by evidence of Wassarman et al. (1976, 1977) that the microvillous and intervillous regions of the oolemma differ in membrane organization and surface character. The microvilli of the mouse egg are relatively stable structures (at least before germinal vesicle breakdown) and do not exhibit membrane flow as do the components of the intervillous regions of the egg surface (Wassarman et al., 1976, 1977). If the microvilli of the oolemma do have a higher membrane viscosity than the intervillous regions, the latter would be expected to be the sites of sperm/egg fusion, based on current notions about the need for fluidity for cell fusion (Poste and Allison, 1973; Ahkong et al., 1975; Poste and Papahadjopoulos, 1976). However, a critical analysis of the character of the surface of eggs has not been performed, and as an approach to this we recommend the quantitative studies of Weiss and Subjeck (1974) and those of Grinnell and colleagues (1976), which show that the microvilli of several cell types have a greater negative-charge density than do the intervillous regions. The microviscosity of different regions of the oolemma could also be assessed by direct means. For example, the microvilli of rat intestinal cells have been shown to have the highest microviscosity of any membrane previously analyzed by fluorescent polarization techniques (Schachter and Shinitzky, 1977) using hydrocarbon fluorophores to measure the molar ratio of cholesterol/phospholipid in membranes and hence the degree of fluidity of the membrane (Inbar and Shinitzky, 1974). Until such approaches are applied to characterize the properties of the oolemma and of the sperm plasmalemma, we will remain ignorant of the features that determine their capacity for fusion.

8. Post-penetration events

Sperm penetration into the ooplasm is soon accompanied by hydration and swelling of the condensed sperm chromatin. This decondensation is preceded by loss of the nuclear envelope, though the mode of this loss has not been elucidated. Thus, an earlier hypothesis that redundant nuclear envelope of the spermatozoon may have a role in accommodating to the increased volume of the male chromatin after it enters the egg is no longer tenable. Interestingly, we now find that the ooplasm of the immature rabbit oocyte does not immediately induce a dissolution of the sperm nuclear envelope; this remains intact for some hours after penetration of spermatozoa into the immature rabbit oocyte (Berrios and Bedford, unpublished observations). The inability of the immature ooplasm to induce chromatin decondensation therefore might relate to its inability to remove the barrier of the nuclear membrane, and not only to an absence of a decondensation factor per se? In the mature egg, however, the decondensing chromatin is not shielded from the ooplasm by any membrane at this stage, and in the succeeding hours, the dispersed sperm chromatin is then reconstituted as a pronucleus, as also is the now dispersed haploid complement of the female chromatin. As this occurs, a pronuclear envelope appears de novo. In electronmicrographs this first resembles a series of flattened vesicles surrounding the faintly defined chromatin that seem to fuse to form long flat cisternae and come to resemble the typical double membrane of the nuclear envelope (Yanagimachi and Noda, 1970a; Bedford, 1972). The origin of the perinuclear vesicles has not been established (Longo, 1973), and in the published studies some disparity exists in the timing of the appearance of the nuclear envelope. That shown in the hamster by Yanagimachi and co-workers sometimes seems to develop even before nuclear decondensation is complete, whereas the nuclear envelope of the rabbit never appears until the chromatin is totally dispersed. Whether this is a real difference or one created by some asynchrony in the in vitro fertilization situation used to study the hamster remains to be determined.

About 4 to 5 hours after penetration, male and female pronuclei become visible in fertilized eggs and then migrate to the center of the egg, to lie next to each other. While commonly believed to be the case, fusion between the membranes of the male and female pronucleus does not occur in mammals, although they can come in very close apposition (Longo and Anderson, 1969; Longo, 1973; Anderson et al., 1975). Rather, the pronuclear envelopes undergo vesiculation, followed by their breakdown. The few mammals studied follow the "Ascaris type" of fertilization (Wilson, 1925) in which the first round of chromosome replication occurs within separate male and female pronuclei, and the "zygote" genome becomes enclosed by a common nuclear envelop only after telophase of the first cleavage division. We have no information on the ultrastructural details of pronuclear association in other vertebrate classes. Polyspermy occurs widely in amphibians (urodeles), reptiles, and birds (Rothschild, 1954), but only one male pronucleus associates with the female pronucleus in these groups. Pronuclei are said to fuse in urodeles (Werner, 1975), but we know of no ultrastructural evidence to support this. As in the rabbit (Longo and Anderson, 1969), there may be apposition and interdigitation of the male and female pronuclei, which could be interpreted as pronuclear fusion in the light microscope. Distinct group difference in the capacity of pronuclear membranes to fuse with one another is a curious phenomenon for which we have no satisfactory explanation.

9. Interaction of spermatozoa with somatic cells

The incorporation of spermatozoa by somatic cells—whether they be epithelial cells of the male excurrent duct and/or female tract, phagocytes, or cell lines maintained in vitro— has interested biologists concerned with such questions, as autoimmunity and telegony, and although controversial (Phillips et al., 1976), has intrigued oncologists exploring the possibility of transfer of genetic information from the sperm to its host cell (Bendich et al., 1976). However, discussion of this area has sometimes seemed to imply also that the interaction of spermatozoa with somatic cells has some homology with fertilization itself and that such a system may be used as a "model" for the study of fertilization. It is unfortunate that the semantics of interaction between spermatozoa and somatic cells—the use of words such as "penetrate" or "enter"—may also connote an active role for the spermatozoon where this is not really intended. It is worth while to reflect that although "penetrate" is used appropriately to describe its passage through the egg vestments, the spermatozoon has a passive role only during its incorporation into the vitellus.

For a variety of reasons, we believe that sperm/somatic cell interaction is not an appropriate or useful model from which extrapolation can be made to fertilization itself. The spermatozoa of several species will adhere readily to somatic cells in vitro, and there is no doubt that they are incorporated eventually by several cell lines and may lack enveloping membranes within the somatic cell (Bendich et al., 1976; Szöllösi, 1976). We know of no evidence, however, that the initial phase of an interaction which culminates in sperm incorporation involves fusion of the sperm plasmalemma with the host cell. In our experience, all early examples of ingestion by phagocytes or by granulosa cells in vivo (Bedford, 1968) reveal that the spermatozoon is encased in a membrane-bound vesicle, and this seems to be so in the early stages of incorporation of spermatozoa by cultured cells (Phillips et al., 1976). The examples provided by Bendich and co-workers (1976) of spermatozoa devoid of limiting membrane within the cytoplasm of cultured cells were prepared 48 hours or more after introduction of spermatozoa, at which time much of their nonkeratinoid element, including the membranes of the spermatozoon, might be expected to have disintegrated. Moreover, the spermatozoon entering the cells may have an intact acrosome, a situation not seen at fertilization. Doubtless, the plasma membranes of spermatozoa and somatic cells can be induced to fuse with each other in the presence of lysolecithin or Sendai virus (Gabara et al., 1973; Lucy, 1975; Goeringer and Dilliplane, 1976), but the value of such systems for elucidation of the mechanisms of fertilization seems doubtful. Furthermore, the ready availability of intact zona-free mammalian ova would seem to make the promotion of other such "models" unnecessary.

10. Concluding remarks

In the sequence of events leading to zygote formation in mammals, membrane fusion is clearly a key feature of the acrosome reaction, of the union of the sperm with the vitellus, of the cortical granule exocytosis which regulates the block to polyspermy, and perhaps also of the formation of the pronuclear envelopes after penetration has occurred. We have focused here almost entirely on mammalian gamete membranes, and have resisted extensive comparison with other membrane systems and fusion phenomena. The discussion serves first to illustrate the fact that our appreciation of the features of mammalian gamete membranes that determine their fusion-related interactions remains for the most part phenomenological and can rarely be considered more precisely in chemical or biochemical terms. There are several reasons for this. For instance, extreme difficulty in obtaining large numbers of mammalian ova prohibits the easy preparation of purified oolemma in sufficient quantities for direct analysis of its composition. Thus, many of the studies of the oolemma have tended to be cytochemical, and this still may be one useful approach to questions about the state and function of the egg microvillus in gamete interaction. Spermatozoa can be obtained in numbers that allow their analysis, but it has often been difficult so far to obtain relatively pure plasmalemmal preparations not contaminated by other sperm membranes. Furthermore, the regionality of their surface properties greatly complicates the functional interpretation of analytical findings where these cannot be attributed to a particular domain, whether it be the acrosomal or the postacrosomal segment of the head or the midpiece of the tail. Thus, again, the cytochemical and ultrastructural approach to the plasmalemma tends to prevail. A fascination with the initial events of fertilization involving penetration of spermatozoa through the zona pellucida, and its specificity, has tended to minimize interest in features that characterize and allow fusion of the spermatozoon with the oolemma. The reasons to believe that this occurs with the intervillous regions over the egg surface through the agency of a restricted region of sperm plasmalemma overlying the equatorial segment are explained, but deeper probing with a variety of techniques will be required before the molecular basis of this interaction becomes evident.

Considerable emphasis has been given in this article to the fact that the surface of the spermatozoon does not remain in the same state after leaving the testis, and seems to be modified by the environment of the excurrent duct and probably also by the female tract as finally it acquires the ability to penetrate the egg (capacitation). Unfortunately, while large numbers of immature spermatozoa can be collected in rete testis fluids, spermatozoa at the stages where these membrane changes are occurring can be recovered as only relatively small populations from the upper regions of the epididymis. The numbers obtainable from the female after capacitation are even smaller, and the spermatozoa exist there as a mixed population; some are dead and others perhaps not yet functionally capacitated. Thus, sampling difficulties allied to physiological differences between species increase the problem of making a sophisticated analysis of the sperm membrane features that relate to the fusion-associated events of fertilization.

Finally, we suggest that many questions about the changing state of the sperm plasmalemma that make it suitable for participation in the acrosome reaction and for fusion with the oolemma are being posed in a conceptual vacuum, since the student of mammalian spermatozoa now studies the surface changes in spermatozoa in the epididymis and female tract without understanding their underlying biological significance or the reasons for their appearance in the evolution of placental mammals. Amphibians and teleosts seem to complete maturation of their spermatozoa within the testis, using the Wolffian duct as a conduit only. Current investigation in our laboratory suggests that surface change is not a common feature of sperm passage through the duct in the several other submammalian species and the prototherian mammal (the echidna) studied, in whom fertilization is internal. Moreover, while numerous investigations indicate the need for capacitation in all the eutherian mammals considered, the studies extant in birds (Howarth, 1970; Howarth and Palmer 1972) suggest that this is not a required feature of sperm/egg interaction in the chicken or turkey; and thus perhaps other submammalian vertebrates do not exhibit a phenomenon comparable to capacitation. Although more information is needed to confirm this theory, it appears possible that eutherian and marsupial mammals alone have adopted a physiological system which requires unusually complex post-testicular changes in the sperm plasmalemma before fertilization can occur. It may be difficult to frame the most logical questions and to interpret analytical data correctly until we comprehend the physiological reasons for such surface modifications in spermatozoa in the male tract of eutherian mammals, and, subsequently, the need for these in the female.

Acknowledgments

We thank Drs. A. R. Bellvé, D. Bennett, G. B. Dooher, P. Higgins, C. F. Millette, M. G. O'Rand, and L. J. Romrell for permission to quote from their unpublished findings, and Drs. L. E. Franklin and P. Higgins for permission to use unpublished photographs. The diagrams were drawn by Ms Jean Jaeckel. Our own unpublished work and that of Dr. H. D. M. Moore, Miguel Berrios and Pa-Nga Viriypanich discussed in this review, have been supported by a grant from the Ford Foundation and by National Institutes of Health grants HD-07257 and HD-09215.

References

Ahkong, Q. F., Fisher, D., Tampion, W. and Lucy, J. A. (1975) Mechanism of cell fusion. Nature (London) 253, 194–195.

Anderson, E., Hoppe, P. C., Whitten, W. K. and Lee, G. S. (1975) In vitro fertilization and early embryogenesis: A cytological analysis. J. Ultrastruct. Res. 50, 231–252.

Aonuma, S., Mayumi, T., Suzuki, K., Noguchi, T., Iwari, M. and Okabe, M. (1973) Studies on sperm capacitation. I. Relationship between a guinea pig sperm-coating antigen and a sperm capacitation phenomenon. J. Reprod. Fetil. 35, 425–432.

Arora, R., Dinakar, N. and Prasad, M. R. N. (1975) Biochemical changes in the spermatozoa and luminal contents of different regions of the epididymis of the rhesus monkey, Macaca mulatta. Contraception 11, 689–699.

Austin, C. R. (1951) Observations on the penetration of sperm into the mammalian egg. Aust. J. Sci. Res. (B) 4, 581–596.

Austin, C. R. (1952) The capacitation of mammalian sperm. Nature (London) 170,326.

Austin, C. R. (1968) Ultrastructure of Fertilization. Holt, Rinehart and Winston, New York.

Baker, T. G. (1972) Oogenesis and ovarian development. In: Reproductive Biology (Balin, H. and Glasser, S., eds.) pp. 398–437, Excerpta Medica, Amsterdam.

Bangham, A. D. (1961) Electrophoretic characteristics of ram and rabbit spermatozoa. Proc. R. Soc. London, Ser. B, 155, 292–305.

Bangham, A. D. (1964) The adhesiveness of leukocytes with special reference to zeta potential. Ann. N.Y. Acad. Sci. 116, 945–949.

Barros, C. (1974) Capacitation of mammalian spermatozoa. In:Physiology and Genetics of Reproduction (Coutinho, E. M. and Fuchs, F., eds.) Part B, pp. 3–24, Plenum Press, New York.

Barros, C. and Franklin, L. E. (1968) Behavior of the gamete membranes during sperm entry into the mammalian egg. J. Cell Biol. 37, C13–C18.

Barros, C. and Herrera, E. (1977) Ultrastructural observations of the incorporation of guinea pig spermatozoa into zona-free hamster oocytes. J. Reprod. Fertil. 49, 47–50.

Barros, C. and Muñoz, G. (1973) Sperm-egg interactions in immature hamster oocytes. J. Exp. Zool. 186, 73–78.

Barros, C. and Muñoz, G. (1974) Sperm penetration through the zona pellucida of immature hamster oocytes. Acta Physiol. Lat. Am. 24, 612–615.

Barros, C. and Yanagimachi, R. (1971) Induction of zona reaction in golden hamster eggs by cortical granule material. Nature (London) 233, 268–269.

Barros, C., Bedford, J. M., Franklin, L. E. and Austin, C. R. (1967) Membrane vesiculation as a feature of the mammalian acrosome reaction. J. Cell Biol. 34, C1–C5.

Barros, C., Fujimoto, M. and Yanagimachi, R. (1973) Failure of zona penetration of hamster spermatozoa after prolonged preincubation in a blood serum fraction. J. Reprod. Fertil. 35, 89–95.

Bavister, B. D., Yanagimachi, R. and Teichman, R. J. (1976) Capacitation of hamster spermatozoa with adrenal gland extracts. Biol. Reprod. 14, 219–221.

Bedford, J. M. (1963) Changes in the electrophoretic properties of rabbit spermatozoa during passage through the epididymis. Nature (London) 200, 1178–1180.

Bedford, J. M. (1965a) Non specific tail-tail agglutination of mammalian spermatozoa. Exp. Cell Res. 38, 654–659.

Bedford, J. M. (1965b) Effect of environment on phagocytosis of rabbit spermatozoa. J. Reprod. Fertil. 9, 249–256.

Bedford, J. M. (1967a) Experimental requirement for capacitation and observations on ultrastructural changes in rabbit spermatozoa during fertilization. J. Reprod. Fertil., Suppl. 2, 35–48.

Bedford, J. M. (1967b) Fertile life of rabbit spermatozoa in rat uterus. Nature (London) 213, 1097–1099.

Bedford, J. M. (1968) Ultrastructural changes in the sperm head during fertilization in the rabbit. Am. J. Anat. 123, 329–358.

Bedford, J. M. (1969a) Limitations of the uterus in the development of the fertilizing ability (capacitation) of spermatozoa. J. Reprod. Fertil., Suppl. 8, 19–26.

Bedford, J. M. (1969b) Morphological aspects of capacitation. In: Schering Symposium on Mechanisms Involved in Contraception (Raspé, G., ed.), Adv. Biosci. 4, 35–50, Pergamon Press, Oxford.

Bedford, J. M. (1970a) Observations on some properties of a potent sperm-head agglutinin in the semen of a fertile rabbit. J. Reprod. Fertil. 22, 193–198.

Bedford, J. M. (1970b) Sperm capacitation and fertilization in mammals. Biol. Reprod. Suppl. 2, 128–158.

Bedford, J. M. (1972) An electron microscopic study of sperm penetration into the rabbit egg after natural mating. Am. J. Anat. 133, 213–254.

Bedford, J. M. (1974a) Biology of primate spermatozoa. In: Contributions to Primatology (Luckett, W. P., ed.). Vol. 3, pp. 97–139, Karger, Basel.

Bedford, J. M. (1974b) Mechanisms involved in penetration of spermatozoa through the vestments of the mammalian egg. In: Physiology and Genetics of Reproduction (Coutinho, E. M. and Fuchs, F., eds.). Part B, pp. 55–68, Plenum Press, New York.

Bedford, J. M. (1975) Maturation, transport, and fate of spermatozoa in the epididymis. In: Handbook of Physiology. Male Reproductive Systems (Greep, R. O. and Astwood, E. B., eds.). Sec. 7, Vol. V, pp. 303–317, American Physiological Society, Washington, D.C.

Bedford, J. M. (1977) Sperm/egg interactions: The specificity of human spermatozoa. Anat. Rec. 188, 477–488.

Bedford, J. M. and Chang, M. C. (1962a) Removal of decapacitation factor from seminal plasma by high-speed centrifugation. Am. J. Physiol. 202, 179–181.

Bedford, J. M. and Chang, M. C. (1962b) Fertilization of rabbit ova in vitro. Nature (London) 193, 898–899.

Bedford, J. M. and Calvin, H. I. (1974) Changes in -S-S- linked structures of the sperm tail during epididymal maturation, with comparative observations in sub-mammalian species. J. Exp. Zool. 187, 181–204.

Bedford, J. M. and Nicander, L. (1971) Ultrastructural changes in the acrosome and sperm membranes during maturation of spermatozoa in the testis and epididymis of the rabbit and monkey. J. Anat. 108, 527–543.

Bedford, J. M. and Millar, R. P. (1978) The character of sperm maturation in the epididymis of the ascrotal hyrax, *Procavia capensis,* and armadillo, *Dasypus novemcintus.* Biol. Reprod. (submitted.)

Bedford, J. M., Cooper, G. W. and Calvin, H. I. (1972) Post-meiotic changes in the nucleus and membranes of mammalian spermatozoa. In: The Genetics of the Spermatozoon (Beatty, R. A. and Gluecksohn-Waelsch, S., eds.) pp. 69–89, University of Edinburgh, Edinburgh.

Bedford, J. M., Calvin, H. and Cooper, G. W. (1973) The maturation of spermatozoa in the human epididymis. J. Reprod. Fertil., Suppl. 18, 199–213.

Belanger, A. M. and Schuetz, A. W. (1975) Precocious induction of activation responses in amphibian oocytes by divalent ionophore A23187. Dev. Biol. 45, 378–381.

Bendich, A., Borenfreund, E., Witkin, S. S., Beju, D. and Higgins, P. J. (1976) Information transfer and sperm uptake by mammalian somatic cells. In: Progress in Nucleic Acid Research and Molecular Biology (Davidson, J. N. and Cohn, W. E., eds.). Vol. 17, pp. 43–75, Academic Press, New York.

Bennett, D. and Boyse, E. A. (1973) Sex ratio in progeny of mice inseminated with sperm treated with H-Y antiserum. Nature (London) 246, 308–309.

Bey, E. (1965) The electrophoretic mobility of sperm cells. In: Cell Electrophoresis (Ambrose, E. J., ed.) pp. 142–151, Little, Brown & Co., Boston.

Biggers, J. D. (1966) Reproduction in male marsupials. Symp. Zool. Soc. (London) 15, 251–280.

Blank, M., Soo, L. and Britten, J. S. (1974) The properties of rabbit sperm membranes in contact with electrode surfaces. J. Membr. Biol. 18, 351–364.

Boettcher, B. (1965) Human ABO blood group antigens on spermatozoa from secretors and non-secretors. J. Reprod. Fertil. 9, 267–268.

Boldt, D. H., Speckart, S. F., Richards, R. L. and Alving, C. R. (1977) Interaction of plant lectins with glycolipids in liposomes. Biochem. Biophys. Res. Commun. 74, 208–214.

Bose, A. R., Kar, A. B. and Dasgupta, P. R. (1966) Sialic acid in the genital organs of the male rat. Curr. Sci. 35, 336–337.

Bose, T. K., Prasad, M. R. N. and Moudgal, N. R. (1975) Changing patterns of sialic acid in the spermatozoa and luminal plasma of the epididymis and vas deferens of the hamster, *Mesocricetus auratus.* Indian J. Exp. Biol. 13, 8–12.

Brackett, B. G. and Oliphant, G. (1975) Capacitation of rabbit spermatozoa *in vitro.* Biol. Reprod. 12, 260–274.

Braden, A. W. H., Austin, C. R. and David, H. A. (1954) The reaction of the zona pellucida to sperm penetration. Austr. J. Biol. Sci., 7, 391–409.

Brummett, A. R. and Dumont, J. N. (1976) Oogenesis in *Xenopus laevis* (Daudin). III. Localization of negative charges on the surface of developing oocytes. J. Ultrastruct. Res. 55, 4–16.

Burgos, M. H. and Fawcett, D. W. (1956) An electron microscope study of spermatid differentiation in the toad, *Bufo arenarum* Hensel. J. Biophys. Biochem. Cytol. 2, 223–240.

Buthala, D. A., Ericsson, R. J. and Chubb, G. T. (1971) Interaction of Sendai virus and rabbit sperm: Transmission and scanning electron microscopy. Biol. Reprod. 5, 325–329.

Calvin. H. I. and Bedford, J. M. (1971) Formation of disulphide bonds in the nucleus and accessory structures of mammalian spermatozoa during maturation in the epididymis. J. Reprod. Fertil., Suppl. 13, 65–75.

Campanella, C. (1975) The site of spermatozoon entrance in the unfertilized egg of *Discoglossus pictus* (Anura): An electron microscope study. Biol. Reprod. 12, 439–447.

Campanella, C. and Andreuccetti, P. (1977) Ultrastructural observations on cortical endoplasmic reticulum and residual cortical granules in the egg of *Xenopus laevis.* Dev. Biol. 56, 1–10.

Chang, M. C. (1951) The fertilizing capacity of spermatozoa deposited into the Fallopian tubes. Nature (London) 168, 697–698.

Chang, M. C. (1955) Development of fertilizing capacity of rabbit spermatozoa in the uterus. Nature (London) 175, 1036–1037.

Chang, M. C. (1957) A detrimental effect of seminal plasma on the fertilizing capacity of sperm. Nature (London) 179, 258–259.

Chien, S., Cooper, G. W., Jan, K.-M., Miller, L. H., Howe, C., Usami, S., and Lalezari, P. (1974) N-Acetylneuraminic acid deficiency in erythrocyte membranes: Biophysical and biochemical correlates. Blood 43, 445–460.

Chulavatnatol, M. and Yindepit, S. (1976) Changes in surface ATPase of rat spermatozoa in transit from the caput to the cauda epididymidis. J. Reprod. Fertil. 48, 91–97.

Colwin, L. H. and Colwin, A. L. (1967) Membrane fusion in relation to sperm-egg association. In: Fertilization (Metz, C. B. and Monroy, A., eds.). Vol. 1, pp. 295–367, Academic Press, New York.

Cooper, G. W. and Bedford, J. M. (1971a) Acquisition of surface charge by the plasma membrane of mammalian spermatozoa during epididymal maturation. Anat. Rec. 169, 300–301.

Cooper, G. W. and Bedford, J. M. (1971b) Charge density change in the vitelline surface following fertilization of the rabbit egg. J. Reprod. Fertil. 25, 431–436.

Cooper, G. W. and Bedford, J. M. (1976) Asymmetry of spermiation and sperm surface charge patterns over the giant acrosome in the musk shrew *Suncus murinus*. J. Cell Biol. 69, 415–428.

Dan, J. C. (1967) Acrosome reaction and lysins. In: Fertilization (Metz, C. B. and Monroy, A., eds.). Vol. 1, pp. 237–294, Academic Press, New York.

Dan, J. C., Hashimoto, S., Kubo, M. and Yonehara, K. (1974) The fine structure of the acrosome trigger. In: The Functional Anatomy of the Spermatozoon (Afzelius, B. A., ed.) pp. 39–45, Pergamon Press, Oxford.

Dandekar, P. and Gordon, M. (1975) Electron microscope evaluation of rabbit eggs exposed to spermatozoa treated with capacitating agents. J. Reprod. Fertil. 44, 143–146.

Darin-Bennett, A., Poulos, A. and White I. G. (1976) The fatty acid composition of the major phosphoglycerides of ram and human spermatozoa. Andrologia 8, 37–45.

Davis, B. K. (1971) Macromolecular inhibitor of fertilization in rabbit seminal plasma. Proc. Natl. Acad. Sci. U.S.A. 68, 951–955.

Delgado, N. M., Huacuja, L., Pancardo, R. Ma., Merchant, H. and Rosado, A. (1976) Changes in the protein conformation of human spermatozoal membranes after treatment with cyclic adenosine 3′, 5′-monophosphate and human follicular fluid. Fertil. Steril. 27, 413–420.

Dukelow, W. R., Chernoff, H. N. and Williams, W. L. (1967) Properties of decapacitation factor and presence in various species. J. Reprod. Fertil. 14, 393–399.

Eager, D. D., Johnson, M. H. and Thurley, K. W. (1976) Ultrastructural studies on the surface membrane of the mouse egg. J. Cell Sci. 22, 345–353.

Eddy, E. M. and Shapiro, B. M. (1976) Changes in the topography of the sea urchin egg after fertilization. J. Cell Biol. 71, 35–48.

Edelman, G. M. and Millette, C. F. (1971) Molecular probes of spermatozoan structures. Proc. Natl. Acad. Sci. U.S.A. 68, 2436–2440.

Elinson, R. P. (1973) Fertilization of frog body cavity eggs: *Rana pipiens* eggs and *Rana clamitans* sperm. Biol. Reprod. 8, 362–368.

Elinson, R. P. (1975) Site of sperm entry and a cortical contraction associated with egg activation in the frog *Rana pipiens*. Dev. Biol. 47, 257–268.

Elinson, R. P. (1977) Fertilization of immature frog eggs: cleavage and development following subsequent activation. J. Embryol. Exp. Morphol. 37, 187–201.

Erickson, R. P., Friend, D. S. and Tennenbaum, D. (1975) Localization of lactate dehydrogenase-X on the surfaces of mouse spermatozoa. Exp. Cell Res. 91, 1–5.

Ericsson, R. J., Buthala, D. A. and Norland, J. F. (1971) Fertilization of rabbit ova *in vitro* by sperm with adsorbed Sendai virus. Science 173, 54–55.

Esbenshade, K. L. and Clegg, E. D. (1976) Electrophoretic characterization of proteins in the plasma membrane of porcine spermatozoa. J. Reprod. Fertil. 47, 333–337.

Fawcett, D. W. (1970) A comparative view of sperm ultrastructure. Biol. Reprod. Suppl. 2, 90–127.

Fawcett, D. W. (1975) The mammalian spermatozoon. Dev. Biol. 44, 394–436.

Fawcett, D. W. and Hollenberg, R. D. (1963) Changes in the acrosome of guinea pig spermatozoa during passage through the epididymis. Z. Zellforsch. Mikrosk. Anat. 60, 276–292.

Fawcett, D. W. and Ito, S., (1965) The fine structure of bat spermatozoa. Am. J. Anat.116, 567–610.

Fellous, M., Gachelin, G., Buc-Caron, M.-H., Dubois, P. and Jacob, F. (1974) Similar location of an early embryonic antigen on mouse and human spermatozoa. Dev. Biol. 41, 331–337.

Fléchon, J.-E. (1975) Ultrastructural and cytochemical modifications of rabbit spermatozoa during epididymal transport. In: The Biology of Spermatozoa (Hafez, E. S. E. and Thibault, C. G., eds.) pp. 36–45, Karger, Basel.

Fléchon, J.-E. and Morstin, J. (1975) Localisation des glycoprotéines et des charges négatives et postives dans le revêtement de surface des spermatozoïdes éjaculés de lapin et de taureau. Ann. Histochim. 20, 291–300.

Fléchon, J.-E., Huneau, D., Solari, A. and Thibault, C. (1975) Réaction corticale et blocage de la polyspermie dan l'oeuf de lapine. Ann. Biol. Anim. Biochim. Biophys. 15, 9–18.

Fournier, S. (1968) Electrophorèse des protéines du tractus génital du Rat. I. Présence dans le sperme épididymaire d'une glycoprotéine migrant vers l'anode à pH=8,5. C. R. Seances Soc. Biol. 162, 568–571.

Frazer, L., Dandekar, P. and Vaidya, R. (1971) In vitro fertilization of tubal rabbit ova partially or totally denuded of follicular cells. Biol. Reprod. 4, 229–233.

Friend, D. S. and Fawcett, D. W. (1974) Membrane differentiations in freeze-fractured mammalian sperm. J. Cell Biol. 63, 641–664.

Gabara, B., Gledhill, B. L., Croce, C. M., Cesarini, J. P. and Koprowski, H. (1973) Ultrastructure of rabbit spermatozoa after treatment with lysolecithin and in the presence of hamster somatic cells. Proc. Soc. Exp. Biol. Med. 143, 1120–1124.

Gall, W. E. and Ohsumi, Y. (1976) Decondensation of sperm nuclei in vitro. Exp. Cell Res. 102, 349–358.

Gasic, G. J., Berwick, L. and Sorrentino, M. (1968) Positive and negative colloidal iron as cell surface electron stains. Lab. Invest. 18, 63–71.

Goeringer, G. C. and Dilliplane, D. (1976) Some aspects of the fusion of avian erythrocytes and bovine spermatozoa in the presence of chemical fusing agents. Anat. Rec. 184, 412.

Gordon, M. (1973) Localization of phosphatase activity on the membranes of the mammalian sperm head. J. Exp. Zool. 185, 111–120.

Gordon, M. and Dandekar, P. V. (1976) Fertilization of rabbit ova pretreated with Concanavalin A (Con A) and wheat germ agglutinin (WGA). Anat. Rec. 184, 413.

Gordon, M. and Dandekar, P. V. (1977) Fine-structural localization of phosphatase activity on the plasma membrane of the rabbit sperm head. J. Reprod. Fertil. 49, 155–156.

Gordon, M., Dandekar, P. V. and Bartoszewicz, W. (1974) Ultrastructural localization of surface receptors for Concanavalin A on rabbit spermatozoa. J. Reprod. Fertil. 36, 211–214.

Gordon, M., Dandekar, P. V. and Bartoszewicz, W. (1975a) The surface coat of epididymal, ejaculated and capacitated sperm. J. Ultrastruct. Res. 50, 199–207.

Gordon, M., Fraser, L. R. and Dandekar, P. V. (1975b) The effect of ruthenium red and Concanavalin A on the vitelline surface of fertilized and unfertilized rabbit ova. Anat. Rec. 181, 95–112.

Gould-Somero, M. and Holland, L. (1975) Fine structural investigation of the insemination response in Urechis caupo. Dev. Biol. 46, 358–369.

Graham, C. F. (1974) The production of parthenogenetic mammalian embryos and their use in biological research. Biol. Rev. 49, 399–422.

Graham, C. F. and Deussen, Z. A. (1974) In vitro activation of mouse eggs. J. Embryol. Exp. Morphol. 31, 497–512.

Grey, R. D., Wolf, D. P. and Hedrick, J. L. (1974) Formation and structure of the fertilization envelope in Xenopus laevis. Dev. Biol. 36, 44–61.

Grinnell, F., Anderson, R. G. W. and Hackenbrock, C. R. (1976) Glutaraldehyde induced alterations of membrane anionic sites. Biochim. Biophys. Acta 426, 772–775.

Gulyas, B. J. (1974) Cortical granules in artificially activated (Parthenogenetic) rabbit eggs. Am. J. Anat. 140, 577–582.

Gulyas, B. J. (1976) Ultrastructural observations on rabbit, hamster and mouse eggs following elec-

118

trical stimulation *in vitro*. Am. J. Anat. 147, 203–218.

Gwatkin, R. B. L. (1976) Fertilization. In: The Cell Surface in Animal Embryogenesis and Development (Poste, G. and Nicolson, G. L., eds.) pp. 1–54, Cell Surface Reviews, Vol. 1, Elsevier/North-Holland, Amsterdam.

Gwatkin, R. B. L. and Williams, D. T. (1974) Heat sensitivity of the cortical granule protease from hamster eggs. J. Reprod. Fertil. 39, 153–155.

Gwatkin, R. B. L. and Williams, D. T. (1977) Receptor activity of the hamster and mouse solubilized zona pellucida before and after the zona reaction. J. Reprod. Fertil. 49, 55–59.

Gwatkin, R. B. L., Rasmusson, G. H. and Williams, D. T. (1976) Induction of the cortical reaction in hamster eggs by membrane-active agents. J. Reprod. Fertil. 47, 299–303.

Hamner, C. E. (1971) Composition of oviductal and uterine fluids. In: Schering Symposium on Intrinsic and Extrinsic Factors in Early Mammalian Development (Raspé, G., ed.), Adv. Biosci. 6, 143–164, Pergamon Press, Oxford.

Hamner, C. E. (1973) Oviducal fluid—composition and physiology. In: Handbook of Physiology. Female Reproductive System (Greep, R. O. and Astwood, E. B., eds.). Sec. 7, Vol. II, Pt. 2, pp. 141–151, American Physiological Society, Washington, D.C.

Hamner, C. E. and Sojka, N. J. (1967) Capacitation of rabbit spermatozoa: Species specificity and organ specificity. Proc. Soc. Exp. Biol. Med. 124, 689–691.

Hanada, A. and Chang, M. C. (1977) Penetration of the zona-free or intact eggs by foreign spermatozoa and the fertilization of deer mouse eggs *in vitro*. J. Exp. Zool. In press.

Harding, H. R., Carrick, F. N. and Shorey, C. D. (1975) Ultrastructural changes in spermatozoa of the brush-tailed possum, *Trichosurus vulpecula* (Marsupialia) during epididymal transit. Part I: The flagellum. Cell Tissue Res. 164, 121–132.

Harding, H. R., Carrick, F. N. and Shorey, C. D. (1976) Ultrastructural changes in spermatozoa of the brush-tailed possum, *Trichosurus vulpecular* (Marsupialia) during epididymal transit. Part II: The acrosome. Cell Tissue Res. 171, 61–73.

Harper, M. J. K. (1970) Factors influencing sperm penetration of rabbit eggs *in vivo*. J. Exp. Zool. 173, 47–62.

Hartmann, J. F. and Gwatkin, R. B. L. (1971) Alteration of sites on the mammalian sperm surface following capacitation. Nature (London) 234, 479–481.

Hartmann, J. F. and Hutchison, C. F. (1974) Nature of the penetration contact interactions between hamster gametes *in vitro*. J. Reprod. Fertil. 36, 49–57.

Hartree, E. F. and Srivastava, P. N. (1965) Chemical composition of the acrosomes of ram spermatozoa. J. Reprod. Fertil. 9, 47–60.

Heard, D. H. and Seaman, G. V. F. (1960) The influence of pH and ionic strength on the electrokinetic stability of the human erythrocyte membrane. J. Gen. Physiol. 43, 635–654.

Hekman, A. and Rümke, P. (1969) The antigens of human seminal plasma (with special reference to lactoferrin as a spermatozoa-coating antigen). In: Protides Biol. Fluids (Peeters, H., ed.). vol. 16, pp. 549–552, Pergamon Press, Oxford.

Henle, W., Henle, G. and Chambers, L. A. (1938) Studies on the antigenic structure of some mammalian spermatozoa. J. Exp. Med. 68, 335–352.

Hjört, T. and Hansen, K. B. (1971) Immunofluorescent studies on human spermatozoa. I. The detection of different spermatozoal antibodies and their occurrence in normal and infertile women. Clin. Exp. Immunol. 8, 9–23.

Hollinger, T. F., Dumont, J. N. and Wallace, R. A. (1976) Calcium-induced cortical granule breakdown in *Xenopus* oocytes. J. Cell Biol. 70, 102a.

Howarth, B. (1970) An examination for sperm capacitation in the fowl. Biol. Reprod. 3, 338–341.

Howarth, B. and Palmer, M. B. (1972) An examination of the need for sperm capacitation in the turkey, *Meleagris gallopavo,* J. Reprod. Fertil. 28, 443–445.

Hunter, R. H. F. (1969) Capacitation in the gold hamster with special reference to the influence of the uterine environment. J. Reprod. Fertil. 20, 223–237.

Hunter, R. H. F. and Léglise, P. C. (1971a) Tubal surgery in the rabbit: Fertilization and polyspermy after resection of the isthmus. Am. J. Anat. 132, 45–52.

Hunter, R. H. F. and Léglise, P. C. (1971b) Polyspermic fertilization following tubal surgery in pigs, with particular reference to the role of the isthmus. J. Reprod. Fertil. 24, 233–246.

Hunter, R. H. F. and Hall, J. P. (1974) Capacitation of boar spermatozoa: synergism between uterine and tubal environments. J. Exp. Zool. 188, 203–214.

Husted, S. (1975) Sperm antibodies in men from infertile couples. Int. J. Fertil. 20, 113–121.

Inbar, M. and Shinitzky, M. (1974) Cholesterol as a bioregulator in the development and inhibition of leukemia. Proc. Natl. Acad. Sci. U.S.A. 71, 4229–4231.

Isojima, S., Koyama, K. and Tsuchiya, K. (1974) The effect on fertility in women of circulating antibodies against human spermatozoa. J. Reprod. Fertil., Suppl. 21, 125–150.

Iwamatsu, T. and Chang, M. C. (1972) Sperm penetration in vitro of mouse oocytes at various times during maturation. J. Reprod. Fertil. 31, 237–247.

Johnson, L. A., Pursel, V. G. and Gerrits, R. J. (1972) Total phospholipid and phospholipid fatty acids of ejaculated and epididymal semen and seminal vesicle fluids of boars. J. Anim. Sci. 35, 398–403.

Johnson, M. H. (1975) The macromolecular organization of membranes and its bearing on events leading up to fertilization. J. Reprod. Fertil. 44, 167–184.

Johnson, M. H. and Edidin, M. (1972) H-2 antigens on mouse spermatozoa. Transplantation 14, 781–786.

Johnson, M. H., Eager, D., Muggleton-Harris, A. and Grave, H. M. (1975) Mosaicism in the organization of concanavalin A receptors on surface membrane of mouse egg. Nature (London) 257, 321–322.

Johnson, W. L. and Hunter, A. G. (1972) Seminal antigens: their alteration in the genital tract of female rabbits and during partial in vitro capacitation with β-amylase and β-glucuronidase. Biol. Reprod. 7, 332–340.

Jones, R. C. (1973) Changes occurring in the head of boar spermatozoa: vesiculation or vacuolation of the acrosome? J. Reprod. Fertil. 33, 113–118.

Katagiri, C. (1974) A high frequency of fertilization in premature and mature coelomic toad eggs after enzymic removal of vitelline membrane. J. Embryol. Exp. Morphol. 31, 573–587.

Katagiri, C. and Moriya, M. (1976) Spermatozoan response to the toad egg matured after removal of germinal vesicle. Dev. Biol. 50, 235–241.

Kemp, N. E. and Istock, N. L. (1967) Cortical changes in growing oocytes and in fertilized or pricked eggs of Rana pipiens. J. Cell Biol. 34, 111–122.

Koehler, J. K. (1972) Human sperm head ultrastructure: a freeze-etching study. J. Ultrastruct. Res. 39, 520–539.

Koehler, J. K. (1976) Changes in antigenic site distribution on rabbit spermatozoa after incubation in "capacitating" media. Biol. Reprod. 15, 444–456.

Koehler, J. K. and Perkins, W. D. (1974) Fine structure observations on the distribution of antigenic sites on guinea pig spermatozoa. J. Cell Biol. 60, 789–795.

Konwinski, M., Vorbrodt, A., Solter, D. and Koprowski, H. (1977) Ultrastructural study of concanavalin-A binding to the surface of preimplantation mouse embryos. J. Exp. Zool. 200, 311–324.

Koo, G. C., Stackpole, C. W., Boyse, E. A., Hämmerling, U. and Lardis, M. P. (1973) Topographical localization of H-Y antigen on mouse spermatozoa by immunoelectronmicroscopy. Proc. Nat. Acad. Sci. U.S.A. 70, 1502–1505.

Kotani, M., Ikenishi, K. and Tanabe, K. (1973) Cortical granules remaining after fertilization in Xenopus laevis. Dev. Biol. 30, 228–232.

Lagunoff, D. (1973) Membrane fusion during mast cell secretion. J. Cell Biol. 57, 252–259.

Laporte, P. (1970) Acides sialiques épididymaires et maturité sexuelle chez le Rat. C. R. Seances Soc. Biol. 164, 2079–2084.

Lawson, D., Raff, M. C., Gomperts, B., Fewtrell, C. and Gilula, N. B. (1977) Molecular events during membrane fusion. J. Cell Biol. 72, 242–259.

Lindahl, P. E. (1973) Activators of the ATP-dependent surface reaction in the apical cell membrane of the bull-sperm head, causing head-to-head association. Exp. Cell Res. 81, 413–431.

Longo, F. J. (1973) Fertilization: A comparative ultrastructural review. Biol. Reprod. 9, 149–215.

Longo, F. J. (1975) Ultrastructural analysis of artificially activated rabbit eggs. J. Exp. Zool. 192, 87–112.

Longo, F. J. and Anderson, E. (1969) Cytological events leading to the formation of the two-cell stage in the rabbit: Association of the maturnally and paternally derived genome. J. Ultrastruct. Res. 29, 86–118.

Lucy, J. A. (1975) Aspects of the fusion of cells *in vitro* without viruses. J. Reprod. Fertil. 44, 193–205.

Mahi, C. A. and Yanagimachi, R. (1976) Maturation and sperm penetration of canine ovarian oocytes *in vitro*. J. Exp. Zool. 196, 189–196.

Mann, T. (1964) The Biochemistry of Semen and of the Male Reproductive Tract. Methuen & Co., London.

Martan, J. and Hruban, Z. (1970) Unusual spermatozoan formations in the epididymis of the flying squirrel *(Glaucomys volans)*. J. Reprod. Fertil. 21, 167–170.

Martan, J., Hruban, Z. and Brown, P. (1971) The formation and development of the cylindrical bodies in the mammalian epididymis. J. Reprod. Fertil. 27, 115–117.

Mathieson, B. J. (1976) Selective expression of surface on differentiated cells of the mouse: Immuno-selection of Y-bearing sperm in an *in vitro* fertilization system and expression of the thymocyte surface markers G_{IX}, TL and LY in tetraparental mice. Ph.D. Thesis, Cornell University, Ithaca, New York.

Mercado, E., Carvajal, G., Reyes, A., and Rosado, A. (1976) Sulfhydryl groups on the spermatozoa membrane. A study with a new fluorescent probe for SH groups. Biol. Reprod. 14, 632–640.

Millette, C. F. and Bellvé, A. R. (1977) Temporal expression of membrane antigens during mouse spermatogenesis. J. Cell Biol. 74, 86–97.

Millonig, G. (1969) Fine structure analysis of the cortical reaction in the sea urchin egg: after fertilization and after electric induction. J. Submicrosc. Cytol. I, 69–84.

Miyamoto, H. and Chang, M. C. (1972) Fertilization *in vitro* of mouse and hamster eggs after the removal of follicular cells. J. Reprod. Fertil. 30, 309–312.

Monroy, A. and Baccetti, B. (1975) Morphological changes of the surface of the egg of *Xenopus laevis* in the course of development. I. Fertilization and early cleavage. J. Ultrastruct. Res. 50, 131–142.

Moore, H. D. M. and Bedford, J. M. (1977) Follicular products are not needed for fertilization but do prevent polyspermy of hamster eggs *in vivo*. 10th Ann. Meet. Soc. Study Reprod., Austin, Texas (Abstract).

Moore, H. D. M. and Bedford, J. M. (1978) Ultrastructure of the equatorial segment of hamster spermatozoa during penetration of oocytes. J. Ultrastruct. Res. In press.

Multamäki, S. (1973) Isolation of pure acrosomes by subcellular fractionation of bull spermatozoa. Int. J. Fertil. 18, 193–205.

Nevo, A. C., Michaeli, I. and Schindler, H. (1961) Eletrophoretic properties of bull and rabbit spermatozoa. Exp. Cell Res. 23, 69–83.

Nicander, L. (1970) Comparative studies on the fine structure of vertebrate spermatozoa. In: Comparative Spermatology (Baccetti, B., ed.) pp. 45–55, Academic Press, New York.

Nicander, L. and Sjödén, I. (1971) An electron microscopical study of the acrosomal complex and its role in fertilization in the river lamprey, *Lampetra fluviatilis*. J. Submicrosc. Cytol. 3, 309–317.

Nicolson, G. L. (1974) The interactions of lectins with animal cell surfaces. Int. Rev. Cytol. 39, 89–190.

Nicolson, G. L. and Yanagimachi, R. (1974) Mobility and the restriction of mobility of plasma membrane lectin-binding components. Science 184, 1294–1296.

Nicolson, G. L., Yanagimachi, R. and Yangimachi, H. (1975) Ultrastructural localization of lectin-binding sites on the zonae pellucidae and plasma membranes of mammalian eggs. J. Cell Biol. 66, 263–274.

Niwa, K. and Chang, M. C. (1975) Fertilization of rat eggs *in vitro* at various times before and after ovulation with special reference to fertilization of ovarian oocytes matured in culture. J. Reprod. Fertil. 43, 435–451.

Noda, Y. D. and Yanagimachi, R. (1976) Electron microscopic observations of guinea pig spermatozoa penetrating eggs *in vitro*. Dev. Growth Differ. 18, 15–23.

O'Connor, C. M. and Smith, L. D. (1976) Inhibition of oocyte maturation by theophylline: Possible mechanism of action. Dev. Biol. 52, 318–322.

Oliphant, G. (1976) Removal of sperm-bound seminal plasma components as a prerequisite to induction of the rabbit acrosome reaction. Fertil. Steril. 27, 28–38.

Oliphant, G. and Brackett, B. G. (1973a) Immunological assessment of surface changes of rabbit sperm undergoing capacitation. Biol. Reprod. 9, 404–414.

Oliphant, G. and Brackett, B. G. (1973b) Capacitation of mouse spermatozoa in media with elevated ionic strength and reversible decapacitation with epididymal extracts. Fertil. Steril. 24, 948–955.

Olson, G. E. and Hamilton, D. W. (1978) Characterization of the surface glyoproteins of rat spermatozoa. Biol. Reprod.—in press.

Olson, G. E., Lifsics, M., Fawcett, D. W. and Hamilton, D. W. (1977) Structural specializations in the flagellar plasma membrane of opossum spermatozoa. J. Ultrastruct. Res. 59, 207–221.

O'Rand, M. G. (1977a) Restriction of a sperm surface antigen's mobility during capacitation. Dev. Biol. 55, 260–270.

O'Rand, M. G. (1977b) The presence of sperm-specific surface autoantigens on the egg following fertilization. J. Exp. Zool. 202, 267–273.

O'Rand, M. G. and Metz, C. B. (1976) Isolation of an "immobilizing antigen" from rabbit sperm membranes. Biol. Reprod. 14, 586–598.

O'Rand, M. G. and Romrell, L. J. (1977) Appearance of cell surface auto- and isoantigens during spermatogenesis in the rabbit. Dev. Biol. 55, 347–358.

Overstreet, J. W. and Bedford, J. M. (1974) Comparison of the penetrability of the egg vestments in follicular oocytes, unfertilized and fertilized ova of the rabbit. Dev. Biol. 41, 185–192.

Overstreet, J. W. and Bedford, J. M. (1975) The penetrability of rabbit ova treated with enzymes or anti-progesterone antibody: A probe into the nature of a mammalian fertilizin. J. Reprod. Fertil. 44, 273–285.

Overstreet, J. W. and Hembree, W. C. (1976) Penetration of the zona pellucida of nonliving human oocytes by human spermatozoa in vitro. Fertil. Steril. 27, 815–831.

Overstreet, J. W. and Cooper, G. W. (1978) The time and location of the acrosome reaction during sperm transport in the female rabbit. J. Reprod. Fertil. Submitted for publication.

Palade, G. E. and Bruns, R. R. (1968) Structural modulations of plasmalemmal vesicles. J. Cell Biol. 37, 633–649.

Parkes, A. S., Rogers, H. J., and Spensley, P. C. (1954) Biological and biochemical aspects of the prevention of fertilization by enzyme inhibitors. Proc. Soc. Study Fertil., 6, 65–80.

Pavlok, A. and MacLaren, A. (1972) The role of cumulus cells and zona pellucida in fertilization of mouse eggs in vitro. J. Reprod. Fertil. 29, 91–97.

Pease, D. C. (1970) Phosphotungstic acid as a specific electron stain for complex carbohydrates. J. Histochem. Cytochem. 18, 455–458.

Peleg, B. A. and Ianconescu, M. (1966) Spermagglutination and spermadsorption due to myxoviruses. Nature (London) 211, 1211–1212.

Pethica, B. A. (1961) The physical chemistry of cell adhesion. Exp. Cell Res., Suppl. 8, 123–140.

Phemister, R. D., Holst, P. A., Spano, J. S. and Hopwood, M. L. (1973) Time of ovulation in the beagle bitch. Biol. Reprod. 8, 74–82.

Phillips, D. M. (1977) Surface of the equatorial segment of the mammalian acrosome. Biol. Reprod. 16, 128–137.

Phillips, S. G., Phillips, D. M., Dev. V. G., Miller, D. A., Van Diggelen, O. P. and Miller, O. J. (1976) Spontaneous cell hybridization of somatic cells present in sperm suspensions. Exp. Cell Res. 98, 429–443.

Pienkowski, M. (1974) Study of the growth regulation of preimplantation mouse embryos using concanavalin A. Proc. Soc. Exp. Biol. Med. 145, 464–469.

Pikó, L. (1969) Gamete structure and sperm entry in mammals. In: Fertilization (Metz, C. and Monroy, A., eds.), vol. 2, pp. 325–403, Academic Press, New York.

Pikó, L. and Tyler, A. (1964) Fine structural studies of sperm penetration in the rat. Proc. 5th Int. Congr. Anim. Reprod., Trento, Italy, vol. 2, pp. 372–377.

Pinto da Silva, P. and Nogueira, M. L. (1977) Membrane fusion during secretion. J. Cell Biol. 73, 161–181.

Plattner, H., (1971) Bull spermatozoa: a reinvestigation by freeze-etching using widely different cryofixation procedures. J. Submicrosc. Cytol. 3, 19–32.

Poirier, G. R. and Spink, G. C. (1971) The ultrastructure of testicular spermatozoa in two species of Rana. J. Ultrastruct. Res. 36, 455–465.

Polge, C. and Dziuk, P. (1965) Recovery of immature eggs penetrated by spermatozoa following induced ovulation in the pig. J. Reprod. Fertil. 9, 357–358.

Poste, G. and Allison, A. C. (1973) Membrane fusion. Biochim. Biophys. Acta 300, 421–465.

Poste, G. and Papahadjopoulos, D. (1976) Lipid vesicles as carriers for introducing materials into cultured cells: Influence of vesicle lipid composition on mechanism(s) of vesicle incorporation into cells. Proc. Nat. Acad. Sci. U.S.A. 73, 1603–1607.

Poulos, A., Darin-Bennett, A., and White, I. G. (1973) The phospholipid-bound fatty acids and aldehydes of mammalian spermatozoa. Comp. Biochem. Physiol. B, 46, 541–549.

Quinn, P. J. (1976) The Molecular Biology of Cell Membranes, pp. 96, University Park Press, Baltimore.

Rajalakshmi, M., Arora, R., Bose, T. K. Dinakar, N., Gupta, G., Thampan, T. N. R. V., Prasad, M. R. N., Ananad Kumar, T. C. and Moudgal, N. R. (1976) Physiology of the epididymis and induction of functional sterility in the male. J. Reprod. Fertil. Suppl. 24, 71–94.

Raz, A. and Goldman, R. (1974) Spontaneous fusion of rat liver lysosomes in vitro. Nature (London) 247, 206–208.

Reed, S. C. and Stanley, H. P. (1972) Fine structure of spermatogenesis in the South African clawed toad Xenopus laevis Dandin. J. Ultrastruct. Res. 41, 277–295.

Roberts, T. K. and Boettcher, B. (1969) Identification of human sperm-coating antigen. J. Reprod. Fertil. 18, 347–350.

de Roeper, A. and Barry, J. M. (1976) Nuclear swelling and chromatin decondensation can occur without cortical granule expulsion in eggs of Xenopus laevis. Exp. Cell Res. 100, 411–414.

Rogers, B. J. and Yanagimachi, R. (1976) Competitive effect of magnesium on the calcium-dependent acrosome reaction in guinea pig spermatozoa. Biol. Reprod. 15, 614–619.

Roomans, G. M. (1975) Calcium binding to the acrosomal membrane of human spermatozoa. Exp. Cell Res. 96, 23–30.

Roomans, G. M. and Afzelius, B. A. (1975) Acrosome vesiculation in human sperm. J. Submicrosc. Cytol. 7, 61–69.

Rosado, A., Velázquez, A. and Lara-Ricalde, R. (1973) Cell polarography. II. Effect of neuraminidase and follicular fluid upon the surface characteristics of human spermatozoa. Fertil. Steril. 24, 349–354.

Ross, M. H. (1976) The Sertoli cell junctional specialization during spermiogenesis and at spermiation. Anat. Rec. 186, 79–104.

Rothschild, L. (1954) Polyspermy. Q. Rev. Biol. 29, 332–342.

Russell, L. and Clermont, Y. (1976) Anchoring device between Sertoli cells and late spermatids in rat seminiferous tubules. Anat. Rec. 186, 79–104.

Russo, J. and Metz, C. B. (1974) The ultrastructural lesions induced by antibody and complement in rabbit spermatozoa. Biol. Reprod. 10, 293–308.

Sandoz, D. (1975) Development of the neck region and the ring during spermiogenesis of Discoglossus pictus (Anuran Amphibia). In: The Functional Anatomy of the Spermatozoon (Afzelius, B. A., ed.), pp. 237–247, Pergamon Press, Oxford.

Satir, B. (1974) Ultrastructural aspects of membrane fusion. J. Supramol. Struct. 2, 529–537.

Satir, B., Schooley, C. and Satir, P. (1973) Membrane fusion in a model system. Mucocyst secretion in Tetrahymena. J. Cell Biol. 56, 153–178.

Schachter, D. and Shinitzky, M. (1977) Fluorescent polarization studies of rat intestinal microvillus membranes. J. Clin. Invest. 59, 536–548.

Schramm, M., Selinger, Z., Salomon, Y., Eytan, E. and Batzri, S. (1973) Pseudopodia formation by secretory granules. Nature (London) New Biol. 240, 203–205.

Skoblina, M. N. (1969) Independence of the cortex maturation from germinal vesicle material during the maturation of amphibian and sturgeon oocytes. Exp. Cell Res. 55, 142–144.

Smith, A. U. (1949) The antigenic relationship of some mammalian spermatozoa. Proc. R. Soc. (London) Ser. B: 136, 472–479.

Smith, L. D. and Ecker, R. E. (1969) Role of the oocyte nucleus in physiological maturation in Rana pipiens. Dev. Biol. 19, 281–309.

Snider, D. R. and Clegg, E. D. (1975) Alteration of phospholipids in porcine spermatozoa during in vivo uterus and oviduct incubation. J. Anim. Sci. 40, 269–274.

Solter, D., Biczysko. W., Graham, C., Pienkowski, M. and Koprowski, H. (1974) Ultrastructure of early development of mouse parthenogenones. J. Exp. Zool. 188, 1–24.

Soupart, P. and Strong, P. A. (1975) Ultrastructural observations on polyspermic penetration of zona pellucida-free human oocytes inseminated *in vitro.* Fertil. Steril. 26, 523–537.

Srivastava, P. N., Munnell, J. F., Yang, C. H. and Foley, C. W. (1974) Sequential release of acrosomal membranes and acrosomal enzymes of ram spermatozoa. J. Reprod. Fertil. 36, 363–372.

Stefanini, M., Oura, C. and Zamboni, L. (1969) Ultrastructure of fertilization in the mouse. 2. Penetration of sperm into the ovum. J. Submicrosc. Cytol. 1, 1–23.

Steinhardt, R. A., Epel, D., Carrol, E. J. and Yanagimachi, R. (1974) Is calcium ionophore a universal activator for unfertilized eggs? Nature (London) 252, 41–43.

Stenuit, K., Geuskens, M., Steinert, G. and Tencer, R. (1977) Concanavalin A binding to oocytes and eggs of *Xenopus laevis.* Exp. Cell Res. 105, 159–168.

Summers, R. G., Talbot, P., Keough, E. M., Hylander, B. L. and Franklin, L. E. (1976) Ionophore A23187 induces acrosome reaction in sea urchin and guinea pig spermatozoa. J. Exp. Zool. 196, 381–385.

Szöllösi, D. (1967) Development of cortical granules and the cortical reaction in rat and hamster eggs. Anat. Rec. 159, 431–446.

Szöllösi, D. (1976) La phagocytose des spermatozoïdes par les cellules du cumulus oophorus de follicules de Veau en culture. C. R. Hebd. Seances Acad. Sci. Ser. D: 283, 801–804.

Szöllösi, D. and Hunter, R. H. F. (1973) Ultrastructural aspects of fertilization in the domestic pig: sperm penetration and pronucleus formation. J. Anat. 116, 181–206.

Talbot, P. (1975) The effect of ions on the acrosome reaction. 8th Ann. Meet. Soc. Study Reprod. Fort Collins, Colorado (Abstract).

Talbot, P. and Franklin, L. E. (1976) Morphology and kinetics of the hamster sperm acrosome reaction. J. Exp. Zool. 198, 163–176.

Talbot, P., Summers, R. G., Hylander, B. L., Keough, E. M. and Franklin, L. E. (1976) The role of calcium in the acrosome reaction: An analysis using ionophore A23187. J. Exp. Zool. 198, 383–392.

Tandler, B. and Poulsen, J. H. (1976) Fusion of the envelope of mucus droplets with the luminal plasma membrane in acinar cells of the cat submandibular gland. J. Cell Biol. 68, 775–781.

Temple-Smith, P. D. and Bedford, J. M. (1976) The features of sperm maturation in the epididymis of a marsupial, the brushtailed possum *Trichosurus vulpecula.* Am. J. Anat. 147, 471–500.

Toullet, F., Voisin, G. A. and Nemirovsky, M. (1973) Histoimmunochemical localization of three guinea-pig spermatozoal autoantigens. Immunology 24, 635–653.

Tyler, A. (1959) Some immunobiological experiments on fertilization and early development in sea urchins. Exp. Cell Res., Suppl. 7, 183–199.

Uehara, T. and Yanagimachi, R. (1977) Activation of hamster eggs by pricking. J. Exp. Zool. 199, 269–274.

Usui, N. and Yanagimachi, R. (1976) Behavior of hamster sperm nuclei incorporated into eggs at various stages of maturation, fertilization, and early development. J. Ultrastruct. Res. 57, 276–288.

Vaidya, R. A., Bedford, J. M., Glass, R. H. and Morris, J. McL. (1969) Evaluation of the removal of tetracycline fluorescence from spermatozoa as a test for capacitation in the rabbit. J. Reprod. Fertil. 19, 483–489.

Vaidya, R. A., Glass, R. H., Dandekar, P. and Johnson, K. (1971) Decrease in the electrophoretic mobility of rabbit spermatozoa following intra-uterine incubation. J. Reprod. Fertil. 24, 299–301.

Van der Stricht, O. (1923) Etude comparée des ovules des mammifères aux différentes périodes de l'ovogenèse, d'après les travaux du Laboratoire d'Histologie et d'Embryologie de l'Université de Gand. Arch. Biol. 33, 229–300.

Velázquez, A. and Rosado, A. (1972) Cell polarography. I. Study of the surface characteristics of human spermatozoa. Fertil. Steril. 23, 562–567.

Vojtíšková, M. Viklický, V., Boubelík, M. and Hattikudur, N. S. (1974) The expression of H-2 and differentiation antigens on mouse spermatozoa. Folia Biol. (Prague) 20, 321–324.

Wassarman, P. M., Albertini, D. F., Josefowicz, W. J. and Letourneau, G. E. (1976) Cytochalasin B-induced pseudo-cleavage of mouse oocytes *in vitro:* Asymmetric localization of mitochondria and

124

microvilli associated with a stage-specific response. J. Cell Sci. 21, 523–535.

Wassarman, P. M., Ukena, T. E., Josefowicz, W. J. and Karnovsky, M. J. (1977) Asymmetrical distribution of microvilli in cytochalasin B-induced pseudocleavage of mouse oocytes. Nature (London) 265, 742–744.

Weil, A. J. (1967) Antigens of the seminal plasma. J. Reprod. Fertil., Suppl. 2, 25–34.

Weiss, L. and Subjeck, J. R. (1974) The densities of colloidal iron hydroxyde particles bound to microvilli and the spaces between them: Studies on glutaraldehyde fixed Ehrlich ascites tumour cells. J. Cell Sci. 14, 215–223.

Werner, G. (1975) Changes in the spermatozoon during fertilization in *Triturus helveticus* Raz. In: The Functional Anatomy of the Spermatozoon (Afzelius, B. A., ed.) pp. 47–56, Pergamon Press, Oxford.

Wilson, E. B. (1925) The Cell in Development and Heredity, Chap. 5, pp. 394–448, Macmillan, New York.

Wolf, D. P. (1974) The cortical response in *Xenopus laevis* ova. Dev. Biol. 40, 102–115.

Wolf, D. P., Inoue, M. and Stark, R. A. (1976) Penetration of zona-free mouse ova. Biol. Reprod. 15, 213–221.

Wooding, F. B. P. (1973) The effect of Triton X-100 on the ultrastructure of ejaculated bovine sperm. J. Ultrastruct. Res. 42, 502–516.

Wooding, F. B. P. (1975) Studies on the mechanism of the Hyamine-induced acrosome reaction in ejaculated bovine spermatozoa. J. Reprod. Fertil. 44, 185–192.

Yamane, J. (1921) Studien über die physikalische und chemische Beschaffenheit des Pferdespermas mit besonderer Berücksichtigung der Physiologie der Spermatozoen. J. Coll. Agric. Hokkaido Imp. Univ. 9, 161–236.

Yanagimachi, R. (1970) The movement of golden hamster spermatozoa before and after capacitation. J. Reprod. Fertil. 23, 193–196.

Yanagimachi, R. (1972) Fertilization of guinea pig eggs *in vitro*. Anat. Rec. 174, 9–20.

Yanagimachi, R. (1973) Behavior and functions of the structural elements of the mammalian sperm head in fertilization. In: The Regulation of Mammalian Reproduction (Segal, S. J., Crozier, R., Corfman, P. A. and Condliffe, P. G., eds.) pp. 215–227, C. C. Thomas, Springfield, Illinois.

Yanagimachi, R. (1975) Acceleration of the acrosome reaction and activation of guinea pig spermatozoa by detergents and other reagents. Biol. Reprod. 13, 519–526.

Yanagimachi, R. and Noda, Y. D. (1970a) Electron microscope studies of sperm incorporation into the golden hamster egg. Am. J. Anat. 128, 429–462.

Yanagimachi, R. and Noda, Y. D. (1970b) Ultrastructural changes in the hamster sperm head during fertilization. J. Ultrastruct. Res. 31, 465–485.

Yanagimachi, R. and Noda, Y. D. (1970c) Physiological changes in the postnuclear cap regions of mammalian spermatozoa: a necessary preliminary to the membrane fusion between sperm and egg cells. J. Ultrastruct. Res. 31, 486–493.

Yanagimachi, R. and Noda, Y. D. (1972) Scanning electron microscopy of golden hamster spermatozoa before and during fertilization. Experientia 28, 69–72.

Yanagimachi, R. and Usui, N. (1974) Calcium dependence of the acrosome reaction and activation of guinea pig spermatozoa. Exp. Cell Res. 89, 161–174.

Yanagimachi, R. and Nicolson, G. L. (1976) Lectin-binding properties of hamster egg zona pellucida and plasma membrane during maturation and preimplantation development. Exp. Cell Res. 100, 249–257.

Yanagimachi, R., Noda, Y. D., Fujimoto, M. and Nicolson, G. L. (1972) The distribution of negative surface charges on mammalian spermatozoa. Am. J. Anat. 135:497–520.

Yanagimachi, R., Nicolson, G. L., Noda, Y. D. and Fujimoto, M. (1973) Electron microscopic observations of the distribution of acidic anionic residues on hamster spermatozoa and eggs before and during fertilization. J. Ultrastruct. Res. 43, 344–353.

Yanagimachi, R., Yanagimachi, H. and Rogers, B. J. (1976) The use of zona-free animal ova as a test-system for the assessment of the fertilizing capacity of human spermatozoa. Biol. Reprod. 15, 471–476.

Zahler, W. L. and Doak, G. A. (1975) Isolation of the outer acrosomal membrane from bull sperm. Biochim. Biophys. Acta 406, 479–488.

Zakai, N., Kulka, R. G., and Loyter, A. (1976) Fusion of human erythrocyte ghosts promoted by the combined action of calcium and phosphate ions. Nature (London) 263, 696–699.

Zakai, N., Kulka, R. G., and Loyter, A. (1977) Membrane ultrastructural changes during calcium phosphate-induced fusion of human erythrocyte ghosts. Proc. Nat. Acad. Sci. U.S.A. 74, 2417–2421.

Zamboni, L. (1970) Ultrastructure of mammalian oocytes and ova. Biol. Reprod., Suppl. 2, 44–63.

Zamboni, L., Patterson, H. and Jones, M. (1976) Loss of cortical granules in mouse ova activated in vivo by electric shock. Am. J. Anat. 147, 95–101.

Myoblast fusion

3

Richard BISCHOFF

Contents

G. Poste & G. L. Nicolson (eds.) Membrane Fusion, pp. 127–179.
© *Elsevier/North-Holland Biomedical Press, 1978.*

1. Introduction

Skeletal muscle fibers are the only permanent multinucleated cells in the vertebrate body. Since myonuclei themselves are postmitotic and do not synthesize DNA or divide, the formation of muscle fibers can occur only by the fusion of individual mononucleated cells, the myoblasts (see Fischman, 1972 for review of evidence in support of multinucleation by fusion). Despite the importance of fusion in myogenesis, there have been relatively few studies concerned directly with elucidating events during fusion. This review focuses on the process of fusion itself, and less emphasis is placed on factors regulating the emergence of fusion-competent cells, although in some cases these phenomena are difficult to separate.

Although fusion of myoblasts is the most dramatic event during normal myogenesis, it is not a necessary condition for subsequent biochemical differentiation and accumulation of muscle-specific proteins. Myoblasts in which fusion has been blocked in a variety of ways can still synthesize normal complements of myosin (Emerson and Beckner, 1975; Moss and Strohman, 1976), creatine kinase (Keller and Nameroff, 1974; Turner et al., 1976a), acetylcholine receptor (Prives and Paterson, 1974) and other cell-specific products, at least for several days. Myoblasts also withdraw from the cell division cycle in the absence of fusion (Cox, 1968; Dienstman and Holtzer, 1975, 1977; Adamo et al., 1976). This illustrates that whatever the significance of myoblast fusion, it does not act as a trigger for other differentiative events.

Most of our information on myoblast fusion has come from studies using cultured cells. Aside from ease of experimental manipulation, the most notable advantage of culture systems is the synchrony of fusion that can be obtained (Fig. 1). Although the rate of fusion in vivo is unknown, it must be exceedingly low since myogenesis extends over many months. On the other hand, the rate of fusion in vitro may reach 7% per hour of the cells plated (O'Neill and Stockdale, 1972b). Fusion can be synchronized further by allowing competent cells to accumulate in low Ca^{2+} medium for several days, a condition that prevents fusion (Shainberg et al., 1969; see section 5.1), then removing the block by adding Ca^{2+}. The fusion rate is greater than 15% per hour under these conditions (van der Bosch et al., 1973).

Finally, a word of caution is needed. Many of the experiments designed to analyze the mechanism of myoblast fusion depend upon visual scoring of the degree of fusion. Since plasma membranes cannot be resolved with the light microscope, indirect clues such as cell shape, size, and arrangement of nuclei must be used to identify multinucleate cells. Although it is obvious that considerable

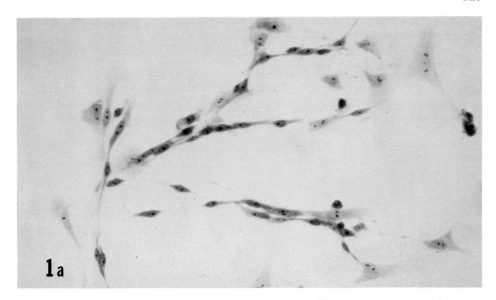

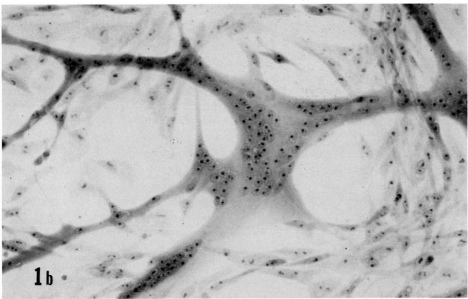

Fig. 1. Monolayer cultures of myogenic cells from 11-day chick embryo. (a) One-day-old culture showing prefusion alignment of small spindle-shaped myoblasts into linear aggregates. (b) Three-day culture containing branching, multinucleated myotubes. Most of the myoblasts have fused by this stage and the remaining mononucleated cells are nonmyogenic. ×330.

fusion has occurred in Fig. 1, it is often difficult to count the number of fused nuclei with accuracy, especially in dense cultures or under conditions in which both the number of myotubes and the number of nuclei per myotube may vary. This problem can be controlled to some extent by optimizing conditions to obtain large differences in fusion (Konigsberg, 1971), by scoring fusion only in sparse cultures (O'Neill and Stockdale, 1972b), or by using a parameter that correlates with fusion but can be measured more accurately (Yeoh and Holtzer, 1977).

2. Specificity of fusion

Cell interactions involving surface determinants display several levels of specificity. When embryonic cells from various tissues are enzymatically dissociated and allowed to reaggregate in culture, initial adhesion is relatively nonspecific and the aggregates consist of a random mixture of cell types. Several hours later as the cells regenerate surface molecules previously lost during dissociation (Roth, 1968), specificity is expressed and sorting out of homotypic subpopulations occurs (Moscona, 1962; Steinberg, 1964). This sorting out exhibits a high specificity for tissue type but a much lower species specificity. For example, chick and mouse liver cells will remain intermixed after reaggregation and form chimeric tissues (Moscona, 1957).

Results of analyses of specificity for myoblast fusion have tended to agree with these earlier studies on cell aggregation in that myoblasts are more likely to fuse with homotypic cells from different species than they are with heterotypic cells. Experiments to measure specificity usually involve initiating monolayer cultures with a mixture of the test cells and allowing fusion to occur. The origin of nuclei is determined either by staining characteristics or by labeling one population of cells with [^{3}H] thymidine prior to mixing. Since myotube nuclei are postmitotic, chromosome markers cannot be used for identification.

Certain precautions must be taken to insure reliable results in experiments with mixtures of myoblasts from different species in which the cells are labeled with [^{3}H] thymidine. Unless 100% labeling of the cells from one species is achieved, it is difficult to determine whether myotubes containing a mixture of labeled and unlabeled nuclei arose from inter- or intraspecific fusions. In addition, cells should be plated to encourage early fusion to avoid dilution of the label beyond detection in rapidly dividing cells. Finally, the mixed cultures should be grown in medium containing unlabeled thymidine to avoid possible artifacts from reutilization of label from dying cells.

In experiments conforming to most of the above criteria (Yaffe and Feldman, 1965; Yaffe, 1969) it was found that calf or rabbit myoblasts readily formed hybrid myotubes with rat cells. Chick myoblasts also fused with rat cells but with a lower frequency than cells from the mammalian species. On the other hand, heterotypic cells such as heart and kidney cells from the same species did not fuse with rat myoblasts. Heterospecific fusion between chick and mouse myoblasts was reported by Maslow (1969). Nuclei were identified by staining characteristics, but the photographs of hybrid myotubes are not convincing.

A wide variety of heterotypic cells have also been tested for their ability to fuse with chick myoblasts (Okazaki and Holtzer, 1965; Holtzer and Bischoff, 1970). Fibroblasts, kidney and liver cells, chondrocytes and dedifferentiated chondrocytes, and smooth and cardiac muscle cells were each labeled with [^{3}H] thymidine and mixed with myogenic cells or added to early cultures containing myotubes. In all cases the labeled cells were excluded from the myotubes.

Evidence that the order in which the test cells are mixed in culture may influence hybridization has come from experiments of Przybylski and Chlebowski (1972). When skeletal myogenic cells were plated onto labeled heart cells, some labeled nuclei were found in myotubes. There was no evidence of heterotypic fusion, however, when labeled heart cells were added to skeletal muscle cultures. Unfortunately, the authors also demonstrated that heart cells could fuse among themselves and, since there was no assurance that 100% of the heart cells were labeled in the mixing experiments, the multinucleate cells with labeled nuclei could have originated entirely from heart cell fusion.

Carlsson and co-workers (1974) used a novel identification technique to confirm the formation of heterospecific myotubes. Fluorescein-labeled antiserum against chick nuclear envelope was used to detect chick nuclei in hybrid myotubes formed in mixed cultures of chick and rat muscle.

3. Prefusion activities

Activities of myogenic cells during the period just prior to fusion are important in bringing fusion-competent cells into contact and facilitating close approximation of their membranes. A minimum sequence of prefusion events would involve collision of motile cells followed by recognition and the formation of stable adhesive contacts. The specificity of myoblast fusion (section 2) could reside in these early events, such as recognition and adhesion, or in membrane rearrangements during the act of fusion itself. Recent experiments have shown that myogenic cells prior to fusion exhibit highly selective interactions that are mediated by peripheral cell surface coat material (Bischoff and Lowe, 1974).

3.1. Myoblast-myotube adhesion

Early observations of muscle differentiating both in vivo and in vitro have repeatedly noted the close association between myoblasts and myotubes (Lewis and Lewis, 1917; Tello, 1922; Capers, 1960). Large numbers of small, basophilic cells lie along the sides of myotubes in 11-day chick embryo muscle (Fig. 2). Myoblast-myotube associations also occur in mammalian muscle in vivo and, according to ultrastructural studies, the great majority of mononucleated

132

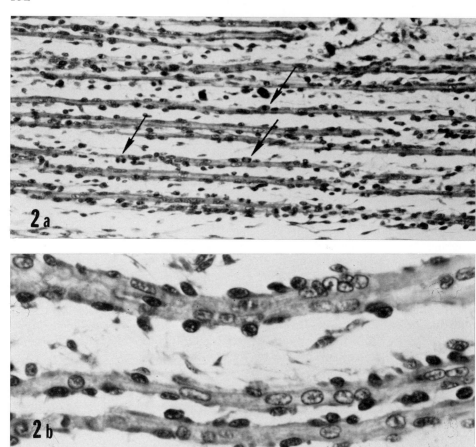

Fig. 2. Sections of 11-day chick embryo breast muscle showing mononucleated cells adhering to myotubes. (a) The majority of single cells in this tissue, including those in mitosis (arrows), are in contact with myotubes. ×315. (b) At higher magnification, the scanty basophilic cytoplasm of the adherent cells and their close contact with the sides of myotubes is evident. ×800.

myogenic cells are found along the walls of myotubes (Kelly and Zacks, 1969; Kelly and Schotland, 1972). These cellular aggregates are often ensheathed by a common basal lamina.

Monolayer cultures also show evidence of adhesion between myoblasts and myotubes (Fig. 3). In addition, the early prefusion alignment of myoblasts into linear aggregates (Fig. 1) is probably an expression of the same phenomenon.

To determine the extent of myoblast-myotube interaction in vitro, time-lapse films were made of chick myogenic cultures in Sykes-Moore chambers (Bischoff and Lowe, 1974). Filming was begun at around 45 hours in vitro when early myotubes are present, and continued for up to 13 hours. Individual myogenic cells were identified (see Bischoff and Lowe, 1974 for discussion of criteria) and

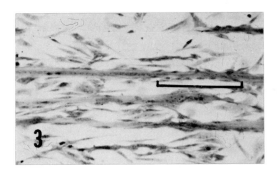

Fig. 3. Two-day-old culture during the period of rapid fusion containing early myotubes and associated mononucleated cells. In several places chains of myoblasts are found lined up along the edge of a myotube (bracket). ×130.

scored throughout the duration of filming for the time spent in contact with the myotubes (Fig. 4). For approximately 170 cell-hours scored, 70% of the time was spent in direct contact with the surface of a myotube. Since myotubes occupy 25% of the surface area in these cultures, the myogenic cells are in contact with myotubes two to three times more often than would be expected on the basis of random collisions. Once collision occurs, the myoblasts tend to remain in contact with myotubes and to develop stable adhesive contacts.

The most impressive aspect of prefusion behavior is the constant and rapid motility of myogenic cells when compared with fibroblasts in the cultures. The fusiform myogenic cells change direction frequently and move equally well in either direction. Although there is no obvious directed movement toward myotubes, once contact is made the myoblast usually aligns itself parallel with the long axis of the myotube and maintains contact for several house (Fig. 4).

There is no evidence of contact inhibition of movement between myogenic cells (Powell, 1973; Bischoff and Lowe, 1974). Despite the tendency to aggregate, individual cells exhibit constant movement and regrouping even after the linear aggregates have formed. Mononucleated cells also migrate extensively while in contact with myotubes. Most of the cells scored for Fig. 4 were in constant movement back and forth along the myotube axis.

Although time-lapse cinematography is useful for studying the dynamic interaction of myogenic cells in the prefusion period, the technique is laborious and unsuited for determing the specificity of myoblast-myotube adhesion. Bischoff and Lowe (1974) developed an assay for adhesion in which [^{3}H] thymidine-labeled cells are added to unlabeled cultures containing myotubes. The cultures are then sacrificed at intervals and the percentage of labeled cells in contact with myotubes is determined from radioautographs (Fig. 5). Myogenic cells rapidly adhere to myotubes in the test cultures. From a random distribution of 25% at the time of initial attachment, within 4 to 6 hours about 60% of the added cells are found in contact with myotubes. Various nonmyogenic cells tested under the same conditions remain randomly distributed and

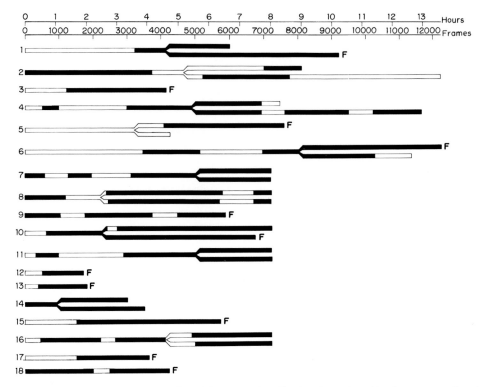

Fig. 4. Cinematographic analysis of myoblast-myotube adhesion. Two-day-old cultures were placed in Sykes-Moore chambers and photographed at 4 sec intervals using an electronic flash unit. Eighteen individual myogenic cells were followed and scored as being either on (filled bars) or off (open bars) the surface of myotubes. Mitoses are indicated by branching of the bars and fusion events are indicated by F. Bars that end without notation represent cells that left the field or were otherwise lost from observation. (From Bischoff and Lowe, 1974.)

show no specific affinity for myotubes (Fig. 5). In fact, myogenic cells in which differentiation has been suppressed by incorporation of 5-bromodeoxyuridine (Stockdale et al., 1964; Bischoff and Holtzer, 1970) seem to show a slight avoidance of myotubes.

3.2. Role of the glycocalyx in mediating adhesion

Cell recognition, adhesion, and fusion probably involve reactions occurring at various strata of the cell periphery. The most external of these layers consists of glycoproteins and glycosaminoglycans loosely attached to the plasma membrane by noncovalent bonds and comprising a sugar-rich coat that has been termed the glycocalyx (Bennett, 1963). Since cells first establish contact through the glycocalyx, the specificity for cell recognition and adhesion might reside in this layer.

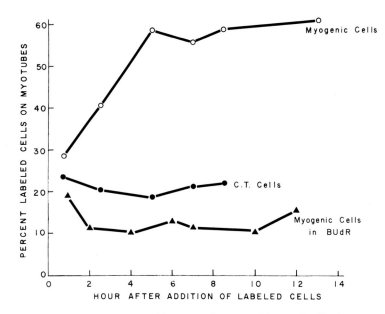

Fig. 5. Radioautographic assay of myoblast-myotube recognition and adhesion. Test cells were labeled with [³H] thymidine and added to a series of 2-day myotube-containing cultures. The cultures were sacrificed as soon as the added cells attached to the substratum and at intervals thereafter. Radioautographs were scored for the percentage of labeled cells adhering to the myotubes. Normal myogenic cells rapidly adhere to myotubes, while cells from chick areolar connective tissue (C.T.) or myogenic cells grown in 5-bromodeoxyuridine do not adhere. (From Bischoff and Lowe, 1974.)

Brief treatment of various cell types with chelating agents such as ethylenediamine tetraacetic acid (EDTA) and ethylenebis (oxyethylenenitrilo) tetraacetic acid (EGTA) has been shown to release macromolecules, including glycocomponents, from the cell surface (Beierle, 1968; Codington et al., 1970; Culp and Black, 1972). This technique was used to investigate the role of loosely bound cell surface components in myoblast recognition and adhesion (Bischoff and Lowe, 1973, 1974).

Gentle treatment with EDTA or EGTA under conditions that do not detach cells or reduce cell viability releases substantial amounts of protein from thoroughly washed myogenic cultures (Fig. 6). The relative amount of solubilized protein is greatest in young cultures. During the period of active fusion, from 2 to 3 days, between 15 and 20% of the total culture protein is EDTA-labile.

Several experiments show that the proteins released by EDTA are derived from the cell surface or culture dish and not from the cell cytoplasm. First, the EDTA treatment does not result in any increase in nigrosin-stainable cells. Second, cultures labeled with [³H]leucine for various periods of time and extracted with EDTA in the presence of cycloheximide show a 20-minute lag in appearance of radioactivity in the EDTA extract (Fig. 7). Label begins to appear in the culture medium about 50 minutes after labeling. These experiments were inter-

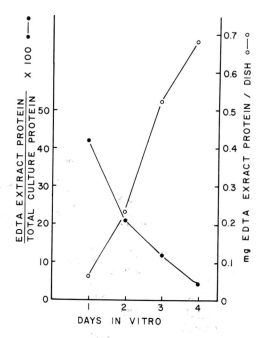

Fig. 6. Removal of surface protein from myogenic cultures by EDTA. Cultures 1–4 days old were washed 4 times with balanced salt solution (BSS) and treated for 10 min at 37 °C with 0.5 mM EDTA in Ca²⁺-free BSS. Protein was measured in both the cells and the EDTA extract.

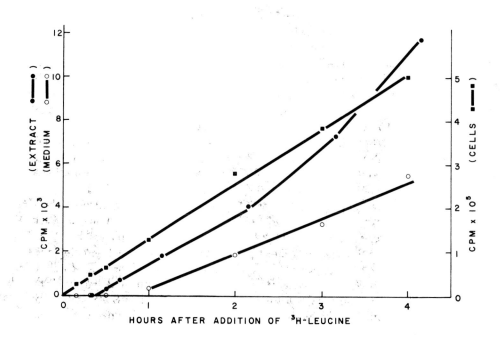

preted as follows: (1) Extraction with EDTA does not cause leakage of cytoplasmic proteins since, if this were the case, label would appear in the extract at the same time as in the cells. (2) At least some of the EDTA-labile proteins are synthesized by the cells. (3) Newly synthesized proteins become susceptible to EDTA release at the cell surface about 20 minutes after addition of label and begin to be shed into the bulk phase of the culture medium about 30 minutes later.

Finally, a more direct demonstration that EDTA removes cell surface material was obtained by electron microscope histochemistry. Cells were stained before and after EDTA extraction with colloidal thorium, a cationic stain for highly acidic groups (Rambourg and Leblond, 1967). EDTA treatment resulted in a substantial reduction in stainable material on the surface of myogenic cells (Fig. 8).

Autoradiographic experiments of the type described earlier have shown that at least some of the EDTA-labile surface material is involved in myoblast-myotube recognition and adhesion (Bischoff and Lowe, 1974). Labeled myogenic cells added to myotube-containing cultures previously treated with EDTA remained in a random distribution for about 6 hours and failed to show any affinity for myotubes (Fig. 9). After this time, the labeled cells adhered to myotubes and reached control levels of adhesion by about 14 hours. That the EDTA might be acting on the culture substratum to interfere with cell attachment or motility was ruled out by time-lapse cinematography.

Further evidence that EDTA removes surface macromolecules involved in cell recognition or binding was obtained by adding EDTA-labile material back to extracted cultures and finding that adhesion was restored (Fig. 10). The proteins were added back to the cultures at their original concentration.

A large number of proteins and glycoproteins are released from the cell surface by chelating agents. Electrophoretic analysis of EDTA-extracted material and used culture medium from cells labeled with [³H]leucine has shown that some of the material is synthesized by the cells and some is tightly adsorbed to the cells from the culture medium (Fig. 11). A rapidly migrating, acidic glycoprotein that also becomes labeled with glucosamine and fucose is especially intriguing since it may represent the material responsible for the reduction in thorium staining of the cell surface following EDTA extraction.

To summarize, the close and almost constant association between myoblasts and myotubes represents a specific adhesion between the two cell types. This adhesion depends on surface components, perhaps glycoproteins or glycosaminoglycans, that can be removed from the cells by chelating agents. Although there is no evidence that a prolonged period of adhesion is necessary for

Fig. 7. (opposite page) Incorporation of [³H] leucine into the cells, culture medium and EDTA extract as a function of time. Two-day-old cultures were incubated in 3 μCi/ml [³H]leucine and treated with EDTA as before (Fig. 6) at various times after labeling. Cycloheximide was added to the EDTA to prevent continued protein synthesis during extraction. Acid-insoluble material was collected separately from the cells, culture medium, and EDTA extract and the radioactivity was determined. (From Bischoff and Lowe, 1974.)

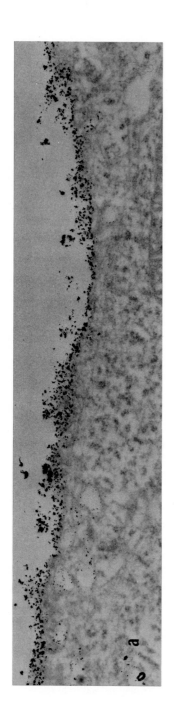

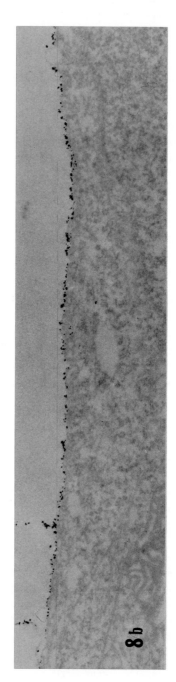

Fig. 8. Effect of EDTA on myotube surface coat material demonstrated by colloidal thorium binding. Two-day-old cultures were fixed in 2.5% glutaraldehyde, postfixed in 2% osmium tetroxide, then treated overnight with 1% colloidal thorium dioxide in 3% acetic acid. Sections were examined without further staining. (a) Control culture treated with BSS. (b) Culture treated with 0.5 mM EDTA for 10 min. (From Bischoff and Lowe, 1974.)

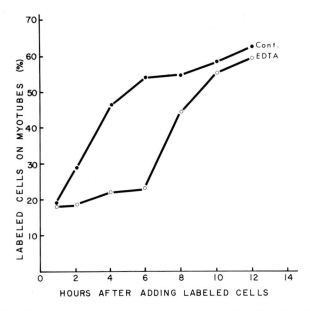

Fig. 9. Effect of EDTA treatment on myoblast-myotube adhesion. Two-day-old cultures were treated with EDTA, then [³H]thymidine-labeled myogenic cells were added to the EDTA-treated cultures and to control cultures treated with BSS. The cultures were sacrificed at intervals and the percentage of labeled cells on myotubes was determined from radioautographs.

fusion, it is tempting to speculate that interactions between the two cells during contact lead to membrane alterations required for fusion.

3.3. Cell coupling and junctions

In view of the rapid mobility of myogenic cells during the prefusion period (Powell, 1973; Bischoff and Lowe, 1974), it would be surprising if there were extensive junctional specializations between cells. Nevertheless, various types of junctions have been reported between myoblasts and myotubes both in vivo and in vitro. The most common type of junction consists of parallel membranes with a ≲ 10 nm gap and aggregations of electron-dense material along the cytoplasmic side of apposed membranes. These are often termed close junctions and have been described in human (Mendell et al., 1972) and chick muscle (Fischman, 1970) in vitro and in chick (Kikuchi and Ashmore, 1976) and rat muscle (Kelly and Zacks, 1969) in vivo.

In a combined ultrastructural and electrophysiological study of myoblast fusion, Rash and Fambrough (1972) found areas of close membrane apposition (3–5 nm) similar to gap junctions in all cells examined that were electrically coupled but not fused. By contrast, only one cell pair that was not coupled possessed a close junction of this type. In addition, there were vesicles bearing remnants of similar junctions in areas apparently undergoing membrane break-

140

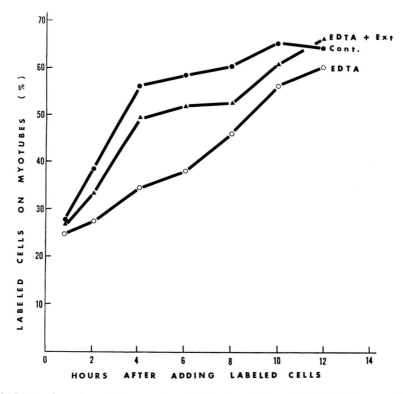

Fig. 10. Restoration of myoblast-myotube adhesion by addition of EDTA-labile material. Macromolecules removed from 2-day myogenic cultures by EDTA were dialyzed, concentrated, and added to extracted cultures at a concentration equivalent to that removed. Labeled cells were added along with the extract, and adhesion to myotubes was measured as before. (From Bischoff and Lowe, 1974.)

down during fusion. The authors suggest that cell coupling via gap junctions may be necessary for fusion, either to transfer information or to act as the fusion initiation site.

Although junctional complexes in myogenic cells have not been studied by lanthanum staining, in view of current knowledge of junction morphology it seems likely that the contact areas described earlier are gap junctions. The presence of electrical coupling is especially convincing (Rash and Fambrough, 1972).

Rash and Staehelin (1974) provided more direct evidence for the existence of gap junctions in a freeze-fracture study of 19-day fetal rat muscle. Small irregular arrays of 8 to 9 nm particles were found on the cytoplasmic leaflet of cells identified as myogenic by their content of myofilaments. Corresponding pits were observed on the complementary leaflet.

No structures resembling desmosones or tight junctions have been described between myogenic cells.

4. Morphology of fusion

4.1. Behavioral aspects

Several studies of myoblast fusion in vitro have used the technique of time-lapse cinematography (Capers, 1960; Cooper and Konigsberg, 1961; Powell, 1973; Bischoff and Lowe, 1974; Bayne and Simpson, 1975). Because of the poor resolution obtained, such studies cannot provide direct information concerning membrane changes during fusion but can illustrate dynamic aspects of myoblast behavior.

All studies agree on the active motility and lack of contact inhibition of myogenic cell movement. Mononucleated cells readily move over and under one another and along the shafts of myotubes (Powell, 1973; Bischoff and Lowe, 1974). Myotubes themselves are relatively sessile. Cells are also constantly regrouping in the prefusion aggregates. It is especially impressive to watch cells come together and appear to fuse only to move apart again during subsequent filming (Capers, 1960; Powell, 1973; Bischoff and Lowe, 1974). Myoblasts often sink into the side of a myotube and displace its cytoplasm without fusing (see Fig. 17-8 in Holtzer, 1970). Such behavior emphasizes the difficulty in scoring fusion with the light microscope in fixed preparations.

The minimum time required for fusion has been measured in time-lapse films. Capers (1960) recorded an interval of about 40 minutes from initial contact to nuclear incorporation. Bischoff and Lowe (1974) found that for cells in active movement along a myotube just prior to fusion there was a minimum interval of 10 minutes between cessation of movement of the myoblast and alignment of the newly fused nucleus with the preexisting myotube nuclei. Alignment of nuclei and the failure of a mononucleated cell to reappear during subsequent filming are good indications that fusion has actually occurred. In several instances there was extensive surface blebbing of the myoblast at the apparent moment of fusion, but this was not a constant finding (Bischoff and Lowe, 1974).

There seems to be no preferential orientation of myoblasts during fusion. Fusion may occur with cells in end-to-end, end-to-side, or side-to-side configurations (Powell, 1973). Likewise, there seems to be no restriction about where fusion may occur along the myotube. We have observed fusion along the myotube shaft as well as at branch points and ends (Bischoff and Lowe, unpublished observations). In view of the observation that the myotube basal lamina first appears along the shaft (Fischman, 1967), fusion in more mature myotubes may be restricted to the ends. There is a report that fusion during myogenesis in vivo occurs only at the ends of the growing fibers (Kitiyakara and Angevine, 1963).

4.2. Ultrastructure

Myoblast fusion under optimal conditions in culture is both rapid and extensive. Up to 70% of all the cells in a standard myogenic culture may fuse within a period of approximately 10 hours (O'Neill and Stockdale, 1972b; Lough and

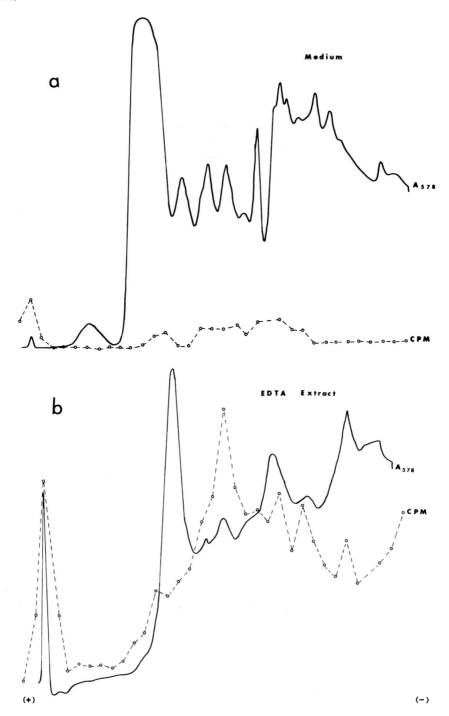

Figure 11. Parts (a) and (b). *Legend on opposite page.*

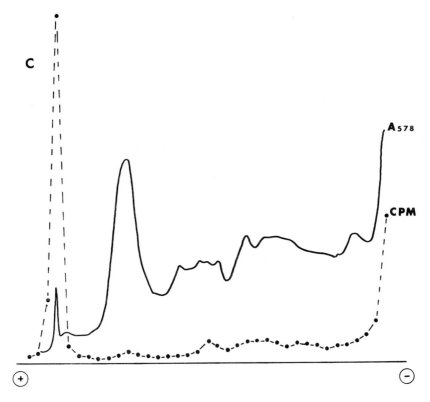

Fig. 11. Polyacrylamide gel electrophoresis of labeled proteins in culture medium and EDTA-extracted material. Two-day-old cultures were labeled for 20 hr with 2 μCi/ml [³H]leucine or 3 μCi/ml [³H]glucosamine then treated with 0.5 mM EDTA for 10 min. The extract and culture medium were dialyzed, concentrated, and fractionated on standard 7% gels. The gels were stained with Coomassie blue and scanned with a densitometer. Parallel gels were sliced and the radioactivity was determined. [³H]leucine-labeled material in (a) culture medium and (b) EDTA extract. [³H]glucosamine-labeled material (c) in the EDTA extract. (From Bischoff and Lowe, 1974.)

Bischoff, 1977). Ca²⁺-induced fusion of competent myoblasts is even more rapid and may exceed 15% of cells per hour (van der Bosch et al., 1973), suggesting that the membrane rearrangements accompanying fusion are rapidly completed. Despite the high frequency of fusion events, there is relatively little unambiguous information on the ultrastructural changes that occur in the membranes of fusing cells.

Ultrastructural examination of early fusion sites could provide important clues about the nature of membrane changes during fusion, such as how much of the cell surface is involved in establishing cytoplasmic continuity. Evidence of dissolution or breakdown of contiguous membranes over a broad area might suggest the secretion of enzymes such as protease or lipase by one or both of the fusing cells. Furthermore, the simultaneous involvement of large areas of membrane would make it more likely that fusion-related biochemical alterations could be identified. If, on the other hand, fusion were restricted to the formation of a small channel of cytoplasmic continuity that subsequently expanded by membrane flow, it would seem doubtful that gross biochemical studies could yield useful information.

A major problem in the ultrastructural identification of early fusion sites is the difficulty in resolving plasma membranes that are not perpendicular to the plane of section. Apparent membrane discontinuities in the region of close apposition between neighboring myoblasts may represent actual areas of cytoplasmic confluence or merely places where the membranes are oblique to the plane of section and therefore appear as amorphous fuzzy material. For example, electron microscopic examination of myogenic cells that have been blocked from fusing by low Ca^{2+} (see section 5.1) reveals numerous examples of cells in close apposition with apparent discontinuities in adjacent membranes (Fig. 12). It is virtually certain, however, that these cells are not fused since no authentic multinucleated cells were found in the cultures and no fusion of Ca^{2+}-blocked cells has been reported by light microscopic studies (Shainberg et al., 1969; van der Bosch et al., 1972; Morris et al., 1976; see Fig. 13).

Among the minimum criteria that should be met in identification of fusion sites are the presence of clearly resolved, U-shaped membranes reflected at either edge of the site and the absence of amorphous material in the area of cytoplasmic confluence.

Several early studies of the ultrastructure of myogenesis during development and regeneration have produced single photographs suggestive of fusion. In general, these images fall into two groups. In some cases the putative fusion site appears to be a small pore between a myoblast and a myotube (Allbrook, 1962; Betz et al., 1966; Kelly and Zacks, 1969). In other instances a larger area of apposing membranes appear to have broken down, leaving a series of small vesicles in the cytoplasmic channel between the two cells (Przybylski and Blumberg, 1966; Larson et al. 1970; Shimada, 1971). Although some of these photographs may depict actual fusion, none of them fulfills the minimum criteria described above the for identification of fusion sites, and hence they are unconvincing.

The first extensive study of the morphology of fusion was carried out by Lipton and Konigsberg (1972), who described seven examples of various stages of myoblast-myotube fusion in cultured quail cells. The investigators used a goniometer (tilting) stage to rule out the possibility of membrane discontinuity arising from an oblique plane of section. Also, each fusion site was analyzed in serial sections. The earliest fusion site consisted of a small cytoplasmic bridge, about 120 nm in diameter, between apposing cells. In more advanced cases, the cytoplasmic bridge was larger but there was no evidence of membrane breakdown such as vesicles or remnants in the region between the reflected membranes. In all cases only a single point of cytoplasmic continuity was found between fusing cells. The authors concluded that fusion is initiated at a single site which subsequently enlarges to provide full cytoplasmic confluence.

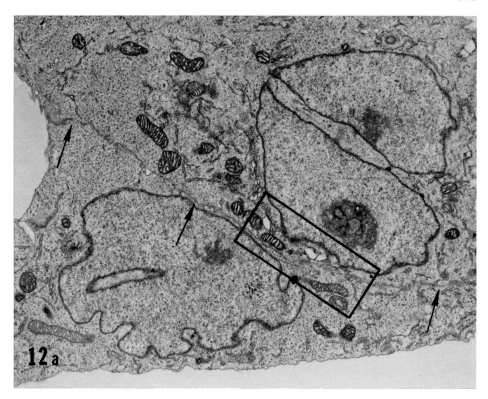

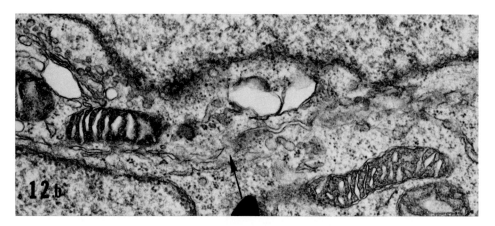

Fig. 12. Close contact of fusion-blocked cells grown for 3 days in medium containing about 160 μM Ca^{2+}. (a) The cell membranes are closely apposed over a large area (arrows). Both cells are mononucleated, but the nuclei are indented. $\times 7700$. (b) At higher magnification there are apparent areas of membrane discontinuity (arrow) between the two cells; however there is no fusion under these conditions and the membrane gaps are an artifact of the plane of section. $\times 30,800$.

Lipton and Konigsberg (1972) also noted the prevalence of coated vesicles in areas of contact between myogenic cells. These vesicles were sometimes confluent with the extracellular space, but it was not determined whether they are involved in exocytosis or endocytosis.

A goniometer stage was also employed in a brief report of myogenesis in vivo (Kikuchi, 1972). Several interesting examples are shown of the effect of specimen tilting in restoring the image of obliquely sectioned membranes. Unfortunately, no actual examples of fusion were found.

A further refinement in the ultrastructural analysis of fusion sites is the use of electrophysiological techniques to determine the onset of fusion in living cells (Rash and Fambrough, 1972). Chick myotubes in culture were impaled with a microelectrode to monitor the membrane potential, then a contiguous myoblast or myotube was depolarized by the iontophoretic application of acetylcholine from another micropipet. The time at which the two cells showed strong electrical coupling was taken as the beginning of fusion and the cultures were fixed for electron microscopic examination at various times after this reference point. A goniometer stage and serial sections were used to analyze the fusion sites.

The most impressive finding of this study is the rapidity with which cytoplasmic confluence proceeds once fusion has begun. In one case, within less than 3 minutes after the onset of electrical coupling, membranes between adjacent myotubes had completely disappeared over an area of about 100 μm^2. The authors calculated that the rate of disappearance of membranes in the contact area was greater than 1 μm^2/sec.

Two small fusion sites were discovered. One consisted of a 4 μm^2 pore between closely apposed cells and the other was a slender cytoplasmic process between cells that appeared separate in the light microscope. Several other more extensive fusion sites were studied and showed evidence of possible surface membrane remnants, usually in the form of flattened cisternae and vesicles, lying in the plane of original cell contact.

The occurrence of cytoplasmic vesicles in the fusion site in this study but not in the Lipton and Konigsberg (1972) study requires some comment. In contrast to the latter study, which involved fusion between undifferentiated myoblasts or between myoblasts and early myotubes, all of the fusion sites studied by Rash and Fambrough (1972) were between differentiated cells, usually two myotubes. This was necessary because of the techniques used to identify fusion: one of the cells must be large enough to record from and the other must be capable of responding to acetylcholine. Early muscle differentiation includes the formation of the sarcoplasmic reticulum and elaborate plasmalemmal invaginations of the developing transverse tubule system (Ezerman and Ishikawa, 1967). Hence, the cytoplasm of myotubes contains many membrane cisternae and vesicles, often in linear arrays, that would complicate the search for plasmalemmal remnants during fusion.

The continuous recording from several myotubes during the initiation of fusion allowed Rash and Fambrough (1972) to comment upon the efficiency of fusion in preserving membrane continuity. Since there was no ionic leakage detected during fusion, the authors concluded that the permeability barrier must remain intact during rearrangement of the lipid bilayers of adjacent cells.

Cultured myogenic cells have also been examined with the scanning electron microscope, but since cytoplasmic confluence would be almost impossible to detect this approach is of limited value in studying fusion. Both mononucleated cells and multinucleated myotubes appear to have an active surface exhibiting numerous microvilli and membrane folds (Shimada, 1972a; Shimada and Fischman, 1975; Bayne and Simpson, 1975). The low radius of microvilli curvature may be important in allowing close contact between cells prior to fusion by minimizing electrostatic repulsive forces (Poste, 1972). Microvilli are also prominent at fusion sites in virus-induced fusion (Hosaka and Koshi, 1968) and sperm egg fusion (Bedford, 1972).

A more promising morphological approach to the study of myoblast fusion is the use of the freeze-fracture technique. This method avoids possible artifacts arising from membrane flow during fixation. Also, the frequency and distribution of intramembranous particles may be studied during myoblast adhesion and fusion. The only freeze-fracture studies published to date are a brief report by Chiquet and co-workers (1975), which noted the numerous filopoida of myogenic cells, and a description of gap junctions on myotubes (Rash and Staehelin, 1974).

5. Fusion inhibitors

Information on the participation of various cellular components in myoblast fusion can be obtained indirectly by the use of specific inhibitors of various aspects of cell function. In experiments of this type, however, it is important to separate the effect of the inhibitor on the process of fusion per se from its effect on the emergence of fusion-competent myoblasts. The latter event depends on the state of differentiation of the cell or its compartment in the myogenic lineage (Holtzer et al., 1975c). The nature of factors governing the emergence of fusion-competent myoblasts is controversial and beyond the scope of this review (O'Neill and Stockdale, 1972a, b; Buckley and Konigsberg, 1974; Dienstman and Holtzer, 1975). All investigators agree, however, that myoblast fusion takes place only during the diploid, or G_1 phase of the cell cycle. Therefore, any inhibitor or culture condition that interferes with the transit of proliferating myogenic cells through the cell cycle and prevents them from reaching G_1 would block the emergence of fusible myoblasts. Experiments designed to avoid or control for any possible interference with proliferation are more likely to provide information about the requirements for fusion itself.

5.1. Calcium withdrawal

In view of the well-known Ca^{2+} requirement in many types of membrane fusion (Poste and Allison, 1973), it is not surprising that myoblast fusion is also Ca^{2+}-

dependent. The phenomenon was first described by Shainberg and associates, (1969) who found that when the Ca^{2+} concentration in the culture medium was reduced by dialysis from 1400 μM to 270 μM, rat myogenic cells failed to fuse but continued to proliferate at near-normal rates. Fusion occurred within 3 to 4 hours when the normal Ca^{2+} concentration was restored to the medium. Subsequently it was found that the same results are obtained whether the Ca^{2+} is removed from the medium or simply complexed by adding EGTA (Patterson and Strohman, 1972).

There is a remarkable similarity in the response of cells from various species to Ca^{2+} concentration in the medium. For example, myogenic cells from calf, rat, and chick all show 50% inhibition of fusion at approximately 500 μM Ca^{2+} (Adamo et al., 1976; Shainberg et al., 1969; Schudt et al., 1973; Merlie and Gros, 1976).

The requirement for Ca^{2+} in myoblast fusion seems to reside in some terminal surface event and not in acquisition of the capacity to fuse or in other aspects of myoblast differentiation. Cells proliferate at a normal rate in low Ca^{2+} medium, and myoblasts withdraw from the cell cycle (Dienstman and Holtzer, 1977; Adamo et al., 1976) as they do in normal medium (Okazaki and Holtzer, 1966; Bischoff and Holtzer, 1969). Synthesis of several muscle-specific proteins such as creatine kinase (Keller and Nameroff, 1974; Morris et al., 1976; Turner et al., 1976b), myokinase (Adamo et al., 1976), myosin (Dienstman and Holtzer, 1975; Emerson and Beckner, 1975; Moss and Strohman, 1976) and acetylcholine receptor (Merlie and Gros, 1976; Patterson and Prives, 1973) takes place at normal rates in mononucleated myoblasts prevented from fusing by low Ca^{2+}. Furthermore, reversal of Ca^{2+} inhibition does not require protein or RNA synthesis (Adamo et al., 1976; Bischoff, unpublished observations).

Fusion of cells initially inhibited with low Ca^{2+} does not appear to require attachment to a substratum. Knudsen and Horowitz (1977) grew monolayer cultures of myoblasts in low Ca^{2+} medium for 2 days, then maintained the cells in suspension during Ca^{2+} restoration and found spherical multinucleate cells formed within a few hours.

Myoblast fusion occurs very rapidly when Ca^{2+} is restored to cells that have been growing in low Ca^{2+} medium for several days (Fig. 13). Within 3 to 4 hours the percentage of nuclei in myotubes approaches that found in control cultures of the same age. Time-lapse cinematography has shown clear evidence of fusion beginning as early as 10 minutes after the addition of Ca^{2+} (Bischoff, unpublished observation). The films also demonstrate that cell motility and cell contact are not impaired by reduced Ca^{2+}. However, whereas normal myogenic cells tend to make relatively stable aggregates that often result in fusion, contacts made by cells in low Ca^{2+} are much more transient.

Schudt, Pette and colleagues examined the requirement for Ca^{2+} in some detail. The basic design for most of their experiments was to grow cells in low Ca^{2+} medium for several days to accumulate fusion-competent myoblasts, then to attempt to reverse the block under various conditions.

As Ca^{2+} concentration is increased during reversal, the optimum fusion rate occurs at about 1400 μM Ca^{2+} and higher concentrations inhibit fusion (van der

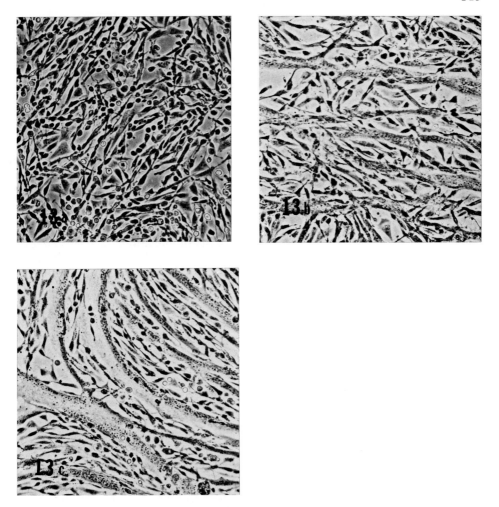

Fig. 13. Inhibition of fusion by low Ca²⁺ medium. (a) Cells were grown for 3 days in medium containing about 160 μM Ca²⁺. All the cells are mononucleated. (b) After 3 days, the Ca²⁺ was increased to 1960 μM and the culture was photographed 4 hr later. Extensive fusion has occurred during the 4 hr. (c) 3-day-old untreated control culture. Phase contrast. ×120.

Bosch et al., 1972; Schudt and Pette, 1975). Excess Mg^{2+} and K^+ also inhibit myoblast fusion under similar conditions (Schudt et al., 1973). The kinetics of reversal as a function of Ca^{2+} concentration suggests that multiple processes are being affected and that Ca^{2+} is not acting simply by adsorption to a protein (van der Bosch et al., 1972).

The requirement for Ca^{2+} in myoblast fusion is specific and other ions such as Mg^{2+}, Zn^{2+}, Mn^{2+}, Ba^{2+}, Cu^{2+}, Cd^{2+}, La^{2+}, and Li^+ cannot substitute, although some fusion occurs with Sr^{2+} (Schudt et al., 1973; Adamo et al., 1976).

To determine whether Ca^{2+} is acting on membrane or cytoplasmic sites, an attempt was made to alter the concentration of free intracellular Ca^{2+} at the time of reversal by the use of the ionophore A23187 (Schudt and Pette, 1975). The ionophore had no effect on fusion rate as a function of Ca^{2+} concentration in the medium under conditions in which there was an enhanced Ca^{2+} flux across the plasma membrane. Although intracellular Ca^{2+} was not measured directly, it appears likely that the Ca^{2+}-sensitive target for fusion resides on the outer surface of the cell. It should be noted that A23187 was found to induce the fusion of hen erythrocytes only after the cell surface was altered by neuraminidase treatment (Ahkong et al., 1975).

If Ca^{2+} acts externally, it may be possible to induce the fusion of plasmalemmal vesicles from fusion-competent myoblasts. The ability to fuse vesicles might lead to simplification of the reaction and eventual identification of the Ca^{2+}-binding component and its role in fusion. An important first step toward this goal has been made by Schudt and colleagues (1976). Membranes were prepared from low Ca^{2+}-blocked myoblasts by the method of Schimmel and co-workers (1973), which produces mostly closed vesicles of high purity by biochemical and ultra-structural criteria (section 8.1, Fig. 21). Addition of 1400 μM Ca^{2+} to such preparations led to aggregation of the vesicles and a shift in the distribution of intramembranous particles, as determined by freeze-fracture, from randomly dispersed to clumped. In addition, the appearance of vesicles in tight apposition and some "dumbbell" shapes suggested that fusion of vesicles was occurring in response to Ca^{2+}. Especially interesting was the observation that, while both Ca^{2+} and Mg^{2+} could induce aggregation and particle clumping, only Ca^{2+} led to subsequent fusion (Schudt et al., 1973). At face value, these experiments suggest that the elements of the fusion reaction survive membrane isolation and respond to Ca^{2+} just as they do in intact living cells. However, additional experiments are necessary before these results can be accepted with confidence. First, fusion was scored by the presence of dumbbell shapes in freeze-fracture preparations and there was no convincing evidence that fusion had actually occurred. Instead, Ca^{2+} could lead to partial constriction of single vesicles. More convincing proof of fusion should include a demonstration that the mean vesicle size increases after Ca^{2+} addition. Furthermore, it would be desirable to show that control vesicles, prepared from fusion-incompetent cells, do not fuse in response to Ca^{2+}

5.2. Cytoskeletal

In view of the role of microfilaments and microtubules in the regulation of surface receptors (Berlin, 1975; Edelman, 1976; Scott et al., 1977), it seems likely that these cytoskeletal elements may also be involved in myoblast fusion. Unfortunately, there have been no systematic studies of the effect of cytoskeletal inhibitors such as cytochalasin B or colchicine on fusion per se. Since these inhibitors may also prevent cell motility and proliferation, it will be difficult to separate direct effects on membrane activities from possible interference with prefusion events.

Myoblasts and immature myotubes are rich in cytoplasmic microtubules

(Ezerman and Ishikawa, 1967; Fischman, 1967; Ishikawa et al., 1968; Warren, 1974). Exposure of myotube-containing cultures to concentrations of colchicine or Colcemid high enough to prevent mitosis (~ 1 μM) leads to the rapid fragmentation of myotubes into structures termed myosacs (Godman and Murray, 1953; Bischoff and Holtzer, 1968; Holtzer et al., 1975b; Fukuda et al., 1976) but does not directly block fusion. Experiments in which cultures were pulse-labeled with [^{3}H] thymidine and then exposed to colchicine at intervals showed that if labeled cells are allowed to complete mitosis, fusion in G_1 is not blocked by 1 μM colchicine (Bischoff and Holtzer, 1968). Ultrastructural observations have shown that cells in 1 μM colchicine contain no detectable microtubules (Bischoff, unpublished observation). Other experiments demonstrated that normal fusion takes place when cells are grown continuously in 0.01 μM colchicine, a concentration that inhibits myotube elongation but does not prevent mitosis or cell proliferation (Bischoff and Holtzer, 1968). The myosacs that form under these conditions incorporate dozens of mononucleated cells and synthesize contractile proteins (Fig. 14). The concentration of microtubules is greatly reduced in such cells but they are not completely absent (Holtzer et al., 1973).

Although more experiments are needed, it appears that, while cells arrested in metaphase by colchicine do not fuse (Okazaki and Holtzer, 1965), a normal complement of cytoplasmic microtubules is not necessary for myoblast fusion.

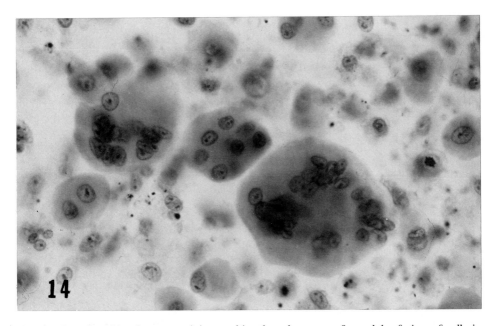

Fig. 14. Four-day-old culture containing multinucleated myosacs formed by fusion of cells in medium containing 0.01 μM colchicine. Although mitosis is not blocked at this concentration, the myosacs fail to elongate. Most mononucleated cells are also rounded and lack normal extensions. ×600.

The possible involvement of microfilaments in fusion is suggested by studies with cytochalasin B, a drug that inhibits cell motility and cytokinesis (Carter, 1967) and is believed to act by interference with the function of 4 to 6 nm microfilaments (Wessells et al., 1971).

Myogenic cells grown continuously in a low concentration of cytochalasin B (0.25–0.5 µg/ml) form differentiated myotubes, but up to 70% of these contain only 2 nuclei (Sanger et al., 1971; Holtzer and Sanger, 1972; Miranda and Godman, 1973). Labeling of nuclei with [³H] thymidine showed that the binucleate myotubes resulted from nuclear replication without cytokinesis and not from fusion (Sanger et al., 1971). Although cell mobility was reduced, this concentration of the drug did not affect cell shape. With time, the frequency of myotubes containing more than two nuclei increased so that some residual fusion, particularly between myotubes, could not be ruled out. Similar results have also been obtained with L_6 myoblasts (Wahrmann et al., 1976), a continuous cell line isolated by Yaffe (1968).

Higher concentrations of cytochalasin B (5–10 µg/ml) result in cessation of movement and produce dramatic effects on cell morphology including contraction of cytoplasm and "arborization" (Sanger et al., 1971; Sanger and Holtzer, 1972; Sanger, 1974; Croop and Holtzer, 1975; Holtzer et al., 1975b). Presumably, there is little or no fusion under these conditions, but the effect of high concentrations of the drug on fusion have not been well described.

To date, experiments have not determined whether cytochalasin B reduces fusion by preventing cell movement and hence opportunity for fusion, or whether the drug acts more directly on some event that occurs after cell contact. In any case, the pleiotropic effects of cytochalasin B (Estensen and Plageman, 1972; Kletzien et al., 1972; Mizel and Wilson, 1972) make it unlikely that this will be a useful inhibitor to probe the function of cortical microfilaments in fusion.

Further work is clearly needed to identify the role, if any, of cytoplasmic fibrous elements in fusion. A promising approach for future studies might be the use of cells blocked from fusing with low Ca^{2+}. Such cultures can be grown to high density before reversal so that interference with motility by cytoskeletal inhibitors would be less of a problem.

5.3 Lectins

Lectins are proteins that can bind to and cross-link specific carbohydrate groups (Sharon and Lis, 1972). This property recommends them as useful reagents to study the role of glycocomponents in myoblast fusion. If glycoproteins or glycolipids are involved in fusion, perhaps a lectin or lectins can be found that will inhibit fusion by binding with the receptors and thus lead eventually to the identification and isolation of the molecules involved.

This strategy was recently used with myogenic cultures in which fusion was blocked with EGTA to reduce the Ca^{2+} concentration (Den et al., 1975). Various lectins were added with Ca^{2+} at the time of reversal in protein-free medium and the degree of fusion was scored 4 hours later. Concanavalin A (Con A) showed a partial inhibition of fusion, which was reversed upon removal of the lectin and prevented by simultaneous addition of a hapten sugar, α-methyl-D-mannoside. However, the inhibition obtained was only about 60% and was independent of lectin concentration. Wheat germ agglutinin (WGA) produced 20 to 30% inhibition of fusion while soy bean agglutinin (SBA) and *Lens culinaris* lectin (LCL) had little or no effect on fusion. Iodinated lectins were used to show that Con A, WGA, and SBA all bound to myogenic cells, although there was no correlation between the number of binding sites and inhibition of fusion.

Inhibition of fusion is also obtained if normal cultures are exposed continuously to lectins (Wollman and Bischoff, 1978). Both Con A and LCL inhibit fusion but cell counts show that Con A is toxic and inhibits proliferation while LCL has only slight effects on cell proliferation (Table 1). Myogenic cells possess surface receptors for the lectins since both Con A and LCL agglutinate cells in a concentration-dependent fashion that is prevented by α-methyl-D-mannoside, a hapten sugar for both lectins. Receptors can also be demonstrated on both myoblasts and myotubes by staining with fluorescein-labeled lectins (Fig. 15) or peroxidase-labeled lectins (Fig. 20).

These early results are promising, but more studies are needed to establish the role of lectin receptors in fusion. The findings that LCL inhibits fusion of cells in normal medium but does not prevent fusion of EGTA-blocked cells upon restoration of Ca^{2+} are not necessarily in conflict. Function of an LCL receptor may be required during the differentiation of fusion capability but not during the process of fusion itself.

Lectins have been implicated in myoblast fusion as endogenous molecules, perhaps located on the cell surface. Teichberg and co-workers (1975) identified and partially characterized a protein from eel electric organ capable of agglutinating trypsin-treated rabbit erythrocytes. The lectin was specific for the β-D-galactoside configuration and agglutination was inhibited by disaccharides

TABLE 1
Inhibition of fusion by lectins[a]

Lectin	Percentage of nuclei in myotubes		Cells per field	
	Exp. 1	Exp. 2	Exp. 1	Exp. 2
None	50	59	165	290
Con A	25 (50)[b]	32 (54)	100 (61)	150 (52)
LCL	24 (48)	37 (63)	140 (85)	270 (93)

[a] Cultures were treated continuously with 30 μg/ml of each lectin from 15 hr after plating until sacrifice at 50 hr.

[b] Numbers in parentheses are percentage of control values.

154

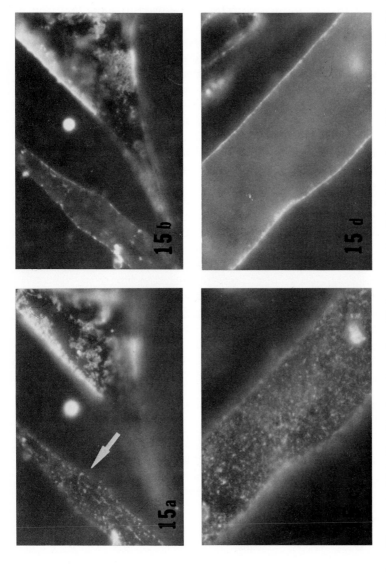

Fig. 15. Binding of fluorescein-labeled lentil lectin to the surface of living myogenic cells in a 3-day culture. Cells were stained for 20 min at 4 °C with 70 μg/ml of the labeled lectin and photographed with epifluorescent optics. (a) and (b) Mononucleated myoblast and fibroblast at different focal planes. Note the finely stippled distribution of lectin receptors on the myoblast (arrow) and the coarse patchy binding to the fibroblast. (c) and (d) Myotube photographed at different focal planes. The distribution of lectin receptors resembles that found on myoblasts. × 1000.

such as lactose and thiodigalactoside. Tests for the presence of lectin activity in saline homogenates of tissues and organs from various species revealed exceptionally high specific agglutinin titers per mg protein in chick embryo muscle in vivo and in vitro.

Gardner and Podleski (1975) first demonstrated that this lectin might be involved in myoblast fusion by showing that 15 mM thiodigalactoside could inhibit fusion of rat L_6 myoblasts, a permanent cell line derived by carcinogen treatment (Yaffe, 1968) and capable of fusion. The hapten sugar did not interfere with cell proliferation and, of several sugars tested, only thiodigalactoside was effective in blocking fusion. Intact cells dissociated by EDTA were also reported to have lectin sites on their surface. More recently, the same authors described a second lectin activity in L_6 myoblasts (Gardner and Podleski, 1976). This agglutinin is present in both soluble and particulate cell fractions and cannot be blocked by thiodigalactoside.

Nowak and co-workers (1976) confirmed the presence of lectin activity in L_6 cells and chick embryo muscle, and showed that the agglutinins from both sources are inhibited by lactose and thiodigalactoside. The specific activity of the lectin reaches a peak during myogenesis that roughly corresponds with the period of myoblast fusion. A comparable developmental curve for lectin activity in chick muscle *in ovo* was reported by Den and colleagues (1976), but these authors were unable to confirm the cell surface location of the lectin or its role in myoblast fusion. High concentrations of thiodigalactoside had no effect on fusion, either in normal chick cultures or in cultures treated with EGTA to accumulate fusion-competent cells. The β-D-galactoside lectin was isolated from acetone powder of embryonic muscle, purified by affinity chromatography and found to be a homodimer with a subunit molecular weight of about 15,000 daltons (Den and Malinzak, 1977).

Experiments in this laboratory with cultured chick muscle have also failed to detect fusion inhibition with the hapten sugar or the presence of hemagglutinating activity on the cell surface (Bischoff, unpublished observations). Disrupted cells have high levels of agglutinin activity but most of it remains in the soluble fraction after high speed centrifugation. We have also found that the appearance of agglutinin activity in cultured muscle is inhibited by growth in 5-bromodeoxyuridine, as are many muscle-specific traits (Stockdale et al., 1964; Bischoff and Holtzer, 1970).

The presence of a muscle-specific protein with carbohydrate binding activity is an intriguing aspect of myogenesis; however, it appears unlikely that the lectin is involved in surface events of myoblast fusion, at least in normal cells. Further work is needed to identify the nature of the lectin and to examine the possibility that more than one lectin with similar specificities may be present in muscle cells.

5.4. Metabolic

Relatively little effort has been made to test the effects of general metabolic and synthetic inhibitors on fusion. A variety of agents can inhibit fusion in randomly growing cultures (Holtzer, 1970) but, as pointed out previously (Holtzer and Bischoff, 1970), it is difficult to determine whether fusion itself is affected or whether the accumulation of fusion-competent cells is blocked. Since energy production and macromolecular synthesis are required for virtually all aspects of cell function, it seems unlikely that their specific role in fusion could be identified in this way.

The only system that would be amenable to studies of this type is the rapid fusion of competent cells prepared by growth in low Ca^{2+} or EGTA containing medium (see section 5.1). The inhibitor could be tested by adding it to the medium when Ca^{2+} is added to induce fusion. This approach was used to show that fusion is an energy-requiring process that can be inhibited by metabolic poisons such as azide, cyanide, fluoride, and dinitrophenol (Den et al., 1975). Complete inhibition of fusion also required the absence of glucose at the time of Ca^{2+} addition. The block was reversible and normal fusion occurred if the inhibitors were removed after 4 hours.

Using a similar technique it has been shown that Ca^{2+}-induced cell fusion does not require protein or RNA synthesis since it is not blocked by cycloheximide, actinomycin D (Adamo et al., 1976) or 5-fluorouridine (Easton and Reich, 1972). Fusion in normal cultures is also unaffected by actinomycin D over short periods of time (Molinaro et al., 1974).

If Ca^{2+}-blocked cells are exposed to 5-bromotubercidin, a reversible inhibitor of both ribosomal and messenger RNA synthesis, for 4 hours prior to adding Ca^{2+}, fusion is delayed for a corresponding time (Easton and Reich, 1972). This suggests that some messenger RNA required for fusion is made in the Ca^{2+}-inhibited cells.

In sum, production of ATP is the only metabolic requirement identified to date for fusion in the Ca^{2+}-induced system. Protein and RNA synthesis are not needed for fusion per se, but may be required earlier for the division or differentiation of fusible cells.

5.5. Enzymes

Digestion or enzymatic alteration of surface components by specific enzymes might provide important clues as to the nature of molecules involved in fusion. The use of enzymes is limited, however, to conditions that do not produce significant cell detachment of fusing cells from the substratum.

Addition of 0.5 $\mu g/ml$ of purified phospholipase C (PLC) to the medium of chick myogenic cultures results in almost complete inhibition of fusion without affecting cell proliferation (Nameroff et al., 1973; Nameroff, 1974). The inhibition is reversible, and normal myotubes form within a few hours after removal of the enzyme. The enzyme produces its effect by acting directly on the cell and not on the culture medium. Curiously, only PLC from *Clostridium welchii* was effective; *B. cereus* PLC did not block fusion although it had enzyme activity.

Since PLC splits the polar head from phospholipids leaving diacylglycerides, it seems possible that the resultant loss of surface charge might alter the binding of some component to the cell surface. Indeed, it was found that PLC releases [^{125}I]-labeled proteins from cells previously labeled with the lactoperoxidase technique (Nameroff, 1974). When this PLC-labile material was added to normal cultures, fusion was inhibited despite the absence of PLC itself. The author postulated that PLC may be removing part of a "lock and key" mechanism necessary for fusion.

The effect of PLC on fusion was confirmed by Schudt and Pette (1976) using the low Ca^{2+} method of synchronizing fusion. These authors surveyed a large variety of enzymes and found that they could be classified into several groups according to their effect on fusion. Proteases (trypsin, collagenase) affect adhesion and fusion to a similar extent and, since they cause cell detachment, a specific effect on fusion could not be tested. Glucosidases (lysozyme, β-galactosidase, β-glucosidase) were active only at high concentration and also caused cell detachment, but some fusion inhibition could be demonstrated at lower concentrations. Phospholipase A and C both showed strong inhibition of fusion without causing detachment. Phospholipase A removes the fatty acid from the 2 position of glycerol to produce lysophosphatides. Thus the action of this enzyme is consistent with the inhibitory effect of exogenous lysolecithin on fusion (Reporter and Norris, 1973; see section 5.6). Finally, enzymes that reduce the surface charge by removing negatively charged groups (hyaluronidase, neuraminidase, acid phosphatase) inhibited fusion at concentrations manyfold lower than those necessary to cause detachment.

5.6. Lipids

The final event in establishing cytoplasmic continuity during fusion is the interaction of the lipid bilayers in membranes of adjacent cells. Assuming that molecular interactions depend on the freedom of movement of membrane lipids, fusion should be inhibited by conditions that reduce membrane fluidity and enhanced by conditions that increase fluidity.

Although studies have demonstrated lateral diffusion in the plane of the membrane of antigens (Edidin and Fambrough, 1973) and acetylcholine receptor (Axelrod et al., 1976) in cultured myotubes, there has been no comparable comparison of membrane fluidity before and after fusion.

Van der Bosch and colleagues (1973) tested the effect on Ca^{2+}-induced myoblast fusion of agents that might be expected to influence membrane fluidity. Fusion was affected dramatically when the temperature was varied at the time of Ca^{2+} addition. The fusion rate increased about 15-fold between 28° and 40°C, the highest temperature tested. The activation energy for fusion calculated from an Arrhenius plot of these data gave a value of 17.8 to 22 kcal per mole for temperatures above 35°C. Similar values have been obtained for chemically-induced fusion of other cell types (Fisher, 1974).

Since Ca^{2+}-induced fusion of competent cells is quite rapid (the fusion rate may exceed 15% of all cells per hour), it seems unlikely that temperature primarily affects energy production. Although inhibitors of ATP synthesis can block Ca^{2+}-induced fusion (Den et al., 1975), they are relatively ineffective in a complete medium such as that used in these experiments.

Cholesterol has a stabilizing effect on membranes and tends to reduce the motion of the hydrophobic tails of the phospholipids (Phillips, 1972). Addition of cholesterol to myogenic cells 4 hours before Ca^{2+}-induced fusion results in a striking inhibition of fusion (van der Bosch et al., 1973). Although the cholesterol was added to the medium in the form of dispersed particles (at 1 mg/ml), there was no evidence that the cholesterol was actually incorporated into the membrane. The fusion inhibition produced by cholesterol could be reversed by increasing the temperature. The activation energy of fusion in the presence of cholesterol was increased to about 140 kcal per mole.

Prives and Shinitzky (1977) tested the effect on fusion of adding various fatty acids to the culture medium of chick myoblasts. Those fatty acids expected to reduce membrane fluidity (stearic and elaidic) delayed fusion while those expected to enhance fluidity (oleic and linoleic) hastened the onset of fusion. The differences were relatively modest, however, and all cultures eventually reached the same degree of fusion after 48 hours in vitro.

Exogenous lysolecithin is capable of promoting fusion of various cells (Poole et al., 1970), and it has been suggested that the endogenous production of lysolecithin might be responsible for spontaneous membrane fusion (Lucy, 1974). Since lysolecithin destabilizes membranes and leads to lysis, its generation in the membrane would have to be localized and highly controlled (Poste and Allison, 1973).

Continuous exposure of rat myogenic cells to medium containing concentrations of lysolecithin (120–160 μg/ml) found to promote fusion in other cells (Ahkong et al., 1972; Gledhill et al., 1972; Koprowski and Croce, 1973) results in inhibition of fusion (Reporter and Norris, 1973). The inhibition is reversible when cells are returned to normal medium. Lysolecithin resulted in some toxicity, as shown by a reduced rate of accumulation of DNA, RNA, and protein in the cultures. Since fusion was only partially blocked, the primary effect may be to retard the production of fusible cells. Although lysolecithin has not been tested in Ca^{2+}-induced fusion, this would be of interest since cells contain increased Ca^{2+} when grown in lysolecithin (Reporter and Norris, 1973).

In view of the unexpected effect of lysolecithin on myoblast fusion, it might be argued that continuous exposure to the lipid selected for cells that were able to remove lysolecithin from the membrane or convert it to lecithin thus neutralizing its fusogenic properties. However, examination of phospholipid metabolism showed greatly reduced turnover in lysolecithin-treated myogenic cells (Reporter and Raveed, 1973).

While phospholipase A might be expected to promote fusion by generating lysolecithin, the enzyme actually inhibits fusion (Schudt and Pette, 1976). Furthermore, Kent and Vagelos (1976) failed to find phospholipase A activity in the membranes of fusing myoblasts. These results make it unlikely that lysolecithin is involved in myoblast fusion.

5.7. Other inhibitors

In addition to agents that may be expected to have a direct effect on membrane structure or metabolism, there are a number of other inhibitors that block muscle differentiation. These include such agents as 5-bromodeoxyuridine (Stockdale et al., 1964; Bischoff and Holtzer, 1969), ethidium bromide (Brunk and Yaffe, 1976), phorbol ester (Cohen et al., 1977) and virus (Holtzer et al., 1975a; Hynes et al., 1976). Since most of these agents appear to act at the nuclear level, they probably affect the generation of fusion-competent cells from precursors and are not likely to provide information about the direct events of fusion.

6. *Environmental factors*

The synthesis and secretion of both surface-bound and soluble macromolecules by myogenic cells (Schubert et al., 1973; Bischoff and Lowe, 1974) suggests that this activity may lead to changes in the microenvironment. Numerous studies have suggested that secretory and metabolic activity of myogenic cells can influence the initiation and rate of myoblast fusion. Myogenic cells may release fusion-promoting substances that accumulate in the immediate vicinity of the cells or remain loosely bound to the cell surface.

In a series of experiments with cultured quail cells, Konigsberg (1971) showed that the onset of fusion could be controlled by varying both the cell density at plating and the volume of the culture medium. High cell density or low medium: cell ratio favored early fusion. Maintaining the cultures on a rotating platform to provide continuous mixing of the medium also delayed fusion, but this effect could be reversed by increasing the cell number. The alteration of the medium by the cells (medium conditioning) was found to be specific for myogenic cells, since other cell types such as heart, liver, and fibroblasts failed to condition medium to accelerate the onset of fusion. A similar reciprocal relationship between cell density and the onset of fusion has also been observed by other investigators (Morris and Cole, 1972; O'Neill and Stockdale, 1972b; Powell, 1973; Doering and Fischman, 1977).

In addition, the onset of fusion occurs earlier in unfed cultures than in cultures in which the medium is replaced after plating (Morris and Cole, 1972; O'Neill and Stockdale, 1972b).

These experiments suggest that myogenic cells produce and secrete into the medium a factor that, when present in sufficient concentration, promotes the fusion of competent cells. Although an alternative explanation that the conditioning involves medium depletion or detoxification has not been ruled out, it seems unlikely in view of the cell specificity involved and the fact that the active component is a macromolecule (Konigsberg, 1971).

Doering and Fischman (1977) have recently attempted to isolate the active material by exposing myogenic cultures to protein-free medium for 24 hours to collect products secreted by the cells. Medium conditioned in this way was found to possess significant fusion-promoting activity in myogenic cultures when compared with fresh medium. Comparable activity was not found with medium conditioned by other cell types or with nonspecific proteins such as albumin and gelatin. Most of the activity was retained by a 10,000 D cut-off ultrafilter and was destroyed by trypsin or boiling.

Although the material studied by Doering and Fischman (1977) promotes fusion, there is no assurance that the activity is comparable to that observed by others since the experimental design is quite different. All earlier studies of the effect of cell density and medium conditioning on fusion were carried out with the cells in a medium enriched with serum and embryo extract (Konigsberg, 1971; Morris and Cole, 1972; O'Neill and Stockdale, 1972b). The activity studied by Doering and Fischman, however, was assayed with the cells in a protein-free medium, a harsh condition that may lead to surface alterations and extensive shedding of macromolecules (Doljanski and Kapeller, 1976). The authors note that the fusion-promoting activity obtained from cells was significantly reduced when tested in a standard enriched medium.

A major weakness in most studies designed to probe the involvement of surface components in myoblast interaction is the method for determining the end point. As discussed earlier (section 1), the onset of fusion is very difficult to measure with any degree of accuracy by light microscope observation. To take an extreme case, fusion of 50 cells into a single hypernucleated myotube would be much easier to detect microscopically than fusion of the same cells into 25 binucleate myotubes. Little is known about factors that regulate the number of nuclei per myotube. The degree of multinuclearity, and hence the relative ease of detecting fusion, could be altered by experimental variables such as cell density, medium mixing, and conditioned medium. Thus, cultures that appear to have quite different numbers of myotubes by microscopic observation may actually have the same amount of fusion.

A novel approach has been designed by Yeoh and Holtzer (1977) in an attempt to confirm previous studies on the effect of cell density and medium conditioning in regulating the onset of fusion. To avoid problems inherent in scoring fusion microscopically, the investigators measured creatine kinase (CK) instead, after first showing that a good correlation exists between the number of fused nuclei and CK activity. Although the kinetics and accumulation of CK and fused nuclei may differ, most studies agree that the initial rise in CK activity is concurrent with, or at most a few hours later than, the onset of fusion determined microscopically (Coleman and Coleman, 1968; Shainberg et al., 1969; Yaffe and Dym, 1973; Turner et al., 1974; Lough and Bischoff, 1977). This method should be especially useful for determining the onset of fusion regardless of the degree of multinuclearity since both large and small myotubes accumulate CK but mononucleated cells do not (Turner et al., 1976a,b).

Using CK activity as an index of fusion, no difference was found in the time of the initial rise in CK when cell density was varied 100-fold (Yeoh and Holtzer, 1977). The effects of conditioned medium were more difficult to analyze since the specific activity of CK was depressed by conditioned medium. Nevertheless, the authors failed to observe an obvious difference in the onset of fusion with conditioned medium prepared by incubating established cultures in protein-free medium as in Doering and Fischman (1977). Although CK activity may be a useful alternative to microscopic observation in determining the onset of fusion, further studies are needed to be confident that there is a good correlation between the two parameters under a variety of culture conditions.

Clearly, environmental factors may have an effect on the onset of fusion and the degree of multinuclearity of the resulting myotubes, but the origin of the factors and their mode of action remains to be determined. Cell density may act by promoting the accumulation of a cell product that is needed in sufficient concentration for the fusion of competent cells. However, cell density also influences proliferation of myogenic cells (Bischoff, 1970; Buckley and Konigsberg, 1974); and since only nonproliferating cells can fuse (Okazaki and Holtzer, 1966; Bischoff and Holtzer, 1969), low density may suppress the emergence of fusion-competent cells by promoting cell replication. Further study of environmental agents must take into account the multiple factors involved in determining whether cells fuse, such as proliferation, motility, and position in the myogenic lineage (Holtzer and Bischoff, 1970; Dienstman and Holtzer, 1975).

7. Surface coats of myogenic cells

The plasma membrane of animal cells contains variable amounts of carbohydrates, most of which are external to the lipid permeability barrier (Winzler, 1970). This peripheral, carbohydrate-rich layer exhibits considerable variation between cell types and within different regions of the same cell and includes a number of elements that may differ widely in structure and function (see Martínez-Palomo, 1970; Rambourg, 1971; Luft, 1971 for general reviews of surface coats).

Myoblast fusion can occur only if the lipid bilayers of adjacent cells are able to come into close enough apposition to permit the molecular rearrangements that establish cytoplasmic confluence. The existence of a cell surface coat or glycocalyx (Bennett, 1963) on myogenic cells prior to fusion would thus appear to be an obstacle that must be overcome to allow fusion. On the other hand, surface coats have been strongly implicated in recognition and adhesion in many cell types (Lilien, 1969; Chipowsky et al., 1973; Lloyd and Cook, 1975) and may be an important determinant of the specificity of myoblast fusion (see section

3.2). Despite the potential importance of cell surface coats during early myogenesis, this topic has received little attention.

7.1. *Ultrastructure and staining of surface coats*

In contrast to embryonic cells, the structure and staining properties of the external surface coats of adult muscle have been well characterized (Bennett, 1963; Rambourg and Leblond, 1967; Luft, 1971; Zacks et al., 1973). Adjacent to the plasma membrane is a thin layer that binds colloidal cationic stains such as iron or thorium. This structure is invisible or, at best, seen in amorphous patches with standard electron microscope processing. External to this layer and about 20 to 30 nm from the plasma membrane is the basal, or external lamina (Zacks et al., 1973). The basal lamina appears as a 40 to 50 nm thick feltwork of fine filaments with standard processing and stains with periodic acid-Schiff and ruthenium red, among other things, but not with colloidal iron or thorium.

In a survey of the ultrastructure of avian myogenesis in vitro, Fischman (1967, 1970) concluded that a glycoprotein surface coat was absent from cells destined to fuse. After fusion, the basal lamina first appeared around the shaft of myotubes at 5 to 6 days in vitro, but the expanded pseudopodial ends remained free of basal lamina.

Shimada (1972b) found a ruthenium-red positive coat on the surface of 12-day chick myogenic cells *in ovo*. The stainable material was lost during trypsin disaggregation but reappeared by the second day in culture. The surface coat was present on early myotubes and mononucleated cells, but the staining was invariably heaviest where the cells were in contact. The free surfaces of cells showed an irregular, patchy distribution. Cultures older than 2 days were not described.

Preliminary observations in this laboratory confirm these findings (Bischoff, unpublished observations). Ruthenium-red staining (Luft, 1971) is light and irregular on mononucleated cells and early myotubes up to 2 days in vitro. The heavier staining in narrow intercellular spaces may represent passive trapping and inefficient washout of the stain complex. As the cultures age staining remains sparse on the mononucleated cells, but the maturing myotubes develop a thick, even sheath of ruthenium-red positive material (Fig. 16). The surface of collagen fibers also stains at periodic intervals with ruthenium red.

The myotube basal lamina is first visible in these cultures with standard processing at 4 days, but there are several differences between the basal lamina and the ruthenium-red stained coat. The basal lamina is separated from the plasma membrane by about 25 nm and, although it follows the contours of the cell quite closely, it does not enter narrow surface pits or nascent transverse (T) tubule openings (Fig. 17). In contrast, the ruthenium-red staining includes the area occupied by the basal lamina but extends up to the plasma membrane and even enters the openings of the T tubules for some distance (Fig. 18). Thus, while the nascent basal lamina stains with ruthenium red, other material at the myotube surface also stains. It may be significant that this material is frequently absent from the surface of the mononucleated myogenic cells.

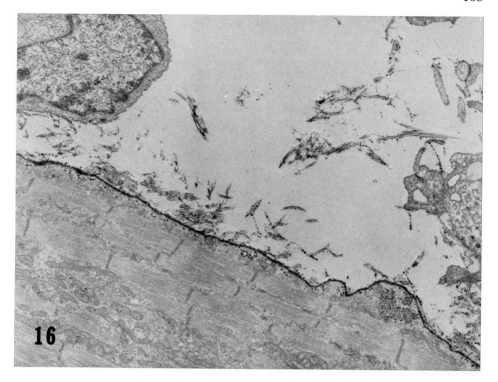

Fig. 16. Four-day-old myogenic culture fixed with glutaraldehyde and osmium tetroxide solutions containing ruthenium red. No counterstains were used. The surface of the myotube and openings of the T-tubules bind the ruthenium red complex, but the mononucleated cells in the culture are largely unstained. Collagen fibrils are also stained. ×8970.

A different picture emerges when myogenic cultures are stained with colloidal thorium (Rambourg and Leblond, 1967). The staining is uniformly present on mononucleated cells and myotubes (Fig. 19) and does not increase during differentiation. Transverse tubule openings and collagen fibrils are unstained, although occasionally some reactive material is present on the inner surface of larger vacuoles which are probably connected with the cell surface.

Colloidal thorium is believed to react with strong acid groups such as carboxyl and sulfate that are present on macromolecules (Hayat, 1975). Although muscle cultures synthesize sulfated glycosaminoglycans (Nameroff and Holtzer, 1967), the cellular origin of these is unknown.

Additional evidence that ruthenium red and thorium stain different surface components comes from the observation that thorium binding is reduced by pretreatment with EDTA, which abolishes myoblast-myotube adhesion (Bischoff and Lowe, 1974, see section 3.2), but ruthenium red binding is unaffected.

Finally, a third staining pattern was obtained with the concanavalin A-peroxidase technique (Bernhard and Avrameas, 1971; Pfenninger, personal

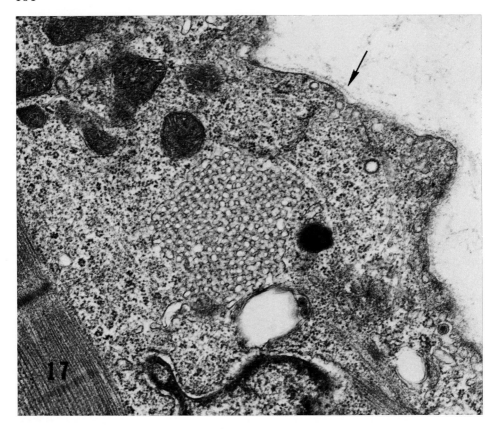

Fig. 17. Surface of a myotube in a 5-day-old culture showing the fine feltwork of the early basal lamina (arrow). The array of smooth membranes in the center is a component of the developing transverse (T)-tubule system. Numerous microtubule profiles and several fuzzy-coated vesicles are also evident. ×30,800.

communication). Concanavalin A (Con A) reacts specifically with sugar groups such as α-D-glucopyranosyl and α-D-mannopyranosyl and staining cannot be attributed to acidic groups.

Con A stains both mononucleated cells and myotubes (Fig. 20) but the staining is not reduced by EDTA treatment. Con A staining is restricted to the outer surface of the myotubes, and the lectin does not react with the T tubule membranes. Con A also fails to stain T tubules in adult muscle (Barchi et al., 1977).

It appears from these studies that, contrary to earlier impressions, the surface of both mononucleated myogenic cells and myotubes is provided with a rich variety of glycocomponents that can be demonstrated with special staining techniques. Although these stains can indicate differences in surface coats during development and between cell types, they are not specific enough to be useful for identification of the reactive material. Further studies are needed to characterize surface coats of myogenic cells and to elucidate their role in cell recognition, adhesion, and fusion.

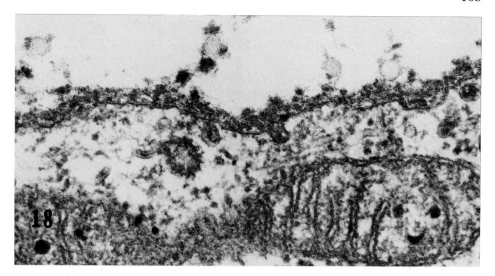

Fig. 18. Surface of a myotube in a 5-day-old culture treated with ruthenium red. The complex binds to the basal lamina, plasma membrane, and openings of the developing T tubules. ×99,000.

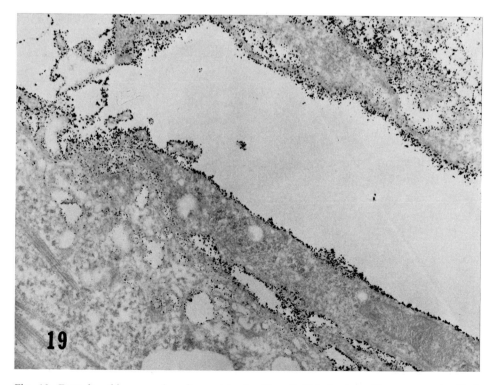

Fig. 19. Four-day-old myogenic culture stained with colloidal thorium. No counterstain. Both the surface of the myotube (lower left) and the adjacent mononucleated cells are heavily coated with the thorium particles. ×18,380.

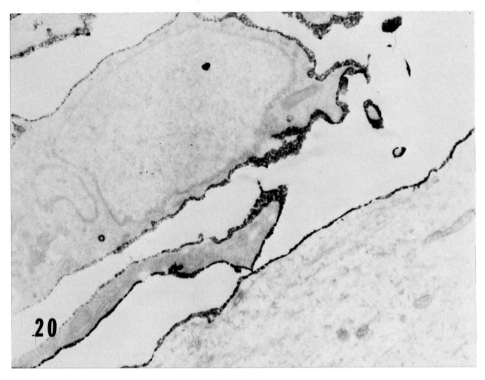

Fig. 20. Myogenic cells in a 3-day culture stained by the concanavalin A-peroxidase technique before fixation. All cells in the culture have lectin binding ites on their surface. ×9920.

7.2. Is the basal lamina a block to fusion?

Of particular importance is the question of whether mononucleated cells can fuse with myotubes ensheathed by a basal lamina. If the basal lamina were an effective block to fusion, its synthesis would be an important element in controlling the number of fibers formed and the degree of multinuclearity per fiber during myogenesis (Bischoff and Lowe, 1974).

Basal lamina material appears to be absent from cells in the few convincing examples of myoblast fusion in vitro (Lipton and Konigsberg, 1972; Rash and Fambrough, 1972). Also, the rate of myoblast fusion is greatly diminished in cultures at 4 or 5 days or about the time a basal lamina forms around the myotubes (Okazaki and Holtzer, 1966). If fresh [³H]thymidine-labeled myogenic cells are added to 5-day cultures, the cells fuse with each other but not with the preexisting myotubes (Bischoff and Holtzer, 1969; Bischoff, 1970). The block to fusion with older myotubes may result from the presence of a basal lamina or from some alteration in the plasma membrane.

Multinucleated myotubes first appear in the chick embryo axial muscles at about 5 days in ovo (Przybylski and Blumberg, 1966) and do not possess a basal lamina. However, basal lamina has been observed on myotomal cells as early as 36 hours (Fig. 10 in Przybylski and Blumberg; Lowe, 1967). Although further study is needed to resolve this point, it is possible that the basal lamina of the somite cells may be related to their epithelial organization at this stage (Hay, 1968). Thus, the early basal lamina material probably does not ensheathe the cells and may be quite different from the basal lamina which eventually surrounds the fibers at a later stage.

A distinct basal lamina first appears at 18 to 20 days of gestation in the rat and ensheathes small groups of cells including both myotubes and mononucleated myogenic cells (Kelly and Zacks, 1969; Kelly and Schotland, 1972). Thus, the internal surfaces of these cells are free of basal lamina and fusion can occur between myotubes and myoblasts within the same fascicle without interference from a continuous basal lamina. At birth in the rat, most fibers occur singly and are completely ensheathed with a basal lamina (Kelly and Schotland, 1972). Although fusion of myoblasts continues at a reduced rate for some time after birth (Moss and Leblond, 1970, 1971), the new myonuclei probably arise from satellite cells, mononucleated cells which lie within the myofiber basal lamina (Mauro, 1961).

Microdissection studies have shown that the basal lamina of mature fibers is a tough, elastic structure resistant to tearing (Bischoff, 1975). It is also resistant to attack by various enzymes such as collagenase, papain, and ficin but can be dissolved by trypsin and pronase (Zacks et al. 1973; Bischoff, 1974). There is no evidence that mononucleated myoblasts can migrate across the basal lamina of single fibers isolated in vitro (Bischoff, 1975). These observations suggest that myoblast-myotube fusion does not occur across the basal lamina. This surface coat may be an important element in regulating fusion.

8. Membrane composition and alterations during myogenesis

Since myoblast fusion is a highly ordered and specific reaction (section 2), it is reasonable to assume that significant membrane alterations occur to facilitate fusion at the appropriate time. Therefore, studies of membrane composition and, particularly, of changes occurring during myogenesis might provide clues to the mechanism of fusion. In view of the large amount of material needed for biochemical analysis, it is not surprising that many of these studies have used established lines of rat myoblasts (Yaffe, 1968).

8.1. Lipids

A method for obtaining plasma membranes in high purity from cultured muscle was developed by Schimmel and colleagues (1973). The final membrane fraction (Fig. 21) from standard 3-day cultures containing a mixture of myotubes and mononucleated cells showed greater than sevenfold enrichment for $[^{125}I]$-α-bungarotoxin binding, a specific marker for myotube plasma membrane

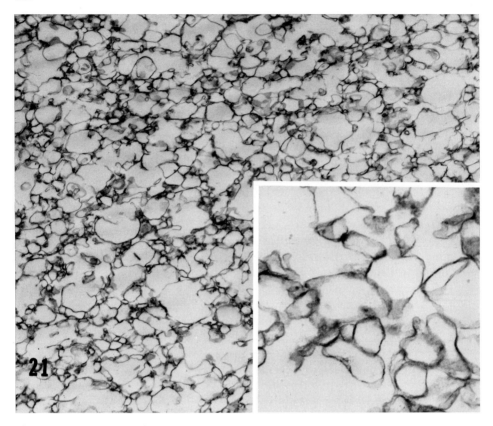

Fig. 21. Plasma membrane fraction from myogenic cells purified according to Schimmel et al. (1973). Most of the membrane exists as closed vesicles, and the preparation is free of contaminating organelles such as ribosomes or mitochondria. ×11,780; inset, ×41,800.

(Hartzell and Fambrough, 1973). The degree of membrane purification and the distribution of marker enzymes among various subcellular fractions was the same in cultures of purified mononucleated cells and myotubes.

This method was subsequently applied to study the lipid composition of plasma membranes before and after fusion using Ca^{2+} induction of fusion (Kent et al., 1974). The cells exhibited an unusually high lipid:protein ratio (about 2.5) but this remained constant during fusion, as did the cholesterol:phospholipid ratio. No difference was found in the fatty acid of phospholipid composition of membranes from mononucleated cells and myotubes. These findings suggest that fusion is not accompanied by alterations in membrane lipids or, if changes do occur, they are highly localized and are not reflected in the composition of total plasma membrane. Alternatively, since normal, mononucleated myogenic cells were not examined in this study, any lipid alterations required for fusion may already have occurred during the growth of the fusion-arrested cells in the low Ca^{2+} medium.

The turnover of phospholipids was found to be reduced severalfold compared with normal fused cultures in cells inhibited from fusing with lysolecithin (Reporter and Raveed, 1973; see section 5.6). Since the authors did not examine turnover in normal unfused cells or in cells blocked by low Ca^{2+}, it cannot be determined whether low turnover is characteristic of unfused cells or is a condition induced by the lysolecithin but unrelated to fusion. Kent and Vagelos (1976) found no difference in the activity of phosphatidic acid phosphatase or phospholipase A, enzymes involved in phospholipid metabolism, in plasma membranes from unfused and fused cells.

Changes in the membrane lipid phase during myogenesis have been detected by monitoring the fluorescence polarization of a lipid probe to provide a measure of membrane fluidity (Prives and Shinitzky, 1977). The membrane viscosity dropped sharply during the first day in culture and then began to increase as myotubes formed. The nadir of viscosity corresponded with the onset of myoblast fusion suggesting that high membrane fluidity may be required for fusion. Further experiments are needed, however, to confirm the correlation between fusion and membrane viscosity. For example, the initial drop in viscosity could result from damage during tissue disaggregation. This could be tested by delaying fusion and determining whether the viscosity change is also delayed.

8.2. Carbohydrates

Studies of the plasma membrane carbohydrate composition during myogenesis have been more promising, as might be expected in view of differences in surface-staining reactions for glycocomponents (see section 7.1).

Purified membranes from rat L_6 myogenic cells showed a decrease in neuraminic acids and amino sugars during differentiation and an increase in glucose and galactose (Winand and Luzzati, 1975). Other sugars, such as mannose and fucose, did not change. The sugar concentrations in these experiments were measured by an isotope dilution method after growth in labeled glucose and need to be confirmed by direct chemical assay. The increase in glucose and galactose may result from the appearance of basal lamina material since a glucose-galactose-hydroxylysine unit was isolated from differentiated cells (Winand and Luzzati, 1975). This disaccharide is linked to hydroxylysine in both basal lamina and collagen (Kornfeld and Kornfeld, 1976).

Rat L_6 cells were also used to measure the distribution of gangliosides, carbohydrate-rich lipids, during myogenesis (Whatley et al., 1976). Cells were labeled with galactose for 48 hours and the gangliosides were extracted and separated by thin-layer chromatography. GM_2 and GM_3 comprised 80 to 90% of the total labeled gangliosides and these showed little change during myogenesis.

Radioactivity in GD_{1a}, however, increased severalfold during the confluent stage just prior to myoblast fusion, then declined when fusion was complete.

Gangliosides are rich in glucose, galactose, and sialic acid, but since no attempt was made to measure the quantities of gangliosides these results cannot be compared with the earlier findings of Winand and Luzzati (1975).

The increase in GD_{1a} during myogenesis is intriguing since endogenous lectin activity also increases simultaneously (see section 5.3), suggesting that the two events may be related. Removal of sialic acid from GD_{1a} would expose galactoside, a specific ligand for the "myoblast lectin."

Although it is not known whether the change in GD_{1a} is related to fusion or to the increase in cell density that accompanies myogenesis, this ganglioside was absent from a nonfusing variant line of L_6 cells (Whatley et al., 1976). Its possible role in fusion could be tested more directly by growing cells at different densities or by blocking fusion with low Ca^{2+}. It has recently been shown that although gangliosides are within the lipid bilayer and beneath the cell surface coat, they are accessible to antibody at the cell surface (Gregson et al., 1977).

8.3. Proteins

Lactoperoxidase-catalyzed iodination of surface proteins of rat L_8 myoblasts before and after fusion failed to exhibit any qualitative differences following SDS-polyacrylamide electrophoresis (Hynes et al., 1976; Chen, 1977). There was, however, a quantitative change. The amount of labeled protein in a high molecular weight band ($\sim 230,000$ D) was substantially increased in fused cells. Similar results were obtained with a galactose oxidase borotritide labeling method, indicating that the material contains galactose. Pretreatment with neuraminidase was necessary to obtain labeling. Other properties of this material, including sensitivity to trypsin and its reduction after viral infection of myoblasts, suggests that it is similar to the large external transformation sensitive (LETS) protein studied in other cells (Hynes, 1973; Yamada and Weston, 1974).

Chen (1977) confirmed these findings, again using L_8 cells, and showed that although the surface quantity of LETS increases during fusion the surface distribution becomes greatly restricted. The pattern of immunofluorescent staining with antibody against LETS protein changes from an even distribution on mononucleated cells in confluent but unfused cultures to highly focal clusters or "hot spots" on myotubes after fusion. The apparent decrease in LETS with immunostaining is hard to reconcile with the increase obtained after iodination and may reflect differences in accessibility of the reactive sites to the different labels.

Because of these contradictory but interesting results, the possible role of LETS in myoblast fusion merits further study. There is also evidence the LETS changes in nonmuscle cells under various conditions also known to affect fusion of myoblasts. For example, LETS is exposed during the G_1 phase of the cell cycle (Hunt et al., 1975) and myoblast fusion occurs only during G_1 (Bischoff and Holtzer, 1969). LETS decreases during viral transformation (Hynes, 1976) or after exposure to phorbol myristate acetate (Blumberg et al., 1976), conditions that also lead to loss of fusion capability (Holtzer et al., 1975a; Hynes et al., 1976; Cohen et al. 1977). Finally, LETS is involved in cell adhesion (Yamada et al., 1976), a necessary precondition for myoblast fusion. Given these correlations, the hypothesis that an increase in LETS is required for fusion is worth testing.

Surprisingly, research to date has detected few changes in membrane composition during myogenesis. The surface of maturing myotubes is significantly different from that of myogenic precursor cells in terms of electrical properties (Fischbach et al., 1971; Kidokoro, 1975; Ritchie and Fambrough, 1975), surface coats, including the basal lamina (see section 7.1), cholinergic sensitivity as a result of specific receptors and enzymes (Fambrough, 1976), and antigenic specificity (Partridge and Smith, 1976). Future research that can identify and focus on specific components is more likely to be productive than studies based on bulk membrane extraction.

9. Summary and conclusions

Myoblast fusion is a specific, highly regulated process that involves several different cell surface components. For the purpose of the following summary, the process will be divided into four stages:

1. Acquisition of fusion competence
2. Contact of competent cells
3. Close apposition of plasma membranes
4. Rearrangement of lipid bilayers

The earliest events, those that confer fusion capability on a cell, are the least well understood. The transition from replicating presumptive myoblast to postmitotic fusion-competent myoblast is linked to the cell cycle and may involve the appearance of new surface properties. Unfortunately, little is known of the membrane characteristics of replicating presumptive myoblasts. That the surface of these cells is already specialized, however, is shown by the ability of dividing myogenic cells to recognize and adhere to myotubes.

Myoblast-myotube recognition and adhesion ensures that fusion-competent cells are brought together and remain in close contact until fusion occurs. The specificity of myoblast fusion probably operates at this level and may reside in the glycocalyx or peripheral surface components. Removal of this material results in loss of adhesion. The lability of the surface coat to EDTA and the fact that the onset of fusion is sensitive to environmental factors such as cell density and medium mixing suggest that the active components are loosely attached to the plasma membrane and are continuously shed into the extracellular space. Conditions that favor retention and a high concentration at the cell surface would tend to facilitate fusion by promoting more extensive cell contact.

Juxtaposition of plasma membranes close enough (a few nanometers) to permit molecular interactions between lipid bilayers has not been described in myogenic cells. Therefore, this event is probably highly localized or exceedingly transient. The glycocalyx presents an obstacle to lipid interaction that must be overcome either by removal with lysosomal enzymes or by penetration by means of a microvillus or fine membrane extension. Although both coated vesicles and microvilli are common in fusing myogenic cells, there is no direct evidence for the involvement of either organelle in fusion. Alternatively, close membrane apposition may occur through the formation of a gap junction between fusing cells. Cytoplasmic confluence may then be a function primarily of the protein molecules comprising the gap junction.

The final stage in fusion, the establishment of cytoplasmic continuity, probably involves rearrangements of membrane lipids and proteins between contiguous cells. Ultrastructural evidence suggests that these changes are restricted to a single small site on the surface of fusing cells. The molecular rearrangements probably require membrane fluidity since agents that promote membrane stability tend to inhibit fusion. The terminal stages of fusion also require a supply of ATP and an appropriate concentration of Ca^{2+}, but the molecular machinery that utilizes these agents is unknown. Once continuity is established, the channel expands very rapidly and the myoblast membrane is incorporated into the surface of the myotube. Fusion is completed.

Acknowledgment

Original contributions from this laboratory were supported by grants from the National Institutes of Health and from the Muscular Dystrophy Association.

References

Adamo, S., Zani, B., Siracusa, G. and Molinaro, M. (1976) Expression of differentiative traits in the absence of cell fusion during myogenesis in culture. Cell differ. 5, 53–67.
Ahkong, Q. F., Cramp, F. C, Fisher, D., Howell, J. I. and Lucy, J. A. (1972) Studies on chemically induced cell fusion. J. Cell Sci. 10, 796–787.
Ahkong, Q. F., Tampion, W. and Lucy, J. A. (1975) Promotion of cell fusion by divalent cation ionophores. Nature (London) 256, 208–209.
Allbrook, D. (1962) An electron microscopic study of regenerating skeletal muscle. J. Anat. 96, 137–152.
Axelrod, D., Ravdin, P., Koppel, D. E., Schlessinger, J., Webb, W. W., Elson, E. L. and Podleski, T. R. (1976) Lateral motion of fluorescently labeled acetylcholine receptors in membranes of developing muscle fibers. Proc. Nat. Acad. Sci. U.S.A. 73, 4594–4598.
Barchi, R. L., Bonilla, E. and Wong, M. (1977) Isolation and characterization of muscle membranes using surface-specific labels. Proc. Nat. Acad. Sci. U.S.A. 74, 34–38.
Bayne, E. K. and Simpson, S. B. Jr. (1975) Lizard myogenesis in vitro. A time-lapse and scanning electron microscopic study. Dev. Biol. 47, 237–256.
Bedford, J. M. (1972) An electron microscopic study of sperm penetration into the rabbit egg after natural mating. Am. J. Anat. 133, 213–254.
Beirle, J. W. (1968) Cell proliferation: enhancement by extracts from cell surfaces of polyoma-virus-transformed cells. Science 161, 798–799.

Bennett, H. S. (1963) Morphological aspects of extracellular polysaccharides. J. Histochem. Cytochem. 11, 14–23.

Berlin, R. D. (1975) Microtubules and the fluidity of the cell surface. Ann. N.Y. Acad. Sci. 253, 445–454.

Bernhard, W. and Avrameas, S. (1971) Ultrastructural visualization of cellular carbohydrate components by means of concanavalin A. Exp. Cell Res. 64, 232–236.

Betz, E. H., Firket, H. and Reznik, M. (1966) Some aspects of muscle regeneration. Int. Rev. Cytol. 19, 203–227.

Bischoff, R. (1970) The myogenic stem cell in development of skeletal muscle. In:Regeneration of Striated Muscle and Myogenesis (Mauro, A., Schafig, S. A. and Milhorat, A. T., eds.) pp. 218–231, Excerpta Medica, Amsterdam.

Bischoff, R. (1974) Enzymatic liberation of myogenic cells from adult rat muscle. Anat. Rec. 180, 645–662.

Bischoff, R. (1975) Regeneration of single skeletal muscle fibers in vitro. Anat. Rec. 182, 215–236.

Bischoff, R. and Holtzer, H. (1968) The effect of mitotic inhibitors on myogenesis in vitro. J. Cell Biol. 36, 111–127.

Bischoff, R. and Holtzer, H. (1969) Mitosis and the processes of differentiation of myogenic cells in vito. J. Cell Biol. 41, 188–200.

Bischoff, R. and Holtzer, H. (1970) Inhibition of myoblast fusion after one round of DNA synthesis in 5-bromodeoxyuridine. J. Cell Biol. 44, 134–150.

Bischoff, R. and Lowe, M. (1973) Role of cell surface proteins in myoblast-myotube interaction in culture. J. Cell Biol. 59,25.

Bischoff, R. and Lowe, M. (1974) Cell surface components and the interaction of myogenic cells. In:Exploratory Concepts in Muscular Dystrophy II. (Milhorat, A. T., ed.) pp. 17–29, Excerpta Medica, Amsterdam.

Blumberg, P. M., Driedger, P. E. and Rossow, P. W. (1976) Effect of a phorbol ester on a transformation-sensitive surface protein of chick fibroblasts. Nature (London) 264, 446–447.

Brunk, C. F. and Yaffe, D. (1976) The reversible inhibition of myoblast fusion by ethidium bromide. Exp. Cell Res. 99, 310–318.

Buckley, P. A. and Konigsberg, I. R. (1974) Myogenic fusion and the duration of the post-mitotic gap. Dev. Biol. 37, 193–212.

Capers, C. R. (1960) Multinucleation of skeletal muscle in vitro. J. Biophys. Biochem. Cytol. 7, 559–566.

Carlsson, S. A., Luger, O., Ringertz, N. R. and Savage, R. E. (1974) Phenotypic expression in chick erythrocyte x rat myoblast hybrids and in chick myoblast x rat myoblast hybrids. Exp. Cell Res. 84, 47–55.

Carter, S. B. (1967) Effects of cytochalasins on mammalian cells. Nature (London) 213, 261–264.

Chen, L. B. (1977) Alteration in cell surface LETS protein during myogenesis. Cell 10, 393–400.

Chipowsky, S., Lee, Y. C. and Roseman, S. (1973) Adhesion of cultured fibroblasts to insoluble analogues of cell-surface carbohydrates. Proc. Nat. Acad. Sci. U.S.A. 70, 2309–2312.

Chiquet, M., Eppenberger, H. M., Moor, H. and Turner, D. C. (1975) Application of freeze-etching to the study of myogenesis in cell culture. Exp. Cell Res. 93, 498–502.

Codington, J. F., Sanford, B. H. and Jeanloz, R. W. (1970) Glycoprotein coat of the TA3 cell. I. Removal of carbohydrate and protein material from viable cells. J. Nat. Cancer Inst. 45, 637–647.

Cohen, R., Pacifici, M., Rubinstein, N., Biehl, J. and Holtzer, H. (1977) Effect of a tumour promoter on myogenesis. Nature (London) 266, 538–539.

Coleman, J. R. and Coleman, A. W. (1968) Muscle differentiation and macromolecular synthesis. J. Cell Physiol. 72, 19–34.

Cooper, W. G. and Konigsberg, I. R. (1961) Dynamics of myogenesis. Anat. Rec. 140, 195–205.

Cox, P. G. (1968) In vitro myogenesis of promuscle cells from the regenerating tail of the lizard, Anolis carolinesis. J. Morphol. 126, 1–18.

Croop, J. and Holtzer, H. (1975) Response of myogenic and fibrogenic cells to cytochalasin B and to colcemid. I. Light microscope observations. J. Cell Biol. 65, 271–285.

Culp, L. A. and Black, P. H. (1972) Release of macromolecules from Balb/C mouse cell lines treated with chelating agents. Biochemistry 11, 2161–2172.

174

Den, H. and Malinzak, D. A. (1977) Isolation and properties of a β-D-galactoside-specific lectin from chick embryo thigh muscle. J. Biol. Chem. 252, 5444–5448.

Den, H., Malinzak, D. A., Keating, H. J. and Rosenberg, A. (1975) Influence of concanavalin A, wheat germ agglutinin and soybean agglutinin on the fusion of myoblasts *in vitro*. J. Cell Biol. 67, 826–834.

Den, H., Malinzak, D. A. and Rosenberg, A. (1976) Lack of evidence for the involvement of a β-D-galactosyl-specific lectin in the fusion of chick myoblasts. Biochem. Biophys. Res. Commun. 69, 621–627.

de Waard, A., Hickman, S. and Kornfeld, S. (1976) Isolation and properties of a β-galactoside binding lectin of calf heart and lung. J. Biol. Chem. 251, 7581–7587.

Dienstman, S. R. and Holtzer, H. (1975) Myogenesis: a cell lineage interpretation. In:Results and Problems in Cell Differentiation. (Reinert, J. and Holtzer, H. eds.), vol. 7, pp. 1–25, Springer-Verlag, Berlin.

Dienstman, S. R. and Holtzer, H. (1977) Skeletal myogenesis. Control of proliferation in a normal cell lineage. Exp. Cell Res. 107, 355–364.

Doering, J. L. and Fischman, D. A. (1977) A fusion-promoting macromolecular factor in muscle conditioned medium. Exp. Cell Res. 105, 437–443.

Doljanski, F. and Kapeller, M. (1976) Cell surface shedding—the phenomenon and its possible significance. J. Theor. Biol. 62, 253–270.

Easton, T. G. and Reich, E. (1972) Muscle differentiation in cell culture. Effects of nucleoside inhibitors and Rous sarcoma virus. J. Biol. Chem. 247, 6420–6431.

Edelman, G. M. (1976) Surface modulation in cell recognition and cell growth. Science 192, 218–226.

Edidin, M. and Fambrough, D. (1973) Fluidity of the surface of cultured muscle fibers. Rapid lateral diffusion of marked surface antigens. J. Cell Biol. 57, 27–37.

Emerson, C. P. and Beckner, S. K. (1975) Activation of myosin synthesis in fusing and mononucleated myoblasts. J. Mol. Biol. 93, 431–447.

Estensen, R. D. and Plagemann, P. A. W. (1972) Cytochalasin B: inhibition of glucose and glucosamine transport. Proc. Nat. Acad. Sci., U.S.A. 69, 1430–1434.

Ezerman, E. B. and Ishikawa, H. (1967) Differentiation of the sarcoplasmic reticulum and T system in developing chick skeletal muscle *in vitro*. J. Cell Biol., 35, 405.

Fambrough, D. M. (1976) Development of cholinergic innervation of skeletal, cardiac, and smooth muscle. In:Biology of Cholinergic Function (Goldberg, A. M. and Hanin, I., eds.) pp. 101–159, Raven Press, New York.

Fischbach, G. D., Nameroff, M., and Nelson, P. (1971) Electrical properties of chick skeletal muscle fibers developing in cell culture. J. Cell Physiol. 78, 289–300.

Fischman, D. A. (1967) An electron microscope study of myofibril formation in embryonic chick skeletal muscle. J. Cell Biol. 32, 557–575.

Fischman, D. A. (1970) The synthesis and assembly of myofibrils in embryonic muscle. Curr. Top. Dev. Biol. 5, 235–280.

Fischman, D. A. (1972) Development of striated muscle. In:The Structure and Function of Muscle. 2nd ed., (Bourne, G. H., ed.) Vol. 1, pp. 75–148, Academic Press, New York.

Fisher, D. (1974) Fluidity and cell fusion. In: Biomembranes, Lipids, Proteins and Receptors, (Burton, R. M. and Packer, L., eds.) pp. 75–93, BI-Science Publ. Div., Webster Groves, Missouri.

Fukuda, J., Henkart, M. P., Fischbach, G. D. and Smith, T. G., Jr. (1976) Physiological and structural properties of colchicine-treated chick skeletal muscle cells grown in tissue culture. Dev. Biol. 49, 395–411.

Gardner, T. K. and Podleski, T. R. (1975) Evidence that a membrane bound lectin mediates fusion of L6 myoblasts. Biochem. Biphys. Res. Commun. 67, 972–978.

Gardner, T. K. and Podleski, T. R. (1976) Evidence that the types and specific activity of lectins control fusion of L6 myoblasts. Biochem. Biophys. Res. Commun. 70, 1142–1149.

Gledhill, B. L., Sawicki, W., Croce, C. M. and Koprowski, H. (1972) DNA synthesis in rabbit spermatozoa after treatment with lysolecithin and fusion with somatic cells. Exp. Cell Res. 73, 33–40.

Godman, G. C. and Murray, M. R. (1953) Influence of colchicine on the form of skeletal muscle in tissue culture. Proc. Soc. Exp. Biol. Med. 84, 668–671.

Gregson, N. A., Kennedy, M. and Leibowitz, S. (1977) Gangliosides as surface antigens on cells isolated from rat cerebellar cortex. Nature (London) 266, 461–462.

Hartzell, H. and Fambrough, D. M. (1973) Acetylcholine receptor production and incorporation into membranes of developing muscle fibers. Dev. Biol. 30, 153–165.

Hay, E. D. (1968) Organization and fine structure of epithelium and mesenchyme in the developing chick embryo. In:Epithelial-Mesenchymal Interactions (Fleischmajer, R. and Billingham, R., eds.) pp. 31–55, Williams and Wilkins, Baltimore.

Hayat, M. A. (1975) Positive Staining for Electron Microscopy, p. 193, Van Nostrand Reinhold, New York.

Holtzer, H. (1970) "Myogenesis". In:Cell Differentiation (Schjeide, O. and de Vellis, J., eds.) pp. 476–503, Van Nostrand Reinhold, New York.

Holtzer, H. and Bischoff, R. (1970) Mitosis and myogenesis. In:The Physiology and Biochemistry of Muscle as a Food (Briskey, E. J., Cassens, R. G. and Marsh, B. B., eds.) pp. 29–51, Univ. of Wisconsin Press, Madison.

Holtzer, H. and Sanger, J. W. (1972) Myogenesis: Old views rethought. In:Research in Muscle Development and the Muscle Spindle, Banker, B., Przybylski, R. J., Van der Muelen, J. P. and Victor, M., eds.) pp. 122–133, Excerpta Medica, Amsterdam.

Holtzer, H., Sanger, J. KW., Ishikawa, H. and Strahs, K. (1973) Selected topics in skeletal myogenesis. Cold Spring Harbor Symp. Quant. Biol. 37, 549–566.

Holtzer, H., Biehl, J., Yeoh, G., Meganathan, R. and Kaji, A. (1975a) Effect of oncogenic virus on muscle differentiation. Proc. Nat. Acad. Sci. U.S.A. 72, 4051–4055.

Holtzer, H., Croop, J., Dienstman, S., Ishikawa, H. and Somlyo, A. P. (1975b) Effects of cytochalasin B and colcemide on myogenic cultures Proc. Nat. Acad. Sci. U.S.A. 72, 513–517.

Holtzer, H., Rubinstein, N., Fellini, S., Yeoh, G., Chi, J., Birnbaum, J. and Okayama, M. (1975c) Lineages, quantal cell cycles, and the generation of cell diversity. Quart. Rev. Biophys. 8, 523–557.

Hosaka, Y. and Koshi, Y. (1968) Electron microscopic study of cell fusion by HVJ virions. Virology 34, 419–434.

Hunt, R. C., Gold, E. and Brown, J. C. (1975) Cell cycle dependent exposure of a high molecular weight protein on the surface of mouse L cells. Biochim. Biophys. Acta. 431, 453–458.

Hynes, R. O. (1973) Alteration of cell surface proteins by viral transformation and by proteolysis. Proc. Nat. Acad. Sci. U.S.A. 70, 3170–3174.

Hynes, R. O. (1976) Cell surface proteins and malignant transformation. Biochim. Biophys. Acta. 453, 73–107.

Hynes, R. O., Martin, G. S., Shearer, M., Critchley, D. R., and Epstein, C. J. (1976) Viral transformation of rat myoblasts: Effects on fusion and surface properties. Dev. Biol. 48, 35–46.

Ishikawa, H., Bischoff, R. and Holtzer, H. (1968) Mitosis and intermediate-sized filaments in developing skeletal muscle. J. Cell Biol. 38, 538–555.

Keller, J. M. and Nameroff, M. (1974) Induction of creatine phosphokinase in cultures of chick skeletal myoblasts without concomitant cell fusion. Differentiation 2, 19–24.

Kelly, A. M. and Schotland, D. L. (1972) The evolution of the "checkerboard" in a rat muscle. In:Research in Muscle Development and the Muscle Spindle (Banker, B. Q., Przybylski, R. J., van der Muelen, J. P. and Victor, M., eds.) pp. 32–48, Excerpta Medica, Amsterdam.

Kelly, A. M. and Zacks, S. I. (1969) The histogenesis of rat intercostal muscle. J. Cell Biol. 42, 135–153.

Kent, C. and Vagelos, P. R. (1976) Phosphatidic acid phosphatase and phospholipase A activities in plasma membranes from fusing muscle cells. Biochim. Biophys. Acta 436, 377–386.

Kent, C., Schimmel, S. D. and Vagelos, P. R. (1974) Lipid composition of plasma membranes from developing chick muscle cells in culture. Biochim. Biophys. Acta 360, 312–321.

Kidokoro, Y. (1975) Developmental changes of membrane electrical properties in a rat skeletal muscle cell line. Gen. Physiol. (London) 244, 129–144.

Kikuchi, T. (1972) Ultrastructural evaluation of the myogenic cell fusion in chick embryo using the goniometer stage of electron microscopy. Tohoku J. Agr. Res. 23, 92–97.

Kikuchi, T. and Ashmore, L. R. (1976) Developmental aspects of the innervation of skeletal muscle fibers in the chick embryo. Cell Tissue. Res. 171, 233–252.

Kitiyakara, A. and Angevine, D. M. (1963) A study of the pattern of post embryonic growth of *M. Gracilis* in mice. Dev. Biol. 8, 322–340.

Kletzien, R. F., Perdue, J. F. and Springer, A. (1972) Cytochalasin A and B. Inhibition of sugar uptake in cultured cells. J. Biol. Chem. 247, 2964–2966.

Knudsen, K. A. and Horowitz, A. F. (1977) Tandem events in myoblast fusion. Dev. Biol. 58, 328–339.

Konigsberg, I. R. (1971) Diffusion-mediated control of myoblast fusion. Dev. Biol. 26, 133–152.

Koprowski, H. and Croce, C. M. (1973) Fusion of somatic and gametic cells with lysolecithin. Meth. Cell Biol. 7, 251–260.

Kornfeld, R. and Kornfeld, S. (1976) Comparative aspects of glycoprotein structure. Ann. Rev. Biochem. 45, 217–237.

Larson, P.F., Jenkison, M. and Hudgson, P. (1970) The morphological development of chick embryo skeletal muscle grown in tissue culture and studied by electron microscopy. J. Neurol. Sci. 10, 385–405.

Lewis, W. H. and Lewis, M. (1917) Behavior of cross-striated muscle in tissue culture. Am. J. Anat. 22, 169–194.

Lilien, J. E. (1969) Toward a molecular explanation for specific cell adhesion. Curr. Top. Dev. Biol. 4, 169–196.

Lipton, B. H. and Konigsberg, I. R. (1972) A fine-structural analysis of the fusion of myogenic cells. J. Cell Biol. 53, 348–364.

Lloyd, C. W. and Cook, G. M. W. (1975) A membrane glycoprotein-containing fraction which promotes cell aggregation. Biochem. Biophys. Res. Commun. 67, 696–700.

Lough, J. and Bischoff, R. (1977) Differentiation of creatine phosphokinase during myogenesis: Quantitative fractionation of isozymes. Dev. Biol. 57, 330–344.

Low, F. N. (1967) Developing boundary (basement) membranes in the chick embryo. Anat. Réc. 159, 231–238.

Lucy, J. A. (1974) Lipids and cell fusion. In:Biomembranes, Lipids, Proteins and Receptors (Burton, R. M. and Packer, L., eds.) pp. 95–116, BI-Science Publ. Div., Webster Groves, Missouri.

Luft, J. H. (1971) Ruthenium red and violet II. Fine structural localization in animal tissues. Anat. Rec. 171, 369–416.

Luft, J. H. (1976) The structure and properties of the cell surface coat. Int. Rev. Cytol., 45 291–382.

Martìnez-Palomo, A. (1970) The surface coats of animal cells. Int. Rev. Cytol. 29, 29–75.

Maslow, D. E. (1969) Cell specificity in the formation of multinucleated striated muscle. Exp. Cell Res. 54, 381–390.

Mauro, A. (1961) Satellite cell of skeletal muscle fibers. J. Biophys. Biochem. Cytol. 9, 493–495.

Mendell, J. R., Roelofs, R. I. and Engel, W. K. (1972) Ultrastructural development of explanted human skeletal muscle in tissue culture. J. Neuropath. Exp. Neurol. 31, 433–446.

Merlie, J. P. and Gros, F. (1976) *In vitro* myogenesis. Expression of muscle specific function in the absence of cell fusion. Exp. Cell Res. 97, 406–412.

Miranda, A. F. and Godman, G. C. (1973) The effects of cytochalasin D on differentiating muscle in culture. Tissue Cell 5, 1–22.

Mizel, S. B. and Wilson, L. (1972) Inhibition of the transport of several hexoses in mammalian cells by cytochalasin B. J. Biol. Chem. 247, 4102–4105.

Molinaro, M., Zani, B., Martinozzi and Monesi, V. (1974) Selective effects of actinomycin D on myosin biosynthesis and myoblast fusion during myogenesis in culture. Exp. Cell Res. 88, 402–405.

Morris, G. E. and Cole, R. J. (1972) Cell fusion and differentiation in cultured chick muscle cells. Exp. Cell Res. 75, 191–199.

Morris, G. E., Piper, M. and Cole, R. (1976) Differential effects of calcium ion concentration on cell fusion, cell division and creatine kinase activity in muscle cell cultures. Exp. Cell Res. 99, 106–114.

Moscona, A. A. (1957) The development *in vitro* of chimeric aggregates of dissociated embryonic chick and mouse cells. Proc. Nat. Acad. Sci. U.S.A. 43, 184–193.

Moscona, A. A. (1962) Analysis of cell recombination in experimental synthesis of tissues *in vitro*. J. Cell. Comp. Physiol. 60, Suppl. 1, 65–80.

Moss, F. P. and Leblond, C. P. (1970) Nature of dividing nuclei in skeletal muscle of growing rats. J. Cell Biol. 44, 459–461.

Moss, F. P. and Leblond, C. P. (1971) Satellite cells as the source of nuclei in muscles of growing rats. Anat. Rec. 170, 421–436.

Moss, P. S. and Strohman, R. C. (1976) Myosin synthesis by fusion-arrested chick embryo myoblasts in cell culture. Dev. Biol. 48, 431–437.

Nameroff, M. (1974) Phospholipase C and the myogenic cell surface. In:Exploratory Concepts in Muscular Dystrophy II (Milhorat, A. T., ed.) pp. 32–34, Excerpta Medica, Amsterdam.

Nameroff, M. and Holtzer, H. (1967) The loss of phenotypic traits by differentiated cells. IV. Changes in polysaccharide production by dividing chondrocytes. Dev. Biol. 16, 250–281.

Nameroff, M., Trotter, J. A., Keller, J. M. and Munar, E. (1973) Inhibition of cellular differentiation by phospholipase C. I. Effects of the enzyme on myogenesis and chondrogenesis in vitro. J. Cell Biol. 58, 107–118.

Nowak, T. P., Haywood, P. L. and Barondes, S. H. (1976) Developmentally regulated lectin in embryonic chick muscle and a myogenic cell line. Biochem. Biophys. Res. Commun. 68, 650–657.

Okazaki, K. and Holtzer, H. (1965) An analysis of myogenesis in vitro using fluorescein-labeled antimyosin. J. Histochem. Cytochem. 13, 726–739.

Okazaki, K. and Holtzer, H. (1966) Myogenesis: fusion, myosin synthesis, and the mitotic cycle. Proc. Nat. Acad. Sci. U.S.A. 56, 1484–1490.

O'Neill, M. and Stockdale, F. E. (1972a) Differentiation without cell division in cultured skeletal muscle. Dev. Biol. 29, 410–418.

O'Neill, M. C. and Stockdale, F. E. (1972b) A kinetic analysis of myogenesis in vitro. J. Cell Biol. 52, 52–65.

Partridge, T. A. and Smith, P. D. (1976) A quantitative test to detect lymphocytes sensitized against the surface of muscle cells. Clin. Exp. Immunol. 25, 139–143.

Paterson, B. and Strohman, R. C. (1972) Myosin synthesis in cultures of differentiating chick embryo skeletal muscle. Dev. Biol. 29, 113–138.

Paterson, B. and Prives, D. (1973) Appearance of acetylcholine receptor in differentiating cultures of embryonic chick breast muscle. J. Cell Biol. 59, 241–245.

Phillips, M. C. (1972) The physical state of phospholipids and cholesterol in monolayers, bilayers and membranes. In:Progress in Surface and Membrane Science (Danielli, J.F., Rosenberg, M. D. and Cadenhead, D. A., eds.) vol. 5, pp. 139–221, Academic Press, New York.

Poole, A. R., Howell, J. I. and Lucy, J. A. (1970) Lysolecithin and cell fusion. Nature (London) 227, 810–812.

Poste, G. (1972) Mechanisms of virus-induced cell fusion. Int. Rev. Cytol. 33, 157–252.

Poste, G. and Allison, A. C. (1973) Membrane fusion. Biochim. Biophys. Acta 300, 421–466.

Powell, J. A. (1973) Development of normal and genetically dystrophic mouse muscle in tissue culture. Exp. Cell Res. 80, 251–264.

Prives, J. M. and Paterson, B. M. (1974) Differentiation of cell membranes in cultures of embryonic chick breast muscle. Proc. Nat. Acad. Sci. U.S.A. 71, 3208.

Prives, J. and Shinitzky, M. (1977) Increased membrane fluidity precedes fusion of muscle cells. Nature (London) 268, 761–763.

Przybylski, R. J. and Blumberg, J. M. (1966) Ultrastructural aspects of myogenesis in the chick. Lab. Invest. 15, 836–863.

Przybylski, R. J. and Chlebowski, J. S. (1972) DNA synthesis, mitosis and fusion of myocardial cells. J. Morphol. 137, 417–431.

Rambourg, A. (1971) Morphological and histochemical aspects of glycoprotein at the surface of animal cells. Int. Rev. Cytol. 31, 57–114.

Rambourg, A. and Leblond, C. P. (1967) Electron microscope observations on the carbohydrate rich cell coat present at the surface of cells in the rat. J. Cell Biol. 32, 27–53.

Rash, J. E. and Fambrough, D. (1972) Ultrastructural and electrophysiological correlates of cell coupling and cytoplasmic fusion during myogenesis in vitro. Dev. Biol. 30, 166–186.

Rash, J. E. and Staehelin, L. A. (1974) Freeze-cleave demonstration of gap junctions between skeletal myogenic cells in vivo. Dev. Biol. 36, 455–461.

Reporter, M. and Norris, G. (1973) Reversible effects of lysolecithin on fusion of cultured rat muscle cells. Differentiation 1, 83–95.

Reporter, M. and Raveed, D. (1973) Plasma membranes: isolation from naturally fused and lysolecithin-treated muscle cells. Science 181, 863–865.

Ritchie, A., and Fambrough, D. M. (1975) Electrophysiological properties of the membrane and acetylcholine receptor in developing rat and chick myotubes. J. Gen. Physiol. 66, 327–341.

Roth, S. (1968) Studies in intercellular adhesive selectivity. Dev. Biol. 18, 602–631.

Sanger, J. W. (1974) The use of cytochalasin B to distinguish myoblasts from fibroblasts in cultures of developing chick striated muscle. Proc. Nat. Acad. Sci. U.S.A. 71, 3621–3625.

Sanger, J. W. and Holtzer, H. (1972) Cytochalasin B.: effects on cell morphology, cell adhesion and mucopolysaccaride synthesis. Proc. Nat. Acad. Sci. U.S.A. 253–257.

Sanger, J. W., Holtzer, S. H. and Holtzer, H. (1971) Effects of cytochalasin B on muscle cells in tissue culture. Nature New Biol. 229, 121–123.

Schimmel, S. D., Kent, C., Bischoff, R. and Vagelos, P. R. (1973) Plasma membrane from cultured cells. Isolation procedure and separation of putative plasma membrane marker enzymes. Proc. Nat. Acad. Sci. U.S.A., 70, 3195–3199.

Schubert, P., Tarikas, H., Humphreys, S., Heinemann, S. and Patrick, J. (1973) Protein synthesis and secretion in a myogenic cell line. Dev. Biol. 33, 18–37.

Schudt, C. and Pette, D. (1975) Influence of the ionophore A 23187 on myogenic cell fusion. FEBS Lett. 59, 36–38.

Schudt, C. and Pette, D. (1976) Influence of monosaccharides, medium factors and enzymatic modification on fusion of myoblasts in vitro. Cytobiologie 13, 74–84.

Schudt, C., van der Bosch, J. and Pette, D. (1973) Inhibition of muscle cell fusion in vitro by Mg^{2+} and K$^+$ ions. FEBS Lett. 32, 296–298.

Schudt, C., Dahl, G. and Gratzl, M. (1976) Calcium-induced fusion of plasma membranes isolated from myoblasts grown in culture. Cytobiologie 13, 211–223.

Scott, R. E., Maercklein, P. B. and Furcht, L. T. (1977) Plasma membrane intramembranous particle topography in 3T3 and SV3T3 cells: The effect of cytochalasin B. J. Cell Sci. 23, 173–192.

Shainberg, A., Yagil, G. and Yaffe, D. (1969) Control of myogenesis in vitro by Ca^{2+} concentration in nutritional medium. Exp. Cell. Res. 58, 163–167.

Sharon, N. and Lis, H. (1972) Lectins: cell agglutinating and sugar-specific proteins. Science 177, 949–959.

Shimada, Y. (1971) Electron microscope observation on the fusion of chick myoblasts in vitro. J. Cell Biol. 48, 128–142.

Shimada, Y. (1972a) Scanning electron microscopy of myogenesis in monolayer culture: A preliminary study. Dev. Biol. 29, 227–233.

Shimada, Y. (1972b) Early stages in the reorganization of dissociated embryonic chick skeletal muscle cells. Z. Anat. Entwick. 138, 255–264.

Shimada, Y. and Fischman, D. A. (1975) Scanning electron microscopy of nerve-muscle contacts in embryonic cell culture. Dev. Biol. 43, 42–61.

Steinberg, M. S. (1964) The problem of adhesive selectivity in cellular interaction. In:Cell Membranes in Development (Locke, M., ed.), pp. 321–366, Academic Press, New York.

Stockdale, F., Okazaki, K. Nameroff, M. and Holtzer, H. (1964) 5-bromodeoxyuridine: effect on myogenesis in vitro. Science 146, 533–535.

Teichberg, V. I., Silman, I., Beitsch, D. D. and Resheff, G. (1975) A β-D-galactoside binding protein from electric organ tissue of Electrophorus electricus. Proc. Nat. Acad. Sci. U.S.A., 72, 1383–1387.

Tello, J. F. (1922) Die entsehung der motorischen und sensiblen nervenigungen. Z. Anat. Entwick. 64, 348–476.

Turner, D. C., Maier, V. and Eppenberger, H. M. (1974) Creatine kinase and aldolase isoenzyme transitions in cultures of chick skeletal muscle cells. Dev. Biol. 37, 63–89.

Turner, D. C., Gmür, R., Siegrist, M., Burckhardt, E. and Eppenberger, H. M. (1976a) Differentiation in cultures derived from embryonic chicken muscle. I. Muscle-specific enzyme changes before fusion in EGTA-synchronized cultures. Dev. Biol. 48, 258–283.

Turner, D. C., Gmür, R., Lebherz, H. G., Siegrist, M., Wallimann, T. and Eppenberger, H. M. (1976b) Differentiation in cultures derived from embryonic chicken muscle. II. Phosphorylase histochemistry and fluorescent antibody staining for creatine kinase and aldolase. Dev. Biol. 48, 284–307.

van der Bosch, J., Schudt, C. and Pette, D. (1972) Quantitative investigation on Ca²⁺ and pH-dependence of muscle cell fusion *in vitro*. Biochem. Biophys. Res. Commun. 48, 326–332.

van der Bosch, J., Schudt, C. and Pette, D. (1973) Influence of temperature cholesterol, dipalmitoyl-lecithin and Ca²⁺ on the rate of muscle cell fusion. Exp. Cell Res. 82, 433–438.

Wahrmann, J. P., Drugeon, G., Delain, E. and Delain, D. (1976) Gene expression during the differentiation of myogenic cells of the L6 line. Biochimie 58, 551–563.

Warren, R. H. (1974) Microtubular organization in elongating myogenic cells. J. Cell Biol. 63, 550–566.

Wessells, N. K., Spooner, B. S., Ash, J. F., Bradley, M. O., Luduena, M. A., Taylor, E. L., Wrenn, J. T. and Yamada, K. M. (1971) Microfilaments in cellular and developmental processes. Science 171, 135–143.

Whatley, R., Ng, S. K. -C, Rogers, J., McMurray, W. C. and Sanwal, B. D. (1976) Developmental changes in gangliosides during myogenesis of a rat myoblast cell line and its drug resistant variants. Biochem. Biophys. Res. Commun. 70, 180–185.

Winand, R. and Luzzati, D. (1975) Cell surface changes during myoblast differentiation: Preparation and carbohydrate composition of plasma membranes. Biochimie 57, 764–771.

Winzler, R. J. (1970) Carbohydrates in cell surfaces. Int. Rev. Cytol. 29, 77–125.

Wollman, Y. and Bischoff, R. (1978) Analysis of cell surface lectin receptors during myogenesis. In preparation.

Yaffe, D. (1968) Retention of differentiation potentialities during prolonged cultivation of myogenic cells. Proc. Nat. Acad. Sci. U.S.A. 61, 447–483.

Yaffe, D. (1969) Cellular aspects of muscle differentiation *in vitro*. Curr. Top. Dev. Biol. 4, 37–77.

Yaffe, D. and Feldman, M. (1965) The formation of hybrid multinucleated muscle fibers from myoblasts of different genetic origin. Dev. Biol. 11, 300–317.

Yaffe, D. and Dym, H. (1973) Gene expression during differentiation of contractile muscle fibers. Cold Spring Harbor. Symp. Quant. Biol. 37, 543–548.

Yamada, K. M. and Weston, J. A. (1974) Isolation of a major cell surface glycoprotein from fibroblasts. Proc. Nat. Acad. Sci. U.S.A. 71, 3492–3496.

Yamada, K. M., Yamada, S. S. and Pastan, I. (1976) Cell surface protein partially restores morphology, adhesiveness, and contact inhibition of movement to transformed fibroblasts. Proc. Nat. Acad. Sci. U.S.A. 73, 1217–1221.

Yeoh, G.C.T. and Holtzer, H. (1977) The effect of cell density, conditioned medium and cytosine arabinoside on myogenesis in primary and secondary cultures. Exp. Cell Res. 104, 63–78.

Zachs, S. I., Sheff, M. F. and Saito, A. (1973) Structure and staining characteristics of myofiber external lamina. J. Histochem. Cytochemistry 21, 703–715.

Macrophage fusion in vivo and in vitro: a review

4

J. M. PAPADIMITRIOU

Contents

G. Poste & G. L. Nicolson (eds.) Membrane Fusion, pp. 181–218.
©Elsevier/North-Holland Biomedical Press, 1978.

1. Historical perspectives

The existence of multinucleate giant cells in both normal and diseased tissues has been recognized for more than a century. Faber (1893) credited Johannes Müller with their first description in 1838 in his treatise on tumors; he thought of them as "mother" cells, their nuclei being "germs" from which new cells could form. Twenty years later Virchow (1858) reported them in human and bovine tuberculous lesions, and suggested that rapid nuclear division without cytokinesis could account for the multinucleate state. Langhans (1868), who did exhaustive studies on the multinucleate giant cells of human tuberculosis and those appearing after the subcutaneous implantation of blood clots in pigeons, suggested that fusion between mononuclear precursors may also operate in the course of multinucleate giant cell formation.

Initially there was much confusion and disagreement regarding the derivation, function, and even the nomenclature of multinucleate giant cells (Virchow, 1858; Langhans, 1868; Faber, 1893; Borrel, 1893; Lambert, 1912). Gradually, it became evident that distinctive types of these cells were associated with defensive or inflammatory reactions (Haythorn, 1929). Yet even in inflammatory cellular exudates distinctions were made between multinucleate giant cells associated with foreign bodies and those (named after Langhans) found chiefly in the infectious (especially tuberculous) granulomas. In the former, the nuclei of the syncytia were believed to be scattered in the central region of the cell, while in the latter they were thought to be arranged in circular or semilunar patterns (Langhans, 1868). It was even suggested that "foreign body" multinucleate giant cells arose by fusion of mononuclear precursors, whereas the "Langhans" type was the result of amitotic nuclear division (Doan et al., 1930). Both these types of multinucleate giant cells, however, may be found in infectious granulomas, and cells possessing a nuclear arrangement intermediate between the two types have also been described frequently (Haythorn, 1929). Eventually the similarity of these two cell types was accepted and a more unified concept of the nature of the multinucleate giant cells of inflamed tissues emerged.

Borrel (1893) was the first to suggest that the multinucleate giant cells seen in inflammatory exudates were fused masses of mononuclear leukocytes. In 1912 Lambert, in a series of in vitro experiments with chicken spleen cells, observed the formation of multinucleate giant cells from wandering mononuclear phagocytes. Awrorow and Tunofejewsky (1914) were the first to demonstrate that multinucleate giant cells could form from circulating leukocytes, and some years later the Lewises (1925, 1926, 1927) reported the transformation of large mononuclear leukocytes from the blood of elasmobranchs, teleosts, amphibia, reptiles, birds, and mammals into multinucleate giant cells. Ebert and Florey (1939), using the rabbit ear window, were able to examine the development in vivo of these polykaryocytes from carbon-labeled mononuclear phagocytes.

More recently, the formation of multinucleate giant cells has been observed in cultures of human (Goldstein, 1954) and avian (Sutton and Weiss, 1966) monocytes. Gillman and Wright (1966) presented data that strongly suggested that the multinucleated giant cells seen in inflammatory granulomas formed from circulating mononuclear phagocytes which they had labeled with tritiated thymidine. Similarly, Spector and Lykke (1966) reported that in both mice and rats, whose circulating monocytes were labeled with carbon and in which inflammatory granulomas were induced with complete Freund's adjuvant, the multinucleate giant cells that formed at the site also contained the cytoplasmic marker. Similar conclusions were also reached when foreign body granulomas were induced in mice whose circulating monocytes were labeled with tritiated thymidine (Mariano and Spector, 1974); in this instance, multinucleate giant cells in the granuloma displayed the nuclear label.

This review examines the general phenomenon of macrophage fusion. The evidence for the fusion of macrophages in vivo and in vitro is discussed and includes some reference to the means employed for the production in vitro of macrophage homokaryoctes and heterokaryocytes. The factors that may be involved in the mechanism of macrophage fusion are considered and compared with those affecting the fusion of other cell types. In addition, the properties of macrophage polykaryocytes forming in inflamed tissues are compared and contrasted with those induced by in vitro techniques. Finally, the functional role of multinucleate giant cells in inflammatory reactions are considered.

2. Macrophage fusion in vivo

2.1. Introduction

Multinucleate giant cells are commonly encountered in inflamed tissues. Classically they are a feature of granulomas, many of which are caused by infective agents; others are induced by physical and chemical agents. In some the precise etiologic factors are unknown or poorly understood (Anderson, 1971). Generally, these cells appear to form where macrophages are plentiful and when the causative agent is persistent and sublethal.

Bacterial infection may result in multinucleate giant cell formation. They are seen frequently in tuberculosis, leprosy, syphilis, lymphogranuloma venereum, actinomycosis and, occasionally, in chronic pyogenic infections. Fungal infections such as aspergillosis, torulosis, and coccidiomycosis are also frequently accompanied by the formation of multinucleate cells in the affected tissue. They may even be induced by such parasitic infestations as amoebiasis.

Multinucleate giant cells may appear in the leukocytic exudates accompanying several viral infections in man and other animals. Infections with the herpes or pox groups of viruses, as well as Newcastle disease virus, respiratory syncytial virus simian paramyxoviruses, canine distemper virus, and those responsible for cat scratch disease, rinderpest, pseudorabies, bovine rhinotracheitis, and visna, are often accompanied by the formation of polykaryocytes in the leukocytic exudate (reviews, Roizman, 1962; Poste, 1972).

As indicated earlier, the implantation of various foreign bodies in man or experimental animals results in multinucleate giant cell formation. In man these cells are commonly seen in asbestosis, gout, in sites of lipid deposition, and in surrounding suture material.

Multinucleate giant cells have been reported in granulomatous diseases of unknown etiology. They occur in such disorders as sarcoidosis, rheumatoid arthritis, regional enteritis, and in giant cell arteritis. The only common feature in all of these unrelated and poorly understood disorders is the establishment of an inflammatory state with a predominantly mononuclear leukocytic exudate. Nevertheless, it is both puzzling and interesting that of the variety of cell types actively involved in the inflammatory response the macrophage readily fuses with others of this kind. Among mammalian differentiated cells, macrophages display a striking tendency for intercellular fusion.

2.2. Direct observations of macrophage fusion

2.2.1. Cinemicrography

Despite the relative ease with which multinucleate giant cells are found in the various pathologic conditions outlined above, few direct observations of the in vivo events leading to their formation are found in the literature. The only report of such observations occurs in the work of Ebert and Florey (1939), who used cinemicrography to investigate the behavior of macrophages labeled with colloidal carbon in the rabbit ear window. Although they observed multinucleate giant cells forming from precursor mononuclear phagocytes, the authors provided little detail about the associated phenomena.

Recently Papadimitriou and Kingston (1976), using cinemicrographic techniques, demonstrated fusion between cells attached to glass coverslips within hours of removal from mice in which they had been implanted. In such a preparation the macrophage population of the leukocytic exudate surrounding the implants adheres preferentially to the surface of the implant, permitting continuous examination of the adherent cells. Despite the close proximity of cells in the explant, fusion was infrequently observed. Repeated intercellular contacts, often lasting for many hours were frequently seen, but in most instances fusion did not occur. These intercellular explorations were invariably performed by ruffling lamellipodia and, on the few occasions that fusion occurred, no distinct pattern of motility preceded the event. Although the initial phase of intercellular contact in such instances was often prolonged, the actual fusion process lasted only a few minutes. After fusion, there was admixture and redistribution of the contributing cytoplasmic contents and the component nuclei occupied a central position in the newly formed syncytium at the periphery of a common centrosphere. The polykaryocytes on the explant exhibited less prominent contact inhibition phenomena than their mononuclear precursors. This resulted in many extensive and prolonged intercellular contacts between polykaryocytes and their immediate neighbors, but again fusion was rarely observed.

The infrequent incidence of macrophage fusion in these cinemicrographic studies may have been caused by technical factors, or by lack of a constant supply of newly emigrated monocytes (Mariano and Spector, 1974), or may have resulted from the absence of some cellular or humoral agènt that might facilitate the process in vivo. Such a "macrophage fusion factor" has been reported in the pulmonary lavage fluid of rabbits immunized by intravenous injections of BCG (Galindo et al., 1974; Parks and Weiser, 1975). As a result of such low rates of fusion in the explants, quantitative assessment was difficult, and it could not be determined whether polykaryocytes, which displayed reduced contact inhibition, would be more prone to use than mononuclears.

2.2.2. Ultrastructural and biochemical changes associated with macrophage fusion
Ultrastructural observations of mononuclear phagocytes in foreign-body granulomas where multinucleate giant cells are common have revealed a wide variety of intercellular contacts. These range from the closely opposed flaps and microvilli of neighboring cells to large tracts of adjacent cell membranes in close apposition. In these regions of apposition the cell surface glycocalyx was only poorly demonstrable with ruthenium red (Papadimitriou and Archer, 1974). Occasionally, flaps of cytoplasm were folded on themselves at the site of contact (Fig. 1). Generally, lysosomal dense bodies occurred frequently near regions of cellular apposition. Breaks in continuity of the cell membranes between two adjacent cells were rarely observed (Fig. 2).

It is unknown which particular type of intercellular contact, if any, is the direct prelude to macrophage fusion. Since animal cells possess a net negative charge on their surface (Weiss and Woodbridge, 1967), contact with microvilli of low curvature radius would be the most effective means of establishing a primary contact between two cells destined to fuse (Pethica, 1961). Such contacts are not only a feature in populations of macrophages where fusion is occurring (Papadimitriou and Archer, 1974), but have also been reported during the course of viral and chemically-induced polykaryocytosis (Poste, 1970; 1972) including macrophage fusion induced by Sendai virus (Schneeberger and Harris, 1966). Unfortunately, the limitations of the techniques did not permit accurate measurements of the exact distance separating the plasma membranes of adjacent cells.

It is also difficult to stain the glycocalyx of adjacent macrophages in regions of close apposition, at least using ruthenium red or the osmium tetroxide—potassium ferrocyanide techniques (Papadimitriou and Archer, 1974). Poste (1970, 1972), using ultrastructural and ellipsometric techniques, has found that unless the glycocalyx of a cell is thinner than 3.5 nm, viral-induced cell fusion cannot occur. However, the glycocalyx on the surface of mononuclear phagocytes and multinucleate giant cells demonstrated by these techniques often measures more than 3.5 nm (Papadimitriou and Archer, 1974). Such measurements, reflect only the dimensions of the deposition zone of the marker material rather than the true thickness of the glycocalyx. The small amounts of marker in areas of close apposition may reflect true thinning of the glycocalyx but may also be artifactual due to poor diffusion of the reagents in such regions.

The formation of cytoplasmic bridges between adjacent cells indicates the breakdown of plasmalemmal continuity (Fig. 2). Such a phenomenon occurs

186

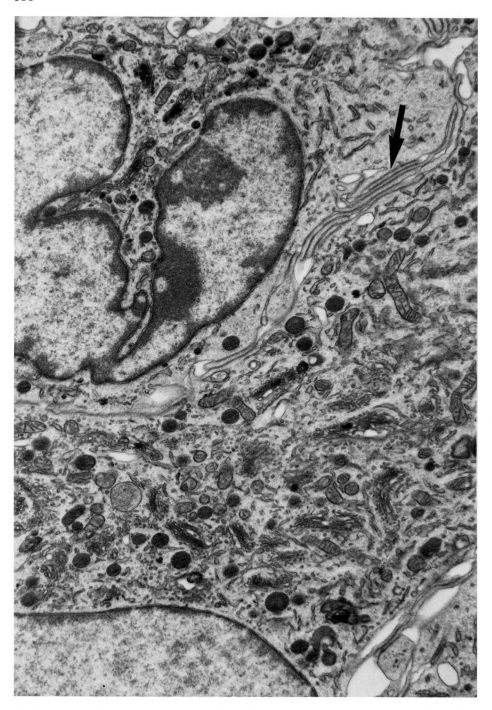

Fig. 1. Electron micrograph showing zone of cytoplasmic apposition between two adjoining cells. In addition to the closeness of the cell membranes there is multilayering of protruding cytoplasmic flaps (arrow). Lysosomal dense bodies are frequent near the area of contact. ×17,000.

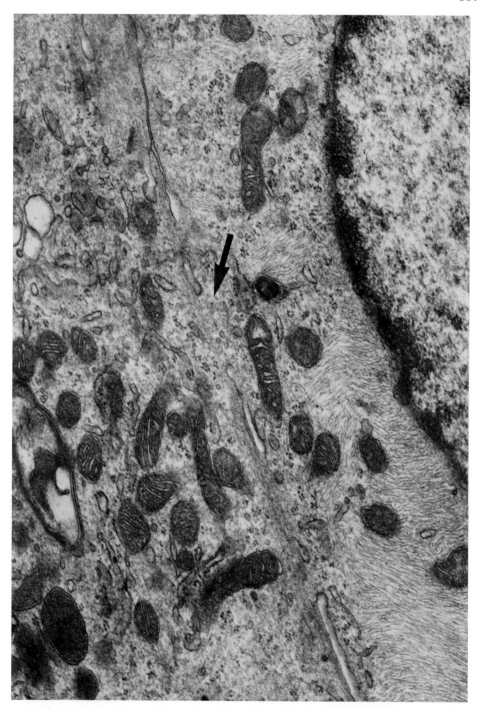

Fig. 2. Electron micrograph showing cytoplasmic continuity (arrow) between two adjacent cells. Small membranous vesicles can be seen in the zone where presumably membrane fusion has occurred. ×41,000.

during the course of the other types of polykaryocytosis (Dales and Siminovitch, 1961; Harris et al., 1966; Schneeberger and Harris, 1966; Meiselman et al., 1967; Hosaka and Koshi, 1968; Svoboda and Dourmashkin, 1969), but superimpositional artifacts of cellular peripheries can result in misleading profiles in the electron micrographs (Hyde et al., 1969) with consequent overestimation of presumed areas of cytoplasmic fusion. These difficulties can be overcome by using an electron microscope equipped with a unicentric tilting stage.

Klenk and Choppin (1969a) have produced evidence suggesting that an increase in relative amounts of phospholipids, especially in phosphatidyl-ethanolamine (Klenk and Choppin, 1969b), may favor cellular fusion, while Rubin (1967) and Howell and Lucy (1969) have hypothesized that the formation and liberation of lysolecithin may produce structural changes in the plasma membrane which would promote cell fusion. Some support for the latter view comes from the finding that treatment of cells in vitro with lysolecithin can induce cell fusion (Howell and Lucy, 1969; Poole et al., 1970; Ahkong et al., 1973).

To obtain an insight into whether a similar mechanism might operate in populations of fusing mononuclear phagocytes, Papadimitriou and Wyche (1976) compared the phospholipid content of peritoneal macrophages and similar cell populations containing multinucleate giant cells induced by the subcutaneous implantation of glass coverslips in mice for a period of two weeks. The results (Table 1) indicate that there is a greater concentration of lysolecithin, phosphatidylethanolamine, lecithin, and, to a lesser degree, sphingomyelin, in cell populations containing multinucleate giant cells. Although they are inconclusive, these results suggest that alterations in phospholipid composition accompany macrophage fusion.

Roizman (1962) has hypothesized that a foreign body implanted in the tissues may affect the orientation of hydrophilic molecules in the cell membranes of adhering macrophages and thus enhance fusion and polykaryocyte formation. Ahkong and his colleagues (1973) have suggested that during the course of chemically- or thermally- induced cell fusion an increase in the degree of membrane fluidity and a reorientation of the polar groups of choline-containing phospholipids occurs. The mode of interaction of the various phospholipids is

TABLE 1

Concentration of various phospholipids in murine peritoneal macrophages and cell population containing multinucleate giant cells

Phospholipid class	Cell populations containing multinucleate giant cells	Peritoneal macrophages
Lysolecithin	135 ± 28 μg/mg soluble protein	43 ± 22 μg/mg soluble protein
Lecithin	413 ± 85 μg/mg soluble protein	94 ± 42 μg/mg soluble protein
Sphingomyelin	124 ± 36 μg/mg soluble protein	58 ± 17 μg/mg soluble protein
Phosphatidylethanolamine	491 ± 64 μg/mg soluble protein	260 ± 73 μg/mg soluble protein

still unresolved though differences in the interaction of different phospholipids has been detected by ultrastructural techniques (Ahkong et al., 1973 and chapter by Lucy in this volume, p. 267).

2.3. Indirect evidence of macrophage fusion

Some authors have attempted to detect the precursors of multinucleate giant cells by labeling circulating monocytes with different agents and then searching for the marker in multinucleate giant cells in variously provoked inflammatory granulomas. Spector and Lykke (1966) labeled circulating monocytes in rats with intravenously injected indian ink and demonstrated the formation of carbon containing multinucleate giant cells at the sites of subcutaneous granulomas subsequently induced by Freund's adjuvant. In such an experiment the possibility must be considered, however, that the cytoplasmic marker may have originated from dead and degenerating cells in the leukocytic exudate and was ingested nonspecifically by the multinucleate giant cells.

Gillman and Wright (1966) and Mariano and Spector (1974) used tritiated thymidine to label the nuclei of circulating monocytes in rats and found that the multinucleate giant cells in foreign body granulomas displayed labeled nuclei. The above data strongly suggest that the mononuclear phagocytes in inflammatory leukocytic exudates are the major precursors of multinucleate giant cells. In such experiments 40 to 70% of syncytial nuclei were labeled (Gillman and Wright, 1966; Mariano and Spector, 1974). In addition the presence in these experiments of binucleate cells, in which only one of the nuclei was labeled (Mariano and Spector, 1974) suggests that cell fusion may start with the involvement of only two cells. Subsequent fusion with fresh mononuclears or with other small polykaryocytes probably results in the formation of large multinucleate giant cells. If macrophages attached to a foreign object are transplanted into the same host but confined within a millipore chamber, then the formation of large polykaryocytes is negligible (Mariano and Spector, 1974). It appears, therefore, that a constant supply of newly emigrated monocytes is necessary for the development of large polykaryocytes.

2.4. The role of the "resident" macrophage

Recent evidence has indicated that monocytes and macrophages derived from monocytes can be distinguished from tissue resident macrophages (Daems and Brederoo, 1971, 1973; Volkman, 1971; Brederoo and Daems, 1972). The cytochemical demonstration of endogenous peroxidase and catalase activities can be used to distinguish these different populations. Although endogenous peroxidatic activity is found in both populations, it is confined to cytoplasmic granules in monocytes and the macrophages arising from them, whereas resident tissue macrophages display the activity within the cisternae of ergastoplasm and in the nuclear envelope (Cotran and Litt, 1970; Daems and Brederoo, 1971, 1973; Brederoo and Daems, 1972).

Using the distribution of endogenous peroxidase as a means of distinguishing resident from exudate macrophages in granulomas induced by foreign body implantation it has been found that in the rat and guinea pig the reaction product is localized within granules and not in the nuclear envelope or ergastoplasm (Papadimitriou, unpublished observations). This granular distribution of reaction product again points to the exudate macrophage as the predominant precursor of multinucleate giant cells. Occasionally, some binucleate cells displayed a distribution of reaction product compatible with that of the resident macrophage. It seems, therefore, that resident macrophages with endogenous peroxidase in the nuclear envelope and ergastoplasm, contribute little, if at all, to the formation of large macrophage syncytia. The meager fusion that resident macrophages display is apparently limited solely to cells of their lineage.

2.5. The efficiency of fusion in macrophage populations

When glass coverslips are implanted in the subcutaneous tissues of mice, the vast majority of multinucleate giant cells are found in the glass adherent cell population and not in the surrounding connective tissue. In the first 2 to 3 days after implantation few multinucleate giant cells are seen (Papadimitriou et al., 1973b; Mariano and Spector, 1974). From the third day onward the multinucleate giant cells increase rapidly in the implant; 5% of all nuclei were found within syncytia on the third day, and up to 45% on the seventh day after implantation (Papadimitriou et al., 1973b). Thus the ability to fuse develops within a few days, and by then almost half of the adherent macrophages have fused into polykaryocytes. The reason for the initial delay before polykaryocytes appear in such a lesion is not understood. Possibly, the initial presence of great numbers of polymorphonuclear leukocytes soon after implantation, or due commital of the exuded mononuclears to active phagocytosis, may prove inhibitory to macrophage fusion (Papadimitriou et al., 1973b), or perhaps even relatively "aged" macrophages are necessary for the formation of macrophage polykaryocytes (Mariano and Spector, 1974).

The continued production of large multinucleate giant cells (syncytia with more than 10 nuclei) in the implant depends on a steady source of circulating monocytes (Papadimitriou et al., 1973b; Mariano and Spector, 1974). In mice which received whole body irradiation, in whom the site of implantation was shielded, only small polykaryocytes (2–4 nuclei) formed (Papadimitriou et al., 1973b). Similar results were obtained by Mariano and Spector (1974), who placed macrophages contained within millipore chambers into irradiated mice and showed that circulating monocytes were required for the ready production of large multinucleate giant cells, even though the cells within the chamber were actively dividing. Possibly, an excess of mononuclear phagocytes is needed to ensure a high frequency of intercellular contacts which in turn ensures intercellular fusion. Alternatively, the chromosomal abnormalities that quickly develop in the macrophage population at the site of the foreign body granuloma (Mariano and

Spector, 1974) could result in macrophages whose potential for fusion is diminished, though definite information on this question is lacking.

2.6. Macrophage fusion in relation to the cell cycle

Macrophages at the site of a foreign body implant synthesize deoxyribonucleic acid (DNA) (Ryan and Spector, 1970; Papadimitriou and Cornelisse, 1975) and, after a 30-minute exposure to tritiated thymidine, 3 to 4% of macrophages on the surface of a foreign body implanted subcutaneously in mice are labeled. In an attempt to determine whether macrophages at a certain phase of the cell cycle are more prone to fuse, the DNA content in individual nuclei of *large* multinucleate giant cells (containing more than 10 nuclei) was assessed by quantitative Feulgen cytophotometry (Papadimitriou and Cornelisse, 1975). Such large syncytia were specifically selected because their component nuclei generally fail to synthesize DNA (Papadimitriou and Cornelisse, 1975). It was found that the vast majority of nuclei in these large multinucleate giant cells contained the diploid amount of DNA (Papadimitriou and Cornelise, 1975), indicating that the optimal time of macrophage fusion is most likely during the G1 phase of the cell cycle. However, exceptions may occur. Mariano and Spector (1974) reported that, after administering colchicine to animals with foreign body granulomas, macrophages in metaphase seemed to be in the process of incorporating multinucleate giant cells. The possibility of close apposition rather than fusion needs to be excluded in such instances. Moreover, cells arrested in metaphase by colchicine may also suffer membrane lesions and may not be comparable to normal metaphase cells.

2.7. Agents influencing macrophage fusion

The exact physicochemical events culminating in cytoplasmic bridge formation and cell fusion are largely unknown. Fusion of two membranes is likely when the molecular interactions within the membranes become no greater than those across the gap separating the two membranes (Poste, 1970, 1972). Nevertheless, some in vitro experiments have shown that a wide range of factors can influence the fusion process (Poste, 1970, 1972; Poste and Reeve, 1972). The possibility that some of these factors may also operate in vivo has been tested by studying the effect of pharmacologic agents on the fusion of macrophages in a foreign body reaction (Papadimitriou and Sforcina, 1975). The various agents were selected because of their effect on ribonucleic acid (RNA) metabolism, protein metabolism, calcium ions, or membrane-associated enzymes such as adenyl cyclase or ATPase (Table 2). The results should be interpreted with caution, since many other variables may operate in the intact animal as opposed to the in vitro situation.

It appears that the process of monocytic fusion in vivo is to some degree dependent on DNA transcription (Papadimitriou et al., 1973b) since treatment with

TABLE 2
Effects of various agents on murine macrophage fusion in foreign body granulomas in vivo

Treatment	Fusion index[a] Mean ± SD
Controls	
Saline-injected mice	30.9 ± 2.6%
Drugs affecting nucleic acid metabolism	
Actinomycin D	20.6 ± 4.2%
Drugs affecting membrane calcium	
Lignocaine	12.3 ± 1.6%
Tetracaine	32.9 ± 9.1%
Trifluoperazine	19.5 ± 2.1%
Diphenhydramine	12.2 ± 7.7%
EDTA	31.8 ± 2.6%
Calcium gluconate	16.7 ± 8.9%
Verapamil	22.6 ± 5.1%
Drugs affecting the adenyl cyclase system	
Isoproterenol	33.4 ± 3.9%
Insulin	41.5 ± 4.8%
Glucagon	16.0 ± 3.3%
Drugs affecting Na^+/K^+ ATPase	
Ouabain	14.3 ± 5.2%
Lysosomal stabilizers	
Mepyramine maleate	16.6 ± 5.6%
Cortisone	24.5 ± 2.3%

[a] Fusion index of a sample is the percentage of the number of nuclei within syncytia divided by the total number of nuclei of the sample.

actinomycin D inhibits macrophage fusion (Table 2), indicating that RNA synthesis is necessary for the process. No inhibition of fusion was observed after treatment with puromycin, suggesting that protein synthesis is not a critical factor in macrophage fusion (Papadimitriou et al., 1973b).

Poste and his collaborators (1970, 1972; Poste and Reeve, 1972) have suggested that changes in the interaction of calcium ions with the plasma membrane are important in fusion. Some circumstantial support for this hypothesis was obtained in experiments in which pharmacologic agents that displace calcium ions from membranes were found to inhibit virus-induced cell fusion (Poste and Reeve, 1972). Drugs that possess this action include local anesthetics and the phenothiazine tranquilizers (Blaustein and Goldman, 1966; Seeman, 1966; Feinstein et al., 1968; Kwant and Seeman, 1969; Ohki, 1970; Papahadjopoulos, 1970). When tested for their effects on macrophage fusion in vivo, most of these agents were found to inhibit the fusion process (Table 2). Verapamil, which also reduces the capacity of membrane-located sites to accumulate calcium (Nayler and Szeto, 1972), was also found to inhibit macrophage fusion in vivo. Not all of this class of agents resulted in reduced fusion, however, (e.g., tetracaine), but this may have been due to the low dosage used to avoid the generalized toxic effect induced by such drugs. These results nevertheless support the role of calcium in cell fusion as enunciated by Poste and his colleagues (Poste, 1970, 1972; Poste and Reeve, 1972) and indicate that similar mechanisms may well operate in different examples of cell fusion.

In vivo treatment with ethylenediaminetetraacetate (EDTA) did not achieve the low serum levels of calcium (Papadimitriou and Sforcina, 1975) which proved effective in the in vitro model of Poste and Reeve (1972), and it is not surprising that little change in macrophage fusion was detected in foreign-body granulomas. On the other hand, repeated injections of calcium gluconate produced elevated serum concentrations of calcium (Papadimitriou and Sforcina, 1975) and induced a significant reduction in macrophage fusion. This finding was interpreted as resulting from the fact that the increased serum concentration of calcium prevented displacement of calcium from the plasma membrane of macrophages, reducing the rate of fusion. This finding is again similar to the effects of calcium on virus-induced cell fusion described by Poste and Reeve (1972).

From the data in Table 2 it appears that the adenyl cyclase system may also be implicated in macrophage fusion (Papadimitriou and Sforcina, 1975). Repeated injection of insulin, which decreases the level of intracellular cyclic adenosine monophosphate (cAMP) (Robinson et al., 1968), enhanced macrophage fusion in sites of foreign body implantation (Papadimitriou and Sforcina, 1975). On the other hand, glucagon, which, decreases intracytoplasmic cAMP (Rodbell et al., 1971) inhibited the process in the same system (Papadimitriou and Sforcina, 1975). Isoproterenol, however, did not produce any significant effect. This may be due to rapid breakdown of the drug and the absence of a sustained effect or to lack of an appropriate receptor on the surface of the mouse macrophage. Overall, the results suggest that macrophage fusion may be associated with low levels of intracellular cAMP. It is interesting to note, however, that the in vitro fusion of myoblasts into myotubes is also inhibited by agents that increase intracellular cAMP (Wahramann et al., 1973), perhaps reflecting a parallel situation.

Inhibition of the sodium-potassium dependent plasma membrane ATPase by ouabain also resulted in a decrease in macrophage fusion (Papadimitriou and Sforcina, 1975) and Norrby (1966) reported that the ATPase inhibitor, phlorizine, reduced in vitro cell fusion by measles virus.

Cortisone and mepyramine maleate have been shown to inhibit macrophage fusion in vivo (Papadimitriou and Sforcina, 1975). Both substances are known to stabilize lysosomes (Allison, 1967), and lysosomal stabilizers have been shown to inhibit virus-induced cell fusion in vitro (Allison, 1967; Poste, 1970, 1972). Conversely, lysosomal labilizers such as anticellular serum (Ptak et al., 1970), lysolecithin (Poole et al., 1970; Ahkong et al., 1973), and hyperbaric oxygen (Stephenson, 1969) promote cell fusion. Poste (1970, 1972) suggested that the secreted lysosomal hydrolases may reduce the thickness of the surface glycocalyx by enzymatic degradation and thereby enhance fusion. Alternatively, phospholipases in lysosomes may induce increased production of lysolecithin (Smith and Winkler, 1969), which may then favor fusion (Howell and Lucy, 1969; Poole et al., 1970; Ahkong et al., 1973). As indicated earlier (section 2.2.2.), some evidence exists that both these processes may operate in macrophage populations in which multinucleate giant cells are forming.

2.8. Miscellaneous factors influencing macrophage fusion

2.8.1. Phagocytosis

Continued phagocytosis restricts the fusion activity of macrophages. If implanted foreign bodies are coated with carrageenan before implantation, the majority of macrophages perusing the vicinity of the foreign body must phagocytize some of the carrageenan layer before they can attach to the surface of the implant and fusion is significantly reduced under these conditions (Papadimitriou et al., 1973a). Possibly, the internalization of surface membrane occurring during phagocytosis removes some surface receptor or component whose presence is essential for fusion. It would appear, therefore, that monocytes and macrophages at a site of inflammation are more likely to fuse with each other if they are not involved in active phagocytosis.

2.8.2. Thymic factors

Bennett and Montes (1973) have suggested that the thymus produces precursors of both epithelioid and multinucleate giant cells. Their hypothesis was based on the observation of increased numbers of epithelioid and multinucleate giant cells in the lymphoid tissues of irradiated F_1 hybrid mice grafted with parental strain thymic cells during lethal graft-versus-host reactions. In some experiments the thymic cells injected into the irradiated F_1 hybrid mice carried the T6 chromosome and permitted the identification of the donor cells. In such instances chromosomal analysis showed that many of the "epithelioid" cells in the lymphoid tissues of the hybrid host were of donor origin; this was not proven for multinucleate giant cells (Bennett and Montes, 1973).

Multinucleate giant cells have been described in granulomatous lesions in athymic children (Di George, 1968), in sarcoidosis (Mitchell et al., 1968), a disorder associated with a selective T-cell dysfunction, and in lethally irradiated thymectomized mice reconstituted with isologous bone marrow in which foreign-body granulomas were produced by the subcutaneous implants of cellophane (Papadimitriou and Spector, 1971). Moreover, a quantitative assessment of multinucleate giant cell formation in foreign-body granulomas in athymic nude mice (Papadimitriou, 1976) showed that the proportion of the total nuclei found within syncytia was the same in both athymic and normal mice, though the mononuclear population at the injury site was reduced in the athymic mice. These data indicate that the thymus does not influence the efficiency of macrophage fusion, nor are the multinucleate giant cells in this type of lesion composed of an actual thymic precursor or a precursor which is under thymic control.

2.8.3. Specific macrophage antisera

Ptak and his colleagues (1970) tested the effect of specific macrophage antisera on the behavior of hamster macrophages grown in vitro. They discovered that the treated macrophages became spherical, displayed reduced phagocytosis, agglutinated with each other, and formed small multinucleate giant cells. After 24 hours of treatment 96% of cells had agglomerated. Antilymphocytic serum, on the other hand, had no effect. The authors postulated that the specific macrophage antiserum produced changes in the macrophage plasma membrane that rendered it more susceptible to fusion, but detailed information is not available.

2.8.4. The role of lymphokines

Gallindo (1972) observed that cultures of alveolar macrophages from the lungs of rabbits previously sensitized with intravenous injections of heat-killed BCG in oil could be induced to fuse and form multinucleate giant cells by adding specific antigen to the culture fluid. Later, it was reported (Gallindo et al., 1974; Parks and Weiser, 1975) that the supernatant fluids from cultures of antigen-stimulated lymph nodes taken from these immunized animals contained a factor(s) which promoted the fusion of normal rabbit alveolar macrophages. This factor was regarded as a lymphokine and was termed "macrophage fusion factor." It was suggested (Parks and Weiser, 1975) that this lymphokine may act most effectively upon more "mature" macrophages. The factor in these experiments may be related to the one described by Godal and colleagues, (1971) in the supernatant fluids obtained from mixed rabbit leukocyte cultures, which also induced monocytes to form multinucleate giant cells in vitro.

The macrophage fusion factor has proved to be a nondialyzable substance, resistant to heating at 80 °C for 30 minutes (Gallindo et al., 1974), suggesting that it is probably different from the macrophage migration inhibition factor (Davis, 1966; Bloom, 1971). Using Sephadex filtration, its molecular weight was shown to be approximately 60,000 (Gallindo et al., 1974), but at present its relationship to other lymphokines is still unknown. If it is a true lymphokine, it is difficult to explain the ready formation of multinucleate giant cells in athymic mice or in children with congenital absence of the thymus (see section 2.8.2) unless the substance is also produced by B lymphocytes. Alternately, it may be that more than one mechanism facilitates the fusion of mononuclear phagocytes.

3. Properties of macrophage polykaryocytes induced in vivo

3.1. Structural characteristics

Cytological descriptions of the structure of multinucleate giant cells in inflammatory reactions have been documented since the early descriptions of Virchow (1858) and Langhans (1868). More recently, Sutton and Weiss (1966) described in detail the ultrastructure of these cells in cultures of avian monocytes, while Dumont and Sheldon (1965) briefly described the architecture of these cells in experimentally-produced tuberculous granulomas of hamsters. Comoglio and his colleagues (1971) studied these syncytia in cultures of chick embryo spleen, and since then their electron microscopic appearances in murine foreign-body reactions have been reported (Papadimitriou, 1973; Papadimitriou, 1974; Papadimitriou and Archer, 1974; Mariano and Spector, 1974). In addition, some investigators have examined the multinucleate giant cells in human tuberculosis and sarcoidosis (Gusek, 1964; Jones-Williams et al., 1970), while Black and Epstein (1974) examined at length the granulomatous lesion in subjects with beryllium and zirconium hypersensitivity.

Although the size and shape of these polykaryocytes varies, the ratio of nuclear to cytoplasmic cross-sectional area is similar to that of mononuclear phagocytes in the same inflammatory site. A crescentic array of nuclei is more frequent if less than 20 nuclei are present in the syncytium. Scanning electron microscopy shows that the multinucleate giant cell is composed of a central elevated mass with irregular surface contours surrounded by one or more relatively smooth-surfaced lamellipodia from which fine filopodia project (Fig. 3). Transmission electron micrographs show the nuclear array dividing the cytoplasm into two regions (Fig. 4). The common centrospheric region enclosed by the majority of nuclei contain most of the mitochondria and lysosomes together with Golgi apparatus, centrioles, vacuoles, and vesicles of varying size. Bundles of filaments 6 to 7 nm in diameter traverse the areas between the organelles. The cytoplasm surrounding the nuclei contains some mitochondria and lysosomes and most of the ergastoplasmic cisternae, ribosomes, lipid globules, glycogen granules, and is also criss-crossed by bundles of filaments. This zone of cytoplasm is dissected by a labyrinthine system of smooth membranes arranged in vesicles, vacuoles, and lamellae. A fine network of 4 to 5 nm thick microfilaments is present at the extreme periphery. The plasma membrane is thrown into folds, flaps, and microvilli, and possesses a thin glycocalyx.

3.2. Cytochemical and biochemical properties

Cytochemical investigations have revealed that multinucleate giant cells seen in inflammatory reactions contain readily detectable activities for a variety of enzymes (Table 3) including those involved in the pentose pathway, glycolysis, and the tricarboxylic acid cycle, and those associated with lysosomes, mitochondria, the Golgi complex, and the plasma membrane (Hodel, 1967; Salthouse and Williams, 1969; Schmalzl and Braunsteiner, 1970; Comoglio et al., 1971; Mariano and Spector, 1974; Papadimitriou and Wyche, 1974). Overall there is a very close resemblance between the cytochemical characteristics of multinucleate giant cells and those of macrophages. The only minor difference is that levels of 5' nucleotidase activity are generally lower in multinucleate giant cells than in mononuclear phagocytes. Although all multinucleate giant cells contained the various enzymatic activities indicated in Table 3, some degree of variation in the

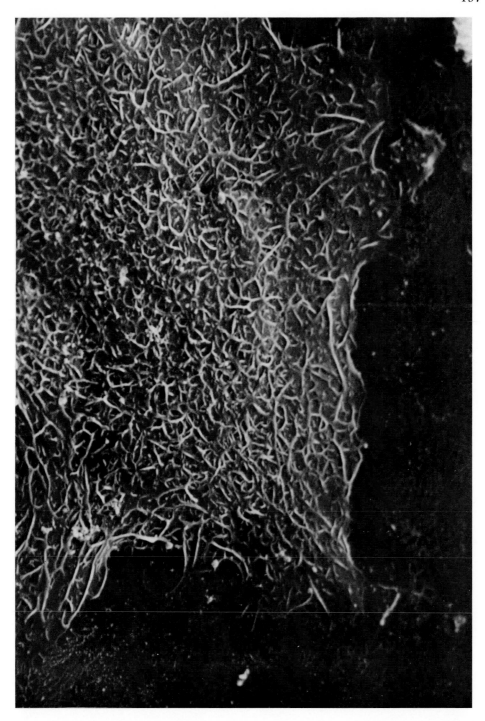

Fig. 3. Scanning electron micrograph of a multinucleate giant cell. A complicated pattern of branching folds is present on the surface. The periphery possesses smoother contours and small filopodia protrude onto the substratum. ×3,200.

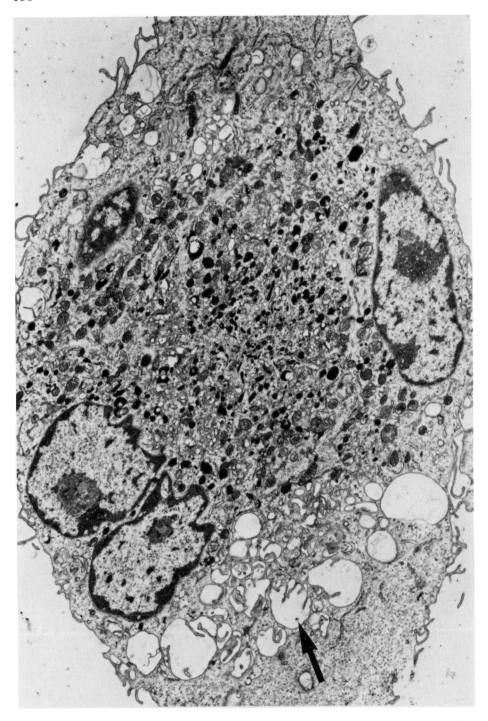

Fig. 4. Transmission electron micrograph of a multinucleate giant cell. The centrosphere of the polykaryocyte contains the majority of the cytoplasmic organelles while the nuclei are situated between it and the peripheral zone of the syncytium. Portions of the vacuolar labyrinth can be seen (arrow). ×7,200.

TABLE 3

Cytochemical enzymatic activities in mouse multinucleate giant cells and macrophages in foreign body granulomas

Enzyme	Intensity of Reaction[a]	
	Multinucleate giant cells	Macrophages
Enzymes of glycolysis		
Lactate dehydrogenase (EC I.1.1.27)	++++	++++
Glycerophosphate dehydrogenase (EC I.1.2.1.)	+++	+++
Glucose-6-phosphatase (EC 3.1.3.9)	+	+
Enzymes of tricarboxylic acid cycle		
Succinate dehydrogenase (EC I.3.99.I)	+++	+++
Isocitrate dehydrogenase (EC.I.1.1.41)	++	++
Malate dehydrogenase (EC I.1.1.37)	++	++
Enzymes of the pentose pathway		
Glucose-6-phosphate dehydrogenase (EC I.1.1.49)	+++	+++
6-Phosphogluconate dehydrogenase (EC I.1.1.44)	++	++
Mitochondrial enzymes		
Cytochrome oxidase (EC I.9.3.I)	+++	+++
3-Hydroxybutyrate dehydrogenase (EC I.1.1.30)	++	++
Oxidoreductases partly associated with mitochondria		
NADH dehydrogenase (EC I.6.2.I)	++++	++++
NADPH dehydrogenase (EC 1.6.2.3. or I.6.99.I)	++++	++++
Alcohol dehydrogenase (EC I.1.1.I.)	+	+
Acid hydrolases		
Acid phosphatase (EC 3.1.3.2.)	++++	++++
Leucine aminopeptidase (EC 3.4.1.I.)	+++	+++
β-Glucuronidase (EC 3.2.1.3)	++	++
Other hydrolases		
Nonspecific esterases: without NaF	+++	+++
Nonspecific esterases: with NaF	—	—
Alkaline phosphatase (EC 3.1.3.I.)	—	—
Thiamine pyrophosphatase	++	++
Adenosine triphosphatase (EC 3.6.1.4)	+++	+++
Peroxidases		
Peroxidase (EC I.II.1.7)	—	—
Transaminases		
Glutamate oxaloacetate transaminase (EC 2.6.1.I.)	+++	+++
Enzymes associated with the cell membrane		
5'-Nucleotidase (EC 3.1.3.5)	++	++++
Sodium-potassium dependent ATPase (EC 3.2.6.I.-)	++	++

[a] Activities are expressed in arbitrary units ranging from + to ++++

concentration of reaction product was observed in syncytia of similar size. It is unclear at this stage whether this represents different stages of functional activity of if it reflects some degenerative change in the polykaryocyte.

Biochemical assays of the enzyme content of cell populations containing multinucleate giant cells show that quantitative differences exist between them and peritoneal macrophages (Tables 3, 4). Populations in which multinucleate giant cells are present also contain macrophages and epithelioid cells, but at the time of collection (2 weeks after implantation) 50% of nuclei on the implant were within syncytial units (Papadimitriou et al., 1973b). Such cell populations contain comparatively lower concentrations of 5' nucleotidase, succinate dehydrogenase, lactate dehydrogenase and aryl hydroxylase, while levels of β glucuronidase and free RNase II are higher than those in peritoneal macrophages (Papadimitriou and Wyche, 1976). LDH 5 was the isoenzyme of lactate dehydrogenase present in the strongest concentrations. This isoenzyme is commonly found in cells that engage in bursts of anaerobic glycolysis (White et al., 1968).

In addition, autoradiographic studies using cells labeled with tritiated thymidine (Ryan and Spector, 1970; Mariano and Spector, 1974; Dreher et al., 1975; Papadimitriou and Cornelisse, 1975), and cytophotometric studies of Feulgen-stained preparations (Papadimitriou and Cornelisse, 1975) have shown that the component nuclei of multinucleate giant cells containing less than 10 nuclei can synthesize DNA and may even successfully divide and produce polyploid and aneuploid daughter cells. DNA synthesis in the component nuclei of the syncytium is generally synchronous, but asynchronous synthesis was also detected in a minority of polykaryons (Mariano and Spector, 1974; Dreher et al., 1975; Papadimitriou and Cornelisse, 1975). Very large polykaryons (with more than 40 component nuclei) did not show evidence of DNA synthesis (Papadimitriou and Cornelisse, 1975).

Similar autoradiographic studies with radioactive uridine demonstrated the ready uptake of the label mainly by the nucleoli and, to a lesser degree, the cyto-

TABLE 4

Intracellular concentrations of some enzymes in murine peritoneal macrophages and in cell populations containing multinucleate giant cell

Enzyme	Peritoneal macrophages	Cell populations containing multinucleate giant cells	Biochemical units
β-Glucuronidase	62.8 ± 5.1	223.9 ± 5.9	μg phenolphalein/mg sol cytoplasmic protein/hr
RNase II (free)	11.1 ± 3.2	23.5 ± 7.2	10^3 units/mg sol protein/hr
Acid phosphatase	41.1 ± 8.0	48.9 ± 5.1	μg naphthol/mg sol. protein/ 30 min
Aryl hydroxylase	22.3 ± 3.6	16.1 ± 2.1	$\mu\mu$M hydroxylated benzo-α-pyrene/mg sol protein/30 min.
Lactate dehydrogenase	336 ± 47	242 ± 28	μM NAD/mg sol protein/min
Succinate dehydrogenase	1.45 ± 0.08	0.85 ± 0.06	μg indophenol/mg sol protein/ min
5'-Nucleotidase	10.2 ± 3.6	3.3 ± 1.8	μM AMP/mg sol protein/min
Glucose-6-phosphate dehydrogenase	2.57 ± 1.1	2.24 ± 0.58	μM NADP/mg sol protein/min

TABLE 5

Concentration of various neutral lipids in murine peritoneal macrophages and cell populations containing multinucleate giant cells

Lipid class	Cell populations containing multinucleate giant cells (μg/mg soluble protein)	Peritoneal macrophages (μg/mg soluble protein)
Free fatty acids	97.3 ± 20.6	36.1 ± 9.8
Cholesterol	52.7 ± 12.2	24.8 ± 7.3
Cholesterol esters	15.1 ± 7.8	63.2 ± 6.5

plasm of these cells (Mariano and Spector, 1974). These results indicate that RNA synthesis occurs and appears to be synchronous within the component nuclei of the polykaryocyte.

As discussed earlier (section 2.1.2), differences in phospholipid concentration exist between cell populations containing multinucleate giant cells and those consisting only of peritoneal macrophages (Table 1) and greater concentrations of lysolecithin, phosphatidylethanolamine, lecithin and, to a lesser degree sphingomyelin, were found in the former. In addition, differences in neutral lipids were detected; cell populations rich in multinucleate giant cells possess greater concentrations of free fatty acids and cholesterol when compared with peritoneal macrophages, while the concentration of cholesterol esters is reduced in comparison (Table 5). These findings may reflect differences in lipid metabolism between the two cell populations. Werb and Cohn (1971) have reported the absence of cholesterol esters in cultured mouse peritoneal macrophages, while in our studies (Papadimitriou and Wyche, 1974, 1976) these lipids were detected in freshly harvested peritoneal macrophages and in cell populations rich in multinucleate giant cells (although the latter contained less than the former). Possibly, in vivo conditions influence lipid metabolism in these cells. The greater concentration of phospholipids in cell populations rich in multinucleate giant cells may be the result of their greater mass of endomembranes (Papadimitriou and Archer, 1974).

3.3. Chromosomal studies

Karyotypic analysis of polyploid and hyperdiploid cells in foreign-body granulomas has revealed chromosomal abnormalities in approximately 25% of the metaphase profiles examined (Mariano and Spector, 1974). Gaps, breaks, rings, and premature chromosome contraction were the commonest changes found. The abnormalities in the hyperdiploid cells reflect the type of chromosomal lesion which can occur in multinucleate giant cells reaching metaphase. Abnormal mitoses were also detected among dividing mononuclears in the same inflammatory lesion. Perhaps this increased frequency of chromosomal abnormality in mononuclear phagocytes and their derivatives may be related to their high content of lysosomal hydrolases (Allison and Paton, 1965).

3.4. The lifespan of multinucleate giant cells

The lifespan of multinucleate giant cells is only 4 to 5 days. This can be demonstrated by transplanting foreign bodies with attached multinucleate giant cells into syngeneic whole-body irradiated recipients (Papadimitriou et al., 1973b). Alternatively, animals in which foreign bodies (e.g., glass coverslips; sheets of melenex) have been implanted can be irradiated while the site of the implant is protected. The results in both instances are similar. Ten days after transplantation few multinucleate giant cells were present in the implant. The largest of the syncytia were the first to be lost, while smaller polykaryocytes possessing 2 to 4 nuclei survived longer. Essentially similar results were obtained by Mariano and Spector (1974), who enclosed freshly explanted foreign bodies and the attached multinucleate giant cells within millipore chambers and reimplanted them into syngeneic recipients.

3.5. The motility of multinucleate giant cells

Cinemicrographic observations of multinucleate giant cells in foreign-body granulomas have shown that these cells are motile, although the degree of translocational movement displayed decreases with increasing size and increasing nuclear content (Papadimitriou and Kingston, 1976). Generally, polykaryocytes are slower than single unfused macrophages, and exhibit speeds of translocation varying from 0.3 to 1.0 μm/min in small syncytia (containing 2–4 nuclei) to 0.04 μm/min with large polykaryocytes (containing more than 15 nuclei). The direction of translocation of small syncytia correlates with the direction of the single, leading lamellipodium. Somewhat larger cells (containing 5–15 nuclei) display loss of directional polarity and possess many lamellipodia, all showing ruffling activity but none of which is dominant. The largest syncytia (possessing more than 20 nuclei), which also show little translocational movement, possess a single large circumferential lamellipodium that extends and retracts rhythmically but induces little positional displacement (Papadimitriou and Kingston, 1976).

Formation and retraction of these very large lamellipodia requires a large reserve of cell membrane material, and Komnick and associates (1972) have suggested that this can be provided by the successive and partial unfolding of cell membrane in regions with a large surface area. Such a source of membrane material in these large polykaryocytes may be provided by the extensive, intracytoplasmic membranous labyrinth which appears to be continuous with the plasma membrane and dissects and traverses the perinuclear zone of the syncytium (Papadimitriou and Archer, 1974). This large reserve of membrane may be utilized during locomotion. On the other hand, it is unlikely that significant amounts of membrane material can be so rapidly synthesized/destroyed to seriously affect cellular movement and, Wolpert and Gingell (1968) have actually produced evidence that the cell membrane is essentially conserved during cell movement.

Differences in contact inhibition of locomotion have also been observed

(Papadimitriou and Kingston, 1976). Generally, contacts between macrophages or small polykaryocytes resulted in prominent contact inhibition, but this was reduced between large polykaryocytes and between large and small polykaryocytes or macrophages. The latter type of behavior would measurably encourage and facilitate surface exploration of adjacent cells and might thus favor the establishment of the cell contact interactions required for fusion.

The energy for locomotion appears to be derived mainly from glycolysis (Papadimitriou and Kingston, 1976), since movement is rapidly inhibited by NaF (but not by NaCN, which interferes with oxidative phosphorylation). Treatment with ethylene glucosamine tetracetate (EGTA), which reduces the concentration of free calcium, also resulted in reduced motility. Similarly, pharmacologic doses of procaine, which has been said to inhibit the release of intracellular calcium into the cytosol (Feinstein and Paimre, 1969), and papaverine, which has been reported to prevent entry of extracellular calcium (Ash et al., 1973), also inhibit the motility of polykaryocytes. Since actin-like proteins have been demonstrated in macrophages (Allison et al., 1971), the available data suggest that a calcium-dependent contractile system of the actomyosin type, similar to that proposed for other nonmyogenic cells (Adelstein and Conti, 1972; Goldman and Knipe, 1972), may also operate in multinucleate giant cells.

3.6. Endocytotic processes

The phagocytic performance of murine multinucleate giant cells of foreign-body granulomas, when tested with suspensions of staphylococci, yeasts, and sheep erythrocytes (treated with glutaraldehyde or with isologous or heterologous antiserum), has been found to be reduced in comparison with mononuclear phagocytes (Papadimitriou et al., 1975). Polykaryocytes containing more than seven nuclei rarely phagocytized yeasts or staphylococci, and similarly the ingestion of sheep erythrocytes treated with heterologous antiserum was also infrequent. Sheep erythrocytes treated with isologous antiserum or glutaraldehyde frequently attached to multinucleate giant cells and some were phagocytized (Figs. 5 and 6). However, when the number of adhering erythrocytes was corrected for the increased size of the polykaryocyte by dividing the number of attached erythrocytes by the number of nuclei in the syncytium, a progressive reduction in phagocytic performance as the nuclear content increased was again observed (Papadimitriou et al., 1975).

These phenomena may be due to the loss of surface receptors involved in binding material to be phagocytosed or to a reduction in the rate of synthesis of the receptor material following fusion. Since fewer sheep erythrocytes treated with either heterologous or isologous antisera attach to the polykaryocytes than would be expected by their size, it is likely that at least the Fc receptor sites are lost or masked. Direct visualization of the Fc receptor for heterologous ferritin-labeled IgG has revealed that the random aggregates of receptor sites were smaller and fewer on multinucleate giant cells than on mononuclear phagocytes (Papadimitriou, 1973), again indicating loss or masking of receptor molecules.

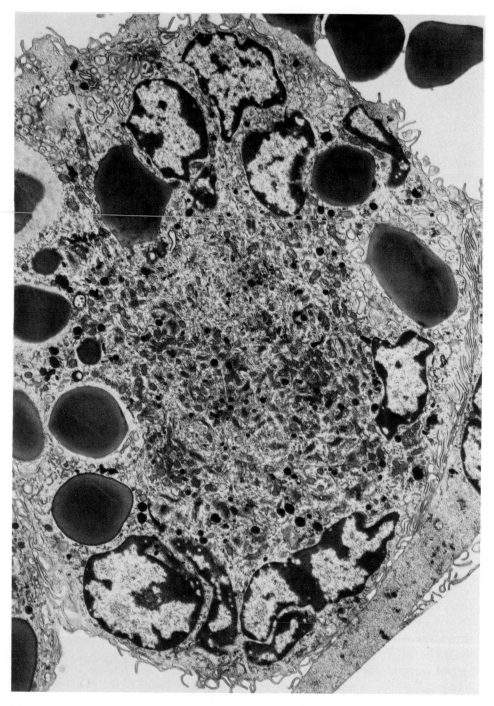

Fig. 5. A multinucleate giant cell which has ingested a number of glutaraldehyde-treated sheep erythrocytes. ×6,000.

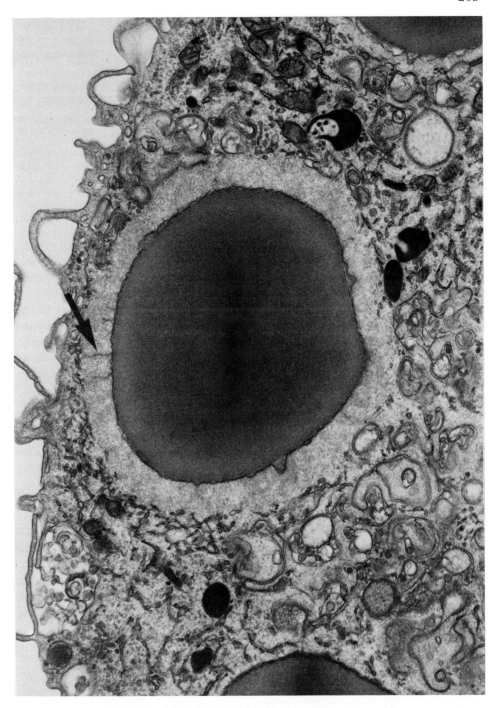

Fig. 6. Electron micrograph showing glutaraldehyde-treated sheep erythrocyte within a phagocytic vacuole in the cytoplasm of a multinucleate giant cell. The phagosome is surrounded by a layer of fine filaments, which is traversed by small tubular channels (arrow). ×23,700.

Since no increase in the number of attached sheep erythrocytes treated with heterologous antiserum occurred after trypsinization of multinucleate giant cells, it is likely that loss rather than masking of the receptor molecules has occurred.

Pinocytosis and micropinocytosis are readily seen in multinucleate giant cells. Colloidal gold and thorotrast are quickly ingested and eventually stored within secondary lysosomes (Fig. 7). At this stage it is not known if the efficiency of these processes is equivalent to those exhibited by macrophages.

4. Macrophage fusion in vitro

4.1. Introduction

Lambert (1912), Weil (1913), Awrorow and Tunofejewski (1914), the Lewises (1925, 1926, 1927), Cohen (1926), and Maximow (1925), observed fusion between macrophages in tissue cultures. Fusion was reported between the large mononuclear phagocytes in the blood of teleosts, elasmobranches, amphibia, reptiles, birds, and mammals. Generally, the cells of amphibia, reptiles, fish, and birds, displayed a greater frequency of fusion in culture than mammalian macrophages. More recently, Goldstein (1954) noted the formation of multinucleate giant cells from human monocytes cultivated on cellophane. Sutton and Weiss (1966) described the ultrastructure of multinucleate giant cells produced in cultures of avian monocytes, and Comoglio and colleagues (1971) studied the formation of multinucleate giant cells in cultures of chick spleen and claimed that they may form either from macrophages or from "reticular" cells. These workers also concluded that the presence of high molecular weight organic compounds was essential for multinucleate giant cell formation. Fusion among chicken macrophages was also accelerated if the cells were maintained in an acidic medium (Franklin, 1958).

Smith and Goldman (1971) have induced multinucleate giant cell formation by treating cultures of macrophages from human colostrum with phytohemagglutinin or concanavalin A. This effect was inhibited by addition of α-methyl-D-mannoside and α-methyl-D-glucoside. In contrast, mouse peritoneal macrophages did not form multinucleate giant cells in response to treatment with these phytomitogens. These results indicate that in some instances agglutination by phytomitogens results in suitable conditions for macrophage fusion but differences exist in the interaction of human colostral macrophages and mouse peritoneal macrophage with these plant agglutinins.

In the last few years a variety of methods have become available for producing somatic cell fusion in vitro. These include inactivated Sendai virus (Okada, 1958; Harris et al., 1966), phospholipase C (Sabban and Loyter, 1974), lysolecithin (Croce et al., 1971), calcium ions at high pH (Peretz et al., 1974), microsurgery (Diacumakos and Tatum, 1972), and high temperature (Ahkong et al., 1973; chapter 6 this volume). The exact mode of action of these agents is still un-

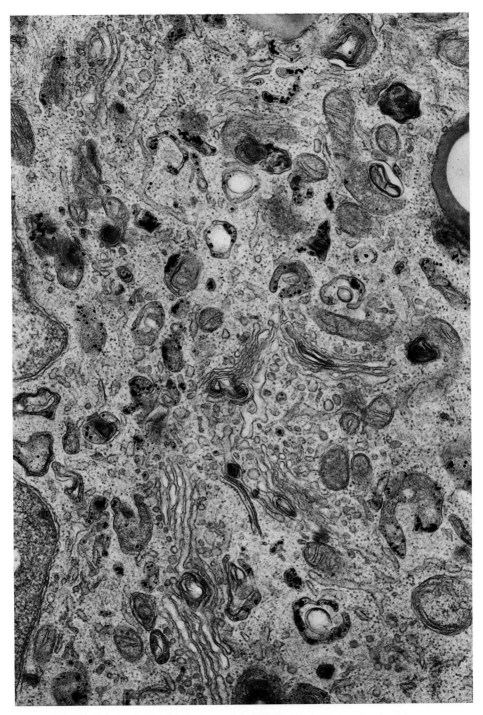

Fig. 7. Electron micrograph showing part of the cytoplasm of a multinucleate giant cell to which colloidal gold was presented. The latter is seen within a variety of vesicles, vacuoles, and secondary lysosomes. ×36,000.

known and has been the subject of recent reviews (Poste, 1970, 1972; Gordon, 1975). Some of these techniques have been used to produce macrophage homokaryocytes and heterokaryocytes (Harris, 1966; Harris et al., 1969; Gordon and Cohn, 1970, 1971a, b, c).

4.2. Macrophage homokaryocytes and heterokaryocytes

4.2.1. Morphology

High yields of macrophage homokaryocytes can be obtained by treatment with UV-irradiated Sendai virus (Harris, 1966; Harris et al., 1969; Gordon and Cohn, 1970). After earlier cultivation, up to 50% can be fused into multinucleate giant cells by this technique. Initially, the nuclei and lipid droplets of such newly formed macrophage homokaryocytes are randomly distributed and a distinct centrosphere is not apparent. One to five hours after fusion the nuclei arrange themselves around a common centrosphere, while the lipid droplets are found peripheral to the annulus of nuclei. This organized pattern is lost after treatment with 10 μg/ml colcemid, and nuclei and lipid globules become randomly scattered throughout the syncytium.

4.2.2. Nucleic acid synthesis

The ability of macrophages to ingest antibody-coated cells, the presence of a highly active ATPase (Cohn and Benson, 1965; Rabinovitch, 1969), the synthesis of significant amounts of acid hydrolases (Cohn and Benson, 1965), and a block at the G_0 phase of the cell cycle (Virolainen and Defendi, 1967; Cohn, 1968; Epifanova and Terskikh, 1969), have made the macrophage a useful tool in the elucidation of nuclear-cytoplasmic relationships in experimentally induced models of cellular fusion (Harris, 1966; Harris et al., 1969; Gordon and Cohn, 1970, 1971a, b, c). Macrophages in culture do not synthesize DNA and appear to be locked in the G_0 phase of the cell cycle (Virolainen and Defendi, 1967; Cohn, 1968; Epifanova and Terskikh, 1969). Virus-induced fusion and formation of rabbit or mouse macrophage homokaryocytes does not modify DNA synthesis in the least (Harris, 1966; Gordon and Cohn, 1970), but DNA synthesis can be induced if heterokaryons are produced in which rapidly proliferating cells are fused with macrophages (Harris, 1966; Harris et al., 1969; Gordon and Cohn, 1970). In this instance the functions of the proliferating cell, irrespective of the number of macrophage nuclei in the syncytium, appear to predominate and DNA synthesis can be initiated in the macrophage nuclei. Rabbit macrophage nuclei begin to synthesize DNA after fusion with HeLa cells (Harris, 1966; Harris et al., 1969), while mouse macrophage nuclei after fusion with mouse melanoma cells initiate both DNA and RNA synthesis (Gordon and Cohn, 1970; 1971a, b), which often proves to be synchronous in all macrophage nuclei (Gordon and Cohn, 1971a). In such heterokaryons the melanocyte nuclei can easily be distinguished, since they are larger, oval-shaped, and contain prominent nucleoli, while the macrophage nuclei are smaller, often bilobed, and possess small nucleoli. Within one hour after fusion, however, the macrophage nuclei swell and their nucleoli become prominent. After several days of cultivation the heterokaryons resemble melanocytes in morphology. The refractile droplets of macrophage origin eventually disappear from the heterokaryons, whereas macrophage homokaryocytes progressively accumulate more. Mouse macrophage-melanoma heterokaryocytes eventually enter mitosis on the second day after fusion, but very few of their progeny progress to a second division, and a third division is rare (Gordon and Cohn, 1970).

Studies with mouse macrophage-melanoma heterokaryocytes have shown that RNA synthesis by heterokaryocytes was essential for subsequent DNA synthesis. Moreover, melanocyte-directed RNA synthesis alone proved adequate in stimulating macrophage DNA synthesis (Gordon and Cohn, 1971a, b). Similar experiments have shown that melanoma proteins also play an important role in the induction of macrophage DNA synthesis (Gordon and Cohn, 1971b, c).

The relationship between the melanoma-cell cycle and macrophage DNA synthesis has been investigated by fusing macrophages with synchronized melanoma cells. If the melanoma cells were in the S phase at the time of fusion, macrophage DNA synthesis occurred 2 hours later, but if the melanoma cells were in G_1 macrophage DNA synthesis was delayed until the melanoma nuclei next entered the S plane (Gordon and Cohn, 1971a, b). These findings agree with observations with other heterokaryons, which have shown that the S phase predominates over G_0 (Rao and Johnson, 1970).

Experiments using actinomycin and cycloheximide have shown that the RNA and proteins required for initiating DNA synthesis in macrophage nuclei in melanoma cell—macrophage heterokaryons were produced during late G_1 as well as the S phase of the melanoma cells. The reason for the 2-hour delay before DNA synthesis begins in the macrophage is unclear. It may be related to the transport of melanoma cell products into the macrophage nucleus or to the physical changes in DNA associated with nuclear swelling, and would thus be related to the heterochromatinic content of the macrophage nucleus (Gordon and Cohn, 1970, 1971a, b). These experiments have provided a useful insight into the mechanisms that control DNA synthesis, not only in the macrophage nucleus but in cell nuclei generally.

4.2.3. Endocytosis and surface characteristics

Macrophage homokaryons remain viable for many days after fusion (Gordon and Cohn, 1970). They continue to phagocytize actively (Gordon and Cohn, 1970), and are richer in surface ATPase and acid phosphatase than their mononuclear precursors (Gordon and Cohn, 1970). On the other hand, heterokaryons produced by fusion of mouse macrophages and mouse melanoma cells display loss of surface ATPase activity, the loss of enzymatic activity being related to the ratio of the nuclei from the two donor cell types (Gordon and Cohn, 1970). Moreover, the enzymatic activity when demonstrable is somewhat patchy in its localization. This loss of activity may be due to the internalization of plasma membrane components or membrane segments without further replacement or to a dilution process resulting from the mixing of the surface components of the two donor cell types, as suggested for other heterokaryons

(Watkins and Grace, 1967; Harris et al., 1969).

When macrophage homokaryons are presented with sheep red blood cells coated with 7s antibody, there is a direct relationship between the number of macrophage nuclei present and the mean number of sheep erythrocytes ingested per cell (Gordon and Cohn, 1970). Heterokaryons formed by fusion of mouse macrophages and mouse melanoma cells initially ingested as many sheep red blood cells as their controls, but this ability disappeared within 12 to 24 hours after fusion and did not reappear (Gordon and Cohn, 1970, 1971c). The number of macrophage nuclei in the heterokaryons influenced the rate of loss of the phagocytic performance. Moreover, this loss of phagocytic function was accelerated in the presence of a high concentration of calf serum that has been shown to increase the rate of pinocytosis (Gordon and Cohn, 1970). The loss of phagocytic activity was virtually complete before the heterokaryons reached their first mitotic division, and none of the daughter cells recovered this property during the subsequent observation period (Gordon and Cohn, 1970).

It appears therefore that macrophage homokaryons are even more actively phagocytic than their mononuclear precursors, whereas fusion with non-phagocytic cells results in a reduced phagocytic performance. The experiments outlined above indicate that alteration in the cell surface properties of the mouse macrophage-mouse melanoma cell heterokaryons also occurs, resulting in loss or masking of the Fc receptor (Gordon and Cohn, 1971c). The latter mechanism seems more likely, since treatment of macrophage-melanoma heterokaryons with trypsin reexposes the masked Fc receptors (Gordon and Cohn, 1971c). Moreover, inhibition of protein synthesis in heterokaryons preserves phagocytic activity presumably by preventing masking of the macrophage receptors. Similarly, inhibition of melanoma RNA synthesis before fusion also blocks the subsequent masking (Gordon and Cohn, 1971c). Failure of the synthesis of the Fc receptor may also be a contributing factor (Gordon, 1974). Finally, different cells vary in their ability to mask the macrophage phagocytic receptor after fusion with macrophages; Ehrlich ascites tumor cells mask the Fc receptor rapidly, primary chick fibroblasts minimally, and embryonic chick erythrocytes not at all (Gordon and Cohn, 1971c; Gordon, 1974).

5. A comparison of in vivo- and in vitro-induced macrophage polykaryocytes

Generally the efficiency of macrophage fusion in both in vivo and in vitro models appears similar, although different means of inducing fusion are involved. Examination of plastic sheet surfaces two weeks after their implantation into the subcutaneous tissue of mice revealed that 50% of all nuclei were within multinucleate cells (Papadimitriou et al., 1973b). Similarly Gordon and Cohn (1971a,b), using Sendai virus, report that 50% of cultured macrophages can be induced to fuse into syncytia varying in size and nuclear content.

Several major differences exist, however, between mouse macrophage homokaryocytes produced in vitro by fusion with inactivated Sendai virus and the multinucleate giant cells which form at the sites of foreign-body reactions in the mouse. First, in vitro-induced macrophage polykaryocytes remain viable for many days (Gordon and Cohn, 1970) while the multinucleate giant cells in murine foreign-body reactions are short-lived, surviving only 4 to 5 days (Papadimitriou et al., 1973b; Mariano and Spector, 1974). Second, virus-induced macrophage polykaryocytes are actively phagocytic (Gordon and Cohn, 1970), while multinucleate giant cells formed in vivo display a progressive loss of phagocytic performance as the number of nuclei in the syncytium increases (Padadimitriou et al., 1975). This reduction of phagocytic performance is accompanied by a simultaneous loss of the Fc receptor on the cell surface (Papadimitriou, 1973; Papadimitriou et al., 1975). Third, small foreign-body multinucleate giant cells are able to synthesize DNA (Mariano and Spector, 1974; Papadimitriou and Cornelisse, 1975) and produce polyploid daughter cells, while macrophage polykaryons produced in vitro do not synthesize DNA. Finally, surface ATPase activity is decreased in multinucleate giant cells (Papadimitriou and Wyche, 1974, 1976), while that in macrophage polykaryons is, if anything, increased (Gordon and Cohn, 1970).

Strangely, the properties of the multinucleate giant cells present in foreign-body reactions are similar to those of mouse macrophage-mouse melanoma cell heterokaryons produced in vitro. The latter syncytia display nuclear DNA synthesis, reduced phagocytic performance, masking of the Fc receptor sites, and decreased surface ATPase activity (Gordon and Cohn, 1970; 1971a, b, c). These similiarities tempt one to suggest that even though circulating monocytes are the major precursors of the multinucleate giant cells of foreign-body reactions they are not necessarily the sole contributors and, possibly, that another non-macrophage cell type donates to the syncytium. In other words, the multinucleate giant cell seen in pathologic reactions may well be a heterokaryon rather than a homokaryon.

At present the identity of this other postulated donor cell is unknown. Judging from the results obtained from virus-induced macrophage heterokaryons (Gordon and Cohn, 1970; 1971a, b, c), it is likely to be a replicating, nonphagocytic cell, with few surface Fc receptor sites and little surface ATPase activity. It is unlikely to be an activated T lymphocyte since multinucleate giant cells form readily in nude athymic mice. Other lymphocytic or connective tissue cells that may also be present in granulomatous inflammatory sites are other possibilities, but there is little evidence to incriminate any of these. Comoglio and co-workers, (1971) have suggested that "reticular" cells may contribute in some instances to the formation of the syncytia, while Mariano and Spector (1974) have proposed that multinucleate giant cells seen at sites of inflammation are formed from the fusion of relatively old macrophages with freshly emigrated monocytes. This latter hypothesis was based on the observation that macrophages enclosed in diffusion chambers and implanted in syngeneic recipients failed to form polykaryons. When the chambers were perforated to allow entry of fresh cells, polykaryocytes formed in the usual way; when tritiated thymidine was used as a cell marker, it was shown that the newly emigrating monocytes fused with those in the diffusion chamber.

6. The role of multinucleate giant cells in inflammatory reactions

As indicated earlier, multinucleate giant cells are frequently found in sites of chronic inflammation but their exact role in such tissue reactions is unknown. Although they form by fusion of exuded monocytes, they differ from macrophages. As a group they display reduced phagocytic efficiency when compared with macrophages (Papadimitriou et al., 1975), and therefore it is unlikely that in this respect their formation would provide any distinct advantage.

Selective release of two lysosomal hydrolases (acid phosphatase and β-glucuronidase) has been demonstrated in cell populations containing multinucleate giant cells (Papadimitriou and Wee, 1976). In this instance, the efficiency of exocytotic enzyme release by cell populations containing multinucleate giant cells is similar to that of peritoneal macrophages (Papadimitriou and Wee, 1976), and a similar proportion of the intracytoplasmic content of the two acid hydrolases tested is released by both these cell populations. Extracellular release of lysosomal enzymes would generally aid in the extracellular lysis of some large, nonphagocytizable material. Thus the envelopment of large, particulate matter by a sheet of multinucleate giant cells releasing lysosomal contents into the engulfed body is merely a modification on a larger scale of the better understood phagocytic events.

Stimulation of endocytosis by such factors as heat-aggregated immunoglobulin increases the extracellular release of lysosomal enzymes from mononuclear phagocytes and their derivatives in foreign-body granulomas (Papadimitriou and Wee, 1976). Since a mixture of cell types derived mainly from circulating monocytes are present in cell populations containing multinucleate giant cells, it becomes difficult to quantitate the relative contribution of each cell type to the total amounts of released hydrolases. Moreover, it is unclear whether variations exist in the efficiency of exocytosis between polykaryocytes of different nuclear content.

Nonselective release of lysosomal hydrolases may also accompany the death of multinucleate cells, which thus results in the release into the immediate microenvironment of relatively large amounts of various enzymes. Since multinucleate giant cells, especially those with many nuclei, have a relatively short lifespan and survive as distinct units for only 4 to 5 days (Papadimitriou et al., 1973b; Mariano and Spector, 1974), death of multinucleate cells will result in the liberation of a variety of lytic enzymes. Mariano and Spector (1974) have suggested that macrophage polykaryon formation may even be an efficient disposal system for "unwanted" macrophages. Such macrophages are the metabolically effete and/or those with chromosomal abnormalities that might even possess neoplastic potential. This concept, however, does not completely explain their presence in some, but not all inflammatory lesions in which macrophages play a predominant role.

Finally, the large multinucleate cells formed in inflammatory lesions may provide a relatively large, physical, protoplasmic barrier between a nonphagocytizable object and the rest of the organism, minimizing and localizing through its physical barrier any injurious effects that a particulate noxious agent may induce.

7. Summary

The multinucleate giant cells found in a variety of inflammatory reactions are short-lived polykaryocytes formed by the fusion of freshly exuded monocytes. Monocytic fusion is influenced by variations in extracellular calcium concentration, cAMP, RNA, sodium-potassium dependent ATPase, and lysosomal stability. These syncytia possess an organized ultrastructure and cytochemical and biochemical profiles suggestive of a range of metabolic activity, including DNA synthesis. The latter, which is more frequent in small polykaryocytes, results in mitotic division and the formation of polyploid daughter cells, some of which may in turn fuse into syncytial units with exuded monocytes. In addition, chromosomal abnormalities are readily detected in polyploid and hyperdiploid cells. Although multinucleate giant cells are motile, they are less active than mononuclear phagocytes, and the larger syncytia display in addition a reduction of contact inhibition. Their phagocytic performance in comparison with that of their mononuclear precursors is also reduced and is accompanied by the loss of some surface receptors. On the other hand, cell populations containing multinucleate giant cells are able to secrete lysosomal enzymes selectively, aiding the extracellular degradation of noningested material.

Macrophage homokaryons produced in vitro by treatment with inactivated Sendai virus differ slightly from multinucleate giant cells formed in vivo. The former do not synthesize DNA, are actively phagocytic, possess more ATPase activity, and survive longer than the latter. DNA synthesis can be induced in these macrophage nuclei if rapidly proliferating cells are also incorporated with the macrophage nuclei. When murine melanoma cells are used, the resulting heterokaryons display DNA synthesis, reduced phagocytic performance, masked surface receptors, and low surface ATPase levels. These characteristics of the macrophage-melanoma heterokaryons tempt one to suggest that a precursor of the multinucleate giant cells of the inflammatory reaction may be a rapidly proliferating, poorly phagocytic cell with little surface ATPase activity.

Acknowledgments

This work was supported by grants from the National Health and Medical Research Council of Australia. I am grateful to Dr. M. N-I. Walters, Professor of Pathology, University of Western Australia, for his help in the prepartion of the manuscript.

References

Adelstein, R. S. and Conti, M. (1972) The characterization of contractile muscle protein from platelets and fibroblasts. Cold Spring Harb. Symp. Quant. Biol. 37, 599–605.

Ahkong, Q. F., Cramp, F. C., Fisher, D., Howell, J. I., Tampion, W., Verrinder, M. and Lucy, J. A. (1973) Chemically-induced and thermally-induced cell fusion : lipid interations. Nature New Biol. 242, 215–217.

214

Allison, A. (1967) Lysosomes in virus infected cells. Perspect. Virol. 5, 29–61.

Allison, A. C. and Paton, G. R. (1965) Chromosome damage in human diploid cells following activation of lysosomal enzymes. Nature (London) 207, 1170–1173.

Allison, A. C., Davies, P. and De Petris, S. (1971) Role of contractile microfilaments in macrophage movement and endocytosis. Nature New Biol. 233, 153–155.

Anderson, W. A. D. (1971) Pathology, C. V. Mosby, St. Louis.

Ash, J. F., Spooner, B. S. and Wessels, N. K. (1973) Effects of papaverine and calcium free medium on salivary gland morphogenesis. Dev. Biol. 33, 463–469.

Awrorow, P. P. and Tunofejewsky, A. (1914) Kultivierungsversuche von leukameschem Blute. Virchows Arch. 216, 184–214.

Bennett, M. and Montes, M. (1973) Graft-versus-host reactions in mice. Am. J. Pathol. 71, 119–134.

Black, M. M. and Epstein, W. L. (1974) Formation of multinucleate giant cells in organized epithelioid cell granulomas. Am. J. Pathol. 74, 263–274.

Blaustein, M. P. and Goldman, D. E. (1966) Action of anionic and cationic nerve blocking agents: experiment and interpretation. Science 153, 429–432.

Bloom, B. R. (1971) In vitro approaches to the mechanism of cell-mediated immune reastions. Adv. Immunol. 13, 102–193.

Borrel, A. (1893) Tuberculose pulmonaire experimentale. Ann. Inst. Past. 7, 593–627.

Brederoo, P. and Daems, W. Th. (1972) Cell coat, worm-like structures, and labyrinths in guinea pig resident and exudate peritoneal macrophages, as demonstrated by an abbreviated fixation procedure for electron microscopy. Z. Zellforsch. 126, 135–156.

Cohen, M. (1926) Observation on the formation of giant cells in turtle blood cultures. Am. J. Pathol. 2, 431–438.

Cohn, Z. A. (1958) The structure and function of monocytes and macrophages. Adv. Immunol. 9, 163–214.

Cohn, Z. A. and Benson, B. (1965) The differentiation of mononuclear phagocytes. Morphology, cytochemistry and biochemistry. J. Exp. Med. 121, 153–170.

Comoglio, P. M., Ottino, G. and Cantino, D. (1971) Experimental study on development and behavior of the multinucleated giant cells "in vitro". J. Reticuloend. Soc. 9, 397–408.

Cotran, R. A. and Litt, M. (1970) Ultrastructural localization of horseradish peroxidase and endogenous peroxidase activity in guinea pig peritoneal macrophages. J. Immunol. 105, 1536–1546.

Croce, C. M., Sawicki, W., Kritchevsky, D. and Koprowski, H. (1971) Induction of homokaryocyte, heterokaryocyte and hybrid formation by lysolecithin. Exp. Cell Res. 67, 427–435.

Daems, W. Th. and Brederoo, P. (1971) The fine structure and peroxidase activity of resident and exudate peritoneal macrophages in the guinea pig. Adv. Exp. Med. Biol. 15, 19–31.

Daems, W. Th. and Brederoo, P. (1973) Electron microscopical studies on the structure, phagocytic properties, and peroxidatic activity of resident and exudate peritoneal macrophages in the guinea pig. Z. Zellforsch. 144, 247–297.

Dales, S. and Siminovitch, L. (1961) The development of vaccinia virus in Earle's L strain cells as examined by electron microscopy. J. Biophys. Biochem. Cytol. 10, 475–503.

Davis, J. R. (1966) Delayed hypersensitivity in vitro: its mediation by cell free substances formed by lymphoid cell-antigen interaction. Proc. Nat. Acad. Sci. U.S.A. 56, 72–77.

Diacumakos, E. G. and Tatum, E. L. (1972) Fusion of mammalian somatic cells by microsurgery. Proc. Nat. Acad. Sci. U.S.A. 69, 2959–2962.

Di George, A. M. (1968) Congenital absence of the thymus and its immunological consequences; concurrence with congenital hypoparathyroidism. In: Birth Defect Original Articles Series, IV, "Immunological Deficiency Diseases in Man" (Bergsma, D. and Good, R. A., eds.), pp. 116–123, National Foundation-March of Dimes, New York.

Doan, C. A., Sabin, F. R. and Forkner, C. E. (1930) The derivation of giant cells with especial reference to those of tuberculosis. J. Exp. Med., Suppl. 3, 52, 89–112.

Dreher, R., Keller, H. U., Hess, M. W. and Cottier, H. (1975) Enhancement of talcum-induced macrophage fusion and giant cell proliferation by delta-hydrocortisone acetate. In: Mononuclear Phagocytes in Immunity, Infection and Pathology (Van Furth, E., ed.) pp. 943–950, Blackwell Scientific Publ., London.

Dumont, A. and Sheldon, H. (1965) Changes in the fine structure of macrophages in experimentally produced tuberculous granulomas in hamsters. Lab. Invest. 14, 2034–2055.

Ebert, R. H. and Florey, H. W. (1939) The extravascular development of the monocyte observed in vivo. Br. J. Exp. Pathol. 20, 342–356.

Epifanova, O. I. and Terskikh, V. V. (1969) On the resting periods in the cell life cycle. Cell Tiss. Kinet. 2, 75–93.

Faber, K. (1893) The part played by giant cells in phagocytosis. J. Pathol. Bacteriol. 1, 349–359.

Feinstein, M. B. and Paimre, M. (1969) Pharmacological action of local anaesthetics on excitation-contraction coupling in striated and smooth muscle. Fed. Proc. 28, 1643–1648.

Feinstein, M. B., Paimre, M. and Lee, M. (1968) Effect of local anaesthetics on excitation-coupling mechanisms. Trans. N.Y. Acad. Sci. 30, 1073–1081.

Franklin, R. M. (1958) Some observations on the formation of giant cells in tissue cultures of chicken macrophages. Z. Naturforsch. 13, 213–214.

Gallindo, B. (1972) Antigen mediated fusion of specifically sensitised rabbit alveolar macrophages. Infect. Immunol. 5, 583–594.

Gallindo, B., Lazdins, J. and Castillo, R. (1974) Fusion of normal alveolar macrophages induced by supernatant fluids from BCG sensitized lymph node cells after elicitation by antigen. Infect. Immunol. 9, 212–216.

Gillman, T. and Wright, L. J. (1966) Probable in vivo origin of multinucleated giant cells from circulating mononuclears. Nature (London) 209, 263–265.

Godal, T., Rees, R. J. W. and Lamvik, J. O. (1971) Lymphocyte mediated modification of blood derived macrophage function in vitro; inhibition of growth of intracellular mycobacteria with lymphokines. Clin. Exp. Immunol. 8, 625–637.

Goldman, R. D. and Knipe, D. M. (1972) The functions of cytoplasmic fibres in non-muscle cell motility. Cold Spring Harb. Symp. Quant. Biol. 37, 523–534.

Goldstein, M. N. (1954) Formation of giant cells from human monocytes cultivated on cellophane. Anat. Rec. 118, 577–591.

Gordon, S. (1974) The expression of macrophage specific functions in fused cells. In: Somatic Cell Hybridization (Davidson R. L. and de la Cruz F. eds.) pp. 271–276, Raven Press, New York.

Gordon, S. (1975) Cell fusion and some subcellular properties of heterokaryons and hybrids. J. Cell Biol. 67, 257–280.

Gordon, S. and Cohn, Z. (1970) Macrophage-melanocyte heterokaryons. I. Preparation and properties. J. Exp. Med. 131, 981–1003.

Gordon, S. and Cohn, Z. (1971a) Macrophage-melanocyte heterokaryons. II. The activation of macrophage DNA synthesis. Studies with inhibitors of RNA synthesis. J. Exp. Med. 133, 321–338.

Gordon, S. and Cohn, Z. (1971b) Macrophage-melanoma cell heterokaryons. III. The activation of macrophage DNA synthesis. Studies with inhibitors of protein synthesis and with synchronized melanoma cells. J. Exp. Med. 134, 935–946.

Gordon, S. and Cohn, Z. (1971c) Macrophage-melanoma cell heterokaryons. IV. Unmasking the macrophage-specific membrane receptor. J. Exp. Med. 134, 947–962.

Gusek, W. (1964) Histologische und Vergleichende Electronmikroscopische Untersuchungsergebuisse zur Zytologie, Histochemese und Structur des Tuberkulosen and Tuberkuloiden granuloms. Med. Welt. 33, 850–866.

Harris, H. (1966) Hybrid cells from mouse and man: A study in genetic regulation. Proc. Roy. Soc. Ser. Biol. Sci. 166, 358–368.

Harris, H., Watkins, J. F., Ford, C. E. and Schoefl, G. I. (1966) Artificial heterokaryons of animal cells from different species. J. Cell Sci. 1, 1–29.

Harris, H., Sidebottom, E., Grace, D. M. and Bramwell, M. E. (1969) The expression of genetic formation: a study with hybrid animal cells. J. Cell Sci. 4, 499–525.

Haythorn, S. R. (1929) Multinucleated giant cells. Arch. Pathol. 7, 651–713.

Hodel, C. (1967) Fermenthistochemische Befunde an Riesenzellen in Talkgranulaomen der Ratte. Pathol. Microbiol. 30, 27–34.

Hosaka, Y. and Koshi, Y. (1968) Electron microscopic study of cell fusion by HJV virions. Virology 34, 419–434.

Howell, J. I. and Lucy, J. A. (1969) Cell fusion induced by lysolecithin. FEBS Lett. 4, 147–150.

Hyde, A., Blondel, B., Matter, A., Cheneval, J. P., Fellouz, B. and Girardier, L. (1969) Homo- and heterocellular junctions in cell cultures: an electrophysiological and morphological study. Progr. Brain Res. 31, 283–311.

Jones-Williams, W., Valerie James, E. M., Erasmus, D. A. and Davies, T. (1970) The fine structure of sarcoid and tuberculous granulomas. Postgrad. Med. J. 46, 496–500.

Klenk, H. D. and Choppin, P. W. (1969a) Chemical composition of the parainfluenza virus SV5. Virology 37, 155–157.

Klenk, H. D. and Choppin, P. W. (1969b) Lipids of plasma membranes of monkey and hamster kidney cells and of parainfluenza virions grown in these cells. Virology 38, 255–268.

Komnick, H., Stockem, W. and Wohlfarth-Botterman, K. E. (1972) Cell motility mechanisms in protoplasmic streaming and amoeboid movement. Int. Rev. Cytol 34, 169–249.

Kwant, W. O. and Seeman, P. (1969) The displacement of membrane calcium by a local anesthetic (chlorpromazine). Biochim. Biophys. Acta 193, 338–349.

Lambert, R. A. (1912) The production of foreign body giant cells in vitro. J. Exp. Med. 15, 510–515.

Langhans, T. (1868) Uber Riezenzellen mit Wandstandigen Kernen in Tuberkeln und die fibrose Form des Tuberkels. Arch. Pathol. Anat. 42, 382–404.

Lewis, M. R. (1925a) Formation of macrophages and epithelioid cells from leukocytes in incubated blood. Am. J. Pathol. 1, 91–100.

Lewis, M. R. (1925b) Origin of phagocytic cells of the lung of the frog. Bull. John Hopkins Hosp. 36, 361–375.

Lewis, M. R. and Lewis, W. H. (1926) Transformation of mononuclear blood cells into macrophages, epithelioid, and giant cells in hanging drop, cultures from lower vertebrates. Cont. Emb. Carn. Inst. Wash. 18, 97–120.

Lewis, W. H. (1927) The formation of giant cells in tissue cultures and their similarity to those in tuberculous lesions. Am. Rev. Tuberc. 15, 616–627.

Lewis, W. H. and Lewis, M. R. (1925) Transformation of white blood cells. J.A.M.A. 84, 798–799.

Mariano, M. and Spector, W. G. (1974) The formation and properties of macrophage polykaryons (inflammatory giant cells). J. Pathol. 113, 1–19.

Maximow, A. A. (1925) The role of the non-granular blood leukocytes in the formation of the tubercle. J. Infect. Dis. 37, 418–429.

Mayhew, E. (1966) Cellular electrophoretic motility and the mitotic cycle. J. Gen. Physiol. 49, 717–725.

Meiselman, N., Kohn, A. and Danon, D. (1967) Electron microscopic study of penetration of Newcastle disease virus into cells leading to formation of polykaryocytes. J. Cell Sci. 2, 71–76.

Mitchell, D. N., Siltzbach, L. E., Sutherland, I. and Hart, P. D. (1968) Sarcoidosis and immunological deficienceis. In: Birth Defects Original Articles, Series IV, Immunological Deficiency Diseases in Man (Bergsma, D. and Good, R. A., eds.) pp. 364–369, National Foundation—March of Dimes, New York.

Nayler, W. G. and Szeto, J. (1972) Effect of verapamil on contractility, oxygen utilization and calcium exchangeability in mammalian heart muscle. Cardiovasc. Res. 6, 120–128.

Norrby, E. (1966) The effect of phlorizin on the multiplication of measles virus. Arch. Ges. Virus Forsch. 18, 333–343.

Ohki, S. (1970) Effect of local anaesthetics on phospholipid bilayers. Biochim. Biophys. Acta 219, 18–27.

Okada, Y. (1958) The fusion of Ehrlich's tumor cells caused by HVJ virus in vitro. Biken J. 1, 103–110.

Papadimitriou, J. M. (1973) Detection of macrophage receptors for heterologous IgG by scanning and transmission electron microscopy. J. Pathol. 110, 213–220.

Papadimitriou, J. M. (1974) Ultrastructural characteristics of murine multinucleate foreign body giant cells. Proc. Eighth Int. Cong. Elect. Microsc., Vol. II, 426–427.

Papadimitriou, J. M. (1976) The influence of the thymus on multinucleate giant cell formation. J. Pathol. 118, 153–156.

Papadimitriou, J. M. and Archer, M. (1974) The morphology of murine foreign body multinucleate giant cells. J. Ultrastruct. Res. 49, 372–386.

Papadimitriou, J. M. and Cornelisse, C. J. (1975) A cytophotometric and autoradiographic study of DNA synthesis in macrophages and multinucleate foreign body giant cells. J. Reticuloend. Soc. 18, 260–270.

Papadimitriou, J. M. and Kingston, K. J. (1976) The locomotory behavior of the multinucleate giant cells of foreign body reactions. J. Pathol. 121, 27–36.

Papadimitriou, J. M. and Sforcina, D. (1975) The effect of drugs on monocytic fusion in vivo. Exp. Cell Res. 91, 233–236.

Papadimitriou, J. M. and Spector, W. G. (1971) The origin, properties and fate of epithelioid cells. J. Pathol. 105, 187–202.

Papadimitriou, J. M. and Wee, S. H. (1976) Selective release of lysosomal enzymes from cell populations containing multinucleate giant cells. J. Pathol. 120, 193–199.

Papadimitriou, J. M. and Wyche, P. A. (1974) An examination of murine foreign body giant cells using cytochemical techniques and thin layer chromatography. J. Pathol. 114, 75–83.

Papadimitriou, J. M. and Wyche, P. A. (1976) A biochemical profile of glass-adherent cell populations containing multinucleated foreign body giant cells. J. Pathol. 119, 239–254.

Papadimitriou, J. M., Finlay-Jones, J-M. and Walters, M. N-I. (1973a) Surface characteristics of macrophages, epithelioid cells and giant cells using scanning electron microscopy. Exp. Cell Res. 76, 353–362.

Papadimitriou, J. M., Sforsina, D. and Papaelias, L. (1973b) Kinetics of multinucleate giant cell formation and their modification by various agents in foreign body reactions. Am. J. Pathol. 73, 349–362.

Papadimitriou, J. M., Robertson, T. A. and Walters, M. N-I. (1975) An analysis of the phagocytic potential of multinucleate foreign body giant cells. Am. J. Pathol. 78, 343–354.

Papahadjopoulos, D. (1970) Phospholipid model membranes. 3. Antagonistic effects of Ca^{2+} and local anaesthetics on the permeability of phosphatidyl serine vesicles. Biochim. Biophys. Acta 211, 467–477.

Parks, D. E. and Weiser, R. S. (1975) The role of phagocytosis and natural lymphokines in the fusion of alveolar macrophages to form Langhans giant cells. J. Reticuloend. Soc. 17, 219–228.

Peretz, H., Toister, Z., Laster, Y. and Loyter, A. (1974) Fusion of intact human erythrocytes and erythrocyte ghosts. J. Cell Biol. 63, 1–11.

Pethica, B. A. (1961) The physical chemistry of cell adhesion. Exp. Cell Res. Suppl. 8, 123–140.

Poole, A. R., Howell, J. I., and Lucy, J. A. (1970) Lysolecithin and cell fusion. Nature (London) 227, 810–814.

Poste, G. (1970) Virus induced polykaryocytosis and the mechanism of cell fusion. Adv. Virus Res. 16, 303–356.

Poste, G. (1972) Mechanisms of virus-induced cell fusion. Int. Rev. Cytol. 33, 157–252.

Poste, G. and Reeve, P. (1972) Inhibition of cell fusion by local anaesthetics and tranquilizers. Exp. Cell Res. 72, 556–560.

Ptak, W., Porwit-Bobr, Z. and Chlap, Z. (1970) Transformation of hamster macrophages into giant cells with antimacrophage serum. Nature (London) 225, 655–657.

Rabinovitch, M. (1969) Uptake of aldehyde treated erythrocytes by L2 cells. Exp. Cell Res. 54, 210–216.

Rao, P. N. and Johnson, R. T. (1970) Mammalian cell fusion: studies on the regulation of DNA synthesis and mitosis. Nature (London) 225, 159–164.

Robinson, G. A., Butcher, R. W. and Sutherland, E. W. (1968) Cyclic AMP. Ann. Rev. Biochem. 37, 149–174.

Rodbell, M., Kraus, M. J., Pohl, S. and Birnbaumer, L. (1971) The glucan sensitive adenyl cyclase system in plasma membranes of rat liver. J. Biol. Chem. 246, 1861–1871.

Roizman, B. (1962) Polykaryocytosis. Cold Spring Harb. Symp. Quant. Biol. 27, 327–342.

Rubin, H. (1967) In: The Specificity of Cell Surfaces (Davis, B. D. and Warren, L., eds.) pp. 181–194, Prentice-Hall, Englewood Cliffs, New Jersey.

Ryan, G. B. and Spector, W. G. (1970) Macrophage turnover in inflamed connective tissue. Proc. Roy. Soc. Ser. Biol. Sci. 175, 269–292.

218

Sabban, E. and Loyter, A. (1974) Fusion of chicken erythrocytes by phospholipase C (clostridium perfringens). Biochim. Biophys. Acta 362, 100–109.

Salthouse, T. N. and Williams, J. A. (1969) Histochemical observations of enzyme activity at suture implant sites. J. Surg. Res. 9, 481–486.

Schmalzl, F. and Braunsteiner, H. (1970) The cytochemistry of monocytes and macrophages. Ser. Haematol. 3, 93–131.

Schneeberger, E. E. and Harris, H. (1966) An ultrastructural study of interspecific cell fusion induced by inactivated Sendai virus. J. Cell Sci. 1, 401–406.

Smith, A. D. and Winkler, H. (1969) Lysosomes and chromaffin granules in the adrenal medulla. In: Lysosomes in Biology and Medicine (Dingle, J. T., and Fell, H. B., eds). Vol. 1, pp. 155–166, Elsevier/North-Holland, Amsterdam.

Smith, C. W. and Goldman, A. S. (1971) Macrophages from human colostrum. Exp. Cell Res. 66, 317–320.

Spector, W. G. and Lykke, A. W. J. (1966) The cellular evolution of inflammatory granulomata. J. Pathol. Bacteriol. 96, 163–177.

Stephenson, N. G. (1969) Effects of increased partial pressures of oxygen, nitrogen and helium on cells in culture. Cell Tiss. Kinet. 2, 225–234.

Sutton, J. S. and Weiss, L. (1966) Transformation of monocytes in tissue culture into macrophages, epithelioid cells and multinucleated giant cells. J. Cell Biol. 28, 303–332.

Svoboda, J. and Dourmashkin, R. (1969) Rescue of Rous sarcoma virus from virogenic mammalian cells associated with chicken cells and trested with Sendai virus. J. Gen. Virol. 4, 532–529.

Virchow, R. (1858) Reizung und Reizbarkeit. Virchows Arch. 14, 1–63.

Virolainen, M. and Defendi, V. (1967) Dependence of macrophage growth in vitro upon interaction with other cell types. Wistar Inst. Symp. Monogr. 7, 67–85.

Volkman, A. (1971) A current perspective of monocytopoiesis. Curr. Top. Pathol. 54, 76–94.

Wahramann, J. P., Winarnd, R. and Luzzati, D. (1973) Effect of cyclic AMP on growth and morphological differentiation of an established myogenic line. Nature New Biol. 245, 112–113.

Watkins, J. F. and Grace, D. M. (1967) Studies on the surface antigens of interspecific mammalian cell heterokaryons. J. Cell Sci. 3, 193–204.

Weil, G. C. (1913) Spontaneous and artificial development of giant cells in vitro. J. Pathol. Bacteriol. 18, 1–7.

Weiss, L. and Woodbridge, R. F. (1967) Some biophysical aspects of cell contacts. Fed. Proc. Fed. Amer. Soc. Exp. Biol. 26, 88–94.

Werb, Z. and Cohn, Z. A. (1971) Cholesterol metabolism in the macrophage. J. Exp. Med. 134, 1545–1569.

White, A., Handler, P. and Smith, E. L. (1968) Principles of Biochemistry, 4th ed., pp. 402, McGraw-Hill, New York.

Wolpert, L. and Gingell, D. (1968) In: Aspects of Cell Motility. Symp. Soc. Exp. Biol. No. 22, (Miller, P. L., ed.) pp. 168–198, Cambridge University Press, London.

Cell fusion in myxomycetes and fungi

5

M. J. CARLILE and G. W. GOODAY

Contents

G. Poste & G. L. Nicolson (eds.) Membrane Fusion, pp. 219–265.
©Elsevier/North-Holland Biomedical Press, 1978.

1. Introduction

A striking feature of both myxomycetes (plasmodial slime molds) and the higher fungi (ascomycetes, basidiomycetes and fungi imperfecti) is that numerous somatic cell fusions are a normal part of the developmental process (Carlile, 1978). When somatic cells that are genetically different come into contact, however, fusion may be prevented or be followed by incompatibility reactions—the heterogenic incompatibility of Esser and Blaich (1973). Sexual fusion also occurs in both groups, but there may be mechanisms that prevent or prematurely terminate the fusion of genetically identical cells—homogenic incompatibility (Esser, 1974a). The sexual processes in fungi are often remarkably complex with elaborate genetical and biochemical control of gametic approach, cell fusion, and subsequent developments. Somatic and sexual fusions in myxomycetes and fungi offer unusual opportunities for the analysis of cell fusion, which will be more fully realized when more is known about the structure of the glycocalyx and plasma membrane in the two groups and of the enzymes involved in the synthesis, modification, and breakdown of the glycocalyx.

Since the myxomycetes and fungi have striking differences as well as important similarities, they will be dealt with separately. Due to the nature of their life cycles we shall discuss sexual fusion before somatic fusion in myxomycetes, and deal with somatic fusion first in fungi.

2. Myxomycetes

2.1. Structure and life cycle

The unique feature of myxomycetes is the plasmodium, a mass of protoplasm that contains many nuclei but is not subdivided into cells. In the best known group, the Physarales, the plasmodium may cover many square centimeters and contain millions of nuclei. Since the nuclei divide synchronously, myxomycete plasmodia offer unusual opportunities for biologists interested in the cell cycle. Plasmodia of one species, *Physarum polycephalum,* are currently being studied in many laboratories and are the main subject of the widely circulated *Physarum Newsletter.* Considerable genetic work has been carried out with a second species, *Didymium iridis,* which has not been grown in pure culture on soluble media and hence is less suitable for physiological and biochemical investigation.

The structure and life cycle of myxomycetes have been described in detail by Ashworth and Dee (1975), Gray and Alexopoulos (1968), and Olive (1975). For convenience, a brief account follows and is summarized in Fig. 1.

A spore germinates to produce a haploid uninucleate amoeba which feeds on bacteria. Repeated binary fission gives rise to an amoeba population. Immersion in water causes the amoebae to turn into flagellate swarmers, which revert to amoebae in the absence of free liquid. Starvation results in encystment of the amoebae, and the presence of further nutrients causes the cysts to germinate to

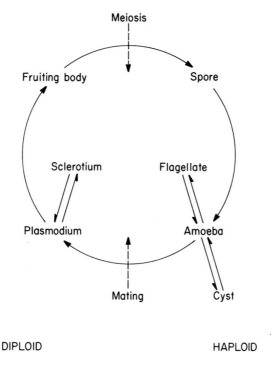

Fig. 1. Life cycle of a myxomycete. Mating and meiosis and the designations haploid and diploid apply to heterothallic and homothallic but not to apogamic strains.

produce amoebae again. Numerous tiny plasmodia may appear in a population of amoebae. Since growth and nuclear division in plasmodia are unaccompanied by cell division, these soon become far larger than an amoeba and are also multinucleate. Fusion (coalescence) of the small plasmodia initiated within an amoeba population occurs, and plasmodia will ingest and digest the remaining amoebae. Finally, as a result of coalescence and cannibalism as well as of more normal feeding, a plasmodium may grow very large and cover an entire petri dish. Such spectacular plasmodia exhibit protoplasmic streaming at rates of up to 1mm/sec in channels ("veins") with reversals in direction at about 1-minute intervals (shuttle streaming), can migrate rapidly, and can surround and digest large objects such as the fruit bodies of toadstools or bits of rotten wood on which they feed. Starvation of a plasmodium may give rise to a sclerotium, a resting phase which, under suitable conditions, will revive to produce a plasmodium again. In the presence of light, however, starvation commonly leads to sporulation, the production of fruiting bodies containing spores. This completes the life cycle of the myxomycete.

The aspects of the myxomycete life cycle that are of interest to students of cell fusion are plasmodium initiation, which may involve sexual fusion between amoebae, and plasmodial coalescence—somatic fusions between plasmodia.

These processes will now be discussed with special reference to the intensively studied *P. polycephalum,* but work on *D. iridis* and other species will be considered where it provides useful additional information.

2.2. *Sexual fusion and plasmodium initiation*

2.2.1. *Physarum polycephalum*
The most widely used plasmodial strain of *P. polycephalum* is the Wisconsin 1 (Wis 1) strain. Dee (1962) germinated spores from Wis 1 plasmodia and obtained amoeba strains by cloning. She found that the amoeba strains could be classified into two mating types. Plasmodia were rarely produced in an amoeba clone, but were readily obtained if amoebae from two strains differing in mating type were mixed. It was established that mating type was controlled by a pair of alleles, later designated mt_1 and mt_2. Genetic recombination was demonstrated, and Dee concluded that the production of plasmodia must involve fusion between amoebae of differing mating type (plasmogamy) and that fusion between nuclei of the two strains (karyogamy) must occur subsequently. The haploid state would be restored by meiosis during sporulation. Later, electron microscopic observation of synaptonemal complexes firmly established that meiosis occurs during spore maturation in *P. polycephalum* (Aldrich, 1967) and other myxomycetes (Aldrich and Mims, 1970; Aldrich and Carroll, 1971), and by determining DNA content per nucleus (Mohberg and Rusch, 1971) and by chromosome counts (Mohberg et al, 1973) that the amoebae of the Wis 1 isolate were haploid and the plasmodia diploid.

Dee (1966) found that an Indiana plasmodial strain behaved similarly to the Wis 1 strain, amoebae derived from the plasmodia being of two mating types which yielded plasmodia on being unified. The mating types were not, however, identical with those obtained from the Wis 1 strain, and plasmodia could be produced by bringing together amoebae of any two of the four mating types so far demonstrated. The Indiana mating types were found to be controlled by genes allelic with those in the Wis 1 strain, and the alleles were designated mt_3 and mt_4. Collins (1975) studied a further four plasmodial strains from different localities and obtained from them amoeba strains carrying eight more mating types controlled by alleles mt_5 to mt_{12}. Collins and Tang (1977) have now obtained two more mating types, alleles mt_{13} and mt_{14}, from a new plasmodial strain from Leningrad. Hence many of the amoeba strains of *P. polycephalum* are *heterothallic,* with mating controlled by multiple alleles at a single locus. Clearly, mating types are numerous, with any new plasmodial strain from nature likely to yield two new mating types.

Meanwhile, Wheals (1970) had studied amoeba progeny of the Colonia plasmodial strain of *P. polycephalum.* It was found that with these amoeba strains plasmodia readily arose within clones, and it was shown that the clonal production of plasmodia was due to an allele at the mating type locus, designated mt_h. Plasmodia were, however, even more readily obtained by mixing mt_h amoebae with amoebae carrying other mating type alleles such as mt_4, and genetic analysis showed that mating between the strains to give diploid plasmodia, followed by meiosis at the time of sporulation must have occurred, a conclusion subsequently substantiated by microdensitometric measurements of nuclear deoxyribonucleic acid (DNA) content (Cooke and Dee, 1974). Since the amoeba strains derived from the Colonia isolate could produce plasmodia by mating with heterothallic strains, the investigators assumed at first the plasmodium production within clones also involved mating, that is, the strains were *homothallic,* capable of producing plasmodia by mating between amoebae that do not differ in mating type. However, it was subsequently shown by measurements on nuclear DNA content that clonally produced plasmodia did not differ from amoebae in ploidy (Cooke and Dee, 1974). One strain (CL) derived from Colonia forms plasmodia with very high frequency within clones. A study of this strain by time-lapse cinefilming showed that neither plasmogamy nor karyogamy occurred in plasmodium initiation—a uninucleate cell becomes binucleate and then multinucleate, either by continued nuclear division or by fusing with other binucleate or multinucleate cells (Anderson et al., 1976). Fusion between amoebae and plasmodia was never observed. The various characteristics of plasmodia as compared with amoebae, such as shuttle streaming, are acquired gradually as the plasmodium increases in size. Hence plasmodium initiation in the CL strain is apogamic, and homothallic initiation of plasmodia in *P. polycephalum* remains unproven.

Adler and Holt (1975) obtained diploid or near-diploid amoeba strains heterozygous with respect to the mating types locus (e.g., carrying both the mt_1 and mt_3 alleles). These strains could be maintained in the amoeboid state, but gave rise to plasmodia within amoeba clones at high frequency, and no ploidy change was involved. Infrequent plasmodium production, without ploidy change, was observed in haploid amoeba clones carrying either the mt_1 or mt_3 allele; such plasmodia, although rarely produced, were normal and vigorous. Hence both haploidy and diploidy, and both the homozygous and heterozygous condition with respect to mating type alleles, are compatible with either the amoeboid or plasmodial state. The frequency of clonal plasmodium initiation, as determined by the proportion of cultures in which plasmodia were produced, was determined for amoeba strains carrying various mating type alleles singly or in combination. It was concluded that the effectiveness of different allele combinations in promoting plasmodium initiation could be characterized as follows:

$$mt_h mt_m \rangle mt_m mt_n \rangle mt_h \rangle \rangle mt_m \text{ or } mt_n$$

where mt_m and mt_n represent different heterothallic mating type alleles. However, since the frequency of plasmodium initiation varies between strains that have the same mating type allele, and since one strain (CL) carrying the mt_h allele shows a very high rate of clonal plasmodium formation (Cooke and Dee, 1974), the finer variations reported above may be due to strain differences. Clearly, the presence in a cell of mt_h or of a pair of heterothallic mating type alleles gives a high probability of plasmodium initiation, whereas with a single heterothallic mating type allele this probability is very low. Adler and Holt (1975) also brought together amoebae carrying both the mt_1 and mt_3 alleles and amoebae with mt_4 alleles and obtained plasmodia carrying all three mating type alleles. Thus it seems that the presence of two mating type alleles does not prevent an amoeba from mating, and it may be possible that an amoeba might undergo two successive matings in a mixture of amoeba strains.

Adler and Holt suggested that mt_h may correspond to two tightly linked, heterothallic mating type alleles. Support for this view has come from work on nonplasmodial forming *(npf)* mutants of apogamic strains carrying the mt_h allele (Anderson and Dee, 1977). These mutants were unable to form plasmodia in amoeba clones but did so, with an interesting exception, if mated with strains carrying heterothallic mating type alleles. Three *npf* genes were recognized. Two of them, *npf* B and *npf* C, could not be separated from each other or from the mating type locus by recombination but were distinguishable by complementation analysis—mixing *npf* B⁻ and *npf* C⁻ amoebae gave diploid hybrid plasmodia, but mixing two *npf* B⁻ strains or two *npf* C⁻ strains did not. It was found that strains carrying a mutation of *npf* B but not *npf* C (i.e., *npf* B⁻ *npf* C⁺) behaved in the same way as strains carrying the mt_2 allele; they would mate with strains carrying the mt_1, mt_3, or mt_4 allele but not the mt_2.

Clearly, the mating type locus has a major role in initiating the plasmodial state in apogamic as well as heterothallic strains. It is possible that occasional cell fusions and nuclear fusions will occur in amoeba clones, although these will not result in any nuclei of changed genotype and will not increase the frequency of plasmodium formation. Cell fusion and nuclear fusion in a population of amoebae of two mating types will, however, result in the production of some amoebae heterozygous with respect to the mating type locus and hence more likely to give rise to plasmodia. If this is so, then the mating type locus does not determine whether plasmogamy or karyogamy occurs; instead, cell and nuclear fusion between amoebae of different strains brings together complementary mating type alleles and thus permits plasmodium initiation to take place. On the other hand, there is no evidence to date that amoebae of *P. polycephalum* of the same mating type can fuse with each other, and it is possible that the "mating type locus" is a cluster of genes controlling several functions related to cell fusion and plasmodium differentiation.

Recently, Mohberg (1977) confirmed by chromosome counts and nuclear DNA measurements that most of the nuclei in Colonia strains are haploid throughout the life cycle. However, she finds that about 2% of the plasmodial nuclei are diploid. The origin of these nuclei is unknown, although fusion between haploid plasmodial nuclei is a possibility. Laffler and Dove (1977) have evidence that these rare diploid nuclei give rise to viable spores, presumably by meiosis. On the other hand, Laane and co-workers (1976) found that spores can arise in Colonia strains by a special form of meiosis ("pseudomeosis") which involves synaptonemal complex formation (and hence possibilities of recombination) but no reduction in ploidy. Further study of diploid nuclei and sporulation in Colonia strains is needed to investigate the intriguing possibility that apogamic plasmodial induction can be followed by events that, in terms of cytology or genetic analysis, might appear as homothallic or parasexual nuclear fusions.

2.2.2. *Didymium iridis and other species*

Collins (1961) obtained from sporangia of a Honduras strain of *Didymium iridis* two amoeba strains of different mating types which yielded plasmodia on being brought together. Studies on a Panama strain yielded amoebae of two more mating types and demonstrated that mating was controlled by multiple alleles at a single locus (Collins, 1963), and examination of a Costa Rica strain brought the number of known mating type alleles to six (Collins and Ling, 1964). Subsequent work resulted in a total of 27 samples from nature being examined (Collins, 1976). Eleven proved to be heterothallic, and from 9 of these a total of 11 mating types were obtained; some mating types were present in more than one sample. Mutation yielded a twelfth mating type. Sixteen samples readily produced plasmodia within clones and were designated "homothallic," although since it was realized that they might be apogamic, perhaps "nonheterothallic" would be a better designation.

Thus *D. iridis*, like *P. polycephalum*, has strains in which plasmodia are readily produced in amoeba clones and strains in which mating is normally needed for plasmodium initiation. As in *P. polycephalum*, plasmodia occasionally arise in clones of heterothallic amoebae (Collins and Ling, 1968). The DNA contents of amoeba and plasmodium nuclei of several heterothallic isolates of *D. iridis* were determined by Collins and Therrien (1976), who concluded that the amoebae were haploid and the plasmodia were diploid, as could be expected where plasmodium initiation results from mating. Therrien and colleagues (1977) examined nuclear DNA contents in six nonheterothallic isolates of *D. iridis;* in two strains the amoebae and plasmodia had similar DNA levels and the investigators concluded that they were apogamic, whereas in four strains the plasmodial levels were double that of the amoebae, a result interpreted as indicating homothallic mating. Aldrich and Carroll (1971) had demonstrated meiosis in a non-heterothallic strain of *D. iridis;* this, however, in view of the work of Laffler and Dove (1977) mentioned earlier, is not proof that plasmodial induction is homothallic.

Ross (1967) studied plasmodium initiation in heterothallic strains of *D. iridis*, showing that plasmodia were initiated by cell fusion between two amoebae followed by nuclear fusion. Plasmodia then developed by growth and by coalescence with other plasmodia. Plasmodia did not fuse with amoebae. Ross and Cummings (1970) harvested and mixed amoebae of different mating types from cultures at various stages of growth and recorded the length of time to plasmodium initiation. Amoebae harvested and mixed near the end of exponential growth fused almost immediately, and under these circumstances unusual behavior—multiple cell fusion and multiple nuclear fusion—was seen. Initially, these multinucleate and polyploid cells behaved more like amoebae than plasmodia, but ultimately plasmodia developed. Ross and associates (1973) developed methods for recording the numbers of plasmodia induced in a population of amoebae of known size. In mixed populations of two mating types up to a dozen plasmodia were initiated per thousand amoebae. Cell fusion or plasmodium initiation was not observed in unmated cultures. Plasmodia numbers reached a maximum about 10 hours after mixing, and then decreased as coalescence between plasmodia occurred.

Plasmodium initiation in *D. nigripes* has been studied by Kerr and Kerr. A cinematographic study (N. Kerr, 1967) of a mutant strain established apogamic plasmodium initiation at a very high frequency and almost all the amoebae turned into plasmodia. Chromosome counts (S. Kerr, 1968) on the mutant and on a wild-type strain showed with both that the amoebae and plasmodia did not differ in ploidy. However, plasmodia contained a small proportion of large nuclei, and the origin of large nuclei both by fusion of small nuclei in interphase and failure of daughter nuclei to separate completely in telophase was observed (S. Kerr, 1970). It was suggested that although plasmodium formation was apogamic, parasexual processes might occur, a view ahead of its time that is currently gaining support from work on *P. polycephalum* (section 2.2.2).

More limited studies of plasmodium initiation have been made in many other myxomycetes. Clark and Collins (1976) carried out a survey on eleven species and summarized the information available on other species. Clearly, many species have both heterothallic strains and strains in which plasmodia readily arise in clones.

2.2.3. Mating type and plasmodium initiation

Genetic studies in *P. polycephalum* suggest that the mating type locus has a major role in controlling plasmodial induction in both heterothallic and apogamic strains, and that it may possibly be a locus that controls differentiation rather than amoeba fusion, the significance of fusion being that it brings together complementary alleles that permit plasmodium initiation to occur. Observational studies on *D. iridis,* on the other hand, suggest that the mating type locus determines whether amoeba fusion will occur (Ross and Cummings, 1970; Ross et al., 1973). The difference may be real, with the mating type locus having evolved a different role in the two species, or may reflect a difference in methods of study. The possibility that the mating type locus in *P. polycephalum* influences amoeba fusion has not been excluded, nor has the role of mating type locus on plasmodium initiation in *D. iridis* been examined so far. It is likely that the problem will be resolved by an approach combining genetic analysis and observational methods such as cine-filming.

The biological significance of the mating types of myxomycetes is that they constitute a system of homogenic incompatibility, preventing the mating of genetically identical amoebae and thus encouraging outbreeding. There are also occasional instances of failure to sporulate or even to produce plasmodia when amoebae from different localities are mixed (e.g., Collins, 1975, 1976) and these could be regarded as due to heterogenic incompatibility—failure to complete mating and genetic recombination as a result of genetic differences. However, it seems unlikely that actual incompatibility mechanisms would exist to combat such improbable matings, and probably failure is due to fortuitous differences between the strains, namely, to genetic incongruity.

2.3. Somatic fusion between plasmodia

Many small plasmodia may be initiated in an amoeba population. These readily fuse with each other (plasmodial coalescence) to yield large plasmodia. Large plasmodia can also fuse with one another (Fig. 2a). Ross (1967) observed that when two small plasmodia happened to come into close proximity there was a surge of protoplasmic movement which brought them into contact, followed by fusion between the plasmodia. The events that occur when plasmodia of the same strain fuse were discussed by Carlile (1973), who also reported that plasmodia of different ages may fail to fuse. Most work on plasmodial fusion has, however, focused on fusion failure due to genetic differences, that is, on somatic incompatibility. These studies will now be discussed.

2.3.1. Physarum polycephalum

Carlile and Dee (1967), working with plasmodial strains derived from the Wis 1 isolate, found two forms of somatic incompatibility. Different strains might fail to fuse (Fig. 2b) or, if fusion did occur, a lethal reaction might destroy much of one or both plasmodia a few hours later. The former reaction can be termed fusion incompatibility, and the latter, lethal reaction, is an example of postfusion incompatibility.

Carlile and Dee (1967) concluded that fusion in the strains being studied was controlled by a pair of alleles at a single locus, later designated (Cooke and Dee, 1975) *fus* A1 and *fus* A2. Fusion occurs when plasmodia, which are diploid, are identical with respect to the fusion locus. For example, plasmodia of genotype *fus* A1A2 will fuse with each other but not with plasmodia that are *fus* A1A1. The alleles are therefore codominant. In a more detailed study Poulter and Dee confirmed (1968) that A1 and A2 are allelic. They also found a pair of fusion alleles in the Indiana strain and reported that these were allelic with those already studied in the Wis 1 strain; this, however, now appears unlikely (Dee, personal communication). Wheals (1970) identified a second fusion locus, *fus* B (Cooke and Dee, 1975), in the Wis 1 strain and found one allele, B2, dominant to the other, B1. Hence, plasmodia of genotype B2B2 and B2B1 would be of the same fusion phenotype and would fuse with each other but not with a B1B1 plasmodium. Adler and Holt (1974) have identified another fusion locus, *fus* C, obtaining the *fus* C1 locus from the Colonia strain and the *fus* C2 from an amoeba strain derived from the Indiana plasmodium. The various fusion loci act independently and phenotypic identity is required at all the *fus* loci for fusion to occur. Collins and Haskins (1972) found four loci, each with a dominant and recessive allele, controlling plasmodial fusion in progeny of an Iowa strain, and Collins (1972) discovered a similar system in the Turtox strain. It is not known which, if any, of the loci reported in the two papers described above are identical with the *fus* A, B, and C loci mentioned earlier. Clearly, however, there are many pairs of alleles at different loci that control fusion in *P. polycephalum,* many of the pairs consist of a dominant and a recessive allele, and at least one pair (*fus* A1 and A2) displays codominance.

Jeffrey and Rusch (1974) carried out gentle homogenization of mixed micro-

228

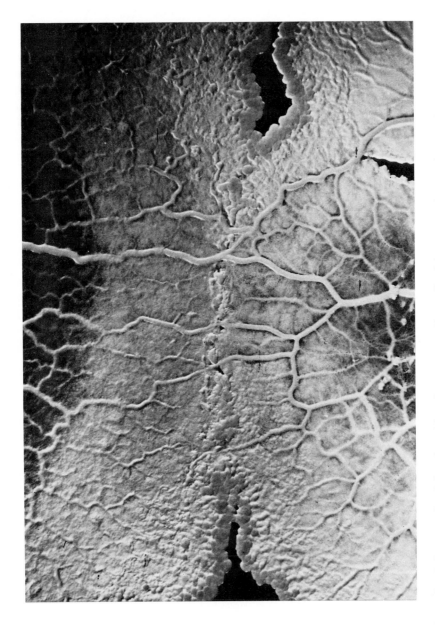

Fig. 2. Plasmodia of *Physarum polycephalum* about 1 hr. after contact (a) Plasmodia of the same strain. Prominent "veins" connect the two plasmodia and the boundary between them is becoming indistinct ×5.6 (b) Plasmodia of strains that differ at the *fus* A locus. Fusion has not occurred. ×7.0 (From Carlile, M. J. and Dee, J. [1967] Nature 215, 832–834.)

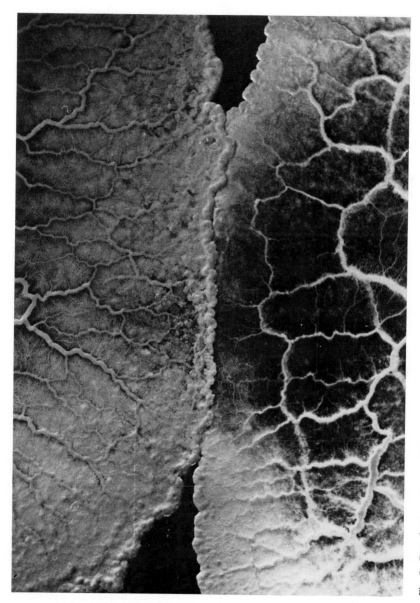

Fig. 2. Part (b). *See caption on facing page.*

plasmodia (tiny plasmodia obtained by growth in shaken liquid culture) of two strains incompatible with respect to fusion and having different nuclear sizes. The procedure yielded nucleated fragments which, under suitable conditions, fused to yield plasmodia in a period of 5 to 10 days. Examination of nuclear size in the reconstructed plasmodia indicated that about one third of the experiments yielded heterokaryotic plasmodia (containing nuclei from both strains). These plasmodia could fuse with each other but not with the parent strains. These experiments suggest that fusion incompatibility might be determined by specificity at the cell surface and that this specificity was destroyed during homogenization, thus permitting indiscriminate fusion among surviving fragments.

If two strains are compatible at all fusion loci, fusion will occur but may be terminated by postfusion reactions. A spectacular postfusion reaction, the lethal reaction, was discovered by Carlile and Dee (1967) in *P. polycephalum.* A detailed study of a unilateral reaction between two strains, 15, which acts as "killer," and 29, a "sensitive," was carried out by Carlile (1972). Strain "29" was destroyed about 5 hours after fusion with strain 15 (Fig. 3).

This lethal effect, however, is only obtained when large plasmodia are fused in

Fig. 3. Lethal interaction between plasmodia of strains 15 (right) and 29 of *Physarum polycephalum* in a 9 cm diameter petri dish. (a) About 5 hr after fusion and 5 min after the first indications of a reaction visible to the naked eye. A small dead area is spreading near the line of fusion between the two plasmodia. (b) About 20 min later. The dead area is rapidly increasing in size. (c) About 17 hr later. All of the area initially occupied by strain 29 and part of that occupied by 15 is dead. (From Carlile, M. J. [1972] J. Gen. Microbiol. 71, 581–590.)

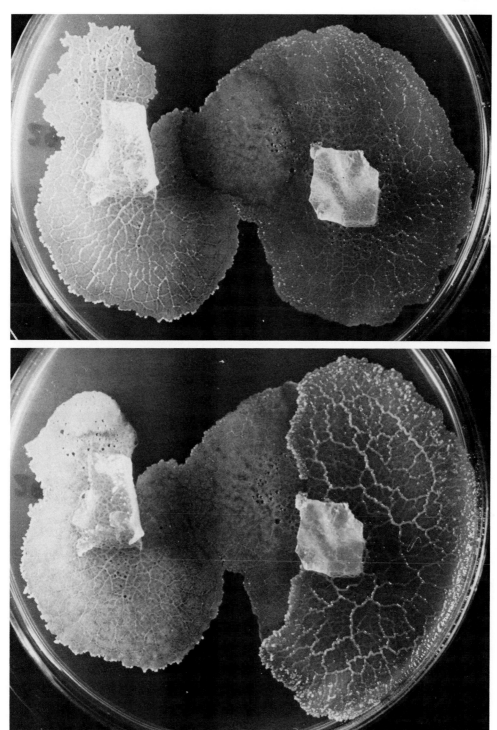

Fig. 3. Part (b): *top;* part (c): *bottom. See caption on facing page.*

certain proportions in the presence of abundant nutrients. Possibly, the "plain agar reaction" obtained when the two strains are fused under starvation conditions is more characteristic of events in nature. No obvious effects are seen, but subsequent tests show that the plasmodium resulting from fusion resembles and behaves as strain 15; the characteristics of strain 29 have vanished. Ultrastructural and autoradiographic studies by Border and Carlile (1974) on the lethal reaction showed that the nuclei of strain 29 were destroyed prior to any other extensive damage, and suggests that under natural conditions the role of the reaction is the elimination of alien nuclei. Genetic studies with the progeny of strain 29 (Carlile, 1976) established control of the lethal reaction in these strains by three genes, *let* (for lethal) A, *let* B, and *let* C and their recessive alleles *let* a, *let* b, and *let* c. A lethal reaction occurs if two plasmodia differ in phenotype with respect to any of the fusion loci, and is directed at the strain that is a homozygous recessive. Thus a strain carrying AA will fuse harmlessly with a strain carrying Aa, but both strains can kill a strain carrying aa. Interaction can be bilateral as well as unilateral. If, for example, plasmodia of phenotypes Ab and aB meet, both will suffer damage. In addition to the three *let* genes established earlier, a fourth must exist to account for the unilateral reaction between strains 15 and 29; indeed, many more probably exist. Adler and Holt (1974) demonstrated a pair of alleles responsible for postfusion lethal effects in hybrids of strains of Colonia and Indiana origin and termed them *kil* A1 and A2; whether the alleles are identical to one of the *let* pairs from Wis 1 progeny is unknown.

2.3.2. Didymium iridis and other species

Collins (1966) showed that somatic incompatibility occurred among plasmodial strains derived from a Honduras isolate of *D. iridis*. Strains fell into several classes, and members of a class were able to fuse with each other but not with those of other classes. Collins and Clark (1968), studying progeny of the Honduras strain, found that somatic incompatibility was controlled by five loci with dominant and recessive alleles at each locus, and plasmodia that were phenotypically identical at all the loci were able to fuse. Ling and Collins (1970) demonstrated a similar system controlling somatic incompatibility in a Panama isolate of *D. iridis*, there being six loci with dominant and recessive alleles. Examination of hybrids between the Honduras and Panama isolates showed at least eleven loci controlling somatic incompatibility in *D. iridis*, but did not establish which of the loci demonstrated in the Panama strain were identical with those already known in the Honduras isolate (Collins and Ling, 1972). Studies aimed at correlating the loci in the two isolates are in progress, and it seems likely that a total of thirteen somatic incompatibility loci will be demonstrated (Clark and Ling, personal communication).

Clark and Collins (1973), studying progeny of the Honduras isolate, found that incompatibility sometimes took the form of failure to fuse. With some combinations of strains fusion occurred but was terminated in a few minutes by a "cytotoxic reaction" which produced a "clear zone" of apparently dead protoplasm. These investigators found that when strains differed at a single incompatibility locus the reaction was unilateral and the damage occurred in the recessive strain. Also, the area destroyed was often large, but with differences at several loci the reaction was more limited. The explanation suggested for this apparently paradoxical result was that a slow, weak incompatibility reaction would permit protoplasmic mixing to continue for a long period and so when damage ultimately occurs a large area is affected. On the other hand, a strong reaction involving differences at several loci will terminate fusion before much mixing has occurred, and hence the area damaged will be small. It could be argued that with an extremely strong reaction there will be virtually no mixing and damage will be impossible to detect. This is actually how Clark and Collins explain instances of nonfusion.

Ling and Ling (1974), studying the effects of the eleven incompatibility loci known in the Panama isolate, distinguish between seven "strong" or "fusion" loci and four "weak" or "clear zone" loci. Differences in phenotype at any fusion locus results in failure of plasmodia to fuse; phenotypic identity at the fusion loci permits fusion and hence possible expression of differences with respect to the clear zone loci. The cytotoxic reaction produced by differences at the clear zone locus is very fast, the effects being visible in 2 to 3 minutes or even 30 seconds after fusion. Ling and Upadhyaya (1974) showed that with single gene differences in the clear zone loci the effects were unilateral, the recessive strain being damaged. They also found that it was sometimes possible to recover viable plasmodia from the apparently dead clear zone, and that the genotype of the surviving material was determined by the clear zone loci in a way that suggested nuclei dominant at these loci tended to survive. Ultrastructural studies have been carried out on clear zones (Upadhyaya and Ling, 1977); however, so much damage occurs so soon that it is unclear which organelles are damaged first.

Few studies have been done on somatic incompatibility in myxomycetes other than *P. polycephalum* and *D. iridis*, but Carlile (1974) has observed both fusion incompatibility and postfusion lethal reactions in *Badhamia utricularis,* and Clark (1977) has demonstrated two loci in *Physarum cinereum* which produce cytotoxic reactions in the recessive phenotype shortly after fusion.

2.3.3. Somatic incompatibility in myxomycetes

Work on *P. polycephalum* and *D. iridis* has demonstrated a wide variety of somatic incompatibility reactions that may occur when plasmodia of different strains meet. Some confusion has resulted, however, as a consequence of authors classifying two types of reactions that look completely different as the same on the basis of a supposed common mechanism. It seems best to classify reactions that are visibly distinguishable, which leads to the recognition of four types of incompatibility, as follows:

1. *Fusion incompatibility.* Two plasmodia do not fuse with each other. The possibility that (difficult to observe) tenuous and transient connections are formed is not excluded, however, and further work may lead to fusion incompatibility being subdivided on the basis of the presence or absence of such connections.

2. *Fast postfusion reactions.* The plasmodia fuse, but this fusion is rapidly ter-

minated by a local cytotoxic effect within seconds or minutes.

3. *Slow lethal reaction.* The plasmodia fuse, and after several hours a lethal reaction causes widespread destruction.

4. *Heterokaryon incompatibility.* The plasmodia fuse, and although no obvious damage is seen at any stage, one of the plasmodial genotypes disappears.

As indicated in the previous section, Clark and Collins (1973) have suggested that what we term "fusion incompatibility" is the result of a "postfusion reaction" so fast and effective as to prevent any observable mixing and hence any damaged region. Possibly, the two forms of incompatibility have a common basis, although in practice it seems wise to discriminate between those instances in which fusion is not observed and those in which dead areas are produced. There are also strong indications that the "slow lethal reaction" and "heterokaryon incompatibility" both have a common genetic basis in differences at the *let* loci and a common ultrastructural basis in the destruction of nuclei of one genotype (Border and Carlile, 1974; Lane and Carlile, to be published). Here again, it may be best to discriminate between the reactions, since not all genetic differences that can cause heterokaryon incompatibility may be capable of provoking a lethal reaction under appropriate conditions.

3. Fungi

The fungi are a diverse collection of organisms containing groups that are phylogenetically distinct. For example, Margulis (1974) splits them between two kingdoms in her five-kingdom classification; those that can form flagellated cells are placed in the Protista and those that cannot are placed in the Fungi. The fungi are all heterotrophic and saprophytic or parasitic. Their characteristic growth form is either as hyphae, cylindrical tubes growing apically and branching subapically to form a ramifying mycelium, or as yeast cells, which lack the polarity of the hyphae, and grow by budding. Many fungi are capable of growing as hyphae or yeast cells depending on environmental conditions, and this dimorphism is an important feature of some parasites. Both growth forms are protected by a rigid cell wall, a major factor in the way of life of the organism and in determining its shape and mode of growth. The major components of hyphae and yeast cell walls are polysaccharides. Cell wall composition varies considerably among different fungi, but Bartnicki-Garcia (1970) shows how they can be classified, according to the occurrence of particular polysaccharides, into groups that closely reflect the phylogenetic relationships of the fungi producing them. For example, as characteristic components the oomycetes have cellulose and β-1,3-glucan, the zygomycetes have chitin and chitosan, and the ascomycetes and basidiomycetes have chitin and glucans.

Except for rare examples, such as the plasmogamy of naked gametes of Allomyces, the fusion of two fungal cells must first involve the coalescence of their walls. In some cases evidence exists for the presence of complementary components on the surfaces of fungal cells that mediate this early stage of cell fusion. The eventual apposition of the two plasma membranes must require the prior localized dissolution of the two walls by specific lytic enzymes, but there is no direct evidence for this hypothesis. The role of the plasma membranes in fungal fusion is proving difficult to elucidate because they are so well protected from investigation by the complex walls; work with artifically prepared protoplasts should help resolve this issue. The fusion of the two membranes is followed by the intermingling of the two cytoplasms and, in the case of sexual fusion, eventually by karyogamy. Nothing is known about the biochemical control of these two final stages, but there is a great deal of genetic information available.

3.1. Life cycles

Burnett (1976) recognizes five basic life cycles among the fungi: *asexual,* with no known sexual stage (e.g., many species of Aspergillus and Penicillium); *haploid,* with meiosis closely following karyogamy (e.g., Mucor); *haploid-diploid,* with a regular alternation of these phases (e.g., Allomyces); *diploid,* with haploid cells occurring briefly as gametes (e.g., Phytophthora); and *haploid-dikaryotic,* with the plasmogamy of two haploid cells being followed by a period in which the pairs of nuclei coexist and divide synchronously until karyogamy occurs. In many ascomycetes, such as Neurospora, the dikaryon has a brief existence just prior to ascus formation; but in most basidiomycetes, such as Coprinus, the dikaryon is the characteristic form and may grow for many generations before karyogamy and meiosis occur in the basidium.

During these life cycles two types of fusion can occur: somatic fusion between vegetative hyphae in many ascomycetes and basidiomycetes, and sexual fusion between haploid sexual hyphae or gametes.

3.2. Somatic fusion

Fusion between vegetative hyphae is very important in the life of many fungi, primarily because it gives rise to three-dimensional networks of mycelium which may be heterokaryotic. With few exceptions, such as Mortierella and Endogone, fusion is unknown among the phycomycetes, but is common in the higher fungi. The proposed sequence of events is as follows: two hyphae, with chemotropic attraction between them, grow towards each other; their walls fuse; their walls dissolve; their plasma membranes fuse; and their cytoplasms intermingle. Related phenomena to be considered here include flocculation or surface fusion, without subsequent wall dissolution and membrane fusion, that occurs between cells of brewer's yeast, and may have been empirically selected for by generations of brewers; the laboratory technique of fusing protoplasts, bypassing the early stages of natural fusion, and in some cases giving rise to fusion unknown in nature; and the surface properties of vegetative cells that have significant interrelationships with other organisms.

3.2.1. Positive autotropism between vegetative hyphae

This phenomenon is common among the ascomycetes and basidiomycetes, where it is a prerequisite for fusions between neighboring hyphae. Many observers, notably Buller (1931, 1933), have described and illustrated how a vegetative hyphal tip will turn toward a neighboring hypha of the same species. This action at a distance must be mediated by chemicals, and has been termed telemorphosis. The agents involved have not been identified, but are probably species-specific macromolecules with a radius of action of about 10 μm, and are unstable (Raper, 1952; Park and Robinson, 1966; Gooday, 1975; Burnett, 1976).

3.2.2. Observations of somatic hyphal fusions

Buller (1931, 1933) made extensive observations of hyphal fusions in vegetative cultures. Among the functions that he lists are the following: to provide a network of mycelium through which nutrients and cytoplasm can pass in any direction; to enable mycelium to resist injury by providing alternative routes; to allow repair of an injury by bridging gaps; and to provide social organization so that several mycelia of the same species can cooperate (e.g., to give a fruiting body). He classifies four types of fusion: hypha-to-hypha fusion, between two hyphal tips attracted from about 15μm apart; hypha-to-peg fusion, between a hyphal tip and a peg, produced as a side branch from an older hypha in response to the advancing hyphal tip; peg-to-peg fusion, with mutual induction of two side branches on adjacent hyphae; and hook-to-peg fusion in the specialized formation of clamp connections in many dikaryotic basidiomycetes. In these last three cases, some observers doubt the formation of pegs, and suggest that a hyphal tip fuses directly with the lateral wall of another hypha without a peg being formed. For example, autoradiographs of wall synthesis during hyphal fusion in young hyphae of *Penicillium claviforme* and *Coprinus cinereus* show no incorporation of radioactivity into lateral walls in response to an approaching tip (S. A. Watkinson, personal communication). However, in view of the formation of a peg in some cases it seems reasonable to suppose that there is always some positive response to the approaching hyphal tip, with some localized preparation of the wall for the impending contact (i.e., the fusion is mutual and not one-sided).

Very little is known about the control of occurrence and frequency of hyphal fusion. The incorporation of 1% activated charcoal greatly increases the rate of heterokaryon formation of *Rhizoctonia solani* (Butler and Bolkan, 1973; Bolkan and Butler, 1974). Increasing the temperature from 28° to 37 °C results in a tenfold increase in fusion frequency for strains of *Coprinus cinereus* (Smythe, 1973), but Ahmad and Miles (1970 a, b) report no differences at 20°, 25° or 30 °C for *Schizophyllum commune*. They do report increased fusion frequencies in the absence of a metabolizable carbon source, and complete inhibition of fusion in the presence of 1 mM copper sulphate. The different strains of *Coprinus* used by Smythe showed widely different fusion frequencies. Bourchier (1957) counted the number of hyphal fusions in cultures of *Corticium vellereum* growing on different media and found very different responses from varying strains, with one strain showing fewer and another strain more fusions when the concentration of nutrients in the medium was increased. Thus, although it appears that hyphal fusion frequencies vary with changing conditions, no consistent pattern in their control has yet emerged.

3.2.3. Vegetative hyphal fusion and somatic incompatibility

The review by Esser and Blaich (1973) on heterogenic incompatibility is an important source of information and references on incompatibility associated with hyphal fusion in fungi. Some more recent developments and resulting perspectives on older work will be discussed below.

3.2.3.1. Neurospora crassa.

Mating in the ascomycete *Neurospora crassa* will occur only between strains carrying a mating type allele *A* and strains with the corresponding allele *a*. The production of heterokaryons by somatic hyphal fusion is limited, however, to pairs of strains identical in mating type (Beadle and Coonradt, 1944). Hence it appears that the mating type gene has a dual role—promoting sexual fusions, and discouraging somatic fusion when the two alleles *A* and *a* are brought together. Newmeyer and co-workers (1973) have confirmed that these two activities are due to a single gene—a thorough test of the possibility that two closely linked genes were involved yielded negative results. Earlier, Newmeyer and Taylor (1967) had obtained partial diploids heterozygous for mating type. These too showed incompatibility, very poor growth taking place until reversion to the homozygous state occurred. A "tolerance" gene has, however, been detected that permits mating type heterozygotes to grow well (Newmeyer, 1970). *Neurospora tetrasperma,* a pseudohomothallic species, has spores carrying two nuclei of different mating type, and a mycelium heterokaryotic with respect to mating type. If these mating type genes are introgressed into an *N. crassa* background, then incompatibility between them is observed, indicating the presence in *N. tetrasperma* of a gene or genes corresponding to the "tolerance" gene (Metzenberg and Ahlgren, 1973). These authors comment on the "brief truce" around the time of mitosis between the mating type alleles of *N. crassa,* during which period they cooperate in promoting sexual fusion and karyogamy.

Garnjobst (1955) studied two other genes, *C* and *D*, with alleles *c* and *d,* which also controlled heterokaryon formation in *N. crassa*. Identity at both alleles was needed for heterokaryons to be produced. Fusion took place between hyphae that differed with respect to these alleles or the mating type locus, but vacuolation and death of protoplasm in the compartments immediately adjacent to the anastomosis soon occurred (Garnjobst and Wilson, 1956). These effects could be produced by microinjection of heterologous but not homologous protoplasm and extracts (Wilson et al., 1961). The active principles in the extracts appear to be proteins (Williams and Wilson, 1966) but this promising approach seems to have been terminated because of technical difficulties (Williams and Wilson, 1968). A further pair of incompatibility alleles, *E* and *e,* were discovered by Wilson and Garnjobst (1966). The genes studied by these workers are now known as *het-C*, *het-D*, and *het-E*. Perkins (1975) showed that, as with the mating type gene, partially diploid strains heterozygous for *het* genes grow poorly until reversion to the homokaryotic state occurs. Mylyk (1975, 1976), using Perkin's methods, has sampled natural populations and identified six new heterokaryon incompatibility genes, designated *het-5* to *het-10*. The somatic incompatibility reactions studied in *N. crassa* are postfusion reactions. However, Wilson and Garnjobst (1966) also observed, but did not study in detail, fusion incompatibility—failure of hyphae to fuse—in some combinations of wild-type strains.

3.2.3.2. Podospora anserina. Extensive studies on incompatibility in the ascomycete *Podospora anserina* have been carried out by Esser's group in Germany and Bernet's in France. Unfortunately the two groups have used different symbols for the genes they studied, but a "French-German dictionary" has been provided by Esser (1974b) in a useful review of the genetics of the fungus.

The incompatibility reactions that occur when two strains of Podospora meet may take several different forms. "Barrage" consists of a zone of dead hyphae separating two colonies of different strains. If the colonies differ in mating type, which should permit perithecium production (sexual sporulation), perithecia may occur on one strain only ("semiincompatibility") or on neither strain ("reciprocal incompatibility"). Although these effects could be considered examples of sexual incompatibility, they are due to genes also responsible for somatic incompatibility and can thus be considered here. Finally, certain combinations of incompatibility genes result in strains which are nonviable or grow very poorly ("self-incompatibility").

Several genes in Podospora can result in postfusion allelic incompatibility. For example, the fusion of a hypha carrying the allele u with one carrying u_1 will result in the process being terminated by cell death. In addition "nonallelic" incompatibility is also known in Podospora, the reaction occurring due to an encounter, in heterokaryon or heterozygote, between alleles of different genes (e.g., a and b). Thus somatic cell fusion, which brings together nuclei carrying alleles a_1 and b, leads to the death of the cell; the presence of these two alleles in an ascospore nucleus will produce a strain of poor viability. That the reaction is nonreciprocal is demonstrated by semiincompatibility in sexual fusion; a male gamete (spermatium) carrying a_1b_1 can fertilize a female carrying ab, but a male ab cannot fertilize a female a_1b_1. Esser (1965) explains that an a_1 nucleus is the killer and b is the "sensitive," and the nuclei in the female receptor organ (trichogyne) are destined to disintegrate in the course of the fertilization process, so that damage to a trichogyne ab nucleus does not matter, whereas damage to an ab spermatium terminates fertilization. A second system of nonallelic incompatibility involves interaction between alleles c_1 and v. Reciprocal incompatibility with respect to sexual sporulation occurs when the two strains are "killers" and "sensitives" with respect to different systems, as for example between a strain carrying a_1b_1cv (killer with respect to a_1, sensitive at v) and abc_1v_1 (killer with respect to c_1, sensitive at b).

Blaich and Esser (1971) have shown that the incompatibility reaction involves the production of aminopeptidases and other catabolic enzymes. Since then much work on both biochemical and genetic aspects of nonallelic incompatibility in Podospora has been carried out by Bernet and co-workers. Bégueret and Bernet (1973) found that with some genes the incompatibility reactions were temperature-dependent and the reaction in a self-incompatible strain could be initiated conveniently by lowering the culture temperature from 32° to 26° C. A sixfold increase in proteolytic enzyme activity accompanied the reaction. Bernet and associates (1973) showed that the genes *mod1* and *mod2* could suppress an incompatibility reaction between two strains, apparently by an effect at the translational stage of proteolytic enzyme formation. Boucherie and Bernet (1974) found a strain that was female sterile (failed to produce protoperithecia) when mutations at both *mod* loci were present. This and a later study (Boucherie et al., 1976) led to the conclusion that proteolytic enzymes involved in both incompatibility and sexual morphogenesis were regulated by the *mod* genes. The biochemical basis of nonreciprocal incompatibility has also been studied (Labarète et al, 1974) and two new "nonallelic" incompatibility systems created (Delettre and Bernet, 1976).

3.2.3.3. Other fungi. A great deal of work has been carried out with Aspergillus on the genetic control of heterokaryon formation by somatic cell fusion by Caten, Jinks and co-workers (e.g., Jinks et al., 1966; Caten, 1971, 1972; Croft and Jinks, 1977). To date these studies have been directed mainly toward establishing whether heterokaryons are readily formed in natural populations, and not to the elucidation of physiological aspects of fusion and incompatibility. Clearly, many loci are involved, and differences between strains of *Aspergillus amstelodami* at some loci prevent heterokaryon formation and at others (e.g., *hetE*) they merely reduce frequency (Caten, 1973). Differences at one locus *(hetB)* will also act on sexual recombination, reducing frequency of outbreeding. The incompatibility loci will also influence the spread of cytoplasmic genetic factors between strains, *hetB* differences totally blocking transmission, *hetE* having no effect, and *hetA* and *hetC* differences restricting rather than wholly preventing spread (Handley and Caten, 1975). Recently Dales and Croft (1977) have obtained heterokaryons between strains of *Aspergillus nidulans* that differ at two incompatibility loci (designated *hetA* and *hetB*) by fusion of protoplasts combined with a selective procedure.

Somatic incompatibility has also been studied in the basidiomycete *Thanatephorus cucumeris* (imperfect stage = *Rhizoctonia solani*). Flentje and Stretton (1964) showed that when hyphae from differing strains met they might either: (1) grow over each other without fusing; (2) attach to each other so firmly that they could not be separated, without cytoplasmic connections forming (i.e., "wallfusion"); (3) fuse, with subsequent death of one or more cells of either or both partners, adjacent to the anastomosis; or (4) fuse without adverse reaction.

It seems that items 1 and 2 are different forms of fusion and item 3 of postfusion incompatibility. McKenzie and co-workers (1969) found that the lethal reaction prevented heterokaryon formation at 5°C but was not wholly effective in so doing at 25° C; both stable and unstable heterokaryons were formed. It appears that at the higher temperature some heterokaryotic hyphae can outgrow the lethal effect and undergo a form of adaptation. A figure showing the lethal effect has been published (Flentje et al., 1970), and it has been concluded that heterokaryon formation in *T. cucumeris* is influenced by heterogenic incompatibility which will restrict sexual and parasexual outbreeding (Stretton and Flentje, 1972). However, in heterokaryon formation in other strains of *T. cucumeris,* heterogenic incompatibility appears to be overruled by a system of homogenic incompatibility resembling that controlling basidiomycete sexuality and promoting outbreeding (Anderson et al., 1972).

3.2.4. Flocculation in yeasts

The ability to flocculate (as opposed to the sexual agglutination discussed later) is an important property to be considered in the use of *Saccharomyces* and other yeasts for fermentations. For the brewer it is useful for the cells of a "bottom yeast" to stick together and sediment as the fermentation is finishing; for the biochemist with a chemostat, flocculation is undesirable (Morris, 1966; Rainbow, 1970; Pringle and Mor, 1975). There is a very wide variation in the degree of flocculation shown by different strains. It is clearly a surface phenomenon, but the molecular mechanisms are uncertain. Surface phosphomannan-protein complexes are implicated in flocculation of *Saccharomyces cerevisiae,* perhaps with calcium ions bridging the cells (Mill, 1964; Lyons and Hough, 1971; Taylor and Orton, 1975). Some analyses suggest that flocculent strains have higher mannan or phosphomannan contents in their walls than nonflocculent strains (Lyons and Hough, 1971; Stewart et al., 1973), and flocculence and phosphomannan-enzyme complexes are lost concomitantly following osmotic shock (Williams and Wiseman, 1973). Stewart and Goring (1976) have further investigated the effect of cations on flocculation. In the more flocculent strains magnesium and manganese ions could induce flocculation, but not as strongly as calcium ions; and in the most flocculent strains the monovalent cations, sodium or potassium, could also induce flocculation. Stewart and Goring suggest that the divalent cations are forming bridges between negative charges on adjacent cells, but that the monovalent cations are acting as "counter ions," and neutralizing these strong negative charges to allow nonionic binding between the cells.

Some strains of *Saccharomyces cerevisiae* appear to require the presence of peptide "inducers" in the growth medium in order to flocculate. (Stewart et al., 1975). These authors point out that the peptides have some similarities, such as a high content of acidic amino acids, to the yeast sex hormone α-factor, which controls sexual agglutination of α and a cells but there is no obvious relationship between sexuality and flocculation. Patel and Ingledew (1975a, b) have found a direct relationship between degree of flocculence and intracellular glycogen content for several strains and mutants of *Saccharomyces* and suggest that glycogen is a "physiological determinant" of flocculation by some unknown mechanism.

Day and colleagues (1975) correlate flocculence of *Saccharomyces* with fimbriae (fibrils seen in electron micrographs protruding from the cell surface). Flocculent strains have dense fimbriation; nonflocculent strains have sparse fimbriation. Strains grown in conditions that inhibit flocculation, or treated with en-zymes such as pronase or α-amylase, which inhibit flocculation, show sparse fimbriation. The fimbriae are about 500 nm long and 5 to 7 nm wide, and Day and co-workers (1975) suggest that they are composed of mannan-protein. Simi-lar fimbriae with roles in sexual agglutination are discussed later.

A related phenomenon is that of "co-flocculation," where two strains are nonflocculent alone, but become flocculent when mixed together (Stewart and Garrison, 1972; Stewart et al., 1973). The mechanism of coflocculation appears to be similar to that of flocculation, in that calcium cross-bridges between anionic surface components are implicated. Genetic studies of brewer's yeast strains are very difficult as many are triploid or aneuploid, and are reluctant to sporulate, but there are at least four flocculation genes (Lewis et al., 1976; Stewart et al., 1976).

3.2.5. Induced fusion of fungal protoplasts

This phenomenon is clearly quite distinct from the hyphal somatic fusions, as removal of the hyphal wall has two effects: it removes a physical and chemical barrier between the plasma membranes of the two cells involved, but also re-moves any recognition sites or enzymic activities in the walls that might be in-volved in the control of "normal" fusions. Thus a comparison between mechanisms of fusion of two hyphae and fusion between corresponding proto-plasts could be fruitful in elucidating the relative roles of wall and membrane in the control of fusion.

Protoplasts can easily be made from yeasts and mycelial fungi by digestion of the cell wall with an appropriate mixture of lytic enzymes (Peberdy et al., 1976). However, as with other protoplasts, they show a very low rate of spontaneous fu-sion. Fusion of fungal protoplasts has been studied by looking for growth after mixing protoplasts of two different auxotrophic mutants and allowing them to regenerate on minimal medium. Ferenczy and colleagues (1974) obtained fre-quencies of complementation of 1.6×10^{-6} per protoplast pair at pH 5.5 and 6×10^{-7} at pH 7 when they mixed protoplasts of two mutants of *Geotrichum candidum* with mannitol or glucose as stabilizer. Vegetative anastomoses or sexual fusions have never been observed in this fungus. The resulting heterokaryons were only stable on minimal medium, as they broke down to yield auxotrophs again on a complete medium. Higher fusion frequencies were obtained with protoplasts of *Asperigillus nidulans,* after centrifuging to form aggregates of about 5 to 20 pro-toplasts (Ferenczy et al., 1975), and with protoplasts of *Phycomyces blakesleeanus* (a frequency of 6×10^{-5}) in a fusion medium containing 0.2 M-$Ca(NO_3)_2$ (Binding and Weber, 1974). Again, *P. blakesleeanus* has never been observed to undergo vegetative anastomoses. Protoplasts of *Saccharomyces cerevisiae* fuse at a higher frequency (ca 1×10^{-4}) than do the intact cells (ca 1×10^{-6}), and this frequency is increased by the addition of the lectin concanavalin A, and sodium nitrite, which together apparently enhance adherence of the protoplasts (Mulony et al., quoted by Crandall, 1976).

Very much higher fusion frequencies can be obtained by use of solutions of polyethylene glycol (Anné and Peberdy, 1975, 1976; Ferenczy et al., 1976). This is a fusigenic agent for plant protoplasts and animal cells and can even give rise to chimeras such as yeast protoplasts fused with hen erythrocytes (Ahkong et al., 1975). Using 30% polyethylene glycol (6000 MW) in the presence of calcium chloride, Anné and Peberdy (1976) have obtained fusion frequencies of up to 0.7% for fusion of protoplasts from pairs of auxotrophic mutants of *Penicillium*, *Aspergillus*, and *Cephalosporium*. Most of these prototrophic progeny were heterokaryons, but some were diploid. For protoplasts of *Penicillium chrysogenum*, Ca^{2+} greatly stimulated fusion, with an optimum of 0.01 M-$CaCl_2$ at pH 7.5 and of 0.6 M at pH 9. Fusion occurred at 4 °C, but increased with rising temperature until there was a sudden drop above 37.5 °C. A pH value of 9.0 was optimum. Maximum fusion frequencies were induced by the polyethylene glycol after a very short incubation time. Anné and Peberdy suggest that these first few minutes are required for the protoplasts to become entirely "coated" with the polyethylene glycol, but that once they are coated fusion can proceed.

Dales and Croft (1977; see also section 3.2.3.3.) have used this technique to fuse protoplasts of *Aspergillus nidulans* strains of different compatibility groups whose hyphae would not fuse naturally. Heterokaryotic and probably diploid progeny were recovered from fusions of two such auxotrophic strains.

F. Kevei and J. F. Peberdy (personal communication) have succeeded in producing interspecific hybrids by inducing fusion between protoplasts of auxotrophic mutants of *Aspergillus nidulans* and *Aspergillus rugulosus* with polyethylene glycol. The resultant heterokaryons regenerated hyphal walls and grew, but only very slowly, to produce irregular colonies. The parental strains readily segregated when the heterokaryons were grown on complete medium. However, after being grown for a long time on minimal medium, vigorously growing sectors tended to appear. These "hybrid" mycelia were characterized by their stability on complete medium and their secretion of a brown pigment into the medium. They produced a poor crop of conidia, some of which were large, uninucleate, and germinable, while others were anucleate and sterile. The cells of the hybrid mycelia contained only about half as many nuclei as either parent, and analysis of protoplasts made from them showed about 10.8 μg DNA per nucleus compared with values of 4.7 and 5.4 from the parental strains. Kevei and Peberdy conclude that the stable hybrid mycelia were diploid. Thus the intermingling of the two parental cytoplasms led to the fusion of the two nuclei. A breakdown of the diploid to yield parental and recombinant sectors was induced by growth in the presence of the fungicide benomyl.

Observations made during the fusion of animal cells and yeast protoplasts (Ahkong et al., 1975) confirmed that the alien plasma membranes fused, producing cytoplasmic bridges between the two cells. The maximum interspecific fusion rate, of about 0.1% of the hen erythrocytes, occurred at hen to yeast cell ratios of 1:1 or 1:10. When ratios of 1:50 or 1:250 were used, very large protoplasts resulted, probably from the fusion of several yeast protoplasts. These giant yeast protoplasts had only a few large intracellular vacuoles, indicating that the vacuolar membranes had also fused.

The technique of inducing protoplast fusion should prove very fruitful both in obtaining new strains and in elucidating relationships and interactions between different cells.

3.2.6. The role of fusion in fungal pathogenesis and symbiosis

Fungi can parasitize other fungi, algae, protozoa, animals, or plants by making contact with and entering their cells, with varying amounts of lysis. In many cases, particularly with the phytopathogens, there is a high degree of specificity, as the pathogen is often highly selective to particular races of the host species. The corollary is that some races are specifically resistant to the pathogen. This very specific phenomenon is the "gene-for-gene host-pathogen system." Albersheim and Anderson-Prouty (1975) suggest that the specificity lies in the cell wall of the pathogen. They view the product of the plant's resistance gene as a plasma membrane receptor that can bind a molecule, the "elicitor," produced by the pathogen; consequently a hypersensitive defense reaction is triggered by the potential host. The pathogen will be virulent when it comes into contact with a host lacking this receptor. These investigators suggest that the elicitors are carbohydrate-containing molecules on the surface of the pathogens that are given a high degree of specificity by glycosyl transferases. If a host plant acquires the ability to recognize and bind the elicitor, and activate a defense system, it becomes resistant. If the then avirulent pathogen can change its surface component, perhaps by acquiring an altered specificity of a glycosyl transferase, it will regain its virulence.

Specific binding between host and parasite has been suggested as a recognition mechanism, for example between conidia of *Cladosporium cucumerium* and host cucumber cells (Raa et al., 1977), and between the chytrid *Rozella allomycis* and its closely related host Allomyces (Held, 1974). The zoospores of Rozellá attach themselves strongly only to the hyphae of Allomyces, not to their contiguous rhizoids and not to hyphae of the closely related Blastocladiella. Held draws an analogy between parasitism and fertilization, suggesting that the specificity of this parasitism arose via a sexual recognition system. Burgeff (1924) made a similar suggestion after observing what appeared to be sex-specific parasitism by members of the Mucorales on saprophytic mucoraceous hosts, but Satina and Blakeslee (1926) investigated this much more fully and could only find a slight tendency for more vigorous parasitism in some cases where host and parasite were of opposite mating types.

The mutualistic symbioses of fungi with algae to form lichens and with higher plants to form mycorrhiza are also very specific, and it is to be expected that surface recognition phenomena between the symbionts play an important role in their establishment and maintenance. To date, however, specific components and mechanisms have not been characterized. There must be physical and biochemical adaptations in the areas of the fungal cells in contact with the algal or plant cells, both to allow the transport of materials between them and to trigger the characteristic morphological development of the fungus. Peveling (1969) has shown for several lichens that the fungal plasma membrane is very convo-

244

luted just underneath where the wall is in contact with an algal cell, implying a transporting role for the increased surface area. Smith (1974) suggests that a very specific physical contact leading to transport between phototroph and heterotroph via localized changes in membrane properties may be essential for stable symbiosis.

3.3. Sexual fusion

Four of the five types of life cycles found in the fungi involve sexual fusion. This may involve morphologically differentiated sexual cells, or gametes, or cells showing no apparent differentiation but with constitutive or inducable surface components involved in sexual fusion.

3.3.1. Fungal sex attractants
There are now many examples of very specific hormones controlling the sexual interactions between two fungal cells that lead to plasmogamy (see reviews by Gooday, 1974, 1975, 1978; Bu'Lock, 1976; Crandall, 1977). In a consideration of cell fusions, these hormones can have two distinct roles: (1) to attract the cells from a distance so they find each other efficiently and come into contact, and (2) to cause a differentiation of the cell surfaces so they become mutually adhesive. Sex hormones are probably very widespread among the fungi, but to date only a few have been fully characterized. The chemical structures of only two of the sex attractants are known.

Sexual fusion of the motile male and female gametes of the chytridiomycete Allomyces is discussed later, but prior to fusion the male gamete is attracted to the female gamete by the specific female metabolite, sirenin, a sesquiterpene (Fig. 4a). Sirenin is active at very low concentrations—the bioassay can detect a 50 picomolar solution—and is rapidly inactivated by the responding male cells (Carlile and Machlis, 1965; Machlis et al., 1968; Machlis, 1973).

In the oomycete water mold, Achlya, the female hyphae produce a steroid hormone, antheridiol, (Fig. 4b) that has a variety of effects on recipient male hyphae, one of which is the chemotropism of the resultant antheridial branches toward the source of the antheridiol, the female cell. As with sirenin, this chemotropic response is very sensitive and specific, and the male cells rapidly metabolize the antheridiol (Barksdale, 1969; Musgrove and Nieuwenhuis, 1975).

In Mucor and its relatives, the sexual hyphae (zygophores) attract each other by complementary volatile chemotropic attractants. Although they are not fully characterized, it appears likely that the two chemicals responsible will prove to be mating-type specific precursors of the sex hormone, trisporic acid (Fig. 4c), that directly causes the production of the zygophores themselves (Mesland et al., 1974; Bu'Lock et al., 1976).

These three examples are from phylogenetically different fungi, but they do have features in common: the attractants are very specific chemicals, are produced in very small amounts, are active in even smaller amounts, and are metabolized by the recipient cell, which would have the effect of accentuating their concentration gradients. For these three hormone systems, there are

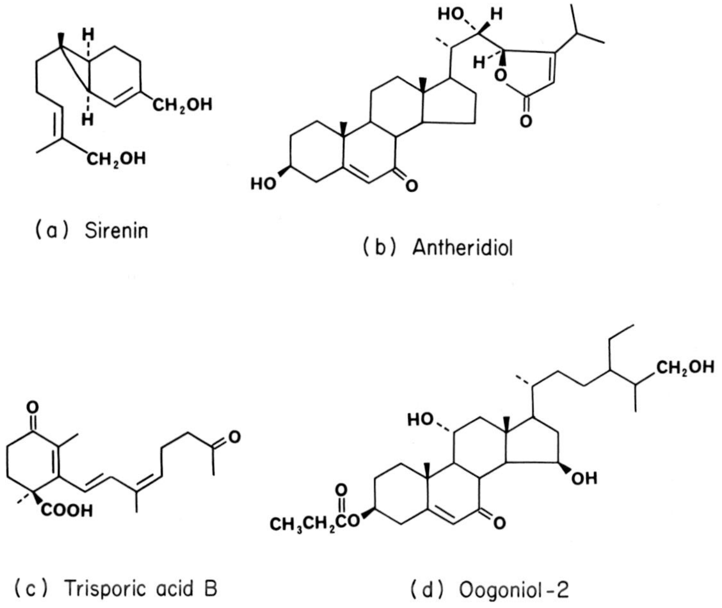

(a) Sirenin

(b) Antheridiol

(c) Trisporic acid B

(d) Oogoniol-2

H_2N-Trp-His-Trp-Leu-Gln-Leu-Lys-Pro-Gly-Gln-Pro-Met-Tyr-COOH

(e) Saccharomyces α-factor 1

Fig. 4. Structures of fungal sex hormones.

homothallic species where the cells producing the hormone and the cells responding to it can be from the same plant, and so are isogenic. Thus the production of, and response to, the hormone can be a phenotypic expression of sexuality. Two adjacent parts of a thallus assume the different chemical attributes of
the two mating types and acquire complementary hormone metabolisms and the
complementary mating surface components discussed further on.

3.3.2. Mechanisms of sexual fusion

Sexual fusion in fungi is specific to species and to mating type, and the assumption must be made that the surfaces of the sexual cells have recognition sites of
the necessary specificity. In some cases (e.g., Achlya, Mucor, and Saccharomyces)
the sexual differentiation involving the acquisition of these sites results from the
action of a sex hormone, in some (e.g., Ustilago) from differentiation triggered
by contact with surface components of the opposite mating types and in others
(e.g., Hansenula) the vegetative cells acquire the sexual agglutinins without any
chemical contact with the opposite mating type. Sexual differentiation can be induced by nutrient starvation, as for example in Allomyces, where the adjacent
male and female gametangia are formed on a hypha of a haploid plant. This has

246

the corollary that sexual differentiation can often be prevented, or reversed (with subsequent loss of ability to fuse with the opposite mating type) by addition of nutrients; for example, antheridial branches of Achlya will revert to vegetative hyphae (Barksdale, 1969).

At present there is very little information about the complete sequence of events leading to successful sexual fusion in any one fungus. Rather, we have a series of examples, each of which highlights one particular aspect of the process.

3.3.2.1. Hormonal control of sexual differentiation. Sexual differentiation of male and female vegetative hyphae to antheridia and oogonia in Achlya results from the action of the two steroid hormones, antheridiol and oogoniol (Fig. 4d), respectively (Barksdale, 1969; McMorris et al., 1975). Thus antheridiol has a dual action, as a morphogen and as an attractant for the resultant sexual hyphae. To date nothing is known of the hormone-induced surface changes that allow the antheridia to fuse with the oogonia.

In the Mucorales, one hormone, trisporic acid, is responsible for the diversion from vegetative growth and asexual reproduction to the formation of the zygophores. On meeting each other, zygophores of opposite mating types, (+) and (−), of *Mucor mucedo* fuse near to their tips, cease elongating, and initiate the developmental sequence leading to the production of a zygospore (Gooday, 1973; Ende, 1978). Vegetative hyphal fusions are very rare in the Mucorales, and the block to fusion is presumably at the cell surface, as viable heterokaryons result following microsurgery (Ootaki, 1973) or premature growth from immature zygospores (Gauger, 1977). Thus trisporic acid must cause each mating type to produce specific sites for sexual fusion on the zygophores. These sites have yet to be characterized, but lectin-binding studies suggest that zygophores have distinctive surface properties as compared to vegetative or asexual hyphae (Jones and Gooday, 1977).

The sexual agglutinability between the two mating types, *a* and *α*, of the yeast *Saccharomyces cerevisiae* is induced, or enhanced, by the complementary hormones, α-factor (Fig. 4e) and *a*-factor, produced by α cells and *a* cells, respectively (Stötzler et al., 1976; Sakurai et al., 1975; Shimoda et al., 1976). *a*/α Diploids do not mate, and do not produce or respond to these hormones.

a-Cells respond to α-factor in several ways: they elongate to become pear-shaped; within one generation they are arrested at the G1 phase of the cell cycle so that DNA replication and further cell division are blocked; they initiate, or greatly increase, the production of *a*-factor; and they gain the property of agglutinating with α-cells. The response of α-cells to *a*-factor is not so well characterized, but is likely to be analogous. The result is that when *a*- and α-cells are mixed, they synchronize their cell cycles and gain sexual agglutinins on their surfaces (Hartwell, 1973; Sena et al., 1973).

Lipke et al., (1976) have investigated the ultrastructural and chemical changes in the walls of *a*-cells responding to α-factor. The wall of the elongating part of the cell has a more diffuse outer surface than that of a vegetative cell, particularly toward the tip, and is much thinner, especially at the tip (Fig. 5). The nucle-

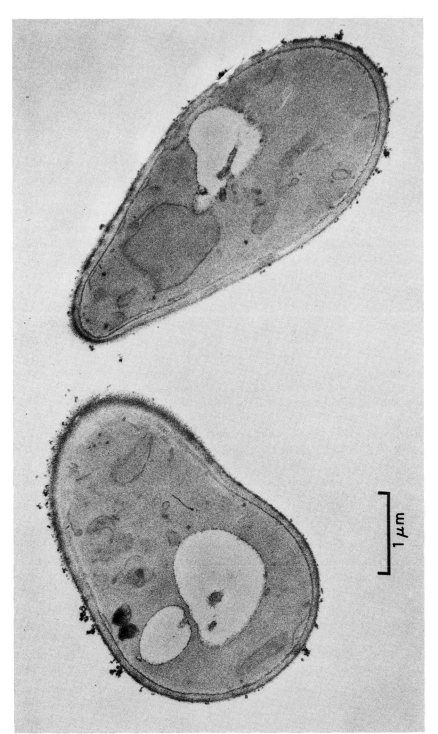

Fig. 5. Electron micrograph of sections of two cells of *Saccharomyces cerevisiae* responding to α-factor, 3 hr after addition of the hormone. Note the diffuse material near the tips of the cell extensions and the presence of the nucleus near the tip of the extension of one cell. Scale represents 1 μm. Reproduced with permission from Lipke, P. N., Taylor, A. and Ballou, C. E., (1976).

us is often seen to have migrated toward the tip. The wall becomes much more susceptible to lysis by the glucanase preparations from *Helix pomatia* and *Arthrobacter luteus,* and the ratio of glucan to mannan in the wall rises from about 1:1 to 1.5:1. Less mannan is produced than in vegetative cells, and it contains an increased proportion of shorter side chains and unsubstituted backbone mannose units. The apical portions of the elongating cells of both α- and a- mating type show a higher affinity for fluorescent concanavalin A, a lectin that especially binds mannans and glucans, and mating pairs of cells show a ring of fluorescence where they have fused (MacKay, 1977).

These changes in the wall are in preparation for mating, and perhaps their diffuse appearance in the electron microscope may reflect the sexual agglutinability of the cells. Cells that have just fused appear to be joined by this material, and have very thin walls at the point of contact, which autolyze as fusion proceeds to plasmogamy (Osumi et al., 1974). Sexual agglutinability of both a- and α-cells is lost by treatment with proteolytic enzymes such as pronase and α-chymotrypsin, but only α-cells are susceptible to trypsin (Shimoda et al., 1975). What appear to be glycoprotein sexual agglutination factors can be released from both mating types by treatment with snail digestive enzyme (Shimoda and Yanagishima, 1975). These factors do not agglutinate cells of opposite mating type, but are specifically adsorbed on their surfaces and inhibit their sexual agglutinability. Thus they behave as univalent factors, analogous to 21-factor of *Hansenula wingei.* Both factors can be purified by binding them onto heat-treated cells of opposite mating type at pH 5.5, and eluting them at pH 9. Molecular-exclusion gel chromatography indicates that they have molecular weights of about one million, and the most purified preparations from a- and α-cells have carbohydrate to protein ratios of about 4:1 and 6:1 respectively.

The roles of the hormones of Saccharomyces and the agglutination factors of Saccharomyces and Hansenula in controlling conjugation in yeasts are examples of widespread phenomena among the fungi. For example, *A* and α cells of the basidiomycete yeast *Rhodosporidium toruloides* show hormonal mutual induction of conjugation tubes which grow toward one another (Abe et al., 1975), nonbudding cells of the chromomycosis fungus *Phialophora dermatitidis* agglutinate in aggregates prior to conjugation (Oujezdsky and Szaniszlo, 1973), as do those of the homothallic Ascomycete, *Schizosaccharomyces pombe* (Calleja, 1970, 1973, 1974; Calleja and Johnson, 1971). Calleja terms this phenomenon "sex-directed flocculation." Only aerated, stationary phase cells of *S. pombe* will flocculate. They can be deflocculated reversibly by extremes of pH, by high temperatures, by detergents such as sodium dodecylsulfate, or by urea and guanidinium chloride, and irreversibly by treatment with proteolytic enzymes.

3.3.2.2. Nature of surface agglutination factors. The most detailed understanding of sex-specific surface components in the fungi is that of the agglutinins of the yeast *Hansenula wingei.* This is an ascomycete, living as a component of fungus gardens of bark beetles in coniferous trees. The haploid cells of the two strains, *5* and *21,* will agglutinate together very strongly in a mass of cells as a preliminary to mating (Crandall and Brock, 1968a; Crandall et al., 1977). Each strain does not agglutinate separately, and neither will agglutinate with the *5 × 21* diploid. Each strain produces a specific surface component, 5-agglutinin and 21-factor, respectively, that is specifically adsorbed by cells of the opposite mating type.

The 5-agglutinin will cause strain *21* cells to agglutinate, and so is multivalent. It can be released from strain *5* cells by treatment with enzymes such as subtilisin or snail digestive juice. It also occurs in the cytoplasm of strain *5* cells (Brock, 1965). It can be purified by affinity chromatography on strain *21* cells. It is a mannan-protein that appears as an aggregate of heterogeneous molecular weight. It is inactivated by treatment with pronase and exo-α-mannanase, and by disulfide-cleaving reagents such as mercaptoethanol (Taylor, 1964a,b, 1965; Taylor and Orton, 1967, 1968; Crandall and Brock, 1968a,b; Yen and Ballou, 1973). Yen and Ballou (1974) have further characterized a preparation released by treatment with subtilisin and shown it to be about 85% carbohydrate, 10% protein, and 5% phosphate. The amino acid composition shows a remarkably high proportion of hydroxyamino acids, with 52% of the residues as serine and 9% as threonine. The molecular weight increases as the concentration decreases, giving a maximum of close to one millon. The mannan chains are highly branched, containing $(1-2)\alpha$, $(1-3)\alpha$ and $(1-6)\alpha$ linkages, and exist as small oligosaccharides directly attached to the protein backbone through the serine and threonine units. This novel structure is distinct from that of the mannan-protein of the wall of *H. wingei*, which has long mannan chains of up to 200 mannose untis attached to the protein via *N*-acylglycosylamine linkages (Ballou, 1976).

The 21-factor is released from strain *21* cells by treatment with trypsin (Crandall and Brock, 1968a). It does not agglutinate strain *5* cells and so is apparently univalent, but it does have the properties required of a surface component responsible for agglutination of strain *21* cells with strain *5* cells: it is only released from strain *21* cells, which concomitantly lose their sexual agglutinability, it neutralizes the activity of 5-agglutinin, and it is adsorbed only by strain *5* cells. It is also a mannan-protein, with 35% carbohydrate, and an apparent molecular weight of about 40,000 (Crandall et al., 1974; Crandall, 1976).

The 21-factor and the 5-factor will form a complex when mixed in vitro (Crandall et al., 1974). The resultant complexes are large, corresponding to the combination of 63, 16, and 6 molecules of 21-factor with one unit of each of the three 5-agglutinin fractions used here of apparent molecular weights of 13×10^5, 5.6×10^5, and 2.2×10^5, respectively. These complexes are soluble, probably as the 21-factor is univalent and so cannot form cross-links. They presumably represent the primary binding mechanism when a strain *5* cell touches a strain *21* cell. Taylor and Orton (1970, 1971), using ^{35}S-labeled 5-agglutinin, found a very high apparent free energy of association for its binding to strain *21* cells. At pH 4, the value is -14.5 kcal/mole, which is higher than most antigen-hapten binding free energies. They suggest that this high value results from the additive effect of several binding sites on each molecule of 5-agglutinin.

The *5 × 21* diploid does not constitutively produce 5-agglutinin or 21-factor

(Crandall and Brock, 1968a). However, certain conditions can initiate the phenotypic production of one or other factor so that the diploid cells will agglutinate with *21* cells or *5* cells. For example, 5-agglutinin production results from the addition of 0.4 mM sodium vanadate or sodium molybdate; 21-agglutinin results from the addition of 1 mM disodium ethylenediaminetetraacetic acid (Na_2 EDTA) (Crandall and Caulton, 1973; 1975). These treatments have no appreciable effect on agglutinability of *5* cells or *21* cells, and so presumably are affecting just the relative expression of the two mating factor genes in the diploid, perhaps via the postulated complementary regulatory genes that are only active in the diploid in repressing each mating factor.

3.3.2.3. The role of fimbriae in sexual fusion. Ustilago violaceae is a plant pathogen, an anther smut of the Caryophyllaceae. It can be grown saprophytically in its haploid yeast form. Mating will occur between yeast cells of the two mating types, a_1 and a_2, to yield dikaryotic cells, which can infect a host plant and much later undergo karyogamy and meiosis to produce basidiospores in the host anthers.

A single locus with the two alleles a_1 and a_2 is the genetic control of mating in Ustilago, but the two mating types show distinct behavior. The a_2 cells can conjugate at any phase of the cell cycle, but for a_1 cells conjugation is restricted to the unbudded G1 phase (Cummins and Day, 1973). Heterozygous $a_1 a_2$ diploids (artificially selected and maintained by pairing two strains that are different auxotrophic and colored mutants) have a_2 mating ability in the S or G2 phase, but in the G1 phase they are active at both alleles and do not mate with a_1 or a_2 cells (Day, 1972; Day and Cummins, 1973). Thus the cell cycle expression controls of the mating type genes are still maintained in the diploids. This phenomenon has been termed "temporal allelic interaction."

Compatible adjacent cells conjugate in pairs by developing conjugation tubes (Poon et al., 1974). These two pegs elongate while in contact with each other, and the two cells become pushed apart. If the cells are separated by up to about 20 μm, conjugation can still be initiated by the a_2 cell producing a peg that grows to the a_1 cell.

An early stage of contact between the mating cells is through long fine hairs, fimbriae (Poon and Day, 1974, 1975; Day, 1976), and a_1 and a_2 cells can be seen by the light microscope, floating in pairs in a water suspension attached by these invisable connections. As they grow the conjugation tubes behave as if they follow the lines of the fimbrial connections between the cells. The fimbriae seem to be essential for conjugation, as cultures of strains that have lost their fimbriae have also lost the ability to conjugate. Ultraviolet light and inhibitors of protein or RNA synthesis will prevent the formation of conjugation tubes by cells that are only in contact via their fimbriae (Day and Cummins, 1974; Cummins and Day, 1974), and so the fimbriae may act as transmitters of regulatory chemicals to allow communication between the two cells. The fimbriae are from 500 to 1500 nm long and 6 to 7 nm wide. The number per cell varies from a few to several hundred. They appear to be proteinaceous, as they are almost completely digested by the proteolytic enzyme pronase. A few of the fimbriae end in a knob about 40 nm in diameter, and these may have a particular role, such as initiating the formation site of the conjugation tube.

Cells can still join in pairs when their fimbriae have been removed by treatment with pronase, and Day and Poon (1975) suggest that another surface component, visible in the electron microscope as amorphous globular material, plays a role in the primary adhesion between adjacent a_1 and a_2 cells. It is produced only in mating cultures, and is susceptible to digestion by α-amylase. Thus in the "courtship" period the cells communicate via this sex-specific amorphous material and also via the fimbriae. Conjugation is continued until the fused apical walls and then the plasma membranes of the two conjugation tubes dissolve to produce a cytoplasmic bridge between the two cells.

Fimbriae may also have a role in the sexual agglutination of *Saccharomyces cerevisiae* (Day et al., 1975) and *Hansenula wingei* (Crandall and Caulton, 1975), as they are seen in electron micrographs of sexually competent cells, and in the fusion of oidia with hyphae of *Psathyrella coprophila*. These oidia and these of some *Copuinus* species do not germinate, and are uninucleate sporelike cells that Kemp (1975a) suggests should be regarded as spermatia. The oidium of *Psathyrella coprophila* has a very distinctive surface ultrastructure; it is covered with filaments about 500 nm long and 10 nm wide (Jurand and Kemp, 1972). These filaments appear analogous to the fimbriae of Ustilago discussed earlier, and presumably play a role in the fusion of the oidium with the chemotropically attracted hypha. Unlike the Ustilago fimbriae they stain with ruthenium red, and so may be of acidic mucopolysaccharide.

3.3.2.4. Membrane fusion of motile gametes.   Allomyces, a member of the chytridiomycetes, produces five types of flagellate cells in its life cycle: male and female gametes, the resultant zygote, and haploid and diploid zoospores. Of these, fusion occurs only between a male and a female gamete. Both gametes are motile by means of one posterior flagellum, but there are many differences between them that must be phenotypic sexual differences since they are both produced by the same haploid plant and must be isogenic. Only the female produces the attractant, sirenin, and only the male (the spermatozoid) is attracted by it (Carlile and Machlis, 1965). The male swims much more actively, is bright orange with an accumulation of α-carotene, is much smaller than the female, and has fewer and smaller mitochondria (Figs. 6a, b). It also appears to lack the "sidebody complex," or "Stüben body," a complex of lipid granules and microbodies of unknown function that is present in the female, and has fewer gamma-like particles (Pommerville and Fuller, 1976).

Hatch (1938) gives us a light microscopic description of fertilization: "They came together smoothly, slid about over each other momentarily, and then fused with a rush. Upon this abrupt mingling of cytoplasms a mad swirling and streaming ensued which resulted in bizarre distortion for the young zygote. It seemed to boil and bubble over its whole surface . . . after seven minutes of this ebullition, the zygote rounded up and swam away . . ." He observed that fusion of young gametes (from 1 to 15 min after release from the gametangia) is "quick, almost instaneous," but it becomes increasingly slower as the gametes age and eventually becomes impossible.

Pommerville and Fuller (1976) have described the fine structure of fertilization in detail. The gametes do not fuse initially at their flagella, as do those, for example, of the alga Chlamydomonas (Mesland, 1976), but they join at a specific region of plasma membrane toward the posterior end of the cell near the flagellum (J. Pommerville, personal communication). Fusion is inhibited by agents that stabilize membranes, such as diphenhydramine and chloroquine, suggesting that membrane fluidity is required, and also by trypsin, implying that a protein or glycoprotein surface is involved. At the "fusion interface" many small cytoplasmic bridges are formed between the fusing cells, a row of vesicles appears (Fig. 6 c, d) and, eventually, the fusion is complete and the resultant binucleate cell becomes spherical. The two nuclei line up, and nuclear fusion follows—first by the two outer nuclear membranes and then separately by the two inner nuclear membranes, to produce a series of bridges between the nuclei analogous to the earlier process of fusion of the cell membranes.

The two gametes are apparently naked, but presumably have some form of glycocalyx. These observations on their fusion strongly suggest that complementary sexual agglutinins are present on their surfaces.

3.3.3. Mating type, incompatibility and sexual fusion

In animals differentiation into male and female has two roles. It permits specialization into static cells with nutrient reserves (ova) and into motile or dispersible cells (spermatozoa). When differentiation extends to the diploid phase, the occurrence of male and female individuals adds a genetic role to the logistic—a certain measure of outbreeding becomes compulsory. In fungi this second function is largely assumed by homogenic incompatibility involving two or more mating types (heterothallism). The distinctness of the two systems of differentiation, into different sexes and into different mating types, is well illustrated in *Neurospora crassa*. This fungus is hermaphrodite, and carries both male and female organs (spermatia and protoperithecia). It is however self-sterile, and for sexual fusion to occur spermatia from a strain of one mating type (e.g., *a*) must reach protoperithecia of the strain of the other type (i.e., *A*). Some fungi carry distinct male and female organs, others have two or more mating types, and still others exhibit both forms of differentiation (Table 1). In a few, such as *Achlya bisexualis*, male and female organs are not usually produced by the same strain, thus resulting in the separate sexes familiar to most biologists.

Some of the lower fungi (chytrids, oomycetes, zygomycetes) and ascomycetes are homothallic (self-fertile), and here the preliminaries to sexual fusion must be controlled by the interaction of male and female structures that do not differ genetically. Thus the female gametangia of the homothallic chytrid Allomyces produce sirenin, and the male gametes respond (section 3.3.1). Where the sexes are separate, as in Achlya, sexual differentiation and gametic approach is controlled by interaction of genetically distinct male and female (section 3.3.2.1) In Mucor, similar controls are exerted through the interaction of two mating types.

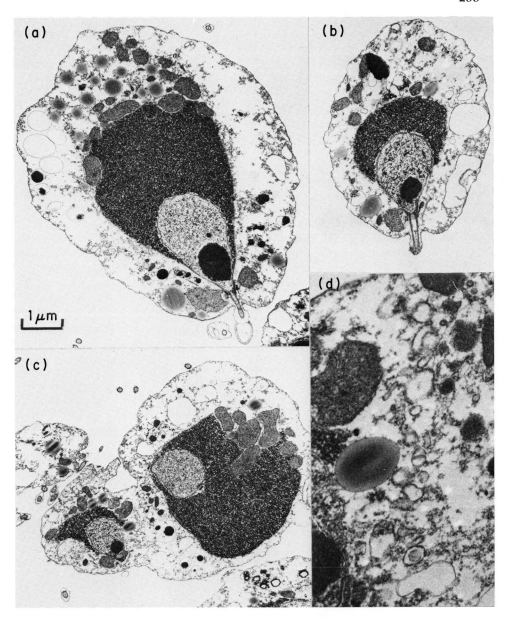

Fig. 6. Electron micrographs of sections of gametes and plasmogamy in *Allomyces macrogynus*. (a) Female gamete. (b) The much smaller male gamete. (c) An early stage of fertilization soon after fusion of gamete plasma membranes. (d) An enlarged view of the fusion interface. Note the formation of cytoplasmic bridges. Scale represents 1 μm in a-c; 0.25 μm in d. (Reproduced with permission of Springer-Verlag from Pommerville and Fuller, 1976.)

TABLE 1
Mating systems and sexual differentiation in fungi and myxomycetes

Mating system	Differentiation into male and female structures		
	Absent	On same strain	On separate strains
1. Self fertility (homothallism)	*Zygorhynchus macrocarpus; Didymium iridis* (some strains)	*Allomyces macrogynous, Aspergillus nidulans*	Not possible[a]
2. Self sterility (heterothallism)			
(a) Two mating types only	*Mucor mucedo; Phycomyces blakesleeanus*	*Neurospora crassa*	*Achlya bisexualis*
(b) Many mating types (homogenic incompatibility)			
(i) Bipolar (many alleles of one factor)	*Coprinus comatus; Physarum polycephalum*	Not known[b]	Not known[b]
(ii) Tetrapolar (many alleles of two factors)	*Coprinus cinereus; Schizophyllum commune*	Not known[b]	Not known[b]
3. Pseudocompatibility (secondary homothallism)	*Agaricus bisporus* (some spores)	*Neurospora tetrasperma*	Not possible[a]

[a] A self-fertile state is clearly not possible where cooperation of separate male and female strains is required.

[b] Differentiation into male and female structures or strains is lacking in myxomycetes and basidiomycetes, groups in which more than two mating types occur.

Attraction occurs, for example, between zygophores of + and − strains, but zygophores of the same strain seem to repel each other. Where there is separate differentiation into male and female and into two mating types, as in *Neurospora crassa,* it is an intriguing problem to elucidate which phases of sexual fusion are under the control of sexual differentiation and which are governed by mating type differences. Some progress has been made in *Neurospora crassa* (section 3.2.3.1; Bistis, 1965).

In basidiomycetes, as in myxomycetes, systems of homogenic incompatibility involving many mating types are common, and differentiation into male and female strains or organs does not occur. Although mating types in myxomycetes are controlled by a series of alleles at a single locus, in most basidiomycetes two unlinked factors, *A* and *B,* are involved. Each of these factors appears to consist of two closely linked loci with many alleles at each. Mating between two strains is completed only if they differ with respect to both factors. The genetics of mating type in basidiomycetes has been reviewed by Koltin and co-workers (1972), Burnett (1975), and Casselton (1978).

Hyphal fusion between basidiomycetes can occur within a strain or between strains whether or not they differ in mating type. If two strains differ in mating type, nuclear migration occurs to establish a stable dikaryon with pairs of nuclei of different mating type coexisting in each hyphal compartment. Karyogamy occurs only in the fruiting body and is immediately followed by meiosis to produce haploid spores. Nuclear migration is controlled by the *B* factor, and will only occur if two strains differ with respect to this factor, as will the fusion of hook cells to the main hypha to give clamp connections (a feature of basidiomycete sexual morphogenesis). Some other aspects of dikaryon morphogenesis—nuclear pairing, conjugate division, and hook cell formation and septation—will occur only if two strains differ with respect to the *A* factor. Thus homogenic sexual incompatibility in basidiomycetes operates mainly after fusion. However, Ahmad and Miles (1970a, b) suggest that the frequency of hyphal fusions in *Schizophyllum commune* is higher when the *A* factor is different in the two hyphae (ca. 3% of contacts of A ≠ B ≠ and A ≠ B =) than when it is common to both (1% for A = B ≠ and A = B =). Smythe (1973) has investigated this situation for *Coprinus cinereus,* and reports that the fusion frequency is consistently lowered only in the case of A ≠ B =. Moreover, there are considerable differences between the strains used here, ranging from less than 3% to 17% for A = B = contacts at 37 °C. Thus the significance of mating type genes in these fusions remains unclear.

In many basidiomycetes, dikaryotization can result from fusion of a hypha with a compatible oidium (section 3.3.2.3). The oidia of Coprinus and Psathyrella are usually uninucleate sporelike cells that are formed on monokaryotic hypha and do not germinate (Kemp, 1975a,b). However, if they are placed near growing hyphae, the hyphae may "home" to them and fuse with them from distances up to 75 μm. Homing is independent of the mating types of hypha and oidium. The hypha may be mono- or dikaryotic, but homing by a monokaryotic hypha is usually stronger. Fusion may be terminated by a lethal reaction, with the hyphal tip becoming vacuolated after about 2 hours. In Coprinus there is heterogenic incompatibility of a type similar to that discussed in section 3.2.3 superimposed on the mating system. By restricting outbreeding, such incompatibility could have a role in speciation.

4. Acrasiales

Virtually no research has been done on cell fusion in the cellular slime molds (acrasiales). Cell fusion does, however, occur in these organisms since a sexual cycle has recently been discovered, and parasexual recombination (which implies the occurrence of somatic cell fusion) has been known for some time. In due course these organisms, especially the intensively studied Dictyostelium discoideum, may be utilized in the investigation of cell fusion. Meanwhile they have already made a contribution to the understanding of two relevant processes, chemotactic aggregation and cell adhesion and sorting. Aggregation is controlled by pulses of the attractant "acrasin," which for Dictyostelium discoideum proves to be cyclic AMP. Adhesion appears to be mediated by interaction between species-specific surface lectins and complementary receptors. Recent reviews on these fascinating organisms are those of Konijn (1975), Loomis (1975), Newell (1975), Frazier (1976), and Sussman and Brackenbury (1976).

5. Discussion

The key events in the sexual process are cell fusion (plasmogamy), nuclear fusion (karyogamy), and meiosis, which involves genetic recombination and restores the haploid condition. In fungi these events may be close together in time, or they may be separated—in basidiomycetes a prolonged dikaryotic phase intervenes between plasmogamy and karyogamy, and in a few fungi there is a long diploid phase between karyogamy and meiosis. Events that may precede and follow sexual fusion are gametic approach controlled by chemotaxis or chemotropism, agglutination between cells of different mating types, fusion and controlled lysis of the glycocalyx, plasma membrane fusion, cytoplasmic mixing, pairing between nuclei of different mating types, and nuclear fusion. The rich morphological diversity of the fungi makes it likely that within the group there are species particularly suited for the study of each of these processes. The myxomycetes are remarkable for the plasticity of their sexual behavior. Within a single species there can be heterothallic, homothallic, and apogamic strains (section 2.2.2), and within single strains there can be a wide variation in nuclear events (Laane and Haugli, 1976; Laane et al., 1976). Genetic and developmental studies on these organisms should throw light on how systems for promoting outbreeding can come into existence and decay.

   Somatic cell fusion is an essential feature of the development of myxomycetes and higher fungi. The heterogenic incompatibility that can terminate cell fusion at various stages should provide opportunities for the genetic dissection of the fusion process. So far little is known about fusion incompatibility. Studies on postfusion incompatibility show similarities in the several myxomycetes and higher fungi studied. The genetics of postfusion heterogenic incompatibility has a formal similarity to DNA restriction and modification in bacteria (Arber, 1974) and perhaps a similar biochemical basis (see Sager and Kitchin, 1975 for a discussion of possible roles for restriction in eukaryotes). As with bacterial restriction, postfusion incompatibility is not 100% effective, and some nuclei or DNA can escape destruction and become adapted (modified?), as implied by the results of Jeffery and Rusch (1974), Ling and Upadhyaya (1974), Dales and Croft (1977), and McKenzie and colleagues (1969) mentioned earlier. The role of somatic heterogenic incompatibility in the life of the organism remains uncertain. Where numerous incompatibility alleles occur, it will severely limit the possibility of somatic fusion between strains and heterokaryon formation and hence parasexual recombination. It has, however, been suggested that its major role is to prevent the spread of infectious cytoplasmic factors including viruses through somatic fusion (Caten, 1972; Handley and Caten, 1975). Some heterogenic incompatibility systems can influence sexual as well as somatic fusion, as in Podospora (section 3.2.3.2), Aspergillus (section 3.2.3.3), Coprinus, and Psathyrella (Kemp 1975a, b). When this occurs, heterogenic incompatibility is likely to promote speciation, as it will also do in completely nonsexual organisms such as the fungi imperfecti.

Perhaps the most urgent need for advancing our understanding of cell fusion in fungi and myxomycetes is a deeper knowledge of the structure, biosynthesis and lysis of the glycocalyx and the plasma membrane. When this has been achieved, these organisms, as a result of their genetic and developmental diversity, will probably make a distinct contribution to our understanding of cell fusion and its consequences.

Acknowledgments

We thank the authors who have sent us reprints and preprints of their work, and Drs. C. E. Ballou and J. Pommerville for photographs.

References

Abe, K., Kusaka, I. and Fukui, S. (1975) Morphological change in the early stages of the mating process of *Rhodosporidium toruloides*. J. Bact. 122, 710–718.

Adler, P. N. and Holt, C. E. (1974) Genetic analysis in the Colonia strain of *Physarum polycephalum*: heterothallic strains that mate with and are partially isogenic to the Colonia strain. Genetics 78, 1051–1062.

Adler, P. N. and Holt C. E. (1975) Mating type and the differentiated state in *Physarum polycephalum*. Dev. Biol. 43, 240–253.

258

Ahkong, Q. F., Howell, J. I., Lucy, J. A., Safwat, F., Davey, M. R. and Cocking, E. C. (1975) Fusion of hen erythrocytes with yeast protoplasts induced by polyethylene glycol. Nature (London) 255, 66–67.

Ahmad, S. S. and Miles, P. G. (1970a) Hyphal fusions in the wood-rotting fungus *Schizophyllum commune* 1. The effects of incompatibility factors. Genet. Res. 15, 19–28.

Ahmad, S. S. and Miles, P. G. (1970b) Hyphal fusions in *Schizophyllum commune* 2. Effects of environmental and chemical factors. Mycologia 62, 1008–1017.

Albersheim, P. and Anderson-Prouty, A. J. (1975) Carbohydrates, proteins, cell surfaces, and the biochemistry of pathogenesis. Ann. Rev. Phytopathol. 26, 31–52.

Aldrich, H. C. (1967) The ultrastructure of meiosis in three species of *Physarum*. Mycologia 59, 127–148.

Aldrich, H. C. and Carroll, G. (1971) Synaptonemal complexes and meiosis in *Didymium iridis:* a reinvestigation. Mycologia 63, 308–316.

Aldrich, H. C. and Mims, C. W. (1970) Synaptonemal complexes and meiosis in Myxomycetes. Am. J. Bot. 57, 935–941.

Anderson, N. A., Stretton, H. M., Groth, J. V. and Flentje, N. T. (1972) Genetics of heterokaryosis in *Thanatephorus cucumeris*. Phytopathology 62, 1057–1065.

Anderson, R. W. and Dee, J. (1977) Isolation and analysis of amoebal-plasmodial transition mutants in the myxomycete *Physarum polycephalum*. Genet. Res. 29, 21–34.

Anderson, R. W., Cooke, D. J. and Dee, J. (1976) Apogamic development of plasmodia in the myxomycete *Physarum polycephalum:* a cinematographic analysis. Protoplasma 89, 29–40.

Anné, J. and Peberdy, J. F. (1975) Conditions for induced fusion of fungal protoplasts in polyethylene glycol solutions. Arch. Microbiol. 105, 201–205.

Anné, J. and Peberdy, J. F. (1976) Induced fusion of fungal protoplasts following treatment with polyethylene glycol. J. Gen. Microbiol. 92, 413–417.

Arber, W. (1974) DNA restriction and modification. Prog. Nucleic Acid Res. Mol. Biol. 4, 1–37.

Ashworth, J. M. and Dee, J. (1975) The Biology of Slime Moulds. Edward Arnold, London.

Ballou, C. E. (1976) Structure and biosynthesis of the mannan component of the yeast cell envelope. Adv. Microbiol. Physiol. 14, 93–158.

Barksdale, A. W. (1969) Sexual hormones of Achlya and other fungi. Science 166, 831–837.

Bartnicki-Garcia, S. (1970) Cell wall composition and other biochemical markers in fungal phylogeny. In: Phytochemical Phylogeny (Harborne, J. B., ed.) pp. 81–103, Academic Press, London.

Beadle, G. W. and Coonradt, V. L. (1944) Heterokaryosis in *Neurospora crassa*. Genetics 29, 291–308.

Bégueret, J. and Bernet, J. (1973) Proteolytic enzymes and protoplasmic incompatibility in *Podospora anserina*. Nature New Biol. 243, 94–96.

Bernet, J., Bégueret, J. and Labarère, J. (1973) Incompatibility in the fungus *Pondospora anserina*. Are the mutations abolishing the incompatibility reaction ribosomal mutations? Mol. Gen. Genet. 124, 35–50.

Binding, H. and Weber, H. J. (1974) The isolation, regeneration and fusion of Phycomyces protoplasts. Mol. Gen. Genet. 135, 273–276.

Bistis, G. M. (1965) The function of the mating-type locus in filamentous Ascomycetes. In: Incompatibility in Fungi (Esser, K. and Raper, J., eds.) pp. 23–31, Springer, Berlin.

Blaich, R. and Esser, K. (1971) The incompatibility relationships between geographical races of *Podospora anserina*. V. Biochemical characterization of heterogenic incompatibility on cellular level. Mol. Gen. Genet. 111, 265–272.

Bolkan, H. A. and Butler, E. E. (1974) Studies on heterokaryosis and virulence of *Rhizoctonia solani*. Phytopathology 64, 513–522.

Border, D. J. and Carlile, M. J. (1974) Somatic incompatibility following plasmodial fusion between two strains of the myxomycete *Physarum polycephalum* J. Gen. Microbiol. 85, 211–219.

Boucherie, H. and Bernet, J. (1974) Protoplasmic incompatibility and female organ formation in *Podospora anserina:* properties of mutations abolishing both processes. Mol. Gen. Genet. 135, 163–174.

Boucherie, M. Bégueret, J. and Bernet, J. (1976) The molecular mechanism of protoplasmic incompatibility and its relationship to the formation of protoperithecia in *Podospora anserina*. J. Gen.

259

Microbiol. 92, 59–66.

Bourchier, R. J. (1957) Variation in cultural conditions and its effect on hyphal fusion in *Corticium vellereum*. Mycologia 49, 20–28.

Brock, T. D. (1965) The purification and characterization of an intracellular sex-specific mannan protein from yeast. Proc. Nat. Acad. Sci., U.S.A. 54, 1104–1112.

Buller, A. H. R. (1931) Researches on Fungi, Vol. 4., Longmans Green, London.

Buller, A. H. R. (1933) Researches on Fungi, Vol. 5., Longmans Green, London.

Bu'Lock, J. D. (1976) Hormones in fungi. In: The Filamentous Fungi (Smith, J. E. and Berry, D. R., eds.), Vol. 2, pp 345–368, Arnold, London.

Bu'Lock, J. D., Jones, B. E. and Winskill, N. (1976) The apocarotenoid system of sex hormones and prohormones in Mucorales. Pure Appl. Chem. 47, 191–202.

Burgeff, H. (1924) Untersuchungen über Sexualität und Parasitismus bei Mucorineen. Bot. Abhandl. 4, 5-135.

Burnett, J. H. (1975) Mycogenetics. Wiley, London.

Burnett, J. H. (1976) Fundamentals of Mycology, 2nd edit. Arnold, London.

Butler, E. E. and Bolkan, H. (1973) A medium for heterokaryon formation in *Rhizoctonia solani*. Phytopathology 63, 542–543.

Calleja, G. B. (1970) Flocculation in *Schizosaccharomyces pombe*. J. Gen. Microbiol. 64, 247–250.

Calleja, G. B. (1973) Role of mitochondria in the sex-directed flocculation of a fission yeast. Arch. Biochem. Biophys. 154, 382–386.

Calleja, G. B. (1974) On the nature of the forces involved in the sex-directed flocculation of a fission yeast. Can. J. Microbiol. 20, 797–803.

Calleja, G. B. and Johnson, B. F. (1971) Flocculation in a fission yeast; an initial step in the conjugation process. Can. J. Microbiol. 17, 1175–1177.

Carlile, M. J. (1972) The lethal interaction following plasmodial fusion between two strains of the myxomycete *Physarum polycephalum*. J. Gen. Microbiol. 71, 581–590.

Carlile, M. J. (1973) Cell fusion and somatic incompatiblity in myxomycetes. Ber. Deutsch. Bot. Ges. 86, 123–139.

Carlile, M. J. (1974) Incompatibility in the myxomycete *Badhamia utricularis*. Trans. Br. Mycol. Soc. 62, 401–402.

Carlile, M. J. (1976) The genetic basis of the incompatibility reaction following plasmodial fusion between different strains of the myxomycete *Physarum polycephalum*. J. Gen. Microbiol. 93, 371–376.

Carlile, M. J. (1978) Bacterial, fungal and slime mould colonies. In: Biology and Systematics of Colonial Organisms (Larwood, G. P. and Rosen, D. R., eds.). Academic Press, London.

Carlile, M. J., and Dee, J. (1967) Plasmodial fusion and lethal interaction between strains in a myxomycete. Nature 215, 832–834.

Carlile, M. J. and Machlis, L. (1965) A comparative study of the chemotaxis of the motile phases of Allomyces. Am. J. Bot. 52, 484–486.

Casselton, L. A. (1978) Dikaryon formation in filamentous fungi. In: The Filamentous Fungi (Smith, J. E. and Berry, D. R., eds.) vol. 3, pp. 275–297, Arnold, London.

Caten, C. E. (1971) Heterokaryon incompatibility in imperfect species of Aspergillus. Heredity 26, 299–312.

Caten, C. E. (1972) Vegetative incompatibility and cytoplasmic infection in fungi. J. Gen. Microbiol. 72, 221–229.

Caten, C. E. (1973) Genetic control of heterokaryon incompatibility and its effect on crossing and cytoplasmic exchange in *Aspergillus amstelodami*. Genetics 74, Suppl. 40.

Clark, J. (1977) Plasmodial incompatibility reactions in the true slime mould *Physarum cinereum*. Mycologia 69, 46–52.

Clark, J. and Collins, O. N. R. (1973) Directional cytotoxic reactions between incompatible plasmodia of *Didymium iridis*. Genetics 73, 247–257.

Clark, J. and Collins, O. N. R. (1976) Studies on the mating systems of eleven species of myxomycetes. Am. J. Bot. 63, 783–789.

Collins, O. N. R. (1961) Heterothallism and homothallism in two myxomycetes. Am. J. Bot. 48, 674–683.

Collins, O. N. R. (1963) Multiple alleles at the incompatibility locus in the myxomycete *Didymium iridis.* Am. J. Bot. 50, 477–480.

Collins, O. N. R. (1966) Plasmodial compatibility in heterothallic and homothallic isolates of *Didymium iridis* Mycologia 58, 362–372.

Collins, O. N. R. (1972) Plasmodial fusion in *Physarum polycephalum:* genetic analysis of a Turtox strain. Mycologia 54, 1130–1137.

Collins, O. N. R. (1975) Mating types in five isolates of *Physarum polycephalum.* Mycologia 67, 98–107.

Collins, O. N. R. (1976) Heterothallism and homothallism: a study of 27 isolates of *Didymium iridis,* a true slime mould. Am. J. Bot. 63, 138–143.

Collins, O. N. R. and Clark, J. (1968) Genetics of plasmodial compatibility and heterokaryosis in *Didymium iridis.* Mycologia 60, 90–103.

Collins, O. N. R. and Haskins, E. F. (1972) Genetics of somatic fusion of *Physarum polycephalum:* the Pp II strain. Genetics 71, 63–71.

Collins, O. N. R. and Ling, H. (1964) Further studies in multiple allelomorph heterothallism in the myxomycete *Didymium iridis.* Am. J. Bot. 51, 315–317.

Collins, O. N. R. and Ling, H. (1968) Clonally produced plasmodia in heterothallic isolates of *Didymium iridis.* Mycologia 60, 858–868.

Collins, O. N. R. and Ling, H. (1972) Genetics of somatic cell fusion in two isolates of *Didymium iridis.* Am. J. Bot. 59, 337–340.

Collins, O. N. R. and Tang, H. C. (1977) New mating types in *Physarum polycephalum.* Mycologia 69, 421–423.

Collins, O. N. R. and Therrien, C. D. (1976) Cytophotometric measurement of nuclear DNA in seven heterothallic isolates of *Didymium iridis,* a myxomycete. Am. J. Bot. 63, 457–462.

Cooke, D. J. and Dee, J. (1974) Plasmodium formation without change in nuclear DNA content in *Physarum polycephalum.* Genet. Res. 23, 307–317.

Cooke, D. J. and Dee, J. (1975) Methods for the isolation and analysis of plasmodial mutants in *Physarum polycephalum.* Genet. Res. 24, 175–187.

Crandall, M. (1976) Mechanisms of fusion in yeast cells. In: Microbial and Plant Protoplasts (Peberdy, J. F., Rose, A. H., Rogers, H. J. and Cocking, E. C., eds.) pp. 161–175, Academic Press, New York.

Crandall, M. (1977) Mating-type interactions in microorganisms. In Receptors and Recognition (Cuatrecasas, P. and Greaves, M. F., eds.) vol. 3, pp. 45–100, Chapman and Hall, London.

Crandall, M. A. and Brock, T. D. (1968a) Molecular basis of mating in the yeast *Hansenula wingei.* Bact. Rev. 32, 139–163.

Crandall, M. A. and Brock, T. D. (1968b) Molecular aspects of specific cell contact. Science 161, 473–475.

Crandall, M. and Caulton, J. H. (1973) Induction of glycoprotein mating factors in diploid yeast of *Hansenula wingei* by vanadium salts or chelating agents. Exp. Cell Res. 82, 159–167.

Crandall, M. and Caulton, J. H. (1975) Induction of haploid glycoprotein mating factors in diploid yeasts. Methods Cell Biol. 12, 185–205.

Crandall, M., Egel, R. and MacKay, V. L. (1977) Physiology of mating in three yeasts. Adv. Micr. Physiol. 15, 307–398.

Crandall, M., Lawrence, L. M. and Saunders, R. M. (1974) Molecular complementarity of yeast glycoprotein mating factors. Proc. Nat. Acad. Sci., U.S.A. 71, 26–29.

Croft, J. H. and Jinks, J. L. (1977) Aspects of the population genetics of *Aspergillus nidulans.* In: Physiology and genetics of Aspergillus (Smith, J. and Pateman, J., eds.) pp. 339–360, Academic Press, London.

Cummins, J. E. and Day, A. W. (1973) Cell cycle regulation of mating type alleles in the smut fungus *Ustilago violaceae.* Nature 245, 259–260.

Cummins, J. E. and Day, A. W. (1974) Transcription and translation of the sex message in the smut fungus *Ustilago violaceae* II. The effects of inhibitors. J. Cell Sci. 16, 49–62.

Dales, R. B. G. and Croft, J. H. (1977) Protoplast fusion and the isolation of heterokaryons and diploids from vegetatively incompatible strains of *Aspergillus nidulans.* FEMS Microbiol. Lett. 1, 201–204.

Day, A. W. (1972) Dominance at a fungus mating type locus. Nature 237, 282–283.

Day, A. W. (1976) Communication through fimbriae during conjugation in a fungus. Nature 262, 583–584.

Day, A. W. and Cummins, J. E. (1973) Temporal allelic interaction, a new kind of dominance. Nature (London), 245, 260–261.

Day, A. W. and Cummins, J. E. (1974) Transcription and translation of the sex message in the smut fungus, *Ustilago violaceae* I. The effect of ultraviolet light. J. Cell Sci. 14, 451–460.

Day, A. W. and Poon, N. H. (1975) Fungal fimbriae. II. Their role in conjugation in *Ustilago violaceae*. Can. J. Microbiol. 21, 547–557.

Day, A. W., Poon, N. H. and Stewart, G. G. (1975) Fungal fimbriae III. The effect on flocculation in Saccharomyces. Can. J. Microbiol. 21, 558–564.

Dee, J. (1962) Recombination in a myxomycete, *Physarum polycephalum*. Genet. Res. 3, 11–23.

Dee, J. (1966) Multiple alleles and other factors affecting plasmodium formation in the true slime mould *Physarum polycephalum*. J. Protozool. 13, 610–616.

Delettre, Y. M. and Bernet, J. (1976) Regulation of proteolytic enzymes in *Podospora anserina:* selection and properties of self-lysing mutants. Mol. Gen. Genet. 144, 191–197.

Ende, H. van den (1978) Sexual morphogenesis in the Phycomycetes. In: The Filamentous Fungi (Smith, J. E. and Berry, D. R., eds.) vol. 3, pp. 257–273, Arnold, London.

Esser, K. (1965) Heterogenic incompatibility. In: Incompatibility in Fungi (Esser, K. and Raper, J., eds.) pp. 6–13, Springer, Berlin.

Esser, K. (1974a) Breeding systems and evolution. In: Evolution in the Microbial World, the 24th Symposium of the Society for General Microbiology (Carlile, M. J. and Skekel, J. J., eds.), pp. 87–104. Cambridge University Press, Cambridge.

Esser, K. (1974b) *Podospora anserina*. In: Handbook of Genetics (King, R. C., ed.) vol. 1, pp. 531–551, Plenum Press, New York.

Esser, K. and Blaich, R. (1973) Heterogenic incompatibility in plants and animals. Adv. Genet. 17 107–152.

Ferenczy, L., Kevei, F. and Zsolt J. (1974) Fusion of fungal protoplasts. Nature (London) 248, 793–794.

Ferenczy, L., Kevei, F. and Szegedi, M. (1975) Increased fusion frequency of *Aspergillus nidulans* protoplasts. Experientia 31, 50–52.

Ferenczy, L., Kevei, F. and Szegedi, M. (1976) Fusion of fungal protoplasts induced by polyethylene glycol. In: Microbial and Plant Protoplasts (Peberdy, J. F., Rose, A. H., Rogers, H. J. and Cocking, E. C., eds.) pp. 177–187, Academic Press, New York.

Flentje N. T. and Stretton, H. M. (1964) Mechanisms of variation in *Thanathephorus cucumeris* and *T. practicolus*. Aust. J. Biol. Sci. 17, 686–704.

Flentje, N. T., Stretton, H. M. and McKenzie, A. R. (1970) Mechanisms of variation in *Rhizoctonia solani*. In: *Rhizoctonia solani,* Biology and Pathology (Parmeter, J. R., ed.) pp. 52–65. Univ. of California Press, Berkeley.

Frazier, W. A. (1976) The role of cell surface components in the morphogenesis of the cellular slime molds. Trends Biochem. Sci. 1, 130–133.

Garnjobst, L. (1955). Further analysis of genetic control of heterokaryosis in *Neurospora crassa*. Am. J. Bot. 42, 444–448.

Garnjobst, L. and Wilson, J. F. (1956) Heterokaryosis and protoplasmic incompatibility in *Neurospora crassa*. Proc. Nat. Acad. Sci., U.S.A. 42, 613–618.

Gauger, W. L. (1977) Meiotic gene segregation in *Rhizopus stolonifer*. J. Gen. Microbiol. 101, 211–217.

Gooday, G. W. (1973) Differentiation in the Mucorales, In: Microbial Differentiation, 23rd Symposium of the Soceity for General Microbiology (Ashworth, J. M. and Smith, J. E., eds.) pp. 269–294. Cambridge University Press.

Gooday, G. W. (1974) Fungal sex hormones. Ann. Rev. Biochem. 43, 35–49.

Gooday, G. W. (1975) Chemotaxis and chemotropism in fungi and algae. In: Primitive Sensory and Communication Systems (Carlile, M. J., ed.) pp. 155–204, Academic Press, London.

Gooday, G. W. (1978) Microbial hormones. In: Companion to Microbiology (Bull, A. T. and Meadow, P. M., eds.) 207–219, Longmans, London.

Gray, W. D. and Alexopoulos, C. J. (1968) Biology of the Myxomycetes. Ronald Press, New York.

Handley, L. and Caten, C. E. (1975) Effect of vegetative incompatibility on cytoplasmic infection in *Aspergillus amstelodami*. Mycovirus Newslett. 3, 6–7.

Hartwell, L. H. (1973) Synchronization of haploid yeast cell cycles, a prelude to conjugation. Exp. Cell Res. 76, 111–117.

262

Hatch, W. R. (1938) Conjugation and zygote germination in *Allomyces arbuscula*. Ann. Bot. N. S. 2, 583–614.

Held, A. A. (1974) Attraction and attachment of zoospores of the parasitic chytrid *Rozella allomycis* in response to host-dependent factors. Arch. Microbiol. 95, 97–114.

Jeffery, W. R. and Rusch, H. P. (1974) Induction of somatic fusion and heterokaryosis in two incompatible strains of *Physarum polycephalum*. Dev. Biol. 39, 331–335.

Jinks, J. L., Caten, C. E., Simchen, G. and Croft, J. H. (1966) Heterokaryon incompatibility and variation in wild populations of *Aspergillus nidulans*. Heredity 21, 227–239.

Jones, B. E. and Gooday, G. W. (1977) Lectin binding to sexual cells in fungi. Biochem. Soc. Trans. 5, 717–719.

Jurand, M. K. and Kemp, R. F. O. (1972) Surface ultrastructure of oidia in the Basidiomycete *Psathyrella coprophila*. J. Gen. Microbiol. 72, 575–579.

Kemp, R. F. O. (1975a) Oidia, plasmogamy and speciation in basidiomycetes. In: The Biology of the Male Gamete (Duckett, J. G. and Racey, P. A., eds.) pp. 57–69, Academic Press, New York.

Kemp, R. F. O. (1975b) Breeding biology of *Coprinus* species in the section *Lanatuli*. Trans. Br. Mycol. Soc. 65, 375–388.

Kerr, N. S. (1967) Plasmodium formation by a minute mutant of the true slime mould, *Didymium nigripes*. Exp. Cell. Res. 45, 646–655.

Kerr, S. (1968) Ploidy level in the true slime mould *Didymium nigripes*. J. Gen. Microbiol. 53, 9–15.

Kerr, S. J. (1970) Nuclear size in plasmodia of the true slime mould *Didymium nigripes*. J. Gen. Microbiol. 63, 347–356.

Koltin, Y., Stamberg, J. and Lemke, P. A. (1972) Genetic structure and evolution of the incompatibility factors in higher fungi. Bact. Rev. 36, 156–171.

Konijn, T. M. (1975) Chemotaxis in the cellular slime moulds. In: Primitive Sensory and Communication Systems (Carlile, M. J., ed.) pp. 101–153, Academic Press, London.

Laane, M. M. and Haugli, F. B. (1976) Nuclear behaviour during meiosis in the myxomycete *Physarum polycephalum*. Norw. J. Bot. 23, 7–21.

Laane, M. M., Haugli, F. B. and Mellem, T. R. (1976) Nuclear behaviour during sporulation and germination in the Colonia strain of *Physarum polycephalum*. Norw. J. Bot. 23 177–189.

Labarère, J., Bégueret, J. and Bernet, J. (1974) Incompatibility in *Podospora anserian:* comparative properties of the antagonistic cytoplasmic factors of a non-allelic system. J. Bact. 120, 854–860.

Laffler, T. G. and Dove, W. F. (1977) Viability of *Physarum polycephalum* spores and ploidy of plasmodial nuclei. J. Bact. 131, 473–476.

Lewis, C. W., Johnston, J. R. and Martin, P. A (1976) The genetics of yeast flocculation. J. Inst. Brewing 82, 158–160.

Ling, H. and Collins, O. N. R. (1970) Control of plasmodial fusion in a Panamanian isolate of *Didymium iridis* Am. J. Bot. 57, 292–298.

Ling, H. and Ling, M. (1974) Genetic control of somatic cell fusion in a myxomycete. Heredity 32, 95–104.

Ling, H. and Upadhyaya, K. C. (1974) Cytoplasmic incompatibility studies in the myxomycete *Didymium iridis:* recovery and nuclear survival in heterokaryons. Am. J. Bot. 61, 598–603.

Lipke, P. N., Taylor, A. and Ballou, C. E. (1976) Morphogenic effects of α-factor on *Saccharomyces cerevisiae a* cells. J. Bacteriol 127, 610–618.

Loomis, W. F. (1975) *Dictyostelium discoideum:* A Developmental System. Academic Press, New York.

Lyons, T. P. and Hough, J. S. (1971) Further evidence for the cross-bridging hypothesis for flocculation of brewer's yeast. J. Inst. Brewing 77, 300–305.

Machlis, L. (1973) The chemotactic activity of various sirenins and analogues and the uptake of sirenin by the sperm of *Allomyces*. Plant. Physiol. 52, 527–530.

Machlis, L., Nutting, W. H. and Rapoport, H. (1968) The structure of sirenin. J. Am. Chem. Soc. 90, 1674–1676.

MacKay, V. L. (1977) Mating-type specific pheromones as mediators of sexual conjugation in yeast. In: Molecular Control of Proliferation and Differentiation, 35th Symposium of Society for Developmental Biology (Papaconstantinou, J., ed.). Academic Press, New York.

Margulis, L. (1974) Five-kingdom classification and the origin and evolution of cells. In: Evolutionary Biology (Dobzhansky, T., Hecht, M. K. and Steere, W. C., eds.) Vol. 7, pp 45–78, Plenum Press, New York.

McKenzie, A. R., Flentje, N. T., Stretton, H. M. and Mayo, M. J. (1969) Heterokaryon formation and genetic recombination within one isolate of *Thanatephorus cucumeris*. Aust. J. Biol. Sci. 22, 895–904.

McMorris, T. C., Seshadri, R., Weihe, G. R., Arsensault, G. P. and Barksdale, A. W. (1975) Structures of oogoniol-1 -2 and -3, steroidal sex hormones of the water mold, *Achlya*, J. Am. Chem. Soc. 97, 2544–2555.

Mesland, D. A. M. (1976) Mating in *Chlamydomonas eugametos*. A scanning electron microscopical study. Arch. Microbiol. 109, 31–35.

Mesland, D. A. M., Huisman, J. G. and Ende, H. van den (1974). Volatile sexual hormones in *Mucor mecedo*. J. Gen. Microbiol. 80, 111–117.

Metzenberg, R. L. and Ahlgren, S. K. (1973) Behaviour of *Neurospora tetrasperma* mating genes intro-gressed into *N. crassa*. Can. J. Genet. Cytol. 15, 571–576.

Mill, P. J. (1964) The nature of the interaction between flocculent cells in the flocculation of *Saccharomyces cerevisiae*. J. Gen. Microbiol. 35, 61–68.

Mohberg, J. (1977) Nuclear DNA content and chromosome numbers throughout the life cycle of the Colonia strain of the myxomycete, *Physarum polycephalum*. J. Cell Sci. 24, 95–108.

Mohberg, J. and Rusch, H. P. (1971) Isolation and DNA content of nuclei of *Physarum polycephalum*. Exp. Cell Res. 66, 305–316.

Mohberg, J., Babcock, K. L., Haugli, F. B. and Rusch, H. P. (1973) Nuclear DNA content and chromosome numbers in the myxomycete *Physarum polycephalum*. Dev. Biol. 34, 228–245.

Morris, E. O. (1966) Aggregation of unicells: yeast. In: The Fungi (Ainsworth, G. C. and Sussman, A. S., eds.) Vol. 2 pp. 63–82, Academic Press, London and New York.

Musgrave, A. and Nieuwenhuis, D. (1975) Metabolism of radioactive antheridiol by *Achlya* species. Arch. Microbiol. 105, 313–317.

Mylyk, O. M. (1975) Heterokaryon incompatibility genes in *Neurospora crassa* detected using duplication-producing chromosome rearrangements. Genetics 80, 107–124.

Mylyk, O. M. (1976) Heteromorphism for heterokaryon incompatibility genes in natural populations of *Neurospora crassa*. Genetics 83, 275–284.

Newell, P. C. (1975) Cellular communication during aggregation of the slime mold Dictyostelium. In: Microbiology—1975 (Schlessinger, D., ed.) pp. 426–433, American Society for Microbiology, Washington, D.C.

Newmeyer, D. (1970) A suppressor of the heterokaryon-incompatibility associated with mating type in *Neurospora crassa*. Can. J. Genet. Cytol. 12, 914–926.

Newmeyer, D. and Taylor, C. W. (1967) A pericentric inversion in *Neurospora*, with unstable duplica-tion progeny. Genetics 56, 771–791.

Newmeyer, D., Howe, H. B. and Galeazzi, D. R. (1973) A search for complexity at the mating-type locus of *Neurospora crassa*. Can J. Genet. Cytol. 15, 577–585.

Olive, L. S. (1975) The Mycetozoans, Academic Press, New York.

Ootaki, T. (1973) A new method for heterokaryon formation in Phycomyces. Mol. Gen. Genet. 121, 49–56.

Osumi, M., Shimoda, C. and Yanagishima, N. (1974) Mating reaction in *Saccharomyces cerevisiae*. V. Changes in the fine structure during the mating reaction. Arch. Microbiol. 97, 27–38.

Oujezdsky, K. B. and Szaniszlo, P. J. (1973) Conjugation in the dimorphic chromomycosis fungus *Phialophora dermatitidis*. J. Bact. 114, 1356–1358.

Park, D. and Robinson, P. M. (1966) Aspects of hyphal morphogenesis in fungi. In: Trends in Plant Morphogenesis (Cutter, E. G., ed.) pp. 27–44, Longmans, London.

Patel, G. B. and Ingledew, W. M. (1975a) The relationship of acid-soluble glycogen to yeast floccu-lation. Can. J. Microbiol. 21, 1608–1613.

Patel, G. B. and Ingledew, W. M. (1975b) Glycogen—a physiological determinant of yeast floccula-tion. Can. J. Microbiol. 21, 1614–1621.

Peberdy, J. F., Rose, A. H., Rogers, H. J. and Cocking, E. C. (1976) Microbial and Plant Protoplasts, Academic Press, London and New York.

Perkins, D. D. (1975) The use of duplication-generating rearrangements for studying hetero-karyon incompatibility genes in Neurospora. Genetics 80, 87–105.

Peveling, E. (1969) Electronenoptische Untersuchungen an Flechten. III. Cytologische Differenzierungen der Pilzzellen im Zusammenhang mit ihrer symbiontischen Lebensweise. Z. Pflanzenphysiol. 61, 151–164.

Pommerville, J. and Fuller, M. S. (1976) The cytology of the gametes and fertilization of *Allomyces macrogynus*. Arch. Microbiol. 109, 21–30.

Poon, N. H. and Day, A. W. (1974) Fimbriae in the fungus *Ustilago violaceae*. Nature 250, 648–649.

Poon, N. H. and Day, A. W. (1975) Fungal fimbriae. I. Structure, origin, and synthesis. Can. J. Microbiol. 21, 537–546.

Poon, N. H., Martin, J. and Day, A. W. (1974) Conjugation in *Ustilago violaceae*. I. Morphology. Can. J. Microbiol. 20, 187–191.

Poulter, R. T. M. and Dee, J. (1968) Segregation between factors controlling fusion between plasmodia of the true slime mould *Physarum polycephalum*. Genet. Res. 12, 71–79.

Pringle, J. R. and Mor, J. R. (1975) Methods for monitoring the growth of yeast cultures and for dealing with the clumping problem. Methods Cell Biol. 11, 131–168.

Raa, J., Robertson, B., Solheim, B. and Tronsmo, A. (1977) Cell surface biochemistry related to specificity of pathogenesis and virulence of microorganisms. In: Cell Wall Biochemistry Related to Specificity in Host-Plant Pathogen Interactions (Solheim, B. and Raa, J., eds.) pp. 11–30, Universitesforlaget, Tromso, Oslo and Bergen.

Rainbow, C. (1970) Brewer's yeasts. In: The Yeasts (Rose, A. H. and Harrison, J. S., eds.) Vol. 3, pp. 147–224, Academic Press, London and New York.

Raper, J. R. (1952) Chemical regulation of sexual processes in the Thallophytes. Bot. Rev. 18, 447–545.

Ross, I. K. (1967) Syngamy and plasmodium formation in the myxomycete *Didymium iridis*. Protoplasma 64, 104–119.

Ross I. K. and Cummings, R. J. (1970) An unusual pattern of multiple cell and nuclear fusions in the heterothallic slime mold *Didymium iridis*. Protoplasma 70, 281–294.

Ross, I. K., Shipley, G. L. and Cummings, R. J. (1973) Sexual and somatic cell fusions in the heterothallic slime mould *Didymium iridis*. 1. Fusion assay, fusion kinetics and cultural parameters. Microbios 7, 149–164.

Sager, R. and Kitchin, R. (1975) Selective silencing of eukaryotic DNA. Science 189, 426–433.

Sakurai, A., Tamura, S., Yanagishima, N. and Shimoda, C. (1975) Isolation of a peptidyl factor controlling sexual agglutination in *Saccharomyces cerevisiae*. Proc. Japan. Acad. 51, 291–294.

Satina, S. and Blakeslee A. F. (1926) The Mucor parasite *Parasitella* in relation to sex. Proc. Nat. Acad. Sci., U.S.A. 12, 202–207.

Sena, E. P., Radin, D. N. and Fogel, S. (1973) Synchronous mating in yeast. Proc. Nat. Acad. Sci., U.S.A. 70, 1373–1377.

Shimoda, C. and Yanagishima, N. (1975) Mating reaction in *Saccharomyces cerevisiae* VIII. Mating-type-specific substances responsible for sexual cell agglutination. Antonie van Leeuwenhoek, 41, 521–532.

Shimoda, C., Kitano, S. and Yanagishima, N. (1975) Mating reaction in *Saccharomyces cerevisiae* VII. Effect of proteolytic enzymes on sexual agglutinability and isolation of crude sex-specific substances responsible for sexual agglutination. Antonie van Leeuwenhoek, 41, 513–519.

Shimoda, C., Yanagishima, N., Sakurai, A. and Tamura, S. (1976) Mating reaction in *Saccharomyces cerevisiae* IX. Regulation of sexual cell agglutinability of *a* type cells by a sex factor produced by α type cells. Arch. Microbiol. 108, 27–33.

Smith, D. C. (1974) Transport from symbiotic algae and symbiotic chloroplasts to host cells. In: Transport at the Cellular Level, 28th Symposium of the Society for Experimental Biology (Sleigh, M. A. and Jennings, D. H., eds.) pp. 485–520. Cambridge University Press.

Smythe, R. (1973) Hyphal fusions in the basidiomycete *Coprinus lagopus sensu* Buller. 1. Some effects of incompatibility factors. Heredity 31, 107–111.

Stewart, G. G., and Garrison, I. F. (1972) Some observations on co-flocculation in *Saccharomyces cerevisiae*. Am. Soc. Brew. Chem. Proc. 1972, 118–131.

Stewart, G. G. and Goring, T . E. (1976) Effect of some monovalent and divalent metal ions on the flocculation of brewers yeast strains. J. Inst. Brewing 82, 341–342.

Stewart, G. G., Garrison, I. F., Goring, T. E., Meleg, M., Pipasts, P. and Russell, I. (1976) Biochemical and genetic studies on yeast flocculation. Kemia-Kemi 3, 465–479.

Stewart, G. G., Russell, I. and Garrison, I. F. (1973) Further studies on flocculation and co-flocculation in brewers yeast strains. Am. Soc. Brew. Chem. Proc. 1973, 100–106.

Stewart, G. G., Russell, I. and Garrison, I. F. (1975) Some considerations of the flocculation charac-teristics of ale and lager yeast strains, J. Inst. Brewing 81, 248–257.

Stötzler, D., Kiltz, H. and Duntze, W. (1976) Primary structure of α-factor peptides from Sac-charomyces cerevisiae. Eur. J. Biochem. 69, 397–400.

Stretton, H. M. and Flentje, N. T. (1972) Interisolate heterokaryosis in Thanatephorus cucumeris. II Be-tween isolates of different pathogenicity. Aust. J. Biol. Sci. 25, 305–318.

Sussman, M. and Brackenbury, R. (1976) Biochemical and molecular-genetic aspects of cellular slime mold development. Ann. Rev. Plant Physiol. 27, 229–265.

Taylor, N. W. (1964a) Specific, soluble factor involved in sexual agglutination of the yeast Hensenula wingei. J. Bact. 87, 863–866.

Taylor, N. W. (1964b) Inactivation of sexual agglutination in Hansenula wingei and Saccharomyces kluyveri by disufide-cleaving agents. J. Bact. 88, 929–936.

Taylor, N. W. (1965) Purification of sexual agglutination factor from the yeast Hansenula wingei by chromatography and gradient sedimentation. Arch. Biochem. Biophys. 111, 181–186.

Taylor N. W. and Orton, W. L. (1967) Sexual agglutination in yeast. V. Small particles of 5-agglutinin. Arch. Biochem. Biophys. 120, 602–608.

Taylor, N. W. and Orton, W. L. (1968) Sexual agglutination in yeast. VII. Significance of the 1.7.S. component from reduced 5-agglutinin. Arch. Biochem. Biophys. 126, 912–921.

Taylor, N. W. and Orton, W. L. (1970) Association constant of the sex-specific agglutinin in the yeast, Hansenula wingei. Biochemistry 9, 2931–2934.

Taylor, N. W. and Orton, W. L. (1971) Cooperation among the active binding sites in the sex-specific agglutinin from the yeast, Hansenula wingei. Biochemistry 10, 2043–2049.

Taylor, N. W. and Orton, W. L. (1975) Calcium and flocculence of Saccharomyces cerevisiae. J. Inst. Brewing 81, 53–57.

Therrien, C. D., Bell, W. R. and Collins, O. N. R. (1977) Nuclear DNA content of myxoamoebae and plasmodia in six non-heterothallic isolates of a myxomycete, Didymium iridis. Am. J. Bot. 64, 286–291.

Upadhyaya, K. C. and Ling, H. (1977) Ultrastructure of clear zones exhibiting cytoplasmic incom-patibility following somatic cell fusion in the myxomycete Didymium iridis. Indian J. Exp. Biol. 14, 652–658.

Wheals, A. E. (1970) A homothallic strain of the myxomycete Physarum polycephalum. Genetics 66, 623–633.

Williams, C. A. and Wilson, J. F. (1966) Cytoplasmic incompatibility reactions in Neurospora crassa. Ann. N. Y. Acad. Sci. 129, 853–863.

Williams C. A. and Wilson, J. F. (1968) Factors of cytoplasmic incompatibility. Neurospora Newslett. 13, 12.

Williams, N. J. and Wiseman, A. (1973) Loss of yeast flocculence after release of extracellular inver-tase (β-fructofuranosidase) and acid phosphatase by a modified osmotic shock procedure that maintains viability. Biochem. Soc. Trans. 1, 1301–1303.

Wilson, J. F. and Garnjobst, L. (1966) A new incompatibility locus in Neurospora crassa. Genetics 53, 621–631.

Wilson, J. F., Garnjobst, L. and Tatum, E. L. (1961) Heterokaryon incompatibility in Neurospora crassa—micro-injection studies. Am. J. Bot. 48, 299–305.

Yen, P. H. and Ballou, C. E. (1973) Composition of a specific intercellular agglutination factor. J. Biol. Chem. 248, 8316–8318.

Yen, P. H. and Ballou, C. E. (1974) Partial characterization of the sexual agglutination factor from Hansenula wingei Y-2340 type 5 cells. Biochemistry 13, 2428–2437.

Mechanisms of chemically induced cell fusion

6

J. A. LUCY

Contents

G. Poste & G. L. Nicolson (eds.) Membrane Fusion, pp. 267–304.
© Elsevier/North-Holland Biomedical Press, 1978.

1. Introduction

The study of chemically induced cell fusion is an area of research that has been developed extensively within the last ten years. There are three main reasons for the present interest in this topic. First, a considerable number of chemicals are now known to induce cells to fuse, and this variety allows investigators to approach the ultrastructural changes and molecular mechanisms involved in cell fusion in various ways. Secondly, studies on the mechanisms of chemically induced cell fusion offer, in view of the comparatively simple systems that can be investigated (Lucy, 1977a), a means of illuminating—at least obliquely—the virtually unknown mechanisms of membrane fusion occurring between cells and within cells in innumerable situations in cell biology and pathology. Finally, chemically induced fusion provides scientists and clinicians with a valuable way of using cell fusion as an easily accessible research technique that would otherwise involve the use of Sendai virus, which has certain inherent practical and theoretical disadvantages. A general review of cell fusion in the latter context has recently been published (Lucy, 1977b).

What features are of primary importance in investigations on mechanisms of chemically induced cell fusion? The subject is sometimes discussed in terms of the behavior of individual chemicals or categories of chemicals that induce cells to fuse. In this chapter a different approach is adopted. Here cell fusion is considered from the viewpoint of membrane structure. The chapter is thus oriented toward an underlying question, namely, what changes occur in membrane lipids, in membrane proteins, and in their interrelationships that permit cells to fuse?

2. Cell fusion and membrane lipids

2.1. Membrane cholesterol

Although the barrier properties and even the existence of the plasma membranes of cells depend on the amphipathic properties of their phospholipid components, plasma membranes also contain relatively high proportions of cholesterol. The role of cholesterol is interesting to consider because although it is clearly important in influencing the fluidity of biological membranes (Chapman, 1973; Kimelberg, 1977), relatively little is known regarding its effects on cell fusion.

In a discussion of membrane fusion in lysosomal digestion, De Duve (1969) asked why endocytosed materials do not appear within the cisternae of the rough-surfaced endoplasmic reticulum of cells. He suggested that fusion between an exoplasmic and an endoplasmic space is not possible, on the assumption that the ability of two membranes to fuse together requires a certain degree of chemical kinship between them. This was consistent with the findings of Thinés-Sempoux (1967), who was an early observer of similarities in phospholipid composition and in cholesterol content between plasma membranes and lysosomal membranes, by comparison with microsomes which have a different lipid composition. De Duve's suggestion would appear, however, not to be supported by the work of Poste and co-workers (1972), who found that the ratio of cholesterol to phospholipid in the plasma membranes of various cell types had no influence on the susceptibility of the cells to fuse on treatment with Sendai virus.

A study has been undertaken in this laboratory on the role on membrane cholesterol in the fusion of hen and guinea pig erythrocytes (Hope et al., 1977a). In this work available techniques were used for decreasing or increasing the cholesterol content of erythrocyte membranes in vitro by incubating the cells with phospholipid liposomes (Bruckdorfer et al., 1968; Bruckdorfer et al., 1969; Grunze and Deuticke, 1974; Cooper et al., 1975). Cell fusion induced by glycerol mono-oleate was unaffected when hen erythrocytes were depleted of cholesterol by treatment with liposomes prepared from pure dipalmitolyphosphatidyl-choline, but depletion of cellular cholesterol decreased cell fusion induced by Sendai virus more than 20-fold. Conversely, it was found that enrichment of hen erythrocytes with cholesterol, by incubation with cholesterol-loaded liposomes of dipalmitoylphosphatidylcholine, increased cell fusion caused by Sendai virus, glycerol mono-oleate and poly(ethylene glycol). Enrichment of erythrocytes from guinea pigs with cholesterol was achieved by feeding the animals a diet containing 1% cholesterol. Again, as with hen erythrocytes, it was observed that virus-induced fusion was enhanced by the presence of additional cholesterol.

The fusion in vitro, induced by glycerol mono-oleate, of erythrocytes from patients with a variety of hepatobiliary disorders has been investigated by Hope and associates (1977b). Although considerable individual variation was found, statistically significant increases in the concentration of cell cholesterol and in the molar ratio of cholesterol to phospholipid were observed in the cells from patients with liver diseases. Furthermore, cell fusion induced by glycerol mono-oleate was more extensive in the erythrocytes of the patients with liver diseases than in erythrocytes obtained from healthy control subjects. Taking data from all of the subjects studied (patients and control subjects), there was a positive correlation between the percentage of polykaryocytosis and both the cholesterol content of the cells and their ratio of cholesterol to phospholipid. Erythrocytes of some of the patients exhibited increased concentrations of phosphatidylcholine and phosphatidylserine, but not of other phospholipid fractions, nor in the mean values for total phospholipid content. A correlation was observed between cell fusion and the cellular concentration of phosphatidylcholine or phosphatidylserine, but not with other phospholipids nor with the total concentration of phospholipid.

These observations show that the effects of membrane cholesterol on cell fusion are complex. However, since the emergence of protein-free areas of lipid bilayer is believed to be important in membrane fusion (see section 4), Hope and colleagues (1977a) have suggested that an enhancement of cell fusion by membrane cholesterol may be due to the sterol acting to facilitate a phase separation of protein-free areas of lipid bilayer. Location of cholesterol in the outer half of the lipid bilayer of erythrocytes (Fisher, 1976) may be important in this respect, as may a vertical displacement of membrane proteins with changes in the content of membrane cholesterol (Borochov and Shinitzky, 1976). It is relevant that the adenosine triphosphatase (ATPase) of rabbit sarcoplasmic reticulum is excluded from solid lipid regions of phospholipid bilayer membranes containing more than 10 mol% of cholesterol (Kleeman and McConnell, 1976); cholesterol-loading of bovine erythrocytes also decreases the surface density of intramembranous particles as shown by freeze-fracture techniques (Deuticke and Ruska, 1976). The results of an immunochemical study with model membranes by Brûlet and McConnell (1976) suggest the possibility that cholesterol has at least two effects on complement fixation in liposome membranes, one related to enhancement of hapten exposure at the membrane surface and the other due to the effects of cholesterol on membrane fluidity.

A change in membrane fluidity may be a more important effect of cholesterol in liposomes, where it has an inhibitory effect on fusion. Thus the fusion of cultured baby hamster kidney cells and of fibroblasts, promoted by liposomes containing phosphatidylserine, was inhibited by the presence of cholesterol (Papahadjopoulos et al., 1973). A similar inhibitory effect of cholesterol, although not on cell fusion per se, was observed by Papahadjopoulous and co-workers (1974) in vesicle fusion occurring between negatively charged phosphatidylserine liposomes. They found that incorporation of equimolar quantities of cholesterol into vesicles composed of lipids in a liquid-crystalline state reduced their ability to fuse with one another. Cholesterol has also been shown to inhibit the fusion of spherical phospholipid membranes of phosphatidylcholine and phosphatidylserine (Breisblatt and Ohki, 1976). In phospholipid systems that are above their transition temperature, the steroid nucleus of cholesterol effectively prevents flexing of the lipid hydrocarbon chains, thereby making them less mobile (Chapman, 1973). Spin-label derivatives of stearic acid (Kroes et al., 1972) and fluorescence probes (Vanderkooi et al., 1974) indicate a reduced fluidity in the membranes of cholesterol-enriched cells. Conversely, a decrease in membrane cholesterol increases the fluidity of biological membranes (Radda and Vanderkooi, 1972; Tanaka and Ohnishi, 1976). A number of studies have been reported that indicate an increase in membrane fluidity favors cell fusion (section 2.2.3), and the inhibitory action of cholesterol on the fusion of membranes in liposome systems may thus be attributed to the sterol acting to decrease membrane fluidity.

2.2. Membrane phospholipids

2.2.1. Introduction

The biomolecular leaflet is a very stable structural organization for amphipathic phospholipids containing hydrophobic fatty acyl chains and polar groups, and it is now generally agreed that the bilayer is probably the primary structural organization for phospholipid molecules in biological membranes (reviews, Singer, 1974; 1977; Nicolson et al., 1977; see also chapter by Papahadjopoulos in this volume). If this premise is accepted, studies on the mechanisms of chemically induced cell fusion can perhaps be reduced to one fundamental question: How is the thermodynamic stability of the bilayers of two closely adjacent plasma membranes disturbed, presumably in a reversible manner, so that their inherent stability is reduced sufficiently to allow the membranes to interact and fuse, but without the perturbation being so great that membrane disintegration occurs? The author thinks that there is no unique answer to this question, and that even if a final common pathway is involved in all instances of membrane/cell fusion, there are still likely to be many differing ways by which the pathway could be initiated.

2.2.2. Bilayer: micelle transitions; lysophosphatidylcholine

The first simple chemical that was investigated in this laboratory in relation to chemically induced cell fusion was lysophosphatidylcholine (Howell and Lucy, 1969). This chemical was studied because it causes bilayers of phosphatidylcholine to break down into globular micelles as a consequence of its wedge shape (Haydon and Taylor, 1963; Bangham and Horne, 1964; Howell et al., 1973). It had previously been postulated that under appropriate circumstances some of the lipids of biological membranes may be organized in globular micelles that are in dynamic equilibrium with the bimolecular leaflet structure (Lucy, 1964, 1968). Also, it was suggested that a primary requirement for the fusion of two membranes is that both of the membranes involved have a relatively high proportion of their phospholipid molecules in the micellar configuration (Lucy, 1969, 1970). Lysophosphatidylcholine was therefore of particular interest in relation to cell fusion since the presence of this molecule in membranes would be expected to lead to a high proportion of membrane lipids being organized in globular micelles, and thus to the facilitation of membrane fusion.

When this hypothesis was tested using hen erythrocytes, the initial experiments with lysophosphatidylcholine in aqueous solution were disappointing because cell lysis occurred without any evidence of cell fusion. It was then found that fixation of the treated cells within 30 seconds of adding lysophosphatidylcholine arrested the damage to membranes at a point where fusion between hen erythrocytes could clearly be seen (Howell and Lucy, 1969). Lysophosphatidylcholine is, nevertheless, extremely damaging to biological membranes. It may well be among the most drastic of fusogenic molecules that can initiate the fusion of a number of different cell types (Poole et al., 1970; Gledhill et al., 1972) and yet allow membranes to retain their integrity, albeit only under special conditions or for a very short time. Different cells show differing sensitivities, however, toward lysophosphatidylcholine as is demonstrated by the work of Halfer and Petrella (1976) on two karyotypically different embryonic cells of *Drosophila melanogaster*. These investigators have also confirmed the earlier report by Becker (1972) that cell fusion of Drosophila cells can be induced by Concanavalin A. Lysophosphatidylcholine acts on the membranes of human erythrocytes to increase their capacity for agglutination; presumably a necessary prerequisite for cell fusion. It has been suggested that insertion of echinocytic agents such as lysophosphatidylcholine into the lipid bilayer of a plasma membrane causes changes in cell shape that result in the redistribution of membrane proteins, bearing surface charge groups, into configurations that favor cell-to-cell interactions (Marikovsky et al., 1976).

It seems likely that the fusogenic action of Sendai virus is localized to the virus particle or at least to within its immediate vicinity when the virus becomes attached to the plasma membrane. This localization may be responsible in part for the ability of the virus to fuse cells without simultaneously causing widespread membrane damage. If the effects of lysophosphatidylcholine on the plasma membranes of cells could be similarly localized, cell fusion induced by this substance might occur in the absence of extensive membrane damage. One way of localizing lysophosphatidylcholine is to include it in the lipid phase of an aqueous emulsion of fat. Since about 10% of the phospholipid of the nascent fat droplets of cow's milk is lysophosphatidylcholine, and the secretion of milk fat has been suggested to depend on budding of the plasma membrane around fat droplets (Keenan et al., 1970; Patton and Keenan, 1975), it seems possible that the presence of lysophosphatidylcholine in the nascent fat droplets of milk may be partially responsible for the membrane fusion involved. With this in mind, emulsions of fat containing lysophosphatidylcholine were prepared and used in experiments on cell fusion (Ahkong et al., 1972). The emulsions were found to induce the formation of relatively stable polykaryocytes and heterokaryons. Treatment of hen erythrocytes with the emulsion yielded homokaryons that generally contained only two or three nuclei, as opposed to the very large multinucleate cells formed on using lysophosphatidylcholine in aqueous solution. This change was presumably a reflection of the localized action of the emulsified lysophosphatidylcholine adhering to the plasma membranes of closely adjacent cells. Mouse fibroblast-hen erythrocyte heterokaryons formed with the aid of the emulsion were also more stable than those produced with lysophosphatidylcholine in solution, but the hybrid cells still had damaged subcellular organelles. Related experiments have been reported by Papahadjopoulos and co-workers (1973) who used negatively charged phospholipid liposomes of phosphatidylcholine and phosphatidylserine, containing lysophosphatidylcholine, to fuse BHK hamster cells, 3T3 and L929 mouse cells in suspension, and monolayer cultures. Martin and MacDonald (1976) reported, however, that they have been unable to obtain significant cell fusion using vesicles consisting of lysophosphatidylcholine and phosphatidylserine. They have instead used positive charged liposomes containing phosphatidylcholine, lysophosphatidylcholine, and stearylamine to induce the fusion of avian erythrocytes and of various cell types in tissue culture. In the author's laboratory stearylamine and, to a greater extent, oleylamine have been found to be potent fusogens of hen erythrocytes. Ultrasonically prepared dispersions of oleylamine and liposomes of phosphatidylcholine containing oleylamine were employed (without lysophosphatidylcholine) to fuse mouse fibroblasts. Oleylamine was used in preference to stearylamine because of its liquidity at incubation temperatures (Bruckdorfer et al., 1974).

Another approach toward the reduction of the toxic effects of lysophosphatidylcholine was developed by Croce and associates (1971), who used the lysophospholipid in the presence of albumin for the fusion of cells in tissue culture. Tissue culture cells and sperm cells have also been fused together using this method (Gledhill et al., 1972). There is no doubt, however, that it is very difficult to overcome the membrane-damaging effects of lysophosphatidylcholine, though the recent work of Martin and MacDonald (1976) referred to above may offer a more successful solution to the problem.

Following the proposal that the formation of globular micelles might be involved in memebrane fusion induced by lysophosphatidylcholine (Lucy, 1970), Breisblatt and Ohki (1975) have suggested that a semimicelle membrane configuration of the hydrocarbon chains of phospholipids may participate in the fusion of spherical bilayers with and without lysophosphatidylcholine. In their experiments, the temperature required for fusion was lowered by 10°C in the presence of lysophosphatidylcholine. For alamethicin-mediated fusion of phosphatidylcholine vesicles, Lau and Chan (1975) have suggested that transient rearrangements in the local structure of the lipid occur, with the formation of an inverted micelle of alamethicin. Martin and McDonald (1976) commented recently that they would agree with the writer's suggestions (Lucy, 1970) regarding possible mechanisms of fusion induced by lysophosphatidylcholine to the extent that membranes that are firmly committed to a compact bilayer structure are not likely to fuse readily. However, they see no reason to invoke a perturbation as extreme as the micellar structure in the fusion process.

Currently, I consider that, although globular micelles may be involved in the fusogenic actions of lysophosphatidylcholine, membrane fusion induced by other lipids (see below) probably occurs by different mechanisms. This opinion is based in part on experiments made with the negative-staining technique of electron microscopy that were designed to investigate interactions between phospholipids and a number of lipids that induce hen erythrocytes to fuse into multinucleated cells. Of the fusogenic compounds studied, only lysophosphatidylcholine was observed to cause bilayers of phosphatidylcholine to become micellar (Howell et al., 1973), (also see section 2.2.4).

The finding by Howell and Lucy (1969) that lysophosphatidylcholine has fusogenic properties was of interest in relation to earlier suggestions that lysophosphatidylcholine might be implicated in a number of instances of membrane fusion occurring in vivo. Thus it had been proposed that lysophosphatidylcholine may be involved in the secretion of catecholamines from both normal and tumor chromaffin cells (Blaschko et al., 1967, 1968), in the release of histamine from mast cells (Högberg and Uvnäs, 1957), and in phagocytosis (Burdzy et al., 1964). In relation to secretions from tumor cells and the possibility that the invasive properties of malignant cells may involve an extracellular release of lysosomal enzymes (Poole and Williams, 1967), attention was drawn by Poole and associates (1970) to a report that some tumors contained a relatively high proportion of lysophosphatidylcholine (Bergelson et al., 1968).

In view of the damaging effects of lysophosphatidylcholine on biological membranes, it seems likely that this substance would be present only transiently in any lysophosphatidylcholine-induced fusion of membranes that may occur in vivo. Thus it is interesting that some cells remove lysophosphatidylcholine as well as producing it by the action of phospholipase A (Mulder et al., 1965; Ferber, 1971; Shier, 1977). Enzymes capable of producing lysophosphatidylcholine by phospholipase activity are known to be associated with plasma membranes (Newkirk and Waite, 1971), mitochondrial membranes (Waite et al., 1969), the endoplasmic reticulum (Bjørnstad, 1966) and lysosomal membranes (Rahman et al., 1970).

Although it is possible that membrane fusion occurring in vivo may involve a bilayer to micelle transition associated with the transient presence in membranes of lysophosphatidylcholine (Suzuki and Matsumoto, 1974), produced and subsequently destroyed by membrane-bound enzymes, there is no experimental support for this concept. Many experiments concerned with the possible involvement of lysophosphatides in virus-induced cell fusion have yielded negative results (Falke et al., 1967; Elsbach et al., 1969; Blough et al., 1973; Pasternak and Micklem, 1974; Parkes and Fox, 1975). Further, lysophosphatidylcholine inhibits the fusion of myoblasts in vitro, subsequent removal of lysophosphatidylcholine allowing cell fusion to proceed (Reporter and Norris, 1973). Reporter and Raveed (1973) have interpreted the inhibition of myoblast fusion by lysophosphatidylcholine as resulting primarily from interference with metabolism and the turnover of new phospholipids. Lysophosphatidylcholine also inhibits the calcium-induced fusion of plasma membrane vesicles isolated from myoblasts grown in tissue culture (Schudt et al., 1976). Although phospholipase A activity has been found in the plasma membranes of fusing muscle cells, Ca^{2+} was not required for enzyme activity even though it is needed for muscle cell fusion (Kent and Vagelos, 1975).

2.2.3. Membrane fluidity

Treatment of hen erythrocytes at 37°C, pH 5.6, with glycerol mono-oleate causes the cells to swell and become spherical within a few minutes; when two or more of these spherical cells make contact, they frequently stick together and many subsequently fuse (Ahkong et al., 1973a). Similar phenomena occur at pH 7.5, although more slowly. Glycerol mono-oleate also causes cell fusion and multinucleate giant-cell formation with mammalian erythrocytes, including those of man, cat, rat, sheep, and rabbit, and interspecies fusion (e.g., between human erythrocytes and hen erythrocytes) occurs (Ahkong et al., 1973b). Our experience in this laboratory has shown that the unsaturated compounds glycerol mono-oleate and oleic acid are among the most effective lipids used to induce the fusion of hen erythrocytes. Oleylamine is also a potent fusogen with erythrocytes (Ahkong and Lucy, unpublished observations). By contrast, glycerol monopalmitate, glycerol monostearate, palmitic acid, and stearic acid were found to be inactive; stearylamine was less active than oleylamine. Two possible explanations for the inactivity of these compounds were proposed: either they did not penetrate into the cell membrane, or they did not induce the required changes in membrane structure (Ahkong et al., 1973a). Electron microscopy of phospholipids codispersed with fusogenic or nonfusogenic lipids, and investigations on mixed surface monolayers of phospholipids and fusogenic or nonfusogenic lipids, indicate that the latter explanation is probably correct (see sections 2.2.4 and 2.2.5).

Because oleic acid caused cell fusion while steric acid was inactive, and because the fusion induced by oleic acid depended heavily on temperature, it was suggested that unsaturated fatty acids and their derivatives may induce erythrocytes to fuse by increasing the proportion of hydrocarbon chains in the membrane that are in a relatively liquid state (Ahkong et al., 1973a). This interpretation of the effects of unsaturated lipids on the erythrocyte membrane was supported by the observation that hen erythrocytes can be fused by heat alone (48°C, on a heated microscope stage) in the absence of exogenously added lipids (Ahkong et al., 1973a). More recently, Kennedy and Rice-Evans (1976) have reported a spectrofluorometric study of the interaction of glycerol mono-oleate with human erythrocyte ghosts. They observed a red-shift in the wavelength of maximum emission of the fluorescent probe l-anilino-8-naphthalene sulfonate (ANS) when the probe interacted with membrane ghosts that had been perturbed by glycerol mono-oleate; a decrease in quantum yield was also found. They suggested that these findings, and related observations with N-phenyl-1-naphthylamine (NPN), indicate that glycerol mono-oleate produces a less ordered structure in the membrane, decreasing the constraint around the environment of the fluorescent probes. These investigators therefore consider that although no data have yet been published on the ability of glycerol mono-oleate to fuse erythrocyte ghosts, their findings suggest that the role of the fusogenic lipid, glycerol mono-oleate, is at least partially a fluidizing one. Kosower and colleagues (1975) have observed that the "membrane mobility agent," A_2C, promotes the fusion of hen erythrocytes, and conclude that it seems reasonable to suppose that some increase in local fluidity in membranes favors cell fusion.

Some thirty different fat-soluble substances (100 μg/ml) were found by Ahkong and co-workers (1973b) to cause the formation of multinucleated cells in experiments in which a suspension of 3×10^8 hen erythrocytes/ml was used. The most effective fusogenic lipids (e.g., glycerol mono-oleate) induced fusion within 5 to 10 minutes at 37°C, but some lipids required about 1 hour to cause a significant incidence of cell fusion (Table 1). This Table shows that fusion induced by saturated fatty acids usually occurred only with acids containing 10 and 14 carbon atoms. It is probably significant that both these medium-chain-length saturated acids and unsaturated C_{16} and C_{18} acids are active, since van Deenen

TABLE 1
Fusion of hen erythrocytes by fat-soluble compounds[a]

Chain length of parent compound	Agent	Ability to induce fusion within 4 hours
C_6	Hexanoic acid (caproic acid)	Negative
C_7	Heptanoic acid (enanthic acid)	Negative
C_8	Octanoic acid (caprylic acid)	Negative
C_9	Nonanoic acid (pelargonic acid)	Negative
C_{10}	Decanoic acid (capric acid)	Positive[c]
C_{11}	Undecanoic acid (hendecanoic acid)	Positive
	Methyl undecanoate	Positive[b]
	Undec-10-enoic acid (undecylenic acid)	Positive[c]
C_{12}	Dodecanoic acid (lauric acid)	Positive
	Methyl dodecanoate	Positive[b]
	Glycerol monolaurate	Positive
	Sucrose monolaurate	Positive
C_{13}	Tridecanoic acid (tridecoic acid)	Positive
	Methyl tridecanoate	Positive[b]
C_{14}	Tetradecanoic acid (myristic acid)	Positive[c]
	Methyl tetradecanoate	Positive[b]
	Tetradec-cis-9-enoic acid (myristoleic acid)	Positive
C_{15}	Pentadecanoic acid	Negative
C_{16}	Hexadecanoic acid (palmitic acid)	Negative
	Glyceryl monopalmitate	Negative
	Methyl hexadecanoate (methyl palmitate)	Positive[b]
	Hexadec-cis-9-enoic acid (palmitoleic acid)	Positive
	Methyl palmitoleate	Positive
C_{18}	Octadecanoic acid (stearic acid)	Negative
	Glyceryl monostearate	Negative
	Ethylene glycol monostearate	Negative
	Propylene glycol monostearate	Negative
	Methyl octadecanoate (methyl stearate)	Positive[b]
	Octadec-cis-9-enoic acid (oleic acid)	Positive
	Octadec-trans-9-enoic acid (elaidic acid)	Positive
	Methyl oleate	Positive
	Glyceryl mono-oleate	Positive
	Glyceryl dioleate	Positive (200 μg/ml)
	Glyceryl trioleate	Negative
	Ethylene glycol mono-oleate	Positive
	Propylene glycol mono-oleate	Positive
	Glyceryl octadec-1-enyl ether (selachyl alcohol)	Positive
	Oleyl alcohol	Positive
	Octadeca-cis,cis-9,12-dienoic acid (linoleic acid)	Positive
	Octadeca-cis,cis,cis-9,12,15-trienoic acid (linolenic acid)	Positive
C_{22}	Docosa-cis-13-enoic acid (erucic acid)	Positive
C_{20}	Retinol (vitamin A_1 alcohol)	Positive
C_{29}	α-Tocopherol (vitamin E alcohol)	Positive (200 μg/ml)
	α-Tocopherol acetate (vitamin E acetate)	Negative

See facing page for notes.

(1969) and his colleagues have shown that these two classes of carboxylic acid behave similarly at an air-water interface and have similar effects on the permeability properties of liposomes.

Interestingly, the *cis*-isomer, octadec-*cis*-9-enoic acid (oleic acid) was noticeably more effective in inducing the fusion of hen erythrocytes than the corresponding *trans*-isomer, octadec-*trans*-9-enoic acid (elaidic acid), although the latter was sufficiently effective to yield a positive result (Table 1). Conversely, both methyl palmitate and methyl sterate were found to have weak fusogenic properties, whereas the parent carboxylic, palmitic, and stearic acids, were inactive (Ahkong et al., 1973b).

The fatty acid components of two bacterial glycolipids and their various derivatives prepared by Dr. T. Ioneda (Institute of Chemistry, University of São Paulo, Brazil), have been investigated by Ahkong (1976) for their ability to fuse hen erythrocytes. Mycolic acids from *C. diphtheriae* were fusogenic. The unsaturated corynomycolenic acid was more effective than the saturated corynomycolic acid, but both mycolic acids were slightly less potent as fusogens than the parent glycolipid. The saturated nocardomycolic acid was not very effective as a fusogen. Interestingly, methyl esters of both saturated and unsaturated mycolic acids from *C. diphtheriae* were fusogenic and more effective than the parent acids; methyl nocardomycolate was also a better fusogen than nocardomycolic acid. The behavior of the methyl esters of mycolic acids therefore paralleled that of the methyl esters of palmitic and steric acids discussed earlier (Table 1).

It is relevant that at concentrations of 0.1 mg/ml the hydrocarbons tetradecane, hexadecane, and octadecane, which have melting points of 5.5°, 20°, and 28°C respectively, have also been found to induce hen erythrocytes to fuse into multinucleate cells (Ahkong et al., 1974).

Fisher (1975) has studied the dependence on concentration of the fusion of hen erythrocytes induced by retinol over a range of temperatures. He measured the time lag, or time to fusion (t_f) ensuing after the addition of fusogen, before fused cells were first observed. Fusion produced by a fixed level of retinol was found to occur earlier as the temperature increased. For example, at a concentration of retinol of 178 μM, the t_f values were > 30 min. at 15°C, 18 min at 27°C, 6.5 min. at 37°C and 3 min at 46°C. Secondly, and conversely, the quantities of retinol produced a t_f value of 5 min: > 700 μM at 15°C, 620 μM at 27°C, 250 μM at 37°C, and 140 μM at 46°C.

Devor and associates (1976) have reported hybridization by centrifugation of membranous vesicles obtained from membranes of *Escherichia coli*. Centrifugation of mixed membrane vesicles at 0°C, in the presence of 20 mM $MgCl_2$ or 25% poly (ethylene glycol) or 8 mM spermine, gave no apparent hybridization.

Note: All lipids were used at a concentration of 100 μg/ml unless otherwise stated. Each lipid was tested at pH 5.6 with erythrocytes at 37°C in Eagle's basal salt solutions buffered with sodium cacodylate containing Dextran 60C (80 mg/ml).

[a] From Ahkong et al., 1973b; reproduced with permission.

[b] Fusion observed after a delay of about 45 min.

[c] Very low incidence of fusion.

However, when the mixture was centrifuged at 42°C (i.e., above the transition temperature of the lipids in both types of vesicles) in the presence of these agents, analyses of lipid phase transitions indicated that membrane hybridization had occurred. Devor and colleagues (1976) concluded that tight association of the membranes under a high centrifugal force resulted in the mixing of different membranes lipids if lipids were in a fluid state, and that polyvalent cations had been added to assure charge neutralization. For various reasons, these investigators thought that membrane fusion was more likely to be involved in the observed membrane hybridization than the transfer or exchange of lipid molecules.

Turning to cell fusion involving phospholipid vesicles, Papahadjopoulos and co-workers (1973) have reported that phospholipid vesicles that are below their transition temperature produce less cell fusion than similar numbers of vesicles containing lipids at or above their transition temperature. Similarly, the fusion of phospholipid vesicles with one another in model systems requires the lipids of the interacting membranes to be in a fluid state, that is at a temperature above the transition temperature (T_c) of the phospholipids present. This topic, which lies outside the scope of the present chapter, is discussed elsewhere (Papahadjopoulos et al., 1976a,b; and the chapter by Papahadjopoulos in this volume).

2.2.4. Interactions between membrane phospholipids and fusogenic lipids in aqueous dispersions

Negative-staining electron microscopy has been used to investigate interactions between phospholipids and a number of substances that induce hen erythrocytes to fuse into multinucleate cells (Howell et al., 1973). Negative staining has often been criticized when applied to the study of lipids on the grounds that as the stain dries, the salts will become concentrated and cause structural changes (Finean and Rumsby, 1963; Miyamoto and Stoeckenius, 1971). The precise molecular interpretation of macromolecular assemblies of lipid molecules observed by negative staining, and of structural changes seen under varying experimental conditions, are therefore problematic. However, as discussed by Howell and associates (1973) a number of observations indicate that at least qualitative structural interpretations can be made with the aid of negative staining since Johnson and Horne (1970) have produced evidence that the negative stain sets rigidly into an amorphous glasslike substance while the specimen is still wet. In addition, hexagonal structures seen in negatively stained aqueous dispersions of phosphatidylcholine, cholesterol, and saponin are also observable in freeze-fractured and deep-etched preparations (Schmitt et al., 1970).

Of the fusogenic compounds investigated by Howell and colleagues (1973), only lysophosphatidylcholine was observed to cause bilayers of phospholipid to become micellar. Unilayered and multi-layered liposomes of phosphatidylcholine, phosphatidylserine, and sphingomyelin gave rise to globular micelles on treatment with lysophosphatidylcholine; the micelles being 7 to 8 nm in diameter. Extracted human red cell lipids, which yielded unilayered vesicles on sonication in aqueous media, also appeared to form micelles on the addition of lysophosphatidylcholine. However, since micelles were not formed when other fusogenic molecules interacted with the phospholipids, it appears that bilayer to micelle transitions—if they feature as a mechanism of cell fusion (see section 2.2.2)—may well be a special case.

Electron microscopy of negatively stained preparations of phosphatidyl-choline, sphingomyelin, phosphatidylserine, extracted red cell lipids, and erythrocyte ghosts, in the presence of fusogenic lipids such as glycerol mono-oleate or oleic acid gave rise to various new vesicular and hexagonal macromolecular assemblies (Howell et al., 1973). By contrast, lipids that are closely related chemically but do not cause erythrocytes to fuse, such as glycerol mono-stearate and steric acid, did not induce these structural changes. This indicated that a primary effect of low-melting lipids (e.g., glycerol mono-oleate and oleic acid) in causing cell fusion is on the organization of the phospholipids of the erythrocyte membrane. Naturally, changes in membrane proteins (section 4) during the fusion process were not excluded by these experiments.

Despite the various effects of cholesterol on the fusion of erythrocytes, discussed in section 2.1., no new structures were seen on treating cholesterol with glycerol mono-oleate in aqueous dispersion. It is also noteworthy, in retrospect, that new structures were not observed in mixtures of phosphatidylethanolamine and glycerol mono-oleate. It may be significant that macromolecular assemblies of phosphatidylethanolamine in aqueous dispersions have a hexagonal appearance before the addition of glycerol mono-oleate, which resembles that given by phosphatidylcholine after its addition. This may be related to findings (section 2.2.5) which indicate that the type of polar group present in membrane phospholipids is of particular importance in their interactions with fusogenic lipids.

2.2.5. Interactions between membrane phospholipids and fusogenic lipids in monomolecular films

The behavior of mixed monolayers of fourteen different lipids with preparations of erythrocyte lipids, purified natural and synthetic phospholipids, cholesterol, and galactosylceramide was investigated by Maggio and Lucy (1975). Mean areas occupied per molecule in mixed films containing lipids that are fusogenic for hen erythrocytes were compared with those for corresponding films containing lipids that are inactive as fusogens. Fusogenic lipids were found to exhibit interactions (negative deviations from the ideality rule for mean molecular areas), which were not shown by nonfusogenic lipids, in mixed monolayers with several species of phospholipid, particularly those containing a choline head group.

Table 2 shows the relationships of the fusogenic ability of a number of fat-soluble compounds to the deviations from ideal behavior that they showed in mixed monolayers with five different synthetic phospholipids. All of the fusogenic compounds in the Table exhibit decreases in mean molecular area per molecule with dipalmitoyl-, distearoyl-, and 1-stearoyl-2-oleylglycerylphosphoryl-choline, whereas the nonfusogenic lipids show virtually ideal mixing with all of these phospholipids except for glycerol monopalmitate at pressures below its phase transition. Similar observations were made in experiments with purified

TABLE 2
Fusogenic properties of fat-soluble compounds and deviations from ideal behavior in mixed monolayers with synthetic phospholipids[a]

Lipids	Fusion ability	Dipalmitoyl-L-3-glycerylphosphoryl-choline	Distearoylglyceryl-phosphorylcholine	1-Stearoyl-2-oleoyl-glycerylphosphoryl-choline	Dipalmitoyl-L-3-glycerylphosphoryl-ethanolamine	Dipalmitoyl-L-3-glycerylphosphoryl-NN-dimethyl-ethanolamine
Glyceryl mono-oleate	Positive	−20	−14	−19	+50	−20
Oleic acid	Positive	−29	−13	−17	+32	−23
Methyl oleate	Positive	−21	−16	−22	0	−20
Palmitoleic acid	Positive	−23	−19	−15	+23	−22
Retinol	Positive	−18	−10	−15	+11	−11
Isostearic acid	Positive	−22	−25	−15	+26	−11
Glyceryl monostearate	Negative	−6	0	0	+50	0
Stearic acid	Negative	+5	+5	−5	+50	0
Glyceryl monopalmitate	Negative	0[b]	−6[b]	−5[b]	+21	0[b]
Palmitic acid	Negative	−6	0	0	+32	0
Glyceryl trioleate	Negative	0[c]	+3[c]	0[c]	+25	0[c]
Methyl stearate	Weak	−28	−23	−18	0	−27
Methyl palmitate	Weak	−33	−16	−10	+8	−35
Glyceryl dioleate	Weak	+9	−19	−7	+11	+18

Note: Maximum percentage deviations in mean molecular area (negative or positive) observed for mixed monolayers are given relative to the corresponding mean molecular areas for ideal films.

[a] From Maggio and Lucy, 1975; reproduced with permission.

[b] Values refer to mixed monolayers with glyceryl monopalmitate at surface pressures above 10 mN · m^{-1}. Below this surface pressure, negative deviations between 10 and 20% were noted.

[c] Immiscible films with glyceryl trioleate.

preparations of egg phosphatidylcholine. Heterogeneity in the hydrophobic chains of phosphatidylcholine, and their degree of unsaturation, had little effect on the interaction of phosphatidylcholine with fusogenic lipids. It was also found that the presence of cholesterol had little effect on the observed pattern of interactions.

As with phosphatidylcholine, all six of the fusogenic lipids studied showed a decrease in mean molecular area when mixed with sphingomyelin from brain tissue, whereas the five nonfusogenic compounds exhibited positive deviations or virtually ideal behavior. By contrast, when fusogenic and nonfusogenic lipids were mixed with galactosylceramide, increases in mean molecular area with respect to ideal behavior were found in every case. Since mixed monolayers of fusogenic lipids with choline-containing phosphatidylcholine and sphingomyelin behaved similarly, and differently from corresponding mixed films containing galactosylceramide, it seems that the choline group participates in the observed interactions. This thesis was supported by the fact that both synthetic and naturally occurring dipalmitoylglycerylphosphorylethanolamine showed either positive deviations or ideal behavior with fusogenic and nonfusogenic lipids. Interestingly, however, in mixed films of dipalmitoyl-NN-dimethyl-phosphatidylethanolamine a differentiation between fusogenic and nonfusogenic compounds was again present (Table 2).

From these experiments it seems that a correlation exists between fusogenic properties in lipid monolayers and negative deviations from ideality in mean molecular area in mixed monolayers with choline-containing phospholipids. Thus it is likely that when a fusogen such as glycerol mono-oleate penetrates into an erythrocyte membrane, the expansion of the membrane will be less than that expected if both phospholipid and fusogenic lipid were to behave as separate, noninteracting entities. The interaction between the two molecular species may well alter the permeability properties of the plasma membrane. Since the choline-containing phospholipids in erythrocyte membranes are asymmetrically distributed in the outer half of the bilayer (Zwaal et al., 1973; Gordesky, 1976), it seems that the outer half of the bilayer will be most susceptible to alterations in structure, and thus possibly in permeability, on treatment with a fusogenic lipid (Maggio and Lucy, 1975).

Clearly, the observations of Maggio and Lucy (1975) show that fusogenic lipids differ from nonfusogenic lipids in their behavior in mixed monolayers with choline-containing phospholipids. At the same time, however, an unsaturated or low-melting hydrophobic chain appears to be required for lipids to be fusogenic. It is therefore surprising that mixed monolayers of fusogenic lipids and choline-containing phospholipids exhibit negative deviations in mean molecular area. Such behavior, often referred to as condensation of mixed monolayers, indicates that the two components of the mixture are less mobile as a result of their molecular interactions in the monolayer (i.e., they occupy a relatively smaller space than would be anticipated). For example, condensation occurs in a mixed monolayer containing cholesterol and a phospholipid that has *cis*-unsaturated chains. The explanation for the condensing effect observed with cholesterol may be, as Shah (1970) has suggested, the result of a molecular cavity being available for cholesterol with a given phospholipid because of the shape of its chains. However, another explanation is that because of its flat shape the cholesterol inhibits some of the freedom of motion of the phospholipid chains when it is present in the monolayer (Chapman and Wallach, 1968). How fusogenic lipids that are not rigid can show condensation effects with saturated and with unsaturated phospholipids in monolayers is not readily apparent. Maggio and Lucy (1976) have suggested that the presence of low-melting molecules of fusogenic lipid, as well as altering the orientation of the phospholipid head group (see below), might cooperatively increase the number of rotational isomers (Traüble and Haynes, 1971) in the phospholipid molecules. This would produce a well-ordered arrangement of molecules in the mixed monolayer, and hence a condensation effect, but the system would still contain disordered hydrocarbon chains (Seelig and Seelig, 1974). Comparable changes occurring in the plasma membrane of a cell treated with a fusogenic lipid may produce an increase in the permeability of its lipid bilayer (Traüble and Haynes, 1971; Marsh, 1974) that allows Ca^{2+} ions to enter the cellular cytoplasm (section 3.2).

Investigations have been undertaken on the behavior of the polar groups of phospholipids in mixed monolayers with fusogenic lipids to obtain a more detailed comparison of the properties of the choline and ethanolamine moieties (Maggio and Lucy, 1976). At pH 5.6 and 10, but not a pH 2, mixed monolayers of glycerol mono-oleate with phosphatidylcholine showed negative deviations from the ideality rule in surface potential per molecule, as well as the negative deviations in mean molecular area already considered. Changes in surface potential were not seen with chemically related but nonfusogenic lipids (e.g., glycerol monostearate), nor were they observed in mixed monolayers with phosphatidylethanolamine. The fact that negative deviations from ideality in surface potential and in surface area, in mixed monolayers of ionizable (e.g., oleic acid) and nonionizable (e.g., glycerol mono-oleate) fusogenic lipids with dipalmitoylglycerylphosphorylcholine, were both decreased on changing the pH of the subphase from pH 5.6 to 2 indicated that the polar head group of the phospholipids is involved in the observed interactions. It was also significant that the differing behavior shown in mixed films by choline-containing phospholipids (negative deviations in mean molecular area) and ethanolamine-containing phospholipids (positive deviations in mean molecular area) was eliminated at pH 2, and that the two species of phospholipid showed almost ideal mixing at this pH with fusogenic and nonfusogenic lipids.

Maggio and Lucy (1976) also observed that the presence in the subphase of divalent cations, including Ca^{2+}, significantly modified the properties of mixed monolayers containing fusogenic lipids and phosphatidylcholine or phosphatidylethanolamine. It was apparent that whether phosphatidylcholine interacts differentially with fusogenic and nonfusogenic lipids, and whether phosphatidylethanolamine behaves in the same way or differently, depends on the surface pressure of the system and also on the degree to which the ion-dipole behavior of the phospholipid is constrained by the influence of an external bivalent metal ion. For a fusogenic lipid to modify the properties of molecules of phosphatidylethanolamine, in which intermolecular ionic interactions are probably greater than in phosphatidylcholine (Phillips et al., 1972), it seems to be necessary to disturb the orientation of the polar moieties of the phospholipid monolayer by adding divalent cations to the subphase. On the basis of these observations it was speculated that in the presence of a fusogenic lipid the properties of the inner half of the lipid bilayer of the erythrocyte membrane (rich in phosphatidylethanolamine) could be altered and behave like the outer half of the bilayer (rich in phosphatidylcholine) if the concentration of cytoplasmic Ca^{2+} were to be raised to about 1 mM as the result of a change in membrane permeability (section 3). This increase in membrane symmetry might lower energy values for the intermixing of the constituence of membranes in adjacent erythrocytes and thus aid the process of cell fusion.

2.2.6. Water-soluble fusogens and phospholipids

Fusogenic properties are not confined to lipid-soluble substances. High concentrations of cryoprotectant molecules such as dimethyl sulfoxide (5 M) and glycerol (1.5–2.5 M) are also fusogenic. On incubation of hen erythrocytes with 5 M dimethyl sulfoxide (35%) for 5 minutes at 37°C the cells became spherical, and binucleate cells were observed after 25 to 30 minutes; the number of fused cells and the number of nuclei per cell increased on further incubation (Ahkong et al., 1975a). Other polyols were investigated for fusogenic properties, and it was found that sorbitol fuses hen erythrocytes; numerous multinucleate cells were present after 2 hours of incubation in 2.5 or 3 M sorbitol. Higher and lower concentrations of sorbitol were less effective. Multinucleate cells were also observed with mannitol, ethylene glycol, and sucrose. Long incubation (30–48 hr) with a high molecular weight dextran (2×10^6) induced the cells to fuse, but a lower molecular weight dextran (Dextran 60C, molecular weight approximately 90,000) was not fusogenic. This latter dextran has, however, been found to be most effective in facilitating chemically induced cell fusion even though it is not itself capable of causing cells to fuse. Its "helper" action is effective both with fusogenic lipids (Ahkong et al., 1973b) and with water-soluble fusogens. For example, Dextran 60C facilitated fusion induced by glycerol (1.5–2.5 M). In its absence most of the glycerol-treated hen erythrocytes became swollen and were lysed, with only a few cells being fused. Similarly, Dextran 60C greatly aided fusion induced by sucrose (Ahkong et al., 1975a). The mechanism of action of the low molecular weight dextran in facilitating cell fusion is unknown, but it may possibly act by causing some diminution in membrane surface potential.

Particularly important among the water-soluble chemicals that induce cell fusion is the polymer, poly(ethylene glycol), which has been used for several years in the fusion of plant protoplasts (Kao and Michayluk, 1974; Wallin et al. 1974; also see the chapter by Power et al., this volume). Since poly(ethylene glycol) is relatively nontoxic, yields an exceptionally high incidence of chemically induced cell fusion, and has a wide range of application, the potential scope for using it as a chemical fusogen is very great. Current developments in studies on the use of poly(ethylene glycol) as a fusogen indicate that it may be more valuable than Sendai virus as an aid in obtaining hybrid cells for use in a wide range of biochemical, genetic, and oncological investigations.

Poly(ethylene glycol) has been shown to fuse mammalian and avian erythrocytes (Hartmann et al., 1974; Ahkong et al., 1975a), with the larger polymer of 6000 MW being more effective than the smaller poly(ethylene glycol) of 1500 MW (Ahkong, 1976; Hartmann et al., 1976). Hartmann and associates (1976) have also shown that pretreatment of hen erythrocytes with a protease preparation greatly enhances the degree of cell fusion observed (to a level of 60–70%) on the subsequent addition of poly(ethylene glycol). These investigators suggest that their results may support the hypothesis that a limited degree of controlled lysis is required for cell fusion (De Boer and Loyter, 1971; Yanovsky and Loyter, 1972).

Using tissue culture techniques, Pontecorvo (1975) found that poly(ethylene glycol) yields a very high incidence of fusion with animal cell lines. He has investigated the production of indefinitely multiplying mammalian somatic cell hybrids using poly(ethylene glycol), and has used the water-soluble fusogen to obtain hybrid cells by the fusion of Chinese hamster HGPRT$^-$ cells (strain WG, clone 1) with mouse TK$^-$ cells (strain 3T3). In addition, human skin fibroblasts have been fused with one another and with human lymphocytes. As might be anticipated from the ability of dimethyl sulfoxide to fuse hen erythrocytes (Ahkong et al., 1975a), this cryoprotectant molecule enhances the fusion of human diploid cells in tissue culture induced by poly(ethylene glycol), the extent of fusion being directly proportional to the concentration of both compounds (Norwood et al., 1976). By contrast with the greater effectiveness of preparations of poly(ethylene glycol) of high molecular weight in inducing the fusion of erythrocytes, polymer molecules of lower molecular weight are more effective in fusing nonerythroid animal cells in monolayer cultures in vitro. Davidson and colleagues (1976) have reported that from a practical viewpoint poly(ethylene glycol)-1000 seems to be the choice for hybridization of cells attached to a substrate. Not only does this polymer, at a concentration of 50%, produce more hybrids than any other preparation of poly(ethylene glycol) at any concentration, but it is also the preparation that is least sensitive to dilution effects. These workers suggest that poly(ethylene glycol)-400 may be useful for inducing fusion in suspensions of cells because its very low viscosity makes it easy to remove after treatment of the cells. In addition, Davidson and co-workers (1976) found that for all poly(ethylene glycol) preparations with molecular weights of between 400 and 6,000, peak hybridization frequencies occurred at polymer concentrations of 50 to 55%, with the higher concentration possibly being more effective with preparations having molecular weights of less than 1000. They suggested that maximum cell hybridization occurs when the concentration of the repeating chemical subunit in poly(ethylene glycol), or the mass of poly(ethylene glycol)/ml, is the same for all preparations of polymer. These various findings with animal cells on interrelationships in tissue culture, between fusogenic properties, molecular weight, and concentration, have been confirmed concomitantly in this laboratory (Goodall, Fisher and Lucy, unpublished observations).

Maul and associates, (1976), using mouse L cells, observed fusion to occur in localized regions only 1 minute after washing the poly(ethylene glycol) off the cells. Contrary to earlier theories (Poste and Allison, 1971), areas of plasma membrane having many microvilli did not seem to fuse. Pontecorvo and colleagues (1977) have also observed that cell fusion occurs very near, or during, the time of treatment of the cells with poly(ethylene glycol). They noted however, that fusion occurs preferentially between or near the tips of elongated cellular processes. This may explain why the proportions of binucleate and multinucleate cells found in sparse monolayers treated with poly(ethylene glycol) were as high as, or even higher, than in confluent cultures. In the latter, there was very extensive side-by-side contact between fibroblasts but not more extensive cell fusion.

The wide range of cells that are susceptible to fusion by poly(ethylene glycol) is one of the most remarkable properties of this substance. In addition to the intra- and interspecific fusion of higher plant protoplasts (see chapter by Power et al., this volume), poly(ethylene glycol) induces the intra- and interspecific fusion of fungal protoplasts (Anné et al., 1976), and of bacterial protoplasts (Fodor and Alföldi, 1976; Schaeffer et al., 1976). Interkingdom fusion has been reported between hen erythrocytes and yeast protoplasts (Ahkong et al., 1975b), HeLa cells and carrot protoplasts (Dudits et al., 1976), and HeLa cells with tobacco protoplasts (Jones et al., 1976).

Although the mechanisms of action of water-soluble fusogens in general, and of poly(ethylene glycol) in particular, have not yet been studied extensively, it is clear that their effects on cell membranes are likely to be complex. It has been suggested (Maggio et al., 1976) that these substances may induce various structural changes in the bulk-phase water (Tanford, 1961) adjacent to membrane surfaces. This may in turn have a variety of consequences on the orientation and hydration of the polar head groups of membrane phospholipids, on the permeability properties of membranes (section 3), and on the spatial orientation and aggregation of membrane proteins (section 4). Maggio and associates (1976) reported a study on the actions on monolayers of dipalmitoylglycerylphosphorylcholine and dipalmitoylglycerylphosphorylethanolamine of poly(ethylene glycol), glycerol, sorbitol, sucrose, and dimethyl sulfoxide, all of which fuse hen erythrocytes (Ahkong et al., 1975a). Small expansion and condensation effects were noted in the surface pressure-area curves for monolayers of the two phospholipids, depending on the particular combinations of phospholipid and water-soluble solute. The most interesting findings, however, were that poly(ethylene glycols), dimethyl sulfoxide, glycerol, and sorbitol all markedly decreased the surface potentials of monolayers of both phospholipids (Fig. 1). Sorbitol had a relatively small effect, but with the other solutes the surface potentials of the phospholipids were decreased by several hundred millivolts. With the preparations of poly(ethylene glycol) that were studied (molecular weights: 1500, 6000, 3×10^5, 5×10^6), lower concentrations of polymer were required to

286

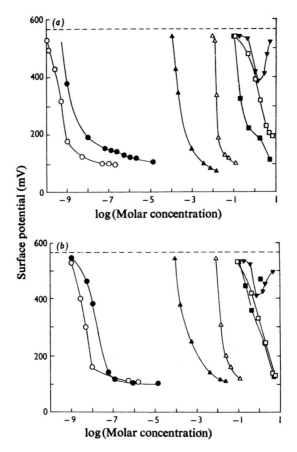

Fig. 1. Surface potentials of phospholipid monolayers on subphases containing differing concentrations of organic solutes. The surface potentials of monolayers of dipalmitoylglycerylphosphorylcholine (a) and dipalmitoylglycerylphosphorylethanolamine (b), each at surface pressures of 35 mN · m^{-1}, were derived from isotherms obtained on subphases of 145 mM-NaCl, pH 5.6, containing poly (ethylene glycol)-5 × 10^6 (o), poly (ethylene glycol)-3 × 10^5 (●), poly (ethylene glycol)-6000 (▲), poly (ethylene glycol)-1500 (△), dimethyl sulfoxide (■), glycerol (□), or sorbitol (▼). The horizontal broken line at 570 mV represents the values of the surface potentials for the phospholipid monolayers, spread at a surface pressure of 35 mN · m^{-1}, on a subphase of 145 mM-NaCl. (Reproduced with permission from Maggio et al., 1976.)

decrease the surface potential with increasing molecular weight. It may be significant that fusogenic lipids give rise only to a rather small decrease in mean surface potential per molecule when they interact with monolayers of phosphatidylcholine (Maggio and Lucy, 1976), whereas poly(ethylene glycol), which is much more effective in inducing cell fusion, causes a large decrease in surface potential. Thus it seems that a decrease in the surface potential of biological membranes may be of major importance in the process of membrane fusion. Conversely, the toxic effects of lipid fusogens might be suggested to result from their causing an undue perturbation of the lipid bilayers of membranes (Maggio and Lucy, 1975), if it is justifiable to extrapolate from phenomena observed in lipid monolayers to biological membranes.

Dimethyl sulfoxide, which enhances the rate of fusion of acidic phospholipid vesicles (Papahadjopoulos et al., 1976a) as well as inducing cell fusion, has been found to produce an increase in the transition temperature, T_c, of acidic dimyristoylphosphatidylglycerol membranes (Lyman et al., 1976). It has therefore been proposed that the fusogenic properties of dimethyl sulfoxide are related to its ability to induce a phase transition in the lipids of membranes, and that its actions may resemble those of Ca^{2+} in promoting vesicle fusion (Papahadjopoulos et al., 1976a,b; see also the chapter by Papahadjopoulos, this volume).

Before leaving the topic of water-soluble fusogens, it is relevant to refer to the behavior of certain nonionic, surface-active, immunological adjuvants that fuse hen erythrocytes (Ahkong et al., 1974). Cell fusion was induced by Arlacel A (mannide mono-oleate), Span 80 (sorbitan mono-oleate), and Span 20 (sorbitan monolaureate). Significantly, fusion was not seen with two related long-chain saturated esters tested; Span 40 (sorbitan monopalmitate) and Span 60 (sorbitan monostearate). This is consistent with the previously reported observations (Ahkong et al., 1973b) that cell fusion occurs with esters of both medium-chain saturated, and longer-chain unsaturated fatty acids, but not in general with esters of palmitic and stearic acids (section 2.2.3 and Table 1). Span 80 and Arlacel A are mixtures of partial esters of a polyol (sorbitol or mannitol) and its anhydrides. When samples of these materials were separated into mono- and dioleate fractions, the mono-oleate fractions of both surfactants induced the fusion of hen erythrocytes. Dioleate fractions were less effective. The behavior of the oleate esters of sorbitol and mannitol therefore parallels that of the mono- and dioleate esters of glycerol (Table 1).

3. Cell fusion and membrane permeability

3.1. Cell swelling

Changes in membrane permeability are intimately associated with chemically induced cell fusion. It would seem likely that many if not all chemicals that induce cell fusion also increase the permeability of the plasma membrane, if only because the decrease in membrane stability that is necessary for membrane fusion to occur may well be accompanied by changes leading to cell swelling and to lysis. This was seen in extreme form on treating hen erythrocytes with lysophosphatidylcholine in aqueous solutions when the cells not only fused to form syncytia within 30 seconds, but the syncytia also disintegrated within 1 minute (Ahkong et al., 1972). In other cases, the decrease in membrane stability appears to be accompanied by an increase in permeability that leads more slowly, but inevitably, to colloid osmotic lysis. Initially, Dextran 60C was included in the incubation medium with fusogenic lipids to combat this process. That the added dextran could inhibit colloid osmotic lysis was shown by the fact that lysis occurred immediately when the resulting multinucleate erythrocytes were resuspended in a medium not containing dextran (Ahkong et al., 1973b).

In my experience, chemically induced cell fusion is preceded by cell swelling. This may play a physical role at the macrolevel to aid cell fusion, as apparently occurs in the fusion of plant protoplasts (Power et al., 1970). With protoplasts, it appears that there must be a sufficient "depth" of cytoplasm at the point of contact of each of the fusing cells before fusion can take place (Power and Cocking, 1971). Alternatively, cell swelling may not only reflect the reduced stability of a cell membrane treated with a fusogen but, as suggested by Ahkong and co-workers (1973a), it may be associated with an exposure of endogenous phospholipids and/or added lipids at certain sites on the cell surface, possibly involving a lateral movement of membrane glycoproteins. Observations made on fibroblasts treated in tissue culture with the fusogenic lipid oleylamine have supported this latter possibility. Thus the appearance in the treated cells, viewed by phase-contrast microscopy, of large, clear cytoplasmic regions suggested a rapid influx of water into the cells, particularly where large protrusions formed. As with erythrocytes, only swollen fibroblasts eventually fused, usually when two adjacent plasma membranes covering the regions of swelling came into contact (Bruckdorfer et al., 1974).

Hen erythrocytes treated with poly(ethylene glycol) initially shrink drastically (Fig. 2a) on treatment with the high concentrations of polymer that are required to induce cell fusion. Presumably this reflects removal of water from the cells by the hydrophilic solute. If poly(ethylene glycol) is used to obtain hybrid cells, care must be exercised to minimize cellular damage due to dehydration by reducing contact of the cells with solutions containing high concentrations of poly(ethylene glycol) to a minimum (ca. 1 min) (Pontecorvo et al., 1977). Erythrocytes that were agglutinated and shrunk by treatment with poly(ethylene glycol)-6000 (400 mg/ml) did not fuse in situ, possibly because polymer molecules between the cells acted as a barrier (Maggio et al., 1976). Extensive cell fusion occurred after removal of most of the poly(ethylene glycol) between swollen, less aggregated cells (Fig. 2b). Once again, cell fusion occurred between swollen cells.

3.2. Calcium ions

The involvement of Ca^{2+} in the fusion of biological membranes seems to be an almost universal phenomenon. For example, Ca^{2+} is necessary in myoblast fusion (Shainberg et al., 1969), in cell fusion induced by liposomes (Papahadjopoules et al., 1973; chapter 14, this volume), and in the fusion of nascent membranes of the protozoan *Echinosphaerium nucleofilum* (Vollet and Roth, 1974). In addition, Ca^{2+} is of major importance in many processes of secretion that involve exocytosis (Carafoli et al., 1975). Nevertheless, it appears that a requirement for Ca^{2+} in cell fusion is not universal. The fusion of hen erythrocytes induced by lysophosphatidylcholine occurs in the absence of exogenous Ca^{2+} (Howell and Lucy, 1969). Also, if low concentrations of virus are used to minimize hemolysis,

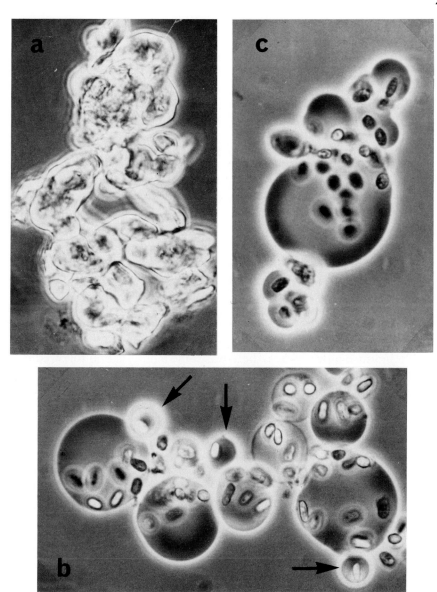

Fig. 2. Morphological changes during the fusion of hen erythrocytes induced by poly(ethylene glycol). (a) Hen erythrocytes (3 × 10⁸ cells/ml), after incubation for 15 min at 37 °C in 1 ml of modified Eagle's basal salt solution (1.8 mM-Ca²⁺) at pH 7.4 (Ahkong et al., 1973b) containing poly(ethylene glycol)-6000 (400 mg/ml), showing extensive cellular aggregation and shrinking. In (b), the aggregated cells were diluted with the (warmed) salt solution (5 ml) free from poly(ethylene glycol), centrifuged (800 g for 5 min), resuspended in the salt solution (1 ml), and incubated at 37 °C for 30 min. Multinucleated and single swollen cells (arrows) are present. The cells in (c) were treated as in (b), except that the salt solution used for dilution and for the second incubation contained EDTA (5 mM). Phase-contrast microscopy: ×1000. (Reproduced with permission from Maggio et al., 1976.)

Sendai virus-induced fusion of human erythrocytes (Peretz et al., 1974) and hen erythrocytes (Hart et al., 1976) occurs in the presence of EDTA and EGTA respectively. With the latter cells, fusion is significantly decreased by the presence of Ca^{2+} even at a concentration of 0.2 mM. This is in marked contrast to the requirement for Ca^{2+} in the fusion of hen erythrocytes by lipid molecules (Ahkong et al., 1973b) and by many water-soluble fusogens (Ahkong et al., 1975a). It is interesting that the general hypothesis for the mechanism of membrane fusion proposed by Poste and Allison (1973) (see also section 4), which involves the activity of a Ca^{2+}-dependent ATPase, would seem not to apply to these instances of cell fusion induced by Sendai virus and by lysophosphatidylcholine.

Lucy and colleagues (1971) noted that hen erythrocytes were fused on the addition of F^- ions to the cells in the presence of Ca^{2+}, and the fusion of human erythrocyte "ghosts" promoted by the combined action of calcium and phosphate ions has been described in detail by Zakai and co-workers (1976). In the latter work, all of the bivalent metal ions that formed precipitates with Ca^{2+} were found to cause the agglutination of erythrocyte ghosts. However, fusion was distinct from agglutination, since Zn^{2+} and Mn^{2+} in phosphate buffer did not promote fusion although they were active in agglutination. To obtain fusion it was necessary to incubate ghosts in phosphate buffer and then add Ca^{2+}. If Ca^{2+} and the phosphate buffer were preincubated for 20 minutes at 37°C before addition of erythrocyte ghosts, calcium phosphate precipitated and, when ghosts were added to such a precipitate, there was some agglutination followed by lysis but no fusion. The inclusion of ATP in the ghosts, which causes internal Ca^{2+} to be pumped out, was found to prevent fusion. Zakai and associates (1976) suggested that these and related results could be explained by assuming that the fusion of ghosts and intact human erythrocytes requires intracellular Ca^{2+}.

The importance of Ca^{2+} in the fusion of plant protoplasts induced by poly-(ethylene glycol) was seen by Kao and Michayluk (1974), who discussed possible interactions of Ca^{2+} and poly(ethylene glycol) at the surfaces of adjacent protoplasts. Similarly, Ferenczy and co-workers (1975) observed a significant effect of Ca^{2+} on the fusion of fungal protoplasts caused by poly(ethylene glycol). Without Ca^{2+}, the complementation of mutant pairs requiring lysine and methionine was negligible, despite good aggregation of the fungal protoplasts at higher concentrations of poly(ethylene glycol). 1 mM calcium chloride was effective in stimulating the fusion process, and the stimulating effect was similar with from 10 to 100 mM $CaCl_2$.

How does Ca^{2+} function in chemically induced cell fusion? The roles of Ca^{2+} are probably many and various, and it is unlikely that Ca^{2+} acts in one way only even in a specific membrane system. The effects of Ca^{2+} are discussed in greater detail in this volume in the chapter by Papahadjopoulos. In the present context, it is interesting to consider the possible actions of Ca^{2+} both outside and inside cells in relation to chemically induced cell fusion.

Outside cells, Ca^{2+} may facilitate cell fusion by bringing them sufficiently close together for fusion to occur. Ca^{2+} may also restore stability to cell membranes subsequent to the fusion process. Ca^{2+} decreases lysis of KB cells, Ehrlich ascites tumor cells, and erythrocytes by Sendai virus (Okada and Murayama, 1966; Yanovsky and Loyter, 1972; Toister and Loyter, 1970), and it has been suggested that Ca^{2+} promotes fusion by stabilizing membranes against viral lysis. Similar considerations may apply in chemically induced cell fusion since, in the case of oleylamine-induced fusion of mouse fibroblasts in tissue culture (Bruckdorfer et al., 1974), significantly more lysis occurred in the absence of Ca^{2+}.

In experiments on avian erythrocytes, Toister and Loyter (1971) found that cell fusion could be induced by Ca^{2+} at pH 10.5. To induce agglutination, lysis, and fusion, the following steps were required: (1) preincubation of the cells at pH 10.5 and 37 °C; (2) addition of the bivalent cation in the cold to cause agglutination; (3) incubation of the agglutinated cells at 37°C. It was proposed that lysis might be caused by saponification of membrane phospholipids at high pH. Such saponification could form lysophospholipid molecules at the point of localization of Ca^{2+} ions, thus promoting lysis. Where the erythrocyte membranes are in contact and lysis is initiated, fusion could also take place, possibly being induced by lysophosphatidylcholine (section 2.2.2).

The spontaneous fusion of myoblasts into myotubes that is important in muscle development is presumably a chemically induced phenomenon, although the chemical inducers involved have not yet been delineated. This fusion process is quantitatively dependent on Ca^{2+} (Van der Bosch et al., 1972), and the fusion of myoblasts themselves (Schudt et al., 1973) and of plasma membrane vesicles isolated from myoblasts grown in culture (Schudt et al., 1976) is inhibited by Mg^{2+} Experiments on the influence of the divalent cation ionophore A23187 on myogenic cell fusion, undertaken by Schudt and Pette (1975), indicate that the site of action of Ca^{2+} in the myoblast system is on the outside surface of the plasma membrane (see also Weidekamm et al., 1976).

With regard to the possible roles of Ca^{2+} inside cells, Poste and Allison (1971), Poste (1972), and Poste and Allison (1973) have drawn attention to the importance of Ca^{2+} in secretion and have proposed a hypothesis for membrane fusion, including cell fusion, which involves the possible participation of a calcium-dependent ATPase, though more recently Poste has suggested that participation of such an enzyme in fusion is by no means a general phenomenon and is at best probably confined to limited examples of fusion (see Papahadjopoulos et al., 1978).

In recent years studies on secretion have been greatly advanced by the use of the bivalent cation ionophores X537A and A23187, which facilitate the movement of Ca^{2+} through membranes (Reed and Lardy, 1972; Pressman, 1973, 1976). These ionophores trigger secretion in the presence of extracellular Ca^{2+} (Foreman et al., 1973; Cochrane and Douglas, 1974; see also chapter by Meldolesi et al., this volume). It is interesting that in direct contrast to the myoblast system, experiments on hen erythrocytes with X537A and A23187 have indicated that Ca^{2+} triggers chemically induced cell fusion by increasing the concentration of cytoplasmic Ca^{2+} (Ahkong et al., 1975c). Bivalent cation ionophores seem to have little aggregating action and, for cell fusion to occur rapidly, it was found necessary to pretreat the hen erythrocytes with neuraminidase and then to aggregate them with dextran before adding ionophore. With cells treated in this way in the presence of Ca^{2+}, limited cell fusion occurred after 3 hours at 37 °C. However, a further incubation at 47 °C for only 10 minutes resulted in extensive fusion, presumably because of the importance of membrane fluidity in cell fusion in hen erythrocytes (section 2.2.3). Cell fusion also occurred with ionophore and Ca^{2+} at 37 °C, albeit more slowly, even with cells that were not pretreated with neuraminidase. Light microscope experiments provided no evidence that either valinomycin, with or without 2,4-dinitrophenol, or 2,4-dinitrophenol alone caused fusion. Increasing the concentration of Ca^{2+} from 1 to 3 mM in the presence of X537A and A23187 allowed cell fusion to occur more rapidly, but this had no effect on cells treated with valinomycin or valinomycin and dinitrophenol (DNP). The actions of both X537A and A23187 with 1 mM Ca^{2+} were inhibited by EGTA and EDTA (2 mM); inhibition by the chelating agents was, however, overcome by an excess of Ca^{2+} (final concentration 3 mM). It was therefore concluded, as with secretory system, that the observed fusogenic action of the two bivalent cation ionophores is mediated by entry of Ca^{2+} into the cells. This would seem to be contrary to the views of Poste and Allison (1973), who have postulated that membranes can exist in two states: (1) the normal Ca^{2+}-associated state, and (2) the "fusion susceptible" state, in which Ca^{2+} has been displaced from the membrane.

What is the probable result of this entry of Ca^{2+} into the erythrocyte cytoplasm? It has been suggested by Ahkong and associates (1975c) that cell swelling, which accompanies fusion induced by ionophores and by lipid fusogens (section 3.1) may result from an inhibition by cytoplasmic Ca^{2+} of membrane-bound Na^+/K^+-dependent ATPase activity. This would lead to a loss of cytoplasmic potassium (Dunn, 1974) and the entry of external water, as postulated for the Ca^{2+}-dependence of the acrosome reaction (Yanagimachi and Usui, 1974). Other consequences of an increased cytoplasmic concentration of Ca^{2+} may be phase separations in membrane phospholipids and the aggregation of intramembranous proteins (section 4).

The entry of Ca^{2+} into cells does not necessarily induce cell fusion. Human erythrocytes, for example, do not fuse with one another at 37 °C or 47 °C on treatment with A23187 in the presence of Ca^{2+} (Allan et al., 1976a, b; Vos et al., 1976), even though the cells rapidly accumulate Ca^{2+} (Edmonson and Li, 1976). Instead, small vesicles bud from human erythrocytes in a process involving the fusion of the plasma membrane with itself (Allan and Michell, 1975). It has been proposed by Allan and co-workers (1976a) that the microvesiculation seems likely to be caused by an increase in the diacylglycerol content of the erythrocyte membrane which they have observed, particularly since they have also found that the microvesicles have a high content of this lipid. They have further suggested that diacylglycerol (Allan et al., 1976b) is probably produced in the inner surface of the membrane bilayer where, since it is fusogenic (cf. Table 1), it could mediate the fusion of cytoplasmic surfaces. As these workers point out, the situation is in contrast to that in which erythrocytes are treated with exogenous phospholipase C, when inward vesiculation (Allan et al., 1975) and cell fusion (De Boer and Loyter, 1971) occur.

It appears probable that the absence of cell-to-cell fusion in human erythrocytes treated with ionophore A23187 is due to entry of exogenous Ca^{2+} into these cells being insufficient, (even after raising the temperature) to cause cell fusion. This is indicated by the fact that human and hen erythrocytes are both fused by treatment with glycerol mono-oleate (Ahkong et al., 1973b), and studies with isotopically labelled Ca^{2+} have shown that Ca^{2+} rapidly enters both species of treated cells prior to the onset of cell fusion (A. M. J. Blow, G. M. Botham and J. A. Lucy, unpublished observations). Erythrocytes treated with water-soluble cryoprotectant molecules that induce fusion also exhibit a similar increase in their permeability to Ca^{2+}; this is not observed in cells treated with nonfusogenic lipid molecules (e.g., glycerol monostearate).

In investigations on the fusion of hen erythrocytes by poly(ethylene glycol), Maggio and colleagues (1976) found that an excess of EDTA inhibited fusion when the chelating agent was present throughout the experiment. EDTA was unable, however, to inhibit cell fusion if it was added to the erythrocytes after the cells had been incubated with polymer in the presence of Ca^{2+}, that is, addition of EDTA when the cells were washed to remove most of the poly(ethylene glycol) was unable to inhibit fusion (Fig. 2c). It was proposed that entry of Ca^{2+} into the erythrocytes during incubation with poly(ethylene glycol), viz. at the stage of cell shrinkage (Fig. 2a), was responsible for cell fusion then becoming independent of exogenous Ca^{2+}. It was also suggested that such an increase in the permeability of the plasma membrane to Ca^{2+} may in turn be related to changes in membrane surface potential induced by poly(ethylene glycol) (cf. section 2.2.6 in view of possibilities for the control of ionic permeability by surface potential (Gingell, 1967; and chapter by Gingell and Ginsberg, this volume). The relationships between membrane phase separations, lipid phase transitions and surface potential that have been considered by Traüble (1977) are also relevant here.

4. Cell fusion and membrane proteins

By comparison with studies on membrane lipids, relatively little work has been published concerning membrane proteins and cell fusion. No doubt this is due in part to the fact that the lipid bilayer component of plasma membranes is primarily responsible for the individual integrity of cells.

p-Chloromercuribenzenesulfonate (0.2mM), which increases the cation permeability of membranes (Dunn, 1974), has been reported to induce the shrinkage and agglutination of hen erythrocytes, which then swell and fuse after 4 hours at 37°C (Hart et al., 1975). Since p-chloromercuribenzenesulfonate traverses the lipid bilayer of cells very slowly, if at all, it was thought that the reagent reacts with a thiol group present in a relatively accessible protein that participates in controlling the concentration of intracellular Ca^{2+}. Hart and co-workers (1975) also found that N-ethylmaleimide (1 mM) inhibited the fusion of hen erythrocytes induced by glycerol mono-oleate, glycerol, poly(ethylene glycol), A23187, and p-chloromercuribenzenesulfonate. N-ethylmaleimide (11 mM) inhibited the fusion of human erythrocytes by glycerol mono-oleate but much less effectively than with hen erythrocytes. It was proposed that the inhibitory action of N-ethylmaleimide in cell fusion is directed toward the thiol groups of protein molecules, possibly located on the cytoplasmic side of the membrane, which respond to an increased concentration of intracellular Ca^{2+} in a process leading to cell fusion. Consistent with this interpretation is the finding that N-ethylmaleimide added to erythrocytes prior to Ca^{2+} inhibits cell fusion, whereas addition of the thiol reagent subsequent to incubation with Ca^{2+} has little effect (Ahkong, 1976).

It is interesting to note that the fusion of hen erythrocytes by Sendai virus was observed to behave quite differently in response to these thiol-blocking agents: p-chloromercuribenzenesulfonate had little action and N-ethylmaleimide (1 mM) more than doubled the fusion index (Hart et al., 1975). This, like the differing effects of Ca^{2+} on the two systems, is another indication that cell fusion induced by Sendai virus and by chemicals apparently proceed by distinctively different pathways. Loyter and associates (1976) have obtained results that illustrate this complexity even further. They found that thiol-blocking reagents are highly active in restoring the fusion susceptibility of human erythrocyte "ghosts" which, unlike the intact cells, are not normally susceptible to fusion by Sendai virus or by glycerol mono-oleate (Peretz et al., 1974). p-chloromercuribenzenesulfonate (0.5–2 mM) and N-ethylmaleimide (1–2 mM) were both effective in this respect. It was proposed that these agents need to react with thiol groups that are located on the inner surface of the erythrocyte membrane to be effective, because the former compound promoted fusion when it was trapped in the ghosts but had no effect when added to resealed ghosts.

Studies on the membrane proteins of human erythrocytes during fusion induced by glycerol mono-oleate have revealed interesting changes (S. J. Quirk and J. A. Lucy, unpublished observations). Polyacrylamide-gel electrophoresis of the proteins of ghosts prepared from fused human erythrocytes showed, by comparison with untreated cells, an increase in polypeptide material of very high molecular weight, a decrease in band 4.1, and an increase in band 4.3, all of which could possibly be ascribed to the effects of an initial entry of Ca^{2+} into cells treated with glycerol mono-oleate. By contrast, a notable loss of band 3 protein appeared to be more specifically associated with cell fusion itself. In addition, the presence of the proteinase inhibitor 1-chloro-4-phenyl-3-L-toluene-p-sulfonamidobutan-2-one (TPCK) partially inhibited both cell fusion and the associated alterations in band 3. These observations indicated that a limited degree of proteolysis of band 3 protein may occur during the fusion of human erythrocytes induced by glycerol mono-oleate that could be specifically associated with the process of cell fusion.

Poste and Allison (1973) suggested that the aggregation of membrane proteins is important in membrane fusion, including cell fusion. This idea was consistent with the movement of intramembranous particles in preparations of erythrocyte membranes treated with Sendai virus (Bächi et al., 1973). Indirect support was also provided by the finding that dimethyl sulfoxide and glycerol induce cell erythrocytes to fuse (Ahkong et al., 1975a), since these chemicals are known to aggregate the intramembranous particles of unfixed T and B mouse lymphocytes (McIntyre et al., 1974). Glycerol-induced aggregation of membrane-intercalated particles has also been seen in *Entamoeba histolytica* (Pinto da Silva and Martínez-Palomo, 1974). Ahkong and associates (1975c) in the context of the possible importance of an increase in the cytoplasmic concentration of Ca^{2+} have pointed out that Ca^{2+} may be expected to aggregate negatively charged spectrin molecules on the cytoplasmic surface; this will in turn aggregate integral membrane proteins (Nicolson and Painter, 1973). Furthermore, the binding of Ca^{2+} to phosphatidylserine molecules in a membrane yields solid aggregates that allow other phospholipids to form fluid clusters (Ohnishi and Ito, 1974; Papahadjopoulos et al., 1974; Jacobson and Papahadjopoulos, 1975; and also see chapter by Papahadjopoulos, this volume). If this should occur at the cytoplasmic surface, where most of the phosphatidylserine is located (Zwaal et al., 1973; Gordesky, 1976; Rothman and Lenard, 1977), it would probably also lead to an aggregation of intramembranous particles (Grant and McConnell, 1974). It is relevant here that Schudt and co-workers (1976) have reported that an aggregation of intramembranous particles occurs in the area of membrane contact during the Ca^{2+}-induced fusion of plasma membranes vesicles isolated from myoblasts grown in vitro.

In addition to proposing that membrane proteins aggregate during membrane fusion, Poste and Allison (1973) also suggested that fusion might proceed by the interdigitation of the aggregated membrane proteins. In my laboratory, we have, by contrast, suggested that fusion proceeds by the intermingling of membrane lipids after the emergence of protein-free areas of lipid bilayer, following protein aggregation (Ahkong et al., 1975a) (Fig. 3). Several electron microscopy observations have been reported in support of the latter view; they are referred to elsewhere (Lucy, 1977b) and in this volume (see chapters by Meldolesi et al. and by Orci and Perrelet), while other workers have described findings supporting the importance of the interdigitation of membrane proteins in membrane fusion (e.g., Weiss et al., 1977). However, most of these investigations do not fall within the scope of the present chapter as they are not concerned with cell fusion per se but with other instances of membrane fusion in phenomena such as exocytosis (Chi et al., 1976; Lawton et al., 1977).

The intramembranous particles of mouse L cells treated with poly(ethylene glycol) have been observed by Maul and colleagues (1976) to aggregate, but a direct correlation with cell fusion could not be made. Vos and associates (1976) have investigated changes in the distribution of intramembranous particles in hen erythrocytes during cell fusion induced by the ionophore A231187. Redistribution of intramembranous particles in the treated cells were observed, which clearly demonstrated that the integral proteins of the treated hen cells were free to move in the lipid bilayer. In addition, absence of cell fusion in experiments with human erythrocytes and A23187 was consistent with an observed lack of

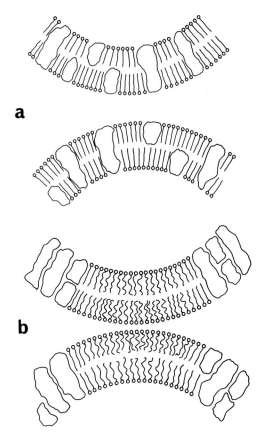

a

b

Fig. 3. Possible relationships between the lipid molecules and intrinsic proteins of biological membranes during membrane fusion: (a) two membranes in close proximity that do not fuse because the proteins of the membranes are randomly arranged in their respective lipid bilayers; (b) membranes in which fusion may now proceed, by the interaction and intermingling of their lipid molecules, after the emergence of protein-free areas of lipid bilayer as a result of aggregation of the intrinsic proteins in both membranes. Compared with (a), the lipid molecules of the membranes that are about to fuse have been perturbed; the lipids of (b) are either more fluid (as illustrated) or, in the extreme case, are rearranged in micellar form (not shown). (Reproduced with permission from Ahkong et al., 1975a.)

movement of intramembranous particles in the treated human cells, and with the well-documented relative immobility of the particles in fresh "ghosts" prepared from human erythrocytes (Elgsaeter et al., 1976). Freeze-fractured erythrocyte membranes have also been investigated recently by Zakai and co-workers (1976) in relation to the fusion of human erythrocytes and their ghosts induced by the combined action of calcium and phosphate ions. They noted an area of close contact of two cells that was almost devoid of intramembranous particles. Further, with a preparation of "fused ghosts," a micrograph revealed two areas studied with tightly packed intramembranous particles connected by a folded smooth area almost devoid of particles. These workers commented that it is tempting to suggest the smooth area is a region of fusion joining two cells.

It seems, however, that electron microscopy will be greatly taxed to provide a conclusive demonstration of fusion occurring either in protein-rich or lipid-rich regions of membranes since the actual instant of membrane fusion is elusive. At best, electron micrographs can probably reveal only the structural situation just before or following the fusion event within a highly dynamic situation.

5. Conclusions

In membrane fusion there are probably many different molecular methods of achieving the same cellular end, and it may therefore be a mistake to attempt to generalize too widely from one membrane system to another on the basis of our limited knowledge of mechanism of fusion. Several paradoxical situations exist at present. Lysophosphatidylcholine induces fusion in a wide variety of cells and in model systems but it inhibits myoblast fusion; Ca^{2+} is essential for many cellular processes involving membrane fusion but it inhibits the fusion of erythrocytes when a low concentration of Sendai virus is used; an elevated concentration of cytoplasmic Ca^{2+} leads to the fusion of erythrocytes but in muscle cells the site of action of Ca^{2+} is apparently on the outside surface of the plasma membrane. An analogy may exist between the gesture of shaking the head from side to side, which is interpreted in the West to mean no, but in India means yes. Nevertheless, no confusion arises when individuals are within their habitual environment. Similar situations may apply to membrane fusion and cell fusion, and it seems necessary and important to obtain more data on the molecular codes that operate in different systems to enable membrane fusion to be treated rationally.

References

Ahkong, Q. F. Y. S. (1976) Studies on the fusion of erythrocytes *in vitro* by non-viral agents. Ph.D. thesis, pp. 1–246, University of London.

Ahkong, Q. F., Cramp, F. C., Fisher, D., Howell, J. I. and Lucy, J. A. (1972) Studies on chemically-induced cell fusion. J. Cell Sci. 10, 769–787.

Ahkong, Q. F., Cramp, F. C., Fisher, D., Howell, J. I., Tampion, W., Verrinder, M. and Lucy, J. A. (1973a) Chemically-induced and thermally-induced cell fusion: lipid-lipid interactions. Nature New Biol. 242, 215–217.

Ahkong, Q. F., Fisher, D., Tampion, W. and Lucy, J. A. (1973b) The fusion of erythrocytes by fatty acids, esters, retinol and α-tocopherol. Biochem. J. 136, 147–155.

Ahkong, Q. F., Fisher, D., Tampion, W. and Lucy, J. A. (1975a) Mechanisms of cell fusion. Nature 253, 194–195.

Ahkong, Q. F., Howell, J. I., Lucy, J. A., Safwat, F., Davey, M. R. and Cocking, E. C. (1975b) Fusion of hen erythrocytes with yeast protoplasts induced by polyethylene glycol. Nature 255, 66–67.

Ahkong, Q. F., Howell, J. I., Tampion, W. and Lucy, J. A. (1974) The fusion of hen erythrocytes by lipid-soluble and surface-active immunological adjuvants. FEBS Lett. 41, 206–210.

298

Ahkong, Q. F., Tampion, W. and Lucy, J. A. (1975c) Promotion of cell fusion by divalent cation ionophores. Nature 256, 208–209.

Allan, D. and Michell, R. H. (1975) Accumulation of 1,2-diacylglycerol in the plasma membrane may lead to echinocyte transformation of erythrocytes. Nature 258, 348–349.

Allan, D., Low, M. G., Finean, J. B. and Michell, R. H. (1975) Changes in lipid metabolism and cell morphology following attack by phospholipase C *(Clostridium perfringens)* on red cells or lymphocytes. Biochim. Biophys. Acta 413, 308–316.

Allan, D., Billah, M. M., Finean, J. B. and Michell, R. H. (1976a) Release of diacylglycerol-enriched vesicles from erythrocytes with increased intracellular [Ca²⁺]. Nature 261, 58–60.

Allan, D., Watts, R. and Michell, R. H. (1976b) Production of 1,2-diacylglycerol and phosphatidate in human erythrocytes treated with calcium ions and ionophore A23187. Biochem. J. 156, 225–232.

Anné, J., Eyssen, H. and De Somer, P. (1976) Somatic hybridisation of *Penicillium roquefortii* with *P. chrysogenum* after protoplast fusion. Nature 262, 719–721.

Bächi, T., Aguet, M. and Howe, C. (1973) Fusion of erythrocytes by Sendai virus studied by immuno-freeze-etching. J. Virol. 11, 1004–1012.

Bangham, A. D. and Horne, R. W. (1964) Negative staining of phospholipids and their structural modification by surface-active agents as observed in the electron microscope. J. Mol. Biol. 8, 660–668.

Becker, J. L. (1972) Fusions *in vitro* de cellules somatiques en culture de *Drosophila melanogaster*, induites par la concanavaline A. C. R. Acad. Sci. (D) (Paris) 275, 2969–2972.

Bergelson, L. D., Dyatloviskaya, E. V., Torkhovskaya, T. I., Sorokina, I. B. and Gorkova, N. P. (1968) Dedifferentiation of phospholipid composition in subcellular particles of cancer cells. FEBS Lett. 2, 87–90.

Bjørnstad, P. (1966) Phospholipase activity in rat liver microsomes studied by the use of endogenous substrates. Biochim. Biophys. Acta 116, 500–510.

Blaschko, H., Firemark, H., Smith, A. D., and Winkler, H. (1967) Lipids of the adrenal medulla. Lysolecithin: a characteristic constituent of chromaffin granules. Biochem. J. 104, 545–549.

Blaschko, H., Jerrome, D. W., Robb-Smith, A. H. T., Smith, A. D. and Winkler, H. (1968) Biochemical and morphological studies on catecholamine storage in human phaechromocytoma. Clin. Sci. 34, 453–465.

Blough, H. A., Gallaher, W. R., and Weinstein, D. B. (1973) Viral lipids-host cell biosynthesis parameters. In: Membrane Mediated Information (Kent, P. W., ed.) Vol. 1, pp. 183–199, Medical and Technical Publishing Co. Ltd., Lancaster, England.

Borochov, H. and Shinitsky, M. (1976) Vertical displacement of membrane proteins mediated by changes in microviscosity. Proc. Nat. Acad. Sci. U.S.A. 73, 4526–4530.

Breisblatt, W. and Ohki, S. (1975) Fusion in phospholipid spherical membranes. I. Effect of temperature and lysolecithin. J. Membrane Biol. 23, 385–401.

Breisblatt, W. and Ohki, S. (1976) Fusion in phospholipid spherical membranes. II. Effect of cholesterol, divalent ions and pH. J. Membrane Biol. 29, 127–146.

Bruckdorfer, K. R., Cramp, F. C., Goodall, A. H., Verrinder, M. and Lucy, J. A. (1974) Fusion of mouse fibroblasts with oleylamine. J. Cell Sci. 15, 185–199.

Bruckdorfer, K. R., Demel, R. A., De Gier, J. and Van Deenen, L. L. M. (1969) The effect of partial replacements of membrane cholesterol by other steroids on the osmotic fragility and glycerol permeability of erythrocytes. Biochim. Biophys. Acta 183, 334–345.

Bruckdorfer, K. R., Edwards, P. A. and Green, C. (1968) Properties of aqueous dispersions of phospholipid and cholesterol. Eur. J. Biochem. 4, 506–511.

Brûlet, P. and McConnell, H. M. (1976) Lateral hapten mobility and immunochemistry of model membranes. Proc. Nat. Acad. Sci. U.S.A. 73, 2977–2981.

Burdzy, Von K., Munder, P. G., Fischer, H. and Westphal, O. (1964) Steigerung der phagocytose von peritoneal makrophagen durch lysolecithin. Z. Naturforsch. 19b, 1118–1120.

Carafoli, E., Clementi, F., Drabikowski, W. and Margreth, A. (Eds.) (1975) Calcium transport in contraction and secretion. Elsevier/North-Holland, Amsterdam.

Chapman, D. (1973) Some recent studies of lipids, lipid-cholesterol and membrane systems. In: Biological Membranes (Chapman, D. and Wallach, D. F. H., eds.) vol. 2, pp. 91–144, Academic Press, London.

Chapman, D. and Wallach, D. F. H. (1968) Recent physical studies of phospholipids and natural membranes. In: Biological Membranes (Chapman, D., ed.) pp. 125–202, Academic Press, London.

Chi, E. Y., Lagunoff, D. and Koehler, J. K. (1976) Freeze-fracture study of mast cell secretion. Proc. Nat. Acad. Sci. U.S.A. 73, 2823–2827.

Cochrane, D. E. and Douglas, W. W. (1974) Calcium-induced extrusion of secretory granules (exocytosis) in mast cells exposed to 48/80 or the ionophores A23187 and X-537A. Proc. Nat. Acad. Sci. U.S.A. 71, 408–412.

Cooper, R. A., Arner, C. A., Wiley, J. S. and Shattil, S. J. (1975) Modification of red cell membrane structure by cholesterol-rich lipid dispersions. A model for the primary spur cell defect. J. Clin. Invest. 55, 115–126.

Croce, C. M., Sawicki, W., Kritchevski, D. and Koprowski, H. (1971) Induction of homokaryocyte, heterokaryocyte and hybrid formation by lysolecithin. Exp. Cell Res. 67, 427–435.

Davidson, R. L., O'Malley, K. A. and Wheeler, T. B. (1976) Polyethylene glycol-induced mammalian cell hybridization: effect of polyethylene glycol molecular weight and concentration. Somat. Cell Genet. 2, 271–280.

DeBoer, E. and Loyter, A. (1971) Formation of polynucleate avian erythrocytes by polylysine and phospholipase C. FEBS Lett. 15, 325–327.

De Duve, C. (1969) The lysosome in retrospect. In: Lysosomes in Biology and Pathology (Dingle, J. T. and Fell, H. B., eds.) vol. 1, pp. 3–40, Elsevier/North-Holland, Amsterdam.

Deuticke, B. and Ruska, C. (1976) Changes of non-electrolyte permeability in cholesterol-loaded erythrocytes. Biochim. Biophys. Acta 433, 638–653.

Devor, K. A., Teather, R. M., Brenner, M., Schwarz, H., Würz, H. and Overath, P. (1976) Membrane hybridization by centrifugation analysed by lipid phase transitions and reconstitution of NADH-oxidase activity. Eur. J. Biochem. 63, 459–467.

Dudits, D., Rasko, I., Hadlacsky, G. and Lima-de-Faria, A. (1976) Fusion of human cells with carrot protoplasts induced by polyethylene glycol. Hereditas 82, 121–124.

Dunn, M. J. (1974) Red blood cell calcium and magnesium: effects upon sodium and potassium transport and cellular morphology. Biochim. Biophys. Acta 352, 97–116.

Edmonson, J. W. and Li, T-K. (1976) The effects of ionophore A23187 on erythrocytes. Relationship of ATP and 2,3 diphosphoglycerate to calcium-binding capacity. Biochim. Biophys. Acta 443, 106–113.

Elgsaeter, A., Shotton, D. M. and Branton, D. (1976) Intramembrane particle aggregation in erythrocyte ghosts. II. The influence of spectrin aggregation. Biochim. Biophys. Acta 426, 101–122.

Elsbach, P., Holmes, K. V. and Choppin, P. W. (1969) Metabolism of lecithin and virus-induced cell fusion. Proc. Soc. Exp. Biol. Med. 130, 903–908.

Falke, D., Schiefer, H-G. and Stoffel, W. (1967) Lipoid-analysis in giant cell growth in herpes hominis virus. Z. Naturforsch. 22b, 1360–1362.

Ferber, E. (1971) Membrane phospholipid metabolism during cell activation and differentiation. In: The Dynamic Structure of Cell Membranes (Wallach, D. F. H. and Fischer, M., eds.) pp. 129–147, Springer-Verlag, Berlin.

Ferenczy, L., Kevei, F. and Szegedi, M. (1975) High-frequency fusion of fungal protoplasts. Experientia 31, 1028–1030.

Finean, J. B. and Rumsby, M. G. (1963) Negatively stained lipoprotein membranes in the validity of the image. Nature 197, 1326–1327.

Fisher, D. (1975) Fluidity and cell fusion. In: Biomembranes—Lipids, Proteins and Receptors (Burton, R. M. and Packer, L., eds) pp. 75–93, BI-Science Publications Divison, Webster Groves, Missouri.

Fisher, K. A. (1976) Analysis of membrane halves: cholesterol. Proc. Nat. Acad. Sci. U.S.A. 73, 173–177.

Fodor, K. and Alföldi, L. (1976) Fusion of protoplasts of Bacillus megaterium. Proc. Nat. Acad. Sci. U.S.A. 73, 2147–2150.

Foreman, J. C., Mongar, J. L. and Gomperts, B. D. (1973) Calcium ionophores and movement of calcium ions following the physiological stimulus to a secretory process. Nature 245, 249–251.

Gingell, D. (1967) Membrane surface potential in relation to a possible mechanism for intercellular interactions and cellular response: a physical basis. J. Theor. Biol. 17, 451–482.

Gledhill, B. L., Sawicki, W., Croce, C. M. and Koprowski, H. (1972) DNA synthesis in rabbit spermatozoa after treatment with lysolecithin and fusion with somatic cells. Exp. Cell Res. 73, 33–40.

Gordesky, S. C. (1976) Phospholipid asymmetry in the human erythrocyte membrane. Trends
Biochem. Sci. 1, 208–211.

Grant, C. W. M. and McConnell, H. M. (1974) Glycophorin in lipid bilayers. Proc. Nat. Acad. Sci.
U.S.A. 71, 4653–4657.

Grunze, M. and Deuticke, B. (1974) Changes of membrane permeability due to extensive cholesterol
depletion in mammalian erythrocytes. Biochim. Biophys. Acta 356, 125–130.

Halfer, C. and Petrella, L. (1976) Cell fusion induced by lysolecithin and concanavalin A in *Drosophila
melanogaster* somatic cells cultured *in vitro*. Exp. Cell Res. 100, 399–404.

Hart, C. A., Ahkong, Q. F., Fisher, D., Hallinan, T., Quirk, S. J. and Lucy, J. A. (1975) Effect of thiol
reagents on virus- and chemically induced fusion of erythrocytes. Biochem. Soc. Trans. 3,
734–736.

Hart, C. A., Fisher, D., Hallinan, T. and Lucy, J. A. (1976) Effects of calcium ions and the bivalent
cation ionophore A23187 on the agglutination and fusion of chicken erythrocytes by Sendai virus.
Biochem. J. 158, 141–145.

Hartmann, J. X., Galla, J. D., Kao, K. N., Gamborg, O. L., Krutman, M., and Garrish, L. M. (1974)
Rapid fusion of blood cells from avian and mammalian species at high frequencies and viability
with polyethylene glycol. Fla. Sci. 37, 31 (Abstr.)

Hartmann, J. X., Galla, J. D., Emma, D. A., Kao, K. N. and Gamborg, O. L. (1976) The fusion of
erythrocytes by treatment with proteolytic enzymes and polyethylene glycol. Can. J. Genet. Cytol.
18, 503–512.

Haydon, D. A. and Taylor, J. (1963) The stability and properties of bimolecular lipid leaflets in aqueous solutions. J. Theor. Biol. 4, 281–296.

Högberg, B. and Uvnäs, B. (1957) The mechanism of the disruption of mast cells produced by compound 48/80. Acta Physiol. Scand. 41, 345–369.

Hope, M. J., Bruckdorfer, K. R., Hart, C. A. and Lucy, J. A. (1977a) Membrane cholesterol and cell
fusion of hen and guinea-pig erythrocytes. Biochem. J. 166, 255–263.

Hope, M. J., Bruckdorfer, K. R., Owen, J. S. and Lucy, J. A. (1977b) Chemically-induced cell fusion
in vitro of erythrocytes from patients with liver diseases. Biochem. Soc. Trans. 5, 1144–1146.

Howell, J. I. and Lucy, J. A. (1969) Cell fusion induced by lysolecithin. FEBS Lett. 4, 147–150.

Howell, J. I., Fisher, D., Goodall, A. H., Verrinder, M. and Lucy, J. A. (1973) Interactions of membrane phospholipids with fusogenic lipids. Biochim. Biophys. Acta 332, 1–10.

Jacobson, K. and Papahadjopoulos, D. (1975) Phase transitions and phase separations in phospholipid membranes induced by changes in temperature, pH, and concentration of bivalent cations. Biochemistry 14, 152–161.

Johnson, M. W. and Horne, R. W. (1970) Some observations on the relative dehydration rates of
negative stains and biological objects. J. Microsc. 91, 197–202.

Jones, C. W., Mastrangelo, I. A., Smith, H. H., Liu, H. Z. and Meck, R. A. (1976) Interkingdom fusion between human (HeLa) cells and tobacco hybrid (GGLL) protoplasts. Science 193, 401–403.

Kao, K. N. and Michayluk, M. R. (1974) A method for high-frequency intergenetic fusion of plant
protoplasts. Planta (Berlin) 115, 355–367.

Keenan, T. W., Morré, D., Olson, D. E., Yunghans, W. N. and Patton, S. (1970) Biochemical and
morphological comparison of plasma membrane and milk fat globule membrane from bovine
mammary gland. J. Cell Biol 44, 80–93.

Kennedy, A. and Rice-Evans, C. (1976) A spectrofluorimetric study of the interaction of glycerol
mono-oleate with human erythrocyte ghosts. FEBS Lett. 69, 45–50.

Kent, C. and Vagelos, P. R. (1975) Phospholipase A activity in plasma membranes from fusing muscle cells. Fed. Proc. 34, 525.

Kleeman, W. and McConnell, H. M. (1976) Interactions of proteins and cholesterol with lipids in
bilayer membranes. Biochim. Biophys. Acta 419, 206–222.

Kimelberg, H. K. (1977) The influence of membrane fluidity on the activity of membrane-bound enzymes. In: Dynamic Aspects of Cell Surface Organization (Poste, G. and Nicolson, G. L., eds.) Cell
Surface Reviews, vol. 3, pp. 205–293. Elsevier/North-Holland, Amsterdam.

Kosower, N. S., Kosower, E. M. and Wegman, P. (1975) Membrane mobility agents. II. Active promoters of cell fusion. Biochim. Biophys. Acta 401, 530–534.

Kroes, J., Ostwald, R. and Keith, A. (1972) Erythrocyte membranes—compression of lipid phases by increased cholesterol content. Biochim. Biophys. Acta 274, 71–74.

Lau, A. L. Y. and Chan, S. I. (1975) Alamecithin-mediated fusion of lecithin vesicles. Proc. Nat. Acad. Sci. U.S.A. 72, 2170–2174.

Lawson, D., Raff, M. C., Gomperts, B., Fewtrell, C. and Gilula, N. B. (1977) Molecular events during membrane fusion. A study of exocytosis in rat peritoneal mast cells. J. Cell Biol. 72, 242–259.

Loyter, A., Kulka, R. G., Zakai, N., Ronn, G., Reichler, Y. and Lalazar, A. (1976) Induction of reversible rearrangement of membrane proteins and phospholipids during fusion of human erythrocyte ghosts. Proc. 6th Eur. Congr. Electron Microscopy, Jerusalem, 2, 33–38.

Lucy, J. A. (1964) Globular lipid micelles and cell membranes. J. Theor. Biol. 7, 360–373.

Lucy, J. A. (1968) Ultrastructure of membranes: Brit. Med. Bull. 24, 127–129.

Lucy, J. A. (1969) Lysosomal membranes. In: Lysosomes in Biology and Pathology (Dingle, J. T. and Fell, H. B., eds.) pp. 313–341, Elsevier/North-Holland, Amsterdam.

Lucy, J. A. (1970) The fusion of biological membranes. Nature (London) 227, 814–817.

Lucy, J. A. (1977a) Cell fusion. Trends Biochem. Sci. 2, 17–20.

Lucy, J. A. (1977b) The membrane of the hen erythrocyte as a model for studies on membrane fusion. In: The Structure of Biological Membranes (Abrahamsson, S., ed.) Nobel Symposium No. 34, pp. 275–291, Plenum Press, New York.

Lucy, J. A., Ahkong, Q. F., Cramp, F. C., Fisher, D. and Howell, J. I. (1971) Cell fusion without viruses. Biochem. J. 124, 46P–47P.

Lyman, G. H., Preisler, H. D. and Papahadjopoulos, D. (1976) Membrane action of DMSO and other chemical inducers of Friend leukaemic cell differentiation. Nature 262, 360–363.

Maggio, B. and Lucy, J. A. (1975) Studies on mixed monolayers of phospholipids and fusogenic lipids. Biochem. J. 149, 597–608.

Maggio, B. and Lucy, J. A. (1976) Polar-group behaviour in mixed monolayers of phospholipids and fusogenic lipids. Biochem. J. 155, 353–364.

Maggio, B., Ahkong, Q. F. and Lucy, J. A. (1976) Poly(ethylene glycol) surface potential and cell fusion. Biochem. J. 158, 647–650.

Marikovsky, Y., Brown, C. S., Weinstein, R. S. and Wortis, H. H. (1976) Effects of lysolecithin on the surface properties of human erythrocytes. Exp. Cell Res. 98, 313–324.

Marsh, D. (1974) Statistical mechanics of the fluidity of phospholipid bilayers and membranes. J. Membr. Biol. 18, 145–162.

Martin, F. J. and MacDonald, R. C. (1976) Lipid vesicle-cell interactions. II. Induction of cell fusion. J. Cell Biol. 70, 506–514.

Maul, G. G., Steplewski, Z., Weibel, J. and Koprowski, H. (1976) Time sequence and morphological evaluation of cells fused by polyethylene glycol 6000. In Vitro 12, 787–796.

McIntyre, J. A., Gilula, N. B. and Karnovsky, M. J. (1974) Cryoprotectant-induced redistribution of intramembranous particles in mouse lymphocytes. J. Cell Biol. 60, 192–203.

Miyamoto, V. K. and Stoeckenius, W. (1971) Preparation and characteristics of lipid vesicles. J. Membr. Biol. 4, 252–269.

Mulder, E., Van den Berg, J. W. O. and Van Deenen, L. L. M. (1965) Metabolism of red cell lipids, II. Conversions of lysophosphoglycerides. Biochim. Biophys. Acta 106, 118–127.

Newkirk, J. J. and Waite, M. (1971) Identification of a phospholipase A_1 in plasma membranes of rat liver. Biochim. Biophys. Acta 225, 224–233.

Nicolson, G. L. and Painter, R. G. (1973) Anionic sites of human erythrocyte membranes. II. Antispectrin-induced transmembrane aggregation of the binding sites for positively charged colloidal particles. J. Cell Biol. 59, 395–406.

Nicolson, G. L., Poste, G. and Ji, T. (1977) The dynamics of cell membrane organization. In: Dynamic Aspects of Cell Surface Organization (G. Poste and G. L. Nicolson, Eds.) Cell Surface Reviews, vol. 3, pp. 1–74. Elsevier/North-Holland, Amsterdam.

Norwood, T. H., Zeigler, C. J. and Martin, G. M. (1976) Dimethyl sulfoxide enhances polyethylene glycol-mediated somatic cell fusion. Somatic Cell Genet. 2, 263–270.

Okada, Y. and Muryayama, F. (1966) Requirement of calcium ions for the cell fusion reaction of animal cells by HVJ. Exp. Cell Res. 44, 527–551.

Ohnishi, S-I. and Ito, T. (1974) Calcium-induced phase separations in phosphatidylserine-phosphatidylcholine membranes. Biochemistry 13, 881–887.

Papahadjopoulos, D., Poste, G. and Schaeffer, B. E. (1973) Fusion of mammalian cells by unilamellar lipid vesicles: influence of lipid surface charge, fluidity and cholesterol. Biochim. Biophys. Acta 323, 23–42.

Papahadjopoulos, D., Poste, G., Schaeffer, B. E. and Vail, W. J. (1974) Membrane fusion and molecular segregation in phospholipid vesicles. Biochim. Biophys. Acta 352, 10–28.

Padahadjopoulos, D., Hui, S., Vail, W. J. and Poste, G. (1976a) Studies on membrane fusion. I. Interactions of pure phospholipid membranes and the effect of myristic acid, lysolecithin, proteins and dimethylsulfoxide. Biochim. Biophys. Acta 448, 245–264.

Papahadjopoulos, D., Vail, W. J., Pangborn, W. A. and Poste, G. (1976b) Studies on membrane fusion. II. Induction of fusion in pure phospholipid membranes by calcium ions and other divalent metals. Biochim. Biophys. Acta 448, 265–283.

Papahadjopoulos, D., Poste, G. and Vail, W. J. (1978) Methods for studying the fusion of natural and model membranes. In: Methods in Membrane Biology (Korn, E., ed.) Plenum Press, New York.

Parkes, J. G. and Fox, C. F. (1975) On the role of lysophosphatides in virus-induced cell fusion and lysis. Biochemistry 14, 3725–3729.

Patton, S. and Keenan, T. W. (1975) The milk fat globule membrane. Biochim. Biophys. Acta 415, 273–309.

Pasternak, C. A. and Micklem, K. J. (1974) The biochemistry of virus-induced cell fusion. Changes in membrane integrity. Biochem. J. 140, 405–411.

Peretz, H., Toister, Z., Laster, Y. and Loyter, A. (1974) Fusion of intact human erythrocytes and erythrocyte ghosts. J. Cell Biol. 63, 1–11.

Phillips, M. C., Finer, E. G. and Hauser, H. (1972) Differences between conformations of lecithin and phosphatidylethanolamine polar groups and their effects on interactions of phospholipid bilayer membranes. Biochim. Biophys. Acta 290, 397–402.

Pinto da Silva, P. and Martínez-Palomo, A. (1974) Induced redistribution of membrane particles, anionic sites and con A receptors in Entamoeba histolytica. Nature 249, 170–171.

Pontecorvo, G. (1975) Production of mammalian somatic cell hybrids by means of polyethylene glycol treatment. Somatic Cell Genet. 1, 397–400.

Pontecorvo, G., Riddle, P. N. and Hales, A. (1977) Time and mode of fusion of human fibroblasts treated with polyethylene glycol (PEG). Nature 265, 257–258.

Poole, A. R. and Williams, D. C. (1967) In vivo effect of an invasive malignant rat tumour on cartilage. Nature (London) 214, 1342–1343.

Poole, A. R., Howell, J. I. and Lucy, J. A. (1970) Lysolecithin and cell fusion. Nature (London) 227, 810–813.

Poste, G. (1972) Mechanisms of virus-induced cell fusion. Int. Rev. Cytol. 33, 157–252.

Poste, G. and Allison, A. C. (1971) Membrane fusion reaction: a theory. J. Theor. Biol. 32, 165–184.

Poste, G. and Allison, A. C. (1973) Membrane fusion. Biochim. Biophys. Acta 300, 421–465.

Poste, G., Reeve, P., Alexander, D. J. and Terry, G. (1972) Effect of plasma membrane lipid composition on cellular susceptibility to virus-induced cell fusion. J. Gen. Virol. 17, 133–136.

Power, J. B. and Cocking, E. C. (1971) Fusion of plant protoplasts. Sci. Prog., Oxford 59, 181–198.

Power, J. B., Cummins, S. E. and Cocking, E. C. (1970) Fusion of isolated plant protoplasts. Nature 225, 1016–1018.

Pressman, B. C. (1973) Properties of ionophores with broad range cation selectivity. Fed. Proc. 32, 1698–1703.

Pressman, B. C. (1976) Biological applications of ionophores. Ann. Rev. Biochem. 45, 501–531.

Radda, G. K. and Vanderkooi, J. (1972) Can fluorescent probes tell us anything about membranes? Biochim. Biophys. Acta 265, 509–549.

Rahman, Y. E., Verhagen, J. and Wiel, D. F. M. v.d. (1970) Evidence of a membrane-bound phospholipase A in rat liver lysosomes. Biochem. Biophys. Res. Commun. 38, 670–677.

Reed, P. W. and Lardy, H. A. (1972) Antibiotic A23187 as a probe for the study of calcium and magnesium function in biological systems. In: The Role of Membranes in Metabolic Regulation (Mehlman, M. A. and Hanson, R. W., eds.) pp. 111–131, Academic Press, New York.

Reporter, M. and Norris, G. (1973) Reversible effects of lysolecithin on fusion of cultured rat muscle cells. Differentiation 1, 83–95.

Reporter, M. and Raveed, D. (1973) Plasma membranes: isolation from naturally fused and lysolecithin-treated muscle cells. Science 181, 863–865.

Rothman, J. E. and Lenard, J. (1977) Membrane asymmetry. Science, 195, 743–753.

Schaeffer, P., Cami, B. and Hotchkiss, R. D. (1976) Fusion of bacterial protoplasts. Proc. Nat. Acad. Sci. U.S.A. 73, 2151–2155.

Schmitt, W. W., Zingsheim, H. P., and Bachmann, L. (1970) Investigation of molecular and micellar solutions by freeze etching. 7th Int. Congress of Electron Microscopy, Grenoble, pp. 455–456.

Schudt, C. and Pette, D. (1975) Influence of the ionophore A23187 on myogenic cell fusion. FEBS Lett. 59, 36–38.

Schudt, C., van der Bosch, J. and Pette, D. (1973) Inhibition of muscle cell fusion in vitro by Mg^{2+} and K^+ ions. FEBS Lett. 32, 296–298.

Schudt, C., Dahl, G. and Gratzl, M. (1976) Calcium-induced fusion of plasma membranes isolated from myoblasts grown in culture. Cytobiologie 13, 211–223.

Seelig, A. and Seelig, J. (1974) The dynamic structure of fatty acyl chains in a phospholipid bilayer measured by deuterium magnetic resonance. Biochemistry 13, 4839–4845.

Shah, D. O. (1970) Surface chemistry of lipids. Adv. Lipid Res. 8, 347–431.

Shainberg, A., Yagil, G. and Yaffe, D. (1969) Control of myogenesis in vitro by Ca^{2+} concentration in nutritional medium. Exp. Cell Res. 58, 163–167.

Shier, W. T. (1977) Inhibition of acyl coenzyme A: lysolecithin acyltransferases by local anesthetics, detergents and inhibitors of cyclic nucleotide phosphodiesterases. Biochem. Biophys. Res. Commun. 75, 186–193.

Singer, S. J. (1974) The molecular organisation of membranes. Ann. Rev. Biochem. 43, 805–833.

Singer, S. J. (1977) The fluid mosaic model of membrane structure. In: The Structure of Biological Membranes. Nobel Symposium No. 34 (Abrahamsson, S., ed.) pp. 443–461, Plenum Press, New York.

Suzucki, Y. and Matsumoto, M. (1974) Acid phospholipase A_1 and A_1 in the cells, and subcellular redistribution of their activities in the cells infected with measles virus. Biochem. Biophys. Rec. Commun. 57, 505–512.

Tanaka, K.-I. and Ohnishi, S.-I. (1976) Heterogeneity in the fluidity of intact erythrocyte membrane and its homogenization upon hemolysis. Biochim. Biophys. Acta 426, 218–231.

Tanford, C. (1961) Physical Chemistry of Macromolecules, pp. 114–200, Wiley, New York.

Thinés-Sempoux, D. (1967) Chemical similarities between the lysosome and plasma membranes. Biochem. J. 105, 20P–21P.

Toister, Z. and Loyter, A. (1970) Virus induced fusion of chicken erythrocytes. Biochem. Biophys. Res. Commun. 41, 1523–1530.

Toister, Z. and Loyter, A. (1971) Ca^{2+}-induced fusion of avian erythrocytes. Biochim. Biophys. Acta 241, 719–724.

Traüble, H. (1977) Membrane electrostatics. In: The Structure of Biological Membranes. Nobel Symposium No. 34 (Abrahamsson, S., ed.), pp. 509–550, Plenum Press, New York.

Traüble, H. and Haynes, D. H. (1971) The volume change in lipid bilayer lamellae at the crystalline-liquid crystalline phase transition. Chem. Phys. Lipids. 7, 324–335.

Van Deenen, L. L. M. (1969) Membrane lipids and lipophilic proteins. In: The Molecular Basis of Membrane Function (Tosteston, D. C., ed.) pp. 47–78, Prentice-Hall, New Jersey.

Van der Bosch, J., Schudt, C. and Pette, D. (1972) Quantitative investigation on Ca^{++}- and pH-dependence of muscle cell fusion in vitro. Biochem. Biophys. Res. Commun. 48, 326–332.

Vanderkooi, J., Fischkoff, S., Chance, B. and Cooper, R. A. (1974) Fluorescent probe analysis of the lipid architecture of natural and experimental cholesterol-rich membranes. Biochemistry 13, 1589–1595.

Vollet, J. J. and Roth, L. E. (1974) Cell fusion by nascent-membrane induction and divalent cation treatment. Cytobiologie 9, 249–262.

Vos, J., Ahkong, Q. F., Botham, G., Quirk, S. J. and Lucy, J. A. (1976) Changes in the distribution of intramembranous particles in hen erythrocytes during cell fusion induced by the bivalent-cation ionophore A23187. Biochem. J. 158, 651–653.

Waite, M., Van Deenen, L. L. M., Ruigrok, T. J. C. and Elbers, P. F. (1969) Relation of mitochondrial phospholipase A activity to mitochondrial swelling. J. Lipid. Res. 10, 599–608.

Wallin, A., Glimelius, K. and Erikkson, T. (1974) The induction of aggregation and fusion of *Daucus carota* protoplasts by polyethylene glycol. Z. Pfl. Physiol. 74, 64–80.

Weidekamm, E., Schudt, C., and Brdiczka, D. (1976) Physical properties of muscle cell membranes during fusion. A fluorescence polarization study with the ionophore A23187. Biochim. Biophys. Acta 443, 169–180.

Weiss, R. L., Goodenough, D. A. and Goodenough, U. W. (1977) Membrane differentiations at sites specialised for cell fusion. J. Cell Biol. 72, 144–160.

Yanagimachi, R. and Usui, N. (1974) Calcium dependence of the acrosome reaction and activation of guinea pig spermatozoa. Exp. Cell Res. 89, 161–174.

Yanovsky, A. and Loyter, A. (1972) The mechanism of cell fusion. I. Energy requirements for virus-induced fusion of Ehrlich ascites tumor cell. J. Biol. Chem. 247, 4021–4028.

Zakai, N., Kulka, R. G. and Loyter, A. (1976) Fusion of human erythrocyte ghosts promoted by the combined action of calcium and phosphate ions. Nature 263, 696–699.

Zwaal, R. F. A., Roelofsen, B. and Colley, C. M. (1973) Localization of red cell membrane constituents. Biochim. Biophys. Acta 300, 159–182.

Virus-induced cell fusion

7

G. POSTE and C. A. PASTERNAK

Contents

G. Poste & G. L. Nicolson (eds.) Membrane Fusion, pp. 305–367.
© *Elsevier/North-Holland Biomedical Press, 1978.*

1. Introduction

The diverse properties shown by the major virus groups in their mode of entry into the cell, the mechanism and site of their intracellular replication, and the way they are assembled and released from the cell creates a complicated pattern of host cell response to virus infection. This diversity reflects the great potential of viruses in the study of cells, since viruses can be selected to induce any one of a wide range of cellular responses. Many questions pertinent to the specificity, multiplication and pathogenicity of viruses have now merged with current research in areas of cell biology, and certain problems in cell biology and virology can begin to be formulated in common terms. This is well illustrated by the phenomenon of virus-induced cell fusion.

Cell fusion is a common cytopathic or cytotoxic response to infection by a wide range of viruses (review, Poste, 1972). In a broader sense, cell fusion represents a special example of fusion between two membranes and information obtained on the mechanism of virus-induced cell fusion has contributed to a better understanding of membrane fusion per se. The striking parallelism between factors regulating cell fusion caused by viruses and those regulating the many other phenomena in which membrane fusion occurs (reviews, Poste and Allison, 1973; Poste and Waterson, 1975; Papahadjopoulos and Poste, 1978) suggests that virus-induced cell fusion represents a valid experimental model for studying the series of membrane alterations that are required to permit membranes to fuse.

A probable example of virus-induced cell fusion was first reported in 1873, when Lüginbuhl described the presence of multinucleate cells at the periphery of smallpox pustules. Subsequently, Unna (1896) and Warthin (1931) observed similar multinucleate cells in tissue lesions produced by varicella and measles viruses, respectively. Similar examples of multinucleate cell formation have been reported in tissues infected with many other viruses (review, Roizman, 1962), and these too probably arise via cell fusion since the viruses in question have all been shown to induce extensive fusion of cells cultured in vitro (reviews, Poste, 1970, 1972). Enders and Peebles (1954), using measles virus, were the first to demonstrate virus-induced cell fusion in a tissue culture system, and this was soon followed by reports of cell fusion in vitro by mumps virus (Henle et al., 1954), human parainfluenza virus (Chanock, 1956), and Sendai (HVJ[1]) virus (Okada, 1958). Subsequently, a large number of viruses have been shown to produce fusion in vitro (see below). In addition to any intrinsic interest that virologists may entertain for virus-induced cell fusion, this phenomenon has also

---

[1]Synonym: Hemagglutinating Virus of Japan.

been widely exploited by investigators in many areas of cell biology, virology, immunology and genetics as a method for producing hybrid cells and hetero-karyons whose novel properties provide valuable insight into numerous aspects of cellular organization (reviews, Harris, 1970; Poste, 1972; Davidson, 1974; Watkins, 1974; Appels and Ringertz, 1975; Gordon, 1975; Zeuthen, 1975; Ringertz and Savage, 1976).

The fusion of cultured cells by viruses has also attracted interest as a possible system for defining the factors involved in membrane fusion (reviews, Poste, 1972; Poste and Allison, 1973). The fusion of cells in vitro by viruses is easily controlled, and the well developed specificities of viruses for particular cell types offers a reliable and highly reproducible method for inducing extensive cell fusion of a wide range of cell types. The experimental study of the mechanism(s) of virus-induced cell fusion is also facilitated by the availability of virus variants that display differing capacities to fuse cells, thus permitting identification of the viral components involved in inducing fusion. Conversely, variation in the susceptibility of different cell types to fusion when challenged by the same virus (reviews, Poste, 1972; Poste and Waterson, 1975) provides useful material for investigating the cellular response(s) during fusion.

The major disadvantage of virus-induced cell fusion as a system for studying membrane fusion is the shortcoming shared by all forms of cell fusion, namely, it is extremely difficult to obtain rapid and synchronized fusion of large number of cells. Although the actual fusion event between any two cells may take place very rapidly (probably on a millisecond time scale, see, Poste and Allison, 1973), the number of cells undergoing fusion at any given time is low, and fusion, assessed in terms of the response of the entire cell population, is extended over several hours. This dictates that it is often difficult to conduct detailed biochemical and biophysical studies on the membrane changes accompanying fusion in such systems because any change(s) occurring in the few cells that are actually fusing at any given time often cannot be detected against the large "background" population of nonfusing cells. Consequently, the study of factors that influence the overall extent of cell fusion by viruses (or any other fusion agent) can generally only provide information on the gross regulation of the fusion process and cannot offer insight into the molecular events responsible for the fusion reaction. However, with the recent isolation of the macromolecular components associated with induction of fusion by certain viruses (see section 4), new opportunities arise where it may be possible to use isolated viral macromolecules to induce rapid fusion of natural and model membranes in cell-free systems.

2. Two kinds of virus-induced cell fusion

It is now recognized that viruses induce cell fusion by two distinct mechanisms. Since some confusion persists in the literature concerning this concept, it is pertinent to outline briefly the differences between these two kinds of cell fusion.

The first type of virus-induced cell fusion, called fusion from without (FFWO), occurs independently of virus replication, is induced equally well by infective and noninfective virus and does not require the synthesis of new virus-specific or host cell-specific material. FFWO takes place only after infection with very high doses of virus and occurs rapidly, being detectable even at the light microscope level as little as 5 minutes after addition of the virus, with maximum fusion occurring 1 to 3 hours after exposure to the virus. The best known example of FFWO is the use of inactivated Sendai virus to induce cell fusion (Okada, 1962; Harris, 1965), but large doses of other RNA-containing lipid-enveloped viruses can induce FFWO including Newcastle disease virus (NDV) (Poste et al., 1972a), mumps (Henle et al., 1954), measles (Cascardo and Karzon, 1965), canine distemper (Rankin et al., 1972), simian virus 5 (SV5) (Holmes and Choppin, 1966), visna virus (Harter and Choppin, 1967) and vesicular stomatitis virus (Nishiyama et al., 1976). Among the DNA-containing viruses, FFWO has been reported only with strains of herpes simplex virus (Tokumaru, 1968) and vaccinia virus (Kaku and Kamahora, 1964; Magee and Miller, 1968).

The second type of virus-induced cell fusion, called fusion from within (FFWI), is characterized by fusion beginning in the later stages of the virus replicative cycle after infection with moderate or low doses of virus. Intracellular replication of the virus and the synthesis of new virus-specific macromolecules are necessary for this type of fusion to occur, and FFWI can be inhibited by a variety of metabolic inhibitors that block the synthesis of the viral components necessary for fusion (review, Poste, 1972). The dependence of fusion from within on virus multiplication and expression of a functional viral genome also dictate that maximum cell fusion coincides with the most intensive phases of virus multiplication and thus occurs several hours, or even days, after infection. However, FFWI does not require the synthesis of complete new virions, since fusion can still occur in cell infected with conditional lethal mutants under restrictive conditions and in cells where the early stages of virus replication have been completed but later stages of virus assembly and release have been blocked by metabolic inhibitors (review, Poste, 1972). FFWI is induced by a large number of DNA-(herpes viruses; poxviruses; retroviruses) and RNA-containing viruses[2] (paramyxoviruses, morbilliviruses, pneumoviruses, coronaviruses, rhabdoviruses) (For a detailed list of individual viruses and references, see Table 1 of the review by Poste, 1972).

The terms "fusion from without" and "fusion from within," introduced by Kohn (1965) and Bratt and Gallaher (1969), respectively, are derived by analogy from the terms "lysis from without" and "lysis from within" used to distinguish between rapid lysis of *Escherichia coli* by high multiplicities of bacteriophage (lysis from without) and lysis occurring after productive infection produced by intracellular accumulation of lysozyme (lysis from within).

In this chapter, we will limit our comments almost exclusively to FFWO induced by lipid enveloped viruses. This type of virus-induced cell fusion offers a

---

[2]International Committee on Taxonomy of Viruses: classification and nomenclature (see Virology, (1976) 71, 371–378).

more convenient system for studying membrane fusion since not only can cell fusion be produced much more rapidly than in FFWI, but induction of FFWO using inactivated viruses devoid of infectivity dictates that virus-cell interactions can be studied without the complication of virus-induced alterations in cellular metabolism created by intracellular virus replication. FFWO is also of greater interest to students of membrane fusion since it actually involves two distinct sequences of membrane fusion; the first involves fusion between the virus envelope and the cellular plasma membrane followed by the second stage of cell-to-cell fusion (see sections 5 and 6 for a full discussion of these two steps).

Most of our present knowledge of events in virus-induced FFWO has come from experimental observations of cell fusion induced by inactivated paramyxoviruses, and Sendai virus and Newcastle disease virus (NDV) in particular. The widespread use of Sendai virus as a tool for fusing cells has been an arbitrary, but successful, decision. This virus can be grown conveniently to high titers in a short time in embryonated eggs, and the fusion activity of such preparations is relatively stable (see below). A second attractive feature, particularly to nonvirologists, is that Sendai virus is considered nonpathogenic for man and is thus easier to handle than certain other viruses with equal or even greater cell fusion activity. For example, virulent strains of NDV are able to induce marked FFWO, fusing up to 60 to 80% of cells in a population within 2 hours (compared with 30–40% for Sendai virus) (Poste et al., 1972a), but their extreme pathogenicity for avian species mandates that they can only be used where strict virus containment facilities are available (and even then usually only under strict licensing supervision from the appropriate national agencies responsible for animal disease control). The range of cell types susceptible to fusion by inactivated Sendai virus and NDV is extremely wide. Susceptibility extends over a considerable range of the vertebrate phylum, and the type of cell used for fusion studies is thus virtually limited only by the requirements and the ingenuity of the investigator.

Detailed technical descriptions of methods for the fusion of cultured cells using inactivated Sendai virus, NDV, and other paramyxoviruses are available in numerous publications (reviews, Okada, 1969; Watkins, 1971; Giles and Ruddle, 1973; Poste, 1973) and will not be discussed here.

3. Optimum conditions for virus-induced cell fusion

The reader is referred to the articles by Okada (1962) and Harris and co-workers (1966) for details of methods for fusing cells in suspension, and to Kohn (1965) and Poste and colleagues (1972a) for fusion of cells in monolayers. Combinations of these two systems have also been described, in which a suspension of cells is added to a monolayer of cells simultaneously with the virus (Davidson, 1969; Klebe et al., 1970; Poste and Reeve, 1972a).

The degree of cell fusion that takes place after treatment with inactivated viruses is related to the concentration of virus used, the number of cells and their inherent susceptibility to fusion and, to a lesser extent, the conditions of incubation during the interaction of virus with the cells. The effect of these factors on the efficiency of cell fusion has been reviewed fully elsewhere (Poste, 1970, 1972; Watkins, 1971; Giles and Ruddle, 1973) and only a few general comments are necessary.

Cell types differ markedly in their susceptibility to virus-induced cell fusion (FFWO and FFWI). This is probably the single most important factor determining both the extent of cell fusion and the size of the polykaryocytes formed. Evidence from many reports (review, Poste, 1970) indicates that cells from established cell lines and malignant cells have a greater fusion capacity than primary and secondary cell strains. Fibroblasts fuse less well than epithelial cells, and both leukocytes and lymphoid cells are relatively resistant to virus-induced cell fusion (Poste, 1970). The reasons for these differences are unclear. The number, availability and types of virus receptors on the cell surface (Poste and Waterson, 1975), differences in plasma membrane lipid composition (Klenk and Choppin, 1969; Poste et al., 1972b), and the nature of the cytoskeletal network of the cells involved may all affect cellular susceptibility to fusion. Detailed information on the mechanism(s) by which these factors and other aspects of cell surface organization can influence the cell fusion reaction is still lacking. Cellular susceptibility to fusion is also influenced by a number of ill-defined factors such as cell passage level, the length of time since subcultivation before exposure to virus and the stage in the cell cycle (Poste, 1972).

Within general limits it can be stated that both the extent of cell fusion and the size of individual polykaryocytes increase with increases in the virus dose. The concentration of virus required to induce cell fusion depends mainly on the fusion susceptibility of the cell(s) in question. For example, some cell types undergo extensive fusion when exposed to relatively low doses of Sendai virus (100–1000 HAU[3]/ml) while other cells routinely require higher virus doses (4000–20,000 HAU/ml) and certain cell types may need to be treated with doses of virus as high as 80,000 HAU/ml before fusion will occur (see Watkins, 1971; Poste, 1972 for examples). Generalization is difficult, however, and a number of cell types may show marked cytotoxicity when exposed to low (<1000 HAU/ml) to intermediate (5000 HAU/ml) doses of inactivated Sendai virus while others show no evidence of injury after incubation with virus concentrations as high as 80,000 HAU/ml for 2 hours (Watkins, 1971; Poste, 1972). Consequently, it may be stated with reasonable confidence that methods which are successful in producing extensive fusion of one cell type may not work with other cells or may induce cytotoxicity. Thus trial-and-error methods are usually required to establish the optimal virus dose, cell numbers, and incubation conditions to obtain efficient fusion of each cell type.

The susceptibility of a wide range of cells to virus-induced cell fusion (FFWO and FFWI) can be increased significantly, however, by pretreatment with phytohemagglutinin (PHA) (Poste et al., 1974, 1976a; Reeve et al., 1974; Sulli-

---

[3]HAU = hemagglutination unit.

van et al., 1975; Yoshida and Ikeuchi, 1975). Commercially available preparations of PHA from Difco Laboratories Inc. (Detroit, Michigan) and the Burroughs Wellcome Co. (Triangle Park, North Carolina and Beckenham, England) are able to induce significant enhancement of virus-induced cell fusion (Poste et all., 1976a). Preparations from both manufacturers are supplied as a lyophilized powder, which can be reconstituted and diluted in phosphate-buffered saline (PBS) or culture medium. Large amounts of PHA with cell fusion-enhancing properties can also be isolated directly from red kidney beans by affinity chromatography on a thyroglobulin-Sepharose column (Poste et al., 1976a).

The mechanism by which PHA enhances virus-induced cell fusion remains to be elucidated. Since PHA conjugated to Sepharose is equally as effective as the free lectin in enhancing cell fusion (Reeve et al., 1974), it is clear that the effect of PHA on the fusion process is mediated entirely at the cell surface and does not require entry of lectin into the cells. The most straightforward explanation for the ability of PHA to enhance virus-induced cell fusion is that agglutination of cells by PHA creates conditions of close cell-to-cell contact that facilitate subsequent fusion. Despite the attractive simplicity of this interpretation, the role of agglutination in promoting cell fusion is confused by the observation that while PHA enhances fusion, other lectins such as concanavalin A (Con A), wheat germ agglutinin, and soybean agglutinin inhibit virus-induced cell fusion (Poste et al., 1974) even though they are equally as effective as PHA in agglutinating cells. A possible explanation for the conflicting effects of different lectins on virus-induced cell fusion concerns the question of whether or not a particular lectin is able to bind to virus-receptor sites on the cell surface. Lectins such as Con A and wheat germ agglutinin that inhibit paramyxovirus-induced cell fusion prevent binding of these viruses to the cell surface, whereas PHA does not alter either the rate or the extent of virus particle binding (Poste et al., 1976a). These results suggest that the enhancement of virus-induced cell fusion by PHA might therefore result merely from its ability to agglutinate cells without affecting the subsequent binding of virus to its receptors on the cell surface. If this interpretation is correct, then other agents that promote cell agglutination without impairing virus binding might also be effective in enhancing virus-induced cell fusion. Thus it is interesting to note that De Boer and Loyter (1971) found that agglutination of erythrocytes by pretreatment with polylysine enhanced their subsequent fusion by inactivated Sendai virus or phospholipase C.

PHA is a mixture of five isolectins, which possess varying degrees of erythroagglutinating, leukoagglutinating, and mitogenic activities (Miller et al., 1973). Ion-exchange chromatography and polyacrylaminde-gel electrophoresis have been used to isolate two major biological fractions, designated L-PHAP and H-PHAP (Miller et al., 1973). The former is a homogeneous single protein with potent leukoagglutinin activity but low erythroagglutinin activity. The H-PHAP fraction is a complex protein mixture with high erythroagglutinin activity but little or no leukoagglutinin activity. Both L-PHAP and H-PHAP possess mitogenic activity. Characterization of the cell fusion-enhancing activity of these various PHA fractions isolated from affinity chromatography-purified PHA indicate that cell fusion-enhancing activity resides largely in the L-PHAP fraction, which contains a single protein corresponding to the most anodal protein of the five PHA proteins resolved by gel electrophoresis (Poste et al., 1976a).

4. Identification of the viral components associated with the induction of membrane fusion

The virological literature on FFWO abounds with proposals for the existence of "fusion factors" within virus particles that can act directly on membranes to render them susceptible to fusion. These have been variously termed cytolysin (Henle et al., 1954), cell wall-destroying enzyme (Zhadanov and Bukrinskaya, 1962), cell fusion substances (Okada et al., 1964), fusion factor (Cascardo and Karzon, 1965), and syncytium-producing toxin (Tokumaru, 1968). In none of these examples has the nature of the fusion factor or its relationship to virus structure and the macromolecular subunits of the virus been defined. The nature of the terms chosen to describe these hypothetical factors implies that they function as lytic enzymes which modify the structural integrity of cellular membranes. Despite extensive effects to identify such factors by fractionation of viruses causing FFWO (for references, see Poste, 1972), fusion of cells by a single molecular (or macromolecular) component isolated from disrupted virus particles has yet to be achieved.

Lucy (1970) and Barbanti-Brodano and co-workers (1971) have proposed that the ability of paramyxoviruses, such as Sendai virus, to induce FFWO results from the presence of lysolecithin within the virus envelope which acts on cell membranes at the site(s) of virus attachment to induce membrane fusion in similar fashion to fusion induced by exogenous lysolecithin (Poole et al., 1970). This hypothesis is unconvincing for several reasons.

First, the lysolecithin content of lipid-enveloped cell-fusing viruses such as Sendai virus is lower than that of other lipid-enveloped viruses such as the influenza viruses which do not induce FFWO (Neurath et al., 1973). Similarly, analysis of lyso-compounds in a number of NDV strains has failed to reveal any correlation between lysolecithin content and cell fusion activity; similar lysolecithin concentrations being found in both fusing and non-fusing strains (Parkes and Fox, 1975). In addition, attempts to identify the formation of lyso-compounds in cell membranes during virus-induced cell fusion (both FFWO and FFWI) have been unsuccessful (Hosaka, 1960; Falke et al., 1967; Elsbach et al., 1969; Pasternak and Micklem, 1974a; Diringer and Rott, 1976). In addition, even assuming that the lysolecithin associated with paramyxovirus particles can gain access to cell membranes the amount of lysolecithin is probably inadequate to induce cell fusion. In egg-grown Sendai virus, 1000 HAU contain approximately 0.1 μg of lysolecithin (calculated from the data of Blough and Lawson, 1968). This dose of virus is sufficient to induce FFWO in a wide range of cell types, yet the induction of cell fusion by addition of exogenous lysolecithin is only achieved at concentrations of 200 μg or more. Even then, the extent of cell fusion induced by exogenous lysolecithin is far less than that produced by the virus and is often accompanied by extensive cell lysis, which is not seen in virus-treated cell populations.

Second, experiments showing that treatment of Sendai virus particles with phospholipase B (lysolecithinase) eliminates their ability to induce FFWO cannot be interpreted (Barbanti-Brodano et al., 1971) as evidence that lysolecithin is responsible for the cell fusion activity. In the experiments described by Barbanti-Brodano and associates, virions were not analyzed for their lysolecithin content so it impossible to tell whether the inhibitory effects of phospholipase treatment on fusion activity were accompanied by removal of lysolecithin from the virus. It seems equally likely that elimination of cell fusion activity by this type of treatment results merely from enzymatic perturbation of the structural integrity of the virus envelope. Similar inhibition of cell fusion activity in lipid-enveloped viruses has been reported after treatment with phospholipases A, C, and D (Kohn, 1965; Tokumaru, 1968; Poste and Allison, 1973). If lysolecithin in the virus particles were functioning as a "fusion factor," then, according to the same logic used by Barbanti-Brodano and colleagues, treatment of virus particles with phospholipase A to generate lysolecithin would be expected to enhance rather than abolish their fusion activity. In fact inhibition of endogenous cellular phospholipase A by more than 95% has no effect on cell fusion (Pasternak and Micklem, 1974a).

In analogous fashion to the proposals for the participation of virus-associated lysolecithin in promoting fashion, Allan and Michell (1976) have proposed that 1,2-diacylglycerol present in Sendai virions might represent the "virus fusion factor" because this glycerolipid has been shown to induce cell fusion when added exogenously to cells in vitro. This proposal at least has the merit that direct measurements of the amount of 1,2-diacylglycerol present in virions were made. Also, studies on the interaction of Sendai virions with [^{32}P]labeled human erythrocytes revealed a marked increase in the labeling of phosphatidate with ^{32}P with the onset of cell fusion, and this was interpreted as a phosphorylation of the diacylglycerol contributed by the virus.

These authors also suggested that during assembly and release of Sendai virus at the cell surface the virus may bud through diacylglycerol-enriched regions of the plasma membrane thus acquiring its capacity to fuse. There is little evidence, however, to support this proposal. Allan and Michell's proposal on virus budding is based on their finding that lipid analysis showed Sendai virions contained 200 times more diacylglycerol than human erythrocyte membranes. Such a comparison is meaningless in the context of virus assembly since viruses do not replicate in erythrocytes, and comparison of the diacylglycerol content in virions and the actual type of host cell in which the virus was grown must be made before any conclusions can be drawn regarding selective enrichment of the virus envelope with diacylglycerol (or any other lipid component). Numerous studies have shown that the lipid composition of Sendai virus and other lipid-enveloped viruses that bud from the cell surface does not differ significantly from that of the plasma membrane from the host cells in which they were grown (review, Rott and Klenk, 1977; but also see Rifkin and Quigley (1974) and Semmel et al. (1975) for dissenting data showing that paramyxoviruses and other enveloped viruses may bud preferentially from regions of the plasma membrane enriched in particular phospholipids).

In addition to the general failure to detect changes in the lipid composition of cells undergoing FFWO, recent evidence indicates that the overall lipid milieu of the plasma membrane is not disturbed during fusion of Sendai virus particles with the plasma membrane. Preliminary experiments by one of the present authors (C.A.P.) in collaboration with R. Evans, G. K. Radda, R. Van Dreele, and J. R. Weir on Lettré cells labeled with fluorescent or spin-labeled fatty acid probes [2-(9-anthroyl) palmitate; 12-(9-anthroyl) stearate; 5-(nitroxide) stearate; or 16-(nitroxide) stearate] indicate that fusion of virus particles with the plasma membrane (assessed by alterations in plasma membrane permeability—see section 6.1.) does not result in any significant change in the rotational freedom of the probe molecules within the membrane. However, at higher virus concentrations, studies with red blood cells have shown that a significant change in the mobility of spin-labeled probes can be detected (see Fig. 10 of Maeda et al., 1977b).

The best evidence concerning the identity of the viral components responsible for cell fusion has come from recent studies on two paramyxoviruses, Sendai virus and NDV. Independent studies in several laboratories over the past five years have established that the ability of these viruses to induce FFWO (and also probably FFWI) is correlated with the presence of a specific glycoprotein in the virus envelope.

Several biological activities of paramyxoviruses are associated with the virus envelope. These include hemagglutinating, hemolytic and cell fusion activities, and the ability to attach and penetrate into susceptible cells. The viron envelope consists of lipids derived from the plasma membrane of the host cell arranged in the form of a bilayer (Rott and Klenk, 1977). The external surface of the lipid bilayer is covered with spikelike projections (Figs. 1, 2) composed of glycoproteins (review, Rott and Klenk, 1977). The inner side of the lipid bilayer of the virus envelope is coated with a nonglycosylated protein, the so-called membrane or matrix (M) protein. The available evidence suggests that the M protein plays a major role in maintaining the structure of the virus envelope and that it serves as the recognition site for the virus nucleocapsid interaction with the putative virus envelope during virus assembly when, before budding, the nucleocapsid aligns beneath areas of the cellular plasma membrane which contain the viral envelope proteins (review, Rott and Klenk, 1977). The virus envelope, in turn, encloses the RNA genome and its associated nucleoprotein (NP) subunit (Fig. 2). For a more detailed discussion of paramyxovirus structure and the synthesis and assembly of viral subunits into intact virions, the reader is referred to recent reviews by Choppin and Compans (1975) and Rott and Klenk (1977). The remainder of this section will be limited to a discussion of the functional properties of the two envelope glycoproteins, since these are involved in the initial attachment of virions to the cell surface and the subsequent series of events that culminate in cell fusion.

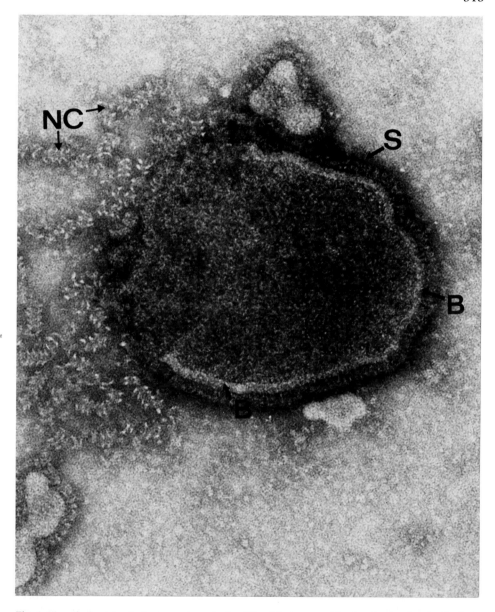

Fig. 1. Negatively stained electron micrograph of Sendai virus particle showing surface "spikes" (S) associated with the underlying lipid bilayer (B) of the envelope. The particle was ruptured at one pole, releasing a portion of the nucelocapsid (NC) with its typical "herringbone" appearance. ×306,000 (Electron micrograph couresty of Dr. June Almeida.)

316

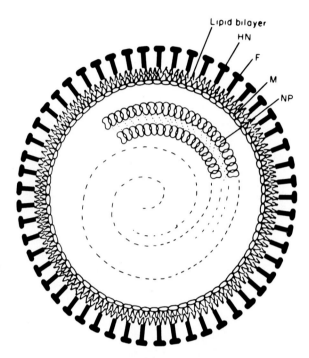

Fig. 2. Schematic representation of the suggested arrangement of structural components in paramyxoviruses. The two envelope glycoproteins, HN and F, are located at the external face of the
lipid bilayer of the virus envelope. The mode of attachment of these glycoproteins to the bilayer and
whether they penetrate the bilayer are unknown. Under certain conditions, another envelope
glycoprotein, F_0, the precursor to F, is present instead of F (see text). The nonglycosylated membrane protein, M, is on the internal surface of the lipid bilayer of the virus envelope. Within the
envelope is the helical ribonucleoprotein containing the protein subunit NP and the viral RNA. In
certain paramyxoviruses, another polypeptide, P, is associated with the nucelocapsid, but has not
been depicted in this figure because details of its arrangement are unknown. (Reproduced with
permission from Choppin and Compans, 1975.)

The envelope glycoproteins of three paramyxoviruses. Sendai, NDV, and Simian virus 5 (SV5), have now been successfully isolated by extraction with
nonionic detergents and the individual glycoproteins separated by sucrose-
density gradient centrifugation and affinity chromatography (Scheid et al., 1972;
Scheid and Choppin, 1973, 1975; Urata and Seto, 1975). In each of these viruses
the larger (69,000-76,000 MW) of the two envelope glycoproteins has been
found to possess both neuraminidase and hemagglutinating activities (Scheid et
al., 1972; Scheid and Choppin, 1973, 1975; Seto et al., 1973; Tozawa et al., 1973;
Nagai and Kleuk, 1977). This protein is referred to as the HN glycoprotein. The
smaller envelope glycoprotein (53,000–56,000 MW), referred to as the F (fusion)
protein, is devoid of neuraminidase and hemagglutinating activities and is involved in virus-induced cell fusion and hemolysis.

The clue that the ability of paramyxoviruses to induce FFWO was associated with the F glycoprotein came from studies in which the structural proteins of virions grown in different host cells were compared. As mentioned earlier, both Sendai virus and NDV exhibit significant differences in their cell fusion activity when grown in different host cells. For example, Sendai virus grown in eggs (Fukai and Suzuki, 1955; Okada, 1958), primary chick embryo fibroblasts (Blair and Robinson, 1968), and primary calf kidney cells (Tozawa et al., 1973) is able to induce FFWO of a wide variety of cell types and also to lyse erythrocytes. However, when the same strain of virus is grown in established mammalian cell lines such as bovine MDBK cells (Scheid and Choppin, 1974, 1975) or mouse L cells (Ishida and Homma, 1961; Homma, 1971) the progeny virions lack cell fusion and hemolytic activities and are also noninfective.

Comparison of the envelope glycoproteins in Sendai virions grown in embryonated eggs and with those of virions grown in MDBK or L cells has revealed that while the content of the HN glycoprotein is similar, virions grown in MDBK and L cells lack the F protein found in egg-grown virus and instead contain large amounts of a higher molecular weight component, F_0 (Fig. 3) (Homma and Ohuchi, 1973; Scheid and Choppin, 1974, 1975). F_0 has since been shown to be a structural precursor for F and posttranslational cleavage of F_0 to F is produced by cellular proteases. Cleavage occurs within the plasma membrane of infected cells during final assembly of new virions (Scheid and Choppin, 1975, 1976). In host cells such as MDBK and L cells, proteolytic cleavage of F_0 to F does not occur and progeny virions are released from infected cells containing F_0 rather than F. Thus the presence of F_0 instead of F in these virions correlates with their inability to induce FFWO.

Similar host-dependent variation in the cleavage of the F_0 precursor to the F protein and a correlation between possession of the F protein and the ability to induce FFWO has been demonstrated for NDV (Samson and Fox, 1973; Hightower et al., 1975; Poste, 1975; Nagai et al., 1976a, b). NDV differs from Sendai virus, however, in that cleavage of F_0 to F occurs in a much wider range of host cells. Infectious virions with cell fusion activity can be obtained not only by growing NDV in eggs but also in a variety of established mammalian cell lines (Reeve and Poste, 1971; Poste et al., 1972a; Alexander et al., 1973; Poste, 1975; Nagai et al., 1976a, b). However, individual strains of NDV differ in this regard. Virulent and mesogenic strains yield infective progeny with cell fusion activity when grown in eggs and a variety of avian and mammalian cells. In contrast, these properties are expressed by avirulent NDV strains only when grown in embryonated eggs or cultured chick chorioallantoic membrane cells, and avirulent strains grown in mammalian cells are usually noninfective and unable to induce cell fusion.

Recent studies by Scheid and Choppin (1977) have shown that the F glycoprotein in Sendai virus, NDV and SV5 is composed of two disulfide-linked polypeptides F_1 (49,000 to 56,000 MW) and F_2 (<10,000 to 16,000 MW). This finding that cleavage of F_0 results in the formation of two polypeptides that remain covalently linked to the virion also sheds some light on the mode of proteolytic cleavage and activation of infectivity, cell fusion and hemolytic activities. Homma and Ohuchi

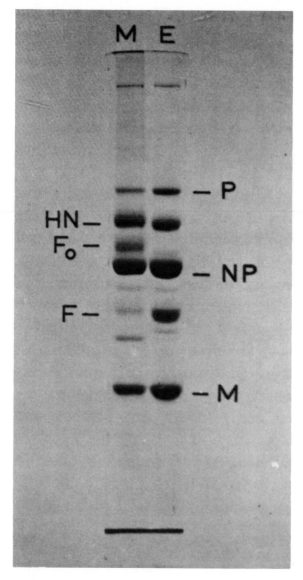

Fig. 3. Polypeptides of Sendai virus grown in embryonated eggs (E) and bovine MDBK cells (M) after SDS-polyacrylamide gel electrophoresis and staining with Coomassie blue. The nonglycosylated proteins, P, NP, and M are identified on the right and the envelope glycoproteins HN, F_0, and F are shown on the left. (Reproduced with permission from Scheid, 1976.)

(1973) had proposed that activation of these properties by cleavage of the F_0 glycoprotein might involve removal of a "masking substance" by proteases. However, Scheid and Choppin's finding that the smaller F_2 fragment is not lost suggests that removal of an inhibitory peptide is probably not involved. It thus seems more likely that proteolytic cleavage results in a more subtle conformational change in the F_0 precursor and that both the F_1 and F_2 chains may be necessary for the functional activity of the F protein. The finding that F_1 and F_2 are disulfide-bonded raises the possibility that inactivation of paramyxovirus infectivity, cell fusion and hemolytic activities by SH-reducing agents (Neurath et al., 1973) results from separation of F_1 and F_2. However, unpublished observations cited by Scheid and Choppin in their 1977 paper indicate that reduction of disulfide bonds between subunits of the HN protein is accompanied by loss of receptor binding and neuraminidase activity and loss of infectivity and cell fusion activities might be due simply to failure of virions to adsorb.

Cleavage of the F_0 precursor to produce F can also be accomplished in vitro. Treatment of Sendai or NDV virions containing the F_0 precursor with 1-10 μg/ml trypsin for 10 minutes at 37 °C converts F_0 to F and this is accompanied by restoration of infectivity and cell fusing and hemolytic activities (Homma and Ohuchi, 1973; Scheid and Choppin, 1974, 1975; Nagai and Klenk, 1977). Such activation cannot be achieved with plasmin, chymotrypsin, or elastase. However, Scheid and Choppin (1976) have isolated a series of Sendai virus mutants in which the F_0 precursor is not cleaved by trypsin but is cleaved by chymotrypsin or elastase. In contrast to wild-type Sendai virus, these mutants do not yield infective virions when grown in eggs. The egg-grown mutant virions contain the uncleaved F_0 precursor and thus lack cell fusion activity. Cleavage of F_0 by treating these virions with chymotrypsin or elastase in vitro restores infectivity and cell fusion activity.

The ability to select for virus mutants in which cleavage of F_0 and activation of cell fusion activities can be induced by different proteases raises the possibility that in the natural host, or during adaptation of a virus to a particular host within the laboratory, variants may emerge that are susceptible to the specific proteases present in the host in question. Differences in the susceptibility of the F_0 glycoprotein to cleavage by the available range of host proteases may explain the observation that whereas Sendai virions grown in MDBK or L cells contain the uncleaved F_0 precursor, are noninfective, and lack cell fusing activities, the related paramyxoviruses SV5 and NDV grown in these cells contain the cleaved F protein, are infective, and display cell fusion activity (Scheid and Choppin, 1974, 1975; Poste, 1975; Nagai et al., 1976a, b).

Additional evidence that the F glycoprotein is necessary for paramyxovirus-induced FFWO has come from studies of virus particles lacking this glycoprotein. Shimizu and Ishida (1975) have reported that the F glycoprotein spikes can be selectively digested from egg-grown Sendai virus by trypsin (20 μg/ml for 1 hr at 35°C), leaving the HN spikes intact on the surface of the virus. F-deficient particles produced by this method exhibit normal hemagglutinating and neuraminidase activities, but their infectivity and cell fusion and hemolytic activities are markedly reduced. Shimizu and Ishida (1975) also reported that the HN spikes could be selectively removed from Sendai virions by a fungal semialkaline protease, producing virions devoid of neuraminidase and hemagglutinating activity that still retained a population of surface spikes which presumably correspond to the F glycoprotein. Unfortunately, the crucial question of whether such virions exhibited cell fusion activity was not tested since HN-deficient particles were unable to adsorb to cells (work discussed on p. 321 suggests, however, that this problem could be overcome by provision of an "artificial virus receptor" on the cell surface by coating cells with plant lectins). The behavior of virions enzymically depleted of the HN glycoprotein is analogous to that of conditional lethal mutants of Sendai virus having a temperature-sensitive mutation affecting HN synthesis (Portner et al., 1975). When grown at nonpermissive temperatures these mutants lack HN. Although the F glycoprotein is present, these virions are unable to induce FFWO since they cannot attach to cells due to the absence of the HN glycoprotein.

Comparable observations have also been made using virus particles reconstituted from isolated Sendai virus subunits (Hosaka and Shimizu, 1972a, b, 1977; Hosaka, 1975). These studies have established that reconstituted particles lacking the F glycoprotein are devoid of cell fusion activity, but the more interesting experiments on the fusion activity of particles containing only the F glycoprotein were again frustrated by failure of the particles to attach to cells.

The attachment of paramyxoviruses to the cell surface via the HN envelope glycoprotein is an essential preliminary step in virus-induced FFWO. Paramyxoviruses are believed to bind to neuraminidase-sensitive sialic acid residues on cell surface glycoproteins in similar fashion to the orthomyxoviruses (influenza viruses) (Gallaher and Howe, 1976; Bächi et al., 1977). Treatment of cells with neuraminidase abolishes their susceptibility to FFWO by Sendai virus (Bächi et al., 1977), presumably by destroying the receptor needed for virus adsorption. Similarly, preinfection of cells with other neuraminidase-containing paramyxoviruses renders them resistant to FFWO when treated subsequently with Sendai virus (Wainberg and Howe, 1973). The loss of cellular susceptibility to FFWO occurs rapidly after infection with the first virus and probably results from cleavage of the receptors for Sendai virus by neuraminidase present in the first virus. The neuraminidase activity associated with paramyxovirus particles does not appear to play a direct role in either virus attachment or subsequent FFWO (e.g., see Micklem and Pasternak, 1977). Indeed, the ability of Sendai virus to induce FFWO has been reported to be enhanced after selective inactivation of viral neuraminidase (Neurath et al., 1972). Also, measles virus, which lacks neuraminidase, is able to induce extensive FFWO (Cascardo and Karzon, 1965). Viral neuraminidase may, however, facilitate the fusion process by promoting structural alterations in the cellular plasma membrane (see p. 345).

The strict requirement for attachment of virus particles to neuraminidase-sensitive receptors as a preliminary to cell fusion is not seen with certain strains of NDV. Bratt and Gallaher (1972) and Poste and Waterson (1975) have shown that virulent strains of NDV are able to attach to and induce FFWO of neuraminidase-treated cells. The nature of the receptors used by NDV on neur-aminidase-treated cells has not been identified. Poste and Waterson (1975) sug-gested that neuraminidase-resistant sialic acid moieties associated with glycolipids (Weiss, 1973) might serve as receptors. Some support for this possibility is pro-vided by recent studies showing that paramyxoviruses can attach to ganglioside receptors in model membranes (Haywood, 1975).

A number of other observations suggest that the initial adsorption of paramyxoviruses to the cell surface prior to cell fusion need not be mediated sol-ely by neuraminidase-sensitive sialoglycoproteins. Horse erythrocytes contain the N-glycolyl form of neuraminic acid instead of N-acetyl neuraminic acid and are not susceptible to agglutination and fusion by Sendai virus (Yamamoto et al., 1974) or certain strains of NDV (Burnet and Lind, 1950). However, after being coated with Con A, horse erythrocytes adsorb Sendai virus and become suscepti-ble to fusion by this virus (Yamamoto et al., 1974; Hosaka and Shimizu, 1977). Similarly, human erythrocytes rendered insusceptible to agglutination and fu-sion by Sendai virus by desialation can be fused by Sendai virus after coating with Con A as an "artificial virus receptor" (Bächi et al., 1977). However, the concen-tration of Con A (and other lectins) used in experiments of this kind is critical. Other studies have shown that pretreatment of cells with Con A prevents virus adsorption and inhibits FFWO induced by paramyxoviruses (Okada and Kim, 1972; Poste et al., 1974).

The creation of 'artificial virus receptors' on the cell surface may offer a po-tentially useful method for studying the functional properties of paramyxovirus envelope glycoproteins. Evaluation of the cell fusion activity of virus particles lacking the HN glycoprotein has been hindered because of the failure of such particles to adsorb to cells. Thus it might be interesting to test whether virus par-ticles lacking HN but containing the F glycoprotein (e.g., naturally occurring *ts* mutants; enzymically depleted particles; or reconstituted particles) could induce agglutination and fusion of lectin-treated erythrocytes. Unpublished data by Shimizu and associates cited by Hosaka and Shimizu (1977) suggests that this ap-proach may be feasible. They reported that HN-deficient Sendai virus particles were able to induce agglutination and FFWO of Con A-treated chick erythrocytes but had no effect on untreated erythrocytes.

Although it is now established that the ability of paramyxoviruses to induce FFWO is correlated with the presence of the F glycoprotein in the virus en-velope, it has not yet been possible to induce cell fusion by adding pure prepara-tions of the isolated F glycoprotein to cell cultures. Expression of the cell fusion activity of the F glycoprotein appears to require the additional presence of lipid(s) (Hosaka and Shimizu, 1972a,b; 1975), presumably in some form of or-dered structural interaction with the F glycoprotein. Fusion was greatest when the viral glycoproteins were mixed with phosphatidylethanolamine, while phos-phatidylcholine and sphingomyelin were relatively ineffective. Interpretation of this work is complicated, however, by the lack of control experiments on the fu-sion activities of the lipids alone. This question is crucial in light of more recent studies showing that lipid vesicles (liposomes) prepared from certain phospholipid(s) can fuse with the cellular plasma membrane and induce cell-to-cell fusion (Poste and Papahadjopoulos, 1976b).

Recent experiments by one of the present authors (G.P.) in collaboration with P. Reeve and H. Bachmeyer suggest that the properties of both the lipid and glycoprotein components of the virus envelope influence the capacity of paramyxoviruses to induce FFWO. Purified envelope glycoproteins isolated from NDV strain Herts grown in chick or BHK cells (virions that contain the F glycoprotein), when incorporated into unilamellar liposomes prepared from either phosphatidylserine (PS), PS and cholesterol (equimolar ratio), or PS and egg phosphatidylcholine (PC) (1:9 molar ratio), induced significant FFWO in BHK cell monolayers. Identical glycoproteins incorporated into liposomes of similar size prepared from PC or PC with distearoylphosphatidylcholine (DSPC) and dipalmitoylphosphatidylcholine (DPPC) were devoid of fusion activity. These marked effects of phospholipid composition in influencing the expression of fusion activity parallel the results of earlier studies on the ability of liposomes lacking incorporated proteins to fuse with other liposomes or with the plasma membrane of cultured cells. For example, liposomes prepared from neutral phospholipids (lecithin) and charged vesicles composed of phospholipid mixtures such as PS/DSPC/DPPC, which are "solid" at 37 °C (i.e., they are below their gel-to-liquid crystalline transition temperature at the experimental temperature), are unable to fuse with other liposomes (Papahajopoulos and Poste, 1975; Martin and Mac-Donald, 1976; Papahadjopoulos et al., 1976a,b) or with cellular membranes (Poste and Papahadjopoulos, 1976a; Poste et al., 1976b, 1978). In contrast, charged liposomes composed of phospholipid mixtures that are "fluid" at 37 °C will readily fuse with similarly charged "fluid" liposomes and with the plasma membrane of cells cultured in vitro (Papahadjopoulos et al., 1976b; Poste and Papahadjopoulos, 1976a,b; Poste et al., 1976b, 1978). It is also interesting that Haywood (1975) was unable to detect fusion of the lipid envelope of Sendai virus with the membrane of neutral lecithin liposomes, but fusion of virions with liposomes was readily observed with lecithin vesicles containing phosphatidylethanolamine.

In contrast to the ability of liposomes of particular lipid compositions containing the F glycoprotein of NDV to induce FFWO, liposomes of identical lipid composition containing the F_0 precursor of NDV (obtained from NDV strain Ulster) were completely devoid of cell fusion activity (Poste, Reeve and Bachmeyer unpublished observations). Liposomes containing glycoproteins from trypsinized NDV strain Ulster virions (in which the F_0 precursor had been converted to F) produced significant FFWO if the liposome had a surface charge and was composed of phospholipids that were "fluid" at 37 °C.

The liposome experiments discussed above are instructive because they indicate that the presence of the F protein alone is not sufficient to perturb the cellular plasma membrane to permit virus envelope-plasma membrane fusion to occur. Thus the properties of the lipid matrix with which the F protein interacts in the virus envelope (or liposome) cannot be dismissed as irrelevant to the fusion process. As discussed further on (p. 355), the possibility exists that the F protein is *not* directly responsible for inducing the membrane perturbation underlying fusion between the virus envelope and the cellular membrane and subsequent cell-to-cell fusion. Rather, the F protein may merely be responsible for ensuring that the virus can bind to a particular class of receptor within the membrane that allows the plasma membrane and the virus envelop to be brought into sufficiently close contact for subsequent fusion to take place.

The liposome experiments described above suggest that the ability of paramyxoviruses to induce FFWO is a property of the intact virus envelope (or at least a structural organization which permits interaction between the F glycoprotein and the lipid bilayer, as in the case of liposomes containing this protein). Thus the activity of any particular component of the virus envelope in causing FFWO almost certainly depends on its organized relationship with other membrane components. This not surprising in view of the many documented examples from membrane reconstitution experiments in which the functional activity of membrane proteins, notably enzymes, is often lost after isolation from membranes but is restored when the protein is allowed to interact with the lipid bilayer of liposomes (Kimelberg, 1977). It seems likely therefore that efforts to isolate a single molecule with FFWO activity from disrupted virus particles are probably unrealistic.

Although viral subunits with cell fusion activity have yet to be isolated, Levitan and Blough (1976) have described an interesting phenomenon in which cell-free extracts from Herpes simplex virus (HSV)-infected BHK cells were found to induce fusion of uninfected BHK cells. This fusion activity was not associated with intact virus particles in the extract. Also, incubation of extracts with antiserum to HSV failed to neutralize the fusion activity, indicating that the factor(s) responsible is not a structural component of the virion. Other experiments using glycosidases and lectins with reactivity to different saccharide residues indicated that the fusion factor(s) is a fucose-containing glycoprotein. However, the question of whether it is a virus-specified nonstructural protein or a host cell (macro)molecule which has undergone a viral-dependent modification remains to be established.

5. Virus-cell interactions in virus-induced cell fusion

There is now a general consensus that FFWO induced by paramyxoviruses and other lipid-enveloped viruses requires fusion of virus particles with the cellular plasma membrane. Differences in opinion remain, however, about whether cell fusion occurs at the same time as fusion of virus particles with the plasma membrane is taking place or whether fusion between the virus envelope and the plasma membrane and cell-to-cell fusion ocur as related, but separate, processes.

Fusion between the virus envelope and the cellular plasma membrane (Figs. 4–6) is the normal mechanism by which paramyxoviruses and many other lipid-enveloped viruses infect cells (Meiselman et al., 1967; Morgan and Howe, 1968; Heine and Schnaitman, 1969; Howe and Morgan, 1969; Bose and Sagik, 1970; Zee and Talens, 1971; Apostolov and Poste, 1972; Bächi and Howe, 1972; Bächi et al., 1973, 1977; Granados, 1973; Wolinsky and Gilden, 1975; Chang and

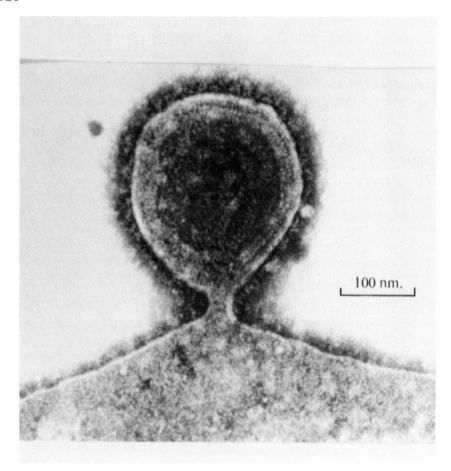

Fig. 4. Negatively stained electron micrograph showing fusion of a Sendai virion with the plasma membrane of a human erythrocyte. ×150,000 (Reproduced with permission from Apostolov and Almeida, 1972.)

Metz, 1976; Okada et al., 1976). The relationship of FFWO to the normal process of virus penetration thus explains the correlation discussed earlier in which paramyxoviruses lacking the F envelope glycoprotein were not only unable to fuse cells but were also noninfective.

Attachment of virus particles to specific receptors on the cell surface is not sufficient to "trigger" the membrane changes needed for fusion between the virus envelope and the cellular plasma membrane. As mentioned in the previous section, Sendai virus particles lacking the F protein adsorb normally to cells via the HN envelope glycoprotein, yet are noninfective and do not induce FFWO. The separate nature of virus attachment and the processes of fusion between virus envelope and the plasma membrane and cell-to-cell fusion is also indicated

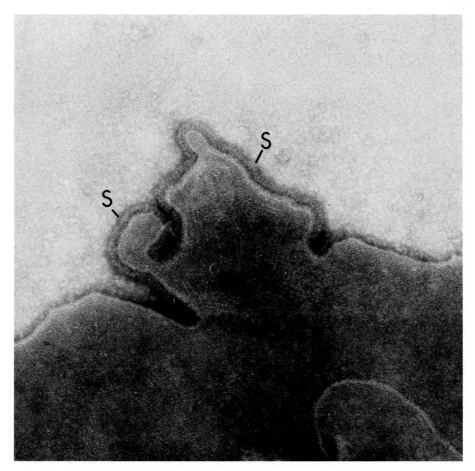

Fig. 5. Negatively stained electron micrograph showing fusion of a Sendai virion with the plasma membrane of a human erythrocyte at a later stage than in Fig. 4, showing complete fusion of the virus with resulting incorporation of the virus envelope (identified by the viral "spikes" [S]) into the cellular plasma membrane. ×132,000. (Electron micrograph courtesy of Dr. K. Apostolov.)

by experiments showing that modification of paramyxoviruses by heat (Okada, 1969), glutaraldehyde (Toister and Loyter, 1972), or antibodies against the F glycoprotein (Seto et al., 1974) does not impair virus attachment but inhibits virus penetration and FFWO.

Virus attachment and the subsequent steps of virus penetration and cell-to-cell fusion can also be distinguished on the basis of their temperature dependence. Virus adsorption, although temperature-independent, is usually done at 2 to 4°C. However, fusion between the virus envelope and the cell membrane and cell-to-cell fusion cannot occur at this temperature (Okada, 1962; Harris and Watkins, 1965; Apostolov and Poste, 1972; Gallaher and Bratt, 1972, 1974; Poste and Allison, 1973; Wainberg and Howe, 1973; Knutton, 1976; Bächi et al.,

326

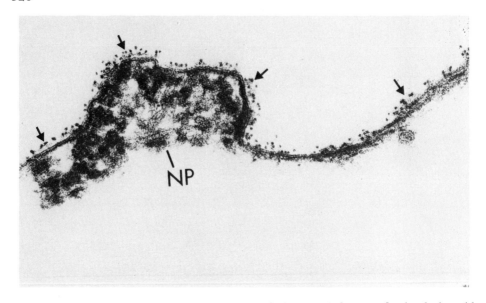

Fig. 6. Thin-section electron micrograph of portion of a human erythrocyte after incubation with Sendai virus for 40 min. at 37°C, showing complete incorporation of the virion into the cell similar to Fig. 5. The presence of viral antigens in the host cell plasma membrane is demonstrated by reactivity to ferritin-labeled antibodies (arrows). Note the dispersal of viral antigens within the plasma membrane from the point at which fusion between virus and cell first occurred. Remnants of nucleoprotein (NP) are still attached to the original site of virus penetration. ×110,000 (Reproduced with permission from Bächi et al., 1977.)

1977). The optimum temperature for these processes is 37 °C; the frequency of both drops rapidly with the reduction in temperature and at temperatures below 18 °C they are completely inhibited (Okada, 1962; Apostolov and Poste, 1972; Gallaher and Bratt, 1972; Pasternak and Micklem, 1973; Okada et al., 1976). This is a general characteristic of most, if not all, examples of membrane fusion (Poste and Allison, 1973) and is probably due to the effect of temperature on the physical state of membrane lipids. Membrane fusion appears to require that phospholipids in the interacting membranes be in a "fluid" state (Papahadjopoulos et al., 1974, 1976a,b; Papahadjopoulos and Poste, 1975; Poste and Papahadjopoulos, 1976a,b; and see also p. 344) and lowering the temperature reduces membrane "fluidity".

Apostolov and Almeida (1972), and more recently Knutton (1978), have proposed that FFWO occurs when a virus particle in close association with the membranes of the two different cells fuses simultaneously with the plasma membrane of both cells (Fig. 7, scheme 2), permitting the virus to act as a connecting "bridge" between the two cells. Progressive enlargement of such bridges results in complete cell fusion.

The virus bridge hypothesis is weakened by the fact that most electronmicroscopic observations of virus-induced FFWO indicate that the fusion of plasma

1. CELL-CELL BRIDGE FORMATION:

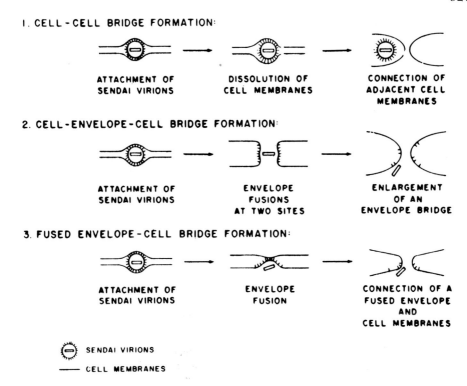

ATTACHMENT OF
SENDAI VIRIONS

DISSOLUTION OF
CELL MEMBRANES

CONNECTION OF
ADJACENT CELL
MEMBRANES

2. CELL-ENVELOPE-CELL BRIDGE FORMATION:

ATTACHMENT OF
SENDAI VIRIONS

ENVELOPE
FUSIONS
AT TWO SITES

ENLARGEMENT
OF AN
ENVELOPE BRIDGE

3. FUSED ENVELOPE-CELL BRIDGE FORMATION:

ATTACHMENT OF
SENDAI VIRIONS

ENVELOPE
FUSION

CONNECTION OF A
FUSED ENVELOPE
AND
CELL MEMBRANES

SENDAI VIRIONS

—— CELL MEMBRANES

Fig. 7. Schematic representation of the three hypotheses proposed to explain cell fusion by Sendai virus and other enveloped paramyxoviruses. (see text for details). (Reproduced with permission from Hosaka and Shimizu, 1977.)

membranes on adjacent cells occurs more commonly at sites where neither intact virus particles nor viral fragments can be recognized (Bächi and Howe, 1972; Bächi et al., 1973, 1977; Cassone et al., 1973; Hosaka and Shimizu, 1974, 1977). In addition, immunoelectronhistochemical studies have shown that viral envelope antigens are not usually present on the membrane over the cytoplasmic bridges formed between cells (Hosaka and Shimizu, 1977) though occasional examples are seen (Fig. 8). While it is possible that viral fragments and antigens have diffused away by the time cytoplasmic bridges are formed, such observations argue strongly against the virus-bridge hypothesis.

Additional evidence against the virus-bridge hypothesis comes from studies showing that the fusion of the envelope of paramyxoviruses with the cellular plasma membrane can be separated experimentally from the process of cell-to-cell fusion. For example, cytochalasin B (Poste and Allison, 1973; Ohki et al, 1975) and tertiary amine local anesthetics (Poste, unpublished observations) do not impair fusion of Sendai virus or NDV with the plasma membrane but both these drugs effectively inhibit the subsequent step of cell-to-cell fusion (Poste and Reeve, 1972b; Pasternak and Micklem 1973, 1974a; Ohki et al., 1975; though see

328

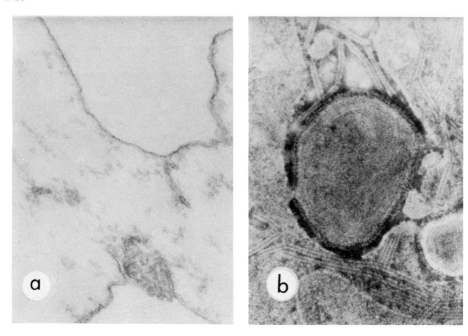

Fig. 8. Thin-section (a) and negatively stained (b) electron micrographs showing points of fusion between two human erythrocytes in which the virus particle has acted as a "bridge" to permit fusion as proposed in scheme 2 of Fig. 7. a = ×57,000; b = ×98,200. (Reproduced with permission from Apostolov and Poste, 1972.)

p. 344 for an alternative explanation of the effect of cytochalasin B). Similarly, Maeda et al. (1977a) have found that incubation of Ehrlich ascites cells with Sendai virus in medium supplemented with high concentrations of saccharides (>0.38 M glucose, mannose or galactose or >0.25 M sucrose) does not affect initial fusion of virus particles with the cellular plasma membrane but blocks cell-to-cell fusion. Although the mechanism underlying this phenomenon remains to be identified, these results indicate that virus envelope-plasma membrane fusion and cell-to-cell fusion are separate events. These results, together with the electronmicroscopic data mentioned above, support the concept that FFWO occurs after the integration of viral envelope proteins into the cell plasma membrane and that cell fusion probably results from an interaction between an area of plasma membrane on one cell containing viral envelope glycoproteins and unmodified plasma membrane on an adjacent cell (Fig. 7, scheme 3).

A sequence of events similar to those in scheme 3 in Figure 7 may also be responsible for virus-induced FFWI. In contrast to FFWO in which viral glycoproteins are inserted into the plasma membrane during virus penetration, modification of the plasma membrane during FFWI would occur when newly synthesized viral glycoproteins are inserted into the plasma membrane during final assembly of new progeny virions at the cell surface. It is known that FFWI does not occur before newly synthesized viral glycoproteins have been inserted into the plasma membrane (Poste, 1975; Rott and Klenk, 1977) and FFWI can be inhibited by ligand-induced cross-linking of the viral glycoproteins within the plasma membrane (Ludwig et al., 1974; Poste et al., 1974).

In the paramyxoviruses, the parallels between FFWO and FFWI can be extended to include a requirement for the presence of the F glycoprotein within the plasma membrane in order for FFWI to occur. FFWI takes place only in those host cells in which cleavage of the F_0 precursor to F occurs. FFWI is not seen during virus replication in host cells in which cleavage of F_0 to F does not take place (Poste, 1975; Nagai et al., 1976a,b). In the latter situation, FFWI can be induced by brief treatment of infected cells with proteases (Nakamura and Homma, 1974; Nagai et al., 1976b).

The role of viral envelope glycoproteins in causing FFWI induced by paramyxoviruses is also indicated by experiments showing that FFWI can be inhibited by agents that block glycosylation of viral proteins. Cultivation of cells infected with NDV (Gallaher et al., 1973) or SV5 (Rott et al., 1975) in the presence of 2-deoxy-2-fluoro-D-glucose or 2-deoxy-2-fluoro-D-mannose and glucosamine prevents complete glycosylation of the newly synthesized virus envelope proteins. Although the partially glycosylated proteins are inserted into the plasma membrane of infected cells, FFWI does not occur. Poste (unpublished observations) has also found that NDV virions released with incompletely glycosylated envelope proteins cannot induce FFWO. Inhibition of FFWI by fluorosugars has also been reported in cells infected with herpesviruses (Ludwig et al., 1974; Knowles and Person, 1976; Schmidt et al., 1976). However, the identity of the (glyco)proteins(s) involved in FFWI induced by these viruses has yet to be established.

If similar modification of the host cell plasma membrane by viral glycoproteins is responsible for both FFWO and FFWI, it is necessary to ask why the capacity of membrane-integrated viral glycoproteins to induce FFWO is lost soon after initial virus penetration even though they persist in the plasma membrane for up to 4 to 6 hours following the initial penetration of the virus via fusion with the plasma membrane (Durand et al., 1975; Kohn, 1975; Okada et al., 1976). Although it might be expected that such membrane-integrated viral glycoproteins would retain the capacity to induce cell fusion, this is clearly not the case. Following the early sequence of FFWO that accompanies initial virus penetration, cells become resistant to fusion until *newly synthesized* envelope glycoproteins are introduced into the plasma membrane, which results in FFWI.

One possible explanation for the rapid decline of FFWO activity following viral penetration is that a "critical concentration" of viral envelope glycoproteins per unit area of plasma membrane might be required to induce the necessary degree of membrane perturbation for fusion to take place. For example, following initial fusion of the virus envelope with the plasma membrane, the viral glycoproteins quickly undergo lateral diffusion within the membrane (Bächi and Howe, 1972; Bächi et al., 1973; Okada et al., 1974 Chang and Metz, 1976), with resulting dilution of their concentration at the original site(s) of penetration.

The possible requirement for a critical concentration of viral envelope glycoproteins within the plasma membrane to induce cell fusion (both FFWO and FFWI) might also account for the finding that antiviral antibodies can enhance virus-induced FFWI in certain situations. For example, using parainfluenza virus type 2 propagated in a cell line in which FFWI did not normally occur, Wainberg and Howe (1972) found that antiviral antibodies added at a stage in infection at which newly synthesized envelope glycoproteins had been inserted into the plasma membrane caused the cells to fuse. A possible interpretation of this observation is that the antibodies served as a ligand to cross-link the viral glycoproteins into "patches" of the necessary size required to induce membrane destabilization and permit fusion to occur. However, as mentioned earlier, in experiments using ligands to cross-link cell surface components both the number of ligand receptors and the valency and concentration of the ligand are crucial in determining the extent of cross-linking and the resulting alteration in cell surface properties. Thus, other studies using multivalent lectin molecules to crosslink surface and the resulting alteration in cell surface properties. Thus, other studies using multivalent lectin molecules to cross-link surface components on paramyxovirus-infected cells have produced marked inhibition of both FFWO and FFWI (Poste et al., 1974; Rott et al., 1975).

Before finishing this discussion of the structural basis of FFWO induced by lipid-enveloped viruses, brief mention should be made of the hypothesis for FFWO proposed by Okada (1962, 1969) since this is often cited in the cell fusion literature even though there is little experimental evidence to support it. Okada's hypothesis proposes that FFWO induced by Sendai virus results from the action of a membrane lytic factor associated with the virus particle. This hypothesis was advanced largely on the basis that Sendai and related viruses causing FFWO were able to lyse erythrocytes. In Okada's scheme, cell fusion is suggested to occur via the fusion of areas of plasma membranes on adjacent cells that have been modified by the viral "lysin." Perturbation of the plasma membrane by the lysin is proposed as occurring in the membrane surrounding attached virus particles, so that fusion occurs solely between plasma membranes and the virus particles do not form part of the bridge (Fig. 7, scheme 1). In this scheme, fusion of the virus envelope with the plasma membrane is not required. Although this hypothesis for FFWO would accomodate the fact that virus antigens are not usually detected on the intercellular bridges (vide supra), there is little other evidence in support of it. Moreover, in this scheme the entire virus is left *outside* fusing cells, thus requiring a quite separate mechanism for viral infection.

No evidence has been obtained for chemical modification of host membranes during FFWO or for the existence of virus-associated molecules with membrane lytic properties. Indeed, even hemolysis does not involve overt lysis of the erythrocyte but results from colloid osmotic swelling of the cell following alterations in plasma membrane permeability caused by fusion of virus particles with the plasma membrane (reviews, Bächi et al., 1977; Hosaka and Shimizu, 1977; also see p. 335).

Although there is a large body of evidence suggesting that the ability of paramyxoviruses to induce hemolysis is closely related to their cell fusion activity and that hemolysis requires fusion of the virus envelope with the erythrocyte plasma membrane (Apostolov and Waterson, 1975), recent work has shown that the cell fusion and hemolytic activities of Sendai virus can be separated. Homma and colleagues (1976) and Shimizu and associates (1976) found that Sendai virions harvested immediately after a one-step growth cycle in eggs are devoid of hemolytic activity yet are fully infective and can induce FFWO. On the other hand, treatments such as freeze-thawing, sonication, prolonged incubation at 37 °C or complement-mediated immune damage to the virus envelope activate the hemolytic activity of such virions. Shimizu and co-workers (1976) and Hosaka and Shimizu (1977) have shown that the envelope of nonhemolytic virions is much less permeable to uranyl acetate (UA) than that of hemolytic virions (so-called UA^+ virions). Hemolysis occurred only after fusion of UA^+ virions with the plasma membrane of erythrocytes. This finding suggests that hemolysis is confined to erythrocytes that have undergone fusion with virus particles which possess permeable (damaged?) envelopes. The leakage of ions through the region of the cellular plasma membrane containing the integrated permeable UA^+ virus envelope then presumably results in colloid-osmotic swelling of the erythrocyte with eventual rupture of the plasma membrane and release of hemoglobin (Apostolov and Waterson, 1975).

6. Mechanisms

6.1. Virus-to-cell and cell-to-cell fusion

The molecular mechanism(s) responsible for fusion between the virus envelope and the plasma membrane, and subsequent fusion between the plasma membranes of adjacent cells in FFWO, is still unknown. In FFWO (and also FFWI) induced by paramyxoviruses, a reasonable assumption can be made that the viral F protein is associated with the process by which membranes are perturbed to render them susceptible to fusion, but details of the exact mechanism(s) involved are unknown.

In common with other examples of membrane fusion (see Chapters by Lucy, Meldolesi et al., and Orci and Perrelet, this volume) freeze-fracture studies have revealed structural alterations in the plasma membranes of cells during FFWO induced by paramyxoviruses. Freeze-etch studies of the plasma membranes of erythrocytes incubated with Sendai virus at 37 °C have revealed redistribution and, in certain instances, clustering, of intramembranous particles during cell-to-cell fusion (Fig. 9) (Bächi et al., 1973, 1977; Zakai et al., 1977). This effect was not induced simply by virus adsorption since attachment of similar numbers of virus particles does not alter membrane structure (Fig. 10). Similarly, incubation of erythrocytes at 37 °C with influenza virus, which cannot fuse with the plasma

332

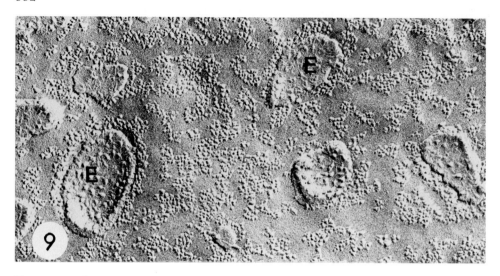

Fig. 9. Freeze-fracture electron micrograph of human erythrocyte after incubation with Sendai virus for 30 min at 37°C showing redistribution of intramembranous particles into clusters (cf. with random distribution of particles shown in Fig. 10). Note the corona of aggregated particles around virus envelope fragments (E). ×944,000. (Reproduced with permission from Bächi et al., 1977.)

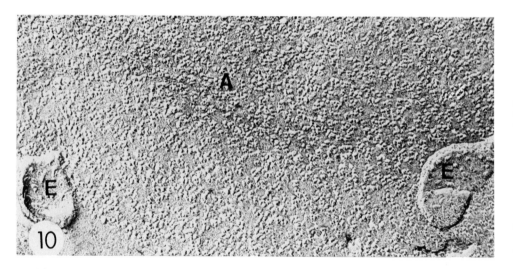

Fig. 10. Freeze-fracture electron micrograph of human erythrocyte after incubation with Sendai virus for 30 min. at 2-4°C showing random distribution of intramembranous particles. Irreversibly bound fragments of virus envelope (E) are also present on the outer Fracture Face (A). (Reproduced with permission from Bächi et al., 1977.)

membrane and does not induce cell-to-cell fusion (Dourmashkin and Tyrrell, 1974; Dales and Pons, 1977), fails to modify membrane ultrastructure (Bächi et al., 1977).

Freeze-fracture studies by Knutton (1976), on the other hand, have shown that fusion of virus particles with the plasma membrane does not induce overt redistribution of intramembranous particles. No clustering was observed, whether cells were fusing or not (Knutton, 1976, 1978). These observations have since been confirmed by Fowler and Branton (1977).

Knutton's work showed the following: freeze-fractured Sendai virus particles examined at 4 °C display numerous intramembranous particles in the concave E face (Fig. 11a) while the convex P face is relatively smooth and contains a complementary arrangement of pits (Fig. 11b). The size of the particles in the E face (approximately 14 nm diameter) is larger than the 9 nm diameter particles seen in the E and P fracture faces of the erythrocyte plasma membrane. On raising the temperature to 37 °C dramatic changes in viral morphology can be detected *before* fusion of the virus envelope and the plasma membrane takes place. First, the surface of the virus changes from a spherical (Fig. 12a) to a highly convoluted form (Figure 12b). These infoldings of the viral surface appear in freeze-fracture replicas as a series of linear ridges. The ridges, which are devoid of intramembranous particles and are some 30 nm wide and up to 500 nm long, are seen only on the E face (Fig. 12c). On the P face they appear as linear grooves (Fig. 12d). Second, the viral "spikes" seen in transmission electron microscopy disappear. At the same time, the viral 14 nm particles also disappear and are replaced by 9 nm particles on both E and P faces (Figs. 12c,d). Such changes indicate considerable reorganization of the spike complexes. In freeze-fracture replicas taken later when fusion of viral envelope with erythrocyte membrane has already occurred, large numbers of 9 nm intramembranous particles are now

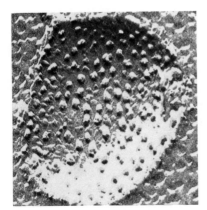 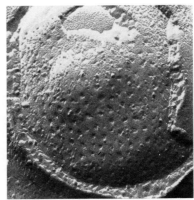

Fig. 11. Freeze-fracture replicas showing fractured Sendai virus particles at 4°C: (a) concave E-fracture face showing numerous 14nm diameter intramembranous particles. ×93,500: (b) convex P fracture face showing pits corresponding to the particles of the E face. ×93,500. (Reproduced with permission of Nature from Knutton, 1976.)

334

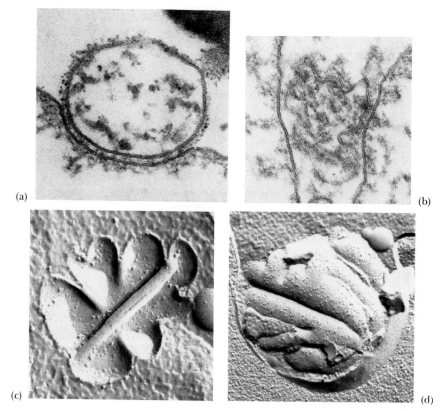

(a)

(b)

(c)

(d)

Fig. 12. Thin-section (a,b) and freeze-fracture (c,d) replicas of Sendai virus particles bound to eryth-rocytes at 4°C (a) and after incubation for 2 min at 37°C (b,c,d). At 37°C, the viral surface becomes invaginated and surface spikes are no longer apparent (b). In freeze-fracture replicas, the invagina-tions are characterized by the presence of 30nm wide, and up to 500nm long, smooth linear ridges on the E face (c), and complementary grooves on the P face (d) of the virus. 9 nm diameter intramem-branous particles are seen on both P and E fracture faces (c and d). (a) ×100,000; (b) ×87,500; (c) ×75,000; and (d) ×45,000. (Reproduced from Knutton (1978) with permission of the Journal of Cell Science.)

seen in the nonridge areas (Figs. 13a,b). The distribution of particles (approxi-mately 1100/μm² on the E face of the erythrocyte plasma membrane and approx-imately 2500/μm² on the P face) suggests that most are of erythrocyte origin. At no stage is a clustering of erythrocyte intramembranous particles apparent. Freeze-fracture replicas such as those of Figure 13, showing both viral ridges and erythrocyte intramembranous particles within the same fracture face, provide strong evidence in confirmation of previous reports documenting direct fusion between erythrocytes and Sendai virus membranes (see p. 323). The engulfment of virions by phagocytosis would not show such mixing. In virus-cell suspensions kept at 0 °C, for example, a viral particle is occasionally seen "embedded" in a "cup-like" depression in the plasma membrane reminiscent of phagocytosis (Fig.

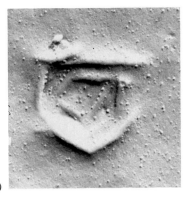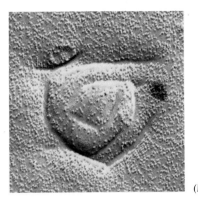

(a)   (b)

Fig. 13. Freeze-fracture replicas showing a later stage during the fusion of the Sendai virus envelope with the erythrocyte membrane. The entire viral envelope has become continuous with the erythrocyte membrane. (a) shows the E face of the erythrocyte; (b) shows the complementary replica of the P face. ×60,000. (Reproduced from Knutton (1978) with permission of the Journal of Cell Science.)

12a), yet no ridge formation and no loss of surface spikes or of 14 nm diameter intramembranous particles is seen until the temperature is raised.

As discussed later in this chapter there is an impressive body of ultrastructural evidence suggesting that membrane fusion may proceed via lipid-lipid interactions in regions where the apposed membranes are depleted of integral membrane proteins. It is therefore perhaps noteworthy that the particle-free ridges on the virus envelope appear to fuse with the cell membrane (Knutton, 1976, 1978).

As yet, the exact sequence of structural events in virus envelope-plasma membrane fusion is unclear. Nor is it known which event is the trigger for membrane fusion. What is obvious is that clustering of intramembranous particles in the cellular plasma membrane (Bächi et al, 1973, 1977) is not a prerequisite for virus-cell fusion, since their distribution on the P and the E face remains unchanged during fusion (Fig. 13). The same is true for HeLa cells (Knutton, unpublished observations). Only occasional aggregates of particles are observed and, since these are found more frequently at high cell concentrations, they may play a part in the establishment of cell-to-cell fusion. Fusion of enveloped paramyxoviruses such as Sendai virus and NDV with the plasma membrane is also accompanied by alterations in plasma membrane permeability which, in turn, promote changes in cellular morphology.

Adsorption of Sendai virions to cells at low external calcium concentrations increases the permeability of the plasma membrane to several compounds, so that potassium leaks out and sodium enters (Fig. 14) (Klemperer, 1960; Fuchs and Giberman, 1973; Pasternak and Micklem, 1974a,b), producing a drop in the resting potential of the plasma membrane (Fuchs et al., 1978). As a result of these alterations, water enters the cell with accompanying morphologic alterations such as swelling (Knutton et al., 1976, 1978; and page 341).

CHANGES IN INTRACELLULAR IONS

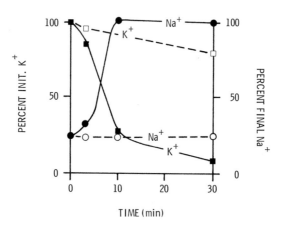

Fig. 14. Cation changes in Lettré cells exposed to Sendai virus. Cells (4.5×10^7/ml) were incubated at 37°C in 2 ml of 0.9% NaCl with or without virus (10^3 HAU/ml) and leakage stopped at various times by addition of 98 ml of 0.25 M sucrose followed by rapid centrifugation. The supernatant medium and the cell pellet were then analyzed by atomic absorption spectrophotometry. ■, Na^+ entry without virus; o, Na^+ entry without virus; □, K^+ leakage without virus.

The perturbation of the permeability barrier in cells incubated with paraoviruses is not due to an inhibition of plasma membrane Na^+/K^+-ATPase (Table 1; Fuchs et al., 1978). Rather, it reflects a general change in plasma membrane integrity, such that the Na^+/K^+-ATPase can no longer maintain ionic asymmetry. In fact, much of the cellular pool of water-soluble intermediates, such as amino acids, sugars and nucleotides leaks out (Pasternak and Micklem, 1973; 1974a,b). This leakage phenomenon is a true permeability change and is not due to cell lysis, since significant release into the medium of lactate dehydrogenase or trichloracetic acid-insoluble proteins labeled with [^{14}C]valine is not detected (Table 2). It is true that Klemperer (1960) noted that NDV-mediated leakage of potassium from HeLa or Krebs 2 ascites tumor cells was accompanied by release of 280 nm absorbing material which was interpreted as being protein. In fact it may have been due to free amino acids, the leakage of which is accelerated by virus (see above). If the material was protein, its loss (which was stimulated only 1.4-fold, in contrast to a 6-fold stimulation of potassium release by virus) may have occurred largely from the outside of cells. Most cultured cells appear to have a loosely bound "coat" of serum-derived proteins (Knox and Pasternak, 1975), and this would not have been labeled under the conditions described in Table 2.

Table 2 also shows that in the presence of Sendai virus (approximately 10^{-6} HAU/cell) under conditions where 65% of the intracellular pool of [^{3}H]metabolites of choline (largely phsophorylcholine) is released into the medium, less than 1% of the cellular phospholipid is released (Pasternak and Micklem, 1973, 1974a). Considerable amounts of water-soluble valine are released independently of vi-

TABLE 1

The effect of Sendai virus on cellular Na^+/K^+-ATPase activity[a]

Enzyme	Activity	
	Untreated control	Virus-treated cells
Mg^{2+}-ATPase	0.31 ± 0.02	0.46 ± 0.01
Na^+/K^+-stimulated Mg^{2+}-ATPase	0.08 ± 0.03	0.09 ± 0.05

[a] Lettré cells (10^8/ml) were incubated with or without Sendai virus (300 HAU/ml) at 37°C and sampled at intervals by centrifuging. The washed pellets were resuspended and disrupted by sonic oscillation (3×10 sec exposure to 100 watts in a Braunsonic 1510 apparatus). Mg^{2+}-ATPase and Na^+/K^+-stimulated Mg^{2+}-ATPase were estimated by the method of Avruch and Wallach (1971). Values are means of three estimations taken over a 20-min incubation period.

TABLE 2

Release of cellular constituents from cells after exposure to Sendai virus[a]

Sample[b]	Percentage of cell-associated radioactivity released into culture medium					
	Untreated controls			Virus-treated cells		
	2 min	6 min	20 min	2 min	6 min	20 min
Lactate dehydrogenase	7	9	12	9	5	7
T.C.A. soluble [^{3}H]	14	12	18	22	42	65
T.C.A. insoluble [^{3}H]	0.4	0.4	1.7	0.4	1.4	0.5
T.C.A. soluble [^{14}C]	65	83	90	71	84	93
T.C.A. insoluble [^{14}C]	4.2	4.3	6.6	3.5	3.0	6.2

[a] Lettré cells (5×10^8) were preincubated with [Me-^{3}H]choline (10 μCi) and [^{14}C]valine (1 μCi) in 20 ml of isotonic saline for 45 min at 37°C, washed and reincubated (at 1×10^7 ml) with or without Sendai virus (80 HAU/ml) at 37°C for the times indicated. Medium and cells were separated by centrifugation, and the lactate dehydrogenase activity (Bergmeyer, 1974) and radioactivity (soluble and insoluble in 5% cold trichloroacetic acid) of each fraction assayed.

[b] Of the total trichloroacetic acid (T.C.A.)-soluble [^{3}H]radioactivity, >90% is in phosphorylcholine (Pasternak and Micklem, 1973). Of the total T. C. A.-insoluble-[^{3}H]radioactivity, >95% is in phosphatidylcholine. T.C.A.-soluble and insoluble-[^{14}C]radioactivity is in free valine and protein, respectively.

rus, presumbably because there is no phosphorylation mechanism to restrict its movement out of cells (Pasternak and Micklem, 1973).

The permeability change seen in cells treated with Sendai virus is thus particularly striking for compounds that are phosphorylated and hence do not normally equilibrate between cells and the extracellular fluid. Two compounds that have proved useful are [³H]- or [¹⁴C]-labeled choline and 2-deoxyglucose. Each compound enters cells by facilitated diffusion and is then rapidly phosphorylated; neither is metabolized further (during short incubations) to any significant amount. Thus more than 90% of the radioactivity associated with cells is in the form of phosphorylcholine or 2-deoxyglucose-6-phosphate. Measurement of the distribution of radioactivity between cells and medium following addition of virus thus provides a simple assay of leakage for both monolayer and suspension cell cultures (Pasternak and Micklem, 1974a,b). The sensitivity of the leakage assay can be further increased by exploiting the fact that the increase in leakage (i.e., passive permeability) is accompanied by a decrease in facilitated diffusion (Pasternak and Micklem, 1974b). Thus by simultaneous exposure of virus-treated cells to a [³H]-labeled compound taken up by facilitated diffusion (e.g., choline or 2-deoxy-glucose at *low* concentration), and to a [¹⁴C]-labeled compound taken up by passive diffusion (e.g., α-methyl glucoside, choline or 2-deoxyglucose at *high* concentration), measurement of the ³H:¹⁴C ratio in the cell pellet permits any change in the permeability properties of the cell to be amplified (see Fig. 15).

It must be emphasized, however, that at low virus concentrations the permeability change occurs only in the absence of Ca^{2+}, being prevented by 1-2 mM Ca^{2+} (Pasternak and Micklem, 1974a). Since this concentration of Ca^{2+} does not prevent virus-cell fusion (as determined morphologically by ridge formation), the permeability change is clearly not a necessary part of the fusion process but is rather a reflection of the fact that the plasma membrane is altered in a general way after fusion with the virus envelope. However, as outlined below, the permeability change *is* a specific outcome of the interaction between the viral envelope and the plasma membrane. Permeability changes of the kind produced by paramyxoviruses are not induced by Ca^{2+}-binding agents such as EDTA or citrate (Pasternak and Micklem, 1974a), or by tertiary amine local anesthetics which destabilize the membrane in a more general manner (Micklem and Pasternak, 1977). How Ca^{2+} prevents the change is not clear, though simple Ca^{2+}-mediated effects on the efficiency of virus adsorption have been excluded (Micklem and Pasternak, 1977).

The permeability alterations induced by Sendai virus and NDV have been detected in all cells tested to date, including established cell lines (HeLa, Lettré, 3T3, SV40-transformed 3T3, HTC, and P815Y), primary cell strains (monkey kidney and human embryo cells) and normal human lymphocytes (Table 3). Although the permeability change is not therefore a cell-specific event, it is virus-specific since only enveloped viruses such as Sendai and NDV which cause FFWO elicit this effect. Thus, cells incubated with influenza (Kunz strain), cytomegalovirus, or adenovirus 5 (which are not known to fuse with the plasma membrane) do not exhibit significant leakage.

TABLE 3
Specificity of virally-induced permeability changes

Virus	Cell	Permeability change[a]
Sendai	Lettré	+
	P815Y	+
	3T3	+
	SV40/3T3	+
	HTC	+
	HeLa[b]	+
	Human embryo fibroblasts[c]	+
	Monkey Kidney[d]	+
	Human lymphocyte[e]	+
Newcastle disease	Lettré	+
Influenza	Monkey kidney[d]	−
Cytomegalovirus	Human embryo fibroblast[c]	−
Adenovirus-5	HeLa[b]	−

[a] Permeability changes were determined by preincubating cells with [³H]choline or 2-deoxy-[³H]glucose and assessing virally stimulated release of [³H]labeled material as described (Pasternak and Micklem, 1974a,b). Sendai virus and Newcastle Disease virus (provided by Dr. D. J. Alexander) were active at <100 HAU/ml (>10^{-5} HAU/cell); Influenza virus (Kunz strain, provided by Professor R. Heath) was inactive at >3000 HAU/ml (3×10^{-3} HAU/cell); cytomegalovirus (provided by Professor H. Stern) was inactive at >10^5 infectious units/ml (>10^{-1} infectious unit/cell); adenovirus-5 (provided by Dr. W. C. Russell) was inactive at >10^9 p.f.u./ml (>100 p.f.u./cell).

[b] Strain sensitive to adenovirus-5.

[c] MRC-5 cells, sensitive to cytomegalovirus.

[d] Flow Labs 0-40305; African green primary monkey kidney cells, sensitive to Kunz influenza virus.

[e] Obtained from normal peripheral blood.

The permeability changes detected in cells treated with Sendai virus or NDV do not result from virally-mediated alterations in the composition and/or turnover of membrane phospholipids (Pasternak and Micklem, 1974a). Also, viral neuraminidase does not appear to be responsible for the permeability changes (Micklem and Pasternak, 1977). In addition, the possibility that fusion of virus particles with the plasma membrane might "activate" membrane-associated proteases causing the observed permeability changes, seems unlikely in view of the observation that exogenously-added trypsin, which certainly modifies cell surface properties in other ways (see Maslow, 1976; Knox and Pasternak, 1977), does not elicit the permeability change (Micklem and Pasternak, 1977). Studies comparing the ability of intact virions and isolated virus envelope glycoproteins to induce permeability changes have produced conflicting results, though differences in the cells used may account for this. Thus, Fuchs and co-workers (1978) have reported that permeability changes in BHK cells treated with Sendai virus are produced only by intact virions, but one of the present authors (C.A.P.) has found that permeability changes can be elicited by isolated Sendai virus envelope proteins alone.

Whatever the cause of fusion of enveloped paramyxoviruses with the plasma membrane and its attendant membrane changes, it is a highly temperature-dependent event. Maximum leakage occurs at 37°C, but declines rapidly below 18-20 °C and is not detectable at 0 °C (Pasternak and Micklem, 1974a, Micklem and Pasternak, 1977). The importance of increased membrane permeability during fusion between the virus envelope and the cellular plasma membrane remains to be established. Certainly, the effect resembles the marked temperature-dependent changes in membrane permeability seen in the fusion of model membranes of defined composition. In these systems the permeability alterations are believed to result from structural instability in the membranes imposed by lateral phase separations of membrane lipids and it has been proposed that this event renders membranes susceptible to fusion (see chapter by Papahadjopoulos, this volume, and section 6.2 below). Moreover, virally-mediated alterations in plasma membrane permeability may be of importance in promoting the structural rearrangements in the plasma membrane required for the second step of cell-to-cell fusion after initial virus-cell fusion has occurred (p. 343).

Alterations in membrane permeability (with accompanying increases in cell volume) are also a prominent feature in cell-to-cell fusion induced by numerous membrane-active amphipathic molecules (see chapter by Lucy, this volume). In view of the highly lytic properties of most of these agents, it is unclear how far the observed permeability changes reflect changes related specifically to fusion, or additional effects created by the type of drastic perturbation of the lipid bilayer induced by these agents (see Micklem and Pasternak, 1977).

The extent to which permeability changes occur in other situations involving the close apposition and/or fusion of natural membranes is not yet clear. In one instance at least, a similar loss of the permeability barrier of the plasma membrane has been observed. This is in the establishment of intercellular "gap" junctions. Here too, monovalent cations and phosphorylated derivatives that do not normally cross the cell membrane can pass between cells, but macromolecules such as protein or RNA cannot (Pitts, 1976). And like the viral situation, transjuctional permeability is inhibited by Ca^{2+} (Rose and Loewenstein, 1975). In a sense, then, a virally modified cell surface (in the absense of Ca^{2+}) resembles one-half of a gap junction. One obvious difference, however, is that the clustering of intramembranous particles which accompanies the establishment of junctions (Pitts, 1976; McNutt, 1977) is not necessary for the virally mediated permeability change. On the other hand, it must be remembered that in most cells the intramembranous particles revealed in freeze-fracture replicas are not the only membrane-embedded proteins present; Con A receptors on 3T3 cells (Pinto da Silva and Martinéz-Palomo, 1975; Micklem et al., 1976), for example, are capable of aggregation without a change in the distribution of the intramembranous particles seen by freeze-fracture electron microscopy. Thus an aggregation or other architectural change in the disposition of some intramembranous proteins leading to the establishment of hydrophilic "channels" across the plasma membrane could still underlie the virally mediated permeability changes.

The permeability change caused by Sendai virus, unlike the establishment of in-

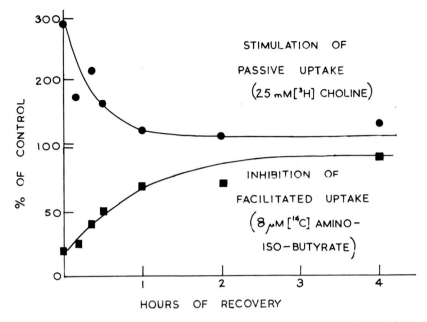

Fig. 15. Recovery of membrane integrity following exposure to virus. HeLa cells (1×10^6 growing as monolayers) were treated with Sendai virus (200 HAU), washed, and incubated in Dulbecco's medium. At the time indicated, the cultures were washed, exposed to [³H]choline (25 mM) and [¹⁴C]amino-isobutyrate (8 μM) and rates of entry of the radiolabeled compounds measured. (Reproduced with permission from Pasternak et al., 1976.)

tercellular junctions, is not stable. If cells that have been exposed to virus are reincubated in media that are conducive to growth, the membrane is "repaired" within a few hours (Fig. 15). Whether this requires the synthesis and insertion, or simply the realignment, of surface membrane components, should not be difficult to determine. Preliminary experiments suggest that protein synthesis (as judged by insensitivity to cycloheximide) is not required (Pasternak, unpublished observations).

As might be expected, changes in plasma permeability produced by paramyxoviruses result in significant changes in cell volume and morphology. In mammalian and avian erythrocytes, this takes the form of simple swelling followed rapidly by lysis (Bächi and Howe, 1972; Baker, 1972; Toister and Loyter, 1973). Similar increases in cell volume are seen in nonerythroid cells. Knutton and colleagues (1976) reported that Lettré cells exposed to Sendai at room temperature (using cell and virus concentrations insufficient to cause cell agglutination) underwent an increase in volume from approximately 700 to 1200 μm^3 with a concomitant loss of surface microvilli (Fig. 16). The swelling can be easily assayed with a Coulter counter, and this has the advantage over the direct assay of permeability changes (which measure the same phenomenon) of being cheaper, faster, and requiring fewer target cells.

The swelling of Lettré and HeLa cells induced by Sendai virus (Knutton et al.,

342

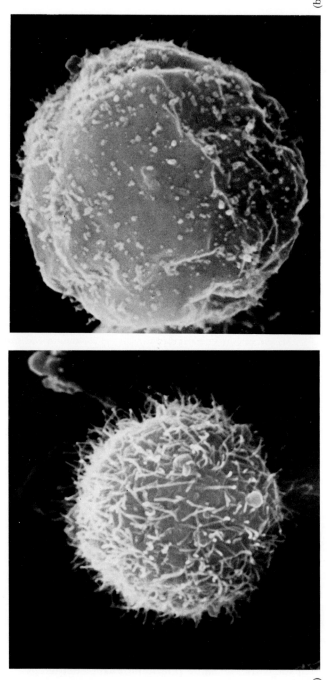

(b)

(a)

Fig. 16. Scanning electron micrographs of (a) control Lettré cells and (b) the same cells after addition of Sendai virus showing marked change in cell surface morphology with loss of microvilli. ×6,000. (Reproduced with permission of Nature from Knuton et al., 1976.)

1976, 1978) shows the same temperature dependence as the permeability changes and fusion of the virus envelope with the plasma membrane (as assayed by freeze-fracture electron microscopy.) Like the permeability changes, swelling is unaffected by cytochalasin B, but is inhibited by Ca^{2+} (1–2 mM), Con A ($8\mu g/$ ml), and ascorbic acid (0.15 M). As might be anticipated, swelling is dependent on the net loss of intracellular potassium. Substitution of potassium chloride for isotonic sodium or choline chloride as the suspending medium for cells reduces Sendai-mediated cell swelling by more than twofold. However, nonvirally mediated swelling induced by suspending cells in 0.06M NaCl is not accompanied by leakage of intracellular Me-[^{3}H]choline or displacement of $^{45}Ca^{2+}$ from the surface of prelabeled cells.

From the data reviewed so far in this section, it is clear that cell fusion (FFWO) induced by paramyxoviruses is preceded by alterations in the permeability of the cellular plasma membrane and also certain structural alterations in membrane organization. As mentioned earlier, freeze-fracture studies have indicated that fusion between adjacent membranes may occur at sites devoid of intramembranous particles. These particle-free zones are generally interpreted as representing membrane regions in which the lipid bilayer is depleted of those integral proteins (or oligomeric protein complexes) that comprise the particles. The ubiquity of this type of particle redistribution during membrane fusion phenomena (for examples, see chapters 6, 11, 12, and 13, this volume) has prompted a number of proposals concerning its functional relevance and the mechanism(s) responsible.

Studies on the fusion of liposomes with the plasma membrane of cultured cells (Poste et al., 1976b), and with other liposomes (see chapter by Papahadjopoulos, this volume), indicate that phospholipid bilayers possess the capacity to fuse spontaneously, provided that the lipid composition and physical state of lipids within the bilayer are appropriate and certain steric and ionic requirements are also met. Fusion between natural membranes might therefore also occur via similar interactions of phospholipid bilayers in adjacent membranes.

Poste and Allison (1973) proposed that the redistribution and aggregation of intramembranous particles seen in virus-induced FFWO and other examples of membrane fusion might serve as a mechanism whereby areas of the membrane could be depleted of integral proteins, with fusion taking place between particle-free (lipid?) areas or via the interdigitation of clustered protein molecules (either mechanism assumes that the particles seen in freeze-fracture replicas are proteins). The ultrastructural data obtained since this proposal was first made generally favor the former view that fusion occurs via lipid-lipid interactions at sites depleted of intramembranous particles (for detailed discussion, see chapters by Lucy and by Orci and Perrelet, this volume). However, even if this view is correct, the interaction of proteins on adjacent membranes may determine the specificity of the fusion reaction by regulating the contact and adhesive interactions between membranes which precede fusion per se (cf., Weiss et al. 1977).

Ahkong and associates (1975a) extended Poste and Allison's hypothesis to include the additional proposal that agents causing membrane fusion promote topographic redistribution of intramembranous particles by increasing the "fluidity" of membrane lipids. While the requirement for a "fluid" membrane for fusion to take place is well documented (Papahadjopoulos et al., 1974, 1976a,b; Poste and Papahadjopoulos, 1976a; Poste et al., 1976b, 1978), there is no evidence to support the view that agents causing cell fusion increase the fluidity of membrane lipids. Actually there is evidence to the contrary. Agents such as lysolecithin, DMSO, and myristic acid, which induce cell fusion, actually decrease the fluidity of phospholipids (Papahadjopoulos et al., 1976a,b). Also, fluorescence polarization studies using the lipid probe 1, 6,-diphenylhexatriene have failed to reveal any significant changes in membrane "microviscosity" during FFWO induced by Sendai virus and NDV (Poste, unpublished observations; and also see p. 314).

An increasing amount of evidence suggests that the mobility and distribution of many integral proteins within the plasma membrane is controlled by cytoplasmic cytoskeletal elements (microfilaments and microtubules) associated with the cytoplasmic face of the membrane (Edelman, 1976; Nicolson, 1976; Poste and Nicolson, 1976, 1977). This raises the question of whether the redistribution of intramembranous particles seen in virus-induced FFWO and other examples of membrane fusion involves participation of cytoskeletal elements. Unfortunately, there is little information available concerning this potentially important question.

Circumstantial evidence that membrane-associated cytoskeletal systems might participate in membrane fusion has come from studies showing that drugs acting on microfilaments and microtubules can alter virus-induced cell fusion. Sendai virus-induced FFWO is inhibited by cytochalasin B (an inhibitor of microfilaments) (Pasternak and Micklem, 1973; Ohki et al., 1975) but is enhanced by drugs such as colchicine that disrupt microtubules (Ohki et al., 1975). However, cytochalasin B does not inhibit the fusion of Sendai virus particles (Poste and Allison, 1973; Ohki et al., 1975) or liposomes (Poste and Papahadjopoulos, 1976a,b) with the plasma membrane. This indicates that cytochalasin B does not impair membrane fusion per se and suggests that the inhibitory effects of this drug on FFWO result from its effect on membrane-associated microfilaments. The recent finding that tertiary amine local anesthetics impair microfilaments in similar fashion to cytochalasin B (Poste et al., 1975a,b) may also explain earlier observations showing that these drugs inhibited virus-induced FFWO and FFWI (Poste and Reeve, 1972b).

An alternative explanation for the effect of cytochalasin B and the participation of microfilaments in cell-to-cell fusion, is the following. The mechanism of the initial fusion event between two cells is the same as that between the virus envelope and the plasma membrane, namely, cell-to-cell contact, followed by Ca^{2+}-mediated, temperature-dependent, destabilization and mixing of membrane components (see below). This permits small cytoplasmic bridges to be formed at focal areas of close cell contact, but the remaining part of each cell remains spherical, in other words, the two fusing cells are in a 'dumb-bell' shape. This raises the possibility that it is the gross morphological change from this shape to that of a single, double-sized sphere, that involves microfilaments and is sensitive to cyhochalasin B. Since the latter shape change is exactly the opposite of that occuring during cytokinesis, which is known to involve a contractile ring of microfilament's (Schroeder, 1968; Tilney and Marsland, 1969), it is perhaps not surprising that it is sensitive to cytochalasin B. Indeed, one of the major actions of cytochalasin B is to inhibit cytokinesis, with resulting formation of binucleate cells (Carter, 1967).

If the above explanation is correct, one would expect to be able to discern cell-to-cell fusion at one or two sites in the presence of cytochalasin B, and this appears to be the case (C.A.P. unpublished observations). One would also expect to be able to observe cell-to-cell fusion at one or two sites with cells such as P815Y that undergo virus-cell fusion (e.g., Table 3), yet form stable polynucleate cells rather poorly. In other words, the cellular specificity of FFWO may be due to an event *subsequent* to that of establishing cytoplasmic bridges.

The permeability changes and resulting swelling that occur when cells are exposed to paramyxoviruses may thus be the driving force for the formation of stable polynucleate cells. In other words, permeability changes are irrelevant to virus-cell fusion *per se,* but play a role in the subsequent change from dumb-bell shape to double-sized sphere. The fact that permeability changes are inhibited by Ca^{2+} at low virus concentrations, but are independent of Ca^{2+} at the higher concentrations of virus required to achieve FFWO, is in accord with this hypothesis. The inhibiting action of higher concentrations of saccharides on FFWO (Maeda et al., 1977a) may then be explicable in terms of an impairment of swelling due to a change in tonicity of the suspending medium.

Irrespective of whether the effect of cytochalasin B on FFWO stems from its action in preventing the topographic redistribution of integral membrane proteins immediately prior to cell-to-cell fusion or the cell shape changes occurring after the establishment of focal areas of cell-to-cell fusion (or both), it seems clear that membrane-associated cytoskeletal elements participate in cell-to-cell fusion leading to the formation of stable polykaryons. However, the mechanisms underlying the changes in cytoskeletal function that lead to alerations in plasma membrane architecture and cell shape in FFWO have yet to be identified.

Treatment of erythrocytes with neuraminidase has been shown to produce redistribution and aggregation of intramembranous particles within the plasma membrane (Elgsaeter and Branton, 1974) and to also increase their susceptibility to chemically induced fusion (Ahkong et al., 1975b). These observations prompt the obvious question as to whether the neuraminidase present in paramyxovirus particles might facilitate structural changes in cell membranes during FFWO. Although the viral enzyme may facilitate some structural rearrangement, it is clear that neuraminidase alone cannot induce the full degree of membrane perturbation required for fusion to occur (Pasternak and Micklem, 1977). As mentioned earlier, neuraminidase activity is associated with the HN glycoprotein in the viral envelope, yet particles containing HN are unable to induce cell fusion unless the F glycoprotein is also present. Also, neuraminidase activity can be selectively inactivated without impairing the cell fusion activity of the virus (Neurath et al., 1972) and vice versa (Calio et al., 1973). Furthermore, lipid-enveloped viruses such as measles and canine distemper lack neuraminidase yet induce extensive cell fusion (p. 320). Conversely, orthomyxoviruses containing neuraminidase do not induce FFWO.

Evidence obtained recently in several laboratories suggests that alterations in the intracellular concentration of free Ca^{2+} can influence the transmembrane control functions mediated by microtubules and microfilaments with resulting changes in cell surface architecture (Poste and Nicolson, 1976; Schreiner and Unanue, 1976, 1977; Vlodavsky and Sachs, 1977), though the mechanisms underlying these Ca^{2+}-mediated effects are unknown.

Studies on the polymerization/depolymerization of microtubules in cell-free systems in vitro have shown that calcium at concentrations $>10^{-5}$ M induces breakdown of microtubules (see Olmstead and Borisy, 1973; Olmstead et al., 1974; Wilson and Bryant, 1974). If a similar response occurs in intact cells, then relatively small increases in intracellular Ca^{2+} might be sufficient to induce microtubule breakdown. Indirect support for this view comes from studies (Poste and Nicolson, 1976; Vlodavsky and Sachs, 1977) showing that the Ca^{2+} ionophore A23187 (which promotes uptake of Ca^{2+} into cells) can mimic certain of the effects of the microtubule disruptive drugs colchicine and vinblastine which bind directly to microtubules. In terms of transmembrane control of plasma membrane architecture, microtubules are presently considered to act as "anchors" restricting the mobility of integral proteins "linked" to them (Edelman, 1976; Nicolson and Poste, 1976). Thus, breakdown of microtubules produced by an increase in intracellular Ca^{2+} would be expected to facilitate the redistribution of membrane proteins since in the absence of "linkage" to microtubule "anchors" they would be free to diffuse within the membrane and/or be subject to mechanically assisted redistribution by microfilaments which remain "linked" to the proteins.

In contrast to the anchorage function performed by microtubules, the microfilaments, which are known to contain actin and myosin, are believed to act as contractile elements that produce redistribution of integral membrane proteins (Nicolson and Poste, 1976; De Petris, 1977; Schreiner and Uhanue, 1977). Assuming that the actomyosin-containing microfilaments in non-muscle cells function in a similar fashion to comparable molecules in skeletal muscle, entry of Ca^{2+} into cells would be expected to induce microfilament contraction with resulting redistribution of membrane proteins since the latter are no longer restrained by microtubules.

Calcium-mediated alterations in the control of cell surface architecture by cytoplasmic cytoskeletal elements thus offers a possible mechanism whereby membrane proteins could be redistributed during viral fusion as described above, as well as by numerous chemical agents (see chapter by Lucy, this volume).

Circumstantial support for the concept that changes in intracellular calcium can induce redistributuon of integral membrane proteins and promote cell fusion is provided by recent studies on erythrocytes. The mobility of integral proteins in the erythrocyte plasma membrane is restrained by their linkage to a network of spectrin molecules on the inner face of the plasma membrane (Nicolson and Painter, 1973; Elgsaeter et al., 1976). The spectrin network can be precipitated by Ca^{2+} (Elgsaeter et al., 1976) and cationic aggregation of spectrin molecules in erythrocyte ghosts is accompanied by redistribution and aggregation of intramembranous particles (Pinto da Silva and Nicolson, 1974; Elgsaeter et al., 1976; Zakai et al., 1977). Similar Ca^{2+}-mediated alterations in transmembrane control of membrane architecture by spectrin might also explain the finding that incubation of erythrocytes with the calcium ionophore A23187 in the presence of calcium phosphate produces aggregation of intramembranous particles in the plasma membrane (Zakai et al., 1977), which is then followed by cell-to-cell fusion (Ahkong et al., 1975b; Zakai et al., 1977).

The role of Ca^{2+} in virus-mediated membrane fusion has also been studied. Initial experiments revealed only slight increases in the uptake of $^{45}Ca^{2+}$ by Lettré cells exposed to Sendai virus (Micklem and Pasternak, 1977). Certainly, equilibration with external Ca^{2+} (as occurs between internal and external Na^+, Fig. 14) does not readily take place (Pasternak and Micklem, 1974a). Nevertheless the increase in uptake of $^{45}Ca^{2+}$ by virus is significant, particularly at low Ca^{2+} concentration (Fig. 17). A similar effect has been observed by Abraham Loyter (personal communication) and by G. Poste, G. Hewlett, P. Reeve, D. J. Alexander, and H. Watkins (unpublished observations) during FFWO induced by Sendai virus and NDV in human erythrocytes, BHK cells, and 3T3 cells. The latter group of investigators have also detected a significant increase in cellular uptake of $^{45}Ca^{2+}$ in FFWI induced by NDV. In this situation, increased $^{45}Ca^{2+}$ uptake begins only with the onset of fusion. The excellent temporal correlation between $^{45}Ca^{2+}$ uptake and the onset of fusion in FFWI is particularly interesting since fusion does not begin until a minimum of 8 hours after initial virus penetration, and the cells have fully recovered from the permeability alterations induced by initial fusion of virus particles with the cell surface. These data suggest that the insertion of virus envelope proteins into the plasma membrane during virus assembly at the plasma membrane may induce permeability changes directly.

How much of the $^{45}Ca^{2+}$ that is taken up by cells is actually within the cells remains to be established. Use of the Ca^{2+}-sensitive protein, aequorin, has also shown an increase in intracellular Ca^{2+}. Until recently, a major drawback to the use of aequorin to monitor changes in Ca^{2+} levels in mammalian cells cultured in vitro has been the problem of introducing sufficient concentrations of the dye into the cells. The microsurgical injection techniques used to introduce this dye into large cells such as invertebrate axons (Baker et al., 1971) and muscle cells (Ashley and Ridgway, 1970) do not lend themselves to work with much smaller mammalian cells. Poste (unpublished observations) has overcome this problem by encapsulating aequorin inside liposomes which are then fused with the plasma membrane of cultured cells with resulting release of the encapsulated dye direct into the cytosol. Using this approach, accurate measurements of intracellular Ca^{2+} in large numbers of cells can be made. Application of this technique to the study of events in FFWO induced by paramyxoviruses has revealed that significant increases in intracellular Ca^{2+} can be detected within as little as 2 minutes after fusion of virus particles with the cell surface. At the time

348

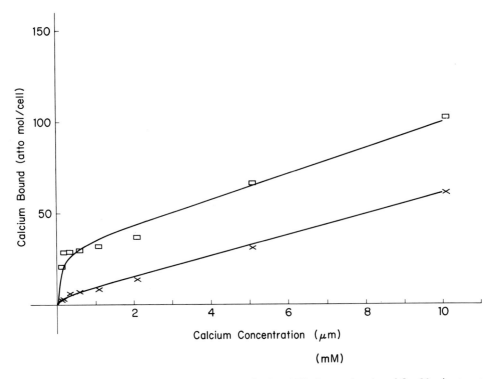

Fig. 17. Uptake of $^{45}Ca^{2+}$ by Lettré cells. Lettré cells (3×10^7/ml) were incubated for 20 minutes at 37°C with $^{45}CaCl_2$ (0.5 μC/ml) and $CaCl_2$ at the indicated concentrations in the presence (□) or absence (×) of Sendai virus (200 HAU/ml). Cells were washed three times in buffer, and radioactivity associated with the pellets measured. Note that this experiment does not distinguish between binding to the cell surface and entry into the cell interior. On the assumption that all the $^{45}Ca^{2+}$ is in the interior of the cell, the intracellular concentration is still <1% of the extracellular concentration under all conditions tested.

cell-to-cell fusion is detected, the intracellular Ca^{2+} concentration has increased from the pre-infection level of 3×10^{-6} M to levels within the range 8×10^{-5} to 4×10^{-5} M.

In short, there is evidence that during viral modification of the cell surface, there is an increase in Ca^{2+} entry into cells. Paramxyoviruses affect not only the *uptake* of $^{45}Ca^{2+}$, but also the *release* of $^{45}Ca^{2+}$ from cells prelabeled with $^{45}Ca^{2+}$; virally stimulated release is not due to displacement of $^{45}Ca^{2+}$ by virus, as the process has the same temperature-dependence as permeability changes (i.e., does not occur at 4°C) (Micklem and Pasternak, 1977). As with uptake, it is not yet clear exactly how much $^{45}Ca^{2+}$ is derived from within cells, and how much from the cell surface. More information is obviously needed concerning the influx and efflux, as well as the binding and release, of Ca^{2+} during virus-induced cell fusion (FFWO and FFWI). For example, it is important to know whether virally stimulated Ca^{2+} movement into and out of cells is indeed a cause of membrane fusion (see section

6.2) or whether it is, like other virally mediated permeability changes (Pasternak and Micklem, 1973, 1974a,b; Fig. 14), a consequence of virus-cell fusion that is important for formation of stable multinucleate cells. That virus-induced cell fusion (both FFWO and FFWI), in common with other examples of membrane fusion, requires Ca^{2+} has been amply documented (see Okada and Murayama, 1966; Poste and Reeve, 1972b; Poste and Allison, 1973; Bächi et al., 1977; Zakai et al., 1977). Sr^{2+} and Ba^{2+} can also substitute for Ca^{2+} with varying degrees of efficiency (Poste and Allison, 1973; Zakai et al., 1977). There are a number of mechanisms by which Ca^{2+} could influence virus-induced cell fusion. These include induction of structural instability and changes in membrane permeability via phase separations in membrane lipids and also alterations in membrane architecture produced by changes in transmembrane cytoskeletal control of protein mobility. These effects are not mutally exclusive and their possible role in membrane fusion will be discussed in the next section.

6.2. Calcium-induced lateral phase separation of membrane phospholipids as a possible mechanism for fusion

Recognition of certain similarities between different forms of membrane fusion (Poste and Allison, 1973; Papahadjopoulos and Poste, 1978) has prompted a number of proposals that there might be a general mechanism common to most, and perhaps all, forms of fusion, though the stimuli responsible for initiating this common reaction almost certainly differ in the highly diverse situations in which membrane fusion occurs.

Earlier proposals for a general mechanism of membrane fusion such as the induction of membrane instability via removal of calcium from membranes (Poste and Allison, 1973) or via an increase in the fluidity of membrane lipids (Ahkong et al., 1975a; Lucy, 1975) do not provide enough detailed information at the molecular level and/or are not consistent with the data now available. To date, the most specific information available concerning the molecular mechanism of membrane fusion has come from recent work on the fusion of model membranes in cell-free systems by Papahadjopoulos and Poste and their colleagues (see chapter 14, this volume). This work has permitted the relative contribution of membrane lipids, proteins, and metal ions in influencing the fusion reaction to be defined in greater detail than has hitherto been possible with natural membranes. The key event in the fusion of model membranes is presently interpreted (Papahadjopoulos and Poste, 1975, 1978; Papahadjopoulos et al., 1977, 1978) as involving the interaction of Ca^{2+} with membrane lipids to produce rigid, crystalline domains of phospholipids created by Ca^{2+}-induced lateral phase separation of acidic phospholipids within a mixed lipid membrane. The formation of such domains is proposed as creating an unstable, highly permeable membrane susceptible to fusion with other similarly perturbed membranes. Although we acknowledge the structural organization of model membranes to be less complex than that of natural membranes, it is instructive to examine whether similar Ca^{2+}-induced phase separation of acidic phospholipids into rigid crystalline domains could serve as a mechanism for fusion of lipid-enveloped viruses with cell membranes and for cell-to-cell fusion induced by viruses.

It is obvious that the close contact (probably less than 15Å separation) of two membranes is an essential prerequisite for fusion. As mentioned earlier, membrane proteins may play an important role in determining whether such contacts can be made, and the high degree of specificity exhibited by most examples of membrane fusion might well be imposed by factors such as protein-protein interactions, which may determine whether the required apposition of membranes can be achieved. However, the finding that cells can be aggregated without concomitant fusion indicates that cell-to-cell contact does not automatically lead to fusion. For example, cells are readily agglutinated by paramyxoviruses containing the F_0 glycoprotein in their envelopes, but neither virus-to-cell nor cell-to-cell fusion occurs. Even when cells are agglutinated by virions containing the F protein (including situations where cell agglutination is augmented by ligands such as PHA or polylysine) fusion does not occur unless specific cations are present. Thus, cell agglutination by virions containing the F glycoprotein can take place in media containing a variety of divalent cations, but virus-to-cell and cell-to-cell fusion are found only in the presence of Ca^{2+}, Ba^{2+}, or Sr^{2+} and cations such as Mg^{2+} at physiologic concentrations do not support fusion (Poste and Reeve, 1972b; Poste and Allison, 1973; Zakai et al., 1976, 1977). Similar cation specificity has been demonstrated in muscle cell fusion (Schudt et al., 1973, 1976) and in the fusion of liposomes (see chapter by Papahadjopoulos, this volume).

These observations suggest that membranes have a very high structural stability and that fusion requires an additional step in which the stability of apposed membranes is altered to render them susceptible to fusion. If recent data on events in the fusion of lipid vesicles can be validly extrapolated to natural membranes, then this additional step might involve a Ca^{2+}-induced lateral phase separation in membrane lipids and formation of rigid crystalline domains of acidic phospholipids within an otherwise "fluid" membrane. (Papahadjopoulos and Poste, 1978; Papahadjopoulos et al., 1977, 1978).

Papahadjopoulos and Poste have proposed that membrane fusion is initiated between closely apposed membranes at the boundaries between crystalline domains of acidic phospholipids and the surrounding noncrystalline lipid. Since these authors have recently presented an extensive review showing how Ca^{2+} mediated lateral phase separation in membrane lipids could serve as a unifying mechanism in all forms of membrane fusion, only an abbreviated summary of this hypothesis will be given here. These investigators propose that membrane fusion is initiated between closely apposed membranes at the boundaries between Ca^{2+}-induced crystalline domains of acidic phospholipids and the surrounding noncrystalline lipid. Such boundaries represent structurally unstable points of high free energy and thus offer focal points for mixing molecules from the apposed membranes (Papahadjopoulos et al., 1977). A further potentially important feature of Ca^{2+}-induced phase changes that may be relevant to fusion concerns the local heat of crystallization which would be expected to be released by interaction of Ca^{2+} with acidic phospholipids. Heat evolved during

Ca²⁺-induced crystallization of lipids such as phosphatidylserine could produce local heating in certain regions of closely apposed membranes which would increase the mixing rate of lipid molecules between membranes, especially at domain boundaries.

Ca^{2+} can therefore be envisaged as playing a central role in membrane fusion. First, Ca^{2+} will promote the close apposition of adjacent membranes by enhancing electrostatic interactions between them. Second, Ca^{2+} will induce destabilization of the apposed membranes by inducing the formation of domains of crystalline acidic phospholipids whose boundaries represent sites at which fusion can occur. Finally, in cell-to-cell fusion, the increased membrane permeability resulting from the establishment of acidic phospholipid domains (Papahadjopoulos et. al., 1977) would permit entry of Ca^{2+} into cells with resulting changes in membrane architecture via Ca^{2+}-mediated alterations in transmembrane control of protein mobility by cytoskeletal elements (section 6.1).

If Ca^{2+}-induced phase separation is indeed the key event underlying membrane fusion, then the fusion susceptibility of any given membrane will be determined by (1) the content of acidic phospholipids in the membrane and their distribution within the bilayer, and (2) by the concentration of Ca^{2+} in the vicinity of the membrane.

The distribution of acidic phospholipids in the two halves (monolayers) of the lipid bilayer would be expected to influence whether a membrane would be able to fuse with other membranes from only one side or from both sides. For example, in the most extreme situation, insufficient acidic phospholipids might be present in either the inner or outer half of the bilayer so that Ca^{2+}-induced phase separation of acidic phospholipids could not occur and the membrane would remain resistant to fusion occurring at that particular side of the membrane.

Since acidic phospholipids vary in their ability to undergo Ca^{2+}-induced phase separation, the fusion susceptibility of any given membrane may be regulated not only by the absolute content of acidic phospholipids, but also by the distribution within the bilayer of those acidic phospholipid species that undergo phase separation in the presence of Ca^{2+}. For example, phosphatidylserine (PS) and phosphatidic acid (PA) undergo lateral phase separation to form rigid crystalline domains in mixed lipid membranes at Ca^{2+} concentrations similar to those found under physiological conditions. In contrast, phosphatidylglycerol and phosphatidylinositol do not undergo lateral phase separation at Ca^{2+} concentrations in the physiologic range. Thus, the content of PS (and also perhaps PA) would be a major factor in determining the fusion susceptibility of a particular membrane, while preferential distribution of PS in either the outer or inner half of the bilayer would further dictate that different sides of the same membrane could vary significantly in their fusion susceptibility.

This point is particularly pertinent to the fusion behavior of the plasma membrane. Studies of plasma membrane lipid composition in mammalian erythrocytes have revealed that PS is present almost exclusively in the inner (cytoplasmic) half of the bilayer, while the outer (external) lipid monolayer is composed predominantly of neutral phospholipids (phosphatidylcholine and sphingomyelin) (Rothman and Lenard, 1977). Although it is not known to what extent similar lipid asymmetry is present in the plasma membrane of other cells, such asymmetry, if widespread, would have major implications for the fusion behavior of the plasma membrane. The scheme outlined above predicts the relative absence of PS in the outer lipid monolayer of the plasma membrane would dictate that Ca^{2+} in the external medium could not induce lateral phase separation of acidic phospholipids into fusion-susceptible domains. The membrane would thus be resistant to fusion at the outer membrane face.

The fusion susceptibility of the two faces of the plasma membrane would change, however, if the distribution of acidic phospholipids within the bilayer were altered and/or the concentration of Ca^{2+} in the cytoplasm was increased. For example, if the content of acidic phospholipids in the outer lipid monolayer of the plasma membrane were to increase as a result of structural rearrangements within the bilayer, then Ca^{2+} in the external medium would induce the formation of crystalline acidic phospholipid domains and the membrane would become susceptible to fusion with other similarly perturbed membranes. This proposal that fusion initiated at the outer face of the plasma membrane might require enrichment of the outer lipid monolayer with acidic phospholipids will be examined further below.

The major assumption made in the scheme above is that the degree of trans-bilayer asymmetry in the distribution of acidic phospholipids is so extreme that the outer face of the plasma membrane is inherently resistant to fusion, and that fusion can occur only when the content of acidic phospholipids in the outer monolayer is increased via some form of redistribution of acidic phospholipids from the inner to the outer monolayer or via the transfer of acidic phospholipids from adjacent membranes by exchange processes. In fact, it has recently been proposed that the physiological role of phospholipid exchange proteins may be to promote membrane fusion (Pasternak, 1977; Quinn and Pasternak, 1978).

An alternative possibility, however, is that although lipid asymmetry may exist, the difference in the distribution of acidic phospholipids between the outer and inner monolayers is not as extreme as that suggested above and that acidic phospholipids are present in the outer monolayer in sufficient amounts to undergo Ca^{2+}-induced lateral phase separation. For example, acidic phospholipids are now known to be present as thin shells surrounding certain integral membrane proteins (Warren et al., 1975a,b; Armitage et al., 1977; Boggs et al., 1977) and it seems likely that at least some of these molecules would be present in the outer monolayer. Consequently, it may well be that sufficient acidic phospholipids are already present in the outer monolayer of the plasma membrane to permit their separation into domains by Ca^{2+} without the need for enrichment of the acidic phospholipid content in the outer monolayer.

Since the Ca^{2+} concentration of the external medium (1-3 mM) is sufficient to induce lateral phase separation of acidic phospholipids, this would mean that the plasma membrane would already be in a "fusion-susceptible" state. At first sight, this is difficult to reconcile with the apparent lack of spontaneous fusion processes involving the plasma membrane. However, the fusion resistance of the plasma membrane could be more apparent than real and that the low frequency of spontaneous fusion is determined by some additional restraint. For example, simple steric factors could be operating to frustrate the establishment of the required initial close contact between the lipid bilayers of the plasma membrane and other membranes. In cell-to-cell fusion, this type of steric hindrance could perhaps be mediated by the various glycoproteins and glycosaminoglycans present in the "cell coat," which would provide a "barrier" to a surface projection of one cell readily achieving direct apposition of its bilayer with that on another cell. It is interesting that enzymic treatment of cells to remove surface glycoproteins and/or glycosaminoglycans renders cells with a fusion-resistant phenotype susceptible to fusion by viruses (Poste, 1972). Similarly, enzymic removal of glycoprotein "coating factors" from the surface of uncapacitated sperm enables them to fuse prematurely without undergoing capacitation (Gwatkin, 1976). Also, the observation (Diacumakos, 1973) that cultured mammalian somatic cells will fuse spontaneously if the cells are merely "brought together" by micromanipulation (in Ca^{2+}-containing medium) suggests that the plasma membrane may indeed be in a fusion-susceptible state.

That most membranes do not fuse under normal conditions is clearly necessary to prevent uncontrolled spontaneous fusion of cells within tissues and also for ensuring that the various membrane-bound organelles retain their structural identity, thus enabling different functional activities to be localized within specific compartments of the cell. In the Ca^{2+}-induced phase separation hypothesis for fusion advanced by Papahadjopoulos and Poste, fusion is considered as not occurring routinely because of one or more of the following conditions: (1) steric hindrance, which frustrates close apposition between two "fusion-susceptible" membranes containing preexisting Ca^{2+}-induced domains of acidic phospholipids; or (2) preexisting asymmetry in the distribution of acidic phospholipids within the membrane, which dictate that insufficient acidic phospholipids are present to undergo phase separation; or (3) the Ca^{2+} concentration in the vicinity of the membrane is insufficient to induce phase separation even if sufficient acidic phospholipids are present. Fusion can be initiated, however, by: (1) removal of steric restraints to the close apposition of two "fusion-susceptible" membranes (i.e., the membranes contain preexisting Ca^{2+}-phase-separated domains); or (2) changes in the distribution of acidic phospholipids within the bilayer that permit the existing concentration of Ca^{2+} in the vicinity of the membrane to induce separation of acidic phospholipids.

In the remainder of this chapter, we will attempt to define how these alternative pathways for generating the formation of domains of acidic phospholipids might operate in virus-to-cell and cell-to-cell fusion. As discussed in several sections of this review, an important functional separation can be made in the case of FFWO induced by paramyxoviruses between the initial fusion of the viral envelope with the cellular plasma membrane and the subsequent step of cell-to-cell fusion. This can be demonstrated experimentally by using cytochalasin B, which permits initial fusion of the virus to take place but blocks subsequent cell-to-cell fusion. This suggests that the latter step requires participation of the membrane-associated microfilament system. As proposed earlier, we envisage that microfilaments are responsible for producing redistribution of integral membrane proteins over a sufficiently large area of the plasma membrane so that the subsequent step of cell-to-cell fusion occurs through the interaction of lipid bilayers on adjacent cells at regions depleted of membrane proteins.

What is the mechanism by which the virus fuses with the plasma membrane? We consider that the envelope of paramyxoviruses could be in a preexisting "fusion-susceptible" state. This proposal is based on recent evidence indicating no pronounced asymmetry in the distribution of acidic phospholipids between the outer and inner monolayers of the virus envelope (Rothman and Lenard, 1977) and data showing that the virus envelope is enriched in acidic phospholipids compared with host cell membranes (Semmel et al., 1975). In the latter studies it was shown that the plasma membranes of chick allantoic cells infected with NDV contain less sphingomyelin and more PS than uninfected cells, while virions released from infected cells contain more PS, PE, and sphingomyelin but less phosphatidylcholine than the plasma membrane of the infected cell. Semmel and colleagues interpreted this data to indicate that virions might bud preferentially from regions of the host cell membrane enriched in these particular components (also see Rifkin and Quigley; 1974, for a similar proposal for enveloped viruses other than paramyxoviruses).

Thus the lack of asymmetry in the distribution of acidic phospholipids in the envelope, together with an increased content of acidic phospholipids, would dictate that the virus envelope would probably contain preexisting domains of acidic phospholipids and would therefore be in an inherently "fusion-susceptible" state (assuming that the Ca^{2+} concentration in the environment is sufficient to induce phase separation of acidic phospholipids). The question remains, however, as to how the virus can render the closely apposed cell plasma membrane fusion susceptible.

Recent data obtained with electron-spin resonance probes (Maeda et al., 1977b) may shed some light on this important question. It was observed that during incubation of HVJ (Sendai) virions with erythrocytes considerable exchange diffusion of phospholipid molecules occurred between the virus envelope and the cellular plasma membrane without accompanying fusion of the two membranes. This was shown clearly by the fact that lipid molecules, but not proteins, were transferred between the two membranes (Maeda et al., 1977b). It is therefore possible that during close contact of the viral envelope with the cellular plasma membrane, the outer monolayer of the plasma membrane is enriched with acidic phospholipids that were originally present in the external monolayer of the virus, Soluble proteins that enhance the exchange of phospholipid molecules have been isolated from various sources, and although some of them are specific for certain head group configurations, others are nonspecific (Wirtz et al., 1976; Bloj and Zilversmit, 1977; Kader, 1977). Recently, Loyter and his colleagues (Zakai et al., 1977; A. Loyter, personal communication) have also shown that the phospholipids exposed at the outer surface of the plasma membrane are altered as a result of exposure to Sendai virus. These investigators reported that following agglutination of human erythrocytes at 37 °C by Sendai virus, there is a marked increase in the susceptibility of plasma membrane lipids to hydrolysis by phospholipase C (from *Bacillus cereus*). Loyter (personal communication) believes that the content of acidic phospholipid accessible at the outer surface of the membrane is increased. Thus, assuming that the lipids rendered accessible to phospholipase C would also be accessible to Ca^{2+}, this enrichment of acidic phospholipids would be expected to facilitate Ca^{2+}-induced phase separation with resulting destabilization of the membrane, which would also become susceptible to fusion.

It is therefore possible to envisage that the content of acidic phospholipids in the outer monolayer of the plasma membrane could be enriched sufficiently to permit Ca^{2+}-induced phase separation to take place either by (1) enhanced exchange diffusion or (2) structural rearrangements within the lipid bilayer, or both. Since Ca^{2+} is present in the extracellular medium at concentrations in the range of 1 to 4mM, phase separations would be expected to occur in both the virus envelope and the cellular plasma membrane and thus initiate fusion.

Alternatively, as mentioned earlier, the plasma membrane may contain preexisting small domains of acidic phospholipids. This suggests that virus particles which fuse with the plasma membrane could interact specifically with these domains. In this situation, both the virus envelope and the cellular plasma membrane are thought to possess preexisting domains of acidic phospholipids, so that fusion between them would take place *automatically* if sufficiently close contact between the bilayers of the two membranes could be achieved with accompanying apposition of domain boundaries. Because of their small size virus particles would be expected to encounter minimum repulsive energy in achieving this type of close contact with the plasma membrane (Poste, 1972).

It is known that the ability of paramyxoviruses to fuse with the plasma membrane is influenced by the nature of the glycoproteins present in the virus envelope (section 5). Virions containing the F glycoprotein are able to fuse with the plasma membrane, but virus particles containing the F_0 glycoprotein cannot. It has previously been assumed that the F protein is *directly responsible for inducing fusion* (i.e., the F glycoprotein is somehow able to induce the membrane perturbation(s) required to destabilize the membrane and permit fusion to occur). It is equally possible, however, that the role of the F and F_0 proteins in determining whether fusion takes place does not involve a direct effect on fusion *per se*, but influences the preceding step of virus attachment which determines the likelihood of achieving close contact between the bilayers of the virus envelope and the plasma membrane required for fusion.

The possible role of virus envelope glycoproteins in virus attachment is indicated by the finding that paramyxoviruses with F_0 glycoprotein bind only to neuraminidase-sensitive moieties usually associated with cell surface glycoproteins, while particles containing the F protein can bind to both the above receptors and to an unidentified class of receptors found on neuraminidase-treated cells (Poste and Waterson, 1975). It has been suggested that the latter receptors are neuraminidase-resistant sialic acid moieties associated with glycolipids, since infective F-containing paramyxoviruses are able to bind to ganglioside-containing lipid vesicles (Haywood, 1975). The ability to attach to glycolipid receptors would have the advantage of bringing the virus envelope into much closer contact with the lipid bilayer than if attachment were to receptors on glycoproteins. Thus, during virus attachment virus particles are envisaged as first interacting with sialic acid receptors associated with glycoproteins, since these are not only more numerous than the glycolipids receptors but also extend further from the bilayer, thus increasing the likelihood of their interaction with incoming virus particles before the glycolipid receptors. Both F- and F_0-containing virions are able to attach to glycoprotein receptors (presumably with equal efficiency). However, to achieve the close apposition of the virus envelope and the plasma membrane needed for fusion to occur, the particles must elute from the glycoprotein receptors (mediated by viral neuraminidase?) and bind to the glycolipid receptors. Once eluted, only virions containing the F glycoprotein might be able to attach to the latter receptors and go on to fuse with the plasma membrane, and virions containing F_0 would be unable to bind to these receptors and thus fail to fuse with the plasma membrane. Support for this proposal comes from recent studies showing that NDV virions containing the F glycoprotein can bind to and fuse with liposomes containing gangliosides, but virions containing F_0 cannot bind or fuse. (G. Poste, unpublished observations).

This proposal that envelope glycoproteins influence virus attachment rather than fusion is a departure from the generally held belief (albeit without experimental proof!) that the F protein is *directly responsible for* fusion of the virus envelope with the plasma membrane. Thus, in our present state of knowledge, it might be more appropriate to say merely that presence of the F protein is "associated with" the ability of the virus to fuse with cells.

Irrespective of whether fusion of enveloped viruses with the plasma membrane requires acidic phospholipid enrichment in the outer monolayer of the plasma membrane or proceeds automatically because both the virus envelope and the plasma membrane contain preexisting domains of acidic phospholipids, fusion of numerous virus particles with the plasma membrane would be expected to increase the overall acidic phospholipid content of the outer monolayer of the plasma membrane. This would create more acidic phospholipid domains which, in turn, would be expected to further increase plasma membrane permeability because of the enhanced permeability of the domain boundaries. As discussed in detail earlier, increased entry of Ca^{2+} into the cytoplasm under these conditions would be expected to produce a number of further changes including: (1) breakdown of the microtubular "anchoring' proteins which would facilitate the redistribution of integral membrane proteins by free diffusion and/or mechanically assisted processes involving the microfilament system; and (2) activation of the actomyosin microfilament system, which would redistribute the "unanchored" membrane proteins. These two events would be expected to produce aggregation of integral proteins and thus create large areas in the bilayer free of integral membrane proteins. Such protein-depleted regions would then act as sites for subsequent fusion with a similarly perturbed plasma membrane on an adjacent cell. This could proceed via: (1) apposition of similar protein-depleted bilayers at sites of simultaneous virus penetration containing acidic phospholipid domain boundaries; or (2) modification of the plasma membrane of an adjacent uninfected cell by viral glycoprotein integrated within the plasma membrane by the same mechanism responsible for initial fusion of virus particles with the plasma membrane. Virus-induced FFWI could also be viewed as analogous to the events just described for FFWO, except that newly synthesized viral F glycoproteins associated with FFWI are inserted into the plasma membrane after transport from the smooth endoplasmic reticulum rather than by penetration from the outside.

Our intention in this final section has been to evaluate critically the extent to which present theories for the mechanism of membrane fusion apply to events in the fusion of enveloped viruses with cellular membranes and to cell-to-cell fusion induced by viruses. We believe that of the different hypotheses advanced in recent years to explain membrane fusion behavior, the proposal that Ca^{2+}-induced phase separations in membrane lipids can trigger membrane instability and initiate the series of membrane changes that culminate in fusion offers the best explanation of events in virus-induced membrane fusion. We recognize that in attempting to explain virus-induced cell fusion in terms of Ca^{2+}-induced changes in membrane properties we have often had to use data obtained in systems of varying complexity, and the validity of certain of the extrapolations made here remains to be established. In addition, considerable speculation must inevitably accompany a number of the issues discussed in this chapter. We hope, however, that by identifying shortcomings in our present knowledge and by focusing attention on unanswered questions this chapter may serve to stimulate further research into this interesting subject.

Acknowledgments

The personal research cited in this article was supported by grants from the Agricultural Research Council of Great Britain, grants CA-13393 and CA-18260 from the National Institutes of Health of the United States of America (G.P.), the Cancer Research Campaign of Great Britain and the Lawson Tait Trust for Medical Research (C.A.P.). We also thank Dr. S. Knutton for permission to cite unpublished observations.

References

Ahkong, Q. F., Fisher, D., Tampion, W. and Lucy, J. A. (1975a) Mechanisms of cell fusion. Nature (London) 253, 194–195.
Ahkong, Q. F., Tampion, W. and Lucy, J. A. (1975b) Promotion of cell fusion by divalent cation ionophores. Nature (London) 256, 208–209.
Alexander, D. J., Hewlett, G., Reeve, P. and Poste, G. (1973) Studies on the cytopathic effects of Newcastle disease virus: the cytopathogenicity of strain Herts 33 in five cell types. J. Gen. Virol. 21, 323–337.

Allan, D. and Michell, R. H. (1976) A possible role for 1, 2-diacylglycerol in fusion of erythrocytes by Sendai virus. Biochem. Soc. Trans. 4, 253.

Apostolov, K. and Almeida, J. D. (1972) Interaction of Sendai virus (HV) with human erythrocytes: a morphological study of hemolysis and cell fusion, J. Gen. Virol. 15, 227–234.

Apostolov, K. and Poste, G. (1972) Interaction of Sendai virus with human erythrocytes: a system for the study of membrane fusion. Microbios 6, 247–261.

Apostolov, K. and Waterson, A. P. (1975) Studies on the mechanism of haemolysis by paramyxoviruses. In:Negative Strand Viruses (Barry, R. D. and Mahy, B. W. J., eds.) pp. 799–822, Academic Press, London.

Appels, R. and Ringertz, N. R. (1975) Chemical and structural changes within chick erythrocyte nuclei introduced into mammalian cells by cell fusion. Curr. Top. Dev. Biol. 9, 137–166.

Armitage, I. M., Shapiro, D. L., Furthmayr, H. and Marchesi, V. T. (1977) [31]NMR evidence for polyphosphoinositide associated with the hydrophobic segment of glycophorin A. Biochem. 16, 1317–1320.

Ashley, C. C. and Ridgway, E. B. (1970) On the relationships between membrane potential, calcium transient and tension in single barnacle muscle fibres. J. Physiol. 209, 105–130.

Avruch, J. and Wallach, D. F. H. (1971) Preparation and properties of plasma membrane and endoplasmic reticulum fragments from isolated rat fat cells. Biochim. Biophys. Acta 233, 334.

Bächi, T. and Howe, C. (1972) Fusion of erythrocytes by Sendai virus studied by electron microscopy. Proc. Soc. Exp. Biol. Med. 141, 141–149.

Bächi, T., Aguet, M. and Howe, C. (1978) Fusion of erythrocytes by Sendai virus studied by immuno-freezing-etching. J. Virol. 11, 1004–1012.

Bächi, T., Deas, J. E. and Howe, C. (1977) Virus-erythrocyte membrane interactions. In:Virus Infection and the Cell Surface (Poste, G. and Nicolson, G. L., eds.), Cell Surface Reviews, vol. 2, pp. 83–127, Elsevier/North-Holland, Amsterdam.

Baker, P. F., Hodgkin, A. L. and Ridgway, E. B. (1971) Depolarization and calcium entry in squid giant axons. J. Physiol. 218, 709–755.

Baker, R. F. (1972) Fusion of human red blood cell membranes. J. Cell Biol. 53, 244–249.

Barbanti-Brodano, G., Possati, L., La Placa, M. and Koprowski, H. (1971) Inactivation of polykaryocytogenic and hemolytic activities of Sendai virus by phospholipase B (lysolecithinase). J. Virol. 8, 796–800.

Bergmeyer, H. U. (1974) Methods in Enzymic Analysis, vol. 1, p. 481, Academic Press, New York.

Blair, C. D. and Robinson, W. S. (1968) Replication of Sendai virus. 1. Composition of the viral RNA and virus specific RNA synthesis with Newcastle disease virus. Virology 35, 537–549.

Bloj, B. and Zilversmit, D. B. (1977) Rat liver proteins capable of transfering PE. J. Biol. Chem. 252, 1613–1619.

Blough, H. A. and Lawson, D. E. M. (1968) The lipids of paramyxoviruses: a comparative study of Sendai and Newcastle disease viruses. Virology 36, 286–292.

Boggs, J., Wood, D. D., Moscarello, M. A. and Papahadjopoulos, D. (1977) Lipid phase separation induced by a hydrophobic protein in PS-PC vesicles. Biochemistry, 16, 2325–2329.

Bose, H. R. and Sagik, B. P. (1970) The virus envelope in cell attachment. J. Gen. Virol. 9, 159–161.

Bratt, M. A. and Gallaher, W. R. (1969) Preliminary analysis of the requirements for fusion from within and fusion from without by Newcastle disease virus. Proc. Nat. Acad. Sci. U.S.A. 64, 536–540.

Bretscher, M. S. (1972) Asymmetrical lipid bilayer structure for biological membranes. Nature (New Biol.) 236, 11–12.

Burnet, F. M. and Lind, P. E. (1950) Hemolysis by Newcastle disease virus. II. General character of hemolytic action. Aust. J. Exp. Biol. Med. 38, 129–150.

Calio, R., Calio, N. and Cassone, A. (1973) Interaction of Sendai virus with human erythrocytes. I. The ultrastructure of untreated and pyridine-treated virions and their relationship with erythrocyte surface during hemagglutination. Boll. Inst. Sieroter. Milan 52, 208–217.

Carter, S. B. (1967) Effects of cytochalasins on mammalian cells. Nature 213, 261–264.

Cascardo, M. R. and Karzon, D. T. (1965) Measles virus giant cell inducing factor (fusion factor). Virology 26, 311–325.

Cassone, A., Calio, R. and Pesce, C. D. (1973) Interaction of Sendai virus with human erythrocytes. II. The fusion reactions. Boll. Inst. Sieroter. Milan 52, 218–223.

Chang, A. and Metz, D. H. (1976) Further investigations on the mode of entry of vaccinia virus into cells. J. Gen. Virol. 32, 275–282.

Chany-Fournier, F., Chany, C. and LaFay, F. (1977) Mechanism of polykaryocyte induction by vesicular stomatitis virus in rat XC cells. J. Gen. Virol. 34, 305–314.

Chanock, R. M. (1956) Association of a new type of cytopathogenic myxovirus with infantile croup. J. Exp. Med. 104, 555–567.

Choppin, F. W. and Compans, R. W. (1975) Reproduction in paramyxo viruses. In: Comprehensive Virology, Vol. V (Franekel-Contrat, H. and Wagner, R. R., eds.) pp. 95–178, Plenum, New York.

Dales, S. and Pons, M. W. (1976) Penetration of influenza examined by means of virus aggregates. Virology 69, 278–286.

Davidson, R. L. (1969) Regulation of melanin synthesis in mammalian cells, as studied by somatic hybridization. III. A method for increasing the frequency of cell fusion. Exp. Cell Res. 55, 424–426.

Davidson, R. L. (1974) Gene expression in somatic cell hybrids. Ann. Rev. Genet. 8, 195–218.

De Boer, E. and Loyter, A. (1971) Formation of polynucleate avian erythrocytes by polylysine and phospholipase C. FEBS Lett. 15, 325–327.

De Petris, S. (1977) Distribution and mobility of plasma membrane components on lymphocytes. In:Dynamic Aspects of Cell Surface Organization (Poste, G. and Nicolson, G. L., eds), Cell Surface Reviews, vol. 3, pp. 643–728. Elsevier/North-Holland, Amsterdam.

Diacumakos, E. G. (1973) Methods for micromanipulation of human somatic cells in culture. In: Methods in Cell Biology. (Prescott, D., ed.), pp. 39–47. Academic Press, New York.

Diringer, H. and Rott, R. (1976) Metabolism of preexisting lipids in baby hamster kidney cells during Newcastle disease virus-induced fusion from within. Eur. J. Biochem. 65, 155–160.

Dourmashkin, R. A. and Tyrrell, D. A. J. (1974) Electron microscopic observations on the entry of influenza virus into susceptible cells. J. Gen. Virol. 24, 129–141.

Durand, D. P., Granlund, D. and Eckhart, T. (1975) Fate of paramyxovirus membrane components following cell infection. In:Negative Strand Viruses (Mahy, B. W. J. and Barry, R. D., eds.) vol. 2, pp. 835–841. Academic Press, London.

Edelman, G. L. (1976) Surface modulation in cell recognition and cell growth. Science 192, 218–226.

Elgsaeter, A. and Branton, D. (1974) Intramembrane particle aggregation in erythrocyte ghosts. I. The effect of protein removal. J. Cell Biol. 63, 1018–1036.

Elgsaeter, A., Shotton, D. M. and Branton, D. (1976) Intramembrane particle aggregation in erythrocyte ghosts. II. The influence of spectrin aggregation. Biochim. Biophys. Acta 426, 101–122.

Elsbach, P., Holmes, K. V. and Choppin, P. W. (1969) Metabolism of lecithin and virus-induced cell fusion. Proc. Soc. Exp. Biol. Med. 130, 903–908.

Enders, J. F. and Peebles, T. C. (1954) Propagation in tissue cultures of cytopathogenic agents from patients with measles. Proc. Soc. Exp. Biol. Med. 86, 277–286.

Falke, D., Schieser, H. G. and Stoffel, W. (1967) Lipoid-analysen bei reisenzellbilund durch herpes virus hominis. Z. Naturforsch B22, 1360–1362.

Fowler, V. and Branton, D. (1977) Lateral mobility of human erythrocyte integral membrane proteins. Nature 268, 23–26.

Fuchs, P. and Giberman, E. (1973) Enhancement of potassium influx in baby hamster kidney cells and chicken erythrocytes during adsorption of parainfluenza (Sendai) virus. FEBS Lett. 31, 127–130.

Fuchs, P., Spiegelstein, M., Haimsohn, M., Gitelman, J. and Kohn, A. (1978) Early changes in the membrane of HeLa cells adsorbing Sendai virus under conditions of fusion. Exp. Cell Res. In press.

Fukai, K. and Suzuki, T. (1955) On the characteristics of a newly isolated hemagglutinating virus from mice. Med. J. Osaka Univ. 6, 1–15.

Gallaher, W. R. and Bratt, M. A. (1972) Temperature-dependent inhibition of fusion from without. J. Virol. 10, 159–161.

Gallaher, W. R. and Bratt, M. A. (1974) Conditional dependence of fusion from within and other cell membrane alterations by Newcastle disease virus. J. Virol. 14, 813–820.

Gallaher, W. R. and Howe, C. (1976) Identification of receptors for animal viruses. Immunol. Commun. 5, 535–552.

Gallaher, W. R., Levitan, D. B. and Blough, H. A. (1973) Effect of 2-deoxy-D-glucose on cell fusion induced by Newcastle disease.

Giles, R. E. and Ruddle, F. H. (1973) Production of Sendai virus for cell fusion. In Vitro 9, 103–107.

Gordon, S. (1975) Cell fusion and some subcellular properties of heterokaryons and hybrids. J. Cell Biol. 67, 257–280.

Granados, R. R. (1973) Entry of an insect poxvirus by fusion of the virus envelope with the host cell membrane. Virology 52, 305–309.

Gwatkin, R. B. L. (1976) Fertilization. In: The Cell Surface in Animal Embryogenesis and Development (Poste, G. and Nicolson, G. L., eds.) Cell Surface Reviews, vol. 1, pp. 1–54, Elsevier-North-Holland, Amsterdam.

Harris, H. (1970) Cell Fusion: The Dunham Lectures. Oxford University Press, Oxford.

Harris, H. and Watkins, J. F. (1965) Hybrid cells derived from mouse and man: artificial heterokaryons of mammalian cells from different species. Nature (London) 205, 640–646.

Harris, H., Watkins, J. F., Ford, C. E. and Schoefl G. I. (1966) Artificial heterokaryons of animal cells from different species. J. Cell Sci. 1, 1-30.

Harter, D. H. and Choppin, P. W. (1967) Cell fusing activity of Visna virus particles. Virology 31, 279-288.

Haywood, A. M. (1975) "Phagocytosis" of Sendai virus by model membranes. J. Gen. Virol. 29, 63–68.

Heine, J. W. and Schnaitman, C. A. (1969) Fusion of vesicular stomatitis virus with the cytoplasmic membrane of L cells. J. Virol. 3, 619–622.

Henle, G., Deinhardt, F. and Girardi, A. (1954) Cytolytic effects of mumps virus in tissue cultures of epithelial cells. Proc. Soc. Exp. Biol. Med. 87, 386–393.

Hightower, L. E., Morrison, T. G. and Bratt, M. A. (1975) Relationships among the polypeptides of Newcastle disease virus. J. Virol. 16, 1599–1607.

Holmes, K. V. and Choppin, P. W. (1966) On the role of the response of the cell membrane in determining virus virulence. Contrasting effects of the parainfluenza virus SV5 in two cell types. J. Exp. Med. 124, 501–520.

Homma, M. (1971) Trypsin action on the growth of Sendai virus in tissue culture cells. I. Restoration of the infectivity for L cells by direct action on trypsin on L cell-borne Sendai virus. J. Virol. 8, 619–629.

Homma, M. (1972) Trypsin action on the growth of Sendi Virus in tissue culture cells. II. Restoration of the hemolytic activity of L cell-borne Sendai virus by trypsin. J. Virol. 9, 829–835.

Homma, M., and Ouchi, M. (1973) Trypsin action on the growth of Sendai virus in tissue culture cells. III. Structural differences of Sendai viruses grown in eggs and tissue culture cells. J. Virol. 12, 1457–1465.

Homma, M. and Tamagawa, S. (1973) Restoration of fusion activity of L-cell-borne Sendai virus by trypsin. J. Gen. Virol. 19, 423–426.

Homma, M., Shimizu, K., Shimizu, Y. K. and Ishida, N. (1976) On the study of Sendai virus hemolysis. I. Complete Sendai virus lacking in hemolytic activity. Virology 71, 41–47.

Hosaka, Y. (1962) Characteristics of growth of HVJ in PS cells. Biken's J. 5, 121–125.

Hosaka, Y. (1968) Isolation and structure of the nucleocapsid of HVJ. Virology 35, 445–457.

Hosaka, Y. (1970) Biological activities of sonically treated Sendai virus. J. Gen. Virol. 8, 43–54.

Hosaka, Y. (1975) Artificial assembly of active envelope particles of HVJ (Sendai virus) In: Negative Strand Viruses (Barry, R. D. and Mahy, B. W. J., eds.) vol. 2, pp. 885–903. Academic Press, London.

Hosaka, Y. and Shimizu, Y. K. (1972a) Artificial assembly of envelope particles of Sendai virus (HVJ). I. Assembly of hemolytic and fusion factors from envelopes solubilized by Nonidet P40. Virology 49, 627–639.

Hosaka, Y. and Shimizu, K. (1972b) Artificial assembly of envelope particles of HVJ (Sendai virus) II. Lipid components for formation of the active hemolysin. Virology 49, 640–646.

Hosaka, Y. and Shimizu, K. (1977) Cell fusion by Sendai virus. In: Virus Infection and the Cell Surface (Poste, G. and Nicolson, G. L. eds.), Cell Surface Reviews, Vol. 2, pp. 129–155, Elsevier/North-Holland, Amsterdam.

Hosaka, Y., Semba, T. and Fukai, K. (1974) Artificial assembly of envelope particles of HVJ (Sendai virus). Fusion activity of envelope particles. J. Gen. Virol. 25, 391–404.

Howe, C. and Bachi, T. (1973) Localization of erythrocyte membrane antigens by immune electron microscopy. Exp. Cell Res. 76, 321–332.

Howe, C. and Morgan, C. (1969) Interaction between Sendai virus and human erythrocytes. J. Virol. 3, 70–81.

Ho Yun-De (1962) Investigations into syncytium formation in cultures of stable cell lines infected with parainfluenza I virus. II. Morphology of syncytium formation in tissue cultures and a hypothesis concerning its mechanism. Acta Virol. 6, 202–206.

Ishida, N. and Homma, M. (1960) A variant Sendai virus, infectious to egg embryos but not to L cells. Tohoku J. Exp. Med. 73, 56–69.

Ishida, N. and Homma, M. (1961) Host controlled variation observed with Sendai virus grown in mouse fibroblast (L) cells. Virology 14, 486–488.

Kader, J. C. (1977) Exchange of phospholipids between membranes. In: Dynamic Aspects of Cell Surface Organization (Poste, G. and Nicolson, G. L., eds.), Cell Surface Reviews, vol. 3, pp. 127–204. Elsevier/North-Holland, Amsterdam.

Kaku, H. and Kamahora, J. (1964) Giant cell formation in L cells infected with active vaccinia virus. Bikens J. 6, 299–315.

Katzman, J. and Wilson, D. E. (1974) Newcastle disease virus-induced plasma membrane damage. J. Gen. Virol. 24, 101–113.

Kimelberg, H. K. (1977) The influence of membrane fluidity on the activity of membrane-bound enzymes. In: Dynamic Aspects of Cell Surface Organization. (Poste, G. and Nicolson, G. L., eds.), Cell Surface Reviews, vol. 3, pp. 205–293. Elsevier/North-Holland, Amsterdam.

Klebe, R. J., Chen, T. R. and Ruddle, F. H. (1970) Controlled production of proliferating somatic cell hybrids. J. Cell Biol. 45, 74–82.

Klemperer, H. G. (1960) Hemolysis and the release of potassium from cells by Newcastle disease virus (NDV). Virology 12, 540–552.

Klenk, H. D. and Choppin, P. W. (1969) Lipids of plasma membranes of monkey and hamster kidney cells and of parainfluenza virions grown in these cells. Virology 38, 255–268.

Knowles, K. W. and Person, S. (1976) Effects of 2-deoxyglucose, glucosamine, and mannose on cell fusion and the glycoproteins of Herpes simplex virus. J. Virol. 18, 644–651.

Knox, P. and Pasternak, C. A. (1975) The binding of serum components to cultured cells. In: Radioimmunoassay in Clinical Biochemistry (Pasternak, C. A., ed.). Heyden and Son, London.

Knox, P. and Pasternak, C. A. (1977) The cell surface and growth in vitro. In: Mammalian Cell Membranes (Jamieson, G. A. and Robinson, D. M., eds.) vol. 5., pp. 237–264. Butterworth and Co., London.

Knutton, S. (1976) Changes in viral envelope structure preceding infection. Nature, 264, 672–673.

Knutton, S. (1978) Studies of membrane fusion. II. Fusion of human erythrocytes by Sendai virus. J. Cell Sci. In press.

Knutton, S., Jackson, D., Graham, J. M., Micklem, K. J. and Pasternak, C. A. (1976) Microvilli and cell swelling. Nature 262, 52–54.

Knutton, S., Jackson, D. and Ford, M. (1978) Studies of membrane fusion. I. Paramyxovirus-induced cell fusion, a scanning electron microscope study. J. Cell Sci. In press.

Kohn, A. (1965) Polykaryocytosis induced by Newcastle disease virus in monolayers of animal cells. Virology 26, 228–245.

Kohn, A. (1975) The fate of protein subunits of parainfluenza (Sendai) virus after adsorption to NIL8 hamster embryo cells. J. Gen. Virol. 29, 179–187.

Kohn, A. and Klibansky, C. (1967) Studies on the inactivation of cell-fusing property of Newcastle disease virus by phospholipase A. Virology 31, 385–388.

Lucy, J. A. (1970) The fusion of biological membranes. Nature (London) 227, 815–817.

Lucy, J. A. (1975) Aspects of the fusion of cells in vitro without viruses. J. Reprod. Fertil. 44, 193–205.

362

Ludwig, H., Becht, H and Rott, R. (1974) Inhibition of herpes virus-induced cell fusion by concanavalin A, antisera and 2-deoxy-D glucose. J. Virol. 14, 307–314.

Lüginbuhl, D. (1873) Der micrococcus der Variola. Arbeiten ans dem Pathologischen Instit. Wurzburg, 1871–72, p. 159.

Maeda, T., Kim, J., Koseki, I., Medada, E., Shiokawa, Y. and Okada, Y. (1977a) Modification of cell membranes with viral envelopes during fusion of cells with HVJ (Sendai virus) Exp. Cell Res. In Press.

Maeda, T., Asano, A., Okada, Y. and Ohnishi, S. I. (1977b) Transmembrane phospholipid motions induced by F. glycoprotein in hemagglutinating virus Japens. J. Virol. 21, 232–241.

Magee, W. E. and Miller, O. V. (1968) Initiation of vaccinia virus infection in actinomycin D-pretreated cells. J. Virol. 2, 678–685.

Martin, F. J. and MacDonald, R. C. (1976) Phospholipid exchange between bilayer membrane vesicles. Biochemistry 15, 321–327.

Maslow, D. E. (1976) In vitro analysis of surface specificity in embryonic cells. In: The Cell Surface in Animal Embryogenesis and Development. (Poste, G. and Nicolson, G. L., eds.), Cell Surface Reviews, vol.1, pp. 697–745. Elsevier/North-Holland, Amsterdam.

Meiselman, N. Kohn, A. and Danon, D. (1967) Electron microscopic study of penetration of Newcastle disease virus into cells leading to formation of polykaryocytes. J. Cell Sci. 2, 71–76.

Micklem, K. J. and Pasternak, C. A. (1977) Surface components involved in virally-mediated membrane changes. Biochem. J. 162, 405–410.

Micklem, K. J., Abra, R. M., Knutton, S., Graham, J. M. and Pasternak, C. A. (1976) The fluidity of normal and virus-transformed cell plasma membrane. Biochem. J. 154, 561–566.

Miller, J. B., Noyes, C., Heinrikson, R., Kingdon, H. S. and Yachnin, S. (1973) Phytohemagglutinin mitogenic proteins. J. Exp. Med. 138, 939–951.

Morgan, C. and Howe, C. (1968) Structure and development of viruses as observed in the electron microscope. IX. Entry of parainfluenza I (Sendai) virus. J. Virol. 2, 1122–1132.

Murayama, F. and Okada, Y. (1965) Calcium effect on cell fusion reaction caused by HVJ. Biken's J. 8, 103–105.

Nagai, Y. and Klenk, H. D. (1977) Activation of precursors to both glycoproteins of Newcastle disease virus by proteolytic cleavage. Virology 77, 125–134.

Nagai, Y., Ogura, H. and Klenk, H. D. (1976a) Studies on the assembly of the envelope of Newcastle disease virus. Virology 69, 523–538.

Nagi, Y., Klenk, H. D. and Rott, R. (1976b) Proteolytic cleavage of the viral glycoproteins and its significance for the virulence of Newcastle disease virus. Virology 72, 494–508.

Nakamura, K. and Homma, M. (1974) Cell fusion by HeLa cells persistently injected with haemadsorption type 2 virus. J. Gen. Virol. 25, 117–124.

Neurath, A. R., Vernon, S. K., Dobkin, M. B. and Rubin, B. A. (1972) Haemolysis by Sendai virus: lack of requirement for neuraminidase. J. Gen. Virol. 16, 245–249.

Neurath, A. R., Vernon, S. K., Hartzell, R. W., Wiener, E. P. and Rubin, B. A. (1973) Haemolysis by Sendai virus: involvement of a virus protein component. J. Gen. Virol. 19, 21–36.

Nicolson, G. L. (1976) Transmembrane control of the receptors on normal and tumor cells. I. Cytoplasmic influence over cell surface components. Biochim. Biophys. Acta 457, 57–108.

Nicolson, G. L. and Painter, R. G. (1973) Anionic sites of human erythrocyte membranes. II. Transmembrane effects of anti-spectrin on the topography of bound positively charged colloidal particles. J. Cell Biol. 59, 395–406.

Nicolson, G. L. and Poste, G. (1976) The cancer cell: Dynamic aspects and modifications in cell-surface organization. New Eng. J. Med. Pt. I: 295, 197–203, Pt. II: 295, 253–258.

Nishiyama, Y., Ito, Y., Shimokata, K., Kimura, Y. and Nagata, I. (1976) Polykaryocyte formation induced by VSV in mouse L cells. J. Gen. Virol. 32, 85–96.

Ohki, K., Nakama, S., Asano, A. and Okada, Y. (1975) Stimulation of virus-induced fusion of Ehrlich ascites tumor cells by 3'5'cyclic AMP. Biochem. Biophys. Res. Commun. 67, 331–337.

Ohuchi, M. and Homma, M. (1976) Trypsin action on the growth of Sendai virus in tissue culture cells. IV. Evidence for activation of Sendai virus by cleavage of a glycoprotein. J. Virol. 18, 1147–1150.

Okada, Y. (1958) The fusion of Ehrlich's tumor cells caused by HVJ virus in vitro. Biken's J. 1, 103–110.

Okada, Y. (1962) Analysis of giant polynuclear cell formation caused by HVJ virus in Ehrlichs ascites tumor cells. 1. Microscopic observation of giant polynuclear cell formation. Exp. Cell Res. 26, 98–107.

Okada, Y. (1969) Factors in fusion of cells by HVJ. Curr. Top. Microbiol. Immunol. 48, 102–156.

Okada, Y. and Kim. J. (1972) Interaction of concanavalin A with enveloped viruses and host cells. Virology 50, 507–515.

Okada, Y., Yamada, K. and Tadokoro, J. (1964) Effect of antiserum on the cell fusion reaction caused by HVJ. Virology 22, 397–409.

Okada, Y., Kim, J., Maeda, Y. and Koseki, I. (1974) Specific movement of cell membranes fused with HVJ (Sendai virus). Proc. Nat. Acad. Sci. U.S.A. 71, 2043–2047.

Okada, Y., Koseki, I., Kim. J., Maeda, Y., Hashimoto, T., Kanno, Y. and Matsui, Y. (1975) Modification of cell membranes with viral envelopes during fusion of cells with HVJ (Sendai virus). Exp. Cell Res. 93, 368–378.

Olmsted, J. B. and Borisy, G. G. (1973) Microtubules. Ann. Rev. Biochem. 42, 507–540.

Olmsted, J. B., Marcum, J. M., Johnson, K. A., Allen, C. and Borisy, G. G. (1974) Microtubule assembly: Some possible regulatory mechanisms. J. Supramol. Struct. 2, 429–450.

Papahadjopoulos, D. and Poste, G. (1975) Calcium-induced phase separation and fusion in phospholipid membranes. Biophys. J. 15, 945–948.

Papahadjopoulos, D. and Poste, G. (1978) Membrane fusion. Int. Rev. Cytol. In press.

Papahadjopoulos, D., Poste, G. and Schaeffer, B. E. (1973) Fusion of mammalian cells by unilamellar lipid vesicles: influence of lipid surface charge, fluidity and cholesterol. Biochim. Biophys. Acta 323, 23–42.

Papahadjopoulos, D., Poste, G., Schaeffer, B. E. and Vail, W. J. (1974) Membrane fusion and molecular segregation in phospholipid vesicles. Biochem. Biophys. Acta 352, 10–28.

Papahadjopoulos, D., Hui, S., Vail, W. J. and Poste, G. (1976a) Studies on membrane fusion I. Interactions of pure phospholipid membranes and the effect of myristic acid, lysolecithin, proteins and DMSO. Biochim. Biophys. Acta 448, 245–264.

Papahadjopoulos, D., Vail, W. J., Pangborn, W. A. and Poste, G. (1976b) Studies on membrane fusion. II. Induction of fusion in pure phospholipid membranes by calcium and other divalent metals. Biochim. Biophys. Acta 448, 265–283.

Papahadjopoulos, D., Vail, W. J., Newton, C., Nir, S., Jacobson, K., Poste, G. and Lazo, R. (1977) Studies on membrane fusion. III. The role of calcium-induced phase changes. Biochim. Biophys. Acta 465, 579–598.

Papahadjopoulos, D., Poste, G. and Vail, W. J. (1978) Methods for studying membrane fusion. In: Methods in Membrane Biology (Korn, E., ed.). Plenum Press. In press.

Parkes. J. G. and Fox, C. F. (1975) On the role of lysophosphatides in virus-induced cell fusion and lysis. Biochemistry 14, 3725–3729.

Pasternak, C. A. (1977) Lipid changes in membranes during growth and development. In: Lipid Metabolism in Mammals (Snyder, F., ed.) pp. 353–383. Plenum, New York.

Pasternak, C. A. and Micklem, K. J. (1973) Permeability changes during cell fusion. J. Membr. Biol. 14, 292–303.

Pasternak, C. A. and Micklem, K. J. (1974a) The biochemistry of virus-induced cell fusion. Changes in membrane integrity. Biochem. J. 140, 405–411.

Pasternak, C. A. and Micklem, K. J. (1974b) Virally-mediated membrane changes: inverse effects on transport and diffusion. Biochem. J., 144, 593–595.

Pasternak, C. A., Strachan, E., Micklem, K. J. and Duncan, J. (1976) New approaches for studying surface membranes. In: Cell Surfaces and Malignancy (Mora, P. T., ed.). U. S. Gov. Printing Office, Washington D.C., Fogarty Intern. Center Proc. 28, 147–151.

Pinto da Silva, P. and Martínez-Palomo, A. (1975). Distribution of membrane particles and gap junctions in normal and transformed 3T3 cells studied in situ, in suspension, and treated with concanavalin A. Proc. Nat. Acad. Sci. U.S.A., 72, 572–576.

Pinto da Silva, P. and Nicolson, G. L. (1974) Freeze-etch localization of concanavalin A receptors to the membrane intercalated particles on human erythrocyte membranes. Biochim. Biophys. Acta 363, 311–319.

Pitts, J. D. (1976) Junctions as channels of direct communication between cells. In: Developmental Biology of Plants and Animals. (Graham, C. F. and Waring, P. F., eds.) pp. 96–110, Blackwell Scientific Publications, Oxford.

Poole, A. R., Howell, J. I. and Lucy. J. A. (1970) Lysolethicin and cell fusion. Nature(London) 227, 810–814.

Portner, A., Scroggs, R. A., Marx, P. A. and Kingsbury, D. W. (1975) A temperature sensitive mutant of Sendai virus with an altered hemagglutinin-neuraminidase polypeptide: consequences for virus assembly and cytopathology. Virology 67, 179–187.

Poste, G. (1970) Virus-induced polykaryocytosis and the mechanism of cell fusion. Adv. Virus. Res. 16, 303–356.

Poste, G. (1972) Mechanisms of virus-induced cell fusion. Int. Rev. Cytol. 33, 157–252.

Poste, G. (1973) Annucleate mammalian cells: applications in cell biology and virology. In:Methods in Cell Biology (Prescott, D. M., ed.) vol. VII, pp. 211–249, Academic Press, New York.

Poste, G. (1975) Interaction of concanavalin A with the surface of virus-infected cells. In: Concanavalin A (Chowdhury, T. K. and Weiss, A. K., eds.) pp. 117–152. Plenum Press, New York.

Poste, G. and Allison, A. C. (1973) Membrane fusion. Biochim. Biophys. Acta 300, 421–465.

Poste, G. and Nicolson, G. L. (1976) Calcium ionophores A23187 and X537A affect cell agglutination by lectins and capping of lymphocyte surface immunoglobulins. Biochim. Biophys. Acta 426, 148–155.

Poste, G. and Nicolson, G. L., Eds. (1977) Dynamic Aspects of Cell Surface Organization. Cell Surface Reviews, vol. 3, Elsevier/North-Holland, Amsterdam.

Poste, G. and Papahadjopoulos, D. (1976a) Lipid vesicles as carriers for introducing materials into cultured cells: Influence of vesicle lipid composition on mechanisms of vesicle incorporation into cells. Proc. Nat. Acad. Sci. U.S.A. 73, 1603–1607.

Poste, G. and Papahadjopoulos, D. (1976b) Fusion of mammalian cells by lipid vesicles. In: Methods in Cell Biology (Prescott, D. M., ed.) vol. 14, pp. 23–32. Academic Press, New York.

Poste, G. and Reeve, P. (1972a) Enucleation of mammalian cells by cytochalasin B. II. Formation of hybrid cells and heterokaryons by fusion of anucleate and nucleated cells. Exp. Cell Res. 73, 287–294.

Poste, G. and Reeve, P. (1972b) Inhibition of cell fusion by local anaesthetics and tranquillizers. Exp. Cell. Res. 72, 556–560.

Poste, G. and Waterson, A. P. (1975) Cell fusion by Newcastle disease virus. In: Negative Strand Viruses (Mahy, B. W. J. and R. D., Barry, eds.) vol. 2, pp. 905–922. Academic Press, London.

Poste, G., Waterson, A. P., Terry, G., Alexander, D. J. and Reeve, P. (1972a) Cell fusion by Newcastle disease virus. J. Gen. Virol. 16, 95–97.

Poste, G., Reeve, P., Alexander, D. J. and Terry, G. (1972b) Effect of plasma membrane lipid composition on cellular susceptibility to virus-induced cell fusion. J. Gen. Virol. 17, 133–136.

Poste, G., Alexander, D. J., Reeve, P. and Hewlett, G. (1974) Modification of Newcastle disease virus release and cytopathogenicity in cells treated with plant lectins. J. Gen. Virol. 23, 255–263.

Poste, G., Papahadjopoulos, D., Jacobson, K. and Vail, W. J. (1975a) Effects of local anesthetics on membrane properties. II. Enhancement of the susceptibility of mammalian cells to agglutination by plant lectins. Biochim. Biophys. Acta 394, 520–539.

Poste, G., Papahadjopoulos, D. and Nicolson, G. L. (1975b) The effect of local anesthetics on transmembrane cytoskeletal control of cell surface mobility and distribution. Proc. Nat. Acad. Sci. U.S.A. 72, 4430–4434.

Poste, G., Alexander, D. J. and Reeve, P. (1976a) Enhancement of virus-induced cell fusion by phytohemagglutinin. In:Methods in Cell Biology (Prescott, D. M., ed.) vol. 14, pp. 1–10. Academic Press, New York.

Poste, G., Papahadjopoulos, D. and Vail, W. J. (1976b) Lipid vesicles as carriers for introducing biologically active materials into cells. In:Methods in Cell Biology (Prescott, D. M., ed.) vol. 14, pp. 33–71. Academic Press, New York.

Poste, G., Porter, C. W. and Papahadjopoulos, D. (1978) Incorporation of lipid vesicles (liposomes) by cultured cells: influence of vesicle size, surface charge and lipid fluidity. Ann. N.Y. Acad. Sci. In press.

Quinn, P. J. and Pasternak, C. A. (1978) Properties of normal cell membranes: lipids. In: Cell Membranes and Envelopes (Blough, H. A. and Tiffany, J. M., eds). Academic Press, London. In press.

Rankin, A. M., Fisher, L. E. and Bussell, R. H. (1972) Cell fusion by canine distemper virus-infected cells. J. Virol. 10, 1179–1183.

Reeve, P. and Poste, G. (1971) Studies on the cytopathogenicity of Newcastle disease virus: relation between virulence, polykaryocytosis and plaque size. J. Gen. Virol. 11, 17–24.

Reeve, P., Hewlett, G., Watkins, H., Alexander, D. J. and Poste, G. (1974) Virus-induced cell fusion enhanced by phytohaemagglutinin. Nature (London) 249, 355–356.

Rifkin, D. B. and Quigley, J. P. (1974) Virus-induced modification of cellular membranes related to viral structure. Ann. Rev. Microbiol., 28, 325–351.

Ringertz, N. R. and Savage, R. E. (1976) Cell Hybrids. Academic Press, New York.

Roizman, B. (1962) Polykaryocytosis. Cold Spring Harbor Symp. Quant. Biol. 27, 327–342.

Roizman, B. (1970) Polykaryocytosis induced by viruses. Proc. Nat. Acad. Sci., (U.S.A.) 48, 228–234.

Rose, B. and Loewenstein, W. R. (1975) Permeability of cell junctions depends on local cytoplasmic calcium activity. Nature 254, 250–252.

Rothman, J. E. and Lenard, J. (1977) Membrane asymmetry. Science 195, 743–753.

Rott, R. and Klenk, H. D. (1977) Structure and assembly of viral envelopes. In:The Cell Surface and Virus Infection (Poste, G. and Nicolson, G. L., eds.), Cell Surface Reviews, vol. 2, pp. 46–81. Elsevier/North-Holland, Amsterdam.

Rott, R., Becht, H., Hammer, G., Klenk, H. D. and Scholtissek, C. (1975) Changes in the surface of the host cell after infection with enveloped viruses. In:Negative Strand Viruses (Mahy, B. W. J. and Barry, R. D., eds.) vol. 2, pp. 843–857. Academic Press, London.

Samson, A. C. R. and Fox, C. F. (1973) Precursor protein for Newcastle disease virus. J. Virol. 12, 579–587.

Scheid, A. (1976) Interactions of paramyxovirus glycoproteins with the cell surface. In:Cell Membrane Receptors for Viruses, Antigens and Antibodies, Polypeptide Hormones and Small Molecules (Beers, R. E., Jr. and Bassett, E. G., eds.) pp. 223–235. Raven Press, New York.

Scheid, A. and Choppin, P. W. (1973) Isolation and purification of the envelope proteins of Newcastle disease virus. J. Virol. 11, 213–271.

Scheid, A. and Choppin, P. W. (1974) Identification of biological activities of paramyxovirus glycoproteins. Activation of cell fusion, hemolysis, and infectivity by proteolytic cleavage of an inactive precursor protein of Sendai virus. Virology 57, 475–490.

Scheid, A. and Choppin, P. W. (1975) Activation of cell fusion and infectivity by proteolytic cleavage of a Sendai virus glycoprotein. In: Proteases and Biological Control (Reich, E., Rifkin, D. B. and Shaw, E., eds.) pp. 645–659, Cold Spring Harbor Laboratory, Cold Spring Harbor, New York.

Scheid, A. and Choppin, P. W. (1976) Protease activation mutants of Sendai virus. Activation of biological properties by specific proteases. Virology 69, 265–277.

Scheid, A. and Choppin, P. W. (1977) Two disulfide-linked polypeptide chains constitute the active F protein of paramyxoviruses. Virology, 80, 54–66.

Scheid, A., Caliguiri, L. A., Compans, R. W. and Choppin, P. W. (1972) Isolation of paramyxovirus glycoproteins. Association of both hemagglutinating and neuraminidase activities with the larger SV5 glycoprotein. Virology 50, 640–652.

Schmidt, M. F. G., Schwarz, R. T. and Ludwig, H. (1976) Fluorosugars inhibit biological properties of different enveloped viruses. J. Virol. 18, 819–823.

Schneeberger, E. E. and Harris, H. (1966) An ultrastructural study of interspecific cell fusion induced by inactivated Sendai virus. J. Cell Sci. 1, 401–406.

Schreiner, G. F. and Unanue, E. R. (1976) Calcium-sensitive modulation of Ig capping: Evidence supporting a cytoplasmic control of ligand-receptor complexes. J. Exp. Med. 143, 15–31.

Schroeder, T. E. (1968) Cytokinesis:filaments in the cleavage furrow. Exp. Cell Res. 53, 272–276.

Schudt, C., Van der Bosch, J. and Pette, C. (1973) Inhibition of muscle cell fusion in vitro by Mg^{2+} and K^{2+} ions. FEBS Lett. 32, 296–298.

Schudt, C., Dahl, G. and Gratzl, M. (1976) Calcium-induced fusion of plasma membranes isolated from myoblasts grown in culture. Cytobiologie 13, 211–223.

Semmel, M., Israel, A. and Audubert, G. (1975) Phospholipids in surface membranes of Newcastle disease virus-infected cells. In: Negative Strand Viruses (Mahy, B. W. J. and Barry, R. D., eds.) vol. 2, pp. 875–883. Academic Press, London.

Seto, J. T., Becht, H. and Rott, R. (1973) Isolation and purification of surface antigens from disrupted paramyxoviruses. Z. Med. Mikrobiol. Immunol. 159, 1–12.

Shimizu, K. and Ishida, N. (1975) The smallest protein of Sendai virus: Its candidate function of binding nucleocapsid to envelope. Virology 67, 427–437.

Shimizu, Y. K., Shimizu, K., Ishida, N. and Homma, M. (1976) On the study of Sendai virus hemolysis. II. Morphological study of envelope fusion and hemolysis. Virology 71, 48–60.

Sullivan, J. L., Barry, D. W., Lucas, S. J. and Albrecht, P. (1975) Measles infection of human mononuclear cells. J. Exp. Med. 142, 773–784.

Tilney, L. G. and Marsland, D. (1969) A fine structural analysis of cleavage induction and furrowing in the eggs of Arbacia Punctualata. J. Cell Biol. 42, 170–184.

Toister, Z. and Loyter, A. (1972) The mechanism of cell fusion. II. Formation of chicken erythrocyte polykaryons. J. Biol. Chem. 248, 422–432.

Toister, Z. and Loyter, A. (1973) The mechanisms of cell fusion. II. Formation of chicken erythrocyte polykaryons. J. Biol. Chem. 248, 422–432.

Tokumaru, T. (1968) The nature of toxins of herpes simplex virus. I. Syncytial giant cell producing components in tissue culture. Arch. Ges. Virusforsch. 24, 104–122.

Tozawa, H., Watanabe, M. and Ishida, N. (1973) Structural components of Sendai virus. Serological and physicochemical characterization of hemagglutinin subunit associated with neuraminidase activity. Virology 55, 242–263.

Unanue, E. R. and Schreiner, G. F. (1977) Structure and function of surface immunoglobulin of lymphocytes. In: Dynamic Aspects of Cell Surface Organization (Poste, G. and Nicolson, G. L., eds.), Cell Surface Reviews, vol. 3, pp. 619–641. Elsevier/North-Holland, Amsterdam.

Unna, P. G. (1896) The Histopathology of the Diseases of the Skin, p. 637. Clay, Edinburgh.

Urata, D. M. and Seto, J. T. (1975) Glycoproteins of Sendai virus: purification and antigenic analysis. Intervirology 6, 108–114.

Vlodavsky, I. and Sachs, L. (1977) Difference in the calcium regulation of Concanavalin A agglutinability and surface microvilli in normal and transformed cells. Exp. Cell Res. 105, 179–189.

Wainberg, M. A. and Howe, C. (1972) Antibody-mediated fusion of FL amnion cells infected with parainfluenza virus type 2. Immunol. Commun. 1, 481–489.

Wainberg, M. A. and Howe, C. (1973) Infection-mediated resistance to cell fusion by inactive Sendai virus. Proc. Soc. Exp. Biol. Med. 142, 981–987.

Wainberg, M. A., Howe, C. and Godman, G. C. (1973) Enhancement of Sendai virus-mediated cell fusion by cupric ions. J. Cell Biol. 57, 388–396.

Warren, G. B., Houslay, M. D., Metcalf, J. C. and Birdsall, N. J. M. (1975a) Cholesterol is excluded from the phospholipid anulus surrounding an active calcium transport protein. Nature 255, 684–687.

Warren, G. B., Metcalf, J. C., Lee, A. G. and Birdsall, N. J. M. (1975b) Mg^{2+} regulates the ATPase activity of a calcium transport protein by interacting with bound phosphatidic acid. FEBS Lett. 50, 261–264.

Warthin, A. S. (1931) Occurrence of numerous large giant cells in the tonsils and pharyngeal mucosa in the prodromal stage of measles. Arch. Pathol. 11, 864.

Watkins, J. F. (1971) Fusion of cells for virus studies and production of cell hybrids. In: Methods in Virology (Maramorosch, K. and Koprowski, H., eds.) vol. 5, pp. 1–32. Academic Press, New York.

Watkins, J. F. (1974) The SV40 rescue problem. Cold Spring Harbor Symp. Quant. Biol. 39, 355–362.

Weiss, L. (1973) Neuraminidase, sialic acids and cell interactions. J. Nat. Cancer Inst. 50, 3–19.

Weiss, R. L., Goodenough, D. A. and Goodenough, V. W. (1977) Membrane differentiations at sites specialized for cell fusion. J. Cell Biol. 22, 144–160.

Wilson, L. and Bryant, J. (1974) Biochemical and pharmacological properties of microtubules. Adv. Cell. Mol. Biol. 3, 21–72.

Wirtz, K. W. A. (1974) Transfer of phospholipids between membranes. Biochim. Biophys. Acta 344, 95–117.

Wirtz, K. W. A., Geurts Van Kessel, W. S. M., Kamp, H. H. and Demel, R. A. (1976) The protein-mediated transfer of phosphatidylcholine between membranes. Eur. J. Biochem. 61, 515–523.

Wolinsky, J. S. and Gilden, D. H. (1975) In vivo studies of parainfluenza 1 (694) virus: mononuclear cell interactions. Arch. Virol. 49, 25–31.

Yamamoto, K., Inoue, D. and Suzuki, K. (1974) Interaction of paramyxovirus with erythrocyte membranes modified by concanavalin A. Nature(London) 250, 511–513.

Yoshida, M. C. and Ikeuchi, T. (1975) Enhancement of Sendai virus-mediated fusion and hybridization of lymphoid cells by addition of phytohemagglutinin. Proc. Jap. Acad. 51, 126–129.

Zakai, N., Kulka, R. G. and Loyter, A. (1976) Fusion of human erythrocyte ghosts promoted by the combined action of calcium and phosphate ions. Nature (London) 263, 696–699.

Zakai, N., Kulka, R. G. and Loyter, A. (1977) Membrane ultrastructural changes during calcium phosphate-induced fusion of human erythrocyte ghosts. Proc. Nat. Acad. Sci. U.S.A. 74, 2411–2421.

Zee, Y. C. and Talens, L. (1971) Entry of infectious bovine rhinotracheitis virus into cells. J. Gen. Virol. 11, 59–63.

Zeuthen, J. (1975) Heterokaryons in the analysis of genes and gene regulation. Humangenetik. 27, 275–301.

Zhadanov. V. M. and Bukrinskaya, A. G. (1962) Studies on the initial stage of virus-cell interaction. Acta Virol. 6, 105–113.

Fusion of plant protoplasts

8

J. B. POWER, P. K. EVANS and E. C. COCKING

Contents

G. Poste & G. L. Nicolson (eds.) Membrane Fusion, pp. 369–385.
© Elsevier/North-Holland Biomedical Press, 1978.

1. Introduction

The plant protoplast consists of that part of the cell which can be plasmolyzed away from the cell wall and thus comprises all the cellular contents bounded by the plasma membrane. The ability to induce fusion of these protoplasts has been an aim ever since they were recognized as a distinct cellular component. To achieve protoplast fusion, conditions must directly alter the nature of the bounding membrane to first permit membrane fusion which will, in most cases, lead to protoplast fusion and cytoplasmic coalescence. Clearly, the fundamental and obvious difference when comparing plant and animal cell fusion is the absolute need to remove the rigid cell wall from plant cells, either mechanically or enzymatically. The development of efficient fusion methods for plant protoplasts has been hampered until recent times because suitable experimental methods for the removal of this wall and for the large scale production and isolation of protoplasts were unavailable.

2. Initial studies on protoplast fusion

The simplest approach to protoplast production, developed by Klerker (1892), was to cut leaf tissue plasmolyzed in 0.4 M potassium nitrate. By chance, the walls of some cells would be cut in such a way that if the protoplasts were not damaged in the process, they could be released into the medium following a controlled deplasmolysis of the system. Protoplasts produced in this way were subsequently used in early attempts at fusion. However, the capacity for protoplasts to undergo fusion was observed initially not in the isolated state, but between subprotoplasts within the plasmolyzed cell. Kuster (1909) plasmolyzed *Allium cepa* roots and *Elodea* leaves in aqueous solutions of calcium nitrate to produce subprotoplasts. On deplasmolysis, the subprotoplasts were observed to fuse and the cells reverted to their original nonplasmolyzed state. This subprotoplast system was used to examine the effect of sugars on membrane fusion, and Kuster concluded that sucrose, for example, was detrimental to the fusion process. Kuster (1910a) also was the first to describe fusion of freely isolated plant protoplasts. This was achieved in two ways. Cells of *Allium* were strongly plasmolyzed in calcium nitrate solutions and, following the method of Klerker, the cut surfaces of cells, still containing their respective protoplasts, were brought into contact. On deplasmolysis the protoplasts expanded, came into contact, and in a very few cases were seen to fuse. Similarly, freely suspended protoplasts could be brought into contact, and if this contact could be maintained for several hours occasional fusion apparently occurred. Although fusion under these conditions was still a completely random and infrequent event, the basic prerequisites for induced fusion were emerging. Protoplasts had to be freed of their cell walls, brought into contact for a finite period of time and upon deplasmolysis they might undergo fusion. The effect of the plasmolyzing solution on the ability of protoplasts to fuse was largely overlooked, however. Kuster (1910b) observed differences, however, in the fusion capacity of subprotoplasts in relation to the plasmolyticum employed. The use of calcium nitrate delayed the fusion of subprotoplasts upon deplasmolysis by the addition of water, as compared to a sucrose-based plasmolyticum. Missbach (1928) showed that the duration of plasmolysis in calcium nitrate, prior to deplasmolysis, also influenced the extent of subprotoplast fusion in *Elodea*. Plasmolysis in excess of three hours was sufficient to prevent subprotoplast fusion. The ability to fuse after plasmolysis was also dependent on the nature of the salts used; potassium nitrate, for example, was preferable as a plasmolyticum since longer periods of plasmolysis could be tolerated without loss of fusion capacity.

Progress was still hampered, however, by the relatively few protoplasts that could be released using mechanical methods, though the careful selection of plant material could produce larger numbers of protoplasts. Kuster (1928), for example, extended his isolation techniques to fruit material of *Solanaceae* species. In maturing fruits of *Atropa* large quantities of "protoplasts" could be released by squashing the fruits but fusion was never observed, probably because a very thin cell wall was still present around the protoplastlike bodies. Vacuoles were, however, seen to fuse. Lorey (1929), using mechanically isolated onion protoplasts, held protoplasts together with glass needles and, avoiding deplasmolysis, reported protoplast fusion in solutions of sucrose or potassium nitrate but not in calcium nitrate. The evidence for induced fusion was thus still scant and often contradictory, but in 1937 Michel introduced some new approaches to the fusion of protoplasts. Pretreatment of plant tissue prior to isolation not only affected the quality of the isolated protoplasts but was shown to influence their fusion potential. Protoplasts still in the intact tissue were subjected to pretreatment with 0.5 M potassium nitrate solution and then mechanically released into viscous solutions of sucrose (starch paste/cane sugar). Using this approach, Michel observed the fusion of subprotoplasts within the cell (autoplastic fusion), protoplasts from different cells of the same species (homoplastic fusion), and protoplast fusion between different species (heteroplastic fusion).

Although fusion was still basically nonreproducible, these early chance observations suggested that there were no barriers to interspecies protoplast fusion, and that if a reliable inducer of fusion could be found interspecies and intergeneric protoplast fusion could be achieved.

Michel (1937) also introduced the concept of using light microscopic markers to monitor fusion, since he reported the fusion of colorless protoplasts (storage root) of *Raphanus* with those possessing anthocyanin in the vacuoles; beetroot root protoplasts were fused with onion root protoplasts, and radish root epidermis protoplasts were fused within horseradish root epidermis protoplasts. Most

372

of the fusion products did not assume a completely spherical shape, suggesting that vacuole fusion was not proceeding quickly, and both the *Allium* and *Raphanus* fusion products failed to survive for more than a few hours. The evidence for vacuole fusion and tonoplast-plasma membrane fusion had, however, been provided by Plowe (1931) who, by inserting a needle into the protoplast, was able to manipulate the protoplast so that withdrawal of the needle brought the tonoplast and the plasma membrane into contact prior to fusion.

Hofmeister (1954) reexamined the evidence for and against protoplast fusion and repeated most of the early fusion experiments described above. No evidence of complete fusion was ever observed using sugars (glucose, sucrose, fructose) as the plasmolyticum, and only occasional fusion was induced by plasmolysis of tissue (onion) with potassium nitrate followed by deplasmolysis. Although the manipulation experiments of the tonoplast/plasma membrane (Plowe, 1931) were reproducible, the previous conclusions that fusion could be induced rather than being a random event could not be confirmed. Hofmeister extended the use of inorganic salts to induce fusion by suspending protoplasts in isotonic solutions of seawater and reported the induction of infrequent fusion of *Allium* protoplasts together with vacuole fusion. In Hofmeister's opinion, seawater and the presence of surface active agents were capable of inducing fusion. He also noted that the use of dilute alkali during deplasmolysis stimulated protoplast adhesion.

Binding (1966), applying the seawater approach for culture, demonstrated the induced fusion of moss protoplasts. Protoplasts of *Funaria hygrometrica, Physcomitrium enrystemum, P. pyriforme* and *Bryum erythorocarpum* were maintained in seawater and underwent cell wall formation. Prior to this, however, they exhibited a weak tendency to fuse. Fusion was rarely complete and viability was low in the fusion products, with the exception of one induced symplasm of *F. hygrometrica* which exhibited cell wall formation.

3. Recent studies on protoplast fusion

The development in the 1960s of enzymatic methods for protoplast isolation has enabled the large-scale production of protoplasts to be undertaken as a reproducible technique. Since protoplast fusion was hitherto an infrequent event, the availability of enzymatically isolated protoplasts in large numbers has created a much broader basis for examining fusion methods. Protoplast isolation methods can be classified into three main groups: mechanical, sequential, enzyme (two-step), and mixed enzyme procedures. The sequential approach involves the production of isolated cells following initial maceration of plant tissues with pectinases which, in turn, are converted into protoplasts by a cellulase treatment (Takebe et al., 1968). In the mixed enzyme approach (Power and Cocking, 1968) plant tissues are plasmolyzed in the presence of mixtures of pectinases and cellulases, thus inducing the concomitant separation and degradation of the cells to release protoplasts directly. This latter method has been extended to many plant species and to tissue from most parts of the plant.

When isolated by the mixed enzyme approach, protoplasts will undergo spontaneous fusion. This results from an expansion of the plasmodesmatal connections, within the symplasm, as the cell walls are degraded, thus releasing homokaryon multinucleate protoplasts. This type of fusion can be observed in up to 10% of the protoplast population. The phenomenon has been reported in protoplast preparations derived from most plant tissues and sometimes from in vitro cultured material. The problem of spontaneous fusion in the protoplast preparations can be eliminated either by using the sequential method for isolation or plasmolyzing the tissue prior to mixed enzyme treatment, thus severing the plasmodesmatal connections before the onset of wall degradation. The process of spontaneous fusion is of little value, however, with respect to the development of methods to bring about induced fusion of protoplasts. Nonetheless, the study of the development of spontaneous fusion bodies in culture provided early clues about the expected behavior of nuclei in multinucleate protoplasts created via fusion (Power and Frearson, 1973). For example, multinucleate spontaneous fusion bodies (homokaryons) were found to be capable of cell wall synthesis and nuclear division was observed to be synchronous.

Initially, the reason for the development of methods to bring about the induced fusion of protoplasts was primarily to examine cell interactions. However, with the advent of culture procedures aimed at the continued survival and division of the plant protoplast, and the prospect of producing whole plant from the regeneration of single protoplasts, it became apparent that fusion of protoplasts also offered one method whereby heterokaryons could possibly be produced between plant species that were sexually incompatible. This therefore afforded fascinating opportunities to produce new hybrids via the process of somatic hybridization, assuming that nuclear fusion would occur. Somatic hybridization thus involves the isolation and fusion of plant protoplasts and their subsequent culture, with the ultimate aim of producing whole plants. The induced fusion of protoplasts is, of course, the key stage in this process, and the prospect of producing new hybrids has provided a potent stimulus to develop methods whereby extensive fusion can be achieved reproducibly.

Using the onion epidermal subprotoplast system described by Yoshida (1961), Power et al. (1970) showed that subprotoplast fusion was enhanced by the use of sodium salts and in particular by sodium nitrate as the plasmolyticum. Mechanically isolated and enzymatically isolated protoplasts held in contact in the presence of sodium nitrate would undergo adhesion and, in some cases, fusion. The extent of sodium nitrate-induced fusion of protoplasts, developed initially using highly cytoplasmic cereal root protoplasts, is variable. It can reach 50%, but for the more commonly used mesophyll protoplasts fusion is usually restricted to less than 1% of the population. To effect fusion, protoplasts are suspended in aqueous solutions of sodium nitrate and allowed to come into contact, often with the aid of centrifugation. The adhesion of protoplasts precedes localized membrane fusion which, often after removal of the sodium nitrate, results in cytoplasmic coalescence and protoplast fusion. Using sodium nitrate, this fusion may take several hours to reach completion, during which cell wall synthesis could be initiated thus restricting further coalescence of the protoplasts (Power and Frearson, 1973).

Using this method fusion has now been induced in many protoplast systems including cereal root protoplasts (Power et al., 1970), mesophyll protoplasts (Carlson et al., 1972; Kameya and Takahashi, 1972; Binding, 1974; Lai and Liu, 1976); mesophyll protoplasts with colorless cultured cell protoplasts (Power et al., 1975), petal protoplasts (Potrykus, 1972), and pollen tetrad protoplasts (Deka et al., 1977).

One of the problems encountered with sodium nitrate-induced fusion is that fusion frequencies, particularly with mesophyll protoplasts, can be low. This is partly due to the physical process of sodium nitrate-induced fusion, which is usually gradual and takes place over a period of many hours, and partly to the structure of mesophyll protoplasts. The presence of chloroplasts within the thin peripheral layer of cytoplasm can prevent the complete coalesence of protoplasts (Power and Cocking, 1971). Despite this, sodium nitrate treatment of mesophyll protoplasts can induce sufficient fusion so that if a powerful selection procedure is available somatic hybrid plants can be recovered (Carlson et al., 1972).

In spite of the problems caused by the structure of mesophyll protoplasts, which hinder rapid fusion, work has concentrated on the fusion of this type of protoplast for two reasons. First, the leaf is a source of many cells that can be readily converted to protoplasts. In some cases yields of up to 7×10^6 protoplasts per gram of leaf material have been obtained (Banks and Evans, 1976). Second, for certain species, mesophyll protoplasts are more amenable to culture than protoplasts derived from other sources and high plating efficiencies can be achieved.

In 1973, Keller and Melchers reported a high frequency of induced fusion in mesophyll protoplasts of tobacco following the treatment of protoplasts at 37 °C in mannitol solutions containing calcium ions buffered at high pH (10.5). This alkali treatment, based on the previously established procedure for the fusion of animals cells (Toister and Loyter, 1973), can induce up to 50% fusion in mesophyll protoplasts. Frequently the treatment of protoplasts with solutions buffered at high pH can adversely affect cell viability, but if a balance is established between fusion frequency and viability and a selection procedure is available, somatic hybrid plants can be produced using this method of fusion. Melchers and Labib (1974) have described the production of somatic hybrids between two light-sensitive varieties of *Nicotiana tabacum* using their high pH and calcium method, and Schieder (1975) has selected a somatic hybrid of two auxotrophic mutants of *Sphaerocarpus donellii* following fusion of protoplasts in calcium nitrate at pH 9.0. Fusion frequencies of 20 to 50% can be achieved for mesophyll protoplasts.

The third, most recent fusion method involves incubation of protoplasts in polyethylene glycol (PEG) solutions. PEG will induce moderate levels of fusion (10–20% of the protoplast population) without adversely affecting the viability. Kao and Michayluk (1974) assessed the induced fusion of mesophyll protoplasts with those of isolates from cultured cells using the light microscope marker method, whereas Wallin and co-workers (1974) used cultured carrot cells and the frequency of multinucleates to evaluate fusion.

Clearly, although they may differ in their merits, these three major methods of fusion, when suitably modified for a particular protoplast system, can bring about significant amounts of fusion. The ultimate assessment of a given fusion procedure must take into account the recovery of somatic hybrids following that procedure. In this respect all three methods have been used successfully to produce somatic hybrid plants. PEG-induced fusion has been used, for example, to produce somatic hybrids between *Nicotiana glauca* and *N. langsdorffii* (Smith et al., 1976), two chlorophyll-deficient mutants of *N. tabacum* (Gleba et al., 1975), *Petunia hybrida* and *P. parodii* (Power et al., 1976, 1977) and albino *P. hybrida* and wild-type *P. parodii* (Cocking et al., 1977).

Certain protoplast systems will undergo high frequencies of fusion (50–90%) without the presence of an obvious inducer (Ito and Maeda, 1973). Unfortunately, this form of protoplast behavior is not typical. In particular, isolated protoplasts of microsporocytes of *Trillium* will fuse upon collision. This fusion potential is short lived, however; if isolation takes longer than 30 minutes this ability to fuse on contact is lost. Thus the change in behavior is related to the period of plasmolysis.

In addition to the three methods of fusion outlined above, other additives have been examined for their ability to induce protoplast fusion but in most cases, while protoplast adhesion occurred fairly readily, fusion failed to occur. Kameya (1975) showed that gelatin and its breakdown products, gelatose and gelatone, would bring about protoplast adhesion without loss of viability. Hartmann and associates (1973) were able to achieve almost the same effect by the use of antibodies prepared against the protoplasts. Antibodies prepared against several different protoplast systems cross-reacted readily to bring about protoplast adhesion, but no fusion ensued. Lysozyme (Potrykus, 1973), salt solutions other than sodium salts (Eriksson, 1971; Kameya and Takahashi, 1972), Concanavalin A (Glimelius et al., 1974; Burgess and Linstead, 1976), polycations (Grout and Coutts, 1974), dextran sulfate, and gelatin (Kameya, 1975) have all been shown to induce adhesion. Fusion was not always observed and fusion frequencies were consistently lower compared with those generally achieved with sodium nitrate, PEG, or the high pH and calcium ion treatments.

4. Mechanisms of fusion

Ahkong and co-workers (1975a) have suggested that in animal cell fusion induced by exogenous chemical agents there is first a perturbation of the bilayer structure of membrane lipids. This results in an increase in the fluidity of the lipid region and, in an extreme case, in micelle formation. These investigators have suggested that such changes in the lipid structure of the plasma membrane would then allow the aggregation of intramembranous protein/glycoprotein particles. When membranes become closely apposed in regions where the protein/ glycoproteins are now absent, intermixing of the disturbed lipid molecules may allow adjacent cells to fuse, though the nature of the driving force for the coalescence of the adjacent bilayers is by no means clear (see chapter by Papahadjopoulos, this volume). These workers have suggested that a requirement for plasma membrane fusion is that the membranes must come sufficiently close together for the intermixing of lipid molecules to be possible and that the topographic arrangement of membrane protein/glycoproteins needs to be selectively altered as well. This may explain why the aggregation of plant protoplasts by ligands such as Con A (Withers, 1973) does not always necessarily result in their fusion.

As discussed by Poste and Allison (1973), membrane fusion probably requires that the membranes must first be brought into apposition at molecular distances (approximately 1.0 nm). It seems likely, therefore, that fusion of plant plasma membranes will be readily possible only if conditions can be devised that permit membranes to come together at distances of 1.0 nm or less. Significantly, isolated plant protoplasts are usually negatively charged at their surfaces, as determined by microelectrophoretic studies (Grout et al., 1972; Pilet and Senn, 1974), and this will cause protoplasts to repel one another in suspension. Normally, therefore, isolated protoplasts do not fuse unless conditions are applied to the system to enable them to come together at separation distances suitable for membrane fusion to take place.

After fusion of the protoplast plasma membranes has been initiated, cytoplasmic mixing occurs more readily in less highly vacuolated and differentiated protoplasts. Thus meristematic protoplasts fuse more readily. Indeed, fusion inducing agents which have a high affinity for water, such as polyethylene glycol (PEG), may enhance cell fusion because they remove water from the vacuoles of differentiated plant protoplasts thereby facilitating cytoplasmic mixing. Most plant protoplasts have a smooth plasma membrane with little tendency for "microvilli" to form. This may explain the apparent inactivity of Sendai virus in inducing their fusion (Withers, 1973), though the possible absence of virus receptors on plant cells may also account for this observation. A few plant protoplasts, such as those from pollen tetrads (Bhojwani and Cocking, 1972), have a more ruffled plasma membrane and this may contribute to the apparently greater susceptibility of these protoplasts to fusion, as well as those from pollen (Ito and Maeda, 1973), where centrifugation is often sufficient to initiate fusion. The induced fusion of protoplasts by salts such as sodium nitrate may also be related to the fact that they reduce the negative charge on the surface of protoplasts. More specifically, high concentrations of Na^+ may result in limited aggregation of the intramembranous protein/glycoprotein particles, with accompanying limited fusion of the plasma membranes (Withers and Cocking, 1972).

The mechanism of action of Ca^{2+} ions at high pH, as used by Keller and Melchers (1973) in inducing protoplast fusion, may relate to the conditions favoring the localized synthesis in the plasma membranes of lysocompounds including the synthesis of lysolecithin. Toister and Loyter (1973) have suggested such a mechanism for the formation of chick erythrocyte polykaryons by Ca^{2+} at pH 10.5. Lucy and his colleagues (1972) have demonstrated that lysolecithin is able to fuse hen erythrocytes in vitro, but the fused cells were highly unstable. Thus there seems to be a highly lytic activity of lysolecithin as well as a marked fusion-inducing capability. Unpublished work from this laboratory has shown that protoplasts treated with very low concentrations of lysolecithin, although they show a tendency to fuse, are highly unstable. They swell, burst, and are pulverized. Poste and Allison (1973) suggest that the type of membrane structure disordering created by lysolecithin that results in membrane fusion is not specific for this molecule since a wide variety of lipophilic and lipolytic agents such as retinol, oleic acid and sucrose (see also Lucy, 1974) have been shown to induce cell fusion. In addition, there is now evidence that lysolecithin formed de novo within membranes is not fusogenic and the ability of lysolecithin to induce fusion is confined to situations where membranes are exposed to exogenous lysolecithin molecules (see the chapter by Papahadjopoulos, this volume).

Poste and Allison (1973) suggested that changes in membrane surface potential may be a factor of fundamental importance in cell fusion. Gingell (1968) has shown that as cells approach and come into contact, the interactions of their electrical double layers cause a marked increase in the negative potential at the outer surface of the apposed membranes that could trigger changes in membrane structure and permeability. This may explain the marked tendency toward formation of large, multinucleate fusion bodies when plant protoplasts are treated with Ca^{2+} at high pH and for the sometimes erratic instability of the fusion products.

The action of PEG in inducing fusion is more difficult to explain. There is no doubt that PEG can induce fusion in membranes from a wide range of organisms including mammalian cells (Pontecorvo, 1975), fungal protoplasts (Anné and Peberdy, 1975), and bacterial protoplasts (Fodor and Alföldi, 1976, Schaeffer et al., 1976). It will also bring about interkingdom fusion between plant and animal cells (Ahkong et al., 1975b; Dudits et al., 1976; Jones et al., 1976). This strongly indicates that PEG may be applied universally as a fusogen. The high efficiency of PEG as a fusion-inducing agent also suggests that it may perturb membranes at several levels, all of which are conducive to fusion. The occasional, pronounced lytic effect of PEG on certain protoplasts may simply be because it is so active at initiating fusion. The most obvious feature of protoplasts exposed to high concentrations of PEG is their tight adhesion to each other. The molecular weight of the PEG influences the degree of protoplast adhesion. Tight adhesion only occurs when the molecular weight exceeds 1000. Wallin and colleagues (1974) used PEG 1540 to induce fusion, while Kao and Michayluk (1974) found that on a per mole basis PEG 6000 was more efficient at causing adhesion than PEG 1540. The presence of divalent cations such as magnesium or calcium in the PEG solution aided adhesion, whereas the presence of potassium appeared to have the reverse effect (Wallin et al., 1974). Attempts to examine protoplasts in the presence of PEG using electron microscopy has proved difficult, probably because of the difficulty of adequate penetration of the fixative under these conditions (Fowke et al., 1975). Tight adhesion of the plasma membrane is restricted to small regions separated by lens-shaped gaps. Fusion of the plasma membranes probably takes place in these small regions, possibly during the dilution of the PEG. Indeed, light microscope observations suggest that fusion occurs while the

PEG is being removed rather than during the period of tight adhesion in the presence of PEG (Kao and Michayluk, 1974).

Recently it has been shown that PEG can decrease the surface potential of lipid monolayers by several hundred millivolts, somewhat similar to the behavior of fusogenic lipids (Maggio et al., 1976). Unlike fusogenic lipids, however, PEG is relatively nontoxic. It has been suggested that PEG acts as a molecular bridge between adjacent membranes, either directly by hydrogen bonding or indirectly by Ca^{2+}, causing a marked clumping together of protoplasts. There is also a marked dehydration effect since concentrated solutions of PEG have a very high affinity for water (Mexal et al., 1975). The molecular mechanisms underlying PEG-induced fusion may eventually be revealed when freeze-fracture techniques are applied to the analysis of changes in the distributions of intramembranous particles. For example, Vos and associates (1976) have observed redistribution of such particles, possibly as a result of interactions between Ca^{2+} and phospholipids, in erythrocytes during fusion induced by the divalent-cation ionophore A23187. Such ionophores (Ahkong et al., 1975c) may also prove useful in the fusion of plant protoplasts. This action may relate to their ability to give rise to protein-free regions of lipid bilayer in the plasma membrane, and their use may be particularly helpful in increasing our knowledge of the mechanisms involved. Similarly, studies on the fusion of lipid vesicles (see chapter by Paphadjopoulos, this volume) may also be helpful in elucidating the mechanisms involved in plant protoplast fusions.

5. Somatic hybrids

In the process of somatic hybridization many factors combine to reduce the frequency of actual hybrid cell formation. Consequently, high fusion frequencies are now considered of secondary importance. Potential hybrid material is lost because only a small proportion of heterokaryons actually lead to true hybrid cell formation even in sexually compatible systems. This could be due partly to the relatively inefficient growth of protoplasts following most fusion treatments. A balance has to be established between fusion frequency and subsequent protoplast viability. Generally, the greater the extent of induced fusion the lower will be the viability. Protoplast systems also vary in their sensitivity to the various fusion agents so that most of the fusion methods have to be adapted to the specific systems under examination.

Using sodium nitrate (Power et al., 1970), Carlson and associates (1972) were able to induce sufficient fusion of mesophyll protoplasts to select somatic hybrids of the genetically tumorous hybrid between *Nicotiana glauca* and *N. langsdorffii*. Also using sodium nitrate to induce fusion, Power and co-workers (1975) were able to isolate callus, following culture under selective growth conditions of fused protoplasts of *Petunia hybrida* and *Parthenocissus tricuspidata*.

These two experiments serve to illustrate not only the absolute need for selection methods in isolating hybrids, but were also the first examples of the likely successes and failures that could follow the fusion and selective culture of two compatible species as compared to two nonrelated, sexually incompatible species.

Using the high pH method of fusion, Melchers and Labib (1974) have selected intraspecific somatic hybrids of *Nicotiana tabacum*. Protoplasts of two light-sensitive, dihaploid lines of *N. tabacum,* known to complement to a normal green, non-light-sensitive wild-type were fused and plated. Calluses developing from the fusion-treated material and grown under the selective light conditions could only develop green shoots if both complementing parental genomes were present.

The third and now most commonly used method of inducing protoplast fusion, polyethylene glycol, seems to provide the right balance between fusion frequency especially for mesophyll protoplasts and cell viability. Gleba and co-workers (1975) produced somatic hybrid plants following fusion of protoplasts isolated from two chlorophyll-deficient varieties of *N. tabacum.* As a result of complementation in the somatic hybrid, photosynthesizing plants could be identified both as true nuclear hybrids and also as cytoplasmic hybrids since the two mutant types used, a plastom-variegated mutant and a codominant genome mutant, permitted two classes of somatic hybrids to be identified.

Smith and colleagues (1976), repeating the original somatic hybridization of *N. glauca* and *N. langsdorffii,* but using PEG as the fusion agent, came to the conclusion that only multiple (triple) fusion events were leading to somatic hybrid cells capable of surviving the selection procedure and regeneration, since none of the somatic hybrids possessed the amphidiploid number of chromosomes. This is in contrast to sodium nitrate-produced somatic hybrids which, although fewer in number, possess a chromosome number equivalent to the summation of the two diploid parents. The extent to which this result is a reflection of the fusion methods used is not clear. It is unlikely that this inconsistency is a direct result of the fusion treatment since Power and associates (1976), again using PEG, were able to produce somatic hybrids of *Petunia hybrida* and *P. parodii* which had a tetraploid chromosome number. In this system, selection was based on naturally occurring differences in protoplast growth in response to nutrient media and drugs. More recently, somatic hybrids of the two *Petunia* species have been produced using nutrient media alone as the mode of selection (Power et al., 1977). Furthermore, following PEG-induced fusion of protoplasts isolated from an albino *P. hybrida* cell suspension with wild-type *P. parodii* mesophyll protoplasts, somatic hybrid callus could be identified. The presence of the *P. parodii* genome in the somatic hybrid corrected the albino deficiency of the *P. hybrida* genome, thus enabling simple visual identification of green somatic hybrid tissues. Unfused or homokaryon material of the potentially green *P. parodii* did not grow beyond the cell colony stage. Somatic hybrid plants with a tetraploid chromosome number were identical in appearance to the somatic hybrids produced using the other methods of selection (Power et al., 1976, 1977). Methods of selection based on the response to media and drug complementation (Power et al., 1976, 1977) or albino complementation (Cocking et al., 1977) may be generally applicable to other pairs of species since neither procedure is dependent upon a prior knowledge of the growth characteristics of the sexual counterparts.

With the possibility of inducing high frequencies of protoplast fusion, particularly with PEG, it becomes possible not only to identify potential hybrid cells but also to physically isolate heterokaryons. This can only happen if visual markers are present in the protoplast population. Kao (1977), using green mesophyll protoplasts of *N. glauca* as a marker for fusion with colorless protoplasts of a soybean cell suspension, were able to isolate heterokaryons of these two species by repeated dilution. It proved possible to culture individual heterokaryons and, in some cases, to induce repeated divisions. Chromosomes of both species were still present in certain cells of some of the cell lines, after several generations. Chromosomes of *N. glauca* were unidirectionally lost but not always completely eliminated from the somatic hybrid cells. Further confirmation of the hybrid nature of these tissues came from the demonstration of specific isozyme patterns alcohol dehydrogenase, and aspartate aminotransferase activity originating from both parents (Wetter, 1977).

Although most interest has centered on the induced fusion of plant protoplasts some attention has also been given to the fusion of animal cells with plant protoplasts, prompted perhaps by the knowledge that fusion methods such as high pH and also PEG used successfully in fusing protoplasts will also fuse animal cells (Pontecorvo, 1975). Dudits and colleagues (1976) demonstrated fusion of human HeLa cells with carrot protoplasts, and Jones and associates (1976) fused HeLa cells with mesophyll protoplasts of the amphiploid hybrid *N. glauca* and *N. langsdorffii*. In both cases PEG was the fusogen, but not surprisingly the development of the fusion products was limited to wall formation and possible nuclear fusion (Dudits et al., 1976) and, in the HeLa-tobacco heterokaryons, to short-term survival of the animal cell nucleus in the other system (Jones et al., 1976). The capacity to produce interkingdom heterokaryons is not surprising since PEG will also induce fusion between hen erythrocytes and yeast protoplasts (Ahkong et al., 1975b).

Doubtless, the reality of somatic hybridization has been demonstrated in a few plant species. Since the capacity to fuse protoplasts is no longer a problem, and the totipotent nature of plant protoplasts is being demonstrated for a growing list of species (Bajaj, 1977), linking these two features should provide a basis for the production of somatic hybrid plants of many species. It is premature to discuss the possible limitations of somatic hybridization, but it has been shown that if adequate selection procedures are developed somatic hybrid plants can be produced at the sexually compatible, interspecific level. The question arises about the likely successes of hybridization of strictly sexually incompatible species. Logically it would appear that the intrageneric sexually incompatible cross should be readily possible via somatic hybridization, and this is actually the area where most of the applied aspects of somatic hybridization will arise, though experimental evidence concerning intergeneric crosses is presently too limited to draw firm conclusions. In the *Petunia-Parthenocissus* system, complete loss of the *Petunia* chromosome component occurs (Power et al., 1975), yet in the soybean-*N. glauca* system, some of the *N. glauca* chromosomes are retained for several generations

(Kao, 1977). Perhaps it will not prove possible to predict the likely successes at the intergeneric crossing level. One factor may be the isolating mechanisms combining to prevent the sexual cross being achieved. Prezygotic incompatability of species may overcome easily if this is due to blocks imposed at pollen germination or pollen tube growth, yet there could still be a postzygotic barrier that may have a more direct effect on somatic hybrid cell formation. Clearly, there are no barriers to interkingdom or interfamilial cell fusion.

The capacity to culture cells and protoplasts of several plant species has enabled microbiological methods and concepts to be applied to higher plants. One of the major barriers to further progress lies in the apparent inability of many of the economically important species, such as the cereals, to culture protoplasts. This is certainly a problem if the development of selection procedures depends initially on growth in the system, but recent evidence suggests that division of one parent in a somatic hybridization scheme may be sufficient. For example, Kao and colleagues (1974) fused nondividing cereal protoplasts with dividing soybean protoplasts and observed more or less synchronous division of nuclei from both species in the heterokaryon. Likewise, in the *P. hybrida-P. parodii* system, selection using differential drug sensitivities or albino complementation relies on the inability of one parent to sustain division at the protoplast/cell level (Cocking et al., 1977; Power et al., 1977).

Somatic hybridization thus offers one approach, along with transformation and uptake studies, to modify plant cells. An increase in genetic variability, the basic aim of plant breeders, can be achieved via somatic hybridization. Somatic hybridization is attractive since whole plants can be recovered following fusion, contrary to the situation that exists in animal cell fusion studies. Loss of this regeneration capacity would be a more general problem, however, if biochemical mutants had to be produced for selection since this would often involve long periods of in vitro cell culture to isolate a suitable mutant. Thus selection methods to date have exploited easily detectable, naturally occurring mutation types such as chlorophyll-deficient mutants (Gleba et al., 1975), albinos (Cocking et al., 1977), and media and drug responses (Power et al., 1976, 1977). If somatic hybridization has a realistic role to play in plant breeding and crop improvement, then the techniques of protoplast isolation, fusion, and hybrid selection must be supported by adequate technology to permit regeneration of the intact plant.

6. Applications of somatic hybridization

Somatic hybridization must generally reflect what the plant breeder is attempting to achieve by using standard plant breeding techniques, but cannot accomplish because of a restriction of available gene pools. In many cases the genetic variability which is sought can be found within the genus. Plant improvement, brought about by combining genotypes of species that are sufficiently divergent to be sexually incompatible, can only be achieved via somatic hybridization.

Plants could thus be modified with respect to disease/host resistance and other similarly desirable characteristics such as flower, fruit, or seed quality and general growth considerations. It is quite likely that the raw somatic hybrid would be incorporated into a conventional breeding programme to transfer the newly acquired characteristics to the established crop or horticultural species. Somatic hybridization is therefore seen as an additional tool for the plant breeder and not as a substitute for conventional breeding techniques.

Since protoplast fusion results in the formation of a cytoplasmic hybrid (cybrid), transfer of cytoplasmically controlled features such as cytoplasmic male sterility could also be achieved via somatic hybridization. Similar procedures could be envisaged for the transfer of certain disease-resistant phenotypes from one species to another. Organelle transfer may be more readily achieved via protoplast fusion rather than attempts to induce direct uptake of isolated organelles. Somatic hybridization may have a role in the modification of vegetatively propagated species where the use of a sexual cycle is either impossible or undesirable. The modification of plants with respect to nitrogen fixation can also be contemplated. However, for some of these possibilities, such as nitrogen fixation (nif), other technologies such as, for example, uptake of plasmids bearing nif genes into plant cells and protoplasts may be more successful (Schell et al., 1976). Clearly, however, somatic hybridization will have a role in the future development and improvement of crop species. Improvements in our knowledge of the mechanisms involved in protoplast fusion will undoubtedly contribute to future progress in this technology since a better understanding of the fusion process will permit hybridization and other modifications in protoplast properties to be engineered more accurately than is possible today.

References

Ahkong, Q. F., Fisher, D. Tampion, W. and Lucy, J. A. (1975a) Mechanisms of cell fusion. Nature 253, 194–195.

Ahkong, Q. F., Howell, J. I., Lucy, J. A., Safwat, F., Davey, M. R. and Cocking, E. C. (1975b) Fusion of hen erythrocytes with yeast protoplasts induced by polyethylene glycol. Nature 255, 66–67.

Ahkong, Q. F., Tampion, W. and Lucy, J. A. (1975c) Promotion of cell fusion by divalent cation ionophores. Nature 256, 208–209.

Anné, J. and Peberdy, J. F. (1975) Conditions for induced fusion of fungal protoplasts in polyethylene glycol solutions. Arch. Microbiol. 105, 201–205.

Bajaj, Y. P. S. (1977) Protoplast isolation, culture and somatic hybridisation. In: Plant Cell, Tissue and Organ Culture (Reinert, J. and Bajaj, Y. P. S., eds.) pp. 467–496, Springer-Verlag, Berlin.

Banks, M. S. and Evans, P. K. (1976) A comparison of the isolation and culture of mesophyll protoplasts from several Nicotiana species and their hybrids. Plant Sci. Lett. 7, 409–416.

Bhojwani, S. S. and Cocking, E. C. (1972) Isolation of protoplasts from pollen tetrads. Nature New Biol. 239, 29–30.

Binding, H. (1966) Regeneration unter verschmelzung nackter laubmoosprotoplasten. Z Pflanzenphysiol. 55, 305–321.

Binding, H. (1974) Fusionversuche mit isolierten protoplasten von Petunia hybrida L. Z. Pflanzenphysiol. 72, 422–426.

Burgess, J. and Linstead, P. J. (1976) Ultrastructural studies of the binding of concanavalin A to the plasmalemma of higher plant protoplasts. Planta (Berlin) 130, 73–79.

Carlson, P. S., Smith, H. H. and Dearing, R. D. (1972) Parasexual interspecific plant hybridisation. Proc. Nat. Acad. Sci. U.S.A. 69, 2292–2294.

Cocking, E. C., George, D., Price-Jones, M. J. and Power, J. B. (1977) Selection procedures for the production of interspecies somatic hybrids of *Petunia hybrida* and *Petunia parodii*. II Albino complementation selection. Plant Sci. Lett. 10, 7–12.

Deka, P. C., Mehra, A. K., Pathak, N. N. and Sen, S. K. (1977). Isolation and fusion studies on protoplasts from pollen tetrads. Experimentia 33, 182–184.

Dudits, D., Raski, I., Hadlaczky, C. Y. and Lima-de-Faria, A. (1976) Fusion of human cells with carrot protoplasts induced by polyethyleneglycol. Hereditas 82, 121–124.

Eriksson, T. (1971) Isolation and fusion of plant protoplasts. In: Les Cultures de Tissus de Plantes, Colloques Internationaux C. N. R. S. No 193, pp 297–302. Strasbourg, France.

Fodor, K. and Alföldi, L. (1976) Fusion of protoplasts of *Bacillus megaterium*. Proc. Nat. Acad. Sci. U.S.A. 73, 2147–2150.

Fowke, L. G., Rennie, P. J., Kirkpatrick, J. W. and Constabel, F. (1975) Ultrastructural characteristics of intergeneric protoplast fusion. Can. J. Bot. 53, 272–278.

Gingell, D. (1968) Computed surface potential changes with membrane interaction. J. Theoret. Biol. 19, 340–346.

Gleba, Y. Y., Butenko, R. G. and Sytnik, K. M. (1975) Fusion of protoplasts and parasexual hybridisation in *Nicotiana tabacum*. Dokl. Akad. Nauk S.S.S.R. 221, 1196–1198.

Glimelius. K. Wallin, A. and Eriksson, T. (1974) Agglutinating effects of concanavalin A on isolated protoplasts of *Daucus carota*. Physiol. Plant. 31, 225–230.

Grout, B. W. W. and Coutts, R. H. A. (1974) Additives for the enhancement of fusion and endocytosis in higher plant protoplasts: an electrophoretic study. Plant Sci. Lett. 2, 397–403.

Grout, B. W. W., Willison, J. H. M. and Cocking, E. C. (1972) Interactions at the surface of plant cell protoplasts: an electrophoretic and freeze-etch study. J. Bioenerg. 4, 311–328.

Hartman, J. X., Kao, K. N., Gamborg, O. L. and Miller, R. A. (1973) Immunological methods for the agglutination of protoplasts from cell suspension cultures of different genera. Planta (Berlin) 112, 45–56.

Hofmeister, L. (1954) Mikrurgische unterschung über die geringe fusionsneigung plasmolysierter, nackter pflanzen-protoplasten. Protoplasma 43, 278–326.

Ito, M. and Maeda, M. (1973) Fusion of meiotic protoplasts in liliaceous plants. Exp. Cell Res. 80, 453–456.

Jones, C. W., Mastrangelo, I. A., Smith, H. H., Liu, H. Z. and Meck, R. A. (1976) Interkingdom fusion between human cells and tobacco hybrid protoplasts. Science 193, 401–403.

Kameya, T. (1973) The effects of gelatin on aggregation of protoplasts from higher plants. Planta (Berlin) 115, 77–82.

Kameya, T. (1975) Induction of hybrids through somatic cell fusion with dextran sulphate and gelatin. Jap. J. Genet. 50, 235–246.

Kameya, T. and Takahashi, N. (1972) The effects of inorganic salts on fusion of protoplasts from roots and leaves of *Brassica* species. Jap. J. Genet. 47, 215–218.

Kao, K. N. (1977) Chromosomal behaviour in somatic hybrids of soybean-*Nicotiana glauca*. Mol. Gen. Genet. 150, 225–230.

Kao, K. N. and Michayluk, M. R. (1974) A method for high-frequency intergeneric fusion of plant protoplasts. Planta (Berlin) 155, 355–367.

Kao, K. N., Constabel, F., Michayluk, M. R. and Gamborg, O. L. (1974) Plant protoplast fusion and growth of intergeneric hybrid cells. Planta (Berlin) 120, 215–227.

Keller, W. A. and Melchers, G. (1973) The effect of high pH and calcium on tobacco leaf protoplast fusion. Z. Naturforsch. 28c, 737–741.

Klerker, J. A. F. (1892) Eine methode zur isolierung lebender protoplasten. Oefvers. Betensk. Akad. Forh. Stokh. 9, 463–471.

Kuster, E. (1909) Uber die verschmelzung nackter protoplasten. Ber. Dtsch. Bot. Ges. 27, 589–598.

Kuster, E. (1910a) Eine method zur gewinnung abnorm grosser protoplasten. Wilhem Rouxs Arch. Entwicklungsmech. Org. 30, 351–355.

Kuster, E. (1910b) Uber veränderungen des plasmaober flache bei plasmolyse. Zeitschr. Bot. 2, 689–717.

Kuster, E. (1928) Uber die gewinnung nackter protoplasten. Protoplasma 3, 223–233.

Lai, K-L. and Liu, L-F. (1976) On the isolation and fusion of rice protoplasts. J. Agr. Assoc. China 93, 10–29.

Lorey, E. (1929) Mikrochirurgische untersuchungen über die viskosität des protoplasmas. Protoplasma 7, 171–178.

Lucy, J. A. (1974) Lipids and membranes. FEBS Lett. 40, S105–111.

Lucy, J. A., Ahkong, Q. F., Cramp, F. C., Fisher, D. and Howell, J. I. (1972) Studies on chemically induced cell fusion. J. Cell Sci. 10, 769–787.

Maggio, B., Ahkong, Q. F. and Lucy, J. A. (1976) Polyethyleneglycol, surface potential and cell fusion. Biochem. J. 158, 647–650.

Melchers, G. and Labib, G. (1974) Somatic hybridisation of plants by fusion of protoplasts. I Selection of light resistant hybrids of 'haploid' light sensitive varieties of tabacco. Mol. Gen. Genet. 135, 277–294.

Mexal, J., Fischer, J. T., Osteryoung, J. and Patrick Reid, C. P. (1975) Oxygen availability in polyethylene glycol solutions and its implications in plant-water relations. Plant Physiol. 35, 20–24.

Michel, W. (1937) Über die experimentelle fusion pflanzlicher protoplasten. Arch. Exp. Zellforsch. Bas. Geweb. 20, 230–252.

Missbach, G. (1928) Versuche zur prüfung der plasmaviskositat. Protoplasma 3, 327–343.

Pilet, P. E. and Senn, A. (1974) Effect du Ca^{++} et du K^+ sur la mobilite electrophoretique des protoplastes. C. R. Acad. Sci. 278, 269–272.

Plowe, J. Q. (1931) Membranes in the plant cell. I. Morphological membranes at protoplasmic surfaces. Protoplasma 12, 196–220.

Pontecorvo, G. (1975) Production of mammalian somatic cell hybrids by means of polyethyleneglycol treatment. Somatic Cell Genet. 1, 397–400.

Poste, G. and Allison, A. C. (1973) Membrane fusion. Biochim. Biophys. Acta 300, 421–465.

Potrykus, I. (1972) Fusion of differentiated protoplasts. Phytomorphology, 22, 91–96.

Potrykus, I. (1973) Transplantation of chloroplasts into protoplasts of Petunia. Z Pflanzenphysiol. 70, 364–366.

Power, J. B. and Cocking, E. C. (1968) A simple method for the isolation of very large numbers of leaf protoplasts by using mixtures of cellulases and pectinases. Biochem. J. 111, 53.

Power, J. B. and Cocking, E. C. (1971) Fusion of plant protoplasts. Sci. Progr. (Oxford) 59, 181–198.

Power, J. B. and Frearson, E. M. (1973) The inter- and intraspecific fusion of plant protoplasts; subsequent developments in culture, with reference to crown gall callus and tobacco and petunia leaf systems. In: Colloques Internationaux C. N. R. S. No 212. Protoplastes et fusion de cellules somatique végétales (Tempe, J., ed.) pp. 409–421, Versailles, France.

Power, J. B., Cummins, S. E. and Cocking, E. C. (1970) Fusion of isolated plant protoplasts. Nature 225, 1016–1018.

Power, J. B., Frearson, E. M., Hayward, C. and Cocking, E. C. (1975) Some consequences of the fusion and selective culture of Petunia and Parthenocissus protoplasts. Plant Sci. Lett. 5, 197–207.

Power, J. B., Frearson, E. M., Hayward, C., George, D., Evans, P. K., Berry, S. F. and Cocking, E. C. (1976) Somatic hybridisation of Petunia hybrida and P parodii. Nature, 263, 500–502.

Power, J. B., Berry, S. F., Frearson, E. M. and Cocking, E. C. (1977) Selection procedures for the production of interspecies somatic hybrids of Petunia hybrida and Petunia parodii. I. Nutrient media and drug sensitivity complementation/selection. Plant Sci. Lett., 10, 1–6.

Schaeffer, P., Cami, B. and Hutchkiss, R. D. (1976) Fusion of bacterial protoplasts. Proc. Nat. Acad. Sci. U.S.A. 73, 2151–2155.

Schell, J., Van Montagu, M., De Picker, A., Da Waele, D., Engler, G., Genetello, C., Hernalsteens, J. P., Holsters, M., Messens, E., Silva, B., Van den Elsacker, S., Van Larebeke, N. and Zaenen, I. (1976) Crown-gall: Bacterial plasmids as oncogenic elements for eucaryotic cells. In: Molecular Biology of Plants (Rubenstein, I., ed.). Academic Press, New York.

Schieder, O. (1975) Selection of a somatic hybrid between auxotrophic mutants of Sphaerocarpos donnellii using the method of protoplast fusion. Z. Pflanzenphysiol. 74, 357–365.

Smith, H. H., Kao, K. N. and Combatti, N. C. (1976) Interspecific hybridisation by protoplast fusion in Nicotiana. Confirmation and extension. J. Hered. 67, 123–128.

Takebe, I., Otsuki, Y. and Aoki, S. (1968) Isolation of tobacco mesophyll cells in intact and active state. Plant Cell Physiol. 9, 115–124.

Toister, Z. and Loyter, A. (1973) The mechanism of cell fusion. II. Formation of chicken erythrocyte polykarons. J. Biol. Chem. 248, 422–432.

Vos, J., Ahkong, Q. F., Botham, G. M., Quirk, S. J. and Lucy, J. A. (1976) Changes in the distribution of intramembraneous particles in hen erythrocytes during cell fusion induced by the bivalent-cation ionophore A 23187. Biochem. J. 158, 651–653.

Wallin, A., Glimelius, K. and Eriksson, T. (1974) The induction of aggregation and fusion of *Daucus carota* protoplasts by polyethylene glycol. Z. Pflanzenphysiol. 74, 64–80.

Wetter, L. R. (1977) Isoenzyme patterns in soybean—*Nicotiana* somatic hybrid cell lines. Mol. Gen. Genet. 150, 231–235.

Withers, L. A. (1973) Plant protoplast fusion: methods and mechanisms. In: Colloques Internationaux C. N. R. S. No. 212. Protoplastes et fusion de cellules somatique végétales (Tempe, J. ed.) pp. 517–545, Versailles, France.

Withers, L. A. and Cocking, E. C. (1972) Fine structural studies on spontaneous and induced fusion of higher plant protoplasts. J. Cell Sci. 11, 59–75.

Yoshida, Y. (1961) Role of nucleus in cytoplasmic activities with special reference to the formation of surface membrane in *Elodea* leaf cells. Plant Cell Physiol. 2, 139–150.

Endocytosis: regulation of membrane interactions

9

Paul J. EDELSON and Zanvil A. COHN

Contents

G. Poste & G. L. Nicolson (eds.) Membrane Fusion, pp. 387–405.
© Elsevier/North-Holland Biomedical Press, 1978.

1. The modern concept of endocytosis

This review has two main goals. First, it intends to provide a brief description of the endocytic process as it is currently understood. Second, it discusses several questions concerning the regulation of this process that are not yet resolved, but for which some recent experimental results are relevant. Several useful reviews of endocytosis and its various aspects include two recent articles which provide much information not presented here (Korn, 1975, Silverstein et al., 1977) as well as an outstanding book with much pertinent material (Holtzman, 1976).

Like diffusion and the various active transport processes, endocytosis is a process for internalizing components of the extracellular environment. However, in contrast to material interiorized by the two other processes, endocytized material is kept segregated from the cell contents, preserving the cytoplasmic integrity. With the exception of certain viral mechanisms for penetration, it is essentially the only physiologic mechanism for the cellular uptake of macromolecules, or particles, and appears to proceed constitutively in nearly every eukaryotic cell examined.

Perhaps a few comments should be made here about the terms popularly used in this area. Endocytosis can be divided into *pinocytosis* (or cell-drinking) and *phagocytosis* (or cell-eating). Phagocytosis generally refers to the adsorptive interiorization of particles larger than 100 nm in diameter. Such adsorption may involve specific interactions between the particle and specific plasma membrane binding sites, as with immunoglobulin G-coated particles bound to macrophage plasma membranes, or it may be a nonspecific physical phenomenon, as occurs in the binding of polystyrene particles. Pinocytosis, on the other hand, refers to the uptake of fluid and to material suspended or dissolved in the fluid and interiorized in the bulk phase.

Adsorptive pinocytosis and phagocytosis typically display classical saturability with increasing loads. Fluid-phase pinocytosis, however, is a low-affinity, high-capacity process which is generally independent of load and not saturable. These characteristics have implications for the selection and validation of suitable marker molecules for the quantitation of bulk-phase pinocytosis. The apparent fluid *volume* endocytized, as opposed to the total quantity of marker interiorized, must be independent of the concentration or total load of the marker used. A concentration dependence would imply that the marker was in fact being interiorized after adsorption to the plasma membrane, rather than free in the fluid phase.

Endocytosis is also strongly temperature-dependent (Steinman et al., 1974) and is completely inhibited at 0°C. This is a particularly useful diagnostic tool for distinguishing internalization of a test marker from its simple adsorption to the cell surface, since no endocytic internalization would be expected at ice temperatures, though surface binding could still proceed. As endocytosis also depends on metabolic energy (Steinman et al., 1974), careful use of inhibitors may help to distinguish binding from actual uptake.

Schematically, endocytosis is a sequence of distortions and reorganizations of plasma membrane. To begin, an area of plasma membrane invaginates to form a pouch on the cytoplasmic side of the cell surface. In the first fusion step in the process, the rim of the pouch seals, closing it with a continuous layer of plasma membrane. As this happens, the pouch itself separates from the plasma membrane, budding into the cytoplasm as a free, closed vesicle, whose contents are derived from the extracellular environment. These initial vesicles have a variety of sizes, but they are on the average about 200 nm in diameter (Steinman et al., 1976). These small pinocytic vesicles then travel centripetally, accumulating in the cell centrosphere. This perinuclear area also contains the Golgi apparatus and the centriole, which is the terminus of the network of cytoplasmic microtubules.

Once formed, and probably throughout the course of their movement toward the nucleus, endocytic vesicles may and do fuse with each other, with primary lysosomes, and with secondary lysosomes. In most cells these fusions are regular and extensive, often occurring within a few minutes of the generation of the endocytic vesicle (Steinman et al., 1976). However, the fusion partner is quite strictly defined, and fusions never involve such membrane-bounded structures as mitochondria, the nucleus, or the endoplasmic reticulum. It is not certain whether fusions may occur between endocytic vesicles and the Golgi lamellae. Some recent information on the control of this fusion process and some potentially important regulatory factors are discussed below.

The lysosomal granules with which these pinosomes and phagosomes may fuse are also divided into two classes. *Lysosomes* are membrane-bounded cytoplasmic organelles containing acid hydrolases. As first generated, whether from the Golgi apparatus itself, or from the Golgi-associated endoplasmic reticulum (GERL), the granules are referred to as *primary lysosomes*. *Secondary lysosomes* are the result of the fusion of these primary lysosomes with incoming endocytic vesicles, and are therefore hybrid organelles with contributions from both internal membrane systems and the plasma membrane.

The morphology of endocytosis was initially established by the work of Lewis on explanted peritoneal cells (Lewis, 1931) and perhaps even earlier by the studies which Metchnikoff carried out on phagocytosis in the starfish (Metchnikoff, 1893). Later, considerable work was done by workers at the Carlsberg Laboratory, using the large amoebas, *Chaos chaos* and *Amoeba proteus* (review, Holter, 1965). More recently, the soil amoeba, *Acanthamoeba castellani,* the mouse peritoneal macrophage, and polymorphonuclear leukocytes of various species have been favorite systems for study. Some work has also been carried out on the endocytic activity of various tissue culture lines, and these are now becoming more popular objects of investigation.

More importantly, quantitative information on rates of basal and stimulated pinocytosis has now been obtained in several systems, and these give a new perspective on the physiologic implications of the process for the economy of the cell. To accurately assess the rate of fluid pinocytosis, it is necessary to use a marker which (1) is completely soluble in the extracellular medium, (2) does not adsorb to the cell surface before being interiorized, (3) does not itself stimulate or inhibit the endocytic process, and (4) is not metabolized so rapidly that its intracellular accumulation over a short period of time (a few hours) is significantly affected. In addition, the assay system must ensure that no contaminating extracellular marker is included in the measurement and that the method can reproducibly detect uptake of about 1 part in 10^6. Radiolabeled compounds have been satisfactorily used for measuring rates of pinocytosis (Wagner et al., 1971; Bowers and Olszewski, 1972) as has the soluble enzyme horseradish peroxidase (Steinman and Cohn, 1972). Phagocytic rates have been assessed in neutrophils with a protein-lipid emulsion (Stossel, 1973), as well as by various particle-counting techniques.

Quantitative studies in the soil amoeba and the mouse macrophage have yielded strikingly similar qualitative conclusions with some interesting quantitative differences. Both cells exhibit a constitutive basal pinocytic rate in culture. For macrophages this rate is about 30 to 50 nl/10^6 cells/hour (Steinman and Cohn, 1972; Edelson et al., 1975), while in Acanthamoeba the rate is 1600 to 2400 nl/10^6 cells/hour (Bowers and Olszewski, 1972). In each case the pinocytic rate is independent of the concentration of marker used for the measurements, and the initial rate of uptake measured is maintained for at least several hours. In both cases no uptake is detected at 0°C. In Acanthamoeba it was estimated that a volume equivalent to 22% of the cell volume was interiorized each hour, while macrophages, because of their smaller volume, interiorize the equivalent of about 25% of their cell volume per hour. Recent studies in the L cell (Steinman et al., 1976) estimate that these cells interiorize about 3% of their cell volume per hour.

Macrophages interiorize about two cell surface equivalents per hour, while L cells interiorize about 50% of their surface area per hour. It is likely that when the pinocytic rate is elevated after physiologic stimulation (Edelson et al., 1975) or upon exposure to the lectin concanavalin A (Edelson and Cohn, 1974a), the surface area and volume turnover rates for the macrophage are still further increased. Bowers and Olszewski (1972) made similar observations of extensive membrane internalization in Acanthamoeba. The implications of these measurements for the economy of plasma membrane are discussed in a later section.

This introduction raises several questions to which the remainder of this review will be devoted. First, what determines the intrinsic pinocytic rate in a given cell type? And, what are the controls and how do they operate in modulating this intrinsic rate? Second, what is the physiologic meaning of the plasma membrane binding sites? Are these sites receptors, whose interactions with specific ligands can have particular physiologic consequences for the cell? Third, how is the membrane organized for endocytosis? Are there membrane areas specialized for the process or is endocytosis an activity characteristic of all portions of the plasma membrane? Fourth, how is phago-lysosome formation controlled, and what accounts for the specificity of the fusion process? Fifth, what determines the fate of the secondary (2°) lysosome membrane, and what is the overall economy of membrane in the endocytic cycle?

2. What determines the intrinsic pinocytic rate?

It is clear that for most cells pinocytosis, whether examined in vitro or in vivo, is an ongoing process. A characteristic pinocytic rate may be measured for cells cultivated even in fairly simple, defined media, and there is no evidence that special extracellular inducers are necessary to stimulate episodic bursts of interiorization. Although the rate at which a cell may pinocytize may be sensitive to a variety of stimulating or inhibiting agents, pinocytosis, at some level, appears to go on continuously. It is also clear that each cell variety studied so far has its own characteristic rate of pinocytosis, and while these may be considerably different among different cell types, the rate is consistent among different samples of the same cell variety.

These observations lead us to the concept of a *basal pinocytic rate* characteristic of each cell type. While this is obviously an idealization, it may be thought of as the level of pinocytic activity a cell would have in the absence of any extracellular influences. Obviously we must be careful not to assume that cells maintained in the usual ways in culture are not being exposed to external influences on their pinocytic activity, yet it seems useful to consider that in the absence of evidence to the contrary such cultivated cells are essentially in a basal state. Clearly, under comparable culture conditions, not only do different cell types have different characteristic rates, but certain manipulations can cause a given cell type to change its characteristic basal rate.

Recently Heiniger and co-workers (1976) demonstrated that the chemical composition of the plasma membrane may be an important influence on the basal pinocytic rate. Kandutsch and Chen (1975) had previously shown that 25-hydroxycholesterol and 7-ketocholesterol are each potent inhibitors of the enzyme 3-hydroxy-3-methylglutaryl-(HMG)-coA reductase, the regulatory enzyme in the sterol synthetic pathway. Thus, administration of either of these compounds to L cells can reduce the sterol content of whole cells to about one-half of their normal levels and may also decrease the sterol content of plasma membrane fractions. Under these conditions, the pinocytic rate is reduced to about 45% of control levels. Either exogenous cholesterol or the sterol precursor, mevalonic acid, can reverse this depression.

Membrane phospholipid composition may also be important in regulating normal endocytic activity. Eileen Mahoney has observed (private communication) that cultured mouse peritoneal macrophages whose saturated fatty acid content has been enriched show striking changes in the way that their ability to ingest antibody-sensitized red blood cells varies with temperature. Phagocytosis by such cells appears to be inhibited at a higher temperature than in cells with a normal lipid composition, as would be expected if the flexibility of the plasma membrane were an important factor in its endocytic behavior.

The results of Heiniger and his colleagues, and Mahoney's results, as well as the striking temperature dependence of endocytosis, would be consistent with a requirement for a reasonably flexible physical state of the plasma membrane in order for endocytosis to proceed. Whether such a requirement would have regulatory implications or would simply play a permissive role in determining endocytic activity is, of course, an open question.

Another interesting, and as yet unexplained, observation is that for certain cell types cell density in culture may affect the "basal" pinocytic rate. L cells (Steinman et al., 1974) and two varieties of epithelial cell lines (Kaplan, 1976), MK (rhesus monkey kidney) and HeLa, increase their pinocytic rate with increasing cell density. This effect is not reproduced by exposing sparse cultures to medium conditioned by confluent cells, and reverses only slowly when dense L-cell cultures are replated at lower cell densities. Thus, this phenomenon does not appear to be caused by the direct action of some diffusible substance accumulated in the extracellular fluid of these cultures. It may, however, represent an indirect nutritional effect of the culture conditions, and thus might not actually represent an effect on the basal pinocytic rate. Not all cells show this density effect. Both 3T3 cells and primary calf embryo fibroblasts show no change in their pinocytic rates when maintained in nongrowing confluent conditions for several days (Steinman et al., 1974). It does not seem likely that this phenomenon is related to a density-dependent inhibition of growth, as L cells in all phases of the cell cycle appear to pinocytize at qualitatively similar rates (Steinman et al., 1974), but its relation to such other density-dependent phenomena as the inhibition of cell motility is not known.

3. Do plasma membrane binding sites function as endocytic receptors?

It is useful to distinguish the notion of a binding site from that of a receptor. A binding site may be considered to be any membrane component which shows a specific, high-affinity, saturable interaction with a specified ligand. When the interaction of a ligand and a binding site has a subsequent physiologic effect on the cell, beyond the mere association of the ligand and site, then the binding site is a receptor. Thus, we can restate the question heading this section and ask what significance the binding of a ligand to the plasma membrane has for the subsequent interiorization of that ligand. Are adsorbed materials simply internalized passively when the bit of membrane to which they are linked is taken in, or does the interaction of the ligand with its binding site generate some intracellular signal specifically to promote the uptake of the complex?

In many examples the formation of a ligand-binding site complex is followed by the interiorization of that complex in a pinocytic or phagocytic vesicle. In some of these cases endocytosis may be incidental to the binding, while in others it is an essential step in the ligand-cell interaction. The reovirus replicative cycle (Silverstein and Dales, 1968) and the regulation of the cholesterol synthetic pathway by low-density lipoproteins (Goldstein and Brown, 1974) are two very closely studied systems in which ligand-binding site complexes are clearly and regularly interiorized in endocytic vesicles. Other examples would be the endocytosis of surface "caps" of antigen-antibody complexes (Taylor et al., 1971), and perhaps the regulation of membrane binding sites by ligands (e.g., Lesniak and Roth, 1976).

Another interesting example of ligand-plasma membrane interaction is offered by a group of bacterial and plant proteins which are toxic to mammalian cells, including Ricin and Abrin (Olsnes et al., 1974), and diphtheria toxin (Collier, 1975).

Each of these proteins is a dimer composed of an A, or enzymatically active, subunit linked by a disulfide bridge to a B, or binding, subunit which is the plasma membrane ligand. The A-subunit acts intracellularly, but its entry into the cells is preceded by a complex binding phase mediated by the B-subunit. Although A-subunits are enzymatically active when exposed to broken cell preparations, the isolated subunits are ineffective against intact cells. The role of the B-subunit appears to be limited to mediating the interaction of the toxin with the plasma membrane. Hybrid molecules containing an A-subunit from either Abrin or Ricin and a B-subunit from the other protein are active toxins (Olsnes et al., 1973). Ricin B-chain may be inactivated with antibody without detoxifying the molecule if the Ricin-antibody complex can still bind to and enter the cell, as for example by interaction with binding sites for the Fc portion of immunoglobulin G (Refsnes and Munthe-Kaas, 1976).

Refsnes and colleagues (1974) have shown that entry of Abrin or Ricin into Ehrlich ascites cells is a temperature-sensitive process which follows the initial attachment of the B-subunit to the plasma membrane. For a brief period of time (up to 10 min after binding), the bound toxin can be neutralized with an antiserum to the A-subunit, but sensitivity to the antitoxin rapidly diminishes when the cells are furthered incubated at 37°C. Refsnes and co-workers interpret these results as reflecting a fairly brief period of toxin binding followed by clearing of the toxins from the plasma membrane by endocytosis. Whether both the A and B-subunits are internalized as a unit, or whether the linking disulfide bond may be cleaved before internalization of the A-subunit is an unresolved question. Additionally, the route by which the A-subunit escapes from the endocytic vesicle and actually reaches the cytoplasm is also unknown, though Nicolson (1974) has suggested that the enzyme may simply be released from broken pinosomes.

Refsnes and associates have called attention to the structural similarity of Abrin and Recin to botulinus and tetanus toxin as well as to nerve growth factor and epidermal growth factor. It is intriguing that many cytoplasmically active molecules occur as sulfhydryl-bonded dimers in which the active subunit is linked to a binding subunit. Adsorptive endocytosis may be an important mechanism for the uptake of a wide variety of hormones, toxins, and other agents, as both Refsnes and co-workers and Boquet and Pappenheimer (1976) point out.

Diphtheria toxin and, perhaps, the analogously active *Pseudomonas aeruginosa* exotoxin (Iglewski and Kabat, 1975) appear to enter cells in a similar way. Extensive studies in Pappenheimer's laboratory have used nontoxic products of mutant genes allelic to the bacteriophage *tox* gene, designated "cross-reacting materials," or crm, to demonstrate that the interaction of diphtheria toxin with toxin-sensitive cells is a saturable process (Gill et al., 1973). The cell surface structure responsible for the toxin sensitivity of human cells appears to be coded for by a gene on chromosome 5 (Creagen et al., 1975) and is presumably the membrane binding site for the toxin. Boquet and Pappenheimer (1976) have recently examined the uptake of radioiodinated toxin by HeLa cells (which are sensitive to the toxin) and mouse L929 cells (which are resistant). In these experiments a specific, saturable component and a nonspecific, apparently nonsaturable component could be demonstrated in HeLa cells, incubated at 30°C with labeled toxin either in the absence or presence of excess unlabeled toxin. Uptake of the variant protein CRM 45, which lacks the COOH terminal 17,000 of the B-chain, shows only a nonspecific component. Uptake of native toxin by mouse L cells shows no specific component, consistent with their lack of the toxin-binding site. Both specific, receptor-mediated uptake and nonspecific uptake are temperature-sensitive processes which are strongly inhibited below 20°C.

Although bulk internalization of toxin in the fluid phase appears insufficient for intoxication, specific uptake by adsorptive pinocytosis followed by escape of the active toxin into the cytoplasm appears compatible with the data presented. Whether bulk-phase uptake is quantitatively insufficient, or whether a specific interaction of the toxin with its membrane-binding site is required for generating an active toxin fragment is unclear, though the experiments of Refsnes and Munthe-Kaas using Ricin suggest that the membrane-binding site does not play a direct role in generating active toxin or in allowing its escape from the endocytic vesicle into the cytoplasm. In addition, we do not know when or how the disulfide bond between the subunits is cleaved, or what the fate of the B-subunit is.

Perhaps the most classic example of ligand-mediated endocytosis is the phenomenon of opsonization. In this case specific antibody of appropriate class, when allowed to coat an otherwise nonphagocytizable particle, permits the particle to be readily ingested by macrophages or neutrophils. This phenomenon is mediated by specific binding sites on the plasma membrane of the phagocyte, which interact strongly with the Fc portion of the IgG molecule, and thus bind the complex to the cell.

One explanation of this phenomenon is that the interaction of the opsonizing antibody with the plasma membrane binding site results in a cytoplasmic signal, or trigger, which directs the cell's attention to the bound particle and stimulates its rapid ingestion. Another possibility is that the bound particle shares the fate of the portion of membrane to which it is bound, and is therefore interiorized passively in the normal cycle of plasma membrane uptake, which accounts for the basal pinocytic activity observed.

The morphology of phagocytosis of such opsonized particles has been carefully examined by Griffin and Silverstein and their colleagues (Griffin et al., 1975). An important observation of theirs was that the initial ligand-binding site interaction is in itself not sufficient for particle uptake over an observation period of an hour or so. However, the initial interaction is followed by subsequent sequential ligand-plasma membrane interactions involving an increasingly greater fraction of the particle surface. Thus the cell goes through a process of "spreading attachment" until its plasma membrane has enveloped essentially the entire particle. This sequential adherence has been referred to by these authors as the "zipper" mechanism of ingestion. An important point this work makes is that the process is not completely determined once the first ligand-membrane interaction is made. Cells prepared so that only a portion of their surface is coated with immunoglobulin will interact with the phagocyte only to the margin of their opsonizing coat (Griffin et al., 1976). Once the plasma membrane extensions have reached the boundary of the sensitized area of the particle, no further interaction takes place. The process of ingestion is stopped in midcourse, and the particles are not interiorized.

These observations are compatible with at least two alternate models of the processes of adsorptive pinocytosis and ligand-mediated ·phagocytosis. In one model, uptake would be specifically triggered by the interaction of the ligand and its binding site. Binding would generate a cytoplasmic signal which would induce the uptake of the ligand-membrane complex. Such an *active* model would imply that membrane ligands could stimulate the rate of endocytosis of the cell to which they bind.

An alternative would be that the interaction of a ligand with its binding site would not specifically stimulate endocytosis. A bound ligand would be interiorized only as a result of the normal endocytic uptake of the area of membrane to which it had been bound. Thus the only role of the binding site would be to attach the ligand to the plasma membrane until it has been *passively* interiorized. In this model no distinction is made between the functions of the various binding sites, so that, given enough time, any bound ligand would be expected to be interiorized. In the "active" model some binding sites might be inert, so that their ligands would not be interiorized, while other binding sites would be endocytic receptors able to promote ingestion of their bound ligands.

Some experimental data support one aspect of the active model. Reaven and Axline (1972) have observed the organization of a subplasmalemmal network of microfilaments beneath an area of plasma membrane adjacent to an adherent polystyrene particle. Other observations on the association of cytoplasmic filaments with forming phagolysosomes have been made by Korn and colleagues (1974). Such a network would be taken as a morphologic consequence of the "signal" generated as a result of particle binding. The meaning of this result is somewhat ambiguous, as polystyrene particles do not appear to interact with specific membrane binding sites but probably interact with the cell as a result of simple physical forces. Thus, this data might also be taken to indicate that there is no need for the mediation of a biologically active receptor for the organization of endocytic activity. Another complication is that the latex particles studied by Reaven and Axline were chosen, for technical convenience, to be so large that they are not ingestible. Changes seen as a result of the interaction of such particles with the cell may only mirror the endocytic process at arm's length.

Another set of observations might also be naturally interpreted in terms of the active model. A plasma membrane binding site for a fragment of the third component of complement (C3b) has been described in the mouse peritoneal macrophage (Bianco et al., 1975). However, the physiology of this binding site is quite different from that for IgG. Erythrocytes sensitized with C3b readily bind to the peritoneal macrophage, but over several hours they fail to be ingested (Griffin et al., 1975). When similar experiments are carried out using inflammatory peritoneal macrophages rather than the resident population used for the initial experiments, the C3b-coated red cells are rapidly ingested. Thus, the C3b binding site fails to mediate ingestion in resident cells, but is quite able to do so in stimulated cells. One explanation for this phenomenon would be that in order for a binding site to mediate ingestion it would need to be coupled to some intracellular process, making it a receptor. Then, the C3b binding site would be a receptor in the inflammatory cell but would be uncoupled in the resident cell. However, it may be unwarranted to assume that the two C3b binding sites are qualitatively and quantitatively identical in the two cell varieties, or that other conditions which may modify endocytic activity of the cell are comparable in the two cell types. Actually, the basal pinocytic rate is 3 to 4 times higher in the inflammatory cell than it is in the resident cell (Edelson et al., 1975). The uptake of C3-sensitized erythrocytes by inflammatory cells might therefore be interpreted as simply reflecting a general enhancement of endocytic activity in these cells. Consistent with this interpretation is the observation that inflammatory cells also show an enhanced uptake of IgG-coated cells, which interact with the Fc binding site. If one argued that the C3 binding site mechanism is different in the inflammatory cell, one would also have to argue that another change has been effected in the Fc binding site of that cell.

Another set of observations that may appear to support the active model are those we have made on the effects of concanavalin A (con A) on the pinocytic rate of mouse peritoneal macrophages (Edelson and Cohn, 1974a). Con A binds to mannosylated glycoproteins on the surface of the macrophage, and stimulates about a three-fold increase in the pinocytic rate of the cells. This seems to indicate that interaction of a ligand with a plasma membrane binding site stimulates the uptake of the complex at a rate greater than that for unfilled sites. However, succinylated con A, which by competition experiments appears to bind to the same sites as the native molecule, does not stimulate endocytosis (Edelson, unpublished observations). Therefore, simple ligand-binding site interaction is not sufficient to promote uptake. This also seems to be true when diphtheria toxin binds to human HeLa cells. Again, ligand-plasma membrane interaction per se does not appear to stimulate fluid uptake (Boquet and Pappenheimer, 1976).

It is possible, as Schekman and Singer (1976) has suggested, that the clustering of membrane binding sites by multivalent ligands specifically promotes membrane interiorization. However, it is also possible that multivalent ligands promote endocytosis by binding disparate membrane areas together long enough for them to fuse, thus forming a closed pinocytic vesicle.

Obviously, resolution of these issues is important to the construction of a more exact model of plasma membrane function. However, this will require the development of methods for assessing not only the bulk uptake of extracellular markers, but also techniques for measuring the rates of internalization of specific membrane components in both free and complexed forms.

4. How is the membrane organized for endocytosis?

There are many examples of cells whose plasma membranes are organized so that distinct functions are alloted to specific membrane regions. There are at least two levels on which the cell can organize its endocytic activity so that only certain components of the plasma membrane are involved. First, the membrane may have specialized endocytic sites that are the only areas which can generate pinocytic or phagocytic vesicles. Such sites may be distributed randomly over the cell surface, or they may be grouped to form an "endocytic face" at one cell pole or discrete "endocytic windows" clustered throughout the plasma membrane.

A second level of control would involve the particular inclusion or exclusion of certain membrane components from a nascent pinosome or phagosome. Thus the endocytic vesicle membrane would not reflect the composition of the plasma membrane as a whole, but would show relative enrichments and depletions of specific components.

Topographic organization of endocytosis and the vectorial transport of materials interiorized at one cell face and discharged at another seems especially characteristic of cells whose main role is protein transport. An excellent example of this sort of arrangement is the transendothelial shuttling of pinocytic vesicles so thoroughly explored by Bruns and Palade (1968) in the capillaries of striated muscle. In this example, waves of vesicles arise at the luminal surface of the endothelial lining cell and cross the cell to the opposite face where they discharge their contents. Similar processes have been described by Fedorko and Hirsch (1971) in mesothelial cells of the mouse omentum, and Rodewald (1972) in the intestinal lining cells of the newborn rat.

It would be of interest to know whether endocytic vesicles in these cells preferentially form from the luminal face rather than the basal surface. One wonders, too, when cells are arranged at the boundary of two very different environments, whether their asymmetric surroundings might not lead to an intracellular polarization of function. Perhaps tissue culture cells growing on a solid substrate are in a similar situation, with the free surface of the cell functionally differentiated from the adherent face. While such adherent cells appear to generate pinocytic vesicles primarily from the free face (Ralph Steinman, personal communication), detailed quantitative information on this question of functionally differentiated surfaces is lacking.

The existence of "endocytic windows" or membrane patches specialized for endocytosis is characteristic of the developing oocyte during the process of interiorization of yolk protein, or vitellogenesis. In this process certain proteins, including the estrogen-dependent proteins lipovitellin and phosvitin, are specifically sequestered in the oocyte, while other normal plasma proteins may accumulate at much lower rates (Wallace, 1972). One of the most interesting aspects of this phenomenon is the occurrence of a distinct morphologic correlate of the specific uptake process. In developing oocytes of both the mosquito and the chicken, Roth and colleagues (Roth et al., 1976) have shown that when the oocyte is actively sequestering protein there is an abundance of plasma membrane pockets, or pits, whose cytoplasmic face shows a spiky pattern of bristles. These coated pits become coated pinocytic vesicles that move toward the centrosphere region where their coats are shed, and the naked vesicles fuse to form a membrane-bounded protein storage compartment, or yolk vesicle.

From studies using ferritin-conjugated IgG (Roth et al., 1976), it appears that the coated pits of chicken oocytes may be the sites at which the binding molecules for the yolk-specific proteins are located. No ferritin conjugate appeared to bind at naked plasma membrane. However, it is still not known whether each of these patches is a mixture of binding sites with different specificities, or whether the clusters have homogeneous ligand specificity. Although it is clear that ligand is not necessary for the formation of the binding site clusters, it is not yet known whether a cluster of sites specifies the location of the underlying membrane coat, whether the location of the coat determines the organization of the sites, or whether coat and sites are inserted into the plasma membrane as a preformed unit.

In addition, the relation of the coated pits to the process of nonadsorptive endocytosis in the oocyte is unclear.

Although coated pits and vesicles are familiar components of a wide variety of other cells, their relation to endocytosis is much more problematic in these. Coated vesicles may occur in several size classes (Friend and Farquhar, 1967), and appear to serve several functions beside endocytosis. In addition, in certain cells that are extremely active endocytically, including the mouse peritoneal macrophage, pinosomes seem unrelated to coated vesicles.

A more subtle level of membrane differentiation with respect to endocytosis is the theoretical possibility that certain components may be selectively included or excluded from endocytic vesicles as they form at the cell surface. Berlin and colleagues have suggested that in neutrophils certain amino acid transport sites are specifically excluded from latex phagosomes (Tsan and Berlin, 1971), while the same cell type specifically enriches certain endocytic vesicles with molecules which can bind Con A or *Ricinus communis* agglutinin (Oliver et al., 1974). They have argued, on the basis of the colchicine-sensitivity of these processes, that these specific rearrangements of membrane components are under the control of cytoplasmic microtubules, perhaps acting together with other structural elements in the cell (Ukena and Berlin, 1972). These interesting ideas cannot yet be fully evaluated, as specificity claims require reliable estimates for the amount of membrane internalized and the original surface area of the cell, and these estimates are not yet readily available for most cells.

There is evidence in other systems that membrane interiorization, in the absence of an interaction of a ligand with a specific binding site, may be nonselective. Hubbard and Cohn (1975) found that the labeled species in the membrane of latex phagolysosomes isolated from enzymatically radioiodinated L cells appeared to resemble the labeling pattern of the plasma membrane quite closely. Obviously, many membrane species would not be identified in such an analysis, and these might show different distributions in plasma membrane and endocytic vesicles. In certain special cases, as for example the interiorization of a lymphocyte "cap," or the clearing of lectin-binding sites from a cell surface (Edelson and Cohn, 1974b), the endocytic vesicles may show a relative enrichment for one or a few membrane components, but a still unresolved question is whether there is a regular division of membrane species into "endocytosis-prone" and "endocytosis-resistant" classes.

5. What determines the fate of endocytic membrane?

The fates of interiorized membrane may be heterogeneous in various situations, but appear in all cases to be highly regulated processes. The generally accepted view is that incoming vesicles—pinosomes and phagosomes—uniformly fuse with lysosomes, as described in the introductory sections of this review. Perhaps the best evidence for this is that in macrophages interiorized proteins are quantitatively degraded with first-order kinetics (Ehrenreich and Cohn, 1967; Steinman and Cohn, 1972; Edelson and Cohn, 1974a,b). As no exocytosis of ingested material can be demonstrated in these cells (Cohn and Benson, 1965), these results imply that under normal circumstances all incoming pinocytic vesicles must deliver their contents to the lysosomal compartment.

However, the fate of pinocytic vesicle membrane may be distinct from the fate of the vesicle contents. Measurements made on the half-life of plasma membrane components, using enzymatic labeling with iodine (Hubbard and Cohn, 1975; Tweto and Doyle, 1976), acetylation (Roberts and Yuan, 1975), or intrinsic labeling (Tweto and Doyle, 1976), or studies of the turnover of specific membrane components (Devreotes and Fambrough, 1975; Edelson and Cohn, 1976a) all indicate that degradation of some or all membrane molecules is a slow process, with a $t^{1/2}$ of 10 to 100 hours, but that the labeled species are ultimately degraded to their constituent amino acids and many of the membrane components appear to be degraded more or less simultaneously. Radioiodination techniques also suggest that there is a rapid degradative component, with a $t^{1/2}$ of a few hours or less, but it is still unclear whether this component has physiologic relevance.

The rate of the slow component of degradation is perhaps two orders of magnitude slower than the rate at which plasma membrane is interiorized by pinocytosis. This discrepancy can be explained if either (1) many, or perhaps most, membrane components are exluded from pinocytic vesicles, or (2) labeled membrane components are interiorized and recycled to the plasma membrane many times without degradation. While it is clear that membrane components can be degraded as the result of their inclusion in an endocytic vesicle (Werb and Cohn, 1972; Hubbard and Cohn, 1975; Edelson and Cohn, 1976b), it remains to be shown that endocytosis is actually the mechanism for steady-state membrane turnover.

6. What regulates phago-lysosome formation?

As mentioned previously, the process of $2°$ lysosome formation is highly controlled and specific. Both the rate of lysosomal fusion and the fusion partner appear to be closely regulated.

Although renal tubular epithelial cells were one of the first cell types in which pinosome fusion was demonstrated (Straus, 1967), it is possible that intracellular fusion is an uncommon event in such transporting epithelia, where most of the pinosomes appear to move across the cell unchanged and deliver their contents exocytotically at the opposite cell face. This special fate for the vesicles may be related to the cell geometry or to a smaller complement of lysosomes in these cells, but other modes of regulation have not been excluded.

In recent years, a number of systems have been described in which there is a failure of lysosomal fusion with endocytic vesicles. Examples of such fusion failures have been described in cells as disparate as Paramecia (Karakashian and Karakashian, 1973) and mouse macrophages, suggesting that such inhibition may be a widespread intracellular mechanism.

Among the most extensively studied examples of fusion failure are the interaction of cultivated mouse peritoneal macrophages with *M. tuberculosis* (Armstrong and D'Arcy Hart, 1971), *Toxoplasma gondii* (Jones and Hirsch, 1972), or with concanavalin A (Edelson and Cohn, 1974a,b), or the interaction of living *Chlamydia psittaci* with mouse L cells (Friis, 1972). D'Arcy Hart and colleagues found that phagocytic vesicles containing living *M. tuberculosis* do not contain histochemically demonstrable acid phosphatase activity. As this enzyme is a constituent of macrophage lysosomes, and can be readily demonstrated in most phagocytic vacuoles surrounding heat-killed organisms, these workers interpreted their observation to mean that live organisms could inhibit the fusion of their enclosing phagosome with lysosomes. Jones and Hirsch made similar observations with another intracellular parasite, *T. gondii*. Again, live but not killed organisms were found to persist in unfused phagocytic vacuoles. In this work, the inferences that were made on the basis of a lack of histochemically demonstrable acid phosphatase in the phagocytic vacuoles were confirmed in cells whose secondary lysosomes were preloaded with the electron-dense marker thorotrast. This marker can be directly viewed in the electron microscope, obviating any problems caused by an artifactual inhibition of lysosomal enzymes in the phagolysosome by the living organism.

Some evidence suggests that this capacity to inhibit phagolysosome formation may be related to surface constituents of the parasite. Antibody-treated parasites, although still entirely viable, no longer inhibit the formation of phagolysosomes (Hirsch et al., 1974; Armstrong and D'Arcy Hart, 1975). Obviously it is of great interest to know precisely which surface structures on the parasite must be reacted with antibody to allow fusion to proceed, and just how such antibodies have their effect on this process.

We have made similar observations on the effects of concanavalin A on pinosome-lysosome fusion (Edelson and Cohn, 1974a,b). As described earlier, con A binds to macrophage plasma membrane and is internalized in pinocytic vesicles whose inner surface is coated with lectin. By several criteria, these pinosomes show a markedly diminished ability to fuse with lysosomes. Lysosomal acid phosphatase cannot be histochemically demonstrated in these pinosomes. Such electron-dense markers as thorotrast or gold, with which secondary lysosomes have been preloaded, are also absent from these vesicles. In addition, exogenous proteins to which the cells have been exposed simultaneously with the con A, and which have been interiorized in the con A pinosomes, are degraded far more slowly than they are in cells that have not been exposed to lectin.

The inhibition of pinolysosome formation depends on the interaction of the lectin with the inner face of the pinosome membrane. When bound lectin is displaced from the membrane by a prolonged postincubation in mannose, each of these indicators of impaired fusion returns to normal. Thus, interactions with glycoproteins on the inner pinosome face may affect the progress of events occurring on the outer or cytoplasmic vesicle surface. One obvious way that such a transmembrane effect could be accomplished would be through a con A-binding glycoprotein which spanned the pinosome membrane, and whose distribution or redistribution was essential for the fusion process. Another possibility is that con A may cross-link heterogeneous membrane species whose sorting out is required for fusion to proceed. Finally, con A binding may affect the disposition of peripheral membrane proteins on the cytoplasmic face that could act as a barrier to the normal interaction of pinosomes with lysosomes. Studies with ferritin-conjugated con A indicate that the lectin is uniformly distributed on the inner face of the pinosome membrane (P. J. Edelson, unpublished observations), suggesting that the lectin is not aggregating normally dispersed membrane components, but any of the other mechanisms is still a lively possibility.

The membrane of an endocytic vesicle may have other fates than incorporation into a hybrid phagolysosome, or persistence as an unfused vesicle. In special cases the membrane appears to be dismantled, allowing its contents to lie free in the cytoplasm. Nogueira and Cohn (1976) have presented morphologic evidence at the ultrastructural level indicating that this is how living *Trypanosome cruzii* escape their phagocytic vacuole and reach the cytoplasm. One could speculate about the role of such a process in the cycle of reovirus replication (Silverstein and Dales, 1968), or in diphtheria intoxication (see above), where material that is initially taken up in endocytic vesicles must eventually gain access to the cyto-plasm.

Undoubtedly the most frequent fate of endocytized membrane must be to re-turn essentially unaltered to the cell surface (Bowers and Olszewki, 1972; Stein-man et al., 1976). However, just how this dissociation of membrane from con-tents is accomplished and whether such recycled membrane might not be subtly changed during its intracytoplasmic stay, are still unresolved questions. In addi-tion, if (as seems reasonable) endocytosis is also the major mechanism for mem-brane degradation, one wonders how the catabolic fraction is determined.

7. Conclusions

Although the general phenomenon of endocytosis in certain specialized cells has been recognized for years, only recently have we begun to appreciate how wide-spread and highly regulated a mechanism it is. In addition, recent work has shifted emphasis from its role in internalizing bulk fluid to its significance for the specific uptake of biologically active macromolecules, viruses, and microor-ganisms through their binding to membrane proteins or lipids. Further, we now recognize endocytosis as an important mechanism for the internalization of plasma membrane itself, and probably for the physiologic renewal and remodel-ing of the cell surface.

Acknowledgments

This work was supported by grants #AI-07012 and CA-23503 from the National Institutes of Health.

References

Armstrong, J. A. and D'Arcy Hart, P. (1971) Response of cultured macrophages to *Mycobacterium tuberculosis,* with observations on fusion of lysosomes with phagosomes. J. Exp. Med. 134, 713–740.

Armstrong, J. A. and D'Arcy Hart, P. (1975) Phago-lysosome interactions in cultured macrophages infected with virulent tubercle bacilli. Reversal of the usual nonfusion pattern and observations on bacterial survival. J. Exp. Med. 142, 1–16.

Bianco, C., Griffin, F. M. and Silverstein, S. C. (1975) Studies of the macrophage complement recep-tor. Alteration of receptor function upon macrophage activation. J. Exp. Med. 141, 1278–1290.

Boquet, P. and Pappenheimer, Jr., A. M. (1976) Interaction of diphtheria toxin with mammalian cell membranes. J. Biol. Chem. 251, 5770–5778.

Bowers, B. and Olszewski, T. E. (1972) Pinocytosis in *Acanthamoeba castellani.* Kinetics and morphol-ogy. J. Cell. Biol. 53, 681–694.

Bruns, R. R. and Palade, G. E. (1968) Studies on blood capillaries. II. Transport of ferritin molecules across the wall of muscle capillaries. J. Cell. Biol. 37, 277–299.

Cohn, Z. A. and Benson, B. (1965) The in vitro differentiation of mononuclear phagocytes. III. The reversibility of granule and hydrolytic enzyme formation and the turnover of granule constituents. J. Exp. Med. 122, 455–466.

Collier, R. J. (1975) Diphtheria toxin: mode of action and structure. Bact. Rev. 39, 54–85.

Creagen, R. P., Chen, S. and Ruddle, F. H. (1975) Genetic analysis of the cell surface: association of human chromosome 5 with sensitivity to diphtheria toxin in mouse-human somatic cell hybrids. Proc. Nat. Acad. Sci. U.S.A. 72, 2237–2241.

Devreotes, P. N. and Fambrough, D. M. (1975) Acetylcholine receptor turnover in membranes of developing muscle fibers. J. Cell. Biol. 65, 353–358.

Edelson, P. J. and Cohn, Z. A. (1974a) Effects of concanavalin A on mouse peritoneal macrophages. I. Stimulation of endocytic activity and inhibition of phago-lysosome formation. J. Exp. Med. 140, 1364–1386.

Edelson, P. J. and Cohn, Z. A. (1974b) Effects of concanavalin A on mouse peritoneal macrophages. II. Metabolism of endocytised proteins and reversibility of the effects by mannose. J. Exp. Med. 140, 1387–1398.

Edelson, P. J. and Cohn, Z. A. (1976a) 5′-nucleotidase activity of mouse peritoneal macrophages. I. Synthesis and degradation in resident and inflammatory populations. J. Exp. Med. 144, 1581–1595.

Edelson, P. J. and Cohn, Z. A. (1976b) 5′-nucleotidase activity of mouse peritoneal macrophages. II. Cellular distribution and effects of endocytosis. J. Exp. Med. 144, 1596–1608.

Edelson, P. J., Zweibel, R. and Cohn, Z. A. (1975) The pinocytic rate of activated macrophages. J. Exp. Med. 142, 1150–1164.

Ehrenreich, B. A. and Cohn, Z. A. (1967) The uptake and digestion of iodinated human serum albumin by macrophages in vitro. J. Exp. Med. 126, 941–958.

Fedorko, M. E. and Hirsch, J. G. (1971) Studies on transport of macromolecules and small particles across mesothelial cells of the mouse omentum. I. Morphologic aspects. Exp. Cell Res. 69, 113–127.

Friend, D. S. and Farquhar, M. G. (1967) Functions of coated vesicles during protein absorption in the rat vas deferens. J. Cell Biol. 35, 357–376.

Friis, R. R. (1972) Interaction of L cells and Chlamydia psittaci: entry of the parasite and host responses to its development. J. Bact. 110, 706–721.

Gill, D. M., Pappenheimer, Jr., A. M. and Uchida, T. (1973) Diphtheria toxin, protein synthesis, and the cell. Fed. Proc. 32, 1508–1515.

Goldstein, J. L. and Brown, M. S. (1974) Binding and degradation of low density lipoproteins by cultured human fibroblasts. Comparison of cells from a normal subject and from a patient with homozygous familial hypercholesterolemia. J. Biol. Chem. 249, 5153–5162.

Griffin, F. M., Bianco, C. and Silverstein, S. C. (1975) Characterization of the macrophage receptor for complement and demonstration of its functional independence from the receptor for the Fc portion of immunoglobulin G. J. Exp. Med. 141, 1269–1277.

Griffin, F. M., Griffin, J. A., and Silverstein, S. C. (1976) Studies on the mechanism of phagocytosis. II. The interaction of macrophages with anti-immunoglobulin IgG coated bone marrow derived lymphocytes. J. Exp. Med. 144, 788–809.

Heiniger, H. J., Kandutsch, A. A. and Chen, H. W. (1976) Depletion of L-cell sterol depresses endocytosis. Nature 263, 515–517.

Hirsch, J. G., Jones, T. C. and Len, L. (1974) Interactions in vitro between Toxoplasma gondii and mouse cells. In: Parasites in the Immunized Host: Mechanisms of Survival. Ciba Foundation Symposium 25 (new series) pp. 205–223, Elsevier, North-Holland, Amsterdam.

Holter, H. (1965) Passage of particles and macromolecules through cell membranes. Symp. Soc. Gen. Microbiol. 15, 89–114.

Holtzman, E. (1976) Lysosomes: A Survey. Continuation of Protoplasmatologia, Cell Biology Monographs. vol. 3. Springer-Verlag, Vienna and New York.

Hubbard, A. L. and Cohn, Z. A. (1975) Externally disposed plasma membrane proteins. II. Metabolic fate of iodinated polypeptides of mouse L cells. J. Cell. Biol. 64, 461–479.

Iglewski, B. H. and Kabat, D. (1975) NAD-dependent inhibition of protein synthesis by Pseudomonas aeruginosa toxin. Proc. Nat. Acad. Sci. U.S.A. 72, 2284–2288.

Jones, T. C. and Hirsch, J. G. (1972) The interaction between Toxoplasma gondii and mammalian cells. II. The absence of lysosomal fusion with phagocytic vacuoles containing living parasites. J. Exp. Med. 136, 1173–1194.

404

Kändutsch, A. A. and Chen, H. W. (1975) Regulation of sterol synthesis in cultured cells by oxygenated derivatives of cholesterol. J. Cell. Physiol. 85, 415–424.

Kaplan, J. (1976) Cell contact induces an increase in pinocytotic rate in cultured epithelial cells. Nature 263, 596–597.

Karakashian, M. W. and Karakashian, S. J. (1973) Intracellular digestion and symbiosis in *Paramecium bursaria.* Exp. Cell Res. 81, 111–119.

Korn, E. D. (1975) Biochemistry of endocytosis. In: Biochemistry of Cell Walls and Membranes (Fox, C. F. ed.) pp. 1–26, University Park Press, Baltimore.

Korn, E. D., Bowers, B., Batzri, S., Simmons, S. R. and Victoria, E. J. (1974) Endocytosis and exocytosis: role of microfilaments and involvement of phospholipids in membrane fusion. J. Supramol. Struc. 2, 517–528.

Lesniak, M. A. and Roth, J. (1976) Regulation of receptor concentration by homologous hormone. Effect of human growth hormone on its receptor in IM-9 lymphocytes. J. Biol. Chem. 251, 3720–3729.

Lewis, W. H. (1931) Pinocytosis. Bull. Johns Hopkins Hosp. 49, 17–28.

Metchnikoff, E. (1893) Comparative Pathology of Inflammation. Lectures at The Pasteur Institute. Paul, London.

Nicolson, G. L. (1974) Ultrastructural analysis of toxin binding and entry into mammalian cells. Nature 251, 628–630.

Nogueira, N. and Cohn, Z. A. (1976) *Trypanosoma cruzi:* Mechanism of entry and intracellular fate in mammalian cells. J. Exp. Med. 143, 1402–1420.

Oliver, J. M., Ukena, T. E. and Berlin, R. D. (1974) Effects of phagocytosis and colchicine on the distribution of lectin binding sites on cell surfaces. Proc. Nat. Acad. Sci. U.S.A. 71, 394–398.

Olsnes, S., Pappenheimer, Jr. A. W., and Meren, R. (1973) Lectins from *Abrin precatorius* and *Ricinus communis.* II. Hybrid toxins and their interaction with chain-specific antibodies. J. Immunol. 113, 842–847.

Olsnes, S., Saltvedt, E. and Pihl, A. (1974) Isolation and comparison of galactose-binding lectins from *Abrin precatorius* and *Ricinus communis.* J. Biol. Chem. 249, 803–810.

Reaven, E. D. and Axline, S. G. (1972) Subplasmalemmal microfilaments and microtubules in resting and phagocytosing cultivated macrophages. J. Cell. Biol. 59, 12–27.

Refsnes, K. and Munthe-Kaas, A. C. (1976) Introduction of B-chain-inactivated ricin into mouse macrophages and rat Kupffer cells via their membrane Fc receptors. J. Exp. Med. 143, 1464–1474.

Refsnes, K., Olsnes, S. and Pihl, A. (1974) On the toxic proteins Abrin and Ricin. Studies of their binding to and entry into Ehrlich ascites cells. J. Biol. Chem. 249, 3557–3562.

Roberts, R. M. and Yuan, B.O-C. (1975) Turnover of plasma membrane polypeptides in non-proliferating cultures of Chinese hamster ovary cells and human skin fibroblasts. Arch. Biochem. Biophys. 171, 234–244.

Rodewald, R. (1972) Intestinal transport of antibodies in the newborn rat. J. Cell. Biol. 58, 189–211.

Roth, T. F., Cutting, J. A. and Atlas, S. B. (1976) Protein transport: a selective membrane mechanism. J. Supramol. Struc. 4, 527–548.

Schekman, R. and Singer, S. J. (1976) Clustering and endocytosis of membrane receptors can be induced in mature erythrocytes of neonatal but not adult humans. Proc. Nat. Acad. Sci. U.S.A. 73, 4075–4079.

Silverstein, S. C. and Dales, S. (1968) The penetration of reovirus RNA and initiation of its genetic function in L-strain fibroblasts. J. Cell. Biol. 36, 197–230.

Silverstein, S. C., Steinman, R. M. and Cohn, Z. A. (1977) Endocytosis. Ann. Rev. Biochem. 46, 669–722.

Steinman, R. M. and Cohn, Z. A. (1972) The interactions of soluble horseradish peroxidase with mouse peritoneal macrophages *in vitro.* J. Cell. Biol. 55, 186–204.

Steinman, R. M., Brodie, S. E. and Cohn, Z. A. (1976) Membrane flow during pinocytosis. A stereological analysis. J. Cell Biol. 68, 665–687.

Steinman, R. M., Silver, J. and Cohn, Z. A. (1974) Pinocytosis in fibroblasts: quantitative studies *in vitro.* J. Cell. Biol. 63, 949–969.

Stossel, T. P. (1973) Quantitative studies of phagocytosis: kinetic effects of cations and heat labile opsonins. J. Cell. Biol. 58, 346–356.

Straus, W. (1967) Lysosomes, phagosomes, and related particles. In: Enzyme Cytology. (Roodyn, D. P., ed.) pp. 239–319, Academic Press, New York.

Taylor, R. B., Duffus, W. P. H., Raff, M. C. and de Petris, S. (1971) Redistribution and pinocytosis of lymphocyte surface immunoglobulin molecules induced by anti-immunoglobulin antibody. Nature New Biol. 233, 225–229.

Tsan, M. F. and Berlin, R. D. (1971) Effects of phagocytosis on membrane transfer of non-electrolytes, J. Exp. Med. 134, 1016–1035.

Tweto, J. and Doyle, D. (1976) Turnover of the plasma membrane proteins of hepatoma tissue culture cells. J. Biol. Chem. 254, 872–882.

Ukena, T. E. and Berlin, R. D. (1972) Effect of colchicine and vinblastine on the topographic separation of membrane functions. J. Exp. Med. 136, 1–7.

Wagner, R., Rosenberg, M. and Estensen, R. (1971) Endocytosis in Chang liver cells. Quantitation by sucrose-^{3}H uptake and inhibition by cytochalasin B. J. Cell. Biol. 50, 804–817.

Wallace, R. A. (1972) The role of protein uptake in vertebrate oocyte growth and yolk formation. In: Oogenesis (Biggers, J. D. and Schuetz, A. W., eds.) pp. 339–359, University Park Press, Baltimore.

Werb, Z. and Cohn, Z. A. (1972) Plasma membrane synthesis in the macrophage following phagocytosis of polystyrene latex particles. J. Biol. Chem. 247, 2439–2446.

Mechanisms of mediator release by inflammatory cells

10

Peter M. HENSON, M.H. GINSBERG
and D.C. MORRISON

Contents

G. Poste & G. L. Nicolson (eds.) Membrane Fusion, pp. 407–508.
© Elsevier/North-Holland Biomedical Press, 1978.

1. Introduction

The secretion of inflammatory mediators from cells has generated increasing interest in recent years. Areas which five years ago we could only indicate as exciting possibilities (Becker and Henson, 1973), such as the role of calcium and membrane depolarization in secretion, are now receiving active and extensive study. Not surprisingly, however, even more questions remain, and at least one of the purposes of this chapter is to point out areas that merit and are ripe for fruitful investigation.

1.1. "Inflammatory" cells and host-defense

In the past, and in this chapter, we have used the term mediator cells, or inflammatory cells. This is meant to describe a group of cells known to have potential for involvement in acute or chronic inflammatory reactions. It does not imply that this is the only role these cells play in vivo. In fact, their physiologic role may be mainly in the area of host defense (but see section 4). It is noteworthy, however, that the processes (intracellular and extracellular mediation mechanisms) involved in protection against infection or against neoplasia are identical to these involved in inflammation. This Janus effect is of particular importance when therapeutic approaches to reduce the inflammatory mechanisms may expose the animal to increasing risk of infection, infestation, and cancer.

1.1.1. Common features of inflammatory cells
The group of cells considered above show functional and mechanistic similarities which support their inclusion into a single group. This comprises one of the main emphases of this chapter. Chief among them is that the cells (and the humoral effectors with which they are associated) constitute the *effector* arm of the immune response. Once the recognition process (antibody or lymphocyte) is complete, these cells act upon the recognized antigen (whether exogenous or endogenous) by reacting with the antibody or complement or by responding to the lymphocyte product. *Each of these cells, therefore, reacts with the Fc region of one or more classes of immunoglobulin and with one or more fragments of complement.*

Other similarities lie in the increasing likelihood that each cell is activated by a precursor serine esterase, that each is controlled (inhibited) by increased levels of cyclic AMP (a point of critical importance for therapy), and that most seem particularly responsive to stimulus-dependent desensitization (see below). Increasingly, each of these cells has been shown to have the potential (although how important this is in vivo is unclear) for chemotaxis and phagocytosis. The possibility of their evolutionary origin from a primitive macrophage-like defense cell cannot be excluded. In other respects, however, secretion of granule constituents from inflammatory cells is similar to secretion from other mammalian and nonmammalian cells (Becker and Henson, 1973, and chapters 11, 12 and 13 in this volume).

1.1.2. Cells discussed in this chapter

The subject has become so large that it is impossible to do justice to the work on all the cells that secrete mediators of acute and chronic inflammatory reactions. We have therefore limited our discussion to four: the mast cell, the basophil, the platelet, and the polymorphonuclear leukocyte (neutrophil). These comprise examples of cells that secrete amine-containing electron-dense granules (mast cells, basophils), those which secrete lysosomes (neutrophils), and those which appear to do both (platelets). An important point of emphasis in this chapter will be the interaction between various inflammatory cells and between these cells and humoral mediation systems.

1.1.3. Omissions and limitations

Of necessity, some cells have been excluded, and even then we have had to impose limitations on our discussion. With apologies therefore, such constrictions will be outlined.

1.1.3.1. Macrophages.

Conspicuously absent is a discussion of macrophage secretion. This is not because we feel it is unimportant. Actually this cell is of critical importance in human inflammatory conditions that commonly present as chronic, self-perpetuating processes. However, the macrophage is an order of magnitude more complex than the other cells described here; it is truly pluripotential. Like the neutrophil, it undergoes chemotaxis, chemokinesis, and phagocytosis and has potent bactericidal properties. However, despite recent evidence for its production of toxic oxygen radicals, the mechanisms of this bactericidal effect are poorly understood. The cell synthesizes new enzymes and granules as well as secreting these to the outside and into phagocytic vacuoles, thus making the study of secretion much more difficult. It is actively pinocytotic and can also reuse its phagolysosomes for discharge into new phagosomes or extracellularly. Secretion is also much slower than in the other cells described here (hours rather than minutes). Pervading the whole field, however, is the main reason we have consciously excluded the macrophage at this time; monocyte and macrophage function are critically dependant upon the state of activation or differentiation of the cell. Since this process of macrophage differentiation (despite continued investigation) is poorly characterized at present, we feel that the information available is too fraught with technical difficulties and inconsistencies for it to be molded into a coherent pattern. Nevertheless, accumulating data on mechanisms of secretion, as yet only in the early descriptive phase, clearly indicate that these mechanisms are very similar, if not identical to those of other cells described here, particularly the neutrophil. We do not intend to

slight the advances made in studies of phagocytosis and of the role of contractile elements by groups such as Silverstein and his collaborators (Griffin et
al., 1976) or Stossel and his colleagues (Stossel, 1977). This subject and that of
the role of macrophages in the immune response has been considered in recent
volumes, such as that edited by Nelson (1976). The subject of macrophage
secretion has also been reviewed recently by Davies and Allison (1976) and
Schorlemmer and colleagues (1977), although the suggestion made by the
latter workers that complement may be involved in stimulation by a wide
variety of agents seems to be premature.

1.1.3.2. Eosinophils. The role of the eosinophil in inflammation is still unclear,
although evidence is accumulating that it is actively cytotoxic for helminth parasites and may have inhibitory potential for other inflammatory mediators. We
know little of its secretory potential or mechanisms due to lack of a specific
grasp of its activation (as was useful for the basophil) and to an inability to
purify the cell adequately. The recent report of Tai and Spry (1977) may
remedy the latter problem and could open the way to an exciting study of the
normal human eosinophil.

1.1.3.3. Lymphocytes and endothelial cells. Lymphocytes produce many mediators,
but the mechanisms of their production and secretion are obscure. In large part
this reflects the difficulties encountered in purifying and identifying the mediators in question, which at the present contain a tremendous array of acronyms.
Once these molecules are identified and characterized, study of their secretion
will follow. Finally, many inflammatory mediators may come from tissue cells,
particularly perhaps from endothelial cells. These are much more difficult to
study, but increasing knowledge of the behavior of these cells in culture is opening the way to adequately characterize them as participants in the inflammatory
process.

1.1.3.4. Additional limitations. Because of the size of the field, we have not attempted to provide an all-inclusive reference list. Where possible, we have
quoted other key reviews and have been guided largely by our own interests and
by an attempt to quote papers as recent as possible. Thus, even in this fast moving area the reader may find this chapter useful to gain access to the literature.
We therefore apologize to those investigators whose studies have not been completely discussed, and hope that our references have been complete enough that
the interested reader will find her/his way to these investigations.

1.2. Secretion of mediators

The rest of this introduction briefly introduces the subjects discussed in more detail for the individual cell types. This will allow a consideration of similarities and
differences and of definitions.

1.2.1. Inflammatory cell activities and mediators

The cells contain membrane-bounded secretory granules containing a wide variety of mediators. A possible similarity between the vasoactive amine-containing granules of the mast cell, the basophil, and the platelet has been considered and may extend to their mode of secretion. Possibly the specific granule of the neutrophil, which is probably not a true lysosome, is related to these. Upon suitable stimulation of the cell, the granules are secreted. This is a noncytotoxic phenomenon which does not involve liberation of cytoplasmic constituents. Cytotoxic reactions of these cells are known and are briefly discussed. Interestingly, in each cell, a transient secretory event often precedes lysis, leading us to coin the term "pseudocytotoxic release."

Chemotaxis has been reported for neutrophils, basophils, eosinophils, macrophages, lymphocytes, and platelets, and phagocytosis for all of these cells and for mast cells. How important or definitive these observations are for some of the cells is at present open to question. However, the data implies an additional commonality between the cell types.

Many different mediators are released, and the specificity and diversity of these cells may lie in this release and the stimuli which activate these mediators. Even here, however, data is accumulating to suggest that prostaglandin intermediates and perhaps oxygen radicals may be liberated from many of these (and other) cells.

1.2.2. Morphology of the secretory process

Basically this involves approach of a secretory granule to the plasma membrane, and membrane fusion with discharge of constituents. Some evidence exists that this exocytosis process occurs with each of the inflammatory cells. However, with platelets (and perhaps even with basophils), an alternative mechanism, involving the opening of a granule-containing space by the pulling apart of two opposed membranes, cannot be completely excluded.

1.2.3. Stimuli for inflammatory cell activation: definition of membrane "receptors"

Each cell has its unique stimuli and, presumably, cell surface receptors. As mentioned earlier, some stimuli act on all or most of these cells. Chief among these are the Fc region of immunoglobulin and complement component fragments, particularly C5a. These are probably the most important in vivo stimuli. However, materials released from cells are also stimuli for inflammatory cells, providing explanations for the many cooperative effects that have been described. Artificial stimuli as, for example, phorbol myristate acetate (PMA), also appear to act on most of these cell types.

A feature of many stimuli is the combination of hydrophobic and basic groups, as for example on a protein (immunoglobulin), a peptide (polymyxin B), and a lipid (platelet activating factor). Another common group of stimuli are serine proteases, one or more of which have now been shown to activate, in some way, all the abovementioned inflammatory cells except the basophil (which may reflect a lack of adequate investigation).

Where particular emphasis has been placed on the interaction of one of these common groups of stimuli with a particular cell type as, for example, with immunoglobulin (IgG) and basophils, or proteases (thrombin) and platelets, similar mechanisms probably exist for the other cells, but they have not been as well studied.

1.2.3.1. Interaction with the cell surface. The relatively few stimuli that have been carefully investigated appear to activate inflammatory cells by interaction with the cell surface. Actually, no good evidence exists for a stimulus acting from within, and even the ionophores presumably act first at the plasma membrane. An exception to this would be the effect of intracellular injection of Ca^{2+} into mast cells, which is hardly physiologic and presumably bypasses the membrane-triggering events.

1.2.3.2. "Receptors". Although the presence of membrane receptors is implied by the comments made above, a working definition of the term receptor is necessary. The term "receptor" is used here to indicate a bifunctional membrane structure (or structures). This must act both as a binding site for the stimulus and, as a result of this binding, as an initiator of the biochemical sequences leading to whatever cell function is being studied. These activities could reside in one molecule, in two molecules that can interact, or in a membrane complex. For none of these stimuli or cells is this point known, so that the receptors at this point remain as activities rather than known structural entities.

1.2.3.3. Cross-linking of receptors. As the interaction of each stimulus with its target cell becomes better understood, the question of valency of "receptor" and of the role of cross-linking, and so-called "patching" and "capping" of such membrane structures is raised. Based on the fluid mosaic theory of membrane structure, cross-linking of receptors by di- or multivalent stimuli can lead to redistribution of receptors into patches, usually visualized by fluorescence microscopy. Generally such aggregation into small patches, distributed around the whole cell, is not under energetic or cytoskeletal control. On the other hand, the association of such patches into one cap, often followed by endocytosis, involves cellular metabolic energy, can usually be inhibited by cytochalasins, and is often modulated by colchicine-sensitive structures, presumably microtubules. This area is discussed in greater detail in the section on neutrophils but probably applies to all the cells. Its role in cell activation is unclear. Considerable evidence exists for the requirement of divalent or multivalent stimuli for activation of most of these cells and this implies that cross-linking of receptors may be an important general mechanism of initial triggering. However, the role of any further redistribution, into patches for example, is not known.

1.2.4. Mechanisms of cell triggering and secretion
Several aspects of the secretory process in inflammatory cells will be discussed. In view of our own intersts, and since this seems to be a common feature of these

cells, we will emphasize the possible role of endogenous proteases (esterases) in the triggering process.

1.2.4.1. Activation of precursor proteases (esterases).   Inhibition of serine proteases abrogates the secretory process in most of these cells. Moreover, where this has been carefully examined, the inhibitor must be present during the stimulation; pretreatment of either stimulus or cell is ineffective. Since this is true of irreversible inhibitors that phosphorylate the active center of these enzymes, it has generally been a working hypothesis that the stimulus activates a precursor serine protease (zymogen) which is involved in cell activation (Becker and Austen, 1966; Becker and Henson, 1973). In many cases the situation is complicated further by the presence of an esterase that is already active (cell-dependent as opposed to stimulus-dependent), which is also required for the cell function.

To date, evidence for the involvement of these activatable enzymes in cell triggering is circumstantial. These enzymes have generally been termed esterases because where their activity has been demonstrated, amino acid esters have been used as substrates or competitive inhibitors (Becker, 1972). In the cell, however, they probably act as endopeptidases (proteases). Where studied, this activatable protease step appears very early in the process of cell triggering (Ranadive and Cochrane, 1971; Henson, 1974a; Kaliner and Austen, 1974), and it is possible to speculate that stimulus-receptor interaction directly induces a conformational change in the zymogen leading to its activation as the first step in the stimulus-secretion coupling process. Perhaps the receptor and the zymogen are identical. This concept is strengthened by the finding that various stimuli for platelets appear to activate different esterases (Henson et al., 1976). Moreover, the precedent for such a process comes from activation of humoral effector sequences, such as the complement and intrinsic coagulation processes, whose first step is also the activation of a serine protease.

It is also relevant that almost all of these inflammatory cells can be activated by the addition of exogenous serine proteases. This is discussed further in the section on platelets. Moreover, the possible role of proteases in cell functions in general has stimulated increasing interest and, where studied, wider application. This role may represent a more general mechanism of cell triggering, even of mitogenesis (Kaplan and Bona, 1974), than was originally believed. This exciting area merits further investigation, particularly to isolate the putative cellular enzymes involved. Of further interest in this regard is the observation that serine protease zymogens can interact with phospholipids (Esmon et al., 1976), which is relevant to a possible membrane site in the cell, and that some proteases may be activated by calcium (Dayton et al., 1976 a,b). This may have significance in light of an early calcium-requiring step associated with esterase activation in mast cells (Kaliner and Austen, 1974).

Finally, the possibilities that inflammatory cells such as mast cells (section 2.3.4), neutrophils (Nakamura et al., 1976), macrophages, and so on may liberate proteases capable of activating more inflammatory cells may explain in part the self-perpetuation of many inflammatory processes and the importance of an adequate plasma protease inhibitor system in vivo.

1.2.4.2. Biochemical processes in stimulus-secretion coupling. The pharmacologic control of secretion in these cells will be considered briefly here, but has been reviewed extensively elsewhere. Similarly, the central role of calcium in these and other secretory processes will be discussed. Each cell requires energy metabolism, but may derive the necessary adenosine triphosphate (ATP) from different pathways. The possible self-regulatory functions (stimulatory or inhibitory) of arachidonic acid metabolism in these cells requires consideration, and is by no means confined to the platelet, although this latter cell has received most study.

1.2.4.3. Cytoskeletal effects. This is a subject of increasing interest. Since both microtubular and contractile microfilamentous structures appear to influence cell membrane functions, including perhaps receptor-stimulus triggering and granule-membrane fusion processes, this subject will receive considerable emphasis for each cell. The cells all exhibit motility and, in most cases, actomyosin has been demonstrated within them. Function and regulation of contractile elements in nonmuscle cells is being intensively investigated, and the reader is directed to reviews by Pollard and Weihing (1974), Allison (1973), and Hitchcock (1977) for further pursuit of this subject.

1.2.4.4. Membrane fusion. This terminal stage of secretion may be one of the least understood, despite the many theories that have been put forward. For the most part the inflammatory cells have not received intensive investigative effort in this area, although what is known has been discussed herein.

2. Mast cells and basophils

In this section we will consider the mechanisms involved in the cellular activation of mast cells and basophilic leukocytes leading to secretion. As several recent review articles have focused upon the activation process (Becker and Henson, 1973; Sullivan and Parker, 1976; Morrison and Henson, 1977; and Ranadive, 1977a,b), particular attention will be devoted to the contribution of the cellular membranes and subcellular structures to the mechanism of secretion. Although mast cells and basophilic leukocytes have distinguishing characteristics which serve to differentiate these two cell types, it is believed that they share sufficient similarities in their secretory response to various stimuli to justify their being considered together in this section. We have also, for the most part, limited discussion to the two most extensively studied experimental systems, the rat peritoneal mast cell and the human basophil.

2.1. Morphologic features of mast cell and basophil secretion

One of the primary distinctions between these two cell types is their location within body tissues; mast cells are located predominantly in the connective tissue surrounding blood vessels whereas basophils circulate in the blood. It is now well recognized that mast cells and basophils represent two distinct cell types (Boseda,

1959; Riley, 1959) although both contain numerous similar (but not identical) storage granules.

2.1.1. Secretion (noncytotoxic release)

Secretion induced by most of the stimuli described in section 2.3 is noncytotoxic in nature. Thus cytoplasmic constituents are not liberated, the cells maintain their ability to exclude Trypan blue (Johnson and Moran, 1969a,b) and intracellular biochemical events are required (see below, and Morrison and Henson, 1977).

The ultrastructural and morphological changes that occur on noncytotoxic stimulation of mast cells and basophils have been examined extensively by a number of investigators over the last ten years using the electron microscope. The results of these examinations provide an elegant morphological description of both intact and degranulated mast cells, and have prompted Morrison and Henson (1977) to observe that there is a remarkable similarity in the appearance of the degranulation process in mast cells despite the enormous diversity of the stimuli used to initiate the secretory process.

This certainly appears to be true in the morphological alterations which accompany secretion from rat peritoneal mast cells initiated by compound 48/80 (Bloom and Haegermark, 1965; Röhlich et al., 1971a,b), antigen (Anderson et al., 1973), polymyxin B sulfate (Lagunoff, 1972a, 1973) and anti-rat Ig or concanavalin A (Lawson et al., 1976). In general, these observers have defined a common morphologic sequence of events which accompany the secretion process.

The initial event appears to be always localized to granules in the periphery of the cell, nearest the cytoplasmic membrane, where a space or halo appears that separates the granule matrix from the surrounding perigranular membrane. The granule matrix changes from a homogeneous electron-dense structure to one of loose internal structure with less electron density. These altered membrane-surrounded granules may then fuse with another altered granule or an intact granule, or may singly approach the cytoplasmic membrane, gradually causing it to bulge. Fusion of the perigranular membrane with the cytoplasmic membrane is preceded by the formation of a pentilaminar structure containing a thickened dense layer where the two unit membranes abut. This membrane fusion may then result in the formation of a thin diaphragm which eventually ruptures, exposing the granule constituents to the external environment. Alternatively, the pentilaminar structure may lead directly to membrane fusion and pore formation without the formation of attenuated intermediates. Nevertheless, the process of fusion of perigranular membranes continues toward the interior of the cell, forming cavities and channels which, by the use of lanthanum, can be shown to be in continuous communication with the external enviornment.

It has recently been reported that the secretory mechanism of rat mast cells, subsequent to the fusion of the perigranular membrane with the cytoplasmic membrane, results in the formation of "blebs," vesicular membranous structures which form principally over altered granules (Lawson et al., 1976). These blebs then pinch off, resulting in the formation of a continuous membrane between the plasma and perigranular membranes. Although such a mechanism, as suggested by the authors, would allow for selective cellular removal of excess membrane lipids, similar morphological structures were previously noted by Bloom and Haegermark (1965) and Röhlich and colleagues (1971a,b) in normal, untreated mast cells. Further, such structures have been suggested by Lagunoff (1973) to be artifacts of the preparatory procedure. Thus precise significance of the observed vesicular structures on the mast cell surface in the degranulation process awaits further experimentation.

Basophil degranulation has been examined by Hastie (1971, 1974, Hastie et al., 1977), who showed an exocytosis of granules from human cells. Previously, the basophil granules appeared to dissolve in situ (see Beneveniste et al., 1972) and the mechanism of release remained unclear. It now appears similar to that in other cells.

Early experiments (Uvnäs and Thon, 1966; Lagunoff, 1966) suggested that histamine and serotonin were bound electrostatically in the mast cell granules and exposure of the matrix to the external ionic environment leads to cation (e.g., Na^+) exchange and elution of the amines (see Uvnäs, 1976; Lagunoff, 1973). No such information is available for basophils although the histamine may be bound to a heparin-like mucopolysaccharide in the granule. Interestingly, Uvnäs has suggested a similar cation exchange for platelet-dense body serotonin (Uvnäs, 1971), though good evidence for this is not yet forthcoming.

2.1.2. Cytotoxic release

The cytotoxic release of mediators from mast cells and basophils generally results from the nonselective destruction of the integrity of the cytoplasmic membrane and the subsequent release of intracellular constituents, both granule- and cell cytoplasm-associated. Loss of membrane integrity may be accomplished by the action of detergents such as n-decylamine, hyperosmotic shock using distilled water, or enzymes that alter membrane lipids, such as phospholipase A (Moran et al., 1962; Morrison and Henson, 1977). Alternatively, cytolysis may be initiated by anti-mast cell antibody with the subsequent generation of the terminal complement components (Valentine et al., 1967). Interestingly, a secretory component may exist in this reaction (Chi et al. 1976), as has also been seen in neutrophil cytotoxic reactions.

2.2. Mast cell activities and mediators

Another characteristic of mast cells and basophils is the remarkable similarity in the types of inflammatory mediators that are released from these cells. These mediators may be conveniently classified on the basis of whether they exist within the cell prior to stimulation (preformed) or whether their activities may be manifest only subsequent to cellular activation processes (newly synthesized). Other recently described functions of these cells (phagocytosis, chemotaxis) will be mentioned briefly and exemplify similarities between all the "mediator" cells.

2.2.1. Preformed mediators

2.2.1.1. Histamine and serotonin. Histamine is one of the earliest and perhaps most well recognized of the mediator substances present in mast cells (Riley and West, 1953) and basophils (Graham et al., 1955; Ishizaka et al., 1972). In mast cells of some species (e.g., the rat), serotonin or 5-OH tryptamine is also present (Benditt et al., 1955). Basophils are the major cellular reservoir of histamine in human blood, but these cells contain less (2.4 $\mu g/10^6$ cells) than that found in rat peritoneal mast cells (40 $\mu g/10^6$ cells) (Sampson and Archer, 1967). Both histamine and serotonin are potent pharmacologic mediators.

2.2.1.2. Chemotactic factors. Molecules selectively chemotactic for specific types of leukocytes are also released from mast cells and basophils. Eosinophil chemotactic factor of anaphylaxis (ECF-A) is released upon antigen challenge of guinea pig and human lung fragments (Kay and Austen, 1971) and exists preformed in preparations of rat peritoneal mast cells (Wasserman et al., 1974), human leukemic basophils (Lewis et al., 1975), and human leukocytes (Czarnetzki et al., 1976). ECF-A has recently been suggested to consist of a mixture of two tetrapeptides which differ in their NH_2-terminal residues (Austen et al., 1976).

The second factor, NCF-A, is preferentially chemotactic for neutrophils and is also distinguished from ECF-A on the basis of its increased relative molecular weight (Austen et al., 1976). This factor, first demonstrated in extracts of human leukemic basophils (Lewis et al., 1975) may also be released from purified rat peritoneal mast cells by rabbit anti-rat light chain antibody (Austen et al., 1976).

2.2.1.3. Heparin. Heparin, a sulfated acidic glycosaminoglycan, is a major constituent of the intracellular granules of rat peritoneal mast cells (Lagunoff et al., 1964), and probably acts there to bind the histamine ionically. It has long been recognized for its potent anticoagulant activity as an inhibitor of thrombin, but its role as a mediator of the inflammatory process following mast cell secretion, or even its ability to be secreted has been questioned. Archer (1961) reported that purified mast cells released heparin only after freezing and thawing, and Lagunoff (1972a) did not find heparin released along with histamine after polymyxin B stimulation. However, studies using compound 48/80 suggested a dose-dependent release of heparin that correlated with the release of histamine (Fillion et al., 1970; Nosál et al., 1970; Slorach, 1971). In addition, Yurt and co-workers (1977) demonstrated a similar correlation using anti F(ab')$_2$ antibody, anti-IgE antibody, or ionophore A23187. At present the reasons for these apparently discrepant results are unclear, but may reflect differences in the preparation of mast cells, the nature of the stimulus, or the assays for heparin.

2.2.2. Newly synthesized mediators

2.2.2.1. SRS-A. One of the classic mediators of inflammation first defined by Feldberg and Kellaway (1938) as slow reacting substance (SRS) causes prolonged contraction of smooth muscle. A similar substance released from guinea pig lung by antigen challenge was termed slow reacting substance of anaphylaxis (SRS-A) by Brocklehurst (1960). SRS-A is released from preparations of human leukocytes (Grant and Lichtenstein, 1974) as well as from tissues rich in mast cells (review, Austen et al., 1976). Interestingly SRS-A usually does not seem to exist preformed in cells and tissues but is generated during the activation process. In this respect Lewis and associates (1974) have shown that stimulation of basophils by antigen under conditions where release is inhibited (e.g., in the presence of 1mM dibutyryl cAMP) does not result in SRS-A generation.

Biochemically, SRS-A has been purified >40,000 times and appears to be an unsaturated low molecular weight acid, possibly containing both hydroxyl and carboxylic acid residues and an active sulfate ester group (Orange et al., 1974).

2.2.2.2. PAF. A second low molecular weight mediator also apparently does not exist preformed within mast cells and basophils. This mediator, termed platelet activating factor (PAF), was originally found in supernatants of rabbit peripheral blood leukocytes after challenge with specific antigen (Henson, 1970d; Siraganian and Osler, 1971; Benveniste et al., 1972). However, more recent experiments have established that a similar mediator is associated with both rabbit and human lung tissue following antigen challenge (Bogart and Stechschulte, 1974; Kravis and Henson, 1975). PAF prepared from rabbit or human basophils appears to have a molecular weight of about 1100 and has chemical properties similar to phospholipids (Benveniste, 1974). Rat PAF has recently been reported to have a lower molecular weight (300–500) and its activity was at least partially destroyed by phospholipase D (Austen et al., 1976). As its name implies, PAF has a number of potent activities on platelets, including initiation of secretion of vasoactive amines, prostaglandins, adenosine diphosphate (ADP) and platelet aggregation (section 3).

2.2.3. Phagocytosis

Although the primary function of the mast cell is in the secretion of mediators, several investigators have suggested that these cells have phagocytic activity in vivo (Padawer and Fruhman, 1967; Padawer, 1971). Fruhman (1973) also showed that mast cell ingestion of zymosan in vitro required a heat-labile factor(s) present in serum and plasma. However, the relative importance of mast cell phagocytosis in vivo as compared to phagocytosis by macrophages or neutrophils is unclear.

2.2.4. Chemotaxis

Human basophils have been reported to respond chemotactically to concentration gradients of the anaphylatoxin C5a (Kay and Austen, 1972; Lett-Brown et al., 1976) and to a factor derived from mitogen-stimulated lymphocytes (Boetcher and Leonard, 1973; Lett-Brown et al., 1976). However, as is the case with phagocytosis by peritoneal mast cells, the basophil chemotactic response has not been studied to the extent of its secretory response, and will not be discussed in detail in this chapter.

2.3. Stimuli for mast cell and basophil secretion

One of the most remarkable features of mast cells (at least those isolated from the rat peritoneal cavity) is the extreme chemical diversity in the molecules which may initiate a noncytotoxic, secretory cellular response. Of additional interest is the fact that molecules which can activate mast cells from one species may be completely ineffective when tested in mast cells from other species (Becker and Henson, 1973). The spectrum of activators that initiate basophil secretion is considerably less broad when compared to mast cells; yet the observation that virtually all molecules which stimulate basophils also act on mast cells offers an additional compelling argument to consider these two cell types together. Since these stimuli have been discussed in detail in a recent review (Morrison and Henson, 1977), they will be considered only briefly here with particular emphasis on cell surface interaction.

2.3.1. Immunoglobulin stimuli

2.3.1.1. IgE immunoglobulin. One of the most prominent features of mast cells and basophils that may actually serve to distinguish these cells from all other mediator cells is the very high affinity with which they bind IgE immunoglobulin. Subsequent interaction of antigen or anti-IgE antibody with this cell-bound immunoglobulin initiates the secretory process. The pioneering experiments of the Ishizakas and their co-workers have contributed significantly to our current understanding of the role of IgE in the activation of mast cells and basophils (review, Ishizaka and Ishizaka, 1975). These authors established that IgE combines with mast cells and basophils through the Fc portion of the immunoglobulin molecule and that binding is responsible for sensitization of the cells to antigen or anti-IgE challenge. They showed further that there are approximately 50,000 IgE binding sites (receptors) per human basophil and that the binding is reversible (association constant 10^9/mole) and noncovalent. Kulezycki and colleagues (1974) and Kulezycki and Metzger (1974) also demonstrated that rat IgE binds reversibly with high affinity to the rat basophilic leukemia cells (association constant 6×10^9 mole); however, the total number of binding sites per cell $(3-10 \times 10^5)$ was somewhat higher than that obtained for the human basophil.

Recently several groups of investigators have explored the biochemical nature of the membrane-localized IgE receptor (Morrison and Henson, 1977) using both rat peritoneal mast cells and rat basophilic leukemia cells. The major approach in these experiments has been to utilize radioiodinated IgE as a marker to localize IgE receptors following disruption of cells by sonication and by the nonionic detergent Nonidet P40. Radioiodinated IgE as was found to be associated with a high molecular weight component of the cell (Conrad et al., 1976; Konig and Ishizaka, 1976), and such preparations could block an IgE-initiated passive, cutaneous anaphylaxis (PCA) reaction. Using radiolabeled cells precipitated with IgE and heterologous anti-IgE, followed by fractionation on SDS polyacrylamide-gel electrophoresis, both Conrad and Froese (1976) and Kulczycki and co-workers (1976) demonstrated a major peak of radioactivity at about 50,000 to 70,000 MW. This peak of radioactivity disappeared if cells were saturated with IgE prior to surface iodination, suggesting that in this circumstance the material was covered up by bound IgE. These authors suggest that the IgE receptor is a glycoprotein on the basis of ability to incorporate exogenous glucosamine, sensitivity to proteases (Bach and Brashler, 1976), and anomalous electrophoretic migration in polyacrylamide gels. More recently, Newman and collaborators (1977) have calculated a molecular weight of 130,000 for the detergent-receptor complex on the basis of sedimentation and diffuse measurements of detergent-solubilized basophilic leukemia IgE receptor. Correction for the presence of bound detergents yielded a value of 77,000 for the isolated receptor. These experiments have also demonstrated that one IgE receptor molecule can bind one IgE molecule, suggesting that the receptor is functionally monovalent (but see below, section 2.5.1).

IgE-mediated activation may be accomplished by interaction with specific antigen to which the IgE molecule is directed (review, Osler et al., 1968), by interaction with anti-IgE antibody (review, Becker and Henson, 1973), or by interaction of cell-bound IgE with concanavalin A (section 2.3.5.1). Each of these mechanisms results in a noncytotoxic, secretory cellular response.

2.3.1.2. IgG_{2A}.   A second immunoglobulin found in the rat, the IgG_{2A} subclass, is also capable of initiating secretion from rat mast cells probably by binding to the IgE receptor (Bach et al., 1971). Initiation of mast cell secretion by either antigen or anti-rat IgG (Kaliner and Austen, 1974) could then be initiated by much the same mechanisms as with membrane-bound IgE.

2.3.1.3. Direct allogeneic anaphylactic degranulation.   A recently published report by Däeron and co-workers (1975) has described a rather novel new mechanism for immunoglobulin-mediated mast cell degranulation in which alloantiserum directed against histocompatability antigens borne by mouse mast cells results in specific and immediate degranulation. The reaction was complement-independent and required an intact Fc. Presumably the Fc region of the attached antibodies (IgE ?) was able to turn around and stimulate the Fc receptors involved in the secretory events initiated by IgG.

2.3.1.4. Anti-rat immunoglobulin and complement.   In addition to the complement-independent mechanisms described above, a complement-dependent immunoglobulin-mediated release of histamine from mast cells has been described that involves heterologous anti-rat immunoglobulin and complement through the fifth component (Austen and Becker, 1966). The secretion has been shown to be noncytotoxic (Johnson and Moran, 1969a), and it has been proposed

(Becker and Henson, 1973; Morrison and Henson, 1977) that the mechanism involves the generation of the anaphylatoxin C5a at the membrane surface (see below).

2.3.2. Complement-derived stimuli

2.3.2.1. Anaphylatoxins C3a, C5a. Another group of immunologic mediators that can initiate secretion from mast cells and basophils may be classified as complement-dependent and immunoglobulin-independent. These are the anaphylatoxin polypeptides C3a and C5a, generated by enzymatic cleavage of the complement components C3 and C5 (review, Müller-Eberhard, 1976). Both anaphylatoxins are highly cationic polypeptides of approximately 8,000 to 10,000 MW. C5a, but not C3a, has a substantial percentage of carbohydrate (30%). Both polypeptides have a COOH terminal arginine residue which appears to be essential for histamine-releasing activity. (This must be contrasted with the action of C5a on neutrophils, section 4). Both of the anaphylatoxin polypeptides, C3a and C5a, have the capacity to initiate histamine release from mast cells (Johnson et al., 1975). Human basophils also secrete histamine in response to low molecular weight molecules generated in serum upon activation of the complement system (Grant et al., 1975; Hook et al., 1975) but this activity appears to be confined to the C5a molecule (Petersson et al., 1975). A similarity with other granulocytes (section 4) is evident. No evidence has been obtained to date for the presence of specific receptor molecules for C5a or C3a. Indirect evidence from desensitization studies (section 2.3.7) in the human basophil, however, suggests that the initial interaction sites of anaphylatoxin and antigen may be distinct. It is possible, though unlikely, that the highly positively charged anaphylatoxins C3a and C5a act by a nonreceptor-specific ionic perturbation of the mast cell membrane.

2.3.2.2. C3b. Although it is well recognized that C3b binding is relatively common for mediator cells (e.g., neutrophils, platelets, lymphocytes, monocytes), the presence of this activity on the surface of rat peritoneal mast cells has only recently been described (Sher, 1976). This report demonstrated that *Shistosoma mansoni* shistosomuli bind C3 and subsequently rosette around rat peritoneal mast cells. Specificity of the requirement for C3 was shown by inhibition with anti-rat C3 antibody but not anti-rat immunoglobulin antibody. Additional studies (Sher and McIntyre, 1977) have shown that, as with neutrophils from some species (reviewed by Henson, 1977), the C3 receptor is sensitive to trypsin and requires Mg^{2+} but not Ca^{2+} for activity. They also show a restricted specificity for C3 from homologous species. However, the functional activity for this C3b binding remains to be elucidated.

2.3.3. Cell-derived stimuli: leukocyte basic proteins.
Another class of mast cell (but not basophil) activator molecules have been shown to be derived from cellular, rather than humoral, mediators. These

molecules are generally low molecular weight proteins, often cationic, which may be isolated from the intracellular granules of rabbit and rat neutrophils (Janoff et al., 1965; Seegers and Janoff, 1966; Ranadive and Cochrane, 1968, 1970; Keller, 1968), human leukocytes (Kelly and White, 1973), and rat eosinophils (Archer and Jackas, 1965).

2.3.4. Proteases, including mast cell proteases.
As first demonstrated by Uvnäs and Antonsson (1963), and since confirmed by numerous investigators (review, Lagunoff et al., 1975), chymotrypsin can initiate mast cell secretion. This property is inhibited by diisopropylflurophosphate, (DFP) which indicates that the active enzyme is required for release. It is interesting to speculate, therefore, on the role of mast cell granule-bound proteases themselves in initiating or enhancing mast cell secretion. Mast cell granules contain an active proteolytic enzyme, termed mast cell chymase by Lagunoff and Benditt (1964), which has many properties similar to α-chymotrypsin. Thus the interesting possibility, previously considered by several investigators, that mast cell chymase itself may participate in the secretory response initiated by other mast cell stimuli, by an interaction in the microenvironment of the mast cell granule, remains a potential mechanism in the secretion process.

2.3.5. Nonimmunologic stimuli.
Of the nonimmunologic stimuli that have been investigated in detail, only two have been shown to have the capacity to initiate secretion from both rat peritoneal mast cells and human basophilic leukocytes.

2.3.5.1. Concanavalin A. The carbohydrate-binding protein concanavalin A (Con A) (Keller, 1973; Hook et al., 1974; Magro, 1974) is an effective stimulus. The consensus is that the lectin binds to the IgE molecules on the cells and therefore initiates secretion by a mechanism identical to anti-IgE antibody (Keller, 1973; Magro, 1974; Morrison and Henson, 1977).

2.3.5.2. Ionophore. The second type of molecules of nonimmunologic origin that may initiate secretion from both mast cells and basophils are the antibiotic ionophores. A23187 (Eli Lilly, Inc.) has been particularly effective for examining the role of calcium in the secretory process. The experiments of Foreman and co-workers (1973) first established that its induction of an influx of calcium into rat peritoneal mast cells could elicit a secretory response of histamine from these cells, and similar results have been obtained by a number of investigators (review, Morrison and Henson, 1977). Similar experiments have been performed using human basophilic leukocytes, and the results of these studies also support the concept that calcium influx into the cell is a sufficient stimulus to initiate the secretion process (Lichtenstein, 1975; Siraganian et al., 1975b). Although ionophore-mediated secretion shares many of the properties of other immunologic and nonimmunologic stimuli, there are some important differences. These will be discussed in more detail in section 2.4.

426

2.3.5.3. *Compound 48/80 and Polymyxin B.* Perhaps one of the most widely utilized of the nonimmunologic stimuli for studying the mechanism of histamine release from rat peritoneal mast cells has been the low molecular weight cationic polymer compound 48/80. This molecule is extremely active in that, assuming a molecular weight of 1300 for the active component (Read and Lenney, 1972), concentrations in the range of 0.5 μM are sufficient to initiate secretion from mast cells. Compound 48/80 shares many of the properties of secretion manifested by antigen as estimated morphologically or biochemically (Johnson and Moran 1969a,b; Bloom and Chakravarty, 1970).

Another cationic peptide which has been widely studied as a model for examining secretion from mast cells is the antibiotic polymyxin B sulfate (Norton and de Beer, 1955). This ten amino acid cyclic peptide contains five diaminobutyric acid residues and is almost as potent a stimulus as compound 48/80. Recent experiments have produced data to suggest that polymyxin B sulfate and compound 48/80 may act by a similar biochemical pathway (Morrison et al., 1975a, 1976) and may compete for the same site on the mast cell membrane (Morrison et al., 1974). Nevertheless, the concept of specific binding site receptor molecules is less clear for these low molecular weight stimuli, both of which are highly positively charged molecules and may also act to perturb the membrane nonspecifically by ionic interactions.

2.3.5.4. *Dextran.* A high molecular weight, neutrally charged polysaccharide, dextran, also initiates secretion from rat peritoneal mast cells (Goth and Knoohuizan, 1962). The molecular weights which appear to be active range from 40,000 (Beraldo et al., 1962) to as high as 2,000,000 (Baxter, 1973), although those of higher molecular weight appear to be more efficient (0.5 μM) than those of lower molecular weight (5–50 μM). However, Baxter and Adamik (1975) found 1.5 nM dextran to be effective. The mechanisms involved in dextran-mediated secretion share a number of properties with those induced by antigen and anti-IgE antibody (Baxter and Adamik, 1974; Foreman and Garland, 1974; Foreman et al., 1977).

2.3.5.5. *Venom-derived activators.* In addition to the cationic polypeptides which are immunologically derived, several similar molecules from nonimmunologic sources are known. Most of these have been isolated from the venom of insects and reptiles (review, Habermann, 1972). They include a mast cell degranulating peptide (MCD peptide) of 22 amino acids (Fredholm and Haegermark, 1967), and Mellitin, a 26-amino acid polypeptide from bee venom, although in this case the release appears to occur by cytotoxic means (Fearn et al., 1964; Habermann, 1972).

More recently, a low molecular weight cationic protein has been isolated and purified from the venom of the cobra, *Naja naja,* which initiates secretion from rat peritoneal mast cells (Morrison et al., 1975b). This cobra venom activator protein (CVA protein) has a molecular weight of approximately 18,000 and is one of the most potent activators of mast cells yet described (concentrations as low as 25 nM initiate significant secretion). CVA protein shares a number of properties with low molecular weight cleavage fragments of C3 (Morrison et al., 1975b; Morrison et al., 1976), suggesting its relationship with immunologically derived activators.

2.3.5.6. ATP.   As was first demonstrated by Keller (1966), exogenous adenosine triphosphate (ATP) in mM concentrations will initiate degranulation from peritoneal mast cells. ADP, AMP, adenosine or other nucleotides are inactive. Substantial evidence exists that the mast cell response to ATP is noncytotoxic (Diamant and Krüger, 1967), although, as in the case with ionophore-mediated secretion, a number of interesting and significant differences exist between the ATP-initiated and immunologically initiated secretory mechanisms (see below). However, as virtually all of the activators require cellular energy to generate a secretory response, the elucidation of the mechanism of ATP-initiated secretion should provide significant information on the mechanisms of release initiated by other activator molecules.

2.3.6. Interaction of stimuli at the cell surface
The development of the concept that an initial interaction which is limited to the cell surface cytoplasmic membrane can stimulate secretion from mast cells and basophils has derived primarily from morphological experiments designed to examine the binding of IgE to cells (Sullivan et al., 1971; Ishizaka and Ishizaka, 1975; Lawson et al., 1975). Thus IgE was demonstrated on the surface of these cells and they could be agglutinated by insolubilized anti-IgE (Renoux et al., 1975).

   This situation is, however, less clear for many of the low molecular weight stimuli, such as the anaphylatoxins, or compound 48/80 and polymyxin B sulfate, where the ability of the cell membrane to provide a permeability barrier is not known. Thus secretion could result from a direct interaction of stimulus with the internalized granules. However, Morrison and co-workers (1975c) demonstrated that insolubilized polymyxin B effectively initiated secretion from rat peritoneal mast cells. As available evidence suggests that compound 48/80 probably acts on the same site as polymyxin B sulfate (Morrison et al., 1974, 1975a), this suggests that the former molecule also acts on the mast cell membrane.

   These experimental data support a primary role for the mast cell and basophil surface cytoplasmic membrane in the initiation of a secretory response.

2.3.7. Desensitization of cells to stimuli
As is true for the other mediator cells discussed in this chapter, mast cells and basophils may, upon interaction with a given suboptimal dose of a given stimulus or with a stimulus under nonsecretory conditions (e.g., 4°C, no Ca^{2+}, etc.), become unresponsive or desensitized to a subsequent exposure to the same or related stimulus but not to unrelated stimuli, when releasing conditions are restored. As pointed out earlier (Becker and Henson, 1973), the relative specificity of the desensitization process argues strongly that desensitization occurs at a stimulus-specific stage in the secretion mechanism. Thus such desensitization experiments may be used to define the relative degrees of similarities of the in-

itiating stimuli in their interaction with the cell. Experiments using the mast cell and basophil (as well as the platelet see section 3) have yielded significant information in this respect. For example, rat mast cells, initially challenged with antigen, are unresponsive to a subsequent challenge with antigen (Schild, 1968) or dextran (Foreman and Garland., 1974) but not to compound 48/80 (Johnson and Moran, 1970). Similarly, mast cells initially desensitized to compound 48/80 are unresponsive to subsequent challenge with polymyxin B sulfate (Morrison et al., 1975a) or cationic band 2 protein (Ranadive and Ruben, 1973) but not to CVA protein or C3a anaphylatoxin (Morrison et al., 1975a); however, the latter two stimuli desensitize to each other. These combined data strongly suggest that (as with the platelet) stimuli-specific desensitization mechanisms play a prominent role in the biochemistry of the secretory response in the mast cell.

Experiments have shown a similar mechanism operating in the human basophil. Thus Grant and colleagues (1975), Petersson and associates (1975), and Siraganian and Hook (1976) have each demonstrated that human basophils challenged with a suboptimal concentration of the anaphylatoxin C5a are desensitized to a subsequent challenge with C5a, but still respond to a secondary challenge with antigen, or vice versa.

The data provide indirect evidence that different mast cell and basophil stimuli initiate distinct biochemical pathways (perhaps through different membrane receptors?) leading to an apparently common secretory response.

2.3.8. Summary

These apparently diverse stimuli actually fall into a number of simplified groups. Thus antigen, anti-IgE, anti-IgG$_{2A}$, Con A, and the allogeneic degranulation system probably all act through the IgE Fc receptor on the cell surface. A second group of stimuli comprise basic peptides from various sources including complement-derived C5a and C3a, (and probably the anti-rat immunoglobulin and complement system), CVA protein, and leukocyte basic proteins. Other stimuli that are also of low molecular weight and cationic include polymyxin B and compound 48/80 which appear to act similarly. Chymotrypsin may represent a class by itself, but parallels are seen with platelets and neutrophils which are also activated by exogenous proteases. Ionophore appears to differ from most stimuli, presumably bypassing the regular initial membrane-triggering event. Finally, and peculiarly, ATP is a stimulus, though the mechanism is unknown.

Dextran appears to act in a manner related to the antigen-IgE system. Dextran sulfate is also a stimulus for macrophage "activation" and induces enzyme release from that cell and it has been suggested that this effect may be mediated through the alternative complement pathway. (Schorlemmer et al., 1977).

2.4. Biochemical mechanisms of mast cell and basophil secretion

In regulation of the cellular responses to external stimuli, mast cells and basophils share a number of biochemical and pharmacologic parameters with each other and with the other mediator cells discussed in this chapter. Because these parameters have recently been reviewed in detail elsewhere (Sullivan and Parker, 1976; Morrison and Henson, 1977; Ranadive, 1977a,b), they will be only briefly summarized here.

2.4.1. General physical parameters

2.4.1.1. Rate. One of the prominent features of the secretory process from mast cells and basophils is the relative rates of the cellular responses to stimuli. Thus the entire cellular response can be manifest within 1 to 2 minutes, as for example, in compound 48/80 (Moran et al., 1962) or antigen-mediated release from mast cells, (Johnson and Moran, 1969b) or C5a anaphylatoxin-mediated release from basophils (Grant et al., 1975; Siraganian and Hook, 1976). Alternatively, responses can be more prolonged, with periods as long as 30 minutes required for full manifestation of the cellular response, as is the case with C3a and C5a (Cochrane and Müller-Eberhard, 1968) and CVA protein (Morrison et al., 1975b) stimulation of mast cells and antigen stimulation of basophils (Lichtenstein and Osler, 1964). It is noteworthy that a stimulus which initiates very rapid secretion from basophils (C5a) induces a slow response in mast cells. Conversely antigen, which initiates a very rapid response in mast cells, elicits a protracted response in basophils. These observations provide additional evidence for a true distinction between these cell types.

2.4.1.2. Concentration of stimulus. For every stimulus thus far examined the rate and the extent of the cellular reponse, as estimated by secretion of histamine or serotonin, is a function of the concentration of stimulus. In most instances, increasing concentrations of stimulus increase both the rate and extent of secretion (Baxter and Adamik, 1975). Supraoptimal concentrations of stimulus (e.g., antigen) can, however, result in a significant decrease in the extent of secretion from both mast cells (Austen et al., 1965) and basophils (Lichtenstein and Osler, 1964).

2.4.1.3. Temperature. All stimuli also demonstrate a temperature dependence of the secretory response, and with at least one stimulus, polymyxin B sulfate, the secretory response of mast cells has been shown to be linear on an Arrhenius plot over the temperature range from 27 °C to 15 °C with a change in slope occurring at that temperature (Lagunoff, 1974). The optimal temperature for secretion, however, varies depending on the stimulus and the type of cell. For example, antigen-mediated release occurs maximally at 37 °C with both basophils (Lichtenstein and Osler, 1964) and mast cells (Mongar and Schild, 1957; Chakravarty, 1960); this is also true with compound 48/80 (Uvnäs and Thon, 1961). Other activators, such as dextran, stimulate maximally at 25 °C. Anaphylatoxin-mediated release from basophils has been reported to occur maximally at 25 °C (Siraganian and Hook, 1976) or at 37 °C (Grant et al., 1975).

2.4.2. Cellular energy
It is well known that noncytotoxic secretion from mast cells and basophils requires cellular energy (review, Austen and Humphrey, 1963; Becker and Henson, 1973). Metabolic energy for secretion from rat mast cells may be generated

either by glycolysis or oxidative phosphorylation. However, antigen-mediated secretion from human basophils requires glycolysis and thus contrasts with mast cells.

Recent experiments of Chi and collaborators (1976) underscore the requirement for cellular ATP in the release of histamine from mast cells. These authors demonstrated that rat peritoneal mast cell cytolysis initiated by rabbit anti-rat mast cell antibody and complement did not result in a concomitant release of histamine. These interesting data may have a significant relation to the degranulation mechanism and bear some resemblance to the pseudocytotoxic secretion seen in neutrophils (section 4.4.6).

2.4.3. Control by cyclic nucleotides
Since cyclic AMP is known to play a vital role in the control of intracellular responses to external stimuli, it is not surprising to find this is also true, in most instances, for basophils and mast cells (review, Sullivan and Parker, 1976; Morrison and Henson, 1977). Stechschulte and Austen (1974) showed that increasing the intracellular levels of cyclic AMP (e.g., by addition of dibutyryl cyclic AMP or isoproterenol) decreased the cellular response of mast cells to antigen. In addition, agents that enhance adenyl cyclase (prostaglandin E_1) or inhibit cyclic AMP phosphodiesterase (theophylline) are also effective inhibitors of mast cell secretion (Sullivan et al., 1975a,b). Similarly, Lichtenstein and Margolis (1968) demonstrated that antigen-induced secretion from human basophils was inhibited by cyclic AMP and agents that increase the levels of this nucleotide in the cell, including histamine itself (e.g., Lichtenstein and Gillespie, 1973). Interestingly, the release from basophils and mast cells initiated by the calcium inophore A23187 does not appear to involve cyclic AMP, suggesting that the cyclic AMP regulatory step in the secretory process precedes a terminal calcium-dependent step. It is, however, important that, unlike calcium influxes, alterations in cAMP levels themselves are not sufficient to initiate secretion from mast cells or basophils.

2.4.4. Divalent cations
It has long been recognized that the divalent cation, calcium, plays a prominent role in mediation of the secretory process from mast cells (review, Foreman and Mongar, 1975). These conclusions have relied primarily upon the results of in vitro experiments where extracellular calcium has been made unavailable either by its exclusion from the incubation buffer or by the use of calcium chelators, such as ethyleneglycoltetraacetic acid (EGTA). Two recent experimental systems, however, have served to strengthen considerably the fundamental role of calcium in the secretory process. Although technically quite distinct, both systems involved the direct translocation of calcium into the cell.

The first of these has utilized a microsystem to physically introduce calcium into the cell. Thus, Kanoe and co-workers (1973) were able to demonstrate extrusion of secretory granules from rat peritoneal mast cells following the direct microinjection of calcium into the cell.

The second experimental system has utilized the ionophore A23187 to generate a calcium flux into the mast cell or basophil and thus initiate degranulation and release of mediators. As demonstrated first by Foreman and associates (1973), A23187 will induce a calcium- *and* energy-dependent secretion of histamine from rat peritoneal mast cells. These authors concluded that calcium entry into the cell was a sufficient stimulus for histamine secretion.

It seems relatively certain that calcium mobilization or entry into the cell plays a prominent role in the secretory mechanism initiated by noncytotoxic stimuli. However, the many potential ways in which calcium may participate in the secretory process, recently discussed by Uvnäs (1976), should not be underestimated in considering the mechanism(s) of degranulation. For example, Ca^{2+} requirements have been shown for at least two steps in antigen-induced release of histamine from sensitized lung (Kaliner and Austen, 1974) or leukocyte cationic protein-induced release from mast cells (Ranadive and Cochrane, 1971). As discussed by Foreman and colleagues (1977), the role of Ca^{2+} in mast cell secretion is also associated in some way with the enhancing effect (or antigen-induced effects) of phosphatidylserine (see below and Morrison and Henson, 1977). Again, this emphasizes the possibility that Ca^{2+} interaction with membrane phospholipids may be important in mediator cell function and regulation.

2.5. Stimulus-receptor triggering at the cell membrane

As mentioned above, it seems likely that most of the stimuli discussed activate the cell by binding to components at the cell membrane. This signal must therefore be transmitted to the intracellular secretory processes of the cell. We will now discuss some of the events that may be involved in this "stimulus secretion coupling" mechanism.

2.5.1. Receptor cross-linking and capping

Available data strongly support the concept that in the mast cell and the basophil, as in the platelet (see section 3) and the lymphocyte (Greaves and Janossy, 1972), an interaction of the stimulus molecule with a membrane-bound receptor is not per se sufficient to evoke a cellular response. An additional requirement of receptor cross-linking by multivalent activator molecules appears to be necessary since only nontoxic di- and multivalent, and not univalent, benzylpenicilloyl haptens could evoke a passive, cutaneous anaphylactic response in guinea pigs (Levine, 1965). A similar requirement for divalent hapten antigens to stimulate human (Siraganian et al., 1975a) and rabbit (Magro and Alexander, 1974) basophils, or for anti-IgE to be divalent to induce release in the human cells (Ishizaka and Ishizaka, 1969), led these investigators to conclude that bridging of membrane-bound antibody is essential for secretion. Furthermore, Taylor and Sheldon (1977) showed that the density of hapten on antigens can profoundly affect the secretory response of mast cells, presumably by the facility with which hapten linked to carrier is able to cross-link receptors. The highly substituted antigen, in addition to being more effective, induced a more rapid response and was more calcium-dependent.

There is also morphological evidence to support the concept of receptor cross-linking. IgE is uniformly distributed in patches around the surface of basophils at 4° but caps at 37° (Sullivan et al., 1971). As expected, the redistribution of IgE required divalent antibody (Becker et al., 1973; Carson and Metzger, 1974). However, there was no correlation between the conditions that were optimal for secretion and those which were necessary for capping. At low concentrations of anti-IgE, release could be initiated in the absence of or prior to redistribution. At high concentrations, redistribution and capping occurred in the absence of secretion. Further, antigen-induced secretion was unaccompanied by gross redistribution of surface IgE. Similar results have been obtained by radioautographic analysis of basophil surface IgE by Ishizaka and Ishizaka (1975). The net conclusion from these studies is that only certain kinds of receptor cross-linking are effective in cell activation. Furthermore, Lawson and collaborators (1975) showed that neither pinocytosis nor capping induced by divalent anti-rat Ig on mast cells were required for triggering and, if clustering was required, less than ten molecules per cluster could induce a secretory response.

These experiments suggest that IgE receptors are mobile and free to migrate in the cell membrane, and recent fluorescence studies by Schlessinger and associates (1976) showed that from 50 to 80% of IgE receptors are mobile in the membrane and that limited cross-linking, sufficient to initiate degranulation, did not influence receptor mobility. Extensive cross-linking, however, abrogated secretion and inhibited the mobility of the receptors.

As at least limited cross-linking of receptors appears to be required for IgE-mediated secretion, the IgE receptors may be associated with each other, that is, they may be either chemically or functionally multivalent as suggested by the autoradiographic studies of Ishizaka and Ishizaka (1975). Thus anti-IgE antibody induced cocapping of occupied and unoccupied IgE receptors. Completely opposite results were obtained by Mendoza and Metzger (1976), who conclude that the receptor for IgE in the mast cell membrane is univalent. Support for a univalent IgE receptor has also come from the result of in vitro studies on the IgE receptor (see section 2.3.1).

2.5.2. Membrane lipids

Until recently, very little experimental evidence existed on the mast cell or basophil lipids during the cellular activation process leading to secretion.

Rat peritoneal mast cells contain approximately equal amounts of phosphatidylethanolamine, phosphatidylcholine, sphigomyelin, and phosphatidylserine or phosphatidylinositol with lesser amounts of lysophosphatidylcholine (Strandberg and Westerberg, 1976). As indicated earlier, Schlessinger and collaborators (1976), using the fluorescent lipid probe 3, 3-dioctadecylindocarbocyanine iodide and fluorescence photobleaching methods, demonstrated that diffusion of membrane lipids was 40-fold greater than the corresponding IgE receptors. Moreover, the experiments of Lagunoff (1974) were consistent with, but not proof that, a phase change in the mast cell membrane lipids could in part regulate the secretory process.

More recently, an acetylenic analogue of arachidonic acid was shown to inhibit con A, anti-IgE, and A23187-mediated release from mast cells (Sullivan and Parker, 1977). Arachidonic acid, (as is also the case with platelets, section 3.4.7) enhanced the release initiated by the above three stimuli and induced secretion by itself, although the evidence suggested that the effect was not mediated by thromboxanes. The possible association of this observation with the suggested stimulatory action of phospholipase A on mast cells (Becker and Henson, 1973; Morrison and Henson, 1977) is still unclear.

2.5.3. The effect of phosphatidylserine on mast cell secretion
An interesting observation on the use of exogenous phospholipids to enhance the secretory response of mast cells to a noncytotoxic stimuli was first made by Goth and co-workers (1971). They showed that low (50 μg/ml) concentrations of phosphatidylserine (but not phosphatidylethanolamine, lecithin, or phosphitidylinositol) would significantly enhance secretion from mast cells stimulated by dextran. Since that time, numerous investigators have demonstrated similar enhancement of release by phosphatidylserine after stimulation with antigen, dextran, con A, and the anaphylatoxins (e.g., Mongar and Svec, 1972; Chakravarty et al., 1973; Garland and Mongar, 1974; Stechschulte and Austen, 1974; Grosman and Diamant, 1975; Johnson et al., 1975; Baxter and Adamik, 1976). Interestingly, whereas most mast cell stimuli are enhanced by this phospholipid, compound 48/80-mediated release is significantly inhibited.

The precise mechanism(s) for the enhancement of secretion is at present unclear, although Foreman and his associates have suggested that the lipid modulates the action of a "calcium gate" in the cell (Foreman et al., 1977). Nevertheless, Sydbom and Uvnäs (1976) have demonstrated that a phospholipid extract of normal rat serum will enhance anaphylactic histamine release from isolated rat mast cells, suggesting a potential role for this effect in vivo. Fractionation of the phospholids has shown that the active element resides in the phosphatidylserine-phosphatidylinositol fraction.

2.5.4. Serine esterases (proteases)
The activation mechanism of mast cells and basophils, as well as the other mediator cells discussed in this chapter, involves the apparent participation of a precursor serine esterase very early in the activation sequence. Thus, antigen-induced secretion from guinea pig lung (Austen and Brocklehurst, 1961) and isolated rat peritoneal mast cells (Becker and Austen, 1966) is inhibited by DFP. Similar inhibition was obtained with antigen stimulation of basophils (Lichtenstein and Osler, 1966; Pruzansky and Patterson, 1975). Since most investigators have been unable to demonstrate inhibition by DFP, if the DFP is removed by washing the cells and the cells are then stimulated with antigen, the presence of an activatable esterase (stimulus-dependent), that is required for secretion and is activated by the stimulus-cell interaction, is implicated (Becker and Henson, 1973). Experiments such as these also tend to exclude as a possible inhibitory mechanism the effect of DFP on glucose metabolism recently shown by Diamant in rat peritoneal mast cells (Diamant, 1976).

434

Using different phosphonate inhibitors, Austen and Becker (1966) presented evidence to suggest that antigen-mediated and complement-dependent non-cytotoxic mediated secretion may involve the same precursor esterase. (This contrasts with the results obtained with the platelet, where different stimuli apparently simulate different precursor esterases (see section 3.4.1).

This same mast cell enzyme was also implicated in secretion induced by neutrophil cationic protein (Ranadive and Cochrane, 1971; Ranadive and Muir, 1972). These investigators also showed clearly that the DFP-inhibitable step came extremely early in the sequence of biochemical events, an observation which was later confirmed for antigen-induced release of histamine from lung by Kaliner and Austen (1974). The effects of compound 48/80, however, are not apparently prevented by DFP (Uvnäs, 1976), perhaps further supporting a different mode of action for this group of stimuli.

2.6. Cytoskeletal elements and secretion

As with the other mediator cells discussed in this chapter, there is evidence to suggest a potential, but as yet undefined, role for microtubules and microfilaments in secretion from mast cells and basophils.

2.6.1. Microtubules
Early evidence that colchicine had a profound effect on the morphological appearance of rat peritoneal mast cells indicated that within 1 hour the cells developed a polymorphic appearance with a peripherally located nucleus and showed increased cell movement (Padawer and Gordon, 1955; Padawer, 1960, 1966). Moreover, the mast cell microtubules (Behnke, 1964; Padawer, 1967a) were absent after colchicine treatment (Padawer, 1967b).

A functional requirement for microtubules for effective mast cell secretion was suggested by the experiments of Gillespie and co-workers (1968) and later by Tolone and collaborators (1974). Colchicine significantly decreased the effectiveness of compound 48/80 and polymyxin B but did not affect release initiated by the cytotoxic agent decylamine. The inhibition was time- and dose-dependent, with 50 to 70% inhibition obtained after 3 hours pretreatment with 5×10^{-4} M colchicine. Deuterium oxide was found to have an opposite effect on secretion (35% D_2O gave an approximate twofold enhancement).

A similar approach has been used to examine microtubule function in preparations of human leukocytes (basophils). Antigen-stimulated secretion was inhibited by pretreatment with 3×10^{-4} M colchicine and, as predicted for an effect on microtubule assembly, pretreatment was more effective in the cold than at 37°C (Levy and Carlton, 1969). Gillespie and Lichtenstein (1972a,b) showed that D_2O enhanced both the rate and extent of antigen-induced histamine release and these effects were partially reversed by colchicine. Data was presented to indicate that microtubule participation occurred in the early, calcium-independent portion of the activation sequence (see below) and a dynamic interplay between AMP levels, calcium concentrations, and microtubule assembly has been suggested to play a regulatory role in the secretory mechanism (Gillespie, 1975).

As with antigen challenge of basophils, both C5a- and ionophore-initiated release were also shown to be inhibited by colchicine (Lichtenstein, 1975; Hook and Siraganian, 1977). Surprisingly, however, whereas ionophore-mediated release (like that induced by antigen) was enhanced by D_2O, C5a secretion was inhibited in a dose-dependent fashion by this agent, (which is presumed to stabilize microtubules). Thus, some questions may be raised about the actual roles of colchicine and D_2O in their interaction with cellular microtubules.

A number of additional experiments have led to the suggestion that microtubules may not play an essential role in the secretory process and that the inhibition observed with colchicine may be the result of effects other than those on colchicine. For example, Orr and associates (1972) found little inhibition of compound 48/80-initiated histamine release from rat peritoneal mast cells even after pretreatment with 10^{-3} M colchicine, a dose significantly higher than that expected to disperse microtubules. Röhlich and co-workers (1971a,b) and Lagunoff (1972a) suggested from morphologic evidence that degranulation does not require movement of granules to the cell surface, and that the inhibition observed with high concentrations of colchicine and vinblastine may be the result of the interactions other than interference with microtubule assembly. In addition, Lagunoff and Chi (1976) found that under treatment conditions where virtually all of the mast cells had the altered morphological appearance due to colchicine and had lost all electron microscopic evidence of intact microtubules, the mast cell secretory response to polymyxin B sulfate remained substantially unaltered. They were unable to correlate morphological changes with inhibition of secretion using colchicine concentration, time of exposure to colchicine, or temperature of incubation. On the basis of these results the authors question the assumption that the inhibition of histamine secretion observed with colchicine is the result of an interference with mast cell microtubule integrity.

Finally, Hastie and collaborators (1977) found no significant changes in the electron micrographic appearance of microtubules during secretion from basophils, and also found it difficult to correlate colchicine effects on microtubules and on secretion. A remarkable similarity exists between the discordance of the data on this subject for mast cells and basophils and that seen for neutrophils (section 4.5.4).

2.6.2. Microfilaments-actomyosin systems
In addition to the potential role played by microtubules, the possible participation in secretion of microfilaments (actomyosin-like contractile elements) present in mast cell cortical cytoplasm has also been suggested.

The microfilaments in the cytoplasm of rat peritoneal mast cells (Röhlich et al., 1971a,b) were shown by the formation of "arrowhead" complexes with heavy meromyosin to be associated with actin filaments (Röhlich, 1975). Many of the microfilaments were shown to be associated with the inner surface of the plasma membrane and occasionally also with the membrane of specific storage granules.

A functional role for microfilaments was suggested by the demonstration that cytochalasin A or B caused significant inhibition or compound 48/80-induced release of histamine from mast cells (Orr et al., 1972). Further, cytochalasin B in vivo strongly inhibited rat subcutaneous mast cell degranulation initiated by antigen. More recently, Schlessinger and associates (1976) found that cytochalasin B markedly inhibited lateral diffusion of IgE receptors on the surface of mast cells. It was concluded that receptor mobility is influenced by phospholipid-independent interactions involving microfilaments.

Douglas and Ueda (1973), however, questioned the mechanism of inhibition of secretion caused by cytochalasin. They were unable to inhibit compound 48/80-initiated secretion from mast cells using concentrations of cytochalasin equivalent to or greater than those used by Orr and colleagues (1972). If, however, they inhibited mast cell oxidative phosphorylation by pretreatment of cells with dinitrophenol, (which by itself had no effect on compound 48/80-induced secretion; see section 2.4.2), then treatment with cytochalasin completely abrogated secretion. These authors proposed that the inhibition occurred not by interference with microfilament integrity but rather by prevention of glucose transport (Mizel and Wilson, 1972; Zigmond and Hirsch, 1972).

The effects of cytochalasin on secretion from human leukocytes by antigen or C5a contrast significantly with those described for the rat peritoneal mast cell. Here (as with neutrophil secretion) the primary effect of cytochalasin was to enhance the release of histamine (Gillespie and Lichtenstein, 1972b; Hook and Siraganian, 1977). Similar results were obtained upon antigen- or anti-IgE induced secretion of histamine from chopped human lung fragments (mast cells) (Orange, 1975). However, release of SRS-A, which does not exist preformed within the cell, was inhibited by cytochalasins.

With the electron photomicrographic evidence of clearly defined microfilament structures within both the basophil (Hastie et al., 1977) and the mast cell, there can be little doubt about the existence of these subcellular elements. As with microtubules, however, the participation of these actomyosin-like contractile elements in the molecular mechanisms of granule exocytosis and secretion is still not clearly defined.

2.7. Membrane fusion

In the process of exocytosis from rat mast cells, fusion of granule-granule as well as granule-plasmalemmal membrane occurs (see section 2.1.1). The mechanisms for this process are as unclear for this cell as for others, and even less is known about basophils because of the problems involved in obtaining purified populations. Nevertheless, several investigators have recently been able to demonstrate directly mobility of intramembranous particles during the secretory process in mast cells. Thus, Chi and co-workers (1976) and Lawson and associates (1976) showed a lack of such particles in the area of the cytoplasmic membrane where perigranular membranes begin to fuse. The former group also demonstrated prominent bulges within seconds of stimulation with polymyxin B sulfate, with each bulge overlying a mast cell granule. The areas of imminent fusion become free of both intramembranous particles (Chi et al., 1976) or of immunoglobulin or con A receptors (Lawson et al., 1976), which is consistent with the recognized processes involved in "fusion" and subsequent "fission," leading to exocytosis. Additional experiments (Lawson et al., 1976) have suggested that the protein-free areas of membrane fusion are then removed from the cell by a process termed *blebbing*, the total process serving to conserve membrane proteins and remove excess lipids from the cell (see section 2.1.1).

2.8. Summary

The mechanisms involved in the secretion of vasoactive amines and other mediators of inflammation from basophils and mast cells are remarkably similar. In addition, the activation processes share properties with the other mediator cells discussed in this chapter and with other types of cells described in this volume.

Thus the generation of a noncytotoxic secretory response is initiated by interaction of the cell with one of a number of specific stimuli, either with a specific receptor localized on the cell membrane or perhaps with nonspecific membrane components. Receptor-stimulus interaction itself is not sufficient to initiate secretion; instead, a cross-linking (but not capping) of the stimulus-receptor complexes is required. Interaction of the stimulus with the receptor eventually results in the desensitization or unresponsiveness of the cell to that or closely related stimuli (e.g., those which act on the same class of receptors) but not to those stimuli which act on different receptors (see section 2.3.7). The biochemical step responsible for desensitization is not shared by different classes of stimuli. As there is strong evidence that one of the early events in the activation process involves the activation of a precursor serine esterase, it has been suggested (Becker and Henson, 1973) that the receptor may itself be the putative precursor serine esterase. To date, however, there is no evidence for this hypothesis.

Subsequent to the interaction of stimulus with receptor, a biochemical role for cyclic AMP, cellular energy, and calcium have been implicated. Experimental data suggest that the sequential requirements in the activation sequence are activation of the precursor serine esterase, the requirement for cellular energy, calcium, and cAMP (review, Ranadive, 1977a,b). Additional calcium and energy steps subsequent to the cAMP requiring step are implied by the calcium and energy requirement for cAMP-independent, ionophore-initiated secretion. The net result of these biochemical events, many of which are presumed to take place either in or associated with the membrane, is that a sufficient biochemical signal to initiate secretion is transferred to the interior of the cell.

The available evidence suggests that the transfer of information to the interior of the cell necessary to initiate the secretory response is manifest very early for those granules which are peripherally localized in the cell. Such granules undergo morphologic changes and eventually fuse with the cytoplasmic membrane. A participatory role of the cytoskeletal elements, microtubules and/or microfilaments in either organizing membrane receptors, aligning receptors with gran-

438

ules, or mobilizing altered granules to the cytoplasmic membrane surface for fusion with the membrane is possible, but is not at present definitive. Continual fusion of perigranular membranes with each other or with cytoplasmic membranes eventually results in a network of altered granules deep within the cell, continuous with the external environment. Release of granular constituents is then realized by a process of ion exchange. One of the most remarkable considerations of this complex series of biophysical and biochemical events is that it may be accomplished by the cell in less than one minute!

3. Platelets

This discussion of the nature of cell surface stimulus interactions and biochemical and morphologic consequences of these interactions in the triggering and modulation of platelet secretion concentrates mainly on platelets from man. Platelets adhere and aggregate at sites of vascular damage to play their critical role in hemostasis. They may also participate in particle clearance from the bloodstream (Copeley and Witte, 1976), inflammation, coagulation and atherogenesis (Mustard and Packham, 1970). The secretion of both newly synthesized and preformed substances from these cells is important in most of these functions.

3.1. Platelet activities and mediators

Only a brief summary of the functions of activated platelets will be undertaken here (reviews, Mustard and Packham, 1970; Weiss, 1975).

3.1.1. Aggregation
This is an early and rapid response to platelet stimulation, and is preceded by a change in platelet shape from discoid to spherical with extension of pseudopodia. The shape change in part requires energy (Behnke, 1970). In the aggregometer, this is detected as an initial decrease in light transmission. In appropriate media (e.g., citrated plasma) certain agents (e.g., epinephrine, serotonin, ADP) directly induce reversible platelet aggregation (Packham et al., 1977), referred to as primary aggregation. Subsequent to this, under specific circumstances (Mustard et al., 1975) "secondary aggregation" occurs with these agents concurrently with the onset of secretion. It has been attributed to the consequences of secreted ADP, but begins before sufficient ADP is released to initiate aggregation de novo (Packham et al., 1973; Charo et al., 1977). Although secondary aggregation probably reflects synergy between small quantities of released ADP and the stimulus (Packham et al., 1973), this point is still unsettled.

3.1.2. Secretion
A number of stimuli directly initiate noncytolytic secretion of platelet constituents, the mechanisms of which form the primary focus of this section.

Others, such as thrombin (rabbit platelets; Kinlough-Rathbone el al., 1975) surface-bound complement (rabbit platelets; Henson, 1970a), polylysine (Guccione et al., 1976), or monosodium urate crystals (Ginsberg et al., 1977b) are capable of initiating both lytic and nonlytic release of platelet constituents.

3.1.3. Coagulation

The platelet plays an important role in the coagulative process (through which numerous mediators are generated). Probably in vivo interaction of the coagulation proteins occurs optimally on the platelet surface. A number of platelet factors may also participate in the process.

3.1.3.1. Platelet factor 3.   This activity, expressed by platelet membrane lipids, acts as a cofactor for two stages in the coagulation sequence (IX + VIII + Ca^{2+} and X + V + Ca^{2+}). The activity is expressed only after cell lysis or as a consequence of noncytotoxic activation by most, but not all, stimuli (Henson and Landes, 1976). It is demonstrated optimally by a mixture of charged (e.g., phosphatidylserine) and neutral (e.g., lecithin) phospholipids. This was discussed in a recent paper by Zwaal and collaborators (1977), who also suggested that the charged lipids are normally on the cytoplasmic surface of the plasma membrane, and that "activation" would allow their mixing with the external, neutral phospholipids, thus allowing interaction with Ca^{2+} and protein coagulation factors on the platelet surface. Presumably, the granule membrane-plasma membrane fusion event might effect such a mixing.

3.1.3.2. Platelet factor 4.   Platelet factor 4 is a protein with heparin-neutralizing activity, which has recently been purified (Handin and Cohen, 1976; Levine and Wohl, 1976). It is released from platelets partly complexed with chondroitin-4-sulfate (Barber et al., 1972), but subsequently binds to heparin with greater avidity (Handin and Cohen, 1976).

3.1.3.3. Procoagulant activity.   In addition to those mentioned above, Walsh (1972a,b,c; Walsh and Lipscomb, 1976) has reported a number of other coagulant activities of washed human platelets. He has suggested that washed platelets may provide a surface for factor XII activation and this activity is enhanced by ADP. In addition, he has described the generation of a factor XII-independent, factor XI-dependent coagulant activity generated in or on platelets by collagen. This activity may account for the absence of bleeding manifestations in patients with Hageman trait. Although other investigators have been unable to repeat these observations (Vecchione and Zucker, 1975) or to demonstrate factor XI on platelets (Schiffman et al., 1977), platelets do contain factor V (Østerud et al., 1977). Moreover, Pinckard and Henson have reported that platelet activating factor (PAF) will initiate coagulative activity in washed rabbit platelets, although the mechanism is still unknown (Pinckard and Henson, 1977). These discrepancies and inconsistencies cannot be explained at present and require further investigation.

3.1.4. Preformed mediators
The large array of potential mediators liberated (secreted) from platelets has been reviewed earlier by Mustard and Packham (1970) and Holmsen and co-workers (1969). Platelets have long been known to secrete both ADP and ATP. The former is a potent platelet-aggregating agent and probably contributes to the secondary wave of aggregation. Platelets also secrete biogenic amines such as serotonin (and in the rabbit, histamine) which profoundly influence both vascular tone and permeability. The serotonin is taken up from the medium (Holmsen et al., 1973) and thus provides, through the use of [³H] serotonin, a useful method for studying amine secretion. These constituents are located in the "dense bodies" (section 3.2).

A number of other released constituents of particular interest include a factor that induces proliferation of smooth muscle cells or fibroblasts in vitro (Rutherford and Ross, 1976) and one that stimulates thymidine incorporation in fibroblasts and glycosaminoglycan synthesis in human synovial cells (Castor, 1977). The biologic role of these materials is unclear at present, but it is exciting to speculate about their potential relevance in the proliferative phenomena associated with atheroclerosis or some forms of arthritis. Platelets also release calcium, potassium, acid hydrolases, mucopolysaccharides, fibrinogen (Holmsen et al., 1969), and collagenase (Chesney et al., 1974).

3.1.5. Prostaglandins and their intermediates
In general, agents that provoke secretion are also capable of stimulating synthesis of arachidonc acid metabolites including prostaglandins and prostaglandin intermediates. They represent an important group of platelet mediators whose full spectrum of activities in vivo is little known. This large group of compounds includes: *thromboxanes* (Hamberg et al., 1975), which stimulate platelets themselves, as well as smooth muscle and probably other cells; *prostaglandins;* and *prostacyclin* (PGI_2), which has potent inhibitory effects. Hydroxyeicosatetranoic acid (HETE) is produced through the lipoxygenase pathway of arachidonic acid breakdown and has been reported to be chemotactic (Goetzl et al., 1977; and section 4). The site and mechanisms of release of these materials are not yet fully defined, although they are intimately associated with other platelet functions (section 3.4.7).

3.2. Morphologic aspects of secretion

Considerable information has come from transmission electron microscopy of platelets during secretion induced by various agents. (For detailed discussion of these data, see reviews of White 1971, 1974). Platelets contain granules that resemble lysosomes in appearance and osmiophilic dense bodies, the latter of which are thought to comprise the storage pool of biogenic amines and adenine nucleotides and may resemble, functionally, the dense granules of basophils and mast cells. A most striking feature of these cells is an interlacing network of channels contiguous with the surrounding milieu, known as the open canalicular system.

Following exposure to stimuli, the platelet may undergo a shape change, consisting of swelling, extension of pseudopods, and assumption of spherical shape (section 3.1.1). Concurrently, platelet granules move away from the surface membrane toward the center of the cell and are tightly surrounded by an ever decreasing ring of peripheral microtubules. White calls this "the contractile wave." Ultimately the granules disappear, leaving a central filamentous mass which appears to contain both actin and myosin (White, 1974).

At first sight, this pattern of secretion appears markedly different from the fusion of granule membranes with the cell surface (exocytosis) observed in the other cell types discussed in this chapter. However, Behnke (1970) and White (1971), using electron-dense tracers, showed that the membranes of the open canalicular system were continuous with the platelet plasma membrane. It remained patent throughout the secretory process, and White (1972a) showed that polycation-stabilized platelet granules were secreted intact into this open canalicular system. Thus the platelets secrete via an extension of the surface membrane which is accessible to soluble stimuli. Insoluble particles may also have access to the surface canalicular system (White, 1972b), and platelet spreading along a surface may bring the open canalicular system in contact with the surface. At present it is not clear whether the platelet granule, like the neutrophil granule, fuses with the plasma membrane at sites in contact with the stimulus.

Platelet secretion may, therefore, proceed along lines similar to that in the other mediator cells. It is possible, however, that in this cell contractile elements serve to pull apart opposing membrane surfaces to "open" a granule-containing space to the canalicular system. This might be mechanistically different from the secretory event in the other cells, but no valid supporting data exists at present.

3.3. Stimuli for platelet activation: stimulus-membrane interactions

A number of important stimuli interact with the cell surface to initiate platelet secretion. This section will discuss several such stimuli with regard to the stimulus structural characteristics required for platelet activation, modes of stimulus-platelet interaction, and evidence for stimulus-specific receptors. Due to space limitations, only selected agents thought to initiate secretion directly will be discussed in detail. Thus agents such as prostaglandin intermediates or calcium ionophores, whose site of action is unknown, will be discussed elsewhere.

3.3.1. Platelet secretion induced by ADP

ADP is primarily an aggregation-promoting agent. However, it has been appreciated that agents such as ADP, epinephrine, and serotonin can induce secretion of platelet constituents in plasma (review, Mustard and Packham, 1970). This secretion appears to require aggregation, leading to the suggestion (Massini and Lüscher, 1971) that ADP-induced secretion is a function of close cell contact (O'Brien and Woodhouse, 1968; Charo et al., 1977).

Aggregation alone, however, is not sufficient for ADP-induced secretion. Moreover, Mustard and colleagues (1975) showed that ADP-induced secretion did not occur in plasma containing normal concentrations of calcium. In washed platelet suspensions, ADP-induced secretion followed aggregation in the presence of (1) lowered calcium concentration, or (2) traces of thrombin, or (3) IgG-containing solutions. It is entirely possible, therefore, that the actual stimulus to the membrane under these conditions is one of those discussed below (e.g., thrombin, IgG) and that the ADP may act as an enhancing agent to increase platelet-stimulus contact. It is of interest that rabbit platelets, which do not respond to aggregated IgG, do not readily undergo ADP-induced secretion.

3.3.2. Interaction of thrombin and other proteases with human platelets

Thrombin, a serine protease generated in the process of coagulation, is a potent platelet stimulus. The activation of platelets by thrombin at sites of hemostasis appears inevitable, and thrombin has been proposed as a critical mediator in platelet hemostatic function. Recently much exciting new data has accumulated concerning the mechanism of action of thrombin and other proteases on human platelets. This has great relevance to the other mediator cells which are also activated by proteases (section 1).

3.3.2.1. Effect of proteases on human platelets.

A variety of proteases stimulate nonlytic secretion by human platelets, possibly (although by no means certainly) through the same substrate. Thus trypsin (which, like thrombin, has a specificity for arginine residues) initiates platelet secretion (Davey and Lüscher, 1967; Martin et al., 1975), but chymotrypsin (which cleaves at aromatic residues) does not. Papain, a sulfhydryl protease showing no structural homology to thrombin but with similar side-chain specificity, also triggers platelet secretion (Davey and Lüscher, 1967; Martin et al., 1975). Data for plasmin, however (which cleaves at lysine residues), has been both positive (Niewiarowski et al., 1973) and negative (Martin et al., 1975).

3.3.2.2. Role of proteolysis in thrombin-induced platelet secretion.

The data cited above suggests the possibility that thrombin activates the platelet through cleavage of a platelet protein substrate with subsequent cell activation (Grette, 1962). In support of this model, thrombin, with its catalytic function blocked by agents such as DFP (Davy and Lüscher, 1967), hirudin (Detwiler and Feinman, 1973a), p-('-nitrophenoxy propoxy)benzamidine benzene sulfonate (Martin et al., 1975), or tosyl lysyl chlormethyl ketone (Mohammed et al., 1976) does not stimulate plateles. Further, Mohammed and co-workers (1976) have reported that proteolytic activities of α, β, and γ thrombin toward fibrinogen correlate with their ability to aggregate platelets. These data show that the catalytic function of thrombin is intimately associated with its effects on platelets and, as a result of their elegant studies of the kinetics of thrombin-initiated platelet secretion, Detwiler and Feinman (1973a,b) proposed that the thrombin platelet interaction could be described by a series of first order enzymatic reactions.

The rate and extent of secretion is related to the concentration of thrombin added (Detwiler and Feinman, 1973a,b). To design a model in which the thrombin substrate was irreversibly cleaved, despite dependence of yield on enzyme concentration, Detwiler and Feinman (1973a) postulated that thrombin was stoichiometrically bound to its hypothetical platelet substrate and was turned over very slowly. However, studies of the binding of thrombin to platelets clearly showed (Tollefsen et al., 1974; Martin et al., 1976) that the bulk of added thrombin remains free even at concentrations well below those inducing maximal stimulation, and that thrombin is reversibly bound at equilibrium. To resolve these conflicts, Martin and colleagues (1975) suggested that thrombin reversibly binds to a receptor which is catalytically but reversibly modified; that is, the cleavage of the receptor is not a requisite for platelet stimulation. This is a fundamental departure from the previously held concept that proteolysis is an essential feature of thrombin action on platelets.

An alternative hypothesis, which preserves a central role for proteolysis, is that the binding of thrombin to its receptor results in proteolysis of the receptor, the rate of receptor cleavage being a function of the concentration of thrombin-receptor complex formed. One may then postulate that the cleavage product decays (or is inactivated) subsequent to its generation. The extent of secretion could then be a function of the concentration of active product generated, namely, the balance between rate of generation and rate of decay. This model predicts that the concentration of thrombin could determine the rate of generation of the hypothetically active cleavage product. The yield would then be a function of net balance between rates of cleavage and inactivation of the hypothetical active cleavage product. Exhaustion of thrombin substrate (receptor) by cleavage would offer a plausible explanation for the desensitization of protease-treated platelets to thrombin (Reimers et al., 1973; Harbury and Schrier, 1974; Okamura and Jamieson, 1976c) while they remain sensitive to other platelet stimuli. Inactivation or decay of activated receptors has been proposed in other cellular systems to account for "desensitization" phenomena (see section 2.3.7). Clearly, although the possibilities are not limited to the two models discussed (e.g., Detwiler et al., 1975; Tollefsen and Majerus, 1976), either would have important implications for strategies to be employed in the search for thrombin substrates in the platelet membrane.

3.3.2.3. Binding of thrombin to the platelet surface. A number of investigators have recently described the binding of thrombin to human platelets (Ganguly 1974; Tollefsen et al., 1974; Martin et al., 1976; Mohammed et al., 1976). Tollefsen and associates (1974) showed by electron microscopic autoradiography that thrombin bound to the platelet surface rather than to cytoplasmic elements, and binding is specific for platelets among blood cells. They calculated 500 high affinity (K_{diss} = 0.21 nM) and 50,000 low affinity sites (K_{diss} = 30 nM), and Martin and collaborators (1976) estimated 300 to 400 high affinity sites (K_{diss} = 1.8-2 nM) and 3300 low affinity (K_{diss} = 120 nM) sites per cell. Both groups agree on the apparent heterogeneity of thrombin-binding sites. There are several possible explanations for this heterogeneity, including negative cooperative binding interactions among equivalent sites, which has been suggested by both kinetic (Detwiler and Feinman, 1973a,b) and binding data (Tollefsen and Majerus, 1976).

Majerus and his colleagues, in a series of elegant experiments, have compared thrombin binding to its effects on platelet secretion. They (Tollefson et al., 1974) noted that K_{diss} for high affinity binding sites was similar to the thrombin concentration initiating half maximal serotonin secretion, indicating a relationship between binding and stimulation. Similarly, structural specificities for both binding and stimulation were indicated by the failure of prothrombin, or activation intermediates 1 or 2, to bind to or stimulate platelets while their proteolytic product, thrombin, does (Tollefson et al., 1975). Bovine thrombin, although antigenically unlike human thrombin, was shown to bind to and stimulate human platelets (Shuman et al., 1976). Finally, Shuman and Majerus (1975) demonstrated that the ionic composition of the suspending buffer media profoundly influences the platelet's avidity for thrombin, and that anion-induced binding changes closely parallel changes in thrombin-induced serotonin secretion. These data provide compelling evidence of the physiologic relevance of thrombin binding to platelets.

3.3.2.4. The search for thrombin substrates. If thrombin initiates platelet secretion by proteolysis, identification of platelet proteins cleaved by thrombin could offer important clues as to the nature of the thrombin "receptor." This problem has been studied by a variety of workers with equivocal results. In one approach, platelet proteins in thrombin-treated or untreated platelets were compared. A number of workers (Salmon and Bounameaux, 1958; Nachman, 1965; Cohen et al., 1969; Baenziger et al., 1971) have demonstrated thrombin-induced deletion of platelet proteins. Most notably, Baenziger and co-workers (1971) demonstrated deletion of an apparent 190,000 MW glycopeptide from thrombin-treated platelets and termed it thrombin-sensitive protein (TSP). TSP was associated with the particulate fraction of lysed platelets, indicating it was either membrane-associated initially or became so after cell lysis. It was deleted only from whole cells, and not membrane fractions, and was released without being cleaved, as judged by SDS gel electrophoresis and amino acid analysis (Baenziger et al., 1971, 1972). Further, its release was inhibitable by PGE_1 (Brodie et al., 1972). Finally, TSP has not been accessible to platelet surface probes (Phillips and Agin, 1974; Okamura and Jamieson, 1976a). Thus, this platelet glycoprotein does not appear to be the platelet membrane thrombin substrate.

A second approach has been to label platelet surface proteins radioactively and to observe changes in the pattern of radioactivity following thrombin treatment. Steiner (1973) and Phillips and Agin (1974) were able to show thrombin-induced subtle deletion in a 118,000 to 121,000 dalton surface glycopeptide from intact platelets using this approach, but only after prolonged thrombin treatment. Hydrolytic products were not demonstrated. Nachman and associates (1973) could demonstrate no such deletion, but iodinated after thrombin treatment rather than before. It has been argued that for thrombin-induced peptide deletion to be significant vis à vis thrombin-induced secretion, it must occur during a similar time course and at similar thrombin levels (Tollefsen and Majerus, 1976; Detwiler and Charo, 1977). However, there is no information concerning the relationship between the amount of the postulated thrombin substrate hydrolyzed (as percent of total present) and secretion. Data indicate that thrombin initiates maximal secretion on occupation of 100 of the 50,000 thrombin-binding sites per cell (Shuman and Majerus, 1975) and that there is equivalence of thrombin-binding sites with slow dissociation of thrombin (Tollefsen and Majerus, 1976). If thrombin were to cleave its receptor, this would only be detectable (i.e., if SDS gel techniques could detect deletion of greater than 5% of a polypeptide) after cleavage of $.05 \times 50,000 = 2500$ sites. This is 25 times the number of sites occupied at maximal secretion. Thus detection of cleavage could reasonably require addition of more enzyme or longer incubation of thrombin with the cells than would secretion. The relationship between polypeptide deletion and thrombin-induced secretion remains uncertain.

In an exciting series of studies, Okamura and Jamieson (1976a,b,c) have described and isolated a platelet glycoprotein (MW 148,000) they termed "glycocalicin." They have marshalled considerable evidence that it is associated with the platelet surface. Based on the observation that trypsinized platelets do not aggregate in response to thrombin and that glycocalicin at 170 μg/ml blocks thrombin-induced platelet aggregation (Okamura and Jamieson, 1976c), these investigators have proposed that glycocalicin is a thrombin receptor. Data concerning the effect of thrombin on platelet glycocalicin in the intact cell that would add considerable weight to this hypothesis is presently unavailable. Further, the ability of chymotrypsin to liberate (proteolytically?) glycocalicin from the cell is inconsistent with its failure to trigger platelet secretion. Nevertheless, this possibility should stimulate new investigation into the nature of the thrombin receptor.

3.3.3. Collagen

The disruption of endothelium following vascular trauma leads to platelet contact with subendothelial structures, platelet adhesion, and secretion. Platelets seem largely reactive toward subendothelial collagenous components (Baumgartner, 1974), lending physiologic significance to collagen-induced platelet activation. Before discussing this important area, it should be noted that collagen is a highly complex material and investigators in this field have not employed any standardized preparations. Thus variations in amino acid composition, carbohydrate content, glycosaminoglycan contamination, and quarternary structure may all contribute to the disparate and occasionally contradictory data reported.

3.3.3.1. Collagen-platelet interaction. The initial binding of collagen to the platelet membrane is viewed as "adhesion." This must be distinguished from subsequent platelet-platelet cohesion or aggregation, usually by use of aggregation inhibitors (Cazenave et al., 1973; Baumgartner, 1974). Subsequent to adhesion, the platelets undergo aggregation (Zucker and Borreli, 1962) and secretion (Hovig, 1963), and synthesize prostaglandins and their intermediates (Smith et al., 1973). It is now known that both the secreted constituents (e.g., ADP) and prostaglandin synthesis (Packham et al., 1977) contribute to collagen-induced aggregation of more platelets.

3.3.3.2. Role of membrane glycosyl transferases in platelet adhesion to collagen. Jamieson and colleagues (1971) were able to correlate inhibition of membrane-bound collagen glycosyl transferases by glucosamine or chlorpromazine with inhibition of platelet adhesion to collagen. This led them to suggest that this surface-bound (Jamieson et al., 1971) enzyme forms a complex with galactosyl residues on incomplete collagen heterosaccharides, thus mediating platelet adhesion. Barber and Jamieson (1971) and Bosmann (1971) observed the presence of such enzymes in platelet membrane preparations. However, the demonstration of a physiologic role for platelet collagen glycosyl transferase in platelet adhesion rests critically on proving that the heterosaccharides of native collagen can interact with the platelet enzyme. Attempts to confirm this hypothesis by chemical modification of collagen have been inconclusive (Puelt et al., 1973; Jaffe and Deykin, 1974a; Harper et al., 1975; but see Brass et al., 1976). Nevertheless, Kang and co-workers (1974) and Chiang and colleagues (1977) have shown that in the case of one collagen from chicken skin the α-chains bind to and stimulate platelets, probably as a direct function of their carbohydrate content. The relevance of this observation to platelet stimulation by native collagen remains to be assessed.

3.3.3.3. Role of collagen quaternary structure. Several workers have appreciated (Zucker and Borreli, 1962; Wilner et al., 1968; Puett et al., 1973) that denatured collagen was ineffective in initating platelet aggregation, indicating that primary structure was usually insufficient. In most cases (Puett et al., 1973; Kang et al., 1974) α- or β-chains alone are also insufficient, suggesting tertiary structural requirements. Subsequently, four groups of investigators (Muggli and Baumgartner, 1973; Brass and Bensusan, 1974; Jaffe and Deykin, 1974b; Simons et al., 1975) have presented convincing data that collagen multimer formation is critical in platelet stimulation. Wilner and associates (1968, 1971) suggested that a rigid array of polar groups (of either charge) was a critical feature of collagen for platelet aggregation. In support of this hypothesis, the simple cationic homopolymer, polylysine (Jenkins et al., 1971), has been shown to aggregate platelets. Further, the urate anion itself is not a platelet stimulus, but when presented to the cell as the uric acid (Mustard et al., 1967) or monosodium urate crystal (Ginsberg et al., 1977b) it initiates platelet secretion.

These data underscore the complexity of platelet-collagen interaction, and it seems likely that both carbohydrate and protein determinants may contribute to this relationship.

3.3.4. Immunoglobulin (immune complexes)
The interaction of immune complexes and platelets has been reviewed by Pfueller and Lüscher (1972a), Becker and Henson (1973), and Osler and Siraganian (1972). Primate, pig, sheep, goat, and ox platelets all adhere directly to immune complexes (Henson, 1969) and are activated by them. Rabbit, dog, mouse, and horse platelets, however, react with complement components associated with the immune complexes (section 3.3.5).

Studies with human platelets, which will be described as an example of the former group, have generally employed heterologous antibody-antigen complexes or nonspecifically aggregated human immunoglobulins. It is assumed (though not proven) that both reagents interact with the same cell surface components to stimulate the cell and are equivalent in effect to human antibody-antigen complexes.

3.3.4.1. Structural requirements for stimulation. Immune complexes initiate aggregation and release of constituents from human platelets in artificial medium and plasma (Humphrey and Jacques, 1955). IgG aggregates (Mueller-Eckhardt and Lüscher, 1968) and IgG-coated particles (Glynn et al., 1965) or surfaces (Packham et al., 1969) have similar effects. The relative contribution of conformational changes in the IgG to that of the polyvalent presentation of IgG in the complexes, aggregated IgG, or adsorbed IgG is unknown at present. However, IgG aggregates of dimer to trimer size are as effective platelet stimuli as those larger than octamers (Ginsberg and Henson, 1977), which implies that if valency alone is critical, bridging on only 2 to 3 "receptors" is sufficient for maximal effect.

Pfueller and Lüscher (1972b) and Henson and Spiegelberg (1973) showed that aggregated IgG of all four subclasses induced aggregation and secretion in washed platelets. The observation of IgG class specificity, activity of Fc fragments, and inactivity of F(ab')$_2$ fragments and effects of tetanus-human antitetanus antibody complexes led to the suggestion that platelets possess human "Fc receptors" (Henson and Spiegelberg, 1973; Israels et al., 1973).

Pfueller and Lüscher (1972c) showed that blockage of free amino groups, or tryptophane or carbohydrate modification of aggregated IgG abolished its ability to stimulate platelets. They proposed that the initial IgG platelet interaction involved both tryptophane residues and carbohydrates while amino groups only mediated subsequent platelet stimulation. They also noted that both IgG and collagen are glycoproteins which stimulate platelets partly by their amino groups, and raised the possibility that they interact with similar sites on the platelet. Moreover, collagen multimerization, like IgG aggregate formation, is critical in platelet stimulation and monomeric IgG blocks platelet stimulation by collagen (Cazenave et al., 1976) as well as by aggregated IgG (Pfueller and Lüscher, 1972b). These intriguing parallels between aggregated IgG and collagen need to be explored.

Aggregated IgG binds to human platelets and 40 to 70 aggregates per cell were required to trigger secretion (Pfueller et al., 1977a). This binding was not prevented by removal of surface glycoproteins and Von Willebrand Factor by proteases (Pfueller et al., 1977b). Interestingly, trypsinization of platelets did not prevent platelet aggregation by collagen (Okamura and Jamieson, 1976c).

3.3.4.2. Possible role for C1 in human platelet activation by Ig.   There is a rough correlation between complement-fixing ability and platelet stimulatory activity of IgG aggregates and immune complexes (Mueller-Eckhardt and Lüscher, 1968). Furthermore, Cl inhibits IgG aggregate-induced cell stimulation (Pfueller and Lüscher, 1972a,b), leading the latter authors to suggest that C1 or a C1-like molecule could be a platelet IgG receptor. In support of this, Wautier and associates (1976a,b) showed that removal of platelet-associated C1q with EDTA inhibited, and subsequent addition of C1q to these cells enhanced, IgG aggregate-induced platelet aggregation. However, the potent effects of EDTA on platelets and the ability of C1q to cross-link particle-bound Ig make this difficult to interpret. Moreover, IgG_4 aggregates (Pfueller and Lüscher, 1972b; Henson and Spiegelberg, 1973), which do not activate complement, stimulate platelet secretion; whereas carbohydrate-modified IgG aggregates (Pfueller and Lüscher, 1972c) and the lipid A constituent of bacterial lipopolysaccharides (Ginsberg and Henson, 1977), which do activate C1, are not potent stimuli of platelet secretion.

3.3.4.3. Stimulation of human platelets by Ig in plasma: the effect of surfaces.   Immune complexes stimulate human platelets in plasma (Humphrey and Jaques, 1955; Movat et al., 1965; Pfueller and Lüscher, 1974), but serum or plasma also has been shown to inhibit the effects of immune complexes on platelets (Mueller-Eckhardt and Lüscher, 1968; Pfueller and Lüscher, 1972a; Henson and Spiegelberg, 1973). This is due partially to the binding of complement components to the immune complexes (Pfueller and Lüscher, 1972b; Henson and Speigelberg, 1973). It is also due to inhibition by the monomeric IgG present in serum (Pfueller and Lüscher, 1972a,b). Thus immune complexes and IgG aggregates initiate platelet aggregation and secretion in fluid phase in plasma only at concentrations far above those expected in vivo (Pfueller and Lüscher, 1974). In contrast, particle-associated immune complexes such as those formed on the surface of a pneumococcus (Zimmerman and Spiegelberg, 1975), particle-bound IgG aggregates (Jobin et al., 1971), or gamma globulin on shunt bifurcations (Evans and Mustard, 1968) are effective platelet stimuli in plasma. Thus in vivo (as with neutrophils, section 4), stimulation of platelet secretion by immune complexes or IgG on particles (e.g., bacteria) or on surfaces may well play an important role in the pathogenesis of tissue injury.

3.3.5. Complement interaction with rabbit platelets
This area will be summarized only briefly and the reader is referred to Becker and Henson (1973) and Osler and Siraganian (1972) for more detailed discussion.

3.3.5.1. Innocent bystander lysis. It has long been recognized that rabbit platelet mediator release could be initiated by immune complexes only in plasma (Humphrey and Jaques, 1955; Barbaro, 1961; Siquiera and Nelson, 1961). The complexes adhere to rabbit platelets in the presence of complement components through C3, presumably by binding to the C3 receptor present on these cells (Henson, 1970a,b, 1974b,c). In the presence of the terminal components of complement, adherent or closely adjacent complexes initiate platelet lysis with mediator release. The alternative pathway of complement activation is involved (Siraganian et al., 1973) and Morrison and co-workers (1975d) found that lipopolysaccharide preparations could only induce this lytic phenomenon, termed "innocent bystander" lysis, if they activated the alternative pathway. These preparations were ineffective if they were classical pathway activators.

3.3.5.2. C3b-induced secretion. The interaction of rabbit platelets with C3 bound to particles or surfaces induces secretion (noncytotoxic) of granule constituents, a process which appears identical to other secretory mechanisms (Henson, 1970a; Henson et al., 1976c). While the stimulation of rabbit platelets has been attributed to C3b, and these cells do not respond to aggregated Ig directly, the possibility of C3 acting as a ligand and allowing previously inactive Ig to effect stimulation cannot be excluded at present. It is interesting that human platelets may also be stimulated by zymosan, through a complement-dependent process, but this does not appear to involve C3b receptors (probably absent from human platelets) and requires fibrinogen and probably other coagulation components (Zucker and Grant, 1974). Its mechanism is unknown.

3.3.5.3. Neutrophil effects. Rabbit platelet secretion by C3b or immune complexes was enhanced in the presence of neutrophils. Part of this apparently involved neutrophil-mediated coagulating processes with generation of thrombin (Henson, 1970b). However, a similar enhancement in the absence of plasma remained unexplained until the recent demonstration of thromboxane release from neutrophils (section 4).

3.3.6. Basophil-derived platelet activating factor (PAF)
This material is another example of an important concept in mediator cell activation, namely, the existence of extensive interaction and cooperation between different types of mediator cells.

PAF has been described from rabbit, rat, and man (reviews, Benveniste, 1977; Henson et al., 1977a), but most work has been performed with the rabbit system. The material is released from basophils (section 2) upon stimulation by a variety of secretory activators and is an extremely potent activator of platelets. It has not been purified as yet but is a lipid, probably a phospholipid, of low molecular weight (about 1000). PAF induces aggregation, secretion, and prostaglandin production from platelets, but not the generation of platelet factor 3 activity (Henson and Landes, 1976). Its action on platelets appears to be direct and does not involve either the liberation of ADP (Henson, 1977) or prostaglandin intermediates (Shaw et al., 1977) although it induces release of these materials. Considerable evidence has been accumulated to suggest the importance of PAF as a mediator of inflammatory process in vivo (Benveniste, 1977; Henson and Pinckard, 1977; Pinckard et al., 1977), but until it has been purified and characterized the biochemical nature of the interaction of this significant mediator with the platelet surface must remain speculative (Henson et al., 1976).

3.3.7. Antiplatelet antibody
A myriad of additional platelet stimuli could be described and some, such as ionophore and thromboxanes (arachidonic acid), are considered below. Among these, however, antiplatelet antibody points up an important concept in platelet stimulation. Several years ago we showed that heterologous antibody against platelets could initiate secretion and that this did not involve the Fc fragment and required divalency (Henson, 1970c). Cross-linking of receptors is implicated. Similar data were recently reported by Coleman and Schreiber (1976), and the system has immediate parallels to antibody stimulation of other cells including macrophages and neutrophils. Moreover, it is suspected that such autoantibodies are an important cause of idiopathic thrombocytopenic purpura (ITP) (Karpatkin et al., 1972).

3.3.8. Desensitization of platelets to stimuli
As with mast cells, incubation of platelets with a variety of stimuli renders them unresponsive to a further addition of that stimulus but normally reactive to unrelated stimuli. Thus pretreatment of platelets with thrombin (section 3.3.2.2) or with PAF (Henson, 1976c) induces a relatively specific state of desensitization. The mechanism of this effect, whether it be exhaustion of membrane substrate, decay of an activatable enzyme or other stimulus-specific step, blockage of receptors and/or inactivation or receptor-bound stimulus, is unknown. Nevertheless, it may serve to limit the extent of secretion in vivo (Henson, 1976a,b,c; Henson and Pinckard, 1977) and provides a useful tool for dissection of the stimulus-secretion coupling process.

3.4. Biochemical mechanisms of platelet activation with particular reference to secretion

This subject is additionally complex in platelets because the products of activation (ADP, serotonin, prostaglandin intermediates, etc.) are themselves potent activators, and it is difficult to dissect primary from secondary effects. Nevertheless, some discussion of implied processes will serve to indicate overall mechanisms and a commonality with other mediator cell types. In some respects this is all the more remarkable in the case of a cell which, after all, is no more than an isolated anucleate fragment of membrane-bounded cytoplasm!

3.4.1. Serine esterases
As discussed at length earlier (3.3.2), exogenous proteases, particularly those with serine in the active center and a specificity for basic amino acids, are potent

platelet stimuli. The question arises, however, as to whether endogenous platelet proteases are involved in stimulation by other activators.

3.4.1.1. Activatable esterases within the cell.   Early experiments (review, Mustard and Packham, 1970; Becker and Henson, 1973) showed inhibition of platelet responses by amino acid esters such as tosylarginine methyl ester (TAME) and by protease inhibitors, and implied the involvement of platelet proteases. More recently, rabbit platelet secretion induced by five stimuli was shown to be inhibited by DFP and, moreover, in the case of four of these, the activation of a precursor esterase was implicated (Henson et al., 1976). Sequencing experiments (Henson, 1974a,b,c) suggested that this step came very early in the stimulus-secretion coupling process, as in mast cells (section 2). Interestingly, use of series of organophosphorus inhibitors (Henson et al., 1976) or amino acid esters (Henson et al., 1977c) suggested that PAF, collagen, C3b, and antiplatelet antibody induced platelet stimulation by different esterases and that each of these was different from thrombin. The simplest hypothesis suggests that the "receptors" for these stimuli are themselves precursor proteases (or are closely linked to them), and that their activation (perhaps by conformational changes as for C1) initiates the stimulation pathway. However, this remains firmly in the realm of speculation and can be no more than a working hypothesis until further characterization of either the "receptors" and/or the enzymes is forthcoming.

3.4.2. Energy metabolism
This subject was discussed in earlier reviews. Platelet secretion (and aggregation) requires ATP generation, which (as in mast cells) may be accomplished by glycolysis or oxidative phosphorylation.

3.4.3. The role of calcium in platelet secretion
As discussed elsewhere in this volume, a calcium flux is implicated in cellular secretory phenomena. In the platelet, Holmsen (1974) has proposed that an increase in intracellular calcium leading to contractile events is the final common pathway of platelet function. As noted below, platelet secretion occurs in the absence of detectable contractile function so that labilization of membranes for fusion is probably also critical in platelet secretion. As discussed by Paphadjopoulos in this volume, calcium may subserve this function as well.

There are several special platelet properties (and problems) which deserve mention. It is well known that there is an absolute requirement for extracellular calcium in the initiation of human platelet aggregation by agents such as ADP. In contrast, chelation of extracellular calcium by EDTA does not abolish secretion initiated by agents such as thrombin (Holmsen et al., 1969) and aggregated gamma globulin (Mueller-Eckhardt and Lüscher, 1968). Further, the human platelet contains large quantities of calcium within platelet-dense bodies (Skaer, 1975) that is secreted (Mürer 1969). Thus net calcium flux is outward during platelet secretion. Finally, platelet phospholipid becomes available during the release reaction and provides platelet factor 3 activity. This coagulant activity may be associated with increased binding of calcium to the platelet surface associated with increases in charged phospholipids in the outer leaflet of the membrane (section 3.1.3.1). The appearance of increased Ca^{2+}-binding activity on activated platelets and its contribution to the apparent platelet calcium influx has been validated experimentally (Massini and Lüscher, 1976). Thus the lack of requirement for extracellular calcium, the presence of a large releasable pool of intracellular calcium, and the generation of new membrane calcium-binding sites during secretion all involve special difficulties in studying the role of calcium in platelet secretion.

Nevertheless, considerable evidence exists linking calcium fluxes to human platelet secretion. Thus the calcium ionophore A23187 directly initiates platelet secretion (Feinman and Detwiler, 1974; Massini and Luscher, 1974; White et al., 1974) even in the absence of external calcium. This indicates that intracellular transfer of calcium alone may trigger secretion. Recently, Charo and collaborators (1976) showed that the calcium antagonist 8-(N,N-diethylamino) acetyl 3,4,5-trimethoxybenzoate (TMB-8) inhibited thrombin and A23187-induced secretion. The inhibition could be decreased by concurrent addition of A23187 and thrombin or by addition of calcium. This further supported a role for calcium in platelet secretion and provides further evidence of the similarity of platelet secretion to secretion by other cells.

3.4.4. Cyclic nucleotides

3.4.4.1. Cyclic AMP. As with the other cells considered here, agents that increase adenylate cyclase activity (e.g., PGE), agents that decrease phosphodiesterase activity (e.g., caffeine), or cAMP analogues (e.g., dibutyryl AMP) increase resting cAMP levels and inhibit platelet aggregation and secretion induced by a number of agents (review, Salzman, 1972). Agents that initiate platelet secretion, such as collagen (Chiang et al., 1976), thrombin (Brodie et al., 1972) and a number of rabbit platelet stimuli (Henson, 1974a; Henson and Oades, 1976), have been shown to inhibit platelet adenyl cyclase or to reduce the levels of cAMP. These data suggest that stimuli could trigger platelet activation by precipitating a fall in intraplatelet cAMP in the resting cell. This critical point remains controversial (Salzman, 1972; Haslam, 1975). Recently Rodan and Feinstein (1976) showed that Ca^{2+} is a potent inhibitor of isolated membrane platelet adenyl cyclase (Ki = 16 μM). In view of the evidence for intraplatelet calcium fluxes during platelet secretion, the possibility arises that changes in adenyl cyclase are secondary to changes in intraplatelet calcium distribution. Thus, while there is good evidence to support a regulatory role for cAMP in platelet secretion, it does not seem as if the drop in cAMP levels per se can trigger secretion or that declines in adenyl cyclase levels are the initiators rather than the result of (and presumably modulators of) the secretory process.

To date, the ability of cAMP to modulate platelet secretion is unexplained. One hypothesis (put forth by a number of workers, e.g., Holmsen, 1975) is that cAMP increases the sequestration of mobilized intraplatelet calcium. Thus cAMP would act to antagonize the postulated increase in cytoplasmic calcium which mediates secretion. In support of this hypothesis, agents known to increase intracellular cAMP inhibited the extent of A23187-induced serotonin secretion in platelet rich plasma (White et al., 1974; Feinstein and Fraser, 1975).

Alternatively, as in other cells, this agent may mediate phosphorylation of contractile proteins involved in secretion. In support of this hypothesis, Booyse and co-workers (1976) have isolated a cAMP-sensitive platelet protein kinase that phosphorylates four endogenous substrates, one of which has a molecular weight of 45,000, similar to actin. These two hypotheses should provide ample stimulation for future work.

3.4.4.2. Cyclic GMP. Considerably less data exists concerning the possible role of cGMP in platelet secretion. In other cells discussed in this chapter, increases in cGMP have been associated with secretion. In the human platelet, increases in platelet cGMP have been observed during collagen-induced secretion and aggregation (Haslem, 1975; Chiang et al., 1976). The enhancing effect of Ca^{2+} on platelet guanyl cyclase activity (Rodan and Feinstein, 1976) makes interpretation of this phenomenon difficult. Similarly, there is little data about the effects of alteration in cellular cGMP level per se on platelet secretion. Thus the role of this cyclic nucleotide as a regulator of secretion is unclear at present.

3.4.5. Microtubules
As noted previously, resting platelets contain a circumferential band of 200 to 250 Å microtubules in close relationship to the submembranous microfilaments. In activated cells, these microtubules move centripetally during the contractile wave and are often found within extended pseudopodia, suggesting a role in platelet contractile function. White (1968) has shown that dissolution of platelet microtubules by treatment with colchicine or vinca alkaloids is associated with loss of discoid shape. Chilling, which leads to loss of microtubules, also leads to loss of discoid shape and extrusion of pseudopodia (White and Krivit, 1967). Finally, D_2O, which stabilizes microtubules, prevents ADP-induced shape change (Le Breton, et al., 1976). These data suggest that microtubules play an important role in maintenance of the discoid shape of the resting platelet.

"Disruption" of microtubules also abrogated the centralization of platelet granules (White, 1969). On the other hand, ADP-induced primary aggregation and clot retraction were unaffected by these doses of colchicine. These data suggested that the microtubule does not provide the motive force of the platelet contractile wave but may serve to orient the contraction. Since this may be involved in secretion, agents disrupting microtubules might affect secretory phenomena. While the second wave of ADP aggregation was inhibited by antimicrotubule agents (White, 1969), the complex relationship between secondary aggregation and secretion (see 3.1.1) makes this data difficult to interpret. Friedman and Detwiler (1975) reported inhibition of thrombin-induced secretion by colchicine at doses above 10^{-3} M, a level in excess of those needed to prevent microtubular association in White's studies. They suggested that microtubule dissolution per se was not the mechanism of the drug effect. Rabbit platelet secretion is also inhibited by colchicine but, as in the abovementioned study, with high concentrations (10^{-4} M) (Henson and Oades, 1976). Concurrent morphologic studies were not performed. As with other cell types, microtubules may have a supporting or even enhancing role, but do not seem to be required for secretion.

3.4.6. Microfilaments

Microfilaments are visible in resting platelets in a submembranous distribution (Zucker-Franklin, 1969). In stimulated cells (White, 1971) microfilaments are observed in pseudopodia and in association with the shrinking circumferential ring of microtubules of the contractile wave. After granule disappearance, the platelet center is occupied by a mass of fused microfilaments. Cytochalasin B inhibits platelet aggregation (Haslam et al., 1975), shape change (White and Krumwiede, 1973), contractile wave (White and Estensen, 1972), and clot retraction (Shepro et al., 1970). The data suggest that microfilaments are implicated in these platelet contractile functions. In contrast, the drug has a variable effect on platelet secretion induced by a variety of agents effecting inhibition or enhancement (White and Estensen, 1972; Friedman and Detweiler, 1975; Haslam et al., 1975; Henson and Oades, 1976). Further, phorbol myristate acetate (PMA) induces platelet secretion without the associated contractile wave (Estenson and White, 1974). Thus inhibition of this effect does not necessarily abolish secretion and secretion can occur in its absence. This indicates that the contractile function may facilitate platelet secretion but is not indispensable to secretion and, presumably, to membrane fusion.

Of these cell types the platelet is a candidate for the secretory mechanism which involves the pulling apart of opposed membranes to open a granule-containing space to the outside (or to the open canalicular system). The contractile elements could produce such an effect. If these are already present and their formation (polymerization) does not have to be initiated as a result of the membrane stimulation, cytochalasin B might have variable effects noticed. This however, remains highly speculative at present.

A platelet actomyosin, termed thrombosthenin, was isolated by Bettex-Galland and Lüscher (1961). Ultrastructurally this material is similar to the platelet microfilaments (Zucker-Franklin et. al., 1967). In sectioned platelets, filaments that form arrowheads with muscle-heavy meromyosin have been demonstrated (Behnke and Kristensen, 1971), and the fused microfilaments of activated cells have been shown to possess Mg^{2+}-ATPase activity (White, 1974). Thus it seems reasonable to suppose that platelet microfilaments contain actomyosin and that platelet actomyosin contributes to aggregation, shape change, contractile wave, and clot retraction. A recent review summarizes novel developments in the chemistry of platelet actomyosin (Detweiler and Charo, 1977).

3.4.7. Prostaglandins

In the past three years there has been an information explosion concerning the interrelationship of prostaglandins and prostaglandin intermediates and various platelet functions. Here we provide a brief and necessarily incomplete synopsis of this exciting area, and refer the reader to the recent volume edited by Silver and colleagues (1977) for a more comprehensive treatment.

3.4.7.1. Prostaglandin metabolites involved in platelet function. Platelets synthesize prostaglandins E_2 and $F_{2\alpha}$, and inhibition of this synthesis by aspirin suggested a relationship to the ability of aspirin to inhibit the second wave of ADP-induced aggregation (Smith and Willis, 1970). PGE_2 and $PGF_{2\alpha}$ per se have little effect on platelets, but intermediates of prostaglandin synthesis such as prostaglandin endoperoxides (PGG_2 and PGH_2) are potent aggregating agents (Hamberg et al., 1974a) and are released from platelets during stimulation by thrombin (Hamberg et al., 1974b), collagen, epinephrine, and arachidonic acid (the substrate for this synthetic process) itself (Smith et al., 1974). Aspirin inhibits arachidonic acid-induced aggregation and endoperoxide synthesis (Smith et al., 1974), but aspirin-like drugs do not inhibit aggregation due to the endoperoxides themselves (Malmsten et al., 1975). More recently, thromboxane A_2, a platelet endoperoxide metabolite, has been shown to be a more potent aggregating agent than the endo-peroxides (e.g., Needleman et al., 1976). It may be the active agent of both arachidonic acid and the endoperoxides.

3.4.7.2. Source of platelet arachidonic acid. The platelet is known to be rich in esterified arachidonic acid (Marcus et al., 1962), possibly on the inner surface of the cell membrane (Derkson and Cohen, 1975), suggesting a ready source for prostaglandin synthesis. Recently, Pickett and Cohen (1976) showed that thrombin treatment of anoxic platelets resulted in accumulation of free arachidonic acid derived from endogenous stores. Adopting a different approach, Bills and associates (1976) and Russell and Deykin (1976) have shown incorporation of radioactive arachidonic acid into platelet phospholipids and subsequent release of the acid followed by oxidative transformation after thrombin stimulation. These studies provide direct evidence for the mobilization of endogenous platelet arachidonate by thrombin.

3.4.7.3. Inhibitors of prostaglandin synthesis. Much of our understanding of the function of prostaglandin intermediates in platelets comes from the effects of various inhibitors of their synthesis. Aspirin, and most of the synthetic nonsteroidal anti-inflammatory drugs, inhibits the second wave of ADP-induced aggregation apparently by preventing synthesis of endoperoxides (Hamberg et al., 1974a) through action on the cyclo-oxygenase enzyme system (Roth et al., 1975). Vane (1976) has proposed that this also represents an important mechanism in the anti-inflammatory effects of aspirin.

Indomethacin also inhibits cyclo-oxygenase but not the platelet lipoxygenase (Bills et al., 1976). Eicosatetranoic acid, however, inhibits both enzymes, providing a means for the simultaneous blockade of endoperoxide and HETE (12.2-hydroxy 5, 8, 10, 14-eicosatetranoic acid) synthesis by platelets.

Mepacrine (quinacrine), an antimalerial and phospholipase inhibitor, has been shown to inhibit arachidonic acid mobilization in tissues other than platelets (Zusman and Keiser, 1977) but does not abrogate the subsequent synthesis of prostaglandins from arachidonic acid. It is of interest that this drug, and more frequently the closely related drug, chloroquine, have been employed with aspirin for synergistic anti-inflammatory effects, perhaps acting to prevent the phospholipase A-induced mobilization of arachidonic acid. However, the mechanisms by which stimuli might induce or enhance phospholipase activity or increase the availability of arachidonic acid containing phospholipids is unknown.

3.4.7.4. Role of prostaglandin intermediates in platelet aggregation. This important area has recently been subjected to lucid analysis by Packham and colleagues (1977). Prostaglandin synthesis appears to induce the secondary aggregation initiated by such agents as ADP, epinephrine, or serotonin. Arachidonic acid itself can stimulate ADP-independent aggregation (Kinlough-Rathbone et al., 1976). Moreover, cyclo-oxygenase inhibitors partially inhibit collagen-induced platelet aggregation (Kinlough-Rathbone et al., 1977). A selective inhibitor of thromboxane A_2 synthesis was found to be an effective antiaggregating agent, suggesting this intermediate to be the mediator of the platelet function (Gryglewski et al., 1976). These data imply that prostaglandin intermediates (thromboxane A_2) may participate (perhaps synergistically) with released ADP in the second wave of ADP-induced aggregation and in collagen-induced aggregation.

3.4.7.5. Role of prostaglandin intermediates in platelet secretion. Platelet secretion initiated in vitro by close cell contact such as is provoked by ADP or by polylysine (Guccione et al., 1976) is inhibited by cyclo-oxygenase inhibitors. This suggests that prostaglandin intermediates play a role in this type of platelet secretion. In contrast, collagen-induced secretion is only partially inhibited and thrombin-induced secretion is inhibited slightly, if at all (Kinlough-Rathbone et al., 1977), by doses of drugs totally abrogating prostaglandin synthesis. This indicates that although thrombin and collagen may induce prostaglandin synthesis and platelet secretion, prostaglandin intermediates are not the only intracellular mediators of this secretory process. Also, two additional agents (superoxide anion (Handin et al., 1977) and PAF (Shaw et al., 1977) that initiate prostaglandin synthesis and platelet secretion have been described. Indomethacin inhibition of prostaglandin synthesis induced by these agents did not abolish platelet secretion. Thus prostaglandin intermediates are not the final common pathway for platelet secretion induced by collagen, thrombin, superoxide anion, or PAF.

A second question is whether prostaglandin metabolites are capable of triggering platelet secretion by themselves. Notably, the majority of studies using these agents are done in citrated platelet rich plasma. Since prostaglandin metabolites are primary aggregating agents (Kinlough-Rathbone et al., 1976) in citrated plasma, they may trigger secretion indirectly by close cell contact. On the other hand, a number of authors have demonstrated prostaglandin intermediate-induced platelet secretion under conditions which would not lead to contact-induced secretion. Thus Malmsten and co-workers (1975) showed endoperoxide-induced secretion from indomethacin-treated platelets and Gerrard and associates (1977) showed endoperoxide-induced secretion in the presence of EDTA (which abrogated aggregation). Further, Kinlough-Rathbone and colleagues (1976, 1977) demonstrated arachidonate-induced secretion in calcium-containing suspensions of washed human platelets, which do not secrete in response to close cell contact. Thus the evidence indicates that prostaglandin intermediates (probably thromboxane A_2) may directly induce platelet secretion.

In conclusion, endoperoxides and thromboxanes are synthesized in response to a variety of platelet stimuli. These intermediates themselves are potent aggregating agents and may directly initiate secretion as well. These agents play an important role in the second wave of ADP-induced platelet aggregation and contribute to secretion occurring in secondary aggregation. They appear to play a synergistic role in the induction of platelet secretion by thrombin and collagen. The full ramifications of these studies in platelet physiology and in modulation of inflammation are just beginning to be appreciated. Finally, though it is not directly involved in the initiation of platelet activation, the potent inhibitor prostacyclin (PGI_2) must be mentioned. This arachidonic acid metabolite may be the most potent inhibitor of platelet activation known. It is generated from blood vessel walls (Moncada et al., 1977) and probably acts on the platelet through effects on cAMP (Best et al., 1977). Presumably it serves as an important negative feedback controlling system in vivo.

3.4.8. Membrane fusion

Even less is known about the interaction of platelet granule membranes with the plasma membrane than about the other mediator cells. In contrast, however, considerable effort is being directed toward study of the platelet surface membrane, and many of its constituent lipids and proteins have been identified and, in some cases, isolated. Thus, there is considerable potential for study of this process which has so far been largely ignored.

4. Neutrophils

4.1. Introduction

Polymorphonuclear neutrophil leukocytes play a central role in the mediation of inflammatory tissue injury, including that associated with bacterial infection, as well as that initiated by intrinsic processes such as the reaction of antigen with antibody, infarction, or even some traumatic injuries. Clinically these cells are associated with acute, severe injury, culminating in tissue necrosis. In addition, it is becoming recognized that neutrophil responses to a stimulus may lead to generation or release of factors promoting or enhancing a subsequent chronic inflammatory reaction involving macrophages.

The neutrophil, a member of the granulocytic series, is produced in the bone marrow, matures there into a cell containing a variety of enzymes and potent "mediators" sequestered in granules (e.g., Bainton and Farquhar, 1968), and circulates in the blood. The granule contents, structure, and so on have recently been discussed in a review by Bainton and collaborators (1976). Two predominant types of granules are present. The azurophil, or primary granule, is a true lysosome and contains acid hydrolases as well as cationic proteins and myeloperoxidase. The specific or secondary granule may not be a lysosome; it is morphologically distinct, especially in the rabbit, and it contains lysozyme, lactoferrin, and alkaline phosphatase (Bainton and Farquhar, 1968; Baggiolini et al., 1970; Nachman et al., 1972).

It is extremely unlikely that significant reactivity of this cell occurs while it is free in the bloodstream (unlike the basophil), and the first step in any neutrophil-mediated event involves margination, localization, and emigration to sites of stimulus production and generation. A thesis will be presented that neutrophils, and probably macrophages, are predominantly reactive with stimuli at surfaces and are optimally effective when crawling over or adherent to tissue surfaces such as basement membranes, collagen, fibrin, or elastic tissue fibrils, cell surfaces, or the surfaces of particles such as bacteria.

Damage to tissue elements, whether cells or supporting structures, clearly involves the release of injurious constituents from neutrophils. It was long thought that this merely comprised the toxic death and cytolysis of the cell, with subsequent release of materials. However in a series of studies extensively discussed in reviews by Becker and Henson (1973), Henson (1974b,c; 1976a,b), Weissmann and co-workers (1973), and Goldstein (1976), it was clearly shown that release of lysosomal enzymes from neutrophils did not generally (in noninfectious circumstances) result in cell lysis but was a secretory phenomenon. Cytotoxic reactions can occur as with urate crystals (Schumacher and Phelps, 1971; Zurier et al., 1973; Sirahama and Cohen, 1974; Ginsberg et al., 1977a), antineutrophil antibody and complement (Hawkins, 1972), Vitamin A, streptolysin D, and staphylococcal leukocidin (Zucker-Franklin, 1965; Dingle, 1969; Woodin and Wieneke, 1970). Even here, however, an early secretory phase has usually been demonstrated before lysis ensues.

Secretion of constituents is central to this theme of injury and will be discussed below with emphasis on the release of lysosomal enzymes since we know next to nothing about the mechanisms of release from neutrophils of other mediators. Extensive documentation of neutrophil secretory processes is described in the abovementioned reviews. Here, we will concentrate on recent work on the mechanisms involved as they relate to membrane-triggering events. It must be emphasized that other neutrophil functions such as chemotaxis, phagocytosis, and oxidative metabolic processes are initiated by similar or identical events and may be integrally associated. Some discussion of these is therefore necessary and appropriate.

4.2. Neutrophil activities and mediators

Sequentially the circulating neutrophil may respond to a focus of inflammation (whether ongoing or potential) by vascular margination, emigration and chemotaxis, adherence to particles or surfaces, engulfment (phagocytosis), de-granulation, and secretion of granule constituents. Other events whose place in the sequence is not yet worked out include (1) activation of oxidative metabolism associated with bactericidal processes and chemiluminescence, (2) release of prostaglandins, (3) synthesis and release of endogenous pyrogen and a variety of other materials such as histaminase (Zeiger et al., 1976) and eosinophil chemotactic factor (König et al., 1976), (4) generation and destruction of kinins (Movat et al., 1973; Wintroub et al., 1974), (5) cytotoxic reactions (e.g., Dean et al., 1974; Gale and Zighelboim, 1974, 1975; Oleske et al., 1977), and (6) generation of procoagulant activity (Niemetz and Morrison, 1977).

4.2.1. Neutrophil adherence and margination

As described below, it is our thesis that adherence of neutrophils to cells or other substrates is an integral part of their activation and function in vivo. Recent data suggest that chemotactic factors induce neutropenia in vivo, and neutrophil aggregation and adherence to surface in vitro (Craddock et al., 1977a; O'Flaherty et al., 1977a,b) and probably in vivo (Lentnek et al., 1976). In addition, we have shown a similar function for platelet activating factor (PAF) from basophils (Henson, unpublished observations). This is of interest because anaphylaxis in the rabbit is associated with immediate but transient neutropenia. Neutropenia is also seen in many other systemic reactions including en-dotoxemia and intravenous complement activation (Ulevitch et al., 1975). One may presume the cells have aggregated and become adherent to vessel walls. However, it must be emphasized that it is extremely difficult experimentally to distinguish between an aggregating and adherence-promoting effect of a given stimulus. The mechanisms of adherence are not known, but adherence to arti-ficial substrates may be Mg^{2+}-dependent (Bryant et al., 1966). The same inves-tigators demonstrated an increased stickiness after phagocytosis, and "activa-tion" of the cell by a variety of stimuli may promote this effect. However, it is important to note that it is not yet possible to distinguish between an increase in intrinsic stickiness or the presence of stimuli that bind to the surface and promote cell adherence by attaching to membrane "receptors." The presence in, and re-lease from, neutrophils of neutrophil-activating substances (cationic proteins?; proteases?) could increase this latter possibility. Neutrophils, in addition, are al-ways isolated from plasma-containing environments and however well washed may have plasma proteins, including immunoglobulin or complement compo-nents, on their surfaces, thus contributing potential adherence-promoting agents to the system. Further consideration of adherence and surface effects on neu-trophil secretion will be included further on.

4.2.2. Chemotaxis and Chemokinesis

Neutrophils are motile cells and exhibit amoeboid movement on surfaces. Stimuli can increase the rate of movement and, if this is nondirectional, the pro-cess is referred to as chemokinesis (Keller et al., 1977). Response to a gradient of the stimulus by directional movement is referred to as chemotaxis. Much energy has been expended on studying this latter phenomenon (Gallin and Quie, 1977). It must be emphasized strongly at the outset that in vitro distinction between chemokinesis and chemotaxis is difficult and must be rigorously pursued for adequate studies of mechanisms of neutrophil "sensing" (Zigmond and Hirsch, 1973). Much confusion in the literature results from lack of effective consideration of this point. Nevertheless, in vivo, where stimulus-induced neutrophil accumulation has been demonstrated (DeShazo et al., 1972; Jones et al., 1977), the nature of the mechanism may be less important than the end result. Moreover, chemotaxis and chemokinesis clearly have certain intrinsic properties in common (see below) and probably can result from identical stimulus-receptor interactions (Becker, 1976b).

4.2.3. Phagocytosis
Phagocytosis is one of the central functions of neutrophils and has recently been reviewed by Stossel (1974, 1975). Since the cell does not undergo pinocytosis, the reaction of neutrophil membrane with the surface of the particle implies a predominantly surface effect previously known as "surface phagocytosis." Some of the potential mechanisms of this process are considered below.

4.2.4. Secretion of granule constituents
This may occur into the vacuole or extracellularly (see below), and is the main focus of this section. The potentially injurious constituents are numerous and have been extensively discussed in earlier reviews.

4.2.5. Oxidative metabolism, production of oxygen radicals, chemiluminescence and bactericidal activity
This complex and highly active field has been discussed in several recent papers by DeChatelet and co-workers (1974), Babior and associates (1973), Johnston and colleagues (1975), Johnston and Lehmeyer (1976), and Goldstein and collaborators (1977a). Stimuli to neutrophils induce a burst of oxidative metabolism of glucose through the hexose monophosphate pathway (Karnovsky, 1962; Becker and Henson, 1973). Classically, this was believed to accompany phagocytosis and was associated with an additional increase in glycolytic activity. This activity appears to power the cell movement, phagocytosis, and secretion (Karnovsky, 1962; Henson and Oades, 1975). More recently, many investigators have shown the metabolic stimulation without phagocytosis, by soluble or surface-bound stimuli (Goldstein et al., 1975b; Henson and Oades, 1975; Johnston et al., 1975). The most exciting advance in this field has been the demonstration that a release of H_2O_2, superoxide anion, singlet oxygen, and hydroxyl radicals accompanies this metabolic burst. Associated with this, and perhaps resulting from singlet oxygen interaction with an unknown "substrate" (Allen et al., 1972), is a burst of chemiluminescence. Significantly, the system is integrally involved with the bactericidal properties of the cell and may be the

most important of many such potential mechanisms in neutrophils (Babior et al., 1973; Johnston et al., 1975).

Upon stimulation, neutrophils consume oxygen and convert it to superoxide anion (O_2^-) (Babior et al., 1973; Johnston et al., 1975). Spontaneous or enzyme-catalyzed dismutation of O_2^- in neutrophils results in H_2O_2 (DeChatelet et al., 1974) which is released (Root et al., 1975). This spontaneous dismutation may generate singlet oxygen, and the interaction of H_2O_2 with O_2^- may form hydroxyl radicals (Johnston et al., 1975). The generation of these materials seems to follow stimulation of the activity of neutrophil NADPH (and/or NADH) oxidase and is deficient in chronic granulomatous disease (CgD) (review, Johnston and McPhail, 1977). The enzyme (s) is probably associated with a granule fraction of the cell (Patriarca et al., 1971; DeChatelet et al., 1974; Curnutte et al., 1975; Segal and Peters, 1977; McPhail et al., unpublished observations). However, Briggs and co-workers (1975) showed a phagocytosis-dependent increase in a membrane-associated NADH oxidase and Goldstein and colleagues (1977a) provided evidence that O_2^- production is a cell surface phenomenon. These discrepant findings require further investigation and reflect the difficulties in working with these radicals, the assay for which are often lacking in specificity. Triggering of the processes leading to O_2^- production and the oxidative metabolic burst certainly occur at the cell membrane, as evidenced by its stimulation by surface-bound stimuli (Gifford and Malawista, 1972; Henson and Oades, 1975; Johnston and Lehmeyer, 1976) and by a requirement for membrane-associated sialic acid (Tsan and McIntyre, 1976).

This is an exciting and important field, but since little is known about the mechanisms involved (particularly of secretion of these materials) it will not be discussed further here except in relation to stimulus-neutrophil interactions in general. It should be emphasized, however, that in vivo these highly reactive oxygen radicals may be extremely potent mediators of cell and tissue injury, either directly or by altering cell membranes to render them more susceptible to attack by other neutrophil constituents (Johnston, R. B., personal communication). If these radicals can kill bacteria so effectively, it is conceivable that they can also damage tissue cells, and it may not be coincidental that so many cells contain superoxide dismutase, which promotes the dismutation of superoxide and prevents the generation of toxic oxygen radicals.

4.2.6. Prostaglandins

Recent data has demonstrated that activation of neutrophils also results in release of prostaglandins (Zurier and Sayadoff, 1975) and their precursors, thromboxanes and endoperoxides (Goldstein et al., 1977b). The suggestion that superoxide dismutase (as well as indomethacin) can inhibit thromboxane production is intriguing. We know little as yet about the mechanisms for this or the cell surface trigger requirements, although some similarities with the well-studied platelet system (section 3) may be implied. They do represent important mediators and should therefore be borne in mind as products of neutrophil stimulation about which we will undoubtedly hear more in the future.

4.2.7. Other mediators
Additional potential mediators from neutrophils will not be discussed further, and the reader is referred to the beginning of this section and to earlier reviews (Becker and Henson, 1973; Goldstein, 1974, 1976; Ranadive, 1977b)

4.3. Morphologic aspects of secretion

4.3.1. Release into phagocytic vacuoles or into the extracellular environment
A prejudice of long standing, deriving from the initial studies of Metchnikov (1891), assumes that the physiologic circumstance of neutrophil secretion involves release of granule constituents only into the phagocytic vacuole, there to kill and/or digest the phagocytized particle (e.g., bacterium or immune complex). Despite its apparent intracellular locale, this process has all the hallmarks of a true secretory reaction, fusion of granule membrane with extracellular membrane (in this case the invaginated vacuole wall), and the biochemical processes usually associated with secretion (see earlier reviews). During this process, however, granule constituents gain access to the outside of the cell. This occurs by a variety of events (Henson, 1973) that could be considered accidental to the overall physiologic purpose: (1) Granules are discharged into vacuoles that have not yet closed off—a process called "regurgitation" while feeding (Weissmann et al., 1972). (2) Granules are liberated into vacuoles that are open to the exterior because of ingestion of additional particles into a vacuole (phagolysosome). (3) More than one cell may attempt to phagocytize one particle, leaving channels to the outside. (4) Occasionally slits may remain after phagocytosis, connecting the vacuole to the outside.

While it is clear that such degranulation is highly significant, the neutrophil may have an additional function as a surface scavenger. Its reaction with stimuli on a surface apparently initiates the same intracellular events that lead to degranulation. Here, however, the granules fuse with a true extracellular membrane and their contents are discharged to the outside of the cell (Henson, 1971abc). We have called this process "frustrated phagocytosis" (but it is more properly granule exocytosis), and have shown it to occur in vivo (Henson, 1974a). Recent studies have demonstrated the increased potency of surface-bound stimuli for both secretion (Henson and Oades, 1975) and superoxide production (Johnston and Lehmeyer, 1976), and chemotactic factors will only stimulate enzyme-release from normal neutrophils if surface bound (Becker et al., 1974). It is interesting in this latter regard that the surface effect can be mimicked by cytochalasin B (Goldstein et al., 1973a,b; and 4.5.5.2). Thus, release of mediators directly against immune reactants complement components and so on on surfaces in the body may play an essential role in removing (digesting) such materials and result in tissue injury only incidentally. Discharge of granule enzymes (e.g., proteases) between a surface and a closely apposed neutrophil membrane could protect such enzymes from inhibitors such as the protease inhibitors in plasma and synovial fluids. The possible ability of some enzymes to adhere to the plasma membrane at the site of secretion (e.g., alkaline phosphatase, Henson, 1971b), would further serve to localize the digestive properties to the required area of the surface.

4.3.2. Sequential discharge or secretion of granules

In the surface-stimulated exocytosis system (Henson, 1971b) we clearly demonstrated a sequential release of first specific granule constituents (e.g., lysosome or alkaline phosphatase) and then azurophil granule contents (acid hydrolases). Bainton (1970, 1973) independently and elegantly showed the same sequential discharge in the degranulation process. The teleological argument suggests release of enzymes acting at neutral or alkaline pH before the pH in the vacuole (or extracellularly?) drops (Jensen and Bainton, 1973) and the medium becomes optimal for the acid hydrolases. Specific granule contents are also released to a greater degree than those from the azurophil granule (Leffell and Spitznagel, 1975). Considerable control of the secretory event is suggested and must be explained biochemically.

The possibility that two different signals are involved in secretion is suggested by the observation that phorbol myristate acetate, the active principal of croton oil, stimulates release only of specific granules (Estensen et al., 1974; Goldstein et al., 1975a; section 4.4.5). Similar dissociation of specific and azurophil granule release was achieved by modulation of calcium concentration (Goldstein et al., 1975b), and ionophore A23187 also induces only release of the specific granule constituents (Goldstein et al., 1974; Wright et al., 1977) or at least a relatively greater amount (Kosa et al., 1975).

An attractive suggestion would consider the granule of the basophil and mast cell, the dense body (serotonin-containing granule) of the platelet, and the specific granule of the neutrophil as similar. Each responds readily to calcium influx (probably whether or not mediated by ionophores) and is discharged rapidly. In the platelet, a hypothesis was proposed of graded magnitudes of response to a stimulus, yielding secretion of first serotonin and then of hydrolases (from different granules) (Holmsen, 1974). This could be also true for the neutrophil; or different pathways could exist for different granules in each cell types. It is relevant that specific and azurophil granule membranes exhibit different compositions (Nachman et al., 1972). Further work is needed to elucidate these processes.

4.3.3. Polarized stimulus-secretion coupling

Discharge of granules occurs only at sites of plasma membrane that are stimulated by the initiating agent (Henson, 1971a,b, 1976b): there does not appear to be a wave of secretion around the cell. This has obvious importance for local discharge of granules into vacuoles or at sites of surface stimulation. It is also significant with regard to consideration of mechanisms of secretion (see below). Cell polarization is also of critical importance in chemotaxis and is briefly discussed in that context.

4.4. Stimuli for neutrophil activation

Emphasis is placed on three groups of stimuli that can clearly induce secretion of both specific and azurophil granule contents if presented to the neutrophil in an appropriate manner. Immunoglobulin and C5a are probably most significant biologically. The synthetic formylated peptides are proving extremely useful as probes for the membrane activation process. The larger fragment of C3 (perhaps C3b) is important as an agent that promotes phagocytosis. Whether it can initiate this process or stimulate secretion is still unclear. It may belong to a class of stimuli that appear to induce the oxidative metabolic burst selectively and to release only specific granule contents. (section 4.4.5). Chemotactic stimuli such as denatured proteins, lipids, arachidonic acid metabolites (e.g., HETE), and so on have been considered in recent reviews (e.g., Wilkinson, 1976; Turner and Lynn, 1977) and will not be discussed further.

4.4.1. Immunoglobulins

Most study of neutrophil stimulation has been with immunologic agents. The cells have Fc "receptors" (review, Henson, 1976b) and can be induced to secrete by aggregated immunoglobulin of all four subclasses of IgG and both subclasses of IgA, but not by IgE, IgM, or IgD. The immunoglobulin must be aggregated or bound to an antigen to stimulate phagocytic uptake and subsequent release of enzymes, or release on surfaces (Hawkins and Peeters, 1971; Henson and Oades, 1975). The larger the particle on which the immunoglobulin is bound, the more effective is the induction of enzyme release, a progression which culminates in presentation of the Ig as a nonphagocytizible surface. It is interesting, however, that soluble aggregates too small to induce enzyme release promote increased oxidative glucose metabolism (Henson and Oades, 1975). We still do not know whether the requirement for aggregation involves a structural alteration of the Fc region or whether it indicates the need for cooperative binding (yielding higher affinity binding for the whole complex) and/or cross-linking of cell surface receptors. The data do not allow a definitive conclusion to be reached, but suggest that the cooperative binding hypothesis plays some role in the effect (Phillips-Quagliata et al., 1969; Metzger et al., 1976; Henson, 1976b).

The nature of the binding site for immunoglobulin Fc on the neutrophil membrane is unknown. However, a number of neutrophil chemotactic stimuli are hydrophobic or even lipid in nature (Wilkinson, 1973, 1976; Turner and Lynn, 1977), and hydrophobicity has also been suggested as a requirement for an opsonic function for molecules (Van Oss and Gillman, 1973). Since the Fc region of immunoglobulin is relatively hydrophobic, the membrane binding sites may be lipoprotein or lipid. Recent data from a number of laboratories (discussion, Anderson and Grey, 1977) on a variety of cell types suggest a phospholipoprotein nature for Fc receptors. It is interesting, however, that mouse macrophage Fc receptors appeared to have little or no associated lipid. Whether this indicates two types of receptors, as suggested recently by Walker (1976) and Grey (1977), or whether the lipid is not essential for the binding properties of the receptor, or whether cells differ in their receptor structure is still unclear. Functional activity for these "receptor" isolates is not yet determined. By contrast, Weissmann (1974) showed that aggregated IgG myelomas could interact with protein free liposomes and disrupt their structure so that contained markers were released. A specificity for IgG_1 and IgG_3 makes this observation particularly interesting since these two subclasses may be most avid in their ability to react with cells (Henson, 1976b). A direct interaction of Fc with lipids is suggested only if IgG is aggregated. Subsequently, preliminary electron spin resonance data was presented to support the Fc-lipid interaction (Schieren et al., 1976).

These observations raise the question of Fc-membrane interaction as it relates to cell activation. The cell and subclass specificity of Fc stimulation suggest strongly the presence of some specific "receptor," perhaps, a lipoprotein. The simultaneous requirement for hydrophobic interactions, possibly with lipid moieties and of nonhydrophobic (e.g., ionic, hydrogen-binding, etc.) through protein or carbohydrate moieties, may be indicated. Trypsin or glycosidases generally do not destroy Fc receptor activity on whole cells (Henson, 1976b; Anderson and Grey, 1977), but this may reflect (1) accessibility, or (2) multiple binding moieties in the "receptor." This latter point is significant in a system where the obligatory use of immune complexes or aggregated immunoglobulin means that quantitative binding data and relative binding affinities are difficult to obtain (but see Arend and Mannix, 1973). Moreover, trypsin apparently destroys solubilized Fc receptor activity (Anderson and Grey, 1977).

The requirement for aggregation of Ig for optimal binding to cells has been discussed earlier. However, there is a similar requirement for neutrophil stimulation. Whether this indicates that receptor cross-linking is important in cell activation is unclear at present. Two lines of preliminary evidence suggest this possibility. Neutrophils bind effectively to immobilized immunoglobulin whether it is aggregated or not (Hawkins, 1971; Henson, 1971a; Henson and Oades, 1975; Lehmeyer and Johnston, 1977). Nevertheless the Ig must be aggregated to stimulate enzyme secretion even when bound to surfaces. Clearly cross-linking does occur, as demonstrated by receptor redistribution (capping), in association with the induction of secretion by complexes (Sajnani et al., 1974, 1976).

Insolubilized immunoglobulin (on surface or particle) is a potent stimulus for the neutrophil and induces secretion of all the mediators described above (section 4.2). The only neutrophil activity that is clearly not stimulated by Ig is chemotaxis, and even this lack may reflect the strong binding of neutrophils to Ig, which could immobilize them on the matrix of the assay system (e.g., the micropore filter) even if the intracellular biochemical processes pertaining to chemotaxis were triggered.

4.4.2. C5a and other complement-derived chemotactic factors
The small fragment of the fifth component of complement has long been known as a chemotactic factor for neutrophils and monocytes (reviews, Gallin and Wolff, 1975; Henson, 1976b; Wilkinson, 1976). An importance advance in our understanding of neutrophil function, however, came from recognition that this

molecule could initiate secretion from neutrophils. It (and other chemotactic factors) also stimulate the respiratory burst in these cells (Goetzl and Austen, 1974) and the release of O_2^- (Goldstein et al., 1975b). To stimulate secretion of enzymes, surface-binding is necessary (Becker and Showell, 1974; Becker et al., 1974) or the cell must be treated with cytochalasin B (Goldstein et al., 1973a,b). These requirements do not hold for superoxide production (see also C3b and Ig stimulation) and suggest some connection between cytochalasin-sensitive structures and the surface effect (see below). Since these observations were made, much work has been devoted to C5a effects on neutrophils. It is noteworthy that C5a is also chemotactic for mononuclear phagocytes and induces them to secrete lysosomal enzymes (Henson and Ryan, unpublished observations).

This factor is one of the most potent neutrophil activators and may be an extremely important stimulus for this cell in vivo. Activation of complement in serum or plasma liberates C3a, C5a, and $\overline{C567}$, all at one point considered to be chemotactic for neutrophils. C3a and C5a are also anaphylatoxins and induce smooth muscle contraction and increased vascular permeability. Recent data has questioned whether C3a anaphylatoxin acts on neutrophils at all (Fernandez et al., 1977). If another small C3 fragment is chemotactic, this has yet to be isolated. Previous C3a preparations may have been contaminated with C5a or such an additional fragment. Similarly, a unique effect of $\overline{C567}$ has not been shown and the action could conceivably result from residual C5a in the complex. This argument is strengthened by the cross-desensitization exhibited by all three chemotactic agents (Becker and Henson, 1973).

It has also been shown that C5a from which the C-terminal arginine has been removed, and which is inactive as an anaphylatoxin (C5ai), is still stimulatory for neutrophils (Fernandez et al., 1977). Most investigators of C5a-induced secretion have actually employed C5ai, since the levels of epsilon amino caproic acid required to totally inhibit the serum carboxypeptidase (which removes the arginine) are higher than those generally used (Goldstein et al., 1973b). It is also highly probable that many or most in vitro effects on neutrophils are mediated by C5ai.

The importance of this stimulus as an in vivo neutrophil activator is manifested by its ability to cause instant neutrophil sequestration upon intravenous injection or intravenous complement activation (Craddock et al., 1977b; O'Flaherty et al., 1977a,b), its presence in inflammatory lesions (e.g., rheumatoid joints; Ward and Zvaifler, 1971), its potent action on neutrophils in vitro, the evidence that complement-derived chemotactic factors are involved in immune complex-initiated neutrophil accumulation in the joint (DeShazo et al., 1972), the sequential accumulation of neutrophils and monocytes into mouse peritoneum after C5ai injection (Henson, unpublished observations), the release of C5-cleaving enzymes from neutrophils (which therefore generate C5a; Ward and Hill, 1970; Goldstein and Weissmann, 1974; Wright and Gallin, 1975), the presence of C5a-activating agents in neutrophils (potent inhibitors are always necessary to limit the effects of potent mediators) (Wright and Gallin, 1975; Bronza and Ward, 1977), and serum (Ward and Ozols, 1976), and so forth.

To date the inability to isolate enough purified, active, radiolabeled C5a to perform binding studies on neutrophils has precluded investigation of the receptor involved. The C5a molecule is a glycoprotein with a peptide fragment of about 9,000 MW. The ability of naturally occurring peptidases to inactivate its neutrophil-activating functions (Boksich and Muller-Eberhard, 1970; Ward and Ozols, 1976) suggests that the peptide moiety is involved. However, a contribution of the carbohydrate has not yet been ruled in or out. C5ai binds to artificial surfaces in vitro and in this form acts as a stimulus for neutrophil secretion (Henson et al., unpublished observations). Unfortunately, since its valency is unknown, we cannot determine whether receptor cross-linking or redistribution occurs.

4.4.3. N-Formylated methionyl peptides

Knowledge of the effects of small peptides on neutrophils has been derived mainly from studies of chemotactic factors. Many species of bacteria liberate potent chemoattractants for neutrophils, and at least some of these were shown to be small peptides with blocked amino groups (Schiffman et al., 1975a). Peptides derived from fibrin (Kay et al., 1973) and collagen (Gallin and Wolff, 1975) are also chemotactic. In addition, two tetrapeptides Val-gly-ser-glu and Ala-gly-ser-glu were shown to attract eosinophils specifically (Austen et al., 1976).

To date, one of the most exciting advances in the study of neutrophil activation is the demonstration of the extremely potent chemotactic potential of N-formylmethionyl peptides by Schiffman and co-workers (1975b). This work was derived from the study of bacterial chemotactic factors, but by providing materials of defined structure has opened the way for an intensive attack on the stimulus-neutrophil interaction mechanism.

Before the advent of these peptides, we showed that chemotactic factors all induced lysosomal enzyme secretion if presented to neutrophils on surfaces (Becker et al., 1974). From this a prediction of the secretory potential of the formylated peptides could be made. This activity was shown by Showell and associates (1976) using cytochalasin B. Furthermore, an exact structure-function correlation between 24 different peptides was found for enzyme secretion and chemotaxis with F-Met-Leu-Phe (FMLP) being the most active. Studies of binding of [³H] peptides revealed similar structure-function relationships (Williams et al., 1977) and demonstrated about 2000 binding sites per cell. However, Schiffmann and colleagues (1977) have suggested a figure of 100,000 per cell.

The possible commonality of the receptors for different chemotactic factors is suggested by competitive studies (Schiffman et al., 1975b; Aswanikumar et al., 1976) in which, for example, the nonchemotactic peptide F-Phe-Met inhibits (in a similar fashion) the responses to F-Met-Phe, bacterial factors and C5a, each of which is also capable of inducing enzyme secretion. However, Musson and Becker (1976), also demonstrated FMLP inhibition of complement-mediated erythrophagocytosis (induced presumably by C3b and/or Ig; see section 4.4.5.2). Whether this is true competition for membrane sites, or involves some intracellular step, needs further elucidation.

4.4.4. Stimuli that apparently selectively induce release of specific granules

A group of diverse stimuli activate neutrophils to cause secretion of specific, but not azurophil, granules. Since all of these also stimulate oxidative metabolism and O_2^- production, the question may again be raised as to whether a graded neutrophil response to a stimulus would involve first metabolic stimulation followed by specific granule secretion and then release of the true lysosomal granule, or whether different pathways are involved.

4.4.4.1. Antineutrophil antibody in the absence of complement. To date, the two stimuli that may only stimulate oxidative metabolism are C3b (but see below, section 4.4.5) and antineutrophil antibody (Rossi et al., 1971). However, the latter stimulus requires a better elucidation of the antigenic determinants involved to be definitive. Moreover its lack of enzyme-releasing effect depends on the report of Hawkins (1972), who unfortunately did not assay for lysozyme or examine the effect of immobilized antibody. Studies of antineutrophil antibodies gain importance from their occurrence and involvement in some cases of idiopathic neutropenia (Boxer et al., 1975) and require further investigation.

4.4.4.2. Ionophore A23187. This ionophore is reported to induce lysozyme, but not acid hydrolase, release in the presence of Ca^{2+} (Goldstein et al., 1975c; Wright et al., 1977), and the latter group actually induced a similar release by manipulation of calcium concentrations alone. The ionophore also stimulates oxidative metabolism (Romeo et al., 1975).

4.4.4.3. Phorbol myristate acetate (PMA). This is a potent stimulus for cells, including neutrophils, and has been shown by many investigators to initiate secretion of specific granules (Estensen et al., 1974; White and Estensen, 1974; Goldstein et al., 1975 a; Wright et al., 1977). As discussed by these authors and in the references in section 4.2.5, PMA is also a widely used stimulus for superoxide production, whose mechanisms of action and binding site in the cell are poorly understood. This stimulus increases intracellular cyclic GMP in neutrophils (Estensen et al., 1973) and (as with other agents that increase cGMP) also induces microtubule formation (Goldstein et al., 1975a). The possible connections between this and the oxidative metabolic burst and specific granule secretion, however, are completely unknown.

4.4.4.4. Concanavalin A. Binding of Con A to neutrophils induces a redistribution of lectin-binding groups (section 4.5.6) and secretion of lysozyme (Hoffstein et al., 1977a). The reports of Hawkins (1974b) and Woodin (1975) that Con A inhibited secretion induced by Ig or leukocidin respectively are not inconsistent, since these investigators examined only release of acid hydrolases and not constituents of the specific granule. Apparently, con A also induces microtubule formation (Hoffstein et al., 1977a) and is an effective but reversible stimulus for the respiratory burst (Romeo et al., 1973), but may inhibit phagocytosis (Berlin, 1972).

4.4.5. C3b

4.4.5.1. C3b "Receptors". The binding of C3b on particles to neutrophils, and the conclusion that C3b "receptors" are present on these cells has been discussed in a previous review (Henson, 1976b). A cautionary note should be included at this point to question whether the C3 fragment which binds to neutrophils is the whole C3b fragment of the molecule or only a portion thereof (Stossel et al., 1975). The C3b receptors described for lymphocytes (Eden et al., 1973; Ross et al., 1973;) appear to be absent from neutrophils.

The C3b-binding site is trypsin-sensitive and is assumed to be protein-containing. However, C3b receptor activity from human erythrocytes was suggested to be a mucopeptide (Nelson and Uhlenbruck, 1967), and that from lymphoid cells to be complex lipoprotein (Dierich and Reisfeld, 1975). A number of investigators, including the present authors, have encountered difficulties in isolating C3b-binding materials from detergent-solubilized membranes, and so far no satisfactory isolation procedure has been devised.

Some evidence of pharmacologic and environmental control of C3 receptor function has been presented. In neutrophils from several species divalent cations are necessary for binding (Lay and Nussenzweig, 1968; Henson, 1969), and binding to rabbit macrophages was inhibited by cytochalasins and colchicine (Atkinson et al., 1977), suggesting microfilament and microtubule involvement (but see below).

4.4.5.2. Is C3b a stimulus for cell activation? C3b on particles promotes their adherence to neutrophils and macrophages (Huber et al., 1968; Lay and Nussenzweig, 1968; Henson, 1969). It was also clearly demonstrated by Gigli and Nelson (1968) that neutrophils phagocytized EAC1423 but not EAC142. This suggested strongly that C3b could stimulate neutrophils. Similar data were reported for stimulation of neutrophil secretion (Henson, 1971a). However, this concept has recently been brought into question by a number of studies that suggest small amounts of IgG present on these particles are actually the stimulus for the surface membrane and that C3b merely acts as a ligand. Thus, erythrocytes coated with IgM antibody and complement (EAIgMC) bound to neutrophils (Schribner and Fahrney, 1976), macrophages (Mantovani et al., 1972; Griffin et al., 1975), and monocytes (Huber et al., 1968) but were not engulfed. EAIgGC were bound and phagocytized. The engulfment of this latter cell into neutrophils was also shown to be inhibitable by Fc fragments (Schribner and Fahrney, 1976) or into macrophages and neutrophils by Fab anti-IgG (Mantovani et al., 1972; Mantovani, 1975). Examination of earlier experiments in this field reveal that the role of complement in phagocytosis was usually studied with IgG antibody or in circumstances (e.g., on zymosan particles) where traces of IgG could not be excluded. Thus conclusive data refuting the ligand hypothesis for C3b are still unavailable. Nevertheless, various questions about its validity and generality remain to be answered.

For example, unlike their unstimulated counterparts, mouse peritoneal exudate macrophages, elicited by thioglycollate, are perfectly capable of engulfing EAIgMC (Griffen et al., 1975). Moreover, in the early, elegant study of Huber and co-workers (1968), uptake of EAIgMC by human monocytes was seen if enough C3 was bound. Similarly, we have shown C3b stimulation of secretion from rabbit platelets, cells which do not react with IgG (Henson, 1969).

Clearly, from both the latter studies much more C3 is required for cell stimulation than for adherence (which only takes a hundred or so molecules). Finally, C3b has been shown to stimulate enzyme release directly from mouse macrophages (Schorlemmer et al., 1977). Thus C3b-induced attachment with IgG-induced cell triggering might easily result if the C3b density on the particle or surface was too low to induce uptake itself. Higher densities however, may themselves stimulate the cell. The elicited macrophages might have a greater sensitivity to a stimulus (known to occur in lipopolysaccharide stimulation of macrophages [Doe et al., 1977]) or an increased C3b receptor density.

In neutrophils, sepharose-bound C3 fragments stimulated superoxide production (oxidative metabolism) but not secretion (even of lysozyme), if IgG was absent. The presence of Ig enhanced the former and *initiated* secretion (Goldstein et al., 1976a). Again, distinction between an absolute or quantitative effect is not possible since no data was presented on the amount of activity of the C3 remaining after boiling (to remove the Ig). The data support the concept of different biochemical pathways stimulated in the neutrophil, and that oxidative metabolism may require a lower stimulus threshold.

This question of C3 stimulation is important only on a conceptual and mechanistic basis, not in practical terms. Complement certainly promotes phagocytosis and secretion, whatever the actual process involved. To answer the question definitively, careful studies with known amounts and densities of surface-bound C3b will be required.

4.4.6. Stimulus for "pseudocytotoxic" reactions

Although a variety of diverse agents induce neutrophil lysis, they initiate a secretory or pseudosecretory phase before lysis ensues. This may be important in vivo in circumstances where release of injurious mediators could precede the cytolytic liberation of intracellular inhibitors (e.g., of proteases; Davies et al., 1971). It has also provided some insight into mechanisms of secretion and lysis (see below). (For references in this area the reader is referred to Hawkins, 1974a: antineutrophil antibody; Woodin and Wieneke 1970: leukocidin; Woodin 1968: streptolysin 0, vitamin A; Zurier et al., 1973, Hoffstein and Weissman, 1975, and Ginsberg et al., 1977a,b: urate crystals.) The urate crystal system is interesting since a similar combination of secretion and lysis is observed with platelets (section 3).

4.5. Biochemical mechanisms of neutrophil stimulation with particular reference to secretion

Clearly a full discussion of this extremely broad area is impossible. Consequently, we concentrate on aspects that relate particularly to the cell membrane, are of special interest, or which serve to compare or contrast different intracellular processes within the neutrophil or within the other mediator cells. We emphasize secretion, but necessarily discuss some of the other neutrophil activities. We also attempt to bring together the data with different stimuli and secretory conditions (i.e., postphagocytic, surface stimulation, cytochalasin-modified) and do not always describe the conditions, only the effective conclusion. Finally, we exercise our option to speculate a little about some of the mechanisms involved in neutrophil activation. Principally this involves a naive oversimplification, but may serve to suggest possible approaches to prove or disprove the hypotheses, thus bringing us closer to a complete understanding of the stimulus-secretion coupling processes within the cell.

4.5.1. Serine esterases (proteases)

4.5.1.1. Activatable esterases within the cell. The studies of Becker and Ward suggested that chemotaxis involved the activation of a serine esterase(s) in the neutrophil as well as the function of a similar preexisting activated enzyme (Becker and Henson, 1973). The data involved a similar approach to that described for mast cells and platelets, and the experiments employed series of organophosphorus inhibitors. The presence of and requirement for preexisting activated esterase activity in chemotaxis (Becker, 1972), erythrophagocytosis (Pearlman et al., 1969; Musson and Becker, 1977), and secretion (Henson, 1974a) made study of the activatable esterase difficult. Nevertheless, clear evidence for the latter has been presented in chemotaxis induced by C5a, $\overline{C567}$, bacterial factors (Becker, 1972; Becker and Henson, 1973) and EAC1423-induced phagocytosis (Musson and Becker, 1977). In the latter, it is unclear whether C3b or Ig is acting as the stimulus for uptake (section 4.4.5.2). This is significant, since one attractive hypothesis for cell triggering involves the activation of such an esterase as an early or initial step in the sequence. Additional preliminary evidence for such an enzyme is coming from studies of [³H]DFP uptake, where we have shown C5a-induced binding of [³H]DFP to neutrophils which had been pretreated with unlabeled DFP. At least 1000 molecules/cell appeared to be bound, but technical difficulties make this an underestimate (Becker and Henson, unpublished observations).

Although the intracellular biochemical sequence in neutrophils has not been explored, it would be attractive to believe that this esterase activation represents an early or initial step at the cell membrane. Certainly cell-associated esterases have been implicated in K^+ uptake (Mandell et al., 1977; and section 5.4) and perhaps in microtubule assembly. Also, it is not known if there are one or more such enzymes (but see Tsung et al., 1977). Some evidence (from phosphonate and ester inhibition profiles) suggested that the same enzyme was involved in C5a-, C3a-, and $\overline{C567}$-induced chemotaxis, but (as mentioned earlier) this may not be surprising. If FMLP, C5a, and bacterial factor act on the same site (section 4.4.3), a similar inhibition would be expected. It is not yet clear whether EAC1423 stimulation (phagocytosis) or C5a stimulation (chemotaxis) have similar structure-function phosphonate inhibition profiles, but examination of the data suggests they may be different—suggesting different esterases. It is important to know whether given stimuli such as C5a or FMLP initiate their various neutrophil activities (chemotaxis, phagocytosis, secretion, oxidative metabolism) via the same esterase and whether IgG and C5a initiate neutrophil stimulation by means of different esterases. We predict that both of these will occur and thus that C5a and IgG have different receptors and esterases, but that one stimulus can induce different cell activities by triggering one receptor and enzyme (section 4.6).

4.5.1.2. Exogenous protease (esterase). As with platelets and mast cells, it might be expected that added esterase would initiate neutrophil activation. We have found that trypsin, if bound to a surface, initiates neutrophil secretion (Henson, 1974a; Henson et al., in preparation). The possibility that the neutrophil proteases themselves could serve as feedback activators has also received support from the observations of neutrophil stimulation by neutrophil products (e.g., Zigmond and Hirsch, 1973; Mandell et al., 1977). In addition, kallikrein, a serine protease, has been shown to stimulate neutrophil chemotaxis and oxidative metabolism (Kaplan et al., 1972; Goetzl and Austen, 1974), although its site of action is also unknown.

4.5.1.3. Mechanisms. It is not known how the esterases (stimulus-dependent, i.e., activable; or cell-dependent, i.e., already active) act to promote neutrophil stimulation. The substrates for the enzymes are unknown, although the activatable esterase, stimulated by C5a, can hydrolyze n-acetyl phenylalanine esters (Becker, 1972). Does this enzyme initiate activation by limited proteolytic cleavage of the next molecule in the stimulus-cell response coupling pathway as occurs in plasma with the coagulation and complement pathways? Does the activated protease destroy the surface-bound stimulus, freeing the receptor for further activation (Aswanikumar et al., 1976)? This latter suggestion is similar to our working hypothesis for neutrophil chemotaxis (see below), but both possibilities are highly speculative.

4.5.2. Energy metabolism
This subject has been considered previously (Karnovsky, 1962; Becker and Henson, 1973; Henson and Oades, 1975) and will not be discussed further except to point out that an ATP-generating system is required for neutrophil phagocytosis, chemotaxis, and secretion, and that these three cases involve the glycolytic pathway. The complicated interrelationships between glycolysis, hexose monophosphate shunt, generation of NADH and NADPH, and the production of H_2O_2 and oxygen radicals are beyond the scope of this review.

4.5.3. Cyclic nucleotides
Extensive studies in this are have revealed an inhibitory action of cyclic AMP on neutrophil secretion, whether added exogenously or produced endogenously (Weissmann et al., 1971b; Becker and Henson, 1973; Henson, 1974a; Goldstein, 1976a, b). Cyclic GMP enhances secretion, as do agents inducing its increase (Ig-

narro, 1973; Ignarro and George, 1974; Zurier et al., 1974). Similar data have been presented for modulation of neutrophil chemotaxis (Gallin and Wolff, 1975). Study of phagocytosis, however, has led to conflicting information. It has been reported that AMP and agents that increase it do not affect uptake (review, Becker and Henson, 1973) or, by contrast, inhibit it (Cox and Karnovsky, 1973). The hexose monophosphate shunt (Henson and Oades, 1975), superoxide production, and chemiluminescene (Lehmeyer and Johnston, 1977) are also inhibited by agents that increase cAMP.

PMA has been shown to increase levels of cGMP in neutrophils (Estensen et al., 1973) and its enhancing effect on C5a-induced azurophil granule release (Goldstein et al., 1975a) is presumed to occur via this process. A series of studies by Goldstein, Weissmann, and Hoffstein (Hoffstein et al., 1977b) and in Chediak-Higashi cells by Oliver (Oliver, 1976) have suggested that increased cGMP is intimately connected with increased microtubule formation (section 4.5.4), that the latter condition is necessary, but not sufficient, for secretion. It would be expected, that C5a would stimulate an increase in cGMP and a decrease in cAMP (as seen in platelets, basophils, etc.). However, direct evidence for this is lacking, although indirect evidence for the Yin-Yang hypothesis is present in abundance (Weissmann et al., 1975).

Some questions remain. In many cases, the levels of drugs that had to be used for inhibitory effects exceeded those required to increase cAMP (Bourne et al., 1973). Since relatively little evidence is available for the effects of natural agents Ig, C5a, and so forth on the cyclic nucleotide levels in neutrophils and on the actions of the nucleotide in the cell, we are far from an understanding of their role in control of secretion.

An explanation for these difficulties may lie in the relative compartmentalization of the nucleotide effects (Ong et al., 1975; Ong and Steiner, 1977) reflected in the local effects of neutrophil stimulation. Thus chemotactic effects are presumably localized to the leading edge of the cell, phagocytosis and secretion to the point of membrane stimulation. This could make interpretation of cyclic nucleotide data extremely difficult, compounded by the recent observation that different modes of providing intracellular cAMP may affect different protein kinases, with perhaps different responsibilities within cells (Byus et al., 1977). It is also intriguing to question the controlling influence of possible (but as yet unknown) cyclic nucleotide-modulated phosphorylation of membrane proteins as suggested for neuronal cells (Greengard, 1976; Nathanson, 1977). While an effect of a neutrophil protein kinase on colchicine-sensitive structures has been suggested (Tsung et al., 1975); the connection between this and secretion is unclear at present.

4.5.4. Microtubules and colchicine-sensitive structures

4.5.4.1. Secretion. As discussed in previous reviews (e.g., Becker and Henson, 1973; Goldstein, 1976; Hoffstein et al., 1977b), secretion into vacuoles is sensitive to colchicine and related drugs and was enhanced by D_2O (Weissmann et al., 1973). Extracellular secretion has also been reported to be inhibited by colchicine (Weissmann et al., 1972), but was less readily affected in other investigations

(Henson, 1972; Hawkins, 1974a). In a series of studies, Hoffstein and associates have shown that increased GMP enhances, and cAMP decreases, microtubule assembly in the neutrophil (Hoffstein et al., 1974, 1977b). Activation of the cell with C5a similarly enhanced microtubule assembly as assessed in the region of the centriole by electron microscopy. A direct effect of microtubules in secretion was originally suggested that disagreed with the data mentioned above, showing an inconsistent inhibitory effect of colchicine. Similarly, inconsistent data have been reported with cells from Chediak-Higashi patients whose defect is probably deficient microtubule assembly (Becker, 1976a,b, Oliver, 1976).

Earlier we suggested (Henson, 1972, 1974a, 1976a), in a naive and unprecise manner, that microtubules might serve to maintain intracellular organization and could thus assist the processes by which granules and vacuoles came into contact with each other. This could be more significant in phagocytosis, where vacuoles are formed in the cytoplasm and more granules are involved than with discharge at the cell surface, where usually only those nearby granules may participate. The recent, elegant quantitative study by Hoffstein and co-workers (1977b) supports and extends this view. A good correlation was seen between inhibition of secretion and microtubule formation, but the former could still occur at concentrations of colchicine that completely prevented microtubule assembly. Thus the exact role for *cytoplasmic* microtubules in secretion is still unknown, but they clearly promote secretion and are increased by stimuli and under conditions where secretion occurs. The possible role of membranous colchicine-sensitive structures is considered below.

4.5.4.2. Chemotaxis.　Chemotaxis may require microtubule function to a greater degree. It is readily inhibited by colchicine (Gallin and Wolff, 1975). The recent exciting studies of Gallin and associates have shown an orientation of the centriole and microtubules toward the leading edge of the cell in chemotaxis (Gallin et al., 1977; Malech and Gallin, 1977) associated with an increase in microtubule assembly similar to that reported for cells in suspension by Hoffstein and colleagues (1974). This orientation is reversed with a reversal of chemotactic gradient. No such orientation occurred in concentrations of colchicine identical to those required for inhibition of microtubule assembly. Here cytoplasmic microtubules appear to play an essential role in cell orientation and directed migration.

4.5.4.3. Phagocytosis.　Colchicine is less effective as an inhibitor of phagocytosis (review, Becker, 1976a,b) but here, as in secretion, total involvement of the cell is not required and local effects may not need microtubule-directed (cytoskeletal) structural control. This does not, however, exclude a role for microtubules or related structures in phagocytosis-associated events (see below).

4.5.4.4. Phagocytosis, cell surface stimulation, "capping," and microtubules and microfilaments.　The studies of Berlin and associates revealed that during phagocytosis by neutrophils, lectin-binding sites (con A) were selectively incorporated into phagocytic vacuoles but membrane transport carrier proteins for purines and nucleotides were selectively excluded from incorporation (Oliver et al., 1974; Berlin et al., 1975). The latter condition could result from active exclusion or from phagocytic uptake occurring only at membrane sites free from these proteins.

Colchicine and diamine were shown to "liberate" the con A sites from linkage with invaginated membrane sites (Berlin et al., 1974; Oliver et al., 1977). This indicated the role of colchicine-sensitive structures in control of cell surface protein distribution and seriously modified the concept of totally freely diffusable proteins in a sea of lipid (Fernandez and Berlin, 1976). Similar control of con A and immunoglobulin-binding sites on lymphocytes has been studied by Edelman and associates (e.g., Edelman, 1976). It is not clear whether these colchicine-sensitive structures are true microtubules that become associated (along with microfilaments) with the interior of the stimulated surface membrane (Reaven and Axline, 1973) or if they represent membrane-associated, assembled tubulin structures different from the morphologically defined microtubules. Recent reports of membrane-associated tubulin (e.g., Bhattachangya and Wolff, 1976) and of tubulin peptides on the outside of neuronal cells (Estridge, 1977) may be relevant.

Incubation of con A with neutrophils results in "patching" of the bound con A and in "capping" (Ryan et al., 1974). If the cells are allowed to move, the cap occurs at the "tail" of the cell; if not, it slowly forms in the central region. It has been suggested that during locomotion the membrane proteins are dragged backward relative to the membrane flow. At 37°C more capping is seen in the presence of colchicine (Mandell et al., 1977) which could be due to the abovementioned "liberation" of receptors or to effects on cell movement (prevented by con A in cells with intact microtubules; Ryan et al., 1974). Chediak-Higashi neutrophils (with a defective microtubule function which can be restored by stimulation of intracellular cGMP; Oliver and Zurier, 1976) show increased con A capping (Oliver et al., 1975). It is of considerable additional interest that a chemotactic factor inhibited capping (Mandell et al., 1977) perhaps also by inducing cGMP and microtubule formation (Hoffstein et al., 1974). Redistribution of submembraneous actin, presumably in microfilaments, into polar aggregates has also been shown during con A-induced capping (Oliver et al., 1977). Similar associations of subplasmalemmal microfilaments and microtubules with con A caps are seen in other cells (Albertini and Anderson, 1977) and are probably important in cell functions (see below).

Con A induces microtubule formation in neutrophils, release of the specific granule enzyme lysozyme (Hoffstein et al., 1977a,b), and activation of oxidative metabolism (see above). The "receptor" redistribution (patching) and its association with microfilaments and microtubules may be important in this and other forms of cell stimulation (section 4.5.5). Thus significance of this con A system lies in the information it may reveal about normal secretory events.

4.5.5. Microfilaments, chemotaxis, and phagocytosis
Neutrophils contain myosin and cortical actin filaments (Senda et al., 1975; Stossel, 1977), and cytochalasins inhibit chemotaxis and phagocytosis (though at very low concentrations an enhancement is seen) and enhance secretion to the out-

side of the cell (reviews, Becker and Henson, 1973; Gallin and Wolff, 1975; Goldstein, 1976; Wilkinson, 1976). It may be presumed that they do this by preventing the gelation of actin (by interfering with actin--actin-binding protein interactions; Hartwig and Stossel, 1976) in the cell. The redistribution of microfilaments into the leading front of the pseudopod in chemotaxis (Gallin et al., 1977), under the con A cap (Oliver et al., 1977) or in pseudopods surrounding a particle during phagocytosis (Stossel, 1977), is prevented by cytochalasin and appears to be essential for these effects. How they act in cell movement, whether in a localized portion of the cell (phagocytosis) or the whole cell (chemotaxis), is still unclear (Allison, 1973; Pollard and Werhing, 1974; Gallin et al., 1977). It may be supposed, however, that colchicine-sensitive cytoskeletal elements are involved to a lesser extent (see below) and that these and the cell membrane itself may be acted upon by the contractile filaments to promote movement (see below).

4.5.5.1. Secretion. It seems possible that the redistribution of microfilaments into pseudopods in phagocytosis, or to the leading edge in chemotaxis, or to the periphery in a cell on a stimulus-coated surface (effectively attempting to phagocytize the whole surface) would remove them from an adjacent area of the membrane. This would allow granules to gain access to the membrane, a process otherwise prevented by subplasmalemmal microfilamentous structures. Cytochalasin B, by effectively disrupting the microfilaments, would allow increased granule-membrane contact, permitting release of granules wherever the plasma membrane was *stimulated* and fusion could occur. Increased secretion would result. Thus neutrophils from a patient with fewer microfilaments (actin dysfunction) also exhibited increased enzyme secretion (Boxer et al., 1974). Moreover, increased mobility of granules has been seen in cytochalasin-treated cells (Henson, unpublished observations).

4.5.5.2. Similarity between surface stimulation of neutrophils and the action of cytochalasin B –an hypothesis. In neutrophils adherent to a surface during chemotaxis, a microfilament network accumulates at the front of the cell and at points of attachment to the filter (Gallin et al., 1977). We suggest that the network does not extend over the whole undersurface of the cell when adherent to surface-bound stimuli, but that it is associated with local membrane movements (as in phagocytic pseudopod formation) and leaves areas of stimulated membrane to which granules can gain access. This would be achieved by microfilament contraction using sites of attachment to the filter (perhaps via transmembrane proteins) as fixed structures to act against, which would free areas of membrane. A soluble stimulus in suspension would not provide a point of binding of membrane to a surface substrate upon which the contractile elements could act, and would not expose the intracellular (cytoplasmic) side of the membrane to the granules. Cytochalasin B, however, by preventing microfilament formation, would allow granules to interact with any "stimulated" portion of the membrane. (A similar hypothesis has been suggested as one alternative by Allison, 1973.)

An alternative or additional hypothesis involves granule interaction with contractile filaments (as may occur in other cells; Østlund et al., 1977), allowing these to be actively pulled to the membrane.

4.5.5.3. Cell contraction. Neutrophils incubated with cytochalasin B, chemotactic factor and cytochalasins, or ionophore A23187, show a calcium-requiring decrease in cell volume (Hsu and Becker, 1975a). The degree of this contraction could not be directly correlated with enzyme release though it may be associated. It is interesting that a similar decrease was seen during stimulation of enzyme release from neutrophils, by ATP in low ionic strength buffer (Becker and Henson, 1975; Hsu and Becker, 1975b). In the latter case an induction of pseudopod formation by ATP was observed (Henson et al., 1975). Whether the volume decrease stimulated by chemotactic factors involved contractile element effects and/or ion efflux with concurrent loss of water is unclear at present, but its occurence in the presence of cytochalasin would suggest the latter.

4.5.6. Cations

As discussed in previous reviews, calcium is required for neutrophil function (Becker and Henson, 1973; Gallin et al., 1977; Naccache et al., 1977). These effects are not discussed in detail since a chapter on calcium in stimulus-secretion coupling is included in this volume. Nevertheless, some exciting new data deserves comment.

4.5.6.1. Calcium. The ionophore A23187, and even calcium itself, can induce the release of lysozyme, but not azurophil granule constituents from neutrophils (section 4.4.4.2). Since calcium in the medium is necessary for rabbit neutrophil secretion (though not always for human secretion), we believe that Ca^{2+} is an absolute requirement for secretion for both types of granule, but that the various mechanisms involved in the different secretory processes are not equally sensitive to Ca^{2+} deprivation or influx. Presumably there is some intracellular compartmentalization and perhaps internal sources of Ca^{2+} may also be involved (Wilkinson, 1975).

Chemotactic factors have been shown to induce uptake of Ca^{2+} (Bouek and Snyderman, 1976) or release of Ca^{2+} (Gallin and Rosenthal, 1974). In a more detailed study, Naccache and colleagues (1977) showed that both influx and efflux was stimulated by FMLP and that the net effect on intracellular Ca^{2+} levels depended on the environmental concentration. An increase in membrane permeability of Ca^{2+}, as well as a mobilization of intracellular pools, was suggested.

Calcium mobilization has been implicated in microtubule assembly (Gallin and Rosenthal, 1974; Hoffstein et al. 1974, 1977b), perhaps in cyclic nucleotide effects, and presumably is involved in contractile-element function. It also probably participates in membrane fusion (see below).

4.5.6.2. Na^+, K^+. External K^+ ions enhance rabbit chemotaxis (Showell and Becker, 1976) and the effect is abolished by ouabain. Recently Dunham and co-workers (1974) showed a stimulus-dependent decrease in K^+ uptake. Naccache and associates (1977) found a stimulus-dependent influx of Na^+ into cells inducted by FMLP and a later and less marked K^+ influx. They suggest that the Na^+ could displace Ca^{2+} from intracellular sites (as shown for Ca^{2+} release from heart mitochondria), leading to an increased available intracellular Ca^{2+} pool.

The role of surface ATPases in neutrophil secretion is unknown, but the work of Becker and associates (Showell et al 1977; Naccache et al., 1977) suggests that they may be of considerable importance. Thus chemotactic factors appear to stimulate Na^+, K^+, ATPase, and the cation flux appears integrally related to cell stimulation, perhaps by local alterations of surface change.

4.5.6.3. Depolarization and charge changes. The abovementioned Na^+ influx could induce membrane depolarization (Naccache et al., 1977) and Gallin and collaborators (1977) have shown chemotactic factor-induced depolarization followed by hyperpolarization in macrophages. They suggest an influx of Na^+ and/or Ca^{2+} for the former and K^+ efflux for the latter, but emphasize that further studies are needed to confirm this. Surface stimulation of neutrophils certainly reduces K^+ influx (Dunham et al., 1974). Gallin and co-workers (1977) also demonstrated a local accumulation of cations (unidentified) at the leading edge of the cell. Previous data had indicated a decrease in negative charge of the whole cell (Gallin et al., 1975). The polarization of the neutrophil response to stimuli is further exemplified by these studies. Such localized charge effects could lead to movement of surface proteins, perhaps by an electrophoretic process (Jaffe, 1977; Poo and Robinson, 1977), a possible mechanism for receptor accumulation at the leading edge of the cell, or of protein movement away from membrane fusion sites.

4.5.7. Membrane fusion
Membrane fusion is discussed elsewhere in this volume and has been reviewed extensively by Poste and Allison (1973). The mechanisms by which granule membranes fuse with cytoplasmic membranes to initiate discharge of their contents into vacuole or extracellular medium are unknown. However, a number of current hypotheses merit comment.

4.5.7.1. A model based on leukocidin effects. Woodin and Wieneke have proposed a model for leukocidin-induced exocytosis (Woodin, 1968; Woodin and Wieneke, 1970). The stimulus is cytotoxic, but in the presence of Ca^{2+} and ATP a secretory stage is seen. The leukocidin induces an increased K^+ efflux (inhibited by tetraethyl ammonium ions), and an influx of Ca^{2+}. A change in the conformation of triphosphoinositide was seen (change in charge). The authors suggest a Ca^{2+} accumulation at the membrane, promoting granule adhesion, followed by exclusion of Ca^{2+} (by its reactivity with phosphate from hydrolyzed ATP), promoting increased permeability at the site and fusion. Speculative evidence for some aspects of this model was provided that could have led to other interpretations.

4.5.7.2. Relevance to current thinking. Some aspects of the abovementioned model formulated 10 or more years ago conform to more current suggestions. Thus Ca^{2+} ions may modulate membrane permeability at sites of membrane-membrane contact with lower concentrations enhancing permeability (Rose et al., 1977). A possible role for Ca^{2+} and phosphate ions in agglutination of red cell ghosts and later in their fusion was suggested by Zakai and co-workers (1976). Calcium-induced displacement of membrane particles at sites of aggregation of chromaffin granules have been reported (Schober et al., 1977), and calcium-induced changes in membrane curvature have been suggested to relate to exocytosis at nerve endings by Hall and Simon (1976). Furthermore, con A induces a rapid phosphatidylinositol turnover in lymphocytes in association with its induction of "transformation" (Schellenberg and Gillespie, 1977). Displacement of Ca^{2+} from membranes, inducing a susceptibility to fusion, is known in other cells (Poste and Allison, 1973).

This could be pieced together to suggest stimulus effects on Na^+, K^+, ATPase, leading to ion flux and changes in local membrane charge. Cation accumulation, perhaps Ca^{2+}, may be associated with binding to membrane phosphatidylinositol. This could induce or be associated with granule attachment to the membrane and, possibly in the presence of PO_4 (from the ATPase function), lead to fusion presumably involving local exclusion of membrane proteins allowing lipid-lipid contact (see Poste and Allison, 1973; Plattner et al., 1977, for various possibilities). It is interesting that neutrophil granules were shown to be disrupted by negatively charged artificial phospholipid spherules, particularly those composed of phosphatidylinositol (Hawiger et al., 1969), and this interaction was prevented if the lipid was previously incubated with positively charged protamine sulfate. Corticosteroids inhibit neutrophil secretion (Wright and Malawista, 1973; Hawkins, 1974a; Goldstein et al., 1976b) and superoxide production (Lehmeyer and Johnston, 1977). Whether this is due to a direct effect on the membrane and on membrane fusion is unknown.

No evidence is available for lysophosphatide-induced fusion, although the generation of prostaglandins and their intermediates during neutrophil stimulation suggest phospholipase A activity that could be associated with lysophosphatide production. Other possibilities exist for the fusion process, but intensive study is required in this cell even to begin to elucidate the mechanisms.

4.6. Summary and speculations

The study of neutrophil functions is proving particularly valuable, mainly because it undergoes phagocytosis and locomotion as well as secretion and can also be obtained in relatively pure populations. Moreover, since it is not actively synthetic, its investigation is somewhat simpler than for the macrophage. We conclude this section with a brief discussion of two unifying hypotheses that may apply to other inflammatory cells, but which (for the reasons mentioned above) may be most effectively studied with neutrophils.

4.6.1. One stimulus may initiate multiple neutrophil activities

An important question for a multifunctional cell is whether it requires different stimuli to activate each function. Earlier we provided preliminary data to suggest that a crude bacterial chemotactic factor could initiate chemotaxis, oxidative metabolism, secretion, and phagocytosis (Henson, 1974b). Similar data were presented for FMLP by Becker (1976a,b), and we (unpublished observations) have obtained comparable results for C5a. Thus it appears that the attachment of such a stimulus to a particle can allow triggering of the cellular events involved in engulfment.

One may assume that a stimulus such as FMLP interacts with only one receptor, and thus that this interaction can trigger multiple pathways in the cell. These are different pathways, as indicated by experimental dissociation of oxidative metabolic events, secretion, chemotaxis, and phagocytosis. At which point these different intracellular pathways diverge is unknown, but these effects seem to be partially controlled by the mode of presentation of cell stimulus. Studies of the role of one or more cell proteases, for example, should reveal much about the activation processes.

4.6.2. A suggested hypothesis for neutrophil activation

For our further studies we have formulated the following possibilities. Through interaction of stimulus with the cell membrane, and probably by way of an activatable protease, either directly and/or through ion fluxes, two types of changes occur: (1) The membrane is altered to become successively reactive to one or more of three types of granules. The first of these is an NADPH-oxidase-containing granule, whose interaction with the plasma membrane leads to the production of superoxide. This suggestion accounts for the discrepant data on surface or granule site for O_2^- production. Subsequent changes allow binding and fusion of specific and then azurophil granules. (2) Formation and action of contractile elements at the site of membrane stimulation. These may serve to pull granules to the membrane, but probably prevent granules from reaching surface membrane fusion sites until they have either disappeared again or have contracted. This contraction would require points of attachment to a particle or surface. It would also induce the pseudopod formation and extension involved in phagocytosis and cell movement. It is further suggested that if the effect of the stimulus on the membrane is transient (because of local desensitization [Becker, 1972], destruction of stimulus [Aswanikumar et al., 1976] or low-affinity binding), relaxation behind the leading edge allows progressive movement of the pseudopod. Around a particle this will lead to engulfment; on a surface this could lead to movement. If a gradient is present, orientation toward the gradient would result; if not, chemokinesis would be the effect.

Thus chemotaxis, chemokinesis, and phagocytosis may be similar and differences among them may be easily explained on the basis of variations between localized (phagocytosis) and generalized effects in the cell. Stimuli for one cell function however, ought to initiate all the effects. Why then does IgG not induce chemotaxis? We suggest that the stimulus here binds firmly to both the cell and membrane substrate (e.g., micropore filter) and does not allow the liberation of the membrane necessary for forward movement.

While these ideas are simplistic, they suggest approaches for experimental verification (or denial).

5. Concluding remarks

In this chapter we have emphasized different aspects of inflammatory cell activation in the discussion of the different cell types. Thus protease effects are best known in platelets, the effects of ion fluxes in neutrophils, morphologic examination of secretion in mast cells, membrane composition in platelets, and so forth. We believe that these cells are fundamentally similar and that studies in one type are directly applicable to another. The present authors and many other groups have found that early suggestions of differences between such cell types later turned out to be trivial and that the underlying mechanisms were the same.

This comment is not meant to detract from differences in functional roles these cells play in vivo. Such differences, however, result mainly from the nature of the stimuli with which the cells interact and/or the contained mediators involved, not from the mechanisms of stimulation or release. Since these cells are integrally involved in host defense mechanisms, as well as in inflammation, a much greater understanding of such mechanisms will be required before we can achieve knowledgeable therapeutic control of the injurious processes without diminishing the protective roles of these cells.

Acknowledgments

We are indebted to Sharon Garland and Vonda Hull for typing the manuscript. Dr. Morrison is a recipient of U.S.P.H.S. Career Development Award AI100081, and we also wish to acknowledge the support of U.S.P.H.S. grants HL16411 and GM24834.

References

Ahkong, Q. F., Tampion, W. and Lucy, J. A. (1975) Promotion of cell fusion by divalent cation ionophores. Nature (London) 256 (5514), 208–209.

Albertini, D. F. and Anderson, E. (1977) Microtubule and microfilament rearrangements during capping of concanavalin A receptors on cultured ovarian granulosa cells. J. Cell. Biol. 73, 111–127.

Allen, R. C., Stjernholm, R. L. and Steele, R. H. (1972) Evidence for the generation of an electronic excitation state(s) in human polymorphonuclear leukocytes and its participation in bactericidal activity. Biochem. Biophys. Res. Commun. 47, 679–684.

Allison, A. C. (1973) The role of microfilaments and microtubules in cell movement, endocytosis and exocytosis. In: Locomotion of Tissue Cells, Ciba Foundation Symposium 14, pp. 109–148. Associated Scientific Publishers, Amsterdam.

Anderson, C. L. and Grey, H. M. (1977) Solubilization and partial characterization of cell membrane Fc receptors. J. Immunol. 118, 819–825.

Anderson, P., Slorach, S. A. and Uvnäs, B. (1973) Sequential exocytosis of storage granules during antigen-induced histamine release from sensitized rat mast cells *in vitro*—an electron microscopic study. Acta Phys. Scand. 88, 359–372.

Archer, G. T. (1961) Release of heparin from the mast cells of the rat. Nature 191, 90.

Archer, G. T. and Jackas, M. (1965) Disruption of mast cells by a component of eosinophil granules. Nature 205, 599–600.

Arend, W. P. and Mannik, M. (1973) The macrophage receptor for IgG: number and affinity of binding sites. J. Immunol. 110, 1455–1463.

Assem, E. S. K., Schild, H. O. and Vickers, M. R. (1972) Stimulation of histamine-forming capacity by antigen in sensitized human leukocytes. Int. Arch. Allergy 42, 343–352.

Aswanikumar, S., Schiffmann, E., Corcoran, B. A. and Wahl, S. M. (1976) Role of a peptidase in phagocyte chemotaxis. Proc. Nat. Acad. Sci. U.S.A. 73, 2439–2442.

Atkinson, J. P., Michael, J. M., Chaplin, H., Jr. and Parker, C. W. (1977) Modulation of macrophage C3b receptor function by cytochalasin-sensitive structures. J. Immunol. 118, 1292–1299.

Austen, K. F. (1973) Biochemical characteristics and pharmacologic modulation of the antigen-induced release of the chemical mediators of immediate hypersensitivity. In: Allergology, Proc. Eighth Intern. Congr. Allergology, pp. 306–324. Elsevier/North-Holland, New York.

Austen, K. F. and Becker, E. L. (1966) Mechanisms of immunologic injury of rat peritoneal mast cells. II. Complement requirement and phosphonate ester inhibition of release of histamine by rabbit anti-rat gamma globulin. J. Exp. Med. 124, 397–416.

Austen, K. F. and Brocklehurst, W. E. (1961) Anaphylaxis in chopped guinea pig lung. I. Effect of peptidase substrates and inhibitors. J. Exp. Med. 113, 521–539.

Austen, K. F. and Humphrey, J. H. (1963) *In vitro* studies of the mechanism of anaphylaxis. Adv. Immunol. 3, 1–96.

Austen, K. F., Bloch, K. J., Baker, A. R. and Arnason, B. G. (1965) Immunological histamine release from rat mast cells *in vitro*. Effect of age of cell donor. Proc. Soc. Exp. Biol. Med. 120, 542–546.

Austen, K. F., Wasserman, S. I. and Goetzl, E. J. (1976) Mast cell derived mediators: structural and functional diversity and regulation of expression. In: Molecular and Biological Aspects of the Acute Allergic Reaction (Johansson, S. G. O., Strandberg, K. and Uvnäs, B., eds.) p. 293–320, Plenum Press, New York.

Babior, B. M., Kipnes, R. S. and Curnutte, J. T. (1973) Biological defense mechanisms: The production by leukocytes of superoxide, a potential bactericidal agent. J. Clin. Invest. 52, 741–744.

Bach, M. K. and Brashler, J. R. (1976) On the nature of the presumed receptor for IgE on mast cells. III. Kinetics of the blocking of the PCA reaction by cell-free particulate preparations from rat peritoneal mast cells and effect of pH and calcium concentration on the reaction. Immunology 30, 101–105.

Bach, M. K., Block, K. J. and Austen, K. F. (1971) IgE and IgGa antibody-mediated release of histamine from rat peritoneal cells. II. Interaction of IgGa and IgE at the target cell. J. Exp. Med. 133, 772–784.

Baenziger, N. L., Brodie, G. N. and Majerus, P. W. (1971) A thrombin-sensitive protein of human platelet membranes. Proc. Nat. Acad. Sci. U.S.A. 68, 240–243.

Baenziger, N. L., Brodie, G. N. and Majerus, P. W. (1972) Isolation and properties of a thrombin-sensitive protein of human platelets. J. Biol. Chem. 247, 2723–2731.

Baggiolini, M., Hirsch, J. G. and DeDuve, C. (1970) Further biochemical and morphological studies of granule fractions from rabbit heterophil leukocytes. J. Cell Biol. 45, 586–597.

Bainton, D. F. (1970) Sequential discharge of polymorphonuclear leukocyte granules during phagocytosis of microorganisms. J. Cell Biol. 47, 11a.

Bainton, D. F. (1973) Sequential degranulation of the two types of polymorphonuclear leukocyte granules during phagocytosis of microorganisms. J. Cell Biol. 58, 249–264.

Bainton, D. F. and Farquhar, M. G. (1968) Differences in enzyme content of azurophil and specific granules of polymorphonuclear leukocytes. II. Cytochemistry and electron microscopy of bone marrow cells. J. Cell Biol. 39, 299–317.

Bainton, D. F., Nichols, B. A. and Farquhar, M. G. (1976) Primary lysosomes of blood leukocytes. In: Lysosomes in Biology and Pathology (Dingle, J. T. and Dean, R. T., eds.) vol. 5, p. 3–32, Elsevier/North-Holland, New York.

Barbaro, J. F. (1961) The release of histamine from rabbit platelets by means of antigen-antibody precipitates. I. The participation of the immune complex in histamine release. J. Immunol. 86, 369–376.

Barber, A. J. and Jamieson, G. A. (1971) Platelet collagen adhesion characterization of collagen glucosyltransferase of plasma membranes of human blood platelets. Biochim. Biophys. Acta 252, 533–545.

Barber, A. J., Käser-Glanzmann, R., Jakábová, M. and Lüscher, E. F. (1972) Characterization of a chondroitin 4-sulfate proteoglycan carrier for heparin neutralizing activity (platelet factor 4) released from human blood platelets. Biochim. Biophys. Acta 286, 312–329.

Baumgartner, H. R. (1974) Morphometric quantitation of adherence of platelets to an artificial surface and components of connective tissue. Thromb. Diath. Haemorrh. Suppl. 60, 39–49.

Baxter, J. H. (1973) The role of Ca²⁺ in mast cell activation, desensitization and histamine release by dextran. J. Immunol. 111, 1470–1473.

Baxter, J. H. and Adamik, R. (1974) Temperature dependence of histamine release from rat mast cells by dextran. Proc. Soc. Exp. Biol. Med. 146, 71–74.

Baxter, J. H. and Adamik, R. (1975) Control of histamine release: Effects of various conditions on rate of release and rate of cell desensitization. J. Immunol. 114, 1034–1041.

Baxter, J. H. and Adamik, R. (1976) Effects of calcium and phosphatidylserine in rat mast cell reaction to dextran. Proc. Soc. Exp. Biol. Med. 152, 266–271.

Becker, E. L. (1972) The relationship of the chemotactic behavior of the complement-derived factors C3a, C5a, and C5̄6̄7̄, and a bacterial chemotactic factor to their ability to activate the proesterase 1 of rabbit polymorphonuclear leukocytes. J. Exp. Med. 135, 376–387.

Becker, E. L. (1976a) Some interrelations of neutrophil chemotaxis, lysosomal enzyme secretion, and phagocytosis as revealed by synthetic peptides. Am. J. Pathol. 85, 385–394.

Becker, E. L. (1976b) Some interrelations among chemotaxis, lysosomal enzyme secretion and phagocytosis by neutrophils. In: Molecular and Biological Aspects of the Acute Allergic Reaction. (Johansson, S. G. O., Strandberg, K. and Uvnäs, B., eds) p. 353–370. Plenum Press, New York.

Becker, E. L. and Austen, K. F. (1966) Mechanisms of immunologic injury of rat peritoneal mast cells. I. The effect of phosphonate inhibitors on the homocytotropic antibody-mediated histamine release and the first component of rat complement. J. Exp. Med. 124, 379–395.

Becker, E. L. and Henson, P. M. (1973) In vitro studies of immunologically induced secretion of mediators from cells and related phenomena. Adv. Immunol. 17, 93–193.

Becker, E. L. and Henson, P. M. (1975) Biochemical characteristics of ATP-induced secretion of lysosomal enzymes from rabbit polymorphonuclear leukocytes. Inflammation 1, 71–84.

Becker, E. L. and Showell, H. J. (1974) The ability of chemotactic factors to induce lysosomal enzyme release. II. The mechanism of release. J. Immunol. 112, 2055–2062.

Becker, K. E., Ishizaka, T., Metzger, H., Ishizaka, K. and Grimley, P. M. (1973) Surface IgE on human basophils during histamine release. J. Exp. Med. 138, 394–409.

Becker, E. L., Showell, H. J., Henson, P. M. and Hsu, L. S. (1974) The ability of chemotactic factors to induce lysosomal enzyme release. I. The characteristics of the release, the importance of surfaces and the relation of enzyme release to chemotactic responsiveness. J. Immunol. 112, 2047–2054.

Behnke, O. (1964) A preliminary report on "microtubules" in undifferentiated and differentiated vertebrate cells. J. Ultrastruct. Res. 11, 139–146.

Behnke, O. (1970) The morphology of platelet membrane systems. Ser. Haematol. 3, 3–16.

Behnke, O., Kristensen, B. I. and Nielsen, L. E. (1971) Electron microscopical observations on ac-

484

tinoid and myosinoid filaments in blood platelets. J. Ultrastruct. Res. 37, 351–369.

Benditt, E. P., Wong, R. L., Arase, M. and Roeper, E. (1955) 5-hydroxytryptamine in mast cells. Proc. Soc. Exp. Biol. Med. 90, 303–304.

Benveniste, J. (1974) Platelet-activating factor, a new mediator of anaphylaxis and immune complex deposition from rabbit and human basophils. Nature (London) 249, 581–582.

Benveniste, J. (1977) Platelet activating factor. Int. Arch. Allergy Appl. Immunol. In press.

Benveniste, J., Henson, P. M. and Cochrane, C. G. (1972) Leukocyte-dependent histamine release from rabbit platelets. The role of IgE, basophils and a platelet-activating factor. J. Exp. Med. 136, 1356–1376.

Beraldo, W. T., Dias da Silva, W. and Lemos Fernandes, A. D. (1962) Inhibitory effects of carbohydrates on histamine release and mast cell disruption by dextran. Brit. J. Pharm. Chemotherapy. 19, 405–413.

Bergendorff, A. and Uvnäs, B. (1973) Storage properties of rat mast cell granules in vitro. Acta Physiol. Scand. 87, 213–222.

Berlin, R. D. (1972) Effect of concanavalin A on phagocytosis. Nature New Biol. 235, 44–45.

Berlin, R. D., Oliver, J. M., Ukena, T. E. and Yin, H. H. (1974) Control of cell surface topography. Nature (London) 247, 45–46.

Berlin, R. D., Oliver, J. M., Ukena, T. E. and Yin, H. H. (1975) The cell surface. New Eng. J. Med. 292, 515–520.

Best, L. C., Martin, T. J., Russell, R. G. G. and Preston, F. E. (1977) Prostacyclin increases cyclic AMP levels and adenylate cyclase activity in platelets. Nature (London) 267, 850–851.

Bettex-Galland, M., and Lüscher, E. F. (1961) Thrombosthenin-a contractile protein from thrombocytes. Its extraction from human blood platelets and some of its properties. Biochim. Biophys. Acta 49, 536–547.

Bhattachargya, B. and Wolff, J. (1976) Polymerisation of membrane tubulin. Nature (London) 264, 576–577.

Bills, T. K., Smith, J. B. and Silver, M. J. (1976) Metabolism of [^{14}C] arachidonic acid by human platelets. Biochim. Biophys. Acta 424, 303–314.

Bloom, G. D. and Chakravarty, N. (1970) Time course of anaphylactic histamine release and morphological changes in rat peritoneal mast cells. Acta Physiol. Scand. 78, 410–419.

Bloom, G. D. and Haegermark, Ö. (1965) A study on morphological changes and histamine release induced by compound 48/80 in rat peritoneal mast cells. Exp. Cell Res. 40, 637–654.

Bloom, G. D., Fredholm, B. and Haegermark, Ö. (1967) Studies on the time course of histamine release and morphological changes induced by histamine liberators in rat peritoneal mast cells. Acta Physiol. Scand. 71, 270–282.

Boetcher, D. A. and Leonard, E. J. (1973) Basophil chemotaxis: augmentation by a factor from stimulated lymphocyte cultures. Immunol. Commun. 2, 421–429.

Bogart, D. B. and Stechschulte, D. J. (1974) Release of PAF from human lung. Clin. Res. (Abstr.) 22, 652A.

Bokisch, V. A. and Müller-Eberhard, H. J. (1970) Anaphylatoxin inactivator of human plasma: its isolation and characterization as a carboxypeptidase. J. Clin. Invest. 49, 2427–2436.

Booyse, F. M., Marr, J., Yang, D-C., Guiliani, D. and Rafelson, M. E., Jr. (1976) Adenosine cyclic 3',5'-monophosphate dependent protein kinase from human platelets. Biochim. Biophys. Acta 422, 60–72.

Bosila, A. A. (1959) The Basophil Leukocyte and its Relationship to the Tissue Mast Cell. Munksgaard, Copenhagen.

Bosmann, H. B. (1971) Platelet adhesiveness and aggregation: the collagen: glycosyl, polypeptide: N-acetylgalactosaminyl and glycoprotein: galactosyl transferases of human platelets. Biochem. Biophys. Res. Commun. 43, 1118–1124.

Boucek, M. M. and Snyderman, R. (1976) Calcium influx requirement for human neutrophil chemotaxis: Inhibition by lanthanum chloride. Science 193, 905–907.

Bourne, H. R., Lehrer, R. I., Lichtenstein, L. M., Weissmann, G. and Zurier, R. (1973) Effects of cholera enterotoxin on adenosine 3',5'-monophosphate and neutrophil function. Comparison with other compounds which stimulate leukocyte adenyl cyclase. J. Clin. Invest. 52, 698–708.

Boxer, L. A., Greenberg, M. S., Boxer, G. J. and Stossel, T. P. (1975) Autoimmune neutropenia. New Eng. J. Med. 293, 748–753.

Boxer, L. A., Hedley-Whyte, E. T. and Stossel, T. P. (1974) Neutrophil actin dysfunction and abnormal neutrophil behavior. New Eng. J. Med. 291, 1093–1099.

Brass, L. F. and Bensusan, H. B. (1974) The role of collagen quaternary structure in the platelet: collagen interaction. J. Clin. Invest. 54, 1480–1487.

Brass, L. F. Faile, D. and Bensusan, H. B. (1976) Direct measurement of the platelet: collagen interaction by affinity chromatography on collagen/sepharose. J. Lab. Clin. Med. 87, 525–534.

Briggs, R. T., Drath, D. B., Karnovsky, M. L. and Karnovsky, M. J. (1975) Localization of NADH oxidase on the surface of human polymorphonuclear leukocytes by a new cytochemical method. J. Cell Biol. 67, 566–586.

Brocklehurst, W. E. (1960) The release of histamine and formation of a slow-reacting substance (SRS-A) during anaphylactic shock. J. Physiol. (London) 151, 416–435.

Brodie, G. N., Baenziger, N. L., Chase, L. R. and Majerus, P. W. (1972) The effects of thrombin on adenyl cyclase activity and a membrane protein from human platelets. J. Clin. Invest. 51, 81–88.

Bronza, J. P. and Ward, P. A. (1977) Chemotactic factor inhibitory activity in rabbit neutrophils. In: Leukocyte Chemotaxis: Methodology, Physiology, and Clinical Applications. (Gallin, J. I. and Quie, P. G., eds.) pp. 229–236, Raven Press, New York.

Bryant, R. E., DesPrez, R. M., VanWay, M. H. and Rogers, D. E. (1966) Studies on human leukocyte motility I. Effects of alterations in pH, electrolyte concentration and phagocytosis on leukocyte migration, adhesiveness and aggregation. J. Exp. Med. 124, 483–499.

Byus, C. V., Klimpel, G. R., Lucas, D. O. and Russell, D. H. (1977) Type I and Type II cyclic AMP-dependent protein kinase as opposite effectors of lymphocyte mitogenesis. Nature (London) 268, 63–64.

Carson, D. A. and Metzger, H. (1974) Interaction of IgE with rat basophilic leukemia cells. IV. Antibody-induced redistribution of IgE receptors. J. Immunol. 113, 1271–1277.

Castor, C. W., Ritchie, J. C., Scott, M. E. and Whitney, S. L. (1977) Connective tissue activation. XI. Stimulation of glycosaminoglycan and DNA formation by a platelet factor. Arthr. Rheum. 20, 859–868.

Caswell, A. H. and Pressman, B. C. (1972) Kinetics of transport of divalent cations across sarcoplasmic reticulum vesicles by ionophores. Biochem. Biophys. Res. Commun. 49, 292–298.

Cazenave, J.-P., Packham, M. A. and Mustard, J. F. (1973) Adherence of platelets to a collagen-coated surface: Development of a quantitative method. J. Lab. Clin. Med. 82, 978–990.

Cazenave, J.-P., Assimeh, S. N., Painter, R. H., Packham, M. A., and Mustard, J. F. (1976) C1q inhibition of the interaction of collagen with human platelets. J. Immunol. 116, 162–163.

Chakravarty, N. (1960) The mechanism of histamine release in anaphylactic reaction in guinea pig and rat. Acta Physiol. Scand. 48, 146–166.

Chakravarty, N. (1968) Further observations on the inhibition of histamine release by 2-deoxyglucose. Acta Physiol. Scand. 72, 425–432.

Chakravarty, N., Goth, A. and Sen, P. (1973) Potentiation of dextran induced histamine release from rat mast cells by phosphatidylserine. Acta Physiol. Scand. 88, 469–480.

Chakravarty, N., Gustafson, G. T. and Pihl, E. (1967) Ultrastructural changes in rat mast cells during anaphylactic histamine release. Acta Pathol. Microbiol. Scand. 71, 233–244.

Charo, I. F., Feinman, R. D. and Detwiler, T. C. (1976) Inhibition of platelet secretion by an antagonist of intracellular calcium. Biochem. Biophys. Res. Commun. 72, 1462–1467.

Charo, I. F., Feinman, R. D. and Detwiler, T. C. (1977) Interrelationships of platelet aggregation and secretion. J. Clin. Invest. 60, 866–873.

Chesney, C. McI., Harper, E. and Colman, R. W. (1974) Human platelet collagenase. J. Clin. Invest. 53, 1647–1654.

Chi, E. Y., Lagunoff, D. and Koehler, J. K. (1976) Freeze-fracture study of mast cell secretion. Proc. Nat. Acad. Sci. U.S.A. 73, 2823–2827.

Chiang, T. M., Beachey, E. H. and Kang, A. H. (1976) Interaction of a chick skin collagen fragment (α1-CB5) with human platelets. Biochemical studies during the aggregation and release reaction. J. Biol. Chem. 250, 6916–6922.

Chiang, T. M., Beachey, E. H. and Kang, A. H. (1977) Binding of collagen α1 chains to human platelets. J. Clin. Invest. 59, 405–411.

Cochrane, C. G. and Müller-Eberhard, H. J. (1968) The derivation of two distinct anaphylatoxin activities from the third and fifth components of human complement. J. Exp. Med. 127, 371–386.

Cohen, I., Bohak, Z., de Vries, A. and Katchalski, E. (1969) Thrombosthenin M. Purification and interaction with thrombin. Eur. J. Biochem. 10, 388–394.

Colman, R. W. and Schreiber, A. D. (1976) Effect of heterologous antibody on human platelets. Blood 48, 119–131.

Conrad, D. H. and Froese, A. (1975) Characterization of an IgE binding component from the surface of normal rat mast cells. Fed. Proc. 34, 1001a.

Conrad, D. H. and Froese, A. (1976) Characterization of the target cell receptor for IgE. II. Polyacrylamide gel analysis of the surface IgE receptor from normal rat mast cells and from rat basophilic leukemia cells. J. Immunol. 116, 319–326.

Conrad, D. H., Berczi, I. and Froese, A. (1976) Characterization of the target cell receptor for IgE. I. Solubilization of IgE receptor complexes from rat mast cells and rat basophilic leukemia cells. Immunochemistry 13, 329–332.

Copley, A. L. and Witte, S. (1976) On physiological microthromboembolization as the primary platelet function: Elimination of invaded particles from the circulation and its pathologic significance. Thrombos. Res. 8, 251–262.

Cox, J. P. and Karnovsky, M. L. (1973) The depression of phagocytosis by exogenous cyclic nucleotides, prostaglandins and theophylline. J. Cell Biol. 59, 480–490.

Craddock, P. R., Hammerschmidt, D., White, J. G., Dalmasso, A. P. and Jacob, H. S. (1977a) Complement (C5a)-induced granulocyte aggregation in vitro. A possible mechanism of complement-mediated leukostasis and leukopenia. J. Clin. Invest. 60, 260–264.

Craddock, P. R., Fehr, J., Dalmasso, A. P., Brigham, K. L. and Jacob, H. S. (1977b) Hemodialysis leukopenia. Pulmonary vascular leukostasis resulting from complement activation by dialyzer cellophane membranes. J. Clin. Invest. 59, 879–888.

Curnutte, J. T., Kipnes, R. S. and Babior, B. M. (1975) Defect in pyridine nucleotide dependent superoxide production by a particulate fraction from the granulocytes of patients with chronic granulomatous disease. New Eng. J. Med. 293, 628–632.

Czarnetzki, B. M., König, W., and Lichtenstein, L. M. (1976) Eosinophil chemotactic factor (ECF). I. Release from polymorphonuclear leukocytes by the calcium ionophore A23187. J. Immunol. 117, 229–234.

Daëron, M., Duc, H. T., Kanellopoulos, J., Le Bouteiller, Ph., Kinsky, R. and Voisin, G. A. (1975) Allogeneic mast cell degranulation induced by histocompatibility antibodies: An in vitro model of transplantation anaphylaxis. Cell Immunol. 20, 133–155.

Da Silva, P. P., and Nogueira, M. L. (1977) Membrane fusion during secretion. A hypothesis based on electron microscope observation of Phytophthora palmivora zoospores during encystment. J. Cell Biol. 73, 161–181.

Davey, M. G. and Lüscher, E. F. (1967) Actions of thrombin and other coagulant and proteolytic enzymes on blood platelets. Nature (London) 216, 857–858.

Davies, P. and Allison, A. C. (1976) The secretion of lysosomal enzymes. In: Lysosomes in Biology and Pathology (Dingle, J. T. and Dean, R. T., eds.) vol. 5, pp. 61–98, Elsevier/North-Holland, New York.

Davies, P., Rita, G. A., Krakauer, K., and Weissmann, G. (1971) Characterization of a neutral protease from lysosomes of rabbit polymorphonuclear leucocytes. Biochem. J. 123, 559–569.

Dayton, W. R., Reville, W. J., Goll, D. E., and Stromer, M. H. (1976a) A Ca^{2+}-activated protease possibly involved in myofibrillar protein turnover. Partial characterization of the purified enzyme. Biochemistry 15, 2159–2167.

Dayton, W. R., Goll, D. E., Zeece, M. G., Robson, R. M. and Reville, W. J. (1976b) A Ca^{2+}-activated protease possibly involved in myofibrillar protein turnover. Purification from porcine muscle. Biochemistry 15, 2150–2158.

Dean, D. A., Wistar, R. and Murrell, K. D. (1974) Combined in vitro effects of rat antibody and neutrophilic leukocytes on schistosomula of Schistosoma Mansoni. Am. J. Trop. Med. Hyg. 23, 420–428.

DeChatelet, L. R., McCall, C. E., McPhail, L. C. and Johnston, R. B., Jr. (1974) Superoxide dismutase activity in leukocytes. J. Clin. Invest. 53, 1197–1201.

Derksen, A. and Cohen, P. (1975) Patterns of fatty acid release from endogenous substrates by human platelet homogenates and membranes. J. Biol. Chem. 250, 9342–9347.

DeShazo, C. V., McGrade, M. T., Henson, P. M. and Cochrane, C. G. (1972) The effect of complement depletion on neutrophil migration in acute immunologic arthritis. J. Immunol. 108, 1414–1419.

Detwiler, T. C. and Charo, I. F. (1977) Research on the biochemical basis of platelet function. Rev. Hemat. In press.

Detwiler, T. C. and Feinman, R. D. (1973a) Kinetics of the thrombin-induced release of calcium (II) by platelets. Biochemistry 12, 282–289.

Detwiler, T. C. and Feinman, R. D. (1973b) Kinetics of the thrombin-induced release of adenosine triphosphate by platelets. Comparison with release of calcium. Biochemistry 12, 2462–2468.

Detwiler, T. C., Martin, B. M. and Feinman, R. D. (1975) Stimulus-response coupling in the thrombin-platelet interaction. In: Ciba Foundation Symposium 35—Biochemistry and Pharmacology of Platelets (Elliott, K. and Knight, J., eds.) pp. 77–100, Elsevier/North-Holland, Amsterdam.

Diamant, B. (1976) Influence of calcium ions and metabolic energy (ATP) on histamine release. In: Molecular and Biological Aspects of the Acute Allergic Reaction (Johansson, S. G. O., Strandberg, K. and Uvnäs, B., eds.) pp. 255–278, Plenum Press, New York.

Diamant, B. and Krüger, P. G. (1967) Histamine release from isolated rat peritoneal mast cells induced by adenosine-5'-triphosphate. Acta Physiol. Scand. 71, 291–302.

Diamant, B. and Krüger, P. G. (1968) Structural changes of isolated rat peritoneal mast cells induced by adenosine triphosphate and adenosine diphosphate. J. Histochem. Cytochem. 16, 707–716.

Dias Da Silva, W. and Lepow, I. H. (1967) Complement as a mediator of inflammation. II. Biological properties of anaphylatoxin prepared with purified components of human complement. J. Exp. Med. 125, 921–946.

Dierich, M. P. and Reisfeld, R. A. (1975) C3 receptors on lymphoid cells: isolation of active membrane fragments and solubilization of receptor complexes. J. Immunol. 114, 1676–1682.

Dingle, J. T. (1969) The extracellular secretion of lysosomal enzymes. In: Lysosomes in Biology and Pathology (Dingle, J. T. and Fell, H. B., eds.) Vol. 2, Chap. 14, pp. 421–435, Elsevier/North-Holland, Amsterdam.

Doe, W. F., Yang, S. T. and Henson, P. M. (1977) Lipopolysaccharide-induced secretion from macrophages: relationship to cytolytic activation. In press.

Douglas, W. W. and Ueda, Y. (1973) Mast cell secretion (histamine release) induced by 48/80: calcium-dependent exocytosis inhibited strongly by cytochalasin only when glycolysis is rate-limiting. J. Physiol. (London) 234, 97P–98P.

Drobis, J. D. and Siraganian, R. P. (1976) Histamine release from cultured human basophils: Lack of histamine resynthesis after antigenic release. J. Immunol. 117, 1049–1053.

Dunham, P. B., Goldstein, I. M. and Weissmann, G. (1974) Potassium and amino acid transport in human leukocytes exposed to phagocytic stimuli. J. Cell Biol. 63, 215–226.

Edelman, G. M. (1976) Surface modulation in cell recognition and cell growth. Some new hypotheses on phenotypic alteration and transmembranous control of cell surface receptors. Science 192, 218–226.

Eden, A., Miller, G. W. and Nussenzweig. V. (1973) Human lymphocytes bear membrane receptors for C3b and C3d. J. Clin. Invest. 52, 3239–3242.

Ehrlich, P. (1877) Beiträge zur kenntnis der anilinfärburgen undihrer verwendung in der mikroskopischen technik. Arch. Mikr. Anat. 13, 263.

Ehrlich, P. (1891) Farbananalytische Untersuchungen zur Histologie und linik des Blutes. Hirschwald, Berlin.

Esmon, C. T., Stenflo, J., Suttie, J. W. and Jackson, C. M. (1976) A new vitamin K-dependent protein. A phospholipid-binding zymogen of a serine esterase. J. Biol. Chem. 251, 3052–3056.

Estensen, R. D. and White, J. G. (1974) Ultrastructural features of the platelet response of phorbol myristate acetate. Am. J. Pathol. 74, 441–452.

488

Estensen, R. D., Hill, H. R., Quie, P. G., Hogan, N. and Goldberg, N. D. (1973) Cyclic GMP and cell movement. Nature (London) 245, 458–460.

Estensen, R. D., White, J. G. and Holmes, B. (1974) Specific degranulation of human polymorphonuclear leukocytes. Nature (London) 248, 347–348.

Estridge, M. (1977) Polypeptides similar to the α and β subunits of tubulin are exposed on the neuronal surface. Nature (London) 268, 60–63.

Evans, G. and Mustard, J. F. (1968) Platelet surface reaction and thrombosis. Surgery 64, 273–280.

Fearn, H. C., Smith, C. and West, G. B. (1964) Capillary permeability responses to snake venoms. J. Pharm. Pharmacol. 16, 79–84.

Feinman, R. D. and Detwiler, T. C. (1974) Platelet secretion induced by divalent cation ionophores. Nature (London) 249, 172–173.

Feinstein, M. B. and Fraser, C. (1975) Human platelet secretion and aggregation induced by calcium ionophores. Inhibition by PGE_1 and dibutyryl cyclic AMP. J. Gen. Physiol. 66, 561–581.

Feldberg, W. and Kellaway, C. H. (1938) Liberation of histamine and formation of lysocithin-like substances by cobra venom. J. Physiol. (London) 94, 187–226.

Fernandez, H. N., Henson, P. M., Otani, A and Hugli, T. E. (1977) Chemotactic response to human C3a and C5a anaphylatoxins I. Evaluation of C3a and C5a leukotaxis $in vitro$ and under simulated $in vivo$ conditions. J. Immunol. In Press.

Fernandez, S. M. and Berlin, R. D. (1976) Cell surface distribution of lectin receptors determined by resonance energy transfer. Nature (London) 264, 411–415.

Fillion, G. M. B., Slorach, S. A. and Uvnäs, B. (1970) The release of histamine, heparin and granule protein from rat mast cells treated with compound 48/80 $in vitro$. Acta Physiol. Scand. 78, 547–560.

Foreman, J. C. and Garland, L. G. (1974) Desensitization in the process of histamine secretion induced by antigen and dextran. J. Physiol. (London) 239, 381–391.

Foreman, J. C. and Mongar, J. L. (1975) Calcium and the control of histamine secretion from mast cells. In: Calcium Transport in Contraction and Secretion. (Carafoli, E., Clementi, F., Drabikowski, W. and Margreth, A., eds.) pp. 175–184. Elsevier/North-Holland, Amsterdam.

Foreman, J. C., Garland, L. G. and Mongar, J. L. (1977) The role of Ca^{2+} in the secretory process-model studies in mast cells. In: Calcium in Biological Systems (Duncan, C. J., ed.) Cambridge Univ. Press, Cambridge. In press.

Foreman, J. C., Mongar, J. L. and Gomperts, B. D. (1973) Calcium ionophores and movement of calcium ions following the physiological stimulus to a secretory process. Nature (London) 245, 249–251.

Fredholm, B. and Haegermark, Ö. (1967) Histamine release from rat mast cells induced by a mast cell degranulating fraction in bee venom. Acta Physiol. Scand. 69, 304–312.

Friedman, F. and Detwiler, T. C. (1975) Stimulus-secretion coupling in platelets. Effects of drugs on secretion of adenosine 5'-triphosphate. Biochemistry 14, 1315–1320.

Fruhman, G. J. (1973) $In vitro$ ingestion of zymosan particles by mast cells. J. Reticuloend. Soc. 13, 424–435.

Gale, R. P. and Zighelboim, J. (1974) Modulation of polymorphonuclear leukocyte-mediated antibody-dependent cellular cytotoxicity. J. Immunol. 113, 1793–1800.

Gale, R. P. and Zighelboim, J. (1975) Polymorphonuclear leukocytes in antibody-dependent cellular cytotoxicity. J. Immunol. 114, 1047–1051.

Gallin, J. I. and Quie, P. G. (1977) (eds.) Leukocyte Chemotaxis: Methodology, Physiology, and Clinical Applications. Raven Press, New York.

Gallin, J. I. and Rosenthal, A. S. (1974) The regulatory role of divalent cations in human granulocyte chemotaxis. Evidence for an association between calcium exchanges and microtubule assembly. J. Cell Biol. 62, 594–609.

Gallin, J. I. and Wolff, S. M. (1975) Leucocyte chemotaxis: physiological considerations and abnormalities. Clin. Haematol. 4, 567–607.

Gallin, J. I., Durocher, J. R. and Kaplan, A. P. (1975) Interaction of leukocyte chemotactic factors with the cell surface. I. Chemotactic factor-induced changes in human granulocyte surface charge. J. Clin. Invest. 55, 967–974.

Gallin, J. I., Gallin, E. K., Malech, H. L. and Cramer, E. B. (1977) Structural and ionic events during leukocyte chemotaxis. Kroc Foundation. In press.

Ganguly, P. (1974) Binding of thrombin to human platelets. Nature (London) 247, 306–307.

Garland, L. G. and Mongar, J. L. (1974) Inhibition by cromoglycate of histamine release from rat peritoneal mast cells induced by mixtures of dextran, phosphatidylserine and calcium ions. Brit. J. Pharmacol. 50, 137–143.

Gerrard, J. M., Townsend, D., Stoddard, S., Witkop, C. J., Jr. and White, J. G. (1977) The influence of prostaglandin G_2 on platelet ultrastructure and platelet secretion. Am. J. Pathol. 86, 99–116.

Gifford, R. H. and Malawista, S. E. (1972) The nitroblue tetrazolium reaction in human granulocytes adherent to a surface. Yale J. Biol. Med. 45, 119–131.

Gigli, I. and Nelson, R. A., Jr. (1968) Complement dependent immune phagocytosis. I. Requirements for C'1, C'4, C'2, C'3. Exp. Cell Res. 51, 45–67.

Gillespie, E. (1975) Microtubules, cyclic AMP, calcium and secretion. Ann N.Y. Acad. Sci. 253, 771–779.

Gillespie, E. and Lichtenstein, L. M. (1972a) Heavy water enhances IgE-mediated histamine release from human leukocytes: Evidence for microtubule involvement. Proc. Soc. Exp. Biol. Med. 140, 1228–1230.

Gillespie, E. and Lichtenstein, L. M. (1972b) Histamine release from human leukocytes: studies with deuterium oxide, colchicine and cytochalasin B. J. Clin. Invest. 51, 2941–2947.

Gillespie, E., Levine, R. J. and Malawista, S. E. (1968) Histamine release from rat peritoneal mast cells: Inhibition by colchicine and potentiation by deuterium oxide. J. Pharmacol. Exp. Ther. 164, 158–165.

Ginsberg, M. H. and Henson, P. M. (1977) Enhancement of platelet response to immune complexes and IgG aggregates by lipid A rich bacterial lipopolysaccharides. J. Exp. Med. 147, 207–218.

Ginsberg, M. H., Kozin, F., Chow, D., May, J. and Skosey, J. L. (1977a) The adsorption of polymorphonuclear leukocyte lysosomal enzyme activity to monosodium urate crystals. Arthr. Rheum. 20, 1538–1542.

Ginsberg, M. H., Kozin, F., O'Malley, M. and McCarty, D. J. (1977b) Release of platelet constitutents by monosodium urate crystals. J. Clin. Invest. 60, 999–1007.

Glynn, M. F., Movat, H. Z., Murphy, E. A. and Mustard, J. F. (1965) Study of platelet adhesiveness and aggregation with latex particles. J. Lab. Clin. Med. 65, 179–201.

Goetzl, E. J. and Austen, K. F. (1974) Stimulation of human neutrophil leukocyte aerobic glucose metabolism by purified chemotactic factors. J. Clin. Invest. 53, 591–599.

Goetzl, E. J., Woods, J. M. and Gorman, R. G. (1977) Stimulation of human eosinophil and neutrophil polymorphonuclear leukocyte chemotaxis and random migration by 12-L-hydroxyl-5,8,10,14-eicosatetraenoic acid. J. Clin. Invest. 59, 179–183.

Goldstein, I. M. (1974) Lysosomal hydrolases and inflammatory materials. In: Mediators of Inflammation, (Weissman, G., ed.) pp. 51–84, Plenum Press, New York.

Goldstein, I. M. (1976) Polymorphonuclear leukocyte lysosomes and immune tissue injury. Progr. Allergy 20, 301–340.

Goldstein, I. M. and Weissmann, G. (1974) Generation of C5-derived lysosomal enzyme-releasing activity (C5a) by lysates of leukocyte lysosomes. J. Immunol. 113, 1583–1588.

Goldstein, I. M., Brai, M., Osler, A. G. and Weissmann, G. (1973a) Lysosomal enzyme release from human leukocytes: mediation by the alternate pathway of complement activation. J. Immunol. 111, 33–37.

Goldstein, I. M., Hoffstein, S., Gallin, J. and Weissmann, G. (1973b) Mechanisms of lysosomal enzyme release from human leukocytes: microtubule assembly and membrane fusion induced by a component of complement. Proc. Nat. Acad. Sci. U.S.A. 70, 2916–2920.

Goldstein, I. M., Horn, J. K., Kaplan, H. B. and Weissmann, G. (1974) Calcium-induced lysozyme secretion from human polymorphonuclear leukocytes. Biochem. Biophys. Res. Commun. 60, 807–812.

Goldstein, I. M., Hoffstein, S. T. and Weissmann, G. (1975a) Mechanisms of lysosomal enzyme release from human polymorphonuclear leukocytes. Effects of phorbol myristate acetate. J. Cell Biol. 66, 647–652.

Goldstein, I. M., Roos, D., Kaplan, H. B. and Weissmann, G. (1975b) Complement and immunoglobulins stimulate superoxide production by human leukocytes independently of phago-cytosis. J. Clin. Invest. 56, 1155–1163.

Goldstein, I. M., Hoffstein, S. T. and Weissmann, G. (1975c) Influence of divalent cations upon complement-mediated enzyme release from human polymorphonuclear leukocytes. J. Immunol. 115, 665–670.

Goldstein, I. M., Kaplan, H. B., Radin, A. and Frosch, M. (1976a) Independent effects of IgG and complement upon human polymorphonuclear leukocyte function. J. Immunol. 117: 1282–1287.

Goldstein, I. M., Roos, D., Weissmann, G. and Kaplan, H. B. (1976b) Influence of corticosteroids on human polymorphonuclear leukocyte function in vitro. Reduction of lysosomal enzyme release and superoxide production. Inflammation 1, 305–315.

Goldstein, I. M., Cerqueira, M., Lind, S. and Kaplan, H. B. (1977a) Evidence that the superoxide-generating system of human leukocytes is associated with the cell surface. J. Clin. Invest. 59, 249–254.

Goldstein, I. M., Malmsten, C. L., Kaplan, H., Samuelsson, B. and Weissmann, G. (1977b) Throm-boxane generation by phagocytosing human leukocytes: inhibition by glucocorticoids and superoxide dismutase. In press.

Goth, A. and Knoohuizen, M. (1962) Tissue thromboplastins and histamine release from mast cells. Life Sci. 1, 459–465.

Goth A., Adams, H. R. and Knoohuizen, M. (1971) Phosphatidylserine: selective enhancer of his-tamine release. Science 173, 1034–1035.

Graham, H. T., Lowry, O. H., Wheelwright, F., Lenz, M. A. and Parish, H. H. (1955) Distribution of histamine among leukocytes and platelets. Blood 10, 467–481.

Grant, J. A. and Lichtenstein, L. M. (1974) Release of slow-reacting substance of anaphylaxis from human leukocytes. J. Immunol. 112, 897–904.

Grant, J. A., Dupree, E., Goldman, A. S., Schultz, D. R. and Jackson, A. L. (1975) Com-plement-mediated release of histamine from human leukocytes. J. Immunol. 114, 1101–1106.

Greaves, M. and Janossy, G. (1972) Elicitation of selective T and B lymphocyte responses by cell sur-face binding ligands. Transplant. Rev. 11, 87–130.

Greengard, P. (1976) Possible role for cyclic nucleotides and phosphorylated membrane proteins in postsynaptic actions of neurotransmitters. Nature (London) 260, 101–108.

Grette, K. (1962) The Mechanism of Thrombin Catalyzed Hemostatic Reactions in Blood Platelets. Norwegian University Press, Oslo.

Grey, H. M., Anderson, C. L., Heusser, C. H., Borthistle, B. K., Von Eschen, K. B. and Chiller, J. M. (1977) Structural and functional heterogeneity of Fc receptors. Proc. Cold Spring Harbor Symp. 61, 315–321.

Griffin, F. M., Jr., Griffin, J. A. and Silverstein, S. C. (1976) Studies on the mechanism of phagocytosis. II. The interaction of macrophages with anti-immunoglobulin IgG-coated bone marrow-derived lymphocytes. J. Exp. Med. 144, 788–809.

Griffin, F. M., Jr., Griffin, J. A., Leider, J. E. and Silverstein, S. C. (1975) Studies on the mechanism of phagocytosis. I. Requirements for circumferential attachment of particle-bound ligands to specific receptors on the macrophage plasma membrane. J. Exp. Med. 142, 1263–1282.

Grosman, N. and Diamant, B. (1975) The influence of phosphatidylserine on the release of his-tamine from isolated rat mast cells induced by different agents. Agents Actions 5, 296–301.

Gryglewski, R. J., Bunting, S., Moncada, S., Flower, R. J. and Vane, J. R. (1976) Arterial walls are protected against deposition of platelet thrombin by a substance (PGX) which they make from prostaglandin endoperoxides. Prostaglandins 12, 685–713.

Guccione, M. A., Packham, M. A., Kinlough-Rathbone, R. L., Perry, D. W. and Mustard, J. F. (1976) Reactions of polylysine with human platelets in plasma and suspensions of washed platelets. Thrombos. Hemostas. 36, 360–375.

Habermann, E. (1972) Bee and wasp venoms. The biochemistry and pharmacology of their peptides and enzymes are reviewed. Science 177, 314–322.

Hall, J. E. and Simon, S. A. (1976) A simple model for calcium-induced exocytosis, Biochim. Biophys. Acta 436, 613–616.

Hamberg, M., Svensson, J. and Samuelsson, B. (1974a) Prostaglandin endoperoxides. A new concept concerning the mode of action and release of prostaglandins. Proc. Nat. Acad. Sci. U.S.A. 71, 3824–3828.

Hamberg, M., Svensson, J., Wakabayashi, T. and Samuelsson, B. (1974b) Isolation and structure of two prostaglandin endoperoxides that cause platelet aggregation. Proc. Nat. Acad. Sci. U.S.A. 71, 345–349.

Hamberg, M., Svensson, J. and Samuelsson, B. (1975) Thromboxanes: a new group of biologically active compounds derived from prostaglandin endoperoxides. Proc. Nat. Acad. Sci. U.S.A. 72, 2994–2998.

Handin, R. I. and Cohen, H. J. (1976) Purification and binding properties of human platelet factor four. J. Biol. Chem. 251, 4273–4282.

Handin, R. I., Karabin, R. and Boxer, G. J. (1977) Enhancement of platelet function by superoxide anion. J. Clin. Invest. 59, 959–965.

Harbury, C. B. and Schrier, S. L. (1974) The effect of a partial release reaction on subsequent platelet function. J. Lab. Clin. Med. 83, 877–886.

Harper, E., Simons, E. R., Chesney, C. M. and Coleman, R. W. (1975) The effect of chemical or enzymatic modifications upon the ability of collagen to form multimers and to initiate platelet aggregation. Thrombos Res. 7, 113–122.

Hartwig, J. H. and Stossel, T. P. (1976) Interactions of actin, myosin and an actin-binding protein of rabbit pulmonary macrophages, III. Effects of cytochalasin B. J. Cell Biol. 71, 295–303.

Haslam, R. J. (1975) Roles of cyclic nucleotides in platelet function, In: Ciba Foundation Symposium 35 — Biochemistry and Pharmacology of Platelets (Elliott, K. and Knight, J., eds.), pp. 121–151. Elsevier/North-Holland, Amsterdam.

Haslam, R. J., Davidson, M. M. L. and McClenaghan, M. D. (1975) Cytochalasin B, the blood platelet release reaction and cyclic GMP. Nature (London) 253, 455–457.

Hastie, R. (1974) A study of the ultrastructure of human basophilic leukocytes. Lab. Invest. 31, 223–231.

Hastie, R., Levy, D. A. and Weiss, L. (1977) The antigen-induced degranulation of basophil leukocytes from atopic subjects studied by electron microscopy. Lab. Invest. 36, 173–182.

Hastie, R. (1971) The antigen-induced degranulation of basophil leucocytes from atopic subjects, studied by phase-contrast microscopy. Clin. Exp. Immunol. 8, 45–61.

Hawiger, J., Collins, R. D., Horn, R. G. and Koenig, M. G. (1969) Interaction of artificial phospholipid membranes with isolated polymorphonuclear leukocytic granules. Nature (London) 222, 276–278.

Hawkins, D. (1971) Biopolymer membrane: A model system for the study of the neutrophilic leukocyte response to immune complexes. J. Immunol. 107, 344–352.

Hawkins, D. (1972) Neutrophilic leukocytes in immunologic reactions: Evidence for the selective release of lysosomal constituents. J. Immunol. 108, 310–317.

Hawkins, D. (1974a) Neutrophilic leukocytes in immunologic reactions in vitro. III. Pharmacologic modulation of lysosomal constituent release. Clin. Immunol. Immunopathol. 2, 141–152.

Hawkins, D. (1974b) The effect of lectins on lysosomal enzyme release from neutrophilic leukocytes. J. Immunol. 113, 1864–1871.

Hawkins, D. and Peeters, S. (1971) The response of polymorphonuclear leukocytes to immune complexes in vitro. Lab. Invest. 24, 483–491.

Henson, P. M. (1969) The adherence of leucocytes and platelets induced by fixed IgG antibody or complement. Immunology 16, 107–121.

Henson, P. M. (1970a) Mechanisms of release of constituents from rabbit platelets by antigen-antibody complexes and complement. I. Lytic and nonlytic reactions. J. Immunol. 105, 476–489.

Henson, P. M. (1970b) Mechanisms of release of constituents from rabbit platelets by antigen-antibody complexes and complement. II. Interactions of platelets with neutrophils. J. Immunol. 105, 490–501.

Henson, P. M. (1970c) Release of vasoactive amines from rabbit platelets induced by antiplatelet antibody in the presence and absence of complement. J. Immunol. 104, 924–934.

Henson, P. M. (1970d) Release of vasoactive amines from rabbit platelets induced to sensitized mononuclear leukocytes and antigen. J. Exp. Med. 131, 287–306.

Henson, P. M. (1971a) The immunologic release of constituents from neutrophil leukocytes. II. Mechanisms of release during phagocytosis, and adherence to non-phagocytosable surfaces. J. Immunol. 107, 1547–1557.

Henson, P. M. (1971b) Interaction of cells with immune complexes. Adherence release of constituents and tissue injury. J. Exp. Med. 134, 114s–135s.

Henson, P. M. (1972) Pathologic mechanisms in neutrophil-mediated injury. Am. J. Pathol. 68, 593–606.

Henson, P. M. (1973) Mechanisms of release of granule enzymes from human neutrophils phagocytosing aggregated immunoglobulin. An electron microscopic study. Arthr. Rheum. 16, 208–216.

Henson, P. M. (1974a) Mechanisms of activation and secretion by platelets and neutrophils. In: Progress in Immunology II. (Brent, L. and Holborow, J., eds.) Vol. 2, pp. 95–105. Elsevier/North-Holland, Amsterdam.

Henson, P. M. (1974b) Mechanisms of mediator release from inflammatory cells. In: Mediators of Inflammation (Weissmann, G., ed.) pp. 9–50, Plenum Press, New York.

Henson, P. M. (1974c) Complement dependent platelet and polymorphonuclear leukocyte reactions. Transplant. Proc. 6, 27–31.

Henson, P. M. (1976a) Secretion of lysosomal enzymes induced by immune complexes and complement. In: Lysosomes in Biology and Pathology (Dingle, J. T. and Dean, R. T., eds.) vol. 5, pp. 99–126, Elsevier/North-Holland Publishing Company, New York.

Henson, P. M. (1976b) Membrane receptors on neutrophils. Immunol. Commun. 5, 757–774.

Henson, P. M. (1976c) Activation and desensitization of platelets by platelet-activating factor (PAF) derived from IgE-sensitized basophils. I. Characteristics of the secretory response. J. Exp. Med. 143, 937–952.

Henson, P. M. (1977) Activation of rabbit platelets by platelet-activating factor (PAF) derived from IgE-sensitized basophils. Characteristics of the aggregation and its dissociation from secretion. J. Clin. Invest. 60, 481–490.

Henson, P. M. and Landes, R. (1976) Activation of platelets by platelet activating factor (PAF) derived from IgE-sensitized basophils. IV. PAF does not activate platelet factor 3. (PF$_3$) Brit. J. Haematol. 34, 269–282.

Henson, P. M. and Oades, Z. G. (1975) Stimulation of human neutrophils by soluble and insoluble immunoglobulin aggregates. Secretion of granule constituents and increased oxidation of glucose. J. Clin. Invest. 56, 1053–1061.

Henson, P. M. and Oades, Z. G. (1976) Activation of platelets by platelet-activating factor (PAF) derived from IgE sensitized basophils. II. The role of serine proteases, cyclic nucleotides and contractile elements in PAF-induced secretion. J. Exp. Med. 143, 953–968.

Henson, P. M. and Pinckard, R. N. (1977) Basophil derived platelet activating factor (PAF) as an *in vivo* mediator of acute allergic reactions: Demonstration of specific desensitization of platelets to PAF during IgE-induced anaphylaxis in the rabbit. J. Immunol. 119, 2179–2184.

Henson, P. M. and Spiegelberg, H. L. (1973) Release of serotonin from human platelets induced by aggregated immunoglobulins of different classes and subclasses. J. Clin. Invest. 52, 1282–1288.

Henson, P. M., Henson, J. E. and Becker, E. L. (1975) Ultrastructural alterations during ATP-induced secretion of lysosomal enzymes from rabbit polymorphonuclear leukocytes. Inflammation 1, 85–91.

Henson, P. M., Gould, D. and Becker, E. L. (1976) Activation of stimulus-specific serine esterases (proteases) in the initiation of platelet secretion. I. Demonstration with organophosphorus inhibitors. J. Exp. Med. 144, 1657–1673.

Henson, P. M., Pinckard, R. N., Shaw, J. O. and Hanahan, D. J. (1977a) Platelet activating factor. Submitted for publication.

Henson, P. M., Zanolari, B. and Ryan, K. (1977b) Manuscript in preparation.

Henson, P. M., Landes, R., and Becker, E. L. (1977c) Activation of stimulus-specific serine esterases (proteases) in the initiation of platelet secretion. II. Demonstration with amino acid esters. Submitted for publication.

Hitchcock, S. E. (1977) Regulation of motility in nonmuscle cells. J. Cell Biol. 74, 1–15.

Hoffstein, S. and Weissmann, G. (1975) Mechanisms of lysosomal enzyme release from leukocytes. IV. Interaction of monosodium urate crystals with dogfish and human leukocytes. Arthr. Rheum. 18, 153–165.

Hoffstein, S., Zurier, R. B. and Weissmann, G. (1974) Mechanisms of lysosomal enzyme release from human leukocytes. III. Quantitative morphologic evidence for an effect of cyclic nucleotides and colchicine on degranulation. Clin. Immunol. Immunopathol. 3, 201–217.

Hoffstein, S., Soberman, R., Goldstein, I. and Weissmann, G. (1977a) Concanavalin A induces microtubule assembly and specific granule discharge in human polymorphonuclear leukocytes. J. Cell Biol. 68, 781–787.

Hoffstein, S., Goldstein, I. M. and Weissmann, G. (1977b) Role of microtubule assembly in lysosomal enzyme secretion from human polymorphonuclear leukocytes: A reevaluation. J. Cell Biol. 73, 242–256.

Högberg, B. and Uvnäs, B. (1960) Further observations on the disruption of rat mesentery mast cells caused by compound 48/80, antigen-antibody reaction, lecithinase A and decylamine. Acta Physiol. Scand. 48, 133–145.

Holmsen, H. (1974) Are platelet shape change, aggregation and release reaction tangible manifestations of one basic platelet function? In: Platelets, Production, Function, Transfusion and Storage. (Baldini, M. G. and Ebbe, S., eds.), pp. 207–220, Grune and Stratton, New York.

Holmsen, H. (1975) Biochemistry of the platelet release reaction. In: Ciba Foundation Symposium 35—Biochemistry and Pharmacology of Platelets (Elliott, K. and Knight, J., eds.) pp. 175–205. Elsevier/North-Holland, Amsterdam.

Holmsen, H., Day, H. J. and Stormorken, H. (1969) The blood platelet release reaction. Scand. J. Haematol. (Suppl.) 8, 1–26.

Holmsen, H., Østvold, A-C. and Day, H. J. (1973) Behaviour of endogenous and newly absorbed serotonin in the platelet release reaction. Biochem. Pharmacol. 22, 2599–2608.

Hook, W. A. and Siraganian, R. P. (1977) Complement-induced histamine release from human basophils. III. Effect of pharmacologic agents. J. Immunol. 118, 679–684.

Hook, W. A., Brown, H. and Oppenheim, J. J. (1974) Histamine release from human leukocytes by concanavalin A and other mitogens. Proc. Soc. Exp. Biol. Med. 147, 659–663.

Hook, W. A., Siraganian, R. P. and Wahl, S. M. (1975) Complement-induced histamine release from human basophils. I. Generation of activity in human serum. J. Immunol. 114, 1185–1190.

Horsfield, G. I. (1965) The effect of compound 48/80 on the rat mast cell. J. Path. Bacteriol. 90, 599–605.

Hovig, T. (1963) Release of a platelet aggregating substance (adenosine diphosphate) from rabbit blood platelets induced by saline "extract" of tendons. Thrombos Diath. Haemorrh. 9, 264–278.

Hsu, L. S. and Becker, E. L. (1975a) Volume decrease of glycerinated polymorphonuclear leukocytes induced by ATP and Ca²⁺. Exp. Cell Res. 91, 469–473.

Hsu, L. S. and Becker, E. L. (1975b) Volume changes induced in rabbit polymorphonuclear leukocytes by chemotactic factor and cytochalasin B. Am. J. Pathol. 81, 1–14.

Huber, H., Polley, M. J., Linscott, W. D., Fudenberg, H. H. and Müller-Eberhard, H. J. (1968) Human monocytes: Distinct receptor sites for the third component of complement and for immunoglobulin G. Science 162, 1281–1283.

Humphrey, J. H. and Jaques, R. (1955) The release of histamine and 5-hydroxytryptamine (serotonin) from platelets by antigen-antibody reactions (in vitro). J. Physiol. (London) 128, 9–27.

Ignarro, L. J. (1973) Neutral protease release from human leukocytes regulated by neurohormones and cyclic nucleotides. Nature New Biol. 245, 151–154.

Ignarro, L. J. and George, W. J. (1974) Hormonal control of lysosomal enzyme release from human neutrophils: Elevation of cyclic nucleotide levels by autonomic neurohormones. Proc. Nat. Acad. Sci. U.S.A. 71, 2027–2031.

Ishizaka, K. and Ishizaka, T. (1969) Immune mechanisms of reversed type reaginic hypersensitivity. J. Immunol. 103, 588–595.

Ishizaka, K., Tomioka, H. and Ishizaka, T. (1970) Mechanisms of passive sensitization. I. Presence of IgE and IgG molecules on human leukocytes. J. Immunol. 105, 1459–1467.

Ishizaka, T. (1976) Functions and development of cell receptors for IgE. In: Molecular and Biological Aspects of the Acute Reaction (Johansson, S. G. O., Strandberg, K. and Uvnäs, B., eds.) pp. 199–216, Plenum Press, New York.

Ishizaka, T. and Ishizaka, K. (1975) Cell surface IgE on human basophil granulocytes, Ann. N.Y. Acad Sci. 254, 462–475.

Ishizaka, T., De Bernardo, R., Tomioka, H., Lichtenstein, L. M. and Ishizaka, K. (1972) Identification of basophil granulocytes as a site of allergic histamine release. J. Immunol. 108, 1000–1008.

Israels, E. D., Nisli, G., Paraskevas, F. and Israels, L. G. (1973) Platelet Fc receptor as a mechanism for Ag-Ab complex-induced platelet injury. Thrombos. Diath. Haemorrh. 29, 434–444.

Jaffe, L. F. (1977) Electrophoresis along cell membranes. Nature (London) 265, 600–602.

Jaffe, R. and Deykin, D. (1974a) Platelet aggregation by different types of collagen. Circulation Suppl. III, 289(a).

Jaffe, R. and Deykin, D. (1974b) Evidence for a structural requirement for the aggregation of platelets by collagen. J. Clin. Invest. 53:875–883.

Jamieson, G. A., Urban, C. L. and Barber, A. J. (1971) Enzymatic basis for platelet: collagen adhesion as the primary step in haemostasis. Nature New Biol. 234, 5–7.

Janoff, A., Schaefer, S., Sherer, J. and Bean, M. A. (1965) Mediators of inflammation in leukocyte lysosomes. II. Mechanism of action of lysosomal cationic protein upon vascular permeability in the rat. J. Exp. Med. 122, 841–851.

Jenkins, C. S. P., Packham, M. A., Kinlough-Rathbone, R. L. and Mustard, J. F. (1971) Interactions of polylysine with platelets. Blood 37, 395–412.

Jensen, M. S. and Bainton, D. F. (1973) Temporal changes in pH within the phagocytic vacuole of the polymorphonuclear neutrophilic leukocyte. J. Cell Biol. 56, 379–388.

Jobin, F., Lapointe, F. and Gagnon, F. (1971) Platelet reactions and immune processes. VI. The effect of immunoglobulins and other plasma proteins on platelet surface interactions. Thrombos. Diath. Haemorrh. 25, 86–97.

Johnson, A. R. and Moran, N. R. (1969a) Release of histamine from rat mast cells: a comparison of the effects of 48/80 and two antigen-antibody systems. Fed. Proc. 28, 1716–1720.

Johnson, A. R. and Moran, N. C. (1969b) Selective release of histamine from rat mast cells by compound 48/80 and antigen. Am. J. Physiol. 216, 453–459.

Johnson, A. R. and Moran, N. C. (1970) Inhibition of the release of histamine from rat mast cells: The effect of cold and adrenergic drugs on release of histamine by compound 48/80 and antigen. J. Pharmacol. Exp. Ther. 175, 632–640.

Johnson, A. R., Hugli, T. E. and Müller-Eberhard, H. J. (1975) Release of histamine from rat mast cells by the complement peptides C3a and C5a. Immunology 28, 1067–1080.

Johnston, R. B., Jr. and Lehmeyer, J. E. (1976) Elaboration of toxic oxygen by-products by neutrophils in a model of immune complex disease. J. Clin. Invest. 57, 836–841.

Johnston, R. B., Jr. and McPhail, L. (1977) The patient with impaired phagocyte function. In: Immunology of Human Infection (Nahmias, A. J. and O'Reilly, R. J., eds.) Plenum, New York, In Press.

Johnston, R. B., Jr., Keele, B. B. Jr., Misra, H. P., Lehmeyer, J. E., Webb, L. S., Baehner, R. L. and Rajagopalan, K. V. (1975) The role of superoxide anion generation in phagocytic bactericidal activity. Studies with normal and chronic granulomatous disease leukocytes. J. Clin. Invest. 55, 1357–1372.

Jones, D. G., Richardson, D. L. and Kay, A. B. (1977) Neutrophil accumulation in vivo following the administration of chemotactic factors. Brit. J. Haematol. 35, 19–24.

Kaliner, M. and Austen, K. F. (1974) Cyclic AMP, ATP and reversed anaphylactic histamine release from rat mast cells. J. Immunol. 112, 664–674.

Kang, A. H., Beachey, E. H. and Katzman, R. L. (1974) Interaction of an active glycopeptide from chick skin collagen (α1-CB5) with human platelets. J. Biol. Chem. 249, 1054–1059.

Kanno, T., Cochrane, D. E. and Douglas, W. W. (1973) Exocytosis (secretory granule extrusion) induced by injection of calcium into mast cells. Can. J. Physiol. Pharmacol. 51, 1001–1004.

Kaplan, A. P., Kay, A. B. and Austen, K. F. (1972) A prealbumin activator of prekallikrein. III. Appearance of chemotactic activity for human neutrophils by the conversion of human prekallikrein to kallikrein. J. Exp. Med. 135, 81–97.

Kaplan, J. G. and Bona, C. (1974) Proteases as mitogens. The effect of trypsin and pronase on mouse and human lymphocytes. Exp. Cell Res. 88, 388–394.

Karnovsky, M. L. (1962) Metabolic basis of phagocytic activity. Physiol. Rev. 42, 143–168.

Karpatkin, S., Strick, N., Karpatkin, M. B. and Siskind, G. W. (1972) Cumulative experience in the detection of antiplatelet antibody in 234 patients with idiopathic thrombocytopenic purpura, systemic lupus erythematosus, and other clinical disorders. Am. J. Med. 52, 776–785.

Kay, A. B. and Austen, K. F. (1971) The IgE-mediated release of an eosinophil leukocyte chemotactic factor from human lung. J. Immunol. 107, 899–902.

Kay, A. B. and Austen, K. F. (1972) Chemotaxis of human basophil leucocyte. Clin. Exp. Immunol. 11, 557–563.

Kay, A. B., Pepper, D. S. and Ewart, M. R. (1973) Generation of chemotactic activity for leukocytes by the action of thrombin on human fibrinogen. Nature New Biol. 243, 56–57.

Keller, H. U., Wilkinson, P. C., Abercrombie, M., Becker, E. L., Hirsch, J. G., Miller, M. E., Scott-Ramsey, W. and Zigmond, S. (1977) A proposal for the definition of terms related to locomotion of leukocytes and other cells. Clin. Exp. Immunol. 27, 377–380.

Keller, R. (1966) Tissue Mast Cells in Immune Reactions (Kallós, P., Goodman, H. C. and Inderbitzen, T., eds.), pp. 38–39. Elsevier/North-Holland, New York.

Keller, R. (1968) Interrelations between different types of cells. II. Histamine-release from the mast cells of various species by cationic polypeptides of polymorphonuclear leukocyte lysosomes and other cationic compounds. Int. Arch. Allergy 34, 139–144.

Keller, R. (1973) Concanavalin A, a model "antigen" for the in vitro detection of cell bound reaginic antibody in the rat. Clin. Exp. Immunol. 13, 139–147.

Kelly, M. T. and White A. (1973) Histamine release induced by human leukocyte lysates: Implication of a specific complement-independent, noncytotoxic reaction. Infect. Immunol. 8, 8–14.

Kinlough-Rathbone, R. L., Chahil, A., Packham, M. A. and Mustard, J. F. (1975) Conditions influencing platelet lysis. Lab. Invest. 32, 352–358.

Kinlough-Rathbone, R. L., Reimers, H. J., Mustard, J. F. and Packham, M. A. (1976) Sodium arachidonate can induce platelet shape change and aggregation which are independent of the release reaction. Science 192, 1011–1012.

Kinlough-Rathbone, R. L., Packham, M. A., Reimers, H. J., Cazenave, J. P. and Mustard, J. F. (1977) Mechanisms of platelet shape change, aggregation, and release induced by collagen, thrombin, or A23187. J. Lab. Clin. Med. 90, 707–719.

Ko, L. and Lagunoff, D. (1976) Depletion of mast cell ATP inhibits complement-dependent cytotoxic histamine release. Exp. Cell Res. 100, 313–321.

König, W. and Ishizaka, K. (1976) Binding of rat IgE with the subcellular components of normal rat mast cells. Immunochemistry 13, 345–353.

König, W., Czarnetzki, B. M. and Lichtenstein, L. M. (1976) Eosinophil chemotactic factor (ECF). II. Release from human polymorphonuclear leukocytes during phagocytosis. J. Immunol. 117, 235–241.

Koza, E. P., Wright, T. E. and Becker, E. L. (1975) Lysosomal enzyme secretion and volume contraction induced in neutrophils by cytochalasin B, chemotactic factor, and A23187. Proc. Soc. Exp. Biol. Med. 149, 476–479.

Kravis, T. C. and Henson, P. M. (1975) IgE-induced release of a platelet activating factor from rabbit lung. J. Immunol. 115, 1677–1681.

Kulczycki, A., Jr., and Metzger, H. (1974) The interaction of IgE with rat basophilic leukemia cells. II. Quantitative aspects of the binding reaction. J. Exp. Med. 140, 1676–1695.

Kulczycki, A., Jr., Isersky, C. and Metzger, H. (1974) The interaction of IgE with rat basophilic leukemia cells. I. Evidence for specific binding of IgE. J. Exp. Med. 139, 600–616.

Kulczycki, A., Jr., McNearney, T. A. and Parker, C. W. (1976) The rat basophilic leukemia cell receptor for IgE. I. Characterization as a glycoprotein. J. Immunol. 117, 661–665.

Lagunoff, D. (1966) Structural aspects of histamine binding: the mast cell granule. In: Mechanisms of Release of Biogenic Amines (Von Euler, U.S., Rosell, S. and Uvnäs, B., eds.) pp. 79–94, Pergamon Press, Oxford.

Lagunoff, D. (1972a) Contributions of electron microscopy to the study of mast cells. J. Invest. Derm. 58, 296–311.

Lagunoff, D. (1972b) The mechanism of histamine release from mast cells. Biochem. Pharmacol. 21, 1889–1896.

Lagunoff, D. (1973) Membrane fusion during mast cell secretion. J. Cell Biol. 57, 252–259.

Lagunoff, D. (1974) Analysis of dye binding sites in mast cell granules. Biochemistry 13, 3982–3986.

Lagunoff, D. and Benditt, E. P. (1964) Proteolytic enzymes of mast cells. Ann. N.Y. Acad. Sci. 103, 185–198.

Lagunoff, D. and Chi, E. Y. (1976) Effect of colchicine on rat mast cells. J. Cell Biol. 71, 182–195.

Lagunoff, D. and Wan, H. (1974) Temperature dependence of mast cell histamine secretion. J. Cell Biol. 61, 809–811.

Lagunoff, D., Phillips, M. T., Iseri, O. A., and Benditt, E. P. (1964) Isolation and preliminary characterization of rat mast cell granules. Lab. Invest. 13, 1331–1344.

Lagunoff, D., Chi, E. Y. and Wan, H. (1975) Effects of chymotrypsin and trypsin on rat peritoneal mast cells. Biochem. Pharm. 24: 1573–1578.

Lawson, D., Fewtrell, C., Gomperts, B. and Raff, M. C. (1975) Anti-immunoglobulin-induced histamine secretion by rat peritoneal mast cells studied by immunoferritin electron microscopy. J. Exp. Med. 142, 391–402.

Lawson, D., Raff, M. C., Gomperts. B., Fewtrell, C., and Gilula, N. B. (1976) Molecular events in membrane fusion occurring during mast cell degranulation. In: Molecular and Biological Aspects of the Acute Allergic Reaction (Johansson, S. G. O., Strandberg, K. and Uvnäs, B., eds.) pp. 279–292, Plenum Press, New York.

Lay, W. H., and Nussenzweig, V. (1968) Receptors for complement on leukocytes. J. Exp. Med. 128, 991–1009.

Le Breton, G. C., Sandler, W. C., and Feinberg, H. (1976) The effect of D_2O and chlortetracycline on ADP-induced platelet shape change and aggregation. Thrombos Res. 8, 477–485.

Leffell, M. S. and Spitznagel, J. K. (1974) Intracellular and extracellular degranulation of human polymorphonuclear azurophil and specific granules induced by immune complexes. Infect. Immun. 10, 1241–1249.

Leffell, M. S. and Spitznagel, J. K. (1975) Fate of human lactoferrin and myeloperoxidase in phagocytizing human neutrophils: Effects of immunoglobulin G subclasses and immune complexes coated on latex beads. Infect. Immun. 12, 813–820.

Lehmeyer, J. E. and Johnston, R. B., Jr. (1977) Effect of anti-inflammatory drugs and agents that elevate intracellular cyclic AMP on the release of toxic oxygen metabolites by phagocytes: Studies in a model of immune complex disease. Clin. Immunol. Immunopathol. In press.

Lentnek, A. L., Schreiber, A. D. and MacGregor, R. R. (1976) The induction of augmented granulocyte adherence by inflammation. Mediation by a plasma factor. J. Clin. Invest. 57, 1098–1103.

Lett-Brown, M. A., Boetcher, D. A. and Leonard, E. J. (1976) Chemotactic responses of normal human basophils to C5a and to lymphocyte-derived chemotactic factor. J. Immunol. 117, 246–252.

Levine, B. B. (1965) The nature of the antigen-antibody complexes which initiate anaphylactic reactions. I. A quantitative comparison of the abilities of nontoxic univalent, toxic univalent, divalent and multivalent benzylpenicilloyl haptens to evoke passive cutaneous anaphylaxis in the guinea pig. J. Immunol. 94, 111–120.

Levine, S. P. and Wohl, H. (1976) Human platelet factor 4: Purification and characterization by affinity chromatography. Purification of human platelet factor 4. J. Biol. Chem. 251, 324–328.

Levy, D. A. and Carlton, J. A. (1969) Influence of temperature on the inhibition by colchicine of allergic histamine release. Proc. Soc. Exp. Biol. Med. 130, 1333–1336.

Lewis, R. A., Goetzl, E. J., Wasserman, S. I., Valone, F. H., Rubin, R. H. and Austen, K. F. (1975) The release of four mediators of immediate hypersensitivity from human leukemic basophils. J. Immunol. 114, 87–92.

Lewis, R. A., Wasserman, S. I., Goetzl, E. J. and Austen, K. F. (1974) Formation of slow-reacting substance of anaphylaxis in human lung tissue and cells before release. J. Exp. Med. 140, 1133–1146.

Lichtenstein, L. M. (1975) The mechanism of basophil histamine release induced by antigen and by the calcium ionophore A23187. J. Immunol. 114, 1692–1699.

Lichtenstein, L. M. and Gillespie, E. (1973) Inhibition of histamine release by histamine controlled by H2 receptor. Nature (London) 244, 287–288.

Lichtenstein, L. M. and Margolis, S. (1968) Histamine release *in vitro*: inhibition by catecholamines and methylxanthines. Science 161, 902–903.

Lichtenstein, L. M. and Osler, A. G. (1964) Studies on the mechanisms of hypersensitivity phenomena. IX. Histamine release from human leukocytes by ragweed pollen antigen. J. Exp. Med. 120, 507–530.

Lichenstein, L. M. and Osler, A. G. (1966) Studies on the mechanisms of hypersensitivity phenomena. XI. The effect of normal human serum on the release of histamine from human leukocytes by ragweed pollen antigen. J. Immunol. 96, 159–168.

Magro, A. M. (1974) Involvement of IgE in Con A-induced histamine release from human basophils *in vitro*. Nature (London) 249, 572–573.

Magro, A. M. and Alexander, A. (1974) *In vitro* studies of histamine release from rabbit leukocytes by divalent haptens. J. Immunol. 112, 1757–1761.

Malech, H. C. and Gallin, J. I. (1977) Functional analysis of human neutrophil migration. II. Functional role of microtubules and microfilaments in orientation and locomotion. In press.

Malmsten, C., Hamberg, M., Svensson, J. and Samuelsson, B. (1975) Physiological role of an endoperoxide in human platelets: Hemostatic defect due to platelet cyclo-oxygenase deficiency. Proc. Nat. Acad. Sci. U.S.A. 72, 1446–1450.

Mandell, B. F., Spilberg, I. and Lichtman, J. (1977) Inhibition of polymorphonuclear leukocyte capping by a chemotactic factor. J. Immunol. 118, 1375–1378.

Mann, P. R. (1969) An electron microscope study of the degranulation of rat peritoneal mast cells brought about by four different agents. Brit. J. Dermatol. 81, 926–936.

Mantovani, B. (1975) Different roles of IgG and complement receptors in phagocytosis by polymorphonuclear leukocytes. J. Immunol. 115, 15–17.

Mantovani, B., Rabinovitch, M. and Nussensweig, V. (1972) Phagocytosis of immune complexes by macrophages. Different roles of the macrophage receptor sites for complement (C3) and for immunoglobulin (IgG). J. Exp. Med. 135, 780–792.

Marcus, A. J., Ullman, H. L., Safier, L. B. and Ballard, H. S. (1962) Platelet phosphatides. Their fatty acid and aldehyde composition and activity in different clotting systems. J. Clin. Invest. 41, 2198–2212.

Martin, B. M., Feinman, R. D. and Detwiler, T. C. (1975) Platelet stimulation by thrombin and other proteases. Biochemistry 14, 1308–1314.

Martin, B. M., Wasiewski, W. W., Fenton II., J. W. and Detwiler, T. C. (1976) Equilibrium binding of thrombin to platelets. Biochemistry 15, 4886–4893.

Massini, P. and Lüscher, E. F. (1971) The induction of the release reaction in human blood platelets by close cell contact. Thrombos. Diath. Haemorrh. 25, 13–20.

Massini, P. and Lüscher, E. F. (1974) Some effects of ionophores for divalent cations on blood platelets. Comparison with the effects of thrombin. Biochim. Biophys. Acta 372, 109–121.

Massini, P. and Lüscher, E. F. (1976) On the significance of the influx of calcium ions into stimulated human blood platelets. Biochim. Biophys. Acta 436, 652–663.

Mendoza, G. and Metzger, H. (1976) Distribution and valency of receptor for IgE on rodent mast cells and related tumor cells. Nature (London) 264, 548–550.

Metchnikoff, E. (1891) Lectures on the Comparative Pathology of Inflammation. Dover Publications, New York (1968).

Metzger, H. (1974) Effect of antigen binding on the properties of antibody. Adv. Immunol. 18, 169–207.

Metzger, H., Budman, D. and Lucky P. (1976) Interaction of IgE with rat basophilic leukemia cells. V. Binding properties of cell free particles. Immunochemistry 13, 417–423.

Mizel, S. B. and Wilson, L. (1972) Inhibition of the transport of several hexoses in mammalian cells by cytochalasin B. J. Biol. Chem. 247, 4102–4105.

Mohammed, S. F., Whitworth, C., Chuang, H. Y. K., Lundblad, R. L. and Mason, R. G. (1976) Multiple active forms of thrombin: Binding to platelets and effects on platelet function. Proc. Nat. Acad. Sci. U.S.A. 73, 1660–1663.

Moncada, S., Higgs, E. A. and Vane, J. R. (1977) Human arterial and venous tissues generate prostacyclin (Prostaglandin X), a potent inhibitor of platelet aggregation. Lancet 1, 18–20.

Mongar, J. L. and Schild, H. O. (1957) Effect of temperature on the anaphylactic reaction. J. Physiol. (London) 135, 320–338.

Mongar, J. L. and Svec, P. (1972) Effect of phospholipids on histamine release. Brit. J. Pharmacol. 44, 327P.

Moran, N. C., Uvnäs, B. and Westerholm, B. (1962) Release of 5-hydroxytryptamine and histamine from rat mast cells. Acta Physiol. Scand. 56, 26–41.

Morrison, D. C. and Henson, P. M. (1977) Release of mediators from mast cells and basophils induced by different stimuli. In: Modern Concepts and Developments in Immediate Hypersensitivity, (Bach, M. K., ed.) pp. 431–502 Academic Press, New York.

Morrison, D. C., Roser, J. F., Henson, P. M. and Cochrane, C. G. (1974) Activation of rat mast cells by low molecular weight stimuli. J. Immunol. 112, 573–582.

Morrison, D. C., Roser, J. F., Cochrane, C. G. and Henson, P. M. (1975a) Two distinct mechanisms for the initiation of mast cell degranulation. Int. Arch. Allergy 49, 172–178.

Morrison, D. C., Roser, J. F., Henson, P. M. and Cochrane, C. G. (1975b) Isolation and characterization of a noncytotoxic mast-cell activator from cobra venom. Inflammation 1, 103–115.

Morrison, D. C., Roser, J. F., Cochrane, C. G. and Henson, P. M. (1975c) The initiation of mast cell degranulation: Activation at the cell membrane. J. Immunol. 114, 966–970.

Morrison, D. C., Henson, P. M. and Kline, L. F. (1975d) Lipopolysaccharide (LPS) initiated lipid A-independent (alternative) and lipid A-dependent (classical) activation of complement in plasma mediated platelet lysis. Fed. Proc 35, 655a.

Morrison, D. C., Henson, P. M., Roser, J. F. and Cochrane, C. G. (1976) Two independent recognition sites for the initiation of histamine release from mast cells. In: Leukocyte Membrane Determinants Regulating Immune Reactivity (Eijsroogel, V. P., Roos, D. and Zeiilemaker, W. P., eds.) p. 551–562 Academic Press, New York.

Movat, H. Z., Mustard, J. F., Taichman, N. S. and Uriuhara, T. (1965) Platelet aggregation and release of ADP, serotonin and histamine associated with phagocytosis of antigen-antibody complexes. Proc. Soc. Exp. Biol. Med. 120, 232–237.

Movat, H. Z., Steinberg, S. G., Habal, F. M. and Ranadive, N. S. (1973) Demonstration of a kinin-generating enzyme in the lysosomes of human polymorphonuclear leukocytes. Lab. Invest. 29, 669–684.

Müller-Eberhard, H. J. (1976) The anaphylatoxins: Formation, structure, function and control. In: Molecular and Biological Aspects of the Acute Allergic Reactions (Johansson, S. G. O., Strandberg, K. and Uvnäs, B., eds.) p. 339–352, Plenum Press, New York.

Mueller-Eckhardt, Ch. and E. F. Lüscher (1968) Immune reactions of human blood platelets. I. A comparative study of the effects on platelets of heterologous antiplatelet antiserum, antigen-antibody complexes, aggregated gamma globulin, and thrombin. Thrombos. Diath. Haemorrh. 20, 155–167.

Muggli, R. and Baumgartner, H. R. (1973) Collagen induced platelet aggregation: Requirement for tropocollagen multimers. Thromb. Res. 3, 715–728.

Mürer, E. H. (1969) Thrombin-induced release of calcium from blood platelets. Science 166, 623.

Musson, R. A. and Becker, E. L. (1976) The inhibitory effect of chemotactic factors on erythrophagocytosis by human neutrophils. J. Immunol. 117, 433–439.

Musson, R. A. and Becker, E. L. (1977) The role of an activatable esterase in immune dependant phagocytosis by human neutrophils. J. Immunol. 118, 1354–1365.

Mustard, J. F. and Packham, M. A. (1970) Factors influencing platelet function: Adhesion, release, and aggregation. Pharmacol. Rev. 22, 97–187.

Mustard, J. F., Glynn, M. F., Nishizawa, E. E. and Packham, M. A. (1967) Platelet-surface interactions: Relationship to thrombosis and hemostasis. Fed. Proc. 26, 106–114.

Mustard, J. F., Perry, D. W., Kinlough-Rathbone, R. L. and Packham, M. A. (1975) Factors responsible for ADP-induced release reaction of human platelets. Am. J. Physiol. 228, 1757–1765.

Naccache, P. H., Showell, H. J., Becker, E. L. and Sha'afi R. I. (1977) Transport of sodium, potassium and calcium across rabbit polymorphonuclear leukocyte membranes: Effect of chemotactic factor. J. Cell Biol. 73, 428–444.

Nachman, R. L. (1965) Immunologic studies of platelet protein. Blood 25, 703–711.

Nachman, R., Hirsch, J. G. and Baggiolini, M. (1972) Studies on isolated membranes of azurophil and specific granules from rabbit polymorphonuclear leukocytes. J. Cell Biol. 54, 133–140.

Nachman, R. L., Hubbard, A. and Ferris, B. (1973) Iodination of the human platelet membrane. Studies of the major surface glycoprotein. J. Biol. Chem. 248, 2928–2936.

Nakamura, S., Yoshinaga, M. and Hayashi, H. (1976) Interaction between lymphocytes and inflammatory exudate cells. II. A proteolytic enzyme released by PMN as a possible mediator for enhancement of thymocyte response. J. Immunol. 117, 1–6.

Nathanson, J. A. (1977) Cyclic nucleotides and nervous system function. Physiol. Rev. 57, 157–256.

Needleman, P., Moncada, S., Bunting, S., Vane, J. R., Hamberg, M. and Samuelsson, B. (1976) Identification of an enzyme in platelet microsomes which generates thromboxane A_2 from prostaglandin endoperoxides. Nature (London) 261, 558–560.

Nelson, D. S. (1976) (Ed.) Immunobiology of the Macrophage. Academic Press. New York and London.

Nelson, D. S. and Uhlenbruck, G. (1967) Studies on the nature of the immune adherence receptor. I. The inhibition of immune adherence by soluble mucoids and mucopeptides and by human erythrocyte ghosts. Vox Sang. 12, 43–67.

Newman, S. A., Rossi, G. and Metzger, H. (1977) Molecular weight and valence of the cell-surface receptor for immunoglobulin E. Proc. Nat. Acad. Sci. U.S.A. 74, 869–872.

Niemetz, J. and Morrison, D. C. (1977) Lipid A as the biologically active moiety in bacterial endotoxin (LPS)-initiated generation of procoagulant activity by peripheral blood leukocytes. Blood. 49, 947–956.

Niewiarowski, S., Senyi, A. F. and Gillies, P. (1973) Plasmin-induced platelet aggregation and platelet release reaction. Effects on hemostasis. J. Clin. Invest. 52, 1647–1659.

Norton, S. and DeBeer, E. J. (1955) Effects of some antibiotics on rat mast cells in vitro. Arch. Int. Pharmacodyn. 102, 352–358.

Nosál, R., Slorach, S. A. and Uvnäs, B. (1970) Quantitative correlation between degranulation and histamine release following exposure of rat mast cells to compound 48/80 in vitro. Acta Physiol. Scand. 80, 215–221.

O'Brien, J. R. and Woodhouse, M. A. (1968) Platelets: their size, shape, and stickiness in vitro: degranulation and propinquity. In: Experimental Biology and Medicine (Hagen, E., Wechsler, W. and Zilliken, F., eds.) vol. 3, pp. 90–102, S. Karger, Basel.

O'Flaherty, J. T., Kreutzer, D. L. and Ward, P. A. (1977a) Neutrophil aggregation and swelling induced by chemotactic agents. J. Immunol. 119, 232–239.

O'Flaherty, J. T., Kreutzer, D. L., Showell, H. J. and Ward, P. A. (1977b) Influence of inhibitors of cellular function on chemotactic factor-induced neutrophil aggregation. J. Immunol. 119, 1751–1756.

Okumura, T. and Jamieson, G. A. (1976a) Platelet glycocalicin. I. Orientation of glycoproteins of the human platelet surface. J. Biol. Chem. 251, 5944–5949.

Okumura, T., Lombart, C. and Jamieson, G. A. (1976b) Platelet Glycocalicin. II. Purification and characterization. J. Biol. Chem. 251, 5950–5955.

Okumura, T. and Jamieson, G. A. (1976c) Platelet Glycocalicin: a single receptor for platelet aggregation induced by thrombin or ristocetin. Thromb. Res. 8, 701–706.

Oleske, J. M., Ashman, R. B., Kohl, S., Shore, S. L., Starr, S. E., Wood, P. and Nahmias, A. J. (1977) Human polymorphonuclear leukocytes as mediators of antibody-dependent cellular cytotoxicity to herpes simplex virus-infected cells. Clin. Exp. Immunol. 27, 446–453.

Oliver, J. M. (1976) Impaired microtubule function correctable by cyclic GMP and cholinergic agonists in the Chediak-Higashi syndrome. Am. J. Pathol. 85, 395–418.

Oliver, J. M. and Zurier, R. B. (1976) Correction of characteristic abnormalities of microtubule function and granule morphology in Chediak-Higashi syndrome with cholinergic agonists. Studies *in vitro* in man and *in vivo* in the beige mouse. J. Clin. Invest. 57, 1239–1247.

Oliver, J. M., Ukena, T. E. and Berlin, R. D. (1974) Effects of phagocytosis and colchicine on the distribution of lectin-binding sites òn cell surfaces. Proc. Nat. Acad. Sci. U.S.A. 71, 394–398.

Oliver, J. M., Zurier, R. B. and Berlin, R. D. (1975) Concanavalin A cap formation on polymorphonuclear leukocytes of normal and beige (Chediak-Higashi) mice. Nature (London) 253, 471–473.

Oliver, J. M., Lalchandani, R. and Becker, E. L. (1977) Actin redistribution during concanavalin A cap formation in rabbit neutrophils. J. Reticuloend. Soc. 21, 359–364.

Olmsted, J. B. and Borisy, G. G. (1973) Microtubules. Ann. Rev. Biochem. 42, 507–540.

Ong, S. H. and Steiner, A. L. (1977) Localization of cyclic GMP and cyclic AMP in cardiac and skeletal muscle: Immunocytochemical demonstration. Science 195, 183–185.

Ong, S. H., Whitley, T. H., Stowe, N. W. and Steiner, A. L. (1975) Immunohistochemical localization of 3':5'-cyclic AMP and a 3':5'-cyclic GMP in rat liver, intestine and testis. Proc. Nat. Acad. Sci U.S.A. 72, 2022–2026.

Orange, R. P. (1975) Dissociation of the immunologic release of histamine and slow-reacting substance of anaphylaxis from human lung using cytochalasins A and B. J. Immunol. 114, 182–186.

Orange, R. P., Murphy, R. C. and Austen, K. F. (1974) Inactivation of slow reacting substance of anaphylaxis (SRS-A) by arylsulfatases. J. Immunol. 113, 316–322.

Orr, T. S. C., Hall, D. E. and Allison, A. C. (1972) Role of contractile microfilaments in the release of histamine from mast cells. Nature (London) 236, 350–351.

Osler, A. G. and Siraganian, R. P. (1972) Immunologic mechanisms of platelet damage. Progr. Allergy 16, 450–498.

Osler, A. G., Lichtenstein, L. M. and Levy, D. A. (1968) *In vitro* studies of human reaginic allergy. Adv. Immunol. 8, 183–231.

Østerud, B., Rapaport, S. I. and Lavine, K. K. (1977) Factor V activity of platelets: Evidence for an activated factor V molecule and for a platelet activator. Blood 49, 819–834.

Østlund, R. E., Leung, J. T. and Kipnis, D. M. (1977) Muscle actin filaments bind pituitary secretory granules *in vitro*. J. Cell Biol. 73, 78–87.

Packham, M. A., Evans, G., Glynn, M. F. and Mustard, J. F. (1969) The effect of plasma proteins on the interaction of platelets with glass surfaces. J. Lab. Clin. Med. 73, 686–697.

Packham, M. A., Guccione, M. A., Chang, P. L. and Mustard, J. F. (1973) Platelet aggregation and release: Effect of low concentrations of thrombin or collagen. Am. J. Physiol. 225, 38–47.

Packham, M. A., Kinlough-Rathbone, R. L., Reimers, H. J., Scott, S. and Mustard, J. F. (1977) Mechanisms of platelet aggregation independent of adenosine diphosphate. In: Prostaglandins in Hematology (Silver, M. J., Smith, J. B. and Kocsis, J. J., eds.), pp. 247–276. Spectrum, New York.

Padawer, J. (1960) Effect of colchicine and related substances on the morphology of peritoneal mast cells. J. Nat. Cancer Inst. 25, 731–747.

Padawer, J. (1966) Induction of cellular movements in mast cells by colchicine treatment. J. Cell Biol. 29, 176–180.

Padawer, J. (1967a) Microtubules in rat peritoneal fluid mast cells. J. Cell Biol. 35(2) Pt. 2), 180a (abstr.)

Padawer, J. (1967b) Fine structure of colchicine treated mast cells. J. Cell Biol. 35(2) Pt. 2), 181a (abstr.).

Padawer, J. (1971) Phagocytosis of particulate substances by mast cells. Lab. Invest. 25, 320–330.

Padawer, J. and Fruhman, G. J. (1967) Phagocytosis of zymosan by mast cells. J. Cell Biol. 35, 181a.

Padawer, J. and Gordon, A. S. (1955) Effects of colchicine on mast cells of the rat. Proc. Soc. Exp. Biol. Med. 88, 522–525.

Paton, W. D. M. (1951) Compound 48/80. A potent histamine liberator. Brit. J. Pharmacol. Chemotherapy. 6, 499–508.

Patriarca, P., Cramer, R., Moncalvo, S., Rossi, F. and Romeo, D. (1971) Enzymatic basis of metabolic stimulation in leucocytes during phagocytosis. The role of activated NADPH oxidase. Arch. Biochem. Biophys. 145, 255–262.

Pearlman, D. S., Ward, P. A. and Becker, E. L. (1969) The requirement of serine esterase function in complement-dependent erythrophagocytosis. J. Exp. Med. 130, 745–764.

Petersson, B. Å., Nilsson, A. and Stålenheim, G. (1975) Induction of histamine release and desensitization in human leukocytes. Effect of anaphylatoxin. J. Immunol. 114, 1581–1584.

Pfueller, S. L. and Lüscher, E. F. (1972a) The effects of immune complexes on blood platelets and their relationship to complement activation. Immunochemistry 9, 1151–1165.

Pfueller, S. L. and Lüscher, E. F. (1972b) The effects of aggregated immunoglobulins on human blood platelets in relation to their complement-fixing abilities. I. Studies of immunoglobulins of different types. J. Immunol. 109, 517–525.

Pfueller, S. L. and Lüscher, E. F. (1972c) The effects of aggregated immunoglobulins on human blood platelets in relation to their complement-fixing abilities. II. Structural requirements of the immunoglobulin. J. Immunol. 109, 526–533.

Pfueller, S. L. and Lüscher, E. F. (1974) Studies of the mechanism of the human platelet release reaction induced by immunologic stimuli. The effect of zymosan. J. Immunol. 112, 1211–1218.

Pfueller, S. L., Jenkins, C. S. P. and Lüscher, E. F. (1977b) A comparative study of the effect of modification of the surface of human platelets on the receptors for aggregated immunoglobulins and for ristocetin—von Willebrand factor. Biochim. Biophys. Acta 465, 614–626.

Pfueller, S. L., Weber, S and Lüscher, E. F. (1977a) Studies of the mechanism of human platelet release reaction induced by immunologic stimuli. III. Relationship between the binding of soluble IgG aggregates to the Fc receptor and cell response in the presence and absence of plasma. J. Immunol. 118, 514–524.

Phillips, D. R. and Agin, P. P. (1974) Thrombin substrates and the proteolytic site of thrombin action on human-platelet plasma membranes. Biochim. Biophys. Acta. 352, 218–227.

Phillips-Quagliata, J. M., Levine, B. B. and Uhr, J. W. (1969) Studies on the mechanism of binding of immune complexes to phagocytes. Nature 22, 1290–1291.

Pickett, W. C. and Cohen, P. (1976) Mechanism of the thrombin-mediated burst in oxygen consumption by human platelets. J. Biol. Chem. 251, 2536–2538.

Pinckard, R. N. and Henson, P. M. (1977) Activation of procoagulant activity in rabbit platelets by basophil-derived platelet activating factor (PAF_B) Fed. Proc. 36, 1329a.

Pinckard, R. N., Halonen, M., Palmer, J. D., Butler, C., Shaw, J. O. and Henson, P. M. (1977) Intravascular aggregation and pulmonary sequestration of platelets during IgE-induced systemic anaphylaxis in the rabbit. Abrogation of lethal anaphylactic shock by platelet depletion. J. Immunol. 119, 2185–2193.

Plattner, H., Reichel, K. and Matt, H. (1977) Bivalent-cation-stimulated ATPase activity at preformed exocytosis sites in Paramecium coincides with membrane-intercalated particle aggregates. Nature (London) 267, 702–704.

Pollard, T. D. and Weihing, R. R. (1974) Actin and myosin and cell movement. CRC Crit. Rev. Biochem. 2, 1.

Poo, M. M. and Robinson, K. R. (1977) Electrophoresis of concanavalin A receptors along embryonic muscle cell membrane. Nature (London) 265, 602–605.

Poste, G. and Allison, A. C., (1973) Membrane fusion. Biochim. Biophys. Acta 300, 421–465.

Pruzansky, J. J. and Patterson, R. (1975) The diisopropylfluorophosphate inhibitable step in antigen-induced histamine release from human leukocytes. J. Immunol. 114, 939–943.

Puett, D., Wasserman, B. K., Ford, J. D. and Cunningham, L. W. (1973) Collagen-mediated platelet aggregation. Effects of collagen modification involving the protein and carbohydrate moieties. J. Clin. Invest. 52, 2495–2506.

Ranadive, N. S. (1977a) Histamine release from mast cells and basophils. In: Inflammation, Immunity and Hypersensitivity, Molecular and Cellular Mechanisms (Movat, H. Z., ed.), Harper & Row, New York.

Ranadive, N. S. (1977b) Tissue injury and inflammation induced by immune complexes. In Inflammation, Immunity and Hypersensitivity: Molecular and Cellular Mechanisms (Movat, H. Z., ed.), Harper & Row, New York.

Ranadive, N. S. and Cochrane, C. G. (1968) Isolation and characterization of permeability factors

502

from rabbit neutrophils. J. Exp. Med. 128, 605–622.

Ranadive, N. S. and Cochrane, C. G. (1970) Basic proteins in rat neutrophils that increase vascular permeability. Clin. Exp. Immunol. 6, 905–911.

Ranadive, N. S. and Cochrane, C. G. (1971) Mechanism of histamine release from mast cells by cationic protein (band 2) from neutrophil lysosomes. J. Immunol. 106, 506–516.

Ranadive, N. S. and Muir, J. D. (1972) Similarity in the mechanism of histamine release induced by cationic protein from neutrophils and by complement dependant Ag-Ab reaction. Int. Arch. Allergy 42, 236–249.

Ranadive, N. S. and Ruben, D. H. (1973) Mechanism of histamine release from rat mast cells by compound 48/80. Comparison with the release induced by cationic protein. Int. Arch. Allergy 44, 745–758.

Read, G. W. and Lenney, J. F. (1972) Molecular weight studies on the active constituents of compound 48/80. J. Med. Chem. 15, 320–323.

Reaven, E. P. and Axline, S. G. (1973) Subplasmalemmal microfilaments and microtubules in resting and phagocytizing cultivated macrophages. J. Cell Biol. 59, 12–27.

Reimers, H. J., Packham, M. A., Kinlough-Rathbone, R. L. and Mustard, J. F. (1973) Effect of repeated treatment of rabbit platelets with low concentrations of thrombin on their function, metabolism and survival. Brit. J. Hematol. 25, 675–689.

Renoux, M. L., De Montis, G. and Marcelli, D. (1975) Binding of human IgE immunoglobulin to rat mast cells. A study by means of three independent techniques. Int. Arch. Allergy 49, 478–490.

Riley, J. F. (1959) The Mast Cells. Livingstone, Edinburgh.

Riley, J. F. and West, G. B. (1953) The presence of histamine in tissue mast cells. J. Physiol. (London). 120, 528–537.

Rocha e Silva, M., Bier, O., and Aronson, M. (1951) Histamine release by anaphylatoxin. Nature (London) 168, 465–466.

Rodan, G. A. and Feinstein, M. B. (1976) Interrelationships between Ca^{2+} and adenylate and guanylate cyclases in the control of platelet secretion and aggregation. Proc. Nat. Acad. Sci. U.S.A. 73, 1829–1833.

Röhlich, P. (1975) Membrane-associated actin filaments in the cortical cytoplasm of the rat mast cell. Exp. Cell. Res. 93, 293–298.

Röhlich, P., Anderson, P. and Uvnäs, B. (1971a) Mast cell degranulation as a process of sequential exocytosis. Acta Biol. Acad. Sci. (Hungary) 22, 197–213

Röhlich, P., Anderson, P. and Uvnäs, B. (1971b) Electron microscope observations on compound 48/80-induced degranulation in rat mast cells. Evidence for sequential exocytosis of storage granules. J. Cell Biol. 51, 465–483.

Romeo, D., Zabucchi, G. and Rossi, F. (1973) Reversible metabolic stimulation of polymorphonuclear leukocytes and macrophages by concanavalin A. Nature New Biol. 243, 111–112.

Romeo, D., Zabucchi, G., Miani, N. and Rossi, F. (1975) Ion movement across leukocyte plasma membrane and excitation of their metabolism. Nature (London) 253, 542–544.

Root, R. K., Metcalf, J., Oshino, N. and Chance, B. (1975) H_2O_2 release from human granulocytes during phagocytosis. 1. Documentation, quantitation, and some regulating factors. J. Clin. Invest. 55, 945–955.

Rose, B., Simpson, I. and Loewenstein, W. R. (1977) Calcium ion produces graded changes in permeability of membrane channels in cell junction. Nature (London) 267, 625–627.

Ross, G. D., Polley, M. J., Rabellino, E. M. and Grey, H. M. (1973) Two different complement receptors on human lymphocytes. One specific for C3b and one specific for C3b inactivator-cleaved C3b. J. Exp. Med. 138, 798–811.

Rossi, F., Zatti, M., Patriarca, P. and Cramer, R. (1971) Effect of specific antibodies on the metabolism of guinea pig polymorphonuclear leukocytes. J. Reticuloend. Soc. 9, 67–85.

Roth, G. J., Stanford, N. and Majerus, P. W. (1975) Acetylation of prostaglandin synthase by aspirin. Proc. Nat. Acad. Sci. U.S.A. 72, 3073–3076.

Russell, F. A., and Deykin, D. (1976) The effect of thrombin on the uptake and transformation of arachidonic acid by human platelets. Am. J. Hematol. 1, 59–70.

Rutherford, R. B. and Ross, R. (1976) Platelet factors stimulate fibroblasts and smooth muscle cells quiescent in plasma serum to proliferate. J. Cell Biol. 69, 196–203.

Ryan, G. B., Borysenko, J. Z. and Karnovsky, M. J. (1974) Factors affecting the redistribution of surface-bound concanavalin A on human polymorphonuclear leukocytes. J. Cell Biol. 62, 351–365.

Sajnani, A. N., Ranadive, N. S. and Movat, H. Z. (1974) The visualization of receptors for the Fc portion of the IgG molecule on human neutrophil leukocytes. Life Sci. 14, 2427–2430.

Sajnani, A. N., Ranadive, N. S. and Movat, H. Z. (1976) Redistribution of immunoglobulin receptors on human neutrophils and its relationship to the release of lysosomal enzymes. Lab. Invest. 35, 143–151.

Salmon, J. and Bounameaux, Y. (1958) Etude des antigènes plaquettaires et, en particulier, du fibrinogène. Trombos. Diath. Haemorrh. 2, 93–110.

Salzman, E. W. (1972) Cyclic AMP and platelet function. New Eng. J. Med. 286, 358–363.

Sampson, D. and Archer, G. T. (1967) Release of histamine from human basophils. Blood 29, 722–736.

Schellenberg, R. R. and Gillespie, E. (1977) Colchicine inhibits phosphatidylinositol turnover induced in lymphocytes by concanavalin A. Nature (London) 265, 741–742.

Schieren, H., Weissmann, G., Seligman, M. and Coleman, P. (1976) Interaction of liposomes with immunoglobulins: an E.S.R. study demonstrating cortisol protection. Fed. Proc. (abstr.) 35, 274.

Schiffman, S., Rimon, A. and Rapaport, S. I. (1977) Factor XI and platelets: Evidence that platelets contain only minimal Factor XI activity. Brit. J. Haematol. 35, 429–436

Schiffmann, E., Showell, H. V., Corcoran, B. A., Ward, P. A., Smith, E. and Becker, E. L. (1975a) The isolation and partial characterization of neutrophil chemotactic factors from Escherichia coli. J. Immunol. 114, 1831–1837.

Schiffmann, E., Corcoran, B. A. and Wahl, S. M. (1975b) N-formylmethionyl peptides as chemoattractants for leucocytes. Proc. Nat. Acad. Sci. U.S.A. 72, 1059–1062.

Schiffmann, E., Corcoran, B. A. and Aswanikumar, S. (1977) Molecular events in the response of neutrophils to synthetic NFMET chemotactic peptides: Demonstration of a specific receptor. In: Leukocyte Chemotaxis: Methodology, Physiology, and Clinical Applications (Gallin, J. I. and Quie, P. G., eds.) pp. 97–111, Raven Press, New York.

Schild, H. O. (1968) Mechanism of anaphylactic histamine release. In: Biochemistry of Acute Allergic Reactions. (Austen, K. F. and Becker, E. L. eds.) pp. 99–118, F. A. Davis, Philadelphia.

Schlessinger, J., Webb. W. W., Elson, E. L. and Metzger, H. (1976) Lateral motion and valence of Fe receptors on rat peritoneal mast cells. Nature (London) 264, 550–552.

Schober, R., Nitsch, L., Rinne, U. and Morris, S. J. (1977) Calcium-induced displacement of membrane-associated particles upon aggregation of chromaffin granules. Science 195, 495–497.

Schorlemmer, H. U., Bitter-Suermann, D. and Allison, A. C. (1977) Complement activation by the alternative pathway and macrophage enzyme secretion in the pathogenesis of chronic inflammation. Immunology 32, 929–940.

Schribner, D. J. and Fahrney, D. (1976) Neutrophil receptors for IgG and complement: their roles in the attachment and ingestion phases of phagocytosis. J. Immunol. 116, 892–897.

Schumacher, H. R. and Phelps, P. (1971) Sequential changes in polymorphonuclear leukocytes after urate crystal phagocytosis. An electron microscopic study. Arth. Rheum. 14, 513–526.

Seegers, W. and Janoff, A. (1966) Mediators of inflammation in leukocyte lysosomes. VI. Partial purification and characterization of a mast cell-rupturing component. J. Exp. Med. 124, 833–849.

Segal, A. W. and Peters, T. J. (1977) Analytical subcellular fractionation of human granulocytes with special reference to the localization of enzymes involved in microbicidal mechanisms. Clin. Sci. Mol. Med. 52, 429–442.

Senda, N., Tamura, H., Shibata, N., Yoshitake, J., Kondo, K. and Tanaka, K. (1975) The mechanism of the movement of leucocytes. Exp. Cell Res. 91, 393–407.

Shaw, J. O., Printz, M., Steward, R., Hirabayshi, K and Henson, P. H. (1977) Relationship of prostaglandin production to secretion in platelets stimulated with platelet activating factor (PAF). Submitted for publication.

504

Shepro, D., Belamarich, F. A., Robblee, L. and Chao, F. C. (1970) Antimotility effect of cytochalasin Bob-served in mammalian clot retraction. J. Cell Biol. 47, 544–547.

Sher, A. (1976) Complement-dependent adherence of mast cells to schistosomula. Nature (London) 263, 334–336.

Sher, A. and McIntyre, S. (1977) Immune adherence receptors on rat mast cells. Fed. Proc. 36, 1264a.

Shirahama, T. and Cohen, A. S. (1974) Ultrastructural evidence for leakage of lysosomal contents after phagocytosis of monosodium urate crystals. A mechanism of gouty inflammation. Am. J. Pathol. 76, 501–512.

Showell, H. J. and Becker, E. L. (1976) The effects of external K^+ and Na^+ on the chemotaxis of rabbit peritoneal neutrophils. J Immunol. 116, 99–105.

Showell, H. J., Freer, R. J., Zigmond, S. H., Schiffmann, E., Aswanikumar S., Corcoran, B. and Becker, E. L. (1976) The structure-activity relations of synthetic peptides as chemotactic factors and inducers of lysosomal enzyme secretion for neutrophils. J. Exp. Med. 143, 1154–1169.

Showell, H. J., Naccache, P. H., Sha'afi, R. I. and Becker, E. L. (1977) The effects of extracellular K^+, Na^+ and Ca^{2+} on lysosomal enzyme secretion from polymorphonuclear leukocytes. J. Immunol. 119, 804–811.

Shuman, M. A. and Majerus, P. W. (1975) The perturbation of thrombin binding to human platelets by anions. J. Clin. Invest. 56, 945–950.

Shuman, M. A., Tollefsen, D. M. and Majerus, P. W. (1976) The binding of human and bovine thrombin ɔ human platelets. Blood 47, 43–54.

Silver, M. K., Smith, J. B. and Kocsis, J. C. (1977) Prostaglandins in Hematology. Spectrum, New York.

Simons, E. R., Chesney, C. M., Colman, R. W., Harper, E. and Samberg, E. (1975) The effect of the conformation of collagen on its ability to aggregate platelets. Thromb. Res. 7, 123–139.

Siqueira, M. and Nelson, R. A., Jr. (1961) Platelet agglutination by immune complexes and its possible role in hypersensitivity. J. Immunol. 86, 516–525.

Siraganian, R. P. and Hook, W. A. (1976) Complement-induced histamine release from human basophils. II. Mechanism of the histamine release reaction. J. Immunol. 116, 639–646.

Siraganian, R. P. and Osler, A. G. (1971) Destruction of rabbit platelets in the allergic response of sensitized leukocytes. I. Evidence for basophil involvement. J. Immunol. 106, 1252–1259.

Siraganian, R. P., Hook, W. A. and Levine, B. B. (1975a) Specific *in vitro* histamine release from basophils by bivalent haptens: Evidence for activation by simple bridging of membrane bound antibody. Immunochemistry 12, 149–157.

Siraganian, R. P., Kulczycki, A. Jr., Mendoza, G. and Metzger, H. (1975b) Ionophore A-23187 induced histamine release from rat mast cells and rat basophil leukemia (RBL-1) cells. J. Immunol. 115, 1599–1602

Siraganian, R. P., Sandberg, A. L., Alexander, A. and Osler, A. G. (1973) Platelet injury due to activation of the alternate complement pathway. J. Immunol. 110, 490–497.

Skaer, R. J. (1975) Elemental composition of platelet dense bodies In: Ciba Foundation Symposium 35: Biochemistry and Pharmacology of Platelets (Elliott, K. and Knight, J., eds.) pp. 239–259, Elsevier/North-Holland, Amsterdam.

Slorach, S. A. (1971) Histamine and heparin release from isolated rat mast cells exposed to compound 48/80. Acta Physiol. Scand. 82, 91–97.

Slorach, S. A. and Uvnäs, B. (1968) Amine formation by rat mast cells *in vitro*. Acta Physiol. Scand. 73, 457–470.

Smith, J. B. and Willis, A. L. (1970) Formation and release of prostaglandins by platelets in response to thrombin. Brit. J. Pharm. 40, 545–546.

Smith, J. B., Ingerman, C., Kocsis, J. J. and Silver, M. J. (1973) Formation of prostaglandins during the aggregation of human blood platelets. J. Clin. Invest. 52, 965–969.

Smith, J. B., Ingerman, C., Kocsis, J. J. and Silver, M. J. (1974) Formation of an intermediate in prostaglandin biosynthesis and its association with the platelet release reaction. J. Clin. Invest. 53, 1468–1472.

Stechschulte, D. J. and Austen, K. F. (1974) Phosphatidylserine enhancement of antigen-induced mediator release from rat mast cells. J. Immunol. 112, 970–978.

Steiner, M. (1973) Effect of thrombin on the platelet membrane. Biochim. Biophys. Acta 323, 653–658.

Stossel, T. P. (1974) Phagocytosis. New Eng. J. Med. 290, 717, 774, 833.

Stossel, T. P. (1975) Phagocytosis, recognition and ingestion. Sem. Hemat. 12, 83–116.

Stossel, T. P. (1977) Contractile proteins in phagocytosis: an example of cell surface-to-cytoplasm communication. Fed. Proc. 36, 2181–2184.

Stossel, T. P., Field, R. J., Gitlin, J. D., Alper, C. A. and Rosen, F. S. (1975) The opsonic fragment of the third component of human complement (C3) J. Exp. Med. 141, 1329–1347.

Strandberg, K. and Westerberg, S. (1976) Composition of phospholipids and phospholipid fatty acids in rat mast cells. Mol. Cell Biochem. 11, 103–107.

Sullivan, T. J. and Parker, C. W. (1976) Pharmacologic modulation of inflammatory mediator release by rat mast cells. Am. J. Path. 85, 437–464.

Sullivan, T. J. and Parker, C. W. (1977) Possible role of arachidonic acid and its metabolites in mediator release from rat mast cells. Fed. Proc. (abstr.) 36, 1216.

Sullivan, A. L., Grimley, P. M. and Metzger, H. (1971) Electron microscopic localization of immunoglobulin E on the surface membrane of human basophils. J. Exp. Med. 134, 1403–1416.

Sullivan, T. J., Parker, K. L., Stenson, W. and Parker, C. W. (1975a) Modulation of cyclic AMP in purified rat mast cells. I. Responses to pharmacologic, metabolic, and physical stimuli. J. Immunol. 114, 1473–1479.

Sullivan, T. J., Parker, K. L., Eisen, S. A. and Parker, C. W. (1975b) Modulation of cyclic AMP in purified rat mast cells. II. Studies on the relationship between intracellular cyclic AMP concentrations and histamine release. J. Immunol. 114, 1480–1485.

Sydbom, A. and Uvnäs, B. (1976) Potentiation of anaphylactic histamine release from isolated rat pleural mast cells by rat serum phospholipids. Acta Physiol. Scand. 97, 222–232.

Tai, P-C. and Spry, J. F. (1977) Purification of normal human eosinophils using the different binding capacities of blood leukocytes for complexed rabbit IgG. Clin. Exp. Immunol. 28, 256–260.

Taylor, W. A. and Sheldon, D. (1977) The effect of antigen concentration and epitope density on rat mast cell degranulation. Fed. Proc. (abstr.) 36, 1216.

Thon, I. L. and Uvnäs, B. (1967) Degranulation and histamine release, Two consecutive steps in the response of rat mast cells to compound 48/80. Acta Physiol. Scand. 71, 303–315.

Tollefsen, D. M. and Majerus, P. W. (1976) Evidence for a single class of thrombin-binding sites on human platelets. Biochemistry 15, 2144–2149.

Tollefsen, D. M., Feagler, J. R. and Majerus. P. W. (1974) The binding of thrombin to the surface of human platelets. J. Biol. Chem. 249, 2646–2651.

Tollefsen, D. M., Jackson, C. M. and Majerus, P. W. (1975) Binding of the products of prothrombin activation to human platelets. J. Clin. Invest. 56, 241–245.

Tolone, G., Bonasera, L. and Parrinello, N. (1974) Histamine release from mast cells: Role of microtubules. Experientia (Basel) 30, 426–427.

Tsan, M-F. and McIntyre, P. A. (1976) The requirement for membrane sialic acid in the stimulation of superoxide production during phagocytosis by human polymorphonuclear leukocytes. J. Exp. Med. 143, 1308–1316.

Tsung, P-K., Kegeles, S. W. and Becker, E. L. (1977) Some differences in intracellular distribution and properties of the chymotrypsin-like esterase activity of human and rabbit neutrophils. Biochim. Biophys. Acta. 499, 212–227.

Tsung, P-K., Sakamoto, T. and Weissmann, G. (1975) Protein kinase and phosphatases from human polymorphonuclear leucocytes. Biochem. J. 145, 437–448.

Turner, S. R. and Lynn, W. S. (1977) Lipid molecules as chemotactic factors. In: Leukocyte Chemotaxis: Methodology, Physiology, and Clinical Applications (Gallin, J. I. and Quie, P. G., eds.) pp. 289–298, Raven Press, New York.

Ulevitch, R. J., Cochrane, C. G., Henson, P. M., Morrison, D. C. and Doe, W. F. (1975) Mediation systems in bacterial lipopolysaccharide-induced hypotension and disseminated intravascular coagulation. I. The role of complement. J. Exp. Med. 142, 1570–1590.

Uvnäs, B. (1964) Release processes in mast cells and their activation by injury. Ann. New York Acad. Sci. 116, 880–890.

Uvnäs, B. (1971) Quantitative correlation between degranulation and histamine release in mast cells. In: Biochemistry of the Acute Allergic Reactions (Austen, K. F., and Becker, E. L., eds.) pp. 175–188, Blackwell, Oxford.

Uvnäs, B. (1976) Modes of action of antigen-antibody reaction and compound 48/80 in histamine release. In: Molecular and Biological Aspects of the Acute Allergic Reaction (Johansson, S. G. O., Strandberg, K. and Uvnäs, B., eds.) p. 217–228, Plenum Press, New York.

Uvnäs B. and Antonsson, J. (1963) Triggering action of phospholipase A and chymotrypsin on degranulation of rat mesentery cells. Biochem. Pharmacol. 12, 867–873.

Uvnäs, B. and Thon, I-L. (1961) Evidence for enzymatic histamine release from isolated rat mast cells. Exp. Cell Res. 23, 45–57.

Uvnäs, B. and Thon, I. L. (1966) A physico-chemical model of histamine release from mast cells. In: Mechanisms of Release of Biogenic Amines (von Euler, U. S., Rosell, S., and Uvnäs, B., eds.) p. 361–370, Pergamon Press, Oxford.

Valentine, M. D., Bloch, K. J. and Austen, K. F. (1967) Mechanisms of immunologic injury of rat peritoneal mast cells. III. Cytotoxic histamine release. J. Immunol. 99, 98–110.

Vane, J. R. (1976) The mode of action of aspirin and similar compounds. J. Allergy Clin. Immunol. 58, 691–712.

Van Oss, C. J. and Gillman, C. F. (1973) Phagocytosis as a surface phenomenon. III. Influence of C1423 on the contact angle and on the phagocytosis of sensitized encapsulated bacteria. Immunol. Commun. 2, 415–419.

Vecchione, J. and Zucker, M. B. (1975) Procoagulant activity of platelets in recalcified plasma. Brit. J. Haematol. 31, 423–428.

Voorhees, A. B., Baker, H. J. and Pulaski, E. J. (1951) Reactions of albino rats to injections of dextran. Proc. Soc. Exp. Biol. Med. 76, 254–256.

Walker, W. S. (1976) Separate Fc-receptors for immunoglobulins (IgG2a) and IgG2b on an established line of mouse macrophages. J. Immunol. 116, 911–914.

Walsh, P. N. (1972a) Albumin density gradient separation and washing of platelets and the study of platelet coagulant activities. Brit. J. Haematol. 22, 205–217.

Walsh, P. N. (1972b) The role of platelets in the contact phase of blood coagulation. Brit. J. Haematol. 22, 237–254.

Walsh, P. N. (1972c) The effect of collagen and kaolin on the intrinsic coagulant activity of platelets. Evidence for an alternative pathway in intrinsic coagulation not requiring factor XII. Brit. J. Haematol. 22, 393–405.

Walsh, P. N. and Lipscomb, M. S. (1976) Comparison of the coagulant activities of platelets and phospholipids. Brit. J. Haematol. 33, 9–18.

Ward, P. A. (1974) The regulation of leukotactic mediators. In: Chemotaxis, its Biology and Biochemistry. (Sorkin, E., ed.) Antibiotics and Chemotherapy 19, pp. 333–337. Karger, Basel.

Ward, P. A. and Hill, J. H. (1970) C5 chemotactic fragments produced by an enzyme in lysosomal granules of neutrophils. J. Immunol. 104, 535–543.

Ward, P. A. and Ozols, J. (1976) Characterization of the protease activity in the chemotactic factor inactivator. J. Clin. Invest. 58, 123–129.

Ward, P. A. and Zvaifler, N. J. (1971) Complement-derived leukotactic factors in inflammatory synovial fluids of humans. J. Clin. Invest. 50, 606–616.

Wasserman, S. I., Goetzl, E. J. and Austen, K. F. (1974) Preformed eosinophil chemotactic factor of anaphylaxis (ECF-A). J. Immunol. 112, 351–358.

Wautier, J. L., Tobelem, G. M., Peltier, A. P. and Caen, J. P. (1976a) C1 and human platelets. II. Detection by immunologic methods and role. Immunology 30, 459–465.

Wautier, J. L., Souchon, H., Solal, L. C., Peltier, A. P. and Caen, J. P. (1976b) C1 and human platelets III. Role of C1 subcomponents in platelet aggregation induced by aggregated IgG. Immunology 31, 595–599.

Weiss, H. J. (1975) Platelet physiology and abnormalities of platelet function. N. Eng. J. Med. 293,

531, 580.

Weissmann, G., Brand, A. and Franklin, E. C. (1974) Interaction of immunoglobulins with liposomes. J. Clin. Invest. 53, 536–543.

Weissmann, G., Zurier, R. B., Spieler, P. J. and Goldstein, I. M. (1971a) Mechanisms of lysosomal enzyme release from leukocytes exposed to immune complexes and other particles. J. Exp. Med. 134, 149s–165s.

Weissmann, G., Dukor, P. and Zurier, R. B. (1971b) Effect of cyclic AMP on release of lysosomal enzymes from phagocytes. Nature New Biol. 231, 131–135.

Weissmann, G., Zurier, R. B. and Hoffstein, S. (1972) Leukocytic proteases and the immunologic release of lysosomal enzymes. Am. J. Pathol. 68, 539–564.

Weissmann, G., Zurier, R. B. and Hoffstein, S. (1973) Lysosomes as secretory organs of inflammation. Agents Actions 3, 370–379.

Weissmann, G., Goldstein, I., Hoffstein, S., Chauvet, G. and Robineaux, R. (1975) Yin/Yang modulation of lysosomal enzyme release from polymorphonuclear leukocytes by cyclic nucleotides. Ann. N.Y. Acad. Sci. 256, 222–232.

Wessells, N. K., Spooner, B. S., Ash, J. F., Bradley, M. O., Luduena, M. A., Taylor, E. L., Wrenn, J. T. and Yamada, K. M. (1971) Microfilaments in cellular and developmental processes. Contractile microfilament machinery of many cell types is reversibly inhibited by cytochalasin B. Science 171, 135–143.

White, J. G. (1968) Effect of colchicine and vinca alkaloids on human platelets. I. Influence on platelet microtubules and contractile function. Am. J. Pathol. 53, 281–291.

White, J. G. (1969) Effects of colchicine and vinca alkaloids on human platelets. III. Influence on primary internal contraction and secondary aggregation. Am. J. Pathol. 54, 467–478.

White, J. G. (1971) Platelet morphology. In: The Circulating Platelet (Johnson, S. A., ed.) pp. 45–121. Academic Press, New York.

White, J. G. (1972a) Exocytosis of secretory organelles from blood platelets incubated with cationic polypeptides. Am. J. Pathol. 69, 41–54.

White, J. G. (1972b) Uptake of latex particles by blood platelets. Phagocytosis or sequestration? Am. J. Pathol. 69, 439–458.

White, J. G. (1974) Electron microscopic studies of platelet secretion. Progr. Hemost. Thromb. 2, 49–98.

White, J. G. and Estensen, R. D. (1972) Degranulation of discoid platelets. Am. J. Pathol. 68, 289–302.

White, J. G. and Estensen, R. D. (1974) Selective labilization of specific granules in polymorphonuclear leukocytes by phorbol myristate acetate. Am. J. Pathol. 75, 45–60.

White, J. G. and Krivit, W. (1967) An ultrastructural basis for the shape changes induced in platelets by chilling. Blood 30, 625–635.

White, J. G. and Krumwiede, M. (1973) Influence of cytochalasin B on the shape change induced in platelets by cold. Blood 41, 823–832.

White, J. G., Rao, G. H. R. and Gerrard, J. M. (1974) Effects of the ionophore A23187 on blood platelets. I. Influence on aggregation and secretion. Am. J. Pathol. 77, 135–150.

Wilkinson, P. C. (1973) Recognition of protein structure in leukocyte chemotaxis. Nature (London) 244, 512–513.

Wilkinson, P. C. (1975) Leucocyte locomotion and chemotaxis. The influence of divalent cations and cation ionophores. Exp. Cell Res. 93, 420–426.

Wilkinson, P. C. (1976) Recognition and response in mononuclear and granular phagocytes. Clin. Exp. Immunol. 25, 355–366.

Williams, L. T., Snyderman, R., Pike, M. C. and Lefkowitz, R. J. (1977) Specific receptor sites for chemotactic peptides on human polymorphonuclear leukocytes. Proc. Nat. Acad. Sci. U.S.A. 74, 1204–1208.

Wilner, G. D., Nossel, H. L. and LeRoy, E. C. (1968) Aggregation of platelets by collagen, J. Clin. Invest. 47, 2616–2621.

Wilner, G. D., Nossel, H. L. and Procupez, T. L. (1971) Aggregation of platelets by collagen: Polar active sites of insoluble human collagen. Am. J. Physiol. 220, 1074–1079.

508

Winquist, G. (1963) Electron microscopy of the basophilic granulocyte. Ann. N.Y. Acad. Sci. 103, 352–375.

Wintroub, B. U., Goetzl, E. J. and Austen, K. F. (1974) A neutrophil-dependent pathway for the generation of a neutral peptide mediator. Partial characterization of components and control by α-1-antitrypsin. J. Exp. Med. 140, 812–824.

Woodin, A. M. (1968) The basis of leucocidin action. In: The Biological Basis of Medicine. (Bittar, E. and Bittar, N., eds.) Vol. 2, pp. 373–395, Academic Press, London.

Woodin, A. M. (1975) Inhibition of secretion in the polymorphonuclear leucocyte by concanavalin A. Exp. Cell Res. 92, 201–210.

Woodin, A. M. and Wieneke, A. A. (1970) Site of protein secretion and calcium accumulation in the polymorphonuclear leucocyte treated with leucocidin. In: Calcium and Cellular Function. (Cuthbert, A. W., ed.) pp. 183–197, Macmillan and Co., Ltd., London.

Wright, D. G. and Gallin, J. I. (1975) Modulation of the inflammatory response by products released from human polymorphonuclear leukocytes during phagocytosis. Generation and inactivation of the chemotactic factor C5a. Inflammation 1, 23–39.

Wright, D. G. and Malawista, S. E. (1973) Mobilization and extracellular release of granular enzymes from human leukocytes during phagocytosis. Inhibition by colchicine and cortisol but not by salicylate. Arthr. Rheum 16, 749–758.

Wright, D. G., Bralove, D. A. and Gallin, J. I. (1977) The differential mobilization of human neutrophil granules. Effects of phorbol myristate acetate and ionophore A23187. Am. J. Pathol. 87, 273–284.

Yurt, R. W., Leid, R. W., Jr., Spragg, J. and Austen, K. F. (1977) Immunologic release of heparin from purified rat peritoneal mast cells. J. Immunol. 118, 1201–1207.

Zakai, N., Kulka, R. G. and Loyter, A. (1976) Fusion of human erythrocyte ghosts promoted by the combined action of calcium and phosphate ions. Nature (London) 263, 696–699.

Zeiger, R. S., Twarog, F. J. and Colten, H. R. (1976) Histaminase release from human granulocytes. J. Exp. Med. 144, 1049–1061.

Zigmond, S. H. and Hirsch, J. G. (1972) Effects of cytochalasin B on polymorphonuclear leukocyte locomotion, phagocytosis and glycolysis. Exp. Cell Res. 73, 383–393.

Zigmond, S. H. and Hirsch, J. G. (1973) Leukocyte locomotion and chemotaxis. New methods for evaluation and demonstration of a cell-derived chemotactic factor. J. Exp. Med. 137, 387–410.

Zimmerman, T. S. and Spiegelberg, H. L. (1975) Pneumococcus-induced serotonin release from human platelets. Identification of the participating plasma/serum factor as immunoglobulin. J. Clin. Invest. 56, 828–834.

Zucker, M. B. and Borrelli, J. (1962) Platelet clumping produced by connective tissue suspensions and by collagen. Proc. Soc. Exp. Biol. Med. 109, 779–787.

Zucker, M. B. and Grant, R. A. (1974) Aggregation and release reaction induced in human blood platelets by zymosan. J. Immunol. 112, 1219–1230.

Zucker-Franklin, D. (1965) Electron microscope study of the degranulation of polymorphonuclear leukocytes following treatment with streptolysin. Am. J. Pathol. 47, 419–433.

Zucker-Franklin, D. (1969) Microfibrils of blood platelets: Their relationship to microtubules and the contractile protein. J. Clin. Invest. 48, 165–175.

Zucker-Franklin, D., Nachman, R. L., and Marcus, A. J. (1967) Ultrastructure of thrombosthenin, the contractile protein of human blood platelets. Science 157, 945–946.

Zurier, R. B. and Sayadoff, D. M. (1975) Release of prostaglandins from human polymorphonuclear leukocytes. Inflammation 1, 93–101.

Zurier, R. B., Hoffstein, S. and Weissmann, G. (1973) Mechanisms of lysosomal enzyme release from human leukocytes. I. Effect of cyclic nucleotides and colchicine. J. Cell Biol. 58, 27–41.

Zurier, R. B., Weissmann, G., Hoffstein, S., Kammerman, S. and Tai, H. H. (1974) Mechanisms of lysosomal enzyme release from human leukocytes. II. Effects of cAMP and cGMP, autonomic agonists, and agents which affect microtubule function. Clin. Invest. 53, 297–309.

Zusman, R. M. and Keiser, H. R. (1977) Prostaglandin E$_2$ biosynthesis in rabbit renomedullary interstitial cells in tissue cultures. J. Biol. Chem. 252, 2069–2071.

Zwaal, R. F. A., Comfurius, P. and van Deenen, L. L. M. (1977) Membrane asymmetry and blood coagulation. Nature (London) 268, 358–360.

Cytoplasmic membranes and the secretory process

11

J. MELDOLESI, N. BORGESE, P. DE CAMILLI and B. CECCARELLI

Contents

G. Poste & G. L. Nicolson (eds.) Membrane Fusion, pp. 509–627.

1. Introduction

In most eukaryotic cells, two different types of secretion are known to occur: (1) the net outward transfer of inorganic ions, small and hydrophobic molecules (molecular secretion); and (2) the quantal release by exocytosis of the packaged content of preformed vesicles or granules (vesicular secretion). The relative importance of these two modes of secretion varies enormously in different cell types. Membrane fusion[1] has a major role in vesicular secretion, and this chapter will be restricted to a review of the latter process and will deal specifically with the interactions of endo- and plasma membranes.

Exocytosis is the final event of vesicular secretion. A number of other complex and interconnected events occurring within the cytoplasm precede this final release process. These include the biosynthesis of secretion products and their transport throughout the cell along a well-identified, membrane-bounded pathway which involves a large portion of the endomembrane system of the cell. Thus both cytoplasmic and surface phenomena will be considered in this article, because vesicular secretion is essentially a unitary process, endowed with a clear finality, in which the structural and dynamic potentialities of endo- and plasma membranes are integrated as a whole. The basic phenomena involved, such as nonrandom interactions and recognition between different membranes, as well as the factors regulating these interactions, appear analogous within and at the cell surface.

A comprehensive description of the properties and functional roles of membranes in secretion is relevant not only for an understanding of the physiology of specifically differentiated secretory cells but also for other cells that are not conventionally considered as secretory cells, yet which retain a minor degree of secretory activity (e.g., as seen in the synthesis and release of their own basement membrane). Actually, the coordinated movement and specific interaction of cytoplasmic membranes occur not only in relation to secretion. A number of other phenomena, such as endocytosis, the biogenesis of lysosomes and peroxisomes, as well as growth and regeneration of organelles and the plasmalemma,

---

[1]In this chapter the term *membrane fusion* will be used as a general term to indicate the process whereby two membranes become continuous. The term *membrane fusion-fission* will be used to specify the events common to the process of establishment of continuity between two interacting membranes and the compartments they delimit (as in exocytosis), and to the process whereby membrane portions pinch off and establish separate compartments (as in endocytosis). As observed in thin-section electronmicrographs, these processes include the focal merging of two morphologically trilaminar membrane segments (of the same membrane or different membranes) to yield a pentalaminar structure (fusion), which is subsequently eliminated (fission).

might be consided as sharing certain functional similarities with secretion. Hence, studies on secretory cells might help in understanding other problems of membrane interactions. In this respect, it should be emphasized that differentiated secretory cells are favorable experimental models because the interactions of their membranes are not only striking in number but are also mainly oriented toward a single function (the intracellular transport and discharge of secretion products), which is carried out according to specific and well-defined programs and can be modified by a variety of controlled experimental manipulations.

In this review, we provide a comprehensive account of intracellular transport operations, as revealed by studies carried out on a variety of different secretory systems (section 2). This is followed by a detailed analysis of the various membranes participating in intracellular transport processes and the discharge of secretion products, including discussion of the chemical composition and structure, and biogenesis and turnover of the various membrane systems. In the last two sections, the dynamic interactions between endomembranes (in particular, the membranes of secretion granules and vesicles) and the plasmalemma will be discussed in relation to the phenomena of exocytosis and secretory granule or vesicle membrane retrieval after exocytosis.

Finally, we would like to emphasize our intentions in preparing this contribution. Due to the enormous literature existing in this field, we have decided to sacrifice completeness in favor of coherence. We have not attempted to provide separate analyses of events in various secretory systems or to pinpoint quantitative deviations and apparent exceptions in different cell types. Furthermore, relatively minor attention is given to a number of specific phenomena which, because of their importance, deserve a much more detailed discussion. One of these is axoplasmic transport, which is considered here only in relation to its possible role in secretion. Another such subject is stimulus-secretion coupling. In this case, due both to the apparent heterogeneity of the mechanisms operating in the different cell systems and to the contradictory results reported in the literature, we were unable to summarize the problem in a short, comprehensive review. Thus the few guidelines that are given may be used to interpret the uncertainties existing at present.

2. Intracellular transport of secretion products

Over the last two decades, it has become clear that the sequence of events leading to the formation of organelles capable of discharging their content by exocytosis starts with the synthesis of specific secretory proteins. This occurs in all cases,[2] regardless of whether the physiologically important components released by the system are the original proteins, products of their posttranslational modification, or different molecules accumulated by the secretory organelles in the course of their maturation or during their entire lifetime (such as neurotransmitters).

---

[2]The only possible exception is the cholinergic vesicle, considered in detail in section 2.1.6.

Hence, to introduce the description of intracellular transport, a short account of the present knowledge about the mechanisms regulating the biosynthesis of secretory proteins is given.

In the mid-1950s, electron microscopists were impressed by the great abundance of bound ribosomes in glandular cells and proposed a specific role for these structures in the biosynthesis of secretory proteins (Palade, 1955). In subsequent years, these early predictions were generally confirmed. It is now recognized that the synthesis of many, and probably all, secretory proteins is carried out on membrane-attached polysomes, whereas a number of soluble proteins of the cytosol are synthesized exclusively by free polysomes (reviews, Rolleston, 1974; Palade, 1975; McIntosh and O'Toole, 1976; Shore and Tata, 1977). The functional significance of the ribosome-membrane interaction has also been studied extensively. Ribosomes are now known to be attached by their large subunits (Sabatini et al., 1966) to specific receptors localized exclusively, and in finite number, in the membrane of the rough-surface endoplasmic reticulum (ER) (Borgese et al., 1974; Mok et al., 1976; Fujita et al., 1977; Kreibich et al., 1978a, b). These receptors allow the transmembrane passage of the growing polypeptide chains. Thus, upon completion, the latter are vectorially discharged within the lumenal space of the ER, where they remain segregated.

Recently a hypothesis (designated as the signal hypothesis) has been proposed to account for the striking selectively of secretory protein synthesis on bound polysomes (Blobel and Dobberstein, 1975a, b). This theory is based on the assumption that all mRNAs coding for the proteins that will be segregated within the ER might contain peculiar base sequences translated as corresponding sequences of amino acids located at the N terminal of the growing chain. The attachment of ribosomes to the ER would be triggered by the interaction of these N-terminal peptides (the signal peptides) with the membrane receptors, and would therefore occur not before, but during, the synthesis of the secretory proteins. Signal peptides would be cleaved from the entire translation products (designated as secretory preproteins) before chain termination, presumably by a microsomal proteolytic enzyme(s).

So far, putative signal peptides have been identified in the translation products corresponding to a considerable number of secretory proteins. Even more important is the observation that for no secretory protein has the existence of signal peptides been excluded. For further information and references on the signal hypothesis, the reader should consult the recent review by Campbell and Blobel (1976). According to an alternative hypothesis, not mutually exclusive with the signal hypothesis, mRNA for secretory proteins is bound directly to the ER membrane (Lande et al., 1975; Sabatini et al., 1976).

The sequence of events that occurs after the release of secretory proteins from the bound ribosomes is very complex. In general, two types of phenomena are known to occur. On the one hand, newly synthesized proteins are transported throughout the cell in order to become available for discharge; concomitantly, the secretion products change in composition and physical arrangement. Based mostly on the work of Palade and his associates, a general model has been developed, according to which all these phenomena take place vectorially within the lumen of a membrane-bounded, functionally continuous secretory pathway. This model is now widely accepted. It is referred to here as the orthodox view of intercellular transport, and is described in its full complexity in section 2.1. below. Alternative interpretations have also been advanced, and these are considered critically in section 2.2.

2.1. The orthodox view

The basic assumption of the orthodox view is that once the secretory proteins are discharged by the bound ribosomes within the lumen of the ER cisternae they can no longer cross the membranes of the secretory pathway, and therefore remain segregated throughout their whole intracellular life and form a totally separate pool with respect to the soluble proteins of the cytosol. Another fundamental aspect of this scheme concerns exocytosis. It is assumed that the membrane of the secretory vesicle, which is programmed to be inserted at the cell surface during exocytosis, is characterized by low permeability properties so that intracellular homeostasis is not disturbed by the exocytotic process. However, the ER membrane, which segregates the newly synthesized proteins, is very permeable and this should preclude its direct involvement in exocytosis. Thus secretion products are assumed to be transported to other membrane-bounded compartments along a specific pathway prior to their discharge.

Although the secretory pathway exhibits considerable heterogeneity in different cell systems, its basic organization appears conserved and includes a number of different and discrete cellular structures aligned in series, through which secretion products travel sequentially: the ER, the Golgi complex, and a class of specific organelles endowed with storage capacity and with the ability to discharge their content by exocytosis (i.e., the secretion granules and vesicles). Evidence for the existence of this general organization is now compelling and appears to be valid for a broad range of secretory systems (Fig. 1) (reviews, Jamieson, 1972, 1973, 1975; Palade, 1975).

Although the picture is clear in its general outline, a number of details are still uncertain or poorly understood. They are considered in the remainder of this section.

2.1.1. What happens in the ER lumen?

ER membranes are highly permeable. Nilsson and co-workers (1973) found that nonelectrolytes with molecular weights lower than 600 are free to diffuse across these membranes, while charged molecules of 90 MW are apparently excluded. Kreibich and colleagues (1974) also found that pyridoxal phosphate (a negatively charged molecule, 247 MW) is able to cross the microsomal membrane. These data suggest that the ionic environment within the ER lumen is probably similar to that of the surrounding cytoplasm. It is generally thought that the segregated protein molecules, which enter the lumen as extended polypeptide chains, acquire their tertiary (or quaternary) structure upon release from ribo-

514

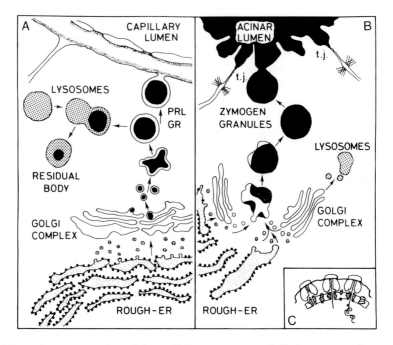

Fig. 1. Schematic representation of intracellular transport and discharge operations, orthodox view. Some existing variations on the theme are exemplified by an endocrine cell (the prolactin-producing cell of the anterior pituitary) shown in (A) and an exocrine cell (the acinar cell of the pancreas) shown in (B). In both systems, the synthesis of secretory proteins on membrane-bound ribosomes is followed by their vectorial discharge and segregation within the lumen of ER cisternae, detailed in (C) (text, section 2.1). Transport from the ER to the Golgi complex is assumed to be carried out by small vesicles (text, section 2.1.2). Two processes occurring in the Golgi complex are illustrated in the figure: the separation of secretion products from lysosomal enzymes (shown only in panel B) and the condensation of secretion products (text, section 2.1.3). The latter process occurs within the lumen of parallel Golgi cisternae in prolactin cells, and in discrete condensing vacuoles in pancreatic acinar cells. In prolactin cells, prolactin granule (PRL GR) discharge occurs by exocytosis localized preferentially at the plasmalemma regions adjacent to the basement membrane (text, section 5.1). Alternatively, secretory granules can be disposed of within the cytoplasm by digestion within lysosomes, which then transform into residual bodies (crinophagy; text, section 2.1.4). In pancreatic acinar cells, exocytosis is restricted to the luminal portion of the plasmalemma, delimited circumferentially by the zonula occludens (tj) (text, section 5.1).

somes and therefore become more bulky and incapable of diffusion back across the membrane.

At least in the liver, some newly synthesized proteins, especially glycoproteins, are apparently not free in the lumenal content but are still membrane bound (Redman and Cherian, 1972; Sauer and Burrow, 1972; Kreibich and Sabatini, 1974). This might be due to the fact that both the glycosylating enzymes and the lipophilic dolichol-phosphate sugar chain donors are located at the inner surface of the ER membrane (Cook, 1977). It may therefore be concluded that the movement of the different segregated proteins within the cisternal space is probably carried out by at least two mechanisms: nonglycosylated proteins might travel through the aqueous phase while glycoproteins might, at least for some time, remain attached to or crawl on the membrane inner surface. In the course of these processes lipoprotein particles are assembled through the binding of increasing amounts of triglycerides and phospholipids (synthesized by ER membrane enzymes) to the apolipoprotein molecules in transit within the cisternae (Glaumann et al., 1975).

A final problem concerns whether or not segregated proteins are distributed homogeneously within the ER system (which is lumenally continuous in its whole). In fractionation studies carried out on liver tissue under pulse-chase conditions, secretory proteins were recovered in both the rough- and smooth-surfaced portions of the ER, with kinetics of appearance and release compatible with a direct participation of the latter in transport as an intermediate compartment between rough ER and Golgi complex (Glaumann and Ericsson, 1970; Peters et al., 1971; Kreibich and Sabatini, 1974; Glaumann et al., 1975).[3] Analogous results were obtained by Olsen and Prockop (1974) studying the distribution of procollagen in fibroblasts. Moreover, three enzyme proteins (ribonuclease in pancreatic acinar cells [Painter et al., 1973]; peroxidase in lacrimal gland acinar cells [Herzog and Miller, 1972; Herzog et al., 1976] and in follicular thyroid cells [Novikoff et al., 1974]; and myeloperoxidase in promyelocytes [Bainton et al., 1971]) have been found by immunoelectron microscopy to be evenly distributed within the ER. However, in plasmocytes differentiating toward secretion, immunoglobulins were first detected in only a few cisternae and later found to spread to most of the others (Leduc et al., 1968). Taken together, these results indicate that even if a certain degree of heterogeneity exists in some cases, in many others the distribution of segregated proteins is probably homogeneous in the whole ER system.

2.1.2. The ER-Golgi boundary

To discuss transport of segregated proteins across the ER-Golgi boundary, it is essential to establish the location of this boundary.

On the basis of classical cytochemical studies, Novikoff and his associates have identified in many cells a region of the endomembrane system, appearing to be spatially and functionally related to the Golgi, which is involved in the synthesis of lysosomes and, in some systems, also in the packaging of secretory proteins. They called this region GERL (Golgi-ER-lysosomes). In their opinion, GERL, although endowed with unique features, is a region of the smooth ER because of its apparent focal membrane continuity with the rest of the ER system (Novikoff et al., 1971, 1974, 1977; Novikoff, 1976; Novikoff and Novikoff, 1977). This assignment can be questioned for several reasons. First, even the concept of GERL itself is vague since it is derived only from the cytochemical localization of a segregated enzyme, acid phosphatase; no information about the features of

---

[3]However, Alexander and co-workers (1976), using an immunocytochemical approach, failed to detect apolipoproteins in the lumen of the smooth ER.

GERL-limiting membranes is available. It is now clear that focal continuity of two membranes cannot be interpreted as an indication of molecular identity or even similarity (sections 3.5, 6). Second, since ER is essentially an homogeneous membrane system the presence of a highly differentiated appendix would be a surprising exception to this general rule. Third, tridimensional reconstructions of the Golgi apparatus carried out on the basis of thick section analyses indicate that the Golgi apparatus is much more irregular, complicated, and ramified than was thought previously and that its interconnected network is not limited to parallel stacks alone (Rambourg et al., 1973; Thiery and Rambourg, 1976). Finally, in some systems (e.g., the exocrine pancreas of the guinea pig), the condensation and packaging of secretion products can occur in membrane-bounded elements (the condensing vacuoles) assigned to GERL by Novikoff (1976, Novikoff et al., 1977) or in "respectable" Golgi stacks, depending on the functional state of the gland (section 2.1.3). In our opinion, a conclusion about the nature of GERL could only be obtained after careful evaluation of the spectrum of biochemical features of its membranes (sections 3.1, 3.2, and 3.5). However, GERL fractions have not been isolated to date.

On the basis of these considerations, the evidence provided in favor of the ER nature of GERL seems inconclusive. In contrast, stronger reasons exist to examine GERL as a component, together with the Golgi *apparatus* (the so-called stacks) of a structurally heterogeneous, but functionally unique, organelle system called the Golgi *complex* (according to the nomenclature used by Palade (1975)).

Two different mechanisms have been proposed to account for the transport of segregated proteins from the ER to the Golgi complex. On the one hand, the results of the Palade and co-workers on the guinea pig pancreas have been interpreted in terms of a physical discontinuity of the two compartments, which would be connected functionally by vesicles containing segregated proteins. These vesicles would originate by budding from transitional elements (partially rough, partially smooth cisternae existing at the ER periphery) and would then unload their content into the cavities of the Golgi complex. Since in the course of this process (which goes on continously in secretory cells) no piling up of membranes is observed in the Golgi area, it has been suggested that the vesicle movement might occur in both directions, like "shuttles" between the two compartments (Jamieson and Palade, 1967a; Jamieson 1972, 1973, 1975; Palade, 1975).

The alternative possibility that has been proposed is based on images (obtained mostly in hepatocytes) showing direct, lumenal continuity of ER cisternae with peripheral large tubules and saccules of the Golgi complex (Claude, 1970; Morré et al., 1971, 1974; Morré, 1977). Recently Thiéry and Rambourg (1976) developed staining procedures by which individual regions of the plasmalemma and endomembrane system can be revealed selectively. The application of these procedures to thick (0.5–1 μ) sections, coupled with sophisticated techniques of sample observation and image reconstruction, seems to confirm the ER-Golgi continuity in several tissues (Thiéry and Rambourg, 1976). According to these findings the previous images obtained in the pancreas might be due to a continuous network of irregularly twisted tubules rather than to discrete vesicles.

At least two firm conclusions can be drawn. First, the continuity (either physical or only functional) of ER and Golgi membranes is unusual because it does not result in the random intermixing of their molecular components. In this respect, it is worth mentioning that in many cell systems the boundary is marked by peculiar beads (Locke and Huie, 1976a, b) of unknown chemical nature which, because of their regular localization, might be endowed with some special physiological significance. Second, in a number of different secretory systems, transport of segregated proteins across the boundary is known to require energy (Jamieson and Palade, 1968b; Howell and Whitfield, 1973; Labrie et al., 1973; Kruse and Bornstein, 1975; Chu et al., 1977; Tartakoff and Vassalli, 1978). Hence, this transport cannot be considered as a simple drainage of the content from one container to the other, but must be thought of as an active process. It has been supposed that fusion-fission of the membrane of the shuttling vesicles with those of the adjacent compartments is the energy-dependent process (Jamieson and Palade, 1968b). This might also explain the case of tubular continuity in that the establishment of the latter might be a dynamic process which undergoes continuous rearrangement and is therefore rapidly disconnected in energy-depleted cells.

2.1.3. Molecular and cellular events occurring in the Golgi complex
The phenomena taking place in the Golgi complex are extremely important. In brief, the secretory proteins transported from the ER as a dilute mixture are modified molecularly, complexed with accumulated ions, redistributed in a specific manner, and finally concentrated and packaged within discrete organelles. Among the molecular modifications of the segregated proteins are the attachment of sugar residues of glycoproteins and the synthesis and sulfation of the carbohydrate chains of protidoglycans. The first of these processes, which occurs by stepwise addition, can be considered as the biochemical hallmark of the Golgi complex, since sugar transferases are used as marker enzymes in cell fractionation studies. As far as protidoglycans are concerned, the synthesis of their sulfated sugar moiety in the Golgi complex and their subsequent transport to granules is now fully demonstrated by cytochemical (Dunn and Spicer, 1969; Olson, 1969), radioautographical (Berg and Young, 1971; Young 1973; Berg and Austin, 1976), and biochemical (Da Prada et al., 1972a; Tartakoff et al., 1974; Blaschke et al., 1976; Giannattasio and Zanini, 1976; Winkler, 1976) evidence obtained in a variety of systems. In this respect it should be mentioned that protidoglycans are released by exocytosis together with other secretion products in cell systems whose major physiological activity is apparently not related to protidoglycan secretion (Berg and Young, 1971; Margolis et al., 1973; Berg and Austin, 1976; Geissler et al., 1977).

Another molecular rearrangement that occurs within the Golgi complex (and possibly also within immature secretion granules) is the processing of some segregated proteins (designated with the prefix pro-), whose size must be reduced by controlled cleavage before release from the cell. This process (often referred to as hormone activation) might appear similar to the processing of secretory preproteins described above in the context of the signal hypothesis. However, this phenomenon is different because: (1) it involves only some of the segregated proteins traveling through the Golgi complex; (2) it is not an obligatory step for protein secretion, since precursor molecules escape cleavage in variable proportion and are released together with the finished peptides; (3) different proteins are cleaved in different positions of their peptide chains; and (4) the existence of precursors of the pro-type has been demonstrated for a limited number of hormones and proteins (reviews, Steiner et al., 1974; Campbell and Blobel, 1976; and also Goldstein, 1976; Morris and Johnson, 1976), while the synthesis of pre-type precursors appears to be a general phenomenon.

A further metabolic event that occurs in the Golgi complex is ion accumulation. This phenomenon is known to occur for Zn^{2+} in pancreatic B cells (Steiner et al., 1974) and for Ca^{2+} in many and possibly all secretory cells. The latter conclusion is based on the following arguments: (1) some Golgi-enriched fractions have been found to bind Ca^{2+} in much higher amounts than true microsomes by an energy-dependent mechanism distinguishable from that of mitochondria (Poisner and Hava, 1970; Selinger et al., 1970; Alonso et al., 1971; Argent et al., 1975; Baumrucker and Keenan, 1975; Moore et al., 1975; Russel and Thorn, 1975; Freedman et al., 1977); (2) in several systems (Clemente and Meldolesi, 1975a; Stoeckel et al., 1975; Ravazzola, 1976; Reith, 1976) calcium was found to be present in considerable amounts in the Galgi apparatus, as revealed by biochemical and/or cytochemical studies; and (3) in many cases calcium was found to be released concomitantly with the secretory proteins (Wallach and Schramm, 1971; Ceccarelli et al., 1975). Furthermore, at least in the exocrine pancreas, evidence exists that the kinetics of discharge under pulse-chase conditions in compatible with its accumulation in the Golgi complex (Ceccarelli et al., 1975). These observations are also important because they suggest that the ionic environment of the Golgi content is different from that of the surrounding cytoplasm, a situation at variance with that discussed earlier for the ER.

Parallel with these biochemical events, the Golgi complex is also involved in the concentration and redistribution of the segregated proteins derived from the ER to yield different types of discrete organelles. Redistribution is a general phenomenon in secretory systems because secretion products must be separated from lysosomal enzymes which always travel the same pathway from the ER to the Golgi complex. In this respect the paper by Kraehenbuhl and collaborators (1977) is of interest. In the exocrine pancreas of the guinea pig [a system which is capable of secreting at least 20 distinct proteins (Sheele, 1975)] these authors have traced, by immunochemistry at the electron microscope level, the distribution of five secretory zymogens, both in the Golgi complex and in the secretory granules. The results indicate a homogeneous distribution for all these proteins, that is, no qualitative regional differences were detected within the gland, within each acinar cell, or within each granule and Golgi element.

On the other hand, enzyme cytochemistry carried out on the exocrine pancreas, as well as on other tissues, indicates that lysosomal enzymes might appear (although in small amounts) in immature secretion granules, where they are located at the extreme periphery of the segregated content. In most mature organelles they are no longer visible (Novikoff and Essner, 1962; Farquhar, 1971; Farquhar et al., 1974; Novikoff 1976). These results indicate that the process of redistribution of segregated proteins is thoroughly controlled. Secretory proteins that must be copacked within discrete organelles might be able, in the ionic environment existing in the Golgi complex, to interact with each other forming supramolecular complexes that would then be segregated in different organelles through fusion-fission of the limiting membranes. Furthermore, additional mechanisms of recognition, probably also mediated by membrane fusion, might operate to correct all possible spills of the redistribution process, as suggested by the disappearance of lysosomal enzymes from secretory granules during maturation.[4]

In this respect, intriguing examples of control are provided by those cells which are capable of discharging more than one population of granules. A classical example is the leukocyte system studied by Bainton and Farquhar (1966, 1970; Bainton, 1973), where azurophil and specific granules are produced from the same Golgi complex at different times and at the opposite faces of the organelle, and then discharged separately. Other examples have also been studied. Thus, the thyroid follicular cell produces B granules (containing thyroglobulin), which originate from Golgi stacks, and A granules (containing peroxidase), which arise from an adjacent tubular network[5] (Novikoff et al., 1974). The intermediate cells of the rat pancreas secrete zymogens (at the cell apex) as well as either glucagon or, more frequently, insulin (at the basolateral surfaces) (Melmed et al., 1972, 1973a,b). Zymogens and insulin are packed within the same Golgi complex, but in different cisternae, whereas glucagon-containing granules arise from an independent dictyosome. Regulation of the two types of discharge appears independent in this system. Another example is somatostatin which is stored together with gastrin in the D cells of Langerhans islets (Erlandsen et al., 1976) and together with calcitonin in thyroid parafollicular cells (Van Norden et al., 1977). Thus the ability of single cells to carry out multiple vesicular secretions concomitantly might be a frequently occurring phenomenon.

The processes of secretion product redistribution and granule formation exhibit clear cytological variability in different systems. In some cases granule maturation does not occur within the Golgi stacks (which seem to be bypassed by the secretion products) but in adjacent structures of the Golgi complex referred to as condensing vacuoles. The latter apparently originate through the successive fusion of many vesicles (some of which are coated[6]), which may be seen all around the condensing vacuoles, as well as in continuity with their limiting

---

[4]It is interesting that a lysosomal enzyme, acid phosphatase, was found in the content of most secretory granules in one case of transplantable hamster insulinoma (Novikoff et al., 1975), which presumably was incapable of full control of the intracellular distribution of segregated proteins. (For a different interpretation of this phenomenon, see Novikoff et al., 1975.)

[5]This network is interpreted by Novikoff and collaborators (1974) as ER. (For a specific discussion of this problem, see section 2.1.2.)

[6]For a detailed discussion of coated vesicles, see section 6.1.

membrane (Jamieson and Palade, 1967a; Herzog and Miller, 1972; Novikoff et al., 1975). Since, as in all spherical organelles, the volume of condensing vacuoles grows faster than their surface, and since their content is simultaneously concentrated, a massive surplus of Golgi membrane must result at this site (~ 50-fold excess has been calculated for the exocrine pancreas by Meldolesi, 1974b) that must be removed, probably as vesicles, either empty or loaded with any mislodged component. In contrast, in many other systems the formation of secretory granules occurs within the Golgi cisternae, usually within the innermost element or toward the edges of the others. Also in these cases there is evidence of intense fusion-fission of vesicles in relation to the process.

This difference in the site of granule assembly may not reflect basically different cytological mechanisms. If pancreatic acinar cells are heavily stimulated for secretion, there is a shift of the zymogen localization and site of granule formation from condensing vacuoles to the parallel cisternae, and the granules produced are much smaller (Jamieson and Palade, 1971b; Völkl et al., 1976). This also occurs if acinar cells are treated with colchicine (unpublished observations, Meldolesi et al.). Thus it seems possible that the apparent bypassing of the piled cisternae is due only to incomplete use of the membrane potentialities of the Golgi complex, which occurs in the unstimulated cell. If needed, the cisternae can be used as well.

The final Golgi phenomenon to be discussed is condensation. For some time extrusion of water was believed to occur through the action of an active pump localized in the membrane of all granules. Later it was recognized that many granules are osmotically inactive (Jamieson and Palade, 1971b; Howell, 1974; Lemay et al., 1974) and that some of them are so stable that they retain their structure even after in vitro removal of the limiting membrane (Anderson et al., 1974; Giannattasio et al., 1975). Furthermore, in many systems identifiable granule cores are seen for some time after exocytosis in the extracellular space, secluded within pocket indentations of the plasmalemma. Thus it is now clear that secretory granules are not membrane-bounded vesicles containing a concentrated solution of secretion products, but organelles whose content is highly structured or even in a solid state. In vitro studies on the stability of isolated granules (Howell and Ewart, 1972; Lemay et al., 1974; Giannattasio et al., 1975; Helman, 1975). and on the reassociation of solubilized components (Pletscher et al., 1974; Rothman et al., 1974) suggest that, at least in some cases, complex ionic interactions among the segregated molecules (which can be extremely heterogeneous; see Ceccarelli et al., 1974; and recent reviews by Uvnäs, 1974a; Dreifuss, 1975; and Winkler, 1976, 1977) might be responsible for their supramolecular arrangement. Since the nature of secretion products contained in different granules vary widely, it is expected that the mechanisms involved in storage might also differ greatly. The available evidence suggests that in some cases a primary role might be played by Ca^{2+}, either alone (Wallach and Schramm, 1970), or in association with nucleotides and acidic proteins (Smith, 1973; Pletscher et al., 1974), or in others by organic polyanions such as protidoglycans (Tartakoff et al., 1974; Uvnäs, 1974a,b).

2.1.4. The secretory granule: function and disposal

Although the granules of various secretory cells differ in their segregated content, they share a number of common features. The peculiarities of the lipid and protein composition of their limiting membranes are discussed specifically in section 3.3. Here we wish to emphasize that these membranes are probably characterized by very low permeability, as indicated by the observation that their incorporation at the cell surface during exocytosis does not lead to gross changes of transmembrane conductance of the plasmalemma. Recently these indirect conclusions were confirmed by detailed studies carried out on the chromaffin granules of the adrenal medulla. The limiting membrane of these organelles was found to exhibit a strikingly low permeability to both monovalent and divalent cations. Furthermore, the internal pH of the granules was low (~ 5.5) and independent of the pH of the surrounding medium (Johnson and Scarpa, 1976). These findings seem to exclude the notion that the content of chromaffin granules be in rapid equilibrium with the cytoplasm, as has been suggested previously (Serk-Hanssen and Christian, 1973; Borowitz et al., 1975). Obviously, this does not mean that chromaffin granules are devoid of any transport activity. On the contrary, throughout their whole lifetime they are capable (Kirschner and Viveros, 1972; Slotkin and Kirschner, 1973; Winkler, 1977) of specific catecholamine uptake[7] probably coupled with the uptake of nucleotides (mostly ATP, of mitochondrial origin) (Kostron et al., 1977a), driven by an ATP-dependent proton translocation (Bashford et al., 1975). Furthermore, a slow accumulation of Ca^{2+}, probably by facilitated diffusion, has been described (Kostron et al., 1977b). Analogous mechanisms might also operate in the vesicles of the adrenergic terminal that have a similar composition, although they contain far less segregated protein (especially chromogranin) (Smith, 1973; Lagerkrantz, 1976; also see section 2.1.5).

In the other secretory granules, ion transport across the limiting membrane has not been thoroughly investigated. The results obtained so far have been limited and sometimes contradictory.

The β-granules of pancreatic islets have been reported to be capable of adjusting their internal calcium concentration in relation to the functional state of the cell, as indicated by cytochemical studies (Herman et al., 1973; Ravazzola et al., 1976). The opposite conclusion was reached by Howell and associates (1975) using radioautography. On the other hand, for many other secretory granules the available evidence strongly suggests that their primary (or even exclusive) function is that of simple storage and segregation organelles (Wallach and Schramm, 1970; Howell and Ewart, 1972; Clemente and Meldolesi, 1975b;

---

[7]The ability to accumulate biogenic amines and nucleotides is not exclusive for chromaffin and other traditional amine-containing granules but is also shared by the granules of pancreatic B cells (Jaim-Etcheverry and Zieher, 1968; Laitner et al., 1975) and thyroid parafollicular cells (Gershon and Nunez, 1973). The physiological significance of these observations is unknown. The phylogenetic and embryological relationships of amine-storing cells have been reviewed recently by Pearse and Polak (1974).

Giannattasio et al., 1975; for alternative views see Rothman, 1975b, and section 2.2).

As already mentioned, the fate of the secretion granules is exocytosis. In some cells, however, a different type of granule elimination has been observed where the need for secretion is suddenly reduced. This phenomenon, originally described by Smith and Farquhar (1966) in the prolactin-producing cells of lactating rats shortly after abrupt separation from their litter, and subsequently observed in other systems (Creutzfeld et al., 1969; Orci et al., 1970; Farquhar, 1971; Habener et al., 1975), has been called crinophagy and consists in the discharge of the granule content within lysosomes by a process analogous to exocytosis (Fig. 1). Crinophagy is not a surprising phenomenon per se since the coordinated fusion of lysosomes and secretion granules is known to occur on other occasions, as for example in leukocytes (Bainton, 1973). However, it is worth mentioning because it represents a further example of fine modulation of membrane fusion in secretory cells.

2.1.5. Modulation and parallelism

So far we have considered the individual steps of intracellular transport separately. To complete the description of the process, we must take into account two specific problems concerning the rate at which transport operations are carried out.

It is well known that many secretory systems are capable of modulating their physiological activity in response to the influence of hormonal, neural, and metabolic signals. Such modulation mostly occurs at the level of discharge and, in many cases, the increase of discharge is accompanied or followed by an increase in synthesis of secretion products (e.g., see Grand and Gross, 1969; Labrie et al., 1973). Intracellular transport is different. The rate of this process, although it varies considerably in different secretory systems, was found to be remarkably constant within each individual system and to remain unchanged even after stimulation in vivo as well as in vitro (Jamieson and Palade, 1971a; Labrie et al., 1973). Recently, however, it was demonstrated that in the exocrine pancreas of rats exhaustively stimulated for several hours by secretagogue infusion, transport operations were also carried out at a considerably faster rate than in unstimulated controls (Bieger et al., 1976). Thus it may be concluded that even if transport is much less sensitive than synthesis and discharge to the action of secretagogues, it is, nevertheless, also a potentially regulated process, at least in the exocrine pancreas.

The second question is whether the rate of transport along the secretory pathway is the same for all exportable proteins released by each single system, that is, whether transport is parallel.[8] Morgan and Peters (1971) reported that in the liver of the rat the time interval between synthesis and release was consider-

---

[8]The same nomenclature is used by Rothman (1975b) in a different context, to indicate the release of different secretory proteins in constant (parallel) or variable (nonparallel transport) proportions relative to each other over time. For a discussion of the work of Rothman, see section 2.2.

ably longer for transferrin than for albumin. Recently the problem was reinvestigated in the exocrine pancreas by Tartakoff and Sheele (1978). Of the secretory proteins analyzed, three were found to be transported more slowly than the others. It is interesting that transferrin and probably the three slow, pancreatic zymogens as well are glycoproteins. It seems likely that the extra time required for their transport might be spent for the biosynthesis of the oligosaccharide chains that occurs in direct association with the ER and Golgi membranes discussed earlier.

2.1.6. Intracellular transport in nerve cells
Intracellular transport in nerve cells is dealt with separately in this chapter because it deviates in some respects from the general scheme given above for other systems. This is due, on the one hand, to the peculiar geometry of neurons and, on the other, to the discrete localization of specific functions in different portions of the cell. In particular, protein synthesis is localized in the cell body (Jones et al., 1975) and transmitter release at the axon terminal. Thus the intracellular transport of some secretion products is necessarily carried out over long distances and can be considered as an important component of axonal transport. The latter includes all the processes of macromolecule exchange that occur in either direction between cell body and axon terminals and play an essential role in the renewal and integrated functionality of nerve cells. (For comprehensive accounts of axonal transport see Dahlstrom, [1971]; Droz, [1975]; Kreutzberg and Schubert [1975]; and Livett [1976], as well as the chapter by Holtzman et al. in the fourth volume of *Cell Surface Reviews*).

The situation for the intracellular transport of secretion products is not the same in all neuronal systems. Thus neurosecretory cells of the hypothalamo-neurohypophysial tract are very similar to regular secretory cells with regard to the synthesis, transport, and packaging of secretion products; the only real difference is the longer time needed for secretory granules to become available for discharge (Dreifuss, 1975; Heap et al., 1975; Pickering et al., 1975; Thorn et al., 1975; Morris et al., 1978).

In contrast, peculiarities clearly exist in both adrenergic and cholinergic neurons. Noradrenaline vesicles have a composition analogous to that of chromaffin granules; they contain segregated proteins, catecholamines, ATP, and divalent cations. Upon exocytosis, these components are all discharged in the expected ratios (Smith et al., 1971; Smith, 1973; Lagerkrantz, 1976). The composition and arrangement of the cholinergic vesicles is less clearly understood. It has been reported repeatedly that ATP is stored together with acetylcholine (ACh) (Dowdall et al., 1974; Silinsky, 1975; Zimmermann and Denston, 1976); for opposite views see Kato et al. (1974); Meunier et al., (1975). Whittaker and co-workers (1974) described the presence of a small acidic peptide, called vesiculin, located within the cholinergic vesicles isolated from the electric organ of *Torpedo marmorata*. However, the significance of this finding is still unclear because so far vesiculin has not been detected in the effluent of stimulated electric organs, nor identified in other cholinergic systems. In addition, the claim of

Musick and Hubbard (1972) of protein discharge from stimulated neuromuscular junctions could be due to leakage from injured muscle cells.

Recently electron-dense particles have been visualized within cholinergic vesicles of different types of synapses in specimens fixed with aldehydes either at low pH (Gray and Paula-Barbosa, 1974) or in the presence of high Ca^{2+} (Boyne et al., 1974; Hillman and Llinas, 1974; Politoff et al., 1974; Benshalom and Flock, 1977). More interestingly, the latter dense particles decrease drastically in number after stimulation (Boyne et al., 1975; Pappas and Rose, 1976). These observations suggest that cholinergic vesicles might also contain a structured internal organization and therefore raise the question as to where and how such an organization is established.

The classical interpretation of these phenomena, based chiefly on radioautographic studies, implies that both cholinergic and adrenergic vesicles are elaborated in the Golgi complex of the perikarya, and then transported down the axon at high rate (a few hundred mm per day-fast axonal transport), together with other discrete structures destined to be incorporated primarily in the terminal, especially in the synaptic membrane. Concomitantly another transport of soluble molecules occurs at a slow rate to renew the axon itself (Dahlstrom, 1971; Bennet et al., 1973; Di Giamberardino et al., 1973; Droz et al., 1973). However, these conclusions are still being debated. Cell fractionation studies suggest that vesicle renewal is relatively independent of fast axonal transport (Carlton and Appel, 1973; Marko and Cuenod, 1973; Kryger-Brevart et al., 1974). Furthermore, one identified component of adrenergic vesicles, dopamine-β-hydroxylase, is apparently transported in two pools with both types of transport, and not by the fast type alone (Brimjoin, 1974).

An alternative mechanism for synaptic vesicle formation within the terminals by budding from smooth ER elements present in the axon had been proposed on the basis of images of conventional electron microscopy and ultrastructural cytochemistry (Tranzer, 1972; Hökfelt, 1973). Recently this hypothesis has been greatly strengthened by tridimensional reconstructions of the endomembrane system of axons. These reconstructions were derived from low- and high-voltage electron micrographs of thick sections coupled to radioautography (Droz, 1975; Droz et al., 1975; Markov et al., 1976). In these studies it was shown convincingly that: (1) the so-called smooth endoplasmic reticulum is a continuous branched system, extending from the perikarya to the terminals, which appears to establish initimate connections with synaptic vesicles; and (2) that a large proportion of the rapidly migrating protein-and glycoprotein-bound radioactivity is associated with this system.

It is probably premature to reach definite conclusions on this issue now. However, the images strongly suggest the possibility that the vesicles, besides their classical origin in the "true" Golgi complex, might also be formed at the periphery by direct budding from the smooth ER membranes (Ceccarelli and Hurlbut, 1975a). Thus one might speculate that one of the major functions of the latter system would be in the transport of segregated molecules to be packaged within the vesicles (Droz, 1975; Holtzman et al., 1977), analogously with the function of the endomembrane system of "typical" secretory cells. If this proves to be the case, however, it is doubtful whether this system should be assigned to the true ER, at least in its terminal portions (section 2.1.2.). Actually, the evidence supporting such an assignment (Holtzman et al., 1977) is very limited and seems to be based mainly on simple morphological analogy. So it seems that alternative possibilities, such as its assignment to the Golgi complex, cannot be completely excluded at present. A more detailed knowledge of some of its properties (such as the chemical composition, possible regional heterogeneity, structural plasticity, and variability in different neurons) is needed before a clear understanding of its nature and functions can be achieved.

A third mechanism that has been proposed to account for vesicle formation is the direct recycling of membrane from the synaptic surface. This is discussed in detail in section 6.2.

2.1.7. Driving forces

It is probably safe to state beforehand that in spite of the large number of reports about the mechanisms regulating intracellular transport, and the driving forces involved, our knowledge of these processes is in fact limited and contradictory.

As a whole, intracellular transport is an energy-dependent process. However, this requirement is limited only to some definite steps. Actually, the vectorial discharge of secretion products and their subsequent flow in the ER cisternal space are apparently unaffected by energy withdrawal. In contrast, the transfer of segregated proteins across the ER-Golgi boundary is an active process. As far as granule formation is concerned, the situation is still unclear. An energy requirement has been described in some systems (Howell, 1972; Howell and Whitfield, 1973; Labrie et al., 1973). However, in the exocrine pancreas, the conversion of condensing vacuoles to mature zymogen granules is apparently energy-independent (Jamieson and Palade, 1971a). Finally, a third energy block coincides with exocytosis and is discussed further in section 5.3. Jamieson and Palade (1968b) proposed membrane fusion-fission as the possible energy-requiring process. This hypothesis has gained wide acceptance even though no direct experimental proof has ever been reported.

In contrast, intracellular transport is not affected by inhibition of protein synthesis (Jamieson and Palade, 1968a, 1971a; Howell and Whitfield, 1973). This observations tends to rule out the simple idea that accumulation of newly synthesized proteins might have a role in the progression of products along the secretory pathway.

The possibility that microtubules might play a role in intracellular transport was proposed originally by Lacy and his associates (Lacy et al., 1968). Later microfilaments also began to be investigated from this standpoint, and both structures have continued to attract much attention. The major limitation of these studies has been the lack of a direct experimental approach. In fact much of the available information has been obtained by employing drugs such as colchicine, vinblastine, and D_2O on the one hand, and cytochalasin B on the other, which are known to interfere with the function of microtubules and microfilaments, respectively. However, the action of these drugs is by no means specific for microtubules and microfilaments, since they all have a wide spectrum of effects. In particular, they can influence the tissue response by metabolic mechanisms, especially interfering with membrane transport. For colchicine, an anticholinergic activity has also been reported (Trifaro et al., 1972). Moreover, it is not yet clear which step(s) of intracellular transport is blocked by antimicrotubule and antimicrofilament drugs. Careful studies have shown that antimicrotubular agents inhibit granule discharge in the liver (Orci et al., 1973c; Stein and Stein, 1973; Redman et al., 1975) and in pancreatic B cells (Malaisse et al., 1976). Other authors have reported, however, that an important inhibitory effect of antimicrotubular drugs is at the ER-Golgi boundary (Feldman et al., 1975; Lohmander et al., 1976; Chu et al., 1977; Rossignol et al., 1977). It should be emphasized that these drugs have a profound effect on the overall geometry of the cell. In particular, the Golgi apparatus is fragmented into many discrete pieces, randomly redistributed throughout the cell. Thus it is possible that the effect on transport is not a specific consequence of microtubule depolymerization but only the result of a disturbance in the controlled interactions of the different cytoplasmic membranes. Also, the widely accepted role of microtubules in rapid axonal transport has been questioned recently (Byers, 1974).

The possible role of the cytoskeleton in the exocytotic process is discussed further in section 5.3.

2.2. An evaluation of alternative views

During recent years a number of experimental observations, which apparently do not fit the constraints of the othodox view, have been reported and have led to the proposal of alternative models. The general feature of many of these models is that secretion products are thought to be stored in multiple pools, one of which would be free in the cytoplasm (soluble pool), and that the release of the different pools is subject to different physiological controls. The data supporting these models derive mainly from experiments designed to study the kinetics of product discharge and the intracellular distribution of secretion products. Thus secretion products are sometimes discharged with multiphasic kinetics, suggesting storage in multiple pools, which are not equally available for exocytosis (Grodsky, 1972; Malaisse et al., 1975). Furthermore, in many systems newly synthesized molecules have been found to be discharged at different times, and with different kinetics with respect to the older ones. (Kopin et al., 1968; Sachs and Haller, 1968; Dunant et al., 1972, 1974; Marchbanks and Israel, 1972; Bennett, 1973; Katz et al., 1973; Molenaar et al., 1973; Norström, 1974). Fractionation studies have revealed that in all systems variable amounts of secretion products are recovered in the final, soluble supernatant obtained by high speed centrifugation of the homogenate. This localization is usually regarded as a redistribution artifact depending on leakage from the membrane bounded organelles. In some cases, however, an opposite interpretation has been envisaged on the basis of the finding that these soluble secretion products are present in different relative proportions compared to their bound counterparts and show different kinetics of biosynthesis and discharge (Marchbanks and Israel, 1972; Norström, 1974; Rothman, 1975b).

Morphological observations, which apparently contrast with the orthodox view, have also been reported. Thus, in extensively stimulated tissues, a lack of correlation has sometimes been observed between the rate of discharge and the number of intracellular granules. In the exocrine pancreas, for example, secretion has been reported to continue after the apparently complete disappearance of granules (Rothman, 1975a). Similarly, in the superior cervical ganglion, Birks (1974) observed a 50% decrease in vesicles while the ACh content remained unchanged. Conversely, under certain conditions secretion of ACh was found to cease after stimulation in sympathetic ganglia and at frog neuromuscular junctions despite the presence of a considerable complement of vesicles in the terminals (Ceccarelli et al., 1972, 1973; Perri et al., 1972; Ceccarelli and Hurlbut, 1975b). Images suggesting intracellular dissolution of granules have often been observed after stimulation of secretion, suggesting that granules release their content into the cytoplasm (Carlsson and Ritzen, 1969; Orci et al., 1969).

Some of these observations are probably due to artifacts. For example, with better fixation techniques, dissolved granules have practically disappeared from the recent literature (Orci 1974; Palade 1975). Furthermore, the persistence of secretion in the absence of typical granules (Rothman, 1975a) can be explained since it has been shown that in heavily stimulated cells granules often decrease in size and density and therefore are not easily identified, especially by the light microscope (Jamieson and Palade, 1971b; Völk et al., 1976). The reports that in some systems individual secretory proteins are present in multiple molecular forms, characterized by different discharge kinetics (Swearingen, 1971; Stachura and Frohman, 1975), are probably also incorrect. In the case of prolactin and growth hormone it was shown that the alleged molecular heterogeneity was due to incomplete sample solubilization. Under correct experimental conditions, molecular heterogeneity was no longer evident and the kinetics of discharge of secretory proteins was entirely consistent with the orthodox view (Zanini et al., 1974).

The differences in the kinetics of discharge of newly synthesized and prestored secretion products may also be artifactual in some cases. Thus the higher specific radioactivity of discharged versus tissue secretion products observed in in vitro experiments carried out under pulse-chase conditions may depend on several factors, such as uneven distributions of oxygen and labeled precursors in the incubated slices. The latter situation is known to favor discharge from the superficial, highly labeled cells, compared with those located in the depth of the tissue (see Farquhar et al., 1975, for a discussion of this problem).

In contrast, other kinetic observations are probably correct and correspond to real differences in the secretion of newly synthesized and prestored secretion products. In our opinion, however, they are not inconsistent with the general lines of the orthodox view. For example, the finding that a small proportion of pulse-labeled secretion products is released from the cell a few minutes after synthesis, showing an intracellular transit time incompatible with storage within granules (McGregor et al., 1975; Rothman, 1976; Slaby and Bryan, 1976), can be explained by assuming that some Golgi-derived vesicles might find their way directly to the plasmalemma, thus bypassing the granule store. This situation would be similar to that of systems such as plasmacytes, which lack secretion granules and thus have a direct, Golgi-to-surface transfer of secretion products (Zagury et al., 1970; Choi et al., 1971; also see section 5.3). In other systems, the early release of newly synthesized products occurs after a time period sufficient for normal intracellular transport. In these cases the phenomenon might be due to simple topographical factors. A good example in this respect is the posterior pituitary, where neurosecretory granules are located in the terminal close to discharge sites immediately after their transport down the neuroaxon. Later on, however, they are displaced to lateral varicosites of the axon, and therefore their discharge is delayed (Sachs and Haller, 1968; Douglas, 1974a; Pickering et al., 1975; Morris et al., 1978). The opposite situation has recently been found by Sharoni and co-workers (1976) in the parotid gland. These investigators demonstrated that the age of granules discharged in vitro depends on the degree of stimulation. When low doses of isoproterenol, a β-adrenergic secretagogue, were used, only old granules, previously accumulated near the luminal plasmalemma, were discharged. At higher doses, the entire granule population was randomly involved, probably because profound infoldings were established at the luminal surface as a consequence of the massive secretory response. In addition, these infoldings made exocytosis possible to the new granules located deeply within the cell. Results on the rat pancreas compatible with this observation had also been reported by Kramer and Poort (1972).

Topographical considerations can be used to explain multiphase discharge kinetics within the framework of the orthodox view. Thus if the time required for the displacement of granules from their original site to a subplasmalemmal location were relatively long, one might observe a first wave of secretion, due to the discharge of granules located close to the plasmalemma and immediately available for discharge, followed by a subsequent wave due to exocytosis of granules previously located more deeply within the cytoplasm (Malaisse et al., 1975). Another explanation for multiphase discharge kinetics is a possible partial inactivation of the excitation mechanism during some stage of the secretory response (section 5.3).

These considerations might also help to explain some of the results obtained by Rothman and his collaborators on the exocrine pancreas. These authors have reported that the relative proportion of the various discharged products (1) differs from that found in isolated granules, (2) changes with time and with the degree of stimulation, and (3) depends on the nature of the secretogogue used. Also, on the basis of cell fractionation experiments, they claim the existence of multiple intracellular pools of secretory zymogens, one or two being particulate and another soluble, and all endowed with unique features. The model proposed (Rothman, 1975b) is somewhat unique because it implies the selective permeability of the granule and plasma membrane to zymogens (secretory proteins, ranging in molecular weight between 13,000 and 70,000) to explain the kinetic heterogeneity and the changes of proportional ratios in the course of discharge. It should be mentioned here that a certain degree of heterogeneity of the pancreatic response can be explained within the framework of the orthodox view, on the one hand, by a rapid and specific modulation of the rate of synthesis of individual zymogens brought about by secretogogue action (Dagorn and Mongeau, 1977) and on the other by the topographic heterogeneity of the gland (Malaisse-Lagae et al., 1975; Putzke and Said, 1975). Furthermore, as pointed out by Sheele and Palade (1975), many of the results reported by Rothman might be due simply to incomplete recovery and activation of individual zymogens. Sheele et al. (1978) have recently reinvestigated the problem of the existence of soluble zymogen pools by means of mixing experiments in which pancreatic juice labeled with [^{3}H] aminoacids was added to the homogenization fluid for [^{14}C]-labeled pancreas. By following the distribution of [^{3}H] and [^{14}C] counts in the individual zymogens purified from the various cell compartments, these authors concluded that the heterogeneity of the soluble and particulate pools previously described by Rothman (1975b) might be accounted for entirely by differential redistribution artifacts (Sheele et al., 1978).

Finally, we will consider the situation of adrenergic and cholinergic terminals. In both these systems it has been reported that the release of neurotransmitters is not random over the entire intraterminal store but involves more readily the fraction recently accumulated as a result of reuptake and neosynthesis (Kopin et al., 1968; Dunant et al., 1972, 1974; Marchbanks and Israel, 1972; Bennett, 1973; Katz et al., 1973). Furthermore, in homogenates of cholinergic tissues, the newly synthesized ACh is recovered mainly in the soluble supernatant (Dunant et al., 1974; Marchbanks, 1975). These findings have fostered the hypothesis that transmitters are released from highly turning over soluble pools, while particulate pools would serve storage functions mainly.

It should be emphasized that the important question is not whether release of transmitters from soluble pools is possible, but whether such a release can be specifically modulated by physiological stimulation. Although the problem is still unresolved, a considerable body of information strongly favors a negative answer to this question. For adrenergic terminals, this conclusion rests on the clear differentiation between Ca^{2+}-dependent release by exocytosis and molecular, Ca^{2+}-independent leakage which can be induced by the administration of drugs (such as tyramine, d-amphetamine, and reserpine), which displace noradrenaline from the granules. Moreover, Thoa and associates (1975) have demonstrated that in adrenergic terminals depleted of noradrenaline by the above-mentioned drugs, secretion elicited by electrical stimulation or high K^+ still occurs by exocytosis, as shown by the release of the secretory protein dopamine-β-hydroxylase.

On the other hand, the extensive work of Ceccarelli and Hurlbut clearly demonstrates that at frog neuromuscular junction the store of quanta of ACh can be exhausted by electrical stimulation without a corresponding decline in the number of synaptic vesicles (Ceccarelli et al., 1972, 1973; Hurlbut and Ceccarelli,

1974; Ceccarelli and Hurlbut, 1975a,b). However, the dissociation between ACh content and number of synaptic vesicles can be accounted for entirely by the fact that the two processes occurring after transmitter discharge, vesicle reformation (section 6.2), and storage of newly synthetized ACh, can operate at different rates. Thus, when secretion of neurotransmitter is induced by stimulation at high frequency, vesicle number and quantal store remain coupled. In contrast, after prolonged stimulation at low frequency, a profound decrease of the quantal store can be observed despite the near-normal complement of vesicles present within the terminals. Finally, a complete dissociation (normal vesicle complement–almost complete exhaustion of the quantal store) can be reached by stimulation at low frequency carried out in the presence of hemicholinium-3, which blocks the uptake of choline by the nerve terminal. These empty vesicles are depleted by electrical stimulation at high frequency as well as by exposure to black widow spider venom (BWSV).

Recent work of two groups of investigators has shown that in frog muscles (Katz and Miledi, 1977) and in mouse diaphragm (Gorio et al., 1978) most of the resting release of ACh may be non-vesicular. Further, Gorio et al. (1978) have demonstrated that in the mouse diaphragm a considerable proportion ($\sim 35\%$ of the total) of ACh is actually located outside the vesicles, not only in the terminals but probably also in nerve trunks, Schwann cells, and so forth. However, the release of this pool cannot be evoked by electrical stimulation, or by high K^+, or by BWSV. Yet this pool is the source of most of the ACh released by resting diaphragms. The remaining 65% is depleted by BWSV, high K^+, and electrical stimulation applied in the presence of hemicholinium-3. The pool is released as quanta, appears to be stored in synaptic vesicles and to participate directly in neuromuscular transmission.

These results are complemented by the important studies of Zimmermann and his associates. In electric organ of *Torpedo* these authors have identified a peculiar population of small synaptic vesicles, rich in newly synthesized ACh and characterized by a low ACh: ATP ratio. These vesicles are fragile, are labeled by extracellular markers (suggesting that they have been in contact with the extracellular space) and are readily discharged on stimulation (Dowdall and Zimmermann, 1974; Zimmermann and Whittaker, 1974; Zimmermann and Denston, 1977a,b). Hence this population of vesicles might be identified tentatively as the tissue counterpart of the readily releasable soluble pool described in the homogenate (Dunant et al., 1974; Marchbanks, 1975).

Most of the available evidence is therefore compatible with the idea that the physiologically modulable release of neurotransmitters is also entirely vesicular.[9] The heterogeneity of the intraterminal store that has been described probably depends on the fact that the various vesicles have different chances to be refilled with the transmitter derived from either direct reuptake (for noradrenaline) or choline reuptake + reacetylation. Thus it is assumed that the vesicles close to the

---

[9]It should be recognized, however, that a clear contradiction to this hypothesis remains with the results reported by Birks (1974).

synaptic membrane (and therefore readier for discharge) would be favored because (1) they are closer to the reuptake sites, and (2) on the average, they may be low in prestored transmitter because they originate mainly from membrane recycling (section 6.2).

3. Composition and organization of the cellular membranes involved in transport and release of secretion products

Since the early days of modern cell biology, the characterization of cellular membranes has been a favorite subject for investigation. Thus cytochemistry of the endomembrane system was already initiated at the end of the 1950s, shortly after the discovery of cytoplasmic compartmentalization in eukaryotes. At approximately the same time, microsomes were identified as membrane-bounded vesicles derived from the fragmentation of the ER and techniques for the isolation of plasma membrane fractions were developed shortly thereafter.

More recently, fractions greatly enriched in Golgi elements and secretion granules have been isolated in many systems and studied in detail, so that now comparative analyses of the membranes participating in intracellular transport and discharge of secretion products are available for at least four secretory tissues: the liver, the exocrine pancreas, the parotid gland, and the adrenal medulla. Some, more fragmentary, evidence exists for other systems as well.

These studies have been carried out with different techniques. Thus the distribution of a variety of enzyme activities has been determined biochemically in the isolated fractions; sometimes these activities have been investigated in more detail by the procedures of classical enzymology. For some enzymes the same problem has been approached by ultrastructural cytochemistry, carried out on the tissue (usually after mild fixation) as well as on isolated fractions. Frequently, lipid analyses have been carried out to reveal the distribution of the various fractions of phospholipids as well as of cholesterol. In some instances the glycolipids and the fatty acid composition of the individual lipid classes have also been investigated. Until the end of the 1960s, the protein composition of membranes remained largely unexplored. Since then some progress has been made, especially by the widespread utilization of high resolution SDS polyacrylamide-gel electrophoresis. Only for a few membrane proteins, however, has the biochemical characterization been pursued to the level of peptide mapping or primary sequence determination. Concomitantly, immunochemical techniques have progressively achieved a major role in the field. Antibodies (sometimes coupled with electron-dense tracers such as ferritin and horseradish peroxidase) have been used for several purposes, such as the purification of antigens and to reveal antigen distribution at the cell surface or in isolated fractions. Other interesting results on the organization of different membranes have been obtained by the use of the plant lectins specific for various sugar residues, by controlled proteolysis and lipolysis and by various techniques of surface labeling. Finally, the recent impetuous development of freeze-fracture studies provides a valuable structural counterpart to many membrane composition studies. Many aspects of the use of these diverse techniques to characterize endomembranes and plasma membranes were reviewed in detail in volumes 3 and 4 of this series.

For a critical understanding of the results obtained so far, the following considerations should be taken into account.

1). Subcellular fractions are never accounted for exclusively by a single class of cell organelles. They always include contaminants in variable proportion. At present, the quantitative evaluation of cross-contamination of subcellular fractions represents an unresolved problem. In most cases this evaluation is carried out by determining in the fraction under study the concentration of markers (most often, enzyme activities) specific for the contaminating organelles. The ratios of marker-specific activities in the fraction versus reference preparations enriched in contaminants (usually considered to be 100% pure) are taken as the quantitative estimation of the degree of contamination. In this procedure several assumptions are made often without experimental confirmation. First, it is assumed that the organelles contaminating the fraction under study are the same as those contained in the corresponding reference fraction. However, it is known that in a population of organelles those which deviate most from average physicochemical features often have a better chance to be distributed as contaminants in heterologous cell fractions. Second, the marker activity is assumed to be due to a single enzyme. Finally, it is assumed that the activity of the marker enzyme remains unchanged from the moment of tissue homogenization to the moment of assay or, at least, that parallel changes occur in all isolated fractions.

2). Fractions are often considered to be fully representative of the corresponding structures present in the tissue or, at least, in the homogenate. However, this is not necessarily the case, especially if one or more of the following conditions apply: (a) the structures under study are heterogeneous; (b) their recovery in the fractions is low; and/or (c) significant changes of important parameters have occurred during fractionation.

3). In many reports published as membrane composition studies, the assays were carried out on fractions of intact organelles, containing not only membranes but also other components, such as segregated proteins and other exportable molecules, ribosomes, proteins adsorbed onto or associated with membrane surfaces, and so on. In other studies the purification of membranes was attempted; however, a rigorous quantitative estimation of the extraction of nonmembrane components was not carried out.

4). Even if one cell type clearly predominates in many secretory tissues, some degree of cellular heterogeneity is present in all cases. Thus, in all isolated fractions, a certain proportion of the organelles probably originates from the minority cell types of the tissue. Usually this source of contamination is totally overlooked. Available evidence indicates, however, that its contribution might not only be relevant in quantitative terms but also variable from fraction to fraction. For example, the morphometric studies by Weibel and co-workers on rat liver indicate that 2.4% of mitochondria, 6.4% of ER, 15.1% of Golgi, 26.5% of plasmalemma, and as much as 32.4% of lysosome membranes present in the tissue are accounted for by nonhepatocytes (Blouin et al., 1977). The contribution to the corresponding isolated fractions remain to be determined.

5). In enzyme cytochemistry special attention is generally focused on experimental limitations and drawbacks. Nevertheless, it is appropriate to recall that a number of problems are a still unresolved and must be considered thoroughly in evaluating the results. These include the effect of tissue processing procedures carried out before the enzyme reaction and the uneven penetration of the reagents throughout the tissue, as well as the diffusion of the reaction products, both before and even after the formation of insoluble precipitates.

6). The available data on the distribution of enzyme activities cannot be extrapolated directly in terms of membrane protein composition because a high degree of enzyme polymorphism has been demonstrated, at least in microsomes and plasmalemma; thus, immunologically different proteins can exhibit the same enzyme activity (Blomberg and Raftell, 1974; Raftell and Blomberg, 1974). Similarly, the distribution of polypeptides, as revealed by polyacrylamide gel electrophoresis, does not yield clear-cut results because co-migration in SDS gels is a criterion insufficient to prove molecular identity and even similarity in size. Furthermore, it is also expected that minor molecular alterations (e.g., limited proteolysis and glycosylation) that are likely to occur within the cell, at least in some cases during the physiological lifetime of individual peptides, might result in considerable changes in migration rates in SDS-polyacrylamide gel electrophoresis.

Current knowledge on the composition and structure of each type of membrane involved in the intracellular transport will be summarized separately in the following pages and then considered together and compared in section 3.5.

3.1. ER membranes

The experimental evidence available in this field is impressive. No other cell fraction has been studied as extensively as liver microsomes, rough- as well as smooth-surfaced (review, De Pierre and Dallner, 1975). Considerable knowledge exists also on microsomal fractions isolated from a number of other secretory systems. Furthermore, microsomes have been subfractioned to yield microsomal membranes and the latter have been analyzed in detail (Meldolesi et al., 1971a; Kreibich et al., 1973; Blackburn et al., 1976; Hortnägl, 1976).

The data on lipid composition are consistent in all systems (Keenan and Morré, 1969; Colbron et al., 1971; Meldolesi et al., 1971b; Morin et al., 1972; Zambrano et al., 1975; Hortnägl, 1976). Relative to many other cytoplasmic membranes, microsomal membranes were found to be characterized by a lower lipid: protein ratio, by high phosphatidylcholine and low sphingomyelin content, and by a very low level of cholesterol. The presence of the latter in liver microsomes is even questioned (Beaufay et al., 1974). The relatively few studies of fatty acid composition have revealed a marked abundance of the unsaturated species.

The protein composition of the ER membranes has turned out to be very complex in all the systems investigated so far. The number of polypeptides separated by SDS-polyacrylamide gel electrophoresis is very high and the spectrum of enzyme activities very large (Meldolesi and Cova, 1972; Amar-Costesec et al., 1974a,b; Beaufay et al., 1974; Kreibich and Sabatini, 1974; Hortnägl, 1976). Furthermore, a large degree of molecular polymorphism of microsomal enzymes has been detected by the combined use of immunological and biochemical techniques. For instance, it is now known that in liver microsomes the activity of cytochrome P_{450} is accounted for by at least six different isoenzymes (Thomas et al., 1976); the existence of a large number of antigenically distinct esterases, nucleoside di-and triphosphatases, and arylamidases (Blomberg and Raftell, 1974; Raftell and Blomberg, 1974) has also been reported.

All these enzymes and the other proteins are thought to be distributed in both the rough and smooth-surfaced portions of the ER (Novikoff and Heus, 1963; Leskes et al., 1971; Amar-Costesec et al., 1974b; Beaufay et al., 1974; Kreibich and Sabatini, 1974). The only identified exception concerns the membrane receptor system specific for ribosome binding (section 2) studied extensively in the liver (Borgese et al., 1974; Mok et al., 1976; Fujita et al., 1977; Kreibich et al., 1978a, b). Two glycoproteins, which apparently belong to that receptor system, were found to be localized exclusively in rough ER membranes (Kreibich et al., 1978a,b). The distribution of the other proteins in liver microsomes, although qualitatively homogeneous, is not exactly the same in quantitative terms, since rough microsomes appear slightly enriched in glucose-6-phosphatase, esterase, and nucleoside diphosphatase, while smooth microsomes are analogously enriched in enzymes of both the NADH- and NADPH-dependent electron transport chains (Amar-Costesec et al.,.1974b; Beaufay et al., 1974) Furthermore, the latter chains were reported to separate partially from each other in microsomes first fragmented into very small vesicles and then resolved by rate or isopycnic gradient centrifugation (Svensson et al., 1972). These findings might be interpreted to indicate that not all microsomal molecular components are individually free for lateral displacement in the plane of the membrane; instead, at least some of them might be arranged in specific supramolecular patches, whose distribution is not entirely random throughout the system.

3.2. Membranes of the Golgi complex

Although the characterization of Golgi membranes has not yet been developed to the point of their microsomal counterpart, a consistent picture has emerged during recent years. Clearly, as shown primarily by cytochemical studies, a large degree of membrane heterogeneity exists within the Golgi complex. Thus, of the membrane-bound enzymes, thiamine pyrophosphatase was found in many systems to be restricted to the innermost (or *trans*) Golgi cisternae (Cheetham et al., 1971; Farquhar et al., 1974; Novikoff, 1976); adenylate cyclase was detected at the cytoplasmic surface of both the very low density lipoprotein (VLDL) vacuoles and the periphery of the flat cisternae (Cheng and Farquhar, 1976a,b). Even more bizarre is the distribution of 5'nucleotidase in the liver Golgi complex. This activity has been detected at the membrane inner surface in VLDL vacuoles and at the outer surface in the *cis* cisternae, where it is limited to the dilated rims and is absent from the centers (Farquhar et al., 1974; Little and Widnell, 1975). Furthermore, Abe and colleagues (1976) have recently demonstrated that the distribution of anionic sites on the surface of Golgi elements is also very heterogeneous. Additional evidence along the same line comes from biochemical studies. The three Golgi subfractions extracted by isopycnic flotation from rat liver microsomes, according to Ehrenreich and co-workers (1973), and enriched primarily in VLDL-containing vacuoles (GFI), *trans* cisternae (GF2) and *cis* cisternae (GF3) respectively, exhibit heterogeneity in relation to the localization of 5'nucleotidase and NADH-dependent electron transport enzymes, while galactosyltransferase (the enzyme generally used as Golgi marker) is distributed more evenly (Bergeron et al., 1973a). A considerable degree of heterogeneity of subfractions isolated from rat liver Golgi preparations was also reported by Ovtracht and collaborators (1973).

A different experimental approach developed by the group at Louvain (Amar-Costesec et al., 1974b; Wibo et al., 1975), and recently applied in our laboratory (Borgese and Meldolesi, unpublished observations), is based on the effect of digitonin (a molecule known to form an insoluble equimolar complex with cholesterol) on the isopycnic density of Golgi elements.[10] When a total Golgi fraction isolated by the method of Ehrenreich and associates (1973) (i.e., GF1+GF2+GF3) was analyzed by this procedure, it was found that the digitonin-induced displacement of the light Golgi elements, as revealed by the distribution of galactosyltransferase activity, was only partial since a considerable proportion of the activity remained in the upper fractions of the gradient (Fig. 2a). Analogous results were obtained by Wibo and co-workers (1975). This finding differs from the situation described for plasma membrane enzymes, which are displaced in bulk by about 0.035 density units, and this might be due to heterogeneous distribution of cholesterol (the molecule responsible for digitonin binding) in Golgi membranes. This hypothesis is also supported by morphological evidence (Fig. 3). In addition, these experiments suggested a heterogeneous localization of Golgi enzymes. Thus the distribution of galactosyltransferase was found to corrrelate well with that of protein in both untreated and digitonin-treated preparations. In contrast, the enzymes of the NADH-dependent electron transport chain were displaced less than galactosyltransferase and their peaks

---

[10]The rationale for this procedure is that cholesterol-rich membranes become heavier as a consequence of digitonin binding and are therefore displaced in the sucrose gradients to a higher buoyant density.

536

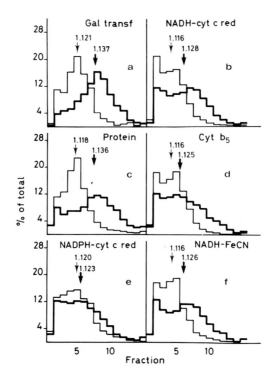

Fig. 2. Effect of digitonin treatment on the distribution of Golgi enzymes on sucrose density gradients. A total rat liver Golgi fraction (GF_1 + GF_2 + GF_3) was prepared according to the method of Ehrenreich et al. (1973), omitting ethanol pretreatment. Half of the sample (~ 20 mg of protein) was incubated at 0°C for 15 min with digitonin (~ 1 mol/mol cholesterol), while the other half was left untreated (Amar-Costessec et al., 1974b). The membrane fractions were then centrifuged into a pellet, resuspended in 0.25 M sucrose, and loaded onto linear sucrose gradients (0.77 M–1.5 M sucrose) with a 0.5 ml 2.5 M sucrose cushion at the bottom. The gradients were centrifuged overnight at 25,000 rpm in the Spinco-Beckmann SW 41 rotor at 3%C. Enzyme activities and protein are plotted as percent of total recovered on the gradient. The heavy line represents the distribution for the digitonin-treated membranes; the light line, the distribution for the untreated membranes. Heavy and light arrows indicate the median densities of the digitonin-treated and untreated samples, respectively. a, galactosyltransferase; b, NADH-cytochrome c reductase; c, protein; d, cytochrome b_5; e, NADPH-cytochrome c reductase; f, NADH-ferricyanide reductase.

were broadened (Fig. 2). This result might be due partly to contamination by microsomes and mitochondrial outer membranes, which are rich in NADH-dependent electron-transport enzymes (in Fig. 2 note that the microsomal marker enzyme NADPH-cytochrome-c-reductase is hardly displaced by digitonin). However, marker enzyme studies, as well as the good correlation between the distribution of galactosyltransferase and total protein on sucrose gradients (Fig. 2), make massive contamination unlikely. Since the specific activities of the NADH-dependent electron transport enzymes in the Golgi fraction were as high as in microsomes, it is probable that most of these activities are indigenous to

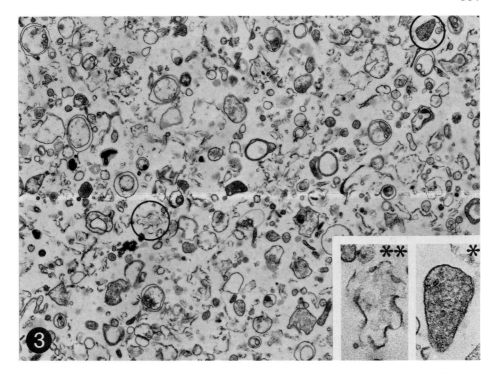

Fig. 3. Electron micrograph of fraction 7 of the gradient shown in Fig. 2, containing digitonin-treated Golgi elements. The field includes numerous very low density lipoproteins containing vacuoles, as identified by their dense particulate content. Some vacuoles are filled with particles while many others appear extracted to a variable extent. Many vacuole bounding membranes appear to be interrupted by discontinuities and give the appearance of dashed lines. Often, at the point of the discontinuities, these membranes appear to undergo sharp bends. These alterations are due to digitonin binding (Amar-Costesec et al., 1974b). Vacuole bounding membranes vary in appearance. The insets show (at higher magnification) the difference between two digitonin-treated VLDL-containing vacuoles: one appears unaffected (*) the other is profoundly modified (**). ×16,200; insets: ×63,000.

Golgi elements. The different effect of digitonin treatment on the distribution of these enzymes compared to galactosyltransferase could be explained if they were preferentially located in cholesterol-poor Golgi elements. Further analysis of the data of Fig. 2 shows that the light fractions of the gradient (containing VLDL vacuoles and light cisternae, as identified by electron microscopy) are enriched in NADH-cytochrome-c-reductase and cytochrome b_5, while a population of heavy Golgi elements (identified as flat cisternae by electron microscopy) contain galactosyltransferase activity but not the NADH-dependent electron transport enzymes. Thus, these data are not incompatible with the immunocytochemical studies of Fowler and colleagues (1976), who have shown that cytochrome b_5 is absent from Golgi stacks (piled flat cisternae) isolated from rat liver.

In this situation, it is safe to conclude that most of the biochemical composition data on Golgi preparations reported so far probably do not reflect the values present in each element but are averages of heterogeneous components. Furthermore, since the yield of isolated Golgi fractions with respect to the homogenate is never quantitative, the possibility exists that these fractions are not representative of the whole complex but are preferentially enriched in one of its portions. We believe that this is the case in the liver, where the fraction isolated according to Morré and associates (1970) is more representative of the parallel stacks, while the fraction isolated according to Ehrenreich and collaborators (1973) (without prior ethanol intoxication of the animals) is enriched comparatively in smaller cisternae, vesicles, and VLDL vacuoles.

Notwithstanding these limitations, certain points of interest have emerged from the composition data available on Golgi fractions. In all systems investigated to date, lipid analyses revealed much higher levels of sphingomyelin, cholesterol, and saturated fatty acids with respect to microsomes (Meldolesi et al. 1971b; Trifaro et al., 1976; Morré, 1977). The same parameters are usually only slightly lower than in the plasma membrane. These data are consistent with the notion that Golgi membranes are less permeable than ER membranes (section 2.1.3). Furthermore, the protein composition, as revealed by SDS-polyacrylamide gel electrophoresis, is very complex (Meldolesi and Cova, 1972; Bergeron et al., 1973b; Meldolesi et al., 1974; Trifaro et al., 1976). The concentration of glycoproteins, especially those containing sialyl residues, is greatly increased. This finding is explained by the specific enrichment of the Golgi membranes in sugar transferases (Wibo et al., 1975; Morré, 1977; Ronzio and Mohrlok, 1977). Several other enzymes have been reported to be present in the Golgi complex (Van Golde et al., 1974; Morré, 1977). In this respect it is interesting to note that, with the exception of NADH-cytochrome c reductase and cytochrome b_5, none of the numerous enzyme activities localized in the ER has been shown conclusively to be present in the Golgi. In contrast, some proteins and enzymes usually believed to be specifically localized in the plasmalemma have been found in the Golgi fractions as well. These include 5'nucleotidase, adenylate cyclase and insulin receptor (Meldolesi et al., 1971c; Bergeron et al., 1973b; Farquhar et al., 1974; Cheng and Fraquhar, 1976a,b; Hodson and Brenchley, 1976; Trifaro et al., 1976). The significance of these overlaps are discussed in section 3.5.

3.3. Membranes of secretory granules and vesicles

In all secretory granule and vesicle membranes analyzed so far, a lipid composition of the plasmalemma type (i.e., high lipid; protein ratio; high levels of cholesterol, sphingomyelin, and saturated fatty acids) has been observed. Glycolipids have been detected in some of these granule membranes, but have not been studied in detail. (Blaschko et al., 1967; White and Hawthorne, 1970; Meldolesi et al., 1971b; Da Prada et al., 1972b; Vilhardt and Hømer, 1972; see also reviews by Meldolesi et al., 1974; Lagerkrantz, 1976; Winkler, 1976, 1977). In addition, chromaffin granules were found to have an usually high level of lysolecithin (Blaschko et al., 1967). The latter is a genuine feature of the membrane and not an artifact resulting from lecithin hydrolysis. The significance of this observation is still unclear. For some time, the high concentration of lyso-lecithin was suspected to play a major role in exocytosis because of the well known ability of lysophospholipids to promote membrane fusion. However, subsequent studies in a number of other systems failed to confirm the generality of high lysolecithin levels in granule membranes (section 5.2).

A large degree of variability has been found in the protein composition of granule membranes of the different secretory systems. Two groups can be identified. The membranes of granules and vesicles isolated from the exocrine pancreas, parotid gland, adrenal medulla, and adrenergic terminals are characterized by a much smaller number of polypeptides with respect to the corresponding microsomal, Golgi, and plasma membranes (McDonald and Ronzio, 1972; Meldolesi and Cova, 1972; Meldolesi et al., 1974; Castle et al., 1975; Wallach et al., 1975; Hörtnagl, 1976; Lagerkrantz, 1976). This finding correlates well with the morphological observation that the density of intramembrane particles (as revealed by freeze-fracture) is low in these secretory organelles (Smith et al., 1973a; De Camilli et al., 1974, 1976; see also Fig. 17A). The same observation has been made about the membranes of other secretory granules (of B cells, mastocytes, principal and mucous cells of the stomach, various pituitary cells) which have not yet been analyzed biochemically (Orci and Perrelet, 1975; Chi et al., 1976; De Camilli unpublished observations; see also Fig. 4). In some systems (exocrine pancreas, parotid gland, adrenal medulla) evidence indicates that the proteins of granule membranes are unique; they are different from their microsomal, Golgi, and plasmalemmal counterparts (Medolesi and Cova, 1972; Meldolesi et al., 1974, Wallach et al., 1975; Hörtnagl, 1976, Trifaro et al., 1976). In zymogen and chromaffin granule membranes the major protein components were found to be glycoproteins (the enzyme dopamine-β-hydroxylase in chromaffin granules; a glycoprotein of unknown function named GP_2 in zymogen granules), which are exposed at the inner membrane surface; other identified proteins (chromomembrin B in chromaffin granules and a minority glycoprotein in zymogen granules) are exposed at the cytoplasmic surface (Hortnägl et al., 1976; König et al., 1976; Winkler, 1976, 1977; Lewis et al., 1977).

The second group includes the specific and azurophilic granules of polymorphonuclear leukocytes, the neurosecretory vesicles of the posterior pituitary, and the synaptic vesicles of brain. In the first group, the membranes of these granules are characterized by a less pronounced molecular specificity. The number of polypeptides present in these membranes is considerable and some of these components, especially glycoproteins, are shared with the plasmalemma (Breckenridge and Morgan, 1972; Morgan et al., 1973; Widlund et al., 1974; Bock and Jorgensen, 1975; Vilhardt et al., 1975; Baggiolini et al., 1977; Bonne et al., 1977; Howe et al., 1977). However, the membranes of this group also exhibit distinct peculiarities, because many of their proteins are not found in other membranes. Furthermore, the distribution of the intramembranous particles in both types of polymorphonuclear leukocyte granule membranes is un-

540

usual since, in contrast to the situation found in most other organelles, the density is higher on the E-than on the P-fracture face (Baggiolini et al., 1977).

On the other hand, studies carried out on other systems, such as the liver and plasmacytes, have failed to reveal the existence of secretory organelles endowed with specific features. In liver, a fraction enriched in VLDL-containing vacuoles (GF1 in the procedure of Ehrenreich et al., 1973) was found to be qualitatively similar to fractions enriched in Golgi cisternae (GF2 and GF3) both in enzyme activities detected (Bergeron et al., 1973a) and in polypeptides revealed by SDS gel electrophoresis (Bergeron et al., 1973b) Hence, in this case, the available evidence suggests that exocytosis is carried out by organelles whose membrane might still retain a Golgi-type composition.

Finally, it should be mentioned that some granule membranes are unusually poor in enzyme activities. This is the case of the pancreas and parotid glands where the existence of not even one enzyme indigenous to the granule membrane has been demonstrated beyond reasonable doubt (Meldolesi et al., 1971c, 1974; Castle et al., 1975; Wallach et al., 1975). In polymorphonuclear leukocytes the only enzyme detected so far in the specific granule membrane is alkaline phosphatase (Bretz and Baggiolini, 1974; Baggiolini et al., 1977). A larger spectrum of enzyme activities (dopamine-β-hydroxylase, phosphatidylinositol kinase, ATPase, cytochrome b_{559}) has been detected in chromaffin granule membranes (Muller and Kirshner, 1975; Hörtnagl, 1976; Winkler, 1976, 1977), in good correlation with the notion that the physiological role of these organelles is not limited to storage and discharge but also includes accumulation of low molecular weight secretion products (section 2.1.4).

3.4. Plasma membrane

One of the major problems to be discussed in regard to plasma membrane composition of secretory cells is regional heterogeneity. A good example of this phenomenon is given by cell junctions. Thus gap junctions are known to be composed of specific, integral membrane protein complexes, intermingled with lipids in discrete, closely packed arrays (Benedetti et al., 1976). An analogous situation probably exists in tight junctions, where the specific integral proteins are arranged into a fibrillar network (Stahelin, 1973).

Fig. 4. Freeze-fracture electron micrographs showing surface views of apical regions of rat stomach cells: (A) mucous cells; (B) principal cells. The fibrillar network of tight junctions (j) separates the luminal from the lateral region of the plasmalemma. The P faces of the luminal surfaces are very low in intramembranous particles and are clearly different from the P faces of the lateral plasma membranes, where particle density is very high. The P faces of granule membranes (arrows) are also characterized by a low density of intramembranous particles. This is a typical feature of the granule membrane, since the density of particles is very high on the P faces of the other intracellular membranes participating in the secretory process (not shown). In contrast, the E-fracture faces of all the cellular membranes are characterized by a very low concentration of intramembranous particles. P*, P face of luminal plasmalemma; P** and E**, P and E faces of the lateral plasmalemma; double arrowheads, E faces of granule membranes. (A) and (B) ×36,500.

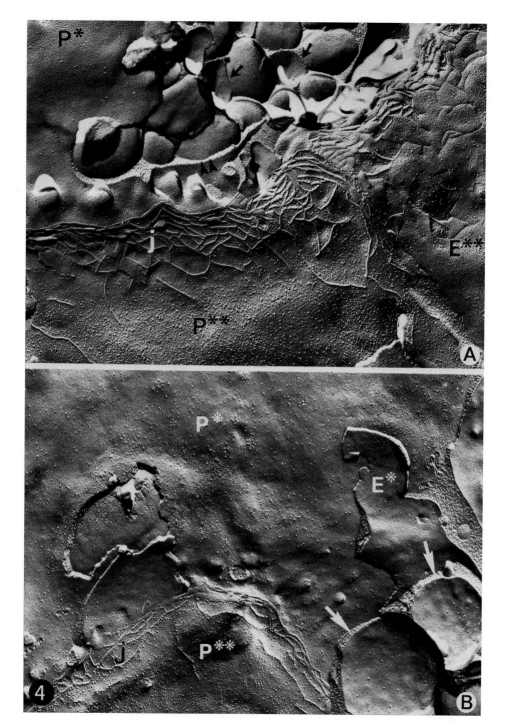

542

In the plasmalemma of epithelial cells forming a continuous lining of a cavity, the region facing the lumen of the cavity (luminal region) is characterized by distinct functional and structural properties. It is separated from the rest of the cell surface (the basolateral membrane) by a continuous circumferential tight junction (zonula occludens). The secretory cells of the exocrine glands and of the thyroid fall into this category, and the release of secretion products in these systems is limited to the luminal region. Cytochemical and freeze-fracture evidence indicates that this luminal region differs from the rest of the plasmalemma not only functionally but also in structure (Figs. 4, 10) and composition (De Camilli et al., 1974, 1976; Farquhar et al., 1974; Orci and Perrelet, 1975).

In the liver, subfractions enriched in plasmalemma fragments originating from three different regions (sinusoidal, canalicular, and lateral; the latter corresponding to the junctional complex and to the adjacent plasmalemma sheets remaining continuous with the junctions during homogenization) have been isolated and analyzed. The results clearly indicate that lipids and most of the proteins are distributed in a relatively homogeneous fashion in the different subfractions but 5'-nucleotidase, alkaline phosphodiesterase and phosphatase, leucine naphthylamidase and Mg^{2+}-ATPase are more concentrated in the canalicular subfraction, whereas hormone-sensitive adenylate cyclase and insulin receptor activity are mostly localized at the sinusoidal plasmalemma. Furthermore, the concentration of glycoproteins is lower in the lateral subfraction, possibly as a consequence of the absence of these molecules from some of the junctions proper (Evans et al., 1973; Wisher and Evans, 1975; Kremmer et al., 1976). Analogous results have been reported by Toda and associates (1975).

Surface heterogeneity has also been reported in another category of secretory cells, the nerve cells, where distinct differences in the polypeptide composition of the synaptic and lateral portions of the plasmalemma have been found (Morgan et al., 1973; Kelly and Cotman, 1977).

In contrast, in many other secretory systems (e.g., endocrine glands, plasmacytes, leukocytes, fibroblasts, etc.) there is no clear indication of cytological heterogeneity of the plasmalemma. However, even in these cases it would be probably naive to conclude that the plasmalemma is entirely homogeneous, since the available evidence clearly indicates that in all cells the distribution of molecular components of the plasmalemma is not random, but controlled by various factors. Among these, underlying membrane associated proteins or structures appear to be important (review by Nicholson et al., 1977 in volume 3, this series). The cellular mechanisms responsible for surface heterogeneity and the functional implications of the latter phenomenon, especially in relation to exocytosis and receptor activation, are discussed in section 5.

On the basis of these considerations it might be concluded that the results of compositional analyses carried out on total membrane fractions represent averages of the features of heterogeneous domains. In the exocrine glands, these results are probably more representative of the composition of the large basolateral surfaces than of the small luminal regions (~5% of the cell surface in pancreatic acinar cells [Bolender, 1974]), which are still unexplored biochemically in most systems.

As summarized in the review article of De Pierre and Karnovsky (1973), plasmalemma fractions isolated from a variety of cell systems share a number of common features. Thus lipid:protein ratios are consistently higher than in microsomes, and the same occurs with the levels of cholesterol, sphingomyelin, glycolipid, and saturated fatty acids. Furthermore, studies of polypeptide and enzyme composition have revealed a high degree of complexity. Finally, it is interesting to recall that plasmalemma markers, such as 5'nucleotidase, alkaline phosphodiesterase, and insulin receptor, are ectoglycoproteins (Evans, 1974; Wisher and Evans, 1975). In contrast, adenylate cyclase is exposed at the cytoplasmic surface (Evans et al., 1973).

3.5. Composition and structure of membranes: comparative evaluation

If one compares the results of the composition studies that we have briefly summarized, the following conclusions can be drawn.

1. The four membrane types involved in transport and discharge of secretion products possess unique biochemical and structural features which can be used as criteria for their identification. This applies not only to microsome and secretion granule membranes, which are essentially homogeneous, at least in qualitative terms, but also to Golgi (identifiable by the distribution of sugar transferases) and plasma membranes (in which the large number of markers compensates for the poor specificity of some of them).

2. As a consequence of refined cell fractionation techniques and of the introduction of immunological techniques in cell biology, it is now clear that the degree of protein overlapping that exists among the different types of membranes is much smaller than was previously thought. The criteria that must be fullfilled to demonstrate the multiple subcellular distribution of membrane markers, particularly of enzyme activities, have become more rigorous over the last few years. Not only cross-contamination, but also the existence of heterogeneous components accounting for the same activity, should be excluded. In at least one case (cytochrome b_5 in microsomes and mitochondrial outer membrane) it has been demonstrated that the same activity in two membrane types is due to two different gene products (Fukushima and Sato, 1973). The same might be true for the microsomal and plasmalemma NADH-dependent flavoproteins, since they display considerable differences in kinetic parameters and acceptor specificities (Crane and Löw, 1976). An important development in the study of the subcellular localization of enzymes has been contributed by the group working at Louvain. By the use of analytical, isopycnic, and velocity gradient centrifugation, in combination with the treatment of subcellular

fractions with agents that selectively modify their buoyant density, microsomal enzymes were classified into five groups, two of which corresponded to ER enzymes, the others presumably contributed by different types of contaminants (Amar-Costesec et al., 1974b; Beaufay et al., 1974). Their results cast doubt on the dual microsome-plasma membrane localization of 5' nucleotidase (Windell, 1972) and nucleotide pyrophosphatase (Bishoff et al., 1975), as well as on the ample overlap between the ER and Golgi membrane enzymes described by several authors (review, Morré, 1977). On the other hand, a higher degree of similarity seems to exist between the membrane of the Golgi complex and the plasmalemma. Although many results are based only on the distribution of enzyme activities and polyacrylamide gel bands, rather than of molecularly identified proteins, the recovery in both the Golgi elements and the plasmalemma of a number of markers, localized at membrane faces that a fusion process would render continous (either at the two cytoplasmic faces, or at the extracellular face of the plasmalemma and at the luminal face of Golgi elements) strongly suggests the existence of a common protein pool shared at least by part of these two compartments, with possible biogenetic and/or functional relationships. In this respect it might be emphasized that this hypothetical pool should involve *macromolecules* and not *entire membranes,* because other components, such as the sugar transferases of the Golgi complex (section 4.1) and $Na^+ + K^+$-ATPase of the plasmalemma, are not shared by the two compartments.

3.   In some systems at least, exocrine as well as endocrine, the membranes of secretory granules have a unique composition and contain neither the membrane proteins of the Golgi nor those of the plasmalemma, and not even those apparently localized in both these compartments. This suggests that these granule membranes are not involved in the transfer of membrane proteins between the Golgi and the cell surface. Thus, with respect to the transport of secretory and membrane proteins, the role of these secretion granules might be totally different: the granule content represents an intermediate compartment of the secretory pathway (section 2), whereas granule membranes constitute a highly specialized portion of the endomembrane system, aligned in parallel to the system of membrane protein exchange which might exist between the Golgi complex and the plasmalemma. Along the same line, one might speculate that liver secretion granules, which apparently have a Golgi-type composition, might be involved in this membrane protein transport (Bergeron, et al., 1973b).

4.   Taken together, the results on membrane composition imply the existence of stringent regulatory mechanisms operating during fusion-fission processes that permit each type of membrane to preserve its molecular individuality. These mechanisms must be extremely precise and efficient if one considers the enormous number of fusions occuring during intracellular transport and discharge.

Further aspects of the biogenetic relationships of the various endomembranes and the plasmalemma are discussed in the following sections of this chapter.

4. Biogenetic relations between membranes involved in secretion

The study of the secretory process led quite some time ago to the suggestion that membrane proteins might be synthesized on membrane-bound ribosomes and transported through the various cell compartments to the plasma membrane along the secretory pathway (Palade, 1959). Because of the suggested parallelism between the transport of secretory material and membrane components, it is relevant also to deal briefly with the issue of membrane biogenesis in this review. The discussion will be limited to the protein components of membranes, since the exchange of lipids, probably occurring between the different membranes within cells (Wirtz, 1974), poses particular problems beyond the scope of this chapter. Moreover, the discussion will focus mainly on the biogenetic relationships between membranes related to each other functionally within the secretory process. Clearly the problem of membrane biogenesis in the semiautonomous organelles is not relevant to the present discussion. For extensive reviews on membrane biogenesis, the reader is referred to volume 4 of this series.

The hypothesis that membrane proteins, synthesized on the rough ER, are subsequently transported via the Golgi complex to the plasma membrane is often referred to, in its various forms, as the "membrane flow" hypothesis (Morré et al., 1971, 1974; Morré, 1977). However, the meaning of this term is not clearly defined, since it has been used to indicate both bulk transfer of membrane from the ER to Golgi to plasma membrane and the selective transfer of single protein molecules from one compartment to another (Morré, 1977). Bulk transfer could involve either one of the following two possibilities: (1) transfer of pieces of the "donor" membrane to the "acceptor" membrane followed by extensive modification of the transferred membrane leading to its transformation into "acceptor" membrane (i.e., membrane differentiation; discussion, Morré et al., 1974; Morré 1977); and/or (2) formation of performed pieces of "acceptor" membrane within the domain of the "donor" membrane and their subsequent transfer to the "acceptor" membrane compartment. Although each of the above two mechanisms of membrane biosynthesis entails fusion-fissions between membranes as the necessary mechanism for transfer of components, they clearly represent radically different modes of membrane biogenesis.

Yet another mechanism for the biogenesis of membranes could involve the direct insertion of proteins, synthesized on free ribosomes, into the appropriate membrane. It must also be emphasized that the various possibilities considered here are not necessarily mutually exclusive. Probably the biogenesis of membranes occurs through the concerted participation of different processes. For instance, it has been proposed (Blobel and Dobberstein, 1975a; Lodish and Small, 1975) that proteins exposed on the *luminal* face of the plasma membrane (which corresponds to the *external* face of the plasma membrane) are made on membrane-bound polysomes, while proteins exposed on the *cytoplasmic* face of membranes are synthesized on free ribosomes and inserted directly into the appropriate membrane. In this way the polar moiety of a protein destined for the luminal side could cross the lipid bilayer during the process of synthesis by a mechanism similar to that used for the vectorial discharge of secretory proteins into the cisternae of the rough ER (section 2). Clearly, it is not necessary to invoke such a mechanism for membrane proteins exposed on the cytoplasmic face, whether they be "integral" or "peripheral" (Singer and Nicolson, 1972).

The experimental evidence favoring the different modes of membrane biosynthesis will now be examined.

4.1. Kinetic studies on membrane biogenesis

Early kinetic experiments (Evans and Gurd, 1971; Franke et al., 1971) suggested that after a pulse with radioactive amino acids ER membrane proteins become rapidly labeled and then travel down the secretory pathway at a rate comparable to that of secretory proteins. Thus Franke and co-workers (1971) carried out in vivo pulse experiments with [14C]guanidino-arginine in the rat liver system, and analyzed the kinetics of incorporation of the radioactive amino acids into the different cell fractions after treatments aimed at extracting the content. They observed that the rough ER membrane fraction was maximally labeled about 10 minutes after the injection, and that thereafter the radioactivity decreased to approximately half the peak value. The radioactivity in smooth ER and Golgi membrane fractions peaked between 15 and 20 minutes and decreased thereafter, while plasma membranes appeared to become labeled more slowly. These data were interpreted to be consistent with the transfer of membrane components from the rough ER to the plasma membrane via the secretory pathway. Moreover, the necessary conclusion which must be derived from these experiments is that large portions of "donor" membrane are transferred to successive "acceptor" membrane compartments within short time intervals (5–10 min), and that membrane therefore must be structures turning over very rapidly.

These studies were criticized, however, on the ground that the results could be accounted for by the presence of small amounts of highly labeled secretory protein contaminating the membrane fractions (Meldolesi, 1974b). That this is possible has been demonstrated in two different systems, the pancreas and the parotid gland. Meldolesi and Cova (1971) performed experiments on well-characterized cell fractions of the guinea pig pancreas. After extraction of the content proteins from microsomes and zymogen granules by a high salt wash, followed by bicarbonate treatment, assays of secretory enzyme activity (chymotrypsinogen and amylase) indicated that the membrane fractions were contaminate with less than 1% of secretory proteins. Pulse-chase experiments on pancreatic slices suggested that, compared to ER membrane proteins, the zymogen granule membrane proteins were labeled with a lag. However, when these fractions were analyzed by SDS-polyacrylamide gel electrophoresis, most of the radioactivity was found to be due to the small amount of contaminating, highly labeled, secretory protein (Meldolesi, 1974a).

Similarly, Amsterdam and collaborators (1971) analyzed rat parotid secretion granule membranes and found that the membrane proteins were labeled at the same rate as the secretory proteins. However, when they carried out their analysis by SDS-polyacrylamide gel electrophoresis (Wallach et al., 1975), these investigators found that only two proline-rich granule membrane proteins were rapidly labeled, whereas all other membrane proteins incorporated very low amounts of radioactivity. Results consistent with this conclusion have also been obtained by Castle and associates (1974) in the rabbit parotid gland.

All the studies carrie̓ ' out over longer time periods (on the order of days), directed at determining the turnover either of individual purified membrane components or total membrane fractions, as analyzed by SDS-polyacrylamide gel electrophoresis, have demonstrated that membrane proteins turn over at very low rates compared to the transit time of exportable proteins within secretory cells, with half-lives ranging from hours to days. These studies have been carried out in a variety of systems such as liver, pancreas, parotid, fibroblasts, and on intracellular as well as surface membranes (reviews, Siekevitz, 1972; Negishi and Omura, 1973; Schimke, 1975; Tweto and Doyle, 1977). It is particularly interesting that in all cases examined secretion granule membrane proteins, whose exclusive fate is to become incorporated into the plasma membrane during the exocytotic process, have half-lives up to one order of magnitude longer than the intracellular transit time of the secretory products they contain (Winkler et al., 1972; Castle et al., 1974; Meldolesi, 1974a; Wallach et al., 1975). That this is probably a general rule is indicated convincingly by the situation of chromaffin granules, where the enzyme dopamine-β-hydroxylase is present both within the content, as an exportable protein, and in a membrane-bound form. The latter form is characterized by considerably lower rates of synthesis and turnover than the former (Gagnon et al., 1976; Geissler et al., 1977; Winkler, 1977).

Thus the membrane fusion-fissions occurring during intracellular transport and exocytosis must be followed by retrieval of membrane by the original compartments (review Palade, 1975; and section 6). One must conclude that whatever the role of the rough ER in membrane biogenesis, in quantitative terms the rate of net transfer of membrane constituents from one compartment to another is necessarily very small compared to the rate of intracellular transport of secretory proteins (discussion, Meldolesi, 1974b). Moreover, the increasing evidence that each type of membrane is characterized by a distinct protein and lipid composition (section 3), makes it highly improbable that one membrane could derive from the transformation of another type of membrane (for a discussion favoring the opposite view, see Morré et al., 1974; Morré, 1977).

Although the turnover studies discussed above, by determining the rates of synthesis of membrane proteins, also give indications of the rates of whatever transfer from one compartment to another does exist, they do not yield information about the site of synthesis of membrane proteins. Dehlinger and Schimke (1971), on the basis of the finding that membrane proteins are degraded at rates proportional to their size in the same way as cytoplasmic proteins, proposed the existence of a cytoplasmic pool of membrane proteins in equilibrium with the membrane. This hypothesis is consistent with the view that membranes grow and turn over by the direct insertion and withdrawal of proteins synthesized by free ribosomes. However, the validity of the correlation between size and degradation rate of membrane proteins has been questioned (Tweto and Doyle, 1977), and experimental evidence for the existence of cytoplasmic pools of membrane proteins is lacking. Thus the results of studies on the turnover of membrane proteins are consistent either with the "direct insertion" hypothesis or with the idea that membrane proteins arrive at the plasma membrane from precursor membranes via fusion-fission. This transport, as discussed earlier, could occur either at a molecular level or as a bulk phenomenon, involving clusters of specific proteins. In the latter case, however, the protein composition of these preformed membrane pieces would have to be extremely modulable, since it is known that the composition of the plasma membrane can change following exposure to hormones or antisera (Gavin et al., 1974; Yu and Cohen, 1974; Hinkle and Tashjian, 1975), and studies on the biosynthesis of erythrocyte plasma membranes also indicate that different membrane proteins are synthesized at different periods during red cell maturation (Koch et al., 1975; Chiang et al., 1976).

In order to distinguish between the various possibilities, it would be important to study the kinetics of synthesis of well-defined membrane proteins in different fractions. However, the difficulty in obtaining pure membrane fractions (section 3), as well as the slow turnover of membrane proteins (entailing extremely low amounts of radioactive amino acids incorporated into membrane proteins in pulse-labeling experiments) (Borgese and Meldolesi, 1976; Borgese et al., 1977), make these experiments technically difficult. Thus relatively few results have been obtained to date. Omura and Kuriyama (1971) studied the biosynthesis of two microsomal enzymes, NADPH-cytochrome c reductase and cytochrome b_5, which they purified from liver rough and smooth microsomal fractions prepared at appropriate times after in vivo administration of labeled amino acid. They found that both enzymes were labeled more rapidly in the rough than in the smooth microsomal fraction, a result suggesting transfer of the newly synthesized proteins from rough to smooth ER. Similar results for cytochrome b_5 had previously been reported by Sargent and Vadlamudi (1968). However, since in these studies the purified enzymes were not analyzed by a high resolution electrophoretic method, it is still possible that a very small amount of secretory protein co-purified with the enzymes, thus giving erroneous results. In any case, the biogenetic relationships between the rough- and smooth-surfaced ER may well be quite different from that between the ER and other membranes. This is because rough and smooth ER constitute a continuous system, both structurally and also as far as enzyme activities and protein composition in general are concerned. (section 3.1). Thus, for transport of membrane proteins from rough to smooth ER to occur, it is sufficient that they travel in the plane of a fluid membrane; membrane fusion-fission processes are not required.

Similar considerations apply to the studies on the appearance of liver microsomal enzyme activities in the newborn rat (Dallner et al., 1966) or in adult rats after either phenobarbital administration (Ernster and Orrenius, 1965; Higgins, 1974) or starvation (Arion and Nordlie, 1965; Jakobsson and Dallner, 1968). These studies have indicated that the enzyme activities in developing rat liver or the induced activities in adult rats reach their maximal values in smooth ER later than in rough ER, suggesting that the enzymes are synthesized on bound ribosomes and then move to the smooth-surfaced membranes.

Studies on the biosynthesis of plasma membrane proteins have yielded diverging results when the analysis was carried out on total membrane fractions, possibly for the reasons discussed above; namely, contamination by secretory or cytoplasmic proteins. For example, Barancik and Lieberman (1971) found that in liver different proteins were inserted into the plasma membrane at varying time intervals after their synthesis, while in HeLa cells Atkinson (1975) found that the [³H]-leucine-labeled proteins appeared at the cell surface immediately following administration of the precursor.

A well-identified plasma membrane component, whose biosynthesis has been followed kinetically, is the Ig molecule present on the surface of B lymphocytes (Vitetta and Uhr, 1975). This protein is an integral membrane protein (Melcher et al., 1975) exposed at the external face of the lymphocyte plasmalemma, and is shed spontaneously from the cell surface (Vitetta and Uhr, 1972). Pulse-labeling experiments have demonstrated an interval of about two hours between the time intracellular Ig was labeled to the time the labeled molecules appeared at the cell surface (Vitetta and Uhr, 1974). Vitetta and Uhr interpreted this result to suggest that surface Ig is synthesized at the rough ER and transported to the cell surface via the secretory route (Vitetta and Uhr, 1975). Analogous studies on B lymphocyte H-2 alloantigen indicate that this protein is also incorporated into membranes intracellularly (Wernet et al., 1973). However, in neither case was the intracellular site of incorporation of these membrane proteins determined.

There is increasing evidence that the incorporation of sugars into plasma membrane glycoproteins occurs predominantly at the Golgi complex (reviews, Cook, 1977; Morré, 1977). Sugar transferases have been located in the Golgi complex, at the luminal membrane surface (section 3.3), and kinetic studies of incorporation of labeled sugars into plasma membrane glycoproteins are consistent with the idea that they arrive at the cell surface after transit through this organelle (Bosmann et al., 1969; Evans and Gurd, 1971; Kawasaki and Yamashima, 1971; Riordan et al., 1974; Atkinson, 1975; Atkinson et al., 1976). In a cell fractionation study on HeLa cells, Atkinson (1975) found that, while [¹⁴C]-amino acid-labeled proteins appeared in the plasma membrane almost immediately after administration of the tracer, [³H]-fucose-labeled glycoproteins appeared with a lag of approximately 15 minutes compared to the incorporation into the total homogenate. Essentially the same result was obtained for the insertion of vesicular stomatitis virus (VSV) envelope glycoproteins (G) and matrix (M) protein into HeLa cell plasma membrane (Atkinson et al., 1976). Detailed cell fractionation studies indicate that core sugars were inserted on to the G protein at the rough ER, while peripheral sugars were attached by Golgi enzymes (Hunt and Summers, 1976a,b).

These results agree with autoradiographic studies carried out with labeled sugars (Bennett and Leblond, 1970; Reith et al., 1970; Bennett et al., 1974; Flickinger et al., 1975; Haddad et al., 1977). The behaviour of newly synthesized glycoproteins was studied in a variety of rat tissues after a single injection of [³H]-fucose (review, Bennett and Leblond, 1977). The label first appeared over the Golgi complex, then over small vesicles between the Golgi complex and the plasma membrane, and finally over the cell surface.

Various authors have presented evidence that the cell possesses sugar transferases at the external surface of the plasma membrane (review, Shur and Roth, 1975). However, this finding has been questioned (Keenan and Morré, 1975) and, given the evidence that incorporation of sugars into proteins occurs predominantly at the Golgi complex, it is unlikely that ectoglycosyltransferases play a major role in plasma membrane glycoprotein biosynthesis. Probably, if they exist, they perform some special function in processes such as cellular recognition, as has been suggested by Roth and co-workers (Roth et al., 1971; Roth and White, 1972; Schachter, 1974).

4.2. Products synthesized by free and membrane-bound ribosomes

Another experimental approach to the problem of the synthesis site of membrane proteins has been to attempt to identify the products synthesized on free and membrane-bound ribosomes. In this type of experiment, after separation of the two populations of ribosomes by density gradient centrifugation, the products synthesized by the polysomes are determined by a variety of methods, such as: (1) in vitro amino acid incorporation followed by characterization of the products; (2) characterization of mRNAs; (3) binding of [¹²⁵I]-labeled antibodies or antigen binding fragments (Fab) to the nascent peptides contained in the ribosomes; and (4) immune precipitation of released nascent chains.

Erroneous interpretation of the data obtained from these experiments can be caused by a variety of artifacts, such as incomplete separation of free and bound ribosomes caused by the adsorption of free polysomes to microsomal membranes (Mechler and Vassalli, 1975a), or detachment of bound ribosomes during homogenization and cell fractionation (O'Toole, 1974). Also, cytoplasmic or membrane proteins may adsorb to the surface of ribosomes and be responsible for [¹²⁵I]-labeled antibody binding (Eschenfeldt and Patterson, 1975).

Presumably because of the various artifacts that can be produced in this kind of experiment, divergent results have been reported for the site of synthesis of NADPH-cytochrome c reductase, an integral flavoprotein exposed on the cytoplasmic face of ER membranes (Depierre and Dallner, 1975). The site of synthesis of this enzyme has been investigated both by its purification from the mixture of products synthesized in vitro by free and bound ribosomes (Ragnotti and Lawford, 1968; Lowe and Hallinan, 1973) and by antibody precipitation of nascent chains (Omura, 1973). While Ragnotti et al. (1969) and Omura (1973) found that the liver enzyme was synthesized on both free and bound ribosomes, Lowe and Hallinan (1973) found that it was produced predominantly by free polysomes. The latter authors attributed the result of Ragnotti et al. (1969) to contamination of their microsomal fraction by free polysomes. Recently, the dual site of synthesis of the enzyme in free and membrane bound polysomes has been confirmed by Harano and Omura (1977a).

In contrast to the situation for NADPH-cytochrome c reductase, two other integral microsomal proteins, cytochrome b_5 (Harano and Omura, 1977a) and cytochrome P_{450} (Negishi et al., 1976), have been reported to by synthesized exclusively by bound ribosomes. These results were obtained both by antibody binding experiments as well as by immune precipitation of nascent chains. Thus, these membrane proteins are synthesized by the same class of polysomes responsible for the synthesis of secretory proteins. However, the two types of proteins do not intermix, apparently because they are separated topologically already during the course of their synthesis (Negishi et al., 1975; Harano and Omura, 1977b).

Bergeron and colleagues (1975) have attempted to identify the site of synthesis of 5'-nucleotidase (for a discussion on the subcellular localization of this enzyme, see section 3.5). After microinjection of mRNAs isolated from mouse liver free and bound ribosomes into *Xenopus laevis* oocytes, the radioactive products synthesized were identified by SDS-polyacrylamide gel electrophoresis of immunoprecipitates obtained with anti-5'-nucleotidase antibodies. It was found that the information for synthesis of the enzyme was contained principally in messengers extracted from bound ribosomes. However, some enzyme was also produced following injection of "free" messengers. Unfortunately, in this study the immunoprecipitates contained large amounts of nonspecifically precipitated proteins responsible for the high background radioactivity present in the gels, making it difficult to quantitatively evaluate the relative amounts of 5'-nucleotidase synthesized by free and bound populations of polysomes.

Elder and Morré (1976) have characterized the products synthesized by the rat liver free and bound ribosomes on the basis of solubility in a high salt-detergent solution and migration rate on SDS-polyacrylamide gels. They found that polypeptides co-purifying with insoluble membrane proteins were synthesized on both free and membrane-bound ribosomes. Although the electrophoretic migration rate of proteins is not a sufficient identification criterion, these studies suggest that hydrophobic membrane proteins can be synthesized by free ribosomes. Analogously, Lodish (1973) reported that reticulocyte free ribosomes could synthesize two erythrocyte membrane proteins subsequently shown to be confined to the cytoplasmic surface of the membrane (Lodish and Small, 1975).

Over the last few years, enveloped virus systems have been found to be convenient models for the investigation of membrane biogenesis, because during infection the cell is mainly engaged in the synthesis of viral proteins, some of which become integrated into the viral membrane. Morrison and Lodish (1975) and Grubman et al. (1975) found that the VSV glycoprotein (G protein), which corresponds to the spikes located on the external surface of the viral envelope, is synthesized exclusively by bound ribosomes, while the M protein, which is found primarily on the inside of the lipid bilayer, is produced by both free and bound ribosomes. The authors suggested that the dual distribution of mRNA coding for the M protein might be artifactual, resulting from the adsorption of free polysomes to membranes during rupture of the cells. Analogous studies in the

Sindbis virus system also showed that the viral capsid protein could be synthesized on free polysomes, while the envelope glycoprotein precursors are synthesized exclusively by the microsomal fraction (Bonatti et al., 1978). Further studies on the VSV system revealed that the unglycosylated precursor G protein (G_1) synthesized by polysomes in a cell-free system could be glycosylated by intracellular membranes obtained from HeLa cells (Toneguzzo and Ghosh, 1977) or dog pancreas (Katz et al., 1977). The glycosylation occurred only if the peptide was synthesized in the presence of membranes, not if membranes were added to the finished peptides, indicating that the protein destined to cross the lipid bilayer must "grow" into the ER membrane in order to be correctly processed (Katz et al., 1977).

In studies on the biosynthesis of VSV envelope proteins, "bound ribosomes" were defined as those recovered in a membrane fraction containing essentially all types of membranes. Thus, these ribosomes were not necessarily associated with the rough ER. It has been reported that cytoplasmic ribosomes can attach to the outer mitochondrial membrane (Kellems et al., 1974) and to the Golgi complex (Elder and Morré, 1976). Furthermore, 70S ribosomes have been found in association with inner mitochondrial membranes (Kuriyama and Luck, 1973) as well as with chloroplast thylakoid membranes (Chua et al., 1973; Margulies and Michaels, 1974; Margulies et al., 1975). It must be mentioned that the interaction between 70S ribosomes and thylakoid membranes appears to be different from that between 80S ribosomes and ER membranes in secretory cells. In order to detect the ribosome-membrane association in chloroplasts of *Chlamydomonas reinhardii,* it was necessary to pretreat the cells with chloramphenicol to prevent polysomal runoff (Chua et al., 1973; Margulies and Michaels, 1974). This was not the case for the 80S ribosome-ER association in secretory cells, which is resistant to puromycin-induced release of nascent chains, both in vivo (Blobel and Potter, 1967; Mechler and Vassalli, 1975b) and in vitro at low ionic strengths (Adelman et al., 1973). Thus the ribosome-membrane association in chloroplasts appears to be more transient than that between 80S ribosomes and ER membranes, and to persist only as long as a determined protein is being synthesized. Chua and collaborators (1973), on the basis of the observation that the association between chloroplast ribosomes and membranes was limited to a time period during which thylakoid membranes are synthesized, suggested that the membrane-bound ribosomes of chloroplasts might be involved in the synthesis of membrane proteins. By analogy, it is conceivable that a transient association between cytoplasmic ribosomes and membranes other than those of the ER plays a role in the direct insertion of proteins into the lipid bilayer.

4.3. Other experimental evidence on membrane biogenesis

Indirect evidence for a role of the Golgi complex in plasma membrane assembly comes from morphological observations. In some systems, Golgi membranes have been observed to increase in thickness from the proximal to the distal face of the dictyosomes, so that at the distal pole they morphologically resemble the plasma membrane (Grove et al., 1968). Yamamoto (1963) observed that in frog sympathetic ganglia Golgi vesicles are characterized by a thickness similar to that of plasma membrane, while Golgi cisternae are thinner and resemble ER membranes. In urinary bladder transitional epithelium, where the cells of the luminal layer have an asymmetrically thickened plasma membrane, Hicks (1966) observed patches of thick membrane in the Golgi cisternae presumably representing pieces of precursor plasma membrane. Analogous observations were made by Mollenhauer and associates (1976) in spermatids in relation to acrosome formation.

On the other hand, there is also morphological evidence suggesting that proteins can be inserted into the plasma membrane directly, without the need for vesicle-mediated transport. Pfenninger and Bunge (1974) studied the morphology of the growth cones of developing nerve fibers by freeze-fracture and found that the growing tip is characterized by a low intramembranous particle density. This increases in the direction of the cell body and with the development of the neuron. The newly formed plasma membrane was interpreted as being derived from particle-poor vesicles present in the axoplasm. Since particle-rich vesicles were not seen, these results can be interpreted to suggest that the protein components lacking in the newly formed plasma membrane are inserted directly at a later stage. However, other interpretations are also possible. The studies of Franke and co-workers (1975) on the nuclear envelope of Acetabularia also suggest that the membrane may increase its surface by the direct insertion of cytoplasmic components without the involvement of vesicle-mediated transport.

Morphological studies on the orientation of membrane glycoproteins have been adduced as support for the concept of plasma membrane biogenesis by the "assembly line" mechanism, that is, by the fusion of ER-derived vesicles. Studies with ferritin-conjugated sugar lectins have demonstrated that, in eukaryotic cells, membrane oligosaccharides are located exclusively at the outer face of the plasma membrane (Nicolson and Singer, 1971) and at the luminal face of rough ER membranes (corresponding to the external face of the plasma membrane) (Hirano et al., 1972). Moreover, the results of these studies indicate that ER glycoproteins possess only core sugar residues, which can be attached by microsomal enzymes, thus suggesting that transport of membrane proteins occurs undirectionally along the secretory pathway (Hirano et al., 1972). These morphological observations with ferritin-conjugated lectins have been confirmed for rough ER by biochemical experiments (Winquist et al., 1974; Rodriguez-Boulan et al., 1975), which have shown that the glycoproteins of rough microsomes become available to concanavalin A only if the vesicles are opened by treatment with detergents. Clearly, however, the study of microsomal membrane glycoproteins is complicated by difficulty in distinguishing between proper membrane constituents and loosely associated secretory proteins (sections 2.1.1 and 3).

Of particular interest is the finding that in two plasma membrane glycoproteins whose orientation in the lipid bilayer has been analyzed, glycophorin (Segrest et al., 1973; Bretscher, 1975) and H-2 antigen (Henning et al., 1976), the oligosaccharide moiety facing the outside of the cell is N-terminal. This is consistent with the idea that plasma membrane apoglycoproteins are synthesized on membrane-bound ribosomes and are inserted into the membrane during the course of their synthesis. It must be stressed that although these findings are consistent with the idea that plasma membrane apoglycoproteins are synthesized at the rough ER, they do not prove this hypothesis; a generally respected pattern of transverse asymmetry common to different types of cellular membranes does not necessarily imply biogenetic relationships among them. Moreover, the rule that carbohydrates are not exposed on the cytoplasmic face of membranes may not be a general one. For example, there is increasing evidence in favor of the presence of protein-bound carbohydrates at the cytoplasmic face of the membranes of secretion granules and lysosomes. This situation has been reported for chromaffin granules (Meyer and Burger, 1976), pancreatic zymogen granules (Lewis et al., 1977), and azurophil and specific granules of polymorphonuclear leukocytes (Feigenson et al., 1975). To explain this orientation of membrane glycoproteins, one must hypothesize either the existence of cytoplasmically oriented sugar transferases or the capacity of membrane proteins to undergo drastic reorientation after glycosylation.

The idea that transport of membrane proteins can occur only in one direction, mainly from ER to plasma membrane, has also been opposed. Evidence has been presented suggesting that in rat liver, a tightly bound, sialic acid containing, microsomal protein is transported to the ER through the cytoplasm after glycosylation in the Golgi complex (Autuori et al., 1975a,b; Elhammer et al., 1975). This claim was based not only on in vivo pulse-labeling experiments but also on the isolation of a sialic acid containing lipoprotein from the high speed supernatant obtained from perfused liver homogenates that could be incorporated into microsomes in vitro. Although the possibility that the sialoglycoprotein present in the microsomal fraction was due to plasma membrane contamination was not excluded, the presence of a lipoglycoprotein in the supernatant fraction, which does not resemble serum lipoproteins by electrophoretic or immunological criteria and might represent a membrane precursor (Svensson et al., 1976), may prove to be important and to subvert dogmas on membrane biogenesis that have gained wide acceptance during recent years unsupported by sufficient experimental evidence.

Finally, the biochemical similarities observed between Golgi and plasma membranes must be mentioned for their relevance to the problem of membrane biogenesis. As discussed in section 3.5, numerous plasma membrane enzymes and hormone receptors have been reported to be present in liver Golgi fractions. It would be interesting to know whether the presence of these enzymes is due to a precursor-product relationship between Golgi and plasmalemma, or whether it is simply a consequence of the intense pinocytotic process occurring at the cell surface (Steinman et al., 1976), which could result in temporary incorporation of plasma membrane portions into the Golgi complex (Gonatas et al., 1975, 1977). Pulse-labeling experiments designed to follow the kinetics of amino acid incorporation into these enzyme molecules in Golgi and plasma membranes must thus be carried out in order to decide between these two possibilities.

4.4. Conclusions

The reader may have found it difficult to make his way through the mass of conflicting evidence presented in this section. Yet the subject could have been treated more coherently if we had only a single model for membrane biogenesis in mind. However, we do not feel that the available evidence justifies a simple, universally applicable scheme. The question that we intended to deal with in this section of the review is what membrane biogenesis has to do with secretion. Clearly, from the evidence presented above, only partial answers can be given. We can say that in secretory cells the rates of membrane biogenesis and intracellular transport processes are of a different order of magnitude and, by analogy with other systems, that at least some plasma membrane proteins are probably inserted directly into the lipid bilayer from the cytoplasm. Comparative studies on the composition of membranes (section 3.5) indicate it is improbable that one type of membrane can be transformed into another type. Therefore, transport of membrane proteins along the secretory pathway, in whatever degree it occurs, must concern well-defined molecules, whether these be distributed homogenously within the domain of the "donor" membrane or clustered into specific patches.

It must be stressed here that although a variety of biochemical and morphological observations suggest that the Golgi complex plays a role in plasma membrane biogenesis, a general answer as to the intracellular route that plasma membrane precursor proteins follow to reach and leave the Golgi complex cannot be given yet. Thus it is possible that at least some membrane proteins do not arrive at the Golgi from the ER, but are inserted directly from the cytoplasm. Also, the mode of transport of membrane proteins from the Golgi to the cell surface is unknown. This could occur along the secretory route, that is, granules or vesicles containing secretory proteins could also carry membrane proteins destined for the plasmalemma, where they would be selectively retained and not retrieved together with the other granule membrane proteins. However, the distinct protein composition of secretory granules found in most systems (sections 3.3; 3.5) suggests that the transport of membrane proteins from Golgi to plasmalemma may occur along a pathway not involving secretory granules, that is, the two phenomena of intracellular transport of secretory and plasma membrane proteins might well be independent.

5. Discharge of secretion products

The final event of vesicular secretion is discharge to the extracellular environment of secretory material prepacked in the cytoplasm within membrane-bounded organelles. On the basis of thin-section images, Palade proposed that discharge occurs following *fusion-fission* (as defined on page 510) of the limiting membrane of granules or vesicles with the plasmalemma (Palade 1959). This process was later termed exocytosis by de Duve (1963). Since then, release of secretory products by exocytosis has been described in a wide variety of systems. Recently this concept has been strongly corroborated and extended by the use of freeze-fracture, a technique that allows direct views of large portions of the internal hydrophobic domain of membranes.

5.1. Topology of exocytosis at the cell surface

In considering the localization of exocytosis at the cell surface, a wide spectrum of situations can be recognized, ranging from the occurrence of exocytosis all over the cell surface, without demonstrable topological restrictions, to situations in which discharge is limited to specific portions of the plasmalemma. In the latter case, two factors should be considered: (1) cellular polarity, the preferential or exclusive localization of the secretory granules or vesicles in one region of the cell; and (2) topological specificity, restriction of exocytosis to well-defined, predetermined areas of the plasmalemma, which can be represented by entire regions or by focal sites. A good example of a secretory cell showing no topological restrictions in exocytosis is the mast cell. The other extreme is the motor neuron, in which synaptic vesicles are found near the synaptic surface of the axon terminal (i.e., high degree of polarity), where discharge is further restricted to focal sites strategically located in close spatial relationship to receptor areas of the muscle cell (see page 562). An analogous degree of polarity exists in all exocrine cells as well as in the follicular cells of the thyroid gland; in these cases, however, the portion of the plasma membrane predetermined for exocytosis is not restricted to focal sites but includes the whole luminal surface, delimited circumferentially by the zonula occludens. On the other hand, in most endocrine systems (e.g., the Langherans islets, the anterior pituitary, the adrenal medulla) exocytosis occurs preferentially, but not exclusively, at the portion of the plasmalemma facing the basal membrane (Farquhar, 1971; Orci et al., 1973a; Benedeczky and Somogyi, 1975; and chapter by Orci and Perrelet, this volume.).

Finally in some protozoa, such as Tetrahymena and Paramecium, no sign of

---

Fig. 5. Low-power electron micrograph showing a portion of resting neuromuscular junction from cutaneous pectoris neuromuscular preparation, soaked for 1 hr in 0.5 mM Ca^{2+} and 4 mM Mg^{2+} and then fixed in OsO_4. The axonal ending (A) is covered by a thin Schwann cell process (Sc), and the terminal contains numerous mitochondria, neurofilaments, elements of smooth endoplasmic reticulum, glycogen (g), and the normal complement of synaptic vesicles (v). Note the electron-dense tufts (active zones) that overlie the presynaptic membrane just opposite the folds in the muscle membrane. × 16,000.

Fig. 6. Freeze-fracture view of a resting neuromuscular junction soaked for 1 hr in Ringer's solution and then fixed with glutaraldehyde and paraformaldehyde. The half interior of the presynaptic membrane, as seen from the outside, is shown in this low power micrograph (P face). The primary features of the prejunctional membrane are the regularly spaced, 90 nm wide, ridges that run perpendicularly to the longitudinal axis of the terminal. The ridges lie in exact register with the postjunctional folds (circle) and are bordered on each side by a double row of large intramembranous particles (inset). Many Schwann cell processes have broken away during fracture (arrows). × 14,000; inset: × 88,000.

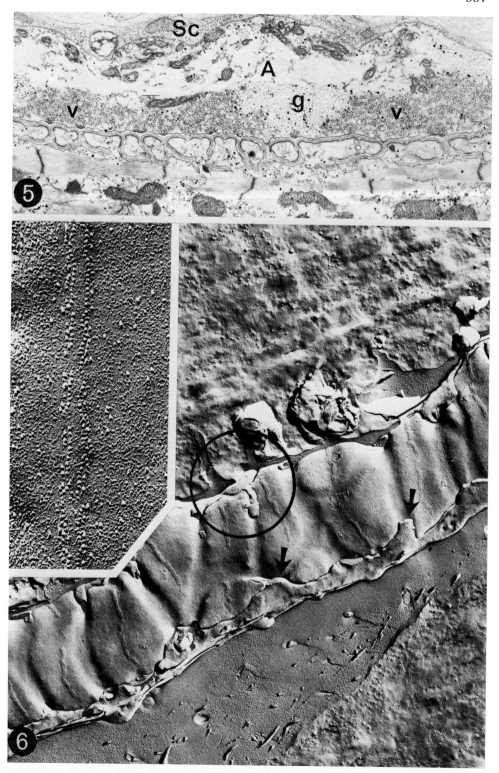

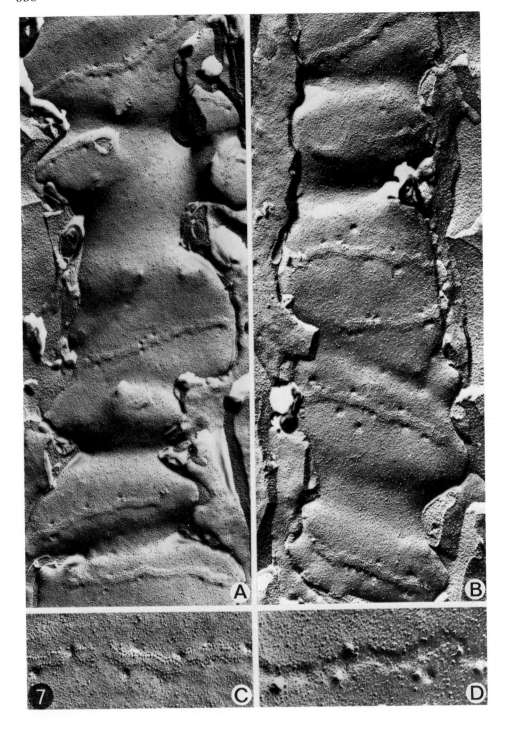

polarity is evident. Discharge occurs, however, at discrete sites of the cortical membrane, topologically identified with respect to cilia and marked by pre-formed particle arrays present in the context of the plasma membrane (Plattner et al., 1973; Satir et al., 1973; and chapter by Allen, this volume).

The mechanisms responsible for establishing and maintaining the specific localization of exocytosis at the cell surface are still poorly understood. In systems characterized by polarity of discharge, the spatial relationship of granules or vesicles to the secretory portions of the plasmalemma can be maintained even after exposure to antimicrotubular drugs, which induce profound alterations in the general architecture of the cell (section 2.1.7) and, at least for some time, after cell dissociation from the tissue by various techniques (Amsterdam and Jamieson, 1974; Herzog et al., 1976). As far as the topological restriction of exocytosis to discrete areas of the plasmalemma is concerned, a major role is apparently played by heterogeneities in the distribution of individual components, either membrane proteins or surface-associated constituents. In many secretory systems the fibrillar network associated with the inner aspect of the plasma membrane is either particularly prominent or arranged according to a peculiar geometry in the secretory region. This observation, coupled with the evidence that drugs affecting the cytoskeleton often interfere in various ways with exocytosis, has prompted speculations about the role of the so called microtubular-microfilamentous system in governing the specific interaction of secretory organelles with the plasmalemma (Berl et al., 1973; Allison and Davies 1974, Malaisse et al., 1975; Palade, 1975; Franke et al., 1976; Holtzman et al., 1977). However, available experimental evidence on this subject is still fragmentary and inconclusive. In contrast, the hypothesis that integral components of membranes may be directly involved in determining the topological specificity of the fusion process is much more soundly based because freeze-fracture studies have demonstrated in some cells the existence of specific arrangements of intramembranous particles located in the plasmalemma either at, or close to, the sites of exocytosis. In other systems discrete arrays were not detected, though in many cases the regions of plasmalemma where exocytosis is known to occur were found to be distinctly different from the nonsecretory regions.

Clear-cut examples of peculiar particle arrays have been described in Protozoa

Fig. 7. Freeze-fracture views of different neuromuscular junctions fixed with glutaraldehyde and paraformaldehyde at the peak of the Black Widow Spider Venom (BWSV) effect. To prevent fibrillation of muscle fibers, the preparations were presoaked for 1 hr in 0.7 mM Ca^{2+} and 4 mg Mg^{2+} and then BWSV was added to the solution. On the P face (A, B and C) many dimples that represent vesicle attachment sites are seen in the immediate vicinity of the double rows of particles. On the complementary E face (D) many protuberances are seen immediately lateral to the furrow. The large intramembranous particles that border the active zone are not clearly seen on this face. Sometimes, however, the large particles leaves imprints on this half of the presynaptic membrane. (C) and (D) are high magnifications of P and E faces, respectively. (C) is from a different neuromuscular junction from the same preparation used for A. (A) ×31,500; (B) ×36,500; (C) ×63,000; (D) ×80,000.

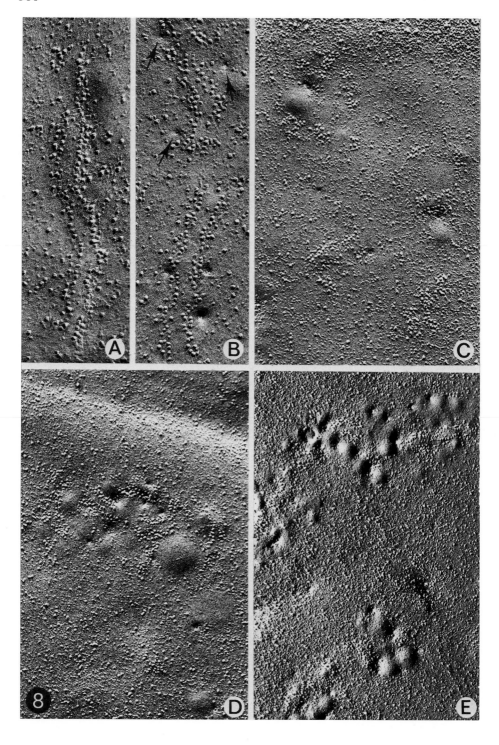

as well as at frog neuromuscular junction. In Tetrahymena, the sites destined for mucocyst discharge are marked on the plasmalemma by a rosette of membrane-intercalated particles, most of which remain associated with the P leaflet after fracture (Fig. 13) (Satir et al., 1972, 1973). In Paramecium, the corresponding structure is a rosette encircled by a double ring of particles (Plattner et al., 1973). Similar particle arrays have been found at sites of exocytosis in other Protozoa (Satir, 1974a). Since these arrays are present in unstimulated cells, they seem to represent rather stable structural elements involved in granule attachment, a process which in protozoa can occur a considerable time before granule discharge. It has been elegantly shown by Beisson and co-workers (1976) that in several Paramecium mutants genetically determined alterations of these arrays are paralleled by the inability of secretory granule (trichocyst) discharge.

The other important example to be considered in the discussion of the role of functional molecules within the plasma membrane, is the frog neuromuscular junction (Dreyer et al., 1973; Heuser et al., 1974; Peper et al., 1974; Akert et al., 1975; Ceccarelli et al., 1976). As shown in Fig. 6, the primary features of the prejunctional membrane are the regularly spaced, 90 nm wide, ridges that run almost perpendicularly to the longitudinal axis of the nerve terminal and are bordered on each side by a double row of large intramembranous particles (P face). These ridges lie in exact register with the postsynaptic infoldings of the underlying muscle fiber. They correspond to the "active zones" described by Couteaux (Fig. 5) and constitute the regions near which fusions of synaptic vesicles were originally seen in thin sections (Couteaux and Pécot-Dechavassine, 1970, 1973; Ceccarelli et al., 1972; Couteaux, 1974; Hurlbut and Ceccarelli, 1974; Ceccarelli and Hurlbut, 1975b).

Freeze-fracture studies of frog neuromuscular junctions during intense secretion of neurotransmitter reveal a striking localization of exocytosis. On the P face, vesicles are seen fused laterally to the ridges in restricted regions of the axolemma immediately adjacent to the double rows of large particles. This exclusive localization of fusions seems to be maintained in all cases, independently of the nature of the stimulus used to evoke secretion. Thus, it has been observed in preparations electrically stimulated in the presence of aldehyde fixative (Heuser et al., 1974) and in junctions in which the spontaneous rate of quantal secretion of ACh was enhanced by exposure to BWSV (Ceccarelli et al., 1976) or

---

Fig. 8. Freeze-fracture views of different neuromuscular junctions fixed with glutaraldehyde and paraformaldehyde at the peak of the (BWSV) effect. The preparations were presoaked for 2 hr in modified Ringer's solution containing no Ca^{2+} and 6 mM Mg^{2+}; BWSV was then added to the same solution. In (D) and (E) 1 mM EGTA was also present throughout the experiment. These micrographs show different degrees of disorganization of the double rows of particles, yet fusions are still seen to occur in regions adjacent to the disorganized rows (arrows in B), or in close relationship to particle arrays clearly representing the remnants of the double rows (C, D, and E). Numerous particle-free bulges are clearly seen in regions of the axolemma near the disorganized rows of particles. They may represent regions of close contact between underlying vesicles and the axolemma. However, images corresponding to fusions are seen always laterally and never within the domain of the particle-free domes. (A) and (B) ×108,000; (C) and (D) ×63,000; (E) ×45,000.

to high K$^+$ solutions (Figs. 7 and 8). No fusions were found in resting terminals (Fig. 6) or when discharge of quanta of ACh in response to nerve stimulation (or to high K$^+$ depolarization) was prevented by soaking the preparation in a solution containing high Mg^{2+} instead of Ca^{2+} (Heuser et al., 1974; Ceccarelli et al., 1978a). BWSV provokes an enormous increase in the frequency of miniature endplate potentials (mepps), which rises in a few minutes to maximum values. The peak release of quanta is maintained for several minutes, then subsides to low levels over $^1/_2$ hour (Longenecker et al., 1970; Ceccarelli et al., 1973). Fig. 7 is a composite of micrographs of prejunctional membranes from terminals that have been fixed at the peak of BWSV effect. Many discharging vesicles that were fixed in the fused state are clearly seen in regions of the axolemma immediately adjacent to the ridges (P face) or furrows (E face).

In contrast to the situation for electrical or high K$^+$ stimulation, the effect of BWSV on mepp frequency is totally independent of the availability of extracellular Ca^{2+} (Longenecker et al., 1970). Therefore, by the use of BWSV in Ca^{2+}-free solutions, information concerning the role of Ca^{2+} in the topological organization of the presynaptic membrane and in the localization of exocytosis could be obtained. Resting muscle bathed for 2 hours or more in Ca^{2+}-free solutions shows fragmentation and profound disorganization of the double rows of particles. However, when BWSV is applied under these condition, fusions are still seen to occur in regions adjacent to the disorganized rows, or in close relationship to particle arrays clearly representing the remnants of the double rows (Fig. 8) (Ceccarelli et al., 1978a,b).

These findings indicate that discharge of ACh at frog neuromuscular junction occurs in localized and definite regions of the axolemma just beside the double row of intramembranous particles bordering the ridges or furrows. However, these particles do not seem to be directly involved in the complex molecular rearrangement of the bilayer structures leading to vesicle fusion and subsequent fission. Rather, it appears they define specific regions of the presynaptic membrane where changes leading to vesicle fusion may occur (see also section 5.2).

In the underlying muscle fiber, the opening of the junctional fold is studded with large intramembranous particles that appear mainly on the P face of postjunctional membrane (Fig. 9) (see also Heuser et al., 1974; Peper et al. 1974; Rash and Ellisman, 1974). The location of this dense concentration of particles in definite areas of the postjunctional membrane, and the correspondence between these areas and the localization of α-bungarotoxin binding sites (Fertuck and Salpeter, 1974; Daniels and Vogel, 1975; Bender et al., 1976; Lentz et al., 1976), suggest that these particles may be involved in ACh reception and/or in ACh signal transduction. Thus, at frog neuromuscular junction, discharge of secretory product and receptor interaction seems not to involve the whole pre- and postjunctional surfaces but only small, discrete areas aligned in register in the two facing membranes. The precise matching between the discharging site and the receptive area probably optimize the coupling between presynaptic secretion and postsynaptic excitation.

In mammals, the organization of neuromuscular junctions does not display

Fig. 9. Freeze-fracture micrograph of a resting frog neuromuscular junction. The fracture has jumped across the synaptic cleft (asterisks) to the muscle membrane. The internal structure of the postsynaptic membrane as seen from the outside of the muscle (P face) is clearly revealed. This region is characterized by clusters of large intramembranous particles that extend into the postsynaptic folds. They correspond to the location of ACh receptors (see text). Note that the basal membrane within the postsynaptic fold is clearly visible. ×63,000.

the regular features of the presynaptic membrane described in the frog. However double rows of intramembranous particles, although shorter and more irregular, are still present and seem to play a critical role in the localization of exocytosis (Rash and Ellisman, 1974). A similar interpretation has been proposed with regard to the presynaptic surface at the central nervous system, where particles appear heterogeneous in size, shape, and distribution (Pfenninger et al., 1972). In other types of mammalian secretory cells particle arrays at sites of granule discharge have either not been found (adrenal medulla: Smith et al., 1973a; Langherans islets: Orci et al., 1973a; parotid gland: De Camilli et al., 1976) or have not been definitively demonstrated. In stimulated neurohypophysis, rosette-like particle aggregates delimiting small pits were observed by Dreifuss and colleagues (1976) on the P face of the plasmalemma of the nerve terminal and interpreted as membrane sites predetermined for exocytosis. However, in the figures these authors published, particle clusters clearly associated with underlying granules were not visible. In other cells (mast cell: Lawson et al., 1977;

564

thyroid: Ishimura et al., 1977) particle aggregates have been observed at early time points after stimulation, but in these cases too no association was found of such clusterings with the discharge of secretory granules. (For further discussion on the possible role of particle arrays in exocytosis, see section 5.2.)

A different type of surface specialization related to exocytosis can be identified in cells where discharge is restricted to the luminal portions of the plasmalemma delimited by zonulae occludentes (exocrine gland and thyroid).[11]

As mentioned earlier (section 3.4), evidence obtained by biochemical and freeze-fracture studies clearly indicates that the secretory surfaces of these cells have a different composition from the rest of the plasmalemma. In a variety of systems that we have examined, such as the exocrine pancreas (De Camilli et al., 1974; see also Fig. 10), parotid (De Camilli et al., 1976), and stomach (principal and mucous cells; see Fig. 4), as well as in the follicular cells of the thyroid (Ishimura et al., 1977), the density of intramembranous particles on the P face is distinctly lower at the luminal than at the basolateral plasmalemma. In both these regions the distribution of particles is essentially homogeneous and no discrete sites predestined for the fusion of individual granules can be identified.

Since in exocrine and thyroid cells other functions [e.g., ion secretion: (Batzri et al., 1973; Petersen and Ueda, 1976)] are known to be carried out at the luminal plasmalemma, it seems probable that the structural specificity of this membrane region is not exclusively related to its ability to participate in exocytosis. This consideration should not be taken to mean that complementary recognition sites, undetectable by available techniques, do not exist in granule and in plasma membranes of all secretory cells. One might speculate that in the plasmalemma of cells which discharge vectorially into a lumen, these hypothetical sites would be localized exclusively at the luminal surface. Thus interesting information might be obtained by studying exocytosis in cells whose topological heterogeneity has been modified experimentally. We found that this can be achieved in pancreatic lobules by in vitro incubation in Ca^{2+}-free Krebs-Ringer bicarbonate medium containing EGTA at low concentrations. This treatment results in the progressive disassembly of tight junctions. The ordered meshwork of junctional fibrils is first disordered and then fragmented into strands of various lengths, which spread in the lateral and luminal portions of the membrane. As soon as the continuity of the beltlike junctional complex is lost, the density of intramembranous particles becomes the same in the luminal and lateral regions (Galli et al., 1976; and Figs. 10 and 11).[12] At this point the readdition of Ca^{2+} in the incu-

---

[11]Although this localization is generally respected, an exception has been reported in the rat exocrine pancreas. In this system, exhaustive in vivo stimulation by prolonged infusion with high doses of caerulein can yield zymogen discharge also at the lateral surface (Lampel and Kern, 1977).

[12]This observation is consistent with the recent results of Maylié-Pfenninger and co-workers (1975) showing that in pancreatic acinar cells dissociated from the tissue by treatments with EDTA and proteolytic enzymes the distribution of glycoproteins (as revealed through the binding of several lectins coupled to ferritin) is homogeneous over the entire cell surface. Furthermore, a rapid redistribution of luminal and lateral plasma membrane components following tight junction disruption has recently been observed in the cells of the epithelium lining the toad bladder by Pisam and Ripoche (1976).

bation medium results in the rapid reformation of tight junctions. However, for at least 2 hours, the density of particles in the luminal plasmalemma remains high (Meldolesi et al., 1977; Ceccarelli et al., 1978b; Fig. 11). These data clearly indicate that at least some integral components are free to redistribute randomly in the plane of the membrane once the zonula occludens (which is apparently the major physical barrier segregating the luminal and lateral plasmalemma) is interrupted.

The last problem to be discussed in relation to the topological specificity of discharge is compound exocytosis (Douglas, 1974b), a phenomenon whereby the membranes of discharged granules acquire the capacity for fusion-fission with other granules. In all secretory systems, granules or vesicles—as long as they are stored in the cytoplasm—are unable to fuse with each other, even though in some cases their mutual apposition is so close as to result in the partial elimination of the adhering membrane layers (Palade, 1975). In many systems, however, during intense stimulation, granules are discharged by fusion of their membranes not only with the original plasmalemma, but also with the membrane of granules previously discharged. Sausagelike infoldings, continuous with the cell surface, are thus formed. In our own experience the most striking examples of compound exocytosis were observed in pancreatic fragments fixed with pyro-antimonate-OsO_4 mixture (Spicer et al., 1968) to reveal cations (Fig. 12).

Figs. 10 and 11. *See pages 566, 567.* Freeze-fracture electron micrographs of lobules of guinea pig exocrine pancreas incubated in vitro. Fig. 10. (A) Control acinar cell showing the P and E fracture faces of the secretory pole of two adjacent cells. The luminal and lateral regions of the plasma membrane are separated by the tight junction (j), which forms a continuous beltlike structure around the lumen (zonula occludens). The density of particles on the P face of the luminal membrane is low compared to the density on the corresponding face of the lateral plasmalemma. The junctional network has a compact organization and is composed of three 3-5 interconnected fibrils running roughly parallel at the luminal side, surrounded by a fibrillar loose meshwork extending abluminally. Figs. 10 (B) and (C) and 11 (A) Changes of plasma membrane architecture induced by incubation in Ca^{2+}-free Krebs-Ringer-bicarbonate (KRB) containing 0.5 mM EGTA. When Ca^{2+} is removed from the incubation medium, the tight junction architecture is progressively disrupted. The peripheral network is first disarranged (Fig. 10B, 10 min after Ca^{2+} removal) and then fragmented to yield smaller strands (arrows), which move in the plane of the membrane (Fig. 10C, 40 min after Ca^{2+} removal); arrowheads mark gap junctions around which tight junction fibrils have wound. 120 min after Ca^{2+} removal (Fig. 11A), the zonula occludens is completely disrupted and the lateral plasmalemma appears at many points (arrow) continuous with the luminal region, distinguishable by the presence of cross-sectioned microvilli. The difference in particle density between the P face of the luminal and lateral membrane has completely disappeared. Figs. 11 (B) and (C) Apical pole of acinar cells of lobules incubated for 2 hr in Ca^{2+}-free KRB + EGTA, then transferred for 15 min in complete KRB medium. These figures show that the reintroduction of Ca^{2+} into the incubation medium results in the rapid reassociation of junctional strands, to yield a continuous zonula occludens around the luminal surface. The latter, however, retains a high density of intramembranous particles on its P face. P** and E**, P and E faces of the lateral plasma membrane; P* and E*, P and E faces of the luminal plasmalemma; L, cross-section of the acinar lumen; m, microvilli; j, tight junctions. Fig. 10 (A) ×71,100, (B) ×27,500; (C) ×22,100; fig. 11 (A) ×35,100, (B) ×45,900; (C) ×43,700. (Figs. 10B and 11A reproduced with permission from Galli et al., 1976.)

566

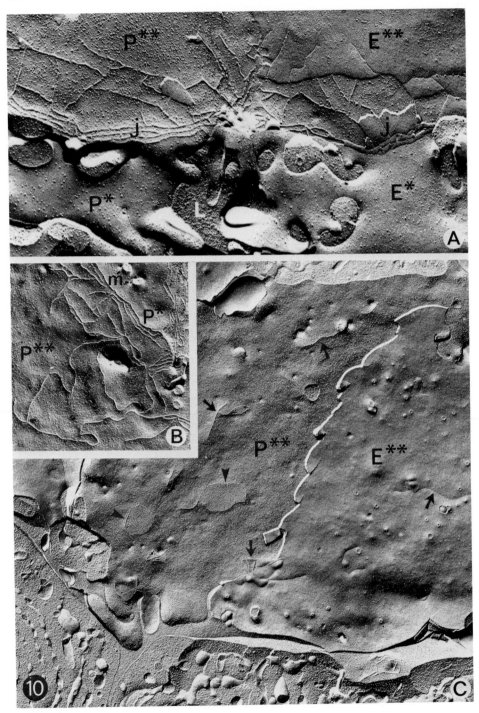

(Captions for Figs. 10 *(above)* and 11 *(facing page)* appear on the preceding page
(p. 565).)

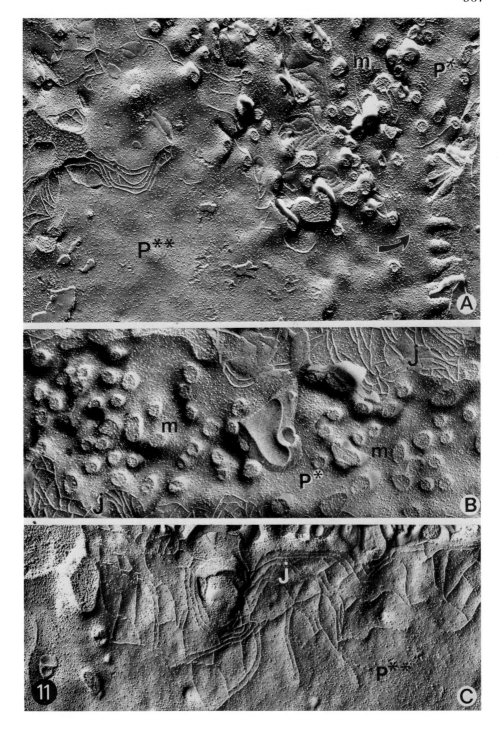

568

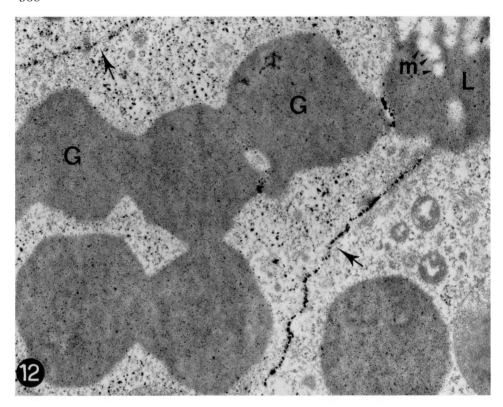

Fig. 12. Compound exocytosis in pancreatic exocrine tissue fixed by immersion in the OsO_4-pyroantimonate mixture of Spicer et al. (1968) to reveal cations. The dense deposits are due to the formation of insoluble precipitates of Na^+ and Ca^{2+} antimonate. Images of compound exocytosis were found in most cells localized at a certain distance from the surface of the block ($\sim 100\mu$), and were encountered far less frequently in more superficial layers. A possible explanation of this finding is that pyroantimonate, diffusing throughout the tissue slightly faster than OsO_4, might have stimulated exocytosis by displacing bound Ca^{2+} in acinar cells (text, section 5.3). L, acinar lumen; m, microvilli, G, fused zymogen granules; arrows indicate the lateral plasma membrane. $\times 28,000$.

To date, a satisfactory explanation of the striking change in fusion ability occurring in the granule membrane after incorporation at the cell surface is not available. Since the phenomenon is very rapid, it probably does not depend on the diffusion of receptors specific for exocytosis from the luminal plasmalemma to the fused granule membrane. It is possible that the environmental conditions existing in the extracellular space might cause a modification of the fused membrane and make it capable of fusion with the membranes of undischarged granules. Thus the environment to which a membrane is exposed should be considered as yet another factor capable of controlling the participation of that membrane in regulated fusion processes.

5.2. Membrane fusion-fission in exocytosis

The molecular events occurring during membrane fusion are still poorly understood. The present knowledge in this field is considered in detail in other chapters of this book. Only the data directly related to exocytosis is reviewed here. Much information on membrane fusion has been derived from the study of artificial systems such as liposomes (see chapter by Papahadjopoulos, this volume). However, it should be emphasized that the mechanisms involved in fusion of artificial membranes are not necessarily relevant to exocytosis and that several significant aspects of this phenomenon cannot even be tackled by this type of experimental approach. These aspects include: (1) the high degree of specificity of the process; and (2) the functional modulation occurring in most systems, which is discussed in detail in section 5.3.

Knowledge concerning the sequence of events occurring during exocytosis has been obtained primarily from morphological studies. On the basis of thin-section images, it was proposed by Palade (1959, 1975) that fusion-fission involves first the close apposition of the partner membranes and then the progressive elimination of membrane layers from the area of contact (Fig. 13), leading to a continuity between the membranes of the granules or vesicles and the plasmalemma. In recent years, new details about fusion-fission in exocytosis have been revealed by the introduction of new morphological techniques, such as freeze-fracture and cytochemistry of specific membrane components, through which regional differences in the structural composition of membranes can be detected.

The data obtained so far appear to be conflicting. In fact, a number of observations have been taken to suggest a key role for intramembranous proteins, whereas other data have been interpreted to support the idea of a primary role for the lipid matrix.

The various examples of particle arrays localized in the plasma membrane at sites of granule discharge have been described earlier (section 5.1). Analogously, structural modifications involving de novo appearance and/or redistribution of particles have been described to occur in the granule membrane before its fusion with the plasma membrane. In Tetrahymena, the membranes of mucocysts located deeply in the cytoplasm are characterized by relatively smooth P and E faces. In contrast, an annulus composed of five to seven rows of closely packed particles appears on the P leaflet of this membrane when it comes within a critical distance from the plasmalemma (Satir et al., 1972, 1973; see also Fig. 14). Similarly, in Paramecium peculiar arrays of particles appear on the apical membrane of trichocysts as they come into close proximity with the plasma membrane (Plattner et al., 1973; Allen and Hausmann, 1976; Beisson et al., 1976; and Chapter by Allen, this volume). Clusters of intramembranous particles have also been observed at the pole of the granule facing the overlying plasma membrane in the neurohypophysis (Dreifuss et al., 1974) and in endocrine pancreas (Berger et al., 1975). These observations appear consistent with the view that clustering

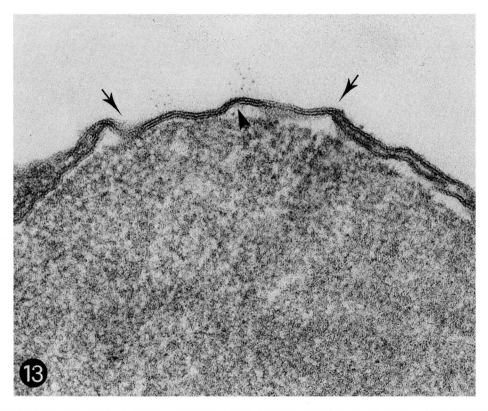

Fig. 13. Electron micrograph of a region of the cell surface in a rat peritoneal mast cell. The figure shows a granule close to discharge. The perigranule membrane has merged with the plasmalemma to yield a pentalaminar structure at the region delimited by the two arrows. At the point indicated by an arrowhead only three layers are visible. This site might correspond to an initial stage in the process of membrane layer elimination leading to exocytosis. × 145,000.

of integral proteins in both interacting membranes is an important event preceding fusion (Poste and Allison, 1973).

Support for this hypothesis derives also from observations on systems of membrane fusion-fission other than exocytosis, as for example, fusion of erythrocytes induced by Sendai virus (Bächi and Howe, 1972, Bächi et al., 1973) and that of gametic cells of *Chlamydomonas reinhardtii* (Weiss et al., 1977). Particle clusters have also been observed in the plasma membrane of endothelial (Smith et al., 1973b; Simionescu et al., 1974) and smooth muscle cells (Orci and Perrelet, 1973) at points of vesicle attachment, which probably correspond to pinocytotic events (Orci and Perrelet, 1973).

Results opposing this view have been reported by Orci and coworkers (1977), by Chi and co-workers (1976) and by Lawson and colleagues (1977). In mast cells (Fig. 15), the latter authors observed that as soon as the granule and plasma membranes come into close proximity, intramembranous particles and three

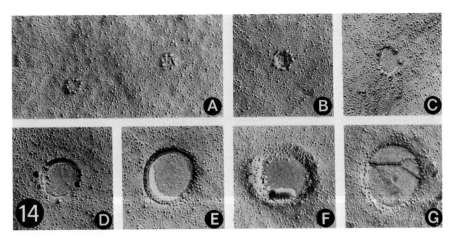

Fig. 14. Freeze-fracture electron micrographs of granule discharge in Tetrahymena. The figure illustrates the sequence of events occurring at the point of interaction between mucocyst membrane and plasmalemma during the fusion-fission process. In the resting protozoon, mature mucocysts are aligned at the periphery of the cytoplasm in close contact with the plasmalemma and bear on their P face an annulus of 5-7 rows of particles. Sites of mucocyst attachment are marked on the plasma membrane by a rosette composed of 1 central and 9 peripheral particles, most of which remain associated with the P face after fracture (A). As fusion is triggered a depression, the fusion pocket, appears within the rosette (B). The particles of the rosette are spread apart progressively while the fusion pocket widens and deepens (C and D). Eventually the particles from the annulus of fracture face P of the mucocyst membrane become visible at the inner edge of the collar of the fusion pocket (E, F, and G), indicating complete continuity between the two P faces of the partner membranes. ×72,000 Reproduced with permission from Satir et al., 1973.)

membrane proteins detectable by electron histochemistry are spread apart from the region of interaction (Lawson et al., 1977). It was suggested, therefore, that most of the membrane proteins are displaced from the point of contact before fusion and that the pentalaminar structure observed in thin sections is composed of two protein-free lipid bilayers. Subsequently, the pinching off of these structures would lead to fission of the fused membranes. Consistent with these observations, we also found that in the parotid gland massive discharge of granules is paralleled by a marked increase in the number of particle-free vesicles (sometimes displaying a multilamellar structure), which can be observed in the lumena by freeze-fracture. In a few cases, particle-free blisters located at the point of fusion of granule membranes with the plasmalemma have also been observed (Fig. 16). These findings lend support to the hypothesis of Ahkong and associates (1975), who proposed that the prerequisite of the fusion process is the presence of protein-free perturbed lipid bilayers and that aggregation of intrinsic membrane proteins may be an important factor only because by this process areas denuded of intramembranous proteins are exposed. Finally, in a recent paper Pinto da Silva and Nogueira (1977) have provided some interesting evidence indicating that, in *Phytophthora palmivora* zoospores, exocytosis is preceded by the

572

Fig. 15. Freeze-fracture electron micrograph of the plasmalemma of rat peritoneal mast cell (P) face. The surface membrane shows numerous bulges clearly related to underlying granules. At the top of the more prominent bulges, areas of plasmalemma showing displacement of intramembranous particles (circles), as well as particle-free blebs (top, single arrows) are visible. The fracture occasionally has exposed E faces of peripheral undischarged granules (*). Cross-section of folds and microvillous projections of the plasma membrane are marked by double arrows. ×27,000.

formation of a hybrid trilaminar diaphragm continuous with the plasmalemma and the secretory granule membrane, resulting from the elimination of the cytoplasmic leaflets of both these membranes. Such a diaphragm, which is assumed to be composed of the lipids and proteins not spanning the membranes, would then rapidly collapse, giving rise to the exocytotic opening.

Considering the data obtained to date on the sequence of events involved in the fusion-fission process of exocytosis, it appears clear that it is impossible to propose a general model consistent with all the available experimental evidence. In this respect, it must be emphasized that the reported morphological observations are not of general value but are restricted to particular systems. Thus, particle-free areas of the plasma membrane, corresponding to granule bulging at the cell surface, have been convincingly shown only in mast cells and in *Phytophthora palmivora* zoospores. They have been observed in other systems, such as the neurohypophysis (Dempsey et al., 1973) and the frog neuromuscular junction (Fig. 8), but only under particular experimental conditions, and their

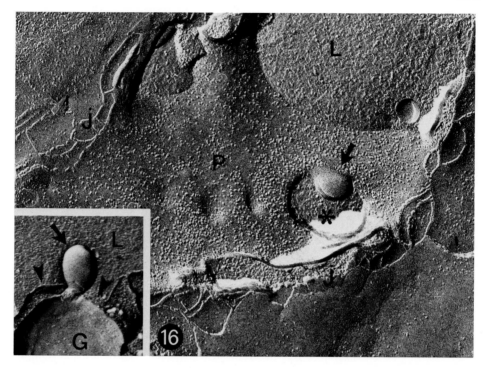

Fig. 16. Freezure-fracture of the apical region of rat parotid acinar cell, 1 min after injection of secretagogue (isoproterenol 0.02 g/kg). The asterisk shows a point where the fracture plane has probably crossed the neck, joining the acinar lumen with the cavity of a granule whose membrane has recently fused with the apical plasma membrane. A particle-free bleb, continuous with the surface membrane at the point of granule fusion, protrudes toward the lumen (arrow). The inset shows another particle-free bleb (arrow), originating at the point where the granule membrane (G) has fused with the plasmalemma (arrowhead). L, cross-sectioned lumen; j, tight junction; P, luminal plasmalemma, P face. ×58,500; inset: ×29,700.

relevance to fusion-fission was not demonstrated. Thus in the neurohyophysis, bulgings were not seen in thin sections, and with freeze-fracture they were most prominent in unstimulated conditions (Dempsey et al., 1973). Moreover, even by freeze-fracture, they were not observed by other authors (Dreifuss et al., 1974; Dreifuss et al., 1976). It may be of interest that the most striking images of such particle free bulgings presented by Dempsey and associates were obtained in tissues soaked by glycerol without prefixation. It is well known that cryoprotectants (McIntyre et al., 1974; Martinez-Palomo et al., 1976) may induce redistribution of intramembranous particles. In the neuromuscular junction, particle-free bulges were seen only after treatment with BWSV in Ca^{2+}-free modified Ringer's solution. In addition, images corresponding to fusions were always observed laterally, and never within, the domain of the particle-free domes (Fig. 8).

The relevance of the observation of particle-free vesicles pinching off from the cell surface in stimulated mast cells (Lawson et al., 1977) and in parotid gland (Fig. 16) should also be questioned. These structures might be due to fixation artifacts occurring at regions where the lipid matrix, highly disarranged by the fusion process, is in an unstable conformation.

Also particle arrays have been clearly demonstrated in only a few cases, namely, in protozoa and at the neuromuscular junction, which represent the best studied examples of systems where exocytosis is restricted to well-defined focal sites of the plasmalemma (section 5.1). Moreover, the particle arrays described in protozoa appear to be different from those of the neuromuscular junction, not only morphologically but also functionally and in relation to the factors responsible for their generation and stabilization. While the rosettes on the plasma membrane of Paramecium and Tetrahymena may play a direct role in fusion-fission, it appears likely that the other particle arrays in these protozoa (annulus of the granule membranes and double ring encircling the rosette on the plasma membrane of Paramecium), as well as those of the neuromuscular junction, do not participate directly. Thus the rosettes in Paramecium and Tetrahymena develop only after the two interacting membranes come into close apposition. Fusion initiates at their center and after discharge they are no longer visible. In contrast, the annulus of the granule membrane of these protozoa, the double ring encircling the rosette on the plasma membrane of Paramecium and the particle rows in neuromuscular junction are still present after exocytosis and are located near but not at the point of fusion. Thus the integral membrane proteins of these arrays are probably necessary for some step preceding fusion-fission itself. Clearly, they must be important in determining the site of exocytosis, but the mechanism by which they play this role is not understood. Possibly, they could be responsible for membrane modifications, or represent energy transducers or specific recognition sites between granule and plasma membranes. An attractive hypothesis suggested by Satir and associates (Satir et al., 1972, 1973; Satir, 1974a) for Tetrahymena is that the particles of the arrays might act as ionophores, triggering exocytosis by increasing the permeability of the membrane to ions and water. Given the increasing evidence for the importance of the free intracellular Ca^{2+} concentration in the regulation of secretion (section 5.3), they might specifically increase Ca^{2+} permeability (Satir, 1976). However, Ceccarelli and colleagues (1978a) found that in the frog neuromuscular junction, after induction of ACh release by BWSV in the absence of extracellular Ca^{2+}, fusion still occurs along the disarrayed particle rows (Fig. 8). This observation makes it highly unlikely that these particles determine the site of exocytosis by allowing a localized influx of extracellular Ca^{2+}. Alternative interpretations of the mechanisms of action of particle arrays have been proposed. In Paramecium, Plattner and coworkers (1977) have found that a bivalent cation-stimulated ATPase coincides with the plasmalemma arrays. In the neuromuscular junction it has been suggested that the particle rows represent anchoring points for cytoplasmic filamentous material involved in the driving of vesicles to the plasmalemma. However, recent results seem to cast some doubt on a possible direct interaction between microfilaments and intramembranous particles (Loor, 1976). In any case, since particle arrays are typical of only a few systems, those processes that might represent general requirements for membrane fusion are likely to be carried out by specific membrane proteins not necessarily arranged in characteristic patterns detectable by freeze-fracture.

The final result of fusion-fission is the establishment of continuity between two lipid bilayers. The necessary rearrangements of membrane lipids leading to this result remain as mysterious today as they were eighteen years ago when fusion-fission was first observed. A popular model was proposed some years ago on the role of lysolecithin (Guttler and Clausen, 1969; Lucy, 1970; Poole et al., 1970; also see chapter by Lucy, this volume). This hypothesis is based on the observed fusogenic ability of lysolecithin in model systems, as well as on the high lysolecithin content of adrenal medulla chromaffin granules (section 3.3). It is proposed that lysolecithin might favor the micellar conformation of phospholipids, thus interrupting the continuity of the lipid bilayer. The transition to the micellar conformation would allow membranes to fuse. Evidence in favor of this idea comes also from the finding that in synaptosomes (Gullis and Rowe, 1975) and blood platelets (Picket et al., 1977) stimulation for secretion is followed by the activation of phospholipase A_2. However, in numerous other systems, evidence supporting this model is lacking (Poste and Allison, 1973; Winkler et al., 1974; Meldolesi, 1974b; and also see chapter by Papahadjopoulos, this volume). Possible membrane modifications following stimulation for secretion will be discussed further in the next section.

In conclusion we feel that exocytosis is such a complex specific event that any model simply indicating a predominant function of proteins or alternatively of lipids, might be misleading. In contrast, fusion-fission of the granule or vesicle membrane with the plasmalemma probably involves an ordered sequence of events, in which both specific membrane proteins and the lipid matrix play a definite specific and controlled role.

5.3. Regulation of secretion and stimulus-secretion coupling

As mentioned in the introduction, only a few problems related to regulation of secretion and stimulus-secretion coupling are discussed in detail here. For more complete accounts, the reader is referred to the paper of Douglas (1968) as well as to more recent review articles and books (Case, 1973; Hubbard, 1973; Douglas, 1974a, 1975a; Rubin, 1974; Berridge, 1975; Carafoli et al., 1975; Lacy, 1975; Rasmussen and Goodman, 1975, 1977; Schramm and Selinger, 1975; Sharp et al., 1975; Case and Goebell, 1976; Coraboef and Rojas, 1976; Ginsborg and Jenkinson, 1976; Scarpa and Carafoli 1978).

With respect to regulation of secretion, the various systems can be grouped into two classes. In systems of the first class (which includes cells with minor or no specialization toward secretion as well as plasmacytes, fibroblasts, macrophages, and hepatocytes), vesicular secretion is not under the direct, short-term control of hormones, neurotransmitters, and metabolic factors. Consequently, discharge of secretion products occurs continuously, approximately at a constant rate. These systems are characterized by the relatively low capacity of their intracellular store of secretion products. This is correlated with the absence of secretory granules, or vesicles, clearly distinguishable on morphological as well as on biochemical grounds from Golgi vesicles and vacuoles (sections 3.3, 3.5). In contrast, the systems of the second class, characterized by the presence (and often by the abundance) of peculiar secretory organelles, are capable of rapidly adjusting their secretory activity in response to environmental stimuli. As mentioned in section 2.1.5, such an adjustment occurs primarily at the level of discharge. In many cases, synthesis is also modified. These two systems will be referred to here as nonhormone-regulated and hormone-regulated, respectively.

On the basis of discharge kinetics it might be assumed that in nonhormone-regulated systems secretory control is localized only at the synthetic level. Studies aimed at detecting other regulatory levels suggest that inhibition of the exocytotic process occurs only as a consequence of drastic alterations in intracellular homeostasis. Thus in fibroblasts, intracellular transport and discharge are known to be energy- and temperature-dependent (Ehrlich and Bornstein, 1972); in fibroblasts and hepatocytes, both these functions are impaired by antimicrotubular drugs (section 2.1.7). Furthermore, Tartakoff and Vassalli (1978) have recently demonstrated that in plasmacytes release of immunoglobulins is unaffected by cyclic nucleotides, hormones, neurotransmitters, and other drugs as well as by incubation in several modified Ringer's solutions (including a Ca^{2+}-free + EGTA Ringer's solution), but is blocked by low temperature. Moreover, depletion of intracellular Ca^{2+} by prolonged incubation with the ionophore A23187 in Ca^{2+}-free conditions, or partial K^+/Na^+ equilibration by treatment with monensin or nigericin result in the shutoff of immunoglobulin discharge and intracellular accumulation of immunoglobulin within Golgi-derived vesicles or vacuoles, respectively. These data suggest that in plasmacytes the discharge of secretion products also depends on the intracellular ionic environment. Preliminary results suggest that the same may occur in both fibroblasts and macrophages (Tartakoff and Vassalli, 1978). Thus one might speculate that in nonhormone-regulated systems, in contrast to the hormone-regulated cells discussed below, control mechanisms of discharge are always adjusted to ongoing activity.

The essential difference between nonhormone-regulated and hormone-regulated systems is that in the latter an evoked discharge, elicited by appropriate stimuli, superimposes from time to time on the slow basal release of secretion products. Relatively little is known about basal release. In mast cells Foreman and Mongar (1972) found that it does not depend on the availability of extracellular Ca^{2+}. In the exocrine pancreas of the guinea pig, Jamieson and Palade (1971b) differentiated basal release from artifactual leakage and showed that the former is blocked at low temperature. In the intact rat pancreas, Kramer and Poort (1972) found that ~20% of newly synthesized zymogens are discharged spontaneously within a few hours, with kinetics that are compatible with a direct Golgi-acinar lumen transport. At the neuromuscular junction basal release could be studied in more detail because release of single quanta of ACh[13] (probably corresponding to exocytosis of individual vesicles; see sections 2.2, 5.1, 6.1) is conveniently revealed by the appearance of mepps at the postsynaptic membrane, whereas adequate stimulation, through the synchronous discharge of multiple quanta, elicits the large end-plate potential which gives rise to the propagated muscle action potential (Fatt and Katz 1952; Del Castillo and Katz 1954a). In this system it has also been demonstrated that the regulation of spontaneous and stimulus-evoked events might be different. In particular, the frequency of mepps is unaffected (in the frog, Fatt and Katz, 1952; Del Castillo and Katz, 1954b) or only slightly decreased (in mammals, Hubbard et al., 1968a) even after prolonged soaking of the terminals in Ca^{2+}-free Ringer's solution. However, a requirement for intracellular Ca^{2+} is suggested by the distinct though transient decrease of mepp frequency that can be observed immediately after a single stimulus at frog neuromuscular junctions soaked in Ca^{2+}-free Ringer's solution, probably as a consequence of a transient increase of Ca^{2+} efflux induced by the nerve depolarization, which temporarily decreases the intraterminal concentration of free Ca^{2+} (Rotschenker et al., 1976)[14]. A similar explanation was suggested by Shimoni and co-workers (1977) to account for the decrease in mepp frequency that occurs when 20 mM K^+ is applied to neuromuscular junctions presoaked in a hypertonic, Ca^{2+}-free solution.

These data suggest that basal release of hormone-regulated systems is analogous in some respects to the release process of nonhormone-regulated cells. For example, in the systems studied so far both types of release were found to be independent of the availability of extracellular Ca^{2+}, but showed dependence on intracellular Ca^{2+}. Whether such analogies are only apparent, or whether the two processes involve the same control mechanisms, remains to be elucidated.

As far as evoked responses are concerned, it is now generally agreed that they are mediated intracellularly by at least two different agents, or second messengers: Ca^{2+} and cyclic AMP (cAMP). In general, these two agents do not act independently from each other but are often strictly interrelated (Berridge, 1975; Rasmussen and Goodman, 1975, 1977; Schramm and Selinger, 1975). In 1974 Goldberg and collaborators introduced the terms mono- and bi-directional to classify cell systems according to their control mechanisms. As further elaborated by Berridge (1975), monodirectional systems are identified as those regulated either by a single stimulant or by multiple stimulants acting in the same direction. Intracellularly, the response might be mediated by either Ca^{2+} or cAMP, but the two second messengers often function in concert, that is, some aspects of the response are controlled by only one messenger, others by both. In contrast, in bidi-

---

[13]Basal release is used here to designate resting spontaneous quantal secretion and does not include nonquantal, molecular leakage (see also section 2.2).

[14]Calcium ions are the link between the voltage change across the nerve membrane and the release of neurotransmitter. In the absence of extracellular Ca^{2+}, the action potential is still able to invade the nerve terminal but fails to cause any significant increase in transmitter release. Under these conditions, the expected effect of the increased Ca^{2+} conductance, provoked by the nerve impulse, is to cause an efflux of Ca^{2+} along the reversed electrochemical gradient.

rectional systems, cells are regulated by two opposite stimulants, the second messengers antagonize each other, and cellular activity is determined by the balance between the two.

Many secretory systems (e.g., some nerve cells, adrenal medulla, endocrine and exocrine pancreas, salivary glands, anterior pituitary) can be classified as monodirectional; others, however (e.g., mast cells and blood platelets) are clearly bidirectional (Berridge, 1975).

During the last two decades the adenylate cyclase-cAMP system has been thoroughly investigated in a variety of cell types, and now there is a general consensus about the general outlines of its organization. Since the information on this subject is not only extensive but is also systematically presented (e.g., in the volumes of the series "Advances in Cyclic Nucleotide Research" Greengard and Robison, eds. 1972; 1973; 1974; 1975a,b; 1976; 1977), it will not be reviewed here. Greater attention will be directed to the role of Ca^{2+} because much more uncertainty exists in this field.

It is now recognized that the concentration of free, ionized Ca^{2+} is several orders of magnitude lower in the cytoplasm of many, and maybe all, cells than in the extracellular fluid (between 10^{-5} and 10^{-7} M intracellularly with respect to 10^{-3} M extracellularly; Baker, 1972). Thus there is a very substantial inward concentration gradient across the membrane, which is further amplified by an electrical gradient, since the interior of cells is negative at rest. Moreover, all cells contain a considerable amount of Ca^{2+} bound to their intracellular structures that is, at least partly, in dynamic equilibrium with the soluble pool of free Ca^{2+}.

An impressive body of evidence indicates that in many secretory cells the evoked release of secretion products is triggered by an increase of the concentration of free Ca^{2+} in the cytoplasm. This concept has been worked out in a number of systems over the last 20 years (discussion, Douglas, 1968). Direct experimental proof was given by Miledi (1973), who obtained an increase in transmitter release from squid giant synapses after intraterminal injection of Ca^{2+}. An analogous result on mast cells was reported shortly thereafter by Kanno and co-workers (1973). More recently, the widespread application of Ca^{2+} ionophores has lent further support to this view (Rasmussen and Goodman, 1975, 1977; Schramm and Selinger, 1975; Scarpa and Carafoli, 1978). However, critical evaluation of the experimental results concerning Ca^{2+}-dependence of various secretory systems reveals a striking heterogeneity, especially in relation to the effect of extracellular Ca^{2+}. These differences can be clearly demonstrated by comparing two systems such as neuromuscular junctions and exocrine pancreas.

It is well known that a nerve impulse arriving at a neuromuscular junction is incapable of releasing the neural transmitter if Ca^{2+} ions are withdrawn from the external medium (review, Ginsborg and Jenkinson, 1976). The quantitative relationship between quantal release of ACh and concentration of external Ca^{2+} and Mg^{2+} was investigated in detail in frog (Jenkinson 1957; Dodge and Rahaminoff, 1967), rat (Hubbard et al., 1968b), and mouse (Cooke et al., 1973) neuromuscular junctions. The general finding is that release rises steeply, in a power relationship, with increasing Ca^{2+} concentration. The effect of Mg^{2+} is to cause a displacement in this power relationship, suggesting simple competition between these two divalent cations. These results, coupled with the work of Katz and Miledi (1969a, 1970) on the giant synapse in the stellate ganglion of the squid, lend strong support to the view that extracellular Ca^{2+} is a primary and immediate requirement for depolarization to evoke transmitter release (Katz and Miledi, 1969b). Thus, based on present evidence, the sequence of events may be described as follows: depolarization opens specific "Ca^{2+} gates" in the presynaptic membrane; this allows Ca^{2+} to flow into the terminal if the membrane potential has not been displaced beyond the Ca^{2+} equilibrium level (Katz and Miledi, 1967).

In pancreatic acinar cells, clear evidence exists that stimulation brought about by either ACh or the peptide hormone pancreozymin-cholecystokinin evokes a secretory response independent of the extracellular Ca^{2+} concentration ranging between 2.5 and 0.1 mM. Moreover, when Ca^{2+} ions are withdrawn from the incubation medium, the response is progressively inhibited only after a delay of several minutes (Case and Clausen, 1973; Matthews et al., 1973). As far as the movements of Ca^{2+} across the plasma membrane are concerned, a marked increase of the efflux was observed after stimulation, indicating a rise in the concentration of free, cytoplasmic Ca^{2+} (Case and Clausen, 1973; Matthews et al., 1973). This rise does not seem to originate from the opening of Ca^{2+} gates because intracellular electrophysiological studies failed to reveal a considerable increase of Ca^{2+} permeability (Nishiyama and Petersen, 1975; Petersen and Ueda, 1975, 1976). However, a moderate increase possibly occurs, since biochemical studies on isolated pancreatic acinar cells indicate that during stimulation the uptake of ^{45}Ca is doubled with respect to the unstimulated controls (Kondo and Schulz, 1976). Taken as a whole, these results suggest that in exocrine pancreas the physiological regulation of the free Ca^{2+} level is controlled primarily by intracellular events, namely, through the distribution of Ca^{2+} from the organelle-bound to the free cytoplasmic pool and vice versa. In this system the role of extracellular Ca^{2+} would be limited to the replenishment of intracellular pools (Case and Clausen, 1973; Matthews et al., 1973; Petersen and Ueda, 1975, 1976). Recent results from studies done by Petersen and Ueda (1975, 1976) suggest that the source of the cytoplasmic Ca^{2+} rise induced by stimulation might be a small pool with high turnover located at the internal surface of the plasmalemma.

Other systems may be considered intermediate between the two examples described above. In relation to extracellular Ca^{2+}, other exocrine glands, such as salivary glands, were found to be independent, although less markedly than the exocrine pancreas (Nielsen and Petersen, 1972). On the other hand, endocrine cells, such as adrenal medulla, more closely resemble the neuromuscular junction (Douglas, 1975a,b). Thus it seems that the behavior of the various systems with respect to the utilization of different Ca^{2+} sources in vesicular secretion varies profoundly. There are, however, at least two reasons to suggest that a clear-cut dualism does not exist. First, as was mentioned earlier, there is a full gamut between strong dependence (as in neuromuscular junction) and considerable independence on extracellular Ca^{2+} (as in exocrine pancreas). Second, in all systems examined to date, secretion can be evoked experimentally by either one of the two mechanisms adequate to increase intracellular free Ca^{2+} concentration, namely, by increasing the influx or by provoking intracellular redistribution. For example, in the neuromuscular junction, Alnaes and Rahaminoff (1975) observed that in the absence of extrancellular Ca^{2+} there is a large increase in mepp frequency following addition of dicoumarol, probably due to the rise of the intracellular concentration of free Ca^{2+} caused by mitochondrial poisoning. In this respect it is interesting that even in the neuromuscular junction (a system with extreme dependence on extracellular Ca^{2+}) a role for intracellular Ca^{2+} in the production of the physiologically evoked epp has been suggested (Rahaminoff and Yaari, 1973; Alnaes and Rahaminoff, 1975; Rahaminoff et al., 1975). According to this hypothesis the secretion of ACh is regulated by the cooperative effect of Ca^{2+} from both intra- and extracellular sources.

We turn now to an examination of the possible mechanisms by which the rise of free Ca^{2+} and/or cAMP concentration is transduced into the secretory response. It is still unclear whether Ca^{2+} acts directly in promoting fusion-fission, or indirectly by activating another process, which in turn would trigger granule discharge. A possible direct action that has been proposed is the shielding of negative charges of granule surfaces, thus permitting their apposition and fusion with the plasmalemma (Dean, 1974a,b; Dean and Matthews, 1974; Matthews and Nordman, 1976). Other mechanisms by which redistribution of Ca^{2+} ions might directly promote membrane fusion have been proposed by Poste and Allison, (1973), Ahkong and co-workers (1975), and by Papahadjópoulos and Poste (see chapter by Papahadjopoulos, this volume). Further evidence supporting direct action of Ca^{2+} comes from observations in Paramecium (Plattner and Fuchs, 1975) and Tetrahymena (Fisher et al., 1976), where a high amount of calcium was detected cytochemically in the trichocyst and the mucocyst membrane regions close to the point of exocytosis. Analogously, the limiting membranes of exocrine pancreas and parotid secretion granules bind large amounts of calcium (Meldolesi et al., 1975). Recently, fusions between secretory granules (Dahl and Gratzl, 1976) or between granules and homologous plasma membrane fractions (Davis and Lazarus, 1976; Milutinovic et al., 1977) have been reported to be inducible in cell-free systems by incubation in appropriate Ca^{2+}-containing solutions. However, since in these studies the demonstration of a true, specific fusion-fission of the membranes participating in exocytosis in the living cell was not provided, no critical evaluation of these results can be given at present. Although there is no question that the decrease of membrane surface charge promotes fusion in a number of experimental conditions (see chapter by Papahadjopoulos, this volume), it is difficult to understand why Ca^{2+}-induced fusion should result specifically in exocytosis (i.e., in the fusion of granule and plasma membranes occurring within the temporal and special restrictions discussed in section 5.1), rather than in generalized fusion of all cellular membranes, since all of them bear net negative charges. Thus, to explain the specificity of exocytosis, other factors should be invoked. Moreover, it should be emphasized that in some systems, such as the parotid gland, exocytosis is triggered preferentially by cAMP through a Ca^{2+}-independent mechanism, whereas an increase of intracellular Ca^{2+} induces only a minor response (Schramm and Selinger, 1975). In this respect, it is interesting that in neuromuscular junction massive exocytosis can be induced by BWSV in the complete absence of extracellular Ca^{2+} (Ceccarelli et al., 1978a,b; see also sections 5.1, 5.2). Although this phenomenon might be attributed to an increase in free intracellular Ca^{2+} concentration, due to a release of Ca^{2+} from bound pools in the terminal (Finkelstein et al., 1976), this interpretation seems unlikely since discharge evoked by BWSV is at least two orders of magnitude larger than that caused by dicoumarol or other drugs which are known to induce release of Ca^{2+} from mitochondria (Alnaes and Rahaminoff, 1975). Thus it appears more likely that the venom exerts its effect bypassing the Ca^{2+} requirement. This conclusion supports the idea that the action of Ca^{2+} in coupling stimulus to secretion is mediated by a subsequent intracellular event.

In recent years a large number of hypotheses have been put forward on the intracellular events directly responsibly for granule discharge, which would be induced by an increased intracellular concentration of second messengers (cAMP or Ca^{2+}). At present, however, none of these ideas appear to be supported by sufficient experimental evidence to justify their general acceptance.

According to one proposal, the second messenger might activate exocytosis through its action on the microtubular-microfilamentous system of the cell (Malaisse et al., 1975). This view rests on the observation that, in some systems, drugs which interfere with microfilaments and microtubules also interfere with exocytosis, and on the well-known involvement of Ca^{2+} in contractile systems. Section 2.1.7 discussed the difficulties in interpreting results obtained with antimicrotubular and antimicrofilamentous drugs, and also mentioned that is is not yet clear whether their main action concerns the exocytotic process itself or some preceding step of intracellular transport. It should be noted here that distribution studies have revealed the presence, in many systems, of considerable amounts of contractile proteins and tubulin, not only adjacent to but apparently also associated with the plasmalemma (Stadler and Franke, 1974; Bhattacharyya and Wolff, 1975, 1976; Rohlich, 1975; Barber and Crumpton, 1976; Scharder and Frerichs, 1976; Tilney and Mooseker, 1976; Zenner and Pfeuffer, 1976), particularly with regions that are important for the discharge of secretion products (Berl et al., 1973; Blitz and Fine, 1974; Westrum and Gray, 1976). Analogous observations have been made recently about the membrane of some secretion granules (Gabbiani et al., 1976; Moore et al., 1976; Puszkin et al., 1977). Of course, these membrane-bound proteins are as good candidates as microtubules and microfilaments to be the target of colchicine and cytochalasin B action, especially because these drugs are known to have profound effects on the molecular topology of the plasmalemma (Wunderlich et al., 1973; Ukena et al., 1974; Yahara and Edelman, 1975). It is interesting that in liver it was reported colchicine inhibits secretory granule discharge at doses that fail to depolymerize microtubules (Redman et al., 1975). For a detailed account on the effect of antimicrotubular and antimicrofilamentous drugs on secretion, the reader is referred to the book edited by Soifer (1975).

Other observations concerning the relation of secretory granules to the microtubule-microfilament system come from time-lapse cinematography of pancreatic B cells grown in culture (Lacy et al., 1975; Malaisse, et al., 1976). It was observed that granules are capable of saltatory movements in fixed directions, apparently predetermined by the arrangement of the cytoskeleton. These movements are increased by stimulation. In the same system it was reported that physiological stimulation with glucose is accompanied by an increase of the rate of tubulin polymerization into microtubules (Pipeleers et al., 1976). Along the same line, it has been reported that stimulation induces the emission of pseudopodia from secretory granules in the parotid gland (Schramm et al., 1972) and in the neurohypophysis (Castel, 1977). Whether or not these observations have relevance in relation to exocytosis is still unclear.

In conclusion, it seems reasonable to assume that tubulin and contractile proteins, in their various modes of organization, play an important role in secretory processes. However, considerable uncertainty remains with regard to the cellular processes in which these proteins are directly involved and whether their activity is in some way modified by increased second messenger concentrations.

Another group of hypotheses implies that the second messenger induces chemical changes in granules and/or plasma membranes. These might include methylation (Diliberto et al., 1976), increase in lysolecithin content through the activation of phospholipase A_2 (Gullis and Rowe, 1975; Picket et al., 1977), or protein phosphorylation. An increase in the rate of phosphorylation of granules or plasma membrane proteins after adequate stimulation has been reported in a large number of secretory tissues, such as the adrenal medulla, (Trifaro and Dworkind, 1971), the pituitary gland, anterior (Labrie et al., 1971) as well as posterior (McKelvy, 1975); the pancreas, exocrine (Lambert et al., 1974; McDonald and Ronzio, 1974) as well as endocrine (Muller et al., 1976). Some of the protein kinases involved in these reactions were found to be dependent on either cAMP or cGMP; others are apparently cyclic nucleotide-independent. The cGMP-dependent phosphorylations are thought to be controlled in turn by intracellular Ca^{2+}, since in many systems it has been demonstrated that the rise in cGMP concentration is elicited by a previous rise in cytoplasmic free Ca^{2+} (Schultz and Hardman, 1975). In secretory tissues, the Ca^{2+}-cGMP relationship has been studied throughly in the exocrine pancreas by Christophe and collaborators (1976) and by Haymovits and Sheele (1976). In the first of these studies, the Ca^{2+} and cGMP increases were reported to occur in sequence, with the kinetics suggesting a possible role of third messenger for the cyclic nucleotide. A phosphorylation directly dependent on Ca^{2+} has been reported recently for specific proteins of synaptosomal membranes (De Lorenzo, 1976; Gordon et al., 1977; Krueger et al., 1977). Although all these data are compatible with the hypothesis that some of the effects of the second messenger are mediated by the phosphorylation of specific proteins, the attempts to demonstrate a link between protein phosphorylation and secretory discharge have never gone beyond temporal correlation. Thus the existence of a cause-effect relationship remains to be demonstrated.

A hypothesis linking stimulus to secretion that was popular some years ago was based on the observation that, in a variety of systems, exposure to cholinergic or adrenergic agents results in a marked breakdown of a single phospholipid type, phosphatidylinositol (PI). This phenomenon was termed the PI effect (review, Hokin, 1968). However, it has recently been shown that the PI effect is not caused by the second messenger; instead, it might be related to the process of receptor activation (review, Michell, 1975).

Yet another hypothesis on the mode of action of the second messenger is that it would activate a membrane-associated ATPase (Woodin and Wieneke, 1964; Rubin, 1970). Recently, a divalent cation-stimulated ATPase has been found by Plattner and coworkers (1977) to coincide with the intramembrane particle arrays present in the plasmalemma of Paramecium at the sites specific for exocytosis. On the other hand, the Ca^{2+}-Mg^{2+} ATPases of various secretory granules (section 3.3; review, Poste and Allison, 1973) may be involved in other functions, such as packaging of the secretory product, rather than in fusion-fission events. Actually, the energy requirements of exocytosis are not yet understood. Thus an energy requirement has been found for exocytosis in both nonhormone regulated and hormone-stimulated systems such as exocrine and endocrine pancreas, parotid gland, and mast cells (Bauduin et al., 1969; Feinstein and Schramm, 1970; Jamieson and Palade, 1971a; Peterson, 1974; Malaisse et al., 1975). Surprisingly, both the basal and the evoked transmitter release at frog neuromuscular junction is increased by uncouplers and electron transport inhibitors (Alnaes and Rahaminoff, 1975). It has been suggested that this effect depends on the release of Ca^{2+} from poisoned mitochondria. However, it is still unclear whether ACh release is really energy-independent or whether it can be carried out at ATP concentrations lower than those required for mitochondrial Ca^{2+} accumulation. As far as the other systems are concerned, it is still not known at which step of the chain of events leading to exocytosis ATP is required, and particularly whether this step comes before or after the rise of free Ca^{2+} in the cytoplasm. Thus the hypothesis of Jamieson and Palade (1971a) that energy is required for membrane fusion still awaits experimental confirmation.

6. Surface remodeling after exocytosis: removal and fate of discharged granule and vesicle membranes

As described in the previous sections, exocytosis results in the insertion of the membrane of granules or vesicles at the surface of secretory cells. This phenomenon is reversible because the cell is endowed with the capacity to remove the excess membranes and to preserve the specificity of its plasmalemma (sections 3.3, 4.1, 4.4). This process does not necessarily imply the removal from the plasmalemma of all components of the inserted granule membrane. In sections 3.5 and 4.4, we discussed the possibility that exocytosis might serve as a general mechanism for transporting membrane macromolecules, or even membrane patches, from intracellular membranes to the plasmalemma. However, in most secretory systems the available evidence indicates that the rate of this process is low compared to that of exocytosis and that an amount of membrane approximately equal to that incorporated in the plasmalemma by exocytosis is eventually removed and somehow disposed of in the cytoplasm. Possible exceptions may be represented by those secretory systems where rapid growth or turnover of the surface membrane is known to occur. In these systems, it has been suggested that the insertion of granule or vesicle membrane into the plasmalemma might contribute to the growth or rapid turnover of the cell surface. For example, in Tetrahymena, an actively dividing protozoon, the surface infoldings corresponding to discharged mucocyst membranes have been found by high voltage and freeze-fracture electron microscopy to undergo a progressive reduction in size coupled with an increase in particle density. This observation has been interpreted to indicate that some constituents of the granule membrane, especially phospholipids, would flow in bulk across the region of fusion and be used to expand the surface of the plasmalemma, whereas integral proteins would remain trapped and eventually be removed (Satir, 1974a,b). Along the same line, in the mammary gland morphological evidence suggests that the membrane of casein granules, which are discharged by exocytosis, might be integrated irreversibly at the cell surface to replace the plasmalemma loss occurring in this system as a consequence of the apocrine secretion of milk fat globules (Franke et al., 1976).

However, since most secretory systems do not have a continuous need for new surface membrane to cope with growth requirements or with massive loss of membrane material to the extracellular space, they must be capable of internalizing excess membrane. In 1959, Palade proposed that discrete patches of intact membrane might pinch off from the cell surface into the cytoplasm, giving rise to vesicles or vacuoles. An alternative model was proposed shortly thereafter by Fawcett (1962). According to this author, the excess membrane might be removed in the form of single molecular constituents. Most of the results obtained over the last 15 years fail to support the latter interpretation, so that at present the view that membranes are recaptured by internalization of intact patches of membrane is widely accepted. In contrast, considerable disagreement exists about the site of the retrieval and the size and nature of the internalized patches as well as about their final fate. These problems are considered in the following sections.

6.1. Membrane retrieval after exocytosis: where, when and how

Over the last few years, evidence has been obtained in a variety of secretory systems (exocrine and endocrine glands, as well as nerve terminals) indicating that the exocytotic process is followed by increased internalization of membrane vesicles or vacuoles. Most of the relevant studies in this field have been carried out by morphological techniques, using extracellular tracers (such as ferritin, horseradish peroxidase, dextran, etc.) which can be detected by electron microscopy. The internalization of these tracers, as a consequence of endocytosis, has generally been found to increase following secretion. Clearly, this increased endocytosis compensates for the incorporation of membrane by exocytosis, thus maintaining the surface area at a constant size. The critical question that must be asked is whether this endocytotic process, in addition to reducing the expanded cell surface, also results in selective removal of the granule membrane constituents from the plasmalemma. This view implies that membrane patches of the granule type would maintain their identity (at least in part), even after incorporation into the plasmalemma. The site of endocytosis is not specified exactly in this hypothesis. It might be speculated that in systems in which exocytosis is followed rapidly by interiorization, the latter process should be localized at, or close to, the site of exocytosis, while in more sluggish systems the granule type patches might undergo limited displacement over the cell surface before interiorization. An alternative view is that, due to the fluidity of membranes, random mixing of membrane components occurs following exocytosis. In this case the plasma membrane would become diluted by granule membrane constituents, and the preservation of plasma membrane specificity would have to be ensured by events subsequent to surface reduction. For example, if "mixed" vesicles containing both plasma and granule membrane constituents were internalized, one might speculate that the granule membrane components could be accumulated selectively within the cytoplasm, while the plasma membrane components would be reinserted at the cell surface. To date, however, clear-cut proof supporting either view has not been obtained, but morphological evidence in a number of systems seems to suggest that the first view is correct, namely, that granule type patches are directly withdrawn from the plasma membrane following exocytosis. The morphological criteria used in drawing this tentative conclusion are the spatial vicinity of the exocytotic and endocytotic events and/or the recognition of granule type patches on the basis of structural peculiarities.

Clear morphological images indicating fast recapture by direct vesiculation of the fused trichocyst membrane have been obtained by Hausmann and Allen (1976) in Paramecium. Analogous models had been proposed previously by Nagasawa and Douglas (Nagasawa et al., 1971; Nagasawa and Douglas, 1972) and by Benedeczy and Smith (1972) for the neurohypophysis and adrenal medulla, where membrane interiorization was suggested to be carried out by coated vesicles that appeared to originate from the exocytotic profiles. Eventually these structures would shed their coat and become regular, clear vesicles (review, Douglas, 1974b). Recently the neurohypophysis has been reinvestigated extensively by other groups. The results obtained are also compatible with the direct recapture of the granule membrane. However, the retrieved organelles appear to be primarily vacuoles of the same size as neurosecretory granules rather than coated vesicles (Nordman et al., 1974, Norman and Morris, 1976;

586

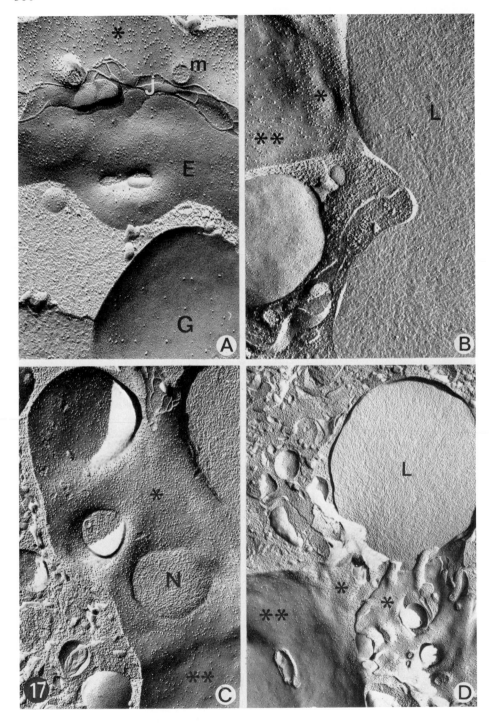

Swann and Pickering, 1976; Theodosis et al., 1976). Analogous results were obtained in pancreatic B cells, another system in which endocytosis seems to follow the exocytotic process rapidly (Orci et al., 1973b).

In exocrine glands, interiorization of granule membranes is carried out at a slower rate compared to the systems described above. Thus, after strong stimulation of the exocrine pancreas and parotid gland, acinar lumena remain enlarged for 2 to 4 hours. Concomitantly, a population of smooth vesicles and small vacuoles accumulates at the apical region of the cells (Amsterdam et al., 1969; Jamieson and Palade, 1971b). Geuze and Kramer (1974) have studied the uptake of extracellular tracers applied at either the basolateral or the luminal surface of exocrine pancreatic cells. They found that following stimulation membrane is withdrawn from the apical surface by means of small, smooth vesicles originating from larger caveolae in continuity with the lumen. Coated vesicles seem to be involved in the recapture process only at a later stage. The relationship of these vesicles to the membrane of discharged granules was not established.

Results strongly supporting the direct retrieval of granule membrane patches have also been obtained by our group in freeze-fracture studies of the parotid gland (De Camilli et al., 1976). In these studies, we took advantage of the fact that the parotid granule and luminal membranes are characterized by greatly different densities of intramembranous particles ($\sim$500/μ^2 and 2000/μ^2, respectively). We found (Fig. 17) that the fusion during exocytosis does not result in the intermixing of the two membrane types, which seem to retain their structural identity, giving rise to a mosaic organization, namely, alternation of patches of high (original luminal plasmalemma) and low (fused granule membrane) particle density. Later in the process the low density patches were redistributed to yield large vacuolar infoldings which were eventually removed. Concomitantly, the acinar lumina reverted to their nonstimulated size.

Fig. 17. Freeze-fracture of the rat parotid gland. Structural modifications of the P face of the luminal plasmalemma after in vivo stimulation with isoproterenol (0.02 g/kg). In unstimulated cells (A) particles are randomly distributed over the P face of the luminal plasma membrane (*). Their density is lower than on the corresponding face of the lateral plasmalemma (not shown; see De Camilli et al., 1976), but it is much higher than on the P face of granule membranes (G). 5 min after isoproterenol injection (B) many granule membranes fused with the plasmalemma can be seen. Particles on fused granule membranes (**) are slightly more concentrated than on membranes of undischarged granules, but their density is distinctly lower than that of the typical luminal plasma membrane (*). Discharge activity continues for 30–40 min after stimulation, leading to a marked increase in size of the lumena, which at this time are bounded by a membrane displaying mosaic organization (i.e., alternating patches of low particle density with patches of high particle density, typical of the resting luminal plasmalemma). As the experiment continues, the low density regions are usually found delimiting large vacuolar infoldings (**) (C, 80 min after isoproterenol injection, and D, 120 min after injection), which eventually disappear (4–6 hr). At this point lumena revert to their original size. *, typical lumenal plasmalemma, P face; **, patches of the luminal membrane originated from fused granule membranes; E, lateral plasmalemma, E face; j, tight junction; m, microvilli; L, cross-section of the acinar lumen; N, cross-section of the neck joining the acinar lumen with the cavity of a vacuolar infolding. (A) ×60,800; (B) ×55,800; (C) ×32,400; (D) ×23,900 (Figures B, C, and D reproduced with permission from De Camilli et al., 1976.)

In nerve terminals, interiorization of vesicles after stimulation has been demonstrated by the use of extracellular tracers or other means in many different systems (Holtzman et al., 1971; Ceccarelli et al., 1972, 1973; Heuser and Reese, 1973; Turner and Harris, 1973; Jorgensen and Mellerup, 1974; Pysh and Wiley, 1974; Ceccarelli and Hurlbut, 1975; Teichberg et al., 1975; Fried and Blaunstein, 1976; see also Fig. 18; Schacher et al., 1976; Zimmermann and Denston, 1977a). However, only at the frog neuromuscular junction has the problem of the localization of vesicle recapture been approached.

Essentially two models have been proposed. Heuser and Reese (1973) have suggested that vesicular membranes are interiorized as coated vesicles from regions of the axolemma relatively distant from the sites at which fusions occur. On the other hand, Hurlbut and Ceccarelli (1974) suggested that uncoated vesicles arise from the axolemma at or near the active zone, where discharge is known to take place. Recently this problem has been reinvestigated by Ceccarelli and colleagues (1978a) by studying the time course of changes in ultrastructure occurring during intense asynchronous release of neurotransmitter induced by BWSV or K^+-enriched solutions.

Solutions with high concentrations of K^+ cause a Ca^{2+}-dependent rapid asynchronous release of quanta of transmitter that is believed to occur as a consequence of an activation of the chain of physiological events involved in transmitter release (Liley 1956; see also section 5.3). When a modified Ringer's solution containing 20 mM K^+ is applied to frog neuromuscular junction, the muscle fibers depolarize by about 40 mV in a few minutes, the mepp amplitude declines by about 50%, and the frequency rapidly increases to a peak level near 500/second in 10 to 15 minutes. Subsequently the mepp frequency subsides slowly, but is still considerable even after 60 minutes. Concomitantly, the mepp amplitude also decreases. When transmitter release is evoked by BWSV, the mepp peak frequency is approximately the same as with 20 mM K^+. However, after 15 to 20 minutes the frequency drops more rapidly and remains at very low levels thereafter. Terminals that had been exposed to either BWSV or high K^+ for 1 hour or more were studied by thin-section electron microscopy (Figs. 19, 20). In agreement with previous observations (Clark et al., 1970, 1972; Frontali et al., 1976), and in good correlation with the depletion of quanta of transmitter observed in electrophysiological experiments, those terminals that had been exposed to BWSV were found to be completely depleted of vesicles (Fig. 19). In addition, both electron microscopy and the direct observation of living preparations showed that BWSV induces a profound swelling of the nerve terminal (Frontali et al., 1976). By contrast, with high K^+, the reduction in number of vesicles was only partial (Fig. 20), despite the fact that in this case the secretion of ACh quanta exceeds the number secreted over the same time in experiments with BWSV. These data clearly indicate that the depletion of vesicles induced by BWSV is due not only to an increased rate of fusion but also to an inhibition of vesicle reformation from the axolemma. Thus it is reasonable to assume that in the 20 mM K^+-treated neuromuscular junctions vesicle attachments to the axolemma occur in relation to both exo- and endocytosis, whereas with BWSV they occur nearly exclusively in relation to exocytosis, since membrane interiorization is greatly impaired.

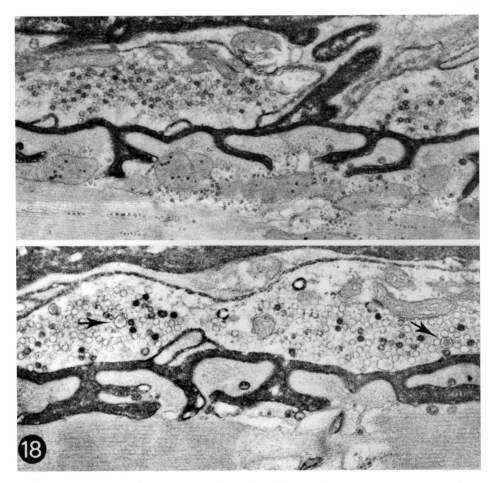

Fig. 18. Electron micrographs of portions of end plates from two different frog preparations stimulated for 2 hr at 2/sec in the presence of horseradish peroxidase and fixed immediately without rest (curare 3μg/ml). Peroxidase was added 0.5 hr before stimulation was begun. Junctional clefts and extracellular spaces between the Schwann cell membrane and the axolemma contain rich deposits of reaction product. Many vesicles contain reaction product, but the proportion of labeled vesicles is different in the two micrographs. Many of the large dense-core vesicles (arrows) that appear in these micrographs show a peripheral layer of reaction product and retain their core. Upper panel ×22,000; lower panel ×38,000. (Reproduced with permission of Raven Press from Hurlbut and Ceccarelli, 1974.)

Freeze fracture of neuromuscular junctions exposed to either BWSV or high K^+ was carried out at the peak of the asynchronous release of quanta to reveal the topological localization of vesicle attachment sites. A striking difference between the two conditions was observed. As mentioned in section 5.1, BWSV induced a dramatic increase of fusions that were localized at the active zone (Fig. 7). Also, with high K^+ many vesicle attachments were seen at the active zone. Many others, however, were located in the portions of the presynaptic axolemma between adjacent active zones, not closely associated with ridges and furrows

590

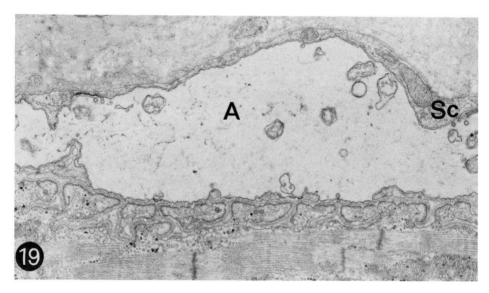

Fig. 19. Electron micrograph showing portion of a neuromuscular junction from a frog cutaneous pectoris neuromuscular preparation fixed 1 hr after the onset of the mepp discharge induced by black widow spider venon. The axonal ending (A) appears to be swollen and the cytoplasmic matrix seems less dense than normal. The terminal contains swollen mitochondria and only a few structures resembling synaptic vesicles. Sc, Schwann cell process. ×18,000.

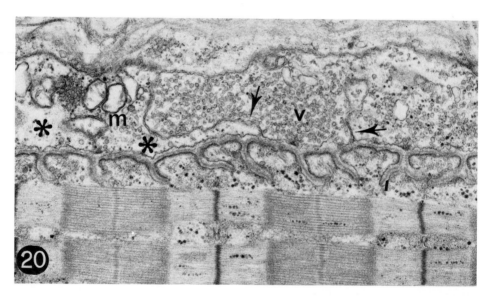

Fig. 20. Electron micrograph of portion of frog neuromuscular junction from a preparation bathed for 1 hr in modified Ringer's solution containing 20 mM K^+ (and normal Ca^{2+}). The mitochondria (m) are swollen, and infoldings (arrows) of the axolemma penetrate deeply into the axoplasm. Some focal depletion in the number of vesicles may have occurred (asterisks), but many synaptic vesicle (v) are still present. ×22,000.

(Fig. 21). Therefore it seems reasonable to conclude that at frog neuromuscular junction exocytosis occurs exclusively at the active zone (see also section 5.1), and specific membrane interiorization can occur at the same place or nearby at the regions of the axolemma located between active zones. These results are in apparent contradiction to those reported by Heuser and associates (1974) who failed to see vesicle attachments in the regions between active zones following a brief period of electrical stimulation. However, since stimulation was carried out in the presence of fixative in these studies, it is likely that specific membrane interiorization was inhibited.

During the above discussion on the site of membrane retrieval, certain points on the rate of this process also emerged. While membrane retrieval in some systems proceeds rapidly after exocytosis, other systems appear to be more sluggish. The idea that exocytosis can be carried out more efficiently than the subsequent endocytosis is supported by observations in frog neuromuscular junction. In this system, prolonged electrical stimulation of secretion at low frequency does not lead to a detectable decrease in the number of vesicles, most of which, however, appear labeled with extracellular tracers, indicating that they have passed through one (or multiple) exo-/endocytosis cycles. In contrast, if transmitter release is increased by applying high frequency stimulation, the rate of vesicle fusion apparently exceeds the rate of reformation. Vesicle-depleted nerve terminals can thus be obtained in which the plasmalemma is characterized by profound surface infoldings (e.g., Figs. 14, 15 in Ceccarelli and Hurlbut, 1975b). These changes revert to normal during a subsequent period of rest (Ceccarelli et al., 1973).

Dissociation of exocytosis from membrane interiorization has also been obtained in the Mauthner fiber-giant fiber synapse of the hatchet fish (Model et al., 1975). In this system, electrical stimulation carried out at physiological temperature produces exocytosis of synaptic vesicles followed rapidly by membrane recapture. Lowering of the temperature to 12 to 14 °C seems to have relatively little effect on discharge. However, retrieval is greatly inhibited and the surface area of the terminals appears very enlarged due to the presence of profound infoldings of the plasmalemma.

Although it is thought that membrane interiorization requires energy (Holtzman, 1976), it is not yet clear at which step this energy is required. Douglas (1974b) proposed that a critical role might be played by coated vesicles, viewed as specific retrieval organelles, because their bristle coats might be able to act as a chemomechanical energy transducer in the process of membrane retrieval. This hypothesis is based on the observation that in various systems coated vesicles have been reported to increase after stimulation and to be labeled by extracellular tracers (Douglas, 1974b).

Recent biochemical and morphological studies (Crowther et al., 1976; Pearse, 1976) demonstrate that the coat isolated from several different tissues is composed of a single, high molecular weight protein, denominated clathrin, arranged according to well-defined geometries. These findings do not support the interpretation put forth in the past that coated vesicles might also arise from fixa-

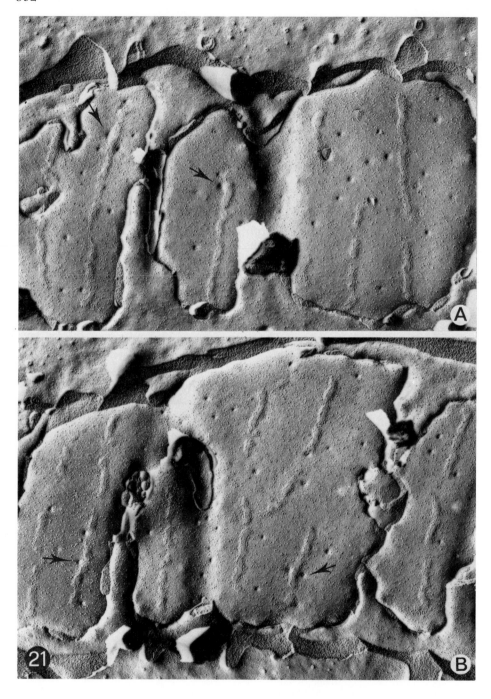

tion artifacts (Ceccarelli et al., 1967; Gray, 1972). However, there is no question that the revelation of coated vesicles in thin sections depends greatly on the techniques of tissue processing and that the number of these organelles appears larger in poorly fixed specimens, especially in nervous tissue (Ceccarelli et al., 1967).

Although the view that coated vesicles may be involved specifically in the interiorization of membrane proteins by endocytosis, proposed originally by Friend and Farquhar (1967), is widely accepted, it should be mentioned that endocytosis per se does not necessarily require coated vesicles. Thus in macrophages, recently studied in detail by Steiman and colleagues (1976; see also chapter by Edelson and Cohn, this volume), endocytosis appears to be carried out by smooth rather than by coated vesicles. In addition, evidence suggesting that coated vesicles might play an important role in membrane retrieval has not been obtained in all systems, or by all investigators. It must be mentioned that coated vesicles have also been observed in secretory cells independently from membrane retrieval. In many systems these vesicles have been found in the Golgi area, where they might participate in the intracellular transport of secretion products (Jamieson and Palade, 1967a,b; Weinstock and Leblond, 1974). In others, bristle coats were found attached to outer surfaces of Golgi cisternae of secretory granules. In the latter case it has been speculated that they might be somehow involved in membrane fusion (Franke et al., 1976; Kartenbeck et al, 1977). Finally, Blitz and coworkers (1977) have recently provided evidence suggesting that coated vesicles might function in the brain as high affinity Ca^{2+} sequestering organelles, analogous in many respects to the sarcoplasmic reticulum of muscle cells. For a discussion on the role of intracellular calcium pools in stimulus-secretion coupling see section 5.3.

In conclusion, it seems fairly well demonstrated that the ability to bind bristle coats is not restricted to a single membrane type but is probably shared by several. It might be speculated that membrane coating really occurs in relation to one or more specific functions. However, the nature of these functions, as well as the specific role of the coat itself, remain obscure.

6.2. The fate of discharged secretory granule membranes

The fate of the membranes of discharged granules and vesicles after interiorization from the cell surface has been the subject of numerous investigations. The results obtained have yielded diverse interpretations. According to many authors, granule membranes would be utilized repeatedly by the cell in successive secretory cycles, whereas others propose that they would be degraded and re-

Fig. 21. Freeze-fracture view of two different portions of neuromuscular junctions (A and B), fixed with glutaraldhyde and paraformaldehyde, 20 min after being exposed to solutions containing 20 mM K^+ (normal Ca^{2+}). Numerous dimples are evident in the P faces throughout the presynaptic membrane. Some are closely associated to the double rows of large particles (arrows), but many occur in regions of the axolemma between the "active zones." (A) and (B) ×31,500.

placed each time by newly synthesized membranes. This section reviews the available experimental evidence and provides a critical evaluation of existing discrepancies.

As was mentioned in section 4, biosynthesis and turnover studies have been carried out on the membrane of secretory granules isolated from a number of systems (exocrine pancreas, parotid gland, and adrenal medulla as well as cultured adrenal medullary cells and adrenergic neurons). These experiments have generally been carried out by exposing secretory cells, in vivo or in vitro, to radioactive precursors without the use of secretagogue drugs. In all these experiments, the rate of synthesis and/or replacement of total proteins of granule membranes, as well as that of individual proteins, was found to be much lower (in most cases by approximately one order of magnitude) than that of the corresponding secretory proteins. In long-term experiments conducted on guinea pig pancreas, it was found that the polypeptides of granule membranes turn over asynchronously, with half-lives between 3 and 13 days, suggesting that the granule membranes are not replaced as a whole, as would be the case if they were digested in lysosomes (Meldolesi, 1974a).

Generally these results have been interpreted as a demonstration that after exocytosis the granule membranes, interiorized by the processes described in the previous section, are not destroyed but are reused many times. However, an alternative interpretation has been proposed by Wallach and co-workers (1975). According to these authors, the slow turnover of granule membranes cannot be taken as proof of membrane recycling because it might be explained equally well by the existence within the cells of a pool of granule-type membranes much larger than that located at the surface of secretory granules. If this were the case, the secretory granule membranes would be characterized by low turnover, even if membranes bounding the granules were destroyed at each secretory cycle. The new membranes, synthesized to replace those destroyed as a consequence of exocytosis, would be diluted by the preexisting, large precursor pool. In our opinion, this interpretation is extremely improbable. It must be emphasized that in order to account for the difference in turnover of membrane and secretory proteins the size of this hypothetical pool would have to be enormous (about ten times the amount of membranes found at the surface of secretory granules). For example, in the exocrine pancreas, this pool would represent at least 25% of the total membranes, corresponding to approximately 40% of the ER membranes (Meldolesi, 1974b). An even larger amount would be necessary in the adrenal medulla and in the parotid gland, where the fraction of total cellular membranes corresponding to granule membranes is proportionally higher. One wonders where such a large membrane mass could be located intracellularly, since it has never been detected by either morphological or cell fractionation studies. The only cellular compartment large enough to contain this pool would be the ER. Were this the case, however, one would expect to find high concentrations of granule membrane markers (such as dopamine-β-hydroxylase in the adrenal medulla) in microsomal fractions. This expectation has not been borne out by experimental data (Hortnägl, 1976). The conclusion is, therefore, that the hypothetical pool probably does not exist. Thus the results of synthesis and turn-over studies are consistent in supporting the hypothesis that the interiorized granule membranes (or at least the macromolecules derived from them) are reutilized repeatedly in granule formation and discharge.[15]

The fate of the granule membranes has also been extensively investigated by morphological techniques. Generally in these studies, carried out in vivo and in vitro, secretory cells were exhaustively stimulated by exposure either to large doses of secretagogue drugs or to appropriate biophysical treatment and exposed to an extracellular tracer. The distribution of this tracer was monitored with time in cytoplasmic organelles during the recovery phase.

The results of some of these studies appear to contradict the biochemical results described above. In exocrine and endocrine secretory cells the retrieved or-ganelles (vesicles and/or vacuoles; see section 6.1), which are labeled by the tracer, first move toward the Golgi area, where they tend to accumulate. From this time on, distinct differences in tracer distribution have been observed by dif-ferent authors. In the adrenal medulla (Abrahams and Holtzman, 1973), the exocrine pancreas (Geuze and Kramer, 1974), the parotid gland (Kalina and Rabinovitch, 1975; Oliver and Hand, 1975), the toad urinary bladder (Masur et al., 1971), and several pituitary cell lines grown in vitro (Tixier-Vidal et al., 1976), the tracer was detected within multivesicular bodies and dense bodies, which usually increase in number after stimulation. In time these organelles were transformed progressively into residual bodies. In other systems, such as the B cells of the pancreas (Orci et al., 1973b), the somatotrophs of the pituitary gland (Pelletier, 1973; Farquhar et al., 1975), the seminal vesicle (Mata, 1976), and the lacrimal acinar cells (Herzog and Farquhar, 1977), tracers were detected not only in multivesicular and dense bodies, but also in Golgi cisternae, and later appeared in newly synthesized secretory granules. In pituitary mammotrophs (Pelletier, 1973), the Golgi cisternae were apparently bypassed by the tracer, which appeared first in the immature and then in the mature secretory granules. Moreover, recent results of Herzog and Farquhar (1977) support the idea that tracer distribution experiments in stimulated secretory cells might be misleading. At variance with previous data (Kalina and Rabinovitch, 1975; Oliver and Hand, 1975) these authors found that, in the parotid gland stimulated with isopro-terenol, an extracellular tracer, dextran (either applied to tissue acini incu-bated in vitro or infused in vivo in the parotid duct) was rapidly accumulated in the stacked cisternae of the Golgi complex and in condensing vacuoles of the

---

[15]The only biochemical experiments that appear to contradict this conclusion are those by De Pot-ter and associates (1972) on splenic nerve terminals. These authors observed that the rate of accumu-lation of dopamine-β-hydroxylase proximally to a nerve ligature (taken as a measure of vesicle turn-over at the nerve terminal) is of the same order of magnitude as the turnover of noradrenaline. However, one wonders whether this experimental approach is adequate to study the problem of membrane reutilization. In fact (as discussed further in section 6.3) there is evidence that vesicle membrane components are transported to the cell body by retrograde axonal flow, where they might be reutilized in the assembly of new secretory vesicles. Thus, not all membrane molecules traveling down the axon in the forward direction are necessarily the product of de novo synthesis.

Golgi region. Concomitantly, there was also an accumulation in lysosomes, which however was distinctly more prominent in isolated acini (when the tracer is available mainly to the basolateral surface) than after intraductal administration.

Nerve cells have also been investigated by these techniques. In neurosecretory terminals of the posterior pituitary, Douglas (1974b) described the transformation of coated vesicles (in his opinion, the retrieval organelles) into smooth, clear vesicles. The latter were thought to join together subsequently in clusters and then to be ferried to the perikarya by retrograde axonal transport.

In sympathetic terminals (Teichberg and Bloom, 1976) and in cultured spinal cord neurons (Teichberg et al., 1975), at least some retrieved vesicles have been reported to fuse into multivesicular bodies, which are then transported to the perikarya. At neuromuscular junctions, the available evidence favors reutilization of the retrieved vesicle membranes. According to Heuser and Reese (1973), the retrieved vesicles would follow a complex intracellular route. After interiorization they would first fuse with large cisternae present in the terminals, from which new synaptic vesicles would originate by budding. A different conclusion was reached by Ceccarelli and collaborators (1972, 1973). According to these authors, uncoated vesicles derive directly from membrane interiorization in the vicinity of the presynaptic membrane and might be available for reuse in transmitter discharge (section 6.1). Observations on other nerve cell systems, which appear consistent with either the first (Model et al., 1975) or the second (Schacher et al., 1976) of these views, have been reported. Recent results by Zimmermann and Denston (1977a,b) represent a strong argument in support of the direct recycling of vesicle membranes. In the *Torpedo* electric organ, these authors have carried out an integrated morphological, neurophysiological, and biochemical study and have shown that the freshly interiorized vesicles, identified by the presence of peculiar biochemical features as well as by the recovery of extracellular tracers, are able to accumulate newly synthesized ACh and to release it on stimulation (sections 2.2, 5). Evidence supporting rapid reutilization of synaptic vesicles was also obtained by McMahan and Yee (1975) in snake motor nerve terminals.

In conclusion, the results of morphological tracer experiments appear quite contradictory. Some of these data seem to correlate well with the results of the synthesis and turnover studies that suggest extensive reutilization of granule membranes. Others, however, are apparently more consistent with the view that all, or at least most, of the membranes of the discharged granules are broken down by lysosomal digestion. However, it should be emphasized that the latter interpretation, although apparently supported by straightforward evidence, is actually based on a number of unproven assumptions. The first and more important of these assumptions implies that extracellular tracers and retrieved membranes remain associated within the cell, so that the distribution of the former necessarily mirrors the distribution of the latter. To date, no experimental evidence suggesting that this is actually the case has ever been reported. In contrast, clear evidence obtained by both morphological and cell fractionation studies indicates that, at least in fibroblasts, fusion of pinocytotic vesicles with lysosomes does not result in the irreversible incorporation of one organelle into the other. Rather, the vesicle content, which includes the extracellular tracers, is released within lysosomes, while the vesicle membrane is able to separate from lysosomes and is probably recycled back to the plasmalemma (Steinman et al., 1976; Schneider et al., 1977; see also chapter by Edelson and Cohn, this volume). At present there is no reason to suppose that the vesicles, interiorized from the cell surface following exocytosis, behave differently from regular pinocytotic vesicles. The recovery of the tracers within lysosomes only indicates that the latter have been in continuity with vesicles interiorized from the cell surface and does not provide any information on the final fate of the retrieved granule membranes. Thus no proof exists that the membrane remnants and lamellar bodies, which are often seen to accumulate within lysosomes together with extracellular tracers, originate from the retrieved granule membranes rather than from other cellular membranes. Again, these considerations emphasize how misleading it can be to extrapolate results concerning the content of organelles to their limiting membranes.

A second unproven assumption concerns the experimental conditions under which tracer experiments are carried out. It is generally believed that exhaustive stimulation of secretory cells has no influence on the final fate of the retrieved membranes. That this may not be the case, however, is suggested by several observations. First it should be emphasized that exhaustive stimulation induces not only the desired massive discharge of secretion products but also a number of unspecific side effects that are expected to be especially severe when drugs such as cholinergic and adrenergic agents or insulin are used at doses known to provoke profound metabolic effects. For instance, in the exocrine pancreas, stimulation with the cholinomimetic agent pilocarpine is known to induce the appearance of autophagosomes (Nevalainen, 1970). That massive stimulation may unspecifically affect secretory cells is further suggested by the following considerations on the exocrine pancreas. In this system, the rate of protein synthesis is hardly affected by the rate of secretion (Jamieson and Palade, 1971b), and the rate of basal secretion is relatively high. In fact, in the unstimulated rat pancreas, the intracellular half-life of secretory proteins is approximately 7 hours (Kramer and Poort, 1972; Christophe et al., 1973). Thus over long enough time periods, the total amount of secreted proteins remains approximately constant whether the system is stimulated or not. Since residual bodies have been reported to remain detectable within cells for long periods of time (up to 90 hr; Smith and Farquhar, 1966), one would also expect to see a large number of these organelles in unstimulated cells if lysosomal digestion were the final fate of all discharged granule membranes. The fact that these structures increase so greatly following massive pharmacological stimulation might suggest that their presence is due to the unphysiological conditions of the experiment.

6.3. Final considerations on granule membrane retrieval

Knowledge about surface remodeling and the fate of the granule membrane has increased considerably in recent years, so that some points now appear to be reasonably well established. Clearly, exocytosis is followed by specific membrane interiorization. The overall result of this process is the maintenance of both the area and the biochemical and structural specificity of the plasmalemma. The capacity to carry out this dimensional and compositional control is not unique to the plasma membrane. It seems to be a general property, common to all membrane types interacting in the course of the intracellular transport of secretion products. The simplest model to explain this process is the direct withdrawal of the heterologous membrane patches from "acceptor" compartments (section 4). In the plasmalemma, a considerable body of evidence exists to indicate that this may well be the case, although direct proof has not yet been obtained.

A second point that appears to be soundly based is that the retrieved membranes are reutilized to a great extent in subsequent secretory cycles, as demonstrated by synthesis and turnover studies as well as by some morphological investigations. Clearly, granule membranes, like all other membranes, undergo turnover. Whether this turnover is due to their lysosomal digestion or whether granule membranes, analogously to the other cytoplasmic membranes, turn over asynchronously molecule by molecule, cannot be determined at present. On the one hand, the appearance of multivesicular and dense bodies following stimulation has been interpreted as an indication of lysosomal involvement. As indicated in the previous section, this interpretation is open to question. The alternative possibility, namely molecular degradation independently of lysosomes, is strongly suggested by our own experiments on guinea pig exocrine pancreas in which the various proteins of zymogen granule membranes were found to turn over at extremely heterogeneous rates (Meldolesi, 1974a). No similar studies have been carried out in other systems so far.

The intracellular route followed by the retrieved granule and vesicle membranes after interiorization in order to be refilled with secretory products probably varies widely in different systems. In the glands, exocrine as well as endocrine, and in other typical secretory cells, this reutilization seems to occur by extensive fusion of the reinternalized membranes either with Golgi membranes or, directly, with other membranes of the granule type that are present in the Golgi area. An analogous process probably occurs in peptidergic and adrenergic neurons. In the latter systems evidence indicates that at least part of the interiorized granule membrane is ferried back to the perikaryon, as shown by the rapid retrograde transport of dopamine β-hydroxylase (Brimjoin and Helland, 1976; Fillenz et al., 1976). To date, the fate of these membranes in the perikaryon has not been investigated. However, the results of Gagnon and co-workers (1976) on adrenergic nerve cell cultures suggest that they might be extensively reutilized. In contrast, cholinergic terminals might have a simpler mechanism of vesicle membrane reutilization. In this system (as discussed in section 2.1.6), at variance with all other secretory cells examined, the secretory vesicles probably do not contain proteins. This might be the reason why the retrieved vesicles appear able to accumulate their secretion product close to the discharge sites without undergoing complex intracellular processing.

Acknowledgements

We wish to thank all the colleagues, students, and technicians of our laboratory for their encouragement and support. We are especially grateful to N. Iezzi and G. Pietrini for their expert technical assistance, to F. Crippa and P. Tinelli who prepared the illustrations, and to Mrs. I. Zambruno for typing the manuscript.

The following colleagues have kindly sent us recent reprints of their papers and preprints of their unpublished work: M. Baggiolini, M. V. L. Bennett, G. Blobel, J. Christophe, J. C. Dagorn, W. W. Douglas, W. J. Evans, W. W. Franke, J. D. Gardner, E. Holtzman, H. F. Kern, N. Kirschner, G. Kreibich, F. Miller, J. J. Nordman, O. H. Petersen, Y. J. Schneider, M. Schramm, I. Schulz, A. M. Tartakoff, E. Weibel, H. Winkler, H. Zimmermann. Special thanks to B. Satir, who provided us with Figure 14, and to W. P. Hurlbut, who critically read the manuscript.

Some of the original work presented was partially supported by a grant from the Muscular Dystrophy Association Inc. of America, awarded to B. Ceccarelli.

References

Abe, A., Moscarello, A. M. and Sturgess, J. M. (1976) The distribution of anionic sites on the surface of the Golgi apparatus. J. Cell Biol. 71, 973–976.

Abrahams, S. J. and Holtzman, E. (1973) Secretion and endocytosis in insulin-stimulated rat adrenal medulla cells. J. Cell Biol. 56, 540–558.

Adelman, M. R., Sabatini, D. D. and Blobel, G. (1973) Ribosome-membrane interactions: non-destructive disassembly of rat liver rough microsomes into ribosomal and membranous components. J. Cell Biol., 56, 206–229

Ahkong, Q. F., Fisher, D., Tampion, W. and Lucy, J. A. (1975) Mechanisms of cell fusion. Nature 253, 194–195.

Akert, K., Peper, K. and Sandri, C. (1975) Structural organization of motor end plate and central synapses. In: Cholinergic Mechanisms (Waser, P. G., ed.) pp. 43–57, Raven Press, New York.

Alexander, C. A., Hamilton, R. L. and Havel, R. J. (1976) Subcellular localization of B apoprotein of plasma lipoproteins in rat liver. J. Cell Biol. 69, 241–263.

Allen, R. D. and Hausmann, K. (1976) Membrane behaviour in exocytotic vesicles. The ultrastructure of Paramecium trichocysts in freeze-fracture preparations. J. Ultrastr. Res. 54, 224–234.

Allison, A. E. and Davies, P. (1974) Interactions of membranes, microfilaments and microtubules in endocytosis and exocytosis. In: Cytopharmacology of Secretion. (Ceccarelli, B., Clementi, F. and Meldolesi, J., eds.) pp. 237–248, Raven Press, New York.

Alnaes, E. and Rahaminoff, R. (1975) On the role of mitochondria in transmitter release from motor nerve terminals. J. Physiol. (London) 248, 285–306.

Alonso, G. L., Bazerque, P. M., Arrigo, D. M. and Tumilasci, O. R. (1971) Adenosine triphosphate-dependent calcium uptake by rat submaxillary gland microsomes. J. Gen. Physiol. 58, 340–350.

Amar-Costesec, A., Beaufay, H., Wibo, M., Thinès-Sempoux, D., Feytmans, E., Robbi, M. and Berthet, J. (1974a) Analytical study of microsomes and isolated subcellular membranes from rat liver. II. Preparation and composition of the microsomal fraction. J. Cell Biol. 61, 201–212.

Amar-Costesec A., Wibo M., Thinès-Sempoux, D., Beaufay, H. and Berthet, J. (1974b) Analytical study of microsomes and isolated subcellular membranes from rat liver. IV. Biochemical, physical and morphological modifications induced by digitonin, EDTA and pyrophosphate. J. Cell Biol. 62, 717–745.

Amsterdam, A. and Jamieson, J. D. (1974) Studies on dispersed pancreatic exocrine cells. I. Dissociation technique and morphologic characteristics of separated cells. J. Cell Biol. 63, 1037–1056.

Amsterdam, A., Ohad, I. and Schramm, M. (1969) Dynamic changes in the ultrastructure of the acinar cell of the rat parotid gland during the secretory cycle. J. Cell Biol. 41, 753–773.

Amsterdam, A., Schramm, M., Ohad, I., Salomon, Y. and Selinger, Z. (1971) Concomitant synthesis of membrane protein and exportable protein of the secretory granules in rat parotid gland. J. Cell Biol. 50, 187–200.

Anderson, P., Rölich, P., Slorach, S. A. and Uvnäs, B. (1974) Morphology and storage properties of rat mast cell granules isolated by different methods. Acta Physiol. Scand. 91, 145–153.

Argent, B. E., Smith, R. K. and Case, M. R. (1975) The distribution of bivalent cation-stimulated adenosine triphosphate hydrolysis and calcium accumulation by subcellular particles. Biochem. Soc. Trans. 3, 713–714.

Arion, W. J. and Nordlie, R. C. (1965) Liver glucose-6-phosphatase and pyrophosphate-glucose phosphotransferase: effects of fasting. Biochem. Biophys. Res. Commun. 20, 606–610.

Atkinson, P. H. (1975) Synthesis and assembly of HeLa cell plasma membrane glycoproteins and proteins. J. Biol. Chem. 250, 2123–2134.

Atkinson, P. H., Moyer, S. A. and Summers, D. F. (1976) Assembly of vesicular stomatitis virus glycoproteins and matrix proteins into HeLa cell plasma membranes. J. Mol. Biol. 102, 613–632.

Autuori, F., Svensson, H. and Dallner, G. (1975a) Biogenesis of microsomal membrane glycoproteins in rat liver. I. Presence of glycoproteins in microsomes and cytosol. J. Cell Biol. 67, 687–699.

Autuori, F., Svensson, H. and Dallner, G. (1975b) Biogenesis of microsomal membrane glycoproteins in rat liver. II. Purification of soluble glycoproteins and their incorporation into microsomal membranes. J. Cell Biol. 67, 700–714.

Bächi, T. and Howe, C. (1972) Fusion of erythrocytes by Sendai virus studied by electron microscopy. Proc. Soc. Exp. Biol. Med. 141, 141–149.

Bächi, T., Aguet, M. and Howe, C. (1973) Fusion of erythrocytes by Sendai virus studied by immuno freeze-etching. J. Virol. 11, 1004–1012.

Baggiolini, M., Bretz, U. and Dewald, B. (1977) Biochemical and structural properties of the vacuolar apparatus of polymorphonuclear leukocytes. In: Movement, Metabolism and Bactericidal Mechanisms of Phagocytes (Rossi, F., Patriarca, P. and Romeo, D., eds.) pp. 89–102. Piccin Medical Books, Padova.

Bainton, D. F. (1973) Sequential degranulation of the two types of polymorphonuclear leukocyte granules during phagocytosis of microorganisms. J. Cell Biol. 58, 249–264.

Bainton, D. F. and Farquhar, M. G. (1966) Origin of granules in polymorphonuclear leukocytes. Two types derived from opposite faces of the Golgi complex in developing granulocytes. J. Cell Biol. 28, 277–301.

Bainton, D. F. and Farquhar, M. G. (1970) Segregation and packaging of granule enzymes in eosinophilic leukocytes. J. Cell Biol. 45, 54–73.

Bainton, D. F., Ullyot, J. L. and Farquhar, M. G. (1971) The development of neutrophilic polymorphonuclear leukocytes in human bone marrow. Origin and content of azurophil and specific granules. J. Exp. Med. 134, 907–934.

Baker, P. F. (1972) Transport and metabolism of calcium ions in nerves. Progr. Biophys. Mol. Biol. 24, 177–224.

Barancik, L. C. and Lieberman, I. (1971) The kinetics of incorporation of protein into liver plasma membrane. Biochem. Biophys. Res. Commun. 44, 1084–1088.

Barber, B. H. and Crumpton, M. J. (1976) Actin associated with purified lymphocyte plasma membrane. FEBS Lett. 66, 215–220.

Bashford, C. L., Radda, G. K. and Ritchie, G. A. (1975) Energy-linked activities of the chromaffin granule membrane. FEBS Lett. 50, 21–24.

Batzri, S., Selinger, Z., Schramm, M. and Rabinovitch, M. R. (1973) Potassium release mediated by the epinephrine α-receptor in rat parotid slices. Properties and relation to enzyme secretion. J. Biol. Chem. 248, 361–368.

Bauduin, H., Colin, M. and Dumont, J. E. (1969) Energy sources for protein synthesis and enzymatic secretion in rat pancreas in vitro. Biochim. Biophys. Acta 174, 722–733.

Baumrucker, C. R. and Keenan, T. W. (1975) Membranes of mammary gland. X. Adenosine triphosphate-dependent calcium accumulation by Golgi apparatus-rich fractions from bovine mammary gland. Exp. Cell Res. 90, 253–260.

Beaufay, H., Amar-Costesec, A., Thinès-Sempoux, D., Wibo, M., Robbi, M. and Berthet, J. (1974) Analytical studies of microsomes and isolated subcellular membranes from rat liver. III. Subfractionation of microsome fraction by isopycnic and differential centrifugation in density gradients. J. Cell Biol. 61, 213–231.

Beisson, J., Lefort-Tran, M., Pouphile, M., Rossignol, M. and Satir, B. (1976) Genetic analysis of membrane differentiation in Paramecium. Freeze-fracture study of the trichocyst cycle in wild-type and mutant strains. J. Cell Biol. 69, 126–143.

Bender, A. N., Ringel, S. P., Engel, W. K., Vogel, Z. and Daniels, M. P. (1976) Immunoperoxidase localization of alpha bungarotoxin: a new approach to myasthenia gravis. Ann. N.Y. Acad. Sci. 274, 20–30.

Benedeczky, I. and Smith, A. D. (1972) Ultrastructural studies of the adrenal medulla of golden hamster: origin and fate of secretory granules. Z. Zellforsch. 124, 367–386.

Benedeczky, I. and Somogyi, P. (1975) Ultrastructure of the adrenal medulla of normal and insulin-treated hamsters. Cell Tissue Res. 162, 541–550.

Benedetti, E. L., Dunia, I., Bentzel, C. J., Vermorken, A. J. M., Kibbelaar, M. and Bloemendal, H. (1976) A portrait of plasma membrane specializations in eye lens epithelium and fibers. Biochim. Biophys. Acta 457, 353–387.

Bennett, G. and Leblond, C. P. (1970) Formation of cell coat material for the whole surface of columnar cells in the rat small intestine as visualized by radioautography with L-fucose-^{3}H. J. Cell Biol. 46, 409–416.

Bennett, G. and Leblond, C. P. (1977) Biosynthesis of glycoproteins present in plasma membrane, lysosomes and secretory material as visualized by radioautography. Histochem. J. 9, 393–418.

Bennett, G., di Giamberardino, L., Koenig, H. L. and Droz, B. (1973) Axonal migration of protein and glycoprotein to nerve endings. II. Radioautographic analysis of the renewal of proteins in nerve endings of chicken ciliary ganglion after intracerebral injection of ^{3}H-fucose and ^{3}H-glucosamine. Brain Res. 60, 129–146.

Bennett, G., Leblond, C. P. and Haddad, A. (1974) Migration of glycoprotein from the Golgi apparatus to the cell surface of various cell types as shown by radioautography after labeled fucose injection into rats. J. Cell Biol. 60, 258–284.

Bennett, M. R. (1973) An electrophysiological analysis of the uptake of noradrenaline at sympathetic nerve terminals. J. Physiol. (London) 229, 533–546.

Benshalom, G. and Flock, A. (1977) Calcium-induced electron density in synaptic vesicles of afferent and efferent synapses on hair cells in the lateral line organ. Brain Res. 121, 173–178.

Berg, N. B. and Austin, B. P. (1976) Intracellular transport of sulfated macromolecules in parotid acinar cells. Cell Tissue Res. 165, 215–226.

Berg, N. B. and Young, R. W. (1971) Sulfate metabolism in pancreatic acinar cells. J. Cell Biol. 50, 469–483.

Berger, W., Dahl, G. and Meissner, H. P. (1975) Structural and functional alterations in fused membranes of secretory granules during exocytosis in pancreatic islet cells of the mouse. Cytobiology 12, 119–139.

Bergeron, J. J. M., Berridge, M. V. and Evans, W. J. (1975) Biogenesis of plasmalemmal glycoproteins. Intracellular site of synthesis of mouse liver plasmalemmal 5′nucleotidase as determined by subcellular location of messenger RNA coding for 5′nucleotidase. Biochim. Biophys. Acta 407, 325–337.

Bergeron, J. J. M., Ehrenreich, J. H., Siekevitz, P. and Palade, G. E. (1973a) Golgi fractions prepared from rat liver homogenates. II. Biochemical characterization. J. Cell Biol. 59, 73–88.

602

Bergeron, J. J. M., Evans, W. H. and Geschwind, I. I. (1973b) Insulin binding to rat liver Golgi fractions. J. Cell Biol. 59, 771–776.

Berl, S., Puszkin, S., Niklas, W. J. (1973) Actomysin-like protein in brain. Science 179, 441–443.

Berridge, M. J. (1975) The interaction of cyclic nucleotides and calcium in the control of cellular activity. Adv. Cyclic Nucleotide. Res. 6, 1–98.

Bhattacharyya, B. and Wolff, J. (1975) Membrane-bound tubulin in brain and thyroid tissue. J. Biol. Chem. 250, 7639–7646.

Bhattacharyya, B. and Wolff, J. (1976) Polymerization of membrane tubulin. Nature 264, 576–577.

Bieger, W., Seybold, J. and Kern, H. F. (1976) Studies on intracellular transport of secretory proteins in the rat exocrine pancreas. V. Kinetic studies on accelerated transport following caerulein infusion invivo. Cell Tissue Res. 170, 203–220.

Birks, R. I. (1974) The relationship of transmitter storage and release to fine structure in a sympathetic ganglion. J. Neurocytol. 3, 133–160.

Bischoff, E., Tran-Thi, T. A. and Decker, K. F. A. (1975) Nucleotide pyrophosphatase of rat liver. A comparative study of the enzymes solubilized and purified from plasma membrane and endoplasmic reticulum. Eur. J. Biochem. 51, 353–362.

Blackburn, G. R., Borneus, M. and Kasper, C. B. (1976) Characterization of the membrane matrix derived from the microsomal fraction of rat hepatocytes. Biochim. Biophys. Acta 436, 387–398.

Blaschke, E., Bergqvist, U. and Uvnäs, B. (1976) Identification of the mucopolysaccharides in catecholamine-containing subcellular particles from various rat, cat and ox tissues. Acta Physiol. Scand. 97, 110–120.

Blaschko, H., Firemark, H., Smith, A. D. and Winkler, H. (1967) Lipids of the adrenal medulla. Lysolecithin, a characteristic constituent of chromaffin granules. Biochem. J. 104, 545–549.

Blitz, A. L. and Fine, R. E. (1974) Muscle-like contractile proteins and tubulin in synaptosomes. Proc. Nat. Acad. Sci. U.S.A. 71, 4472–4476.

Blitz, A. L., Fine, R. E. and Toselli, P. A. (1977) Evidence that coated vesicles isolated from brain are calcium sequestering organelles resembling sarcoplasmic reticulum. J. Cell Biol. 75, 135–147.

Blobel, G. and Dobberstein, B. (1975a) Transfer of proteins across membranes. I. Presence of proteolytically processed and unprocessed nascent immunoglobulin light chains on membrane-bound ribosomes of murine myeloma. J. Cell Biol. 67, 835–851.

Blobel, G. and Dobberstein, B. (1975b) Transfer of proteins across membranes. II. Reconstitution of functional rough microsomes from heterologous components. J. Cell Biol. 67, 852–862.

Blobel, G. and Potter, V. R. (1967) Studies on free and membrane-bound ribosomes in rat liver. II. Interaction of ribosomes and membranes. J. Mol. Biol. 26, 293–301.

Blomberg, F. and Raftell, M. (1974) Enzyme polymorphism in rat liver microsomes and plasma membranes. I. An immunochemical study of multienzyme complexes and other enzyme active antigens. Eur. J. Biochem. 49, 21–31.

Blouin, A., Bolender, R. and Weibel, E. R. (1977) Distribution of organelles and membranes between hepatocytes and non-hepatocytes in the rat liver parenchyma. A stereological study. J. Cell Biol. 72, 441–455.

Bock, E. and Jorgensen, O. S. (1975) Rat brain synaptic vesicles and synaptic membranes compared by cross immunophoresis. FEBS Lett. 52, 37–39.

Bolender, R. P. (1974) Stereological analysis of the guinea pig pancreas. I. Analytical model and quantitative description of non-stimulated pancreatic exocrine cells. J. Cell Biol. 61, 269–287.

Bonatti, S., Cancedda, R., Borgese, N. and Meldolesi, J. (1978) Membrane biogenesis in Sindbis-infected chick embryo fibroblasts I. Determinant role of cellular endomembranes for the synthesis of virus envelope proteins. J. Mol. Biol. Submitted for publication.

Bonne, D., Nicolas, P., Lauber, M., Camier, M., Tixier-Vidal, A. and Cohen, P. (1977) Evidence for an adenylate cyclase activity in neurosecretory granule membranes from bovine neurohypophysis. Eur. J. Biochem. 78, 337–342.

Borgese, N. and Meldolesi, J. (1976) Studies on the biogenesis of rat liver intracellular membranes. J. Cell Biol. 70, 253a.

Borgese, N., Mock, W., Kreibich, G. and Sabatini, D. D. (1974) Ribosomal-membrane interaction. In vitro binding of ribosomes to microsomal membranes. J. Mol. Biol. 88, 559–580.

Borgese, N., De Camilli, P., Brenna, A. and Meldolesi, J. (1977) Immunochemical and freeze-fracture study of membrane biogenesis and interactions. In: Methodological surveys, vol. 6 (Cellular Biochemistry), Membranous Elements and Movement of Molecules (Reid, E., ed.) pp 337–353. Harwood, Chichester.

Borowitz, J. L., Leslie, S. W. and Baugh, L. (1975) Adrenal catecholamine release: possible termination mechanisms. In: Calcium Transport in Contraction and Secretion (Carafoli, E., Clementi, F., Drabikowsky, W. and Margreth, A., eds.) pp. 227–234. Elsevier/North-Holland, Amsterdam.

Bosmann, H. B., Hagopian, A. and Eylar, E. H. (1969) Cellular membranes: The biosynthesis of glycoprotein and glycolipid in HeLa cell membrane. Arch. Biochem. Biophys. 130, 573–583.

Boyne, A. F., Bohan, T. P. and Williams, T. H. (1974) Effects of calcium containing fixation solutions on cholinergic synaptic vesicles. J. Cell Biol. 63, 780–795.

Boyne, A. F., Bohan, T. P., ad Williams, T. H. (1975) Changes in cholinergic sympathetic vesicle populations and the ultrastructure of the nerve terminal membranes of narcine brasiliensis electric organ stimulated in vitro. J. Cell Biol. 67, 814–825.

Breckenridge, W. G. and Morgan, I. G. (1972) Common glycoproteins of synaptic vesicles and the synaptosomal plasma membrane. FEBS Lett. 22, 253–256.

Bretscher, M. S. (1975) C-terminal region of the major erythrocyte sialoglycoproteins is on the cytoplasmic site of the membrane. J. Mol. Biol. 98, 831–835.

Bretz, U. and Baggiolini, M. (1974) Biochemical and morphological characterization of azurophil and specific granules of human neutrophilic polymorphonuclear leukocytes. J. Cell Biol. 63, 251–269.

Brimjoin, S. (1974) Local changes in subcellular distribution of dopamine-β-hydroxilase after blockade of axonal transport. J. Neurochem. 22, 347–358.

Brimjoin, S. and Helland, L. (1976) Rapid retrograde transport of dopamine-β-hydroxilase as examined by the stop flow technique. Brain Res. 102, 217–223.

Burridge, K. and Phillips, J. H. (1975) Association of actin and myosin with secretory granule membranes. Nature 254, 526–529.

Byers, M. R. (1974) Structural correlates of rapid axonal transport: evidence that microtubules may not be involved. Brain Res. 75, 97–113.

Campbell, P. N. and Blobel, G. (1976) The role of organelles in the chemical modification of the primary translation products of secretory proteins. FEBS Lett. 72, 215–226.

Carafoli, E., Clementi, F., Drabikowsky, W. and Margreth, A., (1975) (Eds.) Calcium Transport in Contraction and Secretion. Elsevier/North-Holland, Amsterdam.

Carlsson, S. A. and Ritzen, M. (1969) Mast cells and 5'HT. Intracellular release of 5'-hydroxytriptamine (5'HT) from storage granules during anaphylaxis or treatment with compound 48/80. Acta Physiol. Scand. 77, 449–455.

Carlton, H. C. and Appel, S. H. (1973) The contribution of axoplasmic flow in optic nerve and protein synthesis within the optic tectum to synaptic membrane proteins of the chick optic tectum. J. Neurochem. 20, 1707–1717.

Case, R. M. (1973) Cellular mechanisms controlling pancreatic secretion. Acta Hepatogastroenterol. 20, 435–444.

Case, R. M. and Clausen, T. (1973) The relationship between calcium exchange and enzyme secretion in the isolated rat pancreas. J. Physiol. (London) 235, 75–102.

Case, R. M. and Goebell, H., (Eds.) (1976) Stimulus-Secretion Coupling in the Gastrointestinal Tract. MTP Press Ltd., Lancaster.

Castel, M. (1977) Pseudopodia formation in neurosecretory granules. Cell Tissue Res. 175, 483–498.

Castle, J. D., Jamieson, J. D. and Palade, G. E. (1974) Secretion granules of the rabbit parotid gland. Contamination analysis of the membrane subfractions. J. Cell Biol. 63, 52a.

Castle, J. D., Jamieson, J. D. and Palade, G. E. (1975) Secretion granules of the rabbit parotid gland. Isolation, subfractionation and characterization of the membrane and content subfractions. J. Cell Biol. 64, 182–210.

Ceccarelli, B. and Hurlbut, W. P. (1975a) Transmitter release and the vesicle hypothesis. In: Golgi Centennial Symposium Proceedings (Santini, M., ed.) pp. 529–545. Raven Press, New York.

Ceccarelli, B. and Hurlbut, W. P. (1975b) The effect of prolonged repetitive stimulation in hemicholinium on the frog neuromuscular junction. J. Physiol. (London) 247, 163–188.

Ceccarelli, B., Clemente, F. and Meldolesi, J. (1975) Secretion of calcium in the pancreatic juice. J. Physiol. (London) 245, 617–638.

Ceccarelli, B., Clementi, F. and Meldolesi, J., (Eds.) (1974) Cytopharmacology of Secretion. Raven Press, New York.

Ceccarelli, B., Hurlbut, W. P. and Mauro, A. (1972) Depletion of vesicles from frog neuromuscular junctions by prolonged tetanic stimulation. J. Cell Biol. 54, 30–38.

Ceccarelli, B., Hurlbut, W. P. and Mauro, A. (1973) Turnover of transmitter and synaptic vesicles at the frog neuromuscular junction. J. Cell Biol. 57, 499–524.

Ceccarelli, B., Grohovaz, F. and Hurlbut, W. P. (1978a) Freeze-fracture study of neuromuscular junctions during intense release of neurotransmitter. In preparation.

Ceccarelli, B., Hurlbut, W. P., De Camilli, P. and Meldolesi, J. (1978b) The effect of extracellular calcium on the topological organization of the plasmalemma in two secretory systems. Ann. N.Y. Acad. Sci. In press.

Ceccarelli, B., Peluchetti, D., Grohovaz, F. and Jezzi, N. (1976) Freeze-fracture studies on changes at neuromuscular junctions induced by black widow spider venom. In: Electron Microscopy 1976, vol. II, Biological Sciences (Ben-Shaul, Y., ed.) pp. 376–378. Tal International Publishing Company, Jerusalem.

Ceccarelli, B., Pensa, P., De Giuli, C. and Pelosi, G. (1967) Sur les methodes de fixation pour etudier les terminaisons synaptiques centrales de rat. J. Microscopie 6, 42a.

Cheetham, R. D., Morré, D. J., Pannek, C. and Friend, D. S. (1971) Isolation of a Golgi apparatus-rich fraction from rat liver. IV. Thiamine pyrophosphatase. J. Cell Biol. 49, 899–905.

Cheng, H. and Farquhar, M. G. (1976a) Presence of adenylate cyclase activity in Golgi and other fractions from rat liver. I. Biochemical determination. J. Cell Biol. 70, 660–670.

Cheng, H. and Farquhar, M. G. (1976b) Presence of adenylate cyclase activity in Golgi and other fractions of rat liver. II. Cytochemical localization within Golgi and ER membranes. J. Cell Biol. 70, 671–684.

Chi, E. H., Lagunoff, D. and Koehler, J. K. (1976) Freeze-fracture study of mast cell secretion. Proc. Nat. Acad Sci. U.S.A. 73, 2823–2827.

Chiang, H., Langer, P. J. and Lodish, H. F. (1976) Asynchronous synthesis of erythrocyte membrane proteins. Proc. Nat. Acad. Sci. U.S.A. 73, 3206–3210.

Choi, Y. S., Knopf, P. M. and Lennox, E. S. (1971) Intracellular transport and secretion of an immunoglobulin light chain. Biochemistry 10, 668–679.

Christophe, J. P., Frandsen, E. K., Conlon, T. P., Krishna, G. and Gardner, J. D. (1976) Action of cholecystokinin, cholinergic agents and A23187 on accumulation of guanosine $3',5'$-monophosphate in dispersed guinea pig pancreatic acinar cells. J. Biol. Chem. 251, 4640–4645.

Christophe, J., Vandermeers, A., Vandermeers-Piret, M. C., Rathé, J. and Camus, J. (1973) The relative turnover time in vivo of the intracellular transport of five hydrolases in the pancreas of the rat. Biochim. Biophys. Acta 308, 285–295.

Chu, L. L., Mac Gregor, R. R. and Cohn, D. V. (1977) Energy-dependent intracellular translocation of proparathormone. J. Cell Biol. 72, 1–10.

Chua, N. H., Blobel, G., Siekevitz, P. and Palade, G. E. (1973) Attachment of chloroplast polysomes to thylakoid membranes in Chlamydomonas reinhardtii. Proc. Nat. Acad. Sci. U.S.A. 70, 1554–1558.

Clark, A. W., Mauro, A., Longenecker, H. E. and Hurlbut, W. P. (1970) Fffects on the fine structure of the frog neuromuscular junction. Nature, 225, 703–705.

Clark, A. W., Hurlbut, W. P. and Mauro, A. (1972) Changes in the fine structure of the neuromuscular junction of the frog caused by black widow spider venom. J. Cell Biol. 52, 1–14.

Claude, A. (1970) Growth and differentiation of cytoplasmic membranes in the course of lipoprotein granule synthesis in the hepatic cell. J. Cell Biol. 47, 745–766.

Clemente, F. and Meldolesi, J. (1975a) Calcium and pancreatic secretion. I. Distribution of calcium and magnesium in the acinar cells of the guinea pig pancreas. J. Cell Biol. 65, 88–102.

Clemente, F. and Meldolesi, J. (1975b) Calcium and pancreatic secretion. Dynamics of subcellular calcium pools in the resting and stimulated acinar cells. Brit. J. Pharmacol. 55, 369–379.

Colbron, A., Nachbaur, J. and Vignais, P. M. (1971) Enzymic characterization and lipid composition of rat liver subcellular membranes. Biochim. Biophys. Acta 249, 462–492.

Cook, G. M. W. (1977) Biosynthesis of plasma membrane glycoproteins. In: Cell Surface Reviews, Vol. 4, (Poste, G., and Nicolson, G. L., eds.), pp. 85–136. Elsevier/North-Holland, Amsterdam.

Cooke, J. D., Okamoto, K., Quastel, D. M. G. (1973) The role of calcium in depolarization-secretion coupling at the motor nerve terminal. J. Physiol. (London) 228, 459–497.

Coraboef, E. and Rojas, E. (1976) Colloquium on cell membrane physiology of the endocrine pancreas. J. Physiol. (Paris) 72, 699–832.

Couteaux R. (1974) Remarks on the organization of axonal terminal in relation to secretory process at synapses. In: Cytopharmacology of Secretion (Ceccarelli, B., Clementi, F. and Meldolesi, J., eds.) pp. 369–379. Raven Press, New York.

Couteaux, R. and Pécot-Dechavassine, M. (1970) Vésicules synaptiques et poches au niveau des zones actives de la jonction neuromusculaire. C. R. Acad. Sci. (Paris) D 271, 2346–2349.

Couteaux, R. and Pécot-Dechavassine, M. (1973) Données ultrastructurales et cytochimiques sur le méchanisme de liberation de l'acetylcholine dans la transmission synaptique. Arch. Ital. Biol. 111, 231–262.

Crane, F. L. and Löw, H. (1976) NADH oxidation in liver and fat cell plasma membranes. FEBS Lett. 68, 153–156.

Creutzfeld, W., Creutzfeld, C., Frerichs, H., Perings, E. and Sikinger, K. (1969) The morphological substrate of the inhibition of insulin secretion by diazoxide. Horm. Met. Res. 1, 53–64.

Crowther, R. A., Finch, J. T. and Pearse, B. M. F. (1976) On the structure of coated vesicles. J. Mol. Biol. 103, 785–798.

Dagorn, J. C. and Mongeau, R. (1977) Different action of hormonal stimulation on the biosynthesis of three pancreatic enzymes. Biochim. Biophys. Acta 498, 76–82.

Dahl, G. and Gratzl, M. (1976) Calcium-induced fusion of isolated secretory vesicles from the islets of Langerhans. Cytobiologie 12, 344–355.

Dahlstrom, A. (1971) Axoplasmic transport (with particular respect to adrenergic neurons). Phil. Trans. Roy. Soc. (London) B 261, 325–358.

Dallner, G., Siekevitz, P. and Palade, G. E. (1966) Biogenesis of endoplasmic reticulum membranes. II. Synthesis of constitutive microsomal enzymes in developing rat hepatocytes. J. Cell Biol. 30, 97–117.

Daniels, M. P., and Vogel, Z. (1975) Immunoperoxidase staining of α-bungarotoxin binding sites in muscle end plate shows distribution of acetylcholine receptors. Nature 254, 339–341.

Da Prada, M., Berlepsch, K. and Pletscher, A. (1972a) Storage of biogenic amines in blood platelets and adrenal medulla. Lack of evidence for direct involvement of glycosaminoglycans. N. S. Arch. Pharmakol. 275, 315–322.

Da Prada, M., Pletscher, A. and Tranzer, J. P. (1972b) Lipid composition of membranes of amine-storage organelles. Biochem. J. 127, 681–683.

Davis, B. and Lazarus, N. R. (1976) An in vitro system for studying insulin release by secretory granules-plasma membrane interaction: definition of the system. J. Physiol. (London) 256, 709–730.

Dean, P. M. (1974a) Surface electrostatic-charge measurements on islets and zymogen granules: effect of calcium ions. Diabetology 10, 427–430.

Dean, P. M. (1974b) Neurohypophysial granules: electrokinetic properties and calcium ion binding. Brain Res. 79, 397–403.

Dean, P. M. and Matthews, E. K. (1974) Calcium binding to the chromaffin granule surface. Biochem. J. 142, 637–640.

De Camilli, P., Peluchetti, D. and Meldolesi, J. (1974) Structural difference between lumenal and lateral plasmalemma in pancreatic acinar cells. Nature 248, 245–246.

De Camilli, P., Peluchetti, D. and Meldolesi, J. (1976) Dynamic changes of the lumenal plasmalemma in stimulated parotid acinar cells. A freeze-fracture study. J. Cell Biol. 70, 59–74.

de Duve, C. (1963) Footnote in the paper by Cohn, Z. A., Hirsch, J. G. and Weiner, E. (1963) Lysosomes and Endocytosis. In: Lysosomes. Ciba Found. Symp. (de Reuck, A. V. S., and Cameron, M. P., eds.) p. 126. Churchill J. & A. Ltd., London.

Dehlinger, P. J. and Schimke, R. T. (1971) Size distribution of membrane proteins of rat liver and their relative rates of degradation. J. Biol. Chem. 246, 2574–2583.

Del Castillo, J. and Katz, B. (1954a) Quantal content of the end-plate potential. J. Physiol. (London) 124, 560–573.

Del Castillo, J. and Katz, B. (1954b) The effect of magnesium on the activity of motor nerve endings. J. Physiol. (London) 124, 553–559.

De Lorenzo, R. J. (1976) Ca-dependent phosphorylation of specific synaptosomal fraction proteins: possible role of phosphoprotein in mediating neurotransmitter release. Biochem. Biophys. Res. Commun. 71, 590–597.

Dempsey, G. P., Bullivant, S. and Watkins, W. B. (1973) Ultrastructure of the rat posterior pituitary gland and evidence of hormone release by exocytosis as revelaed by freeze-fracturing. Z. Zellforsch. 143, 465–484.

DePierre, J. W. and Dallner, G. (1975) Structural aspects of the membrane of the endoplasmic reticulum. Biochim. Biophys. Acta 415, 411–472.

DePierre, J. W. and Karnovsky, M. (1973) Plasma membranes of mammalian cells. A review of methods of their characterization and isolation. J. Cell Biol. 50, 275–303.

De Potter, W. P., Chubb, I. W. and de Schaepdryver, A. F. (1972) Pharmacological aspects of peripheral noradrenergic transmission. Arch. Int. Pharmacol. Pharmacodyn. 196, 258–287.

Di Giamberardino, L., Bennett, G., Koening, H. L. and Droz, B. (1973) Axonal migration of protein and glycoprotein to nerve endings. III. Cell fraction analysis of chicken ciliary ganglion after intracerebral injection of labeled precursors of proteins and glycoproteins. Brain Res. 60, 147–159.

Diliberto, E., Viveros, O. H. and Axelrod, J. (1976) Localization of protein carboxymethylase and its endogenous substrate in the adrenal medulla. Methylation of the chromaffin granule membrane. Fed. Proc. 35, 326.

Dodge, F. A. and Rahaminoff, R. (1967) Co-operative action of calcium ions in transmitter release at the neuromuscular junction. J. Physiol. (London) 193, 419–432.

Douglas, W. W. (1968) Stimulus-secretion coupling: the concept and clues from chromaffin and other cells. Brit. J. Pharmacol. 34, 451–474.

Douglas, W. W. (1974a) Mechanisms of release of neurohypophysial hormones: stimulus-secretion coupling. In: Handbook of Physiology, vol. IV, sec. 7, (Knobil, E. and Sawyer, W. H., eds.) pp. 191–224, Am. Physiol. Soc., Washington D.C.

Douglas, W. W. (1974b) Involvement of calcium in exocytosis and the exocytosis-vesiculation sequence. Biochem. Soc. Symp. 39, 1–28.

Douglas, W. W. (1975a) Secretomotor control of adrenal medullary secretion: synaptic, membrane and ionic events in stimulus-secretion coupling. In: Handbook of Physiology, vol. VI, sec. 7, (Blaschko, H., Sayers, G. and Smith, A. D., ed.) pp. 367–388. Am. Physiol. Soc., Washington D.C.

Douglas, W. W. (1975b) Stimulus-secretion coupling in mast cells: regulation of exocytosis by cellular and extracellular calcium. In: Calcium Transport in Contraction and Secretion (Carafoli, E., Clementi, F., Drabikowski, W. and Margreth, A., eds.) pp. 167–175. Elsevier/North-Holland, Amsterdam.

Dowdall, M. J. and Zimmermann, H. (1974) Evidence for heterogeneous pools of acetylcholine in isolated cholinergic synaptic vesicles. Brain Res. 71, 160–166.

Dowdall, M. J., Boyne, A. F. and Whittaker, V. P. (1974) Adenosine triphosphate. A constituent of cholinergic synaptic vesicles. Biochem. J. 140, 1–12.

Dreifuss, J. J. (1975) A review on neurosecretory granules: their contents and mechanisms of release. Ann. N. Y. Acad. Sci. 248, 184–201.

Dreifuss, J. J., Akert, K., Sandri, C. and Moor, H. (1976) Specific arrangements of membrane particles at sites of exo-endocytosis in the freeze-etched neurohypophysis. Cell Tissue Res. 165, 317–325.

Dreifuss, J. J., Nordmann, J. J., Akert, K., Sandri, C. and Moor, H. (1974) Exo-endocytosis in the neurohypophysis as revealed by freeze-fracturing. In: Neurosecretion. The Neuroendocrine Pathway, Sixth Int. Symp. on Neurosecretion, pp. 31–37. Springer-Verlag, Berlin.

Dreyer, F., Peper, K., Akert, K., Sandri, C. and Moor, H. (1973) Ultrastructure of the active zone in the frog neuromuscular junction. Brain Res. 62, 373–381.

Droz, B. (1975) Synaptic machinery and axoplasmic transport: maintenance of neuronal connectivity. In: The Nervous System (Tower, D. B., ed.) vol. I, pp. 111–127, Raven Press, New York.

Droz, B., Koenig, H. J., di Giamberardino, L. (1973) Axonal migration of protein and glycoprotein to nerve endings. I. Radioautographic analysis of the renewal of protein in nerve endings of chick ciliary ganglion after intracerebral injection of ^{3}H-lysine. Brain Res. 60, 93–127.

Droz, B., Rambourg, A. and Koenig, H. L. (1975) The smooth endoplasmic reticulum: structure and role in the renewal of axonal membrane and synaptic vesicles by fast axonal transport. Brain Res. 93, 1–13.

Dunant, Y., Gautron, J., Israël, M., Lesbats, B. and Manaranche, R. (1972) Les compartiments de l'acetylcholine de l'organ electrique de la torpille et leur modifications par la stimulation. J. Neurochem. 19, 1787–2002.

Dunant, Y., Gautron, J., Israël, M., Lesbats, B. and Manaranche, R. (1974) Evolution de la decharge de l'organe éléctrique de la torpille et variations simultanées de l'acetylcholine au cours de la stimulation. J. Neurochem. 23, 635–643.

Dunn, W. B. and Spicer, S. S. (1969) Histochemical demonstration of sulfated mucosubstances and cationic proteins in human granulocytes and platelets. J. Histochem. Cytochem. 17, 668–674.

Ehrenreich, J. H., Bergeron, J. J. M., Siekevitz, P. and Palade, G. E. (1973) Golgi fractions prepared from rat liver homogenates. I. Isolation procedure and morphological characterization. J. Cell Biol. 59, 45–72.

Ehrlich, P. H. and Bornstein, P. (1972) Microtubules in transcellular movement of procollagen. Nature New Biol. 238, 257–260.

Elder, J. H. and Morré, D. J. (1976) Synthesis in vitro of intrinsic membrane proteins by free, membrane-bound and Golgi apparatus-associated polyribosomes from rat liver. J. Biol. Chem. 251, 5054–5068.

Elhammer Å, Svensson, H., Autuori, F. and Dallner, G. (1975) Biogenesis of microsomal membrane glycoproteins in rat liver. III. Release of glycoproteins from the Golgi fraction and their transfer to microsomal membranes. J. Cell Biol. 67, 715–724.

Erlandsen, S. L., Hegre, O. D., Parsons, J. A., Mc Evoy, R. C. and Elde, R. P. (1976) Pancreatic islet cell hormones. Distribution of cell types and evidence for the presence of somatostatin and gastrin within D cells. J. Histochem. Cytochem. 24, 883–896.

Ernster, L. and Orrenius, S. (1965) Substrate-induced synthesis of the hydroxylating enzyme system of liver microsomes. Fed. Proc. 24, 1190–1199.

Eschenfeldt, W. H. and Patterson, R. J. (1975) Do antibody binding techniques identify polysomes synthesizing a specific protein? Biochem. Biophys. Res. Commun. 67, 935–945.

Evans, W. H. (1974) Nucleotide pyrophosphatase, a sialoglycoprotein located in the hepatocyte surface. Nature 250, 391–394.

Evans, W. H. and Gurd, J. W. (1971) Biosynthesis of liver membranes. Incorporation of ^{3}H-leucine into proteins and of ^{14}C glucosamine into proteins and lipids of liver microsomal and plasma-membrane fractions. Biochem. J. 125, 615–624.

Evans, W. H., Bergeron, J. J. M. and Geschwind, I. I. (1973) Distribution of insulin receptor sites among liver plasma membrane subfractions. FEBS Lett. 34, 259–262.

Farquhar, M. G. (1971) Processing of secretory products by cells of the anterior pituitary gland. Mem. Soc. Endocrinol. 19, 79–124.

Farquhar, M. G., Bergeron, J. J. M. and Palade, G. E. (1974) Cytochemistry of Golgi fractions prepared from rat liver. J. Cell Biol. 60, 8–25.

Farquhar, M. G., Skutelky, E. H. and Hopkins, C. R. (1975) Structure and function of the anterior pituitary and dispersed pituitary cells. In vitro studies. In: The Anterior Pituitary (Tixier-Vidal, A. and Farquhar, M. G. eds.) pp. 83–135. Academic Press, New York.

Fatt, P. and Katz, B. (1952) Spontaneous subthreshold activity at motor nerve endings. J. Physiol. (London) 117, 109–128.

Fawcett, D. W. (1962) Physiologically significant specializations of the cell surface. Circulation 26, 1105–1132.

Feigenson, M. E., Schnebeli, H. P. and Baggiolini, M. (1975) Demonstration of ricin binding sites on the outer face of azurophil and specific granules of rabbit polymorphonuclear leukocytes. J. Cell Biol. 66, 183–187.

608

Feinstein, H. and Schramm, M. (1970) Energy production in rat parotid gland. Relation to enzyme secretion and effects of calcium. Eur. J. Biochem. 13, 158–163.

Feldman, G., Maurice, M., Sapin, C. and Benhamou, J. P. (1975) Inhibition by colchicine of fibrinogen translocation in hepatocytes. J. Cell Biol. 6, 237–242.

Fertuck, H. C. and Salpeter, M. M. (1974) Localization of acetylcholine receptor by ^{125}I-labeled α-bungarotoxin binding at mouse motor end-plates. Proc. Nat. Acad. Sci. U.S.A. 71, 1376–1378.

Fillenz, M., Gagnon, C., Stoekel, K. and Thoenen, H. (1976) Selective uptake and retrograde axonal transport of dopamine-β-hydroxylase antibodies in peripheral adrenergic neurons. Brain Res. 114, 293–303.

Finkelstein, A., Rubin, L. L. and Tzeng, M. C. (1976) Black widow spider venom: Effect of purified toxin on lipid bilayers membranes. Science 193, 1009–1011.

Fisher, G., Kaneshiro, E. S. and Peters, P. D. (1976) Divalent cation affinity sites in Paramecium aurelia. J. Cell Biol. 69, 429–440.

Flickinger, C. J. (1975) The relation between the Golgi apparatus, cell surface and cytoplasmic vesicles in Amoeba studied by electron microscopy radioautography. Exp. Cell Res. 96, 189–201.

Foreman, J. C. and Mongar, J. L. (1972) The role of the alkaline earth ions in anaphylactic histamine release. J. Physiol. (London) 244, 753–769.

Fowler, S., Remacle, J., Trouet, A., Beaufay, H., Berthet, J., Wibo, M. and Hauser, P. (1976) Analytical study of microsomes and isolated subcellular membranes from rat liver V. Immunological localization of cytochrome b_5 by electron microscopy: methodology and application to various subcellular fractions. J. Cell Biol. 71, 535–551.

Franke, W. W. and Kartenbeck, J. (1976) Some principles of membrane differentiation. In: Progress in Differentiation Research (Müller-Bérat, N., et al., eds.) pp. 213–243, Elsevier/North-Holland, Amsterdam.

Franke, W. W., Morré, D. J., Deumling, B., Cheetham, R. D., Kartenbeck, J., Jarash, E. D. and Zentgraf, H. W. (1971) Synthesis and turnover of membrane proteins in rat liver: An examination of the membrane flow hypothesis. Z. Naturforsch. 266, 1031–1039.

Franke, W. W., Spring, H., Sheer, U. and Zerban, H. (1975) Growth of the nuclear envelope in the vegetative phase of the green alga Acetabularia. Evidence for the assembly from membrane components synthesized in the cytoplasm. J. Cell Biol. 66, 681–689.

Franke, W. W., Lüder, M. R., Kartenbeck, J., Zerban, H. and Keenan, T. W. (1976) Involvement of vesicle coat material in casein secretion and surface regeneration. J. Cell Biol. 69, 173–195.

Freedman, R. A., Weiser M. M. and Isselbacher K. J. (1977) Calcium translocation by Golgi and lateral-basal membrane vesicles from rat intestine. Decreased in vitamin D-deficient rats. Proc. Natl. Acad. Sci. U.S.A. 74, 3612–3616.

Fried, R. C. and Blaustein, M. P. (1976) Synaptic vesicles recycling in synaptosomes in vitro. Nature 261, 255–256.

Friend, D. S. and Farquhar, M. G. (1967) Functions of coated vesicles during protein absorption in the rat vas deferens. J. Cell Biol. 35, 357–376.

Frontali, N., Ceccarelli, B., Gorio, A., Mauro, A., Siekevitz, P., Tzeng, M. C. and Hurlbut, W. P. (1976) Purification from black widow spider venom of a protein factor causing the depletion of synaptic vesicles at neuromuscular junctions. J. Cell Biol. 68, 462–479.

Fujita, S., Ogata, F., Nakamura, J., Omata, S. and Sugano, H. (1977) Isolation and characterization of membrane proteins responsible for attachment of polyribosomes to rough microsomal fraction of rat liver. Biochem. J. 164, 53–66.

Fukushima, K. and Sato, R. (1973) Purification and characterization of cytochrome b_5-like hemoprotein associated with outer mitochondrial membrane of rat liver. J. Biochem. (Tokyo) 74, 161–173.

Gabbiani, G., Da Prada, M., Richards, G. and Pletscher, A. (1976) Actin associated with membranes of monoamine storage organelles. Proc. Soc. Exp. Biol. Med. 152, 135–138.

Gagnon, C., Schatz, R., Otten, U. and Thoenen, H. (1976) Synthesis, subcellular distribution and turnover of dopamine-β-hydroxilase in organ cultures of sympathetic ganglia and adrenal medullae. J. Neurochem. 27, 1083–1090.

Galli, P., Brenna, A., De Camilli, P. and Meldolesi, J. (1976) Extracellular calcium and the organization of tight junctions in pancreatic acinar cells. Exp. Cell Res. 99, 178–183.

Gavin, J. R., Roth, J., Neville, D. M., DeMeytes, P. and Buell, D. N. (1974) Insulin-dependent regulation of insulin receptor concentrations: a direct demonstration in cell culture. Proc. Nat. Acad. Sci. U.S.A. 71, 84–88.

Geissler, D., Martinek, A., Margolis, R. U., Margolis, R. K., Leeden, R., Konig, P. and Winkler, H. (1977) Composition and biogenesis of complex carbohydrates of adrenal chromaffin granules. Neuroscience 2, 685–693.

Gershon, M. D. and Nunez, E. A. (1973) Subcellular storage organelles for 5'hydroxytryptamine in parafollicular cells of the thyroid gland. The effect of drugs which deplete the amine. J. Cell Biol. 56, 676–689.

Geuze, J. J. and Kramer, M. F. (1974) Function of coated vesicles and multivesicular bodies during membrane regulation in stimulated exocrine pancreas. Cell Tissue Res. 156, 1–20.

Giannattasio, G. and Zanini, A. (1976) Presence of sulfated proteoglycans in prolactin secretory granules isolated from the rat pituitary gland. Biochim. Biophys. Acta 439, 349–357.

Giannattasio, G., Zanini, A. and Meldolesi, J. (1975) Molecular organization of rat prolactin granules. I. In vitro stability of intact and membraneless granules. J. Cell Biol., 64, 246–251.

Ginsborg, B. L. and Jenkinson, D. H. (1976) Transmission of impulses from nerve to muscle. In: Neuromuscular Junction (Zaimis, E. ed.) pp. 229–364, Springer-Verlag, Berlin.

Glaumann, H. and Ericsson, J. L. E. (1970) Evidence for the participation of the Golgi apparatus in the intracellular transport of nascent albumin in the liver cell. J. Cell Biol. 47, 555–571.

Glaumann, H., Bergstrand, A. and Ericsson, J. L. E. (1975) Studies on the synthesis and intracellular transport of lipoprotein particles in rat liver. J. Cell Biol. 64, 356–377.

Goldberg, N. D., Haddox, M. K., Dunham, E., Lopez, C. and Hadden, J. W. (1974) The Yin-Yang hypothesis of biological control: opposing influences of cyclic GMP and cyclic AMP in the regulation of cell proliferation and other biological processes. In: Control of Proliferation in Animal Cells (Clarkson, B. and Baserga, R., eds.) pp. 609–625, Cold Spring Harbor Labs., New York.

Goldstein, A. (1976) Opioid peptides (endorphins) in pituitary and brain. Science 193, 1081–1086.

Gonatas, N. K., Steiber, A., Kim, S. U., Graham, D. and Avrameas, S. (1975) Internalization of neuronal plasma membrane ricin receptors into the Golgi apparatus. Exp. Cell Res. 94, 426–431.

Gonatas, N. K., Kim, S. V., Stieber, A. and Avrameas, S. (1977) Internalization of lectins in neuronal gerl. J. Cell Biol. 73, 1–13.

Gordon, A. S., Davis, C. G. and Diamond, I. (1977) Phosphorylation of membrane proteins at a cholinergic synapse. Proc. Nat. Acad. Sci. U.S.A. 74, 263–267.

Gorio, A., Hurlbut, W. P. and Ceccarelli, B. (1978) Acetylcholine compartments in mouse diaphragm: a comparison of the effects of black widow spider venom, electrical stimulation and high concentration of potassium. J. Cell Biol. In press.

Grand, R. J. and Gross, P. R. (1969) Independent stimulation of secretion and protein synthesis in rat parotid gland. The influence of epinephrine and dibutyryl cyclic AMP. J. Biol. Chem. 244, 5608–5615.

Gray, E. G. (1972) Are the coats of coated vesicles artifacts? J. Neurocytol. 1, 363–382.

Gray, E. G. and Paula-Barbosa, M. (1974) Dense particles within synaptic vesicles fixed with acid aldehyde. J. Neurocytol. 3, 497–512.

Greengard, P. and Robison, G. A., (1972, 1973, 1974, 1975a,b, 1977) (Eds.) Advances in Cyclic Nucleotide Research, vols. 1, 3, 4, 5, 6, 7, 8. Raven Press, New York.

Grodsky, G. M. (1972) A threshold distribution hypothesis for pocket storage of insulin and its mathematical modeling. J. Clin. Invest. 51, 2047–2059.

Grove, S. N., Bracker, C. E. and Morré, D. J. (1968) Cytomembrane differentiation in the endoplasmic reticulum-Golgi apparatus vesicle complex. Science 161, 171–173.

Grubman, M. J., Moyer, S. A., Banerjee, A. K. and Ehrenfeld, E. (1975) Subcellular localization of vesicular stomatitis virus messenger RNA. Biochem. Biophys. Res. Commun. 62, 531–538.

Gullis, R. J. and Rowe, C. E. (1975) The stimulation by transmitter substances and putative transmitter substances of the net activity of phospholipase A_2 of synaptic membranes of cortex of guinea pig brain. Biochem. J. 148, 197–208.

Guttler, T. and Clausen, J. (1969) Changes in the lipid pattern of HeLa cells exposed to immunoglobulin G and complement. Biochem. J. 115, 959–968.

Habener, J. F., Kemper B., and Potts, J. T. (1975) Calcium-dependent intracellular degradation of parathyroid hormone: a possible mechanism for the regulation of hormone stores. Endocrinology 97, 431–441.

Haddad, A., Bennett, G. and Leblond C. P. (1977) Formation and turnover of plasma membrane glycoprotein in kidney tubules of young rats and adult mice as shown by radioautography after an injection of ^{3}H-fucose. Am. J. Anat. 148, 241–274.

Harano, T. and Omura, T. (1977a) Biogenesis of endoplasmic reticulum membrane in rat liver cells. I. Intracellular sites of synthesis of cytochrome b_5 and NADPH-cytochrome c reductase. J. Biochem. 82, 1541–1549.

Harano, T. and Omura, T. (1977b) Biogenesis of endoplasmic reticulum membrane in rat liver cells. II. Discharge of the nascent peptides of NADPH-cytochrome c reductase and cytochrome b_5 on the cytoplasmic site of the endoplasmic reticulum membrane. J. Biochem. 82, 1551–1557.

Haussmann, K. and Allen, R. D. (1976) Membrane behavior in exocytic vesicles. II. Fate of the trichocyst membranes in Paramecium after induced trichocyst discharge. J. Cell Biol. 69, 313–326.

Haymovits, A. and Sheele, G. A. (1976) Cellular cyclic nucleotides and enzyme secretion in the pancreatic acinar cells. Proc. Nat. Acad. Sci. U.S.A. 73, 156–160.

Heap, P. F., Jones, C. W., Morris, J. F., and Pickering, B. T. (1975) Movement of neurosecretory product through the anatomical compartments of the neural lobe of the pituitary gland. An electron microscope autoradiographic study. Cell Tissue Res. 156, 483–497.

Helman, B. (1975) The effects of calcium and divalent ionophores on the stability of isolated insulin secretory granules. FEBS Lett. 54, 343–346.

Henning, R., Milner, R. J., Reske, K., Cunningham, B. A. and Edelman, G. N. (1976) Subunit structure, cell surface orientation and partial amino acid sequences of murine histocompatibility antigens. Proc. Natl. Acad. Sci. U.S.A. 73, 118–122.

Herman, L., Sato, T. and Hales, C. N. (1973) The electron microscopic localization of cations to pancreatic islets of Langerhans and their possible role in insulin secretion. J. Ultrastr. Res. 42, 298–311.

Herzog, V. and Farquhar, M. G. (1977) Uptake of horseradish peroxidase in the Golgi complex of stimulated acinar cells of the parodid gland. Proc. Natl. Acad. Sci. U.S.A. 74, 5073–5079.

Herzog, V. and Miller, F. (1972) The localization of endogenous peroxidase in the lacrimal gland of the rat during postnatal development. J. Cell Biol. 53, 662–680.

Herzog, V., Sies, H. and Miller, F. (1976) Exocytosis in secretory cells of rat lacrimal gland. Peroxidase release from lobules and isolated cells upon cholinergic stimulation. J. Cell Biol. 70, 692–706.

Heuser, J. E. and Reese, T. S. (1973) Evidence for recycling of synaptic vesicle membrane during transmitter release at the frog neuromuscular junction. J. Cell Biol. 57, 315–344.

Heuser, J. E., Reese, T. S. and Landis, D. M. D. (1974) Functional changes in frog neuromuscular junctions studied with freeze-fracture. J. Neurocytol. 3, 109–131.

Hicks, R. M. (1966) The function of the Golgi complex in transitional epithelium. Synthesis of the thick cell membrane. J. Cell Biol. 30, 623–643.

Higgins, J. A. (1974) Studies on the biogenesis of smooth endoplasmic reticulum membranes in hepatocytes of phenobarbital treated rats. II. The site of phospholipid synthesis in the initial phase of membrane proliferation. J. Cell Biol. 62, 635–646.

Hillman, D. E. and Llinas, R. (1974) Calcium containing electron dense structures in the axons of the squid giant synapse. J. Cell Biol. 61, 146–155.

Hinkle, P. M. and Tashjian, A. H. J. (1975) Thyrotropin-releasing hormone regulates the number of its own receptors in the GH_3 strain of pituitary cells in culture. Biochemistry 14, 3845–3851.

Hirano, H., Parkhouse, B., Nicolson, G. L., Lennox, E. S. and Singer, S. J. (1972) Distribution of saccharide residues on membrane fragments from a myeloma-cell homogenate: its implications for membrane biogenesis. Proc. Nat. Acad. Sci. U.S.A. 69, 2945–2949.

Hodson, S. and Brenchley, G. (1976) Similarities of the Golgi apparatus membrane and the plasma membrane in rat liver cells. J. Cell Sci. 20, 167–187.

Hökfelt T. (1973) On the origin of small adrenergic storage vesicles. Evidence for local formation in nerve endings after chronic reserpine treatment. Experientia 29, 580–582.

Hokin, L. E. (1968) Dynamic aspects of phospholipids during protein secretion. Int. Rev. Cytol. 23, 187–241.

Holtzman, E. (1976) Lysosomes, A Survey. Cell Biol. Monographs, Vol. 3, Springer-Verlag, Vienna.

Holtzman, E., Freeman, A. F. and Kashner, L. A. (1971) Stimulation-dependent alterations in peroxidase uptake at lobster neuromuscular junctions. Science 173, 733–736.

Holtzman, E., Schacher, S., Evans, J. and Teichberg, S. (1977) Origin and fate of the membranes of secretion granules and synaptic vesicles. In: Cell Surface Reviews, Vol. 4, (Nicolson, G. L. and Poste, G., eds.) pp. 165–246. Elsevier/North-Holland, Amsterdam.

Hortnägl, H. (1976) Membranes of the adrenal medulla: a comparison of membranes of chromaffin granules with those of endoplasmic reticulum. Neuroscience 1, 9–18.

Howe, P. R. C., Fenwick, E. M., Rostas, J. A. P. and Livett, B. G. (1977) Immunochemical comparison of synaptic plasma membrane and synaptic vesicle membrane antigens. J. Neurocytol. 6, 339–352.

Howell, S. L. (1972) Role of ATP in the intracellular transport of pro-insulin and insulin in the rat pancreatic β-cell. Nature New Biol. 235, 85–86.

Howell, S. L. (1974) The molecular organization of the granule of islets of Langerhans. In: Cytopharmacology of Secretion (Ceccarelli, B., Clementi, F. and Meldolesi, J., eds.) pp. 319–327. Raven Press, New York.

Howell, S. L. and Ewart, R. B. L. (1972) Synthesis and secretion of growth hormone in the rat anterior pituitary. II. Properties of the isolated growth hormone storage granules. J. Cell Sci. 12, 23–31.

Howell, S. L. and Whitfield, M. (1973) Synthesis and secretion of growth hormone in the rat anterior pituitary. I. The intracellular pathway, its time course and metabolic requirements. J. Cell Sci. 12, 1–21.

Howell, S. L., Montague, W. and Tyhurst, M. (1975) Calcium distribution in islets of Langerhans: a study of calcium concentration and of calcium accumulation in B cell organelles. J. Cell Sci. 19, 395–410.

Hubbard, J. I. (1973) Microphysiology of vertebrate neuromuscular transmission. Physiol. Rev. 53, 674–723.

Hubbard, J. I., Jones, S. F. and Landau, E. M. (1968a) On the mechanism by which calcium and magnesium affect the spontaneous release of transmitter from mammalian motor nerve terminals. J. Physiol. (London) 194, 355–380.

Hubbard, J. I., Jones, S. F. and Landau, E. M. (1968b). On the mechanism by which calcium and magnesium affect the release of transmitter by nerve impulses. J. Physiol. (London) 196, 75–86.

Hunt, L. A. and Summers, D. F. (1976a) Association of vesicular stomatitis virus protein with HeLa cell membranes and released virus. J. Virol. 20, 637–645.

Hunt, L. A. and Summers, D. F. (1976b) Glycosylation of vesicular stomatitis virus glycoprotein in virus-infected HeLa cells. J. Virol. 20, 646–657.

Hurlbut, W. P. and Ceccarelli, B. (1974) Transmitter release and recycling of synaptic vesicle membrane at the neuromuscular junction. In: Cytopharmacology of Secretion (Ceccarelli, B., Clementi, F. and Meldolesi, J., eds.) pp. 141–154, Raven Press, New York.

Ishimura, K., Okamoto, H. and Fujita, H. (1977) Freeze-etching observations on the characteristic arrangement of intramembranous particles in the apical plasma membrane of the thyroid follicular cell in TSH-treated mice. Cell Tissue Res. 171, 297–303.

Jacobsson, S. V. and Dallner, G. (1968) Nature of the increase in liver microsomal glucose-6-phosphatase activity during the early stages of alloxan-induced diabetes. Biochim. Biophys. Acta 165, 380–392.

Jaim-Etcheverry, G. and Zieher, L. M. (1968) Electron microscopic cytochemistry of 5-hydroxytryptamine (5HT) in the β-cells of guinea pig endocrine pancreas. Endocrinology 83, 917–923.

Jamieson, J. D. (1972) Transport and discharge of exportable proteins in pancreatic acinar cells: in vitro studies. Curr. Top. Membr. Transport. 3, 273–338.

Jamieson, J. D. (1973) The secretory process in the pancreatic exocrine cell: morphologic and biochemical aspects. In: Handbook Experimental Pharmacology XXXIV (Jorpes, J. and Mutt, V., eds.) pp. 195–217. Springer-Verlag, Berlin.

612

Jamieson, J. D. (1975) Membranes and secretion. In: Cell Membranes: Biochemistry Cell Biology and Pathology (Weissman, G. and Claiborne, R., eds.) pp. 143–152. H. P. Company, New York.

Jamieson, J. D. and Palade, G. E. (1967a) Intracellular transport of secretory proteins in pancreatic acinar cells. I. Role of the peripheral elements of the Golgi complex. J. Cell Biol. 34, 577–596.

Jamieson, J. D. and Palade, G. E. (1967b) Intracellular transport of secretory proteins in pancreatic acinar cells. II. Transport to condensing vacuoles and zymogen granules. J. Cell Biol. 34, 597–615.

Jamieson, J. D. and Palade, G. E. (1968a) Intracellular transport of secretory proteins in the pancreatic exocrine cell. III. Dissociation of intracellular transport from protein synthesis. J. Cell Biol. 39, 580–588.

Jamieson, J. D. and Palade, G. E. (1968b) Intracellular transport of secretory proteins in the pancreatic exocrine cells. IV. Metabolic requirements. J. Cell Biol. 39, 589–601.

Jamieson, J. D. and Palade, G. E. (1971a) Condensing vacuole conversion and zymogen granule discharge in pancreatic exocrine cells: metabolic studies. J. Cell Biol. 48, 503–522.

Jamieson, J. D. and Palade, G. E. (1971b) Synthesis, intracellular transport and discharge of secretory proteins in stimulated pancreatic acinar cells. J. Cell Biol. 50, 135–158.

Jenkinson, D. H. (1957) The nature of antagonism between calcium and magnesium ions at the neuromuscular junction. J. Physiol. (London) 138, 434–444.

Johnson, R. G. and Scarpa, A. (1976) Ion permeability of isolated chromaffin granules. J. Gen. Physiol. 68, 601–631.

Jones, L. R., Mahler, H. R. and Moore, W. J. (1975) Synthesis of membrane protein in slices of rat cerebral cortex. Source of proteins of synaptic plasma membranes. J. Biol. Chem. 250, 973–983.

Jorgensen, O. S. and Mellerup, E. T. (1974) Endocytic formation of rat brain synaptic vesicles. Nature 249, 770–771.

Kalina, M. and Rabinovitch, R. (1975) Exocytosis coupled to endocytosis of ferritin in parotid acinar cells from isoprenaline-stimulated rats. Cell Tissue Res. 163, 373—382.

Kanno, T., Cochrane, D. E. and Douglas, W. W. (1973) Exocytosis (secretory granule extrusion) induced by injection of calcium into mast cells. Can. J. Physiol. Pharmacol. 51, 1001–1004.

Kartenbeck, J., Franke, W. W. and Morré, D. J. (1977) Polygonal coat structures on secretory vesicles of rat hepatocytes. Cytobiologie 14, 284–291.

Kato, A. C., Katz, H. S. and Collier, B. (1974) Absence of adenine nucleotide release from autonomic ganglion. Nature 249, 576–577.

Katz, B. and Miledi, R. (1967) A study of synaptic transmission in the absence of nerve impulses. J. Physiol. (London) 192, 407–436.

Katz, B. and Miledi, R. (1969a) Tetrodotoxin-resistant electric activity in presynaptic terminals. J. Physiol. (London) 203, 459–487.

Katz, B. and Miledi, R. (1969b) Spontaneous and evoked activity of motor nerve endings in calcium Ringer. J. Physiol. (London) 203, 689–706.

Katz, B. and Miledi, R. (1970) Further study of the role of calcium in synaptic transmission. J. Physiol. (London) 207, 789–801.

Katz, B. and Miledi, R. (1977) Transmitter leakage from nerve endings Proc. Roy. Soc. London, B. 196, 59–72.

Katz, F. N., Rothman, J. E., Lingappa, V. R., Blobel, G. and Lodish, H. F. (1977) Membrane assembly in vitro: synthesis, glycosylation and asymmetric insertion of a transmembrane protein. Proc. Natl. Acad. Sci. U.S.A. 74, 3278–3282.

Katz, H. S., Salchmoghaddam, S. and Collier, B. (1973) The accumulation of radioactive acetylcholine by a sympathetic ganglion and by brain: failure to label endogenous stores. J. Neurochem. 20, 569–582.

Kawasaki, T. and Yamashina, I. (1971) Metabolic studies of rat liver plasma membrane using D-(1-^{14}C) glucosamine. Biochim. Biophys. Acta 225, 234–238.

Keenan, T. W. and Morré, J. D. (1969) Phospholipid class and fatty acid composition of Golgi apparatus isolated from rat liver and comparison with other cell fractions. Biochemistry 9, 19–25.

Keenan, T. W. and Morré, D. J. (1975) Glycosyltransferases: do they exist on the surface membrane of mammalian cells? FEBS Lett. 55, 8–13.

Kellems, R. E., Allison, V. F. and Butow, R. A. (1974) Cytoplasmic type 80S ribosomes associated with yeast mitochondria. II. Evidence for the association of cytoplasmic ribosomes with the outer mitochondrial membrane *in situ*. J. Biol. Chem. 249, 3297–3308.

Kelly, P. T. and Cotman, C. W. (1977) Identification of glycoproteins and proteins at synapses in the central nervous system. J. Biol. Chem. 252, 787–793.

Kemper, B., Habener, J. F., Rich, A. and Potts, J. T. (1975) Microtubules and intracellular conversion of proparathyroid hormone to parathyroid hormone. Endocrinology 96, 903–912.

Kirschner, N. (1974) Molecular organization of the chromaffin vesicles of the adrenal medulla. In: Cytopharmacology of Secretion (Ceccarelli, B., Clementi, F. and Meldolesi, J., eds.) pp. 265–272. Raven Press, New York.

Kirschner, N. and Viveros, O. H. (1972) The secretory cycle in the adrenal medulla. Pharmacol. Rev. 24, 385–456.

Koch, P. A., Gartrell, J. E., Gardner, F. H. and Carter, J. R. (1975) Biogenesis of erythrocyte membrane proteins. In vivo studies in anemic rabbits. Biochim. Biophys. Acta 389, 162–176.

Kondo, S. and Schulz, I. (1976) Ca^{++} fluxes in isolated cells of rat pancreas. Effect of secretagogues and different Ca^{++} concentrations. J. Membr. Biol. 29, 185–204.

König, P., Hörtnagl, H., Kostron, H., Sapinsky, H. and Winkler, H. (1976) The arrangement of dopamine β-hydroxylase (E.C.1.14.2.1.) and chromomembrin-B in the membrane of chromaffin granules. J. Neurochem. 27, 1539–1541.

Kopin, I. J., Breese, G. R., Krans, K. R. and Weise, V. R. (1968) Selective release of newly synthesized norepinephrine from the cat spleen during sympathetic stimulation. J. Pharm. Exp. Ther. 161, 271–278.

Kostron, H., Winkler, H., Peer, L. J. and König, P. (1977a) Uptake of adenosine triphosphate by isolated adrenal chromaffin granules: a carrier-mediated transport. Neuroscience 2, 159–166.

Kostron, H., Winkler, H., Geissler, D. and König, P. (1977b) Uptake of calcium by chromaffin granules in vitro. J. Neurochem. 28, 487–493.

Kraehnbuhl, J. P., Racine, L. and Jamieson, J. D. (1977) Immunocytochemical localization of secretory proteins in bovine pancreatic acinar cells. J. Cell Biol. 72, 406–423.

Kramer, M. F. and Poort, C. (1972) Unstimulated secretion of protein from rat exocrine pancreas cells. J. Cell Biol. 52, 147–158.

Kreibich, G. and Sabatini, D. D. (1974) Selective release of contents from microsome vesicles without membrane disassembly. II. Electrophoretic and immunological characterization of microsomal subfractions. J. Cell Biol. 61, 789–807.

Kreibich, G., Debey, P. and Sabatini, D. D. (1973) Selective release of content from microsomal vesicles without membrane disassembly. I. Permeability changes induced by low detergent concentrations. J. Cell Biol. 58, 436–462.

Kreibich, G., Hubbard, A. L. and Sabatini, D. D. (1974) On the spatial arrangement of the proteins in microsomal membranes from rat liver. J. Cell Biol. 60, 616–627.

Kreibich, G., Freinstein, C. M., Pereyra, B. N., Ulrich, B. L. and Sabatini, D. D. (1978a) Proteins of rough microsomal membranes related to ribosome binding. II. Crosslinking of bound ribosomes to specific membrane proteins exposed at the binding sites. J. Cell Biol. In press.

Kreibich, G., Ulrich, B. L. and Sabatini, D. D. (1978b) Proteins of rough microsomal membranes related to ribosome binding. I. Identification of ribophorin I and II, membrane proteins characteristic of rough microsomes. J. Cell Biol. In press.

Kremmer, T., Wisher, M. H. and Evans, W. H. (1976) The lipid composition of plasma membrane subfractions originating from the three major functional domains of the rat hepatocyte cell surface. Biochim. Biophys. Acta 455, 655–664.

Kreutzberg, G. W. and Schubert, P. (1975) The cellular dynamics of intraneuronal transport. In: The Use of Axonal Transport for Studies of Neuronal Connectivity (Cowan, M. W. and Cuenod, M., eds.) pp. 85–112. Elsevier/North-Holland, Amsterdam.

Krueger, B. K., Forn, J. and Greengard, P. (1977) Depolarization-induced phosphorylation of specific proteins, mediated by Ca-ion influx in rat brain synaptosomes. J. Biol. Chem. 252, 2764–2773.

Kruse, N. J. and Bornstein, P. (1975) Metabolic requirements for transcellular movement and secretion of collagen. J. Biol. Chem. 250, 4841–4847.

Kryger-Brevart, V., Weiss, D. G., Mehl, E., Schubert, P. and Kreutzberg, G. W. (1974) Maintenance of synaptic membrane by fast axonal flow. Brain Res. 77, 97–110.

Kuriyama, Y. and Luck, D. J. (1973) Membrane associated ribosomes in mitochondria of Neurospora crassa. J. Cell Biol. 59, 776–784.

Labrie, F., Lemaire, S., Poirier, G., Pelletier, G. and Boucher R. (1971) Adenohypophyseal secretory granules. I. Their phosphorylation and association with protein kinase. J. Biol. Chem. 246, 7311–7317.

Labrie, F., Pelletier, G., Lemay, A., Borgeat, P., Bardeu, N., Dupont, A., Savary, M., Coté, J. and Boucher, R. (1973) Control of protein synthesis in anterior pituitary gland. Acta Endocrinol. Suppl. 180, 301–340.

Lacy, P. E. (1975) Endocrine secretory mechanisms. Am. J. Pathol. 79, 170–188.

Lacy, P. E., Howell, S. L., Young, D. A. and Fink, C. J. (1968) New hypothesis of insulin secretion. Nature 219, 1177–1179.

Lacy, P. E., Finke, E. H. and Codilla, R. C. (1975) Cinemicrographic studies on granule movements in monolayer culture of islet cells. Lab. Invest. 33, 570–576.

Lagerkrantz, H. (1976) On the composition and function of large dense-core vesicles in sympathetic nerves. Neuroscience 1, 81–92.

Laitner, J. W., Sussman, K. E., Vatter, A. E. and Schnaider, F. H. (1975) Adenine nucleotides in the secretory granule fraction of rat islets. Endocrinol. 96, 662–677.

Lambert, M., Camus, J. and Christophe, J. (1974) Phosphorylation of protein components of isolated zymogen granule membranes from rat pancreas. FEBS Lett. 49, 228–232.

Lampel, M. and Kern, H. (1977) Acute interstitial pancreatitis in the rat induced by excessive doses of pancreatic secretagogue. Virch. Arch. Pathol. Anat. Histol. 373, 97–118.

Lande, M. A., Adesnik, M., Sumida, M., Tashiro, Y. and Sabatini, D. D. (1975) Direct association of messenger RNA with microsomal membranes in human diploid fibroblasts. J. Cell Biol. 65, 513–528.

Lawson, D., Raff, M. C., Gomperts, B., Fewtrell, C. and Gilula, N. B. (1977) Molecular events during membrane fusion. A study of exocytosis in rat peritoneal mast cells. J. Cell Biol. 72, 242–259.

Leduc, E. H., Avrameas, S. and Bouteille, M. (1968) Ultrastructural localization of antibody in differentiating plasma cells. J. Exp. Med. 127, 109–118.

Lemay, A., Labrie, F. and Drouin, D. (1974) Stability of secretory granules containing growth hormone and prolactin from bovine anterior pituitary gland. Can. J. Biochem. 52, 327–335.

Lentz, T. L., Rosenthal, J. and Maxarkiewicz, J. E. (1976) Cytochemical localization of acetylcholine receptors by means of peroxidase-labeled α-bungarotoxin. Neurosci. Abstr., 5th Ann. Meet. p. 127.

Leskes, A., Siekevitz, P. and Palade, G. E. (1971) Differentiation of endoplasmic reticulum in hepatocytes. I. Glucose-6-phosphatase distribution in situ. J. Cell Biol. 49, 264–287.

Lewis, D. S., McDonald, R. J., Kronquist, K. E. and Ronzio, R. A. (1977) Purification and partial characterization of integral membrane glycoprotein from zymogen granules of dog pancreas. FEBS Lett. 76, 115–120.

Liley, A. W. (1956) The effects of presynaptic polarization on the spontaneous activity of the mammalian neuromuscular junction. J. Physiol. (London) 134, 427–443.

Little, J. S. and Widnell, C. C. (1975) Evidence for the translocation of 5'-nucleotidase across hepatic membranes in vitro. Proc. Nat. Acad. Sci. U.S.A. 72, 4013–4017.

Livett, B. G. (1976) Axonal transport and neuronal dynamics: contributions to the study of neuronal connectivity. Int. Rev. Physiol., Neurophysiology VI, vol. 10 (Porter, R., ed.) pp. 37–124. University Park Press, Baltimore.

Locke, M. and Huie, P. (1976a) The beads in the Golgi complex-endoplasmic reticulum region. J. Cell Biol. 70, 384–394.

Locke, M. and Huie, P. (1976b) Vertebrate Golgi complexes have beads in a similar position to those found in arthropods. Tissue & Cell 8, 739–743.

Lodish, H. F. (1973) Biosynthesis of reticulocyte membrane proteins by membrane-free polysomes. Proc. Nat. Acad. Sci. U.S.A. 70, 1226–1230.

Lodish, H. F. and Small, B. (1975) Membrane proteins synthesized by rabbit reticulocytes. J. Cell Biol. 65, 51–64.

Lohmander, S., Mokalewsky, S., Madsen, K., Thyberg, J. and Friberg, U. (1976) Influence of colchicine on the synthesis and secretion of proteoglycans and collagen by fetal guinea pig chondrocytes. Exp. Cell Res. 99, 333–345.

Longenecker, H. E., Hurlbut, W. P., Mauro, A. and Clark, A. W. (1970) Effects of black widow spider venom on the frog neuromuscular junction. Effects on end plate potential, miniature end-plate potential and nerve terminal spike. Nature 225, 701–703.

Loor, F. (1976) Cell surface design. Nature 264, 272–273.

Lowe, D. and Hallinan, T. (1973) Preferential synthesis of a membrane-associated protein by free polyribosomes. Biochem. J. 136, 825–828.

Lucy, J. A. (1970) The fusion of biological membranes. Nature 227, 815–817.

Malaisse, W. J., Malaisse-Lagae, F., Van Obberghen, E., Somers, G., Davis, G., Ravazzola, M. and Orci, L. (1975) Role of microtubules in the phasic pattern of insulin release. Ann. N.Y. Acad. Sci. 253, 630–652.

Malaisse, W. J., Davis, G., Somers, G., Van, Obberghen E., Malaisse-Lagae, F., Ravazzola, M., Blondel, B. and Orci, L. (1976) The role of microtubules and microfilaments in secretion. In: Stimulus-Secretion Coupling in the Gastrointestinal Tract (Case, R. M. and Goebell, H., eds.) pp. 65–78, MTP, Lancaster, England.

Malaisse-Lagae, F., Ravazzola, M., Robberecht, P., Vandermeers, A., Malaisse, W. J. and Orci, L. (1975) Exocrine pancreas: evidence for topographic partition of exocrine function. Science 190, 795–797.

Marchbanks, R. M. (1975) The subcellular origin of acetylcholine released at synapses. Int. J. Biochem. 6, 303–312.

Marchbanks, R. M. and Israël, M. (1972) The heterogeneity of bound acetylcholine and synaptic vesicles. Biochem. J. 129, 1049–1061.

Margolis, R. K., Jaanus, S. D. and Margolis, R. U. (1973) Stimulation by acetylcholine of sulfated mucopolysaccharide release from the perfused cat adrenal gland. Mol. Pharmacol. 9, 590–594.

Margulies, N. H. and Michaels, A. (1974) Ribosomes bound to chloroplast membranes in Chlamydomonas reinhardii. J. Cell Biol. 60, 65–77.

Margulies, N. H., Tiffany, H. L. and Michaels, A. (1975) Vectorial discharge of nascent polypeptides attached to chloroplast thylakoid membranes. Biochem. Biophys. Res. Commun. 64, 735–739.

Marko, P. and Cuenod, M. (1973) Contribution of the nerve cell body to renewal of axonal and synaptic glycoproteins in the pigeon visual system. Brain Res. 62, 419–423.

Markov, D., Rambourg, A. and Droz, B. (1976) Smooth endoplasmic reticulum and fast axonal transport. An electron microscope radioautographic study of thick sections after heavy metal impregnation. J. Microsc. Biol. Cell 25, 57–60.

Martínez-Palomo, A., Pinto da Silva, P. and Chavez, B. (1976) Membrane structure of entamoeba histolytica: fine structure of freeze-fractured membranes. J. Ultrastr. Res. 54, 148–158.

Masur, S., Holtzman, E. and Walter, R. (1971) Hormone-stimulated exocytosis in the toad urinary bladder. Some implications for turnover of surface membranes. J. Cell Biol. 52, 211–228.

Mata, L. R. (1976) Dinamics of horseradish peroxidase absorption in the epithelial cells of the hamster seminal vesicle. J. Microsc. Biol. Cell 25, 127–132.

Matthews, E. K. and Nordmann, J. J. (1976) The synaptic vesicle. Calcium ion binding to vesicle membrane and its modification by drug action. Mol. Pharmacol. 12, 778–788.

Matthews, E. K., Petersen, O. H. and Williams, J. A. (1973) Pancreatic acinar cells: acetylcholine-induced membrane depolarization, calcium efflux and amylase release. J. Physiol. (London) 234, 689–701.

Maylié-Pfenninger, M. F., Palade, G. E. and Jamieson, J. D. (1975) Interaction of lectins with the surface of dispersed pancreatic acinar cells. J. Cell Biol. 67, 333a.

McDonald, R. J. and Ronzio, R. A. (1972) Comparative analysis of zymogen granule membrane polypeptides. Biochem. Biophys. Res. Commun. 49, 377–382.

McDonald, R. J. and Ronzio, R. A. (1974) Phosphorylation of a zymogen granule membrane polypeptide from rat pancreas. FEBS Lett. 40, 203–209.

McGregor, R. R., Hamilton, J. W. and Cohn, D. V. (1975) The bypass of tissue hormone stores during secretion of newly synthesized parathyroid hormone. Endocrinology 97, 178–188.

McIntosh, P. R. and O'Toole, K. (1976) The interaction of ribosomes and membranes in animal cells. Biochim. Biophys. Acta 457, 171–212.

McIntyre, J. A., Gilula, N. B., and Karnovsky, M. J. (1974) Cryoprotectant-induced redistribution of intramembranous particles in mouse lyphocytes. J. Cell Biol. 60, 192–204.

McKelvy, J. F. (1975) Phosphorylation of neurosecretory· granules by cAMP-stimulated protein kinase and its implication for transport and release of neurophysin protein. Ann N. Y. Acad. Sci. 248, 80–91.

McMahan, V. J. and Yee, A. (1975) Rapid uptake and release of horseradish peroxidase (HRP) in motor nerve terminals of the snake. Neurosci. Abstr. 5th Ann. Meet., p. 624.

Mechler, B. and Vassalli, P. (1975a) Membrane-bound ribosomes of myeloma cells I. Preparation of free and membrane-bound ribosomal fractions. Assessment of the methods and properties of the ribosomes. J. Cell Biol. 67, 1–15.

Mechler, B. and Vassalli, P. (1975b) Membrane-bound ribosomes in myeloma cells. III. The role of the messenger RNA and the nascent polypeptide chain in the binding of ribosomes to membranes. J. Cell Biol. 67, 25–37.

Melcher, U., Eidels, L. and Uhr, J. W. (1975) Are immunoglobulins integral membrane proteins? Nature 258, 434–435.

Meldolesi, J. (1974a) Dynamics of cytoplasmic membranes in guinea pig pancreatic acinar cells. I. Synthesis and turnover of membrane proteins. 61, 1–13.

Meldolesi, J. (1974b) Secretory mechanisms in pancreatic acinar cells. Role of cytoplasmic membranes. In: Cytopharmacology of Secretion (Ceccarelli, B., Clementi, F. and Meldolesi, J., eds.) pp. 71–86. Raven Press, New York.

Meldolesi, J. and Cova, D. (1971) Synthesis and interactions of cytoplasmic membranes in pancreatic exocrine cells. Biochem. Biophys. Res. Commun. 44, 139–143.

Meldolesi, J. and Cova, D. (1972) Composition of cellular membranes in the pancreas of the guinea pig. IV. Polyacrylamide gel electrophoresis and amino acid composition of membrane proteins. J. Cell Biol. 55, 1–18.

Meldolesi, J., Jamieson, J. D. and Palade, G. E. (1971a) Composition of cellular membranes in the pancreas of the guinea pig. I. Isolation of membrane fractions. J. Cell Biol. 49, 109–129.

Meldolesi, J., Jamieson, J. D. and Palade, G. E. (1971b) Composition of cellular membranes in the pancreas of the guinea pig. II. Lipids. J. Cell Biol. 49, 130–149.

Meldolesi, J., Jamieson, J. D. and Palade, G. E. (1971c) Composition of cellular membranes in the pancreas of the guinea pig. III. Enzymatic activities. J. Cell Biol. 49, 150–158.

Meldolesi, J., De Camilli, P. and Peluchetti, D. (1974) The membrane of secretory granules: structure, composition and turnover. In: Secretory Mechanisms in Exocrine Glands (Thorn, N. and Petersen, O. H., eds.) pp. 137–151. Munksgaard, Copenhagen.

Meldolesi, J., Ramellini, G. and Clemente, F. (1975) Calcium compartmentalization in pancreatic acinar cells. In: Calcium Transport in Contraction and Secretion (Carafoli, E., Clementi, F., Drabikowsky, W. and Margreth, A., eds.) pp. 157–166, Elsevier/North-Holland, Amsterdam.

Meldolesi, J., De Camilli, P. and Brenna, A. (1977) The topology of plasma membrane in pancreatic acinar cells. In: Hormonal Receptors and Digestive Tract Physiology (Inserm Symposium 3) (Bonfils, S., Fromageot, P. and Rosselin, G., eds.) pp. 203–212. Elsevier/North-Holland, Amsterdam.

Melmed, R. N., Benitez, C. and Holt, S. J. (1972) Intermediate cells of the pancreas. I. Ultrastructural characterization. J. Cell Sci. 11, 449–475.

Melmed, R. N., Turner, R. C. and Holt, S. J. (1973a) Intermediate cells of the pancreas. II. Effect of dietary soybean trypsin inhibitor on acinar cell structure and function in the rat. J. Cell Sci. 13, 279–295.

Melmed, R. N., Benitez, C. and Holt, S. J. (1973b) Intermediate cells of the pancreas. III. Selective autophagy and destruction of granules in intermediate cells of the rat pancreas induced by alloxan and streptozotocin. J. Cell Sci. 13, 297–315.

Meunier, F. M., Israël, M. and Lesbats, B. (1975) Release of ATP from stimulated nerve electroplaque junctions. Nature 257, 404–408.

Meyer, D. I. and Burger, M. M. (1976) The chromaffin granule surface. Localization of carbohydrate on the cytoplasmic surface of an intracellular organelle. Biochim. Biophys. Acta 443, 428–436.

Michell, R. H. (1975) Inositol phospholipids and cell surface receptor function. Biochim. Biophys. Acta 415, 81–142.

Miledi, R. (1973) Transmitter release induced by injection of calcium ions into nerve terminals. Proc. Roy. Soc. Ser. B 183, 421–425.

Milutinović, S., Argent, B. E., Schulz, I. and Sachs, G. (1977) Studies on isolated subcellular components of cat pancreas. II. A Ca^{++}-dependent interaction between membranes and zymogen granules of cat pancreas. J. Membr. Biol. 36, 281–295.

Model, P. G., Highstein, S. M. and Bennett, M. V. L. (1975) Depletion of vesicles and fatigue on transmission at a vertebrate central synapse. Brain Res. 98, 209–228.

Mok, W., Freienstein, C., Sabatini, D. D. and Kreibich, G. (1976) Specificity and control of binding of ribosomes to endoplasmic reticulum membranes. J. Cell Biol. 70, 393a.

Molenaar, P. C., Nickolson, V. J. and Polack, R. L. (1973) Preferential release of newly synthetized ^{3}H acetylcholine from rat cerebral cortex slices *in vitro*. Brit. J. Pharmacol. 47, 97–108.

Mollenhauer, H. H., Hass, B. S. and Morré, J. D. (1976) Membrane transformation in Golgi apparatus of rat spermatids. A role for thick cisternae and two classes of coated vesicles in acrosome formation. J. Microsc. Biol. Cell 27, 33–36.

Moore, L., Chen, T., Knapp, H. R. and London, E. J. (1975) Energy-dependent calcium sequestration activity in rat liver microsomes. J. Cell Biol. 250, 4562–4568.

Moore, P. L., Bank, H. L., Brissie, N. T. and Spicer, S. S. (1976) Association of microfilament bundles with lysosomes in polymorphonuclear leukocytes. J. Cell Biol. 71, 659–666.

Morgan, E. H. and Peters, T. (1971) Intracellular aspects of transferrin synthesis and secretion in the rat. J. Biol. Chem. 246, 3508–3511.

Morgan, I. G., Zanetta, J. G., Breckenridge, W. G., Vincendon, G. and Gombos, G. (1973) The chemical structure of synaptic membranes. Brain Res. 62, 405–411.

Morin, F., Tay, S. and Simpkins, H. (1972) A comparative study of the molecular structure of the plasma membranes and the smooth and rough endoplasmic reticulum membranes from rat liver. Biochem. J. 129, 781–788.

Morré, D. J. (1977) The Golgi apparatus and membrane biogenesis. In: Cell Surface Reviews, Vol. 4 (Poste, G. and Nicolson, G. L., eds.) pp. 1–83. Elsevier/North-Holland, Amsterdam.

Morré, D. J., Hamilton, R. L., Mollenhauer, H. H., Mahley, R. W., Cunningham, W. P., Cheetham, R. D. and Lequire, V. S. (1970) Isolation of a Golgi apparatus-rich fraction from rat liver. J. Cell Biol. 44, 484–491.

Morré, D. J., Keenan, T. W. and Mollenhauer, H. H. (1971) Golgi apparatus function in membrane transformations and product compartmentalization: studies with cell fractions isolated from rat liver. In: Advances in Cytopharmacology (Clementi, F. and Ceccarelli, B., eds.) vol. I, pp. 159–182, Raven Press, New York.

Morré, D. J., Keenan, T. W. and Huang, C. N. (1974) Membrane flow and differentiation: origin of Golgi apparatus membranes from endoplasmic reticulum. In: Cytopharmacology of Secretion (Ceccarelli, B., Clementi, F. and Meldolesi, J., eds.) pp. 107–126. Raven Press, New York.

Morris, B. J. and Johnson, C. I. (1976) Isolation of renin granules from rat kidney cortex and evidence for an inactive form of renin (prorenin) in granules and plasma. Endocrinology 98, 1466–1474.

Morris, J. F., Nordman, J. J. and Dyball, R. E. J. (1978) Neurohypophysial secretion. Int. Rev. Exp. Pathol. In press.

Morrison, T. G., Lodish, H. F. (1975) Site of synthesis of membrane and nonmembrane proteins of vesicular stomatitis virus. J. Cell Biol. 250, 6955–6962.

Muller, T. W. and Kirschner, N. (1975) ATPase and phosphatidylinositolkinase activities of adrenal chromaffin vesicles. J. Neurochem. 24, 1155–1162.

Muller, W. A., Amherdt, M. L., Vachon, C. and Renold, A. E. (1976) Phosphorylation reactions in islets of Langerhans. J. Physiol. (Paris) 72, 711–720.

Musick, J. and Hubbard, J. I. (1972) Protein release from mouse notor nerve terminals. Nature 237, 279–281.

Nagasawa, J. and Douglas, W. W. (1972) Thorium dioxide uptake into adrenal medullary cells and the problem of recapture of granule membrane following exocytosis. Brain Res. 37, 141–145.

Nagasawa, J., Douglas, W. W. and Schulz, R. (1971) Micropinocytotic origin of coated and smooth microvesicles (synaptic vesicles) in neurosecretory terminals of posterior pituitary glands demonstrated by incorporation of horseradish peroxidase. Nature 232, 341–342.

Negishi, M. and Omura, T. (1973) Turnover of enzyme proteins in microsomal membranes. Drug Met. Disp. 1, 80–83.

Negishi, M., Sawamura, T., Morimoto, T. and Tashiro, Y. (1975) Localization of nascent NADPH-cytochrome c reductase in rat liver microsomes. Biochim. Biophys. Acta 381, 215–220.

Negishi, M., Fijii-Kurihana, Y., Tashiro, Y. and Imai, Y. (1976) Site of biosynthesis of cytochrome P_{450} in hepatocytes of phenobarbital treated rats. Biochem. Biophys. Res. Commun. 71, 1153–1160.

Nevalainen, T. J. (1970) Effects of pilocarpine stimulation on rat pancreatic acinar cells. Acta Pathol. Microbiol. Scand. (Suppl.) 210.

Nicolson, G. L. and Singer, S. J. (1971) Ferritin-conjugated plant agglutinin as specific saccharide stains for electron microscopy: application to saccharides bound to cell membranes. Proc. Nat. Acad. Sci. U.S.A. 68, 942–945.

Nicolson, G. L. and Singer, S. J. (1974) The distribution and asymmetry of mammalian cell surface saccharides utilizing ferritin conjugated plant agglutinins as specific saccharide stains. Proc. Nat. Acad. Sci. U.S.A. 68, 942–945.

Nicolson, G. L., Poste, G. and Ji, T. H. (1977) The dynamics of cell membrane organization. In: Cell Surface Reviews, Vol. 3 (Poste, G. and Nicolson, G. L., eds.) pp. 1–73. Elsevier/North-Holland, Amsterdam.

Nielsen, S. P. and Petersen, O. H. (1972) Transport of calcium in the perfused submandibular gland of the cat. J. Physiol. (London) 233, 685–697.

Nilsson, R., Peterson, E. and Dallner, G. (1973) Permeability of microsomal membranes isolated from rat liver. J. Cell Biol. 56, 762–776.

Nishiyama, A. and Petersen, O. H. (1975) Pancreatic acinar cells; ionic dependence of acetylcholine induced membrane potential and resistance change. J. Physiol. (London) 244, 431–466.

Nordmann, J. J. and Morris, J. F. (1976) Membrane retrieval at neurosecretory axon endings. Nature 261, 723–725.

Nordmann, J. J., Dreifuss, J. J., Baker, P. F., Ravazzola, M., Malaisse-Lagae, F. and Orci, L. (1974) Secretion-dependent uptake of extracellular fluid by the rat neurohypophysis. Nature 250, 155–157.

Norström, A. (1974) The heterogeneity of the neurohypophysial pool of neurophysin. In: Neurosecretion, the Final Neuroendocrine Pathway (Knowles, F. G. W. and Vollrath, L., eds.) pp. 86–93, Springer-Verlag, Berlin.

Novikoff, A. B. (1976) Endoplasmic reticulum—cytochemist's view. Proc. Nat. Acad. Sci. U.S.A. 73, 2781–2787.

Novikoff, A. B. and Essner, E. (1962) Pathological changes in cytoplasmic organelles. Fed. Proc. 21, 1130–1142.

Novikoff, A. B. and Heus, M. (1963) A microsomal nucleoside diphosphatase. J. Biol. Chem. 238, 710–716.

Novikoff, A. B. and Novikoff, P. M. (1977) Cytochemical contribution to differentiating GERL from Golgi apparatus. Histochem. J. 9, 525–552.

Novikoff, P. M., Novikoff, A. B., Quintana, N. and How, J. J. (1971) Golgi apparatus, GERL and lysosomes of neurons in rat dorsal root ganglia studied by thick section and thin section cytochemistry. J. Cell Biol. 50, 859–886.

Novikoff, A. B., Novikoff, P. M., Ma, M., Shin, W. Y. and Quintana, N. (1974) Cytochemical studies of secretory and other granules associated with the endoplasmic reticulum in rat thyroid epithelial cells. In: Cytopharmacology of Secretion (Ceccarelli, B., Clementi, F. and Meldolesi, J., eds.) pp. 349–368. Raven Press, New York.

Novikoff, A. B., Yam, A. and Novikoff, P. M. (1975) Cytochemical studies of secretory process in transplantable insulinoma of Syrian golden hamster. Proc. Nat. Acad. Sci. U.S.A. 72, 1501–1505.

Novikoff, A. B. Mori, M. Quintana, N. and Yam, A. (1977) Study of the secretory process in the mammalian exocrine pancreas. J. Cell Biol. 75, 148–167.

Oliver, C. and Hand, A. R. (1975) The fate of luminally administered horseradish peroxidase in resting and stimulated rat parotid acinar cells. J. Cell Biol. 67, 316a.

Olsen, B. R. and Prockop, D. (1974) Ferritin-conjugated antibodies used for labeling of organelles involved in the cellular synthesis and transport of procollagen. Proc. Nat. Acad. Sci. U.S.A. 71, 2033–2037.

Olsson, I. (1969) Intracellular transport of glycosaminoglycans in human leukocytes. Exp. Cell Res. 54, 318–324.

Omura, T. (1973) Biogenesis of endoplasmic reticulum membranes in liver cells. Abstr. Ninth. Int. Congr. Biochem., Stockholm, p. 249.

Omura, T. and Kuriyama, Y. (1971) Role of rough and smooth microsomes in the biosynthesis of microsomal membranes. J. Biochem. 69, 651–658.

Orci, L. (1974) A portrait of the pancreatic B-cell. Diabetologia 10, 163–188.

Orci, L. and Perrelet, A. (1973) Membrane-associated particles: increase at sites of pinocytosis demonstrated by freeze-etching. Science 181, 868–869.

Orci, L. and Perrelet, A. (1975) Freeze-etch Histology. A Comparison between Thin Sections and Freeze-etch Replicas. Springer-Verlag, Berlin.

Orci, L., Stauffacher, W., Beaven, D., Lambert, A. E., Renold, A. E. and Rouiller, C. (1969) Ultrastructural events associated with the action of tolbutamide and glibenclamide on pancreatic B cell in vivo and in vitro. Acta Diabetol. Lat. 6 (suppl. 1) 271–299.

Orci, L., Stauffacher, W. E., Dulin, A. E., Renold, A. E. and Rouiller, C. (1970) Ultrastructural changes in A cells exposed to diabetic hyperglycemia. Observations made on pancreas of Chinese hamsters. Diabetologia 6, 199–206.

Orci, L., Malaisse-Lagae, F., Ravazzola, M., Amherdt, M. and Renold, A. E. (1973a) Insulin release by emiocytosis: a demonstration with freeze-etching technique. Science 179, 82–84.

Orci, L., Malaisse-Lagae, F., Ravazzola, M., Amherdt, M. and Renold, A. E. (1973b) Exocytosis-endocytosis coupling in the pancreatic B cell. Science 181, 561–562.

Orci, L., Le Marchand, Y., Singh, A., Assimacopoulos-Jannet, F., Rouiller, C. and Jeanrenaud, B. (1973c) Role of microtubules in lipoprotein secretion by the liver. Nature 244, 30–31.

Orci, L., Perrelet, A. and Friend, D. (1977) Freeze-fracture of membrane fusions during exocytosis in pancreatic B cells. J. Cell Biol. 75, 23–30.

O'Toole, K. (1974) Contamination of free ribosome fractions by detached previously membrane-bound ribosomes. Biochem. J. 305–307.

Ovtracht, L., Morré, D. J., Cheetham, R. D. and Mollenhauer, H. H. (1973) Subfractionation of Golgi apparatus from rat liver: method and morphology. J. Microsc. 18, 87–102.

Painter, G., Tokuyas, K. T. and Singer, S. J. (1973) Immunoferritin localization of intracellular antigens: the use of ultramicrotomy to obtain ultrathin sections suitable for direct immunoferritin staining. Proc. Nat. Acad. Sci. U.S.A. 70, 1649–1653.

Palade, G. E. (1955) A small particulate component of the cytoplasm. J. Biophys. Biochem. Cytol. 1, 59–68.

Palade, G. E. (1959) Functional changes in the structure of cell components. In: Subcellular particles (Hayashi, T., ed.) pp. 64–80. Ronald Press, New York.

Palade, G. E. (1975) Intracellular aspects of the process of protein secretion. Science 189, 347–358.

Pappas, G. D. and Rose, S. (1976) Localization of calcium deposits in the frog neuromuscular junction at rest and following stimulation. Brain Res. 103, 362–365.

Pearse, A. G. E. and Polak, J. M. (1974) Endocrine tumors of neural crest origin: neurolophomas, apudmas and the APUD concept. Med. Biol. 52, 3–22.

Pearse, B. M. F. (1976) Clathrin: a unique protein associated with intracellular transfer of membrane by coated vesicles. Proc. Nat. Acad. Sci. U.S.A. 73, 1225–1259.

Pelletier, G. (1973) Secretion and uptake of peroxidase by rat adenohypophyseal cells. J. Ultrastr. Res. 43, 445–459.

Pelletier, G. (1974) Autoradiographic studies of synthesis and intracellular migration of glycoproteins in rat anterior pituitary gland. J. Cell Biol. 62, 185–197.

Peper, K., Dreyer, F., Sandri, C., Akert, K. and Moore, H. (1974) Structure and ultrastructure of the frog motor endplate. A freeze-etching study. Cell Tissue Res. 149, 437–455.

Perri, V., Sacchi, O., Raviola, E. and Raviola, G. (1972) Evaluation of the number and distribution of

620

synaptic vesicles at cholinergic nerve endings after sustained stimulation. Brain Res. 39, 526–529.

Peters, T., Fleischer, B. and Fleischer, S. (1971) The biosynthesis of rat liver albumin. IV. Apparent passage of albumin through the Golgi apparatus during secretion. J. Biol. Chem. 246, 240–247.

Petersen, O. H. and Ueda, N. (1975) Ca^{++} control of pancreatic enzyme secretion. In: Calcium Transport in Contraction and Secretion. (Carafoli, E., Clementi, F., Drabikowski, W. and Margreth, A., eds.) pp. 147–156, Elsevier/North-Holland, Amsterdam.

Petersen, O. H. and Ueda, N. (1976) Pancreatic acinar cells: the role of calcium in stimulus-secretion coupling. J. Physiol. (London) 254, 583–606.

Peterson, C. (1974) Role of energy metabolism in histamine release. A study on isolated rat mast cells. Acta Physiol. Scand. 92 suppl. 413, 1–34.

Pfenninger, K. H. and Bunge, R. P. (1974) Freeze-fracturing of nerve growth cones and young fibers. A study of developing plasma membrane. J. Cell Biol. 63, 180–196.

Pfenninger, K., Akert, K., Moor, H. and Sandri, C. (1972) The fine structure of freeze-fracture presynaptic membranes. J. Neurocytol. 1, 129–149.

Pickering, B. T., Jones, C. W., Burford, C. D., McPherson, M. A., Swann, R. W., Heap, P. F. and Morris, J. F. (1975) The role of neurophysin proteins: suggestion from the study of their transport and turnover. Ann. N. Y. Acad. Sci. 248, 15–36.

Pickett, W. C., Jesse, R. L. and Cohen, P. (1977) Initiation of phospholipase A$_2$ activity in human platelets by the calcium ionophore A 23187. Biochim. Biophys. Acta 467, 209–221.

Pinto da Silva, P. and Nogueira, M. L. (1977) Membrane fusion during secretion. A hypothesis based on electron microscope observations in *Phytophthora* palmivora zoospores during encystment. J. Cell Biol. 73, 161–182.

Pipeleers, D. G., Pipeleers-Marichal, M. A. and Kipnis, D. M. (1976) Microtubule assembly and the intracellular transport of secretory granules in pancreatic islets. Science 191, 88–90.

Pisam, M. and Ripoche, P. (1976) Redistribution of surface macromolecules in dissociated epithelial cells. J. Cell Biol. 71, 907–920.

Plattner, H. and Fuchs, S. (1975) X-ray microanalysis of calcium binding sites in Paramecium with special reference to exocytosis. Histochemistry 45, 23–48.

Plattner, H., Miller, F. and Bachmann, L. (1973) Membrane specializations in the form of regular membrane-to-membrane attachment sites in Paramecium. A correlated freeze-etching and ultrathin-sectioning analysis. J. Cell. Sci. 13, 687–719.

Plattner, H., Reichel, K. and Matt, H. (1977) Bivalent cation-stimulated ATPase activity at preformed exocytosis sites in Paramecium coincides with membrane-intercalated particle aggregates. Nature 67, 702–704.

Pletscher, A., Da Prada, M., Berneis, K. H., Steffen, H. Lütold, B. and Weder, H. G. (1974) Molecular organization of amine-storage organelles of blood platelets and adrenal medulla. In: Cytopharmacology of Secretion (Ceccarelli, B., Clementi, F. and Meldolesi, J., eds.) pp. 257–264, Raven Press, New York.

Poisner, A. M. and Hava, M. (1970) The role of adenosine triphosphate and adenosine triphosphatase in the release of catecholamines from the adrenal medulla. IV. Adenosine triphosphate-activated uptake of calcium by microsomes and mitochondria. Mol. Pharmacol. 6, 407–415.

Politoff, A. L., Rose, S. and Pappas, G. D. (1974) The calcium binding sites of synaptic vesicles of the frog sartorius neuromuscular junction. J. Cell Biol. 61, 818–823.

Poole, A. R., Howell, J. I. and Lucy, J. A. (1970) Lysolecithin and cell fusion. Nature 227, 810–813.

Poste, G. and Allison, A. C. (1973) Membrane fusion. Biochim. Biophys. Acta 300, 421–465.

Puszkin, S., Schook, W., Puszkin, E., Rouault, C., Ores, C., Schlossberg, J., Kochwa, S. and Rosenfeld, R. E. (1977) Contractile elements in brain tissue: evidence of α-actinin in synaptosomes. In: Contractile Systems in Non-Muscle Tissues (Perry, S. V., Margreth, A. and Adelstein, R. S., eds.) pp. 67–80, Elsevier/North-Holland, Amsterdam.

Putzke, H. P. and Said, F. (1975) Different secretory responses of periinsular and other acini in the rat pancreas after pilocarpin injection. Cell Tissue Res. 161, 133–145.

Pysh, J. J. and Wiley, R. G. (1974) Synaptic vesicle depletion and recovery in cat sympathetic ganglia electrically simulated in vivo. J. Cell. Biol. 60, 365–374.

Raftell, M. and Blomberg, F. (1974) Enzyme polymorphism in rat liver microsomes and plasma membranes II. An immunochemical comparison of enzyme active antigens solubilized by detergents, papain or phospholipases. Eur. J. Biochem. 49, 31–41.

Ragnotti, G., Lawford, G. R. and Campbell, P. N. (1969) Biosynthesis of microsomal nicotinamide adenine dinucleotide phosphate-cytochrome c reductase by membrane-bound and free polysomes from rat liver. Biochem. J. 112, 139–147.

Rahaminoff, R. and Yaari, Y. (1973) Delayed release of transmitter at the frog neuromuscular junction. J. Physiol. (London) 228, 241–257.

Rahaminoff, R., Rahaminoff, H., Binah, O. and Meiri, U. (1975) Control of neurotransmitter release by calcium ions and the role of mitochondria. In: Calcium Transport in Contraction and Secretion (Carafoli, E., Clementi, F., Drabikowsky, W. and Margreth, A., eds.) pp. 253–260, Elsevier/North-Holland, Amsterdam.

Rambourg, A., Marraud, A. and Chretien, M. (1973) Tri-dimensional structure of the forming face of the Golgi apparatus as seen in high voltage electron microscope after osmium impregnation of the small nerve cells in the semiluminar ganglion of the trigeminal nerve. J. Microsc. 97, 49–57.

Rash, J. E. and Ellisman, M. H. (1974) Macromolecular specialization of the neuromuscular junctions and the non-junctional sarcolemma. J. Cell Biol. 63, 567–586.

Rasmussen, H. and Goodman, D. B. P. (1975) Calcium and cyclic AMP as interrelated intracellular messengers. Ann. N. Y. Acad. Sci. 253, 789–796.

Rasmussen, H. and Goodman, D. B. P. (1977) Relationship between calcium and cyclic nucleotides in cell activation. Physiol. Rev. 57, 421–509.

Ravazzola, M. (1976) Intracellular localization of Ca in chromaffin cells of the rat adrenal medulla. Endocrinology 98, 950–953.

Ravazzola, M., Malaisse-Lagae, F., Amherdt, M., Perrelet, A., Malaisse, W. J. and Orci, L. (1976) Patterns of calcium localization in pancreatic endocrine cells. J. Cell Sci. 21, 107–118.

Redman, C. M. and Cherian, M. G. (1972) The secretory pathway of rat liver glycoproteins and albumin. Localization of newly formed protein within the endoplasmic reticulum. J. Cell Biol. 52, 231–245.

Redman, C. M., Banerjee, D., Howell, K. and Palade, G. E. (1975) Colchicine inhibition of plasma protein release from rat hepatocytes. J. Cell Biol. 66, 42–59.

Reith, A., Oftebro, R. and Selielid, R. (1970) Incorporation of ^{3}H-glucosamine in HeLa cells as revealed by light and electron microscope radioautography. Exp. Cell Res. 59, 167–170.

Reith, E. J. (1976) The binding of calcium within the Golgi saccules of the rat odontoblasts. Am. J. Anat. 147, 267–272.

Riordan, J. R., Mitranic, M., Slavic, M. and Moscarello, M. A. (1974) The incorporation of 1- ^{14}C fucose into glycoprotein fraction of liver plasma membranes. FEBS Lett. 47, 248–251.

Rodriguez-Boulan, E. R., Kreibich, G. and Sabatini, D. D. (1975) Spatial localization of glycoprotein in microsomal membranes. Fed. Proc. 34, 582.

Rohlich, P. (1975) Membrane-associated actin filaments in the cortical cytoplasm of the rat mast cell. Exp. Cell Res. 93, 293–298.

Rolleston, F. S. (1974) Membrane-bound and free ribosomes. Subcell. Biochem. 3, 91–117.

Ronzio, R. A. and Mohrlok, S. H. (1977) A possible role of Golgi membrane-associated galactosyltransferases in the formation of zymogen granule glycoproteins. Arch. Biochem. Biophys. 181, 128–136.

Rossignol, B., Keryer, G., Herman, G., Chambaut-Guerin, A. M. and Cahoreau, G. Function of cholinergic and adrenergic receptors in the control of protein secretion in rat salivary glands. In: Hormonal Receptors in Digestive Tract Physiology (Bonfils, S., Fromageot, P. and Rosselin, G., eds.) pp. 311–323. Elsevier/North-Holland, Amsterdam.

Roth, S. and White, D. (1972) Intercellular contact and cell surface galactosyl transferase activity. Proc. Nat. Acad. Sci. U.S.A. 69, 485–489.

Roth, S., McGuire, E. J. and Roseman, S. (1971) Evidence for cell surface glycosyltransferases. Their potential role in cellular recognition. J. Cell Biol. 51, 536–547.

Rothman, S. S. (1975a) Enzyme secretion in the absence of zymogen granules. Am. J. Physiol. 228, 1828–1834.

Rothman, S. S. (1975b) Protein transport by the pancreas. Science 190, 747–753.

Rothman, S. S. (1976) Secretion of new digestive enzymes by pancreas with minimal transit time. Am. J. Physiol. 230, 1499–1503.

Rothman, S. S., Burwen, S. and Liebow, C. (1974) The zymogen granule: intragranular organization and functional significance. In: Cytopharmacology of Secretion (Ceccarelli, B., Clementi, F. and Meldolesi, J., eds.) pp. 341–348. Raven Press, New York.

Rotshenker, S., Erulkar, S. D. and Rahaminoff, R. (1976) Reduction in the frequency of miniature end-plate potentials by nerve stimulation in low calcium solutions. Brain Res. 101, 362–365.

Rubin, R. P. (1970) The role of calcium in the release of neurotransmitter substances and hormones. Pharmacol. Rev. 22, 389–428.

Rubin, R. P. (1974) Calcium and the Secretory Process. Plenum Press, New York.

Russel, J. T. and Thorn, N. A. (1975) Adenosine triphosphate-dependent calcium uptake by subcellular fractions from bovine neurohypophyses. Acta Physiol. Scand. 93, 364–377.

Sabatini, D. D., Tashiro, Y. and Palade, G. E. (1966) On the attachment of ribosomes to microsomal membranes J. Mol. Biol. 19, 503–524.

Sabatini, D. D., Ojakian, G., Lande, M. A., Lewis, J., Mok, W., Adesnik, M. and Kreibich, G. (1976) Structural and functional aspects of the protein synthesizing apparatus in the rough endoplasmic reticulum. In: Control Mechanisms in Development (Meints, R. H. and Davies, E., eds.) pp. 151–180. Plenum Press, New York.

Sachs, H. and Haller, E. W. (1968) Further studies on the neurohypophysis to release vasopressin. Endocrinology 83, 251–262.

Sargent, J. R. and Vadlamudi, B. P. (1968) Characterization and biosynthesis of cytochrome b_5 in rat liver microsomes. Biochem. J. 107, 839–849.

Satir, B. (1974a) Membrane events during the secretory process. In: Transport at the Cell Level (Duncan, J., ed.) pp. 399–418. Cambridge University Press, Cambridge.

Satir, B. (1974b) Ultrastructural aspects of membrane fusion. J. Supramol. Struct. 2, 529–537.

Satir, B. (1976) Genetic control of membrane mosaicism. J. Supramol. Struct. 5, 381–389.

Satir, B., Schooley, C. and Satir, P. (1972) Membrane reorganization during secretion. Nature 235, 53–54.

Satir, B., Schooley, C. and Satir, P. (1973) Membrane fusion in a model system: mucocyst secretion in Tetrahymena. J. Cell Biol. 56, 153–176.

Sauer, L. A. and Burrow, G. N. (1972) The submicrosomal distribution of radioactive proteins released by puromycin from the bound ribosomes of rat liver microsomes labelled in vitro. Biochim. Biophys. Acta 277, 179–187.

Scarpa, A. and Carafoli, E., eds. (1978) Calcium Transport and Cell Function. Ann. N.Y. Acad. Sci. In press.

Schacher, S., Holtzman, E. and Hood, D. C. (1976) Synaptic activity of frog retinal photoreceptors. A peroxidase uptake study. J. Cell Biol. 70, 178–192.

Schachter, H. (1974) The subcellular sites of glycosylation. Biochem. Soc. Symp. 40, 50–71.

Scharder, P. and Frerichs, H. (1976) Cytochalasin B. Evidence for a membrane-directed action on B cells from rat pancreatic islets. Pfl. Arch. Eur. J. Physiol. 364, 95–98.

Schimke, R. T. (1975) Turnover of membrane proteins in animal cells. In: Methods in Membrane Biology (Korn, E. D., ed.) vol. III, pp. 201–236. Plenum Press, New York.

Schneider, Y. J., Tulkens, P. and Trouet, A. (1977) Recycling of fibroblast plasma membrane antigens internalized during endocytosis. Trans. Biochem. Soc. 5, 1164–1166.

Schramm, M. and Selinger, Z. (1975) The functions of cyclic AMP and calcium as alternative second messengers in parotid gland and pancreas. J. Cyclic Nucl. Res. 1, 181–192.

Schramm, M., Selinger, Z., Salomon, Y., Eytan, E. and Batzri, S. (1972) Secretory granule pseudopodia: pseudopodia formation by secretory granules. Nature New Biol. 240, 203–205.

Schultz, G. and Hardman, J. G. (1975) Regulation of cyclic GMP in the ductus deferens of the rat. In: Advances in Cyclic Nucleotide Research (Greengard, P. and Robison, G. A., eds.) vol. 5, pp. 339–355, Raven Press, New York.

Segrest, J. P., Kahane, I., Jackson, R. L. and Marchesi, V. T. (1973) Major glycoprotein of the human erythrocyte membrane: evidence for an amphipathic molecular structure. Arch. Biochem. Biophys. 155, 167–183.

Selinger, Z., Naim, E. and Lasser, M. (1970) ATP-dependent calcium uptake by microsomal preparations from rat parotid and submaxillary glands. Biochim. Biophys. Acta 137, 326–334.

Serk-Hanssen, G. and Christiansen, E. N. (1973) Uptake of calcium in chromaffin granules of bovine adrenal medulla stimulated *in vitro*. Biochim. Biophys. Acta 307, 404–412.

Sharoni, Y., Eimerl, S. and Schramm, M. (1976) Secretion of old versus new exportable protein in rat parotid slices. Control by neurotransmitters. J. Cell Biol. 71, 107–122.

Sharp, G. W. G., Wollheim, C., Muller, W. A., Gutzeit, A., Truehart, P. A., Blondel, B., Orci, L. and Renold, A. E. (1975) Studies on the mechanisms of insulin release. Fed. Proc. 34, 1537–1548.

Sheele, G. A. (1975) Two dimensional analysis of soluble proteins. Characterization of guinea pig exocrine pancreatic proteins. J. Biol. Chem. 250, 5375–5385.

Sheele, G. A. and Palade, G. E. (1975) Studies on the guinea pig pancreas. Parallel discharge of exocrine enzyme activities. J. Biol. Chem. 250, 2660–2670.

Sheele, G. A., Palade, G. E. and Tartakoff, A. M. (1978) Cell fractionation studies of guinea pig pancreas. Molecular redistribution of exocrine proteins during tissue homogenization. J. Cell Biol. In press.

Shimoni, Y., Alnaes, E. and Rahaminoff, R. (1977) Is hyperosmotic neurosecretion from motor endings a calcium-dependent phenomenon? Nature 267, 170–172.

Shore, G. C. and Tata, J. R. (1977) Functions for polyribosome-membrane interactions in protein synthesis. Biochim. Biophys. Acta 472, 197–236.

Shur, B. D. and Roth, S. (1975) Cell surface glycosyltransferases. Biochim. Biophys. Acta 415, 473–512.

Siekevitz, P. (1972) Biological membranes: the dynamics of their organization. Ann. Rev. Physiol. 34, 117–140.

Silinski, E. M. (1975) On the association between transmitter secretion and the release of adenine nucleotides from mammalian motor nerve terminals. J. Physiol (London) 247, 145–162.

Simionescu, N., Simionescu, M. and Palade, G. E. (1974) Morphometric data on the endothelium of blood capillaries. J. Cell Biol. 60, 128–152.

Singer, S. J. and Nicolson, G. L. (1972) The fluid mosaic model of the structure of cell membranes. Science 175, 720–731.

Slaby, F. and Bryan, J. (1976) High uptake of myo-inositol by rat pancreatic tissue *in vitro* stimulates secretion. J. Biol. Chem. 251, 5078–5086.

Slotkin, T. A. and Kirschner, N. (1973) Recovery of rat adrenal amine stores after insulin administration. Mol. Pharmacol. 9, 105–116.

Smith, A. D. (1973) Cellular control of uptake, storage and release of noradrenaline in sympathetic nerves. Biochem. Soc. Symp. 36, 103–131.

Smith, A. D., de Potter, W. P., Moerman, E. J. and de Schaepdryver, A. F. (1971) Release of dopamine-β-hydroxylase and chromogranin A upon stimulation of the splenic nerve. Tissue Cell 2, 547–568.

Smith, R. E. and Farquhar, M. G. (1966) Lysosome function in the regulation of the secretory process in cells of the anterior pituitary gland. J. Cell Biol. 31, 319–347.

Smith, U., Smith, D. S., Winkler, H. and Ryan, J. W. (1973a) Exocytosis in the adrenal medulla demonstrated by freeze-etching, Science 179, 79–82.

Smith, U., Ryan, J. W. and Smith, D. S. (1973b) Freeze-etch studies of the plasma membrane of pulmonary endothelial cells. J. Cell Biol. 56, 492–499.

Soifer, D., ed. (1975) The Biology of Cytoplasmic Microtubules. Ann. N. Y. Acad. Sci. 253.

Spicer, S. S., Hardin, J. H. and Greene, W. B. (1968) Nuclear precipitates in pyroantimonate-osmium tetroxide-fixed tissues. J. Cell Biol. 39, 216–221.

Stachura, M. E. and Frohman, L. A. (1975) Growth hormone: independent release of big and small forms from rat pituitary *in vitro*. Science 187, 447–449.

Stadler, J. and Franke, W. W. (1974) Characterization of the colchicine binding of membrane fractions from rat and mouse liver. J. Cell Biol. 60, 297–303.

Stahelin, L. A. (1973) Structure and function of intercellular junctions. Int. Rev. Cytol. 39, 191–283.

Stein, O. and Stein, Y. (1973) Colchicine-induced inhibition of very low density lipoprotein release by rat liver in vivo. Biochim. Biophys. Acta 306, 142–147.

624

Steiner, D. F., Kemmler, W., Tager, H. S. and Rubinstein, A. H. (1974) Molecular events taking place during intracellular transport of exportable proteins. The conversion of peptide hormone precursors. In: Cytopharmacology of Secretion (Ceccarelli, B., Clementi, F. and Meldolesi, J., eds.) pp. 195–205. Raven Press, New York.

Steinman, R. M., Scott, E. B. and Cohn, Z. A. (1976) Membrane flow during pinocytosis. A stereological analysis. J. Cell Biol. 68, 665–687.

Stoeckel, M. E., Hinderlag-Gertner, C., Dellmann, H. D., Porte, A. and Stutinsky, F. (1975) Subcellular distribution of calcium in mouse hypophysis I. Calcium distribution in the adeno- and neurohypophysis under normal conditions. Cell Tissue Res. 157, 307–322.

Svensson, H., Dallner, G. and Ernster, L. (1972) Investigation on the specificity in membrane breakage occurring during sonication of rough microsomal membranes. Biochim. Biophys. Acta 274, 447–461.

Svensson, H., Elhammer Å., Autuori, F. and Dallner, G. (1976) Biogenesis of microsomal membrane glycoproteins in rat liver. IV. Characteristics of a cytoplasmic lipoprotein having properties of a membrane precursor. Biochim. Biophys. Acta 455, 383–398.

Swann, R. W. and Pickering, B. T. (1976) Incorporation of radioactive precursors into the membrane and contents of the neurosecretory granules of the rat neurohypophysis as a method of studying their fate. J. Endocrin. 68, 95–108.

Swearingen, K. C. (1971) Heterogeneous turnover of adenohypophysial prolactin. Endocrinology 89, 1350–1388.

Tartakoff, A. M. and Scheele, G. (1978) Improvement of the resolution of cell fractionation. J. Cell Biol. Submitted for publication.

Tartakoff, A. M. and Vassalli, P. (1978) Plasma cell immunoglobulin secretion: arrest is accompanied by alterations of the Golgi complex. J. Exp. Med. 146, 1332–1345.

Tartakoff, A. M., Greene, L. J. and Palade, G. E. (1974) Studies on the guinea pig pancreas. Fractionation and partial characterization of exocrine proteins. J. Biol. Chem. 249, 7420–7432.

Teichberg, S. and Bloom, D. (1976) Uptake and fate of horseradish peroxidase in axons and terminals of sympathetic neurons. J. Cell Biol. 70, 285a.

Teichberg, S., Holtzman, E., Crain, S. M. and Peterson, E. R. (1975) Circulation and turnover of synaptic vesicle membranes in cultured fetal mammalian spinal cord neurons. J. Cell Biol. 67, 215–230.

Theodosis, D. T., Dreifuss, J. J., Harris, M. C. and Orci, L. (1976) Secretion-related uptake of horseradish peroxidase in neurohypophysial axons. J. Cell Biol. 70, 294–303.

Thiéry, G. and Rambourg, A. (1976) A new staining technique for studying thick sections in the electron microscope. J. Microsc. Biol. Cell. 26, 103–106.

Thoa, N. B., Wooten, G. F., Axelrod, J. and Kopin, I. J. (1975) On the mechanism of release of norephinephrine from sympathetic nerves induced by depolarizing agents and sympathomimetic drugs. Mol. Pharmacol. 11, 10–18.

Thomas, P. E., Lu, A. Y. H., Ryan, D., West, S. B., Kowalek, J. and Levin, W. (1976) Immunochemical evidence of six forms of rat liver cytochrome P450 obtained using antibodies against purified rat liver cytochromes P450 and P448. Mol. Pharmacol. 12, 746–758.

Thorn, N. A., Russel, J. J. and Vilhardt, H. (1975) Hexosamine, calcium and nemophysin in secretory granules and the role of calcium in hormone release. Ann. Y. Y. Acad. Sci. 248, 202–217.

Tilney, L. G. and Mooseker, M. S. (1976) Actin filament membrane attachment: are membrane particles involved? J. Cell Biol. 71, 402–416.

Tixier-Vidal, A., Picart, R. and Moreau, M. F. (1976) Endocytosis and secretion in cultured anterior pituitary cells. Effect of hypothalamus hormones. J. Microsc. Biol. Cell. 25, 159–172.

Toda, G., Oka, T., Oda, T. and Ikada, Y. (1975) Subfractionation of rat liver plasma membranes. Uneven distribution of plasma membrane-bound enzymes on the liver cell surface. Biochim. Biophys. Acta 413, 52–64.

Toneguzzo, F. and Ghosh, H. P. (1977) Synthesis and glycosylation *in vitro* of glycoprotein of vesicular stomatitis virus. Proc. Natl. Acad. Sci. U.S.A. 74, 1516–1520.

Tranzer, J. P. (1972) A new amine storing compartment in adrenergic axons. Nature New Biol. 237, 57–58.

Trifaro, J. M. and Dworkind, J. (1971) Phosphorylation of membrane components of adrenal chromaffin granules by adenosine triphosphate. Mol. Pharmacol. 7, 52–65.

Trifaro, J. M., Collier, B., Lastowecka, A. and Stern, D. (1972) Inhibition by colchicine and by vinblastine of acetylcholine-induced catecholamine release from an adrenal gland: an anticholinergic action, not an effect upon microtubules. Mol. Pharmacol. 8, 264–272.

Trifaro, J. M., Duer, A. C. and Pinto, J. E. B. (1976) Membranes of the adrenal medulla: a comparison between the membrane of the Golgi apparatus and chromaffin granules. Mol. Pharmacol. 12, 536–545.

Turner, R. T. and Harris, A. B. (1973) Ultrastructure of synaptic vesicle formation using electron dense protein tracer technique. Nature 242, 57–59.

Tweto, J. and Doyle, D. (1977) Turnover of proteins of the eukaryotic cell surface. In: Cell Surface Reviews Vol. 4 (Poste, G. and Nicolson, G. L., eds.) pp. 137–164. Elsevier/North-Holland, Amsterdam.

Ukena, T. E., Borysenko, J. Z., Karnovsky, M. J. and Berlin, R. D. (1974) Effects of colchicine, cytochalasin B and 2-deoxyglucose on the topographical organization of surface-bound concanavalin A in normal and transformed fibroblasts. J. Cell Biol. 61, 70–82.

Unanue, E. R. and Schreiner, F. (1977) Structure and function of surface immunoglobulin of lynphocytes. In: Cell Surface Reviews, Vol. 3 (Poste, G. and Nicolson, G. L., eds.) pp. 619–641. Elsevier/North-Holland, Amsterdam.

Uvnäs, B. (1974a) The molecular basis for the storage and release of histamine in rat mast cell granules. Life Sci. 14, 2355–2366.

Uvnäs, B. (1974b) Histamine storage and release. Fed. Proc. 33, 2172–2176.

Van Golde, L. M. G., Raber, J., Batenburg, J. J., Fleischer, B., Zambrano, F. and Fleischer, S. (1974) Biosynthesis of lipids in Golgi complex and other subcellular fractions from rat liver. Biochim. Biophys. Acta 360, 179–192.

Van Norden, S., Polak, J. M. and Pearse, A. G. E. (1977) Single cell origin of somatostatin and calcitonin in the rat thyroid gland. Histochemistry 53, 243–248.

Vilhardt, H. and Hølmer, G. (1972) Lipid composition of membranes of secretory granules and plasma membranes from bovine neurohypophysis. Acta Endocrinol. 71, 638–648.

Vilhardt, H., Baker, R. V. and Hope, D. P. (1975) Isolation and protein composition of membranes of neurosecretory vesicles and plasma membranes from the neural lobe of the bovine pituitary gland. Biochem. J. 148, 57–65.

Vitetta, E. S. and Uhr, J. W. (1972) Cell surface immunoglobulin. V. Release of cell surface immunoglobulin from murine spleen lymphocytes. J. Exp. Med. 136, 676–694.

Vitetta, E. S. and Uhr, J. W. (1974) Cell surface immunoglobulin. IX. A new method for the study of synthesis, intracellular transport and exteriorization in murine splenocytes. J. Exp. Med. 139, 1599–1620.

Vitetta, E. S. and Uhr, J. W. (1975) Immunoglobulins and alloantigens on the surface of lymphoid cells. Biochim. Biophys. Acta 415, 253–271.

Völkl, A., Bieger, W. and Kern, H. F. (1976) Studies on secretory glycoproteins in the rat exocrine pancrease. I. Fine structure of the Golgi complex and release of fucose-labeled proteins after in vivo stimulation with caerulein. Cell Tissue Res. 175, 227–243.

Wallach, D. and Schramm, M. (1970) Calcium and exportable protein in rat parotid gland. Parallel subcellular distribution and concomitant secretion. Eur. J. Biochem. 21, 433–437.

Wallach, D., Kirschner, N. and Schramm, M. (1975) Non-parallel transport of membrane proteins and content proteins during assembly of the secretory granule in rat parotid gland. Biochim. Biophys. Acta 375, 87–105.

Weinstock, M. and Leblond, C. P. (1974) Synthesis, migration and release of precursor collagen by odontoblasts as visualized by radioautography after ^{3}H-proline administration. J. Cell Biol. 60, 92–127.

Weiss, R. L., Goodenough, D. A. and Goodenough, U. W. (1977) Membrane differentiations at sites specialized for cell fusion. J. Cell Biol. 72, 144–166.

Wernet, D., Vitetta, E. S., Uhr, J. W. and Boyse, E. A. (1973) Synthesis, intracellular distribution and secretion of immunoglobulin and H-2 antigen in murine splenocytes. J. Exp. Med. 138, 847–857.

626

Westrum, L. E. and Gray, E. G. (1976) Microtubules and membrane specializations. Brain Res. 105, 547–550.

White, D. A. and Hawthorne, J. N. (1970) Zymogen secretion and phospholipid metabolism. Phospholipids of zymogen granules. Biochem. J. 120, 533–542.

Whittaker, V. P., Dowdall, M. J., Dowe, J., Facino, G. H. and Scotto, J. (1974) Proteins of cholinergic synaptic vesicles from the electric organ of Torpedo: characterization of a low molecular weight acidic protein. Brain Res. 75, 115–131.

Wibo, M., Godelaine, D. Amar-Costesec, A. and Beaufay, H. (1975) Subcellular location of glycosyltransferases in rat liver. In: The Liver (Preisig, R., Bircher, J. and Paumgartner, G., eds.) pp. 70–83. Editio Cantor Aulendorf, Germany.

Widlund, L., Karlsson, K. A., Wintur, A. and Heilbron, E. (1974) Immunochemical studies on cholinergic synaptic vesicles. J. Neurochem. 22, 455–457.

Widnell, C. C. (1972) Cytochemical localizaton of 5′nucleotidase in subcellular fractions isolated from rat liver. I. The origin of 5′nucleotidase activity in microsomes. J. Cell Biol. 52, 542–557.

Winkler, H. (1976) The composition of adrenal chromaffin granules: an assessment of controversial results. Neuroscience 1, 65–80.

Winkler, H. (1977) The biogenesis of adrenal chromaffin granule. Neuroscience 2, 657–684.

Winkler, H., Schöpf, J. A. L., Hörtnagl, H. and Hörtnagl, H. (1972) Bovine adrenal medulla: subcellular distribution of newly synthesized catecholamines, nucleotides and chromogranins. N. S. Arch. Pharmacol. 273, 43–61.

Winkler, H., Schneider, F. H., Rufener, C., Nakane, P. K. and Hörtnagl, H. (1974) Membranes of the adrenal medulla: their role in exocytosis. In: Cytopharmacology of Secretion (Ceccarelli, B., Clementi, F. and Meldolesi, J., eds.) pp. 127–139. Raven Press, New York.

Winquist, L., Eriksson, L. C. and Dallner, G. (1974) Binding of concanavalin A-sepharose to glycoproteins of liver microsomal membranes. FEBS Lett. 42, 27–31.

Wirtz, K. W. A. (1974) Transfer of phospholipids between membranes. Biochim. Biophys. Acta 344, 95–117.

Wisher, M. H. and Evans, W. H. (1975) Functional polarity of the rat hepatocyte surface membrane. Biochem. J. 146, 375–388.

Woodin, A. M. and Wieneke, A. A. (1964) The participation of calcium, adenosine triphosphate and adenosine triphosphatase in the extrusion of the granule proteins from polymorphonuclear leukocytes. Biochem. J. 90, 498–509.

Wunderlich, F., Müller, R. and Speth, U. (1973) Direct evidence for a colchicine-induced impairment in the mobility of membrane components. Science 182, 1136–1138.

Yahara, I. and Edelman, G. M. (1975) Modulation of lymphocyte receptor mobility by concanavalin A and colchicine. Ann. N. Y. Acad. Sci. 253, 455–469.

Yamamoto, T. (1963) On the thickness of the unit membrane. J. Cell Biol. 17, 413–422.

Young, R. W. (1973) The role of the Golgi complex in sulfate metabolism. J. Cell Biol. 57, 175–189.

Yu, A. and Cohen, E. P. (1974) Studies on the effect of specific antisera on the metabolism of cellular antigens. II. The synthesis and degradation of TL antigens of mouse cells in the presence of TL antiserum. J. Immunol. 112, 1296–1307.

Zagury, D., Uhr, J., Jamieson, J. D. and Palade, G. E. (1970) Immunoglobulin synthesis and secretion. II. Radioautographic studies of sites of addition of carbohydrate moieties and intracellular transport. J. Cell Biol. 46, 52–63.

Zambrano, F., Fleischer, S. and Fleischer, B. (1975) Lipid composition of the Golgi apparatus of rat kidney and liver in comparison with other subcellular fractions. Biochim. Biophys. Acta 380, 357–369.

Zanini, A., Giannattasio, G. and Meldolesi, J. (1974) Studies on in vitro synthesis and secretion of growth hormone and prolactin. II. Evidence against the existence of precursor molecules. Endocrinology 94, 104–111.

Zenner, H. P. and Pfeuffer, T. (1976) Microtubular proteins in pigeon erythrocyte membranes. Eur. J. Biochem. 71, 177–184.

Zimmermann, H. and Denston, C. R. (1976) Adenosine triphosphate in cholinergic vesicles isolated from the electric organ of electrophorus electricus. Brain Res. 111, 365–376.

Zimmermann, H. and Denston, C. R. (1977a) Recycling of synaptic vesicles in the cholinergic synapses of the torpedo electric organ during induced transmitter release. Neuroscience 2, 695–714.

Zimmermann, E. and Denston, C. R. (1977b) Separation of synaptic vesicles of different functional states from the cholinergic synapses of Torpedo electric organ. Neuroscience 2, 715–730.

Zimmermann, H. and Whittaker, V. P. (1974) Effect of electrical stimulation on the yield and composition of synaptic vesicles from the cholinergic synapses of the electric organ of torpedo: a combined biochemical, electrophysiological and morphological study. J. Neurochem. 22, 435–450.

Ultrastructural aspects of exocytotic membrane fusion

12

Lelio ORCI and Alain PERRELET

Contents

G. Poste & G. L. Nicolson (eds.) Membrane Fusion, pp. 629–656.
© *Elsevier/North-Holland Biomedical Press, 1978.*

1. Introduction

Intracellular membrane fusion, including exocytosis, can be considered a by-product of the compartmentalization of the eukaryotic cell cytoplasm by membranes (review, Palade, 1975). Compartmentalization determines a series of membrane-bound spaces in which various substances can be segregated. Consequently, if substances present in two distinct compartments are to mix without being exposed to the cytosol, it can only be done by fusion of their respective enclosing membranes, so that the inner spaces can communicate while at any given time preserving membrane continuity. The reverse mechanism, namely the separation of a single membrane-bound space into two distinct compartments occurs by membrane fission. Membrane fusion and membrane fission allow the intracellular traffic of membrane-bound material (together with that of the respective enclosing membranes) including, in secretory cells, the release of material to the extracellular space. The latter event, known as exocytosis, occurs when the membrane that limits secretory granules fuses with the plasma membrane, a fusion which opens the inside of the granule to the outside of the cell. The ultrastructure of exocytotic membrane fusion has received considerable attention in the last five years from several groups of workers using the freeze-fracture technique to study the morphology of the interacting membranes. The result of such investigation has been the detection of two clear-cut classes of ultrastructural images related to exocytosis. In one class, exemplified by exocytotic membrane fusion in protozoans, the morphological counterparts of membrane proteins detected by freeze-fracture (i.e., intramembranous particles) appear to be directly involved in fusion, whereas in the other class, represented by exocytotic fusion in mammalian secretory cells, lack of involvement of morphologically detectable proteins seems to be the general rule.

   This review has not been conceived as an extensive survey of the literature. Instead, it is an analysis of a limited number of papers that were judged to best convey morphological evidence and concepts underlying exocytotic membrane fusion.

2. Technical approaches to the study of membrane fusion

2.1. Thin sectioning

The importance of membrane fusion as a mechanism whereby a segregated secretory product is released at the cell surface by the merger of the cytoplasmic membrane enclosing the product with the plasma membrane was recognized early (Palade, 1959). However, not until 1968 (Palade and Bruns, 1968) was any serious attempt made to examine membrane fusion (and membrane fission) in the perspective of membrane ultrastructure as seen in thin sections (i.e., the trilaminar "unit" membrane; Robertson 1959, 1964). Using optimal sectioning conditions (sections thinner than 50 nm) and contrast (uranyl acetate and lead staining), which permitted the three layers of the membrane to be clearly resolved, Palade and Bruns (1968) showed that fusion between membrane-limited cytoplasmic vesicles and the plasma membrane in endothelial cells consisted of the following morphological steps: (1) formation of a five-layered diaphragm, which represents the very close apposition of the three-layered vesicle and plasma membranes; (2) reduction of this five-layered diaphragm to a three-layered one by layer elimination (the outer layer of the vesicle membrane and the inner layer of the plasma membrane); (3) formation of a single-layered diaphragm by further layer elimination (the inner layer of the vesicle membrane and the outer layer of the plasma membrane). In this model, the single-layered diaphragm was interpreted to be residual membrane proteins from the eliminated layers rather than an extraneous membrane coat of the endothelial cell; (4) finally, the single-layered diaphragm is eliminated, removing the last barrier between the inside of the vesicle and the extracellular space. At this step, the vesicle either flattens to become an integral part of the surface membrane (operationally, membrane fusion can be considered as the addition of two membranes) or reinvaginates and begins the reverse process of membrane fusion, membrane fission. During the latter process, the invaginating part of the surface membrane progressively acquires a neck connecting the forming vesicle with the outside of the cell via a narrowing stoma. The neck eventually pinches off, resulting in a free cytoplasmic vesicle limited by invaginated surface membrane (operationally membrane fission thus represents the subtraction of one segment of membrane from another).

This careful study provided the important concept of layer elimination during membrane fusion, but left unresolved the question of whether the fusion mechanism involved any chemical or functional differences between the two fusing membranes. Subsequent studies, using conventional thin-section electron microscopy to study membrane fusion in secretory cells, have confirmed the presence of a five-layered (pentalaminar) diaphragm formed by the apposition of the granule-limiting membrane and the plasma membrane as the first morphological event in fusion (Lagunoff, 1973; Satir et al., 1973; Chi et al., 1976; Tandler and Poulsen, 1976; Lawson et al., 1977), which is then followed by a three-layered diaphragm (Tandler and Poulsen, 1976; Pinto da Silva and Nogueira, 1977), though the next step of a single-layered diaphragm was not detected in these studies (except by Chi et al., 1976). Moreover, the uncertainty indicated by Palade and Bruns concerning possible chemical or functional differences between the two fusing membranes remained. In summary, the deceptive simplicity of membrane structure seen in thin sections renders this technique of little use in the further study of membrane fusion if the membrane is considered primarily as a bilayer. However, thin sectioning remains pertinent in exploring structural changes in the "greater membrane" (Revel and Ito, 1967) or "cell boundary" (Orci et al., 1973a) during fusion, that is, the membrane bilayer with its associated polysaccharide coat on the outside (Bennett, 1963; Rambourg, 1971) and microfilamentous network on the inside (Wessels et al., 1971). The usefulness of this approach is illustrated by studies of the distribution of labeled lectin-binding sites on the outer surface of the plasma membrane during its fusion with secretory granules (Lawson et al., 1977) and the detection of microfilaments linking the outer leaflet of secretory granules and the inner leaflet of the plasma membrane in prefusion stages of exocytosis (Malaisse et al., 1972; Franke et al., 1976).

2.2. Freeze-fracture

Given the limitations of the thin-sectioning technique in the study of membrane fine structure, the application of freeze-fracturing to the problem of membrane fusion was undertaken with considerable optimism that it would reveal structural rearrangements in the fusing membranes. As will be detailed in the following sections, these expectations have been fulfilled to a reasonable extent. Freeze-fracturing has the unique property of exposing the internal structure of membranes and permitting—in contrast to thin sectioning—two of the main biochemical components (proteins and phospholipids) to be assigned morphological characteristics and thus be identified within the membrane. Freeze-fracturing splits membranes in the middle of the bilayer to produce two distinct images of the inside of the bilayer: one of these corresponds to a face view of the inner (or protoplasmic) leaflet of the membrane while the other represents a face view of the outer (or exoplasmic) leaflet (Branton, 1966). Conventionally, the protoplasmic leaflet of the plasma membrane has been called a P-fracture face, or P face and the exoplasmic leaflet the E-fracture face, or E face (Branton et al., 1975). In a cytoplasmic vacuole (i.e., secretory granule), the outer leaflet of the limiting membrane (in contact with the cytoplasm) is the protoplasmic P face, and the inner leaflet of the membrane (in contact with the vesicular lumen) is the exoplasmic E face. Morphologically, both fracture faces appear to be smooth regions interrupted by a variable number of globular protrusions. Work with artificial lipid vesicles (Vail et al., 1974) and reconstitution experiments using extracts of biological membranes (Yu and Branton, 1966) have shown that the smooth regions in fracture faces of both artificial and natural membranes correspond to the phospholipid areas of the membrane while the globular protrusions, called intramembranous particles, represent membrane proteins. A wide variation exists in the number of these particles present in fracture faces of different membranes (Branton, 1969; McNutt, 1977), but for a given membrane, the P face is generally richer in particles than the complementary E face. This basic freeze-fracture appearance of membranes, together with considerations of the biochemistry and functional behavior of some of their identifiable components, led Singer and Nicolson (1972) to propose a widely accepted model of the probable arrangement and specific relationships between proteins and lipids, the so-called fluid mosaic model of membrane structure. Specifically, the model suggests that phospholipids are in a "fluid" state at 37°C, permitting lateral (two-dimensional) movement of those protein subunits embedded within the phospholipid matrix. This model correlates well with the freeze-fracture appearance of cellular membranes and the alterations in membrane morphology caused by exocytotic fusion reviewed in this chapter will be interpreted in terms of membrane fluidity and topographic rearrangement of membrane components. As indicated in the introduction, the use of freeze-fracture technique to study membrane fusion has yielded interesting findings in a wide range of different cells, enabling the morphological changes seen in this phenomenon to be classified into two general categories which are described in detail in the following sections.

3. Protozoan membrane fusion

In a paper that aroused considerable interest, since the data presented could be interpreted in terms of the fluid mosaic model of membrane structure, Satir and co-workers (1973) described a specific pattern of membrane organization at the presumptive sites of exocytotic membrane fusion in *Tetrahymena pyriformis*. In this organism, the plasma membrane (P face) contains a circular array of large intramembranous particles (a *rosette*) at sites directly overlying secretory granules (mucocysts). At the tip of mucocysts facing the plasma membrane, the limiting membrane, (P face) contains a corresponding pattern of concentric rows (5–7) of intramembranous particles (an *annulus*). These two differentiations are present in their respective membranes before fusion actually takes place, and after fusion has occurred they delimit the lips of a so-called fusion pocket. This latter structure gradually flattens with the incorporation of the mucocyst membrane into the plasma membrane, progressively separating the large particles of the rosette which eventually become lost in the background particles of the plasma membrane. Although such images strongly suggest a direct involvement of membrane proteins (at least those detectable as intramembranous particles) in the fusion process, and similar sequences have been found in other protozoan species (Plattner, 1976; also see chapter by Allen, this volume), speculations that a rosette mechanism might also occur in exocytosis in mammalian secretory cells did not materialize.

4. Membrane fusion in mammalian secretory cells

Following the initial interesting observations in protozoans, exocytotic membrane fusion has been studied mainly by freeze-fracture technique in a variety of

634

mammalian secretory cells, in both endocrine (B cells of the islet of Langerhans, secretory cells of the neurohypophysis)[1] and exocrine (acinar pancreatic cell) tissues, in single cells where fusion is truly exocytotic (i.e., mast cells, spermatozoa), and also in situations where fusion occurs intracellularly but still represents a form of exocytosis since the internal space of the fused intracellular granules are connected to an exoplasmic space (phagocytic vacuole in the polymorphonuclear leukocyte). Although the main conclusions reached in these studies are essentially similar, each of these systems is discussed separately, with special emphasis on the B cell of the islet of Langerhans. In all of the systems studied to date secretion can be induced in the constitutive cells by various stimuli, greatly increasing the number of exocytotic events which can be studied.

4.1. Mast cells

As seen in freeze-fracture replicas, resting mast cells show two main membrane compartments; one is represented by the plasma membrane, the other by intracellular membranes surrounding the secretory granules. Fracture faces of these two membrane types show a typical random distribution of 6 to 12 nm intramembranous particles with more particles in the P face than in the E face in the plasma membrane, but with approximately the same number in both faces of the secretory granule membrane. In addition, the plasma membranes of both resting and stimulated mast cells have occasional clusters of particles; one type of cluster may contain up to 150 (12 nm) particles, the other being formed by a heterogeneous population of particles (7-12 nm). The latter type of cluster is usually associated with a depression in the membrane or with the base of a microvillus (Chi et al., 1975, 1976; Lawson et al., 1977). Chi and co-workers (1976) studied exocytosis in mast cells stimulated with polymyxin B and found that within seconds after exposure of rat peritoneal mast cells to the drug, bulges form on the surface of the cell. Each bulge appears to overlie a secretory granule and the fracture faces of such bulges (P face), in contrast to nonbulge areas of the plasma membrane, are devoid of intramembranous particles. In some instances there is an increased concentration of particles at the margin of the most prominent, particle-free bulges. Concomitant changes also occur in the fracture faces of secretory granules which normally have an evenly distributed population of particles. In regions of the granule membrane which are closely related to the plasma membrane (probably a stage just preceding fusion), particle-free patches can also be detected. Thin sections of similar areas in stimulated cells reveal either pentalaminar fusions between the granule membranes and the plasma membrane (Lagunoff, 1973; Chi et al., 1976) or a peculiar membrane structure formed by a single dark line with fuzzy material present on both sides (Chi et al.,

---

[1]Another endocrine cell in which exocytosis has been studied with freeze-fracture is the chromaffin cell of the adrenal medulla (Smith et al., 1973b). We have not included this system in the present review since the morphological distinction between exocytotic and endocytotic events is not always evident in the published illustrations.

1976). The fact that particle-free bulges are frequently surrounded by a particle-rich margin in the P face of the plasma membrane suggests that the depletion of particles in the bulges is due to a lateral displacement of these elements rather than their removal from the bulging membrane. This observation was confirmed and extended in a subsequent study of mast cell secretion by Lawson and co-workers (1977). These authors stimulated mast cell secretion by ionophore A23187 or ovalbumin, and examined the topographic distribution of various lectin- and immunoglobulin-binding sites at the cell surface using appropriate ferritin-coupled ligands. In thin-sectioned material the outer surface of regions of the plasma membrane interacting with underlying granule membranes (i.e., involved in a pentalaminar fusion) were devoid of ligand binding sites, which instead were accumulated at the margin of the interacting zone, as were the intramembranous particles seen in freeze-fractured material. Lawson and co-workers (1977) confirmed the absence of intramembranous particles from both fracture faces (P and E) of the interacting plasma and granule membranes and also noted the formation of membrane blebs during degranulation. Blebs appear to be shed from the plasma membrane, but at the time of release the latter retain the collar of intramembranous particles present at the base of each bleb. One possible interpretation of these blebs is that upon pinching off (membrane fission) they expose the content of the secretory granule to the extracellular space and, concomitantly, allow the removal of excess membrane lipid resulting from the fusion process while conserving proteins. As will be seen below, loss of membrane following secretion may be a peculiarity of the single cell status of mast cells since in secretory cells in organized tissues (e.g., pancreas, neurohypophysis, etc.) endocytotic retrieval of secretory granule membranes seems to be the rule after exocytosis (see chapter by Meldolesi et al., this volume). In summary, the mast cell model shows convincingly that the interaction of secretory granules with plasma membrane before and during fusion involves areas of the respective membranes that are depleted of morphologically detectable proteins in the form of intramembranous particles or ligand-binding sites. This sytem also reveals that the changes in the distribution of intramembranous particles are reversible, since in fully open exocytotic channels the distribution of particles is again random (Chi'et al., 1976). Finally, these observations suggest that intramembranous particles do not participate at any stage of membrane fusion. However, as outlined below in a discussion of other systems, the mast cell model is unclear about the mechanism(s) by which fusing membranes mutually recognize one another. It does not explain the mechanisms(s) responsible for particle redistribution (microfilaments?) and offers no insight into events during fusion per se, namely, the opening of the granule to the extracellular space and the accompanying release of the granule content.

4.2. Polymorphonuclear leukocytes

Membrane fusion in polymorphonuclear leukocytes represents a special case of exocytosis, since the fusion of most secretory granule membranes takes place with internalized portions of the plasma membrane (phagocytic vacuoles) rather than with the surface plasma membrane. Another difference found in this cell type concerns the freeze-fracture appearance of the secretory granules. These are of two kinds, azurophilic and specific, differing morphologically in size and shape. The E-fracture faces of both granules have more particles than the corresponding P faces, while the E-fracture faces have similar numbers of particles to the P face of the plasma membrane (Baggiolini et al., 1976). This is unusual, since in most secretory cells studied so far granule membranes contain far less particles than the plasma membrane, as for example, the pancreatic acinar cell (De Camilli et al., 1976), the neurohypophyseal secretory cell (Theodosis et al., 1978) and the B cell of the endocrine pancreas (Orci, 1974). The richly particulated fracture faces of leukocyte secretory granule membranes makes it easy to detect any change in density or distribution occurring during fusion with a phagocytic vacuole. To study exocytosis, polymorphonuclear leukocytes are first challenged with yeast cells to induce phagocytosis, after which fusion of azurophilic and specific granules with the phagocytic vacuoles can be studied by freeze-fracture (Amherdt et al., 1978). As seen in Fig. 1, when a secretory granule is closely applied to a phagocytic vacuole, a well-delimited area devoid of particles appears on the apposed membranes of both organelles. This suggests a clearing of particles from the membranes prior to fusion. No peculiar particle arrangement is observed in any other segment of the membranes involved. Thus the polymorphonuclear leukocyte model again indicates that exocytotic fusion takes place between protein-depleted lipid bilayers.

4.3. Spermatozoa

The acrosome reaction in the spermatozoon is the mechanism whereby the enzymatic content of the acrosome, a large secretory granule packaged in the Golgi region early in spermatozoon development, is released to help the sperm penetrate the layers surrounding the oocyte (see chapter by Bedford and Cooper, this volume). Similar to other processes that permit the discharge of membrane-contained secretory material, the acrosome reaction occurs when the vesicle-limiting membrane (i.e., the acrosomal membrane) fuses with the overlying plasma membrane. The acrosome reaction can be induced in vitro by adding calcium to spermatozoa incubated in a calcium-free, buffered salt solution (Fig. 2). In such a system studied by the freeze-fracture technique, one can observe the following sequence in suitably capacitated cells. Before fusion of the outer acrosomal and plasma membranes, fibrillar intramembranous particles become deleted from the E faces of both membranes (Fig. 3). In addition, there is a clearance of globular particles from limited areas of the P-fracture face of the plasma membrane (Fig. 4). After membrane fusion and consecutive loss of the initially sharp demarcation between the P faces of the plasma and inner acrosomal membranes ("hybrid membrane"), a second series of particle-free areas emerges in the plasma membrane of the postacrosomal region. These clearances may represent an adaptation of the spermatozoon plasma membrane in anticipa-

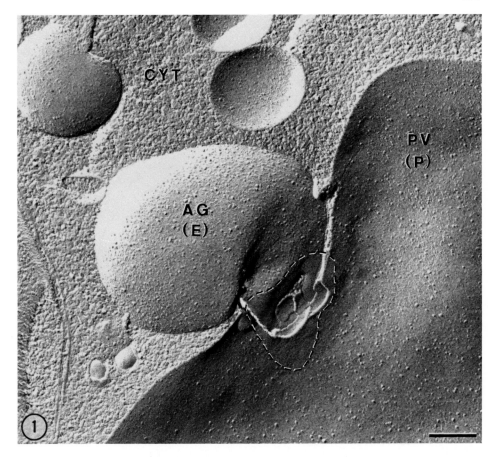

Fig. 1. Rabbit polymorphonuclear leukocyte stimulated to ingest killed yeast cells. This freeze-fracture replica shows part of the cross-fractured cytoplasm (CYT) containing convex AG(E) or concave roundish membrane profiles and part of the membrane delimiting a large phagocytic vacuole. The profile labeled AG(E) represents the E face of an azurophilic granule membrane which, at one of its poles, is in close relationship to the P face of the phagocytic vacuole membrane, PV(P). The closeness of the two membranes is indicated by the presence of a very shallow ridge between them and a patch devoid of intramembranous particles expands on both sides of the ridge. The focal deprivation of particles that are otherwise randomly scattered on the remainder of respective fracture faces is interpreted as a characteristic of incipient fusion between the specific granule and the phagocytic vacuole membranes. Bar = 0.2 μm. ×65,000.

tion of subsequent fusion with the plasma membrane of the oocyte. Further indication that the fusion process takes place in particle-free areas of the membrane can be gained by inducing the acrosome reaction in a medium containing liposomes. Before the addition of calcium, liposomes of varying size are detectable in the vicinity of the plasma membrane in the equatorial region. When calcium is added, however, completely smooth fracture faces of unilamellar liposomes can

638

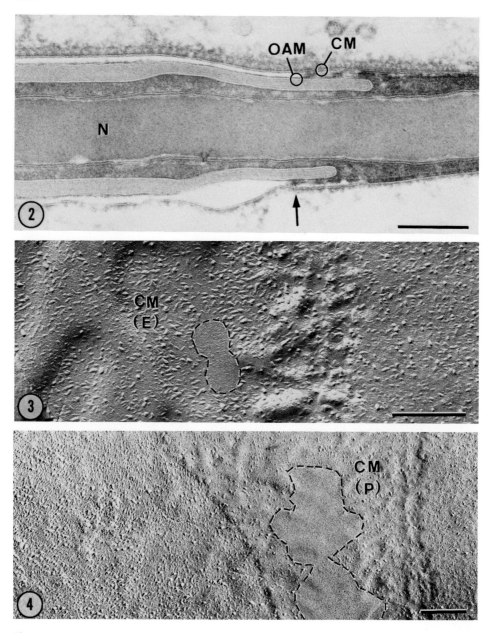

Figs. 2–4. Thin section and freeze-fracture replicas of guinea pig spermatozoa during acrosomal reaction induced in vitro by calcium. In all figures, the tip of head (not visible) is to the left. (Reproduced with permission from Friend et al., 1977.) Detailed legends are given on the opposite page.

be detected attached to or continuous with the hybrid plasma membrane at the point of fusion and in the secondarily cleared postacrosomal region. This may be interpreted as indicating preferential binding of the protein-free liposomes to similarly protein-depleted areas of the plasma membrane, a binding which eventually culminates in fusion of the apposed membranes and total mixing of liposomal and membrane lipids (Friend et al., 1977).

4.4. Acinar cell of the exocrine pancreas

The acinar cell provided the original model in which exocytotic fusion was discovered as the mechanism allowing membrane-bound secretory material to be released to the extracellular space (Palade, 1959). It also yielded convincing thin-section images supporting the layer elimination process in permitting the secretory granule membrane to be incorporated into the plasma membrane following the initial stage of pentalaminar fusion (Palade, 1975; also section 2.1). Although to our knowledge there have not been any publications showing clear-cut redistribution of intramembranous particles at prospective fusion sites in acinar cells[2] (cf. with sections 4.1., 4.3., and 4.6.), the study of this system by freeze-fracture has nonetheless produced interesting data concerning secretory granule and plasma membrane interactions following fusion (De Camilli et al., 1976). The acinar cell shows a distinctive cytoplasmic polarity reflected in the plasma membrane lining each side of the cell. The plasma membrane on lateral and basal poles has a great number of intramembranous particles in its P-fracture face, while the P face at the lumenal pole contains far less particles. The membrane of the secretory granules has still fewer particles than the lumenal plasma membrane. In resting acinar cells, the latter membrane shows a random distribution of particles which is perturbed when exocytotic release is stimulated by isoproterenol: low particulate membrane patches appear in the original lumenal membrane. This mosaic organization of the lumenal plasma membrane in stimu-

---

[2]However, De Camilli et al. (1976) mention the absence of "rosettes, annuli or localized modifications of particle distribution" in the plasma membrane of stimulated acinar cells (also see chapter by Meldolesi et al., this volume).

Fig. 2. Thin section of a sagittally cut spermatozoon showing the equatorial segment. The sperm plasma membrane (CM), with its prominent coat (glycocalyx), appears to be bridged to the outer part of the acrosomal membrane (OAM) by an interrupted dense matrix. Fusion between these two membranes usually takes place in front (arrow) of the compact matrix. Specimen fixed in the presence of tannic acid. N, sperm cell nucleus; Bar = 0.2 μm. ×94,000.

Fig. 3. Freeze-fracture replica of the sperm cell membrane during the acrosomal reaction, showing the formation of a particle-free area (delimited by dashed line) in the E-fracture face of the plasma membrane, CM(E). Note the characteristic fibrillar shape of intramembranous particles outside the depleted area. Bar = 0.2 μm. ×100,000.

Fig. 4. Same as Fig. 3, but showing progressive enlargement (by confluence) of the particle-free area (delimited by dashed line) in the P-fracture face of the plasma membrane, CM(P). Fibrillar particles are also present outside the smooth zone. Bar = 0.2 μm. ×60,000.

lated cells is interpreted as resulting from the addition of poorly particulate secretory granule membrane to the lumenal membrane and the conservation of the identity of the respective membranes after they have fused. The increase in lumenal surface produced by the addition of granule membrane is compensated for by a membrane internalization process (not by shedding the excess membrane; see 4.1.) that seems to involve only the region of low particle density of the mosaic lumenal membrane (i.e., secretory granule membrane). This lends further support to the interpretation given above that upon fusion respective fusing membranes retain their structural identity and do not mix their molecular components (De Camilli et al., 1976). Cytoskeletal elements (microfilaments) have been implicated as possibly being responsible for the maintenance of such membrane mosaicism. It must be noted, however, that observations to date concerning selective removal of membrane patches following exocytotic fusion are restricted to the acinar cell.[3] One reason may be that, in contrast to other systems in which secretory granules fuse individually with the plasma membrane, the acinar cell has the unique characteristic of releasing secretory material by so-called "compound exocytosis" in which "chains" of secretory granules fuse to one another and to the lumenal membrane, resulting in very large, and thus readily identifiable, pockets of infolded membrane (Ichikawa, 1965; Palade, 1975). Another reason is that membrane retrieval after exocytosis has been studied mainly in thin-section preparations using the internalization of electron-dense extracellular markers such as horseradish peroxidase (Orci et al., 1973b; Pelletier, 1973; Theodosis et al., 1976). This technique gives no clue, however, to the identity of the internalizing membrane segments, though future use of ligand binding such as that described earlier for the mast cell model (section 4.1.) should allow interesting correlations to be made with freeze-fracture data (also see the chapter by Meldolesi et al., this volume).

4.5. Secretory cells of the neurohypophysis

Neurohypophyseal cells can be stimulated to release their secretory granules containing oxytocin and vasopressin by various experimental conditions (e.g., electrical stimulation, water deprivation, high potassium concentration). Under stimulatory conditions, thin-section preparations show occasional exocytotic images where continuity between the plasma membrane of the neurosecretory terminal and the granule membrane can be seen in which the granule core is exposed to the extracellular space (Nagasawa et al., 1970; Theodosis et al., 1976). Although these images are infrequent, that exocytosis represents the major mechanism of hormone release in neurohypophyseal cells can be inferred from the fact that stimulation clearly increases endocytotic uptake of surface membrane (Nagasawa et al., 1971; Nordmann et al., 1974; Theodosis et al., 1976; (cf. with section 4.4.). Naturally this system has attracted interest as to the freeze-

[3]With the exception of neurohypophyseal cells (section 4.5), in which membrane to be internalized is characterized by clusters of large particles.

fracture appearance of the respective membranes involved, and the results obtained to date are summarized below (Theodosis et al., 1978) (as in the pancreatic acinar cell, the data obtained pertain mostly to freeze-fracture modifications of membranes after fusion has actually taken place). Images suggesting exocytosis can be detected on both fracture faces of neurosecretory terminal plasma membranes. On the P face these images consist essentially of depressions, often containing adhering granule core material (Fig. 5) and usually free of intramembranous particles, while elsewhere on the membrane the particles are evenly distributed on the fracture face (Fig. 5). Presumably, earlier stages of exocytosis (before fusion) of the kind described in mast cells (see section 4.1) also exist in the neurohypophysis in the form of bulges in the axolemmal P face which are devoid of intramembranous particles (Dempsey et al., 1973; Theodosis et al., 1978). On E faces, completed exocytotic events produce mirror images of those seen on P faces, though the loss of particles in protruding areas of the fracture face is somewhat less impressive because this face normally has a sparse population of particles. Quantitative evaluation reveals that the number of completed exocytotic events (P-face de-

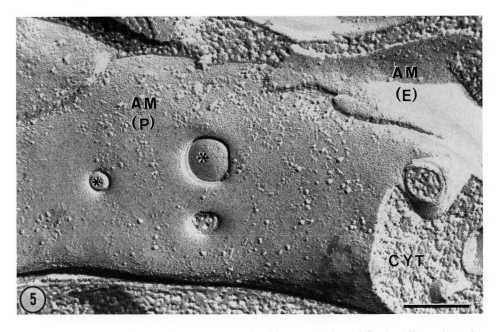

Fig. 5. Dormouse neurohypophysis during stimulated hormone release following hibernation. The freeze-fracture replica shows a longitudinally-fractured neurosecretory axon terminal with part of the cytoplasm (CYT) and of the plasma membrane P face, AM(P), exposed. Part of the plasma membrane E face of a neighboring axon, AM(E), is also seen. On the axonal P face, three completed exocytotic events are observed in face view. Each is surrounded by a wide patch of particle-free membrane. Protruding granule cores are labeled with asterisks. Bar = 0.2 μm. ×92,000. (Unpublished material courtesy of D. Theodosis, Institute of Histology and Embryology, Geneva, Switzerland.)

642

pression with protruding granule core) more than doubles following water deprivation, and is accompanied by a significant decrease in the number of intramembranous particles in both the P and E faces of the plasma membrane (a possible interpretation of the latter data is given in section 4.6). When the secretory granule membrane (E face) is visualized abutting P-fracture faces of the plasma membrane (or, respectively, granule P face abutting axolemmal E faces), a random distribution of intramembranous particles is observed in the granule membrane. It is also noted that particles on the P face of the granule membrane tend to be larger (12 nm) than most particles on the P face of the axolemma. An additional, and potentially significant, finding made in this system is that the P face of neurosecretory terminals contains clusters formed by 6 to 12, large (12 nm) intramembranous particles. The size of these particles contrasts with that of the nonclustered particles (approximately 9 nm), as does their location in frequent association with small depressions of the fracture face (Fig. 6). These clusters, originally believed to be counterparts of the protozoan rosette discussed in section 3, and thus perhaps involved in exocytosis (Dreifuss et al., 1976), can now be reinterpreted as indicative of endocytosis. The latter conclusion is based on two arguments: (1) the number of large particle

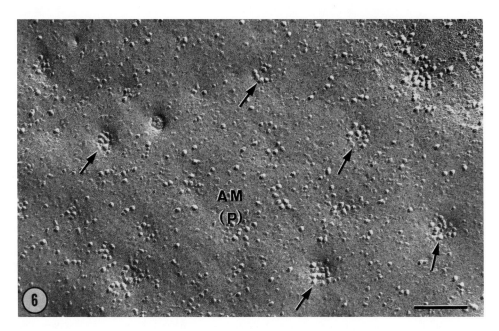

Fig. 6. Freeze-fracture replica of the P face, AM(P), of a neurosecretory axon terminal during stimulation by water deprivation in the rat. The fracture face contains several clusters (arrows) of large particles often associated with a focal depression in the membrane. These clusters probably mark zones of incipient endocytotic activity. Bar = 0.2 μm. ×72,000. (Unpublished material courtesy of D. Theodosis, Institute of Histology and Embryology, Geneva, Switzerland.)

643

clusters associated with small P-face depressions increases significantly following secretory stimulation, a finding in keeping with the known increase of compensatory endocytosis after stimulated hormone release (Nagasawa et al., 1971; Nordmann et al., 1974; Theodosis et al., 1976); and (2) cross-fractured axon terminals contain membrane profiles, usually tubular or cup-shaped, as large or larger than neurosecretory granules whose P faces also contain large (12nm) particles (cup-shaped vacuoles are labeled with peroxidase in experiments tracing membrane retrieval by conventional thin-section electron microscopy [Theodosis et al., 1976]). Although it is not possible to rule out completely that large particles preexist in the membrane and cluster as a result of membrane perturbation following sustained exocytotic release, the evidence favors the view that large particles in the plasma membrane derive from the incorporated secretory granule membrane and that precisely these areas are reinternalized by compensatory endocytosis. Thus the neurohypophysis model again provides evidence for particle clearing at sites of incipient exocytotic membrane fusion and also offers some clues to which parts of the membrane are retrieved following exocytosis (cf. with section 4.4.).

4.6. B cell of the islet of Langerhans

The secretory product of the B cell, insulin, is contained in specific membrane-limited cytoplasmic granules which can be induced to fuse with the plasma membrane by glucose stimulation (Lacy, 1970). In the rat, B cells represent approximately 80% of the total islet cell content, and by means of a mild collagenase digestion it is possible to isolate islets of Langerhans from surrounding exocrine tissue, permitting pure fractions of islets to be obtained for morphological studies. Before fixation and processing for electron microscopy, isolated islets may be kept for variable periods of time in vitro and submitted to different experimental procedures (e.g., incubation in high glucose concentration to stimulate insulin release from B cells). In thin sections of isolated islets stimulated by glucose, B cells (identified from other islet cell types such as A or D cells by the particular shape and electron density of the core of their secretory granules) present typical images of exocytosis. Namely, the continuity of the B cell plasma membrane with the secretory granule membrane can be demonstrated, together with exposure of the core of the granule to the extracellular space (Fig. 7). Earlier stages (e.g., pentalaminar fusion) are rarely documented[4] and, as amply emphasized in the preceding sections, conventional electron microscopy discloses no significant difference between the two membranes involved in fusion. Again

---

[4]Data in the literature indicate that pentalaminar fusions are best observed in cells where secretory granules have large diameters (i.e., large curvature radius), thus offering an extensive area of membrane for interaction with the plasma membrane. In contrast, the small diameter granules (small curvature radius) of the B cell present only narrow interacting zones, and this may explain the rarity of clear-cut pentalaminar images in such cells (and in other cells containing small diameter granules).

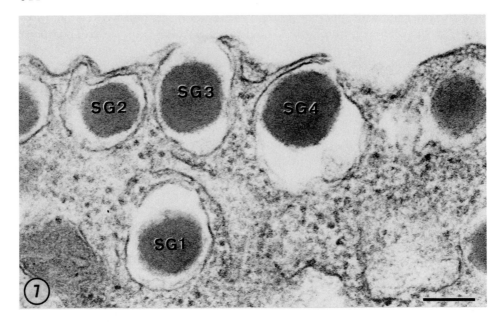

Fig. 7. Thin section of the periphery of a B cell stimulated to release insulin by high glucose concentration. The peripheral cytoplasm contains several secretory granules in various positions with respect to the cell surface. Each granule has a limiting membrane separated from the electron-dense core by a clear halo. SG1 is far from the surface, SG2 close to it, and SG3 and SG4 have their inner space open to the extracellular space due to fusion of the granule-limiting membrane with the plasma membrane. Bar = 0.2 μm. ×70,000.

freeze-fracture, exposing the inner organization of both the granule and plasma membranes, provides new information about the fusion process[5] (Orci, 1974; Orci and Perrelet, 1976; Orci et al., 1977). The protoplasmic (inner) leaflet of the B cell plasma membrane (P face) contains a sizable number of intramembranous particles (approximately 2000/μm^2). Apart from specialized regions of the membrane, such as those containing intercellular junctions, these particles are usually distributed at random and their size varies between 8 and 10 nm (with rare extremes at 13 nm). The exoplasmic (outer) leaflet of the plasma membrane (E face) has fewer particles (approximately 300/μm^2), whose size (8-10 nm) is comparable to that of P face particles. The freeze-fractured secretory granule membrane generally splits to reveal either a convex or a concave

[5]One difficulty encountered in studying B cells by freeze-fracture is the precise identification of this cell. The usual criteria used in conventional, thin-section electron microscopy to distinguish different islet cell types (i.e., specific shape, size, and electron density of secretory granules, distance between the core and the limiting membrane, etc.) have no counterparts in freeze-fracture replicas. Thus one must rely on indirect evidence to distinguish B cells from others. The best evidence so far is that in the rat the center of the islet contains B cells almost exclusively (Ferner, 1952). Freeze-fracture pictures taken in the center of the islet are therefore reasonably interpreted as showing B cells.

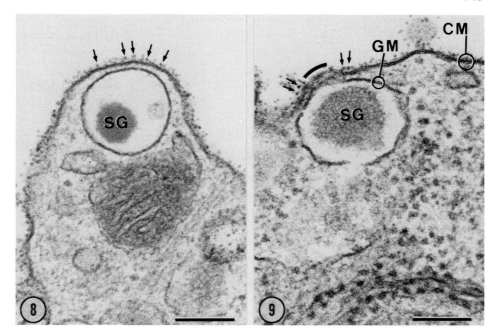

Figs. 8 and 9. Thin sections of glucose-stimulated B cells showing lectin (ricin) binding sites at the cell surface as detected by ricin-ferritin complexes. In Fig. 8 a secretory granule (SG) is seen close to the cell membrane. However, there is no contact between the secretory granule and the plasma membrane and the lectin-ferritin complexes bound to the outer aspect of the cell membrane (arrows) appear as an uninterrupted chain of black dots. In Fig. 9, the secretory granule (GM) and plasma (CM) membranes are closer to one another (with some degree of uncertainty due to a tangential section of the respective membranes), and a stretch of plasma membrane surface devoid of ricin-ferritin complexes (delimited by a black line) appears at the probable contact zone with the secretory granule membrane. Outside this area, ferritin-ricin dots (arrows) are regularly distributed. SG = secretory granule core. Fig. 8: Bar = 0.2 μm. ×81,000. Fig. 9: Bar = 0.2 μm. ×81,000.

spherical face. The convex fracture face corresponds to the inner (exoplasmic) leaflet of the granule's membrane (E face, in relationship with the inside of the vesicle), while the concave fracture face represents the outer (protoplasmic) leaflet relating to the cytoplasmic matrix (P face). The P face of the granule membrane has approximately 250 particles/μm^2 and the E face approximately only 50/μm^2. Particles in both fracture faces are larger (13 nm) than most particles in the plasma membrane. With the different faces of plasma and granule membranes precisely identified (Fig. 10), two conditions favor the observation of membrane interaction during exocytosis: (1) the study of stimulated B cells to increase the chance of detecting exocytotic events; and (2) the examination of replicas showing both the cytoplasm (containing fracture faces of granule membrane) and the plasma membrane (split longitudinally) of the same cell. This second point is significant in trying to correlate a tentative temporal sequence of morphological changes occurring in the fracture faces of granules and the

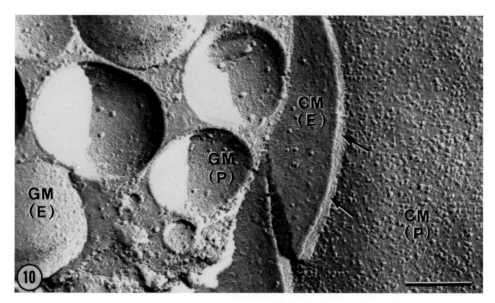

Fig. 10. Freeze-fracture replica of unstimulated B cells showing the membrane faces involved in exocytosis. To the left, a part of the plasma membrane P face, CM(P), is exposed as well as a piece of the E face, CM(E), from the plasma membrane of a neighboring cell. The step separating the two faces (arrows) corresponds to the width of the intercellular space. At right, a cross-fracture of the cytoplasm exposes circular concave and convex profiles. The concave profiles, GM(P), represent the P face of the secretory granule membrane, while the convex ones correspond to the E face of that membrane, GM(E). Note the difference in intramembranous particle content between the respective P and E faces of the plasma and granule membranes. Bar = 0.2 μm. ×91,000. (Reproduced with permission from Orci et al., 1977.)

plasma membrane with the spatial relationships between these two membranes. As detailed in Figs. 11 and 13, the first morphological stage in the tentative sequence is the close relationship between a granule membrane (E face) and the plasma membrane (P face). At the point where no cytoplasmic matrix can be seen separating the two membranes, a round or oval-shaped area appears in the plasma membrane P face. This particle-free area blends, without clear-cut transition, with the normally particulate fracture face. In addition, circumscribed zones without particles in the P face of the plasma membrane can be observed over circular or elliptical bulges which are interpreted to be the result of secretory granules (not visible) adhering to the inner aspect of the plasma membrane (Fig. 11). Occasionally, the secretory granule E face is exposed through an aperture caused by a local removal of the overlying plasma membrane (P face). Given the thinness of the step separating the two membranes, one may assume that the intervening cytoplasm is extremely narrow or absent; in such cases, no particular differentiation of the granule E face is observed (Fig. 12). More advanced stages of exocytosis consist of circular depressions in the plasma membrane P face which often contain protruding spheroids (Fig. 14). The depressions, as well as

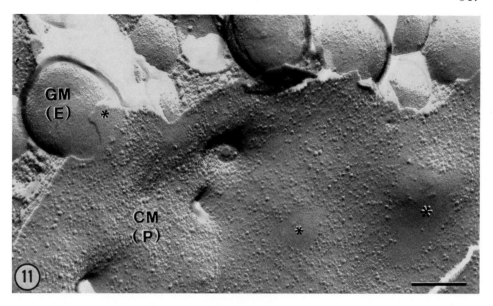

Fig. 11. Freeze-fracture replica of cytoplasmic and plasma membrane fracture faces in stimulated B cells. At left the E face of a secretory granule, GM(E), comes into close relationship with the P face of the plasma membrane, as shown by a thin ridge between the respective faces. In this area of close contact, the plasma membrane P face, CM(P), is devoid of intermembranous particles (asterisk). Particles are also missing in two other limited zones of the membrane (asterisks), and may signal incipient fusion with secretory granules (not visible) situated in the underlying cytoplasm. Bar = 0.2 μm. ×74,000. (Reproduced with permission from Orci et al., 1977).

the lip by which they merge with the membrane face, appear devoid of particles. The central spheroid is equally smooth and is interpreted as the granule core exposed to the extracellular space after incorporation of the granule membrane into the plasma membrane. Finally, images considered as late stage manifestations of exocytosis (i.e., succeeding granule membrane incorporation and granule core dissolution) take the form of circular or slit-shaped lacunae in the plasma membrane P face (Fig. 15). These lacunae are particle-free and may correspond to remnants of the poorly particulated granule membrane. After cessation of stimulation, the plasma membrane again shows a randomly distributed population of particles and no particle-free areas. To date, there is no firm evidence as to whether the normal distribution of particles is restored by addition of new particles in the bare areas or by the selective removal of the latter during the secretion-induced compensatory endocytosis (Orci et al., 1973b), as suggested for the pancreatic acinar cell (De Camilli et al., 1976). The fact that in B cells, as in mast cells (cf. section 4.1), the zone of interaction between a granule and the plasma membrane appears to be deprived of ligand binding sites (Figs. 8, 9), offers a potential way of answering this important question by studying the fate of ligand-deprived areas following stimulation. At any rate, exocytosis induced by a prolonged stimulation of islets with high glucose concentration causes a sig-

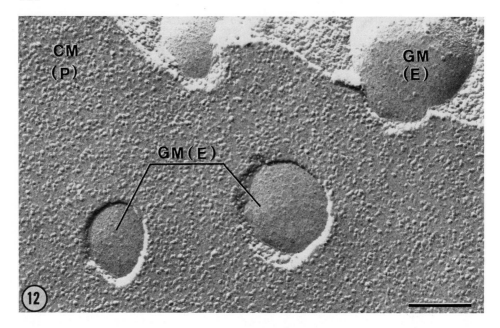

Fig. 12. Freeze-fracture replica of a stimulated B cell showing several E faces of secretory granule membrane, GM(E), both in cross-fractured cytoplasm (upper right) and through "windows" created by the removal of localized areas of the plasma membrane P face, CM(P). The exposed E faces have a characteristic low particle content and do not exhibit any other morphological features. Bar = 0.2 μm. ×84,000. (Reproduced with permission from Orci et al., 1977).

nificant decrease in particle number per μm^2 in the P face of the B cell plasma membrane: 1980 ± 31 (control); 1760 ± 40 (stimulated), p < 0.01. This is comparable to the drop in particle concentration seen in the plasma membranes of stimulated axon terminals of the neurohypophysis reviewed in section 4.5. Both neurohypophyseal and B cells have secretory granule membranes with a low particle content compared to their respective plasma membranes, and probably the decrease observed following stimulation is due to a dilution of the richly particulate plasma membrane by the low particulate membrane of fused secretory granules, and before endocytotic retrieval and resting of the cell reestablish a population of particles at control level. Large (13 nm) particles in the B cell plasma membrane may originate from incorporated granule membrane.

5. Membrane events related to endocytosis

As previously indicated by Palade and Bruns (1968; also see section 3.1) and Bennett (1969), membrane fission—the process by which a single-bilayer structure gives rise to two bilayers—does not differ conceptually from the mechanism

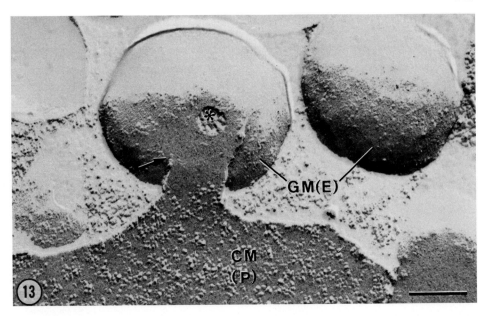

Fig. 13. Freeze-fracture replica of a stimulated B cell revealing two granule membrane E faces, GM(E), and a part of the plasma membrane P face, CM(P). At the arrow, complete fusion between the normally particle-poor E face of the granule membrane and a particle-deprived zone of the plasma membrane can be seen. A small circular depression in the zone of fused membranes (asterisk) may correspond to a forming exocytotic stoma. Bar = 0.2 μm. $\times$81,000.

of membrane fusion (in the latter case two bilayers unite to yield a single bilayer). If membrane fusion is the selective mechanism of exocytosis, membrane fission permits the reverse process, endocytosis. Thus modification of the structure of the plasma membrane at sites of endocytosis may be expected when membranes are examined by the freeze-fracture technique. On P faces of the plasma membrane (cytoplasm not visible) each endocytotic vesicle is revealed by its opening at the membrane surface, which appears as a small (approximately 40 nm in diameter), rounded depression (or "circumvallate papilla," in the terminology of Simionescu et al., 1974) on the fracture face. On E faces, the neck of the vesicle leaves an imprint in the form of a shallow crater delimited by an elevated rim (Orci and Perrelet, 1975). In some cases, the entire vesicle (E face) is exposed by the fracture and is detected·as a smooth, rounded sac above the plane of the plasma membrane E face. In P-face views of endocytotic openings, it is a frequent observation that the intramembranous particles are arranged in a loose necklace figure around the lip of the depression; the fracture face of the depression is usually smooth, but sometimes one or more particles may sit in the center of the depressed area (Orci and Perrelet, 1973; Smith et al., 1973a; Simionescu et al., 1974). Moreover, in cells in which "endocytosis" occur along well-delimited bands of plasma membrane (e.g., smooth muscle: Devine et al., 1971; Muggli

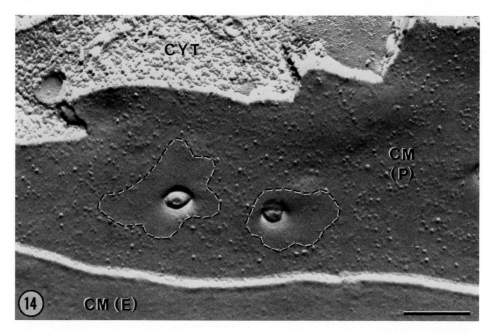

Fig. 14. Freeze-fracture replica of stimulated B cells with cytoplasmic (CYT) and plasma membrane fracture faces. The plasma membrane P face, CM(P), shows two completed exocytotic events, each characterized by a bulging granule core surrounded by a circular patch of particle-deprived membrane (delimited by dashed lines). CM(E) indicates part of the plasma membrane E face of a neighboring islet cell. Bar = 0.2 μm. ×91,000.

and Baumgartner, 1972), the number of intramembranous particles is increased along these bands (both in P and E faces) compared with noninvaginated zones of the membrane (Wells and Wolowyk, 1971; Orci and Perrelet, 1973). More recently, an additional membrane differentiation tentatively related to plasma membrane internalization has been identified in the form of clusters of large particles on P faces of neurophypophyseal axon membrane (Theodosis et al., 1978). At present it is not possible to unequivocally interpret the origin of these nonrandom particle arrangements. One possible cause of necklaces and increased particle concentration around endocytotic openings may be the drifting of particles away from prospective fission zones (similar to the accumulation of particles at the margin of particle-free areas in plasma membranes of secreting mast cells discussed in section 4.1), though in smooth muscle there is no definitive evidence that invaginations will ultimately detach from the plasma membrane. As detailed in the last section, nonrandom arrangements may be considered as evidence of local perturbation of normal protein-lipid relationships in the membrane that should ultimately favor interaction between various membrane components if membrane fission is to occur.

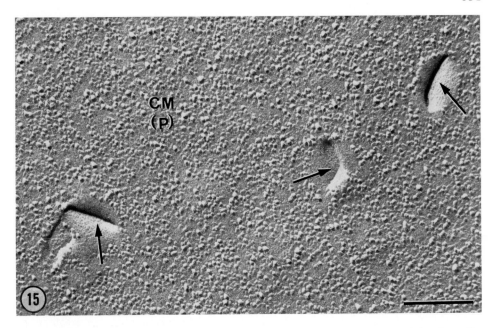

Fig. 15. Freeze-fracture replica of a B cell plasma membrane P face, CM(P), containing three slit-shaped, particle-free depressions (arrows). These depressions are interpreted as remnants of particle-poor granule membrane incorporated into the plasma membrane during exocytotic fusion. At this late stage, granule cores are no longer present due to dissolution in the extracellular space. Bar = 0.2 μm. ×94,000. (Reproduced with permission from Orci et al., 1977.)

6. Summary and conclusions

Morphological observations on a wide range of secretory cells have been re-viewed in this chapter to find out whether a common pattern of membrane or-ganization underlies the mechanism of exocytotic membrane fusion (and, by extension, that of membrane fission). In protozoans (Fig. 16) fusion between a se-cretory granule and the plasma membrane involves a peculiar spatial arrange-ment of intramembranous particles in the respective fusing membranes. In mammalian cells (Fig. 17), fusion is preceded and accompanied by removal of in-tramembranous particles from the membrane zones which are to unite with one another. Although these two processes may appear irreconcilable, they can still fit a general scheme of membrane organization, such as that predicted by Poste and Allison (1973), in which fusion requires the introduction of local *disorder* into membranes. If a random distribution of intramembranous particles is as-sumed to represent an "ordered" relationship between morphological counter-parts of protein and lipid in a membrane, any departure from this randomness may signify disorder, whether it be the focal clearing in a given membrane or the circumscription of a particle-free zone within a rosette of particles. In Poste and

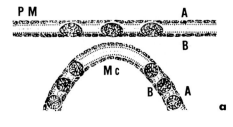

PM   A

B

Mc

B A

a

b

Fig. 16. Interpretive diagram of membrane fusion during mucocyst discharge. (a) Cross-section of plasma membrane (PM) containing large rosette particles, and underlying mucocyst membrane (Mc) containing rows of smaller annulus particles just before fusion. Wavy black lines indicate fractures that split the membranes into A and B halves. (b) Fusion begins when a hinge is formed between surface complexes corresponding to the rosette and the inner edge of the annulus. Wavy black line indicates fracture of the plasma membrane. Seen from face A, a nonetchable depression that corresponds to face B of the mucocyst membrane develops in the center of the rosette. (c) When the rosette particles spread, evolution of the annulus results. Fusion of the A faces is completed. Wavy black line indicates fracture of the plasma membrane. Seen from face A, the depression in the center of the rosette corresponds to the expanding mucocyst content and reorganizing mucocyst membrane face B. This is now etchable. (d) Reorganization is completed around the fusion lip. Fracture line now passes through expanding mucocyst content. A half of the membrane = P face; B half of the membrane = E face. (Reproduced with permission from Satir et al., 1973.)

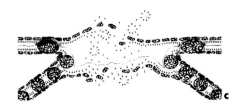

c

d

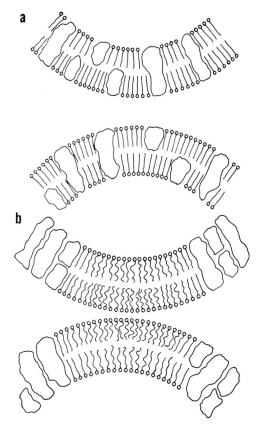

a

b

Fig. 17. Possible relationships between the lipid molecules and the intrinsic proteins of biological membranes during membrane fusion. (a) Two membranes in close proximity which do not fuse because the proteins of the membranes are randomly arranged in their respective lipid bilayers. (b) Membranes in which fusion may now proceed by the interaction and intermingling of their lipid molecules, after the emergence of protein-free areas of lipid bilayer as a result of aggregation of the intrinsic proteins in both membranes. Compared with (a), the lipid molecules of the membranes that are about to fuse have been perturbed; the lipids of (b), are either more fluid (as illustrated) or, in the extreme case, are rearranged in micellar form (not shown). (Reproduced with permission from Ahkong et al., 1975.)

Allison's concept, perturbation or disorder is envisioned essentially as a favorable condition to permit contact and linkage between macromolecular components of the interacting membranes. This logically raises the last question: what are the forces responsible for inducing the local disorder? A considerable amount of experimental evidence recently summarized by Nicolson (1976) points to non-membrane elements of the cytoplasm, the microfilaments (contractile), and the microtubules (skeletal), as the best candidates for controlling movements of individual components of the fluid membrane,[6] especially proteins (these functions

---

[6]Most of the knowledge concerning the cytoskeletal control of membrane components stems from the observation of the behavior of morphologically labeled receptors and binding sites at the cell surface when manipulating microfilaments and/or microtubules experimentally (Yahara and Edelman, 1975; for a recent review, see de Petris, 1977). We are aware of the fact that in nucleated cells (e.g., lymphocytes) the movement of surface receptors is not necessarily linked with that of morphologically detectable proteins as represented by particles in freeze-fracture (Bretscher and Raff, 1975). However, the finding that both ligand-binding sites and intramembranous particles are missing at sites of incipient membrane fusion makes us willing to extend the interpretation of surface receptor labeling experiments to the change in particle distribution detected by freeze-fracture in mammalian secretory cells.

are in addition to the widely assumed role of microtubules and microfilaments in the translocation of entire cytoplasmic organelles from and to various sites of the cell, in promoting cell movements, etc.). Although the exact relationship between microfilaments, microtubules, and membrane proteins is unclear, partly because no morphological technique has been able so far to demonstrate where and how microfilaments and microtubules (or their constituting elements) terminate in the membrane, the presence of a distinctive microfilamentous-microtubular web (review, Nicolson, 1976; Nicolson et al., 1977) at the inner leaflet of the plasma membrane and the recovery of (microfilamentous) actin and myosin from isolated secretory granules (Burridge and Phillips, 1975), indicate that both partners in fusion have the necessary apparatus to induce the required perturbation in membrane organization.

Acknowledgments

We wish to thank Miss Isabelle Bernard for her excellent and tireless secretarial help. Work done in the authors' laboratory was supported by the Swiss National Science Foundation, grants 3.553.75 and 3.120.77.

References

Ahkong, Q. F., Fisher, D., Tampion, W. and Lucy, J. A. (1975) Mechanisms of cell fusion. Nature 253, 194–195.
Amherdt, M., Baggiolini, M., Perrelet, A. and Orci, L. (1978) Membrane fusion during phagocytosis in polymorphonuclear leukocyte. Lab. Invest. In press.
Baggiolini, M., Amherdt, M. and Orci, L. (1976) Unusual membrane fracture faces in polymorphonuclear leukocyte granules. Experientia 32, 1400–1401.
Bennett, H. S. (1963) Morphological aspects of extracellular polysaccharides. J. Histochem. Cytochem. 11, 14–23.
Bennett, H. S. (1969) The cell surface: movements and recombinations. In: Handbook of Molecular Cytology (Lima-de-Faria, A., ed.) pp. 1294–1319. Elsevier/North-Holland, Amsterdam.
Branton, D. (1966) Fracture faces of frozen membranes. Proc. Nat. Acad. Sci. U.S.A. 55, 1048–1056.
Branton, D. (1969) Membrane structure. Ann. Rev. Plant. Physiol. 20, 209–238.
Branton, D., Bullivant, S., Gilula, N. B., Karnovsky, M. J., Moor, H., Mühlethaler, K., Northcote, D. H., Packer, L., Satir, B., Satir, P., Speth, V., Staehelin, L. A., Steere, R. L. and Weinstein, R. S. (1975) Freeze-etching nomenclature. Science 190, 54–56.
Bretscher, M. S. and Raff, M. C. (1975) Mammalian plasma membranes. Nature 258, 43–49.
Burridge, K. and Phillips, J. H. (1975) Association of actin and myosin with secretory granule membranes. Nature 254, 526–529.
Chi, E. Y., Lagunoff, D. and Koehler, J. K. (1975) Electron microscopy of freeze-fractured rat peritoneal mast cells. J. Ultrastr. Res. 51, 46–54.
Chi, E. Y., Lagunoff, D. and Koehler, J. K. (1976) Freeze-fracture study of mast cell secretion. Proc. Nat. Acad. Sci. U.S.A. 73, 2823–2827.
De Camilli, P., Peluchetti, D. and Meldolesi, J. (1976) Dynamic changes of the luminal plasmalemma in stimulated parotid acinar cells. A freeze-fracture study. J. Cell Biol. 70, 59–74.
Dempsey, G. P., Bullivant, S. and Watkins, W. B. (1973) Ultrastructure of the rat posterior pituitary gland and evidence of hormone release by exocytosis as revealed by freeze-fracturing. Z. Zellforsch. Mikrosk. Anat. 143, 465–484.

DePetris, S. (1977) Distribution and mobility of plasma membrane components on lymphocytes. In: Dynamic Aspects of Cell Surface Organization (Poste, G. and Nicolson, G. L., eds.), Cell Surface Reviews, vol. 3, pp. 643–728. Elsevier/North-Holland, Amsterdam.

Devine, C. E., Simpson, F. O. and Bertaud W. S. (1971) Surface features of smooth muscle cells from the mesenteric artery and vas deferens. J. Cell Sci. 8, 427–443.

Dreifuss, J. J., Akert, K., Sandri, C. and Moor H. (1976) Specific arrangement of membrane particles at sites of exo-endocytosis in the freeze-etched neurohypophysis. Cell Tiss. Res. 165, 317–325.

Ferner, H. (1952) Das Inselsystem des Pankreas, p. 119. G. Thieme Verlag, Stuttgart.

Franke, W. W., Lüder, M. R., Kartenbeck, J., Zerban, H. and Keenan, T. W. (1976) Involvement of vesicle coat material in casein secretion and surface regeneration. J. Cell Biol. 69, 173–195.

Friend, D. S., Orci, L., Perrelet, A. and Yanagimachi, R. (1977) Membrane particle changes attending the acrosome reaction in guinea pig spermatozoa. J. Cell Biol. 74, 561–577.

Ichikawa, A. (1965) Fine structural changes in response to hormonal stimulation of the perfused canine pancreas. J. Cell Biol. 24, 369–385.

Lacy, P. E. (1970) Beta cell secretion from the standpoint of a pathologist. Diabetes 19, 895–905.

Lagunoff, D. (1973) Membrane fusion during mast cell secretion. J. Cell Biol. 57, 252–259.

Lawson, D., Raff, M. C., Gomperts, B., Fewtrell, C. and Gilula N. B. (1977) Molecular events during membrane fusion: a study of exocytosis in rat peritoneal mast cells. J. Cell Biol. 72, 242–259.

Malaisse, W. J., Hager, D. L. and Orci, L. (1972) The stimulus-secretion coupling of glucose induced insulin release. The participation of the beta cell web. Diabetes 21 (Suppl. 2), 594–604.

McNutt, N. S. (1977) Freeze-fracture techniques and applications to the structural analysis of the mammalian plasma membrane. In: Dynamic Aspects of Cell Surface Organization (Poste, G. and Nicolson, G. L., eds.) Cell Surface Reviews, vol. 3, pp. 75–126. Elsevier/North-Holland, Amsterdam.

Muggli, R. and Baumgartner, H. R. (1972) Pattern of membrane invaginations at the surface of smooth muscle cells of rabbit arteries. Experientia 28, 1212–1214.

Nagasawa, J., Douglas, W. W. and Schulz, R. A. (1970) Ultrastructural evidence of secretion by exocytosis and of "synaptic vesicle" formation in posterior pituitary gland. Nature 227, 407–409.

Nagasawa, J., Douglas, W. W. and Schulz, R. A. (1971) Micropinocytotic origin of coated and smooth microvesicles ("synaptic vesicles") in neurosecretory terminals of posterior pituitary glands demonstrated by incorporation of horseradish peroxidase. Nature 232, 341–342.

Nicolson, G. L. (1976) Transmembrane control of the receptors on normal and tumor cells. I. Cytoplasmic influence over cell surface components. Biochim. Biophys. Acta 457, 57–108.

Nicolson, G. L., Poste, G. and Ji, T. (1977) The dynamics of cell membrane organization. In: Dynamic Aspects of Cell Surface Organization (Poste, G. and Nicolson, G. L., eds.), Cell Surface Reviews, vol. 3, pp. 1–74, Elsevier/North-Holland, Amsterdam.

Nordmann, J. J., Dreifuss, J. J., Baker, M., Ravazzola, M., Malaisse-Lagae, F. and Orci, L. (1974) Secretion-dependent uptake of extracellular fluid by the rat neurohypophysis. Nature 250, 155–157.

Orci, L. (1974) A portrait of the pancreatic B-cell. Diabetologia 10, 163–187.

Orci, L. and Perrelet A. (1973) Membrane-associated particles: increase at sites of pinocytosis demonstrated by freeze-etching. Science 181, 868–869.

Orci, L. and Perrelet, A. (1975) A comparison between thin section and freeze-etch replicas. In: Freeze-Etch Histology, pp. 22–25. Springer-Verlag, Heidelberg.

Orci, L. and Perrelet, A. (1976) Islet cell membrane perturbation during exocytosis. In: Endocrine Gut and Pancreas (Fujita, T., ed.) pp. 295–299. Elsevier/North Holland, Amsterdam.

Orci, L., Ravazzola, M., Amherdt, M. and Malaisse-Lagae, F. (1973a) The B-cell boundary. In: Intern. Congr. Ser. No. 312. Diabetes (Malaisse, W. J. and Pirart, J., eds.) pp. 104–118. Excerpta Medica, Amsterdam.

Orci, L., Malaisse-Lagae, F., Ravazzola, M., Amherdt, M. and Renold A. E. (1973b). Exocytosis-endocytosis coupling in the pancreatic beta cell. Science 181, 561–562.

Orci, L., Amherdt, M., Malaisse-Lagae, F., Rouiller, C. and Renold, A. E. (1973c) Insulin release by emiocytosis: demonstration with freeze-etching technique. Science 179, 82–84.

Orci, L., Perrelet, A. and Friend, D. S. (1977) Freeze-fracture of membrane fusions during exocytosis in pancreatic B-cells. J. Cell Biol. 75, 23–30.

Palade, G. E. (1959) Functional changes in the structure of cell components. In: Subcellular Particles (Hayashi, T., ed.) pp. 64–83. Ronald Press Company, New York.

Palade, G. E. (1975) Intracellular aspects of the process of protein synthesis. Science 189, 347–358.

Palade, G. E. and Bruns, R. R. (1968) Structural modulations of plasmalemmal vesicles. J. Cell Biol. 37, 633–649.

Pelletier, G. (1973) Secretion and uptake of peroxidase by rat adenohypophyseal cells. J. Ultrastr. Res. 43, 445–459.

Pinto da Silva, P. and Nogueira, M. C. (1977) Membrane fusion during secretion. A hypothesis based on electron microscope observation of *Phytophtora palmivora* zoospore during encystment. J. Cell Biol. 73, 161–181.

Plattner, H. (1974) Intramembraneous changes on cationophore triggered exocytosis in *Paramecium*. Nature 252, 722–724.

Poste, G. and Allison, A. C. (1973) Membrane fusion. Biochim. Biophys. Acta 300, 421–465.

Rambourg, A. (1971) Morphological and histochemical aspects of glycoproteins at the surface of animal cells. Int. Rev. Cytol. 31, 57–114.

Revel, J. P. and Ito, S. (1967) The surface components of cells. In: The Specificity of Cell Surfaces (Davis, E. D. and Warren, L., eds.) pp. 213–234. Prentice-Hall, Englewood Cliffs, New Jersey.

Robertson, J. D. (1959) The ultrastructure of cell membranes and their derivatives. Biochemical Society Symposium, Cambridge, England, vol. 16, pp. 3–43.

Robertson J. D. (1964) Unit membranes: a review with recent new studies of experimental alterations and a new subunit structure in synaptic membranes. In: Cellular Membranes in Development (Locke, M., ed.) pp. 1–81. Academic Press, New York and London.

Satir, B., Schooley, C. and Satir, P. (1973) Membrane fusion in a model system. J. Cell Biol. 56, 153–176.

Simionescu, M., Simionescu, N. and Palade, G. E. (1974) Morphometric data on the endothelium of blood capillaries. J. Cell Biol. 60, 128–152.

Singer, S. J. and Nicolson, G. L. (1972) The fluid mosaic model of the structure of cell membranes. Science 175, 720–731.

Smith, U., Ryan, J. W. and Smith, D. S. (1973a) Freeze-etch studies of the plasma membrane of pulmonary endothelial cells. J. Cell Biol. 56, 492–499.

Smith, U., Smith, D. S., Winckler, H. and Ryan, J. W. (1973b) Exocytosis in the adrenal medulla demonstrated by freeze-etching. Science 179, 79–82.

Tandler, B. and Poulsen, J. H. (1976) Fusion of the envelope of mucous droplets with the luminal plasma membrane in acinar cells of the cat submandibular gland. J. Cell Biol. 68, 775–781.

Theodosis, D. T., Dreifuss, J. J., Harris, C. and Orci, L. (1976) Secretion-related uptake of horseradish peroxidase in neurohypophysial axons. J. Cell Biol. 70, 294–303.

Theodosis, D. T., Dreifuss, J. J. and Orci, L. (1978) A freeze-fracture study of membrane events during neurohypophyseal secretion. J. Cell Biol. In press.

Vail, W. J., Papahadjopoulos, D. and Moscarello, M. A. (1974) Interaction of hydrophobic proteins with liposomes: evidence for particles seen in freeze-fracture as being proteins. Biochim. Biophys. Acta 345, 463–467.

Wells, G. S. and Wolowyk, M. W. (1971) Freeze-etch observations on membrane structure in the smooth muscle of guinea pig. J. Physiol. (London) 218, 11P

Wessels, N. K., Spooner, B. S., Ash, J. F., Bradley, M. O., Luduena, M. A., Taylor, E. L., Wrenn, J. T. and Yamada, K. M. (1971) Microfilaments in cellular and developmental processes. Science 171, 135–143.

Yahara, I. and Edelman, G. M. (1975) Electron microscopic analysis of the modulation of lymphocyte receptor mobility. Exp. Cell Res. 91, 125–142.

Yu, J. and Branton, D. (1976) Reconstitution of intramembrane particles in recombinants of erythrocyte protein Band 3 and lipid: effect of spectrin-actin association. Proc. Nat. Acad. Sci. U.S.A. 73, 3891–3895.

Membranes of ciliates: ultrastructure, biochemistry and fusion

13

Richard D. ALLEN

Contents

G. Poste & G. L. Nicolson (eds.) Membrane Fusion, pp. 657–763.
© *Elsevier/North-Holland Biomedical Press, 1978.*

658

Introduction

The usefulness of ciliated protozoa in biological research has been demonstrated repeatedly since their discovery in the seventeenth century by Leeuwenhoek. These single-celled organisms are easy to grow in the laboratory in mixed cultures and, in recent years, a number of species have been introduced into and maintained in pure culture. The cells undergo a broad spectrum of physiological processes that are required for life in a variety of environments, for ingestion and digestion of a wide variety of nutrients, for both asexual and sexual reproduction, and they also show amazing powers of morphological and biochemical adaptation to environmental changes. Most of the basic phenomena that occur in the cells of higher organisms, plant or animal, can also be found in ciliates. Consequently they have been, and will continue to be, used as models for basic research in cell biology. Information obtained from work with these cells is frequently and directly applicable to problems in higher organisms. In addition, the processes of membrane fusion and recycling can be profitably studied in free-living single-celled ciliates. In these cells the exact sites at which certain membrane fusions will occur are outlined within the plasma membrane. The use of single cells also eliminates the possibility that the membrane phenomena under study have been influenced by neighboring cells of different types.

The ciliates, in particular the genera *Tetrahymena* and *Paramecium,* have been used as models to study the physical and chemical properties of membranes, the turnover of membrane lipids and proteins, cell-to-cell recognition involving the surface coats of membranes and also membrane fusion and recycling. This chapter attempts to review the literature on the membranes of ciliates that has accumulated mainly over the past fifteen years. The morphology of all the membrane systems in ciliates are reviewed in terms of thin-sectioned electron-microscopic studies as well as more recent data obtained using freeze-fracture techniques. All membranes are reviewed since the total membrane population of a cell forms an intercommunicating system which is in constant flow, and between which constant macromolecular exchange is occurring. Biochemical and biophysical studies are reviewed where they are pertinent to this macromolecular exchange. Finally, the dynamic processes of membrane fusion, membrane flow, and membrane recycling are considered.

Nearly all of the illustrations in this chapter are unpublished micrographs taken from my own collection accumulated over this fifteen year period. On occasion I include unpublished observations, interpretations, or reinterpretations that vary from those in the literature if these seem to be justified by my own work.

2. Membrane morphology of ciliates

2.1. Membrane systems in sectioned cells

As in all eukaryotic cells, the protozoa of the phylum Ciliophora possess a number of membranes which have a wide variety of functions and characteristics. This group of mostly free-living, single-celled animals are set apart from most other protozoa at the ultrastructural level by having a pellicle rather than a single membrane that serves as the barrier between their cytoplasm and the external environment. This pellicle consists mainly of membranes, an outer plasma or cell membrane and a closely adherent mosaic of flattened, membrane-limited cisternae called alveoli (Figs. 1–3).*

The plasma membrane is a continuous layer covering all projections (cilia and tentacles) and indentations (parasomal sacs and pellicular pores) of the cell's surface. The alveolar system, on the other hand, is not a continuous covering under the plasma membrane; there are narrow gaps, or septa between the cisternal units of the alveoli where they abut each other (Figs. 2 and 3). These septa usually lie along the rows of cilia, the kineties, and, in some species, alveolar septa also lie along lines paralleling the rows of cilia halfway between two adjacent kineties. Transverse septa often interconnect the adjacent longitudinal septa. Basal bodies and secretory organelles, such as trichocysts and mucocysts, project to the cell's surface through these septal gaps (Figs. 1 and 17). The parasomal sacs, which are small invaginations of the plasma membrane located next to the somatic and oral cilia (Figs. 1, 30 and 34), pass inward penetrating the alveoli to make contact with the underlying ground substance. Similar invaginations of the plasma membrane in ciliates that do not have typical somatic kineties are called pellicular pores. Alveoli are also discontinuous at the specialized openings of the contractile or pulsating vacuoles, and at the cell's anus, termed the cytoproct or cytopyge.

A more extensive gap in this mosaic system occurs in the food vacuole-forming

---

*EDITOR'S NOTE: Because of the numerous and varied citations of the figures in this chapter, they have been arranged together and immediately follow the text beginning on page 725.

area of the cytostome-cytopharynx complex. The alveoli end at the border of this region so that only a single membrane, continuous with the plasma membrane, covers this part of the cytoplasm (Fig. 30). Other more specialized gaps in the alveolar system may be present in other ciliates. One such gap is found in the contractile stalk of peritrich ciliates (Allen, 1973a). The plasma membrane that encloses the myoneme of the stalk of *Vorticella* is subtended by alveoli around only a third of the circumference of the helically wound stalk.

Although at first the alveoli seem to be completely separated at the septa by gaps of 5 to 6 nm, it has been shown in *Paramecium* that they are actually connected by membrane-limited pores (Allen, 1971) (Figs. 2 and 3). Thus the membranes of the alveoli are all continuous with each other via these connecting pores. Such continuities have also been seen in *Tetrahymena* (Fig. 17) and *Coleps* (Allen, 1971) and in other ciliates (Antipa, 1971; Peck, 1971), and may be a common feature of all ciliates that have segmented alveoli.

Internally, ciliates contain the same membrane systems as those found in other eukaryotic cells, as well as some specialized systems not so frequently found. These include double membrane-limited nuclei and mitochondria (Figs. 17 and 42), and single membrane-limited peroxisomes (Figs. 59 and 65), digestive vacuoles (Figs. 42 and 45), and contractile vacuoles (Fig. 54). Ribosome-associated endoplasmic reticulum (ER) is also present (Figs. 2 and 17). This usually assumes the form of irregular tubes and vesicles rather than cisternae. Golgi bodies, in the form of dictyosomes, have been shown in many ciliates (Franke et al., 1971a; Esteve, 1972) but these usually consist of only 2 or 3 stacked cisternae next to a portion of the rough ER (Figs. 58–62). A large number of vesicles of varied size and shape, each with a surrounding membrane, fill the cytoplasm. Many of these are associated with the digestive vacuoles and are probably part of a lysosomal system (Figs. 42 and 45). Others may serve as pools of membrane to be used in the construction of larger organelles (Figs. 33 and 34), (Allen, 1974; Kloetzel, 1974). Accumulations of smooth membrane-limited tubules are associated with the contractile vacuoles of these cells (Fig. 56).

The secretion organelles, such as trichocysts and mucocysts, are all surrounded by membranes that are left behind when their contents are extruded from the cell. Pigment granules and crystals, when present, may also be enclosed by membranes, as is frequently the case for endosymbiotic bacteria and algae. Some pigments and food storage particles, such as lipid, are the only larger cytoplasmic bodies not surrounded by membranes.

Morphologically, the membranes can be separated into two classes based on their thickness and staining characteristics when seen in conventional electron micrographs. One class of membranes has a thickness of 9 to 10 nm, of which one leaflet is frequently thicker and, consequently, appears more electron opaque than the other. In *Paramecium* and *Tetrahymena* the outer leaflet of this class of membranes is thicker than the inner, or cytoplasmic leaflet. The two leaflets are separated by an electron-lucent space. The outer leaflet may be coated by a glycocalyx or glycoprotein material (Nilsson and Behnke, 1971; Wyroba and Przełecka, 1973; Hausmann and Mocikat, 1976), giving the membrane the appearance of being thicker than it really is (Figs. 16 and 34). The membranes that fall into this class are the plasma membrane, food or digestive vacuole membranes, the membranes of a class of vesicles that often have a flattened discoidal shape and accumulate near the oral region, and the membranes of another population of cisternae typically bearing coated pits that lie near the cortex in both the somatic and oral regions (Figs. 2 and 30) close to the proximal ends of the basal bodies and near parasomal sacs (Allen, 1967; Nilsson, 1976). Also, the membranes making up the maturing face of the Golgi bodies or dictyosomes have a thicker membrane than does the rough ER from which they are derived (Franke et al., 1971a; Esteve, 1972).

Other membranes of the cell are thinner, being 7 to 8 nm thick. The trilaminar appearance of these membranes is harder to resolve, particularly when glutaraldehyde fixation precedes osmium tetroxide postfixation. However, the trilaminar nature is clearly revealed in favorable sections. The two electron-opaque leaflets of these membranes are both 2.5 nm thick. A glycocalyx or microfilamentous coat is not usually seen lining these membranes. Most of the membranes within the cytoplasm fall into this category, including the rough and smooth ER, nuclear envelope membranes, the membranes of mitochondria and peroxisomes, the membranes surrounding secretion organelles, and the membranes of the contractile vacuole system. The alveolar membranes may also fit into this category.

Although the most common membranes can be grouped into these two categories, other specialized membranes cannot be classified so easily. One example of such a membrane is found in association with the collecting canals of contractile vacuoles, as for example in *Paramecium* (Fig. 55). These membranes which had been reported to be continuous with the ER (Schneider, 1964), but are now thought to end blindly (McKanna, 1976), have a helically arrayed coating on their cytoplasmic surface (McKanna, 1974, 1976). A similar specialization on surfaces of tubules (Fig. 4) that are continuous with the ER is found in the peritrich, *Opercularia,* and is associated with the contractile microfilamentous system of myonemes (Allen, 1973b). The coating in these membrane tubules, also on the nonluminal surface, has not been studied carefully. The membranes of the ER next to myonemes in peritrichs also have elaborate transmembrane structures (Fig. 5) called linkage complexes (Allen, 1973a,b).

2.2. Membrane systems in freeze-fractured cells

A useful technique that has recently been applied to membrane studies of ciliates is that of fracturing cells that have been quick-frozen with or without prior fixation and with or without prior treatment in a cryoprotectant such as glycerol or DMSO. This procedure splits membranes along their hydrophobic midlines, exposing large areas of their hydrophobic interiors (Branton, 1966). Thus both the fractured face of the protoplasmic leaflet (P face) and the fractured face of the external or luminal leaflet (E face) can be studied (for terminology, see Branton et al., 1975). Membranes fractured in this way typically have a smooth background studded with a varying number of particles. These particles are probably globular proteins or glycoproteins (Branton, 1971; Branton and Deamer, 1972; Packer et al., 1974), while the smooth areas are the hydrophobic surfaces of the two monolayers of phospholipids that make up the matrix of the membrane (Branton, 1971). The greatest advantage of this technique over the averaging techniques of biochemistry is that localized specialization in membranes can be detected.

The two genera of ciliates that have been studied in most detail using this technique are *Tetrahymena* and *Paramecium*. But even here only sketchy information is available on the particle numbers and sizes for the two fracture faces of the various cellular membranes. Figures 6 through 10 give a visual comparison of the difference in particle numbers on the various fracture faces of the pellicular membranes of *Paramecium caudatum*.

The plasma membranes of ciliates have the same relative distribution of particles on the E and P faces as the plasma membranes of other cells; the P face contains more particles than the E face. Most of these particles are randomly distributed in the membrane, though special particle groupings have been observed and some of these have received much attention. For example, rosettes of particles are found on the P face of the plasma membrane (Fig. 6) at the sites of contact with mucocysts in *Tetrahymena* (Satir et al., 1972a, 1973; Wunderlich and Speth, 1972) and with trichocysts in *Paramecium* (Plattner et al., 1973, 1975; Satir, 1974a). These rosettes consist of 8 to 11 particles, 10 to 15 nm in diameter arranged in a circle with a diameter of 60 to 80 nm. In *Paramecium* the rosettes tend to stay with the E face when the cells are frozen without prior fixation (Fig. 7).

A specialization of the P face of the plasma membrane associated with the rosettes in *Paramecium*, but absent in *Tetrahymena*, is a double ring of particles encircling the rosettes (Janisch, 1972; Plattner et al., 1973; Satir, 1974a). These rings, which are actually interrupted where the plasma membrane lies over the septa of the alveolar mosaic system (Fig. 6) (R. D. Allen, unpublished observation), are now considered to be attachment points of the plasma membrane to the margins of the underlying alveolar membranes (Plattner et al., 1973; Beisson et al., 1976; Plattner, 1976). The rings are present whether or not a trichocyst lies under the plasma membrane. However, these rings are flattened into elongated "parentheses" shapes when the trichocysts are absent, and no rosette is found inside the flattened rings under these conditions (Beisson et al., 1976). Formation of the rings does not depend on the presence of trichocysts or of the rosette of particles. Parts of rings or complete rings with rosettes included are sometimes found outside the rows of basal bodies (Fig. 6).

Other modifications of the plasma membrane have also been noted. Rings of particles encircle the ciliary membranes at the bases of all cilia (Figs. 12 and 29). These particles are organized into two rows making up a necklace (Satir et al., 1972b; Wunderlich and Speth, 1972; Sattler and Staehelin, 1974). In addition, only in the GL strain of *Tetrahymena pyriformis* (Wunderlich and Speth, 1972; Sattler and Staehelin, 1974) and in *Paramecium* (Plattner et al., 1973; Satir et al.,

1976) are patches of particles found on the ciliary plasma membrane just distal to the necklace (Figs. 12, 13, and 29). There are probably 9 such patches, each consisting of 3 short rows of particles. The 9 patches are opposite the 9 peripheral doublet microtubules which make up the shaft of the cilium. Linkers, seen in thin sections (Figs. 14, 15 and 30), are present between the membrane and the shaft and are presumably correlated with the patches in the membrane (Sattler, 1974; Satir et al., 1976). Neither the ciliary necklace nor the patches of particles are involved in membrane-to-membrane attachments. The function of these patches or plaques will be considered later in section 3.2.3.

Sattler and Staehelin (1974) have described specializations of particles in the membranes of some oral cilia of *Tetrahymena*. One specialization consists of rows of particles running longitudinally along the axis of the cilium that may be involved in linking the ciliary membrane to 4 to 6 of the peripheral doublets of a ciliary shaft. Other particles apparently reflect sites of bristle attachment to the external leaflet near the tips of certain oral cilia (see also Didier, 1976). Such specializations have never been seen on the somatic cilia.

Rows of particles have also been seen on the plasma membrane covering the oral ribs in *Tetrahymena* (Wunderlich and Speth, 1972; Sattler, 1974). Some of these rows must represent sites of microtubular attachment to the plasma membrane (Sattler, 1974).

Paramecium caudatum also has specializations of the plasma membrane both at the margins of the cytostome-cytopharynx (Fig. 35) and at the cytoproct margins. These consist of 100 to 150 nm wide bands of closely packed particles on the P face of the plasma membrane that are probably involved in linking this membrane firmly to the underlying alveoli which terminate along these lines (Allen, 1976a,c). These bands represent intracellular junctions and are in some ways comparable to the well known intercellular junctions (Satir and Gilula, 1973; Staehelin, 1974) frequently found in multicellular organisms. Rows of particles are also seen on the P face of the cytopharynx membrane of *Paramecium* where microtubules attach to this membrane (Allen, 1976b).

Plattner and collaborators (Plattner et al., 1973, 1975; Plattner and Fuchs, 1975) have summarized a number of less extensive intramembranous particle groups which they feel act as pericellular junctions. They found no correlation between these junctions and calcium-binding sites (Plattner and Fuchs, 1975), nor could they show a transmembrane channel through the center of these intercalated particles by using small tracer molecules (Plattner et al., 1975) of the kind reported to be present in gap junction particles of metazoan cells (Revel and Karnovsky, 1967; Reese et al., 1971). Also, such channels could not be detected in high-resolution freeze-fracture replicas (Plattner et al., 1975). The junctions are found between the cell membrane and the membranes of the alveoli as well as between the cell membrane and the trichocyst tip membranes (Fig. 6), between the trichocyst membrane and adjacent alveolar membranes (Fig. 27), between the plasma membrane and the alveolar membrane surrounding the emerging cilium (described by Plattner et al., 1973, but I was unable to confirm this), and between adjacent alveoli along the inner margin of the alveolar septa (Fig. 10).

Another specialization of the plasma membrane of *Paramecium* has been seen and mentioned only briefly so far (Allen, 1976b). Large rectangular plates of particles are found on the P face of the plasma membrane of *P. caudatum* (Fig. 11). These plates are concentrated on the anterior ventral surface of this cell anterior to the oral opening. Each plate consists of a number of parallel rows of particles. The particles within a row have a center-to-center spacing of 11 nm, while the distance between the midlines of adjacent rows is 25 nm. There does not appear to be a regular arrangement of the plates with regard to the ciliary territories; rather they are scattered randomly over this anterior ventral surface. Their function may be to serve as ion channels, but this is only speculation based on the work on gap junctions (review, Staehelin, 1974) and a possible ATPase activity of small particle arrays in some metazoan cell membranes (Ellisman et al., 1976). Alternatively, they may act as chemoreceptors or be involved in cell union during conjugation. Cell union is a specialized function of this anteroventral surface.

Thus the plasma membrane, although it is homogeneous over most of its area, has regions of local specialization that can be detected only by using the freeze-fracture technique. Isolating plasma membrane biochemically presumably lumps all of these areas together and therefore can yield only the average characteristics of the whole membrane (Branton, 1971). Biochemical studies of isolated membranes must thus be interpreted with this limitation in mind.

The alveoli are flattened or variably expanded cisternae that lie under the plasma membrane and may be bound to it by electron-opaque links (Tokuyasu and Scherbaum, 1965; Franke et al., 1971b). The membrane on the side of an alveolus next to the plasma membrane is called the outer alveolar membrane (OAM), while that next to the cytoplasm is called the inner alveolar membrane (IAM). This latter membrane, under usual conditions of preparation, does not fracture over large areas (Wunderlich and Speth, 1972; Satir et al., 1973; Sattler, 1974; R. D. Allen, unpublished observations) as does the plasma membrane. This indicates that the IAM must contain a higher proportion of intercalated proteins than the plasma membrane (Packer et al., 1974). Generally the OAM contains fewer particles than the IAM and can be found more readily in replicas. The P face of the IAM has more particles than the corresponding E face (Wunderlich and Speth, 1972; Satir et al., 1973; Sattler, 1974), but no particle counts have been reported either in *Tetrahymena* or in *Paramecium*. My own unpublished studies show the IAM of *Paramecium caudatum* to be tightly packed with particles particularly on its E face. At the proximal edge of the septa there is an abrupt transition from a highly particulate membrane to a membrane with far fewer particles (Fig. 9); this latter type of membrane continues throughout the OAM of this organism. The OAM of *Tetrahymena,* according to Wunderlich and Speth (1972), has more particles on its E face than the P face. Sattler (1974), however, reports finding more on the P face. This alveolar membrane thus has large areas containing a high number of particles which are, in turn, continuous with equally large areas containing far fewer particles. Nothing can be said at present about the different functions of these two morphologically distinct regions. However, a possible explanation for the restriction in particle movement from regions of high particle density to low particle density may be found either in the membrane-to-membrane attachments that separate these two regions and which may be able to prevent lateral particle movement, or the particles in the IAM may be anchored to the subtending microfilamentous material of the epiplasm.

Intracellular organelles have received only slight attention in studies using the freeze-fracture technique. Sekiya and co-workers (1975) have published a freeze-fracture study of isolated membranes of *Tetrahymena pyriformis,* strain WH-14. The two membranes of the mitochondria are similar in particle distribution to those of mammalian tissue cells. As could be expected, the microsomal fraction was variable, since such a fraction would contain a heterogeneous group of vesicles and vesiculated smooth and rough endoplasmic reticulum. Wunderlich's group has examined the ER and nuclear membranes in vivo (Wunderlich et al., 1974a, 1975). The P faces of these membranes are uniformly particulate at the normal growth temperature, while the E faces have few particles but numerous pits. The particles in the ER membrane range from 3 to 12 nm, with a mean value close to 7.5 nm (Wunderlich et al., 1975).

Mucocyst and trichocyst membranes have received more study. The membranes of both of these secretory organelles undergo changes during maturation. Particles normally found on the P face of immature organelles apparently migrate to the distal end to form an annulus as they take up their final resting position next to the plasma membrane. This leaves the mature mucocyst membrane almost smooth (Satir et al., 1972a, 1973), while the trichocyst membrane has a smooth area around the distal third of the tip of the trichocyst (Fig. 27) (Allen and Hausmann, 1976). This annular ring of particles may be involved in attaching the trichocyst and mucocyst membrane to the alveolar membrane (Plattner et al., 1973; Allen and Hausmann, 1976) and may also prevent the rupture of this membrane during the discharge of these secretory organelles (Satir, 1974a; Beisson et al., 1976).

During trichocyst maturation the outer surface of the membrane at the distal end of the trichocyst also undergoes a process of differentiation. A collar (Bannister, 1972) is built around that area which undergoes particle depletion (Allen and Hausmann, 1976). The collar imparts strength and integrity to this part of the membrane and remains intact for some time after the trichocyst has been expelled (Hausmann and Allen, 1976).

Digestive vacuole membranes or phagosomes have been receiving attention recently since they show changes in their intramembrane architecture that can be correlated with the digestive cycle. Sattler (1974) saw only a few particles on either fracture face of the digestive vacuole membrane of *Tetrahymena,* while Sekiya and collaborators (1975) report that the particle distribution in digestive vacuoles is like that of plasma membrane. An in vivo study of digestive vacuole membranes of *Tetrahymena pyriformis,* strain GL, was carried out by Batz and Wunderlich (1976). Cells were fed latex beads and then fixed after 5, 30, 60, and 150 minutes following the initial latex feeding. Nascent digestive vacuoles which are continuous with the plasma membrane in the oral region have membranes whose fracture faces bear uniformly distributed 8.5 nm particles in about equal numbers on both P and E faces. Many vesicles of varied sizes and shapes lie close to the nascent digestive vacuole but decrease in number after 60 minutes. Mature digestive vacuoles have particles unequally distributed on the two faces of the membrane; the P face now has numerous particles and the E face has only a few. Smooth areas and indications of fusion with cytoplasmic vesicles or of exocytosis are also seen on the P face. Kitajima and Thompson (1977b) also found an unequal distribution of particles on mature digestive vacuoles of *T. pyriformis,* strain NT-1, but could not confirm the equal numbers on the two leaflets of nascent digestive vacuoles as previously reported (Batz and Wunderlich, 1976). Kitajima and Thompson found 761 ± 219 particles per μm^2 on the P face and 146 ± 85 per μm^2 on the E face of nascent food vacuoles. The number of particles on the E face of mature food vacuoles remains the same but there is an increase to 1625 ± 350 per μm^2 on the P face. Discoidal vesicles have the same number of particles on the two fracture faces as the nascent food vacuoles.

My laboratory has been studying the changes in digestive vacuole membranes of *Paramecium caudatum* (Allen, 1976a,c, 1977). In this organism, as in other ciliates, the digestive vacuole develops from the cytopharynx, a single-membrane-limited region continuous with the pellicle of the buccal cavity. The cytopharynx membrane, although continuous with the plasma membrane, contains an entirely different internal particle morphology than the plasma membrane (Fig. 31). In this membrane the E face is packed with particles while the P face has a more moderate number (Fig. 35). This is the opposite of the plasma membrane, with which it is continuous, since here the E face is quite smooth and the P face contains more particles. The line of change between the plasma membrane and the cytopharyngeal membrane is sharp and coincides with the intracellular junction observed on the P face of the plasma membrane along this terminal line.

The cytopharyngeal membrane to the left of the nascent digestive vacuole is sculptured into ridges and valleys which are aligned at an angle to the terminal line of the alveolus. The ridges of the P face are the lines along which the underlying cytopharyngeal ribbons of microtubules are attached to the cytopharyngeal membrane (Allen, 1974). The valleys are the sites at which vesicles fuse with this membrane. This relationship is reversed on the E face since the ridges are the sites of vesicle fusion and the valleys are sites of ribbon attachment (Fig. 31). Bumps have been seen on the ridges of the E face that must be partially incorporated vesicles (arrows, Fig. 31). Rows of particles that would indicate lines of microtubular-ribbon contact on the P face have been seen, but only in the most favorable replicas. Nascent digestive vacuole membranes, that is, developing digestive vacuoles which have not yet closed and are still connected to the cytopharynx membrane, have identical morphologies to the cytopharyngeal membrane.

A pool of discoidal vesicles lies next to the cytopharyngeal membrane (Figs. 31 and 32). These discoidal vesicles are 60 nm thick and up to 0.5 μm in diameter.

The membranes of these vesicles have the same intramembranous particle distribution and essentially the same particle numbers as the cytopharyngeal membrane; there are more particles on their E face than on the P face. These vesicles are observed to fuse with the cytopharyngeal membranes.

Mature digestive vacuoles that are free in the cytoplasm have a membrane which differs from that of the nascent digestive vacuole. Freeze-fractured membranes of these digestive vacuoles have an entirely smooth E face (Fig. 40); later they have an E face that contains a new population of more prominent particles (Fig. 41). The number of particles on the P face of these vacuoles does not vary much from the nascent digestive vacuole or from the discoidal vesicles described above (also see section 5.2.).

In addition to the intramembrane difference, the replicas reveal sites of pinocytic activity on the digestive vacuole membranes. Both invaginations into the lumen of the vacuole and evaginations into the cytoplasm have been seen (Allen, 1976a). These sites appear as raised or indented circular profiles (Figs. 37 and 39) which are formed by fracturing across the membrane leaflet and the internal contents of the blebs. These circular profiles are regular in size, from 30 to 70 nm in diameter, but their number varies widely from digestive vacuole to digestive vacuole even within the same cell. These changes probably correspond to particular stages of activity within the digestive vacuoles. For example, newly formed digestive vacuoles undergo rapid condensation (Mast, 1947) accompanied by a very irregular vacuole topography with many fractured necks (Fig. 37) (Allen, 1976a). Likewise, old digestive vacuoles ready to be egested show many fractured necks indicating pinocytic activity (Allen, 1976a). These pinocytic invaginations of the old food vacuole membrane have a smooth P face.

Located around or near the cytoplasmic surface of digestive vacuoles at nearly all stages are spherical to irregularly shaped vesicles (not discoidal) with a moderate number of large particles on their E faces (Figs. 43 and 44). Some of these vesicles also appear next to the cytopharynx membrane. Their location suggests that they are the neutral red granules (Mast, 1947; Rosenbaum and Wittner, 1962) that stain positively for acid phosphatase, an indication that they are lysosomes.

Similar freeze-fracture studies applied to other groups of protozoa, and to phagocytic cells in general, will be required to see if the above changes in digestive vacuole membrane morphology is a universal phenomenon in the intracellular digestive process. Such changes suggest that food vacuole membranes may carry intrinsic proteins, which may be enzymatic and may be vital to the various stages of the digestive process. Such membrane changes also indicate that biochemical studies of digestive vacuole membranes isolated at various stages of digestion may lead to a more complete understanding of the digestive process, including sources of enzymes for digestion, the process of movement of digested nutrients from the food vacuole·to the cytoplasm, and possible sites of energy utilization for active transport.

Freeze-fracture studies thus reveal morphologically how membranes change in their intramembranous components with alterations in the physiological activity of the space enclosed by the membrane. Intramembranous particle changes, albeit somewhat different from those in digestive vacuoles, have also been seen previously in chloroplasts (Wang and Packer, 1973; Packer, 1974) and mitochondria (Hackenbrock, 1972; Packer, 1974), and in these organelles have been correlated with changes in functional states of the organelles.

However, the biochemical identity of the individual particles has yet to be determined. Newly formed digestive vacuoles and the discoidal vesicles of *Paramecium* have a glycocalyx lining their luminal surface (Allen, 1974). It might therefore be supposed that the high number of particles on the E face serve as protein attachment sites for the glycosyl groups in the glycocalyx. However, the plasma membrane also has a glycocalyx (Allen, 1971, 1974; Wyroba and Przełecka, 1973; Hausmann and Mocikat, 1976), but the E face of this membrane has only a few particles, making such a hypothesis for the function of particles in digestive vacuole E faces unlikely. The isolation and characterization of these membranes and their various particles will certainly provide new information about the digestive process and presents a major challenge for future research.

The changes observed in membranes that are correlated with physiological changes indicates the necessity for caution in interpreting results of studies on isolated digestive vacuole membranes in which no attempt is made to isolate these membranes in their various physiological states. It is evident that at any given time actively feeding cells will contain digestive vacuoles in all stages of digestion and that on the basis of the above results, their membranes will not be identical at least in protein content. Biochemical isolation studies of even a homogeneous membrane preparation suffer the criticism that the information obtained is not true of any particular site on naturally occurring membranes but can only be an indication of the average properties of the total membrane area, both the protoplasmic and the external leaflet, at the time that the membranes were isolated for study.

As will be discussed in section 4.3.3.2. another important use of the freeze-fracture technique is in determining the temperature at which membrane lipids undergo a phase transition from gel to liquid crystalline. By growing cells at a particular temperature and then lowering it below the phase-transition temperature (Tc) one can, by taking samples at regular temperature intervals, determine the temperature at which smooth patches first appear between the otherwise randomly distributed intramembranous particles. This temperature corresponds roughly to the temperature at which that membrane undergoes a lipid phase transition. This technique has been used in a number of studies on *Tetrahymena* membranes (Wunderlich et al., 1973b, 1974a, 1975; Nozawa et al., 1974; Kasai et al., 1976; Martin et al., 1976) and the results indicate that each membrane has its own phase-transition temperature that can be altered by environmental changes which, in turn, induce fatty acid changes within the membrane. (Fukushima et al., 1976a; Martin et al., 1976; Kitajima and Thompson, 1977a,b).

3. Membrane biochemistry of ciliates

3.1. Lipids

3.1.1. Whole cells

Until very recently nearly all the biochemical studies on ciliate membranes were made on *Tetrahymena*. This cell can be grown easily in pure (axenic) culture and was the first ciliate to be grown on a chemically defined medium. It grows rapidly, having generation times as short as two hours, and reaches population densities of 10^6 cells per milliliter or more (Everhart, 1972). *Tetrahymena* provides an excellent eukaryotic model cell since it has the typical membrane systems normally found in higher animal and plant cells.

Of the total lipid content of exponentially growing *Tetrahymena*, 68% is composed of phospholipids (Erwin and Bloch, 1963) which are a major component of cellular membranes. Two classes of phospholipids, initially identified as phosphatidylethanolamine (PE) and phosphatidylcholine (PC), were reported to make up 54% and 30%, respectively, of the cell phospholipids (Erwin and Bloch, 1963). A third major phospholipid group was later recognized in the PE fraction as glyceride 2-aminoethylphosphonolipid (GAEP)[1] (Thompson, 1967, 1969; Kennedy and Thompson, 1970; Smith et al., 1970; Berger and Hanahan, 1971; Jonah and Erwin, 1971; Nozawa and Thompson, 1971a). Each of these phospholipid groups has subsequently been shown to be a mixture of several closely related lipids; these are listed in the papers and reviews by Berger and associates (1972), Hill (1972), Thompson and Nozawa (1972), Holz and Conner (1973), and in a more recent paper by Ronai and Wunderlich (1975). The relative proportions of the total cellular phospholipid in the various fractions of *Tetrahymena* are given by Baugh and Thompson (1975) as cilia 3%, pellicles 21%, mitochondria 24%, and microsomes 25%, with the remainder in the ciliary supernatant 3%, and postmicrosomal supernatant 24%.

The phospholipids are rich in the polyunsaturated fatty acids linoleic (18:2, $\Delta^{9,\,12}$) and γ-linolenic (18:3, $\Delta^{6,\,9,\,12}$) which are present in largest quantity in the GAEP and PC classes (Jonah and Erwin, 1971). A second isomer of linoleic acid, $\Delta^{6,\,11}$, has also been found to be present in a 3:8 ratio with the $\Delta^{9,\,12}$ isomer in whole cell lipids (Ferguson et al., 1975). The $\Delta^{6,\,11}$ isomer is associated primarily with GAEP (Ferguson et al., 1975) and, when radioactively labeled and fed to cells, is not metabolized but is incorporated unaltered into phospholipids (Koroly and Conner, 1976). GAEP and PC also possess a glyceryl ether group, hexadecyl glycerol, with the ether linkage at the C-1 position (Thompson, 1967; Conner et al., 1968; Berger et al., 1972).

Another unique component of *Tetrahymena* membranes is a neutral lipid, a sterol-like pentacyclic triterpene alcohol, tetrahymenol (Mallory et al., 1963). This neutral lipid may substitute in ciliates (Erwin et al., 1966) for the sterols

---

[1]Though *phosphonolipids* are actually phospholipid analogues, they are included with the phospholipids when these are referred to collectively in this paper.

which are a normal component of many eukaryotic membranes, particularly the plasma membrane.

The lipids of *Paramecium aurelia* are only now beginning to be investigated. Rhoads and colleagues (1977b) have found that the whole cell lipids are composed of 25% neutral lipids and 75% polar forms. The predominant phospholipids are the same as those in *Tetrahymena:* PE, PC, and GAEP. In addition to these, ceramide aminoethylphosphonate (CAEP), phosphatidylserine (PS), phosphatidylinositol (PI), the lyso derivatives of PE and PC, and cardiolipin have been identified. The phosphorus of the phosphonyl molecules comprises 15.7 to 26.5% of the total lipid phosphorus. Andrews and Nelson (1977) confirm the presence of PE, GAEP, and PC in *P. aurelia* as well as cardiolipin, which they report is found only in deciliated cells, not in the ciliary membrane.

The whole cell fatty acids of *P. aurelia* are composed of about 30% polyunsaturated acids (Kaneshiro et al., 1977). These include 18:2 and 18:3, which are found in *Tetrahymena,* but also present are the longer chain acids 20:2, 20:3, 20:4, and 20:5. The major saturated fatty acids are 14:0, 16:0, and 18:0; while a large amount of the monounsaturated acid 18:1, and a smaller amount of 16:1 complete the bulk of the fatty acid complement.

3.1.2. Cilia

When *Tetrahymena* cells are fractionated and the various fractions are analyzed for lipids, it becomes evident that the major phospholipids vary in concentration from fraction to fraction (Nozawa and Thompson, 1971a), as does the neutral lipid tetrahymenol (Thompson et al., 1971).

The cilia-containing fraction which is enriched in membranes is particularly interesting because of its high proportion of phosphonolipids (i.e., GAEP) and much smaller amounts of the choline-containing phospholipid, PC (Kennedy and Thompson, 1970; Smith et al., 1970; Jonah and Erwin, 1971; Nozawa and Thompson, 1971a). The phosphonate content of ciliary lipids is at least twice that of total cellular lipids (Kennedy and Thompson, 1970; Smith et al., 1970). Ciliary membranes also have five to seven times more tetrahymenol than do microsomal membranes (Thompson et al., 1971).

The ciliary membrane fraction of *P. aurelia* has been studied for its lipid content, but this is the only membrane fraction of this cell that has been studied so far. Rhoads and co-workers (1977b) found that the phosphonolipids and phosphatidylserine (PS) are enriched in the ciliary membranes over those in whole cell lipids. The major phospholipids are GAEP (44% of total lipid P), ceramide 2-aminoethylphosphonate (16%), PC (15%), PE (9%), PS (3.5%), PI (1.5%) and four to five sphingosine-containing phospholipids (10%) (Rhoads et al., 1977a). They also found 26 different fatty acids in the total lipids of cilia. Kaneshiro and associates (1977) found that the polyunsaturated fatty acids of ciliary membranes increase with culture age. Cilia always contain a higher percentage of polyunsaturates (71–89%, Rhoads et al., 1977a) than is found in whole cell lipids at a comparable age. The most pronounced change with age occurs in arachidonic acid (20:4$\Delta^{5, 8, 11, 14}$), which increases from 35.9% of the total fatty acid at mid-log phase to 59.2% in early stationary phase (Rhoads et al., 1977a). Over half of this fatty acid is found in GAEP, which contains only the 20:4 (88%) and 20:5 (12%) fatty acids.

3.1.3. Mitochondria

The major lipid component of the mitochondrial fraction is cardiolipin, with PC being the major phospholipid and GAEP being present in only half the concentration of that of whole cells (Jonah and Erwin, 1971). Thus mitochondrial lipids of *Tetrahymena* vary from those of vertebrates (Chapman and Leslie, 1970) since they contain GAEP, which is not found in vertebrate mitochondria, but are alike in that cardiolipin is an important component of mitochondria in both cell types.

Tetrahymenol is the principal neutral lipid of both cilia and mitochondria (Jonah and Erwin, 1971). Mitochondrial fatty acids are more unsaturated than the cilia fraction; mitochondrial phospholipids have a higher concentration of linoleic and γ-linolenic acids than cilia. A high proportion of unsaturated fatty acids is also a feature of vertebrate mitochondria.

3.1.4. Pellicles

Isolated pellicles, which include the plasma membrane without the ciliary membranes and also the inner and outer membranes of the alveoli, have an intermediate amount of phospholipids between that of ciliary membranes and whole cell lipids (Kennedy and Thompson, 1970). This enrichment of the pellicular membranes in phosphonates over whole cells may confer stability on these membranes since the carbon-phosphorus linkage of GAEP is resistant to certain naturally occurring phospholipases (Kennedy and Thompson, 1970). This may be especially important to rumen protozoa where the natural environment contains hydrolytic enzymes.

Tetrahymenol is also enriched in the pellicular membrane fraction over that in the microsomal and mitochondrial fractions (Thompson et al., 1971), but it is present here in about one third to one quarter of the concentration found in ciliary membranes. Alkyl glyceryl ethers, which exist in the phosphonate-containing molecules (Berger and Hanahan, 1971), are also enriched in the surface membranes and may, along with the C-P bonds of the phosphonolipids, give added stability to this membrane. The distribution of tetrahymenol in *Tetrahymena* is similar to the distribution of cholesterol in the plasma membrane of vertebrates (Ashworth and Green, 1966; Coleman and Finean, 1966; Weinstein and Marsh, 1969). Thompson and collaborators (1971) have speculated whether tetrahymenol can determine the physical characteristics and permeability properties of the surface membranes in a manner similar to that of cholesterol in other eukaryotic plasma membranes.

3.1.5. Microsomes

The microsomal fraction consists of all other membranes of the cytoplasm except for the mitochondrial and probably peroxisomal membranes, namely, rough and smooth ER, Golgi apparatus, vesiculated contractile and digestive vacuole membranes, and possibly small lysosomes. This fraction has a relatively low phosphonolipid and glyceryl ether content and a relatively low tetrahymenol content, but higher concentrations of PC and PE than the ciliary membranes (Nozawa and Thompson, 1971a; Thompson et al., 1971).

Ronai and Wunderlich (1975) have subfractionated this microsomal fraction into two smooth membrane fractions and a rough membrane fraction. The lighter (on sucrose gradients) "smooth I" fraction is composed of vesicles 40 to 1000 nm in diameter, the heavier "smooth II" fraction has vesicles ranging from 40 to 500 nm, while the rough fraction has ribosome-studded cisternae and vesicles with diameters up to 350 nm in diameter. All three microsomal fractions have similar phospholipid patterns, with PC, PE, and GAEP being the major phospholipids present. Compared to whole cells, microsomes have less PE, less cardiolipin and more GAEP, CAEP, and lysophosphatidylcholine (LPC). Nearly half of the total fatty acid is constituted of equal amounts of linoleic and γ-linolenic acids. Unsaturated fatty acids make up 65.2%, 69.3%, and 72.7% of the fatty acids in the smooth I, smooth II, and rough fractions, respectively. The neutral lipid tetrahymenol was found only in traces. This is in sharp contrast to its abundant presence in surface membranes.

The source of these three fractions is polymorphic. The rough fraction comes mainly from the rough ER, but the smooth fractions are thought to come from the smooth ER which is composed of cisternae, vesicles, sacs, tubules, and vacuoles of various sizes and shapes and of heterogeneous function (Ronai and Wunderlich, 1975). The lighter smooth I fraction has a higher acid phosphatase activity and may thus be involved to a greater extent in intracellular digestion.

3.1.6. Nuclear membranes

The major lipids found in isolated nuclear membranes are PC (31.0%), PE (26.1%), GAEP (23.3%), and one neutral lipid, tetrahymenol (Nozawa et al., 1975). The PC and PE ratio in the nuclear membranes is higher than in the surface membranes. Small amounts of LPC, lysophosphatidylethanolamine, lyso-2-aminoethylphosphonate and cardiolipin are also present. The nuclear membrane fraction most closely resembles the microsomal fraction, which is to be expected since the ER is continuous with the outer nuclear membrane. However, some differences do exist. The PE concentration is higher in microsomes (34%) than in the nuclear membranes (26.1%), and some cardiolipin is also found in the nuclear membranes but not in microsomes. Compared to mammalian liver nuclear membranes (for references see Franke, 1974), *Tetrahymena* nuclear membranes have a lower content of PC and lack spingomyelin, but have a significant amount of GAEP (Nozawa et al., 1973) which is not present in liver nuclear membranes. Actually these two membrane systems, ER and nuclear, may be even more similar than the above data suggest, but their similarity may be obscured by the heterogeneous nature of the microsomal fraction.

3.1.7. Digestive vacuoles

Weidenbach and Thompson (1974) reported the isolation of intact, young digestive vacuoles from *Tetrahymena* after feeding starved cells ferric oxide. The total surface area of vacuoles formed by these cells within a period of 10 minutes was 387 μm^2, or about 12% of the total cell surface. Lipid analysis of the digestive vacuole membranes and comparison with other fractions from starved cells demonstrated that the phospholipid pattern of digestive vacuole membranes is similar to the phospholipid pattern of the microsomal membranes but varies significantly from the pattern of ciliary membrane phospholipids. The PC and GAEP contents of vacuolar membranes are lower, while the PE content is higher, than their concentrations in whole cell lipids.

3.2. Proteins

3.2.1. Association with membranes

Proteins are now thought to be associated with membranes in two ways: first as integral or intrinsic proteins which have varying degrees of hydrophobic interaction with the lipid bilayer, or second as peripheral or extrinsic proteins attached via electrostatic interactions to the surfaces of the membranes (Singer and Nicolson, 1972; Singer, 1974). Intrinsic proteins are globular, amphipathic molecules which have nonpolar amino acid residues extending into the hydrophobic core of the lipid bilayer and ionic and saccharide hydrophilic groups at the membrane surface. These proteins are tightly bound to the membrane and can only be removed by drastic means such as detergent extraction. Peripheral proteins, bound to the membrane surface, can be removed more easily. These proteins are presumably held to the membrane by relatively weak, noncovalent interactions. Specific examples of these two classes of proteins are listed in the review of Singer (1974).

Only the integral proteins are assumed to be critical to the structural integrity of the membranes. These proteins are grossly heterogeneous in molecular weight and, for the most part, are present in an α-helical conformation, which suggests that the proteins are globular. Any amphipathic protein may potentially be an integral protein and this probably accounts for the highly heterogeneous nature of membrane proteins. Amphipathic proteins may pass all the way through a membrane or may lie in one leaflet of the membrane and be exposed to either the cytoplasmic or external polar environment of the membrane. Consequently, the two surfaces of a membrane are not identical in composition or structure. Such amphipathic proteins would maintain a constant molecular orientation within the membrane; they would be able to move laterally in the plane of the lipid bilayer unless constrained by extrinsic or intrinsic factors, but they would not normally be able to cross from one leaflet of a membrane to the opposite leaflet and would not be able to undergo trans-bilayer "flip-flop."

Interactions between proteins and lipids in the plane of the membrane often play a direct role in membrane functions (reviews, Singer, 1974; Gennis and Jonas, 1977) since certain enzymes require specific phospholipids or other lipids for their activity (review, Kimelberg, 1977).

The proteins of ciliate membranes have not been studied systematically even though the presence of a large number of proteins is indicated from freeze-fracture studies, from in vitro functional studies of various membrane fractions, and from scattered biochemical isolation studies. Proteins of membranes may be structural (having no known enzymatic activity) or they may be enzymatic. Membrane proteins are identified by isolating protein bands using polyacrylamide-gel electrophoresis after membrane isolation and solubilization. The physical characteristics of these membrane proteins can then be compared with known proteins.

3.2.2. Tetrahymena

Baugh and Thompson (1975) have studied the proportion of whole cell proteins in various cell fractions of synchronized cultures of *Tetrahymena*. Cilia contain only 1%, pellicles 14%, mitochondria 26%, and microsomes 14% of the total cellular protein, with the nonmembranous ciliary supernatant (8%) and postmicrosomal supernatant (38%) containing the rest of the protein.

Proteins make up a much higher proportion of the weight of whole cells than of the weight of membranes. Ronai and Wunderlich (1975) show in their microsomal subfractions of strain GL that the weight of proteins in the smooth I and smooth II fractions is about two times the weight of the phospholipids. Baugh and Thompson (1975) report proteins make up three times the weight of lipids in the microsomal fraction of strain WH-14, while in whole cells total protein weight is five times that of total phospholipid. Glucose-6-phosphatase and ATPase have relative specific activities (specific activity of the fraction/ specific activity of whole cells) higher than 1, thus indicating they are enriched in all three microsomal fractions. Acid phosphatase, 5'-nucleotidase, and succinic dehydrogenase have relative specific activities lower than 1 in all three fractions. In most cases these enzymes are distributed unevenly between the three fractions. Glucose-6-phosphatase and ATPase have a higher specific activity in the smooth I fraction than in the smooth II fraction, while ATPase is highest in the rough microsomal fraction.

Many enzymes are known to be associated with membranes either as integral or peripheral proteins and certain enzymes are commonly used as marker enzymes of specific membranes. Similar markers have also been used to identify *Tetrahymena* membrane fractions: succinic dehydrogenase for mitochondrial membranes; 5'-nucleotidase for plasma membranes; NADPH-cytochrome *c* reductase for microsomal membranes; glucose-6-phosphatase for ER membranes; and acid phosphatase for lysosome and digestive vacuole membranes. Franke (1974) has listed the many enzymes that have been localized in the nuclear membranes and microsomes of a number of vertebrate animal cells and tissues. Apparently no attempt has been made to do extensive enzyme localizations of the various membranes of ciliates. It is clear, however, that their membranes must contain a rich assortment of enzymes since they are capable of many of the same metabolic activities as metazoan cells.

There is one report in which cytochemical techniques were found to give positive results for a nonspecific cholinesterase activity in the ciliary membrane and shaft of *Tetrahymena*, strain W (Schuster and Hershenov, 1969). Speculation was presented that this enzyme may form a complex with an ATPase which can control membrane permeability in the cilia. In another study, ATPase reaction product was reported to be present on *Tetrahymena* ciliary membranes (Burnasheva and Jurzina, 1968).

Subbaiah and Thompson (1974) isolated *Tetrahymena* ciliary membranes and compared the structural protein bands obtained on polyacrylamide gels with those of the ciliary axoneme and ciliary matrix. The membrane fraction had four distinct bands, of which three moved slightly slower than the principal protein of the axonemes (tubulin) on urea-containing gels while on sodium dodecyl sulfate (SDS)-containing gels two bands from the membranes had a lower molecular weight than tubulin, another had about the same molecular weight as tubulin, and the fourth band was heavier. Estimated molecular weights were 29,000, 44,000, 53,000, and 63,000 for these four bands.

Structural proteins were also investigated in microsomes of *Tetrahymena*. After the ribosomes had been removed, the proteins of the membrane fraction were run on polyacrylamide gels in SDS. The membranes contained many bands that bore no strong resemblance to the pattern obtained from ciliary membranes (Subbaiah and Thompson, 1974).

The oral apparatus of *Tetrahymena* has been isolated and the proteins analyzed under conditions in which the membranes are retained (Rannestad and Williams, 1971; Wolfe, 1972; Gavin, 1974) as well as under conditions where membranes are lost (Vaudaux, 1976). In the former studies the membranes were not separated from either the basal bodies or the microfilamentous structures of the oral apparatus. Although it might have been possible to identify the membrane proteins by comparing similarly handled oral apparatuses, one sample with membranes and one without membranes, the techniques used in the two types of experiments were different and thus do not allow such a comparison.

3.2.3. Paramecium

The work on membrane proteins in *Paramecium* has been confined to the isolation and characterization of ciliary membranes. Hansma and Kung (1975) have studied *P. aurelia* syngen 4 strain 51s, as well as some membrane-mutant strains. Kitamura and Hiwatashi (1976) have isolated vesicles from cilia of *P. caudatum* which contain the mating reactive proteins in this cell. In *P. aurelia* the ciliary membranes are isolated as 0.1 to 1 μm diameter vesicles and when these are solubilized and the proteins run on SDS gels they contain 12 to 15 bands. A very large band with a high molecular weight (250,000 to 300,000) makes up about three quarters of the total protein. This band is positive for carbohydrates. The minor bands contain proteins with molecular weights of 25,000 to 150,000, with most of these proteins in two bands at 45,000 to 50,000 MW. The latter are probably tubulin. Besides the major band, only one other band contains carbohydrate (determined by the periodate-Schiff stain) and this has a molecular weight of 10,000 or less. The authors suggest that this may be a glycolipid since it stains weakly, or not at all, for protein (Hansma and Kung, 1975).

The major membrane protein is very much like an immobilization-antigen glycoprotein (i-antigen), several of which have been isolated from *P. aurelia* and are discussed in section 3.3.2. The major membrane protein and the i-antigen both have high molecular weights and cannot be distinguished by immunodiffusion (Hansma and Kung, 1975). In addition, both are large single-polypeptide chains. However, some differences are found, such as a slight difference in mobility on Triton X-100 polyacrylamide gels (i-antigen moves ahead) and a difference in solubility, i-antigens being soluble in aqueous solutions without added detergents whereas the major membrane protein of *P. aurelia* is not solubilized by this method. Hansma and Kung (1975) believe that the major membrane protein is a precursor of the i-antigen and is first embedded in the hydrophobic part of the membrane. After minor modification such as the removal of a terminal peptide or the addition of phosphate or carbohydrate prosthetic groups, with the result that the protein becomes more hydrophilic, the major membrane protein presumably will move out of the membrane center to the outer membrane surface.

The protein of the other minor bands may function as "gates," "pumps," "channels," or may have transport functions (Hansma and Kung, 1975). Hansma and Kung (1975) point out that proteins with these functions would only need to be present in very small quantities (ca. 0.1% of the total membrane proteins) to equal the 30 to 80 per μm^2 density of functional sodium channels that have been found in nerves (Keynes, 1972).

Two strains of *P. caudatum* vary in their distribution of ciliary proteins (Kitamura and Hiwatashi, 1976). Using 2 M urea and EDTA, these authors isolated vesicles 100 to 150 nm in diameter that contained 5 to 10% of the total ciliary proteins. They found about 20 bands on SDS gels, thus showing these vesicles have a complex protein composition; proteins from whole cilia have only 30 bands on SDS gels. The factor determining mating activity in *P. caudatum* is tightly bound to these membranes. Four kinds of glycoproteins were identified in the gels and these may determine the mating-type differences (Kitamura and Hiwatashi, 1976).

Electrophysiological studies have shown that the surface membranes of ciliates such as *Paramecium* (reviews, Eckert, 1972; Eckert and Naitoh, 1972; Naitoh, 1974; Eckert et al., 1976) and *Euplotes* (Naitoh and Eckert, 1969; Epstein and Eckert, 1973) are excitable and will depolarize in response to a sudden stimulus such as an increase in sodium or potassium ions in the medium, a mechanical stimulation at the anterior end of the cell, or an electrical current. Calcium ions can pass into the cell when the membrane is depolarized. This increased concentration of Ca^{2+} within the cell causes a reversal of the direction the cilia beat so that the cell begins to swim backward. The calcium ions move down an electrochemical gradient through calcium "gates" within the membrane (Eckert, 1972; Kung et al., 1975; Browning and Nelson, 1976; Eckert et al., 1976). At other sites within the surface membrane postulated calcium "pumps" (Browning and Nelson, 1976) actively remove the calcium from the cell to maintain the internal concentration of free Ca^{2+} within the cell at a level below 10^{-7} M.

Several mutants of *P. aurelia* have been isolated with defective calcium gates and do not react to increased sodium and potassium ion levels as do wild type cells (Kung, 1971; Kung and Eckert, 1972; Kung and Naitoh, 1973; Kung et al.,

1975). These gates may either remain permanently closed (Kung et al., 1975; Browning et al., 1976) or they may be temperature-sensitive, operating in the normal wild-type manner at one temperature but closed to calcium flow at another growth temperature. Other mutants swim backward for extended periods when varied combinations of sodium or potassium ions are present. The duration of backward swimming is correlated with Ca^{2+} influx (Browning et al., 1976) when depolarization is stimulated by Na^+ or K^+.

The site of calcium influx is apparently across the ciliary membrane. Deciliated *P. caudatum* cells are hyperpolarized when mechanically stimulated at the posterior end, as are normal ciliated cells, but when stimulated at the anterior end, they differ from the ciliated cell since they are not depolarized (Ogura and Takahaski, 1976). After cilia regeneration is complete an anterior stimulation will produce depolarization, but it takes several hours for the electrophysiological properties of the membrane to return to normal. The authors conclude (Ogura and Takahashi, 1976) that the Ca^{2+} channels are localized in the ciliary membranes, whereas the K^+ channels which function in hyperpolarization (Naitoh and Eckert, 1973) must be in the plasma membrane.

Depolarization is also eliminated when deciliated *P. caudatum* are stimulated with a pulse of electrical current. Regrowth of the cilia restores the depolarization response (Dunlap and Eckert, 1976).

Several laboratories have been investigating the proximal region of cilia in *P. aurelia* and *P. caudatum,* using cytochemical techniques and x-ray microanalysis of the resulting electron-opaque deposits. Plattner (1975), Plattner and Fuchs (1975), Fisher and co-workers (1976), and Tsuchiya and Takahashi (1976) have all shown electron-opaque deposits at the proximal ends of cilia on the cytoplasmic side of the membrane adjacent to the 9 patches of particles opposite the doublet microtubules. These deposits are enriched in calcium and chlorine over the rest of the cytoplasm and, to a lesser extent, are enriched in phosphorus and sulfur. Plattner and Fuchs (1975) and Tsuchiya (1976) speculate that the deposits are calcium phosphate, the phosphate group coming from the splitting of ATP possibly by an ATPase activity of the membrane-intercalated particles of the patches (Plattner and Fuchs, 1975). Fisher and associates (1976), finding that chlorine is present more frequently along with calcium, believe the deposits may be calcium salts. The calcium is associated with microfilamentous linkers (Figs. 14, 15 and 30), which may be extensions of the intramembranous patches of particles (Figs. 12, 13 and 29) linking the membrane to the microtubular doublets at the base of the axoneme. Sulfur, which is frequently found in the deposits, and at least some of the phosphorus, are probably components of the protein substructure to which the calcium complex is bound (Fisher et al., 1976).

Similar calcium-containing deposits were also found at other membrane-related sites, frequently at locations enriched in intramembranous particles. The most common of these sites was between the plasma membrane and the outer alveolar membrane (Plattner and Fuchs, 1975; Fisher et al., 1976). Deposits were also reported at junctions between the trichocyst and surface membranes (Fisher et al., 1976), on cytoplasmic smooth membranes near parasomal sacs, and on discoidal vesicles in the oral region (Plattner and Fuchs, 1975; Fisher et al., 1976). Plattner and Fuchs (1975), however, failed to find the deposits at the trichocyst-cell membrane junction in intact "resting" trichocysts. In view of the important role calcium plays in exocytosis (Poste and Allison, 1973; Douglas, 1974; Rubin, 1974), the calcium-containing deposits near the parasomal sacs and in the oral region may be related to the presumed endo- or exocytotic activities (Allen, 1967, 1974; Noirot-Timothée, 1968) of these regions. Plattner and Fuchs (1975) also detected calcium-containing deposits on the nonluminal surface of membranes enclosing crystals and within special vacuoles which they called "calcium storing vacuoles." However, these special vacuoles do not show deposits after fixation in calcium-enriched glutaraldehyde.

The accumulated evidence gathered from electrophysiology, electron microscopy and x-ray microanalysis thus argues for the presence of Ca^{2+} gates or channels within the ciliary membrane (Ogura and Takahashi, 1976; Dunlap, 1977). When cilia break away from the cell, the break occurs between the cell body and the ciliary patches (Satir et al., 1976) so that the patches are lost with the cilia. Although these patches are probably not Ca^{2+} gates, they may be Ca^{2+} pumps.

The picture of membrane proteins in the ciliates thus remains very sketchy, and only recently have any attempts been made to look at structural proteins of various cellular fractions. These latter studies were initiated mainly to follow the rates of incorporation of newly synthesized proteins into membranes during membrane biogenesis or as an aid to understanding the basis for membrane mutant alterations.

3.3. Surface and excreted chemicals

3.3.1. Glycocalyx

Electron micrographs reveal the presence of a glycocalyx on the outer surface of the plasma membrane of many ciliates (Figs. 16 and 34). In *P. aurelia* this external membrane coat measures 20 nm in thickness and stains with ruthenium red (Wyroba and Przełecka, 1973), indicating the presence of sugars and anionic groups. Bands of glycoproteins have been detected in ciliary membrane proteins of *Paramecium* which were run on polyacrylamide gels (Hansma and Kung, 1975; Kitamura and Hiwatashi, 1976), but Subbaiah and Thompson (1974) failed to find any glycoprotein bands in SDS gels of cilia, ciliary supernatant, microsomes, or postmicrosomal fractions of *Tetrahymena*, strain WH-14.

The 20 nm surface coat of *P. aurelia* has a homogeneous layer next to the membrane and an external filamentous layer. It covers the entire cell, both pellicle and cilia. Most of the ruthenium red staining is prevented by prior digestion with neuraminidase. For the most part, pronase or trypsin removes the outer filamentous layer. Wyroba and Przełecka (1973) consider this glycocalyx to be composed of glycoprotein with a small amount of sialic acid.

Hausmann and Mocikat (1976) have found that the alcian-blue positive surface coat of the marine ciliate, *Litonotus duplostriatus*, is also composed of two layers, a "ground layer" next to the plasma membrane and a "filamentous" layer. Surfaces of mucocyst membranes that are open to the outside medium do not stain with alcian blue and have no visible coat.

3.3.2. Immobilization antigens

The presence of surface antigens on *Paramecium* and *Tetrahymena* has been known for many years and this topic has been discussed in a number of reviews (Preer, 1969; Gibson, 1970; Finger, 1974). Briefly, in *P. aurelia* a large amount of protein, constituting 30% of the total ciliary proteins (Preer, 1969), covers the entire external cell surface in a 20 to 30 nm layer (Mott, 1963, 1965). The presence of antibodies formed against these antigens causes the cilia to clump together, which immobilizes the cells, and these proteins have consequently been called immobilization antigens (i-antigens). The particular species of i-antigen carried by the cell determines its "serotype," and a large literature has accumulated on the immunology and genetics of serotype determination (S. Allen, 1967; Preer, 1969). However, the function of these i-antigens is still unknown, though they appear to be indispensable to the life of the cell. Changes in environmental conditions, such as changes in temperature, salt concentration or changes in the source of food may cause the cell to replace its existing antigen with another chemically distinct i-antigen.

The i-antigens are easily solubilized in saline or culture fluid (Finger, 1956) and have been shown to consist primarily of protein with small amounts of carbohydrate (Preer, 1959a,b). Although each antigen appears to have a unique amino acid sequence, the antigens have very similar total threonine, leucine, and cysteine contents (Steers, 1965; Reisner et al., 1969b). The i-antigens are very large, having molecular weights ranging from 240,000 to 310,000 (Preer, 1959b; Bishop, 1961; Steers, 1962, 1965; Jones, 1965; Reisner et al., 1969a,b). Although several authors reported that i-antigens were composed of a number of polypeptide chains held together by disulfide bonds, Hansma (1975) has presented evidence that a contaminating protease activated by mercaptoethanol, which was used in the isolation medium of the prior experiments, may have been present in these previous studies. Thus Hansma concluded that the i-antigen of *P. aurelia*, syngen 4, consists of a single-polypeptide chain with a molecular weight of 250,000. Steers and Davis (1977) have confirmed the presence of a protease in the earlier isolation studies of Steers (1962, 1965). These i-antigens would be among the largest naturally occurring single-polypeptide chains known (Reisner et al., 1969a,b).

An immobilization antigen has also been isolated from *P. caudatum* (Koizumi, 1966) and has an even higher molecular weight of 340,000.

Besides the immobilization antigens, Klimetzek (1977b) has identified eight different antigens on the ciliary and cortical membranes of *P. multimicronucleatum*. One of these, the Y antigen, is restricted to ciliary membranes. All eight antigens are found in mating reactive and unreactive cilia of both members of a mating type pair.

3.3.3. Gamones and mating type substances

Other substances that appear to be present on the surface of ciliates, at least on their cilia, are the specific mating type substances (reviews, Preer, 1969; Finger, 1974). In most ciliates these substances are not detectable in the medium but are bound to the cilia. This has been demonstrated in *P. caudatum*, in which cilia isolated from two complementary mating types and mixed together undergo strong agglutination (Takahashi et al., 1974). Mixing the deciliated cells themselves from two mating types does not result in agglutination.

When the membranes are removed from isolated cilia and added to cells of the complementary mating type, a strong self-pairing of these cells occurs (Kitamura and Hiwatashi, 1976). The homotypic pairs not only agglutinate, they also go through the normal conjugation processes. There is no reaction if the membrane-derived vesicles are added to the same mating type cells from which they originate. Separation of the vesicle proteins on polyacrylamide gels indicates a complex mixture of some 20 protein bands (of which 4 are glycoproteins), but the one(s) responsible for the mating reaction have not yet been identified. In *P. caudatum* (Hiwatashi, 1961) and *P. bursaria* (Cohen and Siegel, 1963; Cohen, 1965) only the cilia on the anteroventral surface are active in mating. In most cases the mating type substances have not been isolated but are generally thought to be proteins, probably glycoproteins. Klimetzek (1977a) reports that the mating-reactive substances of *P. multimicronucleatum*, mating type III, forms one major protein band on SDS-polyacrylamide gels from ciliary membranes of cells that are competent to mate but this band is lost when cells enter the unreactive state.

Some ciliates have soluble mating type substances that are liberated into the medium and interact with complementary mating type cells. Such soluble substances are produced by *Euplotes patella* (Kimball, 1939), *Tetrahymena pyriformis*, syngen 7 (Phillips, 1971), and *Blepharisma intermedium* (Miyake, 1968; Miyake and Beyer, 1973). Many details of the mating reaction of this last ciliate have been explained in recent years in the interesting work of Miyake and collaborators.

There are two mating types in *B. intermedium*, types I and II, each of which excretes a specific mating type substance, or gamone. Prior to conjugation mating type I produces gamone I, which reacts with type II cells, transforming them into competent mating type II cells and inducing them to excrete gamone II. Gamone II then attracts (Honda and Miyake, 1975) and reacts with type I cells, transforming them and enhancing the further production of gamone I. After a waiting period of about 2 hours, competent type I and II cells can unite mouth to mouth to form conjugant pairs (Miyake, 1974a,b). Only the oral cilia on the left and right sides of the anterior half of the oral groove are involved in this initial contact and ciliary union (Honda and Miyake, 1976).

Both gamones have been purified and identified. Gamone I is a slightly basic glycoprotein with a molecular weight of about 20,000 (Miyake and Beyer, 1973, 1974) and has been named blepharmone. The carbohydrate component makes up about 5% of the molecule. Glucosamine and mannose are present but there is no detectable sialic acid (Braun and Miyake, 1975). The molecule is highly polar. Gamone II is a lactate molecule (Kubota et al., 1973) called blepharismone. This gamone has been synthesized chemically, and the synthetic gamone induces conjugation (Tokoroyama et al., 1973). The gamones do not play a role in the physical process of agglutination but function as hormone-like substances that induce specific cell types to modify their protein-synthesizing machinery and ciliary surfaces so that cell union will result. Ciliary union seems to require an accumulation of a protein or proteins that will disappear continually unless the gamone is constantly around. Some of these presumably protein factors will be incorporated into the surface glycocalyx of specialized cilia and will function in agglutination.

Once attached to receptor sites, gamones stimulate protein production, which may increase to five times the normal rate. This model system thus postulates the existence on the cell surface of specific receptor sites for the gamones (Miyake, 1974a). The glycoprotein, gamone I, may attach to receptor sites on a type II cell by its polar groups after its sugar portions recognize the specific attachment sites (Braun and Miyake, 1975). Ricci and colleagues (1976) have observed that membrane-active ligands such as the lectins concanavalin A (Con A) and phytohemagglutinin-W (PHA), or antibodies to tubulin, can competitively inhibit pairing of homotypic *B. intermedium* type I cells if gamone II is present in suboptimal concentrations. They conclude that although these ligands do not bind to the same receptor sites as the gamones, they nevertheless inhibit the attachment of gamone to its specific receptor sites. This may result from changes within the cell surface membrane that alter the distribution of receptor-sites or produce a change in the optimal conformation state of the gamone receptors. As in other animal cells, lateral motion of plasma membrane components has been shown in *B. intermedium* after they are reacted with specific ligands (Revoltella et al., 1976). Con A also blocks conjugation in *T. pyriformis* (Ofer et al., 1976).

A requirement for protein synthesis for ciliary union to occur has been demonstrated by Miyake and Honda (1976) in *Blepharisma,* and the union between cells can be disrupted by pronase (Miyake and Beyer, 1973). However, the actual substance that causes the stickiness of cilia has not yet been isolated.

Following ciliary union, a few hours elapse before the surfaces of the two cells fuse to complete conjugation. This fusion is apparently an independent process from that of ciliary union (Honda and Miyake, 1976). Miyake (1974a) views conjugation in *Blepharisma* as a potential model system for the study of cell-to-cell recognition and interaction in tissue formation in developing multicellular organisms.

When the mating types of other species of *Blepharisma,* such as *B. americanum, B. musculus,* and *B. stoltei* were investigated, it was found that they also have two mating types. One (type II) excretes a gamone that has the same activity as blepharismone, while the complementary cells (type I) excrete a gamone that induces type II cells to pair with type I cells (Miyake and Bleyman, 1976). These authors feel that type II gamones from all these species are probably exactly the same molecules as in *B. intermedium* type II cells, while gamone I varies slightly from species to species.

Although mating type substances have not been isolated from other ciliates, there is considerable evidence that they do exist and that they are, in part, proteinaceous and are frequently attached to the surface of ciliary membranes. Inhibitors of protein synthesis have been shown to prevent conjugation in *Euplotes*, *Oxytricha*, *P. aurelia*, and *Tetrahymena* (references, Miyake and Honda, 1976). The compound tunicamycin, which inhibits glycoprotein synthesis, inhibits conjugation in *Tetrahymena pyriformis* (Frisch et al., 1976). Also, enzyme treatments with trypsin and lipase increase the preconjugant pair formation time in starved *Oxytricha bifaria*, trypsin inhibiting the initial phase of contact while lipase prevents cell fusion (Ricci et al., 1975). The same reaction to trypsin and lipase is also seen in *P. multimicronucleatum* (Miyake, 1969, 1974b).

Thus several specific chemicals are involved in the conjugation process. Initially, soluble gamones or surface-attached substances are produced that induce complementary mating types to become competent. In cases where soluble gamones are formed, new protein synthesis results in the formation of compounds that are deposited on the ciliary surface and, when these substances are present, cilia of complementary mating types will stick together. Events then lead to membrane fusion at the surfaces of the mated cells.

3.3.4. Chemoattractants

A recent report by Nohmi and Tawada (1974) shows that the ciliate *Tetrahymena pyriformis* excretes soluble protein into the medium that produces a chemotactic response in *Amoeba proteus*. This negatively charged protein has a molecular weight of 24,000 when calculated from gel-filtration experiments on Sephadex but on SDS-polyacrylamide gels it has a molecular weight of 8,000 which suggests that the molecule is probably composed of 3 subunits. The synthesis site and excretion method of this protein from *Tetrahymena* are unknown.

4. Changes in membrane constituents of ciliates

4.1. Lipid and protein exchange between membranes

The major interest of the laboratories of Thompson and Nozawa over the past decade has been the regulation of the transfer and assembly of newly synthesized lipids and proteins into the various membranes of *Tetrahymena pyriformis* (Thompson, 1972; Thompson and Nozawa, 1972). Thompson and his coworkers in their radioisotope pulse-labeling experiments have made a major contribution to our understanding of biogenesis of the membranes of this eukaryotic model cell.

As reviewed above, various membranes of this ciliate contain different proportions of the major classes of phospholipids, namely PE, PC, and GAEP. Different membranes also contain various protein species as well as different proportions of the same proteins. Consequently, the problem to be solved is how these varied patterns of lipid and protein distribution are achieved from a system in which a presumably common pool of newly synthesized proteins and lipids come from common microsomal enzymes.

Using isotopically labeled precursors of the structural lipids, Thompson (1967) found that nearly all of the [^{14}C]palmitate added to the medium is covalently bound to phospholipids within 5 minutes and subsequent changes in the labeling pattern are due to the redistribution of the labeled lipids formed during this 5-minute period. Microsomes were the first membrane fraction to become highly labeled, taking about 10 minutes to reach maximum activity, while the ciliary membranes and pellicles took up to 6 hours to reach the same specific activity as the microsomal fraction (Nozawa and Thompson, 1971b). The same trends were obtained using [^{14}C]acetate, [^{3}H]hexadecyl glycerol and [^{32}P]i (Thompson, 1972). Thus the conclusion reached by Thompson is that internal membranes are more accessible for lipid addition and exchange than the surface membranes which are known to be morphologically isolated from the endoplasm by the epiplasm and alveoli. (Allen, 1967).

In order to distinguish lipid exchange from net lipid synthesis, cells were placed in an inorganic medium, a condition that stops growth as well as lipid synthesis. After labeling with [^{14}C]palmitate, cells were homogenized and fractionated and it was found that the various fractions contained the same labeling patterns as those obtained for growing cells (Nozawa and Thompson, 1972), indicating that the distribution of lipids results mainly from lipid exchange rather than net lipid synthesis.

To understand this intracellular mobility of lipid molecules, a phospholipid exchange protein, such as that found in rat liver by Wirtz and Zilversmit (1968), was sought. However cycloheximide, which inhibits protein synthesis had no effect on lipid redistribution, nor did the use of tetralylphenoxyisobutyric acid which inhibits 90% of the fatty acid synthesis (Nozawa and Thompson, 1972). Lipid insertion into membranes does not appear to be coupled to the insertion of proteins, and vice versa. This seems to rule out the insertion of newly formed subunits into membranes in the form of vesicles such as proposed by Franke and co-workers (1971a) or Satir (1974a) for plasma membrane formation (see section 5.1). Another conclusion from the above experiments is that the lipid transport process itself is not the mechanism used to maintain a dynamic balance of membrane proteins and lipids (Thompson, 1972).

Additional alterations in lipid composition occur after the incoming lipids reach their membrane destinations. The proportion of radioactive lipids arriving at surface membranes shortly after (1–5 min) tracer administration does not resemble the pattern of bulk lipid distribution in the surface membranes, but resemble that found in the ER where the lipids are synthesized (Thompson et al., 1971). After 30 minutes, however, the distribution of radioactive lipids in the surface membrane is the same as the bulk distribution of lipids in these membranes. Tetrahymenol and the phosphonolipids are now enriched in the surface membranes, and PC and PE are diminished. Thus the mixture of lipids that is first incorporated into these membranes is subsequently altered, increasing the proportion of certain lipids and decreasing others. Thompson (1969) believes the lipids that are retained may be more resistant to the endogenous lipolytic enzymes of the cell surface that degrade PC and PE but not GAEP (Thompson et al., 1971, 1972).

When biogenesis and transport of proteins was followed by using a 5-minute pulse of [^{3}H]leucine followed by a chase of excess nonradioactive leucine, the cilia acquired a low specific activity in their proteins but this activity never reached that of whole cell proteins (Subbaiah and Thompson, 1974). Initially, microsomal protein-specific activity was high but after 30 minutes it declined below whole cell values. Protein activity in other fractions was not measured.

When individual proteins were separated and their radioactivity analyzed, some differences in specific activities of the various proteins were noted (Subbaiah and Thompson, 1974). This may mean that these proteins are moved to the sites of membrane insertion and each is inserted independently of the others. When the ciliary matrix proteins are compared to ciliary membranes and axonemes, label is found first in the matrix and then in the membranes and axonemes; equilibrium is reached in these fractions after 2 hours. The band with highest activity after 5-and 10-minute pulses of radioactive material were components common to both matrix and membrane subfractions. This suggests that new proteins enter cilia first in soluble form (Subbaiah and Thompson, 1974) rather than as performed vesicles. Also, the differences between the time required for the specific activity of lipids to plateau in cilia (6 hr) and that of the specific activity of proteins to plateau (2 hr) suggest that lipids and proteins destined to become part of the ciliary membranes are not transported together through the cell as stable complexes, though they could be transported as loosely bound complexes in which lipids are exchanged with the unlabeled lipids of membranes encountered on the way to the cilia (Subbaiah and Thompson, 1974).

One likely exception to the finding that lipids and proteins are inserted into membranes independently of each other is in the formation of digestive vacuole membranes. Exposing starved *Tetrahymena* cells to ferric oxide resulted in an immediate, rapid formation of digestive vacuoles (Weidenbach and Thompson, 1974). Pretreatment with cycloheximide, which inhibits protein synthesis, does not stop this burst of digestive vacuole formation; thus a reservoir of membrane protein material must exist in the cell from which nascent digestive vacuoles are formed. Also, lipid components of the digestive vacuole membrane do not appear to be formed at the expected increased rate during rapid digestive vacuole formation. Using [^{32}P]orthophosphate, the specific activity of [^{32}P] in vacuoles, which was expected to increase sharply in nascent vacuole membranes, remained about the same as the preexisting vacuole-membrane fractions (Weidenbach and Thompson, 1974).

Electron microscope studies of digestive vacuole formation show vesicle fusion with the nascent vacuole, suggesting that in *Paramecium* a pool of preformed discoidal vesicles serves as a reservoir for digestive vacuole membrane production (Allen, 1974). This topic will be discussed more fully in section 5.2.

4.2. Changes correlated with the cell cycle

When the incorporation of the radioactive lipid precursors [^{32}P]i and [^{14}C]acetate were investigated at various stages of the cell cycle in synchronized cultures of *T. pyriformis* strain GL, a peak of activity was noted at the first stage of cytokinesis when [^{14}C]acetate was used but not when [^{32}P]i was used. Synchronization was produced by the method of using one heat shock treatment per generation (Zeuthen, 1964). Thus increased fatty acid synthesis, but not phospholipid synthesis, is associated with cytokinesis (Baugh and Thompson, 1973). [^{3}H]leucine however, is incorporated into proteins linearly throughout the cell cycle (Baugh and Thompson, 1973).

In a later paper these same authors (Baugh and Thompson, 1975) examined the incorporation of radioactivity into membrane lipids and proteins of various cell fractions in synchronized cultures at specific times in the cell cycle. They showed that phospholipids are incorporated into organelle membranes in a cyclic fashion correlated with specific stages in the division cycle. Mitochondria and microsomes were labeled most rapidly prior to the beginning of furrow formation, and cilia are most rapidly labeled during furrow formation, while the pellicle shows little variation in labeling throughout the cell cycle. Protein incorporation was also cyclical in cilia, mitochondria, and microsomes, but pellicles showed no significant fluctuations in uptake of [^{3}H]proteins during the cell cycle.

Thus, the intracellular distribution of new phospholipids and proteins is closely regulated during the cell cycle. These cyclic radioactive changes may be controlled in part by changes in lipid transport or, in the case of microsomes, by differential retention of newly synthesized phospholipids (Baugh and Thompson, 1975).

The peaks of lipid labeling could not be compared directly with those of protein labeling in the work of Baugh and Thompson (1975), but it was estimated that in cilia and mitochondria the peaks of lipid incorporation follow peaks of protein incorporation by 20 to 30 minutes, while the lipid and protein incorporation peaks coincide more closely in microsomes, exept that there is an additional peak of lipid incorporation just prior to cytokinesis. Baugh and Thompson hypothesize that the production of proteins and their subsequent deployment to particular organelles may induce the movement of those phospholipids that are needed to construct the membrane to the organelle in question. Thus synthesis of specific groups of proteins would initiate membrane expansion. The sequence of protein synthesis would presumably be under the control of regulated gene expression (Baugh and Thompson, 1975).

4.3. Changes induced by physical and chemical alterations of the environment

4.3.1. Ergosterol supplementation
By changing the cell's environment, the chemical composition and fluidity of the membranes of *Tetrahymena* can be altered. The latter alteration can be visualized by changes in the distribution of intramembranous particles. Some of the altera-

tions occur normally in the life cycle of the cell, while others can be induced experimentally and would probably not be encountered in the natural environment of a cell.

One experimentally induced change is the replacement of tetrahymenol in the cell membrane by ergosterol (Conner et al., 1971). The ergosterol appears to replace tetrahymenol in a one-for-one manner, by dilution rather than by metabolic removal. *Tetrahymena* seems to be incapable of catabolizing tetrahymenol (Thompson et al., 1972). Ergosterol-substituted cells become sensitive to antibiotics such as filipin and nystatin (Conner et al., 1971; Nozawa et al., 1975), and when treated with these drugs will lyse within 2 to 3 minutes. Filipin-treated ergosterol-substituted membranes of *Tetrahymena* show large 2.5 to 3.0 nm projections within the freeze-fractured surface membranes, and large surface lesions are apparent in scanning electron micrographs of these cells. The projections in the interior of the membrane are considered by these authors (Nozawa et al., 1975) to be evidence for the incorporation of ergosterol into the membrane's hydrophobic center.

Replacement of tetrahymenol by ergosterol also induces small quantitative changes in the fatty acid composition of the membranes, producing a decrease in the amount of unsaturated fatty acids present and an increase in shorter chain length fatty acids (Ferguson et al., 1971; Mallory and Conner, 1971; Ferguson et al., 1975; Nozawa et al., 1975). The $\Delta^{6, 11}$ isomer of linoleate is also increased in ergosterol-supplemented cells (Ferguson et al., 1975). Changes in the fatty acid composition of the membrane produce changes in membrane fluidity. Higher levels of unsaturated fatty acids will result in greater fluidity of the membrane, while increasing amounts of saturated fatty acids will promote membrane rigidity. Ergosterol-substituted membranes thus become less fluid.

Nozawa and colleagues (1975) have shown that replacement of tetrahymenol by ergosterol causes an increase in PE and a decrease in PC and GAEP, though the total phospholipid quantity remains relatively unchanged. There is also a trend toward increased saturation of hydrocarbon chains in all three phospholipid classes, with PE being the most saturated and GAEP being less saturated. PE may play an important structural and functional role in surface membranes by compensating for the sterol change and thereby regulating membrane fluidity (Nozawa et al., 1975). By analogy to higher animal membranes that contain the sterol cholesterol, Nozawa and collaborators (1975) suggested that tetrahymenol may also regulate membrane fluidity in natural membranes.

4.3.2. Colchicine effects
Colchicine also has a direct effect on the mobility of particles within the membranes of *Tetrahymena* (Wunderlich et al., 1973a). When cells are treated with this alkaloid, aggregation of intramembranous particles upon cooling does not occur within a sharp temperature range, that is, the critical transition temperature range is significantly broadened. Nor does reheating result in complete randomization of the particles. Since the membranes observed were those of the outer alveolar membrane, which are not known to be associated with micro-

tubules, the effect of colchicine on membrane-associated microtubules and, thus, on membrane particle movement, can be ruled out. However, if colchicine could "dissolve" into the apolar membrane regions it might affect membrane fluidity, or it may bind to a membrane protein and consequently alter the mobility of these integral proteins (Wunderlich et al., 1973a).

4.3.3. Temperature shifts

4.3.3.1. Effect on nutritional requirements. Shifts in temperature can induce changes in the nutritional requirements of *T. pyriformis*, mating type II, variety I (Erwin, 1970), as well as changes in membrane fluidity (Nozawa et al., 1974; Wunderlich et al., 1975). At 40°C the strain of *Tetrahymena* studied by Erwin (1970) has an unnatural nutritional requirement for phospholipids. Abnormal cells contain "blebs," elongated cilia, and even lose their typical pyriform shape when pools of phospholipids are exhausted. Growth can be sustained with crude phospholipids, synthetic phospholipids, or with purified ciliate fatty acids. The damage in the cell's membranes is due to an impaired ability to synthesize polyunsaturated fatty acids (Erwin, 1970).

At the ultrastructural level abnormal cells growing at 40°C in a defined medium supplemented with phospholipids which, in turn, have only saturated fatty acids contain all of the normal cortical elements, but there are evident shifts and spatial disorientation of many surface features and structures (Lo et al., 1976). Internal cytoplasmic membranes are also disorganized or reduced. Thus culturing cells under these conditions primarily affects the cell membranes. Other changes, such as the breakdown in basal body rows or the disorder in the surface-associated longitudinal microtubules, may be due to the disorganization of attachment sites for these structures on the inner alveolar membranes (Lo et al., 1976).

4.3.3.2. Effect on membrane fluidity. Shifts in temperature also produce alterations in membrane fluidity. These can be visualized in freeze-fractured membranes by changes in the distribution of intramembranous particles. Normally particles are randomly distributed in the surface membranes of *T. pyriformis*, strain GL, when grown at 28°C but, when cooled to 5°C over a 5-second interval, the particles aggregate into a netlike reticulum (Speth and Wunderlich, 1973). Aggregation begins at 15°C. When cooled slowly from 28°C to 5°C over a 4-minute interval, not only are the particles aggregated but the particle numbers are also reduced. This aggregation is probably the result of the crystallization of lipid domains, with the proteins pushed together in front of the advancing boundary of crystallization (Speth and Wunderlich, 1973; Wunderlich et al., 1973b, 1974a, 1975).

Wunderlich and co-workers (1975) have studied the lipid phase transition in the ER of *Tetrahymena*, strain GL, to confirm that the observed aggregation of particles seen in replicas of freeze-fractured membranes occurs at the same temperature that membrane fatty acids change from a disordered, rapidly moving state to a more ordered phase in which they are less mobile. This change occurs at a specific temperature for each specific membrane and is a function, in part, of the lengths and relative saturation of the fatty acids that are attached to the phospholipids (review, Kimelberg, 1977). By using fluorescent dye probes and electron spin resonance, Wunderlich and associates (1975) observed that the transition temperature for the isolated ER membrane was close to 17 °C, and that the amount of flouresent probe bound to the membrane and the amount of spin label probes in the membranes were also reduced at the same temperature. This is nearly the same temperature at which smooth areas begin to appear in freeze-fractured membranes; the temperature at which smooth areas first appeared was 15 °C (Wunderlich et al., 1975). The conclusion from these experiments is that the lipid molecules undergo a phase transition at the same temperature at which membrane changes first appear in freeze-fractured membranes. Thus the temperature at which replicas of membranes first show smooth areas gives a rough estimate of the bulk transition temperature of that membrane.

Wunderlich and collaborators (1977) have isolated the total lipids from microsomes of *T. pyriformis*, GL, and studied these lipids at different temperatures by x-ray diffraction. The lipids, extracted from cells grown at 28 °C, change from a liquid crystalline state at about 16 °C, while those from cells grown at 18 °C change at about 10 °C. They further determined that those lipid species most responsible for this shift in the phase transition temperature are the ethanolamine-containing phosopholipids.

Using electron spin resonance spectroscopy (ESR) techniques, Nozawa and colleagues (1974) found that the membranes of *Tetrahymena* vary in fluidity in the following order: microsomes > pellicles > cilia. The content of tetrahymenol in these membranes increases in reverse order: cilia > pellicles > microsomes. A tetrahymenol-containing lipid dispersion is more fluid at temperatures below 27 °C and less fluid above 27 °C. Tetrahymenol may thus rigidify membranes at the higher temperature and cause them to be more fluid at the lower temperature (Nozawa et al., 1974). This is, of course, similar to the properties of cholesterol in other cells (Oldfield and Chapman, 1972; Rothman and Engelmann, 1972; Schreier-Muccillo et al., 1973). Thus neutral lipids such as tetrahymenol may not be simply structural components but may also be functional, controlling (in combination with hydrocarbon chains) the fluidity of the membrane and thus, in turn, enzyme activities that are influenced by the physical state of membrane lipids (Nozawa et al., 1974).

Membrane-intercalated proteins may vary in their reaction to being cooled below the lipid phase-transition temperature. Particles within the surface membranes of *Tetrahymena* may only be displaced within the plane of the membrane so that they aggregate into a reticulum (Speth and Wunderlich, 1973). Alternatively, the absolute number of particles may be reduced as seen in the ER membrane, where the number of particles is reduced but there appears to be no aggregation (Wunderlich et al., 1975). In cases where the particle number is reduced Wunderlich and colleagues (1974b, 1975) believe that the particles may be squeezed out of the hydrophobic core of the membrane as were the spin label probes in their study. Nuclear membranes, like surface membranes, undergo particle displacement both normal to, and parallel with, the membrane plane (Wunderlich et al., 1974a).

Changes in membrane fluidity have a pronounced effect on the functional properties of the membranes involved. In the ER in *Tetrahymena*, GL, its glucose-6-phosphatase content undergoes a biphasic decrease in activity when the temperature is lowered to the phase-transition temperature (Wunderlich et al., 1975). Also, there is no incorporation of [^{14}C]uridine into cytoplasmic RNA at temperatures below the phase transition of the nuclear membranes, although labeled RNA continues to accumulate in the nucleus (Wunderlich et al., 1974a). Since the transport of RNA to the cytoplasm is the rate-limiting step at temperatures below 18°C, the membranes of the nuclear envelope are implicated in the RNA transport process (Wunderlich et al., 1974a).

Digestive vacuole membranes are also altered in their physical properties by changes in the temperature at which a ciliate is grown. Kitajima and Thompson (1977b) using *T. pyriformis*, NT-1, find that the digestive vacuole membrane and the surrounding discoidal vesicles undergo a phase separation of their intramembranous protein and lipid components at a temperature of 9°C which is lower than any other membrane in this cell. The plasma membrane undergoes a phase separation at 12°C, while the phase separation of ER occurs at a much higher temperature (24°C), as does that of the outer alveolar membrane (Kitajima and Thompson, 1977a). If these cells are then acclimated to grow at 15°C for 4 hours the phase-separation temperature of digestive vacuole membranes is reduced to the point that phase separation does not occur even at 0°C. Digestive vacuole membranes also undergo a differentiation that parallels the digestive process taking place witin the vacuole. Within 2 to 5 minutes after vacuole formation the phase-separation temperature rises from 9°C to 14°C, during which time no visible fusion with other organelles can be detected. More mature vacuoles have a slightly lower phase-separation temperature of 12°C. Correlated with the sudden rise in the transition temperature is an increase of intramembranous particles on the P face of the newly formed vacuoles; from 761 ± 219 to 1625 ± 350 particles per μ^2. The E face of the nascent digestive vacuole has 146 ± 85 particles per μm^2, and this remains unchanged even in mature vacuoles. These authors also found a correlation between the temperature at which formation ceases and the membrane transition temperature; vacuoles cease to form a few degrees above the transition temperature. They conclude that the changes in the transition temperature are brought about by changes in vacuole membrane lipid composition (Kitajima and Thompson, 1977b) via the transfer of lipids between membranes rather than through dilution of the vacuole lipids as a consequence of fusing with other membranes.

4.3.3.3. Acclimation of membranes. Acclimation of *Tetrahymena* membranes to different temperatures, as for example, 15°C and 35°C, is brought about by alterations in the fatty acid composition of the membranes induced by the growth-temperature shift (Erwin and Bloch, 1963; Wunderlich et al., 1973b). Lower temperatures favor unsaturated fatty acids and more fluid membranes; higher temperatures favor saturated fatty acids and less fluid membranes (Nozawa et al., 1974). *Tetrahymena* cells acclimated to grow at 10°C have their saturated fatty acid content reduced from 25% when growing at 28°C to 13% at 10°C, after 16 hours at the lower temperature. Thus 50% of their saturated fatty acids are replaced by unsaturated fatty acids (Wunderlich et al., 1973b). Monounsaturated fatty acids increase from 16 to 28% of the total fatty acids over a 24-hour period; for the most part fatty acids with two double bonds remain level, though a slight decrease can be detected after 8 hours, while fatty acids with three double bonds increase during the first 4 hours and then remain constant at about 33% of the total fatty acid content of the membranes (Wunderlich et al., 1973b). The absolute number of double bonds increases during the first 8 hours and then remains constant, although the fatty acid pattern continues to change following this 8-hour period (Wunderlich et al., 1973b). Nozawa and co-workers (1974) found that in strain WH-14 the two fatty acids that undergo the most striking changes when the growth temperature is raised from 25°C to 34°C are palmitic (16:0) and palmitoleic (16:1) acids, with the former increasing and the latter decreasing in quantity. Other fatty acids showed less variation. Similar changes occurred in all cell fractions except the cilia. Correlated with these fatty acid changes was a spin-label-determined increase in membrane fluidity in the cilia and pellicular membranes in the 15°C grown cells as compared to 34°C grown cells (Nozawa et al., 1974).

In the thermotolerant strain, NT-1, of *Tetrahymena* the cells will grow normally at 39.5°C and can then be acclimated to growth at 15°C (Fukushima et al., 1976a). Particles of the outer alveolar membrane will initially aggregate, but these aggregated particles will then randomize after 10 hours at the lower temperature. Particles in other internal membranes will aggregate at their own phase-separation temperatures and all membranes will have aggregated particles at 5°C. Although the percentage of the major phospholipid classes within cells usually remains constant during a temperature shift, in *Tetrahymena* an upward temperature shift resulted in a decrease in the concentration of GAEP and a corresponding increase in PE while PC remained constant (Fukushima et al., 1976a). Similar trends were also observed in strain WH-14 (Wunderlich and Ronai, 1975). Glyceryl ethers, of which 1-0-hexadecylglycerol accounts for over 80% of the total, decrease with increasing temperatures while the tetrahymenol to phospholipid ratio remains constant. Plasma membranes and microsomes follow the above pattern, but ciliary membranes have a marked increase in PC and a decrease in lysoPE and lysoGAEP. Accompanying the changes in phospholipid species are changes within the fatty acid composition of the phospholipids. A decrease in growth temperature results in an increase in linoleic (18:2) and γ-linolenic (18:3) acids and a decrease in palmitic acid (16:0). Increases were found in the cilia, the pellicle, and the microsome fractions (Fukushima et al., 1976a). Conner and Stewart (1976) noted a 7% decline in saturated fatty acids when strain WH-14 was shifted downward to 15°C from its normal growth temperature of 35°C. This decline is due mainly to a decrease in palmitic acid, whereas there is an increase of 9% in the unsaturated fatty acids, due principally to an increase of palmitoleic acid. Polyunsaturated acids were reported to be reduced in quantity. Altogether, the number of double bonds remained constant during the temperature shift (Conner and Stewart, 1976).

4.3.4. Lipid precursor supplementation

Tetrahymena, NT-1, cells grown at a constant temperature of 39.5 °C in the presence of high supplemental amounts of partly radioactively labeled linoleic acid (18:2, $\Delta^{9,\,12}$) undergo an increase of polyunsaturated fatty acids in their membranes (Kasai et al., 1976). Polyunsaturated fatty acids in their membranes increased from 35 to 40% of the fatty acid population to over 50%. The major saturated acids (14:0 and 16:0) remained at a normal level but the monounsaturated fatty acids (16:1 and 18:1) decreased significantly. No change in intramembranous particle aggregation in the outer alveolar membrane from that of control cells was noted in the cooled polyunsaturated acid-rich membranes after a 100-minute supplementation. An increase in unsaturation, if it had occurred equally in all cell membranes, should have resulted in increased outer alveolar membrane (OAM) fluidity with a consequent, lowered phase-transition temperature. Extending the experiment to 345 minutes resulted in the appearance of a significant amount of particle aggregation even though the fatty acid composition changed little from what it was at 100 minutes. When other subcellular organelle membranes were investigated, it was found that the ER and nuclear membranes were the first to show the effect of increased membrane fluidity. The phase transition temperature of ER and nuclear membranes had dropped from 24 °C to 18 °C when the cells were fed linoleic acid (18:2) at a constant temperature of 39.5 °C for 100 minutes. By 345 minutes the ER and nuclear membrane transition temperature rose to 21 °C, but the OAM now showed signs of increased fluidity. Thus the flow of newly synthesized lipids was from the ER to the pellicular and ciliary membranes. Newly synthesized fatty acids appeared to be enriched in saturated acids. Thus, by holding the temperature constant and supplementing with high doses of unsaturated fatty acids, which enter the membrane and increase its fluidity, it can be demonstrated that the cell is capable of regulating fatty acid composition in response to increased membrane fluidity by reducing the level of monounsaturated acids and raising the saturated fatty acid level (Lees and Korn, 1966; Kasai et al., 1976). This shows that increases in unsaturated fatty acids are not tied to temperature shifts. The cell is capable of offsetting influxes of unsaturated fatty acids and minimizing changes in membrane fluidity even at constant temperature.

Kasai and collaborators (1976) believe that this evidence favors their interpretation that fluidity changes in the ER membrane are controlled by enzymatic desaturase activity so that the rate of desaturation of fatty acids occurring at the level of the ER increases when the membrane has a suboptimal fluidity and decreases under conditions of superoptimal fluidity, so that as fatty acid desaturation occurs in the ER (Fulco, 1974) the newly formed acids will pass from this organelle to all other membranes of the cell (Nozawa and Thompson, 1971b, 1972) by normal intermembrane lipid exchange. Fukushima and colleagues (1977) observed that a temperature shift from 39.5° to 15°C resulted in an increase in the relative proportion in the phospholipids of [^{14}C]palmitoleate formed from [^{14}C]palmitate. They found that palmityl CoA desaturase is localized in the microsomes and is temperature-dependent, with two breaks in the resulting Arrhenius plot.

This interpretation was supported by another study (Martin et al., 1976), where membrane fluidity and membrane fatty acid distribution in unsupplemented cells following a shift-down in temperature (from 39.5° to 15°C) were compared with the same membrane properties in cells supplemented with large amounts of unsaturated fatty acids during the course of the temperature shift. These supplemented unsaturated acids, linoleate (18:2) and γ-linolenate (18:3), are quickly incorporated into the cell membrane phospholipids; 55% of the supplemented fatty acids are incorporated by the end of the 8-hour acclimation period when cell division, which had been inhibited, begins again. Supplemented cells attain a higher level of unsaturation after they acclimate to the 15°C growth temperature than do the unsupplemented ones. In freeze-fracture replicas these changes are reflected in the pattern changes of smooth area formation in acclimated-cell ER membranes that were treated to show changes in their transition temperatures. Phase-transition alterations in the pellicle membranes were not as extensive or rapid as in the ER, and changes in the ciliary membranes occured last.

Phospholipid composition early in the 8-hour lag phase in both supplemented and control cells was about the same following the downward shifts as that in cells growing at 39.5°C. However, there was some reshuffling of phospholipids in the supplemented cells during the acclimation period; for example, GAEP and PC increased and decreased, respectively, in the microsomes with the reverse occurring in the pellicles during the first hour of the 8-hour period of inhibited division. Although the phospholipid composition of the 15°C grown cells has a much different pattern than that of 39.5°C grown cells (Fukushima et al., 1976a), the alterations in the phospholipids are slower than the rapid adaptation of the phospholipid fatty acids to low temperatures.

Using [^{14}C]acetate it was noted that in unfed cells less myristic (14:0) and palmitic (16:0) fatty acids are synthesized at 15°C than at 39.5°C. Supplemented cells incorporate exogenous linoleic (18:2) and γ-linolenic (18:3) acids at 15°C and consequently do not synthesize as much of these fatty acids. Most of the label becomes incorporated into saturated acids in supplemented cells. Although fatty acid synthesis is altered, the number of double bonds is unchanged between acclimated supplemented cells and acclimated unsupplemented cells. The authors (Fukushima et al., 1976a) conclude that the desaturase activity is decreased in the supplemented cells. If these cells growing at 15°C are rapidly shifted to 39.5°C, the membranes must then have a superoptimal fluidity. Desaturation of linoleate (18:2) is found to be strongly inhibited under these conditions, but desaturation of palmitate (16:0) and oleate (18:1) is inhibited to a lesser extent.

Thus naturally encountered environmental factors other than temperature changes may potentially affect fatty acid saturation. However, changes in oxygen concentration in the culture under normal laboratory conditions were not found to alter the degree of unsaturation of fatty acids in strain NT-1 (Skriver and Thompson, 1976).

Supplementing the culture medium with hexadecyl glycerol, a precursor of the glyceryl ether-containing phospholipids GAEP and PC, resulted in an increased glyceryl ether content in the plasma membrane of NT-1 cells and a decrease in PE. Microsomes of these cells sustained a small increase in GAEP and PC, which was compensated by a decrease in PE. Fatty acid concentrations are altered much more by the hexadecyl glycerol supplementation than are phospholipids. Palmitic acid (16:0), linoleic acid (18:2), and γ-linolenic acid (18:3) are increased while oleic acid (18:1) is decreased, especially in PE. However, the ratio of total saturated to unsaturated acids remains constant. Changes in the proportion of glyceryl ether involvement at the C-1 position of membrane phospholipids may play an additional role in controlling membrane fluidity (Fukushima et al., 1976b).

5. Membrane fusions in the ciliates

5.1 Secretory organelles

Not only do ciliates provide excellent models for studying membrane biogenesis and macromolecular turnover, they also provide unique systems for the study of membrane dynamics in membrane fusion and membrane recycling. Among the first examples of membrane fusion to be studied in detail were the secretion processes of mucocysts in *Tetrahymena* and trichocysts in *Paramecium*. Mucocysts and trichocysts, which have been collectively termed extrusomes (Grell, 1973), are examples of secretory organelles of ciliates and sarcomastigophorans. These organelles have recently been reviewed by Hausmann (1978). The use of freeze-fracture replicas has been particularly helpful in visualizing the sites at which extrusomes fuse with the plasma membrane and in following the membrane-related events of the process of secretion.

Satir has made a thorough study of mucocyst secretion in *Tetrahymena pyriformis,* strain W (Satir et al., 1972a, 1973; Satir, 1974a,b; Satir and Satir, 1974). These organelles are found in rows under the cell surface; some lie in line with the basal body rows while others form anterior-to-posterior lines halfway between the rows of basal bodies (Pitelka, 1961; Allen, 1967) where they lie at the surface along the alveolar septa (Fig. 17). The mucocyst body, which is 900 nm long and 300 nm wide, has a crystalloid structure surrounded by an 8 nm thick membrane (Satir et al., 1973). The origin of mucocysts is generally thought to be from the ER (arrows, Fig. 17) (Tokuyasu and Scherbaum, 1965), but many details of their synthesis in *Tetrahymena* still need to be worked out.

At the surface the mucocysts are oriented so their long axis is perpendicular to the plasma membrane and the tips of the mucocyst come to within 15 nm of the plasma membrane. Electron-opaque material lies between the mucocyst tip and the adjacent plasma membrane (Figs. 17 and 18). Specialized intramembranous particle aggregations are found near this juncture, within both the mucocyst membrane and the plasma membrane (Wunderlich and Speth, 1972; Satir et al., 1972a, 1973). Rosettes of 9 particles, each 15 nm in diameter, are found on the P face of the plasma membrane (Satir et al., 1972a, 1973). Thus the sites where fusion will occur are marked by these rosettes. The rosette itself is about 60 nm across and encloses a 35 nm circle, which is smooth except for the presence of a single central particle within some of the rosettes. Around the tip of the mucocyst membrane, on its P face, is a 68 nm wide annulus of 11 nm intramembranous particles that surrounds a smooth membrane region 61 nm in diameter. Otherwise the mucocyst membrane is quite smooth on both its P and E faces (Satir et al., 1973).

The fusion process as seen in freeze-fracture replicas begins with a depression of the plasma membrane which is surrounded by the rosette (Satir et al., 1972a). The rosette particles spread apart as the diameter of this depression, or fusion pocket grows (Satir et al., 1973). As the pocket deepens the annulus particles from the mucocysts become visible on the P face at the lip of the fused membranes. Initially the complementary E face of the fusion pocket shows a raised area, which widens to about 80 nm before the opening to the exterior is complete. This opening will reach a maximum of at least 245 nm during mucocyst discharge.

From these observations Satir and associates (1973) and Satir (1974a) proposed a model for the structural events leading to, and culminating in, membrane fusion during mucocyst secretion in *Tetrahymena*. The model is divided into three stages. The first stage is the formation of the rosettes within the plasma membrane, which is accomplished by a highly ordered arrangement of intramembranous proteins. The authors postulate that these proteins may contain hydrophilic channels through which ions or polar molecules can pass across the plasma membrane. The second stage in fusion is the formation of the annulus in the mucocyst membrane. The annulus is absent in immature mucocysts which are not close to the plasma membrane. The final stage results in fusion of the complementary leaflets of the two membranes which results in a site-specific rupture at the cell surface whereby an opening is formed into the mucocyst body. Satir and colleagues (1973) also propose that during fusion bonds form between the rosette particles and the particles at the inner edge of the annulus, resulting in the hinging together of the two P-face leaflets. Further on during fusion the rosette particles and the annular particles spread until they become coplanar and the P-face leaflets become completely fused. The two E-face leaflets then reorganize and fuse. During this process some lipids, in the form of micelles, may be lost into the medium. The resulting collar of particles at the open neck may reinforce the membrane and prevent rupture during mucocyst secretion (Satir, 1974a).

At the initial fusion site the two membranes must approach to within a critical distance of each other which, according to Poste and Allison (1973), would have to be within 10 to 15 Å. Within this distance attraction forces will hold the membranes together and any necessary ionic rearrangements or macromolecular conformation changes, which may be required to permit interaction between the phospholipid molecules of the two membranes, can then occur.

As a response to this membrane fusion, molecular rearrangement of the crystalline body of the mucocyst will lead to its enlargement and ejection from the cell (Hausmann, 1972; Hayashi, 1974). During secretion the mucocyst swells within the cell to three times its original volume, thus requiring a 70% increase in mucocyst membrane (Satir et al., 1973). The origin of this increase in membrane remains obscure unless it comes from the ER, to which even mature mucocysts appear to be continuous (Fig. 17).

Paramecium aurelia and *P. caudatum* also have rosettes of particles in the plasma membrane at the sites where trichocysts will be expelled. As in *Tetrahymena* the location of these sites is predictable. In this case there is one trichocyst secretion site halfway between every two consecutive basal body-cilium complexes (Fig. 6) within a ciliary row. At these sites the trichocysts penetrate to the surface through alveolar septa which lie at the tops of the transverse ridges of the pellicle. The rosettes are 80 nm in diameter and are composed of 8 to 10 particles, each 13 nm in diameter (Plattner et al., 1973; Satir, 1974a; Plattner, 1976). In addition to the rosette, *Paramecium* also has a ring composed of one or two circles of 9 nm particles surrounding the rosette (Janisch, 1972; Plattner et al., 1973; Satir, 1974a). These circular rings are 300 to 400 nm in diameter. Plattner has shown that the dimensions of this set of outer rings indicate that the rings coincide with the edge of the underlying alveoli. These rings appear to be involved in binding the plasma membrane to alveolar membranes rather than in the actual processes of membrane fusion. In thin sections links can be seen to traverse the space between these two membranes at the sites of the rings (Plattner et al., 1975).

Other more or less randomly distributed particles are found in the area between the rings and the rosettes, but the space within the rosette is free of particles (Figs. 6 and 19). Curiously, the particles of the 80 nm rosette fracture differently depending on the treatment of the cells prior to freezing. These particles remain with the P face if the cells are fixed before freezing (R. D. Allen, unpublished observations), but they frequently remain with the E face if the cells are frozen without prior fixation (Fig. 7) (Plattner et al., 1973). Fixation may bind these particles to proteins on the cytoplasmic side of the membrane.

The mature trichocyst membrane, like the mucocyst membrane, has an annulus of particles on its P face near its tip (Plattner et al., 1973; Satir, 1974a; Allen and Hausmann, 1976). These particles are about 10 nm in diameter. Upon discharge, the trichocyst membrane fuses with the plasma membrane. During the initial events in this process Plattner and collaborators (1973) reported that the outer rings remain unchanged while the particles of the rosette and those between the rosette and the rings apparently vanish. Unlike the discharge of mucocysts in *Tetrahymena,* where the opening remains long after mucocyst expulsion, the opening after trichocyst discharge is very transient and the plasma membrane quickly closes again over this hole (Plattner et al., 1973; Hausmann and Allen, 1976; Plattner, 1976).

To illustrate the events occurring within the plasma membrane at the sites of secretion, a series of unpublished pictures of the trichocyst-plasma membrane attachment sites in *P. caudatum* is presented. These pictures are organized to show the probable consecutive stages of trichocyst attachment, discharge, and resealing. Figure 19 shows a region near the cell's anterior end where a large number of trichocysts occur. These cells were fixed in low concentrations of glutaraldehyde (0.5%), which allowed some trichocyst discharge before the cells were fixed (Plattner and Fuchs, 1975). Several parentheses-shaped rings (Fig. 19a) indicate unfilled trichocyst sites. Round rings with rosettes at the center (Fig. 19b) are sites at which trichocysts are attached to the plasma membrane. Three rings show central indentations (Fig. 19c) and two show raised areas (Fig. 19d). These are probably indicative of various stages in trichocyst discharge in which the trichocyst membrane first fuses with the plasma membrane and then closes following discharge. One ring, which contains a larger number of particles in random distribution (Fig. 19e), is probably the site of a resealed plasma membrane. The additional particles enclosed by this ring may have come from the annulus of the trichocyst membrane and have been transferred to the resealed plasma membrane during the process of closing.

As was pointed out by Beisson and colleagues (1976), trichocyst membranes develop a very small radius of curvature at their distal tips (Figs. 20 and 22). This tip lies just below the rosette of particles and must insure that the trichocyst membrane can come within the critical hypothetical distance of 10 to 15 Å which will permit membrane fusion (Poste and Allison, 1973). Rows of pits lie on the E face at the end of the trichocyst membrane where the annulus of particles is located. The tip of the membrane is sometimes broken away, leaving the ring of the fractured E-face leaflet (Fig. 21). An early event in trichocyst discharge shows this tip penetrating through the cell membrane and encircled symmetrically by the rings (Fig. 22). The surface membrane appears to curve inward toward the trichocyst, suggesting that membrane fusion has been accomplished. The orifice formed by the fused plasma membrane and the trichocyst membrane must expand beyond the diameter of the ring since the shaft of the trichocyst, as it emerges, is at least 400 nm wide (Fig. 23). A cross-fractured trichocyst tip can be seen within the open secretion site in Fig. 24. Secretion sites, which are thought to be resealing, possess a large number of particles randomly distributed within the rings (Fig. 25), even though a depression is still present in the center of the ring. Sites that are resealed, lack the central depression, and have irregularly shaped or slightly oval rings encircling many particles (Figs. 19e and 26). A few large particles that may have come from the rosette are scattered throughout the randomly distributed smaller particles.

The P face of "resting" trichocysts, those intact trichocysts located at the secretion site, have an annulus of particles at their tip but behind this is a particle-free zone (Fig. 27) which covers one half to one third of the distal end of the trichocyst tip. Allen and Hausmann (1976) reported that the number of particles in the annulus roughly equals the number of particles present within an area of membrane at the proximal end of the trichocyst tip that is the same size as the area of the particle-free zone. Thus the particles may be pushed out of this region toward the tip as the external collar is formed during the final stages of trichocyst maturation (Allen and Hausmann, 1976). The imprint of the microtubulelike collar is visible on the P face of the fractured trichocyst membrane (Plattner et al., 1973; Allen and Hausmann, 1976).

During trichocyst secretion the annulus of particles forms a tightly packed ring at the neck of the opened trichocyst sac at the junction between the plasma membrane and the trichocyst membrane. The particles remain with the P face (Fig. 28), but a prominent ring of pits can be seen on the E face (Fig. 29). The E face of the discharging trichocyst membrane is smooth, as are the "resting" trichocysts (Allen and Hausmann, 1976), and the P face is particulate. The particle-free zone on the P face has disappeared and particles are randomly distributed all the way to the annulus. A small segment of the P face of the plasma membrane can be seen in Fig. 28, while the impression of the double ring of particles on the E face of the plasma membrane is visible distal to the annulus ring in Fig. 29.

As mentioned previously, Satir (1974a) has postulated that the particles of the fusion rosette may function as channels through which ions can enter the cell. Plattner has presented a series of papers, however, that questioned the concept that rosette particles or other secretion-site particles serve as ion channels. By the use of spontaneous incorporation of calcium ionophores into the plasma membrane, he showed that exocytosis of trichocysts can be triggered by promoting the influx of Ca^{2+} into the cell, while simply raising the Ca^{2+} concentration in the medium in the absence of added ionophores had no effect on trichocyst exocytosis (Plattner, 1974). The presence of intramembranous particles above the trichocyst tips does not allow small electron microscopic tracers such as lanthanum hydroxide or heme-peptides to penetrate into the cell or the trichocysts (Plattner, 1974; Plattner et al., 1975), though such a passage has been shown to be possible between some adjacent tissue cells through the particles of gap junctions (Revel and Karnovsky, 1967; Gilula, 1974). Nor could Plattner and associates (1975) demonstrate pores within the various intramembranous particles by using tantalum-tungsten shadowing, which gives better resolution in freeze-fracture replicas than does platinum-carbon shadowing. Thus some circumstantial evidence exists against the contention that the particles serve as hydrophilic channels. Recent cytochemical evidence demonstrates a Ca^{2+} activated ATPase activity in the vicinity of these rosette particles (Plattner et al., 1977).

In whatever manner these rosettes function, they must be essential to trichocyst discharge since studies on *Paramecium* mutants demonstrate that rosettes must be present before normal trichocyst discharge can occur; yet their presence is not necessary for trichocysts to lie next to the plasma membrane (Beisson et al.,

1976). Thermosensitive mutants of *Paramecium tetraurelia* (formerly *P. aurelia*, syngen 4), which form normal membrane-attached trichocysts, will not discharge these organelles upon picric acid stimulation if grown at 27°C. However, if the mutants are grown at 18°C before picric acid stimulation, they will discharge their trichocysts (Beisson et al., 1976). The only obvious difference between the cells grown at 27°C and 18°C is the presence of the rosette of particles at 18°C and their absence at 27°C. Oberg and Satir (1977) have also shown that in secretory mutants picric acid will stimulate release of trichocysts in cells grown at the permissive temperature (18°C) only if $CaCl_2$ is present, and not when $MgCl_2$ is substituted for the calcium. No discharge is seen in the cells grown at 27°C unless the calcium ionophore (A23187) is present along with $CaCl_2$. Satir's group interprets these results to support an ion channel theory, but other explanations are also possible (see Plattner et al., 1977).

Mutants whose trichocysts do not attach to the plasma membrane, or mutants that lack trichocysts altogether, have rings but these are always parentheses-shaped and rings never encircle rosettes. Thus the rings are not an essential feature of exocytosis, but may limit the expansion of the fusion pocket (Beisson et al., 1976).

Bardele (1976) has shown the presence of rosettes of particles on the fractured P face of the plasma membrane next to the distal tips of kinetocysts of a heliozoan, and these rosettes resemble those in ciliates. He also mentions the presence of rosettes of 12 particles with 1 central particle in the tentacle of a suctorian, another ciliate. These rosettes and their surrounding particle-free zones are considered by Bardele to prevent the secretion organelles from being discharged at the wrong time. In a study of two actinopods, whose secretion organelles are similar to those in the heliozoan, Davidson (1976) observed specialized particle groupings on the P face of their plasma membranes adjacent to the tips of the secretion organelles. He points out that these organelles may remain in this position next to the cell membrane for over an hour without undergoing exocytosis. Thus the rosettes must be stable over a relatively long time, and do not appear to arise just before exocytosis. Both of these studies (Bardele, 1976; Davidson, 1976) report the presence of filaments associated with the particle groups on the cytoplasmic side of the cell membrane next to the tip of the secretory vesicle. Also, arrays of filaments are seen on the external cell surface next to these particle groups. Particles at the attachment sites in these cells may be involved in linking the plasma membrane to the mature secretion vesicles and may inhibit membrane fusion until food organisms are recognized at the surface coat of the cell. These attachment zones may themselves be in a crystalline state floating in more liquid membrane material since they can move with the secretion organelles as they pass along the axopods (Bardele, 1976). Bardele has proposed that a lipid phase-shift may be required before fusion can occur between the membranes.

Artificially introduced Ca^{2+} does trigger trichocyst exocytosis (Plattner, 1974). Using low concentrations of glutaraldehyde (0.625%) as a fixative and adding Ca^{2+} in concentrations of 1 to 100 mM before or during fixation, Plattner and

Fuchs (1975) looked for calcium deposits in the trichocysts using x-ray micro-analysis. Most "resting" trichocysts were devoid of electron-opaque Ca^{2+} deposits, but a few contained deposits on the luminal side of the trichocyst tip and body membranes and on the inner lamellar sheath structure of the tip. Discharged trichocysts, however, never had such deposits. When calcium iono-phores are incorporated into the cell membranes, several "resting" trichocysts have electron-opaque calcium deposits, but ghosts of discharged trichocysts are free of calcium deposits. Plattner and Fuchs (1975) believe that calcium enters the trichocysts prior to membrane fusion since the deposits are sometimes present in intact resting trichocysts.

The above experiments support the concept that Ca^{2+} triggers exocytosis, but do not identify the in vivo supply of Ca^{2+}. As Plattner and Fuchs (1975) pointed out, under the experimental conditions used, the low concentration of glutaraldehyde probably alters the permeability properties of the cell membranes, thus permitting an influx of Ca^{2+} from the outside and causing trichocyst discharge before fixation is accomplished. Under natural conditions it is conceivable that the calcium ions could come from internal stores.

The fate of the secretory vesicle membrane following secretion is also important in the membrane economy of the cell. The most probable fate is either incorporation of this membrane into the plasma membrane, as proposed for mucocysts of *Tetrahymena* (Satir, 1974a,b), or a reengulfment of this membrane by endocytosis as shown in trichocysts (Hausmann and Allen, 1976). For the mucocyst or trichocyst membrane to become part of the plasma membrane, it would need to undergo major changes since it is probably neither morphologi-cally (being of the thin-membrane-type) nor biochemically similar to the plasma membrane. The mucocyst membranes of ciliates that have a glycocalyx on the plasma membrane do not bear this coat (Hausmann and Mocikat, 1976). Mem-branes of secretory vesicles of ciliates are usually described as arising from the endoplasmic reticulum (Hausmann, 1977) without passing through the Golgi apparatus, where membranes are differentiated into the thicker plasma mem-brane type and where the glycocalyx is added (Morré et al., 1971). Figure 17 shows connections between mucocyst and ER membranes in *Tetrahymena*. For these vesicle membranes to become part of the plasma membrane, their phos-pholipid and neutral lipid content might need to be altered as well as their intrinsic protein content. A glycocalyx, in ciliates where this coat is found, would also need to be added. This may be possible considering the evidence presented in sections 3 and 4 above, yet evidence from radioisotope labeling does not support the simultaneous insertion of lipids and proteins into the surface mem-branes as large vesicular units (Nozawa and Thompson, 1972; Subbaiah and Thompson, 1974; Baugh and Thompson, 1975).

On the basis of morphological studies, Satir (1974a,b) and Satir and Satir (1974) feel that the mucocyst membrane becomes part of the plasma membrane. However, the only evidence for this view is that open mucocyst sacs range in size from the normal just-secreted size of 800 nm in diameter to far smaller 200 nm diameter sacs. These can be seen either in thick sections prepared for high-voltage electron microscopy or in isolated *Tetrahymena* membranes. Freeze-fracture images are also reported to reveal very shallow, open mucocysts bearing many particles, and these show no evidence of fractured endocytic necks (Satir, 1974b). However, the smaller sacs could result from a comparatively slow endocytic activity or gradual removal of lipids by exchange with other membranes, such as the plasma membrane and alveolar membranes, during which the membrane is slowly reduced in area without endocytic activity. Although Satir (1974a) and Satir and Satir (1974) saw no evidence for reengulfment of mucocyst membranes by endocytosis, Nilsson (1976) has shown a micrograph that could indicate reengulfment of mucocyst membrane by this process, though this one published micrograph does not show continuity between endocytic vesicles and the mucocyst membrane.

Trichocyst membranes of *Paramecium caudatum* are reengulfed first by a rapid (<1 sec) pinching off of the empty sac at the level of the plasma membrane and then by a slower fragmentation of the sac (Hausmann and Allen, 1976) over a period of 5 to 10 minutes. Membrane retrieval by reengulfment also occurs in *P. aurelia* (Plattner, 1976). In *P. caudatum* the collar zone of the trichocyst tip remains intact for the longest period, but this too eventually disintegrates. The final fate of the small vesicular fragments could not be determined. However, no evidence of their incorporation into autophagosomes and subsequent digestion in the lysosomal system was observed. They may possibly be recycled back into the endoplasmic reticulum from where they form originally, or they may be degraded in the course of lipid and protein exchange with other membranes, during which process more subunits are lost from the membrane than are gained, until they eventually disappear as separate entities—a process of "cannibalization" by other cell membranes or "reverse intussusception."

5.2. Digestive vacuole membranes and lysosomes

In sheer quantity of membrane fusion events the digestive or food vacuole system appears to undergo more fusion than any other membrane system within the ciliates. These events include both exocytic and endocytic fusions that are programmed and regulated by the cell to occur in a sequential pattern from the time of formation of food vacuoles until food vacuole defecation and the reengulfment of the membrane of the empty food vacuole has been completed. Light microscopy had initially indicated that food vacuoles undergo a complex series of membrane events. Mast (1947), in his comprehensive study of the digestive process in *Paramecium*, described these events in detail. The food vacuole in this ciliate forms at the posterior dorsal end of the buccal cavity by an invagination of the membrane covering the cytopharynx. Eventually this nascent food vacuole, which becomes quite large, will pinch off from the cytopharynx and begin to move as a separate organelle within the endoplasm. At this time the vacuole begins to diminish in size and within a period of a few minutes it may shrink to one fourth of its original diameter. Neutral red granules continually surround the food vacuole during this early period. The pH of the food vacuole, which is nearly neutral when it is first formed, drops to a very low level as the vacuole shrinks to its smallest diameter. Having reached this minimum diameter, the food vacuole now commences to expand rapidly until it reaches approximately the same diameter as when it first parted from the cytopharynx. The neutral red granules that were close to the food vacuole in its initial formation have now disappeared. The old food vacuole then makes its way to the cell anus (cytoproct) where defecation occurs. The fate of the food vacuole membrane has not been determined in light microscope studies.

Electron microscopy has demonstrated that each of these events in the process of food vacuole formation, shrinking, enlarging, and defecation requires membrane fusion. Recent evidence from thin-section electron microscopy shows that food vacuole membranes are enlarged from a pool of membrane vesicles (Bradbury, 1973, 1974; McKanna, 1973b; Allen, 1974; Kloetzel, 1974; Dass et al., 1976; Nilsson, 1976) that accumulate in the cytoplasm near the cytopharynx—a single membrane-limited part of the cell which invaginates to form food vacuoles (Fig. 30). This is supported by biochemical studies (Weidenbach and Thompson, 1974) and by freeze-fracture studies of the membranes of the pooled vesicles and cytopharyngeal membranes that show identical numbers and an identical distribution of the intercalated particles on the corresponding fractured faces of both membranes (Fig. 31) (Allen, 1976a; Kitajima and Thompson, 1977b).

In *Paramecium*, as well as some other ciliates, such as *Neobursaridium* (Nilsson, 1969) and *Uronema* (Kaneshiro and Holz, 1976), the pool of membranes upon which vacuoles draw assumes the form of flattened vesicles. These vesicles are circular in planar view (Fig. 30). Thus they resemble discs ranging from 0.2 to 0.5 μm in diameter and have a thickness that varies according to the techniques used to fix the cell but which is often about 60 nm. These discoidal vesicles are aligned in ranks and files with their edges next to the cytopharyngeal membrane (Fig. 32), the files being separated by the ribbons of cytopharyngeal microtubules (Fig. 33). Under as yet unknown stimuli, these vesicles fuse edgewise with the cytopharyngeal membrane (Fig. 34). Calcium, which is found to accumulate on these vesicles (Plattner and Fuchs, 1975; Fisher et al., 1976) may be involved in this fusion. Freeze-fracture pictures of the P face of the cytopharyngeal membrane (Fig. 35) do not reveal any obvious particle arrays at the sites at which fusion will occur. However the association of the vesicles with the ribbons of microtubules, which are also set perpendicular to the cytopharyngeal membrane, presumably ensures the proper alignment of vesicles with the membrane so that fusion will be possible. These discoidal vesicles have a glycocalyx on their luminal surface which is morphologically similar to that lining the plasma membrane both in the somatic area and over all the oral region of the cell. Vesicles also have membranes of the thicker type. Once fusion has occurred, membrane flow allows the incorporation of the newly added vesicle into the cytopharynx membrane and thus provides new membrane for the nascent food vacuole (summarized in the drawing, Fig. 36).

At some point the vacuole itself will pinch off where it is connected to the cytopharynx membrane and will now begin its movement through the endoplasm. This pinching in two will leave both the digestive vacuole and the single limiting membrane intact at the posterior dorsal surface of the oral invagination. Thus both exocytosis of vesicles and endocytosis, which results in the release of the food vacuole, will have played a role in forming the new vacuole. The actual mechanisms involved in vacuole release have been considered (Mast, 1947; Nilsson et al., 1973; Allen and Wolf, 1974), but there is no conclusive evidence for any specific mechanism.

Many studies have been directed toward an understanding of the process of food gathering, the recognition of food versus nonfood particles and the parameters that determine how large a vacuole must become before it will be released in the cytoplasm. However, few generalizations can be made from these studies. Recently the necessity for the presence of food vacuole formation to sustain growth in *Tetrahymena* has been considered (Rasmussen and Kludt, 1970; Rasmussen, 1973, 1974; Rasmussen and Modeweg-Hansen, 1973; Hoffmann et al., 1974; Nilsson, 1976). If *Tetrahymena* is grown in a very rich particulate-free growth medium, it seems to be capable of accumulating sufficient nutrients to maintain growth without the requirement for food vacuoles (Rasmussen, 1973, 1974; Hoffmann et al., 1974; Rasmussen and Orias, 1975). Nilsson (1976) believes that even in cells which seem to lack vacuoles, small food vacuoles may still be present. Such experiments, although hardly applicable to cells growing in natural environments, suggest that for sustained growth particulate materials may be necessary to trigger food vacuole formation (Rasmussen and Kludt, 1970; Hoffmann et al., 1974). Indirectly food particles may be necessary for continued exocytosis of the discoidal vesicles, whose addition to the cytopharynx membrane allows the development of nascent food vacuoles. How food particles control food vacuole formation is unknown.

The size attained before food vacuoles can be released is also highly variable. After periods of starvation, the first food vacuole to form in *Paramecium* is usually quite large, up to 60 μm in diameter. If food vacuoles continue to form, they will be smaller than the initial one. Ciliates such as *Paramecium* and *Tetrahymena* are not capable of continued rapid food vacuole formation over a long period of time (Nilsson, 1972, 1976). Following starvation these cells may rapidly form food vacuoles for several minutes, after which they seem to stop and will not make new vacuoles until a period of "rest" has passed (Ricketts, 1971a; Nilsson, 1972; Ricketts and Rappitt, 1974). Food vacuole formation will then commence again but at a slower rate. This may indicate that the pool of preformed membranes has been exhausted during the initial spurt of vacuole production and that, before new food vacuoles can form, new membrane must be either synthesized or existing membrane processed and recycled back to the oral region where it can be reused to form additional vacuoles (Ricketts, 1971b; Allen, 1974; Ricketts and Rappitt, 1975a). The cell is apparently incapable of producing vacuoles at maximum rate indefinitely. Even if they did produce vacuoles at such a rate throughout a complete generation period, the food vacuoles formed would not be able to accumulate enough nutrients to account for the total cell growth during this time (Nilsson, 1976). This suggests that the food vacuoles must be able to concentrate food from the medium (see also Ricketts and Rappitt, 1975b) or that the cell has another method by which it accumulates some of its nutrients. Orias and Rasmussen (1977), working with mutants that do not form food vacuoles, provide evidence that the food vacuoles of *Tetrahymena thermophila* (formerly *T. pyriformis*, syngen 1) are necessary for accumulating iron, copper, folinic acid, and proteins and are necessary for uptake of macromolecular and particulate nutrients. However, extraphagocytic mechanisms result in the uptake of required nutrients when these are supplied in their free form in mutants that do not form food vacuoles.

Mast (1947) observed that when food vacuoles of *Paramecium* are released from the cytopharynx, they quickly pass to the posterior end of the cell where they start to condense and continue condensing until they reach a much smaller diameter than they had when first formed. Similar condensation also occurs in other ciliates (Nilsson, 1976). This process of condensation apparently concentrates the enclosed particulate material and eliminates excess fluid from the vacuole. The fate of the excess membrane around these condensing vacuoles was unknown until recently. Freeze-fracture images of newly formed food vacuoles show channeling and endocytotic events at the surface of these vacuoles (Allen, 1976a, 1977). Condensation is accomplished by an active process of removal of small vesicles from the food vacuole surface until the membrane fits snugly around the mass of food. Vacuoles undergoing condensation show a highly irregular topography (Fig. 39) with many fractured necks, when their E faces (Fig. 37) are seen in planar view, or evaginations into the cytosol when the vacuole is fractured transversely (Fig. 38). The fate of these small endosomes is not known; however, they may become part of the membrane pool from which new food vacuole membrane is derived.

Another observation, dramatically demonstrated in *Paramecium caudatum*, is the change that occurs in membrane particle numbers on the E face of these condensing vacuoles (Allen, 1976a). Initially vacuoles have a highly particulate E face like that of the cytopharyngeal membrane and membranes of the discoidal vesicle pool (Figs. 31, 39). However, within 3 minutes of release, condensation results in vacuoles with smooth-E-face membranes (Fig. 40) and a smooth spherical topography (Allen, 1976a,c). After 10 minutes have elapsed, the food vacuole membrane becomes wavy and a new population of particles appears on the E face which is dominated by a larger particle population (Fig. 41).

These changes in intramembranous particle distribution and numbers can be correlated with the light microscope studies of Mast (1947) and Rosenbaum and Wittner (1962). Mast showed that a steady decrease in volume of the newly formed food vacuole over the first 7 minutes was correlated with a drop in pH of the vacuole contents to as low as pH 1.4. A less drastic drop in pH to a low of 4.0 has been reported in *Tetrahymena* food vacuoles (Nilsson, 1977). In *Paramecium* it is during this time that the bacteria are killed. A rapid increase in size of the vacuole follows this stage. During this volume increase, the pH rises to about neutral and the neutral red granules that had surrounded the vacuole through the early stages disappear. This presents an unresolved paradox: Why is the food vacuole pH increasing at the same time that the neutral red granules, which presumably contain acid hydrolases, are fusing with the vacuoles? Rosenbaum and Wittner (1962) also followed the fate of the neutral red granules, but found neutral red granules to be present around old food vacuoles, even those situated near the cytoproct. Neutral red granules presumably contain hydrolases, since acid phosphatase reaction products show the same pattern of distribution as neutral red granules in starved and feeding cells (Rosenbaum and Wittner, 1962). Esteve (1970) has shown acid phosphatase at the ultrastructural level in dense bodies associated with some food vacuole stages. These bodies are presumably neutral red granules which are lysosomes. In starved cells they are located at the peripheral cortex of the cell, but when the cell feeds they migrate toward the food vacuoles, and if the vacuole contains digestible food they appear to remain close to the vacuole membrane (Rosenbaum and Wittner, 1962). They could not be seen to enter the vacuoles at the light microscope level.

Electron micrographs of some food vacuoles show these bodies lining the surface of food vacuoles (Fig. 42). Freeze-fracture images of these bodies demonstrate that they have a relatively particulate P face and a less particulate E face, with large predominant particles much like the E face of mature food vacuoles (cf. Figs. 41 and 43). Such bodies can sometimes be found to be continuous with the food vacuole membrane (Fig. 44).

From these freeze-fracture studies I conclude that the food vacuole initially decreases in size by endocytosis of small vesicles and then increases in size as a result of fusion with these "neutral red" granules that lie next to the food vacuole and whose disappearance, according to Mast (1947), occurs at the time of vacuole enlargement. This may also explain the initial loss of intramembranous particles from the E face; these particles may be sequestered into the endosomes which are then removed from the vacuole. The later reappearance of another population of particles may be explained by the fusion of vacuoles with the neutral red granules, which have similar particles on their E faces to those on older food vacuole E faces. Another conclusion that might be drawn from the freeze-fracture study is that the smooth E face of the condensed food vacuole membrane indicates an impermeability of this membrane to ions and polar molecules due to its lack of transmembrane proteins. This smooth leaflet may consequently act as a barrier, protecting the cytoplasm of the cell from the very acidic condition of the vacuole during the period that the food organisms are being killed.

Membrane fusions thus play an essential role in the formation, condensation, and later enlargement of food vacuoles and are programmed to occur at specific times during the physiological processes of food intake, killing, and digestion. The events that trigger these fusions are not understood but this intracellular digestive system appears to be well suited to a study in which fusion can be correlated with a natural physiological process within a cell.

One extensive ultrastructural study of the digestive system of a group of closely related peritrichs has been reported. Carasso, Favard, and Goldfischer published a series of papers that described the morphological changes in digestive vacuoles in several peritrichs and correlated these with cytochemical studies. Fauré-Fremiet and co-workers (1962) and Favard and Carasso (1963) have observed 0.2 to 0.6 μm diameter spherical vesicles around food vacuoles in *Epistylis,* which contain the electron-opaque tracer thorium oxide, after cells were exposed to thorium oxide for 2 to 24 hours before being fixed. They concluded that these vesicles formed by endocytosis from the food vacuole membrane. These spherical vesicles were adjacent to other cup-shaped vesicles that also accumulate next to food vacuoles but seemed to be devoid of thorium oxide. Acid phosphatase reaction product, identified by the Gomori technique, is found in *Campanella,* another peritrich, within Golgi cisternae and the small vesicles at the forming face of the dictyosome stacks. A small amount is also found in the ER (Goldfischer et al., 1963). Acid phosphatase has subsequently been shown in food vacuoles (Goldfischer et al., 1967). Carasso and colleagues (1964) believe that the acid phosphatase may pass directly from the ER into the 0.2 to 0.6 μm vesicles. Acid phosphatase may also pass directly from the ER to food vacuoles (Goldfischer et al., 1967). According to these authors (Carasso et al., 1964) the 0.2 to 0.6 μm acid phosphatase-positive vesicles may ultimately flatten out to form the cup-shaped vesicles. A lamellar border is found both on the luminal surface of the food vacuole membrane and within the endosomes (Favard and Carasso, 1964). The cup-shaped vesicles, which contain nucleotidase—as do the food vacuoles (Carasso et al., 1964)—then presumably fuse with the old food vacuoles before the vacuole is defecated.

McKanna (1973b) has interpreted the cup-shaped vesicles of peritrichs to be membranes that move to the cytopharyngeal region, where they are transformed into discoidal shapes and subsequently fuse with the cytopharyngeal membrane to provide for nascent food vacuole formation. He shows examples of pentalaminar membrane structures as an intermediate stage in the fusion process, as has been demonstrated in several other cell types (Palade and Bruns, 1968; Lagunoff, 1973; Tandler and Poulson, 1976). He also believes these cup-shaped vesicles come from young food vacuoles, less than 1 minute old, during the vacuole condensing stage (McKanna, 1973c) rather than originating as the dehydrated products of the 0.2 to 0.6 μm vesicles. McKanna was unable to label these vesicles with ferritin or thorotrast, and so he concluded that these substances must be excluded from the vesicles as they form.

Neither of the above explanations of the origin and fate of these cup-shaped vesicles is completely satisfactory. In a micrograph such as Fig. 45, which shows three food vacuoles of *Vorticella convallaria,* it can be seen that the lower right food vacuole is probably the youngest since the enclosed bacteria show little sign of digestion. As McKanna reported, this is the food vacuole with which most of the cup-shaped vesicles are associated. It also has the smallest diameter, and its food is packed into a discrete mass from which the membrane has apparently begun to separate. The food vacuole in the upper right is older, as indicated by its digested contents, and has enlarged, as indicated by the increased space between the food mass and the membrane. Finally, the left food vacuole appears to have formed from the fusion of two vacuoles since it contains two circular masses of food organisms. These latter two vacuoles have no cup-shaped vesicles fused to their membrane, as has the younger food vacuole (Fig. 45); however, they do have electron-transparent vesicles budding into the cytosol.

Besides those explanations given by previous workers, two additional interpretations that are not mutually exclusive are presented here. Some cup-shaped vesicles may be formed from the larger endosomes that are derived from old vacuole membrane, as Carasso and collaborators (1964) have reported. The cup-shaped vesicles may then be transferred to young vacuoles, where they fuse with these vacuoles (Fig. 47) to provide membrane and a luminal coat to the young vacuoles. Or it may be that all or a part of the cup-shaped vesicles are primary lysosomes formed from the Golgi apparatus (Fig. 58). These lysosomes may fuse with the young food vacuoles (Fig. 47) after the vacuoles have condensed and thus provide the 20 nm electron-opaque coating on the luminal surface of the vacuole membrane, as Fig. 46 might suggest, along with the additional membrane needed for the vacuole to swell. The vesicles budding off the older food vacuoles (Fig. 45) are undoubtedly secondary lysosomes that may retrieve the luminal coat given to the food vacuole by the cup-shaped vesicles. It may be that some of this membrane is transported to the cytopharynx to be used to form new vacuoles. The pool of vacuole precursor membranes that McKanna (1973b) interpreted to be derived from the cup-shaped vesicles did not have the highly organized and dense membrane coat that is typical of cup-shaped vesicles; rather they had a much more transparent lumen. Another likely source of these flattened vesicles is from membrane removed from the vacuole during its early condensation.

This interpretation, which I favor over those already published, would make the cup-shaped vesicles equivalent to primary lysosomes rather than to the discoidal vesicles of *Paramecium*. It would also account for an input of new membrane and Golgi-packaged digestive enzymes into the food vacuoles immediately preceding digestion. This scheme is not inconsistent with the cytochemical evidence presented above, which shows a lack of electron-opaque tracer incorporation from food vacuoles into cup-shaped vesicles, implying that these vesicles are primary lysosomes; nor is it inconsistent with incorporation of such tracers into the 0.2 to 0.6 μm vesicles (presumptive secondary lysosomes) or with the observation that cup-shaped vesicles and the 0.2 to 0.6 μm vesicles both contain acid phosphatase. This last observation would be expected in both primary and secondary lysosomes.

Though the exact sequence of vesicle formation and vesicle incorporation into the food vacuole membranes is still incompletely understood, electron microscope studies have made it clear that these fusion processes are occurring throughout the digestive cycle. As the food vacuole ages and digestion is completed, the vacuole may again start to diminish in size since endocytotic activity again becomes apparent on the surface of vacuole membranes (Fig. 50) (Allen and Wolf, 1974; Allen, 1976a).

We have studied the fate of food vacuole membranes in *Paramecium caudatum* and in *Tetrahymena pyriformis* (Allen and Wolf, 1974; Wolf and Allen, 1974). On the basis of light microscope observations (Foissner, 1972) it was supposed that the food vacuole membrane was sometimes defecated along with the contents. However, timed thin-sectioned studies show that the food vacuole is defecated through a process of fusion of the food vacuole membrane with the plasma membrane at a predetermined site within the pellicle of ciliates, called the cytoproct or cytopyge (Fig. 48). In *Paramecium* this is a long ridge along the posterior suture of the cell. The ridge contains internal fibrous coats on both of its sides and also has microtubules. The top of the ridge is covered with a single membrane which has an external glycocalyx. The pellicular alveoli are separated by a gap of about 140 nm in this area. As the food vacuole approaches this region, it sends out small fingerlike channels (Fig. 50) in the direction of the cytoproct. The ridge flattens out and pulls apart, increasing the single-membrane-limited surface area (Fig. 49). This area increase may be achieved by the fusion and incorporation of discoidal vesicles (Fig. 48) into this membrane. The food vacuole membrane approaches this expanded cytoproct and the membranes fuse. This actual fusion event has not been seen but presumably occurs first between a thin projection of the food vacuole membrane and the cytoproct plasma membrane. Once fusion has been accomplished, the gap will rapidly grow and lengthen so that much of the cytoproct ridge becomes involved in the opening and a relatively large pore is formed. Microtubules seem to be involved in guiding the vacuoles to the cytoproct where defecation is to occur (Allen and Wolf, 1974).

As soon as the cytoproct membrane has been breached, the opening seems to enlarge automatically; there are no restraints limiting the opening except along the lines where the plasma membrane is bound to the underlying alveoli. These intramembranous junctions are seen as bands of intercalated particles on both the P and E faces of the plasma membrane on both sides of the cytoproct ridge (Allen, 1976b). The length of the ridge that becomes involved in the opening is probably determined by the diameter of the vacuole being defecated; the whole length of the cytoproct ridge is not usually involved in one evacuation event.

In *Tetrahymena* the cytoproct is also a special single-membrane-limited slit bordered by the alveoli (Wolf and Allen, 1974). The only differences in this cytoproct from those in *Paramecium* are that the slit is not at the tip of a ridge and the numbers of microtubules surrounding the cytoproct are fewer. Otherwise the approach and fusion of food vacuoles follows essentially the same pattern as that described for *Paramecium*.

Immediately following defecation, the edges of the now empty food vacuole membrane show an extremely rough topography with many channels passing into the cytoplasm (Allen and Wolf, 1974). The cytoplasmic surface of the membrane is coated with a meshwork of microfilaments (Fig. 51) resembling that around pinocytic channels in amoeba (Marshall and Nachmias, 1965). These channels are pinched off and within a few minutes the membrane of the old food vacuole becomes completely reengulfed. Nor is the membrane of the food vacuole added to the outer plasma membrane, although the vacuole membrane is coated with a glycocalyx and it belongs to the thicker type of membrane similar to the plasma membrane (Fig. 51). The closing of the cytoproct seems to be a natural consequence of the removal of all of the food vacuole membrane from between the margins of the cytoproct ridge. The ridge must reform when contact is established between the fibrillar coats at the cytoproct lips.

These reengulfed endosomes of food vacuole membrane are thought to be recycled as vesicles to the cytopharynx region, where the membrane is used to form new food vacuoles (Allen and Wolf, 1974). The basis for this belief is that some of the cytopharyngeal ribbons of microtubules actually pass near the cytoproct region at the time the membrane is being reengulfed (Fig. 52). Discoidal vesicles, which may be the ultimate shape assumed by the endosomes, are found to lie in linear fashion along these ribbons and to be scattered throughout the cytoplasm under the area of endocytosis (arrows, Fig. 52). These vesicles are firmly bound to the ribbons of cytopharyngeal microtubules, as has been shown when the ribbons become highly distorted (Allen, 1975). Figure 53 summarizes our interpretation of the morphological processes involved in vacuole defecation and also illustrates the fusion events involved in the release of the food vacuole contents and the reengulfment of the membrane.

The cycle of membrane flow is completed as the discoidal vesicles are returned to the cytopharynx where they again fuse with the cytopharyngeal membrane (Fig. 34). We do not view this as a closed system of membrane recycling, but one by which membrane will continually enter the system from the fusion of primary lysosomes with the food vacuoles. The total amount of food vacuole membrane in a cell will be divided to the daughter cells during cell division and there may also be other processes that will reduce the total membrane in the food vacuole membrane pool. As already mentioned, the fates of the endosomes from early and late stages of food vacuoles have been subjects of speculation, but intense further work will need to be done to prove which ideas are correct.

5.3. Pulsating vacuole system

The osmoregulatory apparatus of ciliates that is responsible for eliminating excess water is the contractile or pulsating vacuole. Basically this structure is composed of a number of smooth-surfaced anastomosing tubules (Fig. 56) that are contiguous with decorated tubules (Figs. 55 and 56) on their proximal ends (directional terminology is determined by presumptive fluid flow) (McKanna, 1974, 1976) and open into a membrane-bound sac on their distal ends (Fig. 54). This sac may be an ampulla, a collecting canal, or the contractile vacuole that lies close to one or two specialized permanent indentations of the plasma membrane, the contractile vacuole pore(s) (Fig. 54). The sides of these pores have helically arrayed microtubules reenforcing the membrane (Schneider, 1960; Elliott and Bak, 1964; Jurand and Selman, 1969; McKanna, 1973a). The bottom of the pore is covered by the plasma membrane; there are no alveoli here. This pore membrane, which has a circular shape of 0.8 μm in diameter, comes to lie within 25 to 50 nm of the membrane of the contractile vacuole. Systole of the vacuole cycle begins by a fusion of these two membranes, during which an opening is formed into the vacuole. Following expulsion of the contents the pore will close again, at which time the contractile vacuole will pinch off from the plasma membrane. Presumably the vacuole remains as a flattened sac to be refilled during the next cycle.

The membranes of this system of tubules and the vacuole itself are about 7 to 8 nm thick in peritrichs (Carasso et al., 1962; McKanna, 1974) and in *T. pyriformis* (Elliott and Bak, 1964). The tubules are around 20 to 40 nm in diameter and are always attached to the vacuole membrane and its ampullae in *Tetrahymena*.

The excretory apparatus of *Paramecium* is more extensive than that of *Tetrahymena*. Schneider (1959, 1960) has shown that each apparatus consists of a number of radially oriented collecting canals surrounding a central vacuole. The collecting canals swell at their distal ends into ampullae. The ampullae undergo a filling stage and an evacuation stage into the vacuole proper. Schneider (1960) believes the ampullae become continuous with the contractile vacuole during ampullary evacuation and then pinch off from the vacuole prior to refilling of the ampullae. This would require membrane fusion. On the other hand, Patterson and Sleigh (1976) believe the ampullae may always be connected to the vacuole and that a contractile mechanism, operating like a sphincter muscle, exists for closing off the opening. However, since no high resolution studies have shown the details of the relationship between the ampullae and the vacuole, the mechanism of ampullary discharge into the vacuole remains unclear.

Anastomosing tubules join the collecting canals along most of their length. This has recently been shown to occur in rows in both thin sections and freeze-fracture replicas so that the tubules that join the canals have a center-to-center spacing of 80 to 100 nm (Fig. 57) (Hausmann and Allen, 1977). Schneider (1960) reported that these tubules become discontinuous with the collecting canals during discharge of the ampullae; however, recent studies do not support this interpretation (McKanna, 1976; Hausmann and Allen, 1977). Thus the tubules seem to be permanently joined to the collecting canals during both filling and evacuation of the canals and ampullae.

On the outer periphery of the layer of anastomosing tubules that surrounds the collecting canals are fascicles of decorated tubules which McKanna (1974, 1976) has called the "fluid segregation organelles." These tubules, which have a 8 nm thick membrane, are straight and unbranched except that they join to a common, smooth membrane-bound reservoir that leads to the anastomosing tubules. Originally a fluid segregation tubule was thought to be continuous with the endoplasmic reticulum on its proximal end, but McKanna (1974) now feels that they end blindly in the endoplasm. Fluid segregation organelles lie close to mitochondria in the marine ciliate, *Uronema* (Kaneshiro and Holz, 1976).

This whole system of vacuoles, ampullae, and collecting canals with their attached anastomosing tubules is held together by ribbons of microtubules. These microtubules arise at the evacuation pore, pass over the distal hemisphere of the vacuole, and extend out along the ampullae and between the rows of attachments linking tubules and the collecting canal. These microtubular ribbons are joined by bridges to the membranes of the contractile vacuole apparatus and must be of primary importance in maintaining the elongated shape of the collecting canals and in organizing the excretory apparatus so that membrane fusions, where they occur, will do so in an orderly and predetermined fashion (Hausmann and Allen, 1977).

Most of the studies on contractile vacuoles indicate that when the vacuole undergoes systole the contents of the vacuole are forced out by hydrostatic pressure exerted by the cytoplasm. This pressure causes the collapse of the proximal hemisphere of the vacuole into the distal hemisphere (Elliott and Bak, 1964; Organ et al., 1968, 1972; Cameron and Burton, 1969; Patterson, 1976, 1977; Patterson and Sleigh, 1976). The opening at the pore then closes, and the flat, empty contractile vacuole is then capable of being filled again.

However, the problem of what triggers systole and membrane fusion is still not solved. Dunham and Stoner (1969) observed in *Tetrahymena* that the pellicle becomes indented just before systole and that the vacuole becomes rounded as though its wall had contracted. Patterson and Sleigh (1976) and Patterson (1977) describe similar events in *Tetrahymena* and *Paramecium*. Thus, on the basis of light microscope cinematography, it appears that the trigger for systole involves, in part at least, contractile elements around the vacuole and an indentation of the pellicle at the pore. Both these phenomena occur before the culmination of membrane fusion at the pore and evacuation of the vacuole. McKanna (1973a) has shown that in *P. aurelia* the space between the membrane at the bottom of the pore and the membrane of the adjacent contractile vacuole is occupied by unorganized filaments, vesicles, and tubules with a diameter of around 45 nm having an 8.5 nm thick membrane. These vesicles and tubules could possibly enhance the fusion of the two membranes by decreasing the radius of membrane curvature, particularly if they are extensions of the contractile vacuole membrane. No evidence has been reported, however, to suggest they represent such membrane extensions. Also, such tubules may not be a common feature of this zone in other ciliate contractile vacuole pores. The morphological details of this region merit further study since this is a zone in which membrane fusion occurs at regular and controllable intervals.

The plasma membrane of *P. aurelia* is coated with a glycocalyx, even along the sides of the pore and over the outside surface of the membrane at the bottom of the pore (McKanna, 1973a). No coat has been seen on the luminal surface of the contractile vacuole membrane itself. Thus during discharge, two unlike membranes are capable of fusion, but this fusion is temporary and the vacuole membrane will ultimately be released back into the cytoplasm. The amount of mixing (if any) of these two membranes at the fusion site has not been investigated.

5.4. Endomembrane system

The endomembrane system of cells as defined by Morré and Mollenhauer (1974) and Morré (1975) consists of the rough and smooth ER, the nuclear membranes (at least the outer one), the Golgi apparatus, and the various transitional membrane elements derived from the rough ER and the Golgi apparatus such as peroxisomes, lysosomes, and secretory vesicles. The end products formed by the transitional elements include the outer membranes of mitochondria and the plasma membrane (Morré and Mollenhauer, 1974) and probably the digestive vacuole membranes in protozoa. In this system membranes are transferred from one organelle to another by membrane flow and are transformed in transit by a process of membrane differentiation (Morré et al., 1971).

A study of membrane flow in *Tetrahymena pyriformis,* strain GL, was made by Franke and associates (1971a). The rough ER of these ciliates was found to be mostly tubular in shape with some flattened cisternae. Rough ER is frequently associated with mitochondria with which it lies parallel at a distance of 200 Å. A sheet of rough ER underlies the alveoli and epiplasm in *Tetrahymena* as in most ciliates. These authors have also reported bridges between the ER and digestive vacuoles (Franke et al., 1971a,b) and between cisternae of the Golgi apparatus. The rough ER is continuous with smooth-surfaced tubulovesicular profiles. These smooth areas may result from a loss of ribosomes from what was rough ER, but this cannot be determined with certainty. Franke and Kartenbeck (1971) also report that ER tubules are sometimes continuous with the outer mitochondrial membrane. Such continuity would provide a pathway for movement of materials and membrane subunits from the ER synthetic sites to the mitochondria.

As seen in section 4, the major site of biosynthesis of ciliate membrane proteins and phospholipids is in the rough ER. This is true also of other plants and animals (Morré, 1975; Poste and Nicolson, 1977). A specialized region of the ER that lacks ribosomes is the site of small primary transitional vesicle formation (Figs. 59–61). These vesicles, which originate by membrane fusion, when still connected to the ER are spaced at least 40 nm apart along the smooth segment (Franke et al., 1971a). After release they fuse with other vesicles into a fenestrated sheet (Fig. 62) which lies about 0.1 μm from the ER membrane. The next one to three cisternae distal to the fenestrated sheet have a membrane that appears thicker than that of the ER and appears to be of the 9 to 10 nm plasma membrane type. Thus, although the presence of a Golgi apparatus in ciliates was disputed for many years, there should no longer be any question of its presence in *Tetrahymena* and other ciliates (Figs. 58–63). In these cells the Golgi apparatus consists of a number of small stacks of cisternae, called dictyosomes, lying next to a segment of ER which is rough on one membrane and smooth on the membrane next to the dictyosome (Franke et al., 1971a). Dictyosomes can also be formed from the outer nuclear membrane.

Besides being smooth, the ER membrane that undergoes blebbing to form the dictyosome cisternae often appears thicker than the rough ER and has a finely granulofilamentous material in close association with its cytoplasmic surface (Franke et al., 1971a). The identity of this granulofilamentous material is unknown, but Franke and co-workers (1971a) suggest the possibility that it is proteinaceous or lipoproteinaceous and is added to or lost from the membrane during membrane differentiation. Since most primary transition vesicles are "coated," this material may be involved in the formation of the "coat."

The extremities of dictyosomal cisternae are frequently coated with a closed network of hexagons and pentagons made from protein subunits, as is typical of coated pits and coated vesicles (Fig. 58) (Pearse, 1976). Such coated regions are characteristic of membrane transport regions and regions of membrane fusion (Morré and Mollenhauer, 1974; Pearse, 1976). The coated margins of Golgi cisternae probably represent sites of secondary transition-vesicle formation. The content of a cisterna in these coated regions, in peritrichs such as *Vorticella*, appears similar to the cup-shaped vesicles found around young digestive vacuoles, which are often continuous with the vacuole membranes (Figs. 45–47). Whether these cup-shaped vesicles come from the dictyosomes has not been determined but at least some may form this way.

The cisternal membranes bleb at the maturing face to form both "coated" and smooth vesicles. Franke and collaborators (1971a) believe these coated vesicles, which have a thick membrane, may fuse with the parasomal sacs of the plasma membrane thereby contributing to plasma membrane biogenesis (see section 5.6). However, some of these peripheral coated evaginations could represent material being added to the margins of mature Golgi cisternae. Addition of material at the edges of Golgi cisternae has been observed in some algae (Brown et al., 1970). Hausmann (1977) suggests that such transition vesicles in *Pseudomicrothorax dubius* are for transporting polysaccharides in the development of the arms of compound trichocysts.

Since most of the membranes of the endomembrane system lie in an environment that is subject to constant streaming, these membranes must be in a continual state of uniting with other membranes or of fragmentation; both processes require membrane fusion. Some cytoplasmic organelles, such as peroxisomes which are also found in ciliates (Müller, 1969, 1975; Fok and Allen, 1975; Stelly et al., 1975), may form as localized evaginations of the ER (Novikoff and Shin, 1964; de Duve, 1973) and when mature they may separate from the ER. Extrusomes, formed as evaginations of the ER, must also be released from the ER at some point. Primary lysosome formation involves membrane fusion at the dictyosome. The production of secondary lysosomes also involves membrane fusion. In ciliates, most of these processes still require detailed study.

Another unsolved problem is the manner of growth of the alveolar membrane complex. The relationship of these membranes to the endomembrane system is unclear since there are no clues from electron micrographs to suggest how these membranes grow or how they become segmented. Vesicles have never been seen to fuse with, or to form from, the alveolar membranes. This is potentially a very interesting membrane system since the inner membrane, at least, has such a high population of intercalated particles.

5.5. Division of cells and organelles

Membrane fusion must also occur during regular transverse fission of the ciliates, which is the normal mode of asexual division. Tucker (1971) studied fission furrow formation in the ciliate *Nassula*. He found there is a rapid phase followed by a slow phase during furrow formation. The rapid phase is characterized by the presence of a ring of microfilaments around the inside of the pellicle at the forming furrow. Sliding of these microfilaments relative to each other is probably responsible for the rapid phase of furrowing but, when the furrow becomes approximately 5 μm in diameter, it slows abruptly or stops. At this point microtubules are found to develop between the pellicle and the microfilaments. After a quiescent period of about 25 minutes the cleavage constriction suddenly stretches, narrows, and breaks, apparently leaving the outer plasma membranes intact as they fuse during breaking. The mechanisms leading to this final breaking have not been investigated electron microscopically. Consequently, how the fusion of the plasma membrane occurs or what happens to the alveoli during this parting have not been considered. The only change observed in the pellicular membranes at the furrow is a folding or pleating of the inner alveolar membrane and epiplasm. Other ciliates such as *Tetrahymena* (Figs. 64 and 65) presumably develop furrows and divide in a similar manner, but information on this question is still scant.

Each asexual division also involved the division of one or more macronuclei and one or more micronuclei. Most ciliates have micronuclear mitotic divisions within intact nuclear envelopes and amitotic macronuclear divisions also within intact nuclear envelopes. However, in some ciliates the micronuclear envelope breaks down after a new envelope has been reconstituted in mitotic telophase, the new envelope being derived partly from cisternae of the old envelope (e.g., Jenkins, 1967; Raikov, 1973, 1976; Davidson and La Fountain, 1975). Membrane fusion plays a key role in these processes but the process of fusion per se has not been studied.

Membrane fusions of a different sort may also occur during binary fission. At one stage of fission in *Nassula* the three micronuclei and one macronucleus are all linked by membrane bridges, thus making the outer membrane of the nuclear envelope continuous around all 4 nuclei (Tucker, 1967). Later each nucleus elongates up to 100 μm in length. A separation spindle is then formed in each micronucleus. The inner membrane pinches in two around each micronuclear chromatin clump at the end of these 3 dumbbell-shaped nuclei, but the outer membrane continues to enclose the complex of micronuclei, macronuclei, and separation spindle. Finally, the separation spindles and macronucleus divide transversely to form two clumps of nuclei which remain enclosed by the outer nuclear membrane. Golder (1976) has shown that the normal position of the interphase micronucleus in *Woodruffia metabolica* is between the outer and inner nuclear membranes of the macronucleus.

Membrane fusions also play a decisive role in conjugation. During conjugation two cells of complementary mating types come together, agglutinate, and finally their membranes fuse to form channels through which the pronuclei, haploid products of meiosis, can be exchanged from one cell to the other. Elliott and Tremor (1958) first described the ultrastructure of this phenomenon in *Tetrahymena*. They observed small (0.2 μm) pores in the apposed pellicles of the two cells. However, exactly which membranes of the pellicle form the perforated septum between the two cells is not absolutely clear. From published electron micrographs (Elliott and Tremor, 1958; Elliott and Zieg, 1968; Flickinger and Murray, 1974) of this region it appears that only one membrane remains from each member of the pair of cells accompanied by an underlying amorphous layer on its cytoplasmic surface; thus the alveoli are apparently lost from this zone and the plasma membranes of two different cells are capable of fusing to give rise to the cytoplasmic bridges. The pronuclei are exchanged through these bridges.

The situation is clearer in *Paramecium* (Vivier and Andre, 1961; Andre and Vivier, 1962; Schneider, 1963; Jurand and Selman, 1969). Here the pairs of conjugants join together along the right sides of much of their ventral surfaces (a distance of over 80 μm in *P. caudatum*). At the mouth the junction lies to the right of the opening on each cell, leaving the oral region intact. The cilia along this contact zone break away and the trichocysts are lost. The actual cell-to-cell contact is made at the crests of the pellicular ridges. The plasma membranes of the pair fuse at many of these crests and an opening is formed between the cells. In *Paramecium* the alveoli may remain or they may be reduced in the contact zone. Pronuclei have been seen to be fixed in the act of passing through these openings (Inaba et al., 1966) that lie in the zone to the right of the mouth within the junction area, waiting to cross the junction line (Andre and Vivier, 1962).

Conjugating ciliates, which do not feed due to resorption of certain feeding organelles, apparently compensate for the loss of nutrients by forming autophagosomes. *Stylonychia* does not feed for 80 hours while in the conjugation process (Sapra and Kloetzel, 1975); however, it is capable of engulfing and digesting its own mitochondria. Membrane fusion is involved in autophagosome formation and in the eventual fusion of small autophagosomes into larger compound bodies.

In addition to nuclear division the mitochondria presumably are capable of dividing by fission and by a membrane fusion of their outer and inner membranes. However, this process has not so far been captured in electron micrographs.

5.6. Other fusions at the plasma membrane

The surfaces of ciliates have regions other than those in the cytopharynx and at the cytoproct where membrane fusion occurs or may occur. Ciliates can lose their cilia during experimental manipulation and also at some normal stages in the cell's life cycle; for example, conjugating ciliates will lose their cilia in the region of cell attachment (Vivier and Andre, 1961). Blum (1971) has called attention to a zone at which cilia invariably break away from the body. This is at the transition region between the end of the basal body and the beginning of the axoneme, distal to the axosome from which the central axonemal tubules grow (Satir et al., 1976). The membrane at the base of the cilium pinches in two and the membranes fuse so that no opening into the cytoplasm is formed even temporarily. The ciliary necklace is found on the P face of the plasma membrane at sites where cilia have either broken away or not yet formed (Satir et al., 1973,

1976). Only the necklace and some randomly placed particles within the necklace are found in the membrane over these closed ciliary sites. The 9 patches of particles distal to the necklace are lost along with the cilium (Satir et al., 1976).

Another site at which membrane fusion occurs in some peritrich ciliates such as *Vorticella* is at the scopular region where the stalk joins the body of the cell. In *Vorticella* the stalk is contractile and is composed of a bundle of microfilaments that extend in a helical pattern from the proximal to the distal end of the stalk. The myonemal bundle is an extension of the cytoplasm of the body and is surrounded by a plasma membrane and, in part, by the alveoli (Allen, 1973a). Under certain environmental conditions the sessile organism has been reported to shed its stalk, grow a wreath of cilia at its aboral end, and swim away (Mackinnon and Hawes, 1961). Although the details of stalk shedding have not been studied at the fine structural level, this process would require membrane fusion at the site where the stalk is detached from the body.

Another site of potential membrane fusion at the surface of ciliates is present at the regularly placed invaginations of the plasma membrane called parasomal sacs. One of these is found in association with each basal body or pair of basal bodies in the somatic region of *Tetrahymena* and *Paramecium* (Figs. 1, 6, 66). They are also aligned in rows in conjunction with the oral cilia (Figs. 30, 31). In peritrichs and suctorians, which do not have somatic cilia in the normal sessile feeding stage, there are pellicular pores over the body surface that appear to be homologous to the parasomal sacs even though cilia are not present.

These sacs and pores can frequently be seen to be in a state of either endocytosis or exocytosis, that is, enlarged tubular membrane-bound bodies are attached to the plasma membrane at these sites (Fig. 67). Normally the parasomal sacs and pellicular pores have a bristle coat (R. D. Allen, 1967; Franke, 1971) on their cytoplasmic surface that resembles the lining of coated pits. This coat is presumably composed of clathrin protein subunits (Pearse, 1976). During some stage these coated pits may have a spherical aspect and be attached to the plasma membrane at the bottom of the parasomal sac by a narrow neck. Alternatively a larger cisterna, having one or more coated pits along its margin, may be attached at the parasomal sac.

Whether these sites are undergoing endocytosis or exocytosis has not yet been determined. Possibly they are sites of either type of cytotic event, depending on the requirements of the cell at any particular time. Franke and co-workers (1971a) have suggested that they are sites to which Golgi-derived vesicles can fuse with the plasma membrane to allow growth of the cell surface. The Golgi cisternae actually have coated regions and may give rise to coated vesicles or to the thick membrane-bound orthotubular cisternae 30 to 70 nm in diameter that bear coated pits (Franke and Eckert, 1971). Such membranes are known to accumulate in the cytoplasm near the parasomal sacs both in the somatic (Fig. 2) and oral regions (Fig. 30). In studies on other cells, Franke and associates (1976a,b) and Franke and Herth (1974) have documented the role of coated vesicles in providing new membrane for plasma membrane expansion.

On the other hand, these flattened cisternae may arise from the parasomal sacs by invagination and pinching off of these sacs. R. D. Allen (1967) interpreted electron micrographs of inflated parasomal sacs of *Tetrahymena* to be a stage in endocytosis. Noirot-Timothée (1968) using thorotrast to label sites of endocytosis in *Trichodinopsis* showed that electron-opaque material could be found in parasomal sacs, as well as in vesicles within the cytoplasm near the parasomal sacs. Rudzinska (1977) has also reported the uptake of ferritin at these pores in *Tokophrya infusionum*, a suctorian. Evidence for coated vesicles being sites of endocytosis has been presented for many cells (Pearse, 1976). These sites of endocytosis or exocytosis should be studied further since they are doubtless very important in the cell's life cycle. Besides being sites at which the surface membrane area can be increased or resorbed, they may be supplementary sites of specific adsorption and uptake of nutrients from the cell's environment. *Tetrahymena* may be able to use these invaginations as a pathway for nutrient uptake when it is growing in a rich nonparticulate medium, thus bypassing normal digestive vacuole formation.

6. General discussion and conclusions

Ciliated protozoa have some membrane characteristics not commonly found in most other cells, particularly the cells of higher organisms. At the surface of the cell is a plasma membrane, which is universally found in cells, but, unlike other cells, ciliates and a few algae have a plasma membrane that is closely subtended by a mosaic of flattened cisternae. The flattened cisternae intercommunicate with each other by way of pores (Allen, 1971), thus making the alveoli a continuous mosaic system. The plasma membrane seems to be linked to the outer membrane of this mosaic system by bridges that are visible in freeze-fracture replicas as intracellular membrane junctions (Plattner et al., 1973). These junctions are diffused throughout the space between the two adjacent membranes. In addition, more concentrated intracellular junctions join the alveolar membrane to the plasma membrane along lines of termination of alveoli at the cytoproct, cytopharynx, and sites of trichocyst secretion. Alveoli also have junctions with trichocyst tip membranes and between the membranes of adjacent alveoli at the inner margins of septa. All of these junctions appear to cause the membranes of the ciliate pellicle to maintain a certain rigidity. It would appear that the plasma membrane and underlying alveoli probably grow in unison; the incorporation of membranes from secretion vesicles cannot suddenly increase the surface membrane area without the formation of blisters by the newly added membrane.

Many intercalated P-face particles are located in regions of membranes that are known, by thin-sections, to be adjacent to filaments or fibers on the cytoplasmic side of the membrane. Examples of this correspondence of intercalated particles at sites adjacent to cytoplasmic filaments are found at the patches at the proximal end of cilia (Figs. 12–15), at rosettes where the plasma membrane joins the mucocyst or trichocyst membrane (Figs. 17 and 18) and in the extensive zone between the outer alveolar membrane and the plasma membrane (e.g., see Figs. 6 and 48) (Tokuyasu and Scherbaum, 1965; Franke et al., 1971b) at the margins around trichocyst fusion sites and also scattered over the cell surface. Many intercalated particles must be intrinsic proteins to which peripheral proteins of the cytoplasm are attached to the membrane, and many must indicate sites at which membranes are linked together by these peripheral proteins.

Intracellular junctions such as those found in ciliates have not been reported in other cells. One function for the junctions, besides their role in linking membranes, may be to limit the spread of intrinsic proteins. Pellicular membranes contain large numbers of intramembranous particles on the E face of the inner alveolar membrane and on the E face of the cytopharynx. Both regions are contiguous with other membrane surfaces that contain far fewer particles. Also, in both examples the highly particulated regions are "sealed off" from the less particulated contiguous membranes by intracellular membrane junctions as is true between adjacent alveoli at the septal margin next to the inner alveolar membrane (Fig. 10) and at the margin of the cytopharynx between the cytopharynx and surface membranes (Fig. 35). Flow of membrane proteins past these junctions may be limited.

An interesting unsolved problem is the function of the alveoli. These membrane sacs must be important to the cell either to cushion the cell against physical and chemical shock (Pitelka, 1965) and to function as storage sties for calcium (Allen and Eckert, 1969) or to play a role not yet suggested. Thus the function of what is probably the most typically "ciliate" membrane system is not know. Another unique feature of the alveoli is that unlike the plasma membrane there are no sites on these sacs at which membrane fusion is known to occur or even seems to occur; how they grow and their relationship to the endomembrane system of the cell are obscure. The only specialization of these sacs shown to date are the fingerlike projections in *P. caudatum* that extend from the edge of the alveoli at the cytopharynx to the microfilamentous cord that lies in this area (Fig. 36b) (Allen, 1974), but although a function in the control of contractility of this fiber was suggested, no studies have yet been preformed to determine if this fiber is in fact contractile, or if the alveoli play a role in its control.

Ciliates also offer some highly specialized membranes for investigation. These include the fluid segregation organelles of the contractile vacuole complexes (Fig. 55) and other decorated tubules, such as those continuous with the ER of *Opercularia* (Fig. 4) (Allen, 1973b) and the linkage complexes on the ER adjacent to myonemes in peritrichs (Fig. 5) (Allen, 1973a,b). Another membrane specialization is the collar formed around the membranes at the trichocyst tips (Bannister, 1972). The functions of these specializations are again largely unknown. Also, the intramembranous plates scattered on the anterior ventral surface of *Paramecium* (Fig. 11) (Allen, 1976b) have yet to be investigated.

The surface membrane of ciliates appears to differ from most other protozoa and metazoa surface membranes since, although it can fuse with vesicles or form vesicles—whichever occurs—at the parasomal sacs and secretory sites, the membranes added at these sites do not generally flow out onto the cell surface to become part of the plasma membrane. This surface is therefore adapted to fusion between unlike membranes. Thus the thinner membranes of the secretory vesicles that do not have a glycocalyx can fuse with the thicker glycocalyx-lined plasma membrane but they do not flow together; instead, the secretory membrane is resorbed into the cell by endocytosis. Only a very small part of the secretory membrane remains at the cell surface in *Paramecium*. On the other hand, in the cytopharynx where discoidal vesicles become part of the growing, limiting membrane of the food vacuole, the discoidal vesicles have a similar thickness to the cytopharyngeal membrane and also bear a glycocalyx. Since the membranes are alike, this no doubt facilitates their mixing together. Another example of a thin membrane fusing with the thick plasma membrane is at the contractile vacuole pore. Here again, the contractile vacuole membrane does not become part of the surface membrane but is reengulfed by the cell.

The membranes in the endoplasm are also characterized by membranes of two different morphologies. There are thick membranes, 9 to 10 nm in width, which generally have one leaflet more electron-opaque and thicker than the other and a second group of thinner membranes, around 7 to 8 nm wide, whose leaflets are of the same thickness. However, these two classes are not always distinct and the preparatory methods used frequently influence the membrane widths. Generally membranes are thinner at sites of synthesis at the ER, but during membrane flow and differentiation (Morré and Mollenhauer, 1974; Morré, 1975) they become thicker and acquire coats of various kinds. A more systematic approach must be applied in determining thicknesses of ciliate membranes before the thickness of all membranes can be classified. Such a study would provide clarification of the passage of membrane through the cell.

Membrane differentiations are sometimes quite dramatic and do not always occur at the Golgi apparatus. A particularly clear example of this is the differentiation of secretory vesicle membranes as the vesicle comes to lie under the plasma membrane. Both mucocyst and the trichocyst tip membranes (Fig. 27) undergo a clearing of particles from membrane regions that lie near the plasma membrane. These particles apparently move to the distal ends of the vesicle membranes where they form an annulus. In trichocysts a collar of microtubule-like structures is formed around the cytoplasmic surface of the membrane in this particle-free zone. This collar is only present in mature "resting" trichocysts and not in immature trichocysts (Allen and Hausmann, 1976).

The physical and chemical properties of ciliate membranes also present some unique features that have potential use in comparative studies with membrane properties of variously constituted membranes of higher organisms. Membranes of ciliates contain three major classes of polar lipids. Phosphatidylcholine (PC) and phosphatidylethanolamine (PE) are common to other cells, but the third class, a glycerol aminoethylphosphonolipid (GAEP) which contains a C-P bond, is not commonly found in other cells. Ciliates also have a high concentration of hexadecyl glycerol ether (chimyl alcohol) as part of the PC and GAEP classes which have the ether linkage mostly at the C-1 position. In addition the fatty acids of ciliates contain an unusually high concentration of the polyunsaturated acids linoleic (18:2, $\Delta^{9,\ 12}$) and γ-linolenic (18:3, $\Delta^{6,\ 9,\ 12}$), and a second isomeric form of linoleic ($\Delta^{6,\ 11}$) found primarily in GAEP. Finally, the major neutral lipid found in *Tetrahymena* is unique. It is a sterol-like pentacyclic triterpenoid alcohol called tetrahymenol.

The surface membranes of ciliates contain a higher proportion of GAEP than the microsomal membranes and also a higher proportion of tetrahymenol (5-7 times more in ciliary membranes than in microsomes). Microsomes contain higher proportions of PC and PE than do ciliary membranes. The enrichment of the phosphonolipids, the alkyl glycerol ether (hexadecyl glycerol), and the neutral lipids in the surface membranes may impart stability to this membrane (Kennedy and Thompson, 1970; Berger and Hanahan, 1971) while tetrahymenol may help to determine the physical characteristics and permeability properties of the surface membranes (Thompson et al., 1971) and regulate membrane fluidity (Nozawa et al., 1975). These lipids, which are found to be enriched in the surface membranes, seem to be resistant to phospholipases and other degradative enzymes.

So few studies have been done on membrane proteins of ciliates that little general information can be derived. Most of these studies have examined proteins, particularly large glycoproteins, on the exterior of surface membranes in *Paramecium* where very large immobilization antigens are found (250,000–340,000 MW). The function of these i-antigens is unknown, but they are used to determine "serotypes" of a particular species. Intrinsic proteins in ciliary membranes apparently function as calcium gates that are involved in regulating ciliary reversal by allowing calcium to enter the cilia when the membrane is depolarized (Kung et al., 1975; Browning and Nelson, 1976; Eckert et al., 1976; Ogura and Takahashi, 1976). Membrane mutants of *P. aurelia* are now being used to study these important problems (Kung et al., 1975; Browning et al., 1976). Potassium channels involved in hyperpolarization of the cell membrane seem to be located in the plasma membrane on the cell surface rather than in the ciliary membranes (Ogura and Takahashi, 1976). Substances are also found on certain cilia of complementary mating type cells that promote surface stickiness, so that these cilia agglutinate when competent cells touch. This leads to cell aggregation and finally to surface membrane fusions during conjugation. Progress is being made in the identification of gamones (Miyake, 1974a,b), soluble substances released into the cell's environment by different mating types, which cause a series of reactions in opposite mating types that eventually allow mating to occur. Also, ciliary membranes have been isolated and their proteins characterized on polyacrylamide gels (Kitamura and Hiwatashi, 1976). Studies of this nature should lead to the identification of the macromolecules involved in the agglutination reaction of conjugating ciliates.

All membranes within a ciliate are involved in intermembrane exchange of lipids and proteins. However, some membranes are more accessible to exchange than others since a radioactively labeled fatty acid such as palmitate added to the culture medium is found to be incorporated most rapidly into microsomal membranes where it reaches a maximum specific activity within 10 minutes, but it takes 6 hours for ciliary membranes to reach the same specific activity (Nozawa and Thompson, 1971b). This distribution of lipids is due to net lipid exchange rather than to lipid synthesis since starved cells, where lipid synthesis stops, show the same distribution pattern. Also, lipids and proteins are exchanged independently of each other (Nozawa and Thompson, 1972) and lipid insertion is not coupled to protein insertion. In addition, the incorporation of radioactively labeled lipids plateaus in ciliary membranes after 6 hours while proteins plateau after 2 hours (Subbaiah and Thompson, 1974), indicating that both do not enter the membrane at the same rate. Phospholipids can also undergo alterations after they have been incorporated into membranes (Thompson et al., 1971). PE and PC are reduced in the surface membranes of *Tetrahymena* from the levels of their initial incorporation into this membrane, but phosphonolipids that may be resistant to lipolytic enzymes (Thompson et al., 1971, 1972) are not reduced in quantity.

Alterations in the lipid content of membranes can be induced in several ways. Adding high doses of ergosterol to the culture fluid will induce its incorporation into membranes replacing the tetrahymenol (Conner et al., 1971). The cell will compensate for this change by decreasing the unsaturated fatty acid content of the phospholipids and increasing the number of short-chain fatty acids which, taken together, result in a decrease in membrane fluidity (Ferguson et al., 1975). PE is increased and PC and GAEP are decreased (Nozawa et al., 1975). Tetrahymenol in the membrane seems to cause the membrane to be more fluid at temperatures below 27 °C and less fluid at temperatures above 27 °C (Nozawa et al., 1974). Thus tetrahymenol may help regulate those functions of membranes that depend on a particular fluid state of the membrane. Hexadecyl glycerol at the C-1 position of phospholipids may also help to control membrane fluidity (Fukushima et al., 1976b).

The growth temperature affects food vacuoles. Food vacuole formation will cease at a temperature below the phase transition temperature of young food vacuole membranes; this temperature is 9 °C in the NT-1 strain of *Tetrahymena pyriformis* which had been growing at 39.5 °C (Kitajima and Thompson, 1977b). Crystallization of the membrane lipids may prevent fusion of vesicles with the cytopharyngeal membrane or the flow of the vesicles into the nascent food vacuole membrane.

Ciliates are capable of acclimating to growth under various temperatures by changing the lipid makeup of their membranes. The amount of saturated fatty acids will increase with higher temperatures and decrease at lower temperatures, but the number of double bonds seems to remain constant (Conner and Stewart, 1976). The outer alveolar membrane takes 8 to 10 hours to become acclimated to 15 °C in *T. pyriformis*, strain NT-1 (Fukushima et al., 1976a), which is the same time period that division is inhibited in this strain. The phospholipid content of membranes can also be altered as a response to shifts in growth temperature.

When high amounts of linoleic acid are added to the growth medium, the phase transition temperature of cells will change. This is seen first in the ER and nuclear membranes after 100 minutes and in the outer alveolar membrane after 345 minutes (Kasai et al., 1976). Thus the newly synthesized lipids that are enriched in saturated fatty acids apparently move from the ER to the pellicle and the cilia.

These biochemical and biophysical studies demonstrate the amazing capability of cells to adjust their membrane lipid content to compensate for changes in environmental conditions. Proper fluidity of membranes is presumably essential to the proper functioning of membranes.

Ciliate surface membranes, because of the underlying alveoli that limit the movements of the plasma membrane, are almost ideally structured for the study of membrane fusion. Fusion sites, including those of the secretion organelles, contractile vacuole pore, parasomal sacs, and cytoproct, are all permanently outlined or absolutely predictable to within an area of as little as 60 nm in diameter. These sites can be identified in electron micrographs either with thin-section techniques or, of equal importance, in freeze-fracture replicas of membranes.

The usefulness of ciliates in studies of membrane fusion was first recognized and exploited in *Tetrahymena* and *Paramecium* by Satir and coworkers (reviews, Satir, 1974a,b; Satir and Satir, 1974). Plattner (1974), Plattner and associates (1975), and Plattner and Fuchs (1975) have used *P. aurelia* to investigate the trigger for membrane fusion at trichocyst fusion sites. In both of these ciliates the sites at which fusion will occur are outlined by rosettes of intercalated particles with an outer diameter of 60 to 80 nm. Ultrastructural studies of this site at various states of fusion (Figs. 19 and 24–26) have led to an understanding of many of the structural events leading to and culminating in membrane fusion, but little information is presently available on the biochemical events accompanying fusion.

Ciliates also offer other sites for studies of membrane fusion during exocytosis, almost equally predictable, which do not involve rosettes of particles and can be manipulated to enhance the catching of fusion events at specific stages. The contractile vacuole always opens at specialized pores (Fig. 54). Food vacuoles in many ciliates are always egested at specialized sites (Figs. 48–52). Parasomal sacs (Figs. 66 and 67) need further study to see what their role is in the cell, but they certainly are sites of endocytic or exocytic membrane fusion. Finally the growth of food vacuoles, which is highly manipulatable, requires a large number of exocytic fusion events (Fig. 34) that occur at a localized zone of the oral region, the cytopharynx.

Endocytosis is also common in ciliates. The food vacuole is endocytized, the contractile vacuole is pinched off following evacuation, and the membranes of at least some secretory organelles are resorbed (Hausmann and Allen, 1976). However endocytosis, like that in amoeba referred to as "cell drinking" or pinocytosis, also occurs in ciliates. In this case tubular channels surrounded by a microfilamentous network (Fig. 51) invaginate into the cell. These channels are then pinched off from the surface membrane. This is the fate of the membrane of emptied food vacuoles at the cytoproct. This type of pinocytosis thus seems to require a microfilamentous system that is presumably contractile, and the channels formed have a regular diameter. The same type of channeling may occur on the surface of young food vacuoles (Figs. 37 and 38), but this requires investigation. The mechanisms that permit the food vacuole to be pinched off, and the contractile vacuole and secretion vesicle membranes to be released from the surface are unknown.

In all these endocytic and exocytic events in ciliates it is safe to say that the cytoplasm never comes into direct contact with the culture fluid. It is always separated from this fluid by at least one leaflet of the membrane.

Membrane growth and recycling can also be studied profitably using ciliates as a model cell system. Again, the alveoli permit membrane fusion only at specified sites on the plasma membrane. It has been suggested that plasma membrane growth is due to the incorporation of vesicle-derived membrane either from secretory organelles (Satir, 1974a) or from thick-membrane-limited cisternae fusing with parasomal sacs (Franke et al., 1971a). However, the biochemical evidence (Nozawa and Thompson, 1972; Subbaiah and Thompson, 1974) does not support the view that proteins and lipids are inserted together into surface membranes, which would be required if membrane growth were by the incorporation of intact vesicle membranes via fusion processes.

Recycling of membranes probably occurs in the digestive vacuole system of ciliates. The morphological evidence for this has been presented in this review. The biochemical study of Weidenbach and Thompson (1974) also suggests that precursors for these membranes are not synthesized at the time that vacuoles are formed but come from a pool of preformed membrane. They also conclude that these membranes are more like microsomal membranes than surface membranes. However, this latter conclusion should be viewed with caution since the pellicular membrane fraction is presumably composed of only one-third plasma membrane and two-thirds alveolar membranes, which are quite unlike the plasma membrane in freeze-fracture replicas and which would thus mask the true characteristics of the plasma membrane when so much alveolar membrane is also present in the fraction.

Studies of food vacuoles have demonstrated that membranes can undergo structural changes that parallel physiological changes occurring within the membrane-enclosed space (Figs. 37-41). Finally, these studies suggest that membranes which are required to serve as a barrier to ion or polar molecule flow have a leaflet devoid of transmembrane particles located next to the fluid of low pH. Smooth E-face leaflets are also found in trichocyst membranes (Fig. 29) and in the membranes of collecting canals (Fig. 57) of the contractile vacuole system, both of which may act as barriers to the passage of water molecules. On the other hand, membranes with very large numbers of intercalated particles may allow large amounts of ions or polar molecules to pass, as for example, cytopharyngeal membranes (Fig. 31) and the inner alveolar membranes (Fig.8).

It is anticipated that many of the unresolved questions discussed here that relate to membrane dynamics will be answered in the coming years and many of these answers will come from the ciliated protozoa.

Acknowledgments

This review is dedicated to Professor K. R. Porter of the University of Colorado for his encouragement and assistance over a number of years. On several occasions I have been privileged to work in his laboratory under his supervision. The most recent of these stays was a sabbatical spent in Professor Porter's laboratory during which this review was begun. Professor Porter was first to suggest that I work on a problem using the ciliate *Tetrahymena* as a model cell. Since that time I have been using various ciliates as model systems to study a number of problems of basic cell biology.

In preparing the manuscript I had the excellent assistance of Dr. Agnes K. Fok, who helped correct it; Marilyn Allen, who helped in proofreading; and Joanne Muraoka, who typed it. Former students and assistants who contributed to research reported and reviewed here are Marilynn Ueno, David Leaffer, and Robert W. Wolf. Our research has been supported by the U.S. Public Health Service, first at Harvard University under a postdoctoral fellowship (GM-18665) a training grant (2G-707), and a research grant (GM-06637); and later at the University of Hawaii (grant GM-17991).

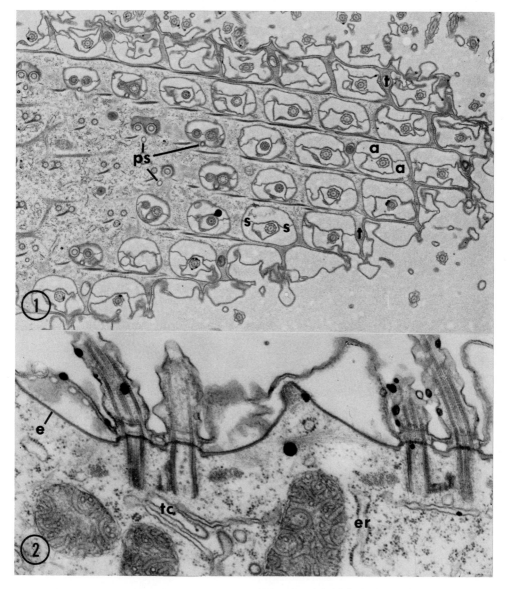

Fig. 1. Tangential section through the cortex showing the typical architecture of the pellicle of *Paramecium caudatum*. Cilia and basal bodies are arranged in rows and arise from the center of a hexagonal or rectangular surface projection. The pellicle consists of a plasma membrane and underlying alveoli (a). The alveoli are swollen here and the septa (s), which separate adjacent alveoli, form lines between the cilia in a row. Trichocysts (t) penetrate to the surface in the transverse ridges that separate cilia. Parasomal sacs (ps) are found next to each pair of basal bodies. ×10,000.

Fig. 2. Two pairs of basal bodies with attached cilia along a kinety, or row of cilia, in *P. caudatum*. A septum between alveoli is seen in planar view (left) and the pores connecting adjacent alveoli are obvious in this septum. A thin epiplasm (e) lies under the inner alveolar membrane. Thick-membrane-bounded cisternae (tc) lie under basal bodies and have localized coated regions. The rough ER membrane (er) is thinner than these thick membranes. The electron opaque deposits, which sometimes occur spontaneously during glutaraldehyde fixation, may be calcium. ×35,000.

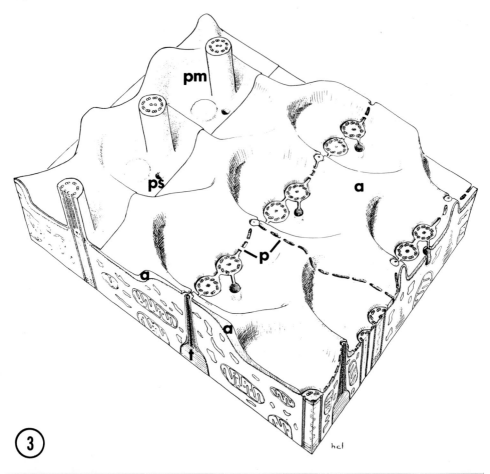

③

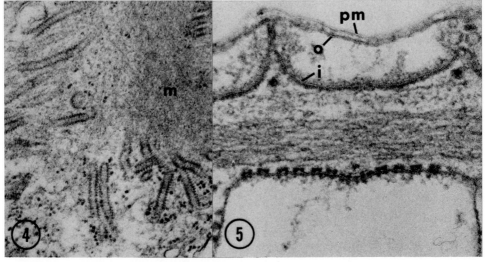

④ ⑤

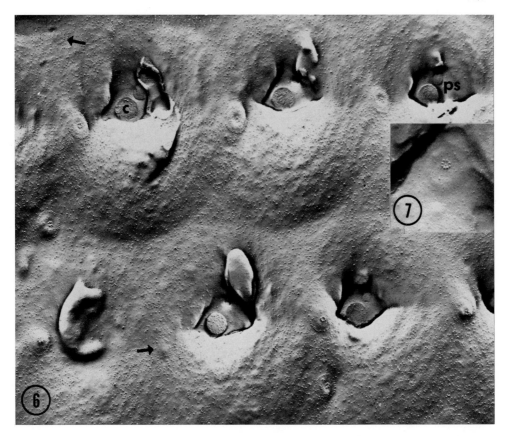

Fig. 3 *(opposite)*. A three-dimensional drawing of the pellicular surface of *P. caudatum*. The plasma membrane (pm) is continuous over the cilia and penetrates to the ectoplasm at the parasomal sacs (ps). The underlying alveoli (a) form a mosaic of cisternae under this membrane and the alveolar units are connected by pores (p) across these septa. Sites of trichocyst (t) and ciliary penetration are also shown. (Drawing courtesy of Mrs. H. C. Lyman.)

Fig. 4 *(opposite)*. Membrane differentiations of the peritrich ciliate *Opercularia*. Tubules are continuous with the endoplasmic reticulum, and lie next to and penetrate into the contractile microfilamentous myonemes (m). Pegs project from the outer surface of the tubule membranes. ×50,000.

Fig. 5 *(opposite)*. In *Vorticella*, the ER adjacent to myonemes has a specialization consisting of a coin-shaped arrangement of transmembrane rails and a central spindle-shaped midpiece. This is fixed to the myoneme and does not separate from it even when the ER vesiculates, as shown in this micrograph. i, inner alveolar membrane; o, outer alveolar membrane; pm, plasma membrane, ×100,000. (From Allen, J. Cell Biol. 56, 1973; reproduced with permission.)

Fig. 6. This freeze-fracture image of the P face of the plasma membrane of *P. caudatum* shows the intercalated particles in this membrane. Particles are randomly arranged except at trichocyst fusion sites, where double rings of particles and central rosettes of large particles are found between fractured cilia (c). Parentheses-shaped rings without rosettes *(arrows)* and circular rings with rosettes can also be identified outside the rows of cilia. Parasomal sacs (ps) lie next to each fractured cilium. ×23,000.

Fig. 7. Normally the rosette in the plasma membrane adjacent to the trichocyst tip remains with the P face, but if the cell is fractured without prior fixation the rosette may remain with the E face of the plasma membrane, as shown here. ×40,000.

728

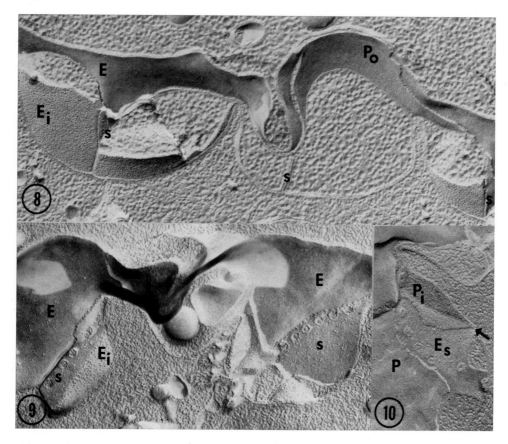

Fig. 8. The pellicle of *P. caudatum* fractured to reveal the E face of the plasma membrane (E), P face of the outer alveolar membrane (P_o), and E face of the inner alveolar membrane (E_i). The differences in relative number of particles on these membranes are evident. The inner alveolar membrane has a particularly large number of particles. Septa (s) of the alveoli are also present. ×30,000.

Fig. 9. The septa (s) between alveoli in *P. caudatum* have pores adjacent to the plasma membrane; the E face (E) of the plasma membrane is shown here. The E face of the inner alveolar membrane (E_i) is densely particulate but this ends abruptly at the bend next to the septum. ×40,000.

Fig. 10. The P face and the E face (E_s) of the septum of adjacent alveoli are shown, as are the P faces of the inner alveolar (P_i) membrane and the plasma membrane (P). Pores in the septum and a row of particles *(arrow)* that separates the P face of the septum from the P face of the inner alveolar membrane are visible. ×40,000.

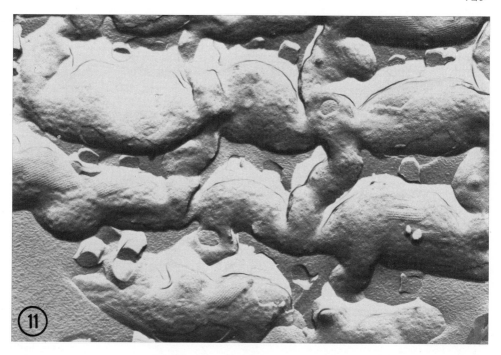

Fig. 11. Plates composed of parallel rows of closely spaced particles are found in the P face of the plasma membrane in the anterior ventral region of *P. caudatum.* The plates are found at the pellicular ridge summits or along the sides of longitudinal ridges. ×22,5000.

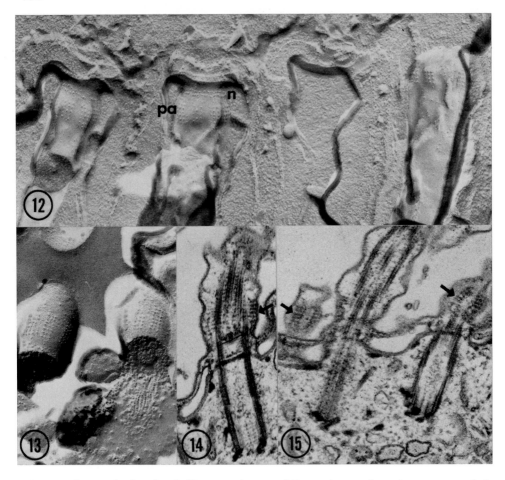

Fig. 12. The proximal ends of ciliary membranes of *Paramecium caudatum* have two-stranded necklaces (n) of particles on their P faces as well as several patches (pa) of particles distal to each necklace. A patch is composed of three short rows of particles. There is probably one patch opposite each doublet of the ciliary axoneme. ×45,000.

Fig. 13. Cells which were fixed and refrigerated for a few days at 4 °C began to swell. The patches of particles at the proximal end of cilia of the oral region are more evident in these swollen cilia, as are the doublet fibers of the axonemes within the cilia. ×50,000.

Fig. 14. Although the intramembranous particles of the patches at the proximal ends of cilia of *P. caudatum* cannot be seen in thin sections, electron-opaque fibers can be seen to pass between the axoneme doublets and ciliary membrane at the level of these patches *(arrow)* just distal to the transition zone between the cilium and basal body. ×50,000.

Fig. 15. The fibers adjacent to the patches are also evident in grazing sections of the ciliary membrane *(arrows)*. The cross-sectioned fibers appear to be arranged in three interlinked rows corresponding to the three rows of intercalated particles within the ciliary membrane at this level. ×50,000.

Fig. 16 *(opposite)*. Surface membranes of many ciliates bear an external glycocalyx. This is especially well developed in *Coleps hirtus*, as shown here. There is a 15 nm electron-opaque layer *(arrow)* lining the membrane of these cilia. In addition long fibers extend beyond this layer. The tips of cilia narrow at the level where the axonemal spokes end, but the peripheral doublets continue on. The central axonemal microtubules extend to the very tip of the cilium *(left cilium)*. ×75,000.

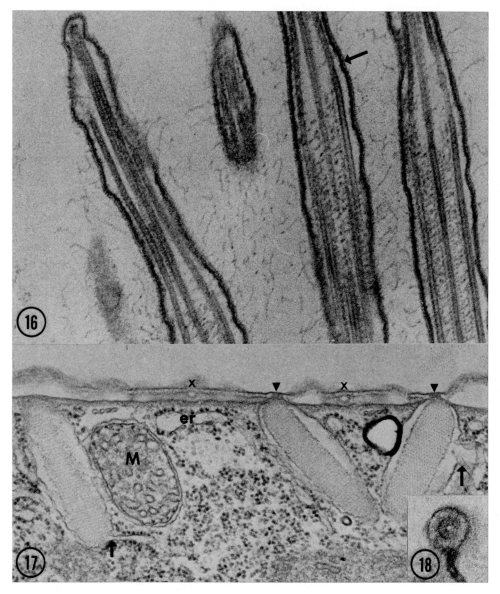

Fig. 17. Mucocysts of *Tetrahymena pyriformis* are aligned along the primary meridians (rows of cilia) as well as secondary meridians as shown in this micrograph. The body of the mucocyst is crystalline and is surrounded by a membrane. This membrane is continuous with the rough endoplasmic reticulum *(arrows)*. Dense material occupies the space between the plasma membrane and the tip of the mucocyst *(arrowheads)*. Pores similar to those in *Paramecium* (x) are seen in the septum between the alveoli along this secondary meridian. M, mitochondrion; er, rough endoplasmic reticulum. ×50,000.

Fig. 18. A tangential section through the point at which the mucocyst meets the plasma membrane shows an electron-opaque disk 65 nm in diameter surrounded by the alveolar margin. An electron-transparent pinhole 10 nm in diameter is located in the center of this disk. The disk is situated at the fusion site of the mucocyst and plasma membranes and has the same diameter as the rosette of intercalated particles at this site. ×100,000. (From Allen, J. Protozool. 14, 1967; reproduced with permission.)

732

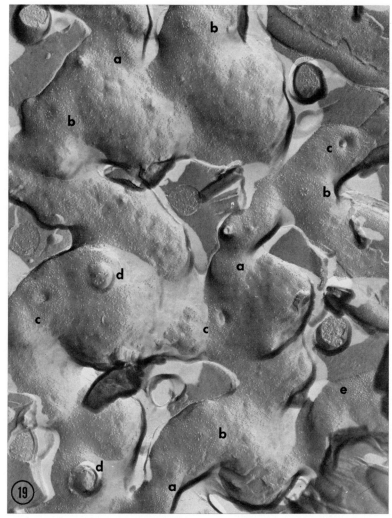

Fig. 19. Many trichocyst fusion sites are found near the anterior end of *Paramecium caudatum*. In freeze-fracture replicas these sites are present at a variety of stages of forming, resting, fusion, and resealing. The fusion sites are indicated by the double rings of particles that identify intermembrane junctions between the alveolar and plasma membranes. Parentheses-shaped rings (a) with an outer width of about 140 nm do not have a trichocyst below them. Resting trichocyst sites (b) have circular rings 300 nm in diameter and an enclosed rosette of 8 particles with an outer diameter of 70-80 nm in the center of the ring. Some of the fusion sites are pulled down in the center (c) while others are pushed out (d). These may represent stages in the processes of fusion and resealing, or the tricho-cysts may have been slightly displaced by the cytoplasmic contents during fixation. Other rings (e) en-close a large number of particles but no rosette; these are probably resealed fusion sites. ×29,600.

Fig. 20 *(opposite)*. The distal end of the E face of the trichocyst membrane has rows of pits corres-ponding to the annulus on the P face (not shown). The pointed tip of the trichocyst lies very close to the plasma membrane. A ring of pits encircles the base of the tip. The ring of pits has about the same diameter (75nm) as the rosette on the plasma membrane. The ring on the P face of the plasma membrane encircling the tip is partially visible. ×40,000.

Fig. 21 *(opposite)*. The E-face leaflet at the point of this trichocyst and possibly the point of the trichocyst itself were fractured away, leaving a circular 63 nm diameter ridge. ×40,000.

Fig. 22 *(opposite)*. The tip of this trichocyst has just penetrated the plasma membrane. The ring of particles is placed symmetrically around the tip and has a slightly swollen diameter of 350 nm. The plasma membrane curves down along the edge of the trichocyst tip, suggesting that fusion with the trichocyst membrane occurred just before fixation was completed. ×40,000.

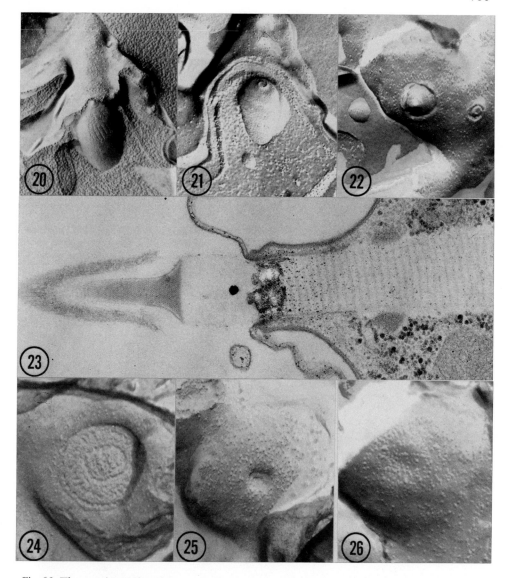

Fig. 23. The opening at the trichocyst fusion site enlarges to a diameter of at least 400 nm. The cap, tip, and shaft of the trichocyst pass through this opening. The ring of particles must be capable of considerable expansion during this process. The electron-opaque material at the pore may be a calcium deposit. Calcium is involved in trichocyst exocytosis. ×37,000.

Fig. 24. Open trichocyst fusion site including a cross-fractured trichocyst cap and tip. The expanded ring is found around the pore (410 nm diam.), and the plasma membrane bends down along the side of the tip. ×60,000.

Fig. 25. Trichocyst fusion site (320 nm diam.) probably in the process of resealing. Many more particles are found within the ring than are found normally at "resting" sites. These particles may come from the annulus of the trichocyst membrane during resealing. ×60,000.

Fig. 26. A fusion site that presumably has resealed after exocytosis encloses many particles, some randomly and others more regularly arranged, as the line of particles shown here. The large particles may be remnants of the rosette. ×60,000.

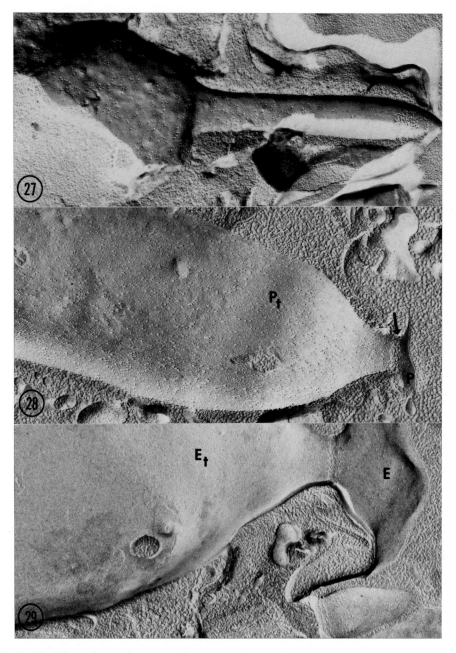

Fig. 27. The P face of the trichocyst membrane, located at the site of secretion, has three different regions based on its particle composition. The body and proximal end of the tip of the trichocyst have randomly arranged particles. Distal to this is a region almost devoid of particles, although some barely exposed particles form lines parallel to the long axis of the tip. Near the distal end of the tip a ring of particles, the annulus, forms a band. ×40,500.

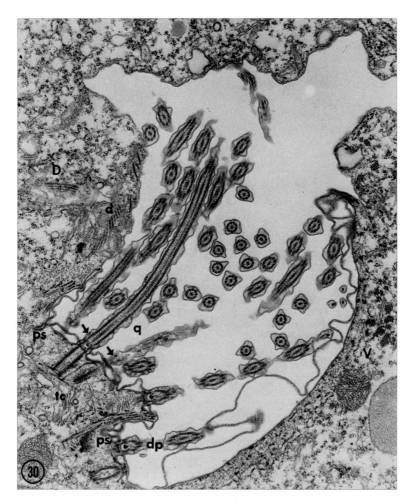

Fig. 28 *(opposite)*. The P face of the trichocyst membrane (P$_t$) after fusion with the plasma membrane (P). The annulus of particles identifies the zone of fusion *(arrow)*. Particles are found on the trichocyst membrane all the way to the annulus; there is no longer a smooth zone. ×40,500.

Fig. 29 *(opposite)*. The E face of the trichocyst membrane (E$_t$) and plasma membrane (E) after fusion of the two membranes. An annulus of pits marks the site of fusion between the trichocyst and plasma membranes; otherwise the E face of the trichocyst membrane is smooth. Pits of the double ring surrounding the fusion site are seen just distal to the annulus. The annulus expands and becomes thinner when the pore opens. A cilium with a necklace and patches of particles is also shown *(lower right)*. ×40,500.

Fig. 30. A cross-section through the distal end of the buccal cavity of *Paramecium caudatum*. The cytopharynx *(left)* is a single membrane-limited region lined with microtubular ribbons and discoidal vesicles (d) *(seen here in planar view)* on the dorsal surface of the buccal cavity (D). The wavy membrane of the nascent food vacuole is toward the top. The floor of the buccal cavity is covered by a pellicle *(lower right)*. Cilia of the quadrulus (q) and dorsal peniculus (dp) occupy the space within the buccal cavity. Thick-membrane-bounded cisternae (tc) with coated pits are found around the basal bodies near the parasomal sacs (ps). Postesophageal bundles of microtubules line the ventral edge (V) of the buccal cavity and nascent food vacuole *(right)*. Arrows indicate the fibers adjacent to patches in the ciliary membrane. ×20,000.

736

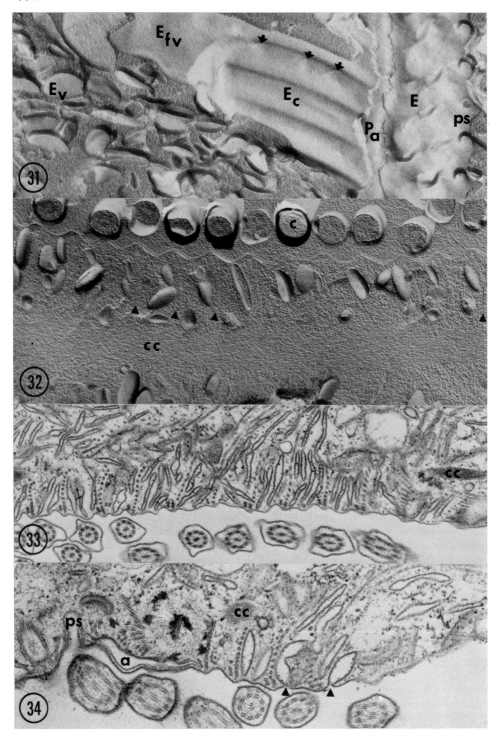

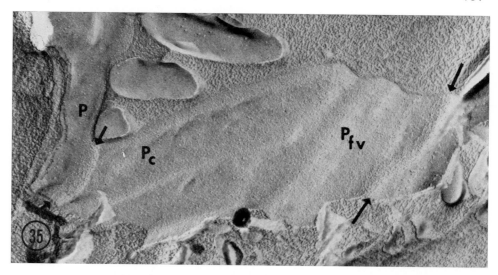

Fig. 31 *(opposite)*. Freeze-fracture replicas of the E face of the cytopharynx (E_c) and plasma membrane (E) that show the ridged nature of the cytopharynx and the densely particulate nature of this membrane. The E face of the nascent food vacuole (E_{fv}) is also particulate. The much less particulate E face of the plasma membrane is also seen. A remnant of the P face of the alveolus (P_a) remains attached to the plasma membrane along this line of juncture between the cytopharyngeal and plasma membranes. Discoidal vesicles with particulate E faces (E_v) similar to those of the cytopharynx membrane are lined up next to the cytopharynx *(lower left)*. Bumps *(arrows)* may indicate sites of vesicle incorporation into the cytopharyngeal membrane. Parasomal sacs in a row (ps), are found along the rows of cilia of the quadrulus. ×30,000.

Fig. 32 *(opposite)*. Cross-section of the wavy topography of the cytopharynx membrane. Discoidal vesicles are oriented perpendicularly to this membrane and are held here by microtubular ribbons barely visible in replicas *(arrowheads)*. The vesicle-free zone separating the vesicles is occupied by the microfilamentous cytostomal cord *(cc)*. The most dorsal ciliary row (c) of the quadrulus extends into the cytopharynx from the buccal cavity and is cross-fractured in this micrograph. ×30,000.

Fig. 33 *(opposite)*. A thin-sectioned view of the cytopharynx in nearly the same plane as the fracture in Fig. 32. The cytopharyngeal microtubular ribbons, discoidal vesicles, microfilamentous cord (cc), and cross-sectioned cilia are all present. ×30,000.

Fig. 34. *(opposite)*. Discoidal vesicles fuse with the cytopharyngeal membrane *(arrowheads)* in a cell stained with ruthenium red. The glycocalyx along the plasma membrane, ciliary membrane, cytopharyngeal membrane, and within the discoidal vesicles is stained. a, alveolus; cc, microfilamentous cord; ps, parasomal sac. ×40,000. (From Allen, J. Cell Biol. 63, 1974; reproduced with permission.

Fig. 35. The P face of the ribbed cytopharynx (P_c) and nascent food vacuole (P_{fv}) of *P. caudatum*. The rough appearance of the face is due mainly to pits. This single-membrane-limited region is continuous with the plasma membrane (P) of the pellicle at both edges. These edges are marked by bands of particles *(between arrows)* that identify lines of intracellular junctions between the plasma membrane and the underlying terminal margins of alveoli. ×40,000.

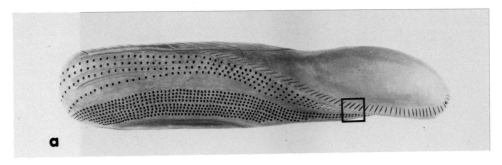

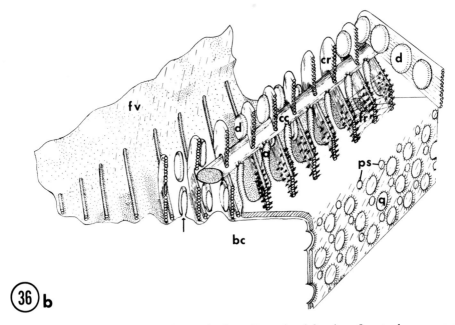

Fig. 36. A clay model of the oral region, and a three-dimensional drawing of a cytopharynx segment of *P. caudatum*. (a) The model shows the dorsal surface of the buccal cavity, cytopharynx *(row of short lines)*, and nascent food vacuole *(dome)* as viewed from inside the cell. The anterior end is toward the left and the left side is toward the bottom. The 12 rows of oral cilia are divided into groups of four: the top four are the quadrulus, the middle four the dorsal peniculus, and the bottom four the ventral peniculus. The box is illustrated in detail in Fig. 36b. (b) The architecture of the cytopharynx is designed to transport vesicles (d) to specific sites of exocytosis *(arrow)*. These vesicles upon fusion with the cytopharyngeal membrane become part of the nascent food vacuole membrane (fv). Cytopharyngeal ribbons (cr) of microtubules guide the vesicles to the growing edge of the cytopharynx and also hold the vesicles in the proper orientation to permit fusion. The roles of the microfilamentous cord (cc) and fingers projecting to this cord from the alveolus (a) are not yet known. bc, buccal cavity; fr, filamentous reticulum; ps, parasomal sacs; q, quadrulus. (From Allen, J. Cell Biol. 63, 1974; reproduced with permission.)

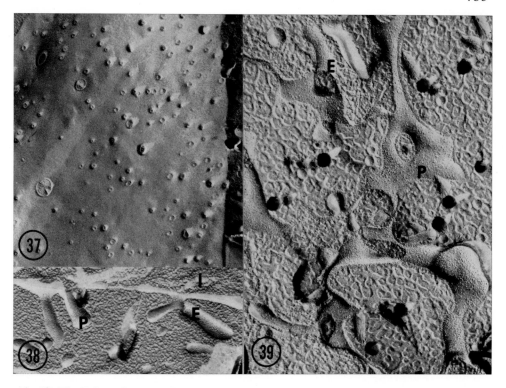

Fig. 37. The E face of a young food vacuole has many raised circular profiles which are fractured necks of endocytic channels. Young vacuoles undergo condensation by a process of endocytosis. ×12,000.

Fig. 38. The channels referred to in Fig. 37 are shown continuous with the food vacuole membrane. The P faces (P) of these channels are relatively smooth while the E face (E) is particulate like the discoidal vesicles and nascent food vacuole membrane. l, lumen of food vacuole. ×30,000.

Fig. 39. The very irregular topography of vacuoles less than 3 min old is illustrated in this replica. The round, opaque balls are latex beads inside the food vacuole lumen. E, E face; P, P face. ×40,000.

740

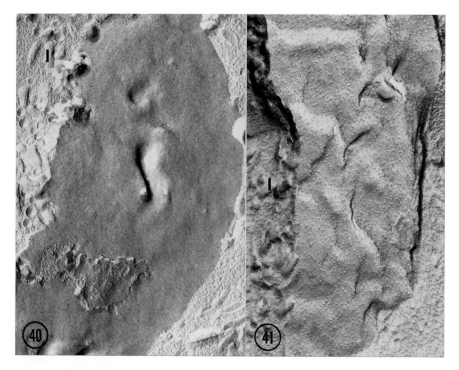

Fig. 40. The E face of a condensed food vacuole is smooth and shows no signs of endocytic activity, a dramatic change from its highly particulate nature just a few minutes earlier (cf. Figs. 31, 39). l, latex beads in food vacuole lumen. ×36,000.

Fig. 41. Following the condensed stage, the food vacuole again enlarges and a new population of prominent particles appears on the E face of the vacuole membrane. l, lumen of food vacuole. ×36,000.

Fig. 42. *(opposite)*. Food vacuoles of *Paramecium caudatum* are associated with neutral red granules (g) through most of their cycle. These elongated granules, which are primary or secondary lysosomes roughly 150 to 350 nm in diameter, are found to accumulate specifically around the vacuoles. They have a thick membrane and an amorphous content with a few electron-opaque grains. Ma, macronucleus. ×21,000.

Fig. 43 *(opposite)*. Organelles of the same size and position as neutral red granules are found around the food vacuoles in freeze-fracture replicas. They are 500–600 nm long and 200–250 nm in diameter. The E face (E) of these organelles has a moderate number of large particles while the P face (P) is populated with smaller particles. ×40,000.

Fig. 44 *(opposite)*. The neutral red granules or lysosomes are rarely seen in continuity with the food vacuole membrane in *P. caudatum*. However, the membrane of the vacuole in this micrograph is in continuity with large evaginations, around 270 nm across, which may represent fusion events with these lysosomes. E, E face; P, P face of food vacuole membrane; l, lumen of food vacuole. ×30,000.

741

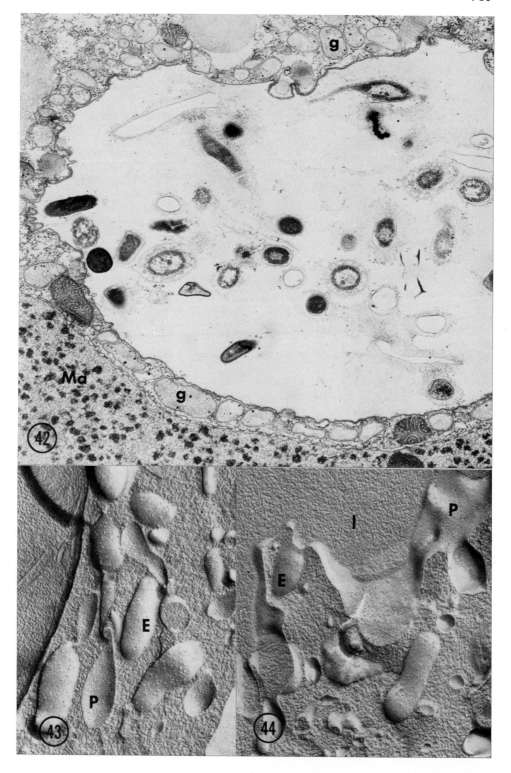

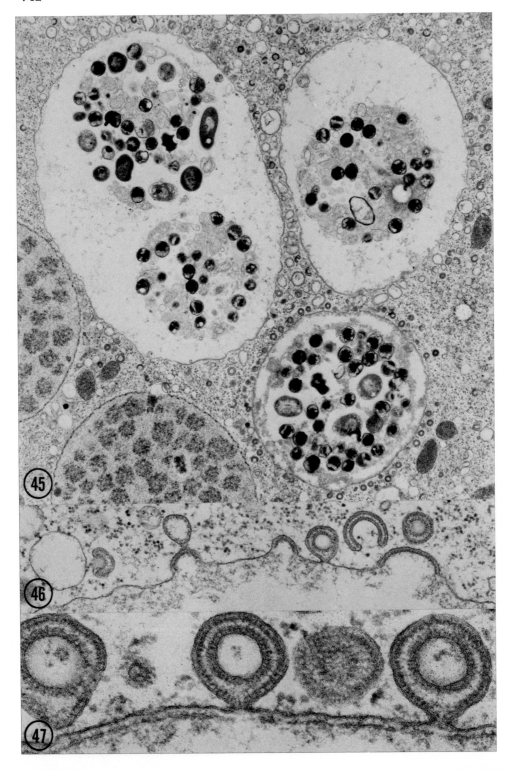

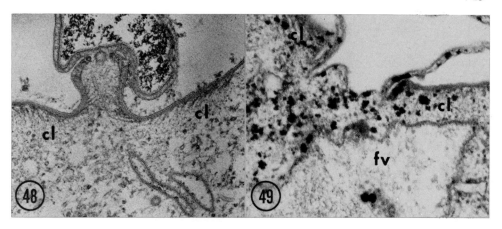

Fig. 45 *(opposite).* Food vacuoles of *Vorticella convallaria* show intense cytotic activity at their membranes. The food vacuole *(lower right)* appears to be in the initial process of enlarging after reaching a maximally condensed stage in which the bacteria are packed into a tight ball. This vacuole is surrounded by a cloud of cup-shaped vesicles, some of them continuous with the vacuole membrane. The older vacuole *(upper right)* shows signs of considerable digestion. It has obviously grown in size, since the membrane is widely separated from the ball of bacteria. Large electron-transparent vesicles with luminal coats on their membranes lie around this vacuole. The left vacuole is apparently the product of two fused vacuoles since it contains two food balls. It has large vesicles in continuity with its membrane *(top)*. (See text for an interpretation of the dynamics of endocytosis and fusion of vesicles with the vacuole membrane.) ×9,000.

Fig. 46 *(opposite).* Coated regions along the luminal surface of the food vacuole appear to be contributed to the vacuole when cup-shaped vesicles fuse with this membrane. Such fusions also provide additional membrane, allowing the vacuole to expand following the condensation phase. ×45,000.

Fig. 47 *(opposite).* Details of the fusion of cup-shaped vesicles with the food vacuole membrane. The structured contents within these cup-shaped cisternae are also shown, as is the coating on the luminal side of the vacuolar membrane. ×100,000.

Fig. 48. A cross-section of the cytoproct, whose lips (cl) have begun to separate prior to food vacuole defecation. This cell was stained with ruthenium red enhancing the tripartite structure of the membranes and staining the glycocalyx, which is continuous over the cytoproct, as well as other external debris. The alveolar membranes do not have a glycocalyx. Discoidal vesicles are present beneath the expanding cytoproct. ×60,000.

Fig. 49. The cytoproct ridge has parted and the space between the lips (cl) is limited by an enlarged single-membrane-limited gap. The food vacuole (fv) membrane, slightly wavy in this micrograph, lies close to this gap. The nature of the electron-opaque stain in this zone is unknown. ×60,000.

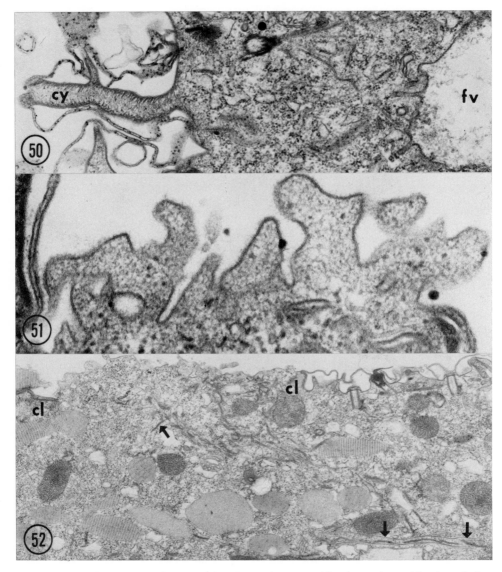

Fig. 50. As the food vacuole (fv) approaches the cytoproct (cy) ridge, it associates with microtubular bundles and sends out small channels from its membrane. ×30,000. (From Allen and Wolf, J. Cell Sci. 14, 1974; reproduced with permission.)

Fig. 51. Reengulfment of the membrane of the defecated food vacuole commences immediately after defecation and continues until the cytoproct is again closed. Microfilaments underlie the endocytosing vacuole membrane. ×75,000.

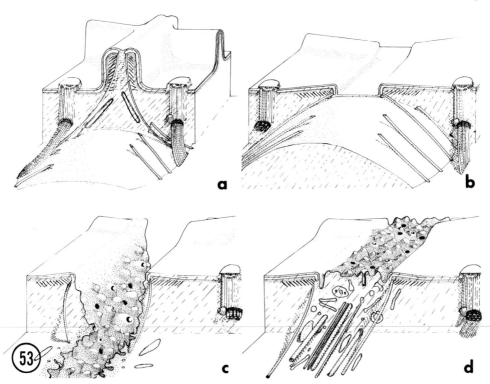

a

b

c

d

Fig. 52 *(opposite).* A cytoproct in the process of closing. The membrane between the two cytoproct lips (cl) is reengulfed in the form of tubules which accumulate under the cytoproct. Cytopharyngeal microtubular ribbons lie near this region and are associated with discoidal vesicles *(lower right).* Arrows indicate a line of discoidal vesicles. ×15,000.

Fig. 53. A series of four drawings illustrates the events in food vacuole defecation and resorption of vacuole membrane. (a) The vacuole is guided to the closed cytoproct by microtubules originating from basal bodies and from the tip of the cytoproct ridge. Vesicles in the area may fuse with the plasma membrane of the cytoproct to increase its area. (b) The ridge spreads apart and exposes the plasma membrane to the approaching food vacuole membrane. (c) Fusion of the two membranes results in defecation of the contents. Rapid endocytosis of the membrane begins. (d) The cytoproct will close as the membrane between the lips of the cytoproct is completely resorbed. This resorption results in a pool of membrane vesicles of various shapes in the vicinity of the cytoproct. (From Allen and Wolf, J. Cell Sci. 14, 1974; reproduced with permission.)

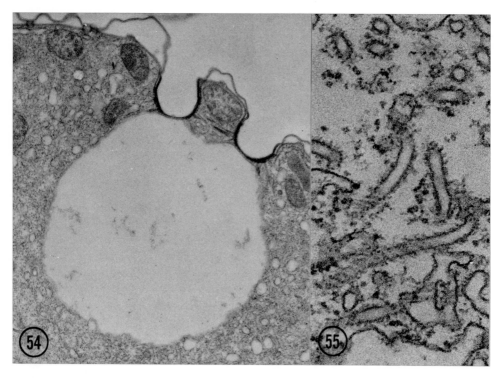

Fig. 54. The contractile or pulsating vacuole of *Tetrahymena pyriformis* lies next to two permanent pores in the pellicle of the cell. Microtubules support the sides of the pores and radiate over the distal hemisphere of the vacuole. Only a small gap separates the vacuole membrane from the membranes at the bottom of the pores. ×15,000. (Photograph courtesy of R. W. Wolf.)

Fig. 55. In *Paramecium caudatum*, tubular fluid segregation organelles are intimately associated with the collecting canals of the contractile vacuole system. These tubules are decorated on their cytoplasmic surface with paired helically-wound pegs. ×75,000.

Fig. 56 *(opposite)*. The collecting canals of *P. caudatum* are continuous with a network of smooth-membrane-limited anastomosing tubules *(arrowhead)*. Microtubules (mt) lie parallel to the longitudinal axis against the membranes of collecting canals. Fluid segregation organelles border the anastomosing tubules *(right)*. ×37,000.

Fig. 57 *(opposite)*. Anastomosing tubules are joined to the collecting canal by long rows of circular necks. These rows are separated by microtubules bordering the collecting canals. The E face of the collecting canal's (E) membrane is smooth while the P face (P) is particulate. The same is true of the anastomosing tubule membranes. ×35,000. (From Hausmann and Allen, Cytobiologie 15, 1977; reproduced with permission.)

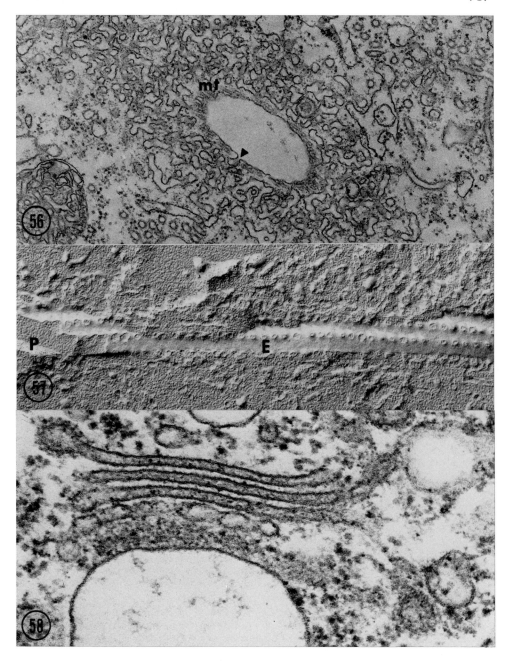

Fig. 58. The Golgi apparatus of ciliates is dissociated into a large number of small dictyosomes, as illustrated here in *Vorticella convallaria*. Smooth areas continuous with the rough ER have amorphous material adjacent to them. Stacks of membrane cisternae lie distal to it. The membrane of these cisternae is thicker than the ER membrane, and the ends of the cisternae frequently have coated blebs. ×100,000.

748

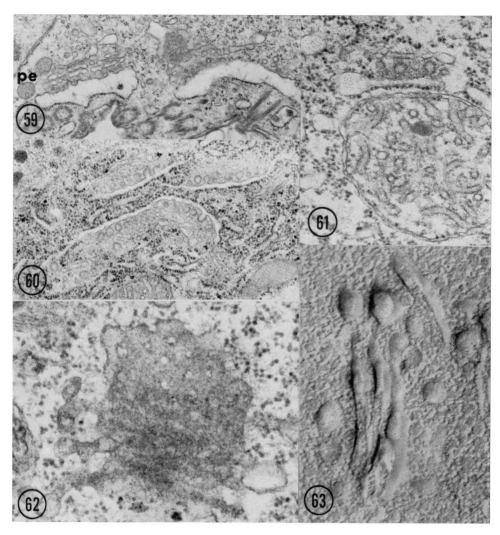

Fig. 59. Dictyosomes of *Opercularia coarctata*. The budding of smooth areas of the rough ER is shown. These small, primary transitional vesicles then fuse to form fenestrated cisternae which lie in the plane of the section to the right of this figure. pe, peroxisome. ×25,000.

Fig. 60. Dictyosomes of *P. caudatum*. Budding of the ER is evident, but the stacks of cisternae are either extremely fenestrated or have not formed at the maturing face of these bodies. ×25,000.

Fig. 61. The tiny dictyosomes of *Tetrahymena pyriformis* were not recognized in this cell for many years. Budding occurs at the ER surface, but only one cisterna may be present at the maturing face. These dictyosomes are frequently formed in close association with an adjacent mitochondrion. ×50,000.

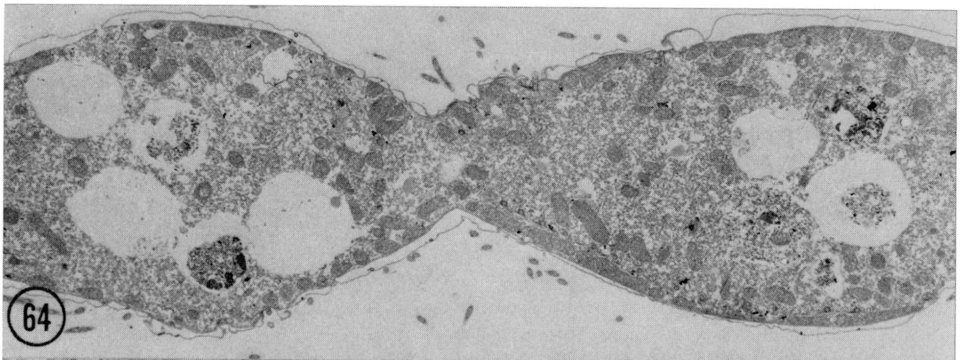

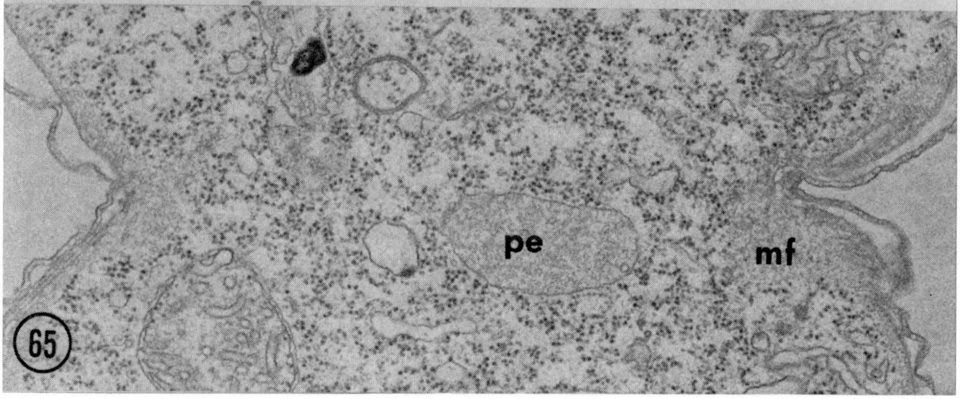

Fig. 62 *(opposite)*. The fenestrated cisterna can be seen in planar view in this dictyosome of *V. convallaria*. ×65,000.
Fig. 63 *(opposite)*. Dictyosomes can also be identified in freeze-fracture replicas of ciliates such as *P. caudatum*. Small vesicles are present at the edges of the cisternae and also seem to lie free in the vicinity of the dictyosome. ×75,000.
Fig. 64. Ciliates divide by transverse fission. A fission furrow constricts the cell around its middle, as illustrated here in *T. pyriformis*. ×2,900.
Fig. 65. The fission furrow of *Tetrahymena pyriformis* involves both the plasma membrane and the alveoli. A ring of microfilaments (mf) surrounds the cytoplasmic surface of the pellicle in this furrow region and produces the furrow by contracting. A peroxisome (pe) is also illustrated. ×32,500.

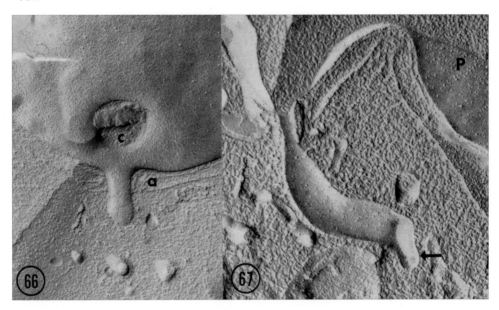

Fig. 66. A parasomal sac, an indentation of the plasma membrane, lies next to the basal body-cilium complex (c) in each pellicular depression in *P. caudatum* as in other ciliates. The E face of this sac, shown here, has no special arrangements of particles. a, alveolus. ×50,000.

Fig. 67. Enlarged parasomal sacs are occasionally found. The P face of this sac in *P. caudatum* is shown to contain particles in approximately the same numbers as found on the P face of the plasma membrane (p). The rounded tip of this sac *(arrow)* is probably a coated pit. ×50,000.

References

Allen, R. D. (1967) Fine structure, reconstruction and possible functions of components of the cortex of *Tetrahymena pyriformis*. J. Protozool. 14, 553–565.

Allen, R. D. (1971) Fine structure of membranous and microfibrillar systems in the cortex of *Paramecium caudatum*. J. Cell Biol. 49, 1–20.

Allen, R. D. (1973a) Structures linking the myonemes, endoplasmic reticulum, and surface membranes in the contractile ciliate *Vorticella*. J. Cell Biol. 56, 559–579.

Allen, R. D. (1973b) Contractility and its control in peritrich ciliates. J. Protozool. 20, 25–36.

Allen, R. D. (1974) Food vacuole membrane growth with microtubule-associated membrane transport in *Paramecium*. J. Cell Biol. 63, 904–922.

Allen, R. D. (1975) Evidence for firm linkages between microtubules and membrane-bounded vesicles. J. Cell Biol. 64, 497–503.

Allen, R. D. (1976a) Freeze-fracture evidence for intramembrane changes accompanying membrane recycling in *Paramecium*. Cytobiologie 12, 254–273.

Allen, R. D. (1976b) The mosaic nature of the plasma membrane of *Paramecium*. J. Protozool. 23, 10A–11A.

Allen, R. D. (1976c) Membranes of digestive vacuoles: topographic and intramembrane particle changes with time. J. Cell Biol. 70, 386a.

Allen, R. D. (1977) Intramembrane alterations that occur in digestive vacuoles of *Paramecium*. 5th Int. Cong. Protozool., (Hutner, S. H., ed.) abstr. 235, The Print Shop, New York.

Allen, R. D. and Eckert, R. (1969) A morphological system in ciliates comparable to the sarcoplasmic reticulum-transverse tubular system in striated muscle. J. Cell Biol. 43, 4a–5a.

Allen, R. D. and Hausmann, K. (1976) Membrane behavior of exocytic vesicles. I. The ultrastructure of *Paramecium* trichocysts in freeze-fracture preparations. J. Ultrastruc. Res. 54, 224–234.

Allen, R. D. and Wolf, R. W. (1974) The cytoproct of *Paramecium caudatum*: structure and function, microtubules, and fate of food vacuole membranes. J. Cell Sci. 14, 611–631.

Allen, S. L. (1967) Chemical genetics of protozoa. In: Chemical Zoology (Kidder, G. W., ed.) pp. 617–694, Academic Press, New York.

André, J. and Vivier, E. (1962) Quelques aspects ultrastructuraux de l'échange micronucléaire lors de la conjugaison chez *Paramecium caudatum*. J. Ultrastruc. Res. 6, 390–406.

Andrews, D. R. and Nelson, D. L. (1977) Phospholipids and fatty acids of the ciliary membrane of *Paramecium*. 5th Int. Cong. Protozool. (Hutner, S. H., ed.) abstr. 315, The Print Shop, New York.

Antipa, G. A. (1971) Structural differentiation in the somatic cortex of a ciliated protozoan, *Conchophthirus curtus* Engelmann 1862. Protistologica 7, 471–501.

Ashworth, L. A. E. and Green, C. (1966) Plasma membranes: phospholipid and sterol content. Science 151, 210–211.

Bannister, L. H. (1972) The structure of trichocysts in *Paramecium caudatum*. J. Cell Sci. 11, 899–929.

Bardele, C. F. (1976) Particle movement of heliozoan axopods associated with lateral displacement of highly ordered membrane domains. Z. Naturforsch. 31c, 190–194.

Batz, W. and Wunderlich, F. (1976) Structural transformation of the phagosomal membrane in *Tetrahymena* cells endocytosing latex beads. Arch. Microbiol. 109, 215–220.

Baugh, L. C. and Thompson, G. A. Jr. (1973) Studies of membrane formation in *Tetrahymena pyriformis*. VI. Variations in the rate of lipid synthesis during division of heat-synchronized cells. Exp. Cell Res. 76, 263–272.

Baugh, L. C. and Thompson, G. A. Jr. (1975) Studies of membrane formation in *Tetrahymena pyriformis*. IX. Variations in intracellular phospholipid and protein deployment during the cell division cycle. Exp. Cell Res. 94, 111–121.

Beisson, J., Lefort-Tran, M., Pouphile, M., Rossignol, B. and Satir, B. (1976) Genetic analysis of membrane differentiation in *Paramecium*. Freeze-fracture study of the trichocyst cycle in wild-type and mutant strains. J. Cell Biol. 69, 126–143.

Berger, H. and Hanahan, D. J. (1971) The isolation of phosphonolipids from *Tetrahymena pyriformis*. Biochim. Biophys. Acta 231, 584–587.

752

Berger, H., Jones, P. and Hanahan, D. J. (1972) Structural studies on lipids of *Tetrahymena pyriformis* W. Biochim. Biophys. Acta 260, 617–629.

Bishop, J. (1961) Purification of an immobilization antigen of *Paramecium aurelia,* variety I. Biochim. Biophys. Acta 50, 471–477.

Blum, J. J. (1971) Existence of a breaking point in cilia and flagella. J. Theor. Biol. 33, 257–263.

Bradbury, P. C. (1973) The fine structure of the cytostome of the apostomatous ciliate *Hyalophysa chattoni.* J. Protozool. 20, 405–414.

Bradbury, P. C. (1974) Stored membranes associated with feeding in apostome trophonts with different diets. Protistologica 10, 533–542.

Branton, D. (1966) Fracture faces of frozen membranes. Proc. Nat. Acad. Sci. U.S.A. 55, 1048–1056.

Branton, D. (1971) Freeze-etching studies of membrane structure. Phil. Trans. Roy. Soc. Lond. B 261, 133–138.

Branton, D. and Deamer, D. W. (1972) Membrane Structure. Protoplasmatologica II/E/1, 1–70.

Branton, D., Bullivant, S., Gilula, N. B., Karnovsky, M. J., Moore, H., Mühlethaler, K., Northcote, D. H., Packer, L., Satir, B., Satir, P., Speth, V., Staehelin, L. A., Steere, R. L. and Weinstein, R. S. (1975). Freeze-etching nomenclature. Science 190, 54–56.

Braun, V. and Miyake, A. (1975) Composition of blepharmone, a conjugation-inducing glycoprotein of the ciliate *Blepharisma.* FEBS Lett. 53, 131–134.

Brown, R. M. Jr., Franke, W. W., Kleinig, H., Falk, H. and Sitte, P. (1970) Scale formation in chrysophycean algae. I. Cellulosic and noncellulosic wall components made by the Golgi apparatus. J. Cell Biol. 45, 246–271.

Browning, J. L. and Nelson, D. L. (1976) Biochemical studies of the excitable membrane of *Paramecium aurelia.* I. $^{45}Ca^{2+}$ fluxes across resting and excited membrane. Biochim. Biophys. Acta 448, 338–351.

Browning, J. L., Nelson, D. L. and Hansma, H. G. (1976) Ca^{2+} influx across the excitable membrane of behavioural mutants of *Paramecium.* Nature 259, 491–494.

Burnasheva, S. A. and Jurzina, G. A. (1968) The ultrastructure and localization of ATPase activity of *Tetrahymena pyriformis* cilia. Repts. 3rd Int. Cong. Histochem. Cytochem. p. 27. Springer, New York.

Cameron, I. L. and Burton, A. L. (1969) On the cycle of the water expulsion vesicle in the ciliate *Tetrahymena pyriformis.* Trans. Am. Microsc. Soc. 88, 386–393.

Carasso, N., Fauré-Fremiet, E. and Favard, P. (1962) Ultrastructure de l'appareil excreteur chez quelques cilies peritriches. J. Microscopie 1, 455–468.

Carasso, N., Favard, P. and Goldfischer, S. (1964) Localisation, a l'echelle des ultrastructures, d'activities de phosphatases en rapport avec les processus digestifs chez un cilie *(Campanella umbellaria).* J. Microscopie 3, 297–322.

Chapman, D. and Leslie, R. B. (1970) Structure and function of phospholipids in membranes. In: Membranes of Mitochondria and Chloroplasts (Racker, E., ed.) pp. 91–126, Van Nostrand Reinhold, New York.

Cohen, L. W. (1965) The basis for the circadian rhythm of mating in *Paramecium bursaria.* Exp. Cell Res. 37, 360–367.

Cohen, L. W. and Siegel, R. W. (1963) The mating-type substances of *P. bursaria.* Genet. Res. 4, 143–150.

Coleman, R. and Finean, J. B. (1966) Preparation and properties of isolated plasma membranes from guinea-pig tissues. Biochim. Biophys. Acta 125, 197–206.

Conner, R. L. and Stewart, B. Y. (1976). The effect of temperature on the fatty acid composition of *Tetrahymena pyriformis* WH-14. J. Protozool. 23, 193–196.

Conner, R. L., Landrey, J. R., Bruns, C. H. and Mallory, F. B. (1968) Cholesterol inhibition of pentacyclic triterpenoid biosynthesis in *Tetrahymena pyriformis.* J. Protozool. 15, 600–605.

Conner, R. L., Mallory, F. B., Landrey, J. R., Ferguson, K. A. Kaneshiro, E. S. and Ray, E. (1971) Ergosterol replacement of tetrahymenol in *Tetrahymena* membranes. Biochem. Biophys. Res. Commun. 44, 995–1000.

Dass, C. M. S., Sapra, G. R. and Kumar, R. (1976) Food vacuole formation and membrane turnover in *Blepharisma musculus v. seshachari* Bhandary. Indian J. Exp. Biol. 14, 535–543.

Davidson, L. A. (1976) Ultrastructure of the membrane attachment sites of the extrusomes of *Ciliophrys marina* and *Heterophrys marina* (Actinopoda). Cell Tissue Res. 170, 353–365.

Davidson, L. and LaFountain, J. R. Jr. (1975) Mitosis and early meiosis in *Tetrahymena pyriformis* and the evolution of mitosis in the phylum Ciliophora. BioSystems 7, 326–336.

de Duve, C. (1973). Biochemical studies on the occurrence, biogenesis and life history of mammalian peroxisomes. J. Histochem. Cytochem. 21, 941–948.

Didier, P. (1976) Organization ultrastructurale des cils des membranelles de quelques ciliés Oligohymenophora. Corps paraxonémiens et coalescence ciliaire. Protistologica 12, 37–47.

Douglas, W. W. (1974) Involvement of calcium in exocytosis and the exocytosis-vesiculation sequence. Biochem. Soc. Symp. 39, 1–28.

Dunham, P. B. and Stoner, L. C. (1969) Indentation of the pellicle of *Tetrahymena* at the contractile vacuole pore before systole. J. Cell Biol. 43, 184–188.

Dunlap, K. (1977) Localization of calcium channels in *Paramecium*. 5th Int. Cong. Protozool. (ed. S. H., Hutner,) abstr. 303, The Print Shop, New York.

Dunlap, K. and Eckert, R. (1976) Calcium channels in the ciliary membranes of *Paramecium*. J. Cell Biol. 70, 245a.

Eckert, R. (1972) Bioelectric control of ciliary activity. Science 176, 473–481.

Eckert, R. and Naitoh, Y. (1972) Bioelectric control of locomotion in the ciliates. J. Protozool. 19, 237–243.

Eckert, R., Naitoh, Y. and Machemer, H. (1976) Calcium in the bioelectric and motor functions of *Paramecium*. Symp. Soc. Exp. Biol. 30, 233–255.

Elliott, A. M. and Bak, I. J. (1964) The contractile vacuole and related structures in *Tetrahymena pyriformis*. J. Protozool. 11, 250–261.

Elliott, A. M. and Tremor, J. W. (1958) The fine structure of the pellicle in the contact area of conjugating *Tetrahymena pyriformis*. J. Biophys. Biochem. Cytol. 4, 839–840.

Elliot, A. M. and Zieg, R. G. (1968) A Golgi apparatus associated with mating in *Tetrahymena pyriformis*. J. Cell Biol. 36, 391–398.

Ellisman, M. H., Rash, J. E., Staehelin, L. A. and Porter, K. R. (1976) Studies of excitable membranes. II. A comparison of specializations of neuromuscular junctions and nonjunctional sarcolemmas of mammalian fast and slow twitch muscle fibers. J. Cell Biol. 68, 752–774.

Epstein, M. and Eckert, R. (1973) Membrane control of ciliary activity in the protozoan *Euplotes*. J. Exp. Biol. 58, 437–462.

Erwin, J. A. (1970) The reversal of temperature-induced cell-surface deformation in *Tetrahymena* by polyunsaturated fatty acids. Biochim. Biophys. Acta 202, 21–34.

Erwin, J. and Bloch, K. (1963) Lipid metabolism of ciliated protozoa. J. Biol. Chem. 238, 1618–1624.

Erwin, J. A., Beach, D. and Holz, G. G., Jr. (1966) Effect of dietary cholesterol on unsaturated fatty acid biosynthesis in a ciliated protozoan. Biochim. Biophys. Acta 125, 614–616.

Esteve, J. C. (1970) Distribution of acid phosphatase in *Paramecium caudatum:* Its relations with the process of digestion. J. Protozool. 17, 24–35.

Esteve, J. C. (1972) L'appareil de Golgi des cilies. Ultrastructure, particulierement chez *Paramecium*. J. Protozool. 19, 609–618.

Everhart, L. P. (1972) Methods with *Tetrahymena*. Methods Cell Physiol. 5, 220–288.

Fauré-Fremiet, E., Favard, P. and Carasso, N. (1962) Étude au microscope électronique des ultrastructures d'*Epistylis anastatica* (Cilié Péritriche). J. Microscopie 1, 287–312.

Favard, P. and Carasso, N. (1963) Mise en evidence d'un processus de micropinocytose interne au niveau des vacuoles digestives d'*Epistylis anastatica* (Cilie Peritriche). J. Microscopie 2, 495–498.

Favard, P. and Carasso, N. (1964) Etude de la pinocytose au niveau des vacuoles digestives de cilies peritriches. J. Microscopie 3, 671–696.

Ferguson, K. A., Conner, R. L. and Mallory, F. B. (1971) The effect of ergosterol on the fatty acid composition of *Tetrahymena pyriformis*. Arch. Biochem. Biophys. 144, 448–450.

Ferguson, K. A., Davis, F. M., Conner, R. L., Landrey, J. R. and Mallory F. B. (1975) Effect of sterol replacement *in vivo* on fatty acid composition of *Tetrahymena*. J. Biol. Chem. 250, 6998–7005.

Finger, I. (1956) Immobilizing and precipitating antigens of *Paramecium*. Biol. Bull. 3, 358–363.

Finger, I. (1974) Surface antigens of *Paramecium aurelia*. In: Paramecium a Current Survey (Van Wagtendonk, W. J., ed.) pp. 131–163, Elsevier/North-Holland, Amsterdam.

Fisher, G., Kaneshiro, E. S. and Peters, P. D. (1976) Divalent cation affinity sites in *Paramecium aurelia*. J. Cell Biol. 69,429–442.

Flickinger, C. J. and Murray, R. L. (1974) Cytoplasmic membrane changes in conjugating *Tetrahymena*. Cell Tissue Res. 153, 357–364.

Foissner, W. (1972) The cytopyge of Ciliata I. Its function, regeneration and morphogenesis in Uronema parduczi. Acta Biol. Acad. Sci. Hung. 23, 161–174.

Fok, A. K. and Allen, R. D. (1975) Cytochemical localization of peroxisomes in *Tetrahymena pyriformis*. J. Histochem. Cytochem. 23, 599–606.

Franke, W. W. (1971) Membrane-microtubule-microfilament-relationships in the ciliate pellicle. Cytobiologie 4, 307–316.

Franke, W. W. (1974) Nuclear envelopes. Structure and biochemistry of the nuclear envelope. Phil. Trans. Roy. Soc. Lond. B 268, 67–93.

Franke, W. W. and Eckert, W. A. (1971) Cytomembrane differentiation in a ciliate, *Tetrahymena pyriformis*. II. Bifacial cisternae and tubular formations. Z. Zellforsch. 122, 244–253.

Franke, W. W. and Herth, W. (1974) Morphological evidence for de novo formation of plasma membrane from coated vesicles in exponentially growing cultured plant cells. Exp. Cell Res. 89, 447–451.

Franke, W. W. and Kartenbeck, J. (1971) Outer mitochondrial membrane continuous with endoplasmic reticulum. Protoplasma 73, 35–41.

Franke, W. W., Eckert, W. A. and Krien, S. (1971a) Cytomembrane differentiation in a ciliate, *Tetrahymena pyriformis*. I. Endoplasmic reticulum and dictyosomal equivalents. Z. Zellforsch. 119, 577–604.

Franke, W. W., Kartenbeck, J., Zentgraf, H., Scheer, U. and Falk, H. (1971b) Membrane-to-membrane cross-bridges. A means to orientation and interaction of membrane faces. J. Cell Biol. 51, 881–888.

Franke, W. W., Kartenbeck, J. and Spring, H. (1976a) Involvement of bristle coat structures in surface membrane formations and membrane interactions during coenocytotomic cleavage in caps in *Acetabularia mediterranea*. J. Cell Biol. 71, 196–206.

Franke, W. W., Lüder, M. R., Kartenbeck, J., Zerban, H. and Keenan, T. W. (1976b) Involvement of vesicle coat material in casein secretion and surface regeneration. J. Cell Biol. 69, 173–195.

Frisch, A., Levkowitz, H. and Loyter, A. (1976) Inhibition of conjugation and cell division in *Tetrahymena pyriformis* by Tunicamycin: A possible requirement of glycoprotein synthesis for induction of conjugation. Biochem. Biophys. Res. Commun. 72, 138–145.

Fukushima, H., Martin, C. E., Iida, H., Kitajima, Y., Thompson, G. A. Jr., and Nozawa, Y. (1976a) Changes in membrane lipid composition during temperature adaptation by a thermotolerant strain of *Tetrahymena pyriformis*. Biochim. Biophys. Acta 431, 165–179.

Fukushima, H., Watanabe, T. and Nozawa, Y. (1976b) Studies on *Tetrahymena* membranes. *In vivo* manipulation of membrane lipids by 1-0-hexadecyl glycerol-feeding in *Tetrahymena pyriformis*. Biochim. Biophys. Acta 436, 249–259.

Fukushima, H., Nagao, S., Yamaguchi, H. and Nozawa, Y. (1977) Palmityl CoA desaturase in *Tetrahymena pyriformis*: its kinetics and temperature dependence. 5th Int. Cong. Protozool. (Hutner, S. H., ed.) abstr. 340, The Print Shop, New York.

Fulco, A. J. (1974) Metabolic alterations of fatty acids. Ann. Rev. Biochem. 43, 215–241.

Gavin, R. H. (1974) The oral apparatus of *Tetrahymena pyriformis*, mating type 1, variety 1. I. Solubilization and electrophoretic separation of oral apparatus proteins. Exp. Cell Res. 85, 212–216.

Gennis, R. B. and Jonas, A. (1977) Protein-lipid interactions. Ann. Rev. Biophys. Bioeng. 6, 195–238.

Gibson, I. (1970) Interacting genetic systems in *Paramecium*. Adv. Morphogen. 8, 159–208.

Gilula, N. B. (1974) Isolation of rat liver gap junctions and characterization of the polypeptides. J. Cell Biol. 63, 111a.

Golder, T. K. (1976) The macro-micronuclear complex of *Woodruffia metabolica*. J. Ultrastruc. Res. 54, 169–175.

Goldfischer, S., Carasso, N. and Favard, P. (1963) The demonstration of acid phosphatase activity by electron microscopy in the ergastoplasm of the ciliate *Campanella umbellaria* L. J. Microscopie 2, 621–628.

Goldfischer, S., Favard, P. and Carasso, N. (1967) The demonstration of acid hydrolase activities in digestive vacuoles of peritrich ciliates by hexazonium pararosanilin procedures. J. Microscopie 6, 867–872.

Grell, K. G. (1973) Protozoology, p. 32, Springer-Verlag, New York.

Hackenbrock, C. R. (1972) States of activity and structure in mitochondrial membranes. Ann. N. Y. Acad. Sci. 195, 492–505.

Hansma, H. G. (1975) The immobilization antigen of *Paramecium aurelia* is a single polypeptide chain. J. Protozool. 22, 257–259.

Hansma, H. G. and Kung, C. (1975) Studies of the cell surface of *Paramecium*. Ciliary membrane proteins and immobilization antigens. Biochem. J. 152, 523–528.

Hausmann, K. (1972) Cytologische studien an trichocysten, IV.-Die feinstruktur ruhender und ausgeschiedener protrichocysten von *Loxophyllum, Tetrahymena, Prorodon* und *Lacrymaria*. Protistologica 8, 401–412.

Hausmann, K. (1977) Development of compound trichocysts in the ciliate *Pseudomicrothorax dubius*. Protoplasma 92, 263–268.

Hausmann, K. (1978) Extrusive organelles in protists. Int. Rev. Cytol. 55. In press.

Hausmann, K. and Allen, R. D. (1976) Membrane behavior of exocytic vesicles. II. Fate of the trichocyst membranes in *Paramecium* after induced trichocyst discharge. J. Cell Biol. 69, 313–326.

Hausmann, K. and Allen, R. D. (1977) Membranes and microtubules of the excretory apparatus of *Paramecium caudatum*. Cytobiologie. 15, 303–320.

Hausmann, K. and Mocikat, K. H. (1976) Direct evidence for a surface coat on the plasma membrane of the ciliate *Litonotus duplostriatus*. Cytobiologie 13, 469–475.

Hayashi, M. (1974) Temperature, chemicals and enzyme effects on the structure of discharged mucocyst from *Tetrahymena pyriformis*. Cytobiologie 9, 460–468.

Hill, D. L. (1972) The Biochemistry and Physiology of Tetrahymena, pp. 46–73, Academic Press, New York.

Hiwatashi, K. (1961) Locality of mating reactivity on the surface of *Paramecium caudatum*. Sci. Rep. Tohoku Univ. Ser. IV (Biol.) 27, 93–99.

Hoffman, E. K., Rasmussen, L. and Zeuthen, E. (1974) Cytochalasin B: aspects of phagocytosis in nutrient uptake in *Tetrahymena*. J. Cell Sci. 15, 403–406.

Holz, G. G. Jr. and Conner, R. L. (1973) The composition, metabolism and role of lipids of *Tetrahymena*. In: Biology of Tetrahymena (Elliott, A. M., ed.) pp. 99–121, Dowden, Hutchinson and Ross, Stroudsburg, Pennsylvania.

Honda, H. and Miyake, A. (1975) Taxis to a conjugation-inducing substance in the ciliate *Blepharisma*. Nature 257, 678–680.

Honda, H. and Miyake, A. (1976) Cell-to-cell contact by locally differentiated surfaces in conjugation of *Blepharisma*. Devel. Biol. 52, 221–230.

Inaba, F., Imamoto, K. and Suganuma, Y. (1966) Electron-microscopic observations on nuclear exchange during conjugation in *Paramecium multimicronucleatum*. Proc. Jap. Acad. 42, 394–398.

Janisch, R. (1972) Pellicle of *Paramecium caudatum* as revealed by freeze-etching. J. Protozool. 19, 470–472.

Jenkins, R. A. (1967) Fine structure of division in ciliate protozoa. I. Micronuclear mitosis in *Blepharisma*. J. Cell Biol. 34, 463–481.

Jonah, M. and Erwin, J. A. (1971) The lipids of membraneous cell organelles isolated from the ciliate *Tetrahymena pyriformis*. Biochim. Biophys. Acta 231, 80–92.

Jones, I. G. (1965) Studies on the characterization and structure of the immobilization antigens of *Paramecium aurelia*. Biochem. J. 96, 17–23.

Jurand, A. and Selman, G. G. (1969) The Anatomy of Paramecium aurelia. St. Martin's Press, New York.

Kaneshiro, E. S. and Holz, G. G. Jr. (1976) Observations on the ultrastructure of *Uronema* spp., marine scuticociliates. J. Protozool. 23, 503–517.

756

Kaneshiro, E. S., Beischel, L., Meyer, K. and Rhoads, D. (1977) Fatty acid composition of *Paramecium aurelia*. 5th Int. Cong. Protozool. (Hunter, S. H., ed.) Abstr. 316, The Print Shop, New York.

Kasai, R., Kitajima, Y., Martin, C. E., Nozawa, Y., Skriver, L. and Thompson, G. A. Jr. (1976) Molecular control of membrane properties during temperature acclimation. Membrane fluidity regulation of fatty acid desaturase action. Biochemistry 15, 5228–5233.

Kennedy, K. E. and Thompson, G. A. Jr. (1970) Phosphonolipids: Localization in surface membranes of *Tetrahymena*. Science 168, 989–991.

Keynes, R. D. (1972) Excitable membrane. Nature 239, 29–32.

Kimball, R. F. (1939) Mating types in *Euplotes*. Am. Naturalist 73, 451–456.

Kimelberg, H. K. (1977) The influence of membrane fluidity on the activity of membrane-bound enzymes. In: Cell Surface Reviews, Vol. 4 (Poste, G. and Nicolson, G., eds.), pp. 205–293. Elsevier/North-Holland, Amsterdam.

Kitajima, Y. and Thompson, G. A. Jr. (1977a) *Tetrahymena* strives to maintain the fluidity interrelationships of all its membranes constant. Electron microscope evidence. J. Cell Biol. 72, 744–755.

Kitajima, Y. and Thompson, G. A. Jr. (1977b) Differentiation of food vacuolar membranes during endocytosis in *Tetrahymena*, J. Cell Biol. 75, 436–445.

Kitamura, A. and Hiwatashi, K. (1976) Mating-reactive membrane vesicles from cilia of *Paramecium caudatum*. J. Cell Biol. 69, 736–740.

Klimetzek, V. (1977a) Isolation and characterization of mating reactive cilia of *Paramecium multimicronucleatum*. 5th Int. Cong. Protozool. (Hutner, S. H., ed.) abstr. 282, The Print Shop, New York.

Klimetzek, V. (1977b) Immunological investigations on isolated cilia of *Paramecium multimicronucleatum*. 5th Int. Cong. Protozool. (Hutner, S. H., ed.) abstr. 283, The Print Shop, New York.

Kloetzel, J. A. (1974) Feeding in ciliated protozoa. I. Pharyngeal disks in *Euplotes:* A source of membrane for food vacuole formation? J. Cell Sci. 15, 379–401.

Koizumi, S. (1966) Serotypes and immobilization antigens in *Paramecium caudatum*. J. Protozool. 13, 73–76.

Koroly, M. J. and Conner, R. L. (1976) Unsaturated fatty acid biosynthesis in *Tetrahymena*. Evidence for two pathways. J. Biol. Chem. 251, 7588–7592.

Kubota, T., Tokoroyama, T., Tsukuda, Y., Koyama, H., and Miyake, A. (1973) Isolation and structure determination of blepharismin, a conjugation initiating gamone in the ciliate *Blepharisma*. Science 179, 400–402.

Kung, C. (1971) Genic mutants with altered system of excitation in *Paramecium aurelia*. I. Phenotypes of the behavioral mutants. Z. Vergl. Physiol. 71, 142–164.

Kung, C. and Eckert, R. (1972) Genetic modification of electric properties in an excitable membrane. Proc. Nat. Acad. Sci. U.S.A. 69, 93–97.

Kung, C. and Naitoh, Y. (1973) Calcium-induced ciliary reversal in the extracted models of "Pawn", a behavioral mutant of *Paramecium*. Science 179, 195–196.

Kung, C., Chang, S., Satow, Y., Houten, J. V. and Hansma, H. (1975) Genetic dissection of behavior in *Paramecium*. Science 188, 898–904.

Lagunoff, D. (1973) Membrane fusion during mast cell secretion. J. Cell Biol. 57, 252–259.

Lees, A. M. and Korn, E. D. (1966) Metabolism of unsaturated fatty acids in protozoa. Biochemistry 5, 1475–1481.

Lo, H. K., Jasper, D. and Erwin, J. A. (1976) Ultrastructural alterations in *Tetrahymena pyriformis* induced by growth on saturated phospholipids at 40.1 °C. Tissue and Cell 8, 19–32.

Mackinnon, D. L. and Hawes, R. S. J. (1961) An Introduction to the Study of Protozoa. University Press, Oxford. p. 321.

Mallory, F. B. and Conner, R. L. (1971) Dehydrogenation and dealkylation of various sterols by *Tetrahymena pyriformis*. Lipids 6, 149–153.

Mallory, F. B., Gordon, J. T. and Conner, R. L. (1963) The isolation of a pentacyclic triterpenoid alcohol from a protozoan. J. Am. Chem. Soc. 85, 1362–1363.

Marshall, J. M. and Nachmias, V. T. (1965) Cell surface and pinocytosis. J. Histochem. Cytochem. 13, 92–104.

Martin, C. E., Hiramitsu, K., Kitajima, Y., Nozawa, Y., Skriver, L. and Thompson, G. A., Jr. (1976)

Molecular control of membrane properties during temperature acclimation. Fatty acid desaturase regulation of membrane fluidity in acclimating *Tetrahymena* cells. Biochemistry 15, 5218–5227.

Mast, S. O. (1947) The food vacuole in *Paramecium*. Biol. Bull. 92, 31–72.

McKanna, J. A. (1973a) Fine structure of the contractile vacuole pore in *Paramecium*. J. Protozool. 20, 631–638.

McKanna, J. A. (1973b) Cyclic membrane flow in the ingestive-digestive system of peritrich protozoans. I. Vesicular fusion at the cytopharynx. J. Cell Sci. 13, 663–675.

McKanna, J. A. (1973c) Cyclic membrane flow in the ingestive-digestive system of peritrich protozoans. II. Cup-shaped coated vesicles. J. Cell Sci. 13, 677–686.

McKanna, J. A. (1974) Permeability modulating membrane coats. I. Fine structure of fluid segregation organelles of peritrich contractile vacuoles. J. Cell Biol. 63, 317–322.

McKanna, J. A. (1976) Fine structure of fluid segregation organelles of *Paramecium* contractile vacuoles. J. Ultrastruc. Res. 54, 1–10.

Miyake, A. (1968) Induction of conjugating union by cell-free fluid in the ciliate *Blepharisma*. Proc. Jap. Acad. 44, 837–841.

Miyake, A. (1969) Mechanism of initiation of sexual reproduction in *Paramecium multimicronucleatum*. Jap. J. Genet. 44 (Suppl.), 388–395.

Miyake, A. (1974a) Conjugation of the ciliate *Blepharisma:* A possible model system for biochemistry of sensory mechanisms. In: Biochemistry of Sensory Functions (Jaenicke, L., ed.) pp. 300–305, Springer-Verlag, Berlin.

Miyake, A. (1974b) Cell interaction in conjugation of ciliates. Curr. Top. Microbiol. Immunol. 64, 49–77.

Miyake, A. and Beyer, J. (1973) Cell interaction by means of soluble factors (gamones) in conjugation of *Blepharisma intermedium*. Exp. Cell Res. 76, 15–24.

Miyake, A. and Beyer, J. (1974) Blepharmone: A conjugation-inducing glycoprotein in the ciliate *Blepharisma*. Science 185, 621–623.

Miyake, A. and Bleyman, L. K. (1976) Gamones and mating types in the genus *Blepharisma* and their possible taxonomic application. Genet. Res. 27, 267–275.

Miyake, A. and Honda, H. (1976) Cell union and protein synthesis in conjugation of *Blepharisma*. Exp. Cell Res. 100, 31–40.

Morré, D. J. (1975) Membrane biogenesis. Ann. Rev. Plant Physiol. 26, 441–481.

Morré, D. J. and Mollenhauer, H. H. (1974) The endomembrane concept: a functional integration of endoplasmic reticulum and Golgi apparatus. In: Dynamic Aspects of Plant Ultrastructure (Robards, A. W., ed.) pp. 84–137. McGraw-Hill, London.

Morré, D. J., Mollenhauer, H. H. and Bracker, C. E. (1971) Origin and continuity of Golgi apparatus. In: Origin and Continuity of Cell Organelles (Reinert, J. and Ursprung, H., eds.) pp. 82–126, Springer-Verlag, New York.

Mott, M. R. (1963) Cytochemical localization of antigens of *Paramecium* by ferritin-conjugated antibody and by counterstaining of the resultant absorbed globulin. J. Roy. Microsc. Soc. 81, 159–162.

Mott, M. R. (1965) Electron microscopy studies on the immobilization antigens of *Paramecium aurelia*. J. Gen. Microsc. 41, 251–261.

Müller, M. (1969) Peroxisomes of protozoa. Ann. N. Y. Acad. Sci. 168, 292–301.

Müller, M. (1975) Biochemistry of protozoan microbodies: peroxisomes, α-glycerophosphate oxidase bodies, hydrogenosomes. Ann. Rev. Microbiol. 29, 467–483.

Naitoh, Y. (1974) Bioelectric basis of behavior of protozoa. Am. Zool. 14, 883–893.

Naitoh, Y. and Eckert, R. (1969) Ciliary orientation: controlled by cell membrane or by intracellular fibrils? Science 166, 1633–1635.

Naitoh, Y. and Eckert, R. (1973) Sensory mechanisms in *Paramecium*. II. Ionic basis of the hyperpolarizing mechanoreceptor potential. J. Exp. Biol. 59, 53–65.

Nilsson, J. R. (1969) The fine structure of *Neobursaridium gigas* (Balech). C. R. Trav. Lab. Carlsberg 37, 49–72.

Nilsson, J. R. (1972) Further studies on vacuole formation in *Tetrahymena pyriformis* GL. C. R. Trav. Lab. Carlsberg 39, 83–110.

Nilsson, J. R. (1976) Physiological and structural studies on *Tetrahymena pyriformis* GL. C. R. Trav. Lab. Carlsberg 40, 215–355.

Nilsson, J. R. (1977) On food vacuoles in *Tetrahymena pyriformis* GL. 5th Int. Cong. Protozool. (Hutner, S. H., ed.) abstr. 310, The Print Shop, New York.

Nilsson, J. R. and Behnke, O. (1971) Studies on a surface coat of *Tetrahymena*. J. Ultrastruc. Res. 36, 542–545.

Nilsson, J. R., Ricketts, T. R. and Zeuthen, E. (1973) Effects of cytochalasin B on cell division and vacuole formation in *Tetrahymena pyriformis* GL. Exp. Cell Res. 79, 456–459.

Nohmi, M. and Tawada, K. (1974) The negatively charged protein extracted from *Tetrahymena pyriformis* as an attractant in *Amoeba proteus* chemotaxis. J. Cell Physiol. 84, 135–140.

Noirot-Timothée, C. (1968) Les sacs parasomaux sont des sites de pinocytose. Etude experimentale a l'aide de thorotrast chez *Trichodinopsis paradoxa* (Ciliata peritricha). C. R. Acad. Sci. Paris, 267, 2334–2336.

Novikoff, A. B. and Shin, W. Y. (1964) The endoplasmic reticulum in the Golgi zone and its relations to microbodies, Golgi apparatus and autophagic vacuoles in rat liver cells. J. Microscopie 3, 187–206.

Nozawa, Y. and Thompson, G. A. Jr. (1971a) Studies of membrane formation in *Tetrahymena pyriformis*. II. Isolation and lipid analysis of cell fractions. J. Cell Biol. 49, 712–721.

Nozawa, Y. and Thompson, G. A. Jr. (1971b) Studies on membrane formation in *Tetrahymena pyriformis*. III. Lipid incorporation into various cellular membranes of logarithmic phase cultures. J. Cell Biol. 49, 722–730.

Nozawa, Y. and Thompson, G. A. Jr. (1972) Studies of membrane formation in *Tetrahymena pyriformis*. V. Lipid incorporation into various cellular membranes of stationary phase cells, starving cells, and cells treated with metabolic inhibitors. Biochim. Biophys. Acta 282, 93–104.

Nozawa, Y., Fukushima, H. and Iida, H. (1973) Isolation and lipid composition of nuclear membranes from macronuclei of *Tetrahymena pyriformis*. Biochim. Biophys. Acta 318, 335–344.

Nozawa, Y., Fukushima, H. and Iida, H. (1975) Studies on *Tetrahymena* membranes. Modification of surface membrane by replacement of tetrahymenol by exogenous ergosterol in *Tetrahymena pyriformis*. Biochim. Biphys. Acta 406, 248–263.

Nozawa, Y., Iida, H., Fukushima, H., Ohki, K. and Ohnishi, S. (1974) Studies on *Tetrahymena* membranes: Temperature-induced alterations in fatty acid composition of various membrane fractions in *Tetrahymena pyriformis* and its effect on membrane fluidity as inferred by spin-label study. Biochim. Biophys. Acta 367, 134–147.

Oberg, S. G. and Satir, B. H. (1977) *Paramecium* fusion rosettes—possible physiological function in trichocyst release. 5th Int. Cong. Protozool. (Hutner, S. H., ed.) abstr. 313, The Print Shop, New York.

Ofer, L., Levkovitz, H. and Loyter, A. (1976) Conjugation in *Tetrahymena pyriformis*. The effect of polylysine, concanavalin A, and bivalent metals on the conjugation process. J. Cell Biol. 70, 287–293.

Ogura, A. and Takahashi, K. (1976) Artificial deciliation causes loss of calcium-dependent responses in *Paramecium*. Nature 264, 170–172.

Oldfield, E. and Chapman, D. (1972) Dynamics of lipid in membranes: heterogeneity and role of cholesterol. FEBS Lett. 23, 285–297.

Organ, A. E., Bovee, E. C., Jahn, T. L., Wigg, D. and Fonseca, J. R. (1968) The mechanism of the nephridial apparatus of *Paramecium multimicronucleatum*. I. Expulsion of water from the vesicle. J. Cell Biol. 37, 139–145.

Organ, A. E., Bovee, E. C. and Jahn, T. L. (1972) The mechanism of the water expulsion vesicle of the ciliate *Tetrahymena pyriformis*. J. Cell Biol. 55, 644–652.

Orias, E. and Rasmussen, L. (1977) *Tetrahymena* food vacuoles: What are they good for? 5th Int. Cong. Protozool. (Hutner, S. H., ed.) abstr. 230, The Print Shop, New York.

Packer, L. (1974) Membrane structure in relation to function of energy-transducing organelles. Ann. N. Y. Acad. Sci. 227, 166–174.

Packer, L., Mehard, C. W., Meissner, G., Zahler, W. L. and Fleischer, S. (1974) The structural role of lipids in mitochondrial and sarcoplasmic reticulum membranes. Freeze-fracture electron micros-copy studies. Biochim. Biophys. Acta 363, 159–181.

Palade, G. E. and Bruns, R. R. (1968) Structural modulations of plasmalemmal vesicles. J. Cell Biol. 37, 633–649.

Patterson, D. J. (1976) Observations on the contractile vacuole complex of Blepharisma americanum Suzuki, 1954 (Ciliophora, Heterotrichida). Arch. Protistenk. 118, 235–242.

Patterson, D. J. (1977) On the behaviour of contractile vacuoles and associated structures of Paramecium caudatum (Ehrbg.) Protistologica 13, 205–212.

Patterson, D. J. and Sleigh, M. A. (1976) Behavior of the contractile vacuole of Tetrahymena pyriformis W: a redescription with comments on the terminology. J. Protozool. 23, 410–417.

Pearse, B. M. F. (1976) Clathrin: a unique protein associated with intracellular transfer of membrane by coated vesicles. Proc. Nat. Acad. Sci. U.S.A. 73, 1255–1259.

Peck, R. K. (1971) Fine structure, morphogenesis and interrelationships within representatives of three ciliated protozoan genera, Ph. D. Thesis, University of Illinois at Urbana-Champaign.

Phillips, R. B. (1971) Induction of competence for mating in Tetrahymena by cell-free fluids. J. Protozool. 18, 163–165.

Pitelka, D. R. (1961) Fine structure of the silverline and fibrillar systems of three tetrahymenid ciliates. J. Protozool. 8, 75–89.

Pitelka, D. R. (1965) New observations on cortical ultrastructure in Paramecium. J. Microscopie 4, 373–394.

Plattner, H. (1974) Intramembraneous changes on cationophore-triggered exocytosis in Paramecium. Nature 252, 722–724.

Plattner, H. (1975) Ciliary granule plaques: membrane-intercalated particle aggregates associated Ca^{2+}-binding sites in Paramecium. J. Cell Sci. 18, 257–269.

Plattner, H. (1976) Membrane disruption, fusion and resealing in the course of exocytosis in Paramecium cells. Exp. Cell Res. 103, 431–435.

Plattner, H. and Fuchs, S. (1975) X-ray microanalysis of calcium binding sites in Paramecium with special reference to exocytosis. Histochemistry 45, 23–47.

Plattner, H., Miller, F. and Bachmann, L. (1973) Membrane specializations in the form of regular membrane-to-membrane attachment sites in Paramecium. A correlated freeze-etching and ultrathin-sectioning analysis. J. Cell Sci. 13, 687–719.

Plattner, H., Reichel, K. and Matt, H. (1977) Bivalent-cation-stimulated ATPase activity at pre-formed exocytosis sites in Paramecium coincides with membrane-intercalated particle aggregates. Nature 267, 702–704.

Plattner, H., Wolfram, D., Bachmann, L. and Wachter, E. (1975) Tracer and freeze-etching analysis of intra-cellular membrane-junctions in Paramecium. Histochemistry 45, 1–21.

Poste, G. and Allison, A. C. (1973) Membrane fusion. Biochim. Biophys. Acta 300, 421–465.

Poste, G. and Nicolson, G. L., eds. (1977) Dynamic aspects of cell surface organization. Cell Surface Reviews, Vol. 3. Elsevier/North-Holland, Amsterdam.

Preer, J. R., Jr. (1959a) Studies on the immobilization antigens of Paramecium. II. Isolation. J. Immunol. 83, 378–384.

Preer, J. R., Jr. (1959b) Studies on the immobilization antigens of Paramecium. III. Properties. J. Immunol. 83, 385–391.

Preer, J. R., Jr. (1969) Genetics of the protozoa. In: Research in Protozoology (Chen, T. T., ed.) Vol. 3, pp. 129–278, Pergamon Press, New York.

Raikov, I. B. (1973) Mitose intranucleaire acentrique due micronoyau de Loxodes magnus, Cilie Holotriche. Etude ultrastructurale. C. R. Acad. Sci. Paris 276, 2385–2388.

Raikov, I. B. (1976) Fine structure of the nuclear apparatus of the ciliate Loxodes magnus. II. Micronuclear mitosis. Tsitiologica 18, 144–151 (in Russian).

Rannestad, J. and Williams, N. E. (1971) The synthesis of microtubules and other proteins of the oral apparatus in Tetrahymena pyriformis. J. Cell Biol. 50, 709–720.

Rasmussen, L. (1973) On the role of food vacuole formation in the uptake of dissolved nutrients by Tetrahymena. Exp. Cell Res. 82, 192–196.

Rasmussen, L. (1974) Food vacuole membrane in nutrient uptake by *Tetrahymena*. Nature 250, 157–158.

Rasmussen, L. and Kludt, T. A. 1970) Particulate material as a prerequisite for rapid cell multiplication in *Tetrahymena* cultures. Exp. Cell Res. 59, 457–463.

Rasmussen, L. and Modeweg-Hansen, L. (1973) Cell multiplication in *Tetrahymena* cultures after addition of particulate material. J. Cell Sci. 12, 275–286.

Rasmussen, L. and Orias, E. (1975) *Tetrahymena:* Growth without phagocytosis. Science 190, 464–465.

Reese, T. S., Bennett, M. V. L. and Feder, N. (1971) Cell-to-cell movement of peroxidases injected into the septate axon of crayfish. Anat. Rec. 169, 409.

Reisner, A. H., Rowe, J. and Macindoe, H. M. (1969a) The largest known monomeric globular proteins. Biochim. Biophys. Acta 188, 196–206.

Reisner, A. H., Rowe, J. and Sleigh, R. W. (1969b) Concerning the tertiary structure of the soluble surface proteins of *Paramecium*. Biochemistry 8, 4637–4644.

Revel, J. P., and Karnovsky, M. J. (1967) Hexagonal array of subunits in intercellular junctions of the mouse heart and liver. J. Cell Biol. 33, C7–C12.

Revoltella, R., Ricci, N., Esposito, F. and Nobili, R. (1976) Cell surface control in *Blepharisma intermedium* Bhandary (Protozoa Ciliata) 1. Distinctive membrane receptor-sites in mating type I cells for various interacting ligands. Monitore Zool. Ital. (N.S) 10, 279–292.

Rhoads, D., Beischel, L., Meyer, K. and Kaneshiro, E. S. (1977a) Lipid composition of *Paramecium aurelia* cilia. J. Cell Biol. 75, 223a.

Rhoads, D., Meyer, K. and Kaneshiro, E. S. (1977b) Lipids of *Paramecium aurelia*. 5th Int. Cong. Protozool. (Hutner S. H., ed.) abstr. 436, The Print Shop, New York.

Ricci, N., Esposito, F. and Nobili, R. (1975) Conjugation in *Oxytricha bifaria:* Cell interaction. J. Exp. Zool. 192, 343–348.

Ricci, N., Esposito, F., Nobili, R. and Revoltella, R. (1976) Cell surface control in *Blepharisma intermedium* (Protozoa Ciliata). Blocking of gamone II-induced homotypic pairing by different membrane interacting ligands. J. Cell Physiol. 88, 363–370.

Ricketts, T. R. (1971a) Periodicity of endocytosis in *Tetrahymena pyriformis*. Protoplasma 73, 387–396.

Ricketts, T. R. (1971b) Endocytosis in *Tetrahymena pyriformis*. The selectivity of uptake of particles and the adaptive increase in cellular acid phosphatase activity. Exp. Cell Res. 66, 49–58.

Ricketts, T. R. and Rappitt, A. F. (1974) Determination of the volume and surface area of *Tetrahymena pyriformis,* and their relationship to endocytosis. J. Protozool. 21, 549–551.

Ricketts, T. R. and Rappitt, A. F. (1975a) The effect of puromycin and cycloheximide on vacuole formation and exocytosis in *Tetrahymena pyriformis* GL-9. Arch. Microbiol. 102, 1–8.

Ricketts, T. R. and Rappitt, A. F. (1975b) A radioisotopic and morphological study of the uptake of materials into food vacuoles by *Tetrahymena pyriformis* GL-9. Protoplasma 86, 321–337.

Ronai, A. and Wunderlich, F. (1975) Membranes of *Tetrahymena*. IV. Isolation and characterization of temperature-sensitive smooth and rough microsomal subfractions. J. Membrane Biol. 24, 381–399.

Rosenbaum, R. M. and Wittner, M. (1962) The activity of intracytoplasmic enzymes associated with feeding and digestion in *Paramecium caudatum*. The possible relationship to neutral red granules. Arch. Protistenk. 106, 223–240.

Rothman, J. E. and Engelmann, D. M. (1972) Molecular mechanism for the interaction of phospholipid with cholesterol. Nature New Biol. 237, 42–44.

Rubin, R. P. (1974) Calcium and the Secretory Process. Plenum Press, New York.

Rudzinska, M. A. (1977) The role of the "pits" in *Tokophrya infusionum*. 5th Int. Cong. Protozool. (Hutner, S. H., ed.) abstr. 238, The Print Shop, New York.

Sapra, G. R. and Kloetzel, J. A. (1975) Programmed autophagocytosis accompanying conjugation in the ciliate *Stylonychia mytilus*. Devel. Biol. 42, 84–94.

Satir, B. (1974a) Membrane events during the secretory process. Symp. Soc. Exp. Biol. 28, 399–418.

Satir, B. (1974b) Ultrastructural aspects of membrane fusion. J. Supramol. Struc. 2, 529–537.

Satir, P. and Gilula, N. B. (1973) The fine structure of membrane and intercellular communication in insects. Ann. Rev. Entomol. 18, 143–166.

Satir, P. and Satir, B. (1974) Design and function of site specific particle arrays in the cell membrane.

In: Control of Proliferation in Animal Cells (Clarkson, B. and Baserga, R., eds.) pp. 233–249, Cold Spring Harbor, New York.

Satir, B., Sale, W. S. and Satir, P. (1976) Membrane renewal after dibucaine deciliation of *Tetrahymena*. Exp. Cell Res. 97, 83–91.

Satir, B., Schooley, C. and Satir, P. (1972a) Membrane reorganization during secretion in *Tetrahymena*. Nature 235, 53–54.

Satir, B., Schooley, C. and Satir, P. (1972b) The ciliary necklace in *Tetrahymena*. Acta Protozool. 11, 291–294.

Satir, B., Schooley, C. and Satir, P. (1973) Membrane fusion in a model system. Mucocyst secretion in *Tetrahymena*. J. Cell Biol. 56, 153–176.

Sattler, C. A. (1974) Membrane specializations in *Tetrahymena pyriformis:* An ultrastructural study of the pellicle and ciliary membranes of the somatic and oral regions including a reconstruction of the buccal cavity. Ph. D. Thesis, University of Colorado, Boulder.

Sattler, C. A. and Staehelin, L. A. (1974) Ciliary membrane differentiations in *Tetrahymena pyriformis*. *Tetrahymena* has four types of cilia. J. Cell Biol. 62, 473–490.

Schneider, L. (1959) Die feinstruktur des nephridialplasma von *Paramecium*. Zool. Anz., 23 (Suppl.), 457–470.

Schneider, L. (1960) Elektronenmikroskopische Untersuchungen über das Nephridialsystem von *Paramecium*. J. Protozool. 7, 75–90.

Schneider, L. (1963) Elektronenmikroskopische untersuchungen der konjugation von *Paramecium*. I. Die auflösung und neubildung der zellmembran bei den konjuganten. (Zugleich ein Beitrag zur morphogenese cytoplasmatischer Membranen.) Protoplasma 56, 109–140.

Schneider, L. (1964) Elektronenmikroskopische untersuchungen an den ernährungsorganellen von *Paramecium*. I. Der cytopharynx. Z. Zellforsch. 62, 198–224.

Schreier-Muccillo, S., Marsh, D., Dugas, H., Schneider, H. and Smith, I. C. P. (1973) A spin probe study of the influence of cholesterol on motion and orientation of phospholipids in oriented multibilayers and vesicles. Chem. Phys. Lipid 10, 11–27.

Schuster, F. L. and Hershenov, B. (1969) Ultrastructural localization of cholinesterase in the pellicle of *Tetrahymena pyriformis* (W). Exp. Cell Res. 55, 385–392.

Sekiya, T., Kitajima, Y. and Nozawa, Y. (1975) Freeze-fracture studies of various membrane components isolated from *Tetrahymena pyriformis* cells. J. Elec. Microsc. 24, 155–165.

Singer, S. J. (1974) The molecular organization of membranes. Ann. Rev. Biochem. 43, 805–833.

Singer, S. J. and Nicolson, G. L. (1972) The fluid mosaic model of the structure of cell membranes. Science 175, 720–731.

Skriver, L. and Thompson, G. A., Jr. (1976) Environmental effects on *Tetrahymena* membranes. Temperature-induced changes in membrane fatty acid unsaturation are independent of the molecular oxygen concentration. Biochim. Biophys. Acta 431, 180–188.

Smith, J. D., Snyder, W. R. and Law, J. H. (1970) Phosphonolipids in *Tetrahymena* cilia. Biochem. Biophys. Res. Commun. 39, 1163–1169.

Speth, V. and Wunderlich, F. (1973) Membranes of *Tetrahymena*. II. Direct visualization of reversible transitions in biomembrane structure induced by temperature. Biochim. Biophys. Acta 291, 621–628.

Staehelin, L. A. (1974) Structure and function of intercellular junctions. Int. Rev. Cytol. 39, 191–282.

Steers, E., Jr. (1962) A comparison of the tryptic peptides obtained from immobilization antigens of *Paramecium aurelia*. Proc. Nat. Acad. Sci. U.S.A. 48, 867–874.

Steers, E., Jr. (1965) Amino acid composition and quaternary structure of an immobilizing antigen from *Paramecium aurelia*. Biochemistry 4, 1896–1901.

Steers, E., Jr. and David, R. H., Jr. (1977) A reexamination of the structure of the immobilization antigen from *Paramecium aurelia*. Comp. Biochem. Physiol. 56B, 195–199.

Stelly, N., Balmefrezol, M. and Adoutte, A. (1975) Diaminobenzidine reactivity of mitochondria and peroxisomes in *Tetrahymena* and in wild-type and cytochrome oxidase-deficient *Paramecium*. J. Histochem. Cytochem. 23, 686–696.

Subbaiah, P. V. and Thompson, G. A., Jr. (1974) Studies of membrane formation in *Tetrahymena pyriformis*. The biosynthesis of proteins and their assembly into membranes of growing cells. J. Biol. Chem. 249, 1302–1310.

762

Takahashi, M., Takeuchi, N. and Hiwatashi, K. (1974) Mating agglutination of cilia detached from complementary mating types of *Paramecium.* Exp. Cell Res. 87, 415–417.

Tandler, B. and Poulsen, J. H. (1976) Fusion of the envelope of mucous droplets with the luminal plasma membrane in the acinar cells of the cat submandibular gland. J. Cell Biol. 68, 775–781.

Thompson, G. A., Jr. (1967) Studies of membrane formation in *Tetrahymena pyriformis.* I. Rates of phospholipid biosynthesis. Biochemistry, 6, 2015–2021.

Thompson, G. A., Jr. (1969) The metabolism of 2-aminoethylphosphonate lipids in *Tetrahymena pyriformis.* Biochim. Biophys. Acta 176, 330–338.

Thompson, G. A., Jr. (1972) *Tetrahymena pyriformis* as a model system for membrane studies. J. Protozool. 19, 231–236.

Thompson, G. A., Jr., and Nozawa, Y. (1972). Lipids of protozoa: Phospholipids and neutral lipids. Ann. Rev. Microbiol. 26, 249–278.

Thompson, G. A., Jr., Bambery, R. J. and Nozawa, Y. (1971) Further studies of the lipid composition and biochemical properties of *Tetrahymena pyriformis* membrane systems. Biochemistry 10, 4441–4447.

Thompson, G. A., Jr., Bambery, R. J. and Nozawa, Y. (1972) Environmentally produced alterations of the tetrahymenol:phospholipid ratio in *Tetrahymena pyriformis* membranes. Biochim. Biophys. Acta 260, 630–638.

Tokoroyama, T., Horii, S. and Kubota, T. (1973) Synthesis of blepharismone, a conjugation inducing gamone in ciliate *Blepharisma.* Proc. Jap. Acad. 49, 461–463.

Tokuyasu, K. and Scherbaum, O. H. (1965) Ultrastructure of mucocysts and pellicle of *Tetrahymena pyriformis.* J. Cell Biol. 27, 67–81.

Tsuchiya, T. (1976) Electron microscopy and electron probe analysis of the Ca^{2+}-binding sites in the cilia of *Paramecium caudatum.* Experientia 32, 1176–1177.

Tsuchiya, T. and Takahasi, K. (1976) Localization of possible calcium-binding sites in the cilia of *Paramecium caudatum.* J. Protozool. 23, 523–526.

Tucker, J. B. (1967) Changes in nuclear structure during binary fission in the ciliate *Nassula.* J. Cell Sci. 2, 481–498.

Tucker, J. B. (1971) Microtubules and a contractile ring of microfilaments associated with a cleavage furrow. J. Cell Sci. 8, 557–571.

Vaudaux, P. (1976) Isolation and identification of specific cortical proteins in *Tetrahymena pyriformis* strain GL. J. Protozool. 23, 458–464.

Vivier, E. and Andre, J. (1961) Donnees structurales et ultrastructurales nouvelles sur la conjugaison de *Paramecium caudatum.* J. Protozool. 8, 416–426.

Wang, A. Y. I. and Packer, L. (1973). Mobility of membrane particles in chloroplasts. Biochim. Biophys. Acta 305, 488–492.

Weidenbach, A. L. S. and Thompson, G. A., Jr. (1974). Studies of membrane formation in *Tetrahymena.* VIII. On the origin of membranes surrounding food vacuoles. J. Protozool. 21, 745–751.

Weinstein, D. B. and Marsh, J. B. (1969) Membrane of animal cells. IV. Lipid of the L cell and its surface membrane. J. Biol. Chem. 244, 4103–4111.

Wirtz, K. W. A. and Zilversmit, D. B. (1968) Exchange of phospholipids between liver mitochondria and microsomes *in vitro.* J. Biol. Chem. 243, 3596–3602.

Wolf, R. W. and Allen, R. D. (1974) The cytoproct of *Tetrahymena pyriformis:* microtubules, food vacuole membrane and endocytosis. J. Protozool. 21, 425.

Wolfe, J. (1972) Basal body fine structure and chemistry. Adv. Cell Molec. Biol. 2, 151–188.

Wunderlich, F. and Ronai, A. (1975) Adaptive lowering of the lipid clustering temperature within *Tetrahymena* membrane. FEBS Lett. 55, 237–241.

Wunderlich, F. and Speth, V. (1972) Membranes in *Tetrahymena.* I. Cortical pattern. J. Ultrastruc. Res. 41, 258–269.

Wunderlich, F., Müller, R. and Speth, V. (1973a) Direct evidence for a colchicine-induced impairment in the mobility of membrane components. Science 182, 1136–1138.

Wunderlich, F., Speth, V., Batz, W. and Kleinig, H. (1973b) Membranes of *Tetrahymena.* III. The effect of temperature on membrane core structures and fatty acid composition of *Tetrahymena* cells. Biochim. Biophys. Acta 298, 39–49.

Wunderlich, F., Batz, W., Speth, V. and Wallach, D. F. H. (1974a) Reversible, thermotropic alteration of nuclear membrane structure and nucleocytoplasmic RNA transport in *Tetrahymena*. J. Cell
Biol. 61, 633–640.

Wunderlich, F., Wallach, D. F. H., Speth, V. and Fischer, H. (1974b) Differential effects of temperature on the nuclear and plasma membranes of lymphoid cells. A study by freeze-etch electron
microscopy. Biochim. Biophys. Acta 373, 34–43.

Wunderlich, F., Ronai, A., Speth, V., Seelig, J. and Blume, A. (1975) Thermotropic lipid clustering
in *Tetrahymena* membranes. Biochemistry 14, 3730–3735.

Wunderlich, F., Mahler, P. H., Ronai, A. and Kreutz, W. (1977) Thermotropic membrane transitions in *Tetrahymena*. 5th Int. Cong. Protozool. (Hutner, S. H., ed.) abstr. 428, The Print Shop,
New York.

Wyroba, E. and Przełecka, A. (1973) Studies on the surface coat of *Paramecium aurelia*. I. Ruthenium
red staining and enzyme treatment. Z. Zellforsch. 143, 343–353.

Zeuthen, E. (1964) The temperature-induced division synchrony in *Tetrahymena*. In: Synchrony in
Cell Division and Growth (Zeuthen, E., ed.) pp. 99–158. Interscience Publishers, New York.

Calcium-induced phase changes and fusion in natural and model membranes

14

Demetrios PAPAHADJOPOULOS

Contents

G. Poste & G. L. Nicolson (eds.) Membrane Fusion, pp. 765–790.
© *Elsevier/North-Holland Biomedical Press, 1978.*

1. Introduction

Studies on membrane fusion with model membranes such as lipid vesicles and black lipid films provide a unique opportunity for investigating the mechanism of membrane fusion at the molecular level. Phospholipid vesicles and black lipid films are stable structures with a limited permeability to small ions and large molecules (Thompson and Henn, 1970; Papahadjopoulos and Kimelberg, 1973; Bangham et al., 1974). Their chemical composition can vary considerably from pure one-component systems to mixtures of various phospholipids differing in head-group or hydrocarbon chain configuration. Thus questions relating to head-group specificity, bilayer fluidity, phase transitions and separations, for example, can be studied in great detail. In addition, other membrane components such as cholesterol, glycolipids, and various proteins or glycoproteins can be incorporated into the phospholipid bilayer, and their effects on fusion studied under controlled conditions. Finally, membrane-active molecules and drugs that are known to alter the fusion of natural membranes can be added externally to a well-defined system and their influence on the mechanism of fusion assessed in detail. The results from these simple systems can then be correlated with the evidence on fusion with natural membranes.

In recent years several laboratories have begun to study membrane fusion with model membranes. The methods and systems that have been used to detect fusion vary widely. The evidence for fusion has been based on observations such as the transfer of labeled lipid molecules between vesicles (Taupin and McConnell, 1972; Maeda and Ohnishi, 1974; Papahadjopoulos et al., 1974), changes in nuclear magnetic resonance (NMR) spectra of sonicated vesicles (Prestegard and Fellmeth, 1974; Lau and Chan, 1975), increase in vesicle size (Papahadjopoulos et al., 1974; Poznansky and Weglicki, 1974; Prestegard and Fellmeth, 1974; Lau and Chan, 1975), increase in optical density of vesicle suspensions (Lansman and Haynes, 1975; Martin and MacDonald, 1976a), mixing of lipid molecules from two vesicle populations (Papahadjopoulos et al., 1974; Martin and MacDonald, 1976a), transfer of fluorescent probes or ionophores between vesicles and black lipid films (Pohl et al., 1973; Moore, 1976), mixing and reactivity of two membrane proteins incorporated initially in different lipid vesicles (Miller and Racker, 1976), optical microscopic observations on spherical bilayer black lipid films (Breisblatt and Ohki, 1976), and electrical conductance measurements between apposed hemispherical bilayer black lipid films (Liberman and Nenashev, 1972; Neher, 1974). The lack of a consistent definition for the phenomenon of membrane fusion has led to disagreement about whether a particular experimental observation provides conclusive evidence for fusion. For example, transfer of spin-labeled lipid molecules (Maeda and Ohnishi, 1974; Van Der Bosch and McConnell, 1975) and fluorescent probes (Pohl et al., 1973) from one type of membrane to another could be due to fusion but could equally well result from molecular exchange or diffusion of single molecules or clusters of molecules (Martin and MacDonald, 1976a; Papahadjopoulos et al., 1976a). Similar alternative interpretations can be offered to explain evidence involving changes in proton magnetic resonance spectra (for references, see above) and also for evidence on increase in vesicle size, and even for evidence indicating complete mixing of the lipid molecules between two different vesicle populations.

The most rigorous and incontrovertible evidence for fusion between two vesicles would involve a demonstration of stoichiometric mixing of the vesicle membrane components as well as mixing of the vesicle contents enclosed within their interior aqueous space. Positive evidence for either of the above mixing reactions would strongly indicate fusion since the occurrence could not be explained by any of the alternative mechanisms proposed to date. Stoichiometric (one to one) mixing of vesicle lipids has been unequivocally demonstrated for calcium-induced fusion of acidic phospholipid vesicles (Papahadjopoulos et al., 1974, 1976b). Evidence for mixing water-soluble vesicle contents has not been reported so far except for an early observation (Taupin and McConnell, 1972) involving a small degree of reactivity of spin-labeled compounds trapped within lecithin vesicles. However, lack of evidence for mixing water-soluble vesicle contents cannot be taken as an indication that fusion has not occurred, since the vesicles may become leaky during fusion and their contents would thus be released into the outside medium. Such an occurrence could be particularly important for the fusion of small, sonicated lipid vesicles. Release of vesicle contents has been observed to occur during calcium-induced fusion of unilamellar phosphatidylserine vesicles (Papahadjopoulos et al., 1977).

For a more detailed, critical survey of the advantages and disadvantages of particular techniques in monitoring fusion of lipid vesicles, the reader is referred to the recent review by Papahadjopoulos and co-workers (1978).

2. Molecular exchange and fusion between pure phospholipid membranes

Fusion (coalescence) of oil droplets in water is driven by the high interfacial tension of the system and results in an overall decrease of the oil-water interphase, thus minimizing the free energy of the system. Since it is generally accepted that phospholipid vesicles and black lipid films are stable structures with low interfacial tension (Thompson and Henn, 1970; Bangham, 1972), it would be surprising to find that they tend to fuse extensively. A critical evaluation of the studies published recently on this subject now permits reasonable description of the ability of pure phospholipid membranes to fuse.

An early report on sonicated lecithin vesicles indicated a small but measurable degree of fusion (Taupin and McConnell, 1972). A later study with sonicated dimyristoyl phosphatidylcholine (DMPC) vesicles revealed that extensive and rapid fusion occurred only at the temperature region of the expected transition (Tc) from solid to liquid crystalline (Prestegard and Fellmeth, 1974). The occurrence of fusion in this case was concluded from the observed increase in vesicle size. It was reported subsequently, however, (Kantor and Prestegard, 1975) that this increase in vesicle size could occur only in the presence of myristic acid, which was present as an impurity in original DMPC preparations. Other studies using similar methodology have indicated that the observed changes in proton magnetic resonance spectra and increase in vesicle size could be the result of incomplete "annealing" of the vesicles (Lawaczek et al., 1975, 1976).

Evidence for a slow, time-dependent increase in the size of sonicated DMPC and also dipalmitoyl phosphatidylcholine (DPPC) vesicles incubated at temperatures near or above the Tc has also been obtained more recently in several laboratories with chromatographically pure lipids under conditions of preparation that would ensure adequate annealing of the vesicles (Martin and MacDonald, 1976a; Papahadjopoulos et al., 1976a; Suurkuusk et al., 1976). The authors of the latter three studies were in general agreement that the existing evidence is not definitive enough concerning the mechanism of the observed increase in vesicle size. Although fusion is not ruled out, other processes such as molecular diffusion could also account for the observed phenomena.

The interaction of vesicles composed of pure phospholipids was also studied by mixing their lipid components. Using unsonicated lecithin vesicles (Papahadjopoulos et al., 1974, 1976a), it was observed that no appreciable mixing occurred at temperature a few degrees above the phase transition of the vesicles, indicating the absence of fusion even after long incubation times (20 hr). More recently, a study with sonicated lecithin vesicles observed considerable mixing between two populations of lecithin vesicles (Martin and MacDonald, 1976a). However, the kinetics of mixing indicated a process of gradual transfer of the more soluble molecules to the more stable vesicles. This appears to be compatible with a diffusion process rather than true fusion. Similar observations and conclusions were reported in another study involving the interaction of multilamellar phosphatidylglycerol (PG) vesicles suspended in 0.1 M NaCl at pH 7.4 (Papahadjopoulos et al., 1976a).

The interaction of pure phospholipid vesicles has also been studied by measuring the transfer of radioactive lipid from small sonicated vesicles to large multilamellar vesicles (Papahadjopoulos et al., 1974). It was observed that no transfer above background occurred between egg lecithin vesicles or brain phosphatidylserine (PS) vesicles or several mixtures of these lipids, even during incubation at 37 °C for 20 hours in 0.1 M NaCl at pH 7.4.

Fusion of spherical bilayer black lipid films has also been studied recently by direct optical observations (Breisblatt and Ohki, 1975, 1976). These large spherical bilayers are allowed to float in a density gradient, and the percentage of fusion (one larger sphere) is scored among spheres that have formed a common interface. Although the percentage of fusion events was low at low temperature, a large increase in fusion was seen at higher temperatures (30 to 40 °C) for both lecithin and PS bilayers. Although this system is interesting, it is not possible at present to exclude the problems of lipid impurities and mechanical instability. Enhancement of fusion by lysolecithin and a decrease of fusion by cholesterol has also been described in this system (Breisblatt and Ohki, 1976).

In conclusion, the studies to date with pure phospholipid vesicles suspended in water or monovalent salt solutions reveal that fusion between fluid vesicles above their transition temperatures is very slow, if it occurs at all. This appears to be true for both sonicated and nonsonicated neutral or negatively charged vesicles. However, extensive growth in the size of sonicated vesicles at temperatures close to the Tc has now been observed in several laboratories. The interpretation of this phenomenon has varied from a mechanism involving diffusion (Martin and MacDonald, 1976a) to incomplete annealing (Lawaczek et al., 1975) to a myristic acid impurity (Kantor and Prestegard, 1975). The refractoriness of pure phospholipid vesicle membranes to fusion could reflect the innate stability of these systems. Furthermore, the existence of large repulsive forces at the interface of even neutral (lecithin) membranes (Le Neveu et al., 1976)—possibly due to structured water (Marcelja and Radic, 1976)—dictate that close contact between colliding membranes cannot be achieved.

3. The role of lysolecithin, fatty acids, and DMSO in fusion of phospholipid membranes

The role of lysolecithin in membrane fusion has been of considerable interest since the demonstration by Howell and Lucy (1969) that this compound can induce fusion of erythrocytes. Since then, this method has been used in several laboratories for the production of cell hybrids (Croce et al., 1971), though its usefulness is limited by the problem of cell lysis. The possible role of lyso compounds in membrane fusion under physiological conditions has been discussed in detail elsewhere (Poste and Allison, 1973). The most interesting proposal in this respect involves the possible production of endogenous lysolecithin by the action of calcium-activated phospholipase(s), which would make these membranes susceptible to fusion (Lucy, 1975). Although this possibility cannot be totally excluded, its potential importance as a major mechanism in natural membrane fusion has been diminished greatly by several recent observations. First, phospholipid vesicles can cause extensive cell fusion without lysolecithin and without appreciable cell lysis (Papahadjopoulos et al., 1974; Papahadjopoulos et al., 1978). Indeed, neutral (lecithin) vesicles containing up to 20% lysolecithin are unable to induce cell fusion (Martin and MacDonald, 1976b). Second, calcium alone can induce fusion of phospholipid membranes without any concomitant enzymatic hydrolysis (Papahadjopoulos et al., 1974, 1976b). Third, extensive hydrolysis of intact red cells by phospholipases can occur without inducing cell lysis, indicating that endogenous lysolecithin per se is not lytic for cell membranes (Zwaal et al., 1973). It has also been shown that the lytic effects of lysolecithin on model membranes are reduced substantially in the presence of cholesterol, which may be related to the ability of lysolecithin to form a stoichiometric complex with cholesterol which has a lamellar rather than micellar configuration (Inoue and Kitagawa, 1974; Rand et al., 1975).

The effect of exogenous lysolecithin on the thermotropic properties and fusion of phospholipid vesicles has been studied recently in detail (Papahadjopoulos et al., 1976a). It was shown that addition of lysolecithin produced an increase in the Tc of individual phospholipids, even when added at a 20% molar ratio. This unexpected result does not support the impression derived from the literature that lysolecithin is able to increase the fluidity of membrane lipids (Howell et al., 1972; Ahkong et al., 1973). Fluidization of the lipid bilayer following addition of lysolecithin would be expected to result in a decrease in the Tc of the membranes. If lysolecithin is added to two populations of preformed multilamellar lecithin vesicles (Papahadjopoulos et al., 1976a), considerable enhancement of mixing of the lipid components is observed. However, the kinetics of mixing indicate a gradual enrichment of each of the lecithins with the other, which suggests an exchange diffusion process rather than true fusion. True fusion should have resulted in the formation of a third component with an intermediate melting point, reflecting the formation of unique mixture with the stoichiometry of the initial two populations of lecithins. It was observed in the same study, however, that lysolecithin induces the formation of such a third component when added to two populations of PG vesicles.

It was concluded from this data that lysolecithin can produce some fusion, at least with acidic phospholipid membranes, but the initial effect on lecithin membranes is to enhance the rate of molecular mixing probably via exchange diffusion rather than fusion. It was also concluded that at relatively low concentrations of lysolecithin, before solubilization of the phospholipid membranes becomes pronounced (up to 20% mole ratio), its effect on lipid bilayers is not fluidization but rather an apparent "stabilization," as evidenced by an increase in the Tc.

As mentioned earlier, myristic acid has been shown to enhance the rate of growth in size of sonicated DMPC vesicles, possibly indicating an effect on vesicle fusion (Kantor and Prestegard, 1975). It is not clear, however, whether this effect is due to improper annealing (Lawaczek et al., 1975), or related to structural fluctuations induced by the bilayer phase transition (Kantor and Prestegard, 1975) or to the properties of the sonicated vesicles containing myristic acid. To provide insight into this problem Papahadjopoulos and collaborators (1976a) studied the effect of myristic acid on the rate of mixing of the lipid components between two lecithin populations, DMPC and DPPC. They first established that the presence of myristic acid (premixed with the phospholipid in a 10% mole ratio before vesicle formation) produced an increase in the Tc and a broadening of the main transition, while the premelt transition was not apparent at all. These effects were similar to those observed with lysolecithin and discussed above. When such multilamellar vesicles (each composed of myristic acid mixed with one of the two lecithins) were incubated together, no appreciable mixing was observed. The conditions included incubation for 4 hours at

50°C at pH from 4.5 to 8.5 in 0.1 M NaCl salt solution (Papahadjopoulos et al., 1976a). This type of experiment was also performed with sonicated preparations of the same lipid. Incubation of the two populations for 3 hours and 20 hours at 50°C produced a gradual shift of the two endothermic peaks toward an intermediate temperature. This result demonstrates extensive mixing between the two populations of vesicles, but again the shift of the two peaks was consistent with the kinetics of exchange diffusion rather than fusion.

It can be concluded from the above results that myristic acid has a considerable effect on phospholipid mixing between bilayer membranes of different vesicles. However, the extent of its effect and the type of mixing observed depend mainly on the condition of the phospholipid bilayers. For example, the acid had no effect (no enhancement of mixing) on large multilamellar lecithin vesicles, but induced extensive exchange diffusion of lecithin molecules between sonicated vesicle preparations at temperatures above their Tc. The effect of myristic acid on the growth of sonicated DMPC vesicles at a temperature identical with their Tc could thus be related to the structural fluctuations that the bilayer is undergoing at the Tc. Since this phenomenon depends on the phase transition, but is greatly enhanced by myristic acid at the Tc, it could be concluded that myristic acid is providing the electrostatic attractive forces necessary to bring two vesicles into close apposition. Fusion or molecular exchange can then occur at points of structural discontinuities induced by the phase transition. This interpretation is based on recent observations concerning the existence of large repulsive forces between lecithin bilayers (Le Neveu et al., 1976) and the importance of phase changes during calcium-induced fusion events (Papahadjopoulos et al., 1977), which will be discussed in detail below.

Dimethylsulfoxide (DMSO) has been shown to produce cell fusion (Ahkong et al., 1975) and also to induce particle redistribution in cell membranes (McIntyre et al., 1974) and drastic changes in the phase transition characteristics of phospholipid membranes (Lyman et al., 1976a,b). Interest in the possible effects of DMSO on membranes is augmented because of its well known use as a carrier to increase drug absorption and as an inducer of leukemic cell differentiation (Preisler et al., 1973). Its effect on the ability of phospholipid membranes to fuse was examined recently (Papahadjopoulos et al., 1976a). This study involved the addition of high concentrations of DMSO into suspensions of two populations of multilamellar vesicles composed of pure phospholipids. It was found that 30% DMSO has no appreciable effect on the mixing of lecithins (incubation for 20 hr at 45°C) as detected by differential scanning calorimetry. However, under similar conditions, incubation of two populations of PG vesicles, produced a high degree of mixing (~50%) following incubation for 1.5 hours at 45°C. Furthermore, the characteristics of this mixing indicated that it was due to true fusion rather than to exchange diffusion (Papahadjopoulos et al., 1976a).

The effect of DMSO in promoting fusion between PG but not PC (lecithin) vesicles seems to be related to its ability to induce a new phase transition with a Tc at much higher temperatures (Lyman et al., 1976a). This remarkable effect of DMSO in stabilizing acidic phospholipid membranes in the frozen (solid) state was also shown recently to be correlated with the ability of this and other compounds to induce leukemic cell differentiation (Lyman et al., 1976b). Although further studies with DMSO are needed before its mechanism of action is understood, its effect in promoting both fusion and a large increase in the Tc is reminiscent of the effect of Ca^{2+}, which will be discussed later. A synergistic effect between Ca^{2+} and DMSO has already been noted (Lyman et al., 1976a) and the phenomenon is now being investigated in this laboratory in more detail.

4. The role of proteins and polypeptides in fusion of phospholipid membranes

The phospholipid vesicle system presents excellent opportunities for studying the possible role of specific proteins involved in natural membrane fusion phenomena such as secretion, fusion of viruses with cells and the many other examples discussed in this volume. Although it is possible that the key step in the mechanism of membrane fusion involves only calcium and acidic phospholipids (see below), it is conceivable that specific proteins may play an important role in conferring specificity or in organizing the lipid bilayer in microdomains that may or may not favor fusion. It is also possible that specific proteins may induce fusion between membranes not otherwise susceptible to fusion. The reports that have appeared to date indicate that proteins or peptides may play a significant role, although the ones used so far are not known to be involved in natural membrane fusion phenomena.

Alamethicin is a cyclic oligopeptide (Payne et al., 1970) that is known to form voltage-dependent ion channels in black lipid films (Mueller and Rudin, 1968). Its interaction with phospholipid vesicles had been studied before (Hauser et al., 1970), and its effect in broadening the lipid proton resonance spectra had been interpreted as immobilization of the hydrocarbon chains by the peptide. More recently, evidence has been presented which was interpreted as indicating that alamethicin induces fusion of lecithin vesicles at temperatures well above the Tc (Lau and Chan, 1975). This evidence was based partly on the observed increase in vesicle size, the translocation of external ions and alamethicin molecules into the interior of the vesicles during incubation, and NMR data indicating that alamethicin interacts primarily with the polar head-group region of the lecithin vesicles (Lau and Chan, 1974, 1976). It is not clear, however, from the existing evidence whether these authors have excluded the possibility that the observed phenomena are due to an increased diffusion of lecithin molecules from smaller to larger vesicles, which is mediated by alamethicin.

Concanavalin A (con A) is a protein that has been used extensively to characterize cell surface phenomena because of its strong binding to specific glycoproteins (review, Nicolson, 1974). A recent study with lecithin vesicles observed that the presence of con A induces a large increase in the transfer rate of spin-labeled lipid molecules between vesicles (Van Der Bosch and McConnell, 1975). Moreover, it was noted that this effect was pronounced only at the phospholipid transition temperature. It was concluded that con A caused fusion between lecithin vesicles, although again a rigorous exclusion of diffusion mechanisms was not attempted. This phenomenon is interesting although the mechanism is unclear at present. In some respects the effect of con A is reminiscent of the effect of myristic acid discussed earlier.

The effect of a hydrophobic membrane protein on the fusion of lecithin vesicles has also been reported recently (Papahadjopoulos et al., 1975a,b). The protein was a purified major component of myelin proteolipid (Folch and Lees, 1951; Gagnon et al., 1971), and its interaction with phospholipid vesicles had already been studied in some detail (Papahadjopoulos et al., 1975b,c). Briefly, it had been shown that this protein can be incorporated into phospholipid bilayers (up to 50% by weight) without affecting the Tc of the main endothermic transition. The resultant lipoprotein membranes have an intermediate buoyant density and contain intramembrane particles in freeze-fracture electron micrographs (Vail et al., 1974; Papahadjopoulos et al., 1975c). Low-angle x-ray diffraction of such membranes indicated that the incorporation of the protein did not increase the thickness compared to that of the pure lipid, indicating that a good part of the protein must be embedded within the lipid bilayer (Rand et al., 1975).

When two populations of vesicles composed of either DMPC or DPPC, and each containing 15% of the proteolipid apoprotein, were prepared separately and incubated together for 0.5 and 1.5 hours at 45 °C, a large degree of mixing was observed (Papahadjopoulos et al., 1976a). Under similar conditions, no mixing was observed in the absence of protein with the same two lipids as multilamellar vesicles. The kinetics of mixing indicated that increasing amounts of DPPC were mixing with all of the available DMPC. Such mixing was observed only when both vesicle populations contained the protein. It was concluded that the data are best explained by a mechanism involving fusion, assuming a decreasing propensity for fusion with the increasing content of DPPC (Papahadjopoulos et al., 1976a).

It is still too early to generalize about the possible effects of various proteins on phospholipid membrane fusion. Existing evidence indicates that proteins and oligopeptides enhance molecular mixing between vesicles, and that this effect could be related to fusion. Very little can be said concerning the mechanism(s) involved in such interactions.

5. The role of divalent metals (calcium and magnesium) in fusion of vesicle membranes

Particular attention has been focused on the role of calcium ions (Ca^{2+}) in regulating the many cellular processes in which membrane fusion occurs (Poste and Allison, 1973; Rubin, 1975; and various chapters in this volume). Although the involvement of Ca^{2+} in such phenomena may include participation in diverse reactions with several cellular components, its interaction with membrane phospholipids seemed a reasonable area for further investigation. The reason for this optimism was based on observations made several years ago that Ca^{2+} could induce abrupt changes in the permeability of PS vesicles (Papahadjopoulos and Bangham, 1966). Later it was shown that this effect was related to the Ca^{2+} asymmetry across the PS membrane (Papahadjopoulos and Ohki, 1969), since such membranes are very stable and impermeable in the absence of Ca^{2+} or when Ca^{2+} is present on both sides of the membrane. Other details of the interaction of PS with Ca^{2+}, which have been reviewed elsewhere (Papahadjopoulos et al., 1977), include the formation of a stoichiometric (2:1, PS:Ca^{2+} mole ratio) complex at lmM Ca^{2+} in the presence of 100 mM NaCl, pH 7.4 (Bangham and Papahadjopoulos, 1966), the condensation of the area per molecule in PS monolayers (Bangham and Papahadjopoulos, 1966; Papahadjopoulos, 1968), the crystallization of the acyl chains, and disappearance of the normal PS endothermic transition (Jacobson and Papahadjopoulos, 1975).

The first report on the role of Ca^{2+} in the fusion of phospholipid vesicles (Papahadjopoulos et al., 1974) presented evidence showing that incubation of PS vesicles in Ca^{2+}-containing solutions induced an increase in the vesicle size which could not be reversed by excess ethylenediaminetetraacetic acid (EDTA). It was later demonstrated by freeze-fracture electron microscopy that the addition of Ca^{2+} to sonicated PS vesicles (1–10 mM) induces the formation of large, spiral (cochleated) lipid cylinders (Papahadjopoulos et al., 1975a) which become large, closed spherical vesicles following addition of excess EDTA. Calcium-induced fusion was also demonstrated with vesicles containing mixtures of PS and PC, as shown by mixing the lipid components (Papahadjopoulos et al., 1974). The same technique has been used more recently to demonstrate fusion between PG vesicles (at 10 mM Ca^{2+}), and phosphatidic acid (PA) vesicles at 0.2 mM Ca^{2+} (Papahadjopoulos et al., 1976b). Evidence on mixing two populations of multilamellar PG vesicles is shown in Fig. 1. It can be seen that at 5 mM Ca^{2+} the vesicles aggregate but no mixing is observed. However, at 7.5 mM Ca^{2+}, aggregation is followed by the formation of a new component with a melting point identical to the equimolar mixture of the two PG species. This third component becomes the predominant membrane present at 10 mM Ca^{2+} with only very small amounts of the two initial pure PG membranes remaining (Fig. 1, curve d).

One of the most remarkable aspects of these interactions is the specificity for divalent metals. For example, it was observed with a PG system similar to that shown in Fig. 1 that addition of Mg^{2+} (up to 20 mM) does not produce mixing of lipids from the two vesicle populations even though it produces aggregation of the vesicles (Papahadjopoulos et al., 1975b). Similarly, incubation of PS vesicles in the presence of 2 to 5 mM Mg^{2+} at 37 °C produces slow aggregation of intact vesicles (Papahadjopoulos et al., 1977) instead of the cochleated cylinders produced by 1 to 10 mM Ca^{2+} under similar conditions. Thus it appears that Mg^{2+} does not induce fusion of either PS or PG vesicles under the conditions used for Ca^{2+}-induced fusion. On the other hand, both Ca^{2+} and Mg^{2+} are equally effective in inducing fusion of PA vesicles (Papahadjopoulos et al., 1976b). Earlier studies on the binding of Ca^{2+} and Mg^{2+} to PS and PA vesicles indicated similar

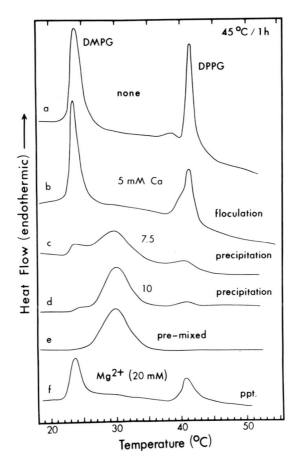

Fig. 1. Effects of different Ca^{2+} concentrations on the mixing of preformed vesicles of dimyristoyl phosphatidylglycerol (DMPG) and dipalmitoyl phosphatidylglycerol (DPPG). The phospholipids were suspended separately by shaking in 100 mM NaCl buffer, pH 7.4 at 45 °C at a concentration of 5 μmol/ml, and then equilibrated for 30 min at 45 °C. Aliquots from each suspension containing approximatley 2 μmol of each lipid were mixed, diluted to a total concentration of 1 μmol/ml with 100 mM NaCl buffer pH 7.4, and incubated at 45 °C for 1 hr under the following conditions: (a) no additions; (b) 5 mM CaCl$_2$, 1 hr, then 6 mM EDTA, 0.5 hr at 45 °C; (c) 7.5 mM CaCl$_2$, 1 hr, then 8.5 mM EDTA, 0.5 hr at 45 °C; (d) 10 mM CaCl$_2$, 1 hr, then 12 mM EDTA, 0.5 hr, 45 °C; (e) equimolar mixture of the two lipids suspended in 100 mM NaCl buffer; (f) 20 mM MgCl$_2$, 1h, then 25 mM EDTA, 0.5 h. (Reproduced with permission from Papahadjopoulos et al., 1976b.)

binding constants (Hendrickson and Fullington, 1965; Abramson et al., 1966), although considerable differences were noted between Ca^{2+} and Mg^{2+} in their effects on the condensation of PS monolayers and the decrease in surface potential (Papahadjopoulos, 1968). Because of the differences between Ca^{2+} and Mg^{2+} discussed above, it can be concluded that the effects of Ca^{2+} on fusion of PS and PG vesicles must be due to specific binding to the head-group charges (Papahad-

jopoulos, 1968; Papahadjopoulos et al., 1977) rather than to simple neutralization and screening of the negative charges relating to double-layer electrostatics
(McLaughlin et al., 1971) since the latter effect would not be expected to exhibit
specificity between bivalent metal ions.

The following general conclusions have been drawn concerning the characteristics of Ca^{2+}-induced fusion of phospholipid vesicles (Papahadjopoulos et al.,
1976b): (1) the vesicles should carry a net negative surface charge, and the
bilayers must be fluid (above the Tc) at the experimental temperature; (2) there
is a definite "threshold" concentration at which Ca^{2+} becomes effective. This
threshold varies with different phospholipids suspended in 100 mM NaCl at pH
7.4, from 0.2 mM for PA, to 0.1 mM for PS, to 10 mM for PG. All these vesicles
become leaky to their contents at approximately the same threshold concentrations of Ca^{2+}; (3) although fusion is accompanied by vesicle aggregation, it is
clear that aggregation as such is not enough to induce fusion; and (4) Ca^{2+}-
induced fusion is accompanied by a drastic change in the thermotropic properties of the vesicle membranes, resulting in crystallization of the acyl chains and a
shift of the phase transition to a much higher temperature. It has been proposed
recently that such phase changes are a key event for the mechanism of Ca^{2+}-
induced fusion (Papahadjopoulos et al., 1977, 1978).

6. The role of lipid phase changes in fusion of phospholipid membranes

The correlation between Ca^{2+}-induced phase changes and membrane fusion is illustrated in Fig. 2. Here, the addition of 1 mM Ca^{2+} to PG vesicles produces only
a shift of the two peaks to slightly higher temperatures, as would be expected
from considerations of simple double-layer electrostatics (Träuble and Eibl,
1974). Addition of EDTA reverses the process, giving back the original components without any mixing, or fusion (Fig. 2, curve c). When the Ca^{2+} concentration is raised to 10 mM, both transitions disappear completely (curve d) and, as
shown earlier, they now appear at much higher temperature (Verkleij et al.,
1974). Thus the bilayers have undergone an isothermal phase transition under
these conditions. Addition of excess EDTA to the mixture (Fig. 2 curve e) indicates that the two populations of PG vesicles are now completely mixed, presumably as a result of fusion.

The same phenomenon of an isothermic phase transition occurs during the interaction of PS with Ca^{2+}, in which case crystallization of the acyl chains is obtained at 1 mM Ca^{2+} (Jacobson and Papahadjopoulos, 1975). A recent study on
the mechanism of calcium-induced fusion of phospholipid (PS) vesicles
(Papahadjopoulos et al., 1977) indicates that the phase change from a fluid to a
solid state is a key event during fusion induced by divalent cations. For example,
it was shown that fusion by Ca^{2+} is inhibited at 0°C, a temperature at which the
bilayers would be crystallized before Ca^{2+} is added. In this case the vesicles were
observed only to aggregate and not fuse. Similarly, addition of Mg^{2+} up to 10

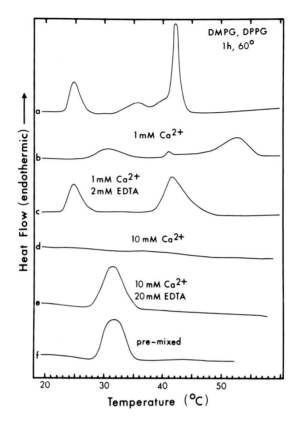

Fig. 2. Effects of Ca²⁺ concentrations on the thermotropic phase transitions and mixing of pre-formed vesicles of dimyristoyl phosphatidylglycerol (DMPG) and dipalmitoyl phosphatidylglycerol (DPPG). Phospholipids were prepared as described in Fig. 1, except that after mixing the samples were incubated at 60°C for 1 hr under the following conditions: (a) no other additions; (b) 1 mM $CaCl_2$; (c) 1 mM $CaCl_2$ for 1 hr, then 2 mM EDTA and additional incubation for 0.5 hr at 60°C; (d) 10 mM $CaCl_2$; (e) 10 mM $CaCl_2$ for 1 hr, followed by 20 mM EDTA for 0.5 hr; and (f) equimolar quantities of DMPG and DPPG mixed in chloroform and suspended in 100 mM NaCl buffer at 45°C as above. Following each of these incubations the suspensions were centrifuged at 100,000 × g for 15 min at 20°C. The wet pellets were then transferred to the sample pans of the differential scanning calorimeter and analyzed within 1–2 hr. (Reproduced with permission from Papahad-jopoulos et al., 1976b.)

mM at 37°C induced immediate aggregation of vesicles with only a minimal degree of fusion. The Tc was increased from 7° to 18°C but the vesicles remained fluid at the experimental temperature, 37°C. Furthermore, it was shown that decreasing the temperature from 37° to 12°C greatly enhanced the degree of fusion induced by 10 mM Mg^{2+}. In this latter case the PS vesicles are fluid initially (Tc = 7°C in 100 mM NaCl, pH 7.4) but become solid following addition of Mg^{2+} (Tc = 18°C in 100 mM NaCl/5 mM $MgCl_2$, pH 7.4). It was therefore concluded that the difference in effectiveness between Ca^{2+} and Mg^{2+} in inducing fusion of PS vesicles at 22° to 37°C is due to the fact that Ca^{2+} induces a phase transition (crystallization) while Mg^{2+} does not.

At this point it is relevant to ask whether Ca^{2+} can induce similar phase transitions in mixed phospholipid membranes, and whether such transitions could account for the fusion of PS/PC vesicles mentioned earlier. This question has been studied independently in two laboratories using different techniques, one involving electron spin resonance probes (Ohnishi and Ito, 1973, 1974; Ito and Ohnishi, 1974) and the other involving differential scanning calorimetry (DSC) and x-ray diffraction (Papahadjopoulos et al., 1974; Jacobson and Papahadjopoulos, 1975). The evidence indicates that Ca^{2+} can induce a lateral phase separation resulting in the formation of separate domains of PS-Ca^{2+} and PC.

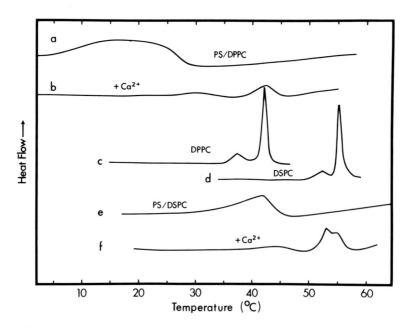

Fig. 3. Differential scanning calorimetry thermogram of mixed lipid membranes before and after addition of Ca^{2+}. Phospholipid dispersions were made in 100 mM NaCl buffer, pH 7.4, at concentrations of 4 μmoles/ml. (a) Phosphatidylserine/dipalmitoyl phosphatidylcholine (PS/DPPC) (2:1) mixture, dispersed without sonication for 30 min at 37°C, (b) PS/DPPC (2:1) mixture, dispersed without sonication for 30 min at 37°C, then incubated for 30 min with Ca^{2+} (10 mM) at 37°C (c) pure DPPC nonsonicated dispersions dispersed at 42°C, (d) pure distearoyl phosphatidylcholine (DSPC), nonsonicated, dispersed at 58°C, (e) PS/DSPC (2:1) mixture dispersed without sonication at 42°C and (f) PS/DSPC (2:1) mixture, dispersed by sonication for 1 hr at 42°C, then incubated for 1 hr with Ca^{2+} (10 mM) at 42°C. If EDTA (equimolar to Ca^{2+}) is added after the above Ca^{2+} treatment, the calorimeter scan of the sample is identical to that shown in e. All the above samples were centrifuged at room temperature for 10 min at 100,000 × g and the pellets transferred into the calorimeter sample pans with a Pasteur pipette. Amount of phospholipid per sample was approximately 1 μmole. (Reproduced with permission from Jacobson and Papahadjopoulos, 1975.)

Fig. 3 shows data obtained by DSC which indicate the appearance of a pure PC component from vesicles composed of a PS/PC mixture following the addition of Ca^{2+} to the sonicated vesicles. Similar phase separation has been induced by Ca^{2+} with PA/PC mixtures (Galla and Sackmann, 1975) but not with PG/PC mixtures (Verkleij et al., 1974; Van Dijck et al., 1975).

It has been pointed out (Papahadjopoulos et al., 1974, 1977) that the occurrence of a phase separation in PS/PC vesicles correlates with fusion of the same vesicles. Thus both phenomena are observed only above a threshold concentration of Ca^{2+}, the magnitude of which depends on the initial molar ratio of PS:PC; neither phenomenon is observed with ratios of PS:PC less than 1:1, and neither can be induced by Mg^{2+} even at high concentrations. Recently the same conditions as those above were also found to be effective in inducing fusion of PS/PC vesicles by following the mixing of two membrane-embedded proteins initially reconstituted in separate vesicles (Miller and Racker, 1976). This latter study established that calcium was more effective in inducing fusion between PS/phosphatidylethanolamine (PE) vesicles compared with PS/PC vesicles. It is not known, however, whether this effect is related to a greater effectiveness of Ca^{2+} in inducing phase separations in a PS/PE membrane system.

It is evident from the experiments discussed above that a strong correlation exists between aggregation of vesicles, an increase in their permeability, the onset of phase changes and membrane fusion. All these events are initiated in a highly cooperative fashion by increasing Ca^{2+} (and only in certain cases by Mg^{2+}) concentrations. The Ca^{2+} concentrations needed for these effects vary widely with different phospholipids or mixtures, generally requiring a high surface charge density (negative). Charge neutralization, which generally leads to aggregation of vesicles, is not sufficient per se to induce fusion. The additional key event required for fusion of lipid bilayer membranes seems to be the phase transition from fluid to crystalline (i.e., condensed acyl chain packing) that is induced by Ca^{2+} isothermally at concentrations above a threshold. Recently it has been proposed (Papahadjopoulos et al., 1977) that the addition of Ca^{2+} induces a transient unstable state during which the lipid bilayer is highly susceptible to fusion, but which becomes stable again after Ca^{2+} equilibration or removal.

The molecular events responsible for creating this unstable state were discussed in detail recently (Papahadjopoulos et al., 1977) and could involve the following events: (1) Ca^{2+} asymmetry across the phospholipid bilayer membrane; (2) the establishment of structural discontinuities (phase boundaries) between two different phospholipid domains (solid and fluid) in the plane of the membrane; and (3) transient local release of heat liberated by the exothermic crystallization of the phospholipid acyl chain. Fig. 4 summarizes diagramatically the concepts and proposals outlined above. In this diagram, A represents a phospholipid vesicle in its resting (stable) fluid state at temperatures above the Tc of the bilayer. The vesicle is composed either of pure acidic phospholipids (PS, PA, PG) or a mixture of acidic and neutral lipids (PS, PC). Addition of Ca^{2+} in the external aqueous space can produce a phase transition (B) or a phase separation (B'), thus inducing complete (B) or partial (B') crystallization of the outer mono-

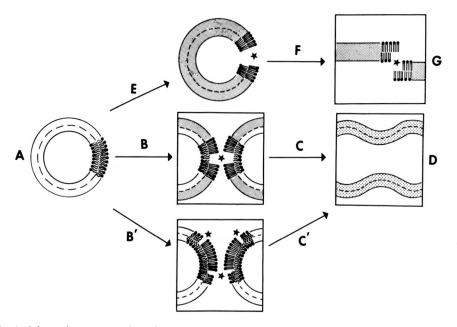

Fig. 4. Schematic representation of proposed events in fusion between phospholipid vesicles resulting from Ca²⁺-induced phase changes. The stippled areas of the vesicle membranes indicate crystallized domains; the broken lines indicate the position of the bilayer midplanes; and the asterisks indicate structural defects, domain boundaries, or regions of transient hydrocarbon-water contact. As indicated in the text, these transient energy states may be generated by the mismatch of monolayer regions that have undergone a Ca²⁺-induced phase transition (separation) with the remaining regions of the vesicle. Phospholipid molecules in fluid domains are represented by "wavy" acyl chains and in solid domains by straight acyl chains. (Reproduced with permission from Papahadjopoulos et al., 1977.)

layer of the vesicle membrane. This molecular condensation may well create structural defects with resulting exposure of the hydrocarbon chains to water. The instability of such structures would make them susceptible to fusion that would occur at such points of hydrocarbon-water contact (C and C'). The fused vesicle (D) would be a much more stable structure as a result of Ca²⁺ having gained access to the inner compartment and possible in-out molecular rearrangements during the fusion events might create vesicles less susceptible to further fusion.

7. The possible role of calcium-induced phase changes in triggering fusion of natural membranes

Despite the bewildering diversity of the many examples of membrane fusion phenomena discussed elsewhere in this volume, certain similarities warrant consideration of the possibility that there is an underlying key mechanism common to most if not all forms of membrane fusion, though the stimulus(i) responsible for triggering this key reaction will vary in different examples of membrane fusion.

Earlier proposals for the mechanism of membrane fusion, such as the induction of membrane instability via removal of calcium from membranes (Poste and Allison, 1973) or via an increase in the fluidity of membrane lipids (Ahkong et al., 1975; Lucy, 1975), do not provide enough detailed information at the molecular level and/or are not consistent with the data now available. To date the most detailed information available concerning the molecular mechanism(s) of membrane fusion has come from recent work on the fusion of model membranes in cell-free systems reviewed in this chapter. These studies have permitted the relative contribution of membrane lipids, proteins, and metal ions in influencing the fusion reaction to be defined in greater detail than has hitherto been possible with natural membranes. The key event in the fusion of model membranes is presently interpreted (Papahadjopoulos et al., 1977) as involving the interaction of Ca^{2+} with membrane lipids to induce a close apposition of two membranes and to create rigid, crystalline domains of phospholipids following Ca^{2+}-induced lateral phase separations of acidic phospholipids within a mixed lipid membrane. The formation of such domains envisaged to produce an unstable, highly permeable membrane susceptible to fusion with other similarly perturbed membranes. Although the structural organization of model membranes is acknowledged to be considerably less complex than that of natural membranes, it is instructive to examine whether similar Ca^{2+}-induced phase separation of acidic phospholipids into rigid crystalline domains might provide the mechanism underlying fusion occurring in natural membranes.

In summarizing the work discussed above on the fusion of phospholipid vesicles, the following general points can be made regarding the factors that influence fusion (Papahadjopoulos et al., 1976a,b, 1977, 1978).

1. In the absence of Ca^{2+}, purified phospholipid membranes in the form of closed vesicles are highly stable at temperatures both above and below the gel-to-liquid crystalline transition temperature of the lipids in the vesicle membrane. Such vesicles do not fuse or undergo changes in size and composition via the exchange of phospholipid molecules between adjacent vesicles.
2. Modification of the structural organization of the lipid bilayer of such vesicles by the introduction of exogenous amphipathic molecules such as lysolecithin or myristic acid (at temperatures when the vesicles are fluid, $T > T_c$) can increase the rate of phospholipid exchange between vesicles but fusion probably does not occur in the absence of Ca^{2+}.
3. Aggregation and fusion of vesicles prepared from pure acidic phospholipids (PS, PA, PG) can be induced simply by the addition of Ca^{2+} at temperatures where the vesicle lipids are "fluid." The concentration of Ca^{2+} required to induce fusion ranges from 10^{-2} M to 10^{-4} M, depending

on the phospholipid head group involved. However, in contrast to the high susceptibility of vesicles containing acidic phospholipids to Ca^{2+}-induced fusion, vesicles composed of neutral phsopholipids (lecithins) do not fuse under these conditions.

4. The fusion of vesicles containing acidic phospholipids (PS, PG) is highly specific for Ca^{2+}. Mg^{2+} causes rapid vesicle aggregation but fusion only occurs under special conditions.

5. The presence of proteins and other amphipathic molecules can increase phospholipid exchange between vesicles and perhaps induce fusion, but this phenomenon is enhanced at a temperature where the lipid bilayers are undergoing a phase transition.

Several general conclusions can be drawn from these observations (Papahadjopoulos et al., 1978).

First, it is obvious that the close apposition of two membranes is an essential prerequisite for fusion. However, the finding that vesicles can be aggregated without concomitant fusion indicates that the close apposition of two lipid bilayers does not automatically lead to fusion even when the lipids in both membranes are in a fluid state. The phospholipid bilayer is thus viewed as having very high structural stability. Consequently, fusion of apposed membranes *requires an additional step* in which the stability of the membranes is altered to render them susceptible to fusion. Papahadjopoulos and associates (1977, 1978) consider that this additional step, *and the crucial event in fusion,* is a Ca^{2+}-induced lateral phase separation which results in the formation of rigid crystalline domains of acidic phospholipids within a mixed lipid membrane. It is proposed that membrane fusion is initiated between closely apposed membranes at the boundaries between crystalline domains of acidic phospholipids and the surrounding noncrystalline lipid. Such boundaries represent structurally unstable points of high free energy and thus offer focal points for mixing of molecules from the apposed membranes.

Ca^{2+} is therefore envisaged as playing a dual role in membrane fusion. First, Ca^{2+} will promote the close apposition of adjacent membranes by enhancing electrostatic interactions between them. Second, Ca^{2+} will induce destabilization of the apposed membranes by inducing the formation of domains of crystalline acidic phospholipids whose boundaries represent sites at which fusion can occur. If Ca^{2+}-induced phase separation of acidic phospholipids is actually the key event underlying membrane fusion, then the fusion susceptibility of membranes will be determined by (a) the content of acidic phospholipids in the membrane and also by their distribution within the bilayer (see below); and (b) by the concentration of Ca^{2+} in the vicinity of the membrane.

The distribution of acidic phospholipids in the two halves (monolayers) of the lipid bilayer would be expected to influence whether a cellular membrane would be able to fuse with other membranes from only one side or from both sides. For example, in the most extreme situation, insufficient acidic phospholipids might be present in either the inner or outer half of the bilayer so that Ca^{2+}-induced phase separation of acidic phospholipids could not occur and the membrane would remain resistant to fusion.

Since acidic phospholipids vary in their ability to undergo Ca^{2+}-induced phase separation, the fusion susceptibility of any given membrane may be regulated not only by the absolute content of acidic phospholipids but also by the specific distribution within the bilayer of those acidic phospholipid species that undergo phase separation in the presence of Ca^{2+}. For example, PS and PA undergo lateral phase separation to form rigid crystalline domains in mixed lipid membranes at Ca^{2+} concentrations similar to those found under physiological conditions. In contrast, PG (and perhaps PI) does not undergo lateral phase separation at Ca^{2+} concentrations in the physiologic range. Thus the content of PS (and also perhaps of PA) would be a major factor in determining the fusion susceptibility of a particular membrane, while the preferential distribution of PS in either the outer or inner half of the bilayer would further dictate that different sides of the same membrane could vary significantly in their fusion susceptibility.

This point is particularly pertinent to the fusion behavior of the plasma membrane. Studies of plasma membrane lipid composition in mammalian erythrocytes have revealed that PS is present almost exclusively in the inner (cytoplasmic) half of the bilayer, while the outer (external) lipid monolayer is composed predominantly of neutral phospholipids (PC and sphingomyelin) (reviews, Bergelson and Barsukov, 1977; Rothman and Lenard, 1977). While it is not known whether similar lipid asymmetry is present in the plasma membrane of other cells, this type of asymmetry in the distribution of acidic phospholipids, if widespread, has major implications for the fusion behavior of the plasma membrane. The hypothesis outlined above which predicts the relative absence of PS in the outer lipid monolayer of the plasma membrane would dictate that Ca^{2+} in the external medium could not induce lateral phase separation of acidic phospholipids into fusion-susceptible domains. The membrane would thus be resistant to fusion initiated at the outer membrane face such as cell-to-cell fusion or the fusion of lipid-enveloped viruses or lipid vesicles with the cellular plasma membrane (see chapter by Poste and Pasternak, this volume).

In contrast, the inner monolayer of the plasma membrane with its higher content of PS would be expected to form such domains when exposed to the necessary concentration of Ca^{2+} and thus be highly susceptible to fusion. Under normal conditions, however, the concentration of Ca^{2+} in the cytoplasm ($<10^{-5}$ M) is at least two orders of magnitude (and often more) lower than that in the external medium (10^{-3} M). This lower intracellular calcium concentration would thus be insufficient to induce the lipid phase separation required for fusion. Although Mg^{2+} is present intracellularly at higher concentrations (10^{-3} M), Mg^{2+} cannot substitute for Ca^{2+} in inducing phase separation and the membrane therefore remains resistant to fusion.

The fusion susceptibility of the two faces of the plasma membrane would change if the distribution of acidic phospholipids within the bilayer were altered and/or the concentration of Ca^{2+} in the cytoplasm was increased. For example, if the content of acidic phospholipids in the outer lipid monolayer of the plasma membrane were to increase as a result of structural rearrangements within the bilayer, then Ca^{2+} in the external medium would induce the formation of crystalline acidic phospholipid domains and the membrane would become susceptible to fusion with other similarly perturbed membranes. This proposal, that fusion initiated at the outer face of the plasma membrane might require enrichment of the outer lipid monolayer with acidic phospholipids, is particularly pertinent in relation to the possible mechanism underlying cell-to-cell fusion induced by lysolecithin and other amphipathic molecules and also to the fusion of lipid-enveloped viruses and lipid vesicles with the plasma membrane (review, Papahadjopoulos et al., 1978 and chapter by Poste and Pasternak, this volume).

The major assumption made above is that the degree of transbilayer asymmetry in the distribution of acidic phospholipids is so extreme that the outer face of the plasma membrane is inherently resistant to fusion and that fusion can occur only when the content of acidic phospholipids in the outer monolayer is increased via some form of redistribution of acidic phospholipids from the inner to the outer monolayer, or via the transfer of acidic phospholipids from adjacent membranes by exchange processes.

An alternative possibility, however, is that although lipid asymmetry may exist, the difference in the distribution of acidic phospholipids between the outer and inner monolayers is not as extreme as that suggested above and that acidic phospholipids are actually present in the outer monolayer in sufficient amounts to undergo Ca^{2+}-induced lateral phase separation. For example, acidic phospholipids are now known to be present as an annulus, or shell, around certain integral membrane proteins (Warren et al., 1975a,b; Armitage et al., 1977; Boggs et al., 1977), and it seems likely that at least some of these molecules would be present in the outer monolayer. Consequently, it may well be that sufficient acidic phospholipids are already present in the outer monolayer of the plasma membrane to permit their separation into domains by Ca^{2+} without the need for enrichment of the acidic phospholipid content in the outer monolayer.

Since the Ca^{2+} concentration of the external medium is sufficient to induce lateral phase separation of acidic phospholipids, this would mean that the plasma membrane would be in a "fusion-susceptible" state. At first sight this reasoning is difficult to reconcile with the apparent lack of spontaneous fusion processes involving the plasma membrane. However, the fusion resistance of the plasma membrane could be more apparent than real and the low frequency of spontaneous fusion is determined by some additional restraint. For example, simple steric factors could be operating to frustrate the establishment of the required initial close contact between the lipid bilayers of the plasma membrane and other membranes. In cell-to-cell fusion this type of steric hindrance could perhaps be mediated by the various glycoproteins and glycosaminoglycans present in the "cell coat," which would provide a "barrier" to a surface projection by one cell readily achieving direct apposition of its bilayer with that on another cell. It is of interest that enzymic treatment of cells to remove surface glycoproteins and/or glycosaminoglycans renders cells with a fusion-resistant phenotype susceptible to fusion by viruses (review, Poste, 1972). Similarly, enzymic removal of glycoprotein "coating factors" from the surface of uncapacitated sperm enables them to fuse prematurely with or without undergoing capacitation (review, Gwatkin, 1976). Also, the observation (Diacumakos, 1973) that cultured mammalian somatic cells will fuse spontaneously if the cells are merely "brought together" by micromanipulation (in Ca^{2+}-containing medium) suggests that the plasma membrane may indeed be in a fusion-susceptible state.

In intracellular fusion events involving membrane-bound organelles, either with each other or with the inner face of the plasma membrane (e.g., exocytosis), changes in intracellular Ca^{2+} concentration alone would be sufficient to trigger fusion *without* the need for redistribution of acidic phospholipids within the bilayer. The inner face of the plasma membrane and the outer face of the membranes surrounding the various organelles that undergo fusion could contain a high percentage of acidic phospholipids such as PS (Bretscher, 1972; Gordesky and Marinetti, 1973; Zwaal et al., 1973; Marinetti and Love, 1974; Nilsson and Dallner, 1977; Rothman and Lenard, 1977) and, when exposed to the necessary concentration of Ca^{2+}, would be expected to undergo *immediate* lateral phase separation to create fusion-susceptible crystalline domains of acidic phospholipids. Such membranes are not continually undergoing fusion because the intracellular Ca^{2+} concentration is too low for phase separation and domain formation to take place. However, under conditions where the Ca^{2+} concentration in the vicinity of such membranes is raised to a level sufficient to trigger phase separation, they will immediately become fusion-susceptible. This scheme is compatible with the very rapid fusion of secretory granules with each other and with the plasma membrane induced in secretory cells by treatment with the Ca^{2+} ionophore A23187 which increases intracellular Ca^{2+} levels (Cochrane and Douglas, 1974; Eimerl et al., 1974; Garcia et al., 1975; Williams and Chandler, 1975; Gwatkin, 1976). Similarly, *immediate* fusion of isolated secretory granules in cell-free systems occurs following addition of Ca^{2+} to the solution surrounding the granules (Dahl and Gratzl, 1976; Vacquier, 1976; Schober et al., 1977).

It is proposed therefore that the crucial event responsible for triggering most, and probably all, membrane fusion phenomena is Ca^{2+}-induced separation of acidic phospholipids such as PS into rigid crystalline domains with fusion occurring between domain boundaries on closely apposed membranes (Papahadjopoulos et al., 1977, 1978; Papahadjopoulos and Poste, 1978). Under normal conditions most membranes do not fuse. Clearly, this is necessary to prevent uncontrolled spontaneous fusion of different cells within tissues and also for ensuring that the various membrane-bound organelles retain their structural identity, enabling different functional activities to be localized within specific compartments of the cell. It is proposed that membrane fusion does not occur under normal conditions because of: (1) steric hindrance, which frustrates close apposition between two "fusion-susceptible" membranes containing Ca^{2+}; or (2) preexisting asymmetry in the distribution of acidic phospholipids within the membrane, which dictates that Ca^{2+} cannot induce fusion susceptible "domains" of acidic phospholipids; or (3) an insufficient Ca^{2+} concentration in the vicinity of the membrane. The latter two factors constitute a "forbidding asymmetry" which ensures that acidic phospholipids cannot be segregated into rigid domains to render membranes unstable and susceptible to fusion. Fusion can, however, be initiated by: (1) removal of steric restraints to the close apposition two "fusion-susceptible" membranes (i.e., both membranes contain preexisting Ca^{2+}-acidic phospholipid domains); or (2) changes in the distribution of acidic phospholipids within the bilayer that permit the existing (extracellular) concentration of Ca^{2+} in the vicinity of the membrane to induce separation of acidic phospholipids (this is proposed as the mechanism operating in certain fusion events initiated at the outer face of the plasma membrane, such as cell-to-cell fusion, or fusion of lipid-enveloped viruses, or vesicles with the plasma membrane); or (3) changes in the Ca^{2+} concentration (intracellular) in the vicinity of membranes that contain sufficient preexisting amounts of acidic phospholipids to undergo immediate phase separation to create the fusion-susceptible acidic phospholipid domains (this is proposed as the mechanism underlying fusion of intracellular organelles with each other and/or with the inner face of the plasma membrane following influx of Ca^{2+} into the cell).

This proposal for the induction of membrane fusion via Ca^{2+}-induced phase separation of acidic phosopholipids offers a single mechanism which can explain both extracellular fusion (cell-to-cell; enveloped virus-to-cell; lipid vesicle-to-cell) as well as intracellular fusions (endocytosis; exocytosis). However, this chapter has merely presented an abbreviated outline of how Ca^{2+}-induced phase separations in membrane lipids could explain different classes of membrane fusion behavior. A fuller discussion of this hypothesis, together with a detailed analysis of its applicability to membrane fusion behavior in cellular and subcellular situations, has been presented recently in two extensive reviews (Papahadjopoulos and Poste, 1978; Papahadjopoulos et al., 1978).

References

Abramson, M. B., Katzman, R., Gregor, H. and Curci, R. (1966) The reaction of cations with aqueous dispersions of phosphatidic acid. Determination of stability constants. Biochemistry 5, 2207–2213.

Ahkong, Q. F., Cramp, F. C., Fisher, D., Howell, J. I. and Lucy, J. A. (1972) Studies on chemically induced cell fusion. J. Cell. Sci. 10. 769–787.

Ahkong, Q. F., Fisher, D., Tampion, W. and Lucy, J. A. (1973) The fusion of erythrocytes by fatty acids, esters, retinol and α-tocopherol. Biochem. J. 136, 147–155.

Ahkong, Q. F., Fisher, D., Tampion, W. and Lucy, J. A. (1975) Mechanisms of cell fusion. Nature 253, 194–195.

Armitage, I. M., Shapiro, D. L., Furthmayr, H. and Marchesi, V. T. (1977) [31]NMR evidence for polyphosphoinositide associated with the hydrophobic segment of glycophorin A. Biochemistry 16, 1317–1320.

Bangham, A. D. (1972) Lipid bilayers and biomembranes. Ann. Rev. Biochem. 41, 753–776.

Bangham, A. D. and Papahadjopoulos, D. (1966) Biophysical properties of phospholipids. I. Interaction of phosphatidylserine monolayers with metal ions. Biochim. Biophys. Acta 126, 181–184.

Bangham, A. D., Hill, M. W. and Miller, N. G. A. (1974) Preparation and use of liposomes as models of biological membranes. In: Methods in Membrane Biology (Korn, E. D., ed.) vol. I, pp. 1–68. Plenum Press, New York.

Bergelson, L. D. and Barsukov, L. I. (1977) Topological asymmetry of phospholipids in membranes. Science 197, 224–230.

Boggs, J., Moscarello, M. A. and Papahadjopoulos, D. (1977) Lipid phase separation induced by a hydrophobic protein in PS-PC vesicles. Biochemistry 16, 2325–2329.

Breisblatt, W. and Ohki, S. (1975) Fusion in phospholipid spherical membranes. I. Effect of temperature and lysolecithin. J. Membr. Biol. 23, 385–401.

Breisblatt, W. and Ohki, S. (1976) Fusion in phospholipid spherical membranes: effect of cholesterol, divalent ions and pH. J. Membr. Biol. 29, 127–146.

Bretscher, M. S. (1972) Asymmetrical lipid bilayer structure for biological membranes. Nature New Biol. 236, 11–12.

Cochrane, D. E. and Douglas, W. W. (1974) Calcium-induced extrusion of secretory granules (exocytosis) in mast cells exposed to 48/80 or the ionophores A-23187 and X-537A. Proc. Nat. Acad. Sci. U.S.A. 71, 408–412.

Croce, C. M., Sawicki, W., Kritchevsky, D. and Koprowski, H. (1971) Induction of homokaryocyte, heterokaryocyte and hybrid formation by lysolecithin. Exp. Cell Res. 67, 427–435.

Dahl, G. and Gratzl, M. (1976) Calcium-induced fusion of isolated secretory vesicles from the islet of Langerhans. Cytobiologie 12, 344–355.

Diacumakos, E. G. (1973) Methods for micromanipulation of human somatic cells in culture. In: Methods in Cell Biology (Prescott, D. M., ed.) vol. VII, pp. 287–311, Academic Press, New York.

Eimerl, S., Savion, N., Heichal, O. and Selinger, Z. (1974) Induction of enzyme secretion in rat pancreatic slices using the ionophore A23187 and calcium. J. Biol. Chem. 249, 3991–3993.

Folch, J. and Lees, M. (1951) Proteolipids, a new type of tissue lipoproteins. J. Biol. Chem. 191, 807–817.

Gagnon, J., Finch, P. R., Wood, D. D. and Moscarello, M. A. (1971) Isolation of a highly purified myelin protein. Biochemistry 10, 4756–4763.

Galla, H.-J. and Sackmann, E. (1975) Chemically induced lipid phase separation in model membranes containing charged lipids: a spin label study. Biochim. Biophys. Acta 401, 509–529.

Garcia, A. G., Kirpekar, S. M. and Prat, J. C. (1975) A calcium ionophore stimulating the secretion of catecholamines from the cat adrenal. J. Physiol. 244, 253–262.

Gordesky, S. E. and Marinetti, G. V. (1973) The asymmetric arrangement of phospholipids in the human erythrocyte membrane. Biochem. Biophys. Res. Commun. 50, 1027–1031.

Gwatkin, R. B. L. (1976) Fertilization. In: The Cell Surface in Animal Embryogenesis and Development (Poste, G. and Nicolson, G. L., eds.), Cell Surface Reviews, vol. 1, pp. 1–54. Elsevier/North-Holland, Amsterdam.

Hauser, H., Finer, E. G. and Chapman, D. (1970) Nuclear magnetic resonance studies of the polypeptide alamethicin and its interaction with phospholipids. J. Mol. Biol. 53, 419–433.

Hendrickson, H. S. and Fullington, J. G. (1965) Stabilities of metal complexes of phospholipids: Ca(II), Mg(II) and Ni(II) complexes of phosphatidylserine and triphosphoinositide. Biochemistry 4, 1599–1605.

Howell, J. I. and Lucy, J. A. (1969) Cell fusion induced by lysolecithin. FEBS Lett. 4, 147–150.

Howell, J. I., Ahkong, Q. F., Cramp, F. C., Fisher, D., Tampion, W. and Lucy, J. A. (1972) Membrane fluidity and membrane fusion. Biochem. J. 130, 44P–45P.

Inoue, K. and Kitagawa, T. (1974) Effect of exogenous lysolecithin on liposomal membranes: Its relation to membrane fluidity. Biochim. Biophys. Acta 363, 361–372.

Ito, T. and Ohnishi, S.-I. (1974) Ca^{2+} induced lateral phase separations in phosphatidic acid-phosphatidylcholine membranes. Biochim. Biophys. Acta 352, 29–37.

Jacobson, K. and Papahadjopoulos, D. (1975) Phase transitions and phase separations in phospholipid vesicles, induced by changes in temperature, pH and concentration in bivalent metals. Biochemistry 14, 152–161.

Kantor, H. L. and Prestegard, J. (1975) Fusion of fatty acid containing lecithin vesicles. Biochemistry 14, 1790–1794.

Lansman, J. and Haynes, D. A. (1975) Kinetics of a Ca^{2+}-triggered membrane aggregation reaction of phospholipid membranes. Biochim. Biophys. Acta 394, 335–347.

788

Lau, A. L. Y. and Chan, S. I. (1974) Nuclear magnetic resonance studies of the interaction of alamethicin and lecithin bilayers. Biochemistry 13, 4942–4948.

Lau, A. L. Y. and Chan, S. I. (1975) Alamethicin-mediated fusion of lecithin vesicles. Proc. Nat. Acad. Sci. U.S.A. 72, 2170–2174.

Lau, A. L. Y. and Chan, S. I. (1976) Voltage-induced formation of alamethicin pores in lecithin bilayer vesicles. Biochemistry 15, 2551–2555.

Lawaczek, R., Kainosho, M., Girardet, J.-L. and Chan, S. I. (1975) Effects of structural defects in sonicated phospholipid vesicle on fusion and ion permeability. Nature 256, 584–586.

Lawaczek. R., Kainasho, M. and Chan, S. I. (1976) The formation and annealing of structural defects in lipid bilayer vesicles. Biochim. Biophys. Acta 443, 313–330.

Le Neveu, D. M., Rand, R. P. and Parsegian, V. A. (1976) Measurement of forces between lecithin bilayers. Nature 259, 601–603.

Liberman, E. A. and Nenashev, V. A. (1972) Kinetics of adhesion and surface electrical conductivity of bimolecular phospholipid membranes. Biofizika 17 (2), 231–238.

Lucy, J. A. (1975) Aspects of the fusion of cells in vitro without viruses. J. Reprod. Fertil. 44, 193–205.

Lyman, G. H., Papahadjopoulos, D. and Preisler, H. D. (1976a) Phospholipid membrane stabilization by dimethylsulfoxide and other inducers of Friend leukemic cell differentiation. Biochim. Biophys. Acta 448, 460–473.

Lyman, G. H., Preisler, H. D. and Papahadjopoulos, D. (1976b) Membrane action of DMSO and other chemical inducers of Friend leukemic cell differentiation. Nature 262, 360–363.

Maeda, T. and Ohnishi, S.-I. (1974) Membrane fusion transfer of phospholipid molecules between phospholipid bilayer membranes. Biochem. Biophys. Res. Commun. 60, 1509–1516.

Marcelja, S. and Radic, N. (1976) Repulsion of interfaces due to boundary water. Chem. Phys. Lett. 42, 199–230.

Marinetti, G. V. and Love, R. (1974) Extent of cross-linking of amino-phospholipid neighbors in the erythrocyte membrane as influenced by the concentration of difluorodinitrobenzene. Biochem. Biophys. Res. Commun. 61, 30–37.

Martin, F. J. and MacDonald, R. C. (1976a) Phospholipid exchange between bilayer membrane vesicles. Biochemistry 15, 321–327.

Martin, F. J. and MacDonald, R. C. (1976b) Lipid vesicle-cell interactions. II. Induction of cell fusion. J. Cell Biol. 70, 506–514.

McIntyre, J. A., Gilula, N. B. and Karnovsky, M. J. (1974) Cryoprotectant-induced redistribution of intramembranous particles in mouse lymphocytes. J. Cell Biol. 60, 192–203.

McLaughlin, S. G. A., Szabo, G. and Eisenman, G. (1971) Divalent ions and the surface potential of charged phospholipid. J. Gen. Physiol. 58, 667–687.

Miller, C. and Racker, E. (1976) Fusion of phospholipid vesicles reconstituted with cytochrome C oxidase and mitochondrial hydrophobic protein. J. Membr. Biol. 26, 319–333.

Moore, M. R. (1976) Fusion of liposomes containing conductance probes with black lipid films. Biochim. Biophys. Acta 426, 765–771.

Mueller, P. and Rudin, R. O. (1968) Alamethicin, a cyclopeptide, can induce action potentials in biomolecular lipid membranes. Nature 217, 713–719.

Neher, E. (1974) Asymmetric membranes resulting from the fusion of two black lipid membranes. Biochim. Biophys. Acta 373, 327–336.

Nicolson, G. L. (1974) The interaction of lectins with animal cell surfaces. Int. Rev. Cytol. 39, 89–190.

Nilsson, O. S. and Dallner, G. (1977) Transverse asymmetry of phospholipids in subcellular membranes of rat liver. Biochim. Biophys. Acta 464, 453–458.

Ohnishi, S.-I. and Ito, T. (1973) Clustering of lecithin molecules in phosphatidylserine membranes induced by calcium ion binding to phosphatidylserine. Biochem. Biophys. Res. Commun. 51, 132–138.

Ohnishi, S.-I. and Ito, T. (1974) Calcium-induced phase separations in phosphatidylserine-phosphatidylcholine membranes. Biochemistry 13, 881–887.

Papahadjopoulos, D. (1968) Surface properties of acidic phospholipids: Interaction of monolayers

Papahadjopoulos, D. and Kimelberg, H. K. (1973) Phospholipid vesicles (liposomes) as models for biological membranes: their properties and interactions with cholesterol and proteins. In: Progress in Surface Science (Davison, S. G., ed.) vol. 4, part 2, pp. 141–232. Pergamon Press, Oxford, England.

Papahadjopoulos, D. and Ohki, S. (1969) Stability of asymmetric phospholipid bilayers. Science 164, 1075–1077.

Papahadjopoulos, D. and Poste, G. (1975) Calcium-induced phase separation and fusion in phospholipid membranes. Biophys. J. 15, 945–948.

Papahadjopoulos, D. and Poste, G. (1978) Membrane fusion. Int. Rev. Cytol. In press.

Papahadjopoulos, D., Poste, G., Schaeffer, B. E. and Vail, W. J. (1974) Membrane fusion and molecular segregation in phospholipid vesicles. Biochim. Biophys. Acta 352, 10–28.

Papahadjopoulos, D., Vail, W. J., Jacobson, K. and Poste, G. (1975a) Cochleate lipid cylinders: formation by fusion of unilamellar lipid vesicles. Biochim. Biophys. Acta 394, 483–491.

Papahadjopoulos, D., Moscarello, M., Eylar, E. H. and Isac, T. (1975b) Effects of proteins on thermotropic phase transitions of phospholipid membranes. Biochim. Biophys. Acta 401, 317–335.

Papahadjopoulos, D., Vail, W. J. and Moscarello, M. A. (1975c) Interaction of a purified hydrophobic protein from myelin with phospholipid membranes: studies on ultrastructure, phase transitions and permeability. J. Membr. Biol. 22, 143–184.

Papahadjopoulos, D., Hui, S., Vail, W. J. and Poste, G. (1976a) Studies on membrane fusion. I. Interactions of pure phospholipid membranes and the effect of myristic acid, lysolecithin, proteins and DMSO. Biochim. Biophys. Acta 448, 245–264.

Papahadjopoulos, D., Vail, W. J., Pangborn, W. A. and Poste, G. (1976b) Studies on membrane fusion. II. Induction of fusion in pure phospholipid membranes by calcium and other divalent metals. Biochim. Biophys. Acta 448, 265–283.

Papahadjopoulos, D., Vail, W. J., Newton, C., Nir, S., Jacobson, K., Poste, G. and Lazo, R. (1977) Studies on membrane fusion. III. The role of calcium-induced phase changes. Biochim. Biophys. Acta 465, 579–598.

Papahadjopoulos, D., Poste, G. and Vail, W. J. (1978) Studies on membrane fusion with natural and model membranes. In: Methods in Membrane Biology (Korn, E., ed.). Plenum Press, New York. In press.

Payne, J. W., Jakes, R. and Hartley, B. S. (1970) The primary structure of alamethicin. Biochem. J. 117, 757–766.

Pohl, G. W., Stack, G. and Trissl, H. W. (1973) Interaction of liposomes with black lipid membranes. Biochim. Biophys. Acta 318, 478–481.

Poste, G. (1972) Mechanisms of virus-induced cell fusion. Int. Rev. Cytol. 33, 157–252.

Poste, G. and Allison, A. C. (1973) Membrane fusion. Biochim. Biophys. Acta 300, 421–465.

Poznansky, M. J. and Weglicki, W. B. (1974) Lysophospholipid induced volume changes in lysosomes and in lysosomal lipid dispersions. Biochem. Biophys. Res. Commun. 58, 1016–1021.

Preisler, H. D., Housman, D., Scher, W. and Friend, C. (1973) Effects of 5-bromo-2'-deoxyuridine on production of globin messenger RNA in dimethyl sulfoxide-stimulated Friend leukemia cells. Proc. Nat. Acad. Sci. U.S.A. 70, 2956–2959.

Prestegard, J. H. and Fellmeth, B. (1974) Fusion of dimyristoyl lecithin vesicles as studied by proton magnetic resonance spectroscopy. Biochemistry 13, 1122–1126.

Prestegard, J. H. and Fellmeth, B. (1975) Fusion of fatty acid containing lecithin vesicles. Biochemistry 14, 1790–1794.

Rand, R. P., Pangborn, W. A., Purdon, A. D. and Tinker, D. O. (1975) Lysolecithin and cholesterol interact stoichiometrically forming bimolecular lamellar structures in the presence of excess water, or lysolecithin or cholesterol. Can. J. Biochem. 53, 189–195.

Rothman, J. E. and Lenard, J. (1977) Membrane asymmetry. Science 195, 743–753.

Rubin, R. P. (1975) Calcium and the Secretory Process. Plenum Press, New York.

Schober, R., Nitsch, C., Rinne, V. and Morris, S. J. (1977) Calcium-induced displacement of membrane-associated particles upon aggregation of chromatin granules. Science 195, 495–497.

Suurkuusk, J., Lentz, B. R., Berenholz, Y., Biltonen, R. L. and Thompson, T. E. (1976) A calorimetric and fluorescent probe study of the gel-liquid crystalline phase transition in small, single-lamellar dipalmitoyl phosphatidylcholine vesicles. Biochemistry 15, 1393–1401.

Taupin, C. and McConnell, H. M. (1972) Membrane fusion. In: Mitochondria-Biomembranes. Fed. Eur. Biochem. Soc. Symp. 8th Mtg., Amsterdam, (van den Burgh, S. G., Borst, P., van Deenen, L. L. M., Riemeroma, J. C., Slater, E. C. and Tager, J. M., eds.) pp. 219–220, Elsevier/North-Holland, Amsterdam.

Thompson, T. E. and Henn, F. A. (1970) Experimental phospholipid model membranes. In: Membranes of Mitochondria and Chloroplasts (Racker, E., ed.) pp. 1–52, Van Nostrand Reinhold Co., New York.

Träuble, H. and Eibl, H. (1974) Electrostatic effects on lipid phase transitions: membrane structure and ionic environment. Proc. Nat. Acad. Sci. U.S.A. 71, 214–218.

Vacquier, V. D. (1976) Isolated cortical granules: a model system for studying membrane fusion and calcium-mediated exocytosis. J. Supramol. Struc. 5, 27–35.

Vail, W. J., Papahadjopoulos, D. and Moscarello, M. A. (1974) Interaction of a hydrophobic protein with liposomes. Evidence for particles seen in freeze fracture as being proteins. Biochim. Biophys. Acta 345, 463–467.

Van der Bosch, J. and McConnell, H. M. (1975) Fusion of dipalmitoyl phosphatidylcholine vesicle membranes induced by concanavalin A. Proc. Nat. Acad. Sci. U.S.A. 72, 4409–4413.

Van Dijck, P. W. M., Ververgaert, P. H. J. Th., Verkleij, A. J., Van Deenen, L. L. M. and De Gier, J. (1975) Influence of Ca^{2+} and Mg^{2+} on thermotropic behaviour and permeability properties of liposomes. Biochim. Biophys. Acta 406, 465–478.

Verkleij, A. J., deKruyff, B., Ververgaert, P. H. J. Th., Tocanne, J. F. and Van Deenen, L. L. M. (1974) The influence of pH, Ca^{2+} and protein on the thermotropic behavior of the negatively charged phospholipid, phosphatidylglycerol. Biochim. Biophys. Acta 339, 432–437.

Warren, G. B., Houslay, M. D., Metcalf, J. C. and Birdsall, N. J. M. (1975a) Cholesterol is excluded from the phospholipid annulus surrounding an active calcium transport protein. Nature 255, 684–687.

Warren, G. B., Metcalf, J. C., Lee, A. G. and Birdsall, N. J. M. (1975b) Mg^{2+} regulates the ATPase activity of a calcium transport protein by interacting with bound phosphatidic acid. FEBS Lett. 50, 261–264.

Williams, J. A. and Chandler, D. (1975) Ca^{2+} and pancreatic amylase release. Fed. Proc. 228, 1729–1732.

Zwaal, R. F. A., Reolofsen, B. and Colley, C. M. (1973) Localization of red cell membrane constituents. Biochim. Biophys. Acta 300, 159–182.

Problems in the physical interpretation of membrane interaction and fusion

15

David GINGELL and Lionel GINSBERG

Contents

G. Poste & G. L. Nicolson (eds.) Membrane Fusion, pp. 791–833.
© *Elsevier/North-Holland Biomedical Press, 1978.*

1. Introduction

Membranes hate edges. All our ideas of membrane transformations are based on this fact. Cellular and artificial membranes appear to form closed bags of various shapes. Holes in the bags are not tolerated; they are either repaired spontaneously or lead immediately to membrane disruption, into smaller bags. Therefore an important and intriguing question is how membranes can join and fuse in a smooth process, preserving their contents from inadvertent release while passing through what appears to be an unstable intermediate topographic form.

The experimental study of membrane fusion has been conducted in three ways, using (1) natural membrane fusion occurring under physiological conditions, (2) mutual interactions between artificial bilayer lipid membranes, and (3) the interaction between artificial lipid vesicles and natural membranes. From the point of view of physical analysis, interaction between artificial vesicles (liposomes) offers a system where components are sufficiently well defined for quantitative study, but it is perhaps the least satisfactory of the three categories regarding physiological relevance. There are two major objections leveled against artificial vesicles: first, despite assertions to the contrary (Papahadjopoulos et al., 1977) these vesicles have not been clearly shown to undergo mutual fusion. Second, they can represent thermodynamically metastable forms following sonication procedures to form small unilamellar vesicles (Gershfeld, 1976).

Although there may be better evidence for the fusion of artificial vesicles with natural membranes (Huang and Pagano, 1975; Pagano and Huang, 1975; Martin and MacDonald, 1976; Poste and Papahadjopoulos, 1976), these interactions do not readily lend themselves to physical analysis. The use of lipid vesicle interactions to parallel biological fusion is supported by recent ultrastructural observations on membrane interactions in exocytotic secretion (Satir et al., 1973; Lawson et al., 1977), which suggest that fusion takes place at a region of lipid-lipid apposition. From this area of the membrane intercalated macromolecules are largely excluded. This is important since the opposite result, that fusion occurs at regions of macromolecular aggregation, would cast doubt on all fusion studies using lipid-only systems. Consequently, we have drawn upon material from artificial and natural fusion systems.

In the following sections we discuss the evidence for vesicle fusion and give an energetic analysis of vesicle interactions. The problem of the stimulus for fusion is discussed in answer to the question of how the apposition of two membranes can trigger the profound topological change which characterizes biological fusion. Finally, we consider the topological changes in the bilayer and discuss the mechanism of macromolecular dispersal from the zone of fusion in biological membranes.

2. Theoretical and experimental aspects of vesicle fusion

2.1. Experimental evidence for bilayer lipid vesicle fusion

A major experimental problem in the analysis of artificial membrane fusion is that the methods of detection of the putative "fusion event" do not clearly distinguish between genuine fusion (where the membrane-bound internal compartments of the separate vesicles become confluent, in a process that transfers the contents of the separate vesicles to the resulting vesicle) and molecular mixing of the two participating membranes as a result of some other process. Thus, for example, nuclear magnetic resonance (NMR) studies, which claim to demonstrate vesicle fusion (Prestegard and Fellmeth, 1974; Lau and Chan, 1974), demonstrate an increase in the size of vesicles in a suspension under a variety of experimental conditions but fail to show whether this is the result of membrane fusion or the diffusion of individual phospholipid molecules from one vesicle to another (the latter may not even be mediated via adhesion of the vesicles; discussion by Papahadjopoulos et al., 1976a). Similarly, transfer of a conductance probe from a liposome to a black lipid film (Moore, 1976) does not prove that the former has fused with the latter.

The most convincing evidence to date for fusion in an artificial system has been obtained via the technique of differential scanning calorimetry (Papahadjopoulos et al., 1974, 1976a, 1976b, 1977 and chapter by Papahadjopoulos, this volume). By means of this technique it has been possible to distinguish molecular mixing of membranes as a result of diffusion of individual components from other processes occurring in vesicle populations (Papahadjopoulos et al., 1976a). The conclusion drawn from these studies is that vesicles composed of acidic phospholipids (e.g., phosphatidylglycerol, phosphatidic acid and phosphatidylserine) are able to fuse in the presence of divalent cations, the nature and concentration of the divalent cation required for fusion being dependent on the particular phospholipid employed. Surprisingly, neither adhesion nor fusion between small vesicles of phosphatidylcholine appears to take place. Since they carry no net charge and possess a zero zeta potential they might be expected to flocculate. Possibly, the repulsion of uncertain etiology between egg lecithin membranes described by Le Neveu and co-workers (1976) prevents adhesion of the vesicles. But if this is so, it is unclear how small phosphatidylcholine vesicles can adhere to cell membranes (Pagano et al., 1974).

The problem with the differential scanning technique, however, is that it fails to distinguish between genuine fusion and a process whereby vesicles rupture on addition of divalent cation, the resulting membrane fragments "reannealing" with one another to form larger structures. This alternative has been suggested by Papahadjopoulos and associates (1975) on the basis of freeze-fracture evidence, and is supported by the finding that asymmetric calcium can destabilize and lyse artificial membranes (Papahadjopoulos and Ohki, 1969). The notion that vesicle rupture is brought about by asymmetric calcium is consistent with the expected large area difference between the two halves of a bilayer that would be produced by addition of calcium to one side (effecting a fluid $\rightarrow$ gel phase transition in that monolayer while the other half of the membrane remains fluid). Clearly, if a process of rupturing ("cracking") and reannealing is occurring in vesicle populations exposed to divalent cations, the usefulness of this system as a paradigm for biological membrane fusion may be severely reduced. We have attempted to resolve the question of whether acidic phospholipid vesicles in the presence of divalent cations can fuse rather than rupture by determining whether adhesive contact of these liposomes is an energetically feasible event. The results of our calculations are outlined in section 2.2, using phosphatidylserine vesicles as a specific example. This is one of the most clearly documented experimental systems, and it has been claimed that phosphatidylserine vesicles fail to interact in the absence of calcium but that fusion occurs when there are concentrations of this cation exceeding 1 mM in the aqueous medium (Papahadjopoulos and Poste, 1975).

2.2. Energetic considerations in liposome interaction

The likelihood of two vesicles being able to come into close contact under conditions in which fusion of artificial membranes has been claimed can be estimated by applying the DLVO (Derjaguin, Landau, Verwey, Overbeek) theory of colloid particle interaction. Physical modeling of this kind has scarcely been attempted in any detail for membrane systems in which fusion follows contact, the work of Dean (1975) on chromaffin granule secretion being a rare exception.

2.2.1. Attraction between vesicles

We consider the approach of two identical liposomes as illustrated in Fig. 1. The vesicles are of equal radius (a) and are both bounded by a lipid bilayer membrane of thickness $t = 40$ Å. At the outer and inner surfaces of the boundary membranes, the lipid headgroups occupy a region of thickness $\delta = 8$ Å; thus the thickness of each membrane's hydrocarbon interior is t-2δ. The minimum distance of separation of the spheres is $2d$. For this geometry, we write the following equation for the van der Waals attraction energy G^{ed} between the two thin lipid shells (Verwey and Overbeek, 1948):

$$G^{ed} = -\frac{A}{3}\left[\left\{\frac{1}{s^2 - 4} + \frac{1}{s^2} + \frac{1}{2}\ln\left(\frac{s^2 - 4}{s^2}\right)\right\} - \left\{\frac{1}{\bar{s}^2 - 4} + \frac{1}{\bar{s}^2} + \frac{1}{2}\ln\left(\frac{\bar{s}^2 - 4}{\bar{s}^2}\right)\right\}\right], \tag{1}$$

where

$$s = 2 + \left[(2d + \delta)/\left(a - \frac{\delta}{2}\right)\right],$$

$$\bar{s} = 2 + \frac{2d + 2t - \delta}{a - t + \frac{\delta}{2}}.$$

and A, the Hamaker coefficient is assumed to maintain a value of either 3×10^{-14} erg or 6×10^{-14} erg independent of interaction distance. (Haydon and Taylor, 1968; Gingell and Parsegian, 1972)

2.2.2. Repulsion between vesicles: linear and exact methods

As the spheres approach, they also experience an electrostatic repulsion as a result of the presence of charges of like sign on their surfaces, which arise from ionization of acidic moieties in the lipid headgroup region. We have estimated the energy of repulsion G^{es} between the spheres (Fig. 1) in two ways, both of which involve the device of Derjaguin (1934) whereby it is possible to calculate the interaction energy of spheres if the interaction of infinitely large planes having the same surface properties is known. The first (linear) method, described in Appendix A, involves approximations that restrict the validity of the results to potentials less than about 25 mV. In this model the magnitude of the potential at the vesicle surfaces is considered to be a function of interaction distance, the pKa of the surface ionogenic group and the degree of counterion penetrability of the phospholipid headgroups. The second (nonlinear) method is based on an "exact" solution valid for any potential. In this case the surface potential is assumed to remain constant. The interaction energy was obtained using equation (A8) to integrate nonlinear parallel plane energies obtained from the tabulation of Devereux and de Bruyn (1963). Constant potential results were checked against values for constant surface charge density calculated by an "exact" method (Ohshima [1975] equations 45, 50, 52, using equation 57 from Ohshima [1974]). The constant charge and constant potential curves for spheres with surface potentials at infinite separation of −50 mV differ by only 5% at 12 Å and 9% at 8 Å, when the Debye length equals 8 Å. Comparison of a linear constant surface charge equation without provision for surface counterion penetrability with a nonlinear equation showed the linear method grossly overestimated the interaction energy. We also learned from the linear method that the surface potential rise, which is overestimated by this method, was insufficient to cause association of the ionogenic group, even for a surface potential at infinite separation $\psi_0^\infty = -60$ mV. Therefore use of a nonlinear equation based on constant charge or con-

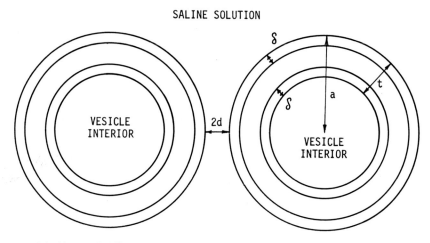

Fig. 1. Model of interacting liposomes.

stant potential entails insignificant error from failure to consider association. The only remaining difference is that the nonlinear model has no provision for surface counterion penetrability. We corrected for the resulting overestimate of the repulsive energy as follows. Linearized equations for sphere-sphere interactions at constant pKa were employed, in one case including a surface penetrability factor, in the other omitting it, giving a higher repulsion. The ratio (r) of the higher to lower energies was identical for each potential and sphere radius treated. We then reduced each corresponding nonlinear interaction energy by $r = 1.2$. Since linear equations will overestimate r, this procedure provides a maximal correction, resulting in a minimal estimate of the repulsive energy. Thus, we have upper and lower bounds to the repulsion.

2.2.3. Interaction energy

Having obtained the electrostatic and electrodynamic contributions to the total free energy of interaction of the vesicles (G^t), we can now write, for any given distance of separation,

$$G^t = G^{es} + G^{ed}. \tag{2}$$

We follow Dean (1975) in comparing G^t with the kinetic energy of translational motion of the spheres, kT, and arrive at the following expression for the fraction of particles (p) in a suspension having sufficient kinetic energy to come to a given distance of separation $2d$ from one another, where the interaction energy is $G^t(2d)$:

$$p(2d) = \exp(-G^t(2d)/kT). \tag{3}$$

This is valid if the distance of separation of the particles is greater than, or equal to , that at which a maximum free energy of interaction is located, which is the case in all our calculations. Fraction p may be computed for varying values of separation, particle radius and membrane surface potential ψ_0 (the last parameter being for membranes at infinite separation). The results of these computations are illustrated in Fig. 2. Interaction takes place in 145 mM NaCl solution at pH 7.4, at 37 °C, where the pKa of the ionizable group at the membrane surface is assumed to be 4.5 and with a Hamaker coefficient of 3×10^{-14} erg. The minimum distance of separation for the spheres in Fig. 2 was chosen to be one Debye length (8Å). This value was selected as a compromise between the need to obtain p for small separations of the vesicles (membrane fusion requiring close contact of the surfaces concerned) and the risk of applying the theory of electrical double layer interaction to distances of separation where even a "rigorous" treatment is inadequate. The latter point is clear from measurements of the forces between bilayers of phosphatidylcholine; a short-range force of repulsion was demonstrated that did not seem to show the exponential decay behavior with distance expected of classical electrostatic repulsion (Le Neveu et al., 1976).

The curves in Figs. 2a (linearized electrostatics) and 2b (nonlinear method)

797

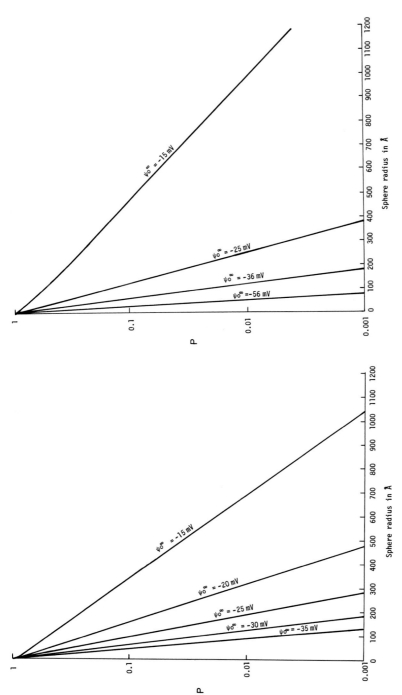

Fig. 2. Probability p of vesicle approach to 8Å as a function of vesicle radius for various surface potentials, ψ_0^∞. Curve (a) approximate (linearized) electrostatic repulsion; (b), exact (nonlinear) electrostatic repulsion.

show that at low potentials (~ -15 mV) the probabilities of adhesion predicted by the two electrostatic models are similar for vesicles of 100 Å radius, but as vesicle size or potential increases the inadequacy of the linear model becomes marked. Since linearization magnifies the repulsive interaction energy, it artifactually diminishes the probability of close approach.

The results of our calculations are best examined in the light of experimental work on vesicle fusion. In a review of much recent experimental work, Papahadjopoulos and Poste (1975) claim that phosphatidylserine (PS) liposomes will not fuse in the absence of calcium but that pronounced fusion occurs when low concentrations of this divalent cation ($\geqslant 1$ mM) are present in the suspending medium. The electrophoretic zeta potential (ζ) of PS liposomes in the absence of calcium is -55.2 mV in 145 mM-NaCl-Tris at pH 7.4. In the presence of 1 mM-CaCl$_2$, $\zeta =, -36$ mV (Papahadjopoulos, 1968, Fig. 5). Initially we expected to find that the reduction in zeta potential by 1 mM-Ca^{2+} would effect an increase in the probability of adhesion and thus explain its potentiation of fusion. Although it actually increases the probability, the latter is still small. Probabilities of approach to 8 Å at -36 mV and -55 mV for 100 Å and 200 Å radius vesicles are shown in Table 1. The attractive energy constant is set at $3-6 \times 10^{-14}/$ erg and values are given for counterion-penetrable and impenetrable lipid headgroup regions. Clearly, the probabilities are all small and fall sharply with increasing vesicle size. The table confirms that PS vesicles cannot approach close enough to fuse in the absence of calcium. If the zeta potential value of -36 mV were a reliable measure of surface potential, one could conclude that the addition of calcium would cause a small but possibly significant aggregation (and then fusion) of the smallest vesicles from a completely nonadherent state. However, the ζ-potential measurement of -36 mV probably seriously underestimates the surface potential of a highly charged surface such as pure PS. The depression of the ζ-potential at high surface charge densities when compared with theoretical predictions for ψ_0 based on Gouy-Chapman theory is clearly demonstrated for cationic surfactants in the work of Carroll and Haydon (1975), and similar results were obtained by MacDonald and Bangham (1972) for phospholipid systems. Thus it is likely that the ζ-potential measurements for PS in the absence of calcium (-55.2 mV) and in its presence (-36 mV for 1 mM-Ca^{2+}) are both minimal values, "true" membrane surface potentials exceeding them considerably (i.e., true potentials more negative). This theme is developed in section 2.2.4 and Appendix B, where an attempt is made to determine theoretical values for potentials at the surface of a vesicle made from an acidic phospholipid, (using PS as a specific example) via a refined form of the Gouy-Chapman equation (Parsegian and Gingell, 1973). If the measured ζ-potentials underestimate the surface potential the probabilities of approach to 8 Å (and thence of fusion) would be even less than we have calculated. It is quite clear, however, that 1 mM-Ca^{2+} does not reduce the surface charge density to zero (as implied by Papahadjopoulos et al., 1977), and we conclude that the addition of 1 mM-Ca^{2+} is not expected to reduce the potential energy barrier to close-approach to a value where the vast majority of vesicles can adhere and fuse. Only a tiny fraction of the smallest, with a radius of ~ 100 Å, may do so.

TABLE 1
Probability of approach to 8 Å

Attractive energy coefficient	$\zeta = -36$ mV				$\zeta = -55$ mV			
	Penetrable surface		Impenetrable surface		Penetrable surface		Impenetrable surface	
	$a = 100$ Å	$a = 200$ Å	$a = 100$ Å	$a = 200$ Å	$a = 100$ Å	$a = 200$ Å	$a = 100$ Å	$a = 200$ Å
3×10^{-14} erg	4.4×10^{-2}	2.2×10^{-3}	2.0×10^{-2}	6.0×10^{-4}	8.0×10^{-4}	7.2×10^{-7}	1.9×10^{-4}	3.9×10^{-8}
6×10^{-14} erg	5.0×10^{-2}	3.5×10^{-3}	2.7×10^{-2}	9.5×10^{-4}	9.5×10^{-4}	1.1×10^{-6}	2.2×10^{-4}	6.1×10^{-8}

2.2.4. *Ionization of phosphatidylserine*

In the previous section we used experimental ζ-potentials in the calculation of adhesion probabilities for PS vesicles, in an attempt to rationalize their reported fusion. Here we shall consider the ionization of PS bilayers to estimate the true surface potentials of the vesicles. Details of the protracted analysis of the ionization model are given in Appendix B, and here we shall simply quote the results as they relate to the previous section.

The ionization model suggests that PS vesicles in the presence of 1 mM-Ca^{2+} have a surface potential of -73 mV. In the absence of calcium the value is -80 mV. These values exceed the electrophoretically determined ζ-potentials of -36 mV and -55 mV, respectively. In section 2.2.3 we concluded from ζ-potential data that in the absence of calcium vesicles will never come into close contact (Fig. 2) and that the probability of fusion is minute in the presence of calcium. The higher surface potentials predicted on the basis of ionization data suggest that these probabilities will be even smaller. Thus the conclusion to be drawn from this description of the PS surface is that the changes seen when calcium is added to a suspension of PS vesicles are likely to be the result of rupture and reformation of the vesicles by the divalent cation, rather than genuine membrane fusion. Fragments of ruptured vesicles might "reanneal" with one another to produce large membranous structures such as the "cochleate" cylinders described by Papahadjopoulos and co-workers (1975). This process may be a reflection of some of the events of biological membrane fusion on a rather gross scale. It is not, however, a satisfactory model of how cells fuse by localized molecular reorganization at a point of adhesive contact.

3. *The triggering event in membrane fusion*

Although the status of charged lipid vesicles as models for fusion may be problematical, one is still faced with the fact of fusion of natural membranes in vivo. The nature of the event or events that trigger the transformation is crucial to the fusion problem. Here we consider how lipid membrane apposition may initiate fusion.

The triggering process must lie at the heart of the important distinction between the processes of membrane adhesion and fusion. The problem may be rephrased as the question of how a membrane can "know" it has been approached. Any factor that automatically changes when membranes come very close together is a possible candidate for the trigger. In our present state of ignorance we have a fairly free hand to explore several promising possibilities. First we shall briefly describe a variety of triggering pathways shown in Fig. 3 and then analyze some of them in greater detail. Our analysis embraces those situations where calcium is required for fusion and where the interacting membranes carry a significant fraction of charged phospholipids. A calcium requirement is well documented in secretion (review, Douglas, 1968, 1974; Foreman et al, 1973; Poste and Allison, 1973). Raised intracellular calcium has been shown to induce

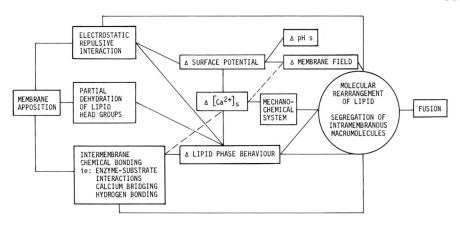

Fig. 3. Interrelationships between proposed fusion-triggering events. Solid lines indicate arrows pointing to the right; broken lines, the reverse. Δ signifies "a change in."

fusion of red cells (Ahkong et al., 1975), and divalent cations are essential to the putative fusion of liposomes (see references to section 2.1. Charged phospholipids, including PS, are found predominantly on the inner leaflet of the plasmalemma (Zwaal et al., 1973), where they may play a role in exocytosis. Their requirement for liposome fusion was discussed in section 2.1. On the outer surface of the plasmalemma, glycolipids (Cook and Stoddart, 1973) may play a role comparable to the inner charged phospholipids. The decision to consider triggering as a lipid-lipid membrane phenomenon is based mainly on the experimental findings that fusion of natural membranes, in the few cases examined so far, seems to involve interactions between areas of lipid bilayers from which membrane intercalated macromolecules are wholly or partially excluded (section 5). It is also conceivable that despite the low probabilities of adhesion calculated in section 2, charged bilayer vesicles can undergo true fusion. While we would not claim that membrane fusion is explicable solely as an interaction between charged lipids, it is consistent with experimental results and provides a useful working hypothesis. Within this framework two candidates for important roles in the fusion process are a change in lipid phase behavior and a change in electrostatic surface potential. Consider the implications of electrostatic interaction between similar negatively charged membranes. When such surfaces approach to within about 10 Å in physiological solutions, there is a predicted increase in the negative surface potential (Gingell, 1967). This effect has been verified by Ninham and Parsegian (1971), using a more sophisticated method of calculation. The rise in potential is maximal when no association of surface ionogenic groups occurs, for example when the pKa is low. One of the consequences of a potential change is a change in the concentrations of small diffusible ions near the interface. Calcium would be more concentrated at closely interacting membrane surfaces than at the same surfaces when separated. Hydrogen ions would also be concentrated but to a lesser degree. A change in the surface calcium concentration might have several effects including lipid condensation (Papahadjopoulos, 1968), which produces a change in transition temperature for the liquid-gel transformation. It has also been shown that calcium can cause a two-dimensional phase separation in lipid bilayers composed of mixtures of negatively charged and uncharged lipids (Ohnishi and Ito, 1974; Papahadjopoulos et al., 1974; Galla and Sackmann, 1975; Ito et al., 1975). The charged lipids are believed to form discrete clusters. Calcium might also be involved in the activation of surface enzymes in natural membranes, as might a fall in surface pH. The role of such biochemical events in fusion is, of course, unknown. Another result of surface potential change is that the electrostatic field across the membrane is altered. This could cause reorientation of molecular dipoles either in membrane proteins or at the headgroup region of phospholipids. A more direct effect of electrostatic interaction might also be significant: the energy of pushing charged membranes together depends on the size of the surface charge density, at infinite separation. Since the latter is greater in the gel than in the fluid state, apposition would tend to drive membranes toward a fluid state, that is, the transition temperature would change. However, this effect would be reduced if significant charge neutralization due to calcium binding occurs during membrane apposition.

The displacement of intramembranous particles (IMPs) might occur by virtue of several of the mechanisms mentioned. A calcium-induced phase separation due to membrane contact could result in the particles being preferentially concentrated in one phase. A change in transition behavior at the contact region, which is not primarily mediated by calcium, might have similar effects. Speth and Wunderlich (1973) have described reversible particle clustering as a result of temperature changes in *Tetrahymena* membranes. Particle segregation might be a secondary event, perhaps following enzyme activation or a change in membrane field. In exocytosis it is also possible that intramembranous particle displacement may be due to their mechanical linkage to cytoskeletal elements in the cytoplasm, whose structural organization and contraction is regulated by intracellular calcium (see p. 346 this volume). In secretion an increase in calcium concentration below the plasma membrane occurs due to calcium entry into the cell (Douglas, 1974). Alternatively, particle dispersal might be caused by lateral electrostatic displacement from the contact zone under the influence of an approaching negatively charged surface, without the involvement of phase changes or cytoplasmic strings.

There are several other ways in which contact might lead to fusion, apart from those involving electrostatics. For example, intramembranous calcium binding may convert to intermembranous binding. Some dehydration of the lipid headgroup region may occur, although in the case of lecithin (Le Neveu et al., 1976) there is apparently some strongly associated water which may not be readily displaced on interaction. More specific chemical interactions, such as enzyme-substrate associations, may be involved in physiological fusion. Weiss and associates (1977b) carried out an ultrastructural analysis of *Chlamydomonas* gamete fusion and tentatively suggested that an IMP-associated lipase may play a role. There is little data relating to mechanisms of chemical specificity in the fusion reaction, but Scheid and Choppin (1974) have demonstrated a requirement for a specific virus envelope protein in the case of Sendai virus-induced cell fusion (also see chapter 7, this volume). Although there can evidently be great selectivity in fusion reactions, there is no a priori reason to suppose that this is determined by fusogenic factors and could equally well be determined by factors operating at the cell contact and adhesion stages preceding fusion.

We shall select some of the above possibilities for a more detailed discussion. Because our ignorance of possible chemical interactions precludes such analysis, we shall restrict ourselves to the more amenable electrostatic events shown in Fig. 3.

3.1. Changes in small ion concentrations

The potential rise that occurs when membranes come into contact causes a redistribution of small ions in accordance with the Boltzmann relation

$$C_i^S = C_i^B e^{-ze\psi_0/kT} \tag{4}$$

where C_i^S and C_i^B are surface and bulk concentrations of ion species i of valency z and ψ_0 is the surface potential with respect to that in the bulk phase. For H$^+$ ions, when $\psi_0 < 0$, $C_{H^+}^S > C_{H^+}^B$. The increase in [H$^+$] at the surface can reciprocally reduce the surface charge by causing association of ionogenic acidic groups, as discussed in the earlier section on interaction energy, with the result that the potential change is itself decreased. However, in physiological media, this reduction is not serious where the pKa of the ionogenic group is below ~4.5 and potentials are less than ~100 mV. Since the concentrating effect is exponential in the valency, calcium is more strongly attracted to negatively charged membranes than H$^+$. It is possible to calculate ion concentration changes following the methods of Parsegian and Gingell (1973) for membranes coming into contact. Consider, for simplicity, membranes composed of charged phospholipid mixed with uncharged phospholipid interacting in 145 mM NaCl containing 1 mM Ca^{2+}. Allowing for penetrability of the headgroup region, as described in Appendix B, using an effective thickness (h) for the region of the phospholipid headgroups that is penetrable to counterions, we have at membrane contact

$$\frac{1}{APC}\left(\frac{K_a}{K_a + 10^{-pHx}}\right) + nh\left[\frac{1}{x} - (1-\eta)x - \eta x^2\right] = 0, \tag{5}$$

where APC is area per ionizable charge group, n is concentration of monovalent anions, K_a is the dissociation constant of the ionogenic group, η is the fraction of positive charge in bulk solution carried by divalent cations, and $x \equiv \exp(-e\psi_0^c/kT)$; for $\psi_0^c < 0$; $x > 1$, where ψ_0^c refers to surface potential at contact.

At infinite separation, $x \equiv \exp(-e\psi_0^\infty/kT)$ and

$$\sqrt{\frac{2\pi e^2}{\epsilon kTn}}\left\{\frac{1}{APC}\left(\frac{K_a}{K_a + 10^{-pHx}}\right) + nh\left[\frac{1}{x} - (1-\eta)x - \eta x^2\right]\right\} - \sqrt{\frac{1}{x} + \frac{\eta}{2}}(x-1) = 0, \tag{6}$$

where ϵ is the bulk dielectric constant of the aqueous phase.

Numerical solution of equations (5) and (6) for x gives the change in surface potential in the penetrable region at contact, from which ion concentration changes in that region can be obtained by Boltzmann's relation. Following this procedure for 0.1 and 0.2 mole fraction PS in lecithin with $h = 4\text{Å}$, as discussed in Appendix B, we obtain the values shown in Table 2 for two cases; first, where the lipids are in a liquid state (65Å^2 per PS headgroup) and second in a gel state (48Å^2 per PS headgroup).

Clearly, substantial changes in surface potential are predicted, and therefore high concentrations of calcium in the penetrable region should be found on contact. Although 100 mM Ca^{2+} may seem unreasonable in a region of effective thickness 4 Å, it is readily calculated that this represents only 2 calcium ions in the 8 Å thick hydrophilic penetrable region existing between 100 phospholipid molecules on each membrane. Thus in terms of volume, the space occupied by calcium counterions is negligible even at contact. Such large changes in calcium concentration induced by membrane contact obviously represent the basis for a plausible fusion trigger. While we feel that this is a very promising result, there is little evidence at present upon which to base a mechanism for fusion given such a triggering event. It might be interesting to study the effects of such high calcium concentrations on the stability of isolated membranes before speculating further.

In contrast to divalent cation accumulation at contact, the predicted change in surface pH is small. It may, however, be sufficient to activate enzymes in the membrane or to initiate other biochemical events. As noted earlier, the [H$^+$] change will have little effect on the density of ionized charge, even at contact, unless a significant fraction of groups of pKa > 4.5 are present.

TABLE 2

1 mM bulk Ca^{2+} 48 Å^2 per PS headgroup

Mole fraction PS	mV			mM				
	ψ_0^∞	ψ_0^c	$\Delta\psi_0$	$[Ca^{2+}]_s^\infty$	$[Ca^{2+}]_s^c$	$(pH_s^+)^\infty$	$(pH_s^+)^c$	ΔpH_s^+
						(bulk pH = 7.4)		
0.1	−23	−44	−21	6.2	33	7.0	6.6	0.4
0.2	−40	−60	−20	24	116	6.7	6.4	0.3

0 mM bulk Ca^{2+} 65 Å^2 per PS headgroup

Mole fraction PS	ψ_0^∞	ψ_0^c	$\Delta\psi_0$	$(pH_s^+)^\infty$	$(pH_s^+)^c$	ΔpH_s^+
0.1	−18	−38	−20	7.1	6.7	0.4
0.2	−32	−54	−22	6.9	6.5	0.4

3.2. Changes in membrane field

Consider the potential profile through a symmetrical charged phospholipid membrane in symmetrical monovalent electrolyte in the absence of a resting potential, Fig. 4a. For simplicity, we draw a straight line between the surface potentials on either side to indicate the membrane potential profile. If calcium is added to one side, the double layer potential will get smaller due to Debye screening and also to adsorption and the field will increase (Fig. 4b). For a gel membrane of 0.2 mole fraction PS, reduction of the potential on one side to half the initial value by adding calcium would change the field from zero to 10^5 V/cm. In this context it is interesting that asymmetric calcium has been implicated in membrane instability and rupture (Papahadjopoulos and Ohki, 1969). In lipid membranes such a large field change might affect the conformation of the polar regions of phospholipids, which account for about 40% of the bilayer thickness. For membranes with intercalated proteins, a torque would be exerted on the macromolecules as a result of their dipole moment. Now consider Fig. 4c, which represents

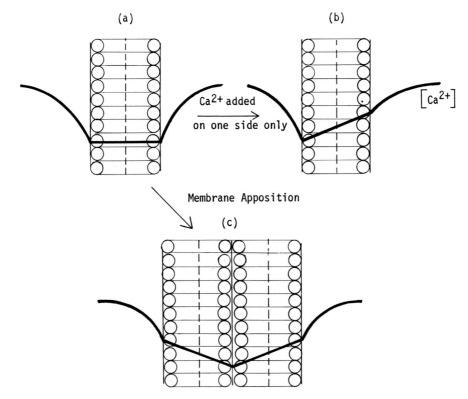

Fig. 4. Electrostatic field across symmetrical membranes containing charged lipids. (a) isolated membrane with identical solutions bathing both faces; (b) calcium added to one side only; (c) contact between two similar membranes.

the field change due to membrane apposition for membranes composed of 0.2 mole fraction PS in lecithin. Table 2 shows that the change in potential on apposition in 1 mM bulk calcium is -20 mV. This represents a change in membrane field from 0 to 10^5 V/cm. The significant point here may be that membrane apposition may bring about very large changes in membrane field, comparable with those calculated for isolated charged membranes made unstable by having divalent cations on one side only. The latter process may represent an isolated "half-fusion" event. However, a mere increase in electrostatic field may be insufficient to act as a trigger. Contact in the absence of calcium would cause similar surface potential changes (Table 2) without initiating fusion. But if the charge density were too high for contact to occur without calcium the potential rise would obviously not occur. It is not possible to carry this further without knowing more about the conditions under which vesicles of various mole fractions PS/lecithin can come into adhesive contact in the absence of calcium. There is, however, a clear case where a change in field is not sufficient to cause fusion. Positive and negative vesicles adhere on mixing but do not fuse (Korn et al., 1974). Simple electrostatic analysis shows that at contact the field in each membrane increases to a high value as the potential at their contacting faces falls to zero. In this system strong electrostatic attraction between headgroups of contiguous phospholipid membranes may restrain the molecules from undergoing rearrangements necessary for fusion.

3.3. Membrane interaction and transition temperature

We have considered the question of whether close approach of two membranes might provide the stimulus for fusion by a change in lipid transition temperature. Such an effect does not appear to have been analyzed previously. Our calculations suggest that there may be a very small tendency toward fluidization, but increased calcium binding during interaction may reverse this effect.

Appendix C (p. 828) shows that the change in transition temperature due to electrostatic interaction as membranes come into contact is given by

$$T_t^c = T_t^\infty + \frac{g_\lambda - g_\gamma}{S_\lambda^\infty - S_\gamma^\infty}. \tag{7}$$

T_t^∞ is the transition temperature at infinite separation, T_t^c is that at contact, g_λ and g_γ are the repulsive electrostatic free energies of interaction for membranes in the fluid and gel states, respectively, coming from infinite separation to contact. The difference in these terms is due to the different surface charge densities in the two states. S_λ^∞ and S_γ^∞ refer to entropies in these two states. This difference in entropy is given for dimyristoyl lecithin by Phillips and associates (1969) as 22.4 cal/degree mole, which is equivalent to 0.27 erg/degree cm² for each half of a bilayer.* Values of g_λ and g_γ were calculated after Parsegian and Gingell (1973) from

---

*We have assumed that it is necessary to melt all the molecules in one half of the bilayer. In mixed membranes of charged and neutral lipids a larger ΔT_t may result if only a proportion of the lipid has to melt.

$$g = ze(\psi_0^c - \psi_0^\infty)/APC, \qquad\qquad (8)$$

where g is interaction free energy per surface. This is a legitimate procedure since APC remains practically constant when pKa < 4.5 for interaction potentials in the range considered.

Values of ψ_0^c (at contact) and ψ_0^∞ (infinite separation) were obtained for membranes containing 20% PS as follows. In the presence of calcium we use $\psi_0^\infty = -29$ mV from the theoretical curve of surface potential based on the 10% PS electrophoretic result (see Appendix B and Fig. 10, curve c). The potential at contact was calculated from equation 5 using a value of 372 Å²/charge, representing a charge density twice that corresponding to the experimentally derived ζ-potential for 10% PS in lecithin, $\zeta = -15.6$ mV, (see p. 827). These values give $g_\gamma = 0.93$ erg/cm². For g_λ we used values of ψ_0^c and ψ_0^∞ calculated at 90°C, which is above the estimated transition temperature of PS in the presence of calcium (Papahadjopoulos et al., 1977). The area per charge, $APC = 500$ Å², was calculated for 20% PS in the manner outlined above. This procedure gave $g_\lambda = 0.78$, resulting in $\Delta T_t = -0.5$°C.

We also calculated the change in transition temperature for membrane interaction in the absence of calcium, which involves simpler assumptions. The potentials ψ_0^∞, ψ_0^c for gel and liquid were calculated from equations 5 and 6 on the basis of 48 Å² and 65 Å² per PS headgroup, respectively. The value of $\psi_0^\infty = -32$ mV for the liquid state corresponds with curve c of Fig. 9. Values of g_λ and g_γ derived from these potentials were substituted into equation 7 giving $\Delta T_t = -0.85$°C.

These calculations do not take into account possible strong calcium adsorption at contact due to the rise we predicted in surface calcium concentration (Table 2). In principle, the resulting reduction in charge could be built into equation (5) using a calcium binding constant. However, besides the uncertainty attached to a binding constant for high calcium concentrations, the interaction energy cannot be obtained from equation 8 because the area per charge becomes a function of interaction. Preliminary calculations suggest that a rise in transition temperature of several degrees might occur if calcium binding reduces the surface charge to near zero at contact. However, in the case of PS vesicles, which are already in a gel state in 1 mM Ca²⁺ before contact, a rise in transition temperature would have no effect on fusion. In physiological conditions, for biological membranes in a fluid state, a gelation at the contact site might have important consequences; but it must be remembered that our calculations refer to membranes with significant fractions of charged phospholipids. If the data obtained on lipid asymmetry in the erythrocyte plasma membrane also applies to other cells, then charged phospholipids are present on the inner membrane leaflet and sparse on the outer one.

4. Structural changes in lipid bilayers during fusion

In the previous section we discussed several ways that membrane contact might initiate fusion. Now we examine the gross molecular reorganization of the lipid bilayers that must occur before fusion is completed. This reorganization leads to the formation of an aqueous channel through the contact zone. Few models have been proposed to describe the nature of the molecular rearrangements involved in this process. Here we consider and extend some of these models in the light of recent experimental observations.

One pertinent question in this context is whether two apposed bilayers are converted into a single membrane as an intermediate stage in the fusion process. The observations of Neher (1974) on fusion of hemispherical black lipid films support this suggestion. In this system, the area of contact of the two hemispheres is transformed into a single bimolecular leaflet assumed to contain interspersed "lenses" and "micelles," as shown by measurements of specific capacitance and unit conductance in the presence of gramicidin A. Subsequently this transient bilayer formed in the contact zone tends to rupture, completing the fusion process. Electron microscopic observations on calcium-induced aggregation of isolated chromaffin granules (Edwards et al., 1974) provide further support for the notion that two contacting membranes initially form a single "trilaminar" structure during fusion. Neither Lagunoff (1973) nor Lawson and co-workers (1977), however, were able to demonstrate stages in fusion beyond an initial pentalaminar "bi-membrane" structure in mast cell degranulation. Thus this question remains unresolved, particularly since the vast area of membrane contact in the system described by Neher (1974) may not be relevant to membrane fusion in vivo.

A major model which dispenses completely with bimolecular leaflets (paired or single) at the point of fusion is that of Lucy (1970). In this model, the phospholipid molecules of the two fusing membranes assume a globular micellar configuration instead of their original lamellar orientation prior to contact and fusion. The micelles of the two membranes then intermingle at the point of close approach and fusion is completed when lamellar membranes re-form. This idea is based in part on the observation that lysolecithin, a compound known to form micelles when dispersed in water, induces cell fusion (Poole et al., 1970). The action of lysolecithin in membrane fusion can, however, be explained in terms of mechanisms other than that of destabilization and micelle formation within the interacting membranes, as shown schematically in Fig. 5. The mechanism shown in this diagram involves (a) the simultaneous association of a lysolecithin micelle with the external surfaces of two cell membranes. This event is energetically feasible because of the small electrostatic repulsion between the cell surface and such a small particle, composed of lipids with headgroups carrying no net charge. This is followed by (b) a molecular reorganization resulting in the formation of an amphiphilic "stalk" connecting the two membranes. Subsequent stages would involve lateral diffusion of the lysolecithin molecules from the stalk and their replacement by indigenous membrane lipid molecules, and an increase in girth of the stalk, resulting finally in apposition of the cytoplasmic halves (monolayers) of the two membranes followed by rupture.

As an alternative to the membrane fusion model proposed by Lucy (1970), we may consider the scheme put forward by Lau and Chan (1974) to explain putative fusion of lecithin vesicles in the presence of the ionophore alamethicin. This

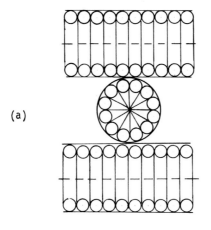

(a)

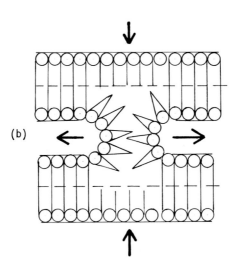

(b)

Fig. 5. Model for lysolecithin-induced membrane fusion. (a), lysolecithin micelle bridging two membranes; (b) formation of an amphiphilic stalk connecting the membranes. Arrows indicate directions of thinning and expansion that lead to completion of fusion.

scheme can be adapted to provide a far more general mechanism for biological membrane fusion, summarized in Fig. 6. At the region of contact (a) between two membranes, the formation of an "inverted micelle" (b) is predicted to occur, leading finally to the rearrangement shown in (c), at which stage fusion is complete. Inverted micelles of the type shown in Fig. 6b are thought to correspond to the hexagonal (II) and cubic phases described in numerous x-ray diffraction studies. Occurrence of these phases is widespread and includes that formed with phosphatidyléthanolamine at elevated temperature in lipid excess (Reiss-Husson, 1967) and the hexagonal (II) phase demonstrated for the case of car-

810

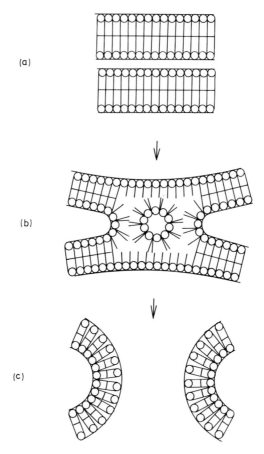

Fig 6. Hypothetical scheme for fusion. (a) apposition of bilayers; (b) formation of an inverted micelle; (c) completed process.

diolipin in the presence of calcium (Deamer et al., 1970; Rand and Sengupta, 1972; Costello and Gulik-Krzywicki, 1976). In a recent series of experiments on membrane fusion, Papahadjopoulos and collaborators (1976b) demonstrated a hexagonal (II) phase in aqueous dispersions of phosphatidic acid treated with magnesium. This phase is only present at pH = 6.0, the lipid molecules retaining a lamellar arrangement at pH = 8.0. Failure to demonstrate a hexagonal phase for phosphatidylserine in the presence of high concentrations (100 mM) of calcium (Rand and Sengupta, 1972) does not weaken the model discussed here, despite the presumed importance of this lipid in membrane fusion (Papahadjopoulos et al., 1974, 1977; Papahadjopoulos and Poste, 1975). Hexagonal phases might form even with PS under the special conditions prevalent when membranes come into contact (e.g., possible dehydration, changes in membrane electrostatic field, calcium concentration increases, etc; see section 3).

The "inverted micelle" model is useful when considering the experiments of Neher (1974), described earlier in this section: formation of a single bilayer from two black lipid films could result from coalescence of the contacting halves (monolayers) of the two membranes into a large number of inverted micelles (unlike the single structure shown in Fig. 6), which could then aggregate into larger lenses or possibly diffuse within the membrane away from the site of contact, leaving predominantly a single bilayer in that zone. In biological membranes, where the fusion process may be more complex, possibly involving activation of various enzymes (e.g., phospholipases) that can alter the membrane lipid composition in the region of fusion, many lipids and lipid derivatives could be implicated in the construction of inverted micelles, including diacylglycerol, which has been shown to be fusogenic (Ahkong et al., 1973; Allan et al., 1976). The main weakness of the inverted micelle model is that it is difficult to see how such marked molecular rearrangements could occur in membranes that have been rendered "solid" by the presence of calcium. (Papahadjopoulos and Poste [1975] state that calcium brings about liquid → gel phase transitions for many acidic phospholipids, yet both these components are thought to be required in fusion.) This criticism could, however, be leveled at all the models for membrane fusion proposed so far.

5. Fusion of natural membranes

5.1. Behavior of membrane macromolecules in fusion

So far we have restricted our discussion to interactions between simple lipid membranes. Such systems lack proteins and glycoproteins found in natural membranes, which must be included in any complete account of fusion. It is implicit in the study of membranes composed of lipid alone that they are a simplified paradigm for natural membranes. Indirect evidence supporting this idea has recently been obtained in a freeze-fracture study of mast cell secretion by Lawson and colleagues (1977). These authors found that IMPs were practically absent from both P and E faces of granule and plasma membranes at the region of close apposition where fusion was occurring. A surrounding ring of increased particle density suggests that the particles are centrifugally displaced, together with the intervening cytoplasm, as the membranes come into contact. In addition, surface binding sites for ferritin-conjugated lectins, concanavalin A, and phytohemagglutinin, as well as surface immunoglobulins are not found on either membrane in the region where they are interacting but are found elsewhere. In contrast it is instructive that a cholera toxin label binds both at contact sites and outside them, since the binding site is a small ganglioside rather than a macromolecule. These results suggest that at the region of fusion all proteins and glycoproteins are laterally displaced, leaving two lipid bilayers which then fuse. Less extensive studies by Schober and associates (1977), showing that chromaffin granules incubated in 10 mM Ca^{2+} formed mutual attachment sites free of IMPs and sometimes fused, support this interpretation.

These results complement the work of Satir and co-workers (1973) on muco-cyst secretion by *Tetrahymena,* which is characterized by the formation of a bald region at the apex of the secretory vesicle where it is closely apposed to the plas-malemma. This bald region is surrounded by an annulus of clustered ~80 Å diameter IMPs, clearly seen on the P face of the membrane. The overlying plasmalemma shows a distinctive rosette arrangement of large (~150 Å diam.) P-face particles which form a ~600 Å diameter ring with a central particle. The rosette and annulus come into "contact" and the particle-free zone widens from ~600 Å to ~2000 Å, whereupon rupture and mucous discharge occur. In-crease in vesicle size immediately before fusion indicates that the water perme-ability of the apposition region dramatically increases before fusion occurs. The function of the particles that surround the ~ 2000 Å bald regions would seem to be to limit the extent of the membrane region lost when rupture takes place. This sequence of events can be contrasted with that described by Weiss and co-workers (1977a) for contractile vacuole-plasmalemma interaction in *Chlamydo-monas.* The plasmalemmal contact region shows an apparently widening, central particle aggregate about 1800 Å in diameter in a field of reduced particle den-sity. The limited data available suggest that the underlying vacuole membrane shows a ringlike disposition of particles at some stage in the process. In this sys-tem, however, fusion does not seem to occur, and it is postulated that the particle aggregate forms a hydrophilic patch for water transfer. (This suggestion and that of Satir et al. [1973], concerning the influx of water into the mucocyst at a stage when an annulus of closely apposed particles has formed, are interesting in relation to the hypothesis [Gingell, 1973, 1976] that IMP aggregation can automatically cause a local increase in permeability.) Fusion between $mt(+)$ and $mt(-)$ gametes of *Chlamydomonas* has also been studied by Weiss and associates (1977b). In this process it is far less clear that fusion takes place at particle-free zones than in the preceding examples. Although mating structures seen in both P and E faces of $mt(+)$ strain cell membranes (which form the fertilization tube) are practically devoid of large particular inclusions, that seen in the E face of $mt(-)$ cells after activation by flagellar contact with $mt(+)$ cells shows a dense cluster of central particles. In this case it may be that fusion occurs between a region with strongly clustered particles on one membrane and an almost particle-free zone on another.

5.2. Possible mechanisms for particle displacement

Although it is perhaps too early to conclude that displacement of membrane glycoproteins and proteins at sites of fusion is an invariant feature of membrane fusion, an analysis of possible mechanisms responsible for IMP migration is worthwhile. Three such mechanisms are discussed here.

5.2.1. Direct electrostatic displacement

Direct electrostatic displacement may occur if there is close apposition be-tween charged membranes, provided the intramembranous particles are kinetically associated with charge-bearing groups of the cell surface (discussion, Gingell, 1976). This is the case in the normal human red cell, but probably not in lymphocytes where agents that cross-link surface groups in the patching reaction apparently do not cause corresponding IMP clustering (Karnovsky and Unanue, 1973; Pinto da Silva and Martinez-Palomo, 1975; Yahara and Edelman, 1975). Clearly, however, not all particle migrations occurring during fusion can be ascribed to contact-mediated electrostatic displacement. The close membrane association (<50 Å) that would be necessary to cause such displacement of IMPs does not take place in the genesis of the particle-free zone of the fertilization tube of *Chlamydomonas* (Weiss et al., 1977b). Outside the realm of membrane fusion, particle migrations can take place in reconstituted membranes, as shown by the imaginative work of Chen and Hubbell (1973). Clearly, some other driving force must be sought in such cases. However, the argument for electrostatic particle displacement where close contact does occur is greatly strengthened by the novel observations of Poo and Robinson (1977) that concanavalin A receptors can be impelled to diffuse across muscle cell membranes by electric fields in the order of 1 mV/cell diameter. This phenomenon has been treated theoretically by Jaffe (1977). Since macromolecular surface receptors are characteristically free to undergo translational diffusion in the plane of the membrane (reviewed, Edidin, 1974; Poste and Nicolson, 1977), it is not unreasonable to anticipate that their lateral motion away from a point of contact could be induced by electrostatic forces. The operation of such forces between cells and charged surfaces has been demonstrated experimentally by Gingell and Fornes (1975), Gingell and Todd (1975), and Gingell and collaborators (1977).

These findings may be summarized as follows. IMPs are displaced in the plane of the membrane prior to fusion; most surface receptors are not normally linked mechanically to IMPs; and electrostatic forces will tend to displace surface receptors. How can these incongruent features be reconciled? It is tempting to speculate that the decision to fuse on contact might be determined by a prior chemical event in one or both membranes that mechanically links charged diffusible surface macromolecules with IMPs, thus providing a mechanism for electrostatic repulsion of IMPs from the region of contact that leads to fusion. Such a mechanism might be switched on by metabolic events, priming the membrane so that it becomes susceptible to fusion. A diagramatic illustration of this idea is shown in Fig. 7. In Fig. 7a charged glycoproteins of the cell surface are shown in a "normal" state kinetically separated from IMPs; in Fig. 7b, a priming reaction has linked them. Repulsion of the linked units occurs as the membranes come together leading to lipid-lipid contact (Fig. 7c) and then fusion. Alternatively, the stage shown in Fig. 7b might lead to the formation of a gap junction by desialylation of the hydrophilic moiety of the glycoproteins (Fig. 7e). Reduction in mutual electrostatic repulsion might then lead to aggregation of the remaining IMP-glycoprotein complexes, (Fig. 7f), under the influence of van der Waals attractive forces (Gingell, 1976). Gap junctions are characterized by clustered IMPs (McNutt and Weinstein, 1973), and chemical analysis has failed to detect sugars in them (Gilula, 1978). Contact between "unprimed" membranes (Fig. 7d) would lead to lateral

814

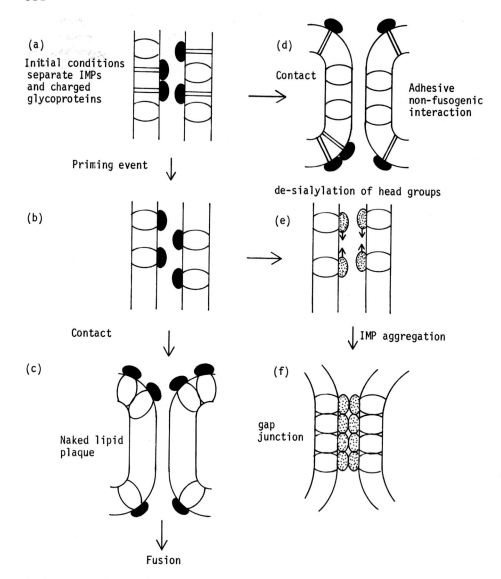

Fig. 7. Hypothesis for intercellular interactions, leading to fusion or junction formation. The path followed depends on the state of association of membrane macromolecules. (See text for details.)

repulsion of surface glycoproteins without concomitant motion of IMPs, resulting in adhesive but not fusogenic contact.

If membrane apposition causes IMP displacement, creating a bald spot, we can ask whether the energy of apposition is sufficient to drive the process. Suppose the surface energy of the lipid interface is 2 erg/cm² (Requena and Haydon,

1975). To create a bald spot of 500 Å radius in a bilayer would require 3×10^{-10} erg. This can be compared with the calculated interaction energy between model membranes (Parsegian and Gingell, 1973). For secondary minimum interaction the range is 10^{-3} to 10^{-4} erg/cm², which is fairly independent of the minutiae of the model. Clearly, secondary minimum contact could not drive IMP dispersal. On the other hand, primary contact, whose energy must be of the order of the surface energy of the individual membranes, would be expected to power such a process. To get through the repulsive energy barrier and come into primary contact, cells almost certainly need an energetic process such as filopod protrusion. In this case the energy of IMP dispersal would be provided by cellular metabolism. Metabolism could provide the necessary energy because the energy previously calculated for making a bald spot could be supplied by 600 molecules of ATP, assuming 100% efficiency. The repulsive energy barrier for the interaction between secretory vesicles and the inside of a plasmalemma may be reduced sufficiently for primary contact to occur by the influx of calcium (Dean, 1975). The energy barrier to primary contact may be artificially reduced by micelles such as lysolecithin (section 4), which will experience minimal electrostatic repulsion by virtue of their small radius of curvature and lack of formal charge. A localized contact will form, leading to cell-cell fusion without even producing a large bald region.

5.2.2. *Displacement by lateral phase separations*
An alternative conceptual approach to the production of a bald patch by electrostatic repulsion is that a lipid phase change may exclude intercalated macromolecules from the contact zone. Such exclusion has been observed in freeze-fracture studies of *Tetrahymena* membranes (Speth and Wunderlich, 1973). The portions of membrane proteins that penetrate the lipid bilayer may not readily be incorporated into the crystal structure of the lipid in the gel state and will be excluded. The preferential gelling of specific phospholipids may thus freeze out the proteins into more fluid regions (though it is possible that they would segregate into the gel phase). In the case of PS mixed with lecithin, Ohnishi and Ito (1974) have shown that calcium causes preferential clustering and gelling of PS leaving lecithin in the liquid state. Thus it is possible, as Ahkong and co-workers (1975) suggest, that the action of calcium might be to sequester membrane proteins away from the site of incipient fusion. This implies that the region of membrane contact would be occupied by lipid in the gel state, a notion that is perhaps counterintuitive and disagrees with some published conclusions (Papahadjopoulos et al., 1973). Nevertheless, the calorimetric studies of Jacobson and Papahadjopoulos (1975) show that calcium gels PS membranes under conditions where they argue that fusion occurs. It is possible that entry of calcium in secretion (Katz and Miledi, 1967; Miledi, 1973; Douglas, 1974) may thus cause PS aggregates to form, resulting in phase-sensitive IMP sequestration and a naked lipid patch at the contact region. In this context it is interesting that PS is found predominantly in the inner leaflet of the plasma membrane of the erythrocyte

816

(Zwaal et al., 1973) and presumably on the outer surface of secretory vesicles. It is less simple to rationalize intercellular fusion in this way because of the dearth of charged phospholipids on the outer surfaces of cells, though it is possible that charged glycolipids (Cook and Stoddart, 1973) in the outer leaflet may respond to high local calcium concentrations at contact (section 3) in an analogous way. Even the gelling of such relatively minor membrane components might cause localized macromolecular exclusion and provide a site for intercellular fusion.

5.2.3. Membrane-microfilament interactions
The third category of processes that may cause IMP exclusion from the region of presumptive fusion involves cytoplasmic contractile elements. It has been suggested (Allison, 1971) that cytoplasmic proteins may be capable of exerting traction on membrane macromolecules. They may therefore be capable of causing lateral motion of IMPs. Possible mechanical linkages, between membrane elements and contractile proteins in relation to cell movement (Bray, 1973) and endocytosis (Gingell 1973, 1976) have been proposed. In this field conjecture is more abundant than hard facts. Nevertheless, a most interesting relationship between axial actin filaments and the membranes of microvilli has been described by Mooseker and Tilney (1975b). The filaments appear to send out projections that make periodic contact with the plasmalemma. The demonstration of the Z-disc protein α-actinin in the dense material of the tips of the microvilli (Schollmeyer et al. 1974) and the muscle-like relationship between actin filament polarity and putative insertion via α-actinin heighten the interest of this work, although the authors have been unable to demonstrate connection of multiple filament projections with the apparently correspondingly aligned IMPs by freeze-fracture techniques (Mooseker and Tilney, 1975a). Tilney and Detmers (1975) have also shown that an actin-like protein is associated with spectrin in the red cell membrane, and have argued that the two proteins serve to anchor the IMPs. Evidence from studies on ligand-induced capping of surface receptors (de Petris, 1975) may also perhaps be interpreted as suggesting that cytochalasin B-sensitive filaments are "linked" to certain classes of receptors. However, it is difficult to disentangle receptor "linkage" from a general requirement of cell motility for capping to take place. Also, since such receptors are probably not associated kinetically with IMPs (Karnovsky and Unanue, 1973; Pinto da Silva and Martinez-Palomo, 1975) these studies are of limited interest here.

The calcium influx that occurs in secretion might induce contraction of filamentous proteins attached to IMPs, causing them to drag away IMPs, creating bald patches in the plasmalemma. Whether a similar process might occur in the secretory vesicle surface is problematic. If it does, adhesion and fusion would occur when a random collision brought bald patches on vesicle and plasmalemma into apposition. If the formation of bald patches is an event that occurs before contact, one would expect to see them at regions where no contact had yet formed as well as at contact sites. The available freeze-fracture evidence seems inadequate to decide this question.

6. Summary

We have not attempted to give a balanced review of the literature relating to physical aspects of membrane fusion. Instead, we have tried to analyze what seemed to be some of the better documented facts of fusion, primarily studies on the fusion of small vesicles which alone may be sufficiently well defined to allow some attempt at quantitative analysis. Only bilayer vesicles of negatively charged lipids seem to undergo topological changes suggesting fusion. This requires divalent cations. Our energetic analysis confirms that vesicles of phosphatidylserine in the absence of calcium should never approach to within 8 Å and should therefore not fuse. However, the addition of calcium does not reduce the energy of repulsion sufficiently for any but a fraction of the smallest vesicles to approach to 8 Å. We feel, therefore, that further experimental evidence is required to show that fusion is not actually proceeding via cracking and reannealing of small vesicles as opposed to direct fusion. The vital experiment showing transfer of contents between vesicles without spillage has never to our knowledge been successfully performed—possibly because it simply doesn't happen that way. It is most important to resolve this question, since the rationale of this field of study has been based on the assumption that small vesicles represent a legitimate paradigm for the fusion of natural membranes.

The nature of a triggering event in fusion, based on calculated interactions between charged lipid membranes has been discussed. We do not feel that the validity of these results depends on the interpretation of interactions between small lipid vesicles. Possible triggering stimuli include a large increase in calcium concentration at membrane surfaces brought about by the increased electrostatic potential at close membrane apposition. Another contender is the vast change in electric field through the lipid, which is predicted on contact. We also investigated the possibility that apposition might automatically induce a change in lipid transition temperature, but the calculated effect is not encouragingly large.

The nature of the fusion event has also been examined and a tentative model based on inverted micelles has been developed. In natural membranes it appears that the essential event is the removal of intramembranous particles from the contact zone. This may take place under the influence of direct electrostatic repulsion, by lipid phase changes that exclude protein from the contact zone or, possibly, in some cases by contractile proteins which drag the particles centrifugally.

Appendix A

Electrostatic energy of sphere-sphere interaction: linear approximation

Consider the approach of two identical, infinitely large parallel planar membranes in symmetrical monovalent electrolyte. These membranes are made from

818

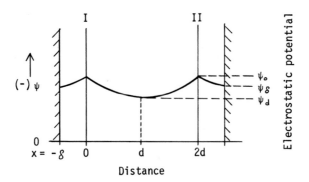

Fig. 8. Electrostatic potential profile between interacting lipid bilayers.

the same acidic phospholipid as the liposomes in Fig. 1, and their interacting surfaces are negatively charged as a result of the dissociation of the acidic moieties in the lipid headgroup region. Electrical double layers in the aqueous phase are associated with these charged surfaces and, as the membranes approach, the double layers will interact to yield an electrical potential profile of the type shown in Fig. 8. This interaction model follows the scheme of Gingell (1967). Formal charges are considered to reside in planes I and II only. Behind the charged plane on each membrane there is a region of thickness, δ (the headgroup region), which is assumed to be partially penetrable to ions in solution. The degree of penetrability is taken to be constant in a direction normal to the surface and is defined by a factor $\omega = 1/2$, which represents the reduction in available space for counterions inside the charged plane. At any separation, $2d$, between planes I and II, the potential at the inner limit of the penetrable region ψ_δ is lower than that, ψ_0, at the charged plane as a result of counterion penetration behind the charged plane. The potential at the midpoint between the interacting surfaces is termed ψ_d

If the surface potential ψ_0^∞ at infinite separation of the membranes is known, the surface charge density σ can readily be obtained by substituting into equation 17 from Gingell (1967). Thus

$$\sigma = \frac{\epsilon \kappa \psi_0^\infty}{4\pi}[(1 - \omega)^{\frac{1}{2}} \tanh \kappa(1 - \omega)^{\frac{1}{2}} \delta + 1]. \tag{A1}$$

where ϵ is the bulk dielectric constant of the aqueous phase and κ is the Debye-Hückel parameter defined by

$$\kappa = \sqrt{\frac{8\pi n e^2}{\epsilon kT}},$$

where k is Boltzmann's constant, T is the absolute temperature, e is the electronic charge, and n is the electrolyte concentration in the bulk aqueous phase.

With σ now known, ψ_0 can be defined as a function of interaction distance, again using equation 17 from Gingell (1967):

$$\psi_0 = \frac{4\pi\sigma}{\epsilon\kappa}[(1 - \omega)^{\frac{1}{2}} \tanh \kappa(1 - \omega)^{\frac{1}{2}} \delta + \tanh \kappa d]^{-1}. \qquad \text{(A2)}$$

The surface potentials of the interacting membranes are related to the midpoint potential ψ_d by

$$\psi_d = \psi_0 \operatorname{sech} \kappa d, \qquad \text{(A3)}$$

combining (A2) and (A3):

$$\psi_d = \frac{4\pi\sigma}{\epsilon\kappa} \cdot \frac{\operatorname{sech} \kappa d}{\tau + \tanh \kappa d} \qquad \text{(A4)}$$

where

$$\tau \equiv (1 - \omega)^{\frac{1}{2}} \tanh \kappa (1 - \omega)^{\frac{1}{2}} \delta.$$

The electrostatic repulsive pressure P between the two membranes is in general related to ψ_d by

$$P = 2nkT\left(\cosh\frac{e\psi_d}{kT} - 1\right). \qquad \text{(A5)}$$

Thus, combining (A4) and (A5):

$$P = 2nkT\left[\cosh\left\{\frac{\beta}{(\tau + 1)e^\phi + (\tau - 1)e^{-\phi}}\right\} - 1\right]. \qquad \text{(A6)}$$

where

$$\phi \equiv \kappa d,$$

and

$$\beta \equiv 8\pi\sigma e/kT\kappa\epsilon.$$

Using equation (A6), the change in electrostatic free energy $G_p{}^{es}$ for a single membrane, on bringing the two planar surfaces from infinite separation to a distance of separation, $2d$, may be expressed as follows:

$$G_p{}^{es} = \frac{2nkT}{\kappa} \int_\infty^\phi \cosh\left\{\frac{\beta}{(\tau + 1)e^\phi + (\tau - 1)e^{-\phi}}\right\} - 1 \cdot d\phi. \qquad \text{(A7)}$$

Derjaguin's method shows that the repulsive energy $G_s{}^{es}$ for two spheres having the same surface properties as the planes under consideration is given by the following further integration:

where

$$G_s{}^{es} = \frac{2\pi a}{\kappa} \int_\infty^\xi G_p{}^{es} (\xi) d\xi, \qquad \text{(A8)}$$

$$\xi \equiv 2\kappa d, \ a = \text{sphere radius.}$$

The integrations in equations (A7) and (A8) were carried out numerically, allowing $G_s{}^{es}$ to be obtained as a function of d, a, and σ. Accuracy of integration was assessed by substituting trial functions which were analytically soluble. Surface charge density was obtained from membrane surface potential at infinite separation via equation (A1). This allowed estimation of G_s^{es} for experimentally deter-mined liposome surface potentials (electrokinetic potentials). The use of an equation essentially designed to describe planar interactions (equation A1) in the determination of σ from the electrokinetic potentials of spheres is justifiable in this system as κa is large at all times.

Implicit in the model developed thus far is the assumption of constant surface charge density, σ, on membrane interaction. In liposomes made from acidic phospholipids, however, the surface charges result from ionization of, for ex-ample, carboxyl and phosphate groups in the lipid headgroup region. The de-gree of ionization of these moieties will be altered by membrane interaction, as ψ_0 increases when the distance between the surfaces decreases, leading to an increase in the surface hydrogen ion concentration (via Boltzmann's relation-ship) which will in turn cause reassociation of the acidic groups. To correct for this variation in σ, we adopted the following procedure.

At infinite separation of the membranes, the surface charge density σ^∞ is given by equation A1:

$$\sigma^\infty = \frac{\epsilon \kappa \psi_0^\infty}{4\pi} (\tau + 1)$$

using the definition of τ following equation (A4).

The degree of dissociation of the surface ionizable groups at infinite separa-tion α^∞, derived in Appendix B (equations B1 to B5), is

$$\alpha^\infty = \frac{K_a}{K_a + 10^{-pH}e^{-e\psi_0^\infty/kT}}, \tag{A9}$$

where K_a is the acid dissociation constant of the single ionizable group under consideration in this model (a value of $10^{-4.5}$ was used throughout this analysis, corresponding to the mean of values quoted by numerous authors for the dis-sociation of the carboxyl group of PS). 10^{-pH} refers to the hydrogen ion concen-tration in the bulk phase.

Combining equations (A1) and (A9) gives the maximum surface ionizable group charge density, q:

$$q = \frac{\sigma^\infty}{\alpha^\infty}. \tag{A10}$$

For any given membrane separation distance ($2d$) at which the membrane surface potentials are ψ_0, we write the following expressions for surface charge density:

$$\sigma = \frac{\epsilon \kappa \psi_0}{4\pi} (\tau + \tanh \kappa d). \tag{A11}$$

and

$$\sigma = q\alpha = \frac{qK_a}{K_a + 10^{-pH}e^{-e\psi_0/kT}}. \tag{A12}$$

Excluding σ between equations (A11) and (A12) yields a single expression which can readily be solved numerically for ψ_0. Thus it is possible to calculate membrane surface potential for any interaction distance, this potential including a correction for reassociation of ionizable groups. The charge density corresponding to the corrected surface potential was then used in the expressions for sphere repulsion energy. In practice this correction was usually very small due to the low pKa value of the carboxyl group.

A weakness of the vesicle interaction energy model is that, in solutions containing calcium, it fails to take into account the dependence of calcium binding on distance of separation of the vesicles (as the vesicles approach, their surface potentials will increase and thus the surface calcium concentration increases, (via Boltzmann's relationship), leading to further calcium binding). This effect is difficult to quantify because of insufficient data despite several attempts made by various authors to obtain values for calcium binding constant(s) (by a variety of techniques, e.g., interpretation of electrophoresis results [Barton, 1968] and fluorescence measurements [Haynes, 1974]). Clearly, however, a significant reduction in membrane surface charge density will only be achieved as a result of increased divalent cation binding if the vesicles are allowed to approach to within one Debye length (8 Å) of each other since membrane surface potential is only markedly increased at these small distances of separation. The small probabilities of approach (section 2.2.3) imply that such augmented Ca^{2+} binding will rarely occur.

Appendix B. Model for phosphatidylserine ionization

B.1. Membrane surface potential in the absence of calcium

To obtain theoretical estimates of surface potential ψ_0 for PS membranes, it is necessary to consider the ionization characteristics of the individual dissociable groups in the headgroup region of the lipid. Each PS molecule has one basic (amino) and two acidic (phosphate and carboxyl) groups in its headgroup region. For these groups, we write the following expressions for the degree of ionization, α (cf. Appendix A, equation A9):

$$\alpha_1 = \frac{K_{a_1}}{K_{a_1} + [H^+]} \tag{B1}$$

$$\alpha_2 = \frac{K_{a_2}}{K_{a_2} + [H^+]} \tag{B2}$$

$$\alpha_3 = \frac{K_b}{K_b + [OH^-]} \tag{B3}$$

where α_1 and K_{a_1} are the degree of ionization and acidic dissociation constant, respectively, for the phosphate moiety of PS, α_2 and K_{a_2} refer to similar parameters for the carboxyl group, and α_3 and K_b are the degree of ionization and

822

basic dissociation constant for the amino group. The ion concentrations (hydrogen and hydroxyl) in equations (B1), (B2), and (B3) will be surface concentrations for the case of PS molecules in lamellar array in a membrane. These surface concentrations are given by a Boltzmann factor as follows:

$$[H^+]_S = [H^+]_B x. \tag{B4}$$

and

$$[OH^-]_S = [OH^-]_B \frac{1}{x}, \tag{B5}$$

where

$$x \equiv \exp(-e\psi_0/kT).$$

$[H^+]_S$ and $[H^+]_B$ are the surface and bulk hydrogen ion concentrations, respectively, $[OH^-]_S$ and $[OH^-]_B$ refer to hydroxyl ion concentrations, e is the electronic charge, k is Boltzmann's constant, and T is the absolute temperature.

Combining equations (B1), (B2), and (B3) with (B4) and (B5) yields an expression for the total net negative charge density, σ, at the surface of a PS vesicle in the absence of ion adsorption from the aqueous phase:

$$\sigma = q\left(\frac{K_{a_1}}{K_{a_1} + [H^+]_B x} + \frac{K_{a_2}}{K_{a_2} + [H^+]_B x} - \frac{K_b}{K_b + [OH^-]_B \frac{1}{x}} \right). \tag{B6}$$

In equation (B6), q is the electronic charge multiplied by the common density of each of the three dissociable group types. This, multiplied by the degree of ionization α, gives the actual surface charge density, σ.

Another expression for membrane surface charge density in terms of potential can be obtained from the Gouy-Chapman theory. Here we use a form of the Gouy-Chapman equation modified for the case of surface charge occupying a region of finite thickness, that is, charges not all coplanar (Parsegian and Gingell, 1973; Parsegian, 1974):

$$\sigma = \frac{e}{Q}\left[\sqrt{\frac{1}{x}(x-1)} - Qnh\left(\frac{1}{x} - x\right) \right], \tag{B7}$$

where

$$Q \equiv \sqrt{2\pi e^2/\epsilon kTn}.$$

and ϵ is the bulk dielectric constant of the aqueous phase, n is the concentration of the symmetrical monovalent electrolyte, and h denotes the thickness of the surface layer on the membrane, which is penetrable to ions in solution and contains the surface charges. A value of 4 Å was used throughout these calculations for this thickness; the PS headgroup has a length of approximately 8 Å (denoted δ in section 2), and it is assumed that about half of its volume is penetrable to ions in solution.

Equation (B7) applies to the case of a single isolated membrane, that is, at infinite separation from any other membrane. It retains the nonlinear form of the underlying Poisson-Boltzmann equation, unlike the equations in Appendix A. When combined with equation (B6), excluding σ between the two expressions, the result is a single, transcendental algebraic equation that may be solved numerically for ψ_0 after inserting known values for the dissociation constants, ion concentrations, and so forth. The numerical solution may be simplified by considering the values of some of the constants concerned: the following dissociation constants for the ionizable groups of PS were obtained from the combined data of Abramson and co-workers (1964), Bangham and Papahadjopoulos (1966), Papahadjopoulos (1968), and Hauser and collaborators (1976):

K_b (amino group) $= 10^{-4}$

K_{a_1} (phosphate) $> 10^{-1.2}$ (the isoelectric point for PS vesicles $=$ pH 1.2 in 0.1 M NaCl, under which condition the phosphate group still appears to be ionized)

and K_{a_2} (carboxyl) $\approx 10^{-4.5}$ (estimates vary between $10^{-3.4}$ and $10^{-5.2}$).

Using these figures, it can be seen that the second term in the argument in parentheses in equation (B6) (i.e., the term relating to the ionization of the carboxyl group) will start to change significantly from 1 at a level of ψ_0 lower than that which will bring about significant changes in the first and third terms. Thus it is legitimate to write the following simplification for equation (B6) at "low" potentials, where ψ_0 lies between zero and -300 mV:

$$\sigma = \frac{qK_{a_2}}{K_{a_2} + [\mathrm{H^+}]_B x} \qquad \textbf{(B8)}$$

Combining (B8) with (B7) yields a relatively simple expression to be solved numerically as follows:

$$e\left[\sqrt{\frac{1}{x}}\,(x-1) - Qnh\left(\frac{1}{x} - x\right)\right](K_{a_2} + [\mathrm{H^+}]_B x) - QqK_{a_2} = 0. \qquad \textbf{(B9)}$$

The value of q was obtained by assuming that each lipid headgroup in a 100% PS membrane occupies an area of 65 Å^2 and by assigning one electronic charge to each headgroup, yielding $q \approx 74,000$ e.s.u./cm^2. The solution to equation (B9) is then $\psi_0 = -80$ mV for vesicles in 145 mM NaCl at pH 7.4. This value may be compared to the figure of -55.2 mV quoted in section 2.2.3 as the electrokinetic potential for PS membranes in the absence of calcium (Papahadjopoulos, 1968). Clearly, at these high surface charge densities, ζ is an underestimate of ψ_0, in agreement with the observations of numerous authors (e.g., MacDonald and Bangham, 1972; Carroll and Haydon, 1975). To investigate the relationship between membrane surface charge density and potential in more detail, theoretical values for ψ_0 were calculated for lower values of q than that quoted above (using equation B9). This is equivalent to determining membrane surface potential for PS bilayers in which the acidic phospholipid is diluted in a neutral dipolar lipid such as a lecithin. The results of these calculations are shown in Fig. 9 (curve a). On this graph, the theoretical curve may be compared with an experimentally derived curve of electrokinetic potentials for PS diluted in egg yolk lecithin to varying extents (curve b). Electrokinetic potential measurements were carried out using standard cylindrical microelectrophoresis apparatus (Bangham et al., 1958). The results obtained agree qualitatively with those of Barton

824

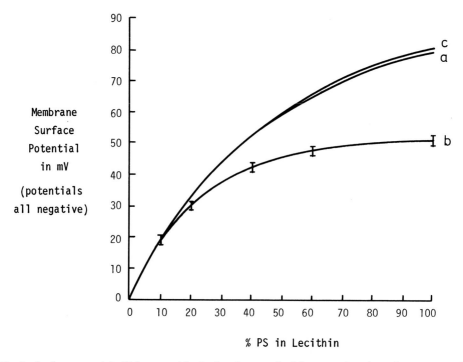

Fig. 9. Surface potential of bilayer vesicles in the absence of calcium as a function of percentage of PS in lecithin. Curve (a) theoretical; (b) measured ζ-potential; (c) constructed curve, based on the 10% point of curve (b). Error bars represent standard errors.

(1968), although the latter author did not take into account membrane surface penetrability and the distinctive ionization characteristics of PS in his theoretical analysis. From the curves in Fig. 9 it is again clear that the electrokinetic potential is an underestimate of the "true" membrane surface potential at high surface charge densities. This underlines the improbability of PS vesicle fusion in the absence of calcium. In section 2 it was shown that vesicle contact would be an improbable event if surface potential = -55 mV; it would be far more unlikely if $\psi_0 = -80$ mV, thus theory is consistent with the experimental observations on lack of fusion in the absence of Ca^{2+}. The close similarity of ζ to ψ_0 at *low* surface charge densities (e.g., 10% PS in lecithin) suggests that our model for the PS surface in the absence of calcium may well be a valid approximation. This view is supported by the similarity of curve a (the theoretical curve of ψ_0 against percent of PS) to curve c, the latter being an extrapolation of the experimental result for 10% PS in lecithin to higher charge densities. To obtain curve c, the charge density corresponding to ζ for 10% PS in lecithin was ascertained using equation B7; this equation was then solved to obtain ψ_0 for multiples of the charge density of 10% PS in lecithin.

B.2. Presence of calcium

A theoretical value for the surface potential (ψ_0*) of PS membranes when the bulk aqueous phase contains 1 mM Ca^{2+} cannot be determined simply by modifying equation (B7) to include divalent cations in the medium because calcium appears to bind specifically to the surface of vesicles made from acidic phospholipids; thus the form of equation (B6) will change too. The extent to which calcium binds to PS has been determined by labeling experiments in which the activity per unit weight of dried phospholipid (using ^{45}Ca) is obtained (Bangham and Papahadjopoulos, 1966; Papahadjopoulos, 1968; Papahadjopoulos et al., 1977). When the bulk concentration of calcium is 1 mM, one *equivalent* of calcium appears to be bound to each PS headgroup. This would seem to be sufficient binding to bring about complete surface charge neutralization from simple stoichiometric considerations. However, PS vesicles in the presence of 1 mM Ca^{2+} retain an electrokinetic potential of approximately -36 mV. To explain this apparent contradiction, Papahadjopoulos (1968) has suggested that the binding of one equivalent of calcium per PS headgroup prevents ionization of the molecule's amino group by forming a coordination complex. If this is the case, the relevant form of equation (B6) in the presence of calcium is:

$$\sigma = q^*\left(\frac{K_{a_1}^*}{K_{a_1}^* + [\text{H}^+]_B x^*} + \frac{K_{a_2}^*}{K_{a_2}^* + [\text{H}^+]_B x^*} - 1\right). \tag{B10}$$

The asterisks indicate the changed values of the various terms (q, K_a, and ψ_0) on calcium binding. The -1 relates to the binding of one equivalent of calcium per headgroup.

If amino group ionization were still permitted in the presence of calcium, one would be obliged to write:

$$\sigma = q^*\left(\frac{K_{a_1}^*}{K_{a_1}^* + [\text{H}^+]_B x^*} + \frac{K_{a_2}^*}{K_{a_2}^* + [\text{H}^+]_B x^*} - \frac{K_b^*}{K_b^* + [\text{OH}^-]_B \frac{1}{x^*}} - 1\right). \tag{B11}$$

The form of equation (B11) reinforces the view that calcium binding prevents amino group ionization, for at "low" ψ_0* (the condition when calcium is bound) the argument in parentheses ≈ 0 or is negative, yielding a value for σ that cannot be reconciled with the experimental ζ potential. Equation (B10) is thus confirmed and may be simplified in a way similar to the treatment of equation (B6) (the equation for membrane surface charge density in the absence of calcium), yielding an expression analogous to equation (B8):

$$\sigma = \frac{q^* K_{a_2}^*}{K_{a_2}^* + [\text{H}^+]_B x^*}. \tag{B12}$$

The modified Gouy-Chapman equation relating membrane surface charge to surface potential in the presence of divalent cations may be written (Parsegian and Gingell, 1973) as follows:

$$\sigma = \frac{e}{Q}\left[\sqrt{\frac{1}{x^*} + \frac{\eta}{2}} (x^* - 1) - Qnh\left(\frac{1}{x^*} - (1 - \eta)x^* - \eta(x^*)^2\right)\right], \qquad \textbf{(B13)}$$

where η denotes the fraction of the positive counterion charge carried by divalent cations, and n is the concentration of monovalent anions in the medium. Combining equations (B12) and (B13) yields a single expression that may be solved numerically for ψ_0^* in a similar way to the solution of equation (B9) for ψ_0:

$$e\left[\sqrt{\frac{1}{x^*} + \frac{\eta}{2}} (x^* - 1) - Qnh\left(\frac{1}{x^*} - (1 - \eta)x^* - \eta(x^*)^2\right)\right]\left(K_{a_2}^* + [\text{H}^+]_Bx^*\right) - Qq^*K_{a_2}^* = 0 \quad \textbf{(B14)}$$

The value of q^* in equation (B14) was deduced by assuming that the area per lipid headgroup in a 100% PS membrane in the presence of 1 mM Ca^{2+} is 48 Å^2 (calcium brings about a condensation of PS membranes; thus the area per headgroup is less than the 65 Å^2 quoted for the case of no divalent cation). This gives $q^* \approx 10^5$ e.s.u./cm^2. Clearly, the numerical solution to equation (B14) will then yield a value for $|\psi_0^*|$ greatly in excess of the electrokinetic potential quoted for PS vesicles in the presence of 1 mM Ca^{2+} ($|\zeta| = 36$ mV), unless the values of some of the "constants in the equation (e.g., Ka_2 and h) are changed by calcium binding. For ψ_0^* to be of similar magnitude to ζ, it would be necessary for either the thickness, h, of the headgroup penetrable region, or pKa_2 to increase on calcium binding. However, divalent cation adsorption probably acts to contract rather than expand the surface penetrable region of the membrane. Furthermore, no change in the pKa of the carboxyl group of PS is seen when 1 mM Ca^{2+} is added to the aqueous medium (unpublished measurements of ζ-potential against pH for PS membranes in the presence of 1 mM Ca^{2+}, using both 100% PS membranes and 20% PS diluted in lecithin). Thus it would appear that the "true" potential at the surface of a PS membrane when the bulk aqueous phase contains 1 mM Ca^{2+} considerably exceeds the quoted electrokinetic potential. This will only be false if the stoichiometry of calcium binding is not what we have modelled on the basis of the labeling studies quoted above (Papahadjopoulos, 1968). Recent experimental work suggests that more than one equivalent of calcium may be able to bind to each PS headgroup under different conditions of ionic strength from those used by Papahadjopoulos (Hauser et al., 1976). We investigated in more detail the extent of calcium binding under the present experimental conditions (aqueous phase containing 145 mM NaCl and 1 mM CaCl$_2$ at pH 7.4). Theoretical and experimental curves of membrane surface potential were plotted against percent of PS in lecithin for membranes in the presence of calcium (compare with the curves in Fig. 9 that were obtained for membranes in the absence of divalent cations). Fig. 10 shows the results of these experiments. These curves show that ζ diverges from ψ_0^* at high surface charge densities, as was the case in the absence of calcium. This divergence, however, is also apparent

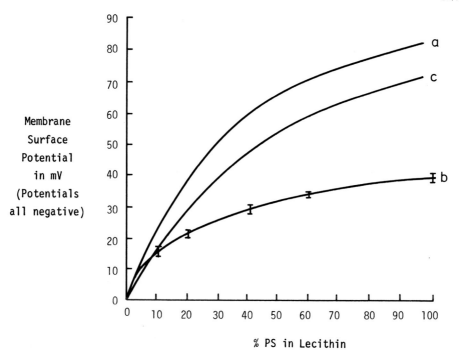

Fig. 10. Same as Fig. 9, except that calcium is present.

at lower charge densities. Thus, at 10% PS in lecithin, $\zeta = -15.6$ mV whereas $\psi_0^* = -23$ mV (in the absence of calcium, $\zeta = \psi_0$ at a corresponding dilution of the acidic phospholipid). It is surprising that there should be such a difference between theoretical predictions and experimental results at a charge density where ζ would be expected to be an accurate measure of membrane surface potential. One cause of this difference might be that the extent of calcium binding exceeds the value we have incorporated into our theoretical model, as suggested above. If it is assumed that $\zeta = -15.6$ mV is the "correct" value for the surface potential of 10% PS-in-lecithin membranes in 1 mM Ca^{2+}, it is possible to work backward and deduce a figure for the "correct" amount of calcium bound under these conditions, as follows. From equation (B13), a surface potential of -15.6 mV corresponds to an area of 743 Å^2 per electronic charge. On the average, one PS molecule will be found per 480 Å^2 of membrane surface (as the PS is diluted to 10% in lecithin). Thus the number of electronic charges per PS headgroup ≈ 0.65. This suggests that 0.35 equivalents of calcium are bound to each PS headgroup, in addition to the one equivalent bound that prevents ionization of the headgroup's amino moiety when the bulk calcium concentration is 1 mM.

If it is true that 35% of the charge on each PS headgroup is neutralized as a result of calcium binding under these experimental conditions, a membrane composed solely of PS will have a surface potential of -73 mV (reading from curve c,

Fig. 10, and assuming that the degree of calcium binding remains constant as the percent of PS in lecithin increases). This value is still much greater in magnitude than the electrokinetic potential for identical experimental conditions (-36 mV). It is very difficult to reconcile such a high membrane surface potential (-73 mV) with the idea that PS vesicles are able to fuse in 1 mM Ca^{2+} (Papahadjopoulos and Poste, 1975; Papahadjopoulos et al., 1975, 1977).

Appendix C.

Effect of vesicle interaction and contact on the transition temperature of membrane phospholipids

Recent work has indicated that the fluidity of an isolated membrane can be altered at a constant temperature merely by changing its ionic environment and hence the electrostatic free energy of the system (Träuble and Eibl, 1974). This concept can be applied to the case of two charged phospholipid membranes approaching each other in a salt solution, where the interaction of their accompanying electrical double layers will effect a change in the interfacial free energy of each membrane and hence its fluid-gel phase transition temperature, T_t.

Consider first the free energy, G, of an isolated system of two membranes at a temperature well below any transition temperature. The energy of this system at close approach of the membranes will differ from that at infinite separation by a quantity equal to the integral of the force needed to bring the surfaces together, as follows:

$$G_\gamma^c = G_\gamma^\infty + g_\gamma, \tag{C1}$$

where c refers to membrane contact, ∞ refers to infinite separation, γ indicates the gel state of the lipid below T_t (the temperature of the system being sufficiently low to ensure that the membranes remain in the gel phase at all distances of separation), and g is the electrostatic free energy (the integral of the force needed to bring the membranes to contact from infinity).

The analogous free energy expression for membranes brought from ∞ to contact, where the whole process is conducted at an elevated temperature so that the membranes are always in a fluid state may be written:

$$G_\lambda^c = G_\lambda^\infty + g_\lambda. \tag{C2}$$

where λ indicates the fluid state.

At the transition temperature, T_t^c, for membranes in contact (signifying that the transition temperature is a function of separation distance), the molar free energies of the gel and fluid states will be equal at constant pressure:

$$G_\gamma^c = G_\lambda^c,$$

Therefore
$$\tag{C3}$$
$$G_\gamma^\infty + g_\gamma = G_\lambda^\infty + g_\lambda,$$

that is:

$$H_\gamma^\infty - T_t^c S_\gamma^\infty + g_\gamma = H_\lambda^\infty - T_t^c S_\lambda^\infty + g_\lambda, \qquad \text{(C4)}$$

where H = enthalpy, S = entropy. So that:

$$T_t^c = \frac{H_\lambda^\infty - H_\gamma^\infty}{S_\lambda^\infty - S_\gamma^\infty} + \frac{g_\lambda - g_\gamma}{S_\lambda^\infty - S_\gamma^\infty}. \qquad \text{(C5)}$$

At the transition temperature for membranes at infinite separation (T_t^∞):

$$H_\gamma^\infty - T_t^\infty S_\gamma^\infty = H_\lambda^\infty - T_t^\infty S_\lambda^\infty, \qquad \text{(C6)}$$

therefore

$$T_t^\infty = \frac{H_\lambda^\infty - H_\gamma^\infty}{S_\lambda^\infty - S_\gamma^\infty} \qquad \text{(C7)}$$

Substituting Equation (C7) into (C5) yields:

$$T_t^c = T_t^\infty + \frac{g_\lambda - g_\gamma}{S_\lambda^\infty - S_\gamma^\infty}. \qquad \text{(C8)}$$

As the entropy of the fluid state exceeds that of the gel and $g_\gamma > g_\lambda$ (these electrostatic energy terms are positive and the surface charge density in the gel state exceeds that of the fluid), it can be seen that $T_t^c < T_t^\infty$, that is, the transition temperature falls as the membranes approach, implying that membranes in the gel state would tend to become more fluid.

Acknowledgments

We would like to thank Dr V. Adrian Parsegian for showing us the elegant relation between transition temperature and interaction energy, and Dr. A. D. Bangham who kindly invited us to use his microelectrophoresis equipment. L. G. is grateful to the Medical Research Council of Great Britain for financial support.

References

Abramson, M. B., Katzman, R. and Gregor, H. P. (1964) Aqueous dispersions of phosphatidylserine. Ionic properties. J. Biol. Chem. 239, 70–76.

Ahkong, Q. F., Fisher, D., Tampion, W. and Lucy, J. A. (1973) The fusion of erythrocytes by fatty acids, esters, retinol and α-tocopherol. Biochem. J. 136, 147–155.

Ahkong, Q. F., Tampion, W. and Lucy, J. A. (1975) Promotion of cell fusion by divalent cation ionophores. Nature 256, 208–209.

Allan, D., Billah, M. M., Finean, J. B. and Michell, R. H. (1976) Release of diacylglycerol-enriched vesicles from erythrocytes with increased intracellular [Ca²⁺]. Nature 261, 58–60.

Allison, A. C. (1971) Analogies between triggering mechanisms in immune and other cellular reactions. In: Cell Interactions. Proc. Third Lepetit Colloq. London (Silvestri, L. G., ed.) pp. 156–161, Elsevier/North-Holland, Amsterdam.

830

Bangham, A. D. and Papahadjopoulos, D. (1966) Biophysical properties of phospholipids. I. Interaction of phosphatidylserine monolayers with metal ions. Biochim. Biophys. Acta 126, 181–184.

Bangham, A. D., Flemans, R., Heard, D. H. and Seaman, G. V. F. (1958) An apparatus for microelectrophoresis of small particles. Nature 182, 642–644.

Barton, P. G. (1968) The influence of surface charge density of phosphatides on the binding of some cations. J. Biol. Chem. 243, 3884–3890.

Bray, D. (1973) Model for membrane movements in the neural growth cone. Nature 244, 93–96.

Carroll, B. J. and Haydon, D. A. (1975) Electrokinetic and surface potentials at liquid interfaces. J. Chem. Soc. (Faraday Trans. I) 71, 361–377.

Chen, Y. S. and Hubbell, W. L. (1973) Temperature and light-dependent structural changes in rhodopsin-lipid membranes. Exp. Eye Res. 17, 517–532.

Cook, G. M. W. and Stoddart, R. W. (1973) Surface Carbohydrates of the Eukaryotic Cell, chap. 3. Academic Press, London.

Costello, M. J. and Gulik-Krzywicki, T. (1976) Correlated x-ray diffraction and freeze-fracture studies on membrane model systems. Perturbations induced by freeze-fracture preparative procedures. Biochim. Biophys. Acta 455, 412–432.

Deamer, D. W., Leonard, R., Tardieu, A. and Branton, D. (1970) Lamellar and hexagonal lipid phases visualized by freeze-etching. Biochim. Biophys. Acta 219, 47–60.

Dean, P. M. (1975) Exocytosis modelling: an electrostatic function for calcium in stimulus-secretion coupling. J. Theor. Biol. 54, 289–308.

de Petris, S. (1975) Concanavalin A receptors, immunoglobulins and Θ antigen of the lymphocyte surface. Interactions with concanavalin A and with cytoplasmic structures. J. Cell Biol. 65, 123–146.

Derjaguin, B. V. (1934) Untersuchungen über die reibung und adhäsion, IV. Theorie des anhaftens kleiner teilchen. Kolloid-Z. 69, 155–164.

Devereux, O. F. and de Bruyn, P. L. (1963) Interaction of Plane-Parallel Double Layers. M. I. T. Press, Cambridge, Massachusetts.

Douglas, W. W. (1968) Stimulus-secretion coupling: the concept and clues from chromaffin and other cells. Brit. J. Pharm. 34, 451–474.

Douglas, W. W. (1974) Involvement of calcium in exocytosis and the exocytosis-vesiculation sequence; Biochem. Soc. Symp. 39, 1–28.

Edidin, M. (1974) Rotational and translational diffusion in membranes. Ann. Rev. Biophys. 3, 179–201.

Edwards, W., Phillips, J. H. and Morris, S. J. (1974) Structural changes in chromaffin granules induced by divalent cations. Biochim. Biophys. Acta 356, 164–173.

Foreman, J. C., Mongar, J. L. and Gomperts, B. D. (1973) Calcium ionophores and movement of calcium ions following the physiological stimulus to a secretory process. Nature 245, 249–251.

Galla, H-J. and Sackman, E. (1975) Chemically induced lipid phase separation in model membranes containing charged lipids: a spin label study. Biochim. Biophys. Acta 401, 509–529.

Gershfeld, N. L. (1976) Physical chemistry of lipid films at fluid interfaces. Ann. Rev. Phys. Chem. 27, 349–368.

Gilula, N. B. (1978) Structure of intercellular junctions. In: Intercellular Junctions and Synapses in Development (Feldman, J., Gilula, N. B. and Pitts, J., eds.) pp. 1–22 Chapman and Hall Ltd. London.

Gingell, D. (1967) Membrane surface potential in relation to a possible mechanism for intercellular interactions and cellular responses: a physical basis. J. Theor. Biol. 17, 451–482.

Gingell, D. (1973) Membrane permeability change by aggregation of mobile glycoprotein units. J. Theor. Biol. 38, 677–679.

Gingell, D. (1976) Electrostatic control of membrane permeability via intramembranous particle aggregation. In: Mammalian Cell Membranes (Jamieson, G. A. and Robinson, D. M., eds.) pp. 198–223, Butterworths, London.

Gingell, D. and Fornes, J. A. (1975) Demonstration of intermolecular forces in cell adhesion using a new electrochemical technique. Nature 256, 210–211.

Gingell, D. and Parsegian, V. A. (1972) Computation of van der Waals interactions in aqueous systems using reflectivity data. J. Theor. Biol. 36, 41–52.

Gingell, D. and Todd, I. (1975) Adhesion of red blood cells to charged interfaces between immiscible liquids. A new method. J. Cell Sci. 18, 227–239.

Gingell, D., Parsegian, V. A. and Todd I. (1977) Long-range attraction between red cells and a hydrocarbon surface. Nature 268, 767–769.

Hauser, H., Darke, A. and Phillips, M. C. (1976) Ion-binding to phospholipids. Eur. J. Biochem. 62, 335–344.

Haydon, D. A. and Taylor, J. L. (1968) Contact angles for thin lipid films and the determination of London-van der Waals forces. Nature 217, 739–740.

Haynes, D. H. (1974) 1-Anilino-8-naphthalenesulfonate: a fluorescent indicator of ion binding and electrostatic potential on the membrane surface. J. Membr. Biol. 17, 341–366.

Huang, L. and Pagano, R. E. (1975) Interaction of phospholipid vesicles with cultured mammalian cells. I. Characteristics of uptake. J. Cell Biol. 67, 38–48.

Ito, I., Ohnishi, S.-I., Ishinaga, M. and Kito, M. (1975) Synthesis of a new phosphatidylserine spin-label and calcium-induced lateral phase separation in phosphatidylserine-phosphatidylcholine membranes. Biochemistry 14, 3064–3069.

Jacobson, K. and Papahadjopoulos, D. (1975) Phase transitions and phase separations in phospholipid induced by changes in temperature, pH and concentration of bivalent cations. Biochemistry 14, 152–161.

Jaffe, L. F. (1977) Electrophoresis along cell membranes. Nature 265, 600–602.

Karnovsky, M. J. and Unanue, E. R. (1973) Mapping and migration of lymphocyte surface macromolecules. Fed. Proc. 32, 55–59.

Katz, B. and Miledi, R. (1967) A study of synaptic transmission in the absence of nerve impulses. J. Physiol. 192, 407–436.

Korn, E. D., Bowers, B., Batzri, S., Simmons, S. R. and Victoria, E. J. (1974) Endocytosis and exocytosis: role of microfilaments and involvement of phospholipids in membrane fusion. J. Supramol. Str. 2, 517–528.

Lagunoff, D. (1973) Membrane fusion during mast cell secretion. J. Cell Biol. 57, 252–259.

Lau, A. L. Y. and Chan, S. I. (1974) Nuclear magnetic resonance studies of the interaction of alamethicin with lecithin bilayers. Biochemistry 13, 4942–4948.

Lawson, D., Raff, M. C., Gomperts, B., Fewtrell, C. and Gilula, N. B. (1977) Molecular events during membrane fusion. A study of exocytosis in rat peritoneal mast cells. J. Cell Biol. 72, 242–259.

Le Neveu, D. M., Rand, R. P. and Parsegian, V. A. (1976) Measurement of forces between lecithin bilayers. Nature 259, 601–603.

Lucy, J. A. (1970) The fusion of biological membranes. Nature 227, 814–817.

MacDonald, R. C. and Bangham, A. D. (1972) Comparison of double layer potentials in lipid monolayers and lipid bilayer membranes. J. Membr. Biol. 7, 29–53.

Martin, F. J. and MacDonald, R. C. (1976) Lipid vesicle-cell interactions. III. Introduction of a new antigenic determinant into erythrocyte membranes. J. Cell Biol. 70, 515–526.

McNutt, S. and Weinstein, R. S. (1973) Membrane ultrastructure at mammalian intercellular junctions. Progr. Biophys. Mol. Biol. 26, 45–101.

Miledi, R. (1973) Transmitter release induced by injection of calcium ions into nerve terminals. Proc. Roy. Soc. Lond. B 183, 421–425.

Moore, M. (1976) Fusion of liposomes containing conductance probes with black lipid films. Biochim. Biophys. Acta 426, 765–771.

Mooseker, M. S. and Tilney, L. G. (1975a) Actin filament-membrane attachment: are membrane particles involved? J. Cell Biol. 67, 293a (abstr).

Mooseker, M. S. and Tilney, L. G. (1975b) Organization of an actin filament-membrane complex, filament polarity and membrane attachment in the microvilli of intestinal epithelial cells. J. Cell Biol. 67, 725–743.

Neher, E. (1974) Asymmetric membranes resulting from the fusion of two black lipid bilayers. Biochim. Biophys. Acta 373, 327–336.

832

Nicolson, G. L. and Painter, R. G. (1973) Anionic sites of human erythrocyte membranes. II. Antispectrin-induced transmembrane aggregation of the binding sites for positively charged colloidal particles. J. Cell Biol. 59, 395–406.

Ninham, B. and Parsegian, V. A. (1971) Electrostatic potential between surfaces bearing ionizable groups in ionic equilibrium with physiological saline solution. J. Theor. Biol. 31, 405–428.

Ohnishi, S-I. and Ito, T. (1974) Calcium-induced phase separations in phosphatidylserine-phosphatidylcholine membranes. Biochemistry 13, 881–887.

Ohshima, H. (1974) Diffuse double layer interaction between two parallel plates with constant surface charge density in an electrolyte solution. II. The interaction between dissimilar plates. Colloid Polymer Sci. 252, 257–267.

Ohshima, H. (1975) Diffuse double layer interaction between two parallel plates with constant surface charge density in an electrolyte solution. III. Potential energy of double layer interaction. Colloid Polymer Sci. 253, 150–157.

Pagano, R. E. and Huang, L. (1975) Interaction of phospholipid vesicles with cultured mammalian cells. II. Studies of mechanism. J. Cell Biol. 67, 49–60.

Pagano, R. E., Huang, L. and Wey, C. (1974) Interaction of phospholipid vesicles with cultured mammalian cells. Nature 252, 166–167.

Papahadjopoulos, D. (1968) Surface properties of acidic phospholipids: interaction of monolayers and hydrated liquid crystals with uni- and bi-valent metal ions. Biochim. Biophys. Acta 163, 240–254.

Papahadjopoulos, D. and Ohki, S. (1969) Stability of asymmetric phospholipid membranes. Science 164, 1075–1077.

Papahadjopoulos, D. and Poste, G. (1975) Calcium-induced phase separation and fusion in phospholipid membranes. Biophys. J. 15, 945–948.

Papahadjopoulos, D., Poste, G. and Schaeffer, B. E. (1973) Fusion of mammalian cells by unilamellar lipid vesicles: influence of lipid surface charge, fluidity and cholesterol. Biochim. Biophys. Acta 323, 23–42.

Papahadjopoulos, D., Poste, G., Schaeffer, B. E. and Vail, W. J. (1974) Membrane fusion and molecular segregation in phospholipid vesicles. Biochim. Biophys. Acta 352, 10–28.

Papahadjopoulos, D., Vail, W. J., Jacobson, K. and Poste, G. (1975) Cochleate lipid cylinders: formation by fusion of unilamellar lipid vesicles. Biochim. Biophys. Acta 394, 483–491.

Papahadjopoulos, D., Hui, S., Vail, W. J. and Poste, G. (1976a) Studies of membrane fusion. I. Interactions of pure phospholipid membranes and the effect of myristic acid, lysolecithin, proteins and dimethylsulfoxide. Biochim. Biophys. Acta 448, 245–264.

Papahadjopoulos, D., Vail, W. J. Pangborn, W. A. and Poste, G. (1976b) Studies on membrane fusion. II. Induction of fusion in pure phospholipid membranes by calcium ions and other divalent metals. Biochim. Biophys. Acta 448, 265–283.

Papahadjopoulos, D., Vail, W. J., Newton, C., Nir, S., Jacobson, K., Poste, G. and Lazo, R. (1977) Studies on membrane fusion. III. The role of calcium-induced phase changes. Biochim. Biophys. Acta 465, 579–598.

Parsegian, V. A. (1974) Possible modulation of reactions on the cell surface by changes in electrostatic potential that accompany cell contact. Ann. N.Y. Acad. Sci. 238, 362–371.

Parsegian, V. A. and Gingell, D. (1973) A physical force model of biological membrane interaction. In: Recent Advances in Adhesion (Lee, L-H, ed.) pp. 153–192, Gordon and Breach, New York.

Phillips, M. C., Williams, R. M. and Chapman, D. (1969) On the nature of hydrocarbon chain motions in lipid liquid crystals. Chem. Phys. Lipids 3, 234–244.

Pinto da Silva, P. and Martínez-Palomo, A. (1975) Distribution of intramembranous particles and gap junctions in normal and transformed 3T3 cells studies in situ, in suspension, and treated with concanavalin A. Proc. Nat. Acad. Sci. U.S.A. 72, 572–576.

Poo, M.-M. and Robinson, K. R. (1977) Electrophoresis of concanavalin A receptors along embryonic muscle cell membrane. Nature 265, 602–605.

Poole, A. R., Howell, J. I. and Lucy, J. A. (1970) Lysolecithin and cell fusion. Nature 227, 810–813.

Poste, G. and Allison, A. C. (1973) Membrane fusion. Biochim. Biophys. Acta 300, 421–465.

Poste, G. and Nicolson, G. L., eds. (1977) Dynamic Aspects of Cell Surface Organization. Cell Surface Reviews, Vol. 3. Elsevier/North-Holland, Amsterdam.

Poste, G. and Papahadjopoulos, D. (1976) Lipid vesicles as carriers for introducing materials into cultured cells: influence of vesicle lipid composition on mechanism(s) of vesicle incorporation into cells. Proc. Nat. Acad. Sci. U.S.A. 73, 1603–1607.

Prestegard, J. H. and Fellmeth, B. (1974) Fusion of dimyristoyllecithin vesicles as studied by proton magnetic resonance spectroscopy. Biochemistry 13, 1122–1126.

Rand, R. P. and Sengupta, S. (1972) Cardiolipin forms hexagonal structures with divalent cations. Biochim. Biophys. Acta 255, 484–492.

Reiss-Husson, F. (1967) Structure des phases liquide-cristallines de différents phospholipides, monoglycérides, sphingolipides, anhydres ou en présénce d'eau. J. Mol. Biol. 25, 363–382.

Requena, J. and Haydon, D. A. (1975) Van der Waals forces in oil-water systems from the study of thin lipid films. II. The dependence of the van der Waals free energy of thinning on film composition and structure. Proc. Roy. Soc. Lond. A. 347, 161–177.

Satir, B., Schooley, C. and Satir, P. (1973) Membrane fusion in a model system: mucocyst secretion in Tetrahymena. J. Cell Biol. 56, 153–176.

Scheid, A. and Choppin, P. W. (1974) Identification of biological activities of paramyxovirus glycoproteins. Activation of cell fusion, hemolysis and infectivity by proteolytic cleavage of an inactive precursor protein of Sendai virus. Virology 57, 475–490.

Schober, R., Nitsch, C., Rinne, U. and Morris, S. J. (1977) Calcium-induced displacement of membrane-associated particles upon aggregation of chromaffin granules. Science, 195, 495–497.

Schollmeyer, J. V., Goll, D. E., Tilney, L. G., Mooseker, M., Robson, R. and Stromer, M. (1974) Localization of α-actinin in non-muscle material. J. Cell Biol. 63, 304a (abstr.).

Speth, V. and Wunderlich, F. (1973) Membranes of Tetrahymena. II. Direct visualization of reversible transitions in biomembrane structure induced by temperature. Biochim. Biophys. Acta 291, 621–628.

Tilney, L. G. and Detmers, P. (1975) Actin in erythrocyte ghosts and its association with spectrin. Evidence for a non-filamentous form of these two molecules in situ. J. Cell Biol. 66, 508–520.

Träuble, H. and Eibl, H. (1974) Electrostatic effects on lipid phase transitions: membrane structure and ionic environment. Proc. Nat. Acad. Sci. 71, 214–219.

Verwey, E. J. W. and Overbeek, J. Th. G. (1948) Theory of the Stability of Lyophobic Colloids. Elsevier/North-Holland, Amsterdam.

Weiss, R. L., Goodenough, D. A. and Goodenough U. W. (1977a) Membrane particle arrays associated with the basal body and with contractile vacuole secretion in Chlamydomonas. J. Cell Biol. 72, 133–143.

Weiss, R. L. Goodenough, D. A. and Goodenough, U. W. (1977b) Membrane differentiation at sites specialized for cell fusion. J. Cell Biol. 72, 144–160.

Yahara, I., Edelman, G. M. (1975) Electron microscopic analysis of the modulation of lymphocyte receptor mobility. Exp. Cell Res. 91, 125–142.

Zwaal, R. F. A., Roelofsen, B. and Colley, C. M. (1973) Localization of red cell membrane constituents. Biochim. Biophys. Acta 300, 159–182.

Subject Index

A23187
— aggregation of intramembranous particles by, 347
— cell fusion and, 291–93, 295, 378
— cortical granule reaction, 28, 36, 47
— intracellular Ca^{2+} and, 785
— lymphocyte immunoglobulin synthesis and, 576
— mast cell exocytosis, 420, 425, 427, 430–31, 433, 636
— mucocyst exocytosis, 697–710
— myoblast fusion and, 150
— neutrophil secretion and, 463, 468, 477
— platelet secretion and, 452
— sperm acrosome reaction, 10, 36, 97
— synaptic vesicle exocytosis, 578
— trichocyst exocytosis, 697–710
A_2C (*see* Membrane mobility agents)
Abrin
— toxic effects, 393–97
Acanthomoeba castellani,
— endocytosis in, 389–90
Acetabularia
— nuclear envelope, 553
Acetylcholine receptor, 128, 148, 157
Achyla (*see* Fungi)
Acrasiales (*see also Dictyostelium discoideum*)
— fusion, 256
Acrosomal process (*see* Sperm, invertebrate)
Acrosome reaction (*see also* Fertilization; Sperm) 66–67, 70–79, 86–98, 292, 636–39
Actin
— invertebrate eggs, 39–42
— invertebrate sperm, 6, 8, 12, 14, 30, 43
— microfilaments and, 346, 435
— secretory granules, 654
— thrombosthenin, 454
α-actinin, 816

Actinomycin D
— macrophage fusion and, 191
— myoblast fusion and, 156
Adenyl cyclase
— platelet secretion and, 452
ADP
— platelets and, 438, 441–42, 446, 450–51, 453, 455
Adrenal medulla (*see also* Chromaffin granules; Exocytosis)
— secretion
—— calcium and, 273, 479, 575–85
—— recycling of secretory granules in, 595–97
—— ultrastructure, 556–75
Aequorin, 7, 47, 347
Agglutinins (*see* Lectins)
Alamethicin
— liposome fusion, 273, 772
Alcian blue
— glycocalyx and, 678
Allium Sp.
— protoplasts, fusion of, 370
Allocentrotus fragilis, 4
Allomyces (*see* Fungi)
Aminopeptidase, 238
Ammonia
— fertilization and, 37
Amoeba proteus
— endocytosis, 389
Anaphylaxis, 420
Anesthetics (*see* Local anesthetics)
Antheridiol, 244–45
Antigens
— blood group, 85
— capping of, 399, 415, 432, 475
— H-2, 74, 549, 553
— H-4Y antigen, sperm, 71
— immobilization (ciliates), 675–76, 679
— sperm, mammalian, 71, 74, 84–86, 96

838

Cell fusion *(cont'd)*
— cell cycle and, 128, 147, 246, 251
— cell swelling and, 287–88
— chemically induced, 268–304
— cholesterol and, 158, 268–270, 281–82
— cinematography, 133, 137, 141, 148, 184, 223, 226
— concanavalin A, 18–19, 26, 100, 152–55, 163, 206, 241, 272, 310, 375–76
— cyclic nucleotides, 193
— dextrans, 283, 287, 292
— diacylglycerol, 292–93
— dimethylsulfoxide, 283–87, 295
— EDTA and, 290, 292–93
— erythrocytes, 269–287, 290–297
— fat soluble chemical fusogens, 271–297
— fertilization *(see separate entry)*
— fungi *(see separate entry)*
— gametes *(see Fertilization)*
— genetic control of,
—— fungi, 237–240, 252–256
—— myxomycetes, 222–25, 227–33
— glycerol mono-oleate, 269, 274–79, 293–97
— glycocalyx and, 134–39, 142–43, 161–67, 184–85, 193, 257, 353
— heterokaryons, 272, 274
— hyperbaric oxygen, 193, 206
— hyperthermia and, 188, 275, 278, 292
— insulin and, 193
— interkingdom, 285
— lectins and, 18–19, 26, 100, 152–55, 163, 206, 241, 272, 310–12, 375–76
— liposomes and, 270, 272, 278, 287
— local anesthetics and, 191–92
— lysolecithin and, 188, 193, 271–74, 278–79, 287, 312–13, 330
— lysosomal stabilizers and, 193
— lysosomes and, 185, 193, 196–201, 212
— macrophages *(see separate entry)*
— membrane cholesterol and, 268–70, 281–82
— membrane fluidity and, 268–70, 274–79
— membrane lipids and, 188, 193, 201
— membrane mobility agents, 275
— microfilaments, 30, 43, 327, 344–49, 356

— microsurgical, 206, 353, 785
— microvilli, 108–09, 147, 185, 285
— myoblasts *(see separate entry)*
— myxomycetes *(see separate entry)*
— oleylamine, 272–73, 275–79, 291–97
— permeability changes, 146–47, 331, 335–45
— phenothiazine tranquillizers and, 192, 252
— phospholipases and, 143, 156–58, 169, 193, 206, 311
— plant protoplasts *(see separate entry under Protoplasts)*
— poly (ethylene glycol), 283–88, 290, 294–97
— proteases, 284
— protease inhibitors, 294
— redistribution of membrane proteins, 288, 294–97, 331–36, 343–49, 813–15
— saturated fatty acids, 274–78
— sorbital, 283–87
— span (20, 40, 60, 80), 287
— spontaneous, 353, 785
— steroids, 193
— surfactants, 287
— temperature-dependence, 157–58, 188, 239, 242, 275, 278, 292, 325–26, 333
— thiol-blocking agents, 293–95
— unsaturated fatty acids, 274–78
— valinomycin, 292
— virus-induced *(see separate entry)*
— water-soluble chemical fusogens, 283–87, 290, 292, 294–97
— Zn^{2+} and, 149
Cell junctions
— gap junctions,
—— myoblasts, 139–40, 147
—— permeability of, 340, 663–64, 697, 813
—— structure, 140, 540, 663–64, 697, 813–15
Cell motility
— fungal gametes, 251–52
Cell permeability
— virus-induced fusion and, 331, 335–45
Cell wall
— fungi, 235, 240–41, 247–51
Cellulases

862